To my Dear, Wonderful Wife Sheila

Contents

Preface

With the publication of the fourth edition, this text moves into the fourth decade of its existence. In the spirit of the times, the first edition introduced a number of "firsts" in an introductory engineering mechanics textbook. These "firsts" included

a) the first treatment of space mechanics
b) the first use of the control volume for linear momentum considerations of fluids
c) the first introduction to the concept of the tensor

Users of the earlier editions will be glad to know that the 4th edition continues with the same approach to engineering mechanics. The goal has always been to aim toward working problems as soon as possible from first principles. Thus, examples are carefully chosen during the development of a series of related areas to instill continuity in the evolving theory and then, after these areas have been carefully discussed with rigor, come the problems. Furthermore at the ends of each chapter, there are many problems that have not been arranged by text section. The instructor is encouraged as soon as he/she is well along in the chapter to use these problems. The instructors manual will indicate the nature of each of these problems as well as the degree of difficulty. The text is not chopped up into many methodologies each with an abbreviated discussion followed by many examples for using the specific methodology and finally a set of problems carefully tailored for the methodology. The nature of the format in this and preceding editions is more than ever first to discourage excessive mapping of homework problems from the examples. And second, it is to lessen the memorization of specific, specialized methodologies in lieu of absorbing basic principles.

A new feature in the fourth edition is a series of starred sections called "Looking Ahead . . . ." These are simplified discussions of topics that appear in later engineering courses and tie in directly or indirectly to the topic under study. For instance, after discussing free body diagrams, there is a short "Looking Ahead" section in which the concept and use of the control volume is presented as well as the system concepts that appear in fluid mechanics and thermodynamics. In the chapter on virtual work for particles and rigid bodies, there is a simplified discussion of the displacement methods and force methods for deformable bodies that will show up later in solids courses. After finding the forces for simple trusses, there is a "Looking Ahead" section discussing briefly what has to be done to get displacements. There are quite a few others in the text. It has been found that many students find these interesting and later when they come across these topics in other courses or work, they report that the connections so formed coming out of their sophomore mechanics courses have been most valuable.

Over 400 new problems have been added to the fourth edition equally divided between the statics and dynamics books. A complete word-processed solutions manual accompanies the text. The illustrations needed for problem statement and solution are taken as enlargements from the text. Generally, each problem is on a separate page. The instructor will be able conveniently to select problems in order to post solutions or to form transparencies as desired.

Another important new feature of the fourth edition is an organization that allows one to go directly to the three dimensional chapter on dynamics of rigid bodies (Chapter 18) and then to easily return to plane motion (Chapter 16). Or one can go the opposite way. Footnotes indicate how this can be done, and complimentary problems are noted in the Solutions Manual.

Another change is Chapter 16 on plane motion. It has been reworked with the aim of attaining greater rigor and clarity particularly in the solving of problems.

There has also been an increase in the coverage and problems for hydrostatics as well as examples and problems that will preview problems coming in the solids course that utilize principles from statics.

It should also be noted that the notation used has been chosen to correspond to that which will be used in more advanced courses in order to improve continuity with upper division courses. Thus, for moments and products of inertia I use I_{xx}, I_{yy}, I_{xz} *etc. rather than* I_x, I_y, P_{xy} etc. The same notation is used for second moments and product second moments of area to emphasize the direct relation between these and the preceding quantities. Experience indicates that there need be no difficulty on the student's part in distinguishing between these quantities; the context of the discussion suffices for this purpose. The concept of the tensor is presented in a way that for years we have found to be readily understood by sophomores even when presented in large classes. This saves time and makes for continuity in all mechanics courses, particularly in the solid mechanics course. For bending moment, shear force, and stress use is made of a common convention for the sign—namely the convention involving the normal to the area element and the direction of the quantity involved be it bending moment, shear force or stress component. All this and indeed other steps taken in the book will make for smooth transition to upper division course work.

In overall summary, two main goals have been pursued in this edition. They are

1. To encourage working problems from first principles and thus to minimize excessive mapping from examples and to discourage rote learning of specific methodologies for solving various and sundry kinds of specific problems.
2. To "open-end" the material to later course work in other engineering sciences with the view toward making smoother transitions and to provide for greater continuity. Also, the purpose is to engage the interest and curiosity of students for further study of mechanics.

During the 13 years after the third edition, I have been teaching sophomore mechanics to very large classes at SUNY, Buffalo, and, after that, to regular sections of students at The George Washington University, the latter involving an international student body with very diverse backgrounds. During this time, I have been working on improving the clarity and strength of this book under classroom conditions giving it the most severe test as a text. I believe the fourth edition as a result will be a distinct improvement over the previous editions and will offer a real choice for schools desiring a more mature treatment of engineering mechanics.

I believe sophomore mechanics is probably the most important course taken by engineers in that much of the later curriculum depends heavily on this course. And for **all** engineering programs, this is usually the first real engineering course where students can and must be creative and inventive in solving problems. Their old habits of mapping and rote learning of specific problem methodologies will not suffice and they must learn to see mechanics as an integral science. The student must "bite the bullet" and work in the way he/she will have to work later in the curriculum and even later when getting out of school altogether. No other subject so richly involves mathematics, physics, computers, and down to earth common sense simultaneously in such an interesting and challenging way. We should take maximum advantage of the students exposure to this beautiful subject to get him/her on the right track now so as to be ready for upper division work.

At this stage of my career, I will risk impropriety by presenting now an extended section of acknowledgments. I want to give thanks to SUNY at Buffalo where I spent 31 happy years and where I wrote many of my books. And I want to salute the thousands (about 5000) of fine students who took my courses during this long stretch. I wish to thank my eminent friend and colleague Professor Shahid Ahmad who among other things taught the sophomore mechanics sequence with me and who continues to teach it. He gave me a very thorough review of the fourth edition with many valuable suggestions. I thank him profusely. I want particularly to thank Professor Michael Symans, from Washington State University, Pullman for his superb contributions to the entire manuscript. I came to The George Washington University at the invitation of my longtime friend and former Buffalo colleague Dean Gideon Frieder and the faculty in the Civil, Mechanical and Environmental Engineering Department. Here, I came back into contact with two well-known scholars that I knew from the early days of my career, namely Professor Hal Liebowitz (president-elect of the National Academy of Engineering) and Professor Ali Cambel (author of recent well-received book on chaos). I must give profound thanks to the chairman of my new department at G.W., Professor Sharam Sarkani. He has allowed me to play a vital role in the academic program of the department. I will be able to continue my writing at full speed as a result. I shall always be grateful to him. Let me not forget the two dear ladies in the front office of the department. Mrs. Zephra Coles in her decisive efficient way took care of all my needs even before I was aware of them. And Ms. Joyce Jeffress was no less helpful and always had a humorous comment to make.

I was extremely fortunate in having the following professors as reviewers.

Professor Shahid Ahmad, SUNY at Buffalo
Professor Ravinder Chona, Texas A&M University
Professor Bruce H. Karnopp. University of Michigan

Professor Richard F. Keltie, North Carolina State University
Professor Stephen Malkin, University of Massachusetts
Professor Sudhakar Nair, Illinois Institute of Technology
Professor Jonathan Wickert, Carnegie Mellon University

I wish to thank these gentlemen for their valuable assistance and encouragement.

I have two people left. One is my good friend Professor Bob Jones from V.P.I. who assisted me in the third edition with several hundred excellent statics problems and who went over the entire manuscript with me with able assistance and advice. I continue to benefit in the new edition from his input of the third edition. And now, finally, the most important person of all, my dear wife Sheila. She has put up all these years with the author of this book, an absent-minded, hopeless workaholic. Whatever I have accomplished of any value in a long and ongoing career, I owe to her.

About the Author

Irving Shames presently serves as a Professor in the Department of Civil, Mechanical, and Environmental Engineering at The George Washington University. Prior to this appointment Professor Shames was a Distinguished Teaching Professor and Faculty Professor at The State University of New York—Buffalo, where he spent 31 years.

Professor Shames has written up to this point in time 10 textbooks. His first book *Engineering Mechanics, Statics and Dynamics* was originally published in 1958. All of the books written by Professor Shames have been characterized by innovations that have become mainstays of how engineering principles are taught to students. *Engineering Mechanics, Statics and Dynamics* was the first widely used Mechanics book based on vector principles. It ushered in the almost universal use of vector principles in teaching engineering mechanics courses today.

Other textbooks written by Professor Shames include:

- *Mechanics of Deformable Solids*, Prentice-Hall, Inc.
- *Mechanics of Fluids*, McGraw-Hill
- *Introduction to Solid Mechanics*, Prentice-Hall, Inc.
- *Introduction to Statics*, Prentice-Hall, Inc.
- *Solid Mechanics—A Variational Approach* (with C.L. Dym), McGraw-Hill
- *Energy and Finite Element Methods in Structural Mechanics*, (with C.L. Dym), Hemisphere Corp., of Taylor and Francis
- *Elastic and Inelastic Stress Analysis* (with F. Cozzarelli), Prentice-Hall, Inc.

In recent years, Professor Shames has expanded his teaching activities and has held two summer faculty workshops in mechanics sponsored by the State of New York, and one national workshop sponsored by the National Science Foundation. The programs involved the integration both conceptually and pedagogically of mechanics from the sophomore year on through graduate school.

Dynamics

CHAPTER 11

Kinematics of a Particle—Simple Relative Motion

11.1 Introduction

Kinematics is that phase of mechanics concerned with the study of the motion of particles and rigid bodies without consideration of what has caused the motion. We can consider kinematics as the geometry of motion. Once kinematics is mastered, we can smoothly proceed to the relations between the factors causing the motion and the motion itself. The latter area of study is called *dynamics*. Dynamics can be conveniently separated into the following divisions, most of which we shall study in this text:

1. Dynamics of a single particle. (You will remember from our chapters on statics that a particle is an idealization having no volume but having mass.)
2. Dynamics of a system of particles. This follows division 1 logically and forms the basis for the motion of continuous media such as fluid flow and rigid-body motion.
3. Dynamics of a rigid body. A large portion of this text is concerned with this important part of mechanics.
4. Dynamics of a system of rigid bodies.
5. Dynamics of a continuous deformable medium.

Clearly, from our opening statements, the particle plays a vital role in the study of dynamics. What is the connection between the particle, which is a completely hypothetical concept, and the finite bodies encountered in physical problems? Briefly the relation is this: In many problems, the size and shape of a body are not relevant in the discussion of certain aspects of its motion; only the mass of the object is significant for such computations. For example, in towing a truck up a hill, as shown in Fig. 11.1, we would only be concerned

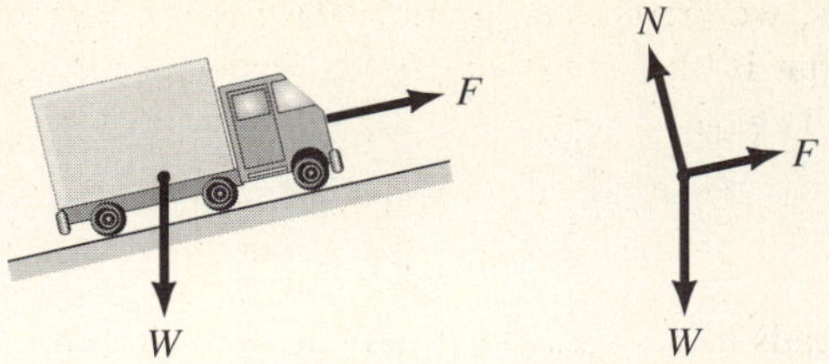

Figure 11.1. Truck considered as a particle.

with the mass of the truck and not with its shape or size (if we neglect forces from the wind, etc., and the rotational effects of the wheels). The truck can just as well be considered a particle in computing the necessary towing force.

We can present this relationship more precisely in the following manner. As will be learned in the next chapter (Section 12.10), the equation of motion of the center of mass of any body can be formed by:

1. Concentrating the entire mass at the mass center of the body.
2. Applying the total resultant force acting on the body to this hypothetical particle.

When the motion of the mass center characterizes all we need to know about the motion of the body, we employ the particle concept (i.e., we find the motion of the mass center). Thus, if all points of a body have the same velocity at any time t (this is called *translatory motion*), we need only know the motion of the mass center to fully characterize the motion. (This was the case for the truck, where the rotational inertia of the wheels was neglected.) If, additionally, the size of a body is small compared to its trajectory (as in planetary motion, for example), the motion of the center of mass is all that might be needed, and so again we can use the particle concept for such bodies.

Part A: General Notions

11.2 Differentiation of a Vector with Respect to Time

In the study of statics, we dealt with vector quantities. We found it convenient to incorporate the directional nature of these quantities in a certain notation and set of operations. We called the totality of these very useful formulations "vector algebra." We shall again expand our thinking from scalars to vectors—this time for the operations of differentiation and integration with respect to any scalar variable t (such as time).

For scalars, we are concerned only with the variation in magnitude of some quantity that is changing with time. The scalar definition of the time derivative, then, is given as

$$\frac{df(t)}{dt} = \lim_{\Delta t \to 0} \left[\frac{f(t + \Delta t) - f(t)}{\Delta t} \right] \tag{11.1}$$

This operation leads to another function of time, which can once more be differentiated in this manner. The process can be repeated again and again, for suitable functions, to give higher derivatives.

In the case of a vector, the variation in time may be a change in magnitude, a change in direction, or both. The formal definition of the derivative of a vector $\boldsymbol{F}$ with respect to time has the same form as Eq. 11.1:

$$\frac{d\boldsymbol{F}}{dt} = \lim_{\Delta t \to 0} \left[\frac{\boldsymbol{F}(t + \Delta t) - \boldsymbol{F}(t)}{\Delta t} \right] \tag{11.2}$$

If $\boldsymbol{F}$ has no change in direction during the time interval, this operation differs little from the scalar case. However, when $\boldsymbol{F}$ changes in direction, we find for the derivative of $\boldsymbol{F}$ a new vector, having a magnitude as well as a direction, that is different from $\boldsymbol{F}$ itself. This directional consideration can be somewhat troublesome.

Let us consider the rate of change of the position vector for a reference *xyz* of a particle with respect to time; this rate is defined as the *velocity vector*, $\boldsymbol{V}$, of the particle relative to *xyz*. Following the definition given by Eq. 11.2, we have

$$\frac{d\boldsymbol{r}}{dt} = \lim_{\Delta t \to 0} \left[\frac{\boldsymbol{r}(t + \Delta t) - \boldsymbol{r}(t)}{\Delta t} \right]$$

The position vectors given in brackets are shown in Fig. 11.2. The subtraction

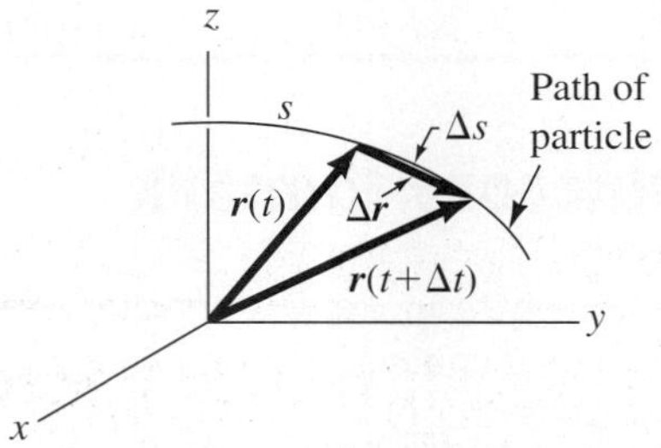

Figure 11.2. Particle at times t and $t + \Delta t$.

between the two vectors gives rise to the displacement vector $\Delta\boldsymbol{r}$, which is shown as a chord connecting two points Δs apart along the trajectory of the particle. Hence, we can say (using the chain rule) that

$$\frac{d\boldsymbol{r}}{dt} = \lim_{\Delta t \to 0} \left(\frac{\Delta\boldsymbol{r}}{\Delta t} \right) = \lim_{\Delta t \to 0} \left(\frac{\Delta\boldsymbol{r}}{\Delta s} \frac{\Delta s}{\Delta t} \right)$$

where we have multiplied and divided by Δs in the last expression. As Δt goes to zero, the direction of $\Delta \boldsymbol{r}$ approaches tangency to the trajectory at position $\boldsymbol{r}(t)$ and approaches Δs in magnitude. Consequently, in the limit, $\Delta \boldsymbol{r}/\Delta s$ becomes a unit vector $\boldsymbol{\epsilon}_t$, tangent to the trajectory. That is

$$\frac{\Delta \boldsymbol{r}}{\Delta s} \to \frac{\Delta s \boldsymbol{\epsilon}_t}{\Delta s} = \boldsymbol{\epsilon}_t \qquad \therefore \frac{d\boldsymbol{r}}{ds} = \boldsymbol{\epsilon}_t \tag{11.3}$$

We can then say

$$\frac{d\boldsymbol{r}}{dt} = \boldsymbol{V} = \lim_{\Delta t \to 0}\left[\left(\frac{\Delta s}{\Delta t}\right)\left(\frac{\Delta \boldsymbol{r}}{\Delta s}\right)\right] = \frac{ds}{dt}\boldsymbol{\epsilon}_t \tag{11.4}$$

Therefore, $d\boldsymbol{r}/dt$ leads to a vector having a magnitude equal to the speed of the particle and a direction tangent to the trajectory. Keep in mind that there can be any angle between the position vector and the velocity vector. Students seem to want to limit this angle to 90°, which actually restricts you to a circular path. The acceleration vector of a particle can then be given as

$$\boldsymbol{a} = \frac{d\boldsymbol{V}}{dt} = \frac{d^2\boldsymbol{r}}{dt^2} \tag{11.5}$$

The differentiation and integration of vectors $\boldsymbol{r}$, $\boldsymbol{V}$, and $\boldsymbol{a}$ will concern us throughout the text.

Part B: Velocity and Acceleration Calculations

11.3 Introductory Remark

As you know from statics, we can express a vector in many ways. For instance, we can use rectangular components, or, as we will shortly explain, we can use cylindrical components. In evaluating derivatives of vectors with respect to time, we must proceed in accordance with the manner in which the vector has been expressed. In Part B of this chapter, we will therefore examine certain differentiation processes that are used extensively in mechanics. Other differentiation processes will be examined later at appropriate times.

We have already carried out a derivative operation in Section 11.2 directly on the vector $\boldsymbol{r}$. You will see in Section 11.5 that the approach used gives the derivative in terms of *path variables*. This approach will be one of several that we shall now examine with some care.

11.4 Rectangular Components

Consider first the case where the position vector $\boldsymbol{r}$ of a moving particle is expressed for a given reference in terms of rectangular components in the following manner:

$$\boldsymbol{r}(t) = x(t)\boldsymbol{i} + y(t)\boldsymbol{j} + z(t)\boldsymbol{k} \tag{11.6}$$

where $x(t)$, $y(t)$, and $z(t)$ are scalar functions of time. The unit vectors $\boldsymbol{i}, \boldsymbol{j}$, and $\boldsymbol{k}$ are fixed in magnitude and direction at all times, and so we can obtain $d\boldsymbol{r}/dt$ in the following straightforward manner:

$$\frac{d\boldsymbol{r}}{dt} = \boldsymbol{V}(t) = \frac{dx(t)}{dt}\boldsymbol{i} + \frac{dy(t)}{dt}\boldsymbol{j} + \frac{dz(t)}{dt}\boldsymbol{k} = \dot{x}(t)\boldsymbol{i} + \dot{y}(t)\boldsymbol{j} + \dot{z}(t)\boldsymbol{k} \tag{11.7}$$

A second differentiation with respect to time leads to the acceleration vector:

$$\frac{d^2\boldsymbol{r}}{dt^2} = \boldsymbol{a} = \ddot{x}(t)\boldsymbol{i} + \ddot{y}(t)\boldsymbol{j} + \ddot{z}(t)\boldsymbol{k} \tag{11.8}$$

By such a procedure, we have formulated velocity and acceleration vectors in terms of components parallel to the coordinate axes.

Up to this point, we have formulated the rectangular velocity components and the rectangular acceleration components, respectively, by differentiating the position vector once and twice with respect to time. Quite often, we know the acceleration vector of a particle as a function of time in the form

$$\boldsymbol{a}(t) = \ddot{x}(t)\boldsymbol{i} + \ddot{y}(t)\boldsymbol{j} + \ddot{z}(t)\boldsymbol{k} \tag{11.9}$$

and wish to have for this particle the velocity vector or the position vector or any of their components at any time. We then integrate the time function $\ddot{x}(t)$, $\ddot{y}(t)$, and $\ddot{z}(t)$, remembering to include a constant of integration for each integration. For example, consider $\ddot{x}(t)$. Integrating once, we obtain the velocity component $V_x(t)$ as follows:

$$V_x(t) = \int \ddot{x}(t)\,dt + C_1 \tag{11.10}$$

where C_1 is the constant of integration. Knowing V_x at some time t_0, we can determine C_1 by substituting t_0 and $(V_x)_0$ into the equation above and determining C_1. Similarly, for $x(t)$ we obtain from the above:

$$x(t) = \int\left[\int \ddot{x}(t)\,dt\right]dt + C_1 t + C_2 \tag{11.11}$$

where C_2 is the second constant of integration. Knowing x at some time t, we can determine C_2 from Eq. 11.11. The same procedure involving additional constants applies to the other acceleration components.

We now illustrate the procedures described above in the following series of examples.

Example 11.1

Pins A and B must always remain in the vertical slot of yoke C, which moves to the right at a constant speed of 2 m/s in Fig. 11.3. Furthermore, the pins cannot leave the elliptic slot. (a) What is the speed at which the pins approach each other when the yoke slot is at $x = 1.5$ m? (b) What is the rate of change of speed toward each other when the yoke slot is at $x = 1.5$ m?

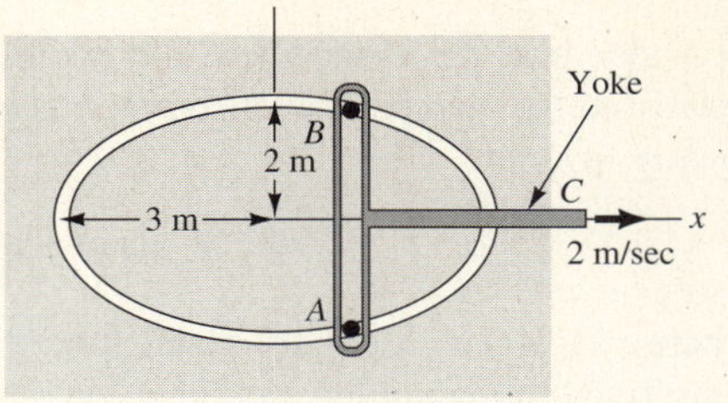

Figure 11.3. Pin slides in slot and yoke.

The equation of the elliptic path in which the pins must move is seen by inspection to be

$$\frac{x^2}{3^2} + \frac{y^2}{2^2} = 1 \qquad \text{(a)}$$

Clearly, if coordinates (x, y) are to represent the coordinates of pin B, they must be time functions such that for any time t the values $x(t)$ and $y(t)$ satisfy Eq. (a). Also, $\dot{x}(t)$ and $\dot{y}(t)$ must be such that pin B moves at all times in the elliptic path. We can satisfy these requirements by first differentiating Eq. (a) with respect to time. Canceling the factor 2, we obtain

$$\frac{x\dot{x}}{3^2} + \frac{y\dot{y}}{2^2} = 0 \qquad \text{(b)}$$

Now $x(t)$, $y(t)$, $\dot{x}(t)$, and $\dot{y}(t)$ must satisfy Eq. (b) for all values of t to ensure that B remains in the elliptic path.

We can now proceed to solve part (a) of this problem. We know that pin B must have a velocity $\dot{x} = 2$ m/s because of the yoke. Furthermore, when $x = 1.5$ m, we know from Eq. (a) that

$$\frac{1.5^2}{3^2} + \frac{y^2}{2^2} = 1 \qquad \text{(c)}$$

$$\therefore\ y = 1.732 \text{ m}$$

Now going to Eq. (b), we can solve for $\dot{y}$ at the instant of interest.

$$\frac{(1.5)(2)}{3^2} + \frac{(1.732)(\dot{y})}{2^2} = 0$$

Therefore,

$$\dot{y} = -0.77 \text{ m/s}$$

Thus, pin B moves downward with a speed of 0.77 m/s. Clearly, pin A must move upward with the same speed of 0.77 m/s. The pins approach each other at the instant of interest at a speed of 1.54 m/s.

Example 11.1 (Continued)

To get the acceleration $\ddot{y}$ of pin B, we first differentiate Eq. (b) with respect to time.

$$\frac{x\ddot{x} + \dot{x}^2}{3^2} + \frac{y\ddot{y} + \dot{y}^2}{2^2} = 0 \tag{d}$$

The accelerations $\ddot{x}$ and $\ddot{y}$ must satisfy the equation above. Since the yoke moves at constant speed, we can say immediately that $\ddot{x} = 0$. And using for x, y, $\dot{x}$, and $\dot{y}$ known quantities for the configuration of interest, we can solve for $\ddot{y}$ from Eq. (d). Thus,

$$\frac{0 + 2^2}{3^2} + \frac{1.732\ddot{y} + 0.77^2}{2^2} = 0$$

Therefore,

$$\ddot{y} = -1.37 \text{ m/s}^2$$

Pin B must be accelerating downward at a rate of 1.37 m/s^2 while pin A accelerates upward at the same rate. The pins accelerate toward each other, then, at a rate of 2.74 m/s^2 at the configuration of interest.

In the motion of particles near the earth's surface, such as the motion of shells or ballistic missiles, we can often simplify the problem by neglecting air resistance and taking the acceleration of gravity g as constant (9.81 m/s^2). For such a case (see Fig. 11.4), we know immediately that $\ddot{y}(t) = -g$ and $\ddot{x}(t) = \ddot{z}(t) = 0$. On integrating these accelerations, we can often determine for the particle useful information as to velocities or positions at certain times of interest in the problem. We illustrate this procedure in the following examples.

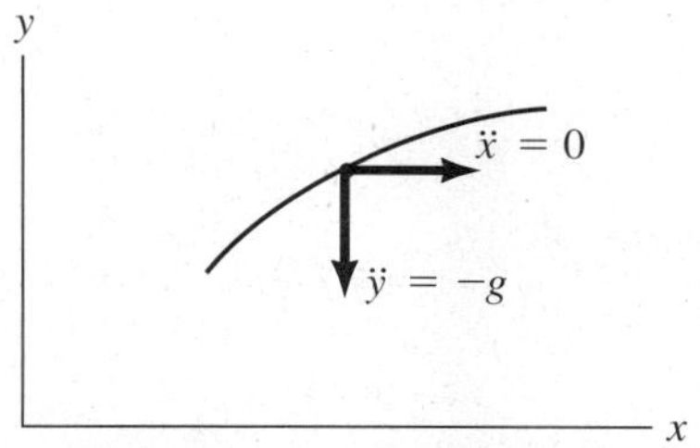

Figure 11.4. Simple ballistic motion of a shell.

Example 11.2

Ballistics Problem 1. A shell is fired from a hill 150 m above a plain. The angle α of firing (see Fig. 11.5) is 15° above the horizontal, and the muzzle velocity V_0 is 900 m/s. At what horizontal distance, d, will the shell hit the plain if we neglect friction of the air? What is the maximum height of the shell above the plain? Finally, determine the trajectory of the shell [i.e., find $y = f(x)$].

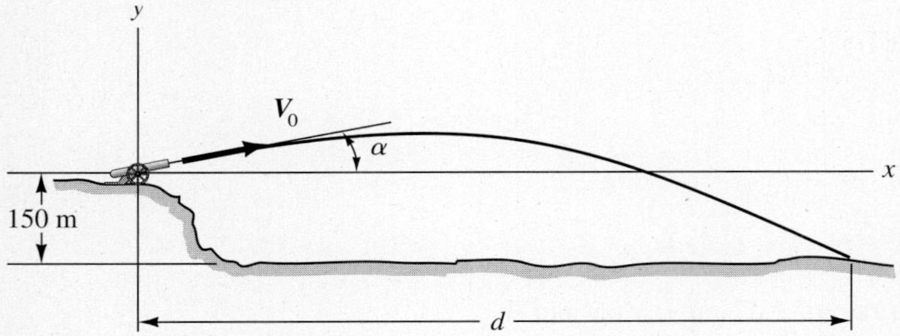

Figure 11.5 Ballistics problem: find d.

We know immediately that

$$\ddot{y}(t) = \frac{dV_y}{dt} = -9.81 \text{ m/s}^2 \tag{a}$$

$$\ddot{x}(t) = \frac{dV_x}{dt} = 0 \tag{b}$$

We need not bother with $\ddot{z}(t)$, since the motion is coplanar with $\dot{z}(t) = z = 0$ at all times. We next separate the velocity variables from the time variables by bringing dt to the right sides of the previous equations. Thus

$$dV_y = -9.81\, dt$$
$$dV_x = 0\, dt$$

Integrating the above equations, we get

$$V_y(t) = -9.81t + C_1$$
$$V_x(t) = C_2$$

We shall take $t = 0$ at the instant the cannon is fired. At this instant, we know V_y and V_x and can determine C_1 and C_2. Thus,

$$V_y(0) = 900 \sin 15° = (-9.81)(0) + C_1$$

Example 11.2 (Continued)

Therefore,

$$C_1 = V_y(0) = 233 \text{ m/s}$$

Also

$$V_x(0) = 900 \cos 15° = C_2$$

Therefore,

$$C_2 = V_x(0) = 869 \text{ m/s}$$

We can give the velocity components of the shell now as follows:

$$V_y(t) = \frac{dy}{dt} = -9.81t + 233 \text{ m/s} \qquad \text{(c)}$$
$$V_x(t) = \frac{dx}{dt} = 869 \text{ m/s} \qquad \text{(d)}$$

Thus, the horizontal velocity is constant. Separating the position and time variables and then integrating, we get the x and y coordinates of the shell.

$$y(t) = -9.81\frac{t^2}{2} + 233t + C_3 \qquad \text{(e)}$$
$$x(t) = 869t + C_4 \qquad \text{(f)}$$

When $t = 0$, $y = x = 0$. Thus, from Eqs. (e) and (f), we clearly see that $C_3 = C_4 = 0$. The coordinates of the shell are then

$$y(t) = -4.905t^2 + 223t \qquad \text{(g)}$$
$$x(t) = 869t \qquad \text{(h)}$$

To determine *distance d*, first find the time t for the impact of the shell on the plain. That is, set $y = -150$ in Eq. (g) and solve for the time t. Thus,

$$-150 = -4.905t^2 + 233t$$

Example 11.2 (Continued)

Therefore,

$$4.905t^2 - 233t - 150 = 0$$

Using the quadratic formula, we get for t:

$$t = 46.8 \text{ s}$$

Substituting this value of t into Eq. (h), we get

$$d = (869)(46.8) = \boxed{40.67 \text{ km}}$$

To get the *maximum height* y_{max}, first find the time t when $V_y = 0$. Thus, from Eq. (c) we get

$$0 = -9.81t + 233$$

Therefore,

$$t = 23.75 \text{ s}$$

Now substitute $t = 23.75$ s into Eq. (g). This gives us y_{max}.

$$y_{max} = -(4.905)(23.75)^2 + 233(23.75)$$

$$\therefore \boxed{y_{max} = 2.77 \text{ km}}$$

Finally, to get the *trajectory* of the shell (i.e., y as a function of x), solve for t in Eq. (h) and substitute this into Eq. (g). We then have

$$y_{max} = -(4.905)\left(\frac{x}{869}\right)^2 + 233\left(\frac{x}{869}\right)$$

Therefore,

$$\boxed{y = -6.495 \times 10^{-6} x^2 + 0.268x}$$

Clearly, the trajectory is that of a *parabola*.

Example 11.3

Ballistics Problem 2. A gun emplacement is shown on a cliff in Fig. 11.6. The muzzle velocity of the gun is 1000 m/s. At what angle α must the gun point in order to hit target A shown in the diagram? Neglect friction.

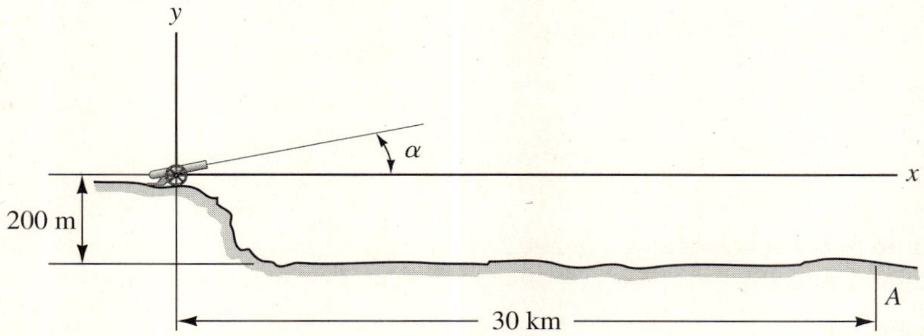

Figure 11.6. Find α to hit A.

Newton's law for the shell is given as follows for a reference xy having its origin at the gun.

$$\ddot{y}(t) = -9.81$$
$$\ddot{x}(t) = 0$$

Integrating, we get

$$\dot{y}(t) = V_y(t) = -9.81t + C_1 \quad \text{(a)}$$
$$\dot{x}(t) = V_x(t) = C_2 \quad \text{(b)}$$

When $t = 0$, we have $\dot{y} = 1000 \sin \alpha$ and $\dot{x} = 1000 \cos \alpha$. Applying these conditions to Eqs. (a) and (b), we solve for C_1 and C_2. Thus,

$$1000 \sin \alpha = 0 + C_1$$

Therefore,

$$C_1 = 1000 \sin \alpha$$

Also,

$$1000 \cos \alpha = C_2$$

Therefore,

$$C_2 = 1000 \cos \alpha$$

Hence, we have

$$\dot{y}(t) = -9.81t + 1000 \sin \alpha$$
$$\dot{x}(t) = 1000 \cos \alpha$$

Integrating again, we get

$$y(t) = -9.81\frac{t^2}{2} + 1000 \sin \alpha\, t + C_3$$
$$x(t) = 1000 \cos \alpha\, t + C_4$$

Example 11.3 (Continued)

When $t = 0$, $x = y = 0$. Hence, it is clear that $C_3 = C_4 = 0$. Thus, we have

$$y = -4.095t^2 + 1000 \sin \alpha\, t \tag{c}$$

$$x = 1000 \cos \alpha\, t \tag{d}$$

To get the *trajectory*, we solve for t in Eq. (d) and substitute into Eq. (c).

$$\begin{aligned} y &= -4.095 \frac{x^2}{(1000 \cos \alpha)^2} + 1000 \sin \alpha \frac{x}{(1000 \cos \alpha)} \\ &= -4.095 \times 10^{-6} \frac{x^2}{\cos^2 \alpha} + x \tan \alpha \end{aligned} \tag{e}$$

where we have replaced $\sin \alpha/\cos \alpha$ by $\tan \alpha$. When $x = 30$ km, $y = -200$ m. Hence, we have on substituting these data into Eq. (e):

$$-200 = -4.095 \times 10^{-6} \frac{(30 \times 10^3)^2}{\cos^2 \alpha} + 30 \times 10^3 \tan \alpha$$

Replace $1/\cos^2 \alpha$ by $\sec^2 \alpha = \left(1 + \tan^2 \alpha\right)$:

$$-200 = -4.095 \times 10^{-6} (30 \times 10^3)^2 (1 + \tan^2 \alpha) + 30 \times 10^3 \tan \alpha$$

Therefore,

$$\tan^2 \alpha - 6.796 \tan \alpha + 0.955 = 0 \tag{f}$$

Using the quadratic formula, we find the following angles:

$$\alpha_1 = 8.17^\circ$$
$$\alpha_2 = 81.44^\circ$$

There are thus *two* possible firing angles that will permit the shell to hit the target, as shown in Fig. 11.7.

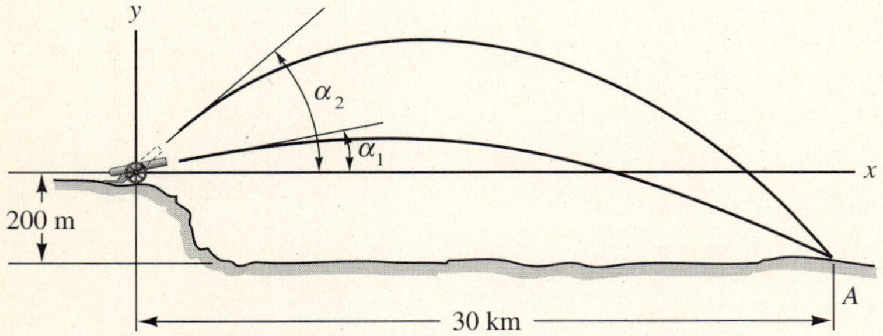

Figure 11.7. Two firing angles are possible.

Example 11.4

The engine room of a freighter is on fire. A fire-fighting tugboat has drawn alongside and is directing a stream of water to enter the stack of the freighter as shown in Fig. 11.8. If the initial speed of the jet of water is 21 m/s, is there a value of α of the issuing jet of water that will do the job? If so, what should α be?

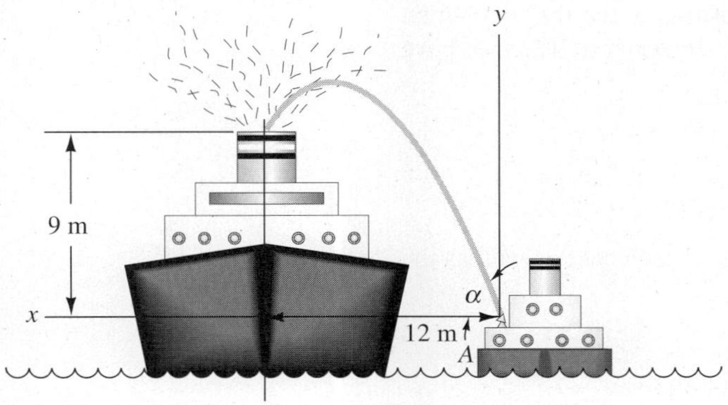

Figure 11.8. Fire-fighting tugboat directing a jet of water into the stack of a freighter.

Consider a particle within the stream of water. Neglecting friction, **Newton's law** for the particle is given as follows:

$$\ddot{y} = -9.81 \text{ m/s}^2 \qquad \ddot{x} = 0 \text{ m/s}^2$$

Integrating twice, and using initial conditions at A, we get

$$\dot{y} = -9.81t + 21\sin\alpha \text{ m/s} \quad \text{(a)} \qquad \dot{x} = 21\cos\alpha \text{ m/s} \quad \text{(c)}$$

$$y = -4.905t^2 + 21\sin\alpha\, t \text{ m} \quad \text{(b)} \qquad x = 21\cos\alpha\, t \text{ m} \quad \text{(d)}$$

Solve for t from Eq. (d) and substitute into Eq. (b) to get

$$y = -4.905\left[\frac{x}{21\cos\alpha}\right]^2 + 21\sin\alpha\left[\frac{x}{21\cos\alpha}\right]$$

Replace $\cos^2\alpha$ by $1/(1+\tan^2\alpha)$ and $(\sin\alpha/\cos\alpha)$ by $\tan\alpha$ in the previous equation and then substitute the coordinates of point B at the stack where the water is supposed to reach. That is, set $x = 12$ m and $y = 9$ m. We then get

$$9 = -(11.12\times 10^{-3})(12^2)(1+\tan^2\alpha) + 12\tan\alpha$$

$$\therefore\ \tan^2\alpha - 7.45\tan\alpha + 6.62 = 0$$

Using the quadratic formula we get

$$\tan\alpha = \frac{7.45 \pm \sqrt{7.45^2 - (4)(6.62)}}{2} = 1.031;\ 6.419$$

Example 11.4 (Continued)

We thus have two angles for α, each of which will theoretically cause the stream to go to point B of the stack. These angles are

$$\alpha_1 = 45.87^\circ \qquad \alpha_2 = 81.15^\circ$$

Does one, none, or both angles above yield a stream of water that will come down at B so as to enter the stack? We can determine this by finding the maximum value of y and locating the position x for this maximum value. To do this, we set $\dot{y} = 0$ and solve for t using each α. Thus we have

$$0 = -9.81t + 21\sin\begin{Bmatrix} 45.87^\circ \\ 81.15^\circ \end{Bmatrix}$$

$$\therefore t = \begin{Bmatrix} 1.536 \\ 2.115 \end{Bmatrix} \text{ s}$$

Now get the position x for maximum elevation for each case as well as the elevation maximum, y_{max}.

For $\alpha = 45.87^\circ$:

$$x = (21)(\cos 45.87^\circ)(1.536) = 22.46 \text{ m}$$

$$y_{max} = -(4.905)(1.536)^2 + (21)(\sin 45.87^\circ)(1.536) = 11.58 \text{ m}$$

For $\alpha = 81.37^\circ$:

$$x = (21)(\cos 81.15^\circ)(2.115) = 6.83 \text{ m}$$

$$y_{max} = -(4.905)(2.115)^2 + (21)(\sin 81.15^\circ)(2.115) = 21.95 \text{ m}$$

A sketch of the two possible trajectories is shown in Fig. 11.9. Clearly the shallow trajectory will hit the side of the stack and is unacceptable, while the high trajectory will deposit water inside the stack and is thus the desired trajectory. Thus,

$$\boxed{\alpha = 81.15^\circ}$$

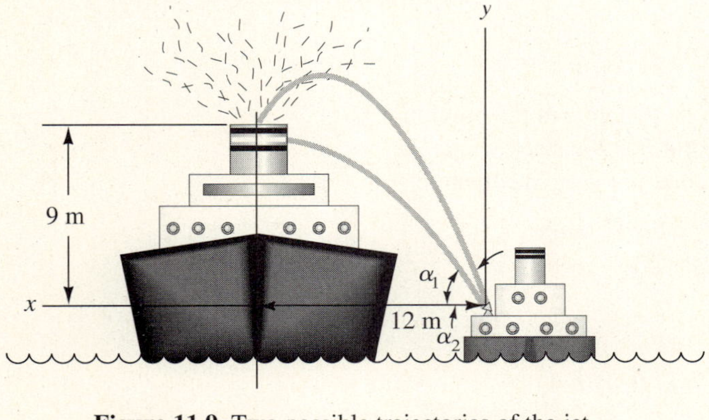

Figure 11.9. Two possible trajectories of the jet.

We do not always know the variation of the position vector with time in the form of Eq. 11.6. Furthermore, it may be that the components of velocity and acceleration that we desire are not those parallel to a fixed Cartesian reference. The evaluation of $\boldsymbol{V}$ and $\boldsymbol{a}$ for certain other circumstances will be considered in the following sections.

11.5 Velocity and Acceleration in Terms of Path Variables

We have formulated velocity and acceleration for the case where the rectangular coordinates of a particle are known as functions of time. We now explore another approach in which the formulations are carried out in terms of the path variables of the particle, that is, in terms of geometrical parameters of the path and the speed and the rate of change of speed of the particle along the path. These results are particularly useful when a particle moves along a path that we know apriori (such as the case of a roller coaster).

As a matter of fact, in Section 11.2 (Eq. 11.4) we expressed the velocity vector in terms of path variables in the following form:

$$\boldsymbol{V} = \frac{ds}{dt}\boldsymbol{\epsilon}_t \tag{11.12}$$

where ds/dt represents the speed along the path and $\boldsymbol{\epsilon}_t = d\boldsymbol{r}/ds$ (see Eq. 11.3) is the unit vector tangent to the path (and hence collinear with the velocity vector). The acceleration becomes

$$\frac{d\boldsymbol{V}}{dt} = \boldsymbol{a} = \frac{d^2s}{dt^2}\boldsymbol{\epsilon}_t + \frac{ds}{dt}\frac{d\boldsymbol{\epsilon}_t}{dt} \tag{11.13}$$

Replace $d\boldsymbol{\epsilon}_t/dt$ in this expression by $(d\boldsymbol{\epsilon}_t/ds)(ds/dt)$, the validity of which is assured by the chain rule of differentiation. We then have

$$\boldsymbol{a} = \frac{d^2s}{dt^2}\boldsymbol{\epsilon}_t + \left(\frac{ds}{dt}\right)^2\frac{d\boldsymbol{\epsilon}_t}{ds} \tag{11.14}$$

Before proceeding further, let us consider the unit vector $\boldsymbol{\epsilon}_t$ at two positions that are Δs apart along the path of the particle as shown in Fig. 11.10. If Δs is small enough, the unit vectors $\boldsymbol{\epsilon}_t(s)$ and $\boldsymbol{\epsilon}_t(s + \Delta s)$ can be considered to intersect and thus to form a plane. If $\Delta s \to 0$, these unit vectors then form a *limiting plane*, which we shall call the *osculating plane*.[1] The plane will have an orientation that depends on the position s on the path of the particle. The osculating plane at $\boldsymbol{r}(t)$ is illustrated in Fig. 11.10. Having defined the osculating plane, let us continue discussion of Eq. 11.14.

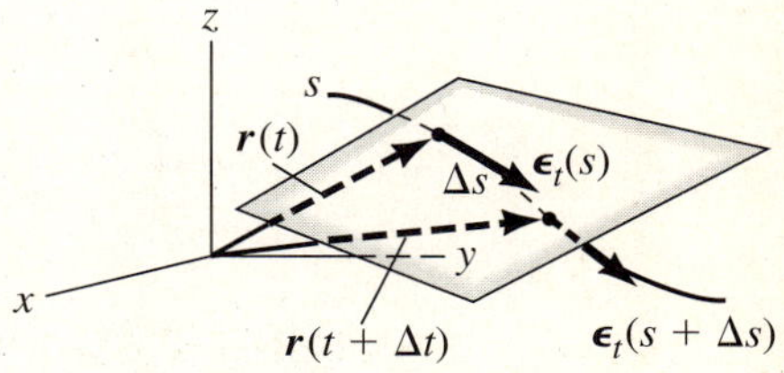

Figure 11.10. Osculating plane at $\boldsymbol{r}(t)$.

[1]From the definition, it should be apparent that the osculating plane at position s along a curve is actually *tangent* to the curve at position s. Since osculate means to kiss, the plane "kisses" the curve, as it were, at s.

Since we have not formally carried out the differentiation of a vector with respect to a spatial coordinate, we shall carry out the derivative $d\boldsymbol{\epsilon}_t/ds$ needed in Eq. 11.14 from the basic definition. Thus,

$$\frac{d\boldsymbol{\epsilon}_t}{ds} = \lim_{\Delta s \to 0}\left[\frac{\boldsymbol{\epsilon}_t(s + \Delta s) - \boldsymbol{\epsilon}_t(s)}{\Delta s}\right] = \lim_{\Delta s \to 0}\left(\frac{\Delta \boldsymbol{\epsilon}_t}{\Delta s}\right) \tag{11.15}$$

The vectors $\boldsymbol{\epsilon}_t(s)$ and $\boldsymbol{\epsilon}_t(s + \Delta s)$ are shown in Fig. 11.11(a) along the path and are also shown (enlarged) with $\Delta\boldsymbol{\epsilon}_t$ as a vector triangle in Fig. 11.11(b). As pointed out earlier, for small enough Δs the lines of action of the unit vectors $\boldsymbol{\epsilon}_t(s)$ and $\boldsymbol{\epsilon}_t(s + \Delta s)$ will intersect to form a plane as shown in Fig. 11.11(a). Now in this plane, draw normal lines to the aforementioned vectors at the respective positions s and $s + \Delta s$. These lines will intersect at some point O, as shown in the diagram. Next, consider what happens to the plane and to point O as $\Delta s \to 0$. Clearly, the limiting plane is our osculating plane at s [see Fig. 11.11(c)]. Furthermore, the limiting position arrived at for point O is *in the osculating plane* and is called the *center of curvature* for the path at s. The distance between O and s is denoted as R and is called the *radius of curvature*. Finally, the vector $\Delta\boldsymbol{\epsilon}_t$ (see Fig. 11.11 (b)), in the limit as $\Delta s \to 0$, ends up in the osculating plane normal to the path at s and directed toward the center of curvature. The unit vector collinear with the limiting vector for $\Delta\boldsymbol{\epsilon}_t$ is denoted as $\boldsymbol{\epsilon}_n$ and is called the *principal normal vector*.

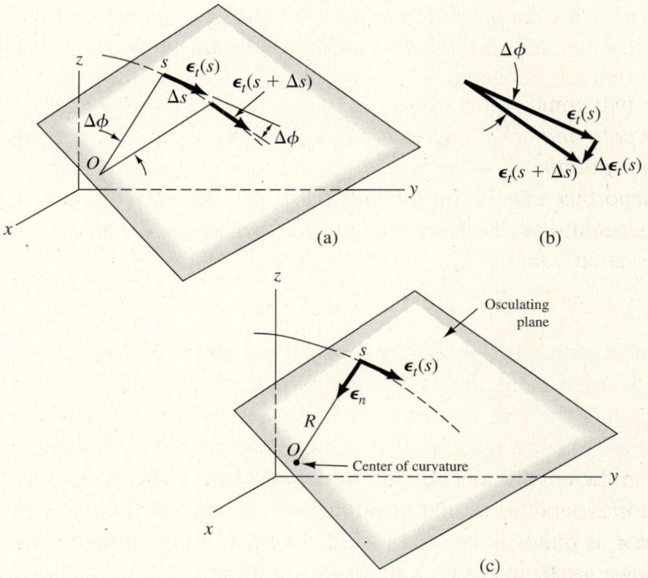

Figure 11.11. Development of the osculating plane and the center of curvature.

With the limiting *direction* of $\Delta\boldsymbol{\epsilon}_t$ established, we next evaluate the *magnitude* of $\Delta\boldsymbol{\epsilon}_t$ as an approximate value that becomes correct as $\Delta s \to 0$. Observing the vector triangle in Fig. 11.11(b), we can accordingly say:

$$|\Delta\boldsymbol{\epsilon}_t| \approx |\boldsymbol{\epsilon}_t|\Delta\phi = \Delta\phi \tag{11.16}$$

Next, we note in Fig. 11.11(a) that the lines from point O to the points s and $s + \Delta s$ along the trajectory form the same angle $\Delta\phi$ as is between the vectors $\boldsymbol{\epsilon}_t(s)$ and $\boldsymbol{\epsilon}_t(s + \Delta s)$ in the vector triangle, and so we can say:

$$\Delta\phi = \frac{\Delta s}{Os} \approx \frac{\Delta s}{R} \tag{11.17}$$

Hence, we have for Eq. 11.16:

$$|\Delta\boldsymbol{\epsilon}_t| \approx \frac{\Delta s}{R} \tag{11.18}$$

We thus have the magnitude of $\Delta\boldsymbol{\epsilon}_t$ established in an approximate manner. Using $\boldsymbol{\epsilon}_n$, the principal normal at s, to approximate the direction of $\Delta\boldsymbol{\epsilon}_t$ we can write

$$\Delta\boldsymbol{\epsilon}_t \approx \frac{\Delta s}{R}\boldsymbol{\epsilon}_n$$

If we use this result in the limiting process of Eq. 11.15 (where it becomes exact), the evaluation of $d\boldsymbol{\epsilon}_t/ds$ becomes

$$\frac{d\boldsymbol{\epsilon}_t}{ds} = \lim_{\Delta s \to 0}\left(\frac{\Delta\boldsymbol{\epsilon}_t}{\Delta s}\right) = \lim_{\Delta s \to 0}\left[\frac{(\Delta s/R)\boldsymbol{\epsilon}_n}{\Delta s}\right] = \frac{\boldsymbol{\epsilon}_n}{R} \tag{11.19}$$

When we substitute Eq. 11.19 into Eq. 11.14, the acceleration vector becomes

$$\boldsymbol{a} = \frac{d^2 s}{dt^2}\boldsymbol{\epsilon}_t + \frac{(ds/dt)^2}{R}\boldsymbol{\epsilon}_n \tag{11.20}$$

We thus have two components of acceleration: *one component in a direction tangent to the path and one component in the osculating plane at right angles to the path and pointing toward the center of curvature*. These components are of great importance in certain problems.

For the special case of a *plane curve*, we learned in analytic geometry that the radius of curvature R is given by the relation

$$R = \frac{\left[1 + \left(\dfrac{dy}{dx}\right)^2\right]^{3/2}}{\left|\dfrac{d^2 y}{dx^2}\right|} \tag{11.21}$$

Furthermore, in the case of a plane curve, the osculating plane at every point clearly must correspond to the plane of the curve, and the computation of unit vectors $\boldsymbol{\epsilon}_n$ and $\boldsymbol{\epsilon}_t$ is quite simple, as will be illustrated in Example 11.5.

How do we get the principal normal vector $\boldsymbol{\epsilon}_n$, the radius of curvature R, and the direction of the osculating plane for a three-dimensional curve? One procedure is to evaluate $\boldsymbol{\epsilon}_t$ as a function of s and then differentiate this vector with respect to s. Accordingly, from Eq. 11.19 we can then determine $\boldsymbol{\epsilon}_n$ as well as R. We establish the direction of the osculating plane by taking the cross product $\boldsymbol{\epsilon}_n \times \boldsymbol{\epsilon}_t$, to get a unit vector normal to the osculating plane. This vector is called the *binormal vector*. This is illustrated in starred problem 11.7.

Example 11.5

A particle is moving along a circular path in the xy plane (Fig. 11.12). When the particle crosses the x axis, it has an acceleration along the path of 1.5 m/s^2 and is moving with the speed of 6 m/s in the negative y direction. What is the total acceleration of the particle?

Clearly, the osculating plane must be the plane of the path. Hence, R is 0.6 m, as is shown in the diagram. We need simply to employ Eq. 11.20 for the desired result. Thus,

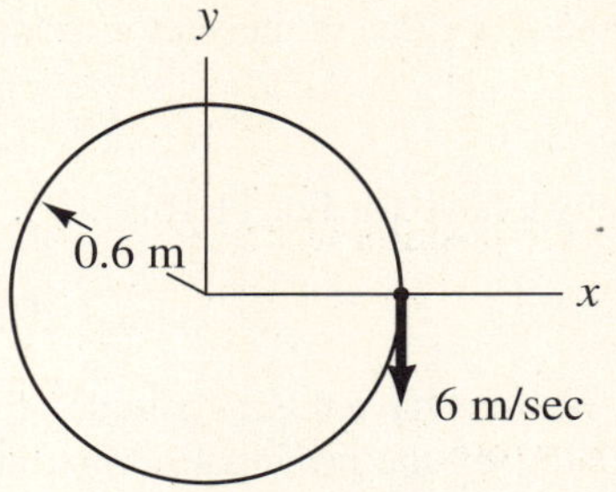

Figure 11.12. Particle on circular path.

$$\boldsymbol{a} = 1.5\boldsymbol{\epsilon}_t + \frac{6^2}{0.6}\boldsymbol{\epsilon}_n \text{ m/s}^2$$

For the xy reference, the acceleration is

$$\boldsymbol{a} = -1.5\boldsymbol{j} - 60\boldsymbol{i} \text{ m/s}^2$$

Example 11.6

A particle is moving in the xy plane along a parabolic path given as $y = 1.22\sqrt{x}$ (see Fig. 11.13) with x and y in metres. At position A, the particle has a speed of 3 m/s and has a rate of change of speed of 3 m/s^2 along the path. What is the acceleration vector of the particle at this position?

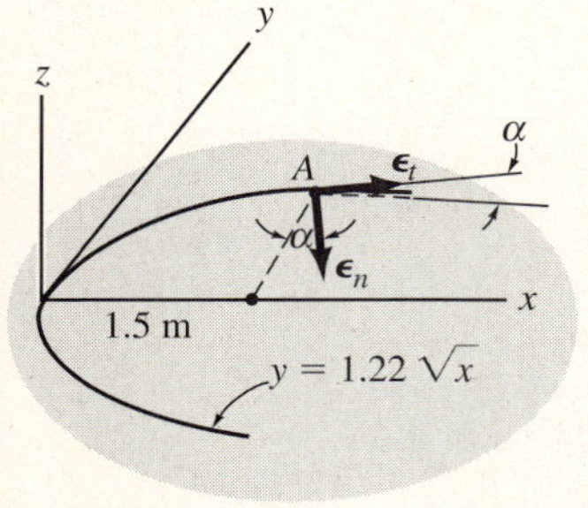

Figure 11.13. Particle on a parabolic path.

We first find $\boldsymbol{\epsilon}_t$ by noting from the diagram that

$$\boldsymbol{\epsilon}_t = \cos\alpha\,\boldsymbol{i} + \sin\alpha\,\boldsymbol{j} \qquad \text{(a)}$$

Example 11.6 (Continued)

where

$$\tan\alpha = \frac{dy}{dx} = \frac{d}{dx}\left(1.22\sqrt{x}\right) = \frac{0.61}{\sqrt{x}} \tag{b}$$

At the position of interest ($x = 1.5$ m) we have

$$\tan\alpha = \frac{0.61}{\sqrt{1.5}} = \frac{1}{2}$$

Therefore,

$$\alpha = 26.5°$$

Hence,

$$\boldsymbol{\epsilon}_t = 0.895\boldsymbol{i} + 0.446\boldsymbol{j} \tag{c}$$

As for $\boldsymbol{\epsilon}_n$, we see from the diagram that

$$\boldsymbol{\epsilon}_n = \sin\alpha\,\boldsymbol{i} - \cos\alpha\,\boldsymbol{j}$$

Therefore,

$$\boldsymbol{\epsilon}_n = 0.446\boldsymbol{i} - 0.895\boldsymbol{j} \tag{d}$$

Next, employing Eq. 11.21, we can find R. We shall need the following results for this step:

$$\frac{dy}{dx} = 0.61x^{-1/2} \tag{e}$$

$$\frac{d^2y}{dx^2} = -0.305x^{-3/2} \tag{f}$$

Substituting Eqs. (e) and (f) into Eq. 11.21, we have for R:

$$R = \frac{\left[1 + \left(0.61x^{-1/2}\right)^2\right]^{3/2}}{0.305x^{-3/2}} \tag{g}$$

At the position of interest, $x = 1.5$, we get

$$R = 8.40 \text{ m} \tag{h}$$

We can now give the desired acceleration vector. Thus, from Eq. 11.20, we have

$$\boldsymbol{a} = 3(.895\boldsymbol{i} + .446\boldsymbol{j}) + \frac{9}{8.40}(.446\boldsymbol{i} - .895\boldsymbol{j}) \tag{i}$$

$$\therefore \quad \boldsymbol{a} = 3.16\boldsymbol{i} + .379\boldsymbol{j} \text{ m/s}^2$$

*Example 11.7

A particle is made to move along a spiral path, as is shown in Fig. 11.14.

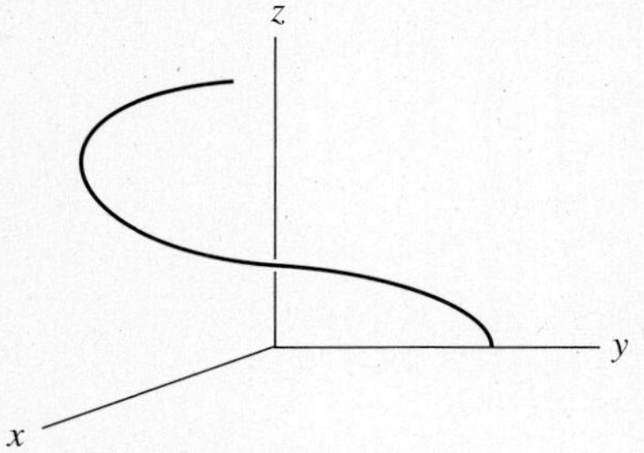

Figure 11.14.

The equations representing the *path* are given parametrically in terms of the variable τ in the following manner:

$$\begin{aligned} x_p &= A \sin \eta\tau \\ y_p &= A \cos \eta\tau \quad (A,\ \eta,\ C \text{ are known constants}) \\ z_p &= C\tau \end{aligned} \tag{a}$$

where the subscript p is to remind the reader that these relations refer to a fixed path. When the particle is at the xy plane ($z = 0$), it has a speed of V_0 m/s and a rate of change of speed of N m/s^2. What is the acceleration of the particle at this position?

To answer this, we must ascertain $\boldsymbol{\epsilon}_t$, $\boldsymbol{\epsilon}_n$, and R. To get $\boldsymbol{\epsilon}_t$ we write:

$$\boldsymbol{\epsilon}_t = \frac{d\boldsymbol{r}_p}{ds} = \frac{dx_p}{ds}\boldsymbol{i} + \frac{dy_p}{ds}\boldsymbol{j} + \frac{dz_p}{ds}\boldsymbol{k} \tag{b}$$

But:

$$\frac{dx_p}{ds} = \frac{dx_p}{d\tau}\frac{d\tau}{ds} \quad \text{and} \quad \frac{dy_p}{ds} = \frac{dy_p}{d\tau}\frac{d\tau}{ds}, \quad \text{etc.}$$

Solving for $dx_p/d\tau$, $dy_p/d\tau$, and $dz_p/d\tau$ from Eq. (a), we can express Eq. (b) as:

$$\boldsymbol{\epsilon}_t = (A\eta \cos \eta\tau \boldsymbol{i} - A\eta \sin \eta\tau \boldsymbol{j} + C\boldsymbol{k})\frac{d\tau}{ds} \tag{c}$$

But:

$$ds = \sqrt{(dx_p)^2 + (dy_p)^2 + (dx_p)^2} \tag{d}$$

Solving for the differentials dx_p, dy_p, and dz_p from Eq. (a) and substituting into Eq. (d), we get:

$$ds = \left[(A\eta \cos \eta\tau)^2 + (A\eta \sin \eta\tau)^2 + C^2\right]^{1/2} d\tau \tag{e}$$

Solving for $d\tau/ds$ from the above equation, we have:

$$\frac{d\tau}{ds} = \frac{1}{\left[(A\eta)^2(\cos^2 \eta\tau + \sin^2 \eta\tau) + C^2\right]^{1/2}} = \frac{1}{(A^2\eta^2 + C^2)^{1/2}} \tag{f}$$

Example 11.7 (Continued)

in which we replaced ($\cos^2 \eta\tau + \sin^2 \eta\tau$) by unity. Returning to Eq. (c), we can thus say:

$$\boldsymbol{\epsilon}_t = \frac{1}{(A^2\eta^2 + C^2)^{1/2}}\left[A\eta(\cos\eta\tau\,\boldsymbol{i} - \sin\eta\tau\,\boldsymbol{j}) + C\boldsymbol{k}\right] \tag{g}$$

To get $\boldsymbol{\epsilon}_n$ and R we employ Eq. 11.19, but in the following manner:

$$\boldsymbol{\epsilon}_n = R\frac{d\boldsymbol{\epsilon}_t}{ds} = R\frac{d\boldsymbol{\epsilon}_t/d\tau}{ds/d\tau} = \frac{R}{(A^2\eta^2 + C^2)^{1/2}}\frac{d\boldsymbol{\epsilon}_t}{d\tau} \tag{h}$$

in which we have replaced $ds/d\tau$ using Eq. (f). We can now employ Eq. (g) to find $d\boldsymbol{\epsilon}_t/d\tau$:

$$\frac{d\boldsymbol{\epsilon}_t}{d\tau} = -\frac{A\eta^2}{(A^2\eta^2 + C^2)^{1/2}}(\sin\eta\tau\,\boldsymbol{i} + \cos\eta\tau\,\boldsymbol{j}) \tag{i}$$

When we substitute this relation for $d\boldsymbol{\epsilon}_t/d\tau$ in Eq. (h), the principal normal vector $\boldsymbol{\epsilon}_n$ becomes:

$$\boldsymbol{\epsilon}_n = -\frac{RA\eta^2}{A^2\eta^2 + C^2}(\sin\eta\tau\,\boldsymbol{i} + \cos\eta\tau\,\boldsymbol{j}) \tag{j}$$

If we take the magnitude of each side, we can solve for R:

$$R = \frac{A^2\eta^2 + C^2}{A\eta^2} \tag{k}$$

We now have $\boldsymbol{\epsilon}_t$ and $\boldsymbol{\epsilon}_n$ at any point of the curve in terms of the parameter τ. As the particle goes through the xy plane, this means that the z coordinate of the position of the particle is zero and z_p of the path corresponding to the position of the particle is zero. When we note the last of Eqs. (a), it is clear, that τ must be zero for this position. Thus $\boldsymbol{\epsilon}_n$ and $\boldsymbol{\epsilon}_t$ for the point of interest are:

$$\boldsymbol{\epsilon}_t = \frac{1}{(A^2\eta^2 + C^2)^{1/2}}(A\eta\,\boldsymbol{i} + C\boldsymbol{k}) \tag{l}$$

$$\boldsymbol{\epsilon}_n = -\frac{RA\eta^2}{A^2\eta^2 + C^2}\boldsymbol{j} = -\boldsymbol{j} \tag{m}$$

where we have used Eq. (k) to replace R in Eq. (m). We can now express the acceleration vector using Eq. 11.20. Thus:

$$\boldsymbol{a} = \frac{N}{(A^2\eta^2 + C^2)^{1/2}}(A\eta\,\boldsymbol{i} + C\boldsymbol{k}) - \frac{V_0^2 A\eta^2}{A^2\eta^2 + C^2}\boldsymbol{j} \tag{n}$$

The direction of the osculating plane can be found by taking the cross product of $\boldsymbol{\epsilon}_t$ and $\boldsymbol{\epsilon}_n$.

PROBLEMS

11.1. A mass is supported by four springs. The mass is given a vibratory movement in the horizontal (x) direction and simultaneously a vibratory movement in the vertical (y) direction. These motions are given as follows:

$$x = 2 \sin 2t \text{ mm}$$
$$y = 2 \cos (2t + 0.3) \text{ mm}$$

What is the value of the acceleration vector at $t = 4$ s? How many g's of acceleration does this correspond to?

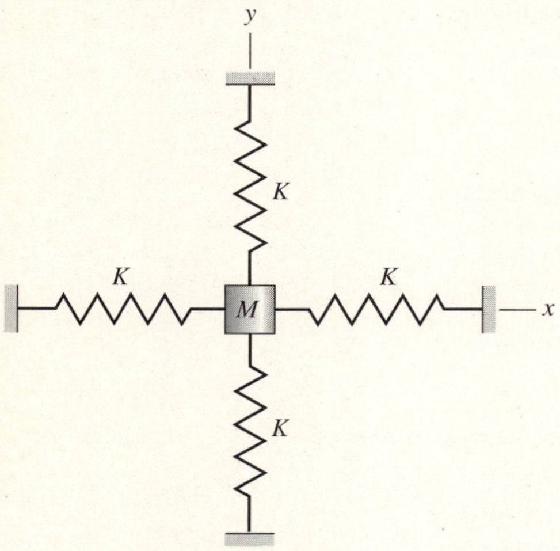

Figure P.11.1.

11.2. A particle moves along a plane circular path of radius r equal to 0.3 m. The position OA is given as a function of time as follows:

$$\theta = 6 \sin 5t \text{ rad}$$

where t is in seconds. What are the rectangular components of velocity for the particle at time $t = 0.2$ s?

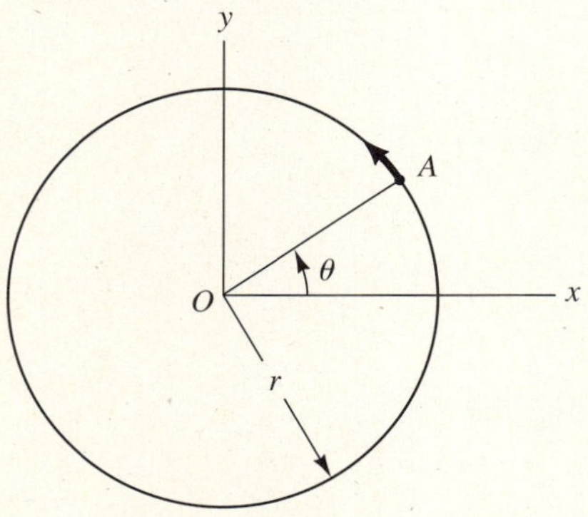

Figure P.11.2.

11.3. A particle with an initial position vector $\boldsymbol{r} = 5\boldsymbol{i} + 6\boldsymbol{j} + \boldsymbol{k}$ m has an acceleration imposed on it, given as

$$\boldsymbol{a} = 6t\boldsymbol{i} + 5t^2\boldsymbol{j} + 10\boldsymbol{k} \text{ m/s}^2$$

If the particle has zero velocity initially, what are the acceleration, velocity, and position of the particle when $t = 10$ s?

11.4. The position of a particle at times $t = 10$ s, $t = 5$ s, and $t = 2$ s is known to be, respectively:

$$\boldsymbol{r}(10) = 10\boldsymbol{i} + 5\boldsymbol{j} - 10\boldsymbol{k} \text{ m}$$
$$\boldsymbol{r}(5) = 3\boldsymbol{i} + 2\boldsymbol{j} + 5\boldsymbol{k} \text{ m}$$
$$\boldsymbol{r}(2) = 8\boldsymbol{i} - 20\boldsymbol{j} + 10\boldsymbol{k} \text{ m}$$

What is the acceleration of the particle at time $t = 5$ s if the acceleration vector has the form

$$\boldsymbol{a} = C_1 t\boldsymbol{i} + C_2 t^2\boldsymbol{j} + C_3 \ln t\boldsymbol{k} \text{ m/s}^2$$

where C_1, C_2, and C_3 are constants and t is in seconds?

11.5. A highly idealized diagram is shown of an *accelerometer*, a device for measuring the acceleration component of motion along a certain direction—in this case the indicated x direction. A mass B is constrained in the accelerometer case so that it can only move against linear springs in the x direction. When the accelerometer case accelerates in this direction, the mass assumes a displaced position, shown dashed, at a distance δ from its original position. This configuration is such that the force in the springs gives the mass B the acceleration corresponding to that of the accelerometer case. The shift δ of the mass in the case is picked up by an electrical sensor device and is plotted as a function of time. The damping fluid present eliminates extraneous oscillations of the mass. If a plot of a_x versus time has the form shown, what is the speed of the body after 10 s, 30 s, and 45 s? The acceleration a_x is measured in g's—i.e., in units of 9.81 m/s^2. Assume that the body starts from rest at $x = 0$.

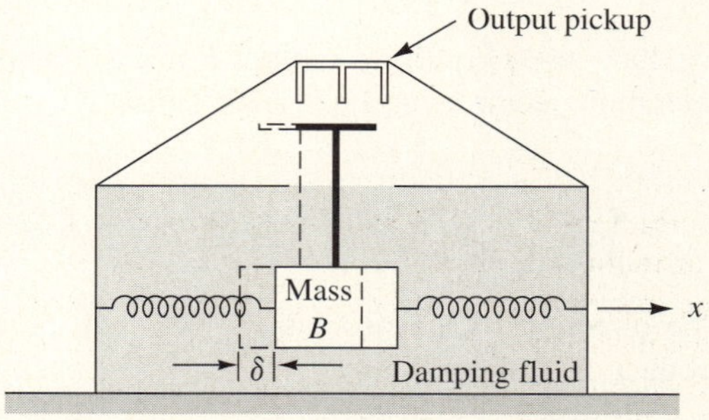

Figure P.11.5 *(Continued)*

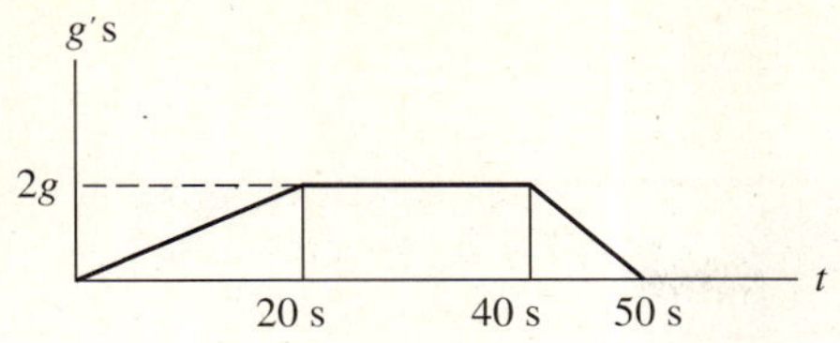

Figure P.11.5.

11.6. The position vector of a particle is given as

$$\boldsymbol{r} = 6t\boldsymbol{i} + (5t + 10)\boldsymbol{j} + 6t^2\boldsymbol{k} \text{ m}$$

What is the acceleration of the particle at $t = 3$ s? What distance has been traveled by the particle during this time? [*Hint:* Let $dr = \sqrt{dx^2 + dy^2 + dz^2}$ and divide and multiply by dt in second half of problem. Look up integration form $\int \sqrt{a^2 + t^2}\,dt$ in Appendix I.]

11.7. In Example 11.1, what is the acceleration vector for pin B if the yoke C is accelerating at the rate of 3 m/s^2 at the instant of interest?

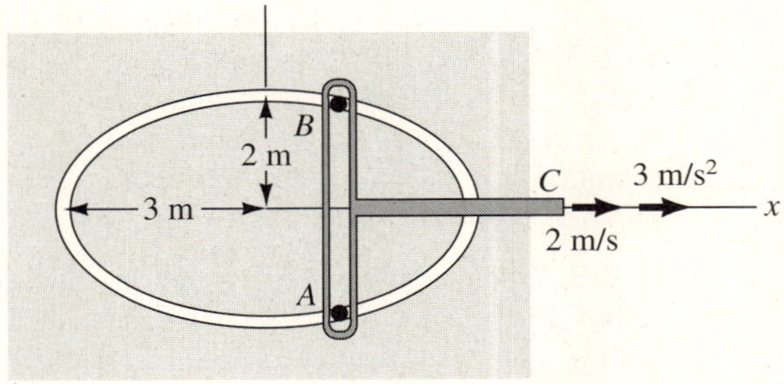

Figure P.11.7.

11.8. Particles A and B are confined to always be in a circular groove of radius 1.5 m. At the same time, these particles must also be in a slot that has the shape of a parabola. The slot is shown dashed at time $t = 0$. If the slot moves to the right at a constant speed of 1 m/s, what are the speed and rate of change of speed of particles toward each other at $t = 1$ s?

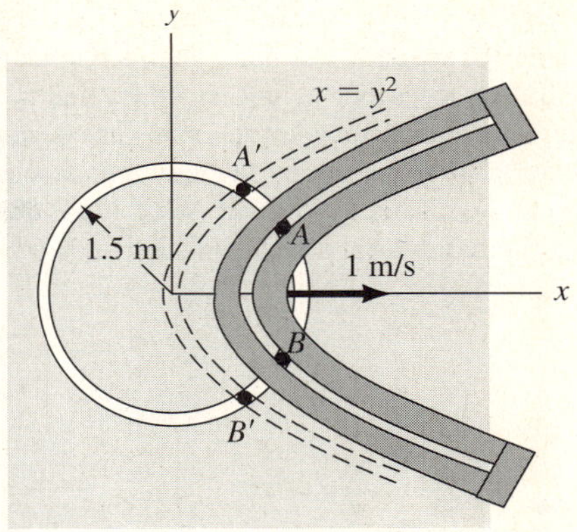

Figure P.11.8.

11.9. The face of a cathode ray tube is shown. An electron is made to move in the horizontal (x) direction due to electric fields in the cathode tube with the following motion:

$$x = A \sin \omega t \text{ mm}$$

Also, the electron is made to move in the vertical direction with the following motion:

$$y = A \sin\ (\omega t + \alpha) \text{ mm}$$

Show that for $\alpha = \pi/2$, the trajectory on the screen is that of a *circle* of radius A mm. If $\alpha = \pi$, show that the trajectory is that of a *straight line* inclined at $-45°$ to the xy axes. Finally, give the formulations for the directions of velocity and acceleration of the electron in the xy plane.

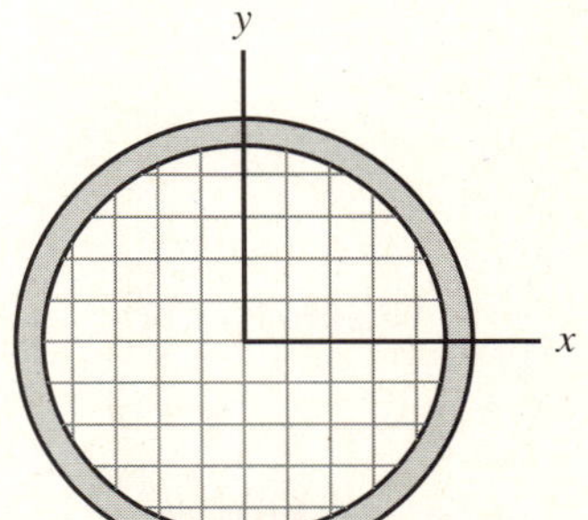

Figure P.11.9.

11.10. A yoke A moves to the right at a speed $V = 2$ m/s and a rate of change of speed $\dot{V} = .6$ m/s^2 when the yoke is at a position $d = 0.27$ m from the y axis. A pin is constrained to move inside a slot in the yoke and is forced by a spring in the slot to slide on a parabolic surface. What are the velocity and acceleration vectors for the pin at the instant of interest? What is the acceleration normal to the parabolic surface at the position shown?

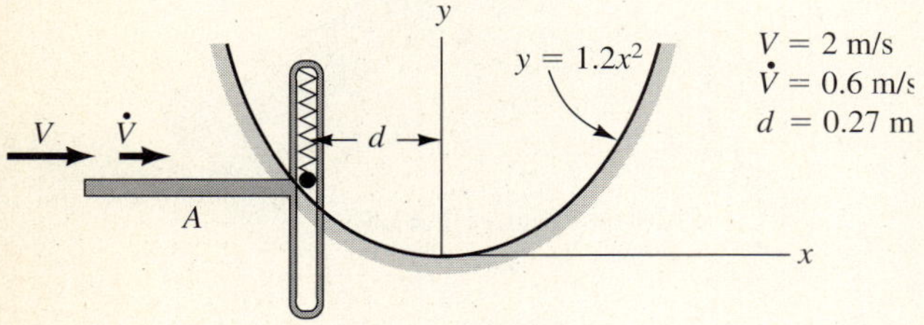

Figure P.11.10.

11.11. A flexible inextensible cord restrains mass M. Both pins A acting on top of the cord move downward at a constant speed V_2 while pin B acting on the bottom of the cord moves upward at a constant speed V_1. The cord is free to slide along the pins without friction. Starting from the horizontal orientation of the cord, what is the velocity of the mass M as a function of time?

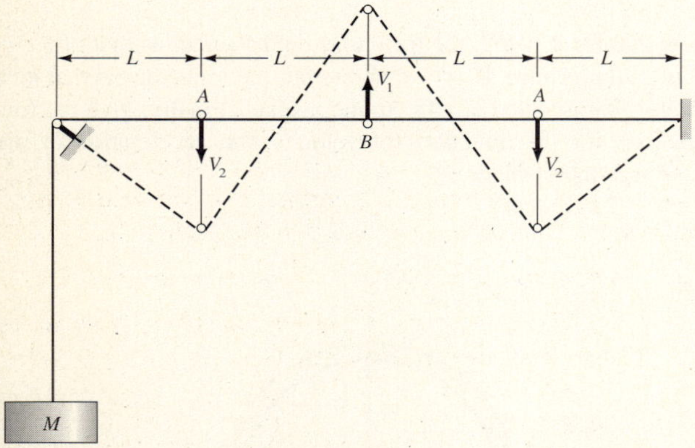

Figure P.11.11.

11.12. Mass M is held by an inextensible cord. What is the velocity of M as a function of α; the time t; the constant velocities V_A and V_C; and the distance h? Disc G is free to turn.

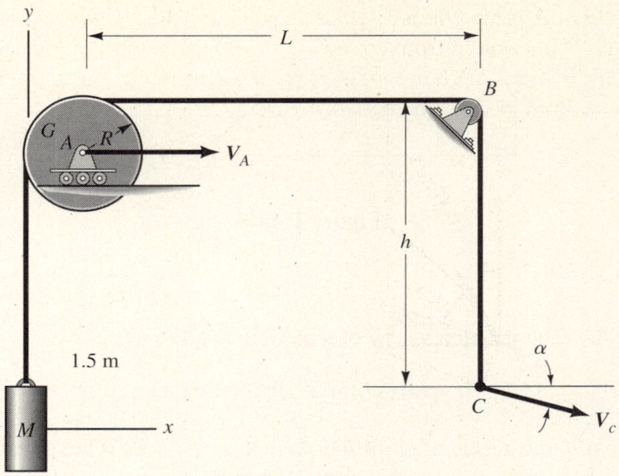

Figure P.11.12.

11.13. A stunt motorcyclist is to attempt a "jump" over a deep chasm. The distance between jump-off point and landing is 100 m. As technical advisor to this stunt man, what minimum speed do you tell him to exceed at the jump-off point A? The cycle is highly streamlined to minimize wind resistance. Give the result in km/hr.

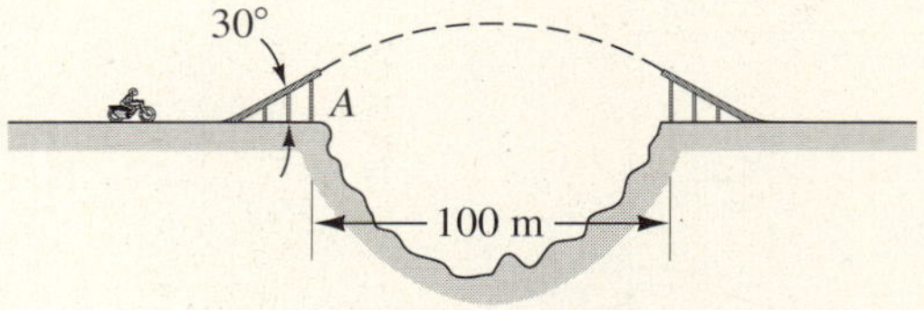

Figure P.11.13.

11.14. A charged particle is shot at time $t = 0$ at an angle of 45° with a speed 3 m/s. If an electric field is such that the body has an acceleration $-60t^2\boldsymbol{j}$ m/s^2, what is the equation for the trajectory? What is the value of d for impact?

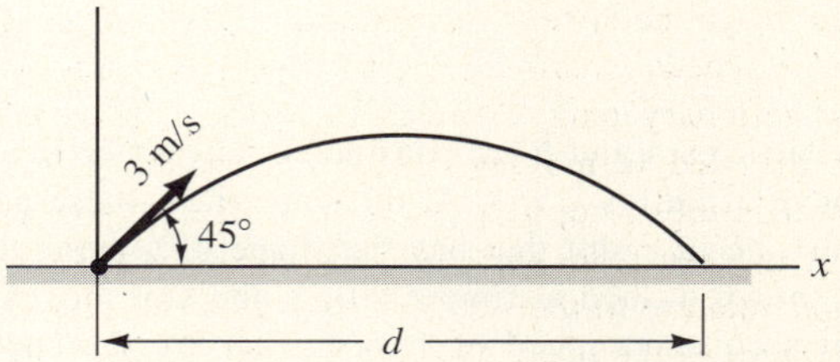

Figure P.11.14.

11.15. A projectile is fired at a speed of 1000 m/s at an angle ϵ of 40° measured from an inclined surface, which is at an angle ϕ of 20° from the horizontal. If we neglect friction, at what distance along the incline does the projectile hit the incline?

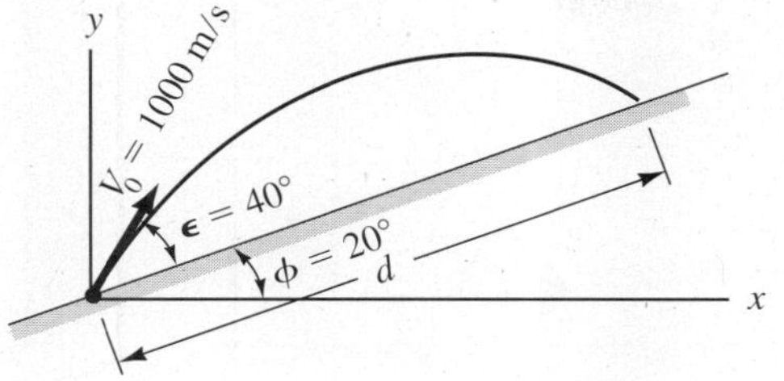

Figure P.11.15.

11.16. Grain is being blown into an open train container at a speed V_0 of 6 m/s. What should the minimum and maximum elevations d be to ensure that all the grain gets into the train? Neglect friction and winds.

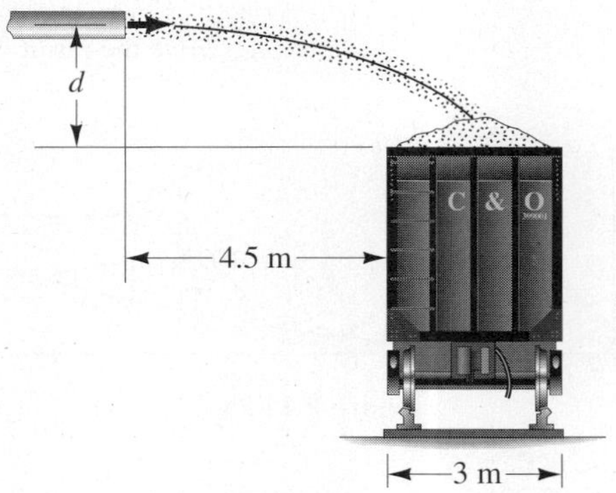

Figure P.11.16.

11.17. A rocket-powered test sled slides over rails. This test sled is used for experimentation on the ability of man to undergo large persistent accelerations. To brake the sled from high speeds, small scoops are lowered to deflect water from a stationary tank of water placed near the end of the run. If the sled is moving at a speed of 100 km/h at the instant of interest, compute h and d of the deflected stream of water as seen from the sled. Assume no loss in speed of the water relative to the scoop. Consider the sled as an inertial reference at the instant of interest and attach xy reference to the sled.

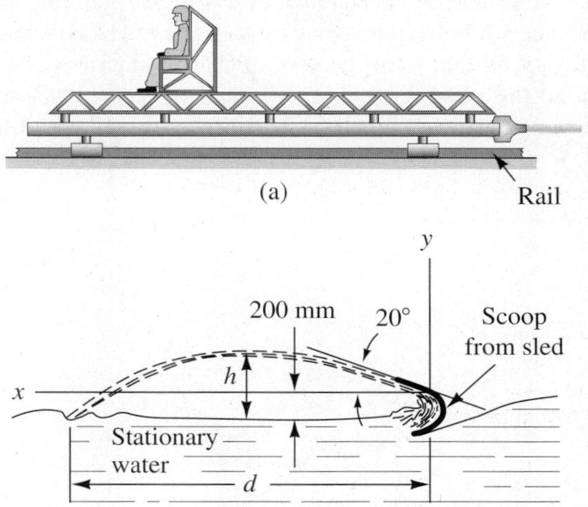

Figure P.11.17.

11.18. In the previous problem, the vane has a velocity given relative to the ground reference XY as

$$V = -5t^2 + 27.8 \text{ m/s}$$

What is the distance δ between the vane and the position of impact of the water that left the vane at time $t = 0$. Use the trajectory of the preceding problem, which relates x and y for a reference xy attached to the vane and moving to the left at $t = 0$ at a speed of 100 km/h = 27.8 m/s. The trajectory of the water after leaving the vane at $t = 0$ is

$$y = -7.2 \times 10^{-3} x^2 + 0.364x$$

with x in meters.

11.19. A fighter-bomber is moving at a constant speed of 500 m/s when it fires its cannon at a target at B. The cannon has a muzzle velocity of 1000 m/s (relative to the gun barrel).

(a) Determine the distance d. Use reference shown.

(b) What is the horizontal distance between the plane and position B at the time of impact?

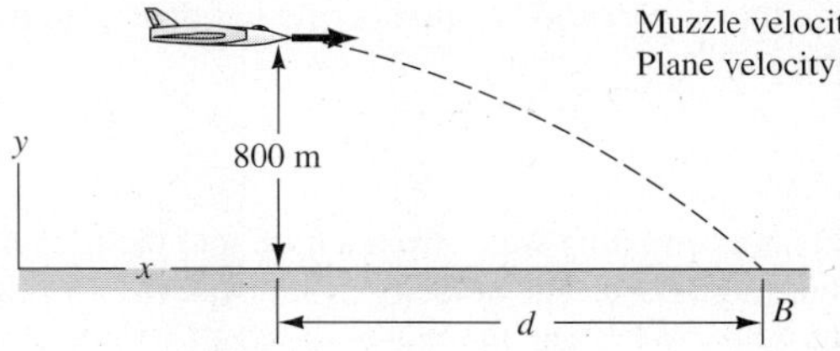

Figure P.11.19.

11.20. A golfer has the bad luck of having his golf ball strike a nearby tree while having a shallow trajectory. The ball bounces off at a speed that is 60 percent of the preimpact speed. If it moves in the same plane as the initial trajectory, compute the distance d at which the ball hits the ground with respect to the tee at A.

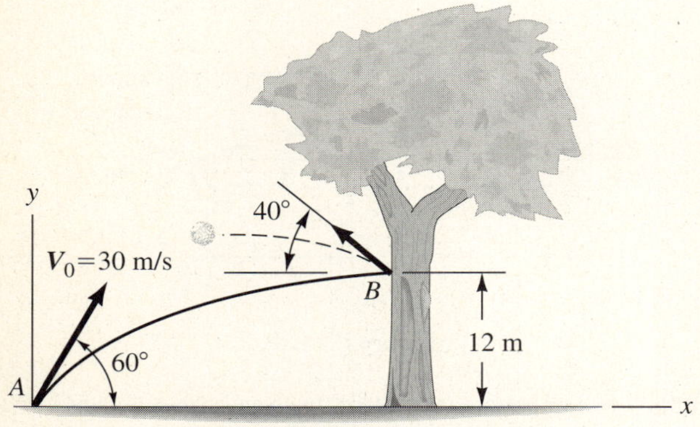

Figure P.11.20.

11.21. What angle α will result in the longest distance d at impact? The muzzle velocity is V_0. The surface is flat.

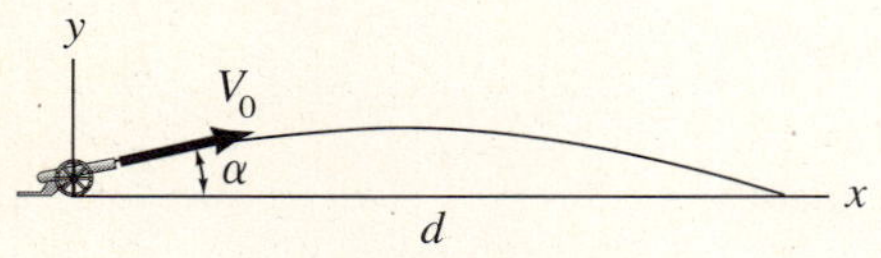

Figure P.11.21.

11.22. A sportsman in a valley is trying to shoot a deer on a hill. He quickly estimates the distance of the deer along his line of sight as 500 m and the height of the hill as 100 m. His gun has a muzzle velocity of 1000 m/s. If he has no graduated sight, how many feet above the deer should he aim his rifle in order to hit it? (Neglect friction.)

11.23. A fireman is directing water from a hose into the broken window of a burning house. The velocity of the water is 15 m/s as it leaves the hose. What are the angles α needed to do the job?

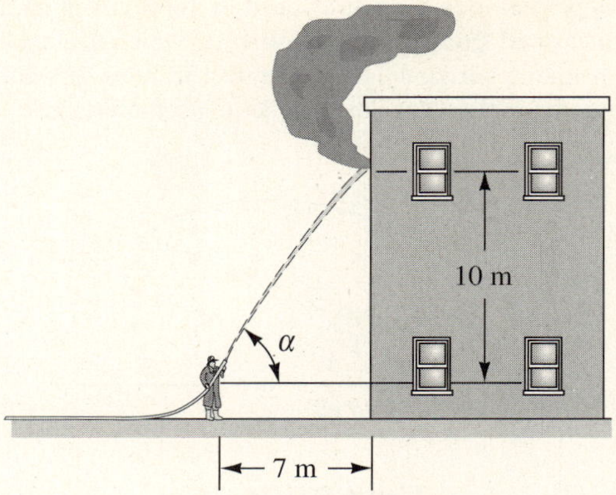

Figure P.11.23.

11.24. A long range gun is shown for which the muzzle velocity is 1000 m/s. If we neglect friction, at what position $\bar{x}$, $\bar{y}$ does the shell hit the ground?

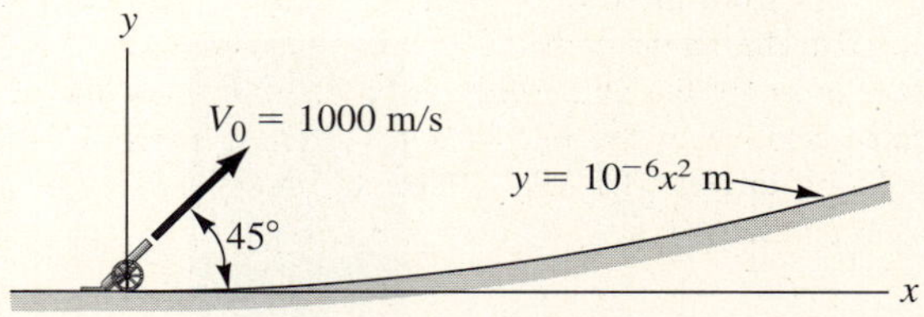

Figure P.11.24.

11.25. An archer in a Jeep is chasing a deer. The Jeep moves at 15 m/s and the deer moves at 7.5 m/s along the same direction. At what inclination must the arrow be shot if the deer is 100 m ahead of the Jeep and if the initial speed of the arrow is 60 m/s relative to archer? (Neglect friction.)

11.26. A fighter plane is directly over an antiaircraft gun at time $t = 0$. The plane has a speed V_1 of 140 m/s. A shell is fired at $t = 0$ in an attempt to hit the plane. If the muzzle velocity V_0 is 1000 m/s, how many metres d should the gun be aimed ahead of the plane to hit it? What is the time of impact?

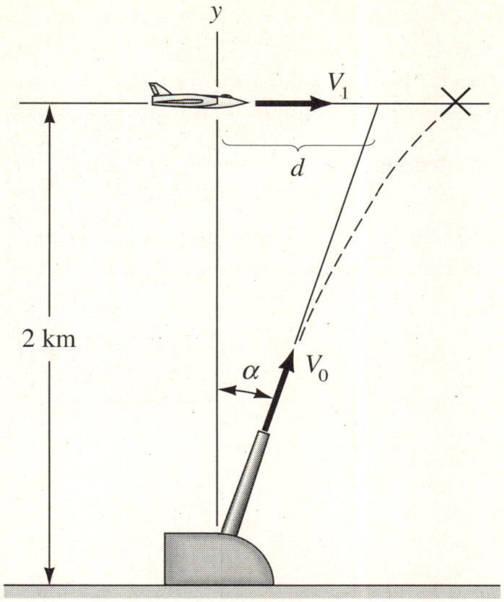

Figure P.11.26.

11.27. A destroyer is making a run at full speed of 20 m/s. When abreast of a missile site target, it fires two shells. The target is 12 km from the destroyer. If the muzzle velocity is 400 m/s, what is the angle of firing α with the horizontal that the computer must set the guns? Also, what angle β must the turret be rotated relative to the line of sight at the instant of firing? [*Hint:* To hit target, what must V_y of the shell be? Result: $\alpha = 23.7°$ and $\beta = 3.26°$.]

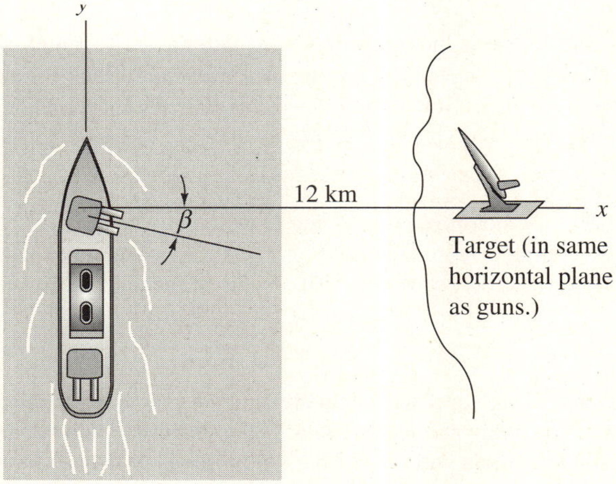

Figure P.11.27.

11.28. In the preceding problem, a second smaller destroyer is firing at the target as shown in the diagram. The data of the preceding problem applies with the following additional data. Due to strong wind and current, the destroyer has a drift velocity of 1.7 m/s in a northeast direction in addition to its full speed of 20 m/s. Form two simultaneous transcendental equations for α and β and verify that $\alpha = 21.39°$ and that $\beta = 10.727°$.

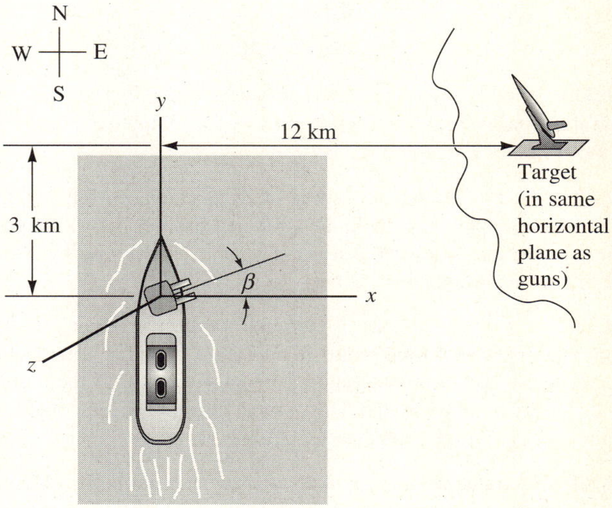

Figure P.11.28.

***11.29.** A Jeep with an archer is moving at a speed of 15 m/s. At 100-m distance and moving at right angles to the Jeep is a deer running at a speed of 7.5 m/s. If the initial speed of the arrow shot by the archer to bag the deer is 66.7 m/s, what inclination α must the shot have with the horizontal and what angle β must the vertical projection of the shot onto the ground have relative to the line AB?

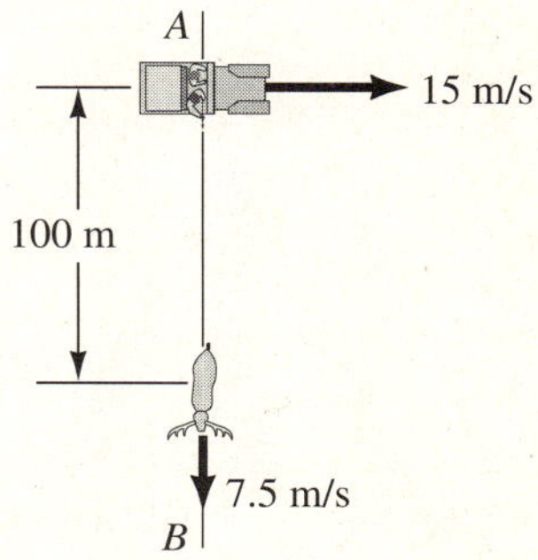

Figure P.11.29.

11.30. A particle moves with a constant speed of 5 m/s along the path. Compute the acceleration at points 1, 2, and 3.

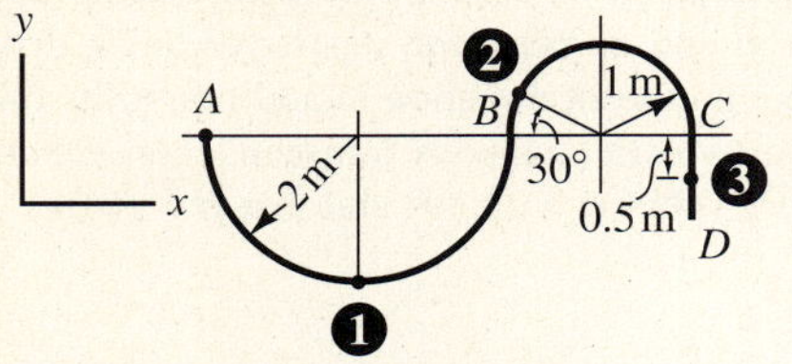

Figure P.11.30.

11.31. If, in Problem 11.30 the speed is 5 m/s only at point A, and it increases 5 m/s for each foot traveled, compute the acceleration at points 1, 2, and 3.

11.32. A car is moving at a speed of 25 m/s along a highway. At a curve in the highway, the radius of curvature is 1.3 km. What is the acceleration of the car? To decrease this acceleration by 30%, what must its speed be?

11.33. A high-speed train is running at 30 m/s. It goes into a curve having a minimum radius of curvature of 2 km. What is the acceleration that sitting passengers are subjected to? If the radius of curvature were to be doubled, at what constant speed could the train then go with the same acceleration?

11.34. An amusement park ride consists of a cockpit in which a passenger is strapped in a seated position. The cockpit rotates about A with angular speed ω. The average person's head is 3 m from the axis of rotation at A. We know that if a person's head is subjected to an acceleration of 3 g's or more in a direction from shoulders to head for any length of time, he/she will be uncomfortable and perhaps black out. What, then, is the maximum value of ω in r/min to prevent these effects, using a safety factor of 3? [*Hint:* You will soon learn that the speed in a circular path is $R\omega$.]

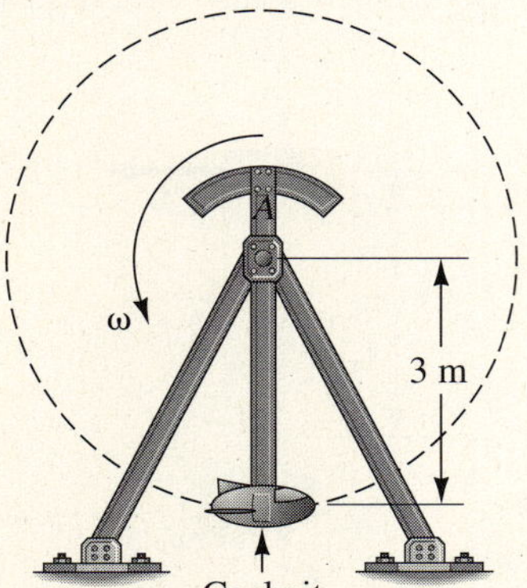

Figure P.11.34.

11.35. A motorcyclist is moving along a circular path having a radius of curvature of 400 m. He is increasing his speed along the path at the rate of 1.4 m/s^2. If he enters the curve at a speed of 14 m/s, what is his total acceleration after traveling for 10 s along this path?

11.36. What is the direction of the normal vector and the value of the radius of curvature at a position a of the curve?

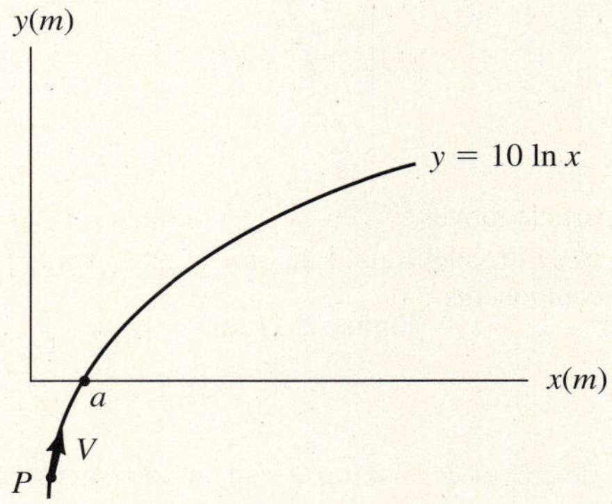

Figure P.11.36.

11.37. A particle P moves with constant speed V along the curve $y = 10 \ln x$ m. At what position x does the particle have the maximum acceleration? What is the value of this acceleration if $V = 1$ m/s?

11.38. A car is moving along a circularly curved road of radius 600 m so as to merge with traffic on a highway. If the car accelerates at a constant rate of 3 m/s^2, what will be the total acceleration of the car when it is going 25 m/s?

11.39. A motorcycle is moving along a circular flat road and is accelerating at a uniform rate of 5 m/s^2. At what speed will the total acceleration be 6 m/s^2? The radius of the path of the motorcycle is 220 m.

11.40. A fighter plane is in a diving maneuver along a trajectory that is approximately a parabola. If the maximum number of g's that the pilot can withstand is 5 g's from shoulder to head from the dynamics of the plane, what is the maximum allowable speed for this maneuver when the plane reaches A?

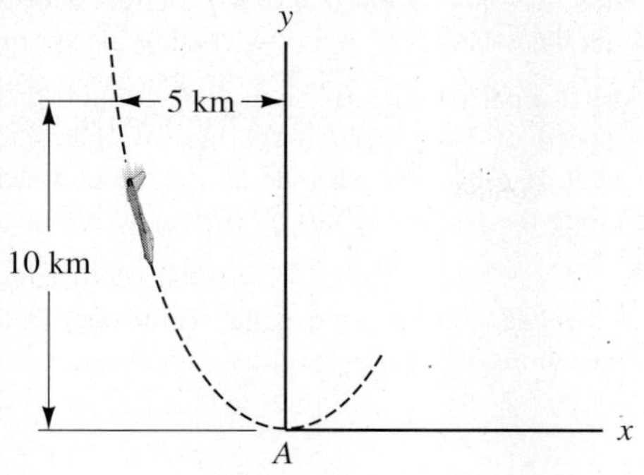

Figure P.11.40.

11.41. A particle moves with a constant speed of 3 m/s along the path. What is the acceleration $\boldsymbol{a}$ at position $x = 1.5$ m? Give the rectangular components of $\boldsymbol{a}$.

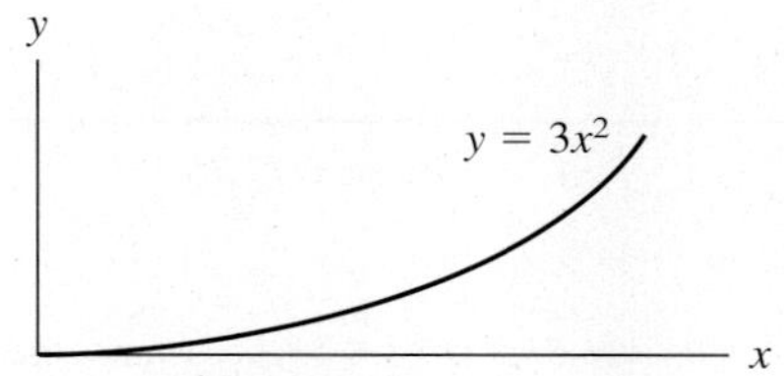

Figure P.11.41.

11.42. A particle moves along a sinusoidal path. If the particle has a speed of 3 m/s and a rate of change of speed of 1.5 m/s² at *A*, what is the *magnitude* of the acceleration? What is the magnitude and direction of the acceleration of the particle at *B*, if it has a speed of 6 m/s and a rate of change of speed of 1 m/s² at this point?

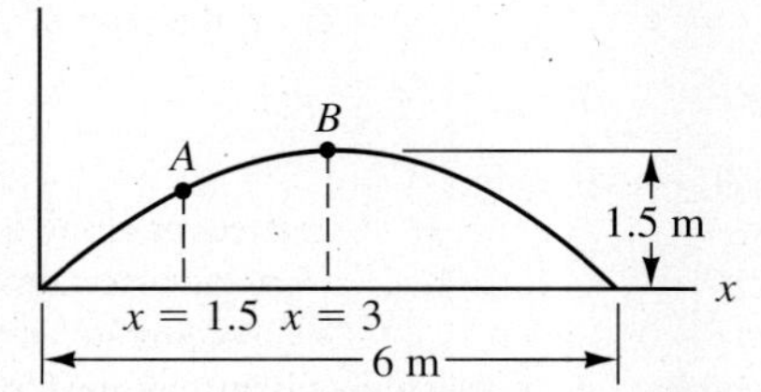

Figure P.11.42.

11.43. A passenger plane is moving at a constant speed of 55 m/s in a holding pattern at a constant elevation. At the instant of interest, the angle β between the velocity vector and the x axis is 30°. The vector is known through on-board gyroscopic instrumentation to be changing at the rate $\dot{\beta}$ of −5°/s. What is the radius of curvature of the path at this instant? $\left[\textit{Hint}: \boldsymbol{a} = \frac{d(V\boldsymbol{\epsilon}_t)}{dt} = \frac{V^2}{R}\boldsymbol{\epsilon}_n.\right]$

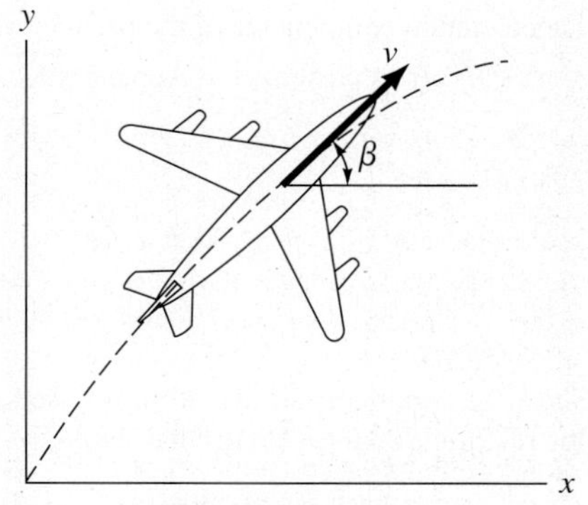

Figure P.11.43.

11.44. At what position along the ellipse shown does the normal vector have a set of direction cosines (0.707, 0.707, 0)? Recall that the equation for an ellipse in the position shown is $x^2/a^2 + y^2/b^2 = 1$.

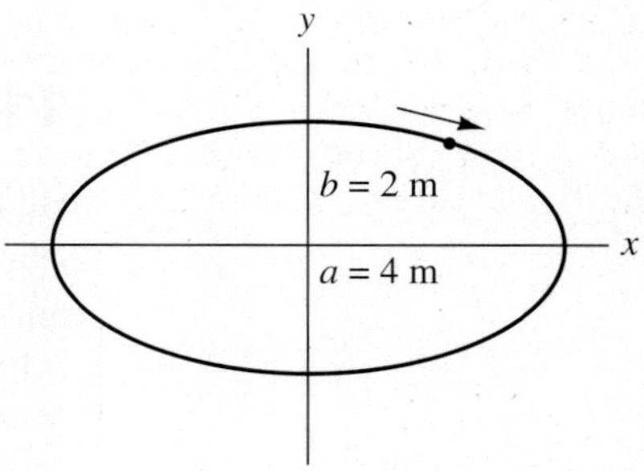

Figure P.11.44.

11.45. A particle moves along a path given as

$$y = 3x^2 \text{ m}$$

The projection of the particle along the x axis varies as $\sqrt{0.2t^2}$ m (where t is in seconds) starting at the origin at $t = 0$. What are the acceleration components normal and tangential to the path at $t =$ 2 s? What is the radius of curvature at this point?

11.46. A particle moves along a path $y^2 = 10x$ with x and y in metres. The distance traversed along this path starting from the origin is given by S such that

$$S = \frac{t}{2} + \frac{t^2}{100} \text{ m}$$

where t is measured in units of seconds. What are the normal and tangential acceleration components of the particle when $y = 10$ m? [*Hint:* $\int \sqrt{y^2 + a^2}\, dy$ is presented in Appendix I.] Also, note that

$$ds = \sqrt{dx^2 + dy^2} = \sqrt{\left(\frac{dx}{dy}\right)^2 + 1}\, dy.$$

11.47. Show by arguments similar to those used in the text for deriving the relation $d\boldsymbol{\epsilon}_t/ds = (1/R)\boldsymbol{\epsilon}_n$ that $d\boldsymbol{\epsilon}_n/ds = -(1/R)\boldsymbol{\epsilon}_t$.

11.48. **(a)** For coplanar paths in the xy plane, find the formula for $\dot{\boldsymbol{a}}$, that is, the "jerk."

(b) If a particle moves on a plane circular path of radius 5 m at a speed of 5 m/s, and if the rate of change of speed is 2 m/s^2, what is $\dot{\boldsymbol{a}}$ for the particle if the second derivative of its speed along the path is 10 m/s^3? [*Hint:* Use the result of Problem 11.47.]

11.49. A beebee gun shoots a pellet as shown. Determine the radius of curvature of the trajectory as a function of x.

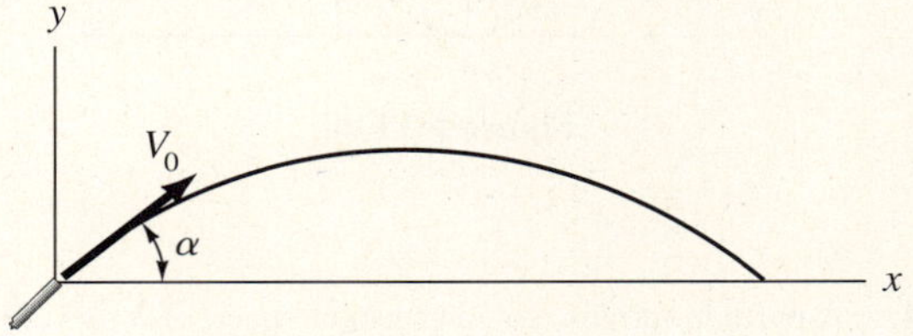

$V_0 = 30$ m/s
$\alpha = 40°$

Figure P.11.49.

11.6 Cylindrical Coordinates

The final method we shall consider for evaluating the velocity and acceleration of a particle brings us back to considering coordinates of the particle as time functions, as we did at the outset of this study. Now we shall employ cylindrical coordinates, and we shall evaluate velocity and acceleration with components having certain directions that are associated with the cylindrical coordinates of the particle. Thus, particle P in Fig. 11.15 is located by specifying cylindrical coordinates θ, $\bar{r}$, and z.[2] The transformation equations between Cartesian and cylindrical coordinates are

$$x = \bar{r}\cos\theta, \qquad \bar{r} = (x^2 + y^2)^{1/2}$$
$$y = \bar{r}\sin\theta, \qquad \theta = \tan^{-1}\frac{y}{x} \tag{11.22}$$

[2]The notation $\bar{r}$ is used to distinguish it from r, which, according to previous definitions in statics, is the magnitude of $\boldsymbol{r}$, the position vector.

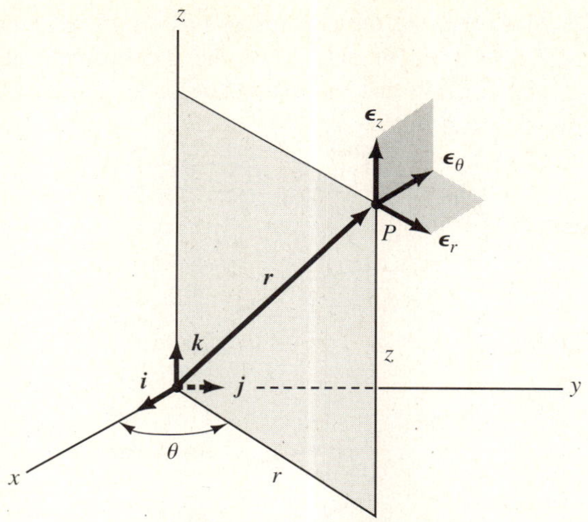

Figure 11.15. Cylindrical coordinates.

Unit vectors are associated with these coordinates and are given as:

$\boldsymbol{\epsilon}_z$, which is parallel to the z axis and, for practical purposes, is the same as $\boldsymbol{k}$. This is considered to be the *axial direction*.

$\boldsymbol{\epsilon}_{\bar{r}}$, which is normal to the z axis, pointing out from the axis, and is identified as the *radial direction* from z.

$\boldsymbol{\epsilon}_\theta$, which is normal to the plane formed by $\boldsymbol{\epsilon}_z$ and $\boldsymbol{\epsilon}_{\bar{r}}$ and has a sense in accordance with the right-hand-screw rule for the permutation z, $\bar{r}$, θ. We call this the *transverse direction*.

Note that $\boldsymbol{\epsilon}_{\bar{r}}$ and $\boldsymbol{\epsilon}_\theta$ will change direction as the particle moves relative to the xyz reference. Thus, these unit vectors are generally *functions of time*, whereas $\boldsymbol{\epsilon}_z$ is a constant vector.

Using previously developed concepts, we can express the velocity and acceleration of the particle relative to the xyz reference *in terms of components always in the transverse, radial, and axial directions and can use cylindrical coordinates exclusively in the process*. This information is most useful, for instance, in turbomachine studies (i.e., for centrifugal pumps, compressors, jet engines, etc.), where, if we take the z axis as the axis of rotation, the axial components of fluid acceleration are used for thrust computation while the transverse components are important for torque considerations. It is these components that are meaningful for such computations and not components parallel to some xyz reference.

The position vector $\boldsymbol{r}$ of the particle determines the direction of the unit vectors $\boldsymbol{\epsilon}_{\bar{r}}$ and $\boldsymbol{\epsilon}_\theta$ at any time t and can be expressed as

$$\boldsymbol{r} = \bar{r}\boldsymbol{\epsilon}_{\bar{r}} + z\boldsymbol{\epsilon}_z \tag{11.23}$$

To get the desired velocity, we differentiate $\boldsymbol{r}$ with respect to time:

$$\frac{d\boldsymbol{r}}{dt} = \boldsymbol{V} = \bar{r}\dot{\boldsymbol{\epsilon}}_{\bar{r}} + \dot{\bar{r}}\boldsymbol{\epsilon}_{\bar{r}} + \dot{z}\boldsymbol{\epsilon}_z$$

Our task here is to evaluate $\dot{\boldsymbol{\epsilon}}_{\bar{r}}$. On consulting Fig. 11.16, we see clearly that changes in direction of $\boldsymbol{\epsilon}_{\bar{r}}$ occur only when the θ coordinate of the particle changes. Hence, remembering that the magnitude of $\boldsymbol{\epsilon}_{\bar{r}}$ is always constant, we have for $\dot{\boldsymbol{\epsilon}}_{\bar{r}}$ using the chain rule:

$$\dot{\boldsymbol{\epsilon}}_{\bar{r}} = \frac{d\boldsymbol{\epsilon}_{\bar{r}}}{dt} = \frac{d\boldsymbol{\epsilon}_{\bar{r}}}{d\theta}\frac{d\theta}{dt} = \frac{d\boldsymbol{\epsilon}_{\bar{r}}}{d\theta}\dot{\theta} \tag{11.24}$$

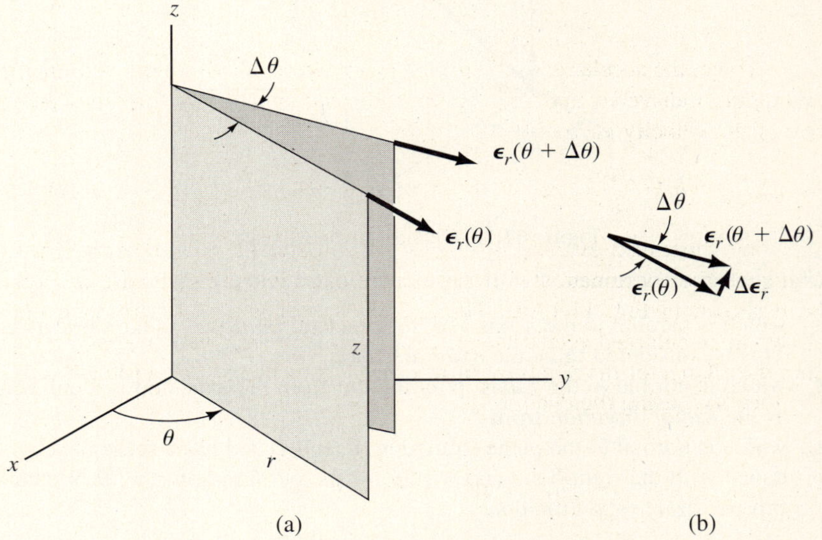

Figure 11.16. Change of unit vector $\boldsymbol{\epsilon}_{\bar{r}}$.

To evaluate $d\boldsymbol{\epsilon}_{\bar{r}}/d\theta$, we have shown in Fig. 11.16(a) the vector $\boldsymbol{\epsilon}_{\bar{r}}$ for a given $\bar{r}$ and z at positions corresponding to θ and $(\theta + \Delta\theta)$. In Fig. 11.16(b), furthermore, we have formed an enlarged vector triangle from these vectors and, in this way, we have shown the vector $\Delta\boldsymbol{\epsilon}_{\bar{r}}$ (i.e., the change in $\boldsymbol{\epsilon}_{\bar{r}}$ during a change in the coordinate θ). From the vector triangle, we see that

$$\left|\Delta\boldsymbol{\epsilon}_{\bar{r}}\right| \approx \left|\boldsymbol{\epsilon}_{\bar{r}}\right|\Delta\theta = \Delta\theta \tag{11.25}$$

Furthermore, as $\Delta\theta \to 0$ we see, on consulting Fig. 11.16, that the *direction* of $\Delta\boldsymbol{\epsilon}_{\bar{r}}$ approaches that of the unit vector $\boldsymbol{\epsilon}_{\theta}$ and so we can approximate $\Delta\boldsymbol{\epsilon}_{\bar{r}}$ as

$$\Delta\boldsymbol{\epsilon}_{\bar{r}} \approx \left|\Delta\boldsymbol{\epsilon}_{\bar{r}}\right|\boldsymbol{\epsilon}_{\theta} \approx \Delta\theta\,\boldsymbol{\epsilon}_{\theta} \tag{11.26}$$

where we have used Eq. 11.25 in the last step. Going back to Eq. 11.24, we utilize the preceding result to write

$$\frac{d\boldsymbol{\epsilon}_{\bar{r}}}{dt} = \left(\frac{d\boldsymbol{\epsilon}_{\bar{r}}}{d\theta}\right)\dot{\theta} \approx \left(\frac{\Delta\boldsymbol{\epsilon}_{\bar{r}}}{\Delta\theta}\right)\dot{\theta} \approx \left(\frac{(\Delta\theta)\boldsymbol{\epsilon}_{\theta}}{\Delta\theta}\right)\dot{\theta} = \dot{\theta}\boldsymbol{\epsilon}_{\theta} \tag{11.27}$$

In the limit, as $\Delta\theta \to 0$, all the previously made approximations become exact statements and we accordingly have

$$\frac{d\boldsymbol{\epsilon}_{\bar{r}}}{dt} = \dot{\theta}\boldsymbol{\epsilon}_{\theta} \tag{11.28}$$

The velocity of particle P is, then,

$$\boldsymbol{V} = \dot{\bar{r}}\boldsymbol{\epsilon}_{\bar{r}} + \bar{r}\dot{\theta}\boldsymbol{\epsilon}_{\theta} + \dot{z}\boldsymbol{\epsilon}_{z} \tag{11.29}$$

To get the acceleration relative to xyz in terms of cylindrical coordinates and radial, transverse, and axial components, we simply take the time derivative of the velocity vector above:

$$\boldsymbol{a} = \frac{d\boldsymbol{V}}{dt} = \ddot{\bar{r}}\boldsymbol{\epsilon}_{\bar{r}} + \dot{\bar{r}}\dot{\boldsymbol{\epsilon}}_{\bar{r}} + \dot{\bar{r}}\dot{\theta}\boldsymbol{\epsilon}_{\theta} + \bar{r}\ddot{\theta}\boldsymbol{\epsilon}_{\theta} + \bar{r}\dot{\theta}\dot{\boldsymbol{\epsilon}}_{\theta} + \ddot{z}\boldsymbol{\epsilon}_{z} \tag{11.30}$$

We must next evaluate $\dot{\boldsymbol{\epsilon}}_{\theta}$. Like $\boldsymbol{\epsilon}_{\bar{r}}$, the vector $\boldsymbol{\epsilon}_{\theta}$ can vary only when a change in the coordinate θ causes a change in direction of this vector, as has been shown in Fig. 11.17(a). The vectors $\boldsymbol{\epsilon}_{\theta}(\theta)$ and $\boldsymbol{\epsilon}_{\theta}(\theta + \Delta\theta)$ have been shown in an enlarged vector triangle in Fig. 11.17(b) and here we have shown $\Delta\boldsymbol{\epsilon}_{\theta}$, the change of the vector $\boldsymbol{\epsilon}_{\theta}$ as a result of the change in coordinate θ. We can then say, using the chain rule,

$$\dot{\boldsymbol{\epsilon}}_{\theta} = \frac{d\boldsymbol{\epsilon}_{\theta}}{dt} = \frac{d\boldsymbol{\epsilon}_{\theta}}{d\theta}\,\dot{\theta} \approx \frac{\Delta\boldsymbol{\epsilon}_{\theta}}{\Delta\theta}\,\dot{\theta} \tag{11.31}$$

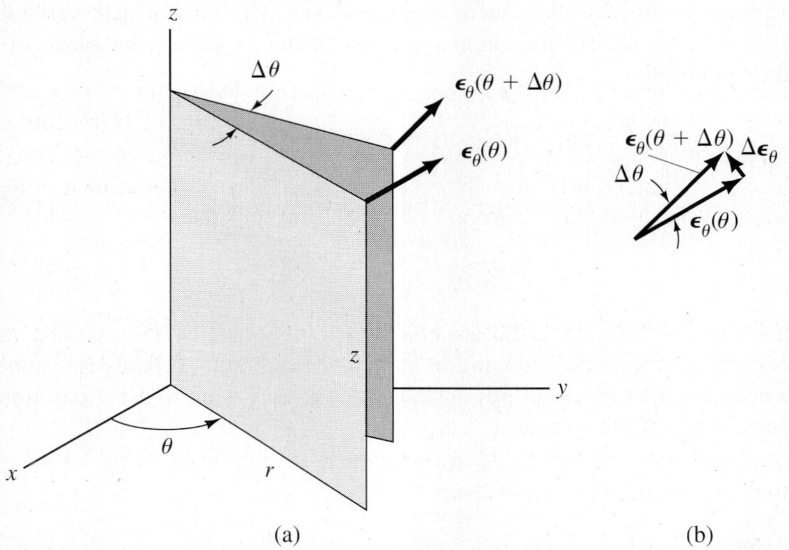

Figure 11.17. Change of unit vector $\boldsymbol{\epsilon}_{\theta}$.

As $\Delta\theta \to 0$, the direction of $\Delta\boldsymbol{\epsilon}_\theta$ becomes that of $-\boldsymbol{\epsilon}_{\bar{r}}$ and the magnitude of $\Delta\boldsymbol{\epsilon}_\theta$, on consulting the vector triangle, clearly approaches $|\boldsymbol{\epsilon}_\theta|\Delta\theta = \Delta\theta$. Thus, the vector $\Delta\boldsymbol{\epsilon}_\theta$ becomes approximately $-\Delta\theta\ \boldsymbol{\epsilon}_{\bar{r}}$. In the limit, we then get for Eq. 11.31:

$$\dot{\boldsymbol{\epsilon}}_\theta = -\dot{\theta}\boldsymbol{\epsilon}_{\bar{r}} \tag{11.32}$$

Using Eqs. 11.28 and 11.32, we find that Eq. 11.30 now becomes

$$\boldsymbol{a} = \ddot{\bar{r}}\boldsymbol{\epsilon}_{\bar{r}} + \dot{\bar{r}}\dot{\theta}\boldsymbol{\epsilon}_\theta + \dot{\bar{r}}\dot{\theta}\boldsymbol{\epsilon}_\theta + \bar{r}\ddot{\theta}\boldsymbol{\epsilon}_\theta - \bar{r}\dot{\theta}^2\boldsymbol{\epsilon}_{\bar{r}} + \ddot{z}\boldsymbol{\epsilon}_z$$

Collecting components, we write

$$\boldsymbol{a} = \left(\ddot{\bar{r}} - \bar{r}\dot{\theta}^2\right)\boldsymbol{\epsilon}_{\bar{r}} + \left(\bar{r}\ddot{\theta} + 2\dot{\bar{r}}\dot{\theta}\right)\boldsymbol{\epsilon}_\theta + \ddot{z}\boldsymbol{\epsilon}_z \tag{11.33}$$

Thus, we have accomplished the desired task. A similar procedure can be followed to reach corresponding formulations for spherical coordinates. By now you should be able to produce the preceding equations readily from the foregoing basic principles.

For motion in a *circle* in the *xy* plane, note that $\dot{\bar{r}} = \dot{z} = 0$, and $\bar{r} = r$. We get the following simplifications:

$$\boldsymbol{V} = r\dot{\theta}\boldsymbol{\epsilon}_\theta \tag{11.34a}$$

$$\boldsymbol{a} = r\ddot{\theta}\boldsymbol{\epsilon}_\theta - r\dot{\theta}^2\boldsymbol{\epsilon}_r \tag{11.34b}$$

Furthermore, the unit vector $\boldsymbol{\epsilon}_\theta$ is tangent to the path, and the unit vector $\boldsymbol{\epsilon}_r$ is normal to the path and points away from the center. Therefore, when we compare Eq. 11.34b with those stemming from considerations of path variables (Section 11.5), clearly for circular motion in the *xy* coordinate plane of a right-hand triad:

$$\left|r\ddot{\theta}\right| = \left|\frac{d^2s}{dt^2}\right| \qquad \left|r\dot{\theta}^2\right| = \left|\frac{V^2}{r}\right| \qquad \begin{cases} \boldsymbol{\epsilon}_\theta = \boldsymbol{\epsilon}_t & \text{(for counterclockwise motion as seen from } +z)^3 \\ \boldsymbol{\epsilon}_\theta = -\boldsymbol{\epsilon}_t & \text{(for clockwise motion as seen from } +z) \\ \boldsymbol{\epsilon}_r = -\boldsymbol{\epsilon}_n & \end{cases} \tag{11.35}$$

Thus, Eqs. 11.34b and 11.20 are equally useful for quickly expressing the acceleration of a particle moving in a circular path. You probably remember these formulas from earlier physics courses and may want to use them in the ensuing work of this chapter.

[3]The sense of $\boldsymbol{\epsilon}_t$ is that of the velocity of the particle, whereas the sense of $\boldsymbol{\epsilon}_\theta$ is determined by the reference *xyz*. For this reason a multiplicity of relations between these unit vectors exists.

Example 11.8

A *towing tank* is a device used for evaluating the drag and stability of ship hulls. Scaled models are moved by a rig along the water at carefully controlled speeds and attitudes while measurements are being made. Usually, the water is contained in a long narrow tank with the rig moving overhead along the length of the tank. However, another useful setup consists of a rotating radial arm (see Fig. 11.18), which gives the model a transverse motion. A radial motion along the arm is another degree of freedom possible for the model in this system.

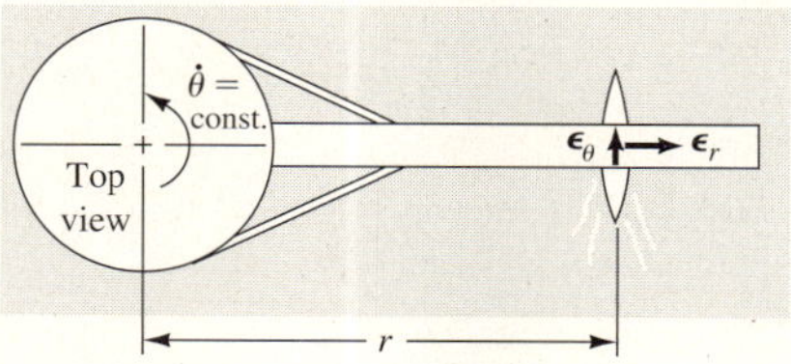

Figure 11.18. Circular towing tank.

Consider the case where a model is being moved out radially so that in one revolution of the main beam it has gone, at constant speed relative to the main beam, from position $\bar{r} = 3.3$ m to $\bar{r} = 4$ m. The angular speed of the beam is 3 r/min. What is the acceleration of the hull model relative to the water when $\bar{r} = 4$ m?

In order to find the radial speed of the model, note that one revolution of the arm corresponds to a time τ evaluated as:

$$\tau = \frac{1}{\frac{3}{60}} = 20 \text{ s}$$

Hence, we can say for $\dot{\bar{r}}$:

$$\dot{\bar{r}} = \frac{4 - 3.3}{\tau} = 35 \text{ mm/s}$$

We can now readily describe the motion of the system at the instant of interest with cylindrical coordinates as follows:

$$\bar{r} = 4 \text{ m}, \qquad \dot{\theta} = 3\left(\frac{2\pi}{60}\right) = 0.314 \text{ rad/s}$$
$$\dot{\bar{r}} = 35 \text{ mm/s}, \qquad \ddot{\theta} = 0$$
$$\ddot{\bar{r}} = 0, \qquad \ddot{z} = 0$$

Using Eq. 11.33, we may now evaluate the acceleration vector,

$$\begin{aligned} \boldsymbol{a} &= \left[0 - (4)(.314)^2\right]\boldsymbol{\epsilon}_{\bar{r}} + \left[0 + (2)(.035)(.314)\right]\boldsymbol{\epsilon}_\theta + [0]\boldsymbol{\epsilon}_z \\ &= -.394\boldsymbol{\epsilon}_{\bar{r}} + .022\boldsymbol{\epsilon}_\theta \text{ m/s}^2 \end{aligned}$$

Finally,

$$|\boldsymbol{a}| = 0.395 \text{ m/s}^2$$

Note that we could have used notation r instead of $\bar{r}$ here. (Why?)

Example 11.9

A firetruck has a telescoping boom holding a firefighter as shown in Fig. 11.19. At time t, the boom is extending at the rate of 6 m/s and increasing its rate of extension at .3 m/s^2. Also at time t, $r = 10$ m and $\beta = 30°$. If a velocity component of the firefighter in the vertical direction is to be 3.3 m/sec at this instant, what should $\dot{\beta}$ be? Also, if at this instant the vertical acceleration of the firefighter is to be 1.7 m/s^2, what should $\ddot{\beta}$ be? Note that the motion of the firefighter is that of joint A.

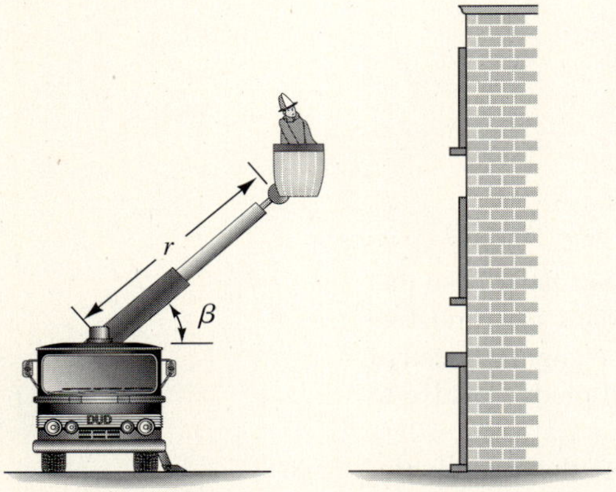

Figure 11.19. A firetruck with a telescoping boom.

We first insert stationary reference xyz at the base of the boom with the boom in the xy plane as shown in Fig. 11.20. Clearly, the boom is rotating

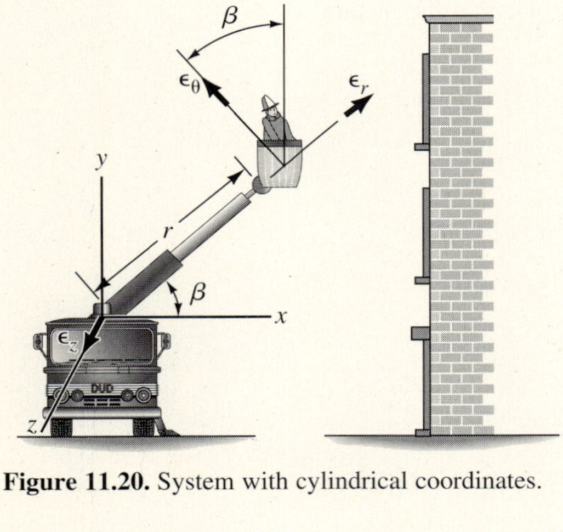

Figure 11.20. System with cylindrical coordinates.

Example 11.9 (Continued)

about the z axis and so this axis is the *axial* direction for the system. Furthermore, the boom being normal to the z axis must then be in the *radial* direction. It should be obvious that there is no motion of the firefighter in the axial direction. We now give the velocity vector as follows, noting that $\theta \equiv \beta$ here.

$$\begin{aligned} \boldsymbol{V} &= \dot{\bar{r}}\boldsymbol{\epsilon}_{\bar{r}} + \bar{r}\dot{\theta}\boldsymbol{\epsilon}_\theta + \dot{z}\boldsymbol{\epsilon}_z \\ &= (0.6)\boldsymbol{\epsilon}_{\bar{r}} + (10)(\dot{\beta})\boldsymbol{\epsilon}_\theta + 0\boldsymbol{\epsilon}_z \text{ m/s} \end{aligned}$$

We require that $\boldsymbol{V} \bullet \boldsymbol{j} = 3.3$ m/s. We thus have for this purpose

$$3.3 = (0.6)(\boldsymbol{\epsilon}_{\bar{r}} \bullet \text{j}) + (10)(\dot{\beta})(\boldsymbol{\epsilon}_\theta \bullet \text{j}) = 0.6 \sin\beta + 10\dot{\beta}\cos\beta$$

where we have used the fact that the dot product between two unit vectors is simply the cosine of the angle between the unit vectors. Noting that at the instant of interest $\beta = 30°$, we can readily solve for $\dot{\beta}$. We get

$$\dot{\beta} = 0.346 \text{ rad/s}$$

Now we write the equation for the acceleration.

$$\begin{aligned} \boldsymbol{a} &= \left(\ddot{\bar{r}} - \bar{r}\dot{\theta}^2\right)\boldsymbol{\epsilon}_{\bar{r}} + \left(\bar{r}\ddot{\theta} + 2\dot{\bar{r}}\dot{\theta}\right)\boldsymbol{\epsilon}_\theta + 0\boldsymbol{\epsilon}_z \\ &= \left(0.3 - (10)(0.346)^2\right)\boldsymbol{\epsilon}_{\bar{r}} + \left(10\ddot{\beta} + (2)(0.6)(0.346)\right)\boldsymbol{\epsilon}_\theta \\ &= -0.900\boldsymbol{\epsilon}_{\bar{r}} + \left(10\ddot{\beta} + 0.416\right)\boldsymbol{\epsilon}_\theta \text{ m/s}^2 \end{aligned}$$

We require that $\boldsymbol{a} \bullet \boldsymbol{j} = 1.7$ m/s^2. Hence

$$\begin{aligned} 1.7 &= -0.900(\boldsymbol{\epsilon}_{\bar{r}} \bullet \text{j}) + \left(10\ddot{\beta} + 0.416\right)(\boldsymbol{\epsilon}_\theta \bullet \text{j}) \\ &= -0.900 \sin 30° + \left(10\ddot{\beta} + 0.416\right)\cos 30° \end{aligned}$$

We then have the following desired information:

$$\ddot{\beta} = 0.207 \text{ rad/s}^2$$

Again, as in the preceding example, we could have used r instead of $\bar{r}$.

PROBLEMS

11.50. A car is moving along a circular track of radius 40 m. The position S along the path is given as

$$S = 3t^2 + \frac{t^3}{6} \text{ m}$$

The time t is given in seconds. What are the angular velocity and angular acceleration of the car at $t = 5$ s?

11.51. A point P fixed on a rotating plate has an acceleration in the x direction of -10 m/s^2. If r for the point is 1 m, what is the angular acceleration of the plate? The angular speed at the instant of interest is 2 rad/s counterclockwise.

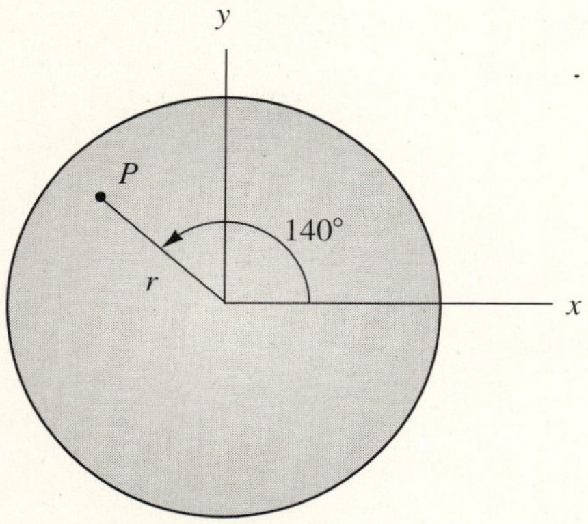

Figure P.11.51.

11.52. A flat disc A with a rubber surface is driven by bevel gears having diameters $D_1 = 200$ mm and $D_2 = 75$ mm. A second rubber disc B of diameter $D_3 = 50$ mm is turned by the friction contact with A. We thus have a *friction drive* system. At the instant of interest, $\omega = 5$ rad/s and $\dot{\omega} = 3$ rad/s^2. If wheel B is moved downward at a speed $V_B = 75$ mm/s at the instant of interest, what is the rotational speed Ω and the rate of change of rotational speed $\dot{\Omega}$ of the small disc B? Slipping between B and A occurs only in the radial direction of disc A. The distance r is 100 mm at the instant of interest.

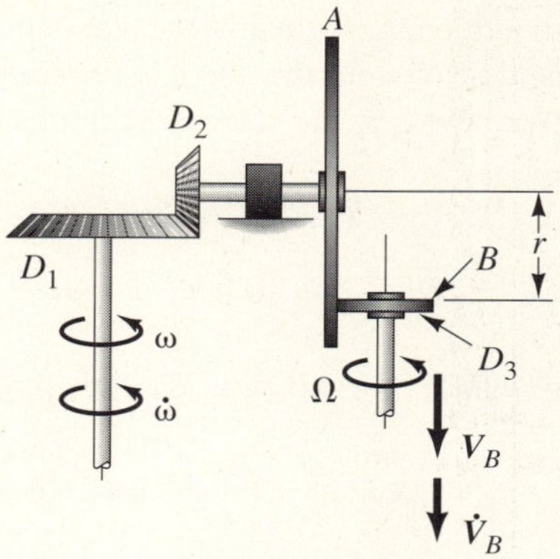

Figure P.11.52.

11.53. What are the velocity and acceleration components in the axial, transverse, and radial directions for a particle moving relative to xyz in the following way:

$$\bar{r} = 10e^{-2t} \text{ m} \qquad \theta = 0.2t \text{ rad} \qquad z = 0.6t \text{ m}$$

with t in seconds. Make a rough sketch of the early portion of the path of the particle.

11.54. A vertical member rotates in accordance with:

$$\omega = 3\sin(0.1t) \text{ rad/s}$$

with t in seconds. Attached to CD is a system of rods HI and FG each of length 200 mm and pinned together at their midpoints K. Also, GA and IA of length 100 mm, are pinned together as shown. At the end of A is a stylus which scribes a curve on plate J. The angle β of the system is given as

$$\beta = 1.3 - \frac{t}{10} \text{ rad}$$

with t in seconds. What are the radial and transverse velocity and acceleration components of the stylus at time $t = 5$ s about axis N–N? (*Note:* Pin F is fixed but pin H moves vertically in a slot as shown.)

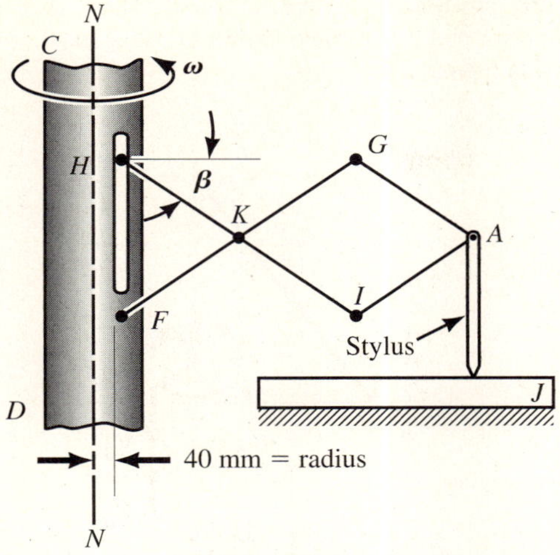

Figure P.11.54.

11.55. A particle moves with a constant speed of 1.5 m/s along a straight line having direction cosines $l = 0.5$, $m = 0.3$. What are the cylindrical coordinates when $|\boldsymbol{r}| = 6$ m? What are the axial and transverse velocities of the particle at this position?

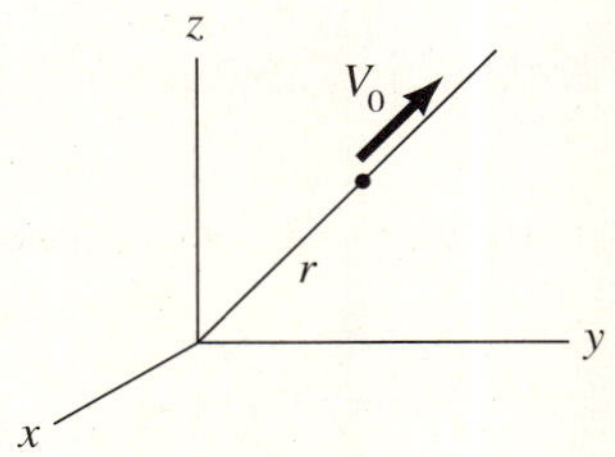

Figure P.11.55.

11.56. A wheel is rotating at time t with an angular speed ω of 5 rad/s. At this instant, the wheel also has a rate of change of angular speed of 2 rad/s². A body B is moving along a spoke at this instant with a speed of 3 m/s relative to the spoke and is increasing in its speed at the rate of 1.6 m/s². These data are given when the spoke, on which B is moving, is vertical and when B is 0.6 m from the center of the wheel, as shown in the diagram. What are the velocity and acceleration of B at this instant relative to the fixed reference xyz?

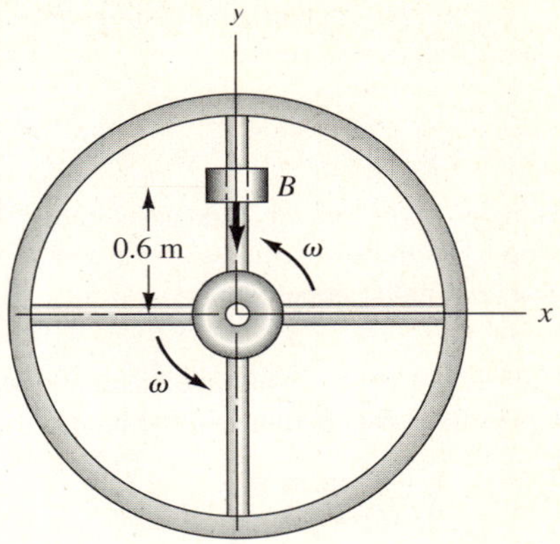

Figure P.11.56.

11.57. A plane is shown in a dive-bombing mission. It has at the instant of interest a speed of 135 m/s and is increasing its speed downward at a rate of 22 m/s². The propeller is rotating at 150 r/min and has a diameter of 4 m. What is the velocity of the tip of the propeller shown at A and its acceleration at the instant of interest? Use cylindrical velocity components.

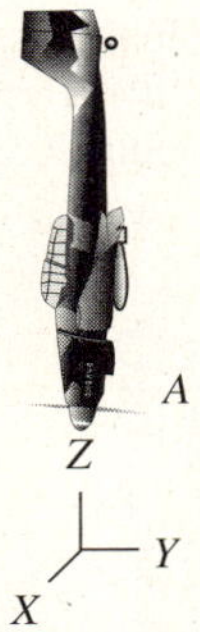

Figure P.11.57.

11.58. The motion of a particle relative to a reference xyz is given as follows:

$$\bar{r} = 0.2 \sinh t \text{ m} \qquad \theta = 0.5 \sin \pi t \text{ rad} \qquad z = 6t^2 \text{ m}$$

with t in seconds. What are the magnitudes of the velocity and acceleration vectors at time $t = 2$ s? Note that sinh 2 = 3.6269 and cosh 2 = 3.7622.

11.59. Given the following cylindrical coordinates for the motion of a particle:

$$\bar{r} = 20 \text{ m} \qquad \theta = 2\pi t \text{ rad} \qquad z = 5t \text{ m}$$

with t in seconds. Sketch the path. What is this curve? Determine the velocity and acceleration vectors.

11.60. A grain of plutonium is being tracked in a turbulent atmosphere. Relative to reference xyz, the displacement components are

$$x = 6t \text{ m} \qquad y = 10t \text{ m} \qquad z = t^3 + 10 \text{ m}$$

Express the position, velocity, and acceleration vectors of the particle using cylindrical coordinates with components in the axial, transverse, and radial directions.

11.61. The motion of a particle in cylindrical coordinates is given by the following parametric equations:

$$\bar{r} = 3 \sin \pi t \text{ m}$$
$$\theta = 6t + 3t^2 \text{ rad}$$
$$z = 5 \cos \pi t + 3 \text{ m} \quad (t \text{ in seconds})$$

Determine the velocity and acceleration of the particle at $t = 0.35$ s.

11.62. A flyball governor has the following data at the instant of interest:

$$\omega = 0.2 \text{ rad/s} \qquad \dot{\omega} = 40 \text{ mrad/s}^2 \qquad \alpha = 45°$$
$$\dot{\alpha} = 5 \text{ rad/s} \qquad \ddot{\alpha} = 0.2 \text{ rad/s}^2$$

If at this instant, the arms are in the xz plane, give the velocity and acceleration vectors of the spheres using cylindrical coordinates for the axial, transverse, and radial directions.

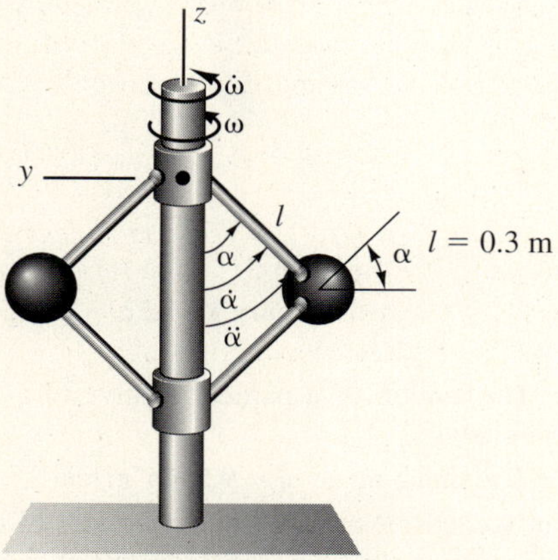

Figure P.11.62.

11.63. A stray tomahawk missile is being tracked. It is moving at a constant speed of 220 m/s along a straight path having direction cosines $l = 0.23$ and $m = 0.64$. When

$$\boldsymbol{r} = 10\boldsymbol{i} + 6\boldsymbol{j} + 8\boldsymbol{k} \text{ km}$$

express the position in cylindrical coordinates. Also, give the velocity vector and the acceleration vector using axial, transverse, and radial components.

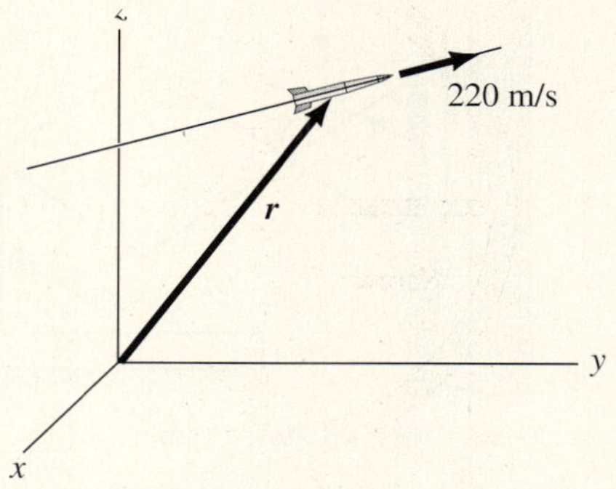

Figure P.11.63.

11.64. A wheel of diameter 600 mm is rotated at a speed of 2 rad/s and is increasing its rotational speed at the rate of 3 rad/s². It advances along a screw having a pitch of 12 mm. What is the acceleration of elements on the rim in terms of cylindrical components?

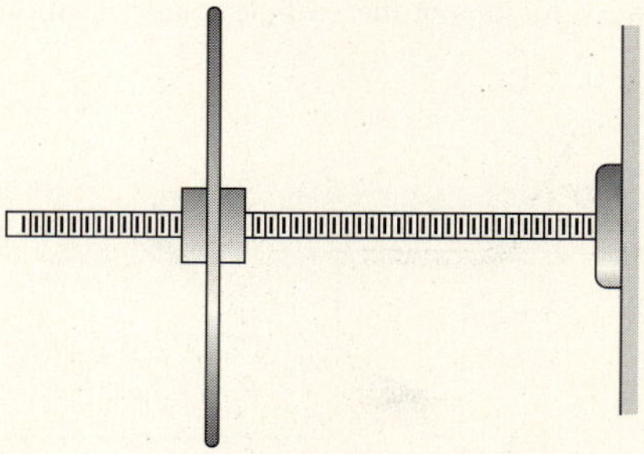

Figure P.11.64.

11.65. A wheel with a threaded hub is rotating at an angular speed of 80 r/min on a right-hand screw having a pitch of 12 mm. At the instant of interest, the rate of change of angular speed is 20 r/min/s. A sleeve A is advancing along a spoke at this instant with a speed of 1.5 m/s and a rate of increase of speed of 1.5 m/s². The sleeve is 0.6 m from the centerline at O at the instant of interest. What are the velocity vector and the acceleration vector of the sleeve at this instant? Use cylindrical coordinates. (See footnote for definition of pitch.)

[4]Pitch is the distance advanced along a screw for one revolution.

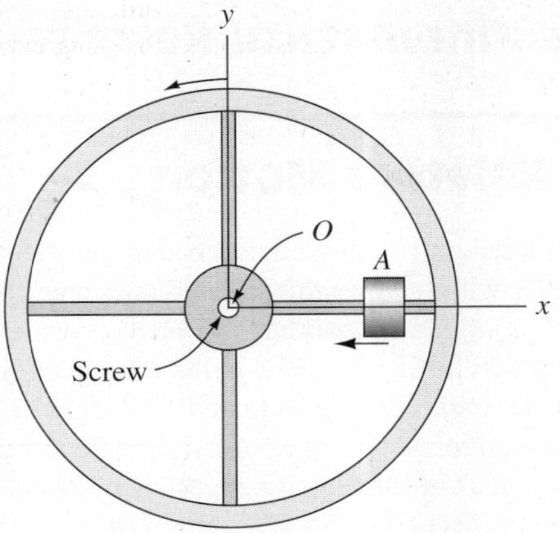

Figure P.11.65.

11.66. A simple garden sprinkler is shown. Water enters at the base and leaves at the end at a speed of 3 m/s as seen from the rotor of the sprinkler. Furthermore, it leaves upward relative to the rotor at an angle of 60° as shown in the diagram. The rotor has an angular speed ω of 2 rad/s. As seen from the ground, what are the axial, transverse, and radial velocity and acceleration components of the water just as it leaves the rotor?

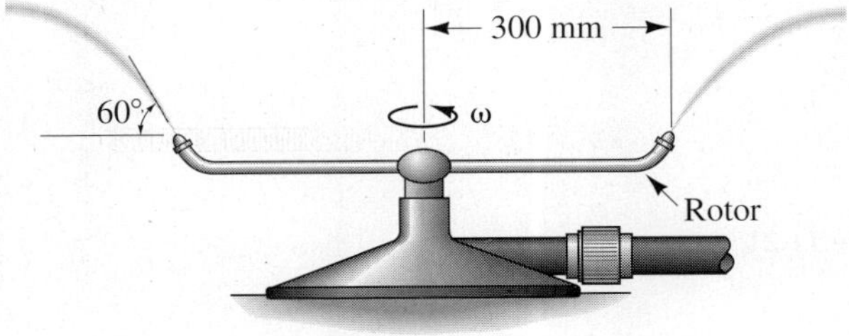

Figure P.11.66.

11.67. The acceleration of gravity on the surface of Mars is 0.385 times the acceleration of gravity on earth. The radius R of Mars is about 0.532 times that of the earth. What is the time of flight of one cycle for a satellite in a circular parking orbit 1.3 Mm from the surface of Mars? [*Note:* $GM = gR^2$.]

11.68. A threaded rod rotates with angular position $\theta = 0.315t^2$ rad. On the rod is a nut which rotates relative to the rod at the rate $\omega = 0.4t$ rad/s. When $t = 0$, the nut is at a distance 0.6 m from A. What is the velocity and acceleration of the nut at $t = 10$ s? The thread has a pitch of 5 mm (see footnote 4). Give results in radial and transverse directions.

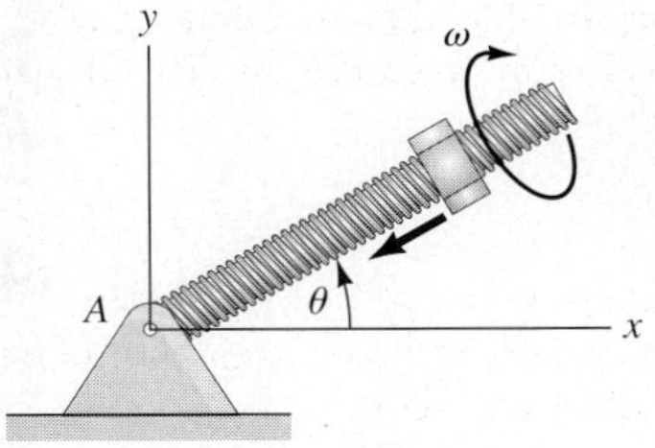

Figure P.11.68.

11.69. Underwater cable is being laid from an ocean-going ship. The cable is unwound from a large spool A at the rear of the ship. The cable must be laid so that is *not dragged* on the ocean bottom. If the ship is moving at a speed of 1.5 m/s, what is the necessary angular speed ω of the spool A when the cable is coming off at a radius of 3.2 m? What is the average rate of change of ω for the spool required for proper operation? The cable has a diameter of 150 mm.

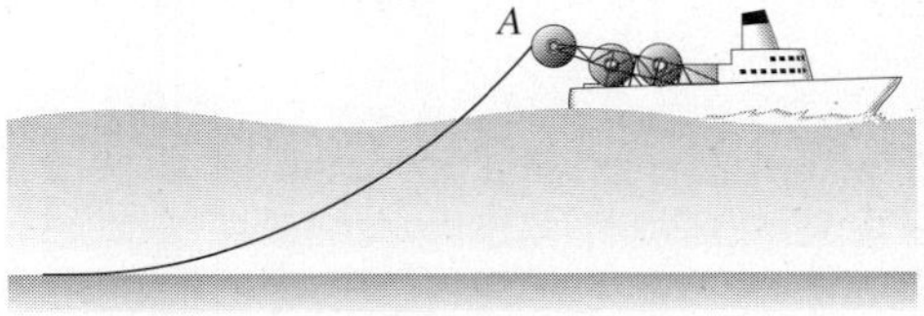

Figure P.11.69.

11.70. A variable diameter drum is rotated by a motor at a constant speed ω of 10 r/min. A rope of diameter d of 12 mm wraps around this drum and pulls up a weight W. It is desired that the velocity of the weight's *upward* movement be given as

$$\dot{X} = 0.12 + \frac{t^2}{26 \times 10^3} \text{ m/s}$$

where for $t = 0$ the rope is just about to start wrapping around the drum at $Z = 0$. What should the radius $\bar{r}$ of the drum be as a function of Z to accomplish this? What are the velocity components $\dot{Y}$ and $\dot{Z}$ of the weight W when $t = 100$ s?

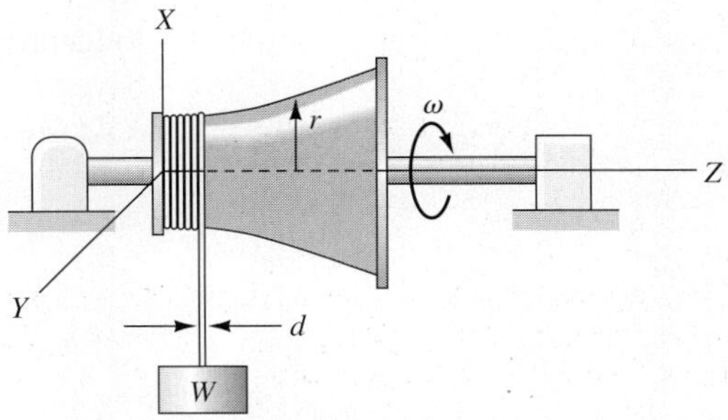

Figure P.11.70.

Part C: Simple Kinematical Relations and Applications

11.7 Simple Relative Motion

Up to now, we have considered only a single reference in our kinematical considerations. There are times when two or more references may be profitably employed in describing the motion of a particle. We shall consider in this section a very simple case that will fulfill our needs in the early portion of the text.

As a first step, consider two references *xyz* and *XYZ* (Fig. 11.21) moving in such a way that the direction of the axes of *xyz* always retain the same orientation relative to *XYZ* such as has been suggested by the dashed references giving successive positions of *xyz*. Such a motion of *xyz* relative to *XYZ* is called *translation*.

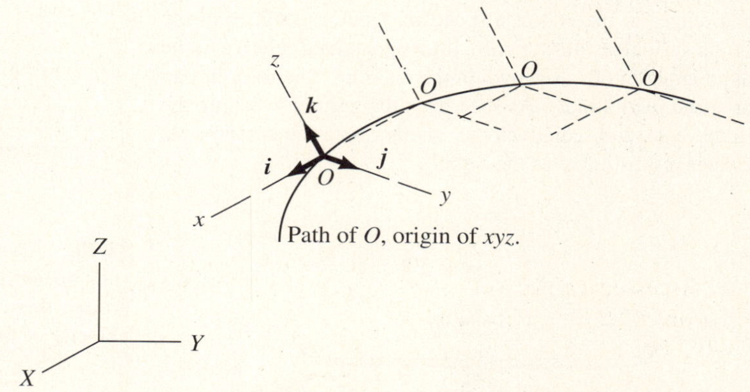

Figure 11.21. Axes *xyz* are translating relative to *XYZ*.

Suppose now that we have a vector $\boldsymbol{A}(t)$ which varies with time. Now in the general case, the time variation of $\boldsymbol{A}$ will depend on from which reference we are observing the time variation. For this reason, we often include subscripts to identify the reference relative to which the time variation is taken. Thus, we have $(d\boldsymbol{A}/dt)_{xyz}$ and $(d\boldsymbol{A}/dt)_{XYZ}$ as time derivatives of $\boldsymbol{A}$ as seen from the *xyz* and *XYZ* axes, respectively. How are these derivatives related for axes *xyz* and *XYZ* that are translating relative to each other? For this purpose, consider $(d\boldsymbol{A}/dt)_{XYZ}$. We will decompose $\boldsymbol{A}$ into components parallel to the *xyz* axes and so we have

$$\left(\frac{d\boldsymbol{A}}{dt}\right)_{XYZ} = \left[\frac{d}{dt}(A_x\boldsymbol{i} + A_y\boldsymbol{j} + A_z\boldsymbol{k}\right]_{XYZ} \tag{11.36}$$

where A_x, A_y, and A_z are the scalar components of $\boldsymbol{A}$ along the *xyz* axes. Because *xyz* translates relative to *XYZ* (see Fig. 11.21), the unit vectors of *xyz*, which we have denoted as $\boldsymbol{i}, \boldsymbol{j}$, and $\boldsymbol{k}$, are *constant vectors* as seen from XYZ. That is, whereas these vectors may change their lines of action, they *do not* change *magnitude* and *direction* as seen from XYZ and are thus constant vectors as seen from XYZ. We then have, for the equation above:

$$\left(\frac{d\boldsymbol{A}}{dt}\right)_{XYZ} = \left(\frac{dA_x}{dt}\right)_{XYZ} \boldsymbol{i} + \left(\frac{dA_y}{dt}\right)_{XYZ} \boldsymbol{j} + \left(\frac{dA_z}{dt}\right)_{XYZ} \boldsymbol{k} \qquad (11.37)$$

But A_x, A_y, and A_z are *scalars* and a time derivative of a scalar, as you may remember from the calculus, is not dependent on a reference of observation.[5] We could readily replace $(dA_x/dt)_{XYZ}$ by $(dA_x/dt)_{xyz}$, etc., with no change in meaning—or we could leave off the subscripts entirely for these terms. Thus, we can say now:

$$\left(\frac{d\boldsymbol{A}}{dt}\right)_{XYZ} = \left(\frac{dA_x}{dt}\right)\boldsymbol{i} + \left(\frac{dA_y}{dt}\right)\boldsymbol{j} + \left(\frac{dA_z}{dt}\right)\boldsymbol{k} \qquad (11.38)$$

Now consider $(d\boldsymbol{A}/dt)_{xyz}$. Again, decomposing $\boldsymbol{A}$ into components along the xyz axes and noting that $\boldsymbol{i}, \boldsymbol{j}$, and $\boldsymbol{k}$ are constant vectors as seen from xyz, we can conclude that

$$\begin{aligned}\left(\frac{d\boldsymbol{A}}{dt}\right)_{xyz} &= \left(\frac{dA_x}{dt}\right)_{xyz} \boldsymbol{i} + \left(\frac{dA_y}{dt}\right)_{xyz} \boldsymbol{j} + \left(\frac{dA_z}{dt}\right)_{xyz} \boldsymbol{k} \\ &= \left(\frac{dA_x}{dt}\right)\boldsymbol{i} + \left(\frac{dA_y}{dt}\right)\boldsymbol{j} + \left(\frac{dA_z}{dt}\right)\boldsymbol{k} \end{aligned} \qquad (11.39)$$

where as discussed earlier we have dropped the xyz subscripts. Observing Eqs. 11.38 and 11.39, we conclude that

$$\left(\frac{d\boldsymbol{A}}{dt}\right)_{XYZ} = \left(\frac{d\boldsymbol{A}}{dt}\right)_{xyz} \qquad (11.40)$$

We can conclude that

$$\left(\frac{d}{dt}\right)_{XYZ} = \left(\frac{d}{dt}\right)_{xyz} \qquad (11.41)$$

That is, the *time derivative of a vector is the same for all reference axes that are translating relative to each other.*

Note in the discussion that the fact that the unit vectors of xyz were *constant* relative to XYZ resulted in the simple relation 11.41. If xyz were *rotating* relative to XYZ, the unit vectors of xyz would not be constant as seen from XYZ and a more complex relationship would exist between $(d\boldsymbol{A}/dt)_{XYZ}$ and $(d\boldsymbol{A}/dt)_{xyz}$. We shall develop this relationship later in the text.

[5]Clearly, the time variation of the temperature $T(x,y,z)$, a scalar, at any position in the classroom does not depend on the motion of an observer in the classroom who might be interested in the temperature at a particular position at a particular time.

11.8 Motion of a Particle Relative to a Pair of Translating Axes

A pair of references *xyz* and *XYZ* are shown now in Fig. 11.22 moving in translation relative to each other. The *velocity vector* of any particle *P* depends on the reference from which the motion is observed. More precisely, we say that the velocity of particle *P* relative to reference *XYZ* is the time rate of change of the position vector $\boldsymbol{r}$ for this reference, where this rate of change is viewed from the *XYZ* reference. This can be stated mathematically as

$$\boldsymbol{V}_{XYZ} = \left(\frac{d\boldsymbol{r}}{dt}\right)_{XYZ} \tag{11.42}$$

Similarly, for the velocity of particle *P* as seen from reference *xyz*, we have

$$\boldsymbol{V}_{xyz} = \left(\frac{d\boldsymbol{\rho}}{dt}\right)_{xyz} \tag{11.43}$$

where we now use position vector $\boldsymbol{\rho}$ for reference *xyz* and view the change from the *xyz* reference (see Fig. 11.22). By the same token, $(d\boldsymbol{R}/dt)_{XYZ}$ is the velocity of the origin of the *xyz* reference as seen from *XYZ*. Since all points of the *xyz* reference have the same velocity relative to *XYZ* at any time *t* for this case (translation of *xyz*), we can say that $(d\boldsymbol{R}/dt)_{XYZ}$ is the velocity of reference *xyz* as seen from *XYZ*.

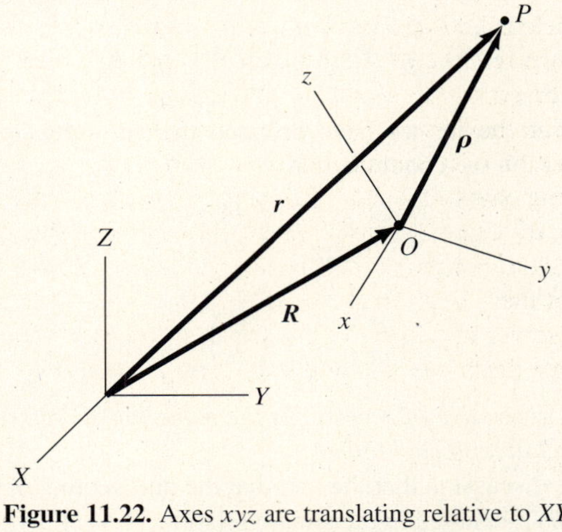

Figure 11.22. Axes *xyz* are translating relative to *XYZ*.

From Fig. 11.22 we can relate position vectors $\boldsymbol{\rho}$ and $\boldsymbol{r}$ by the equation

$$\boldsymbol{r} = \boldsymbol{R} + \boldsymbol{\rho} \tag{11.44}$$

Now take the time rate of change of these vectors as seen from *XYZ*. We get

$$\left(\frac{d\boldsymbol{r}}{dt}\right)_{XYZ} = \left(\frac{d\boldsymbol{R}}{dt}\right)_{XYZ} + \left(\frac{d\boldsymbol{\rho}}{dt}\right)_{XYZ} \tag{11.45}$$

The term on the left side of this equation is $\boldsymbol{V}_{XYZ}$, as indicated earlier, and we shall use the notation $\dot{\boldsymbol{R}}$ for $(d\boldsymbol{R}/dt)_{XYZ}$. We can replace the last term by the derivative $(d\boldsymbol{\rho}/dt)_{xyz}$ in accordance with Eq. 11.41 since the axes are in translation relative to each other. But $(d\boldsymbol{\rho}/dt)_{xyz}$ is simply $\boldsymbol{V}_{xyz}$, the velocity of P relative to xyz. Thus, we have

$$\boldsymbol{V}_{XYZ} = \boldsymbol{V}_{xyz} + \dot{\boldsymbol{R}} \tag{11.46}$$

By the same reasoning, we can show that the acceleration of particle P is related to references XYZ and xyz as follows[6]:

$$\boldsymbol{a}_{XYZ} = \boldsymbol{a}_{xyz} + \ddot{\boldsymbol{R}} \tag{11.47}$$

Equations 11.46 and 11.47 convey the physically simple picture that the motion of a particle relative to XYZ is the sum of the motion of the particle relative to xyz plus the motion of xyz relative to XYZ.

It must be kept clearly in mind that the equations which we have developed apply only to references which have a *translatory* motion relative to each other. In Chapter 15 we shall consider references which have arbitrary motion relative to each other. (Since a reference is a rigid system, we shall need to examine at that time the kinematics of rigid bodies in order to develop these general considerations of relative motion.) The equations presented here will then be special cases.

How can we make use of multiple references? In many problems the motion of a particle is known relative to a given rigid body, and the motion of this body is known relative to the ground or other convenient reference. We can fix a reference xyz to the body, and if the body is in translation relative to the ground, we can then employ the given relations presented in this section to express the motion of the particle relative to the ground.

If, in ensuing chapters, we talk about the "motion of particles relative to a point," such as, for example, the center of mass of the system, then it will be understood that this motion is relative to a *hypothetical reference* moving with the center of mass in a *translatory manner* or, in other words, relative to a nonrotating observer moving with the center of mass.[7]

We illustrate these remarks in the following examples.

[6]As you no doubt will anticipate, the acceleration of a particle as seen from reference XYZ is

$$\boldsymbol{a}_{XYZ} = \left(\frac{d\boldsymbol{V}_{XYZ}}{dt}\right)_{XYZ}$$

Similarly, we have for $\boldsymbol{a}_{xyz}$,

$$\boldsymbol{a}_{xyz} = \left(\frac{d\boldsymbol{V}_{xyz}}{dt}\right)_{xyz}$$

[7]Using a point to convey information about relative motion of a particle only allows you to convey information as to how far or how near the particle is to the point and also as to the speed and rate of change of speed of the particle toward or away from the point. The important information regarding *direction* is entirely left out, requiring a reference frame in order to give this kind of information.

Example 11.10

A jet airliner is shown in Fig. 11.23 flying at a speed of 270 m/s in a translatory manner relative to the ground reference *XYZ*. At the instant of interest, a downdraft causes the plane to accelerate downward at a rate of 22 m/s². While this is happening, the pilot cuts back on the throttle so that the plane is decelerating in the *Y* direction at the rate of 14 m/s². Thus, the plane has an acceleration given as

$$\boldsymbol{a} = -22\boldsymbol{k} - 14\boldsymbol{j} \text{ m/s}^2 \tag{a}$$

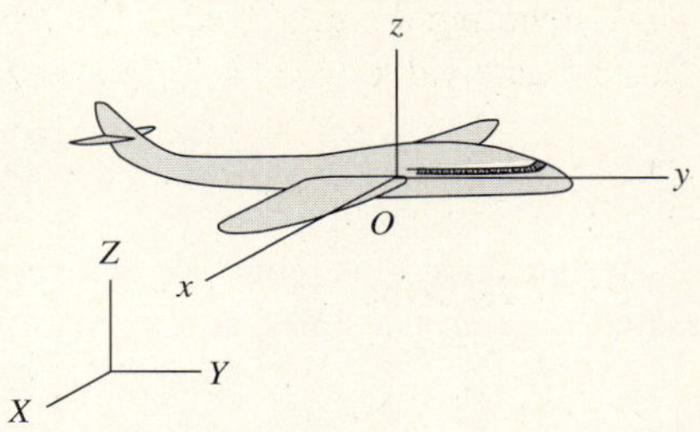

Figure 11.23. Plane translates relative to *XYZ*.

while maintaining a translatory attitude. While this is happening, a solenoid is operated to close a valve gate that weighs 2.2 N. What is the force on the valve gate from the plane at the instant when the valve gate is moving downward relative to the airplane at a speed of 3 m/s and accelerating downward relative to the plane at a rate of 4.9 m/s²?

We must find the acceleration of the valve relative to the ground reference *XYZ*, which may be taken in the problem to be an *inertial reference*. This information will permit us to use the familiar form of Newton's law. It will be convenient in this undertaking to *fix* a reference *xyz*, having the same unit vectors as reference *XYZ*, to the airplane at any convenient location (see Fig. 11.23). Thus

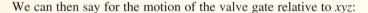

Fix *xyz* to the plane
Fix *XYZ* to the ground

We can then say for the motion of the valve gate relative to *xyz*:

$$\boldsymbol{a}_{xyz} = -4.9\boldsymbol{k} \text{ m/s}^2 \tag{b}$$

The acceleration of *O*, the origin of *xyz* relative to *XYZ*, is

$$\ddot{\boldsymbol{R}} = -22\boldsymbol{k} - 14\boldsymbol{j} \text{ m/s}^2 \tag{c}$$

Since the references are translating relative to each other, we can employ Eq. 11.47 to get $\boldsymbol{a}_{XYZ}$, the acceleration of the valve gate relative to inertial space. Thus,

$$\begin{aligned}\boldsymbol{a}_{XYZ} &= (-22\boldsymbol{k} - 14\boldsymbol{j}) + (-4.9\boldsymbol{k})\\ &= -14\boldsymbol{j} - 26.9\boldsymbol{k} \text{ m/s}^2\end{aligned}$$

We can now employ **Newton's law** in the form

$$\boldsymbol{F} = m\boldsymbol{a}_{XYZ} \tag{d}$$

Example 11.10 (Continued)

Thus, denoting the total force from the airplane as $\boldsymbol{F}_{\text{plane}}$, and remembering that the gate valve weighs 2.2 N, we have

$$\boldsymbol{F}_{\text{plane}} - 2.2\boldsymbol{k} = \frac{2.2}{g}(-14\boldsymbol{j} - 26.9\boldsymbol{k}) \quad \text{(e)}$$

where $-2.2\boldsymbol{k}$ is the force of gravity. Solving for $\boldsymbol{F}_{\text{plane}}$ we get

$$\boldsymbol{F}_{\text{plane}} = -3.140\boldsymbol{j} - 3.833\boldsymbol{k}\ \text{N} \quad \text{(f)}$$

Example 11.11

The freighter in Fig. 11.24 is moving at a steady speed V_1 of 4.5 m/s relative to the water. The freighter is 200 m long at the waterline with point A at midship. A stalking submerged submarine fires a torpedo when the submarine and freighter are at the positions shown in the diagram. The torpedo maintains a steady speed V_2 of 12 m/s relative to the water. Will the torpedo hit the freighter?

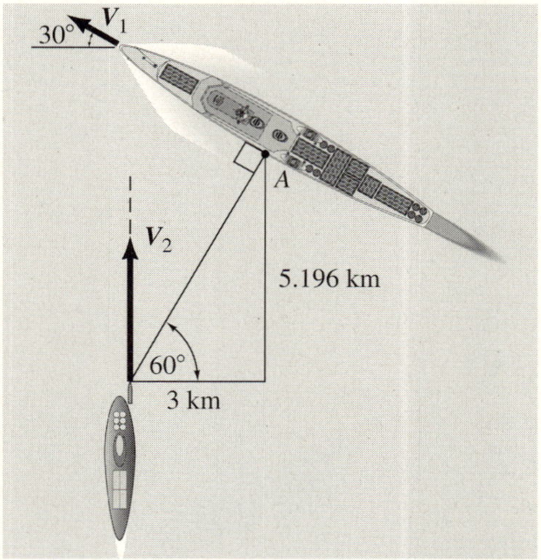

$V_1 = 4.5$ m/s
$V_2 = 12$ m/s
Freighter length = 200 m

Figure 11.24. A torpedo is fired toward a freighter. Does it hit or miss?

A key feature in solving this problem (and others like it) is that we can readily tell whether there is a hit or a miss and, if there is a hit, exactly where this takes place. This is done by simply observing the torpedo from a vantage point of the freighter. The torpedo velocity *relative to the freighter* (i.e., the motion seen by an on-board observer) will point directly to the position of potential contact with the freighter or will indicate a miss.

Example 11.11 (Continued)

We accordingly make the following reference fixes:

Fix xyz to the freighter
Fix XYZ to the water

This is shown in Fig. 11.25. The velocity of xyz, and, hence the freighter, relative to XYZ (i.e., $\dot{\boldsymbol{R}}$) is $(-4.5 \cos 30°\boldsymbol{i} + 4.5 \sin 30°\boldsymbol{j})$ m/s. The velocity of the torpedo relative to XYZ is $12\boldsymbol{j}$ m/s. We can then say

$$\boldsymbol{V}_{XYZ} = \boldsymbol{V}_{xyz} + \dot{\boldsymbol{R}}$$

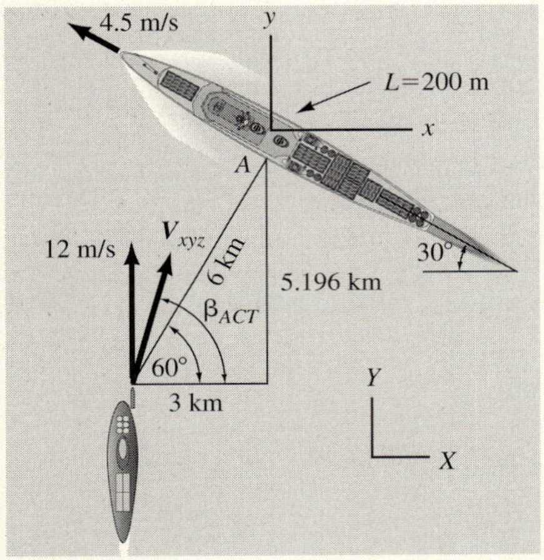

Figure 11.25. Velocity vectors and references for the engagement.

Hence,

$$12\boldsymbol{j} = \boldsymbol{V}_{xyz} - 4.5\cos 30°\boldsymbol{i} + 4.5\sin 30°\boldsymbol{j}$$

$$\therefore \boldsymbol{V}_{xyz} = 3.90\boldsymbol{i} + 9.75\boldsymbol{j} \text{ m/s} \qquad \text{(a)}$$

To just miss the freighter, the velocity vector of the torpedo relative to the freighter, $\boldsymbol{V}_{xyz}$, must have a course such that this vector forms an angle β_0 with the horizontal axis given as (see Fig. 11.26)

$$\beta_0 = \alpha + 60° = \tan^{-1}\frac{100}{6000} + 60° = 60.95° \qquad \text{(b)}$$

Now go back to Eq. (a) to obtain the actual angle, β_{ACT} (see Fig. 11.25), for the actual relative velocity vector $\boldsymbol{V}_{xyz}$.

$$\beta_{ACT} = \tan^{-1}\frac{\left(\boldsymbol{V}_{xyz}\right)_y}{\left(\boldsymbol{V}_{xyz}\right)_x} = \tan^{-1}\frac{9.75}{3.90} = 68.20°$$

Thus we may all relax; the torpedo just misses the freighter since $\beta_{ACT} > \beta_0$.

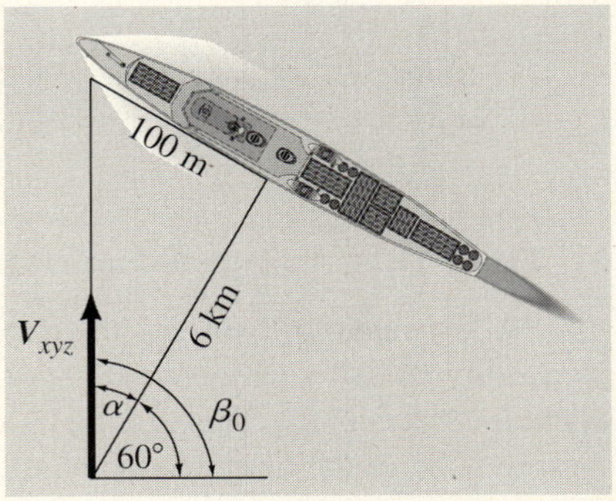

Figure 11.26. Relative velocity vector $\boldsymbol{V}_{xyz}$ for just missing the freighter.

PROBLEMS

11.71. Two wheels rotate about stationary axes each at the same angular velocity, $\dot{\theta} = 5$ rad/s. A particle A moves along the spoke of the larger wheel at the speed V_1 of 1.5 m/s relative to the spoke and at the instant shown is decelerating at the rate of 1 m/s^2 relative to the spoke. What are the velocity and acceleration of particle A as seen by an observer on the hub of the smaller wheel? What are the velocity and acceleration of particle A as seen by an observer on the hub of the smaller wheel if the axis of the larger wheel moves at the instant of interest to the left with a speed of 3 m/s while decelerating at the rate of 0.7 m/s^2? Both wheels maintain equal angular speeds.

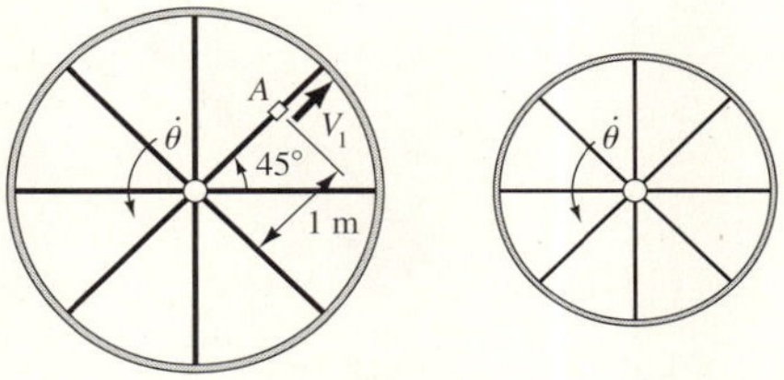

Figure P.11.71.

11.72. Four particles of equal mass undergo coplanar motion in the xy plane with the following velocities:

$$V_1 = 2 \text{ m/s}$$
$$V_2 = 3 \text{ m/s}$$
$$V_3 = 2 \text{ m/s}$$
$$V_4 = 5 \text{ m/s}$$

We showed in Section 8.3 that the velocity of the center of mass can be found as follows:

$$\left(\sum_i m_i\right)\mathbf{V}_c = \sum_i m_i \mathbf{V}_i$$

where $\mathbf{V}_c$ is the velocity of the center of mass. What are the velocities of the particles relative to the center of mass?

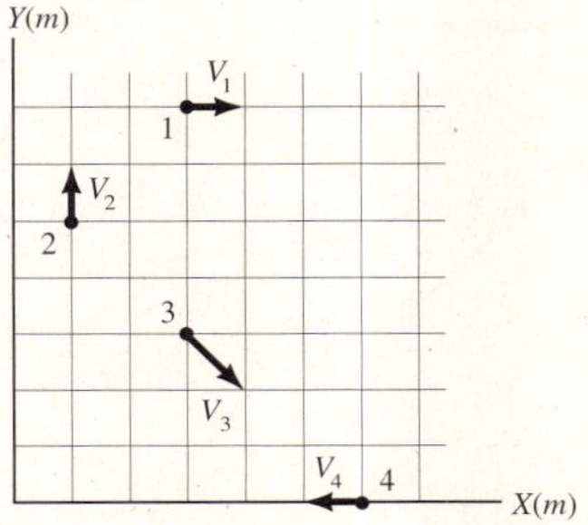

Figure P.11.72.

11.73. A sled, used by researchers to test man's ability to perform during large accelerations over extended periods of time, is powered by a small rocket engine in the rear and slides on lubricated tracks. If the sled is accelerating at $6g$, what force does the man need to exert on a 86 g body to give it an acceleration relative to the sled of

$$9\boldsymbol{i} + 6\boldsymbol{j} \text{ m/s}^2$$

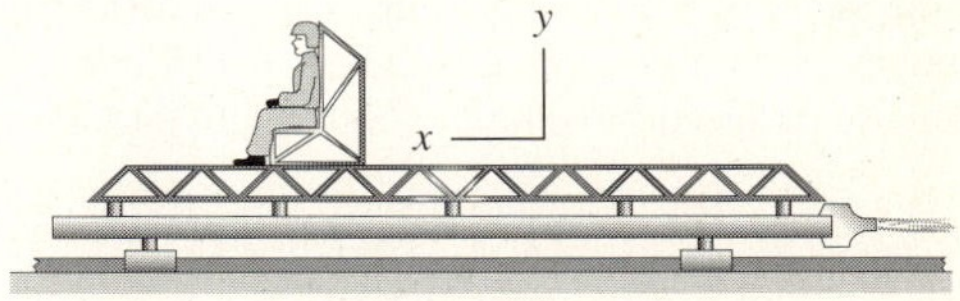

Figure P.11.73.

11.74. On the sled of Problem 11.73 is a device (see the diagram) on which mass M rotates about a horizontal axis at an angular speed ω of 5×10^3 r/min. If the inclination θ of the arm BM is maintained at 30° with the vertical plane C–C, what is the total force on the mass M at the instant it is in its uppermost position? The sled is undergoing an acceleration of $5g$. Take M as having a mass of 0.15 kg.

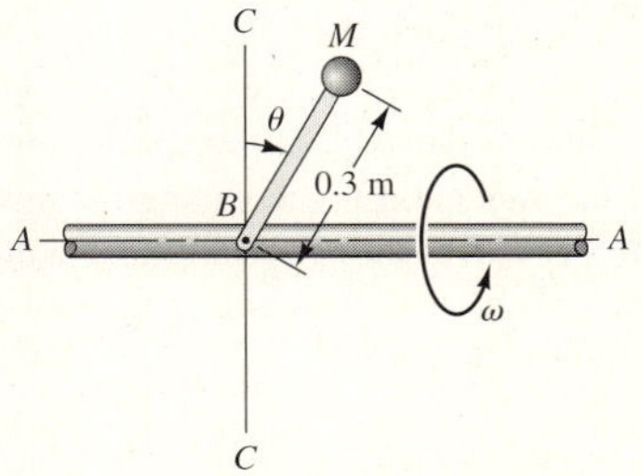

Figure P.11.74.

11.75. A vehicle, wherein a mass M of 0.5 kg rotates with an angular speed ω equal to 5 rad/s, moves with a speed V given as $V = 1.5 \sin \Omega t$ m/s relative to the ground with t in seconds. When $t = 1$ s, the rod AM is in the position shown. At this instant, what is the dynamic force exerted by the mass M along the axis of rod AM if $\Omega = 3$ rad/s?

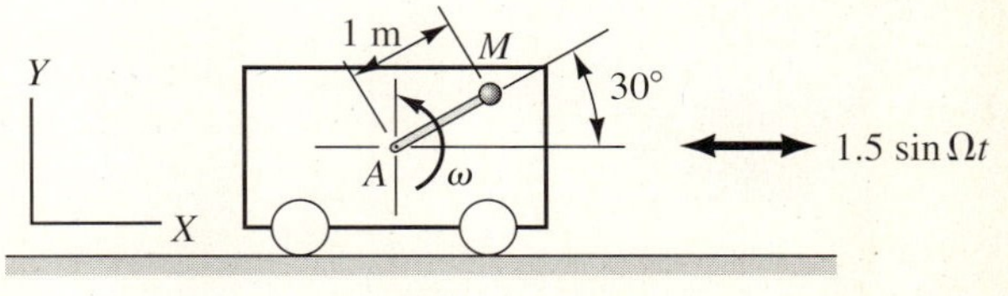

Figure P.11.75.

11.76. In Problem 11.75, what is the frequency of oscillation, Ω, of the vehicle and the value of ω if, at the instant shown, there is a force on the mass M given as

$$\boldsymbol{F} = 110\boldsymbol{i} - 160\boldsymbol{j} \text{ N}$$

11.77. A cockpit C is used to carry a worker for service work on road lighting systems. The cockpit is moved always in a translatory manner relative to the ground. If the angular speed ω of arm AB is 16 mrad/s when $\theta = 30°$, what are the velocity and acceleration of any point in the cockpit body relative to the truck? At this instant, what are the velocity and acceleration, relative to the truck, of a particle moving with a horizontal speed V of 0.15 m/s and with a rate of increase of speed of 6 mm/s², both relative to the cockpit?

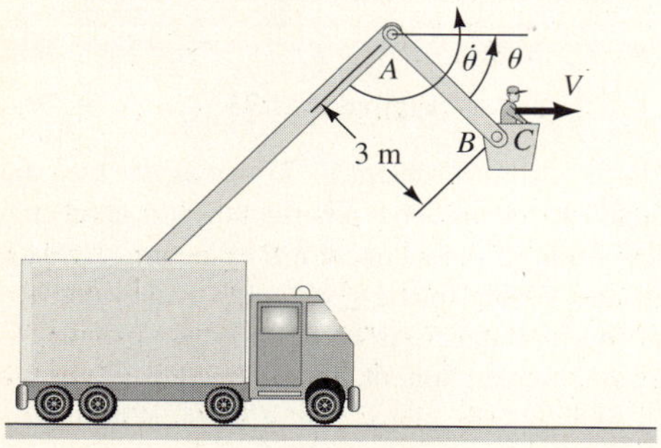

Figure P.11.77.

11.78. A Ferris wheel rotates at the instant of interest with an angular speed $\dot{\theta}$ of 0.5 rad/s and is increasing its angular speed at the rate of 0.1 rad/s². A ball is thrown from the ground to an occupant at A. The ball arrives at the instant of interest with a speed relative to the ground given as

$$\boldsymbol{V}_{XYZ} = -3\boldsymbol{j} - 0.6\boldsymbol{k} \text{ m/s}$$

What are the velocity and the acceleration of the ball relative to the occupant at seat A provided that this seat is not "swinging?" The radius of the wheel is 6 m.

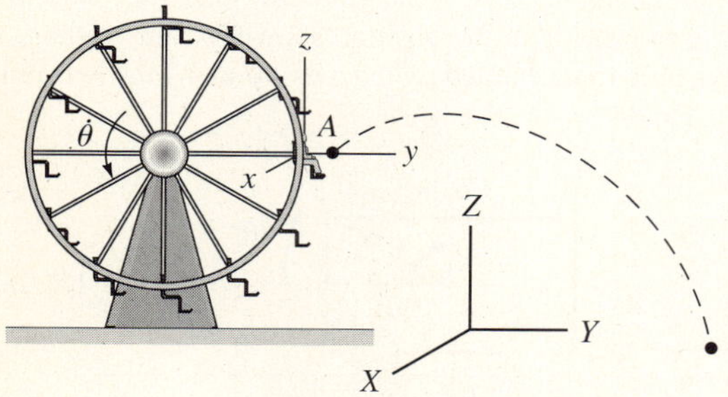

Figure P.11.78.

11.79. A rocket moves at a speed of 700 m/s and accelerates at a rate of $5g$ relative to the ground reference XYZ. The products of combustion at A leave the rocket at a speed of 1.7 km/s relative to the rocket and are accelerating at the rate of 30 m/s² relative to the rocket. What are the speed and acceleration of an element of the combustion products as seen from the ground? The rocket moves along a straight-line path whose direction cosines for the XYZ reference are $l = 0.6$ and $m = 0.6$.

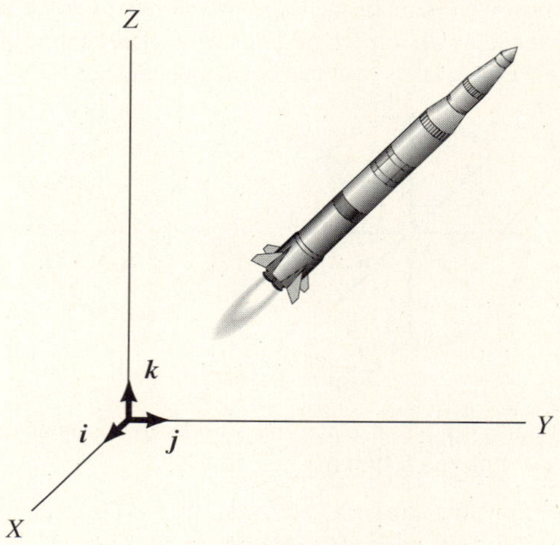

Figure P.11.79.

11.80. Two 10 m international-class sailboats are racing. They are on different tacks. If there is no change in course, will A hit B? If so, where from the center of B does this happen?

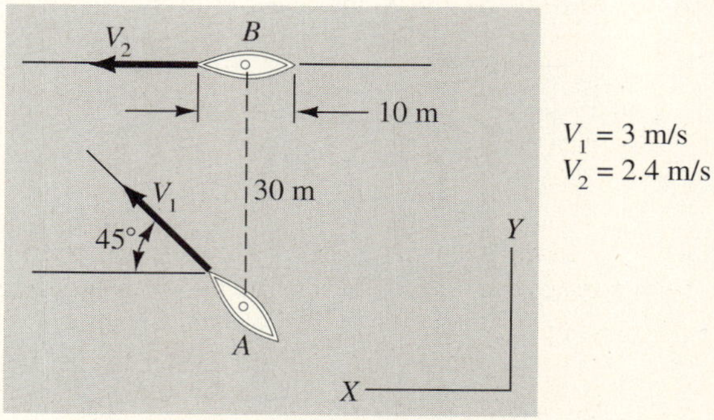

Figure P.11.80.

11.81. A train is moving at a speed of 2.8 m/s. What speed should car A have to just barely miss the front of the train? How long does it take to reach this position? Use a multireference approach only.

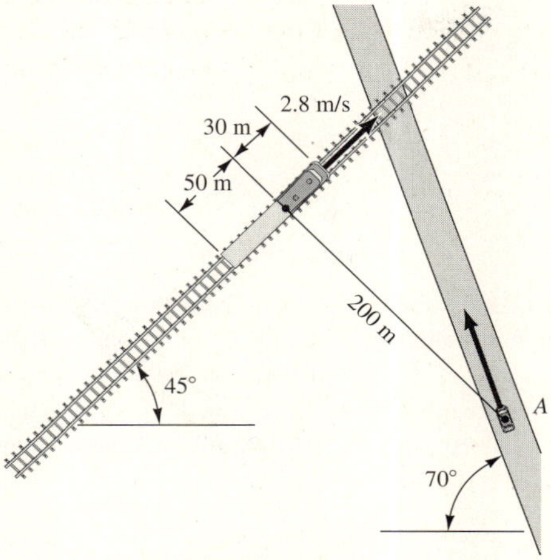

Figure P.11.81.

11.82. A Tomahawk missile is being tested for its effect on a naval vessel. A destroyer is towing an old expendable naval frigate at a speed of 8 m/s. The missile is shown at time t moving along a straight line at a constant speed of 200 m/s, the guidance system having been shut off to avoid an accident involving the towing destroyer. Does the missile hit the target and if so where does the impact occur? The missile moves at a constant elevation of 3 m above the surface of the water.

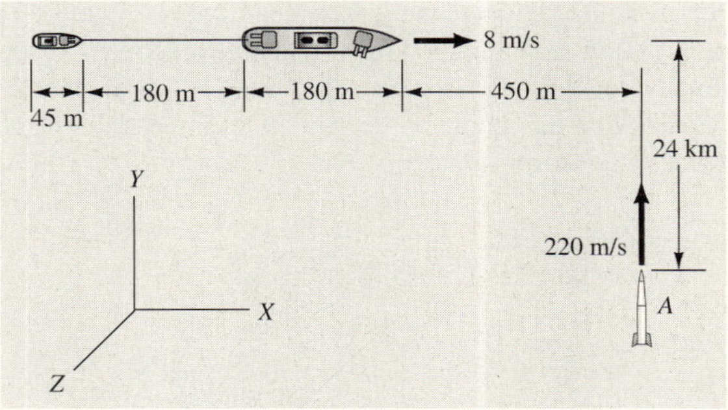

Figure P.11.82.

11.83. On a windy day, a hot air balloon is moving in a translatory manner relative to the ground with the following acceleration:

$$\boldsymbol{a} = 2\boldsymbol{i} - 5\boldsymbol{j} + 3\boldsymbol{k} \text{ m/s}^2$$

Simultaneously, a man in the balloon basket is swinging a small device for measuring the dew point. The device of mass 5 kg is connected to a massless rod. At the instant shown, $\omega = 2$ rad/s and $\dot{\omega} = 3$ rad/s^2 both relative to the balloon. What force does the rod exert on the device at this instant? Give the result in vector and scalar form. The rod is in a horizontal position (see elevation view) at the instant shown and has a length of 0.5 m.

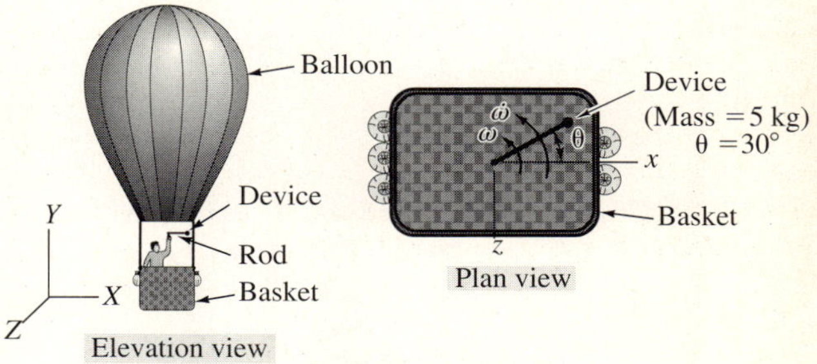

Figure P.11.83.

11.84. A submarine is moving at a constant horizontal speed of 7.7 m/s below the surface of the ocean. At the same time, the sub is descending downward by discharging air with an acceleration of $0.023g$ while remaining horizontal. In the submarine, a flyball governor operates with weights having a mass of 500 g each. The governor is rotating with speed ω of 5 rad/s. If at time t, $\theta = 30°$, $\dot{\theta} = 0.2$ rad/s, and $\ddot{\theta} = 1$ rad/s^2, what is the force developed on the support of the governor system as a result solely of the motion of the weights at this instant? [*Hint:* What is the *acceleration* of the *center of mass* of the spheres relative to *inertial* space?]

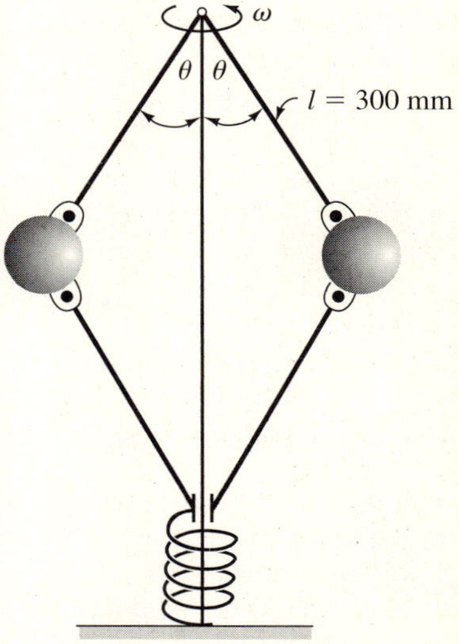

Figure P.11.84.

11.85. A fighter plane is landing and has the following acceleration relative to the ground while moving in a translatory manner:

$$\boldsymbol{a} = -0.2g\boldsymbol{k} - 0.1g\boldsymbol{j} \text{ m/s}^2$$

The wheels are being let down as shown. What is the dynamic force acting at the center of the wheel at the instant shown? The following data apply:

$$\omega = 0.3 \text{ rad/s} \qquad \dot{\omega} = 0.4 \text{ rad/s}^2$$

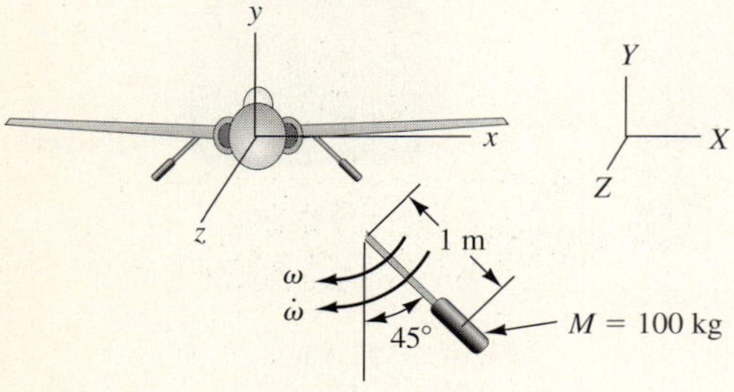

Figure P.11.85.

11.86. In a steam turbine, steam is expended through a stationary nozzle at a speed V_0 of 3 km/s at an angle of 30°. The steam impinges on a series of blades mounted all around the periphery of a cylinder, which is rotating at a speed Ω of 5×10^2 r/min. The steam impinges on the blades at a radial distance of 1.20 m from the axis of rotation of the cylinder. What angle α should the left side of the blades have for the steam to enter the region between blades most smoothly? [*Hint:* Let *xyz* move with a blade in a translatory manner relative to the ground (*xyz* is thus not entirely fixed to the blade and hence does not rotate).]

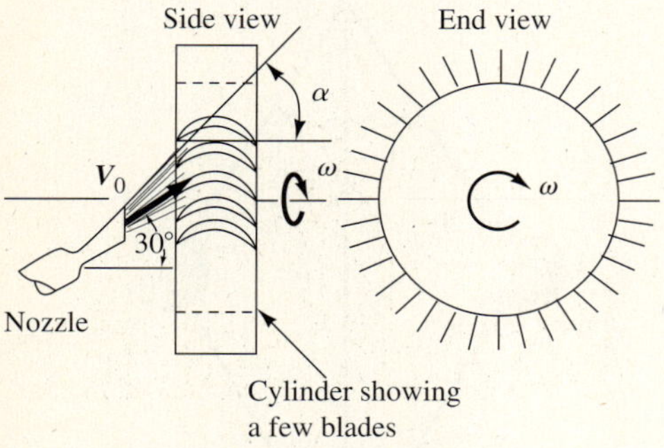

Figure P.11.86.

11.87. A sailboat moves at a speed V_0 of 3.6 m/s relative to a stationary reference *XY*. The wind is moving uniformly at a speed V_1 relative to *XY* in the direction shown. On top of the mast is a direction vane responding to the wind relative to the boat. If this vane points in a direction of 170° from the *X* axis, what is the velocity V_1 of the wind?

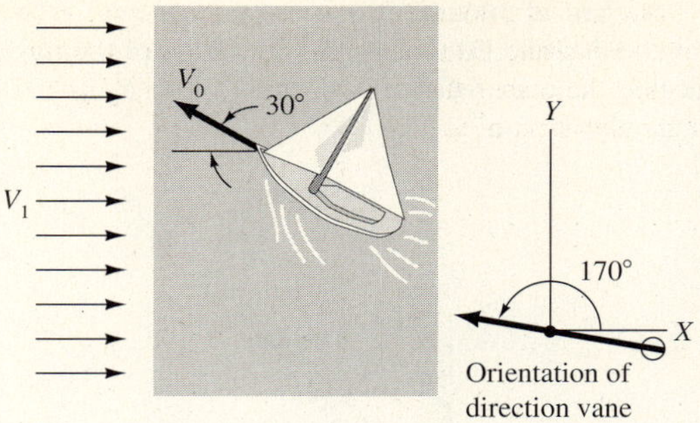

Figure P.11.87.

11.88. A boat is about to depart from point *A* on the shore of a river that has a uniform velocity V_0 of 1.5 m/s. If the boat can move at the rate of 4.5 m/s relative to the water, and if we want to move along a straight path from *A* to *B*, how long will it take to go from *A* to *B*? At what angle β should the boat be aimed relative to the water?

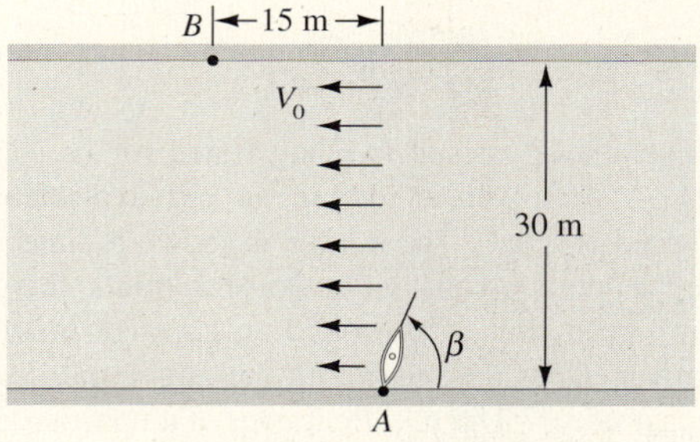

Figure P.11.88.

11.89. A jet passenger plane is moving at a speed V_0 of 240 m/s. A storm region extending 4 km in width is reported 15 km due East of its position. The region is moving NW at a speed V_1 of 30 m/s. At what maximum angle α from due N can the plane fly to just miss the storm front?

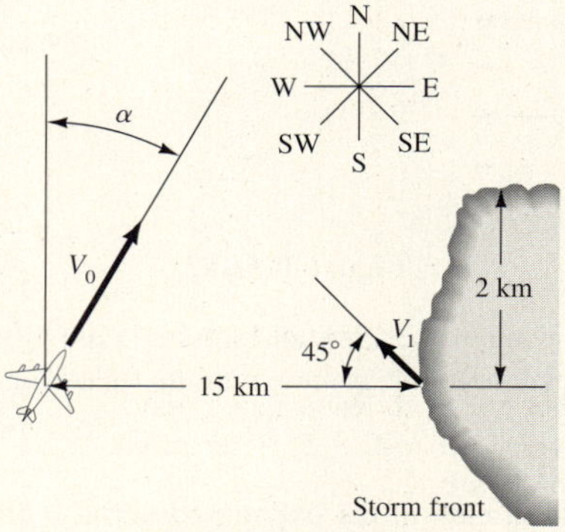

Figure P.11.89.

11.90. Mass M of 3 kg rotates about point O in an accelerating rocket in the xy plane. At the instant shown, what is the force from the rod onto the mass? Include the effects of gravity if $g = 7.00$ m/s^2 at the elevation of the rocket.

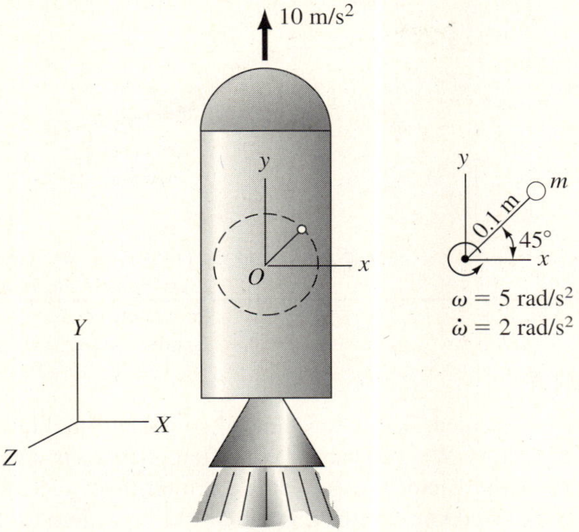

Figure P.11.90.

11.91. A light plane is approaching a runway in a cross-wind. This cross-wind has a uniform speed V_0 of 15 m/s. The plane has a velocity component V_1 parallel to the ground of 32 m/s relative to the wind at an angle β of 30°. The rate of descent is such that the plane will touch down somewhere along A–A. Will this touchdown occur on the runway or off the runway for the data given?

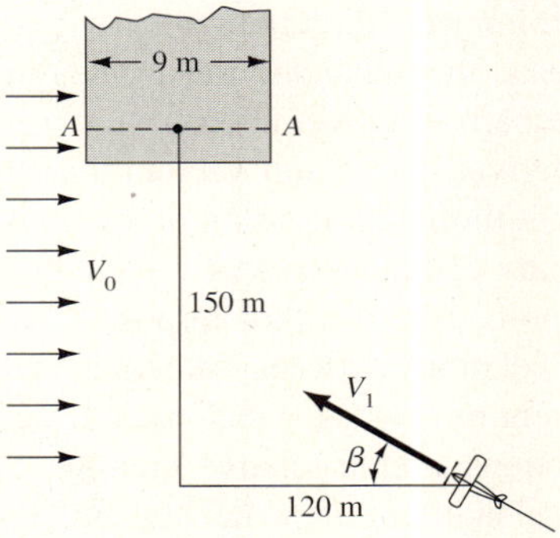

Figure P.11.91.

11.92. A helicopter is shown moving relative to the ground with the following motion:

$$V = 36i + 19j + 5.5k \text{ m/s}$$
$$a = 2.8i + 4.4j + 1.9k \text{ m/s}$$

The helicopter blade is rotating relative to the helicopter in the following manner at the instant of interest:

$$\omega_1 = 100 \text{ r/min} \qquad \dot{\omega}_1 = 10.8 \text{ rad/s}^2$$

The blade is 10 m long. What is the velocity and the acceleration of the tip B relative to the ground reference XYZ? Give your results in meters and seconds. The blade is parallel to the X axis at the instant of interest.

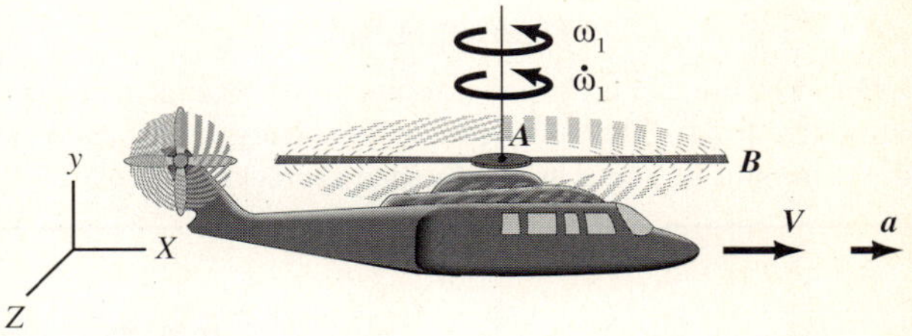

Figure P.11.92.

11.93. A destroyer in rough seas has the following translational acceleration as seen from inertial reference XYZ when it is firing its main battery in the YZ plane:

$$a = 5j + 2k \text{ m/s}^2$$

What must ω_1 and $\dot{\omega}_1$ of the gun barrel be relative to the ship at this instant so that tip A of the barrel has zero acceleration relative to XYZ?

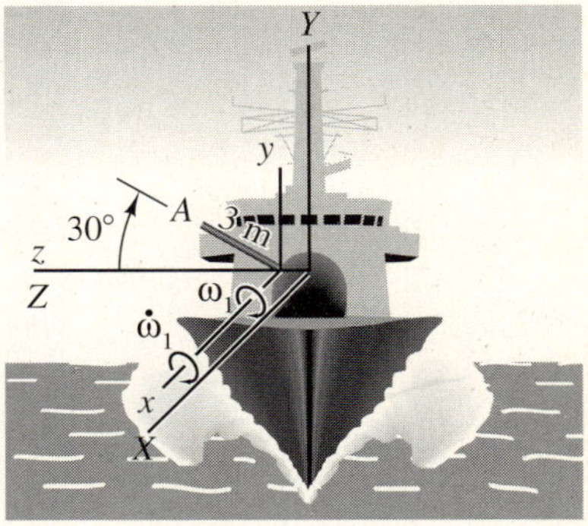

Figure P.11.93.

11.94. A small elevator E in an ocean-going vessel has the following motion relative to the ship:

$$a_{Elev.} = 0.2gk \text{ m/s}^2$$

The ship has the following motion relative to nearby land:

$$a_{Ship} = 0.2i + 3j + 0.6k \text{ m/s}^2$$

If the weight of the elevator including passengers is 8 kN, what is the force on the ship from the elevator?

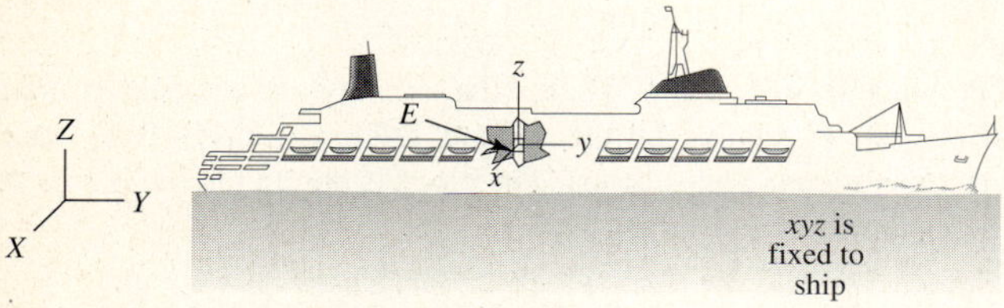

Figure P.11.94.

11.95. You are a court expert consultant. At what distance from point C, the front of the moving train, do you testify that car A collides with train B? Use a multireference approach.

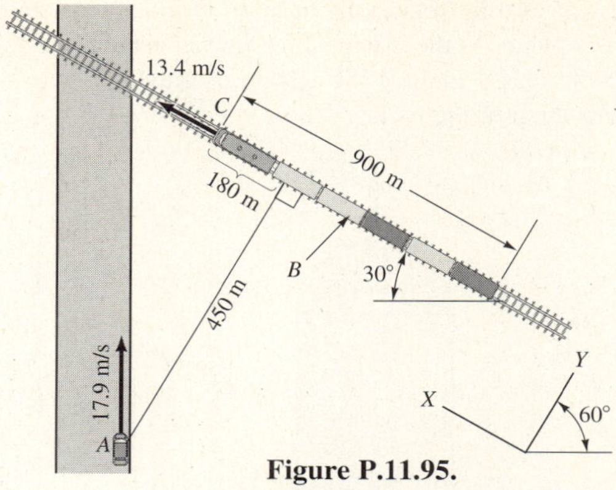

Figure P.11.95.

11.9 Closure

In this chapter, we have presented, first, a few general comments on differentiation and integration of vectors. We then carried out differentiations in a variety of ways. In the first case, the vector $\boldsymbol{r}$ was expressed in terms of rectangular scalar components and the fixed unit vectors $\boldsymbol{i}$, $\boldsymbol{j}$, and $\boldsymbol{k}$. The procedure for finding $\dot{\boldsymbol{r}}$ and $\ddot{\boldsymbol{r}}$ in terms of rectangular scalar components is straightforward and involves only the familiar differentiation operations of scalar calculus. We next considered the kinematics of a particle moving along some given path. Here, we obtained $\dot{\boldsymbol{r}}$ and $\ddot{\boldsymbol{r}}$ in terms of speeds and rates of changes of speeds of the particle along the path with component directions no longer fixed in space but instead related at each point along the path to the geometry of the path. For this reason, we brought in certain concepts of differential geometry such as the osculating plane, the normal vector, etc. Finally, we computed $\dot{\boldsymbol{r}}$ and $\ddot{\boldsymbol{r}}$ in terms of cylindrical coordinates with component directions always in the radial, transverse, and axial directions. Clearly, the radial and transverse directions are not fixed in space and change as the particle moves about.

In carrying out various derivatives of unit vectors that are not fixed in space, such as $\boldsymbol{\epsilon}_{\bar{r}}$ and $\boldsymbol{\epsilon}_{\theta}$, we went through a limiting process in arriving at the desired results. Later, in the study of kinematics of a rigid body, we present simple straightforward formal procedures for this purpose.

We next investigated the relations between velocities and accelerations of a particle, as seen from different references, which are translating relative to each other. We called such motions simple relative motion. Later, when we undertake rigid-body motion, we shall consider the case involving references moving arbitrarily relative to each other. It is vital to remember that we must measure $\boldsymbol{a}$ relative to an *inertial reference* when we employ **Newton's law** in the form $\boldsymbol{F} = m\boldsymbol{a}$. We may at times find it convenient to employ two references in this connection where one reference is the inertial reference needed for the desired acceleration vector. This situation is illustrated in Example 11.10.

In Chapter 12, we shall consider the *dynamics* of motion of a particle. We shall then have ample opportunity to employ the kinematics of Chapter 11.

PROBLEMS

11.96. A particle at position (3, 4, 6) m at time $t_0 = 1$ s is given a constant acceleration having the value $6\boldsymbol{i} + 3\boldsymbol{j}$ m/s^2. If the velocity at the time t_0 is $16\boldsymbol{i} + 20\boldsymbol{j} + 5\boldsymbol{k}$ m/s, what is the velocity of the particle 20 s later? Also give the position of the particle.

11.97. A pin is confined to slide in a circular slot of radius 6 m. The pin must also slide in a straight slot that moves to the right at a constant speed, V, of 3 m/s while maintaining a constant angle of 30° with the horizontal. What are the velocity and acceleration of the pin A at the instant shown?

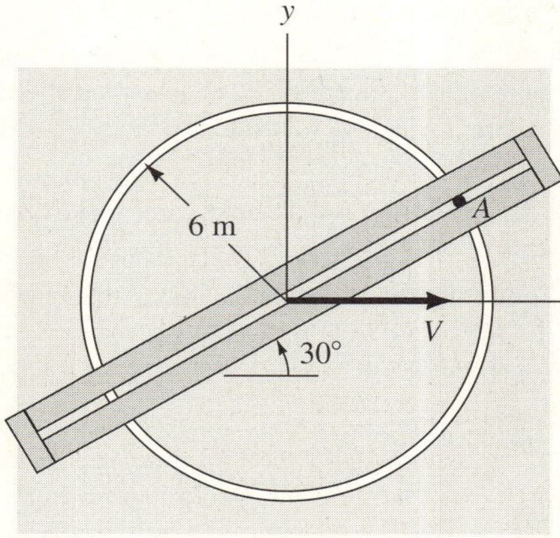

Figure P.11.97.

11.98. A freighter is moving in a river at a speed of 1.5 m/s relative to the water. A small boat A is moving relative to the water at a speed of 1.5 m/s in a direction as shown in the diagram. The river is moving at a uniform speed of 0.3 m/s relative to the ground. Will the boat hit the freighter and, if so, where will the impact occur?

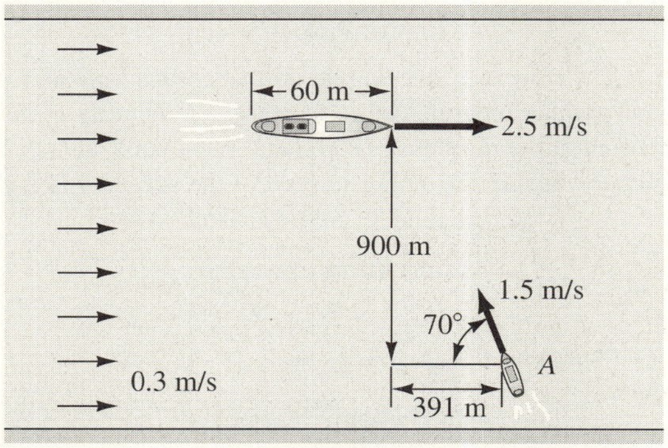

Figure P.11.98.

11.99. A light line attached to a streamlined weight A is "shot" by a line rifle from a small boat C to a large boat D in heavy seas. The weight must travel a distance of 18 m horizontally and reach the larger boat's deck, which is 6 m higher than the deck of boat C. If the angle α of firing is 40°, what minimum velocity V_0 is needed? At the instant of firing, boat C is dipping down into the water at a speed of 1.5 m/s. Assume that the larger boat remains essentially fixed at constant level.

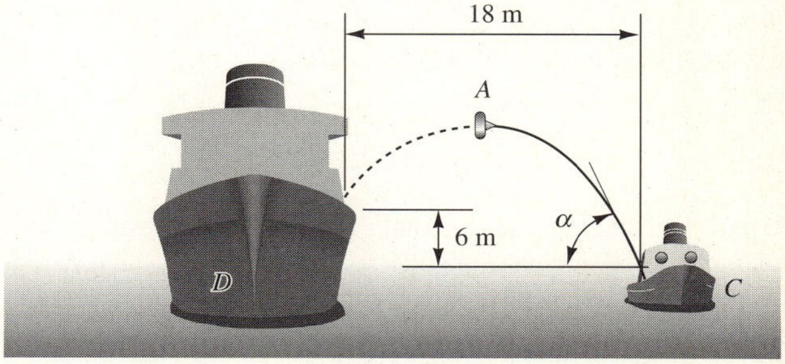

Figure P.11.99.

11.100. A projectile is fired at an angle of 60° as shown. At what elevation y does it strike the hill whose equation has been estimated as $y = 10^{-5}x^2$ m? Neglect air friction and take the muzzle velocity as 1000 m/s.

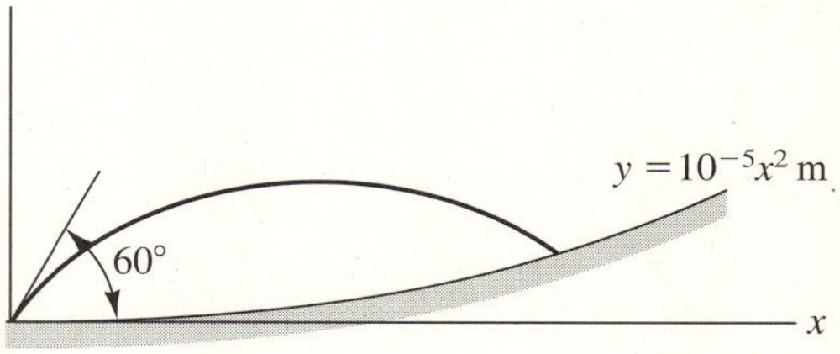

Figure P.11.100.

11.101. A proposed space laboratory, in order to simulate gravity, rotates relative to an inertial reference XYZ at a rate ω_1. For occupant A in the living quarters to be comfortable, what should the approximate value of ω_1 be? Clearly, at the center, there is close to zero gravity for zero-g experiments. A conveyor connects the living quarters with the zero-g laboratory. At the instant of interest, a package D has a speed of 5 m/s and a rate of change of speed of 3 m/s^2 relative to the space station, both toward the laboratory. What are the axial, transverse, and radial velocity and acceleration components at the instant of interest relative to the inertial reference? What are the rectangular components of the acceleration vector?

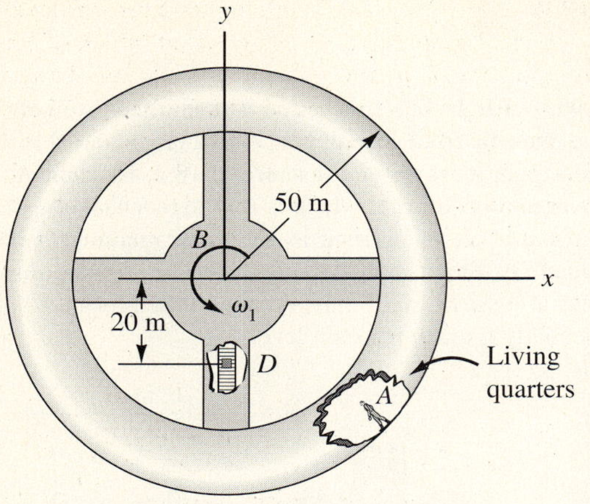

Figure P.11.101.

11.102. A yoke AB rotates about pin A. When the angle α is equal to $\pi/6$ rad, it has an angular speed $\dot{\alpha} = \pi$ rad/s and a rate of change of angular speed $\ddot{\alpha} = 0.3\pi$ rad/s^2. A slot in the yoke constrains a pin C to move with the slot while a spring forces the pin to slide on the parabolic surface. What are the velocity and acceleration vectors at the instant of interest?

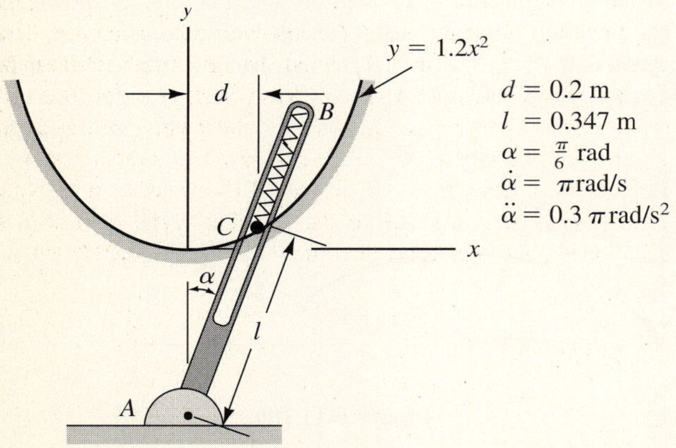

Figure P.11.102.

11.103. In Problem 11.17, what is the change of radius of curvature of a stream of water as it goes out from a circular vane of radius 0.30 m into free flight? What and where is the minimum radius of curvature?

11.104. A batter has hit a home run ball that just clears the fence. It has a line drive trajectory. A fan attempts to catch the ball. Neglecting friction, what is the speed of the ball as the fan attempts to catch it?

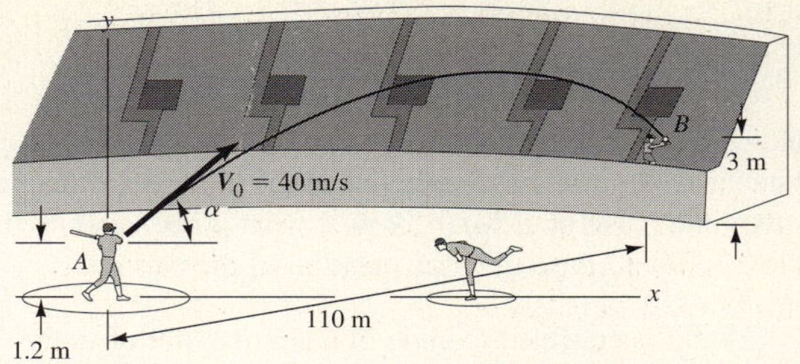

Figure P.11.104.

11.105. A platform is rotating relative to a stationary reference xyz with angular speed $\omega_1 = 1$ rad/s and a rate of change of angular speed $\dot{\omega}_1 = 0.2$ rad/s^2 at the instant of interest. Also at this instant, the platform is being raised at a speed $V_z = 0.47$ m/s with a rate of change of speed of $\dot{V}_z = 0.26$ m/s^2. A rod and mass A rotate on the platform with an angular speed relative to the platform equal to $\omega_2 = 0.4$ rad/s. At the instant of interest, the angle α is 30°. If $r = 30$ mm,

(a) determine the radial, axial, and transverse acceleration components of A;

(b) determine the rectangular components of the acceleration of A.

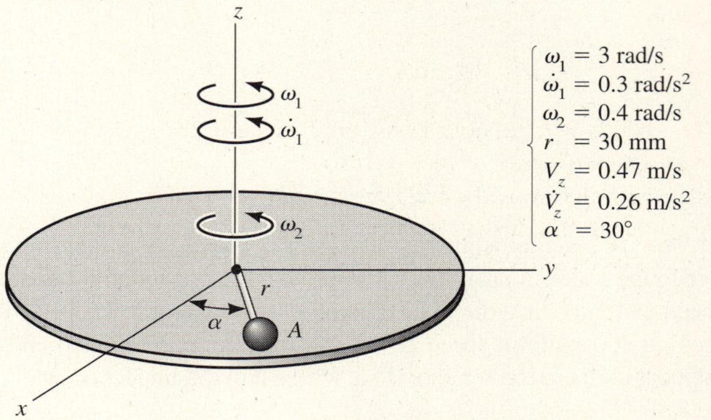

Figure P.11.105.

11.106. The flow of water into an ordinary water sprinkler is 25 L/s initially and is programmed to increase continuously to 50 L/s. The exit area of the nozzles is 1960 mm^2. What is the area of lawn that will be watered? The angular speed of the sprinkler is 2 rad/s.

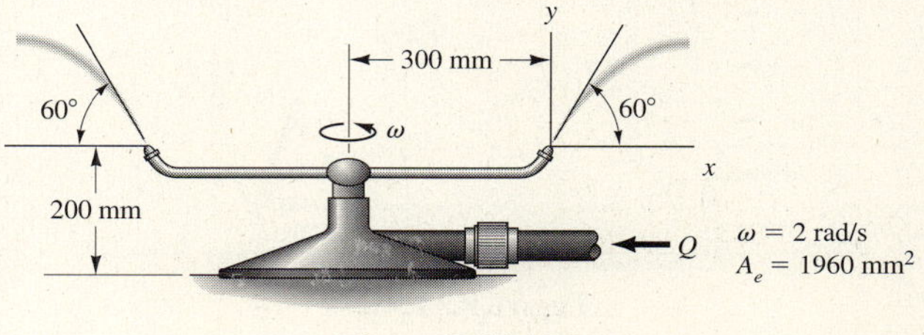

Figure P.11.106.

11.107. Pilots of fighter planes wear special suits designed to prevent blackouts during a severe maneuver. These suits tend to keep the blood from draining out of the head when the head is accelerated in a direction from shoulders to head. With this suit, a flier can take 5g's of acceleration in the aforementioned direction. If a flier is diving at a speed of 280 m/s, what is the minimum radius of curvature that he can manage at pullout without suffering bad physiological effects?

11.108. A particle moves with constant speed of 1.5 m/s along a path given as $x = y^2 - \ln y$ m. Give the acceleration vector of the particle in terms of rectangular components when the particle is at position $y = 3$ m. Do the problem by using path coordinate techniques and then by Cartesian-component techniques. How many g's of acceleration is the particle subject to?

11.109. A submarine is moving in a translatory manner with the following velocity and acceleration relative to an inertial reference:

$$\boldsymbol{V} = 3\boldsymbol{i} + 3.8\boldsymbol{j} + \boldsymbol{k} \text{ m/s} \qquad \boldsymbol{a} = 0.1\boldsymbol{i} - 12.35\boldsymbol{j} + 0.27\boldsymbol{k} \text{ m/s}^2$$

A device inside the submarine consists of an arm and a mass at the end of the arm. At the instant of interest, the arm is rotating in a vertical plane with the following angular speed and angular acceleration:

$$\omega = 10 \text{ rad/s} \qquad \dot{\omega} = 3 \text{ rad/s}^2$$

The arm is vertical at this instant. The mass at the end of the rod may be considered to be a particle having a mass of 5 kg. What are the velocity and acceleration vectors for the motion of the particle at this instant relative to the inertial reference? What must be the force vector from the arm onto the particle at this instant?

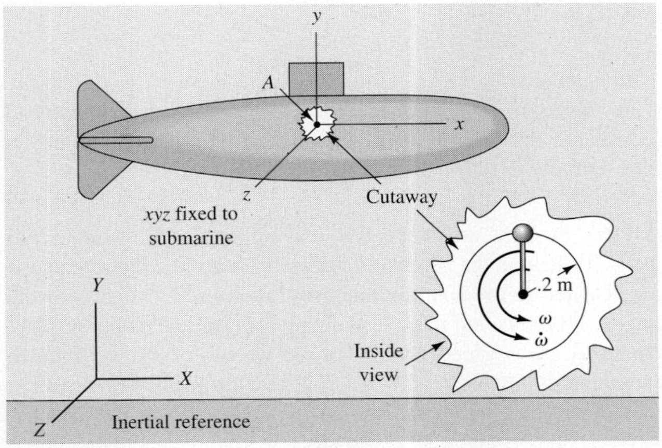

Figure P.11.109.

11.110. A mechanical "arm" for handling radioactive materials is shown. The distance $\bar{r}$ can be varied by telescoping action of the arm. The arm can be rotated about the vertical axis A–A. Finally, the arm can be raised or lowered by a worm gear drive (not shown). What is the velocity and acceleration of the object C if the end of the arm moves out radially at a rate of 0.3 m/s while the arm turns at a speed ω of 2 rad/s? Finally, the arm is raised at a rate of 0.6 m/s. The distance $\bar{r}$ at the instant of interest is 1.5 m. What is the acceleration in the direction $\boldsymbol{\epsilon} = 0.8\boldsymbol{i} + 0.6\boldsymbol{j}$?

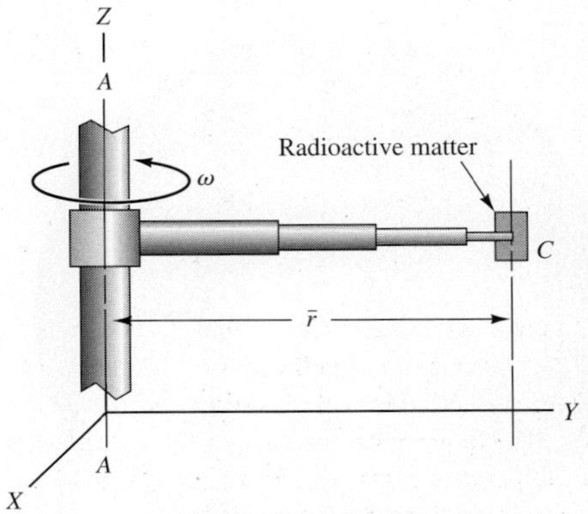

Figure P.11.110.

11.111. A top-section view of a water sprinkler is shown. Water enters at the center from below and then goes through four passageways in an impeller. The impeller is rotating at constant speed ω of 8 r/min. As seen from the impeller, the water leaves at a speed of 3 m/s at an angle of 30° relative to r. What is the velocity and acceleration as seen from the ground of the water as it leaves the impeller and becomes free of the impeller? Give results in the radial, axial, and transverse directions. Use one reference only.

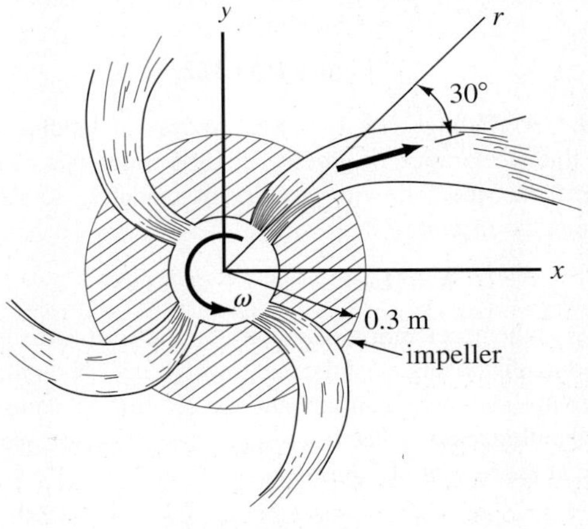

Figure P.11.111.

11.112. A luggage dispenser at an airport resembles a pyramid with six flat segments as sides as shown in the diagram. The system rotates with an angular speed ω of 2 r/min. Luggage is dropped from above and slides down the faces to be picked up by travelers at the base.

A piece of luggage is shown on a face. It has just been dropped at the position indicated. It has at this instant zero velocity as seen from the rotating face but has at this instant and thereafter an acceleration of $0.2g$ along the face. What is the total acceleration, as seen from the ground, of the luggage as it reaches the base at B? Use one reference only.

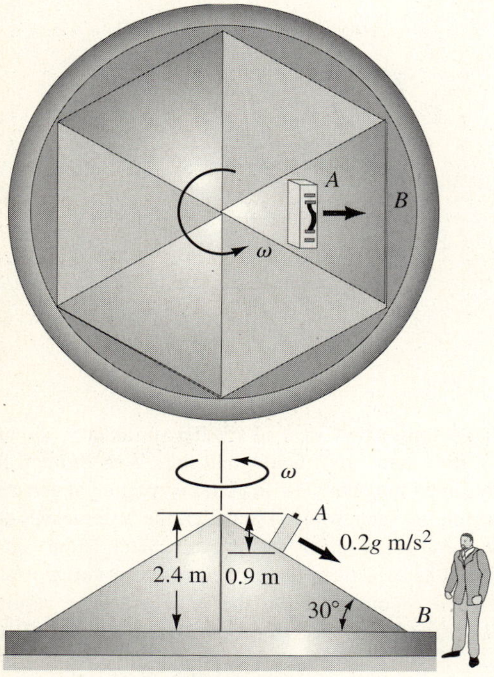

Figure P.11.112.

11.113. A landing craft is in the process of landing on Mars, where the acceleration of gravity is 0.385 times that of the earth. The craft has the following acceleration relative to the landing surface at the instant of interest:

$$\boldsymbol{a} = 0.2g\boldsymbol{i} + 0.4g\boldsymbol{j} - 2g\boldsymbol{k} \text{ m/s}^2$$

where g is the acceleration of gravity on the earth. At this instant, an astronaut is raising a hand camera weighing 3 N on the earth. If he is giving the camera an upward acceleration of 3 m/s^2 relative to the landing craft, what force must the astronaut exert on the camera at the instant of interest?

[7]It is estimated that it would cost $100 billion to land a man on Mars.

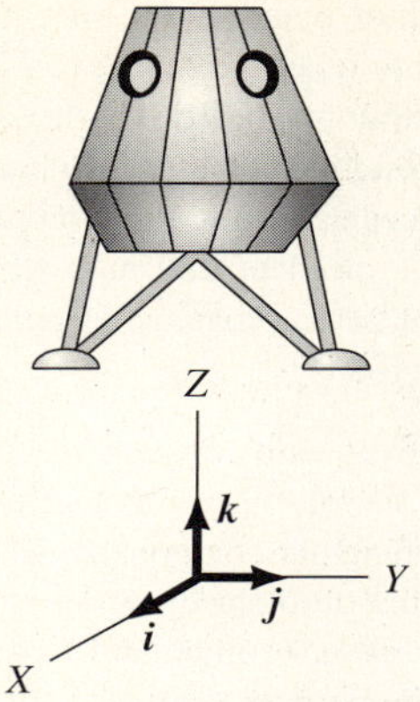

Figure P.11.113.

11.114. A jet of water has a speed at the nozzle of 20 m/s. At what position does it hit the parabolic hill? What is its speed at that point? Do not include friction.

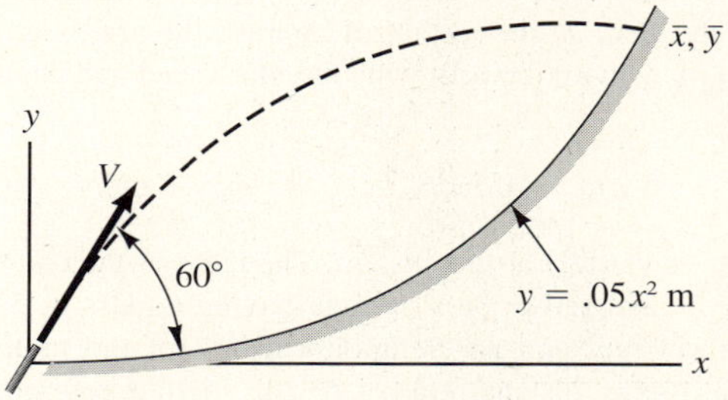

Figure P.11.114.

11.115. A particle moves along a circular path of diameter 10 m such that

$$V = 3t + 6 \text{ m/s}$$

When it has traveled a distance of 15 m, the velocity starts to decrease at the rate of 0.2 m/s^2. What is the acceleration at t = 1.3 s and at t = 18 s?

11.116. A light attack boat is leaving an engagement at full speed. To help in the process, a battery of four 50-caliber machine guns is fired to the rear continuously. The muzzle velocity of the guns is 1000 m/s and the rate of firing is 3000 rounds per minute. The guns are oriented parallel to the water in order to achieve maximum thrust. Neglecting friction, how far from the rear of the boat does each bullet hit the water and what is the spacing between successive bullets from any one gun as the bullets hit the water? The boat is moving at a constant speed of 23 m/s.

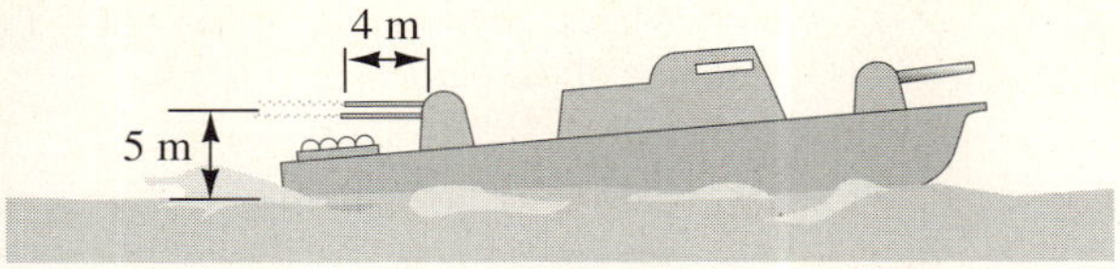

Figure P.11.116.

***11.117.** A particle has a variable velocity $V(t)$ along a helix wrapped around a cylinder of radius e. The helix makes a constant angle α with plane A perpendicular to the z axis. Express the acceleration $\boldsymbol{a}$ of the particle using cylindrical coordinates. Next, express $\boldsymbol{\epsilon}_t$ using cylindrical unit vectors and note that the sum of the transverse and axial components of $\boldsymbol{a}$ (just computed) can be given simply as $\dot{V}\boldsymbol{\epsilon}_t$. Next, express the acceleration of the particle using path coordinates. Finally, noting that $\boldsymbol{\epsilon}_n = -\boldsymbol{\epsilon}_{\bar{r}}$, show that the radius of curvature is given as $R = \dfrac{e}{\cos^2 \alpha}$.

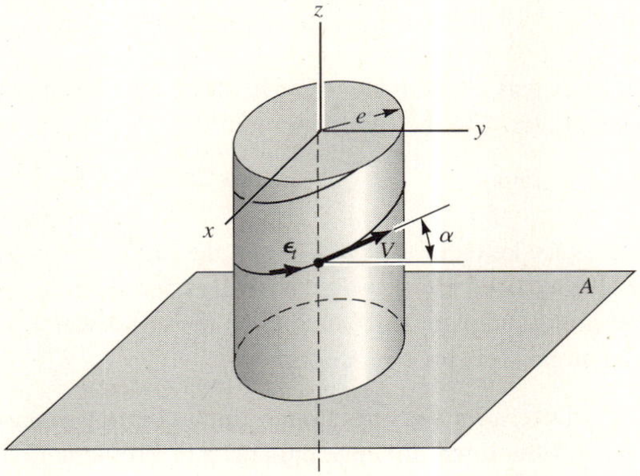

Figure P.11.117.

11.118. An eagle is diving at a constant speed of 12 m/s to catch a 3-m snake that is moving at a constant speed of 5.6 m/s. What should α be so that the eagle hits the small head of the snake? The eagle and the snake are moving in a vertical plane.

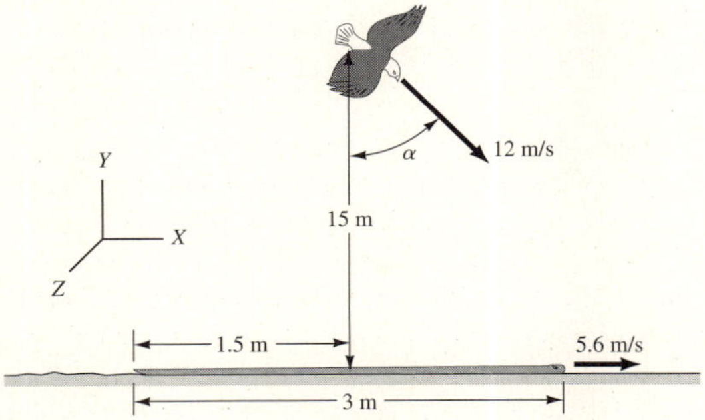

Figure P.11.118.

11.119. A tube, most of whose centerline is that of an ellipse given as

$$\frac{z^2}{1.8^2} + \frac{\bar{r}^2}{.32^2} = 1$$

has a cross-sectional diameter $D = 100$ mm. The tube has the following rotational motion at the instant of interest:

$$\omega = 0.15 \text{ rad/s} \qquad \dot{\omega} = 36 \text{ mrad/s}^2$$

Water is flowing through the tube at the following rate at the instant of interest:

$$Q = 0.18 \text{ L/s} \qquad \dot{Q} = 25 \text{ mL/s}^2$$

The tube is in the vertical plane at the instant of interest. What is the acceleration of the water particles at the centerline of the tube at point C using cylindrical coordinates and cylindrical components? Assume over the cross-section of the tube that the water velocity and acceleration are uniform.

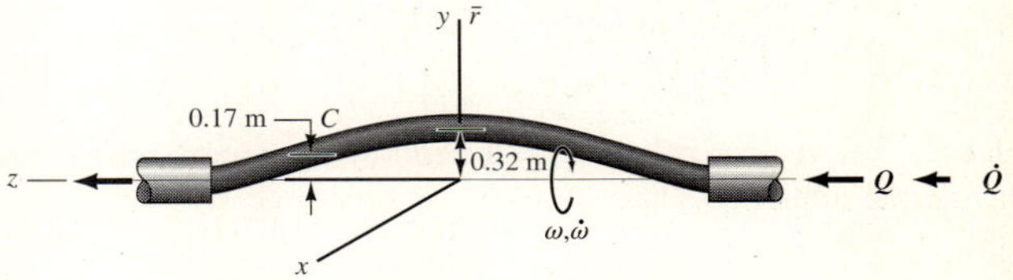

Figure P.11.119.

11.120. A World War I fighter plane is in level flight moving at a speed of 16 m/s. At time t_0 it has an acceleration given as:

$$\boldsymbol{a} = 0.2g\boldsymbol{i} - 0.3g\boldsymbol{j} + 2g\boldsymbol{k} \text{ m/s}^2$$

Also at this time, the co-pilot is raising a camera upward with an acceleration of $0.1g$ relative to the plane. If the camera has a mass of 10 g, what force must the co-pilot exert on the camera to give it the desired motion at time t_0? Note that the plane never rotates during this action. Take $g = 9.81 \text{ m/s}^2$.

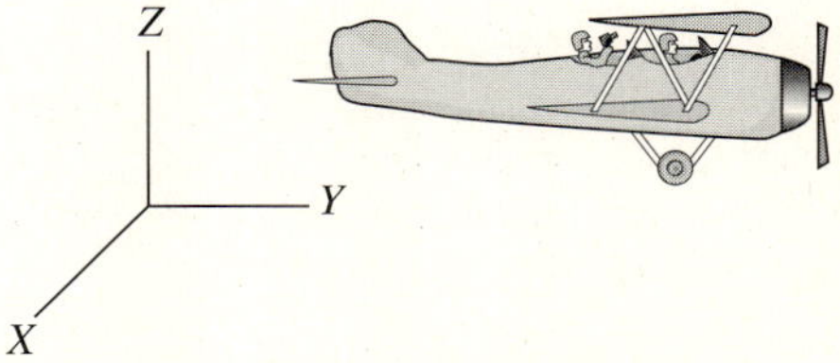

Figure P.11.120.

11.121. A space device has a velocity of $0.50\boldsymbol{i} + 0.20\boldsymbol{j}$ m/s and an acceleration of $0.20\boldsymbol{i} + 0.30\boldsymbol{k}$ m/s^2, both relative to the ground reference XYZ. Rod CD has an angular motion relative to the space device equal to $\omega = 2$ rad/s and $\dot{\omega} = 3$ rad/s^2. What are the velocity and the acceleration of mass M relative to the ground? CD is in the vertical zy plane at the instant of interest. Reference xyz shown is fixed to rod AB. CD is at 60° from AB.

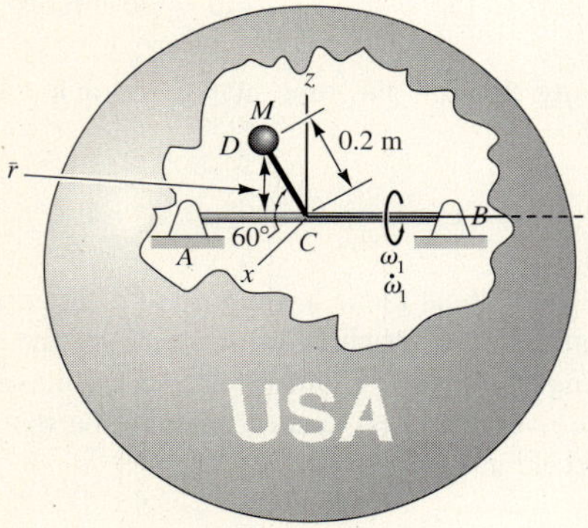

Figure P.11.121.

11.122. A weather balloon has the following motion relative to the inertial reference XYZ:

$$\boldsymbol{V} = 150\boldsymbol{i} + 200\boldsymbol{j} + 60\boldsymbol{k} \text{ m/s}$$
$$\boldsymbol{a} = 20\boldsymbol{i} - 40\boldsymbol{j} + 38\boldsymbol{k} \text{ m/s}^2$$

A light rod at A is connected to particle D and is rotating relative to the balloon at the instant of interest with angular speed $\omega_1 = 5$ rad/s and rate of change of speed $\dot{\omega}_1 = 2$ rad/s^2. What is the velocity of the particle at D relative to XYZ at the instant of interest? The distance $AD = 0.2$ m. What is the tensile force on the rod in the direction of AD? The mass of the particle at D is 1 kg.

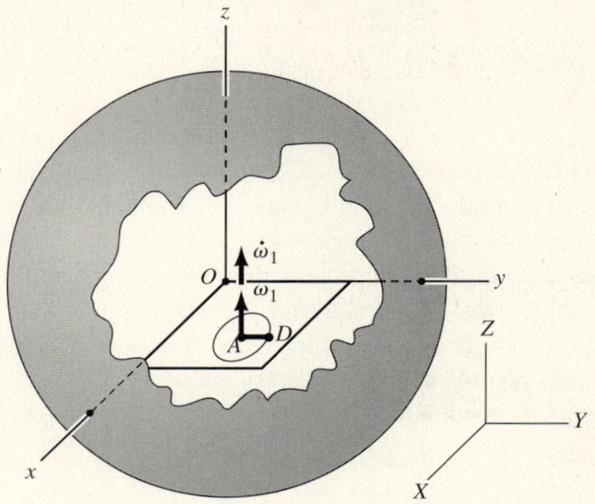

Figure P.11.122.

***11.123.** A particle is made to move along a path given in terms of the parameter τ in the following manner:

$$(x)_p = -\sin 2\tau \qquad (y)_p = \cos 2\tau \qquad (z)_p = e^{-\tau}$$

Give a simple sketch of the path. When the particle is at an elevation $z = 1$, the speed along the path is 5 m/s and the rate of change of speed along the path is 10 m/s^2. Find the acceleration vector at $z = 1$ in components for path coordinates.

***11.124.** Determine the direction of the osculating plane at position $z = 1$ for the three-dimensional curve of Problem 11.123. We got the following results from the previous problem:

$$\boldsymbol{\epsilon}_n = R(0.08\boldsymbol{i} - 0.80\boldsymbol{j} + 0.16\boldsymbol{k})$$
$$\boldsymbol{\epsilon}_t = 0.894\boldsymbol{i} - 0.447\boldsymbol{k}$$
$$R = 0.8198 \text{ m}$$

CHAPTER 12

Particle Dynamics

12.1 Introduction

In Chapter 11, we examined the geometry of motion—the kinematics of motion. In particular, we considered various kinds of coordinate systems: rectangular coordinates, cylindrical coordinates, and path coordinates. In this chapter, we shall consider Newton's law for the three coordinate systems mentioned above, as applied to the motion of a particle.

Before embarking on this study, we shall review notions concerning units of mass presented earlier in Chapter 1. Recall that a pound mass (lbm) is the amount of matter attracted by gravity at a specified location on the earth's surface by a force of 1 pound (lbf). A slug, on the other hand, is the amount of matter that will accelerate relative to an inertial reference at the rate of 1 ft/s^2 when acted on by a force of 1 lbf. Note that the slug is defined via Newton's law, and therefore the slug is the proper unit to be used in Newton's law. The relation between the pound mass (lbm) and the slug is

$$M\text{ (slugs)} = \frac{M\text{ (lbm)}}{32.2} \tag{12.1}$$

Note also that the weight of a body in pounds force near the earth's surface will numerically equal the mass of the body in pounds mass. It is vital in using Newton's law that the mass of the body in pounds mass be properly converted into slugs via Eq. 12.1.

In SI units, recall that a kilogram is the mass that accelerates relative to an inertial reference at the rate of 1 metre/s^2 when acted on by a force of 1 newton (which is about one-fifth of a pound force). If the weight W of a body is given in terms of newtons, we must divide by 9.81 to get the mass in kilograms needed for Newton's law. That is,

$$M\text{ (kg)} = \frac{W\text{ (N)}}{9.81} \tag{12.2}$$

We are now ready to consider Newton's law in rectangular coordinates.

Part A: Rectangular Coordinates; Rectilinear Translation

12.2 Newton's Law for Rectangular Coordinates

In rectangular coordinates, we can express Newton's law as follows:

$$\begin{aligned} F_x &= ma_x = m\frac{dV_x}{dt} = m\frac{d^2x}{dt^2} \\ F_y &= ma_y = m\frac{dV_y}{dt} = m\frac{d^2y}{dt^2} \\ F_z &= ma_z = m\frac{dV_z}{dt} = m\frac{d^2z}{dt^2} \end{aligned} \tag{12.3}$$

If the motion is known relative to an inertial reference, we can easily solve for the rectangular components of the resultant force on the particle. The equations to be solved are just algebraic equations. The *inverse* of this problem, wherein the forces are known over a time interval and the motion is desired during this interval, is not so simple. For the inverse case, we must get involved generally with integration procedures.

In the next section, we shall consider situations in which the resultant force on a particle has the same direction and line of action at all times. The resulting motion is then confined to a straight line and is usually called *rectilinear translation*.

12.3 Rectilinear Translation

For rectilinear translation, we may consider the line of action of the motion to be collinear with one axis of a rectilinear coordinate system. Newton's law is then one of the equations of the set 12.3. We shall use the x axis to coincide with the line of action of the motion. The resultant force F (we shall not bother with the x subscript here) can be a constant, a function of time, a function of speed, a function of position, or any combination of these. At this time, we shall examine some of these cases, leaving others to Chapter 19, where, with the aid of the students' knowledge of differential equations,[1] we shall be more prepared to consider them.

Case 1. Force Is a Function of Time or a Constant. A particle of mass m acted on by a time-varying force $F(t)$ is shown in Fig. 12.1. The plane on which the body moves is frictionless. The force of gravity is equal and opposite

[1]Most students studying dynamics will concurrently be taking a course in differential equations.

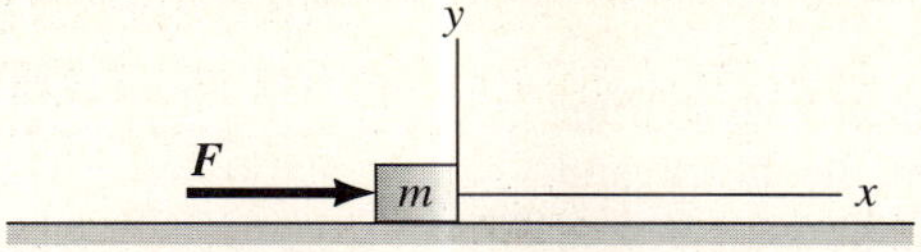

Figure 12.1. Rectilinear translation.

to the normal force from the plane so that $F(t)$ is the resultant force acting on the mass. Newton's law can then be given as follows:

$$F(t) = m\frac{d^2x}{dt^2}$$

Therefore,

$$\frac{d^2x}{dt^2} = \frac{F(t)}{m} \tag{12.4}$$

Knowing the acceleration in the x direction, we can readily solve for $F(t)$.

The inverse problem, where we know $F(t)$ and wish to determine the motion, requires integration. For this operation, the function $F(t)$ must be piecewise continuous.[2] To integrate, we rewrite Eq. 12.4 as follows:

$$\frac{d}{dt}\left(\frac{dx}{dt}\right) = \frac{F(t)}{m}$$

$$d\left(\frac{dx}{dt}\right) = \frac{F(t)}{m}\,dt$$

Now integrating both sides we get

$$\frac{dx}{dt} = V = \int \frac{F(t)}{m}\,dt + C_1 \tag{12.5}$$

where C_1 is a constant of integration. Integrating once again after bringing dt from the left side of the equation to the right side, we get

$$x = \int\left[\int \frac{F(t)}{m}\,dt\right]dt + C_1t + C_2 \tag{12.6}$$

We have thus found the velocity of the particle and its position as functions of time to within two arbitrary constants. These constants can be readily determined by having the solutions yield a certain velocity and position at given times. Usually, these conditions are specified at time $t = 0$ and are then termed initial conditions. That is, when $t = 0$,

$$V = V_0 \quad \text{and} \quad x = x_0 \tag{12.7}$$

These equations can be satisfied by substituting the initial conditions into Eqs. 12.5 and 12.6 and solving for the constants C_1 and C_2.

Although the preceding discussion centered about a force that is a function of time, the procedures apply directly to a force that is a constant. The following examples illustrate the procedures set forth.

[2]That is, the function has only a finite number of finite discontinuities.

Example 12.1

A 45 kg body is initially stationary on a 45° incline as shown in Fig. 12.2(a). The coefficient of dynamic friction μ_d between the block and incline is 0.5. What distance along the incline must the weight slide before it reaches a speed of 12 m/s?

A free-body diagram is shown in Fig. 12.2(b). Since the acceleration is zero in the direction normal to the incline, we have from **equilibrium** that

$$45g\cos 45° = N = 312.2 \text{ N} \qquad \text{(a)}$$

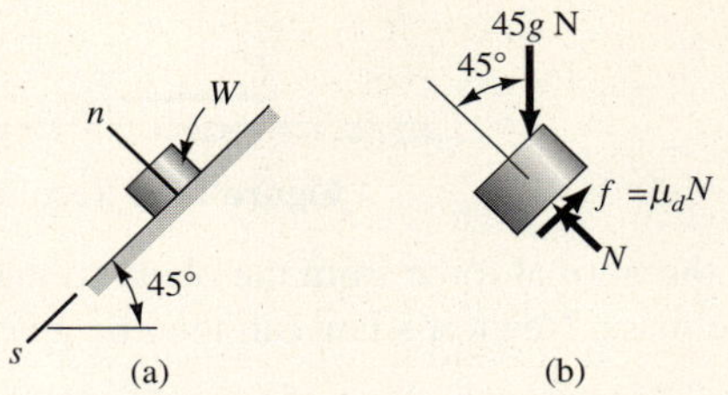

Figure 12.2. Body slides on an incline.

Now applying **Newton's law** in a direction along the incline, we have

$$45\frac{d^2s}{dt^2} = 45g\sin 45° - \mu_d N$$

Therefore,

$$\frac{d^2s}{dt^2} = 3.468 \qquad \text{(b)}$$

Rewriting Eq. (b) we have

$$d\left(\frac{ds}{dt}\right) = 3.468\,dt$$

Integrating, we get

$$\frac{ds}{dt} = 3.468t + C_1 \qquad \text{(c)}$$

$$s = 3.468\frac{t^2}{2} + C_1 t + C_2 \qquad \text{(d)}$$

When $t = 0$, $s = ds/dt = 0$, and thus $C_1 = C_2 = 0$. When $ds/dt = 12$ m/s, we have for t from Eq. (c) the result

$$12 = 3.468t$$

Therefore,

$$t = 3.46 \text{ s}$$

Substituting this value of t in Eq. (d), we can get the distance traveled to reach the speed of 12 m/s as follows:

$$s = 3.468\frac{(3.46)^2}{2} = \boxed{20.76 \text{ m}}$$

Example 12.2

A charged particle is shown in Fig. 12.3 at time $t = 0$ between large parallel condenser plates separated by a distance d in a vacuum. A time-varying voltage V (notation not to be confused with velocity) given as

$$V = 6 \sin \omega t \tag{a}$$

is applied to the plates. What is the motion of the particle if it has a charge q coulombs and if we do not consider gravity?

As we learned in physics, the electric field E becomes for this case

$$E = \frac{V}{d} \tag{b}$$

The force on the particle is qE and the resulting motion is that of rectilinear translation. Using **Newton's law** we accordingly have

$$\frac{d^2x}{dt^2} = q\frac{6 \sin \omega t}{md} \tag{c}$$

Rewriting Eq. (c), we have

$$d\left(\frac{dx}{dt}\right) = q\frac{6 \sin \omega t}{md}\,dt$$

Integrating, we get

$$\frac{dx}{dt} = -\frac{6q}{\omega md}\cos \omega t + C_1 \tag{d}$$

$$x = -\frac{6q}{\omega^2 md}\sin \omega t + C_1 t + C_2 \tag{e}$$

Applying the initial conditions $x = b$ and $dx/dt = 0$ when $t = 0$, we see that $C_1 = 6q/m\omega d$ and $C_2 = b$. Thus, we get

$$x = -\frac{6q}{\omega^2 md}\sin \omega t + \frac{6q}{m\omega d}t + b$$

The motion of the charged particle will be that of sinusoidal oscillation in which the center of the oscillation drifts from left to right.

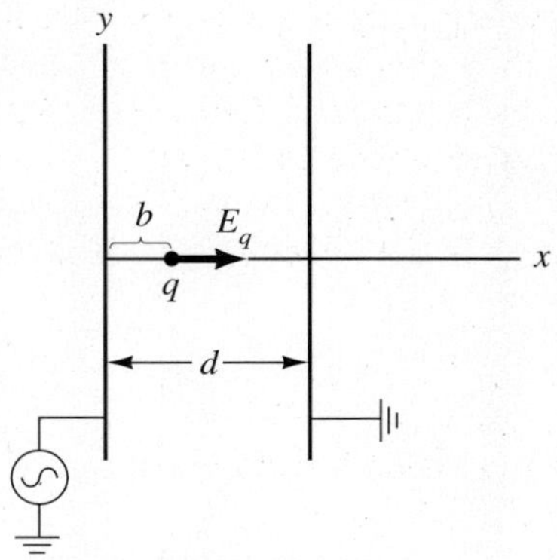

Figure 12.3. Charged particle between condenser plates.

Case 2. Force Is a Function of Speed. We next consider the case where the resultant force on the particle depends only on the value of the speed of the particle. An example of such a force is the aerodynamic drag force on an airplane or missile.

We can express *Newton's law* in the following form:

$$\frac{dV}{dt} = \frac{F(V)}{m} \tag{12.8}$$

where $F(V)$ is a piecewise continuous function representing the force in the positive x direction. If we rearrange the equation in the following manner (this is called *separation* of *variables*):

$$\frac{dV}{F(V)} = \frac{1}{m}\,dt$$

we can integrate to obtain

$$\int \frac{dV}{F(V)} = \frac{1}{m}t + C_1 \tag{12.9}$$

The result will give t as a function of V. However, we will generally prefer to solve for V in terms of t. The result will then have the form

$$V = H(t, C_1)$$

where H is a function of t and the constant of integration C_1. A second integration may now be performed by first replacing V by dx/dt and bringing dt over to the right side of the equation. We then get on integration

$$x = \int H(t, C_1)\,dt + C_2 \tag{12.10}$$

The constants of integration are determined from the initial conditions of the problem.

Example 12.3

A high-speed land racer (Fig. 12.4) is moving at a speed of 100 m/s. The resistance to motion of the vehicle is primarily due to aerodynamic drag, which for this speed can be approximated as $0.2V^2$ N with V in m/s. If the vehicle has a mass of 4 Mg, what distance will it coast before its speed is reduced to 70 m/s?

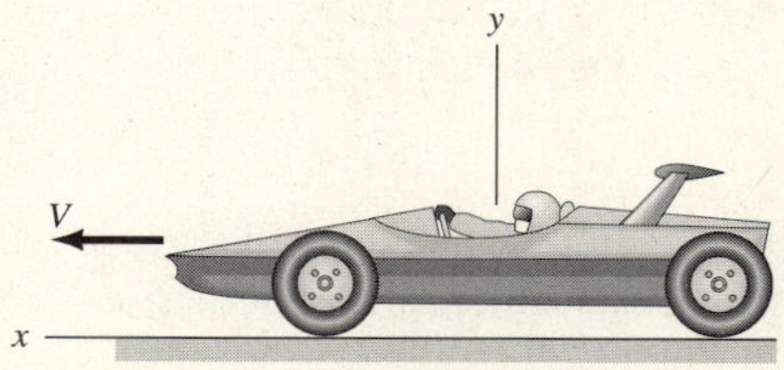

Figure 12.4. High-speed racer.

We have, using **Newton's law** for this case,

$$\frac{dV}{dt} = -\frac{0.2V^2}{4\times 10^3} = -5\times 10^{-5}V^2 \tag{a}$$

Separating the variables, we get

$$\frac{dV}{V^2} = -5\times 10^{-5}\,dt \tag{b}$$

Example 12.3 (Continued)

Integrating, we have

$$-\frac{1}{V} = -5 \times 10^{-5}t + C_1 \tag{c}$$

Taking $t = 0$ when $V = 100$, we get $C_1 = -1/100$. Replacing V by dx/dt, we have next

$$\frac{1}{V} = \frac{dt}{dx} = 5 \times 10^{-5}t + \frac{1}{100} \tag{d}$$

Separating variables once again, we get

$$\frac{dt}{5 \times 10^{-5}t + (1/100)} = dx$$

To integrate, we perform a change of variable. Thus

$$\eta = 5 \times 10^{-5}t + (1/100)$$
$$\therefore\ d\eta = 5 \times 10^{-5}\,dt$$

We then have as a replacement for our equation

$$\frac{d\eta}{\eta} = 5 \times 10^{-5}\,dx$$

Now integrating and replacing η, we get

$$\ln\left(5 \times 10^{-5}t + \frac{1}{100}\right) = 5 \times 10^{-5}x + C_2$$

When $t = 0$, we take $x = 0$ and so $C_2 = \ln(1/100)$. We then have on combining the logarithmic terms:

$$\ln\left(5 \times 10^{-3}t + 1\right) = 5 \times 10^{-5}x \tag{e}$$

Substitute $V = 70$ in Eq. (d); solve for t. We get $t = 85.7$ s. Finally, find x for this time from Eq. (e). Thus,

$$\ln\left[(5 \times 10^{-3})(85.7) + 1\right] = 5 \times 10^{-5}x$$

Therefore,

$$\boxed{x = 7.13 \text{ km}}$$

The distance traveled is then 7.13 km.

Example 12.4

A conveyor is inclined 20° from the horizontal as shown in Fig. 12.5. As a result of spillage of oil on the belt, there is a viscous friction force between body D and the belt. This force equals 1.5 N per unit relative velocity between body D and the belt. The belt moves at a constant speed V_B up the conveyor while initially body D has a speed $(V_D)_0 = 0.6$ m/s relative to the ground in a direction down the conveyor. What speed V_B^* should the belt have in order for body D to be able to eventually approach a zero velocity relative to the ground? For belt speed V_B^*, and for the given initial speed of body D, namely $(V_D)_0 = 0.6$ m/s, determine the time when body D attains a speed of 0.3 m/s relative to the ground. The mass of D is 2.5 kg.

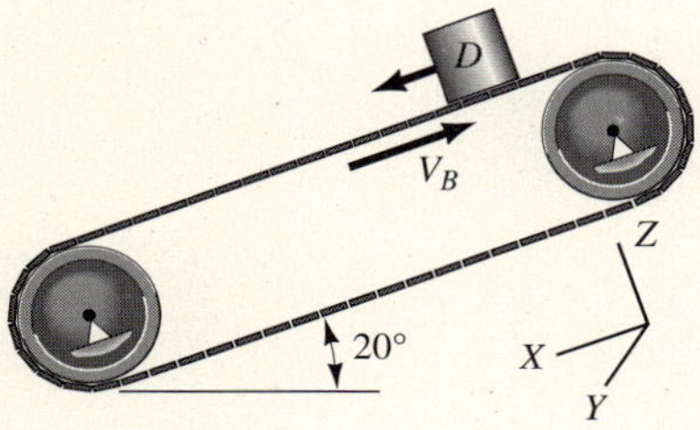

Figure 12.5. A body slides down a conveyor belt wet with oil.

We begin by assigning axes for the problem as follows (see Fig. 12.6):

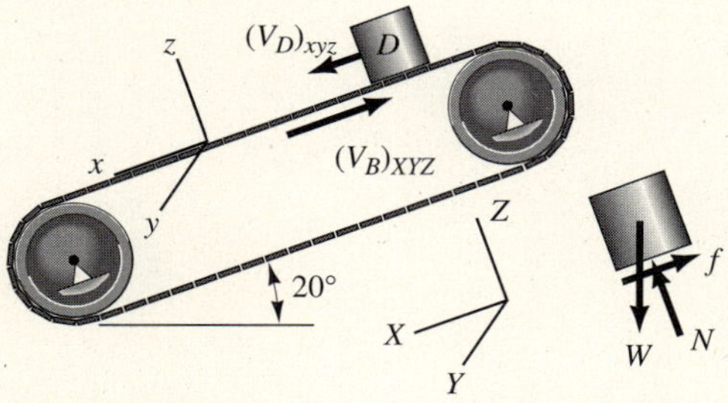

Figure 12.6. Friction force f is 0.1 times the relative velocity between body D and the belt.

Fix xyz to the belt
Fix XYZ to the ground

From **kinematics** we can say

$$(\boldsymbol{V}_D)_{XYZ} = (\boldsymbol{V}_D)_{xyz} + \dot{\boldsymbol{R}} \qquad \text{Note that } \dot{\boldsymbol{R}} = -(V_B)_{XYZ}\boldsymbol{i}$$

$$\therefore (\boldsymbol{V}_D)_{xyz} = (\boldsymbol{V}_D)_{XYZ} - \dot{\boldsymbol{R}} = \left[V_D - (-V_B)\right]_{XYZ}\boldsymbol{i} = (V_D + V_B)_{XYZ}\boldsymbol{i}$$

Example 12.4 (Continued)

For the friction force f we have

$$f = -(0.1)(V_D)_{xyz} = -(0.5)(V_D + V_B)_{XYZ}\boldsymbol{i}$$

We may now use **Newton's law** for body D in the x direction. Since all velocities from here on will be relative to the ground, we can dispense with the reference subscripts. Thus

$$2.5\frac{dV_D}{dt} = -(1.5)(V_D + V_B) + 2.5g\sin 20° \qquad \text{(a)}$$

When body D attains a theoretical permanent zero velocity relative to the ground, V_D and $\frac{dV_D}{dt}$ are equal to zero. This gives us (noting that V_B now becomes V_B^*)

$$0 = -(1.5)(0 + V_B^*) + 25g\sin 20° \quad \therefore \quad V_B^* = 5.59 \text{ m/s}$$

Now determine the time for body D to attain a velocity of 0.3 m/s relative to the ground for a belt speed of 5.59 m/s. For this we go back to Eq. (a).

$$2.5\frac{dV_D}{dt} = (-1.5)(V_D + 5.59) + 2.5g\sin 20° = -1.5V_D$$

$$\frac{dV_D}{V_D} = -\left(\frac{1.5}{2.5}\right)dt = -0.6dt$$

$$\therefore \ln V_D = -0.6t + C_1$$

When

$$t = 0, \quad V_D = 0.6 \text{ m/s}, \quad \therefore C_1 = \ln 0.6$$

Hence, on combining log terms[3]

$$\ln\left(\frac{V_D}{0.6}\right) = -0.6t$$

Set $V_D = 0.3$. Solve for t.

$$t = -\frac{1}{0.6}\ln(0.5) = \quad 1.155 \text{ s}$$

[3]Note from this equation that $V_D = 0.6e^{-0.6t}$ and that $\dot{V}_D = -0.36e^{-0.6t}$ and so we see that as t approaches infinity both of these quantities approach zero. Thus, theoretically body D could approach a permanent zero velocity relative to the ground.

Case 3. Force Is a Function of Position. As the final case of this series, we now consider the rectilinear motion of a body under the action of a force that is expressible as a function of position. Perhaps the simplest example of such a case is the frictionless mass–spring system shown in Fig. 12.7. The body is shown at a position where the spring is unstrained. The horizontal force from the spring at all positions of the body clearly will be a function of position x.

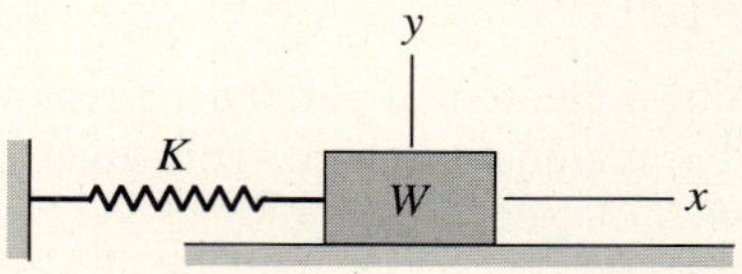

Figure 12.7. Mass–spring system.

Newton's law for position-dependent forces can be given as

$$m\frac{dV}{dt} = F(x) \tag{12.11}$$

We cannot separate the variables for this form of the equation as in previous cases since there are three variables (V, t, and x). However, by using the chain rule of differentiation, we can change the left side of the equation to a more desirable form in the following manner:

$$m\frac{dV}{dt} = m\frac{dV}{dx}\frac{dx}{dt} = mV\frac{dV}{dx}$$

We can now separate the variables in Eq. 12.11 as follows:

$$mV\,dV = F(x)\,dx$$

Integrating, we get

$$\frac{mV^2}{2} = \int F(x)\,dx + C_1 \tag{12.12}$$

Solving for V and using dx/dt in its place, we get

$$\frac{dx}{dt} = \left[\frac{2}{m}\int F(x)\,dx + C_1\right]^{1/2}$$

Separating variables and integrating again, we get

$$t = \int \frac{dx}{\left[\frac{2}{m}\int F(x)\,dx + C_1\right]^{1/2}} + C_2 \tag{12.13}$$

For a given $F(x)$, V and x can accordingly be evaluated as functions of time from Eqs. 12.12 and 12.13. The constants of integration C_1 and C_2 are determined from the initial conditions.

A very common force that occurs in many problems is the *linear restoring force*. Such a force occurs when a body W is constrained by a linear spring (see Fig. 12.7). The force from such a spring will be proportional to x measured from a position of W corresponding to the *undeformed configuration* of the system. Consequently, the force will have a magnitude of $|Kx|$, where K, called the *spring constant*, is the force needed on the spring per unit elongation or compression of the spring. Furthermore, when x has a positive value, the spring force points in the negative direction, and when x is negative, the spring force points in the positive direction. That is, it always points toward the position $x = 0$ for which the spring is undeformed. The spring force is for this reason called a *restoring* force and must be expressed as $-Kx$ to give the proper direction for all values of x.

For a *nonlinear* spring, K will not be constant but will be a function of the elongation or shortening of the spring. The spring force is then given as

$$F_{\text{spring}} = -\int_0^x K(x)\,dx \tag{12.14}$$

In the following example and in the homework problems, we examine certain limited aspects of spring–mass systems to illustrate the formulations of case 3 and to familiarize us with springs in dynamic systems. A more complete study of spring–mass systems will be made in Chapter 19. The motion of such systems, we shall later learn, centers about some stationary point. That is, the motion is *vibratory* in nature. We shall study vibrations in Chapter 19, wherein time-dependent and velocity-dependent forces are present simultaneously with the linear restoring force. We are deferring this topic so as to make maximal use of your course in differential equations that you are most likely studying concurrently with dynamics. It is important to understand, however, that even though we defer vibration studies until later, such studies are not something apart from the general particle dynamics undertaken in this chapter.

Example 12.5

A cart A (see Fig. 12.8) having a mass of 200 kg is held on an incline so as to just touch an undeformed spring whose spring constant K is 50 N/mm. If body A is released very slowly, what distance down the incline must A move to reach an equilibrium configuration? If body A is released suddenly, what is its speed when it reaches the aforementioned equilibrium configuration for a slow release?

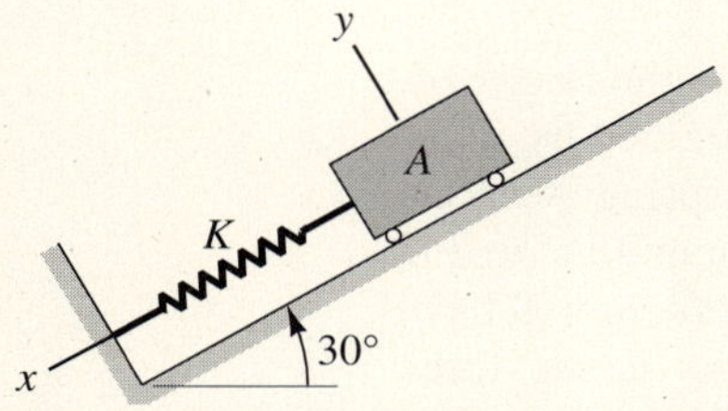

Figure 12.8. Cart–spring system.

As a first step, we have shown a free body of the vehicle in Fig. 12.9. To do the first part of the problem, all we need do is utilize the

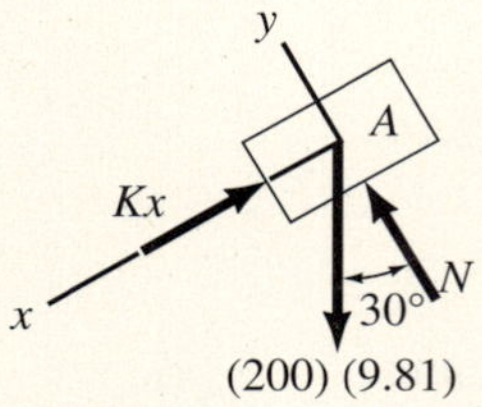

Figure 12.9. Free-body diagram of cart.

definition of the spring constant. Thus, if δ represents the compression of the spring, we can say:

$$K = \frac{F}{\delta}$$

Example 12.5 (Continued)

Therefore,

$$\delta = \frac{F}{K} = \frac{(200)(9.81)\sin 30^\circ}{50}$$

$$\delta = 19.62 \text{ mm}$$

Thus, the spring will be compressed 0.01962 m by the cart if it is allowed to move down the incline very slowly.

For the case of the quick release, we use **Newton's law**. Thus, using x in metres so that K is (50)(1000) N/m:

$$200\ddot{x} = (200)(9.81)\sin 30^\circ - (50)(1000)(x)$$

Therefore,

$$\ddot{x} = 4.905 - 250x$$

Rewriting $\ddot{x}$, we have

$$V\frac{dV}{dx} = 4.905 - 250x$$

Separating variables and integrating,

$$\frac{V^2}{2} = 4.905x - 125x^2 + C_1$$

To determine the constant of integration C_1, we set $x = 0$ when $V = 0$. Clearly, $C_1 = 0$. As a final step, we set $x = 0.01962$ m and solve for V.

$$V = \left\{2\left[(4.905)(0.01962) - (125)(0.01962)^2\right]\right\}^{1/2}$$

$$V = 0.310 \text{ m/s}$$

The following example illustrates an interesting device used by the U.S. Navy to test small devices for high, prolonged acceleration. Hopefully, the length of the problem will not intimidate you. Use is made of the gas laws presented in your elementary chemistry courses.

Example 12.6

An *air gun* is used to test the ability of small devices to withstand high prolonged accelerations. A "floating piston" A (Fig. 12.10), on which the device to be tested is mounted, is held at position C while region D is filled with highly compressed air. Region E is initially at atmospheric pressure but is entirely sealed from the outside. When "fired," a quick-release mechanism releases the piston and it accelerates rapidly toward the other end of the gun, where the trapped air in E "cushions" the motion so that the piston will begin eventually to return. However, as it starts back, the high pressure developed in E is released through valve F and the piston only returns a short distance.

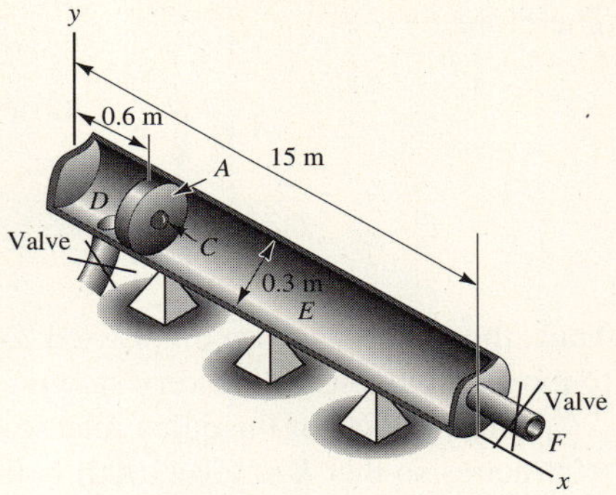

Figure 12.10. Air gun.

Suppose that the piston and its test specimen have a combined mass of 1 kg and the pressure initially in the chamber D is 7 MPa (above atmosphere). Compute the speed of the piston at the halfway point of the air gun if we make the simple assumption that the air in D expands according to pv = constant and the air in E is compressed also according to pv = constant.[4] Note that v is the specific volume (i.e., the volume per unit mass). Take v of this fluid at D to be initially 13×10^{-3} m^3/kg and v in E to be initially 0.82 m^3/kg. Neglect the inertia of the air.

The force on the piston results from the pressures on each face, and we can show that this force is a function of x (see Fig. 12.10 for reference axes). Thus, examining the pressure p_D first for region D, we have, from initial conditions,

$$(p_D v_D)_0 = (7\times10^6 + 0.1\times10^6)(13\times10^{-3}) = 92.3\times10^3 \qquad \text{(a)}$$

Furthermore, the mass of air D given as M_D is determined from initial data as

$$M_D = \frac{(V_D)_0}{(v_D)_0} = \frac{(0.6)\left(\frac{\pi}{4}\right)(0.3)^2}{13\times10^{-3}} = 3.26 \text{ kg} \qquad \text{(b)}$$

where $(V_D)_0$ is the volume of the air in D initially. Noting that pv = const. and then using the right side of Eq. (a) for $p_D v_D$ as well as the first part of Eq. (b) for v_D, we can determine p_D at any position x of the piston:

$$p_D = \frac{92.3\times10^3}{v_D} = \frac{92.3\times10^3}{V_D/M_D} = \frac{92.3\times10^3}{(\pi/4)(0.3)^2(x)/3.26}$$

Therefore,

$$p_D = \frac{4.257\times10^6}{x} \qquad \text{(c)}$$

[4]You should recall from your earlier work in physics and chemistry that we are using here the isothermal form of the equation of state for a perfect gas. Two factors of caution should be pointed out relative to the use of this expression. First, at the high pressures involved in part of the expansion, the perfect gas model is only an approximation for the gas, and so the equation of state of a perfect gas that gives us pv = constant is only approximate. Furthermore, the assumption of isothermal expansion gives only an approximation of the actual process. Perhaps a better approximation is to assume an adiabatic expansion (i.e, no heat transfer). This is done in Problem 12.130.

Example 12.6 (Continued)

We can similarly get p_E as a function of x for region E. Thus,

$$(p_E v_E)_0 = (0.1\times10^6)(0.82) = 82\times10^3$$

and

$$M_E = \frac{(V_E)_0}{(v_E)_0} = \frac{(14.4)\left(\frac{\pi}{4}\right)(0.3)^2}{0.82} = 1.24 \text{ kg}$$

Hence, at position x of the piston

$$p_E = \frac{82\times10^3}{v_E} = \frac{82\times10^3}{V_E/M_E} = \frac{82\times10^3}{(\pi/4)(0.3)^2(15-x)/1.24}$$

Therefore,

$$p_E = \frac{1.438\times10^6}{15-x}$$

Now we can write **Newton's law** for this case. Noting that V without subscripts is velocity and not volume,

$$MV\frac{dV}{dx} = \frac{\pi(0.3)^2}{4}(p_D - p_E) = \frac{\pi}{4}(0.3)^2\left(\frac{4.257\times10^6}{x} - \frac{1.438\times10^6}{15-x}\right) \quad \text{(d)}$$

where M is the mass of piston and load. Separating variables and integrating, we get

$$\frac{MV^2}{2} = \frac{\pi}{4}(0.3)^2[4.257\times10^6 \ln x + 1.438\times10^6 \ln(15-x)] + C_1 \quad \text{(e)}$$

To get the constant C_1, set $V = 0$ when $x = 0.6$ m. Hence,

$$C_1 = -\frac{\pi}{4}(0.3)^2(4.257\times10^6 \ln 0.6 + 1.438\times10^6 \ln 14.4)$$

Therefore,

$$C_1 = -117.4\times10^3$$

Substituting C_1 in Eq. (e), we get

$$V = \left(\frac{2}{M}\right)^{1/2}$$
$$\left\{\frac{\pi}{4}(0.3)^2\left[4.257\times10^6 \ln x + 1.438\times10^6 \ln(15-x)\right] - 117.4\times10^3\right\}^{1/2}$$

We may rewrite this as follows noting that $M = 1$ kg:

$$V = 1.414\left[300.9\times10^3 \ln x + 101.6\times10^3 \ln(15-x) - 117.4\times10^3\right]^{1/2}$$

At $x = 7.5$ m, we then have for V the desired result:

$$V = 1.414\left(300.9\times10^3 \ln 7.5 + 101.6\times10^3 \ln 7.5 - 117.4\times10^3\right)^{1/2}$$

$$V = 1.178 \text{ km/s}$$

*Example 12.7

A light stiff rod is pinned at A and is constrained by two linear springs, $K_1 = 1000$ N/m and $K_2 = 1200$ N/m. The springs are unstretched when the rod is horizontal. At the right end of the rod, a mass $M = 5$ kg is attached. If the rod is rotated 12° *clockwise* from a horizontal configuration and then released, what is the speed of the mass when the rod returns to a position corresponding to the *static equilibrium* position with mass M attached?

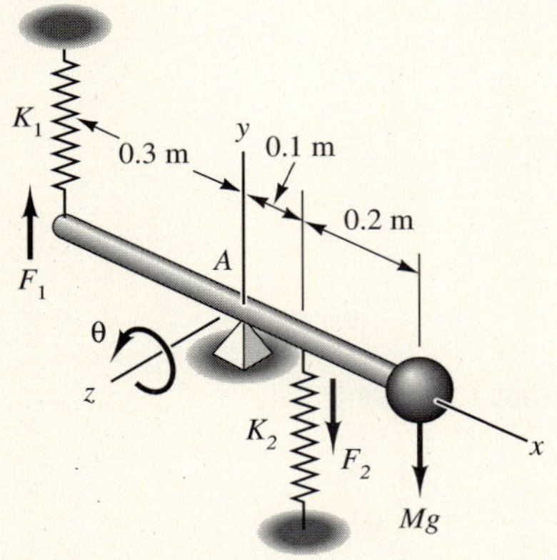

Figure 12.11. Two linear springs and a particle on a weightless rigid rod. Spring forces shown for positive θ.

A free-body diagram of the system for *positive* θ is shown in Fig. 12.12(a) and a free-body diagram of the particle M is shown in

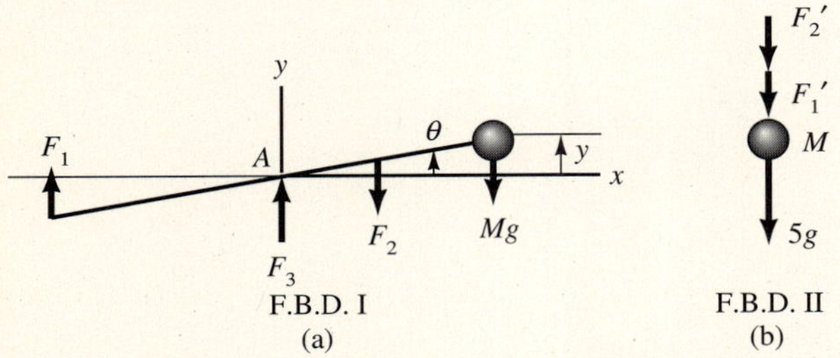

Figure 12.12. Free-body diagrams of the system and the particle for positive θ.

Fig. 12.12(b). The spring forces F_1 and F_2 on the rod are given as follows for small positive rotations θ:

$$F_1 = (0.3)(\theta)(K_1) = 300\,\theta \text{ N}$$
$$F_2 = -(0.1)(\theta)(K_2) = -120\,\theta \text{ N} \qquad \text{(a)}$$

where θ is in radians. In the first free body, we will think of the rod as a massless perfectly rigid lever as studied in high school or perhaps even

Example 12.7 (Continued)

earlier. Then we can say for the forces on the second of our free bodies stemming from the springs[5]

$$F_1' \quad (\text{from } F_1) = -F_1 = -300\,\theta\,\text{N}$$
$$F_2' \quad (\text{from } F_2) = \frac{0.1}{0.3}(F_2) = -40\,\theta\,\text{N}$$

We can now give **Newton's law** for M as follows using y for the vertical coordinate of the particle:

$$5\ddot{y} = -5g - 300\,\theta - 40\,\theta$$
$$\therefore\ \ddot{y} = -g - \frac{340}{5}\,\theta \qquad \text{(b)}$$

Next, from **kinematics**, we can say for small rotation

$$\theta = \frac{y}{0.3}$$

Now going back to Eq. (b) we replace $\ddot{y}$ by

$$\frac{d\dot{y}}{dt} = \left(\frac{d\dot{y}}{dy}\right)\left(\frac{dy}{dt}\right) = \dot{y}\left(\frac{d\dot{y}}{dy}\right)$$

in order to be able to separate variables. Also replace θ by $y/0.3$. We then may say

$$\dot{y}\,d\dot{y} = (-227y - g)\,dy$$

Integrating

$$\frac{\dot{y}^2}{2} = (-227)\left(\frac{y^2}{2}\right) - gy + C_1 \qquad \text{(c)}$$

When $\theta = -12° = -\left(\frac{12}{360}\right)(2\pi)$ rad $= -\ 2.094$ rad, $\dot{\theta} = \dot{y} = 0.$ We can then solve for the constant of integration using $y = 0.3\ \theta.$

$$C_1 = (227)\left[\frac{[(0.3)(-0.2094)]^2}{2}\right] + (9.81)(0.3)(-0.2094) = -0.1683$$

Hence,

$$\dot{y}^2 = 2\left[(-227)\left(\frac{y^2}{2}\right) - gy - 0.1683\right] \qquad \text{(d)}$$

[5]Note that a positive θ gives negative values for F_1' and F_2' on M and vice versa. It is for this reason that we require the minus signs.

Example 12.7 (Continued)

For the static **equilibrium** configuration of the rod, we require from Fig. 12.12(a)

$$\underline{\sum M_A = 0:}$$

$$\therefore -(Mg)(0.3) - F_1(0.3) - F_2(0.1) = 0$$

Substituting values from Eqs. (a) and noting that we are only using the magnitudes of the forces above for the required negative moments we get

$$-(5)(9.81)(0.3) - (300\,\theta_{Eq})(0.3) - (120\,\theta_{Eq})(0.1) = 0$$

Solving for θ_{Eq}

$$\theta_{Eq} = -0.1443 \text{ rad}$$

Hence

$$y_{Eq} = (0.3)(-0.1443) = -0.04328 \text{ m/s}$$

Now go to Eq. (d) and substitute y_{Eq}. We get

$$\dot{y}_{Eq} = \sqrt{2}\left[(-113.5)(-0.04328)^2 - (9.81)(-0.04328) - 0.1683\right]^{1/2}$$

The desired result is then

$$\dot{y}_{Eq} = 0.968 \text{ m/s}$$

12.4 A Comment

In Part A, we have considered only rectilinear motions of particles. Actually in Chapter 11, we considered the coplanar motion of particles having a constant acceleration of gravity in the minus y direction and zero acceleration in the x direction. These were the *ballistic* problems. We treated them earlier in Chapter 11 because the considerations were primarily kinematic in nature. In the present chapter, they correspond to the coplanar motion of a particle having a constant force in the minus y direction along with an initial velocity component in this direction, plus a zero force in the x direction, with a possible initial velocity component in this direction. Therefore, in the context of Chapter 12 we would have integrated two scalar equations of Newton's law in rectangular components (Eqs. 12.3) for a single particle. The resulting motion is sometimes called *curvilinear* translation.

PROBLEMS

12.1. A particle of mass 15 kg is moving in a constant force field given as

$$\boldsymbol{F} = 3\boldsymbol{i} + 10\boldsymbol{j} - 5\boldsymbol{k} \text{ lb}$$

The particle starts from rest at position (3, 5, –4). What is the position and velocity of the particle at time $t = 8$ s? What is the position when the particle is moving at a speed of 6 m/s?

12.2. A particle of mass m is moving in a constant force field given as

$$\boldsymbol{F} = 2m\boldsymbol{i} - 12m\boldsymbol{j} \text{ N}$$

Give the vector equation for $\boldsymbol{r}(t)$ of the particle if, at time $t = 0$, it has a velocity $\boldsymbol{V}_0$ given as

$$\boldsymbol{V}_0 = 6\boldsymbol{i} + 12\boldsymbol{j} + 3\boldsymbol{k} \text{ m/s}$$

Also, at time $t = 0$, it has a position given as

$$\boldsymbol{r}_0 = 3\boldsymbol{i} + 2\boldsymbol{j} + 4\boldsymbol{k} \text{ m}$$

What are coordinates of the body at the instant that the body reaches its maximum height, y_{max}?

12.3. A block is permitted to slide down an inclined surface. The coefficient of friction is 0.05. If the velocity of the block is 9 m/s on reaching the bottom of the incline, how far up was it released and how many seconds has it traveled?

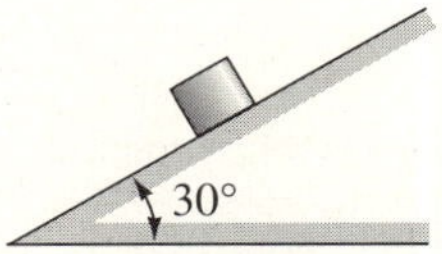

Figure P.12.3.

12.4. An arrow is shot upward with an initial speed of 24 m/s. How high up does it go and how long does it take to reach the maximum elevation if we neglect friction?

12.5. A mass D at $t = 0$ is moving to the left at a speed of 0.6 m/s relative to the ground on a belt that is moving at constant speed to the right at 1.6 m/s. If there is coulombic friction present with $\mu_d = 0.3$, how long does it take before the speed of D relative to the belt is 0.3 m/s to the left?

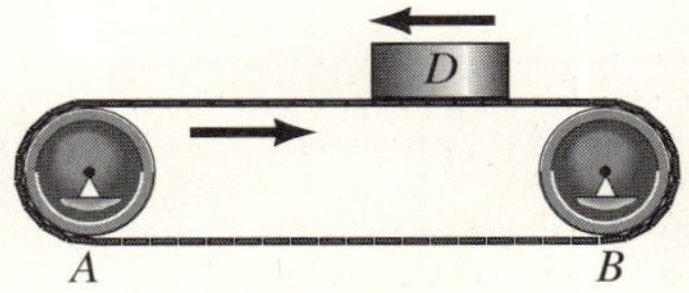

Figure P.12.5.

12.6. Do Problem 12.5 with the belt system inclined 15° with the horizontal so that end B is above end A.

12.7. A drag racer can develop a torque of 270 N m on each of the rear wheels. If we assume that this maximum torque is maintained and that there is no wind friction, what is the time to travel 400 m from a standing start? What is the speed of the vehicle at the 400 m mark? The weight of the racer and the driver combined is 7 kN. For simplicity, neglect the rotational effects of the wheels.

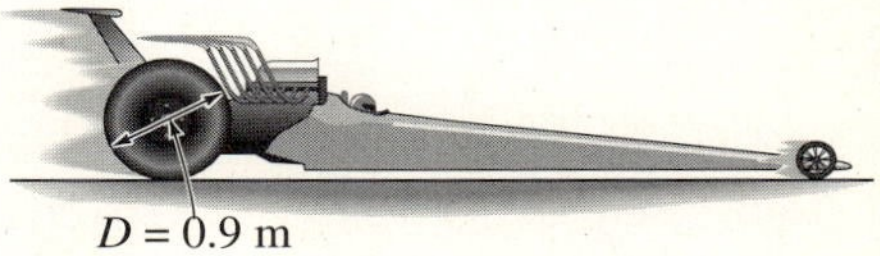

Figure P.12.7.

12.8. A truck is moving down a 10° incline. The driver strongly applies his brakes to avoid a collision and the truck decelerates at the steady rate of 1 m/s². If the static coefficient of friction μ_s between the load W and the truck trailer is 0.3, will the load slide or remain stationary relative to the truck trailer? The weight of W is 4.5 kN and it is not held to the truck by cables.

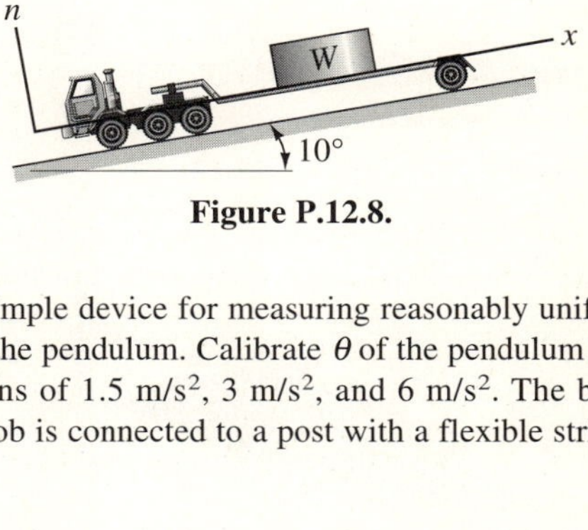

Figure P.12.8.

12.9. A simple device for measuring reasonably uniform accelerations is the pendulum. Calibrate θ of the pendulum for vehicle accelerations of 1.5 m/s², 3 m/s², and 6 m/s². The bob weighs 5 N. The bob is connected to a post with a flexible string.

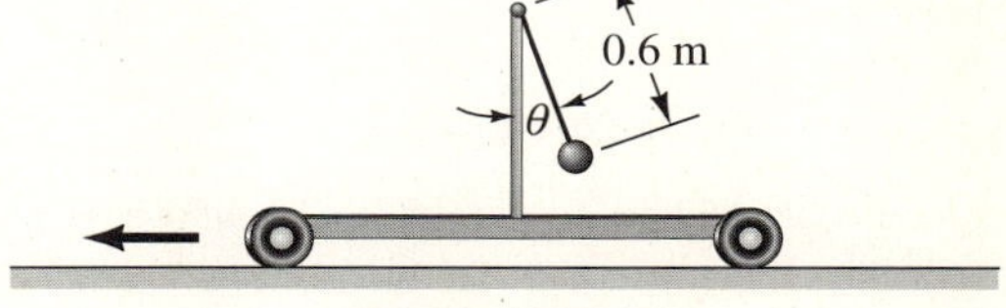

Figure P.12.9.

12.10. A piston is being moved through a cylinder. The piston is moved at a constant speed V_p of 0.6 m/s relative to the ground by a force F. The cylinder is free to move along the ground on small wheels. There is a coulombic friction force between the piston and the cylinder such that $\mu_d = 0.3$. What distance d must the piston move relative to the ground to advance 10 mm along the cylinder if the cylinder is stationary at the outset? The piston has a mass of 2.5 kg and the cylinder has a mass 5 kg.

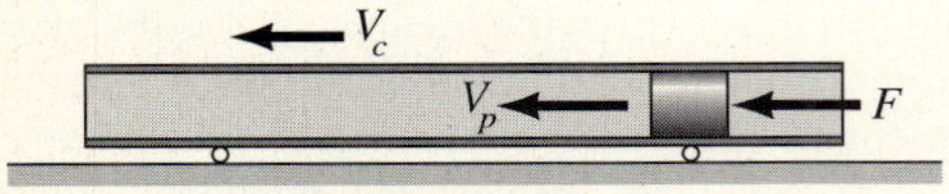

Figure P.12.10.

12.11. A force F of 5 kN is suddenly applied to mass A. What is the speed after A has moved 0.1 m? Mass B is a triangular block of uniform thickness.

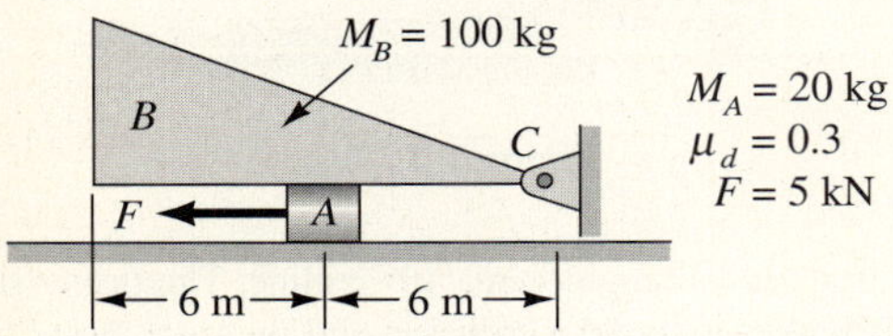

Figure P.12.11.

12.12. A fighter plane is moving on the ground at a speed of 100 m/s when the pilot deploys the braking parachute. How far does the plane move to get down to a speed of 60 m/s? The plane has a mass of 8 Mg. The drag is $27.5V^2$ with V in m/s.

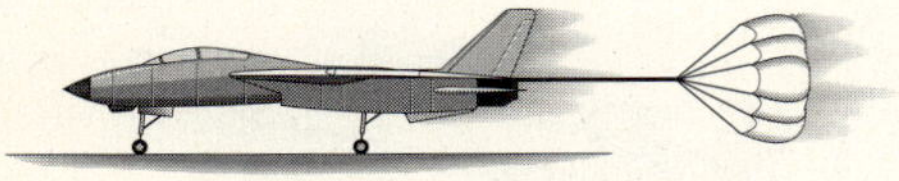

Figure P.12.12.

12.13. Blocks A and B are initially stationary. How far does A move along B if A moves 0.2 m relative to the ground?

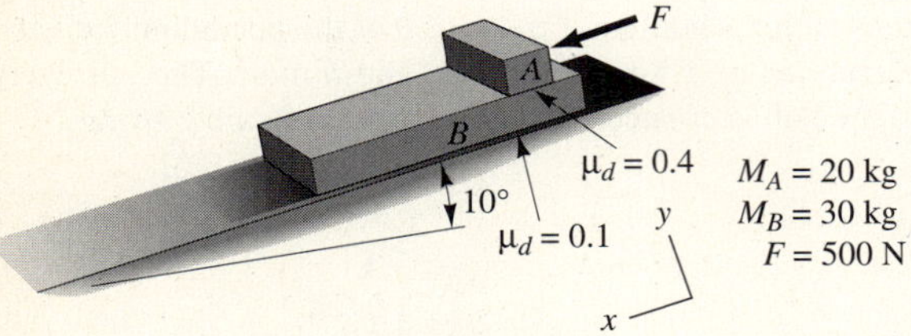

Figure P.12.13.

12.14. A 30 N block at the position shown has a force $F = 100$ N applied suddenly. What is its velocity after moving 1 m? Also, how far does the block move before stopping? Member AB weighs 200 N.

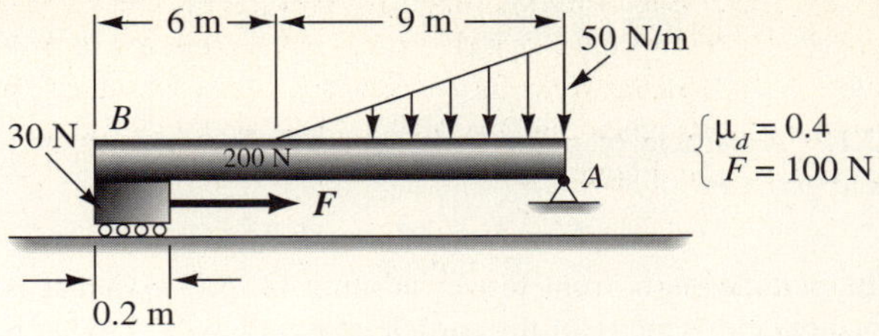

Figure P.12.14.

12.15. A block B of mass M is being pulled up an incline by a force F. If μ_d is 0.3, at what angle α will the force F cause the maximum steady acceleration?

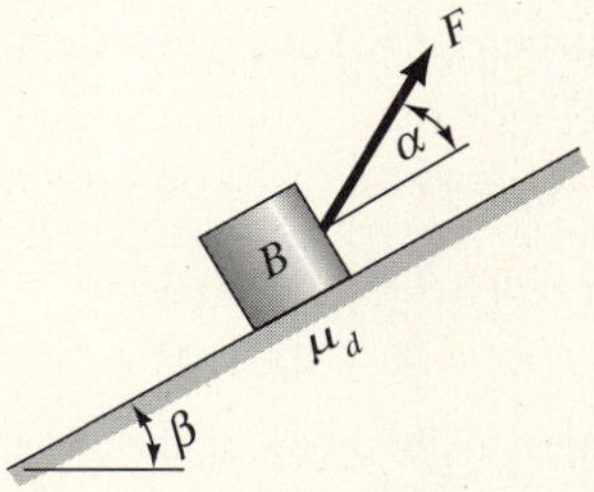

Figure P.12.15.

12.16. A 10 kN force is applied to body B whose mass is 15 kg. Body A has a mass of 20 kg. What is the speed of B after it moves 3 m? Take $\mu_d = 0.28$. The center of mass of body A is at its geometric center.

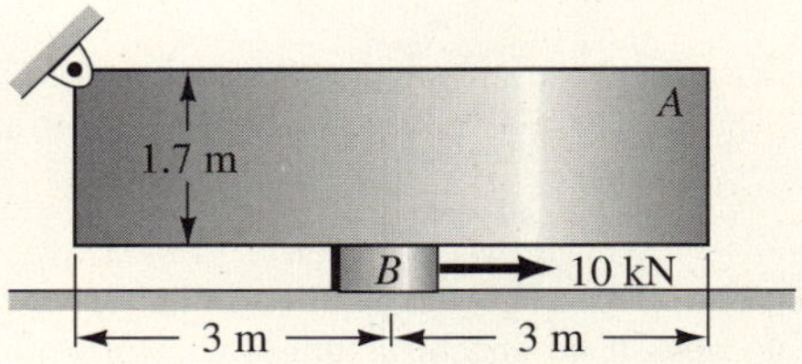

Figure P.12.16.

12.17. A constant force F is applied to the body A when it is in the position shown. What should F be if A is to attain a velocity of 2 m/s after moving 1 m? The spring is unstretched at the position shown.

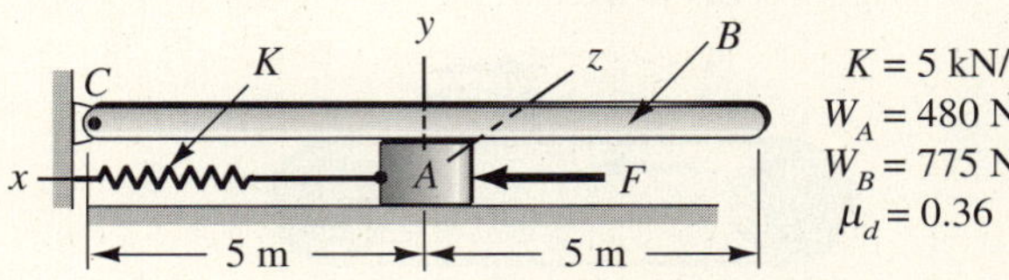

Figure P.12.17.

12.18. Two slow moving steam roller vehicles are moving in opposite directions on a straight path. They start at A and B at the time $t = 0$. How far from point A do they pass each other? What are their speeds when this happens? [*Hint:* Show that the time for this is 1.5 hours.] Note t is in hours.

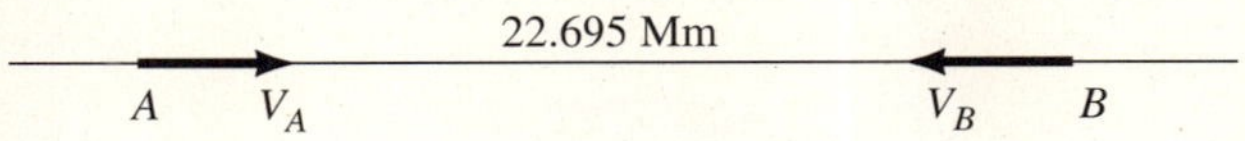

$$V_A = 6t + \sqrt{3t} + 3 \text{ km/h} \quad V_B = 5 + t^{2/3} + 0.5t^{1/3} \text{ km/h}$$

Figure P.12.18.

12.19. As you learned in chemistry, the *coefficient of viscosity* μ is a measure, roughly speaking, of the "stickiness" of a fluid. To measure this property for a highly viscous liquid-like oil, we let a small sphere of metal of radius R descend in a container of the liquid. From fluid mechanics, we know that a drag force will be developed from the oil given by the formula

$$F = 6\pi\mu VR$$

This relation is called *Stoke's law*. The other forces acting on the sphere are its weight (take the density of the sphere as ρ_{Sphere}) and the buoyant force, which is the weight of the oil displaced (take the density of the oil as ρ_{Oil}). The sphere will reach a constant velocity called the *terminal velocity* denoted as $V_{\text{Term.}}$. Show that

$$\mu = \frac{2}{9}\frac{gR^2}{V_{\text{Term.}}}\left(\rho_{\text{Sphere}} - \rho_{\text{Oil}}\right)$$

12.20. A force F is applied to a system of light pulleys to pull body A. If F is 10 kN and A has a mass of 5 Mg, what is the speed of A after 1 s starting from rest?

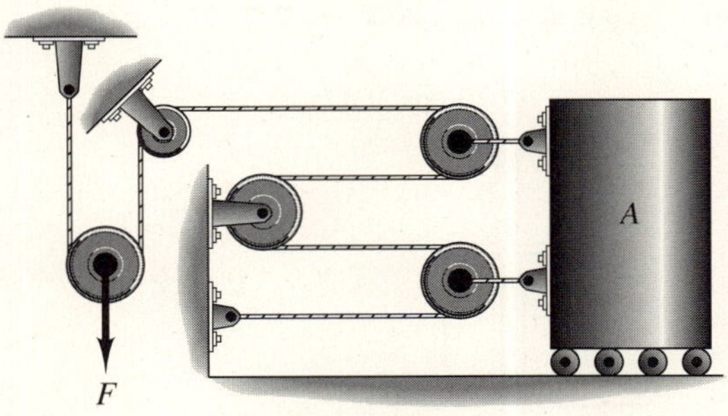

Figure P.12.20.

12.21. A force represented as shown acts on a body having a mass of 15 kg. What is the position and velocity at $t = 30$ s if the body starts from rest at $t = 0$?

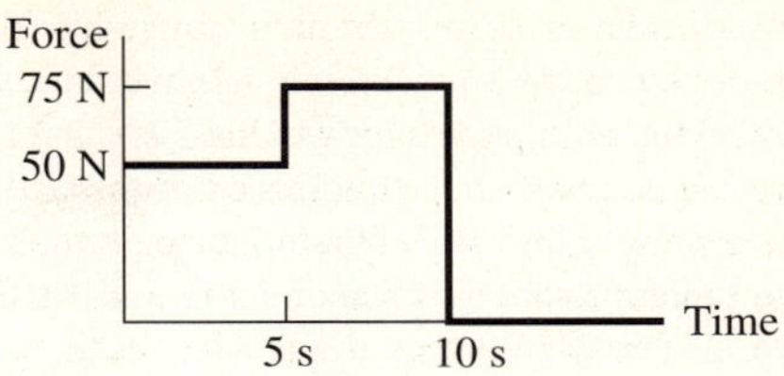

Figure P.12.21.

12.22. A body of mass 1 kg is acted on by a force as shown in the diagram. If the velocity of the body is zero at $t = 0$, what is the velocity and distance traversed when $t = 1$ min? The force acts for only 45 s.

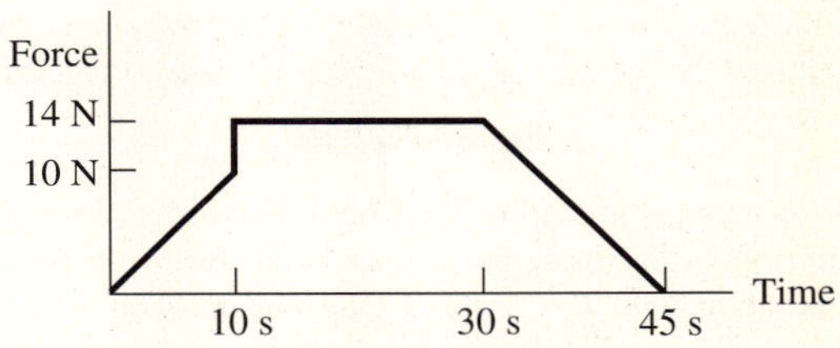

Figure P.12.22.

12.23. Three coupled streetcars are moving down an incline at a speed of 6 m/s when the brakes are applied for a panic stop. All the wheels lock except for car B, where due to a malfunction all the brakes on the front end of the car do not operate. How far does the system move and what are the forces in the couplings between the cars? Each streetcar weighs 220 kN and the coefficient of dynamic friction μ_d between wheel and rail is 0.30. Weight is equally distributed on the wheels.

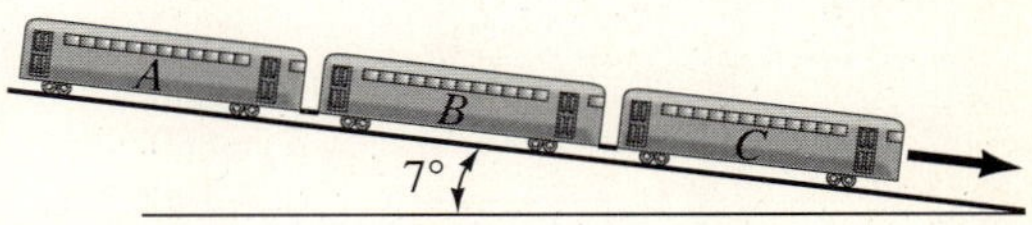

Figure P.12.23.

12.24. A body having a mass of 14 kg is acted on by a force given by

$$F = 140t^2 + e^{-t} \text{ N}$$

If the velocity is 3 m/s at $t = 0$, what is the body's velocity and the distance traveled when $t = 2$ s?

12.25. A body of mass 10 kg is acted on by a force in the x direction, given by the relation $F = 10 \sin 6t$ N. If the body has a velocity of 3 m/s when $t = 0$ and is at position $x = 0$ at that instant, what is the position reached by the body from the origin at $t = 4$ s? Sketch the displacement-versus-time curve.

12.26. A water skier is shown dangling from a kite that is towed via a light nylon cord by a powerboat at a constant speed of 14 m/s. The powerboat with passenger weighs 3 kN and the man and kite together weigh 1.2 kN. If we neglect the mass of the cable, we can take it as a straight line as shown in the diagram. The horizontal drag from the air on the kite plus man is estimated from fluid mechanics to be 360 N. What is the tension in the cable? If the cable suddenly snaps, what is the instantaneous horizontal relative acceleration between the kite system and the powerboat?

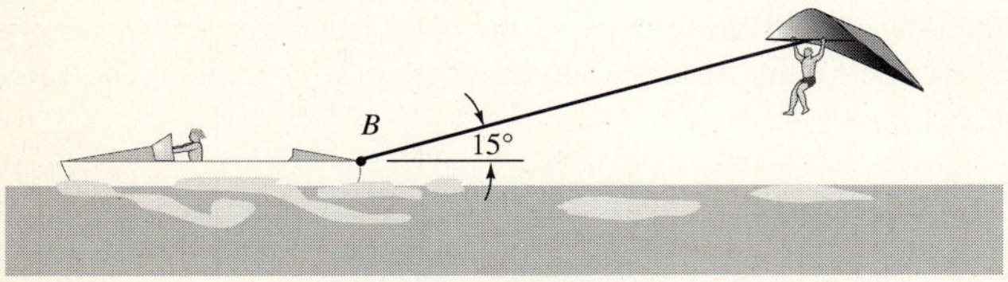

Figure P.12.26.

12.27. A mass M is held by stiff light telescoping rods that can elongate or shorten freely but cannot bend. Each rod is pin connected at the ends A, B, C, and D. The system is on a horizontal, frictionless surface. Two linear springs having spring constants $K_1 = 880$ N/m and $K_2 = 1400$ N/m are connected to the rods as shown in the diagram. If mass $M = 3$ kg is moved 3 mm to the right and is released from rest, what is the equation for the velocity in the x direction as a function of x? What is the speed of the mass when it returns to the vertical position of the rods?

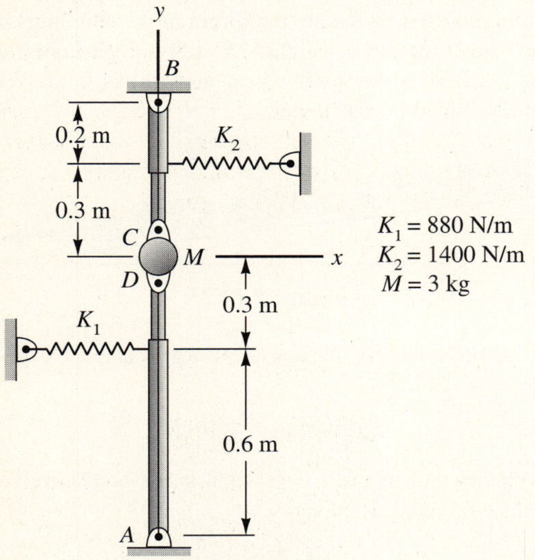

Figure P.12.27.

12.28. A force given as 22 sin $3t$ N acts on a mass of 15 kg. What is the position of the mass at $t = 10$ s? Determine the total distance traveled. Assume the motion started from rest.

12.29. A block A of mass 500 kg is pulled by a force of 10 kN as shown. A second block B of mass 200 kg rests on small frictionless rollers on top of block A. A wall prevents block B from moving to the left. What is the speed of block A after 1 s starting from a stationary position? The coefficient of friction μ_d is 0.4 between A and the horizontal surface.

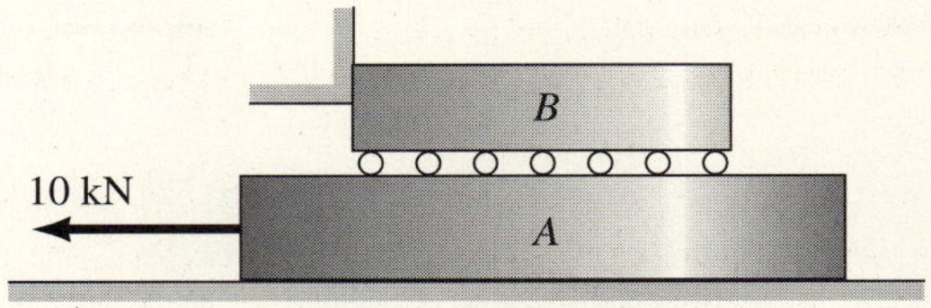

Figure P.12.29.

12.30. Block B weighing 500 N rests on block A, which weighs 300 N. The dynamic coefficient of friction between contact surfaces is 0.4. At wall C there are rollers whose friction we can neglect. What is the acceleration of body A when a force F of 5 kN is applied?

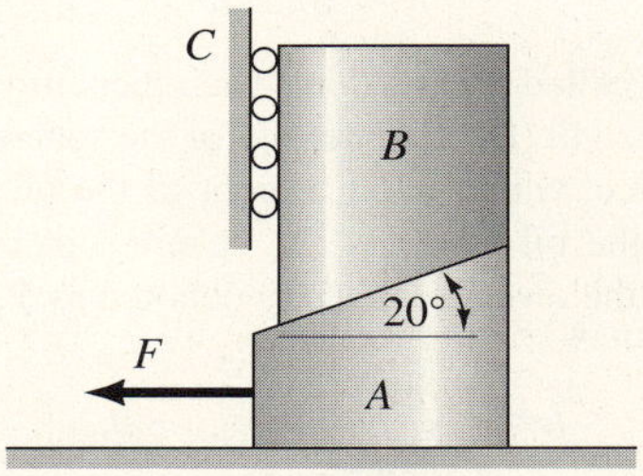

Figure P.12.30.

12.31. A body A of mass 0.5 kg is forced to move by the device shown. What total force is exerted on the body at time $t = 6$ s? What is the maximum total force on the body, and when is the first time this force is developed after $t = 0$?

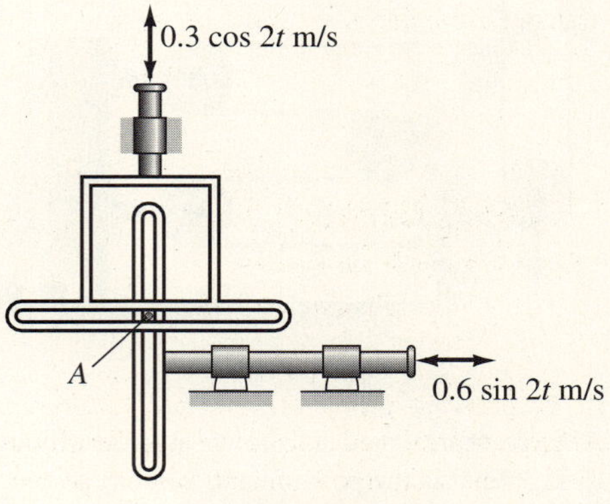

Figure P.12.31.

12.32. Do Problem 12.10 for the case where there is viscous friction between piston and cylinder given as 150 N/m/s of relative speed. Also, what is the maximum distance l the piston can advance relative to the cylinder?

12.33. The high-speed aerodynamic drag on a car is $0.98V^2$ N with V in m/s. If the initial speed is 50 m/s, how far will the car move before its speed is reduced to 30 m/s? The mass of the car is 900 kg.

12.34. A block slides on a film of oil. The resistance to motion of the block is proportional to the speed of the block relative to the incline at the rate of 7.5 N/m/s. If the block is released from rest, what is the *terminal speed*? What is the distance moved after 10 s?

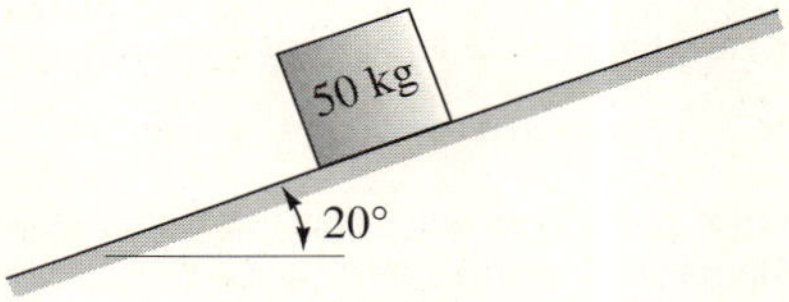

Figure P.12.34.

12.35. When you study fluid mechanics, you will learn that the drag D on a body when moving through a fluid with mass density ρ is given as $\frac{1}{2}C_D\rho V^2 A$ where V is the velocity of the body relative to the fluid; A is the frontal area of the object; and C_D is the so-called *coefficient of drag* usually determined by experiment.

A racing plane on landing is moving at a speed of 100 m/s when a braking parachute is deployed. This parachute has a frontal area of 30 m^2 and a $C_D = 1.2$. The plane has a frontal area of 20 m^2 and a $C_D = 0.4$. If the plane and parachute have a combined mass of 8 Mg, how long does it take to go from 100 m/s to 60 m/s by just coasting? Take $\rho = 1.2475$ kg/m^3 and neglect rolling resistance from the tires. There is no wind.

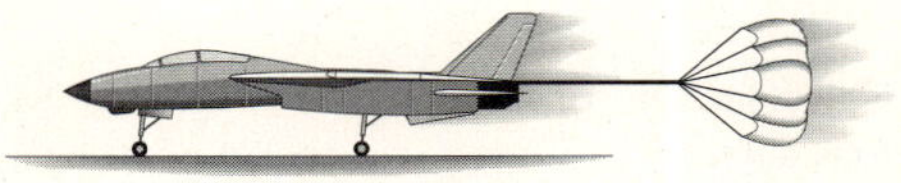

Figure P.12.35.

12.36. In the previous problem, what is the largest frontal area of the braking parachute if the maximum deceleration of the plane is to be 5g's when at a speed of 100 m/s the parachute is first deployed?

12.37. Mass B is on small rollers and moves down the incline. It is connected to a linear spring, which at the position shown is stretched from its undeformed length of 2 m to a length of 5 m. What is the speed of B after it moves 1 m? Use Newton's law as well as the x coordinate shown in the diagram.

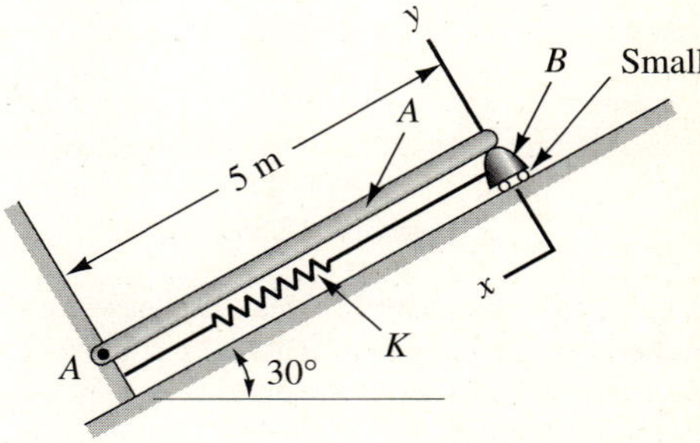

$M_A = 40$ kg
$M_B = 20$ kg
$\mu_d = 0.2$
$l_0 = 2$ m (unstretched length of spring)
$K = 20$ N/m

Figure P.12.37.

12.38. A wedge of wood having a density of 600 kg/m^3 is forced into the water by a 650 N force. The wedge is 0.6 m in width.

(a) What is the depth d?

(b) What is the speed of the wedge when it has moved upward 0.15 m after releasing the 650 N force assuming the wedge does not turn as it rises? Recall, a buoyant force equals the weight of the volume displaced (Archimedes).

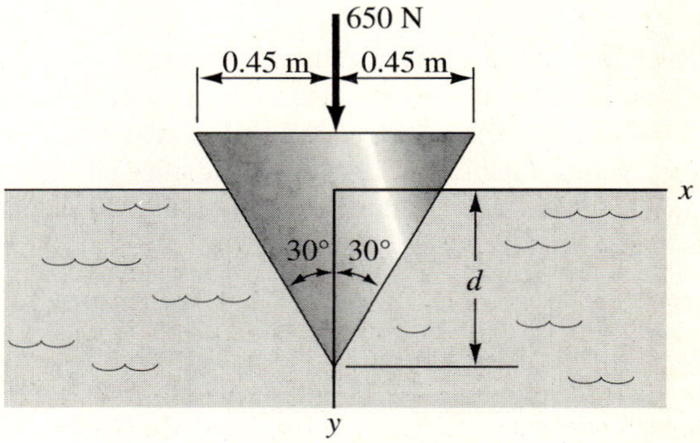

Figure P.12.38.

12.39. A poison dart gun is shown. The cross-sectional area inside the tube is 650 mm^2. The dart being blown weighs 1 N. The dart gun bore has a viscous resistance given as 44 N per unit velocity in m/s. The hunter applies a constant pressure p at the mouth of the gun. Express the relation between p, V (velocity), and t. What constant pressure p is needed to cause the dart to reach a speed of 20 m/s in 2 s? Assume the dart gun is long enough.

Figure P.12.39.

12.40. Using the diagram for Problem 12.5, assume that there is a lubricant between the body D of mass 2 kg and the belt such that there is a viscous friction force given as 1.5 N per unit relative velocity between the body and the belt. The belt moves at a uniform speed of 1.5 m/s to the right and initially the body has a speed to the left of 0.6 m/s relative to ground. At what time later does the body have a zero instantaneous velocity relative to the ground?

12.41. In Problem 12.40 assume that the belt system is inclined 20° from the horizontal with end B above end A. What minimum belt speed is required so that a body of mass M moving downward will come to a permanent halt relative to the ground? For this belt speed, how long does it take for the body to slow down to half of its initial speed of 0.6 m/s relative to the ground?

12.42. One of the largest of the supertankers in the world today is the *S.S. Globtik London*, having a weight when fully loaded of 4.24 GN. The thrust needed to keep this ship moving at 5 m/s is 50 kN. If the drag on the ship from the water is proportional to the speed, how long will it take for this ship to slow down from 5 m/s to 2.5 m/s after the engines are shut down? (The answer may make you wonder about the safety of such ships.)

12.43. A cantilever beam is shown. It is observed that the vertical deflection of the end A is directly proportional to a vertical tip load F provided that this load is not too excessive. A body B of mass 200 kg, when attached to the end of the beam with F removed, causes a deflection of 5 mm there after all motion has ceased. What is the speed of this body if it is attached suddenly to the beam and has descended 3 mm?

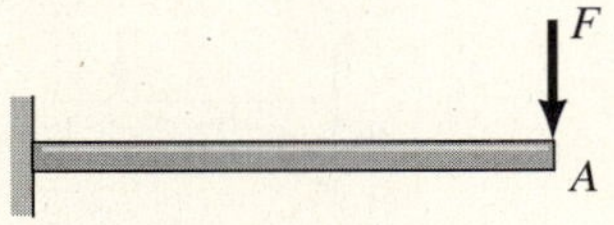

Figure P.12.43.

12.44. The spring shown is nonlinear. That is, K is not a constant, but is a function of the extension of the spring. If $K = 0.4x + 0.6$ N/mm. with x measured in mm, what is the speed of the mass when $x = 0$ after it is released from a state of rest at a position 75 mm from the equilibrium position? The mass of the body is 15 kg.

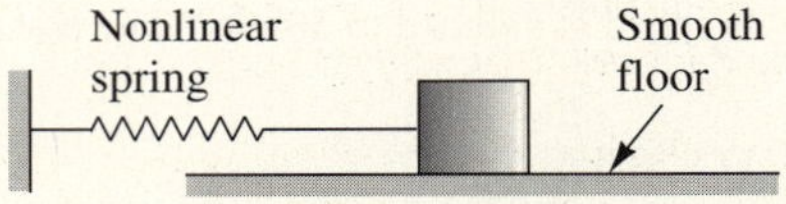

Figure P.12.44.

12.45. A particle of mass m is subject to the following force field:

$$\boldsymbol{F} = m\boldsymbol{i} + 4m\boldsymbol{j} + 16m\boldsymbol{k} \text{ N}$$

In addition, it is subjected to a frictional force $\boldsymbol{f}$ given as

$$\boldsymbol{f} = -m\dot{x}\boldsymbol{i} - m\dot{y}\boldsymbol{j} + 2m\dot{z}\boldsymbol{k} \text{ N}$$

The particle is stationary at the origin at time $t = 0$. What is the position of the particle at time $t = 1$ s?

12.46. A beebee is shot vertically upward with an initial velocity of 36 m/s. If the air resistance is $0.67 \times 10^{-3}V^2$ N, how much time elapses for the projectile to reach its maximum elevation? How high does it go? The beebee weighs 0.24 N.

12.47. If in the previous problem, the beebee has reached a maximum height of 28.3 m, what is the speed when it returns to the ground, assuming it does not reach its terminal velocity? If it has reached the terminal velocity, what is your answer?

12.48. A rocket weighing 25 kN is fired vertically from a test stand on the ground. A constant thrust of 100 kN is developed for 20 seconds. If just as an exercise, we do not take into account the amount of fuel burned, and if we neglect air resistance, how high up does this hypothetical rocket go? Note that neglecting fuel consumption is a serious error! In the next problem we will investigate the case of the variable mass problem.

***12.49.** Calculate the velocity after 20 seconds for the case where there is a *decrease* of mass of a rocket of 50 kg/s as a result of exhaust combustion products leaving the rocket at a speed of 1.8 km/s relative to the rocket. At the outset the rocket weighs 25 kN. [*Hint:* Start with Newton's law in the form $\boldsymbol{F} = (d/dt)(m\boldsymbol{V})$ where $\boldsymbol{F}$ is the weight, a variable that decreases as fuel is burned.] The first term on the right side of this equation is $m(dV/dt)$ where m is the instantaneous mass of the rocket and unburned fuel. Now there is a force on the 50 kg/s of combustion products being expelled from the rocket at a speed relative to the rocket of 1.8 km/s. The rate of change of linear momentum associated with this force clearly must be $(dm/dt)(1800)$. The *reaction* to this force for this momentum change is on the rocket in the direction of flight of the rocket and must be added to $m(dV/dt)$. The force exerted by the exhaust gases on the rocket is a propulsive force and is called the *thrust* of the rocket. Again, neglect drag of the atmosphere since it will be small at the outset because of low velocity and small later because of the thinness of the atmosphere.

12.50. We start with a cylindrical tank with diameter 15 m containing water up to a depth of 3 m. Initially the solid movable cylindrical piston A having a diameter of 6 m and a centerline colinear with the centerline of the tank is positioned so that its top is flush with the bottom of the tank. Now the cylinder is moved upward so that the following data apply at the instant of interest assuming the free surface of the water remains flat;

$$h_2 = 0.6 \text{ m} \qquad \dot{h}_2 = 1.5 \text{ m/s} \qquad \ddot{h}_2 = 0.9 \text{ m/s}^2$$

What is the external force from the *ground* support on the water needed for this condition not including the force required to support the dead weight of the water?

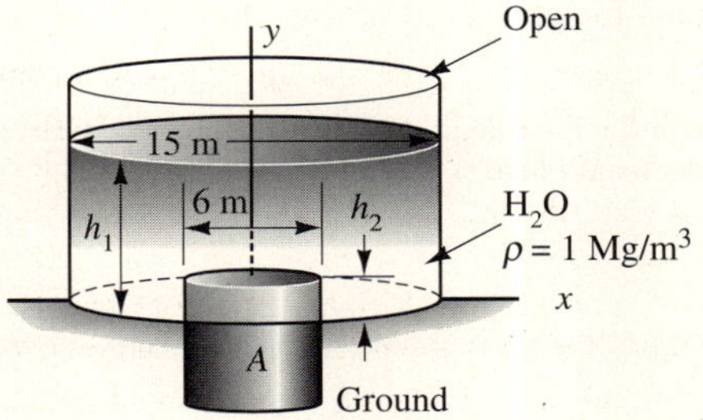

Figure P.12.50.

12.51. A sleeve slides downward along a pipe on which there is dry friction with $\mu_d = 0.35$. A wire having a constant tension of 80 N is attached to the sleeve and moves with it always retaining the same angle α with the horizontal. If the sleeve weighs 60 N, what should α be for the sleeve to move for 10 seconds before stopping after starting downward with an initial speed of 5 m/s?

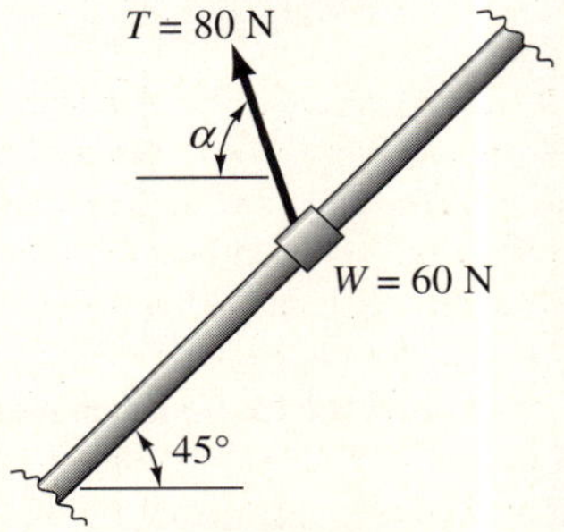

Figure P.12.51.

12.52. An electron having a charge of $-e$ coulombs is moving between two parallel plates in a vacuum with an impressed voltage E. If at $t = 0$, the electron has a velocity V_0 at an angle α_0 with the horizontal in the xy plane, what will be the trajectory equation taking the initial conditions to be at the origin of xy? Show that

$$y = \frac{eE}{2m} + \frac{x^2}{(V_0 \cos \alpha_0)^2} + x \tan \alpha_0$$

where m is the mass of the electron. Note we have neglected gravity here since it is very small compared with the electrostatic force.

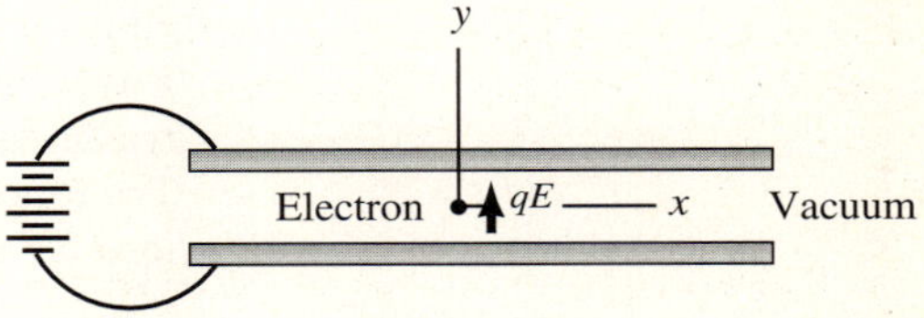

Figure P.12.52.

12.53. A system of light pulleys and inextensible wire connects bodies A, B, and C as shown. If the coefficient of friction between C and the support is 0.4, what is the acceleration of each body? Take M_A as 100 kg, M_B as 300 kg, and M_C as 80 kg.

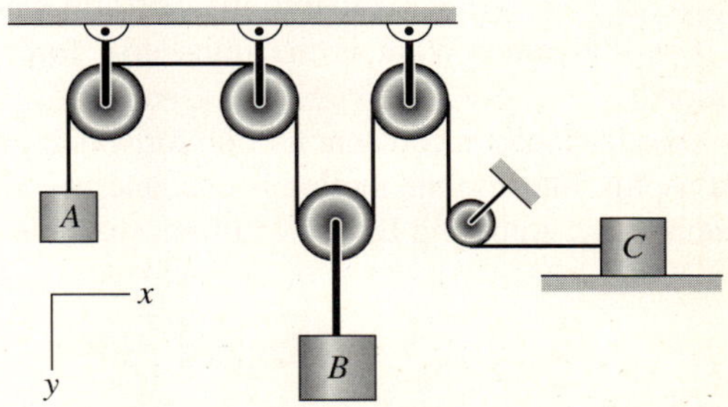

Figure P.12.53.

Part B: Cylindrical Coordinates; Central Force Motion

12.5 Newton's Law for Cylindrical Coordinates

In cylindrical coordinates we can express Newton's law as follows:

$$F_{\bar{r}} = m(\ddot{\bar{r}} - \bar{r}\dot{\theta}^2) \tag{12.15a}$$

$$F_\theta = m(\bar{r}\ddot{\theta} - 2\dot{\bar{r}}\dot{\theta}) \tag{12.15b}$$

$$F_z = m\ddot{z} \tag{12.15c}$$

If the motion is known, it is a simple matter to ascertain the force components using Eqs. 12.15. The inverse problem of determining the motion given the forces is particularly difficult in this case. The reason for this difficulty, as you may have already learned in your differential equations course, is that Eqs. 12.15a and 12.15b are *nonlinear*[6] for all force functions. For this reason, we cannot present integration procedures as in Part A of this chapter. The following example will serve to illustrate the kind of problem we are able to solve with the methods thus far presented in this chapter.

[6]A differential equation is nonlinear if the dependent variable and its derivatives form powers greater than unity or form products anywhere in the equation.

Example 12.8

A platform shown in Fig. 12.13 has a constant angular velocity ω equal to 5 rad/s. A mass B of 2 kg slides in a frictionless chute attached to the platform. The mass is connected via a light inextensible cable to a linear spring having a spring constant K of 20 N/m. A swivel connector at A allows the cable to turn freely relative to the spring. The spring is unstretched when the mass B is at the center C of the platform. If the mass B is released at $r = 200$ mm from a stationary position relative to the platform, what is its speed relative to the platform when it has moved to position $r = 400$ mm? What is the transverse force on the body B at this position?

We have here a coplanar motion for which cylindrical coordinates are most useful. Because the motion is coplanar, we can use r instead of $\bar{r}$ with no ambiguity. Applying Eq. 12.15a first, we have

$$-20r = 2(\ddot{r} - 25r)$$

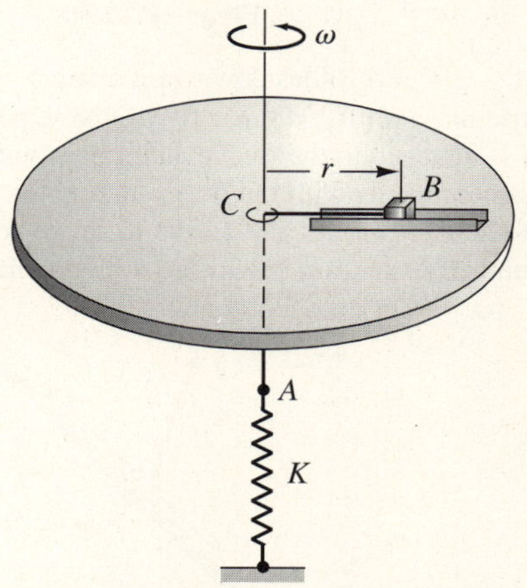

Figure 12.13. Slider on rotating platform.

Example 12.8 (Continued)

Therefore,

$$\ddot{r} = 15r \qquad \text{(a)}$$

As in Example 12.5, we can replace $\ddot{r}$ so as to allow for a separation of variables.

$$\ddot{r} \equiv \frac{dV_r}{dt} \equiv \frac{dV_r}{dr}\frac{dr}{dt} \equiv V_r \frac{dV_r}{dr} = 15r$$

Therefore,

$$V_r \, dV_r = 15r \, dr$$

Integrating, we get

$$\frac{V_r^2}{2} = \frac{15r^2}{2} + C_1 \qquad \text{(b)}$$

To determine C_1, note that, when $r = 0.20m$, $V_r = 0$. Hence,

$$C_1 = -\frac{0.600}{2}$$

Equation (b) then becomes

$$V_r^2 = 15r^2 - 0.600 \qquad \text{(c)}$$

When $r = 0.40$ m, we get for V_r from Eq. (c):

$$V_r = 1.342 \text{ m/s} \qquad \text{(d)}$$

This is the desired velocity relative to the platform.

To get the transverse force F_θ, go to Eq. 12.15b. Substituting the known data into the equation, we have

$$F_\theta = 2[(0.40)(0) + (2)(1.342)(5)]$$

$$F_\theta = 26.84 \text{ N}$$

This is the transverse force on the mass B.

Although you will be asked to solve problems similar to the preceding example, the main use of cylindrical coordinates in Part B of this chapter will be for gravitational central force motion. We shall first present the basic physics underlying this motion expressing certain salient characteristics of the motion, and then we shall arrive at a point where we can effectively employ cylindrical coordinates to describe the motion.

12.6 Central Force Motion—An Introduction

At this time, we shall consider the motion of a particle on which the resultant force is always *directed toward some point fixed in inertial space*. Such forces are termed *central forces* and the resulting motion of this particle is called *central force motion*. A simple example of this is the case of a space vehicle moving with its engine off in the vicinity of a large planet (see Fig. 12.14).

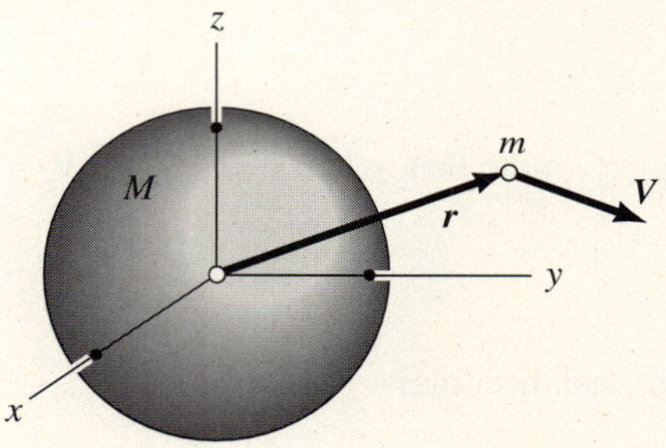

Figure 12.14. Body m moving about a planet.

The space vehicle is very small compared to the planet and may be considered to be a particle. Away from the planet's atmosphere, this vehicle will experience no frictional forces, and, if no other astronautical bodies are reasonably close, the only force acting on the vehicle will be the gravitational attraction of the fixed planet.[7] This force is directed toward the center of the planet and, from the gravitational law, is given as

$$\boldsymbol{F} = -G\frac{M_{\text{planet}}m_{\text{body}}}{r^2}\hat{\boldsymbol{r}} \tag{12.16}$$

In the ensuing problems for this chapter and also for Chapter 14, we shall need to compute the quantity GM in the equation above. For this purpose, note that, for any particle of mass m at the surface of any planet of mass M and radius R, by the law of gravitation:

$$W = mg = \frac{GMm}{R^2}$$

[7]We are neglecting drag developed from collisions of the space vehicle with solar dust particles.

where g is the acceleration of gravity at the surface of the planet. Solving for GM, we get

$$GM = gR^2 \tag{12.17}$$

Thus, knowing g and R for a planet, it is a simple matter to find GM needed for orbit calculations around this planet.

As pointed out earlier, the motion of a space vehicle with power off is an important example of a central force motion—more precisely a *gravitational* central force motion. The vehicle is usually launched from a planet and accelerated to a high speed outside the planet's atmosphere by multistage rockets (see Fig. 12.15). The velocity at the final instant of powered flight is called the *burnout* velocity. After burnout, the vehicle undergoes gravitational central force motion. Depending on the position and velocity at burnout, the vehicle can go into an orbit around the earth (elliptic and circular orbits are possible), or it can depart from the earth's influence on a parabolic or a hyperbolic trajectory. In all cases, the motion must be coplanar.

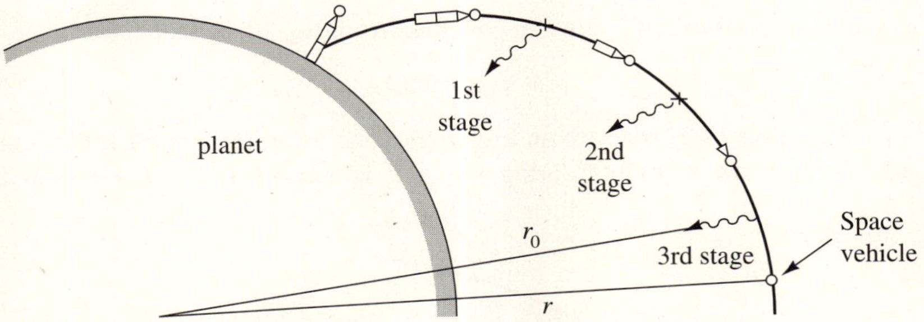

Figure 12.15. Launching a space vehicle.

In the following sections, we shall make a careful detailed study of gravitational central force trajectories. Those who do not have the time for such a detailed study of the trajectories can still make many useful and interesting calculations in Chapter 14 using energy and momentum methods that we shall soon undertake.

*12.7 Gravitational Central Force Motion

For gravitational central force motion, we shall employ an inertial reference xy in the plane of the trajectory with the origin of the reference taken at the point P toward which the central force is directed (see Fig. 12.16). We shall use cylindrical coordinates r and θ for describing the motion. Because $z = 0$ at all times, these coordinates are also called polar coordinates. Since the motion is coplanar in plane xy, we can delete the overbar used previously for r with no danger of ambiguity.

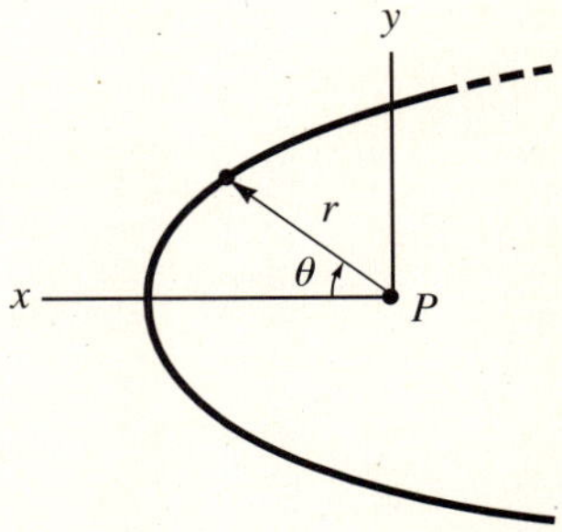

Figure 12.16. xy is inertial reference in plane of the trajectory.

Let us consider *Newton's law* for a body of mass m, which is moving near a star of mass M:

$$m\frac{d\mathbf{V}}{dt} = -G\frac{Mm}{r^2}\hat{\boldsymbol{r}} \tag{12.18}$$

Canceling m and using cylindrical coordinates and components, we can express the equation above in the following manner:

$$(\ddot{r} - r\dot{\theta}^2)\boldsymbol{\epsilon}_r + (r\ddot{\theta} + 2\dot{r}\dot{\theta})\boldsymbol{\epsilon}_\theta = -\frac{GM}{r^2}\hat{\boldsymbol{r}} \tag{12.19}$$

Since $\boldsymbol{\epsilon}_r$ and $\hat{\boldsymbol{r}}$ are identical vectors, the scalar equations of the preceding equation become

$$\ddot{r} - r\dot{\theta}^2 = -GM/r^2 \tag{12.20a}$$

$$r\ddot{\theta} + 2\dot{r}\dot{\theta} = 0 \tag{12.20b}$$

Equation 12.20b can be expressed in the form

$$\frac{1}{r}\frac{d}{dt}(r^2\dot{\theta}) = 0 \tag{12.21}$$

as you can readily verify. We can conclude from Eq. 12.21 that

$$r^2\dot{\theta} = \text{constant} = C \tag{12.22}$$

Equation 12.22 leads to an important conclusion. To establish this, consider the area swept out by r during a time dt, which in Fig. 12.17 is the shaded

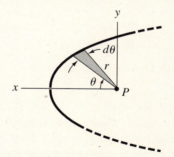

Figure 12.17. Particle sweeps out area.

area. By considering this area to be that of a triangle, we can express it as

$$dA = \frac{r^2\,d\theta}{2}$$

Dividing through by dt, we have

$$\frac{dA}{dt} = \frac{r^2\dot{\theta}}{2}$$

Now dA/dt is the rate at which area is being swept out by r; it is called *areal velocity*. And, since $r^2\dot{\theta}$ is a constant for each gravitational central force motion (see Eq. 12.22), we can conclude that the areal velocity is a constant for each gravitational central force motion. (This is Kepler's second law.) This means that when r is decreased, $\dot{\theta}$ must increase, etc. The constant, understand, will be different for each different trajectory.

In order to determine the general trajectory, we replace the independent variable t of Eq. 12.20a. Consider first the time derivatives of r:

$$\dot{r} = \frac{dr}{dt} = \frac{dr}{d\theta}\frac{d\theta}{dt} = \frac{d\theta}{dt}\frac{dr}{d\theta} = \frac{C}{r^2}\frac{dr}{d\theta} \tag{12.23}$$

where we have used Eq. 12.22 to replace $d\theta/dt$. Next, consider $\ddot{r}$ in a similar manner:

$$\ddot{r} = \frac{d\dot{r}}{dt} = \frac{d}{dt}\left(\frac{C}{r^2}\frac{dr}{d\theta}\right) = \frac{d}{d\theta}\left(\frac{C}{r^2}\frac{dr}{d\theta}\right)\frac{d\theta}{dt} \tag{12.24}$$

Again, using Eq. 12.22 to replace $d\theta/dt$, we get

$$\ddot{r} = \left[\frac{d}{d\theta}\left(\frac{C}{r^2}\frac{dr}{d\theta}\right)\right]\frac{C}{r^2} \tag{12.25}$$

For convenience, we now introduce a new dependent variable, $u = 1/r$, into the right side of this equation

$$\begin{aligned}\ddot{r} &= \left[\frac{d}{d\theta}\left(Cu^2\,\frac{d(1/u)}{d\theta}\right)\right]Cu^2 \\ &= \left\{\frac{d}{d\theta}\left[Cu^2\left(-\frac{1}{u^2}\right)\frac{du}{d\theta}\right]\right\}Cu^2 \\ &= -C^2u^2\,\frac{d^2u}{d\theta^2}\end{aligned}$$

By replacing $\ddot{r}$ in this form in Eq. 12.20a and $\dot{\theta}^2$ in the form C^2u^4 from Eq. 12.22, and finally, r by $1/u$, we get

$$-C^2u^2\,\frac{d^2u}{d\theta^2} - C^2u^3 = -GMu^2$$

Canceling terms and dividing through by C^2, we have

$$\frac{d^2u}{d\theta^2} + u = \frac{GM}{C^2} \tag{12.26}$$

This is a simple differential equation that you may have already studied in your differential equations course. Specifically, it is a second-order differential equation with constant coefficients and a constant driving function GM/C^2. We want to find the most general function $u(\theta)$, which when substituted into the differential equation satisfies the differential equation—i.e., renders it an identity. The theory of differential equations indicates that this general solution is composed of two parts. They are:

1. The general solution of the differential equation with the right side of the differential equation set equal to zero and hence given as

$$\frac{d^2u}{d\theta^2} + u = 0 \qquad (12.27)$$

   This solution is called the *complementary* (or *homogeneous*) solution, u_c.

2. *Any* solution u_p that satisfies the full differential equation. This part is called the *particular solution.*

The desired general solution is then the sum of the complementary and particular solutions. It is a simple matter to show by substitution that the function $A \sin \theta$ satisfies Eq. 12.27 for any value of A. This is similarly true for $B \cos \theta$ for any value of B. The theory of differential equations tells us that there are two independent functions for the solution of Eq. 12.27. The general complementary solution is then

$$u_c = A \sin \theta + B \cos \theta \qquad (12.28)$$

where A and B are arbitrary constants of integration. Considering the full differential equation (Eq. 12.26), we see by inspection that a particular solution is

$$u_p = \frac{GM}{C^2} \qquad (12.29)$$

The general solution to the differential equation (Eq. 12.26) is then

$$u = \frac{GM}{C^2} + A \sin \theta + B \cos \theta \qquad (12.30)$$

By simple trigonometric considerations, we can put the complementary solution in the equivalent form, $D \cos (\theta - \beta)$, where D and β are then the constants of integration.[8] We then have as an alternative formulation for u (= $1/r$):

$$u = \frac{1}{r} = \frac{GM}{C^2} + D \cos(\theta - \beta) \qquad (12.31)$$

You may possibly recognize this equation as the general *conic equation* in polar coordinates with the focus at the origin. In your analytic geo-

[8]By expanding $[D \cos (\theta - \beta)]$ as $[(D \cos \beta) \cos \theta + (D \sin \beta) \sin \theta]$, we see, since D and β are arbitrary, that $[D \cos (\theta - \beta)]$ is equivalent to $[A \sin \theta + B \cos \theta]$, where A and B are arbitrary.

metry class, you probably saw the following form for the general conic equation.[9]

$$\frac{1}{r} = \frac{1}{\epsilon p} + \frac{1}{p}\cos(\theta - \beta) \tag{12.37}$$

where ϵ is the *eccentricity*, p is the distance from the *focus to the directrix*, and β is the angle between the x axis and the axis of symmetry of the conic section.

Comparing Eqs. 12.31 and 12.37, we see that

$$p = \frac{1}{D} \tag{12.38a}$$

$$\epsilon = \frac{DC^2}{GM} \tag{12.38b}$$

[9]A *conic section* is the locus of all points whose distance from a *fixed point* has a *constant ratio* to the distance from a *fixed line*. The fixed point is called the *focus* (or focal point) and the line is termed the *directrix*. In Fig. 12.18 we have shown point P, a directrix DD, and a focus O. For a conic section to be traced by P, it must move in a manner that keeps the ratio $r/\overline{DP}$, called the *eccentricity*, a fixed number. Clearly, for every acceptable position P, there will be a mirror image position P' (see the diagram) about a line normal to the directrix and going through the focal point O. Thus, the conic section will be *symmetrical* about axis OC.

Using the letter ϵ to represent the eccentricity, we can say:

$$\frac{r}{\overline{DP}} \equiv \epsilon = \frac{r}{p + r\cos\eta} \tag{12.32}$$

where p is the distance from the focus to the directrix. Replacing $\cos\eta$ by $-\cos(\theta - \beta)$, where β (see Fig. 12.18) is the angle between the x axis and the axis of symmetry, we then get

$$\frac{r}{p - r\cos(\theta - \beta)} = \epsilon \tag{12.33}$$

Now, rearranging the terms in the equation, we arrive at a standard formulation for conic sections:

$$\frac{1}{r} = \frac{1}{\epsilon p} + \frac{1}{p}\cos(\theta - \beta) \tag{12.34}$$

To understand the significance of the eccentricity ϵ, let us consider conic sections in terms of a reference xy, where x is the axis of symmetry (i.e., consider $\beta = 0$ in preceding formulations and refer to Fig. 12.19). Equation 12.34 can be expressed for these rectangular coordinates in the following manner:

$$\frac{1}{\sqrt{x^2 + y^2}} = \frac{1}{\epsilon p} + \frac{1}{p}\frac{x}{\sqrt{x^2 + y^2}} \tag{12.35}$$

Simple algebraic manipulation permits us to put the preceding equation into the following form:

$$(1 - \epsilon^2)x^2 + y^2 + 2p\epsilon^2 x - \epsilon^2 p^2 = 0 \tag{12.36}$$

If $\epsilon > 1$, the coefficients of x^2 and y^2 are different in sign and unequal in value. The equation then represents a *hyperbola*.

If $\epsilon = 1$, only one of the squared terms remains and we have a *parabola*.

If $\epsilon < 1$, the coefficients of the squared terms are unequal but have the same sign. The curve is that of an *ellipse*.

If $\epsilon = 0$, clearly we have a *circle* since the coefficients of the squared terms are equal in value and sign.

In Appendix III, we discuss in more detail the particular case of the ellipse.

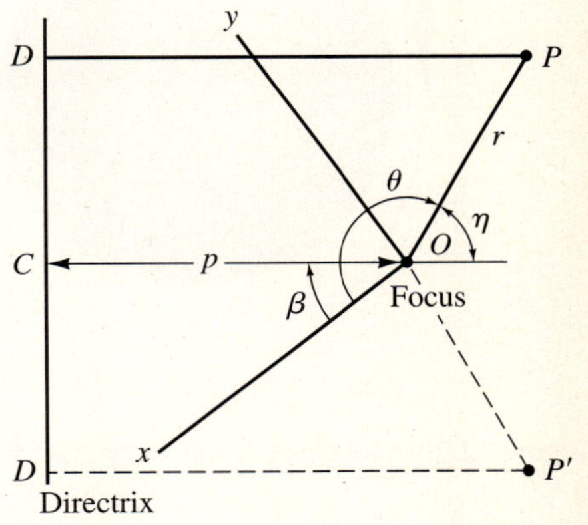

Figure 12.18. $r/\overline{DP} = \epsilon \equiv$ constant for conic section.

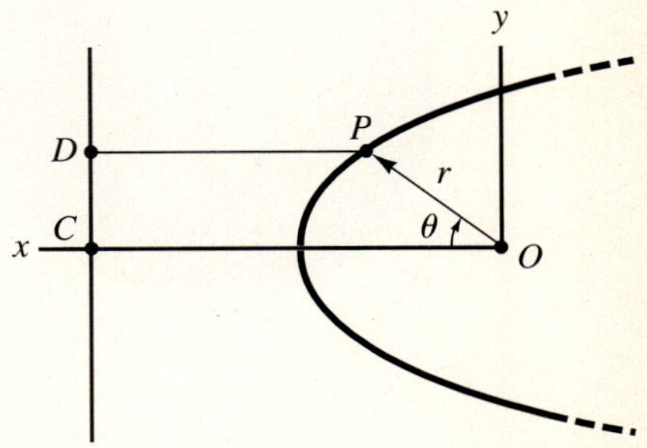

Figure 12.19. Case for $\beta = 0$; x axis is axis of symmetry.

From our knowledge of conic sections, we can then say that if

$$\frac{DC^2}{GM} > 1, \text{ the trajectory is a hyperbola} \tag{12.39a}$$

$$\frac{DC^2}{GM} = 1, \text{ the trajectory is a parabola} \tag{12.39b}$$

$$\frac{DC^2}{GM} < 1, \text{ the trajectory is an ellipse} \tag{12.39c}$$

$$\frac{DC^2}{GM} = 0, \text{ the trajectory is a circle} \tag{12.39d}$$

Clearly, DC^2/GM, the eccentricity, is an extremely important quantity. We shall next look into the practical applications of the preceding general theory to problems in space mechanics.

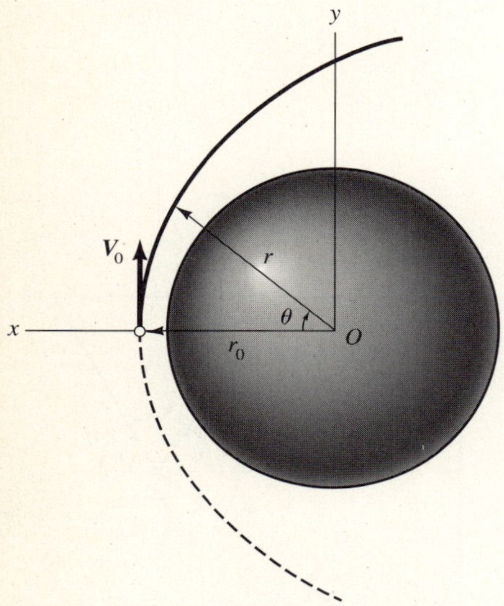

Figure 12.20. Launching at axis of symmetry.

*12.8 Applications to Space Mechanics

We shall now employ the theory set forth in the previous section to study the motion of space vehicles—a problem of great present-day interest. We shall assume that at the end of powered flight the position $\boldsymbol{r}_0$ and velocity $\boldsymbol{V}_0$ of the vehicle are known from rocket calculations. The reference employed will be an inertial reference at the center of the planet and so the reference will translate with the planet relative to the "fixed stars." Accordingly, the earth will rotate one cycle per day for such a reference. We know that the trajectory of the body will form a plane fixed in inertial space and so, for convenience, we take the *xy* plane of the reference to be the plane of the trajectory. It is the usual practice to choose the *x* axis to be the axis of symmetry for the trajectory. If there is a *zero radial velocity* component at "burnout," then the launching clearly occurs at a position along the axis of symmetry of the trajectory (i.e., along the *x* axis). This case has been shown in Fig. 12.20, wherein the subscript 0 denotes launch data. If, on the other hand, a radial component $(V_r)_0$ is present at burnout, then the launch condition occurs at some position θ_0 from the *x* axis, as shown in Fig. 12.21. We generally do not know θ_0 a priori, since its value depends on the equation of the trajectory. Finally, the angle α shown in the diagram will be called the *launching angle* in the ensuing discussion.

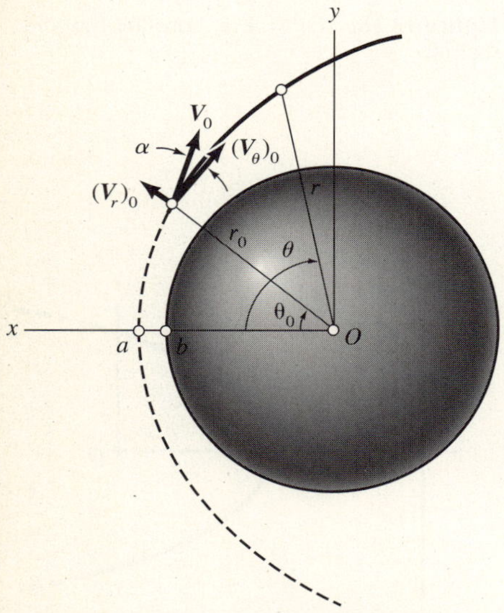

Figure 12.21. Burnout with radial velocity present.

Since the *x* axis has been chosen to be the axis of symmetry, the equation of motion of the vehicle after powered flight is given in terms of arbitrary constants C and D by Eq. 12.31 with the angle β set equal to zero. Thus, we have

$$\frac{1}{r} = \frac{GM}{C^2} + D\cos\theta \tag{12.40}$$

The problem is to find the constants C and D from launching data. We shall illustrate this step in the examples following this section. Note that when these constants are evaluated, the value of the eccentricity $\epsilon = DC^2/GM$ is then available so that we can state immediately the general characteristics of the trajectory.

Furthermore, if the vehicle goes into orbit, we can readily compute the *orbital time* τ for one cycle around a planet. We know from the theory that the aerial velocity is constant and given as

$$\frac{dA}{dt} = \frac{r^2\dot{\theta}}{2} = \text{constant} \tag{12.41}$$

But $r^2\dot{\theta}$ equals the constant C in accordance with Eq. 12.22. Hence,

$$dA = \frac{C}{2}\,dt \tag{12.42}$$

The area swept out for one cycle is the area of an ellipse given as πab, where a and b are the semimajor and semiminor diameters of the ellipse, respectively. Hence, we have on integrating Eq. 12.42:

$$A = \pi ab = \int_0^\tau \frac{C}{2}\,dt = \frac{C}{2}\tau$$

Therefore,

$$\tau = \frac{2\pi ab}{C} \tag{12.43}$$

We have shown in Appendix III that

$$a = \frac{\epsilon p}{1-\epsilon^2} \tag{12.44a}$$

$$b = a(1-\epsilon^2)^{1/2} \tag{12.44b}$$

Replacing p by $1/D$ in accordance with Eq. 12.38a, we then get

$$a = \frac{\epsilon}{D(1-\epsilon^2)} \tag{12.45a}$$

$$b = a(1-\epsilon^2)^{1/2} = \frac{\epsilon}{D(1-\epsilon^2)^{1/2}} \tag{12.45b}$$

Thus, we can get the orbital time τ quite easily once the constants of the trajectory, D and C, are evaluated.

To illustrate many of the previous general remarks in a most simple manner, we now examine the special case where, as shown in Fig. 12.22, various launchings (i.e., burnout conditions) are made from a given point a such that the launching angle $\alpha = 0$. Clearly $(V_r)_0 = 0$ for these cases and the launching axis corresponds to the axis of symmetry of the various trajectories. Only V_0 will be varied in this discussion.

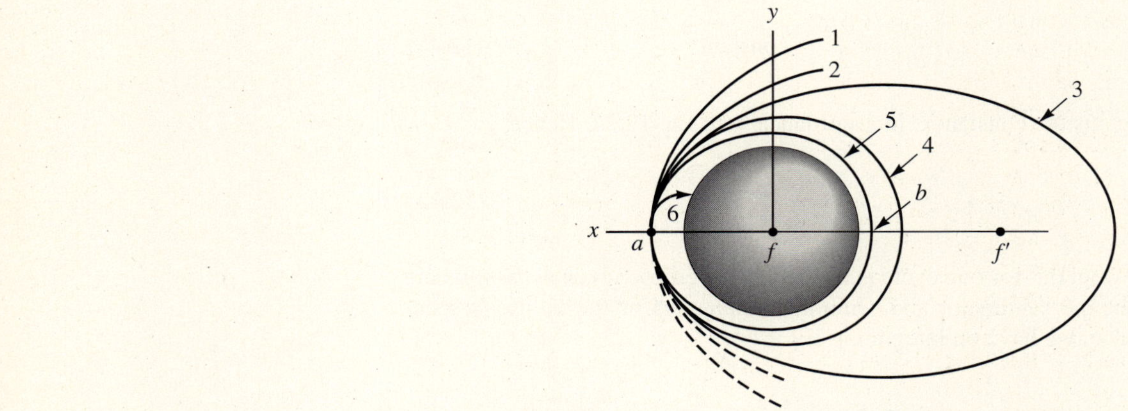

Figure 12.22. Various launchings from the earth or some other planet.

The constants C and D are readily available for these trajectories. Thus, we have from Eq. 12.22:

$$C = r^2\dot{\theta} = rV_\theta = r_0 V_0 \tag{12.46}$$

And from Eq. 12.40, setting $r = r_0$ when $\theta = \theta_0 = 0$, we get, on solving for D:

$$D = \frac{1}{r_0} - \frac{GM}{C^2} = \frac{1}{r_0} - \frac{GM}{r_0^2 V_0^2} \tag{12.47}$$

Since C and D above, for a given r_0, depend only on V_0, we conclude that the eccentricity here is dependent only on V_0 for a given r_0.

If V_0 is so large that DC^2/GM exceeds unity, the vehicle will have the trajectory of a hyperbola (curve 1) and will eventually leave the influence of the earth. If V_0 is decreased to a value such that the eccentricity is unity, the trajectory becomes a parabola (curve 2). Since a further decrease in the value of V_0 will cause the vehicle to orbit, curve 2 is the limiting trajectory with our launching conditions for outer-space flight. The launching velocity for this case is accordingly called the *escape velocity* and is denoted as $(V_0)_E$. We can solve for $(V_0)_E$ for this launching by substituting for C and D from Eqs. 12.46 and 12.47 into the equation $DC^2/GM = 1$. We get

$$(V_0)_E = \sqrt{\frac{2GM}{r_0}} \tag{12.48}$$

a result that is correct for more general launching conditions (i.e., for cases where launching angle $\alpha \neq 0$). Thus, launching a vehicle with a speed equaling or exceeding the value above for a given r_0 will cause the vehicle to leave the earth until such time as the vehicle is influenced by other astronomical bodies or by its own propulsion system. If V_0 is less than the escape velocity, the vehicle will move in the trajectory of an ellipse (curve 3). The closest point to the earth is called *perigee*; the farthest point is called *apogee*. Clearly, these points lie along the axis of symmetry. Such an orbiting vehicle is often called a space satellite. (Kepler, in his famous first law of planetary motion, explained the motion of planets about the sun in this same manner.) One focus for the aforementioned conic curves is at the center of the planet. Another focus f' now moves in from infinity for the satellite trajectories. As the launching speed is decreased, f' moves toward f. When the foci coincide, the trajectory is clearly a circle and, as pointed out earlier (see Eq. 12.39d), the eccentricity ϵ is zero. Accordingly, the constant D must be zero (the constant C clearly will not be zero) and, from Eq. 12.47, the speed for a *circular* orbit $(V_0)_C$ is

$$(V_0)_C = \sqrt{\frac{GM}{r_0}} \tag{12.49}$$

For launching velocities less than the preceding value for a given r_0, the eccentricity becomes negative and the focus f' moves to the left of the earth's center. Again, the trajectory is that of an ellipse (curve 5). However, the satellite will now come closer to the earth at position b, which now becomes the perigee, than at the launching position, which up to now had been the minimum distance from the earth.[10] If friction is encountered, the satellite will slow up, spiral in toward the atmosphere, and either burn up or crash. If V_0 is small enough, the satellite will not go into even a temporary orbit but will plummet to the earth (curve 6). However, for a reasonably accurate description of this trajectory, we must consider friction from the earth's atmosphere. Since this type of force is a function of the velocity of the satellite and is not a central force, we cannot use the results here in such situations for other than approximate calculations.

[10]Note that with the positive x axis going through perigee, r is *minimum* when $\theta = 0$. From Eq. 12.40, we can conclude for this case (θ is measured here from perigee) that, to minimize r, the constant D must be positive. The eccentricity must then be positive for θ measured from perigee. If the positive x axis goes through apogee, then r is *maximum* when $\theta = 0$. From Eq. 12.40 we can conclude that D must be negative for this case (θ is here measured from apogee). Thus, the eccentricity is negative for θ measured from apogee. This is clearly the case for curve 5.

Example 12.9

The first American satellite, the Vanguard, was launched at a velocity of 8.05 km/s at an altitude of 640 km (see Fig. 12.23). If the "burnout" velocity of the last stage is parallel to the earth's surface, compute the maximum altitude from the earth's surface that the Vanguard satellite will reach. Consider the earth to be perfectly spherical with a radius of 6.37 Mm (r_0 is therefore 7.01 Mm).

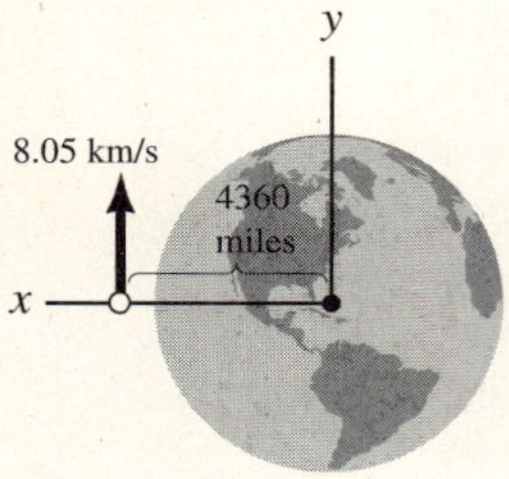

Figure 12.23. Launching of the Vanguard satellite.

We must now compute the quantities GM, C, and D from the initial data and other known data. To determine GM, we employ Eq. 12.17 and in terms of units of miles and hours we get

$$\begin{aligned} GM &= (9.81)(6.37\times10^6)^2 \\ &= 398.1\times10^{12}\ \text{m}^3/\text{s}^2 \end{aligned} \tag{a}$$

The constant C is readily determined directly from initial data as

$$\begin{aligned} C = r_0V_0 &= (7.01\times10^6)(8.05\times10^3) \\ &= 56.43\times10^9\,\text{m}^2/\text{s} \end{aligned} \tag{b}$$

Finally, the constant D is available from Eq. 12.47:

$$\begin{aligned} D &= \frac{1}{r_0} - \frac{GM}{C^2} = \frac{1}{(7.01\times10^6)} - \frac{398.1\times10^{12}}{(56.43\times10^9)^2} \\ &= 17.7\times10^{-9}\ \text{m}^{-1} \end{aligned} \tag{c}$$

The eccentricity DC^2/GM can now be computed as

$$\epsilon = \frac{DC^2}{GM} = \frac{(17.7\times10^{-9})(56.43\times10^9)^2}{398.1\times10^{12}} = 0.1416 \tag{d}$$

The Vanguard will thus definitely not escape into outer space.

The trajectory of this motion is formed from Eq. 12.40:

$$\frac{1}{r} = \frac{398.1\times10^{12}}{(56.43\times10^9)^2} + 17.7\times10^{-9}\cos\theta$$

Example 12.9 (Continued)

Therefore,

$$\frac{1}{r} = 125.0 \times 10^{-9} + 17.7 \times 10^{-9} \cos\theta \qquad \text{(e)}$$

We can compute the maximum distance from the earth's surface by setting $\theta = \pi$ in the equation above:

$$\frac{1}{r_{\text{max}}} = (125.0 - 17.7) \times 10^{-9} = 107.3 \times 10^{-9} \text{ m}^{-1}$$

Therefore,

$$r_{\text{max}} = 9.32 \text{ Mm} \qquad \text{(f)}$$

By subtracting 6.37 Mm from this result, we find that the highest point in the trajectory is 2.95 Mm from the earth's surface.

Example 12.10

In Example 12.9, first compute the escape velocity and then the velocity for a circular orbit at burnout.

Using Eq. 12.48, we have for the escape velocity:

$$(V_0)_E = \sqrt{\frac{2GM}{r_0}} = \left[\frac{2(398.1 \times 10^{12})}{7.01 \times 10^6}\right]^{1/2}$$

$$(V_0)_E = 10.66 \text{ km/s}$$

For a circular orbit, we have from Eq. 12.49:

$$(V_0)_C = \sqrt{\frac{GM}{r_0}} = 7.536 \text{ km/s}$$

Thus, the Vanguard is almost in a circular orbit.

Example 12.11

Determine the orbital time in Example 12.9 for the Vanguard satellite.

We employ Eqs. 12.44 for the semimajor and semiminor axes of the elliptic orbit. Thus, recalling that $p = 1/D$ we have

$$a = \frac{\epsilon}{D(1-\epsilon^2)} = \frac{0.1416}{(17.7\times10^{-9})(1-0.1416^2)}$$
$$= 8.164 \text{ Mm}$$
$$b = a(1-\epsilon^2)^{1/2} = 8.164\times10^6(1-0.1416^2)^{1/2}$$
$$= 8.082 \text{ Mm}$$

Therefore, from Eq. 12.43 we have for the orbital time:

$$\tau = \frac{\pi ab}{C/2} = \frac{(\pi)(8.164\times10^6)(8.082\times10^6)}{56.43\times10^9/2}$$

$$\tau = 7.347 \text{ ks} = 122.4 \text{ min}$$

Example 12.12

A space vehicle is in a circular "parking" orbit around the planet Venus, 320 km above the surface of this planet. The radius of Venus is 6.16 Mm, and the escape velocity at the surface is 10.26 km/s. A retro-rocket is fired to slow the vehicle so that it will come within 32 km of the planet. If we consider that the rocket changes the speed of the vehicle over a comparatively short distance of its travel, what is this change of speed? What is the speed of the vehicle at its closest position to the surface of Venus?

We show the vehicle in a circular parking orbit in Fig. 12.24. We shall consider that the retro-rockets are fired at position A so as to establish a new elliptic orbit with apogee at A and perigee at B.

As a first step, we shall compute GM using the escape-velocity equation 12.48. Thus, we have

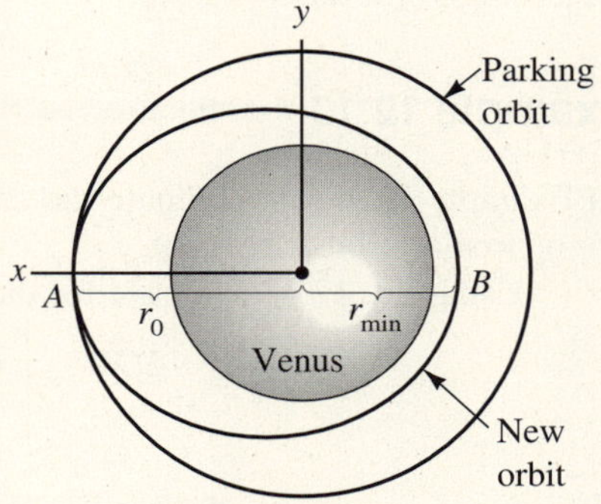

Figure 12.24. Change of orbit.

$$V_E = \sqrt{\frac{2GM}{R}}$$

Therefore,

$$GM = \frac{V_E^2 R}{2} = (10.26\times10^3)^2\left(\frac{6.16\times10^6}{2}\right)$$
$$= 324.2\times10^{12} \text{ m}^3/\text{s}^2$$

Example 12.12 (Continued)

The equation for the *new elliptic orbit* is given as

$$\frac{1}{r} = \frac{GM}{C^2} + D\cos\theta \qquad \text{(a)}$$

Note that when

$$\theta = 0, \qquad r = r_0 = 6.480 \text{ Mm} \qquad \text{(b)}$$

$$\theta = \pi, \qquad r = r_{min} = 6.192 \text{ Mm} \qquad \text{(c)}$$

To determine the constant C, we subject Eq. (a) to the conditions (b) and (c). Thus,

$$\frac{1}{6.48\times10^6} = \frac{324.2\times10^{12}}{C^2} + D \qquad \text{(d)}$$

$$\frac{1}{6.192\times10^6} = \frac{324.2\times10^{12}}{C^2} - D \qquad \text{(e)}$$

Adding these equations, we eliminate D and can solve for C. Thus,

$$\frac{648.4\times10^{12}}{C^2} = \frac{1}{6.480\times10^6} + \frac{1}{6.192\times10^6}$$

Therefore,

$$C = 45.31 \times 10^9 \text{ m}^2/\text{s}$$

Accordingly, for the new orbit,

$$r_0V_0 = 45.31\times10^9$$

Therefore,

$$V_0 = 6.992 \text{ km/s}$$

For the *circular* parking orbit the velocity V_c is

$$V_c = \sqrt{\frac{GM}{r_0}} = \sqrt{\frac{324.2\times10^{12}}{6.480\times10^6}}$$
$$= 7.073 \text{ km/s}$$

The change in velocity that the retro-rocket must induce is then

$$\Delta V = 6.992 - 7.073 = \boxed{-81 \text{ m/s}}$$

The velocity at the perigee at B is easily computed since

$$r_BV_B = C = 45.31\times10^9$$

Therefore,

$$V_B = 45.31 \times 10^9/6.192 \times 10^6 = \boxed{7.317 \text{ km/s}}$$

Now let us consider more general launching conditions where the launching angle α is not zero (see Fig. 12.25). The constant C is still easily evaluated (see Eq. (12.46)) in terms of launching data as $r_0(V_\theta)_0$. To get D, we write Eq. 12.40 for launching conditions. Thus,

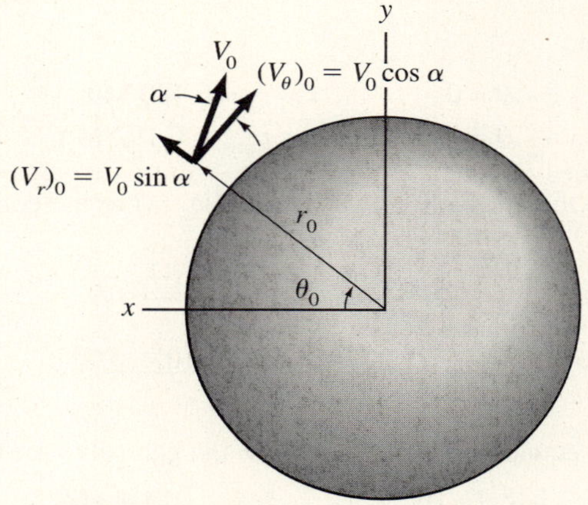

Figure 12.25. Launch with radial velocity.

$$\frac{1}{r_0} = \frac{GM}{C^2} + D\cos\theta_0 \tag{12.50}$$

The value of θ_0 is not yet known. Thus, we have two unknown quantities in this equation, namely D and θ_0. Differentiating Eq. 12.40 with respect to time and solving for $\dot{r}$, we get

$$\dot{r} = Dr^2\dot{\theta}\sin\theta = DC\sin\theta \tag{12.51}$$

Noting that $\dot{r}$ is equal to V_r and submitting the preceding equation to launching conditions, we then form a second equation for the evaluation of the unknown constants D and θ_0. Thus,

$$(V_r)_0 = DC\sin\theta_0 \tag{12.52}$$

Rearranging Eq. 12.50, we have

$$\frac{1}{r_0} - \frac{GM}{C^2} = D\cos\theta_0 \tag{12.53}$$

Divide both sides of Eq. 12.52 by C. Now, squaring Eqs. 12.52 and 12.53, adding terms, and using the fact that $\sin^2\theta_0 + \cos^2\theta_0 = 1$, we get for the constant D the result :[11]

$$D = \left\{\left(\frac{1}{r_0} - \frac{GM}{C^2}\right)^2 + \left[\frac{(V_r)_0}{C}\right]^2\right\}^{1/2} \tag{12.54}$$

[11]The student has the option of formulating Eqs. 12.52 and 12.53 for each problem and finding D from these equations, or he/she can use Eq. 12.54 directly. In some of the homework problems we shall ask you to do both.

Having taken the positive root for D, we note (see footnote #10 on page 547) that θ is to be measured from perigee. The eccentricity is

$$\epsilon = \frac{C^2}{GM}\left\{\left(\frac{1}{r_0} - \frac{GM}{C^2}\right)^2 + \left[\frac{(V_r)_0}{C}\right]^2\right\}^{1/2} \tag{12.55}$$

First, bringing C^2 into the bracket and then replacing C by $r_0(V_\theta)_0$ in the entire equation, we get the eccentricity conveniently in terms of launching data:

$$\epsilon = \frac{r_0(V_\theta)_0}{GM}\left\{(V_r)_0^2 + \left[(V_\theta)_0 - \frac{GM}{r_0(V_\theta)_0}\right]^2\right\}^{1/2} \tag{12.56}$$

One can show, using the preceding formulations, that the equation for the escape velocity developed earlier, namely

$$V_E = \sqrt{\frac{2GM}{r}}$$

is valid for any launching angle α. Remember that V_E in this equation is measured from a reference xyz at the center of the planet translating in inertial space. The velocity attainable by a rocket system relative to the planet's surface does not depend on the position of firing on the earth, but depends primarily on the rocket system and trajectory of flight. However, the velocity attainable by a rocket system relative to the aforementioned reference xyz *does* depend on the position of firing on the planet's surface. This position, accordingly, is important in determining whether an escape velocity can be reached. The extreme situations of a launching at the equator and at the North Pole are shown in Fig. 12.26 and should clarify this point. Note that the motion of the planet's surface adds to the final vehicle velocity at the equator, but that no such gain is achieved at the North Pole.

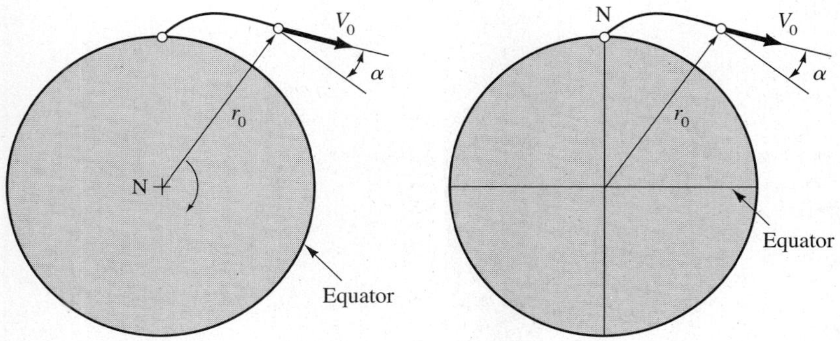

Figure 12.26. Launching at equator and North Pole.

Example 12.13

Suppose that the Vanguard satellite in Example 12.9 is off course by an angle $\alpha = 5°$ at the time of launching but otherwise has the same initial data. Determine whether the satellite goes into orbit. If so, determine the maximum and minimum distances from the earth's surface.

The initial data for the launching are

$$r_0 = 7.01 \text{ Mm}, \qquad V_0 = 8.05 \text{ km/s}$$

Hence,

$$(V_r)_0 = (8.05\times 10^3)\sin\alpha = (8.05\times 10^3)(0.0872)$$
$$= 701.6 \text{ m/s}$$

$$(V_\theta)_0 = (8.05\times 10^3)\cos\alpha = (8.05\times 10^3)(0.996)$$
$$= 8019.4 \text{ m/s}$$

To determine whether we have an orbit, we would have to show first that the eccentricity ϵ is less than unity. This condition would preclude the possibility of an escape from the earth. Furthermore, we must be sure that the perigee of the orbit is far enough from the earth's surface to ensure a reasonably permanent orbit. Actually, for both questions we need only calculate r for $\theta = 0$ and $\theta = \pi$. An infinite value of one of the r's will mean that we have an escape condition, and a value not sufficiently large will mean a crash or a decaying orbit due to atmospheric friction.

Using the value of GM as 398.1×10^{12} m^3/s^2 from Example 12.9 and using Eq. 12.54 for the constant D, we can express the trajectory of the satellite (Eq. 12.40) as

$$\frac{1}{r} = \frac{398.1\times 10^{12}}{[(7.01\times 10^6)(8019.4)]^2} + \left\{\left[\frac{1}{7.01\times 10^6} - \frac{398.1\times 10^{12}}{[7.01\times 10^6)(8019.4)]^2}\right]^2 + \left[\frac{701.6}{(7.01\times 10^6)(8019.4)}\right]^2\right\}^{1/2} \cos\theta$$

Example 12.13 (Continued)

Therefore,

$$\frac{1}{r} = 125.97 \times 10^{-9} + 20.83 \times 10^{-9} \cos\theta \ \text{m}^{-1} \qquad \text{(a)}$$

Set $\theta = 0$:

$$\frac{1}{r(0)} = 125.97 \times 10^{-9} + 20.83 \times 10^{-9} \text{m}^{-1}$$

Hence,

$$r(0) = 6.81 \text{ Mm}$$

Thus, after being launched at a position 640 km above the earth's surface, the satellite comes within 440 km of the earth as a result of a 5° change in the launching angle. This satellite, therefore, must be launched almost parallel to the earth if it is to attain a reasonably permanent orbit.

Now, setting $\theta = \pi$, we get

$$\frac{1}{r(\pi)} = 125.97 \times 10^{-9} - 20.83 \times 10^{-9}$$

Hence,

$$r_{max} = 9.51 \text{ Mm}$$

Obviously, the maximum distance from the earth's surface is 3.14 Mm.

PROBLEMS

12.54. A device used at amusement parks consists of a circular room that is made to revolve about its axis of symmetry. People stand up against the wall, as shown in the diagram. After the whole room has been brought up to speed, the floor is lowered. What minimum angular speed is required to ensure that a person will not slip down the wall when the floor is lowered? Take $\mu_s = 0.3$.

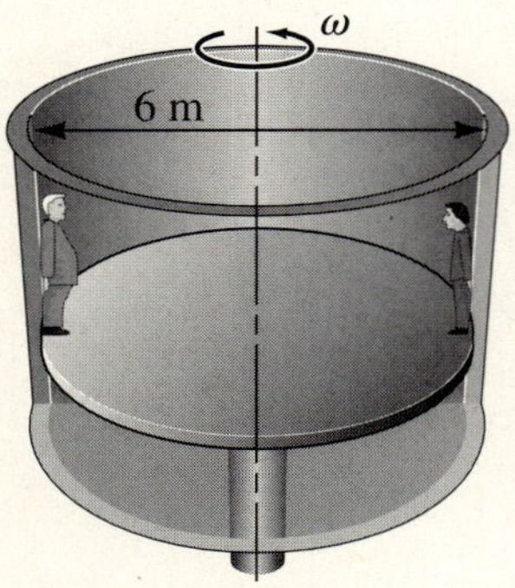

Figure P.12.54.

12.55. A flywheel is rotating at a speed of $\omega = 10$ rad/s and has at this instant a rate of change of speed $\dot{\omega}$ of 5 rad/s^2. A solenoid at this instant moves a valve toward the centerline of the flywheel at a speed of 1.5 m/s and is decelerating at the rate of 0.6 m/s^2. The valve has a mass of 1 kg and is 0.3 m from the axis of rotation at the time of interest. What is the total force on the valve?

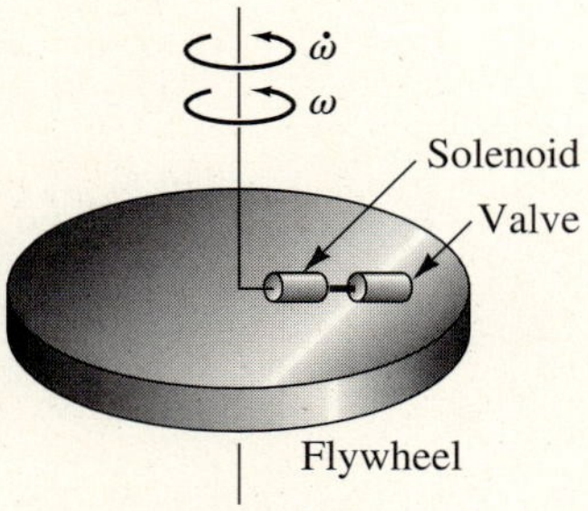

Figure P.12.55.

12.56. A conical pendulum of length l is shown. The pendulum is made to rotate at a constant angular speed of ω about the vertical axis. Compute the tension in the cord if the pendulum bob has weight W. What is the distance of the plane of the trajectory of the bob from the support at O?

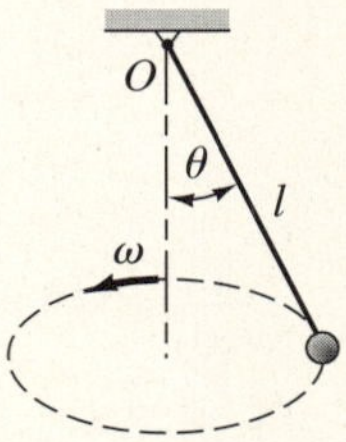

Figure P.12.56.

12.57. A shaft AB rotates at an angular velocity of 100 r/min. A body E of mass 10 kg can move without friction along rod CD fixed to AB. If the body E is to remain stationary relative to CD at any position along CD, how must the spring constant K vary? The distance r_0 from the axis is the unstretched length of the spring.

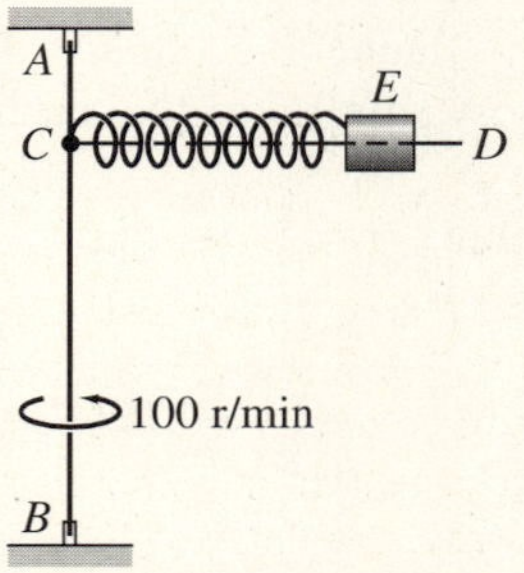

Figure P.12.57.

12.58. A device consists of three small masses, three weightless rods, and a linear spring with $K = 200$ N/m. The system is rotating in the given fixed configuration at a constant speed $\omega = 10$ rad/s in a horizontal plane. The following data apply:

$M_A = 2$ kg $\qquad M_B = 3$ kg $\qquad M_C = 2$ kg

If the spring is stretched by an amount 25 mm, determine the total force components acting on the mass at C and the tensile force in member DA. [*Hint:* Consider a single particle, then a system of particles; DA and DB are pin connected.]

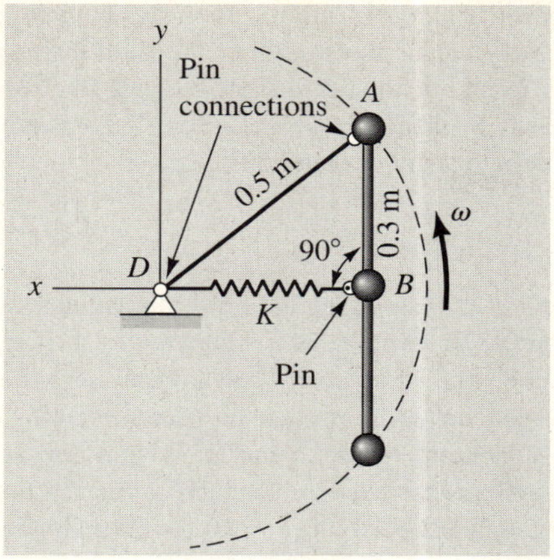

Figure P.12.58.

12.59. In the preceding problem consider that member *DA* is welded to the sphere at *A*. Now, at the instant of interest, there is also an angular acceleration of the system having the value of 0.28 rad/s^2 counterclockwise. What are the force components acting on particle *C*, and what are the force components from rod *AD* acting on particle *A*? See hint given in the preceding problem.

12.60. A device called a *flyball governor* is used to regulate the speed of such devices as steam engines and turbines. As the governor is made to rotate through a system of gears by the device to be controlled, the balls will attain a configuration given by the angle θ, which is dependent on both the angular speed ω of the governor and the force *P* acting on the collar bearing at *A*. The up-and-down motion of the bearing at *A* in response to a change in ω is then used to open or close a valve to regulate the speed of the device. Find the angular velocity required to maintain the configuration of the flyball governor for $\theta = 30°$. Neglect friction.

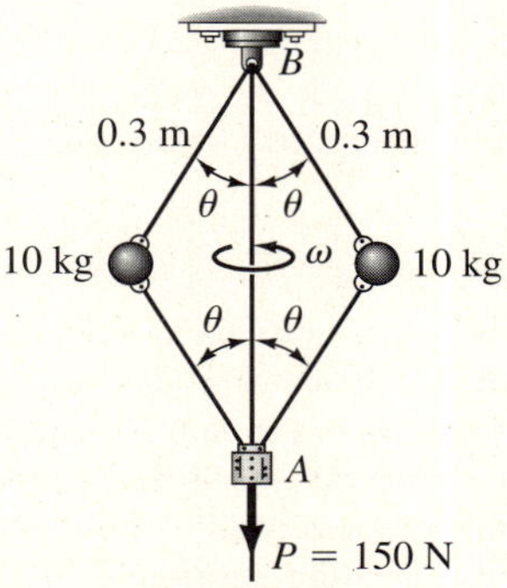

Figure P.12.60.

12.61. A platform rotates at 2 rad/s. A body *C* weighing 450 N rests on the platform and is connected by a flexible weightless cord to a mass weighing 225 N, which is prevented from swinging out by part of the platform. For what range of values of *x* will bodies *C* and *B* remain stationary relative to the platform? The static coefficient of friction for all surfaces is 0.4.

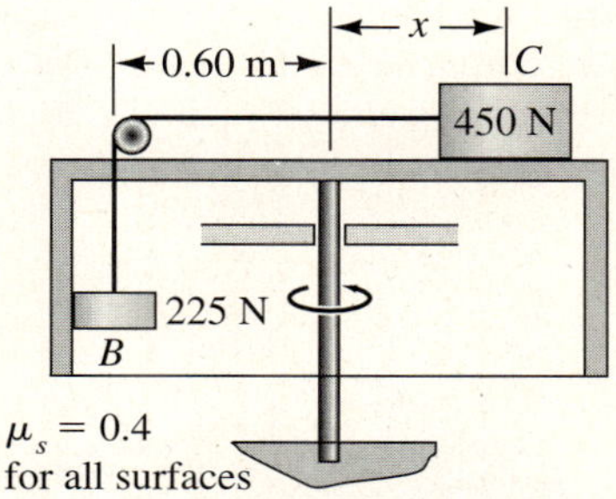

Figure P.12.61.

12.62. A particle moves under gravitational influence about a body *M*, the center of which can be taken as the origin of an inertial reference. The mass of the particle is 750 kg. At time *t*, the particle is at a position 7.2 Mm from the center of *M* with direction cosines $l = 0.5$, $m = -0.5$, $n = 0.707$. The particle is moving at a speed of 7.6 km/s along the direction $\boldsymbol{\epsilon}_t = 0.8\boldsymbol{i} + 0.2\boldsymbol{j} + 0.566\boldsymbol{k}$. What is the direction of the normal to the plane of the trajectory?

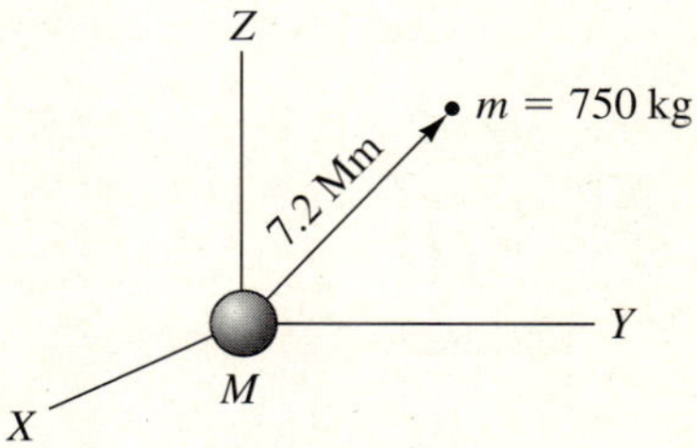

Figure P.12.62.

12.63. If the position of the particle in Problem 12.62 were to reach a distance of 6.9 Mm from the center of body *M*, what would the transverse velocity V_θ of the particle be?

12.64. Use Eqs. 12.38b and 12.40 to show that if the eccentricity is zero, the trajectory must be that of a circle.

12.65. A satellite has at one time during its flight around the earth a radial component of velocity 900 m/s and a transverse component of 7.36 km/s. If the satellite is at a distance of 7.04 Mm from the center of the earth, what is its areal velocity?

12.66. Compute the escape velocity at a position 8 Mm from the center of the earth. What speed is needed to maintain a circular orbit at that distance from the earth's center? Derive the equation for the speed needed for a circular orbit directly from Newton's law without using information about eccentricities, etc.

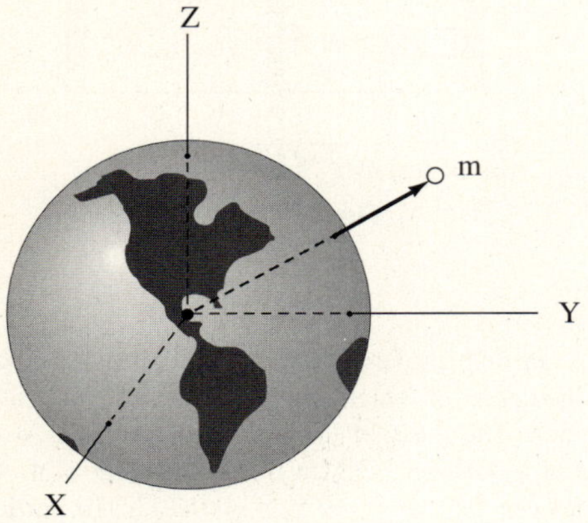

Figure P.12.66.

12.67. A small sphere is swinging in the *xy* plane at the end of a thin light rod while in a tank of water. If we neglect the buoyant force (see next problem) on the sphere and on the rod, determine the angular acceleration $\ddot{\theta}$ and the tensile force T on the rod for the following conditions, which include viscous drag D on the sphere only.

$$M = 20g \quad L = 3.3 \text{ m} \quad \theta = 65°$$
$$\dot{\theta} = 35°/\text{s} \quad D = 1.5 \times 10^{-3} V^2 \text{ N}$$

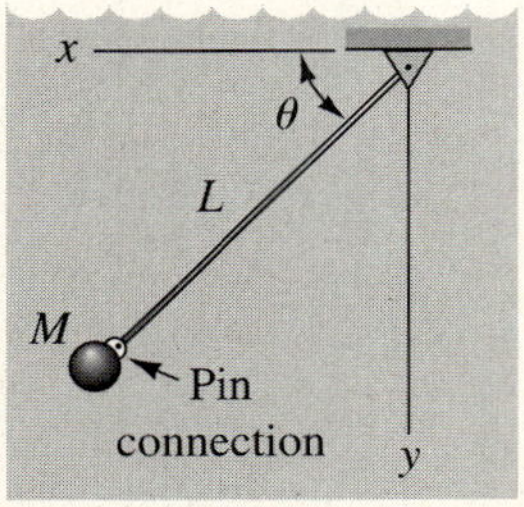

Figure P.12.67.

12.68. As you learned in high school, the buoyant force on a submerged body equals the weight of the fluid displaced. Do the preceding problem but now include the buoyant force on the sphere whose diameter we shall take to be 10 mm. The volume of a sphere is $\frac{4}{3}\pi r^3$ and the density of water is 1 Mg.

12.69. A mass M is swinging around a vertical axis at the end of a weightless cord of length L. M is supported by a frictionless platform that can be moved vertically upward from its lowest position, where it just touches M. Formulate an equation giving the tension T in the cord in terms of M, L, and ω. What is the value of θ at which the platform first ceases to touch M as the platform is moved down from its highest position ($\theta = 0$) for the following data:

$$\omega = 8 \text{ rad/s} \qquad M = 3.1 \text{ kg} \qquad L = 0.42 \text{ m}$$

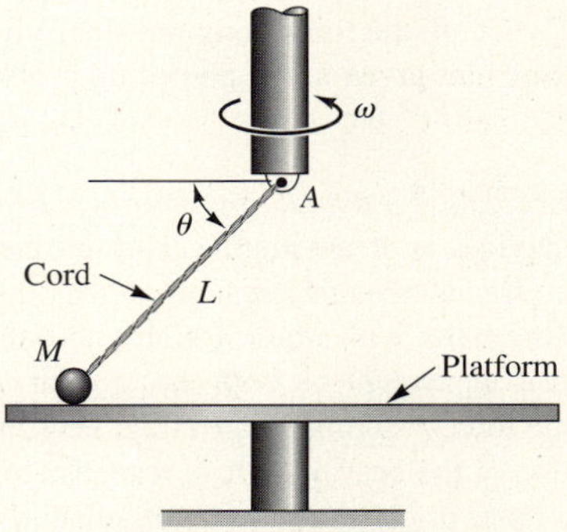

Figure P.12.69.

12.70. A satellite is in a circular (parking) orbit around the earth, whose radius is 6.37 Mm. If the velocity of the satellite is 6.9 km/s, how much time is needed for one complete orbit?

12.71. A space vehicle is in a circular orbit around the earth in a plane corresponding to the equator and moving so as to remain above a specific position on the earth's surface. This means that the satellite has the same angular speed as the earth, making it useful as a communications satellite. What should the radius of the orbit be? The radius of the earth is 6.37 Mm. Work in units of miles and hours.

12.72. Consider a satellite of mass m in a circular orbit around the earth at a radius R_0 from the center of the earth. Using the universal law of gravitation (Eq. 1.11) with M as the mass of the earth and using *Newton's law* in a direction normal to the path, show that

$$V_{Circ.} = \sqrt{\frac{GM}{R_0}}$$

for a circular orbit. Now at the earth's surface use the universal gravitational law again and the weight of the body, to show that $GM = gR^2_{Earth}$, where g is the acceleration of gravity.

12.73. The acceleration of gravity on the planet Mars is about 0.385 times the acceleration of gravity on earth, and the radius of Mars is about 0.532 times that of the earth. What is the escape velocity from Mars at a position 160 km from the surface of the planet?

12.74. In 1971 Mariner 9 was placed in orbit around Mars with an eccentricity of 0.5. At the lowest point in the orbit, Mariner 9 is 320 km from the surface of Mars.

(a) Compute the maximum velocity of the space vehicle relative to the center of Mars.

(b) Compute the time of one cycle.

Use the data in Problem 12.73 for Mars.

12.75. A man is in orbit around the earth in a space-shuttle vehicle. At his lowest possible position, he is moving with a speed of 8.3 km/s at an altitude of 320 km. When he wants to come back to earth, he fires a retro-rocket straight ahead when he is at the aforementioned lowest position and slows himself down. If he wishes subsequently to get within 80 km from the earth's surface during the first cycle after firing his retro-rocket, what must his decrease in velocity be? (Neglect air resistance.)

12.76. The Pioneer 10 space vehicle approaches the planet Jupiter with a trajectory having an eccentricity of 3. The vehicle comes to within 1.6 Mm of the surface of Jupiter. What is the speed of the vehicle at this instant? The acceleration of gravity of Jupiter is 27.67 m/s^2 at the surface and the radius is 69.83 Mm.

12.77. If the moon has a motion about the earth that has an eccentricity of 0.0549 and a period of 27.3 days, what is the closest distance of the moon to the earth in its trajectory?

12.78. The satellite Hyperion about the planet Saturn has a motion with an eccentricity known to be 0.1043. At its closest distance from Saturn, Hyperion is 1.485 Gm away (measured from center to center). What is the period of Hyperion about Saturn? The acceleration of gravity of Saturn is 13.93 m/s^2 at its surface. The radius of Saturn is 57.6 Mm.

12.79. Two satellite stations, each in a circular orbit around the earth, are shown. A small vehicle is shot out of the station at A tangential to the trajectory in order to "hit" station B when it is at a position E 120° from the x axis as shown in the diagram. What is the velocity of the vehicle relative to station A when it leaves? The circular orbits are 320 km and 640 km, respectively, from the earth's surface.

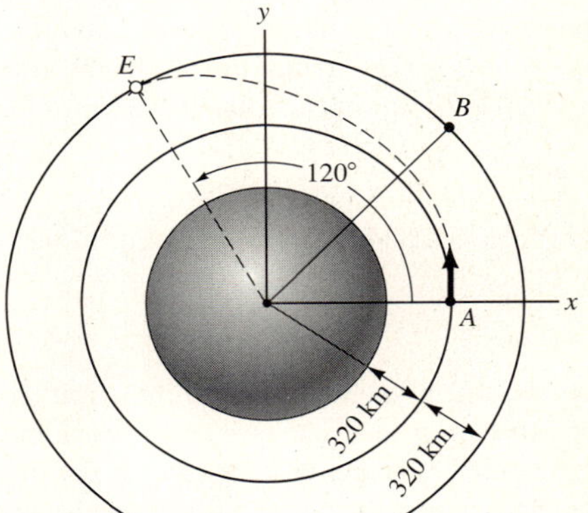

Figure P.12.79.

12.80. In Problem 12.79, determine the total velocity of the vehicle as it arrives at E as seen by an observer in the satellite B. The values of C and D for the vehicle from Problem 12.79 are 52.44×10^9 m^2/s and 4.582×10^{-9} m^{-1}, respectively.

12.81. The Viking I space probe is approaching Mars. When it is 80.65 Mm from the center of Mars, it has a speed of 4.48 km/s with a component (V_r) toward the center of Mars of 4.39 km/s. Does Viking I crash into Mars, go into orbit, or have one pass in the vicinity of Mars? If there is no crash, how close to Mars does it come? The acceleration of gravity on the surface of Mars is 4.13 m/s^2, and its radius is 3.4 Mm. Do not use formula for D as given by Eq. 12.54, but work from the trajectory equations.

12.82. A meteor is moving at a speed of 9 km/s relative to the center of the earth when it is 560 km from the surface of the earth. At that time, the meteor has a radial velocity component of 1.8 km/s toward the center of the earth. How close does it come to the earth's surface? Do this problem without the aid of Eq. 12.54.

12.83. Do Problem 12.82 with the aid of Eq. 12.54.

12.84. The moon's radius is about 0.272 times that of the earth, and its acceleration of gravity at the surface is 0.165 times that of the earth at the earth's surface. A space vehicle approaches the moon with a velocity component toward the center of the moon of 880 m/s and a transverse component of 220 m/s relative to the center of the moon. The vehicle is 3.2 Mm from the center of the moon when it has these velocity components. Will the vehicle go into orbit around the moon if we consider only the gravitational effect of the moon on the vehicle? If it goes into orbit, how close will it come to the surface of the moon? If not, does it collide with the moon? Do this problem without the aid of Eq. 12.54.

12.85. Do Problem 12.84 with the aid of Eq. 12.54.

12.86. Assume that a satellite is placed into orbit about a planet that has the same mass and diameter as the earth but no atmosphere. At the minimum height of its trajectory, the satellite has an elevation of 645 km from the planet's surface and a velocity of 8.3 km/s. To observe the planet more closely, we send down a smaller satellite from the main body to within 16 km of this planet at perigee. The "subsatellite" is given a velocity component toward the center of the planet when the main satellite is at its lowest position. What is this radial velocity, and what is the eccentricity of the trajectory of the subsatellite? What is a better way to get closer to the planet?

12.87. Suppose that you are on a planet having no atmosphere. This planet rotates once every 6 hr about its axis relative to an inertial reference *XYZ* at its center. The planet has a radius of 1.6 Mm, and the acceleration of gravity at the surface is 7 m/s². A bullet is fired by a man at the equator in a direction normal to the surface of the planet as seen by this man. The muzzle velocity of the gun is 1.5 km/s. What is the eccentricity of the trajectory and the maximum height h of the bullet above the surface of the planet?

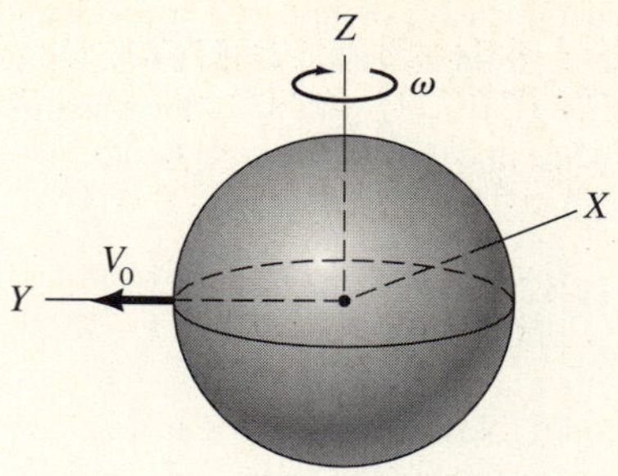

Figure P.12.87.

***12.88.** A satellite is launched at A. We wish to determine the time required, Δt, to get to position B. Show that for this calculation we can employ the formulation

$$\Delta t = \frac{1}{C}\int_{\theta_0}^{\theta_B} r^2\, d\theta$$

For integration purposes, show that the formulation above becomes

$$\Delta t = \frac{1}{C}\int_{\theta_0}^{\theta_B} \frac{d\theta}{[(GM/C^2) + D\cos\theta]^2}$$

Carry out the integration.

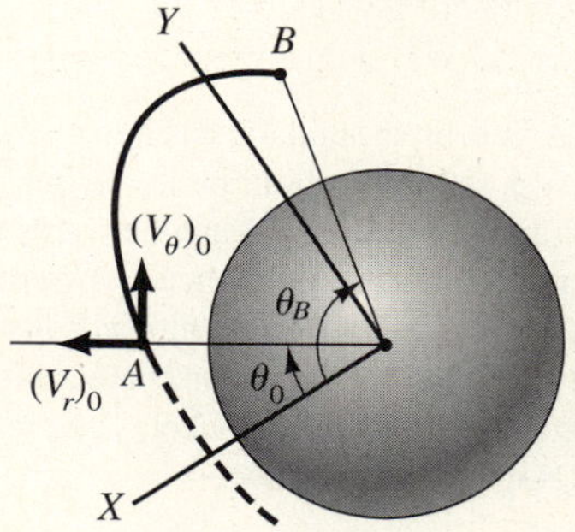

Figure P.12.88.

***12.89.** A satellite is launched at a speed of 9 km/s relative to the earth's center at an altitude of 550 km above the earth's surface. The guidance system has malfunctioned, and the satellite has a direction 20° up from the tangent plane to the earth's surface. Will the satellite go into orbit? Give the time required for one cycle if it goes into orbit or the time it takes before it strikes the earth after firing. Neglect friction in both cases. (See Problem 12.88 before doing this problem.)

Part C: Path Variables

12.9 Newton's Law for Path Variables

We can express Newton's law for path variables as follows:

$$F_t = m\frac{d^2s}{dt^2} \tag{12.57a}$$

$$F_n = m\frac{(ds/dt)^2}{R} \tag{12.57b}$$

Notice that the second of these equations is always nonlinear, as discussed in Section 12.5.[12] This condition results from both the squared term and the radius of curvature R. It is therefore difficult to integrate this differential equation. Accordingly, we shall be restricted to reasonably simple cases. We now illustrate the use of the preceding equations.

[12]Equation 12.57a could also be nonlinear, depending on the nature of the function F_t.

Example 12.14

A portion of a roller coaster that one finds in an amusement park is shown in Fig. 12.27(a). The portion of the track shown is coplanar. The curve from A to the right on which the vehicle moves is that of a parabola, given as

$$(y - 30)^2 = 30x \tag{a}$$

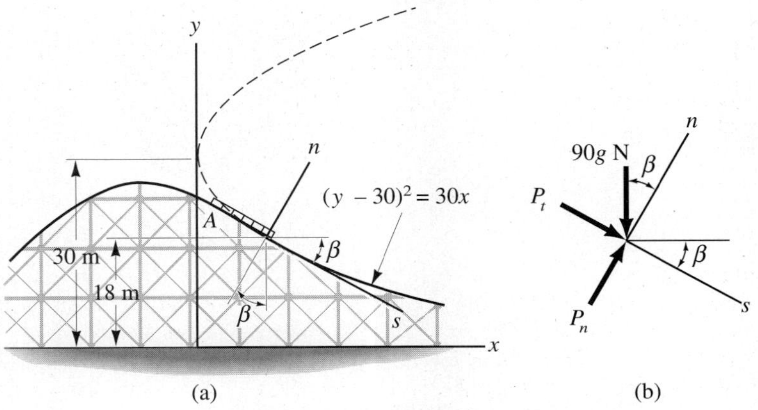

Figure 12.27. Roller coaster trajectory.

Example 12.14 (Continued)

with x and y in metres. If the train of cars is moving at a speed of 12 m/s when the front car is 18 m above the ground, what is the total normal force exerted by a 90 kg occupant of the front car on the seat and floor of the car?

Since we require only the force F normal to the path, we need only be concerned with a_n. Thus, we have

$$a_n = \frac{(ds/dt)^2}{R} = \frac{12^2}{R} \tag{b}$$

We can compute R from analytic geometry as follows:

$$R = \frac{\left[1 + (dy/dx)^2\right]^{3/2}}{\left|d^2y/dx^2\right|} \tag{c}$$

wherein from Eq. (a) we have

$$\frac{dy}{dx} = \frac{15}{y-30} \tag{d}$$

$$\frac{d^2y}{dx^2} = -\frac{15}{(y-30)^2}\frac{dy}{dx} = -\frac{225}{(y-30)^3}$$

Substituting into Eq. (c), we have

$$R = \frac{\left[1 + \left(\frac{15}{y-30}\right)^2\right]^{3/2}}{\left|225/(y-30)^3\right|}$$

At the position of interest, we get

$$R = \frac{\left[1 + \left(\frac{225}{144}\right)\right]^{3/2}}{(225/1728)} = 31.5 \text{ m} \tag{f}$$

Example 12.14 (Continued)

Accordingly, we now have for F_n, as required by **Newton's law**:

$$F_n = Ma_n = 90\left(\frac{144}{31.5}\right) = 411.4 \text{ N} \qquad \text{(g)}$$

Note that F_n is the *total* force component normal to the trajectory needed on the occupant for maintaining his motion on the given trajectory. This force component comes from the action of gravity and the forces from the seat and floor of the car. These forces have been shown in Fig. 12.27(b), where P_n and P_t are the normal and tangential force components from the car acting on the occupant. The resultant of this force system must, accordingly, have a component along n equal to 411.4 N. Thus,

$$-90g\cos\beta + P_n = 411.4 \qquad \text{(h)}$$

To get β, note with the help of Eq. (d) that

$$\tan^{-1}\left(\frac{dy}{dx}\right)_{y=18} = \tan^{-1}\frac{15}{-12} = -51.3° \qquad \text{(i)}$$

Therefore,

$$\beta = 51.3°$$

Substituting into Eq. (h) and solving for P_n, we get

$$P_n = 90g\cos 51.3° + 411.4$$

$$\boxed{P_n = 963 \text{ N}}$$

This is the force component *from* the vehicle *onto* the passenger. The reaction to this force is the force component *from* the passenger onto the vehicle.

Part D: A System of Particles

12.10 The General Motion of a System of Particles

Let us examine a system of n particles (Fig. 12.28) that has interactions between the particles for which *Newton's third law* of motion (action equals reaction) applies. *Newton's second law* for any particle (let us say the ith particle) is then

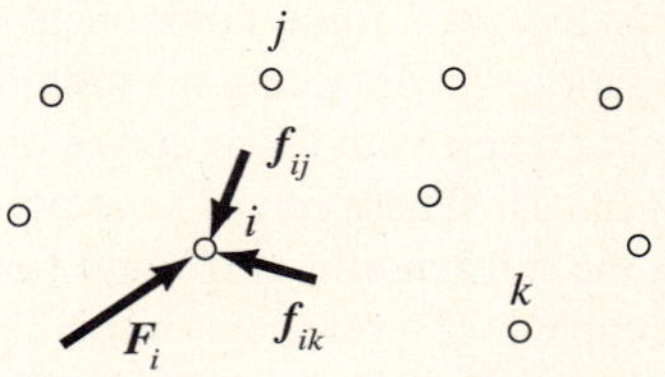

Figure 12.28. Forces on ith particle of the system.

$$m_i \frac{d^2 \boldsymbol{r}_i}{dt^2} = \boldsymbol{F}_i + \sum_{\substack{j=1 \\ i \neq j}}^{n} \boldsymbol{f}_{ij} \tag{12.58}$$

where $\boldsymbol{f}_{ij}$ is the force on particle i from particle j and is thus considered an *internal* force for the system of particles. Clearly, the $j = i$ term of the summation must be deleted since the ith particle cannot exert force on itself. The force $\boldsymbol{F}_i$ represents the resultant force on the ith particle from the forces *external* to the system of particles.

If these equations are added for all n particles, we have

$$\sum_{i=1}^{n} m_i \frac{d^2 \boldsymbol{r}_i}{dt^2} = \sum_{i=1}^{n} \boldsymbol{F}_i + \sum_{i=1}^{n} \sum_{j=1}^{n} \boldsymbol{f}_{ij} \tag{12.59}$$

Carrying out the double summation and excluding terms with repeated indexes, such as $\boldsymbol{f}_{11}$, $\boldsymbol{f}_{22}$, etc., we find that for each term with any one set of indexes there will be a term with the reverse of these indexes present. For example, for the force $\boldsymbol{f}_{12}$, a force $\boldsymbol{f}_{21}$ will exist. Considering the meaning of the indexes, we see that $\boldsymbol{f}_{ij}$ and $\boldsymbol{f}_{ji}$ represent action and reaction forces between a pair of particles. Thus, as a result of *Newton's third law*, the double summation in Eq. 12.59 should add up to zero. *Newton's second law* for a system of particles then becomes:

$$\boldsymbol{F} = \sum_{i=1}^{n} m_i \frac{d^2 \boldsymbol{r}_i}{dt^2} = \frac{d^2}{dt^2} \sum_{i=1}^{n} m_i \boldsymbol{r}_i \tag{12.60}$$

where $\boldsymbol{F}$ now represents the vector sum of all the *external* forces acting on *all* the particles of the system.

To make further useful simplifications, we use the first moment of mass of a system of n particles about a fixed point A in inertial space given as

$$\text{first moment vector} \equiv \sum_{i=1}^{n} m_i \boldsymbol{r}_i$$

where $\boldsymbol{r}_i$ represents the position vector from the point A to the ith particle (Fig. 12.29). As explained in Chapter 8, we can find a position, called the

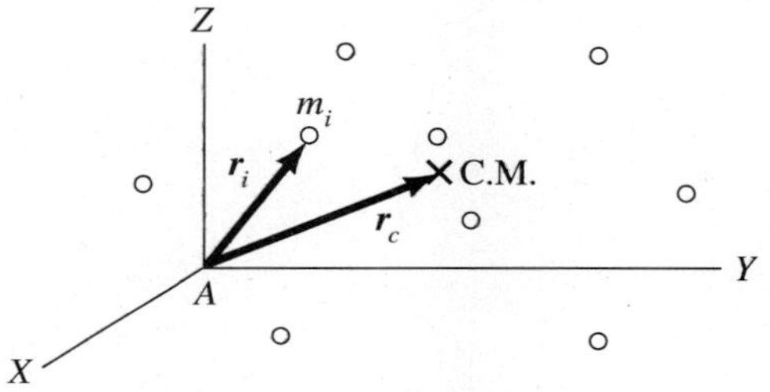

Figure 12.29. Center of mass of system.

center of mass of the system, with position vector $\boldsymbol{r}_c$, where the entire mass of the system of particles can be concentrated to give the correct first moment. Thus,

$$\boldsymbol{r}_c \sum_{i=1}^{n} m_i = \sum_{i=1}^{n} m_i \boldsymbol{r}_i$$

Therefore,

$$\boldsymbol{r}_c = \frac{\sum m_i \boldsymbol{r}_i}{\sum m_i} = \frac{\sum m_i \boldsymbol{r}_i}{M} \tag{12.61}$$

Let us reconsider Newton's law using the center-of-mass concept. To do this, replace $\sum m_i \boldsymbol{r}_i$ by $M\boldsymbol{r}_c$ in Eq. 12.60. Thus,

$$\boldsymbol{F} = \frac{d^2}{dt^2}(M\boldsymbol{r}_c) = M\frac{d^2\boldsymbol{r}_c}{dt^2} \tag{12.62}$$

We see that *the center of mass of any aggregate of particles has a motion that can be computed by methods already set forth, since this is a problem involving a single hypothetical particle of mass M*. You will recall that we have alluded to this important relationship several times earlier to justify the use of the particle concept in the analysis of many dynamics problems. We must realize for such an undertaking that $\boldsymbol{F}$ is the total *external* force acting on *all* the particles.

Example 12.15

Three charged particles in a vacuum are shown in Fig. 12.30. Particle 1 has a mass of 10 mg and a charge of 4 mC (coulomb) and is at the origin at the instant of interest. Particles 2 and 3 each have a mass of 20 mg and a charge of 50 μC and are located, respectively, at the instant

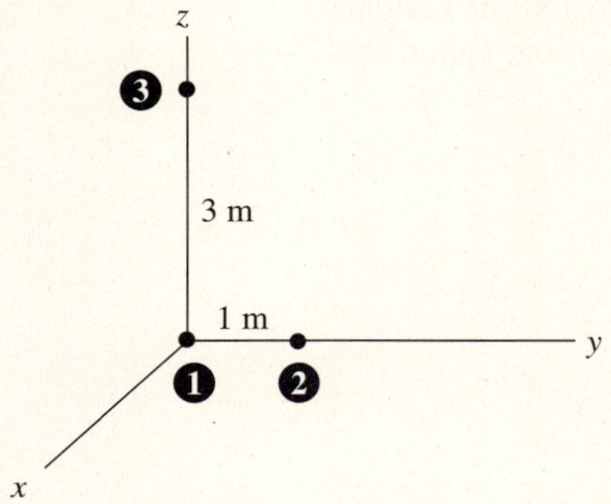

Figure 12.30. Charged particles in field ***E***.

of interest 1 m along the y axis and 3 m along the z axis. An electric field $\boldsymbol{E}$ given as

$$\boldsymbol{E} = 2x\boldsymbol{i} + 3z\boldsymbol{j} + 3(y + z^2)\boldsymbol{k} \text{ N/C} \quad \text{(a)}$$

is imposed from the outside. Compute: (a) the position of the center of mass for the system, (b) the acceleration of the center of mass, and (c) the acceleration of particle 1.

To get the position of the center of mass, we merely equate moments of the masses about the origin with that of a particle having a mass equal to the sum of masses of the system. Thus,

$$(10+20+20)\times 10^{-6}\boldsymbol{r}_c = (20\times 10^{-6})\boldsymbol{j} + (20\times 10^{-6})3\boldsymbol{k}$$

Therefore,

$$\boldsymbol{r}_c = 0.4\boldsymbol{j} + 1.2\boldsymbol{k} \text{ m} \quad \text{(b)}$$

To get the acceleration of the mass center, we must find the sum of the *external* forces acting on the particles. Two external forces act on each particle: the force of gravity and the electrostatic force from the external field. Recall from physics that this electrostatic force is given as $q\boldsymbol{E}$, where q is the charge on the particle. Hence, the total external force for each particle is given as follows:

$$\boldsymbol{F}_1 = -(9.81)(10\times 10^{-6})\boldsymbol{k} + \boldsymbol{0} \text{ N} \quad \text{(c)}$$
$$\boldsymbol{F}_2 = -(9.81)(20\times 10^{-6})\boldsymbol{k} + (50\times 10^{-6})(3\boldsymbol{k}) \text{ N} \quad \text{(d)}$$
$$\boldsymbol{F}_3 = -(9.81)(20\times 10^{-6})\boldsymbol{k} + (50\times 10^{-6})(9\boldsymbol{j} + 27\boldsymbol{k}) \text{ N} \quad \text{(e)}$$

Example 12.15 (Continued)

The sum of these forces F_T is

$$F_T = 450 \times 10^{-6} j + 1009.5 \times 10^{-6} k \text{ N} \tag{f}$$

Accordingly, we have for $\ddot{r}_c$:

$$\ddot{r}_c = \frac{450 \times 10^{-6} j + 1009.5 \times 10^{-6} k}{50 \times 10^{-6}}$$

$$\ddot{r}_c = 9j + 20.2k \text{ m/s}^2 \tag{g}$$

Finally, to get the acceleration of particle 1, we must include the coulombic forces from particles 2 and 3. As you learned in physics, this force is given between two particles a and b with charges q_a and q_b as follows:

$$f_{\text{coul}} = -\frac{q_a q_b}{4\pi\epsilon_0 r^2} \hat{r}$$

where $\hat{r}$ is the unit vector between the particles, and ϵ_0 is the dielectric constant equal to 8.854 ρ F/m (farad per metre) for a vacuum. Note that the coulombic force is repulsive between like charges. The total coulombic force F_C from particles 2 and 3 is

$$F_C = -\frac{(40 \times 10^{-6})(50 \times 10^{-6})}{(4\pi\epsilon_0)(1^2)} j - \frac{(40 \times 10^{-6})(50 \times 10^{-6})}{(4\pi\epsilon_0)(3^2)} k \tag{h}$$
$$= -18j - 2k \text{ N}$$

The total force acting on particle 1 is then

$$(F_1)_T = \underbrace{-(9.81)(10 \times 10^{-6})k}_{\text{from weight}} + \underbrace{\mathbf{0}}_{\text{from external field}} + \underbrace{(-18j - 2k)}_{\text{from internal field}} \text{ N} \tag{i}$$

Clearly, the internal field dominates here. **Newton's law** then gives us

$$\ddot{r}_1 = \frac{-18j - 2k}{10 \times 10^{-6}}$$

$$\ddot{r}_1 = -1.8 \times 10^6 j - 0.2 \times 10^6 k \text{ m/s}^2 \tag{j}$$

We see here from Eqs. (g) and (j) that although the particles tend to "scramble" away from each other due to very strong internal coulombic forces, the center of mass accelerates slowly by comparison.

Example 12.16

A young man is standing in a canoe awaiting a young lady (Fig. 12.31). The man has a mass of 72 kg and, as shown, is positioned near the end of the canoe, which has a mass of 96 kg. When the young lady appears, he quickly scrambles forward to greet her, but when he has moved 6 m to the forward end of the canoe, he finds (not having studied mechanics) that he cannot reach her. How far is the tip of the canoe from the dock after our hero has made the 6 m dash? The canoe is in no way tied to the dock and there are no water currents. Neglect friction from the water on the canoe.

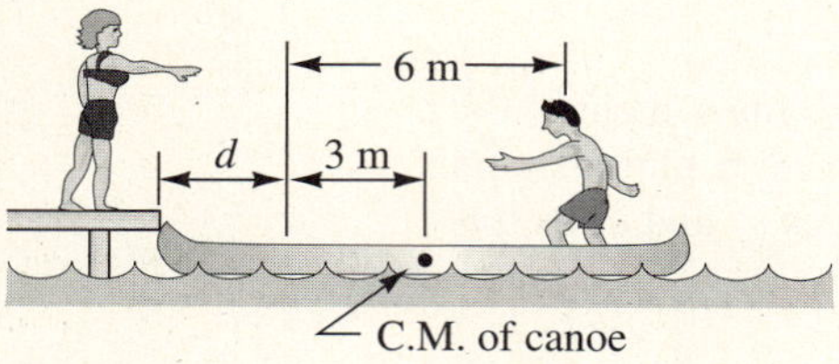

Figure 12.31. Man in canoe awaits his date.

The center of mass of the man plus the canoe cannot change position during this action since there is no net external force acting on this system during this action. Hence the first moment of mass about any fixed position must remain constant during this action. In Fig. 12.32 we have shown the man in the forward position and we choose the position at the tip of the dock to equate moment of mass at the beginning of the action and just when the man has moved the 6 m. We then can say, noting that we are denoting the unspecified distance between the tip of the canoe and the forward position of the man as d as shown in Figs. 12.31 and 12.32,

$$96(d+3)+72(d+6)=72(x+d)+96(x+d+3)$$

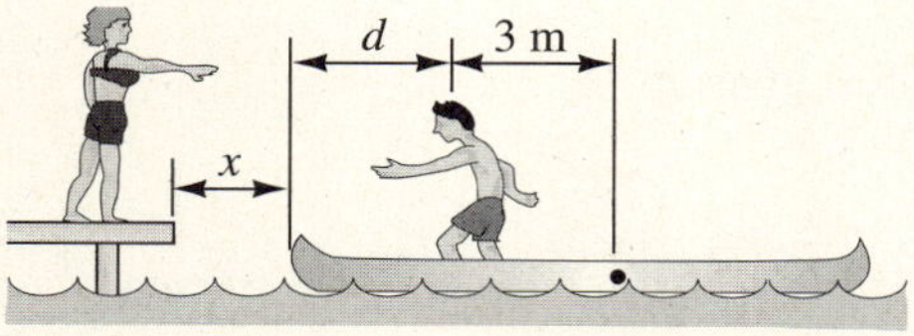

Figure 12.32. Man rushes forward to greet his date.

Canceling terms where possible we then have

$$168x = 432 \quad \therefore \quad \boxed{x = 2.57 \text{ m}}$$

12.90. A warrior of old is turning a sling in a vertical plane. A rock of mass 0.3 kg is held in the sling prior to releasing it against an enemy. What is the minimum speed ω to hold the rock in the sling?

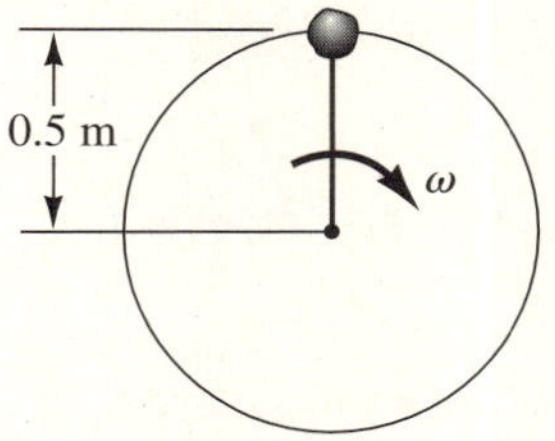

Figure P.12.90.

12.91. A car is traveling at a speed of 25 m/s along a banked highway having a radius of curvature of 150 m. At what angle should the road be banked in order that a zero friction force is needed for the car to go around this curve?

12.92. A car weighing 20 kN is moving at a speed V of 16 m/s on a road having a vertical radius of curvature of 200 m as shown. At the instant shown, what is the maximum deceleration possible from the brakes along the road for the vehicle if the coefficient of dynamic friction between tires and the road is 0.55?

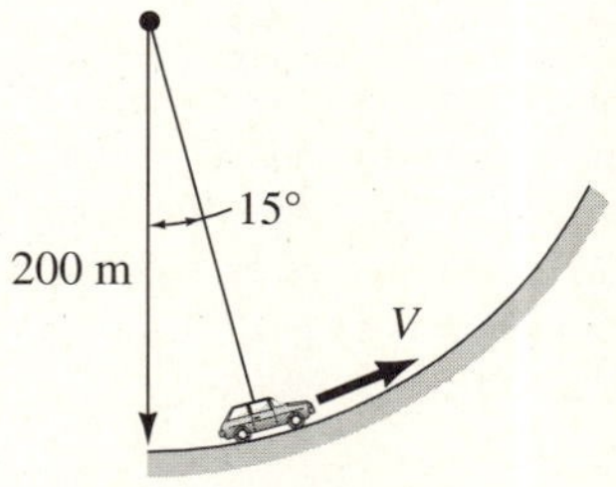

Figure P.12.92.

12.93. A particle moves at uniform speed of 1 m/s along a plane sinusoidal path given as

$$y = 5 \sin \pi x \text{ m}$$

What is the position between $x = 0$ and $x = 1$ m for the maximum force normal to the curve? What is this force if the mass of the particle is 1 kg?

12.94. A catenary curve is formed by the cable of a suspension bridge. The equation of this curve relative to the axes shown can be given as

$$y = \frac{a}{2}(e^{ax} + e^{-ax}) = a \cosh ax$$

with x and y in metres. A small one-passenger vehicle is designed to move along the catenary to facilitate repair and painting of the bridge. Consider that the vehicle moves at uniform speed of 3 m/s along the curve. If the vehicle and passenger have a combined mass of 110 kg what is the force normal to the curve as a function of position x?

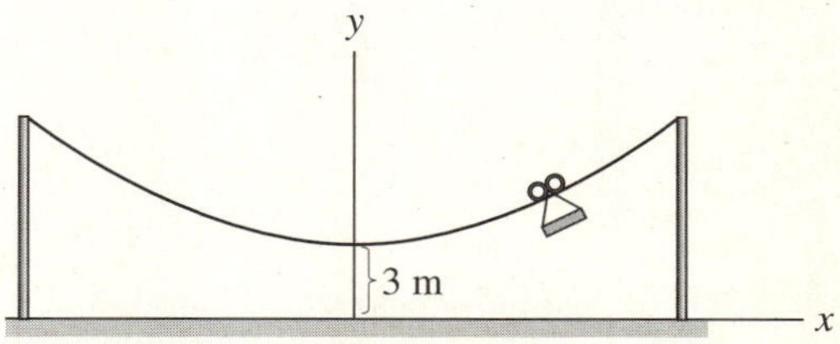

Figure P.12.94.

12.95. A rod CD rotates with shaft G–G at an angular speed ω of 300 r/min. A sleeve A of mass 500 g slides on CD. If no friction is present between A and CD, what is the distance S for no relative motion between A and CD?

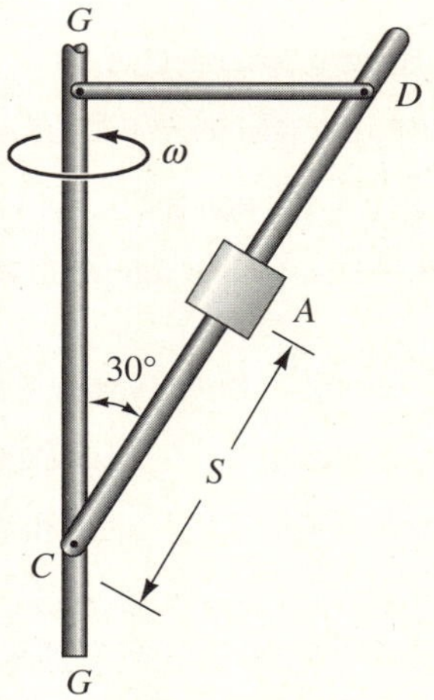

Figure P.12.95.

12.96. In Problem 12.95, what is the range of values for S for which A will remain stationary relative to CD if there is coulombic friction between A and CD such that $\mu_s = 0.4$?

12.97. A circular rod EB rotates at constant angular speed ω of 50 r/min. A sleeve A of mass 1 kg slides on the circular rod. At what position θ will sleeve A remain stationary relative to the rod EB if there is no friction?

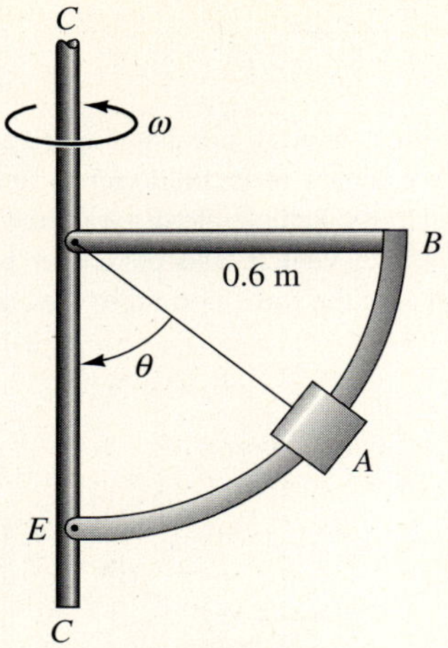

Figure P.12.97.

12.98. In Problem 12.97 assume that there is coulombic friction between A and EB with $\mu_s = 0.3$. Show that the minimum value of θ for which the sleeve will remain stationary relative to the rod is 75.45°.

12.99. The following data for a system of particles are given at time $t = 0$:

$$M_1 = 50 \text{ kg at position } (1, 1.3, -3) \text{ m}$$

$$M_2 = 25 \text{ kg at position } (-0.6, 1.3, -2.6) \text{ m}$$

$$M_3 = 5 \text{ kg at position } (-2.6, 5.3, 1) \text{ m}$$

The particles are acted on by the following respective external forces:

$$F_1 = 50\boldsymbol{j} + 10t\boldsymbol{k} \text{ N (particle 1)}$$

$$F_2 = 50\boldsymbol{k} \text{ N} \qquad \text{(particle 2)}$$

$$F_3 = 5t^2\boldsymbol{i} \text{ N} \qquad \text{(particle 3)}$$

What is the velocity of M_1 relative to the mass center after 5 sec, assuming that at $t = 0$, the particles are at rest?

***12.100.** Given the following force field:

$$\boldsymbol{F} = -0.6x\boldsymbol{i} + 0.9\boldsymbol{j} - 0.32\boldsymbol{k} \text{ N/kg}$$

what is the force on any particle in the field per unit mass of the particle. If we have two particles initially stationary in the field with position vectors

$$\boldsymbol{r}_1 = 0.9\boldsymbol{i} + 0.6\boldsymbol{j} \text{ m}$$

$$\boldsymbol{r}_2 = 1.2\boldsymbol{i} - 0.6\boldsymbol{j} + 1.2\boldsymbol{k} \text{ m}$$

what is the velocity of each particle relative to the center of mass of the system after 2 s have elapsed? Each particle has a mass of 3 g.

12.101. A stationary uniform block of ice is acted on by forces that maintain constant magnitude and direction at all times. If

$$F_1 = (25g) \text{ N}$$

$$F_2 = (10g) \text{ N}$$

$$F_3 = (15g) \text{ N}$$

what is the velocity of the center of mass of the block after 10 sec? Neglect friction. The density of ice is 900 kg/m³.

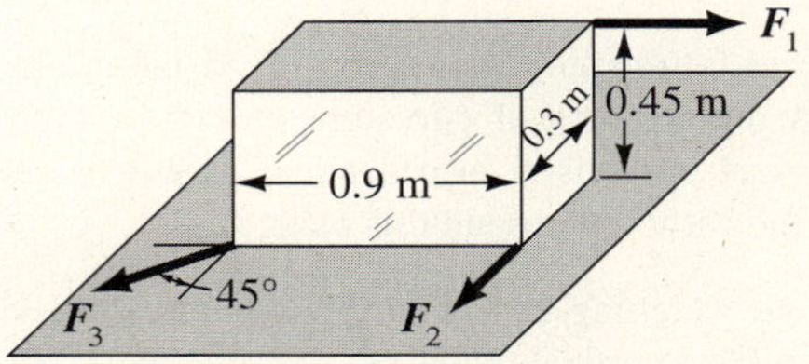

Figure P.12.101.

12.102. A space vehicle decelerates downward (Z direction) at 4.48 m/s² while moving in a translatory manner relative to inertial space. Inside the vehicle is a rod BC rotating in the plane of the paper at a rate of 50 rad/s relative to the vehicle. Two masses rotate at the rate of 20 rad/s around BC on rod EF. The masses are each 300 mm from C. Determine the force transmitted at C between BC and EF if the mass of each of the rotating bodies is 5 kg and the mass of rod EF is 1 kg. BC is in the vertical position at the time of interest. Neglect gravity.

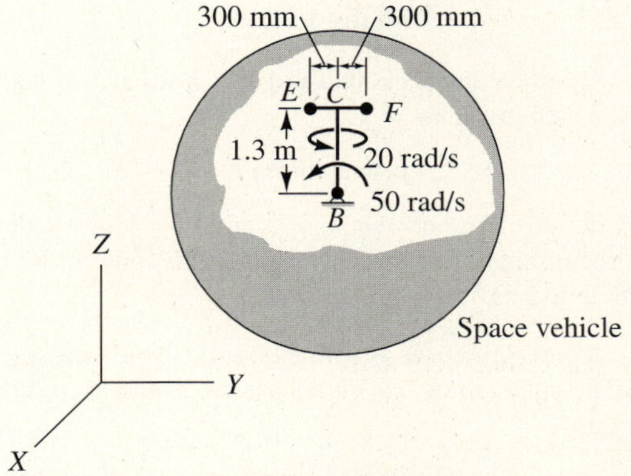

Figure P.12.102.

12.103. Two men climb aboard a barge at A to shift a load with the aid of a fork lift. The barge has a mass of 20 Mg and is 10 m long. The load consists of four containers each with a mass of 1.3 Mg and each having a length of 1 m. The men shift the containers to the opposite end of the barge, put the fork lift where they found it, and prepare to step off the barge at A, where they came on. If the barge has not been constrained and if we neglect water friction, currents, wind, and so on, how far has the barge shifted its position? The fork lift has a mass of 1 Mg.

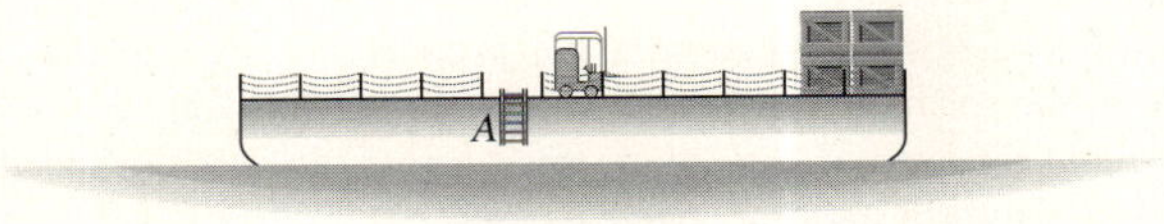

Figure P.12.103.

12.104. An astronaut on a space walk pulls a mass A of 100 kg toward him and shortens the distance d by 5 m. If the astronaut weighs 660 N on earth, how far does the mass A move from its original position? Neglect the mass of the cord.

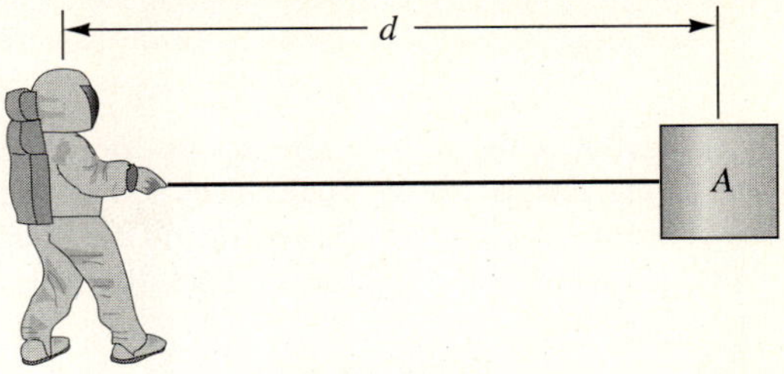

Figure P.12.104.

12.105. Two identical adjacent tanks are each 3 m long, 1.5 m high, and 1.5 m wide. Originally, the left tank is completely full of water while the right tank is empty. Water is pumped by an internal pump from the left tank to the right tank. At the instant of interest, the rate of flow Q is 0.56 m^3/s, while $\dot{Q}$ is 0.14 m^3/s^2. What horizontal force on the tanks is needed at this instant from the foundation? Assume that the water surface in the tanks remains horizontal. The density of water is 1 Mg/m^3.

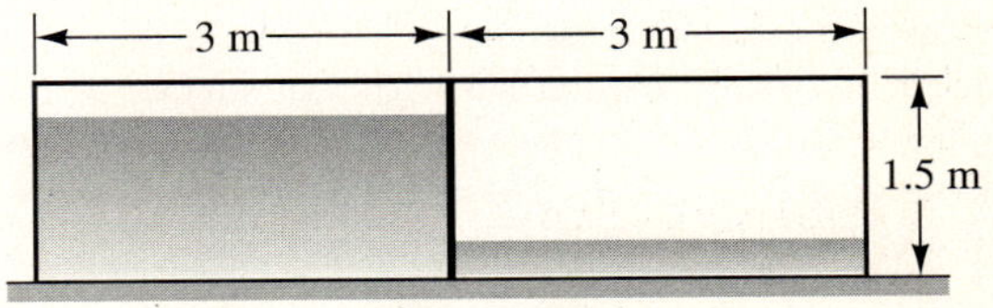

Figure P.12.105.

12.11 Closure

In this chapter, we integrated Newton's law for various coordinate systems. Also, with the aid of the mass center concept, we formulated Newton's law for any aggregate of particles. In the next two chapters, we shall present alternative procedures for more efficient treatment of certain classes of dynamics problems for particles. You will note that, since the new concepts are all derived from Newton's law, whatever problems can be solved by these new methods could also be solved by the methods we have already presented. A separate and thorough study of these topics is warranted by the gain in insight into dynamics and the greater facility in solving problems that can be achieved by examining these alternative methods and their accompanying concepts. As in this chapter, we will make certain generalizations applicable to any aggregate of particles.

PROBLEMS

12.106. A block A of mass 10 kg rests on a second block B of mass 8 kg. A force F equal to 100 N pulls block A. The coefficient of friction between A and B is 0.5; between B and the ground, 0.1. What is the speed of block A relative to block B in 0.1 s if the system starts from rest?

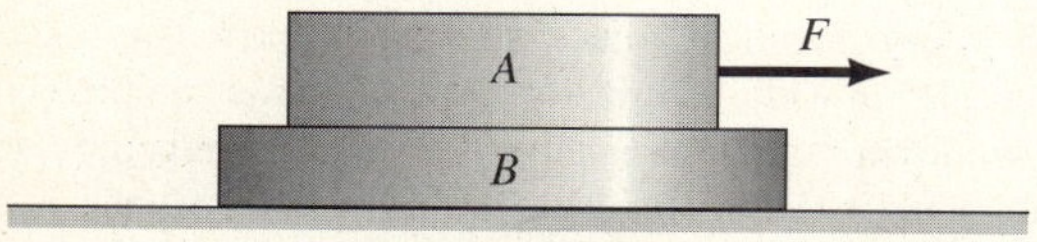

Figure P.12.106.

***12.107.** A block B slides from A to F along a rectangular chute where there is coulombic friction on the faces of the chute. The coefficient of dynamic friction is 0.4. The bottom face of the chute is parallel to face $EACF$ (a plane surface) and the other two faces are perpendicular to $EACF$. The body weighs 25 N. How long does it take B to go from A to F starting from rest?

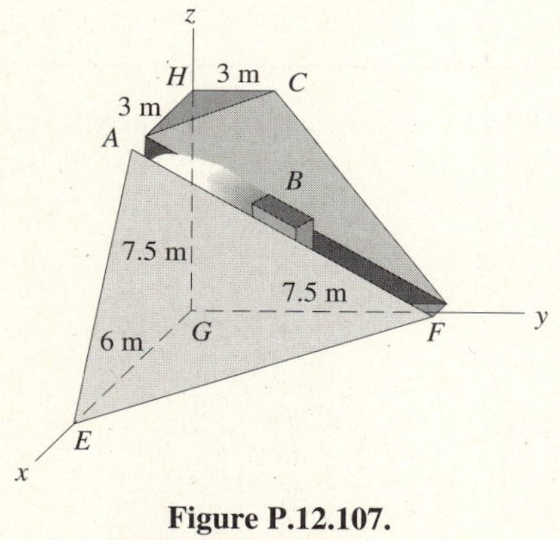

Figure P.12.107.

12.108. A tugboat is pushing a barge at a steady speed of 4 m/s. The thrust from the tugboat needed for this motion is 3.6 kN. The barge with load weighs 900 kN. If the water resistance to the barge is proportional to the speed of the barge, how long will it take the barge to slow to 2.5 m/s after the tugboat ceases to push?

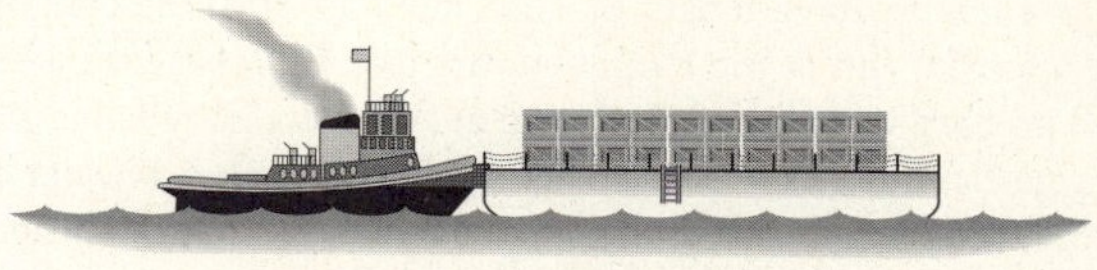

Figure P.12.108.

12.109. A spring requires a force x^2 N for a deflection of x mm, where x is the deflection of the spring from the undeformed geometry. Because the deflection is not proportional to x to the first power, the spring is called a *nonlinear* spring. If a 100 kg block is suddenly released on the undeformed spring, what is the speed of the block after it has descended 10 mm?

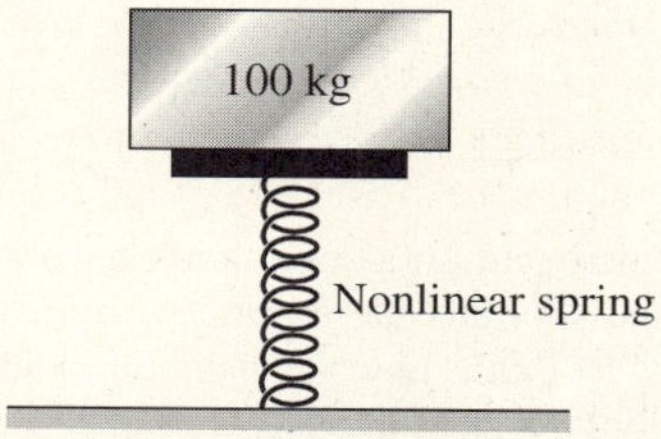

Figure P.12.109.

12.110. A horizontal platform is rotating at a constant angular speed ω of 5 rad/s. Fixed to the platform is a frictionless chute in which two identical masses each of 2 kg are constrained by a pair of linear springs each of spring constant K = 250 N/m. If the unstretched length l_0 of each of the springs is 0.18 m, show that at steady state the angle θ must have the value 36.87°. Springs are fixed to the platform at A.

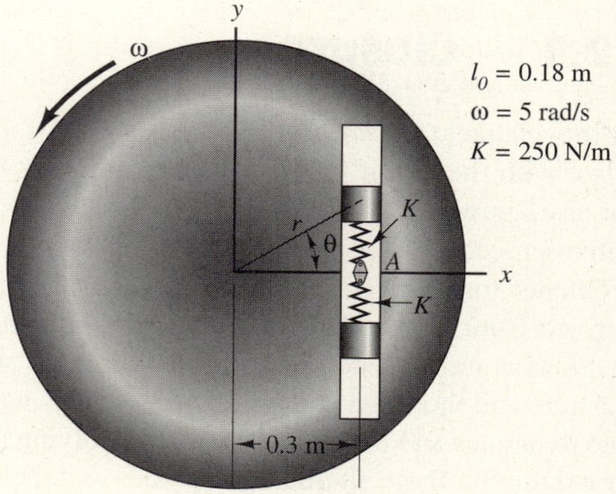

Figure P.12.110.

12.111. What is the velocity and altitude of a communications satellite that remains in the same position above the equator relative to the earth's surface?

12.112. A satellite is launched and attains a velocity of 8.5 km/s relative to the center of the earth at a distance of 385 km from the earth's surface. The satellite has been guided into a path that is parallel to the earth's surface at burnout.

(a) What kind of trajectory will it have?
(b) What is its farthest position from the earth's surface?
(c) If it is in orbit, compute the time it takes to go from the minimum point (perigee) to the maximum point (apogee) from the earth's surface.
(d) What is the minimum escape velocity for this position of launching?

12.113. A rocket system is capable of giving a satellite a velocity of 9.8 km/s relative to the earth's surface at an elevation of 320 km above the earth's surface. What would be its maximum distance h from the surface of the earth if it were launched (1) from the North Pole region or (2) from the equator, utilizing the spin of the earth as an aid?

12.114. A space vehicle is to change from a circular parking orbit 320 km above the surface of Venus to one that is 1620 km above this surface. This motion will be accomplished by two firings of the rocket system of the vehicle. The first firing causes the vehicle to attain an apogee that is 1620 km above the surface of Venus. At this apogee, a second firing is accomplished so as to achieve the desired circular orbit. What is the change in speed demanded for each firing if the thrust is maintained in each instance over a small portion of the trajectory of the vehicle? Neglect friction. The radius of Venus is 6.16 Mm, and the escape velocity at the surface is 10.26 km/s^2.

12.115. Weights A and B are held by light pulleys. If released from rest, what is the speed of each weight after 1 s? Weight A is 50 N and weight B is 200 N.

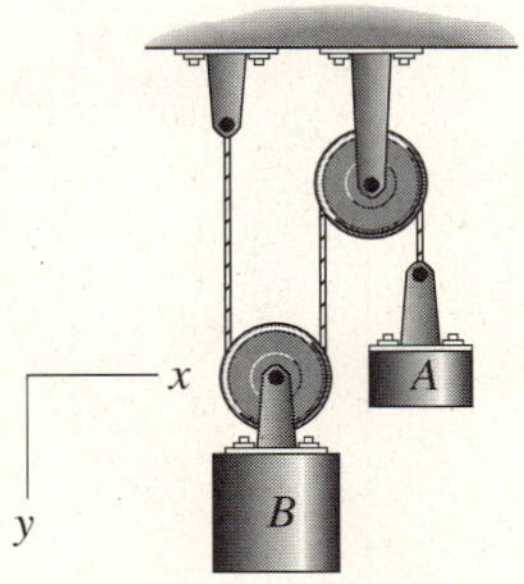

Figure P.12.115.

12.116. The following data are given for the flyball governor (read Problem 12.60 for details on how the governor works):

$$l = 275 \text{ mm}$$
$$D = 50 \text{ mm}$$
$$\omega = 300 \text{ r/min}$$
$$\theta = 45°$$

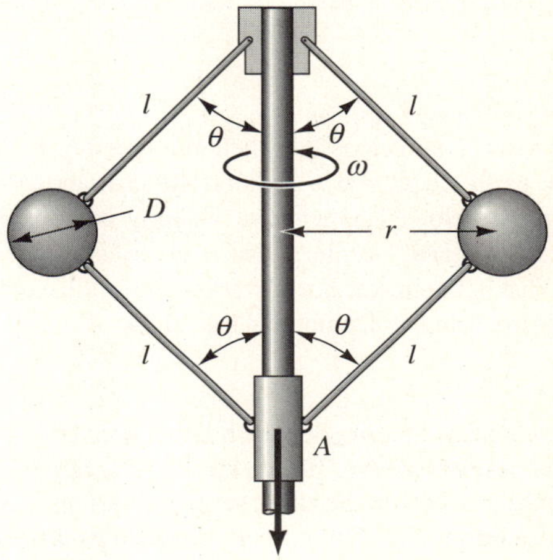

Figure P.12.116.

What is the force P acting on frictionless collar A if each ball has a mass of 1 kg and we neglect the weight of all other moving members of the system?

12.117. A spy satellite to observe the United States is put into a circular orbit about the North and South Poles. The satellite is to make 10 cycles/day (24 h). What must be the distance from the surface of the earth for this satellite?

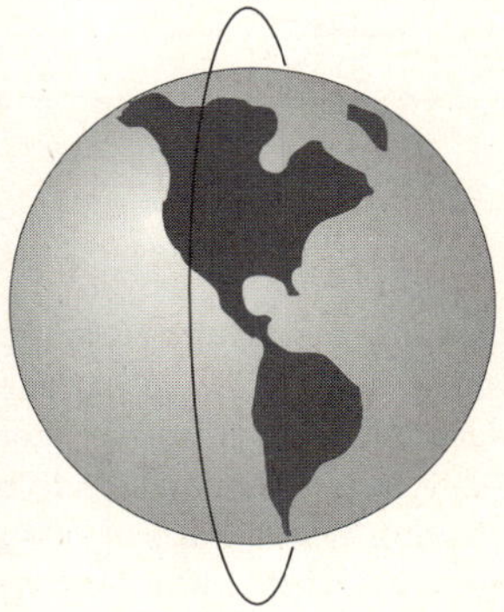

Figure P.12.117.

12.118. A skylab is in a circular orbit about the earth at a distance of 500 km above the earth's surface. A space shuttle has rendezvoused with the skylab and now, wishing to depart, decouples and fires its rockets to move more slowly than the skylab. If the rockets are fired over a short time interval, what should the relative speed between the space shuttle and skylab be at the end of rocket fire if the space shuttle is to come as close as 100 km to the earth's surface in subsequent ballistic (rocket motors off) flight?

12.119. A space vehicle is launched at a speed of 8.5 km/s relative to the earth's center at a position 400 km above the earth's surface. If the vehicle has a radial velocity component of 1.34 km/s toward the earth's center, what is the eccentricity of the trajectory? What is the maximum elevation above the earth's surface reached by the vehicle? Do not use Eq. 12.54.

12.120. A skier is moving down a hill at a speed of 14 m/s when he is at the position shown. If the skier weighs 800 N, what total force do his skis exert on the snow surface? Assume that the coefficient of friction is 0.1. The hill can be taken as a parabolic surface.

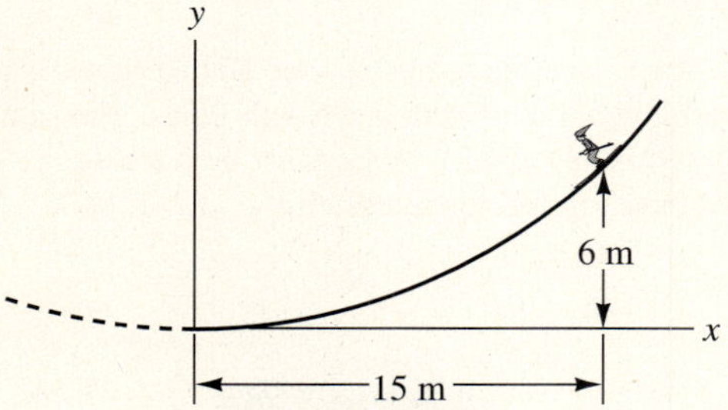

Figure P.12.120.

12.121. A submarine is moving at constant speed of 8.2 m/s below the surface of the ocean. The sub is at the same time descending downward while remaining horizontal with an acceleration of $0.023g$. In the submarine a flyball governor operates with weights having a mass each of 500 g. The governor is rotating with speed ω of 5 rad/s. If at time t, $\theta = 30°$, $\dot{\theta} = 0.2$ rad/s, and $\ddot{\theta} = 1$ rad/s^2, what is the force developed on the support of governor system as a result solely of the motion of the weights at this instant?

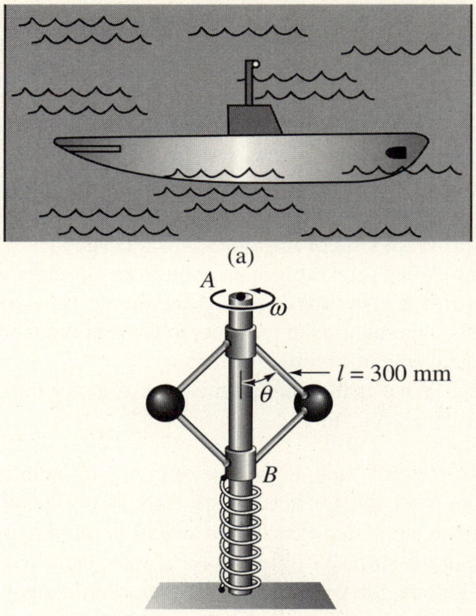

Figure P.12.121.

12.122. A mass spring system is shown. Write two simultaneous differential equations describing the motion of the mass. The spring has an unstretched length r_0. Consider that the spring does not bend and only changes length. Neglect all masses except that of the particle. If you restrict the motion to small rotations, how can you simplify the equations? Consider that the motion is confined to the xy plane.

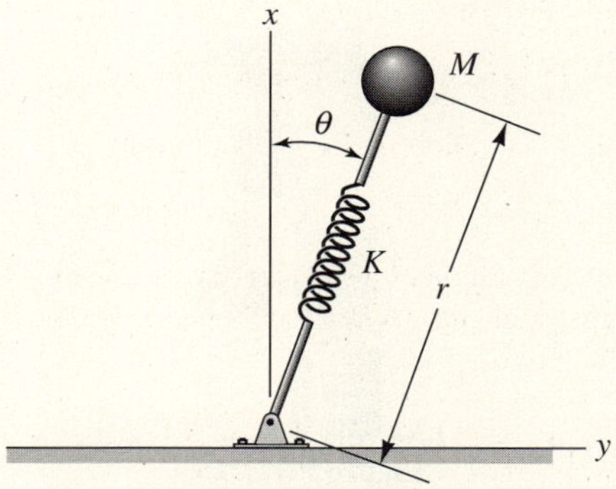

Figure P.12.122.

12.123. Three bodies have the following weights and positions at time t:

$$W_1 = 50 \text{ N}, \quad x_1 = 1.8 \text{ m}, \; y_1 = 3 \text{ m}, \; z_1 = 3 \text{ m}$$

$$W_2 = 25 \text{ N}, \quad x_2 = 1.5 \text{ m}, \; y_2 = 1.8 \text{ m}, \; z_2 = 0$$

$$W_3 = 40 \text{ N}, \quad x_3 = 0, \; y_3 = -1.2 \text{ m}, \; z_3 = 0$$

Determine the position vector of the center of mass at time t. Determine the velocity of the center of mass if the bodies have the following velocities:

$$V_1 = 1.8\boldsymbol{i} + 0.9\boldsymbol{j} \text{ m/s}$$

$$V_2 = 3\boldsymbol{i} - 0.9\boldsymbol{k} \text{ m/s}$$

$$V_3 = 1.8\boldsymbol{k} \text{ m/s}$$

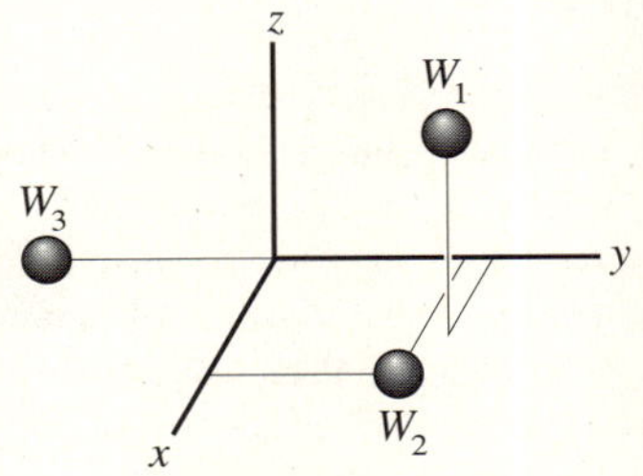

Figure P.12.123.

12.124. In Problem 12.123, the following external forces act on the respective particles:

$$\boldsymbol{F}_1 = 6t\boldsymbol{i} + 3\boldsymbol{j} - 10\boldsymbol{k} \text{ lb} \quad \text{(particle 1)}$$

$$\boldsymbol{F}_2 = 15\boldsymbol{i} - 3\boldsymbol{j} \text{ lb} \quad \text{(particle 2)}$$

$$\boldsymbol{F}_3 = \boldsymbol{0} \text{ lb} \quad \text{(particle 3)}$$

What is the acceleration of the center of mass, and what is its position after 10 s from that given initially? From Problem 12.123 at $t = 0$:

$$\boldsymbol{r}_C = 1.109\boldsymbol{i} + 1.278\boldsymbol{j} + 1.304\boldsymbol{k} \text{ m}$$

$$V_C = 1.435\boldsymbol{i} + 0.391\boldsymbol{j} + 0.430\boldsymbol{k} \text{ m/s}$$

12.125. A small body M of mass 1 kg slides along a wire from A to B. There is coulombic friction between the mass M and the wire. The dynamic coefficient of friction is 0.4. How long does it take to go from A to B?

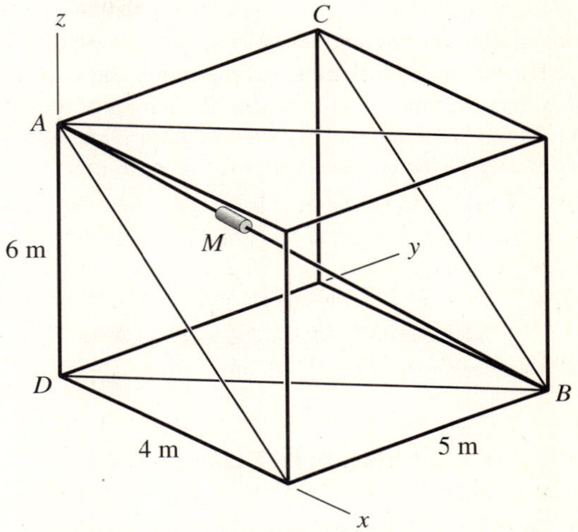

Figure P.12.125.

12.126. For $M = 15$ kg and $K = 1.8$ N/mm, what is the speed at $x = 25$ mm if a force of 25 N in the x direction is applied suddenly to the mass–spring system and then maintained constant? Neglect the mass of the spring and friction.

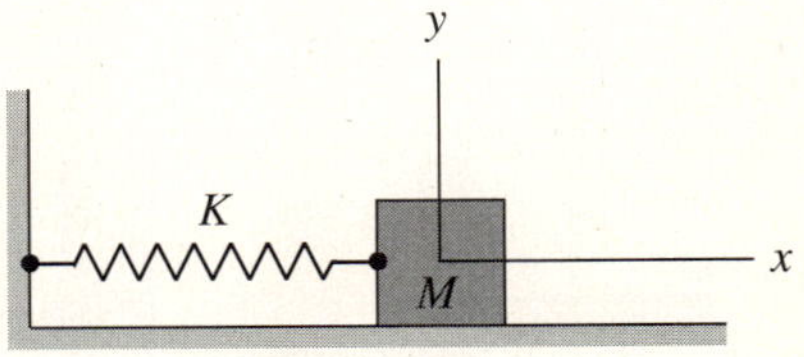

Figure P.12.126.

12.127. A rod B of mass 500 kg rests on a block A of mass 50 kg. A force F of 10 kN is applied suddenly to block A at the position shown. If the coefficient of friction μ_d is 0.4 for all contact surfaces, what is the speed of A when it has moved 3 m to the end of the rod?

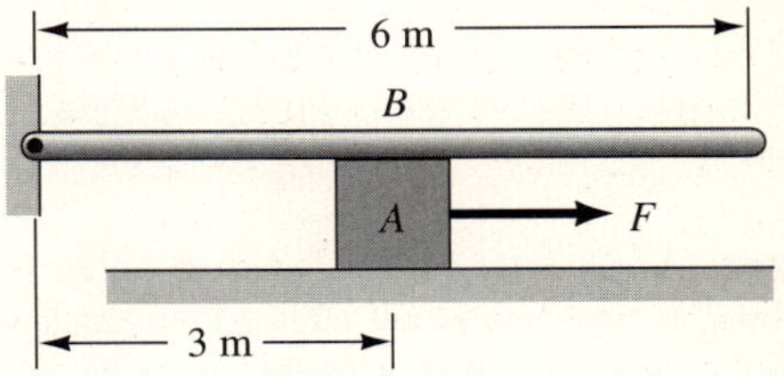

Figure P.12.127.

12.128. A simply supported beam is shown. You will learn in your course on strength of materials that a vertical force F applied at the center causes a deflection δ at the center given as

$$\delta = \frac{1}{48}\frac{FL^3}{EI}$$

If a mass of 100 kg, fastened to the beam at its midpoint, is suddenly released, what will its speed be when the deflection is 3 mm? Neglect the mass of the beam. The length of the beam, L, is 6 m. Young's modulus E is 207 kN/mm^2, and the second moment of area of the cross-section I is 8.3×10^6 mm^4.

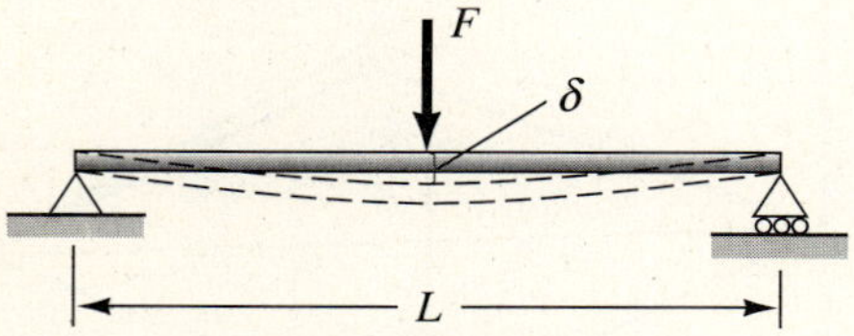

Figure P.12.128.

12.129. A piston is shown maintaining air at a pressure of 55 kPa above that of the atmosphere. If the piston is allowed to accelerate to the left, what is the speed of the piston after it moves 75 mm? The piston assembly has a mass of 1.4 kg. Assume that the air expands *adiabatically* (i.e., with no heat transfer). This means that at all times pV^k = constant, where V is the volume of the gas and k is a constant which for air equals 1.4. Neglect the inertial effects of the air.

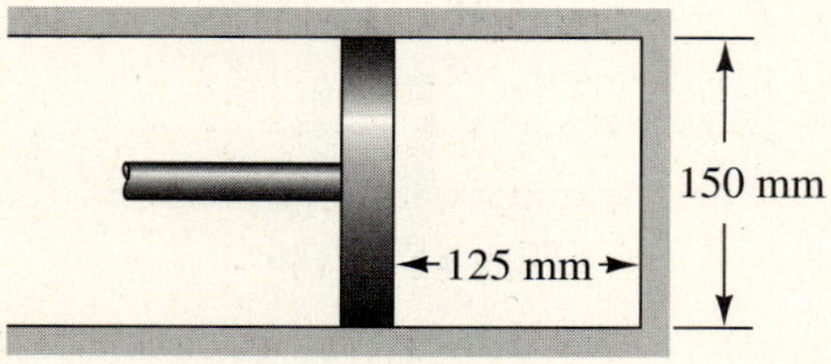

Figure P.12.129.

***12.130.** In Example 12.6 assume that there are adiabatic expansions and compressions of the gases (i.e., that pv^k = constant with k = 1.4). Compare the results for the speed of the piston. Explain why your result should be higher or lower than for the isothermal case.

12.131. Body A and body B are connected by an inextensible cord as shown. If both bodies are released simultaneously, what distance do they move in 0.5 s? Take M_A = 25 kg and M_B = 35 kg. The coefficient of friction μ_d is 0.3.

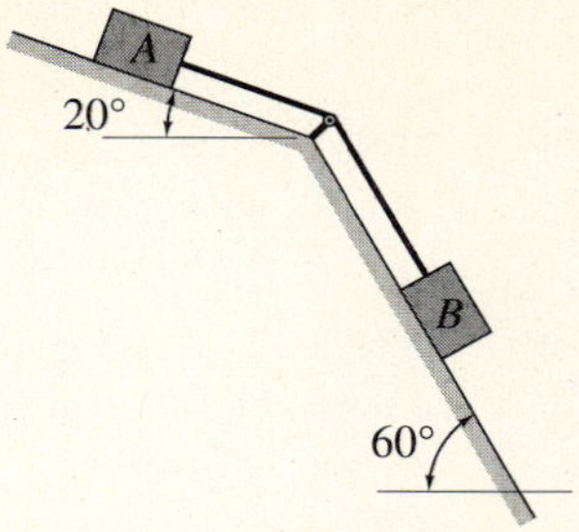

Figure P.12.131.

12.132. A force F of 2 kN is exerted on body C. If μ_d for all surface contacts is 0.2, what is the speed of C after it moves 1 m? The body C is initially stationary at the position shown, when the force F is applied. Solve using Newton's law. The following are the masses of the three bodies involved.

$$M_A = 100 \text{ kg} \qquad M_B = 80 \text{ kg} \qquad M_C = 50 \text{ kg}$$

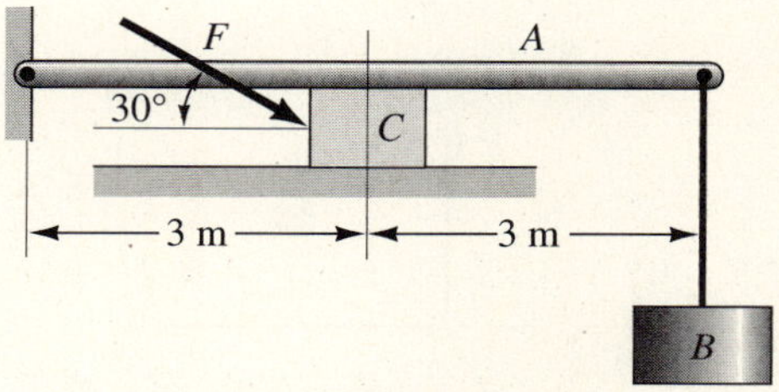

Figure P.12.132.

12.133. A constant force F of 5 kN acts on block A. If we do not have friction anywhere, what is the acceleration of block A?

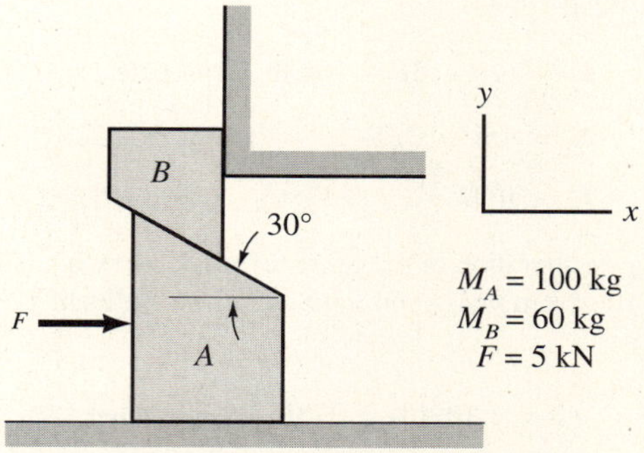

Figure P.12.133.

12.134. The system shown is released from rest. What distance does the body C drop in 2 s? The cable is inextensible. The coefficient of dynamic friction μ_d is 0.4 for contact surfaces of bodies A and B.

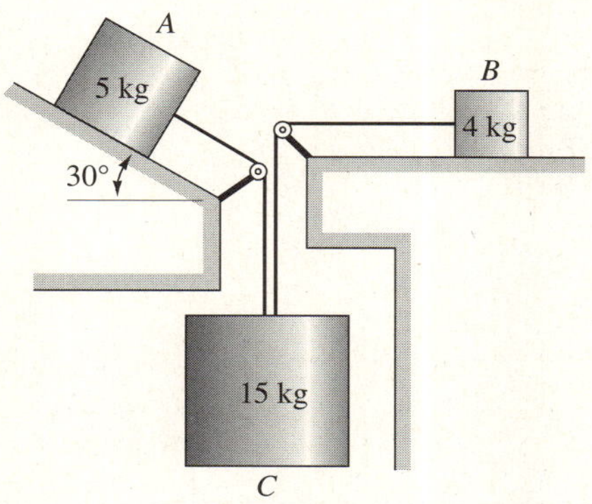

Figure P.12.134.

12.135. Do Problem 12.134 for the case where there is viscous damping for the contact surfaces of bodies A and B given as $7.5V$ N, with V in m/s.

12.136. Two bodies A and B are shown having masses of 40 kg and 30 kg, respectively. The cables are inextensible. Neglecting the inertia of the cable and pulleys at C and D, what is the speed of the block B 1 s after the system has been released from rest? The dynamic coefficient of friction μ_d for the contact surface of body A is 0.3. [*Hint:* From your earlier work in physics, recall that pulley D is instantaneously rotating about point a and hence point c moves at a speed that is twice that of point b.]

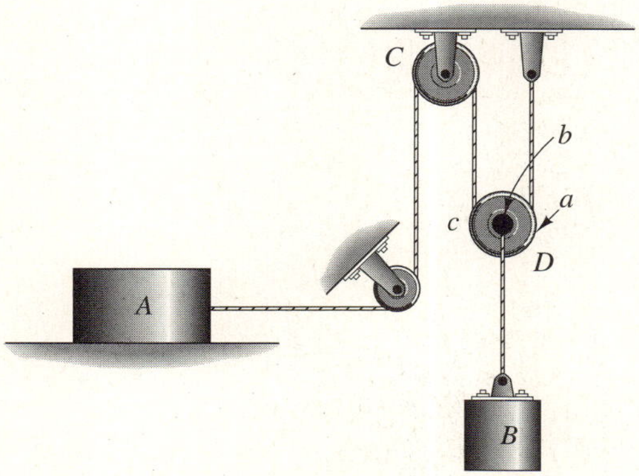

Figure P.12.136.

12.137. Bodies A, B, and C have weights, respectively, of 500 N, 1000 N, and 750 N. If released from rest, what are the respective speeds of the bodies after 1 s? Neglect the weight of pulleys.

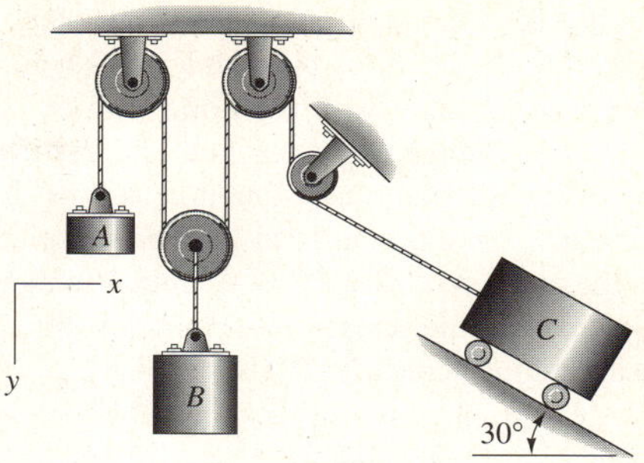

Figure P.12.137.

12.138. A car is moving at a constant speed of 18 m/s on a road part of which ($A \to B$) is parabolic and part of which ($C \to D$) is circular with a radius of 3 km. If the car has an anti-lock braking system and the static coefficient of friction μ_s between the road and the tires is 0.6, what is the maximum deceleration possible at the $x = 2$ km position and at the $x = 10$ km position? The total vehicle weight is 12 kN.

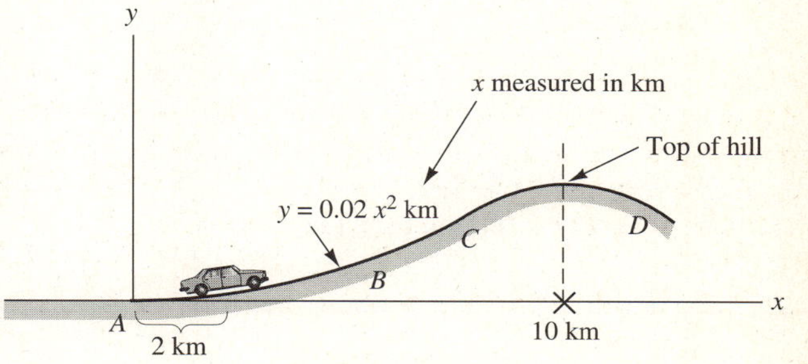

Figure P.12.138.

12.139. A mass of 3 kg is moving along a vertically oriented parabolic rod whose equation is $y = 3.4x^2$. A linear spring with $K = 550$ N/m connects to the mass and is unstretched when the mass is at the bottom of the rod having an unstretched length $l_0 = 1$ m. When the spring centerline is 30° from the vertical, as shown in the diagram, the mass is moving at 2.8 m/s. At this instant, what is the force component on the rod directed normal to the rod?

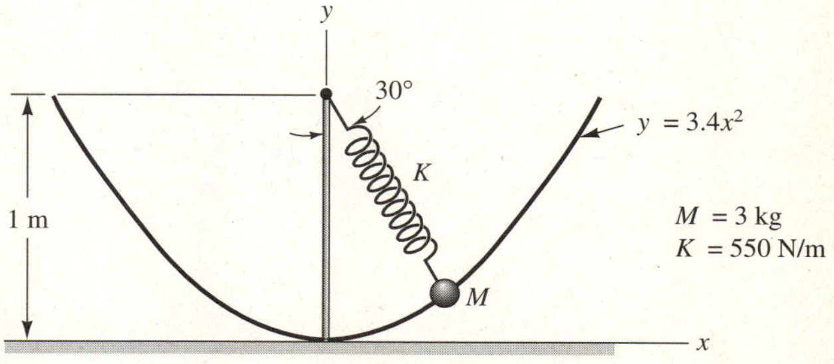

Figure P.12.139.

12.140. A heated cathode gives off electrons which are attracted to the positive anode. Some go through a small hole and enter the parallel plates at an angle with the horizontal of $\alpha_0 = 0$ and a velocity of V_0. Determine the horizontal and vertical motions of the electron inside the plates as a function of time. Letting $x = l$, find the time that the electron is in the parallel plate region and then obtain the exit vertical velocity. Assuming straight-line motion until the electron hits the screen, show that the vertical position of impact, assuming the screen is flat, is

$$y_{\text{Impact}} = \frac{eElL}{mV_0^2} + \frac{eEl^2}{2mV_0^2}$$

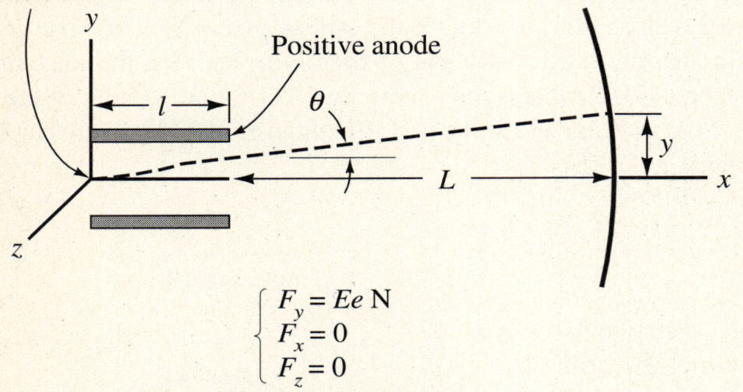

Figure P.12.140.

***12.141.** A weightless cord supports two identical masses each of weight W. The cord is being pulled at a constant speed V_0 by a force F. Formulate an equation for F in terms of V_0, L, l, W, and h. Determine F for the following conditions:

$V_0 = 2.2$ m/s $W = 40$ N

$L = 3.3$ m $h = 0.16$ m

$l = 0.26$ m

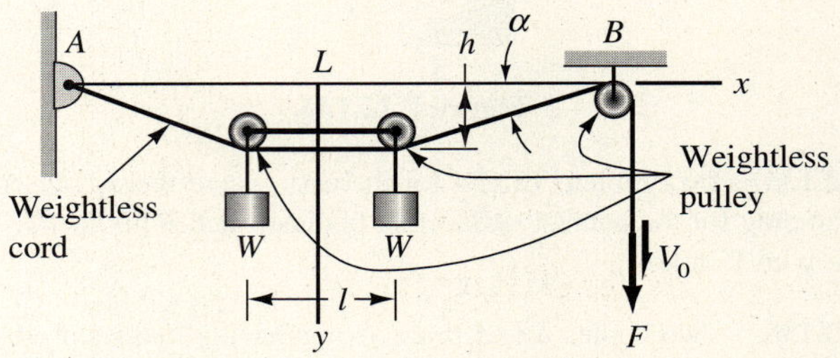

Figure P.12.141.

CHAPTER 13

Energy Methods for Particles

Part A: Analysis for a Single Particle

13.1 Introduction

In Chapter 12, we integrated the differential equation derived from Newton's law to yield velocity and position as functions of time. At this time, we shall present an alternative procedure, that of the method of energy, and we shall see that certain classes of problems can be more easily handled by this method in that we shall not need to integrate a differential equation.

To set forth the basic equation underlying this approach, we start with *Newton's law* for a particle moving relative to an inertial reference, as shown in Fig. 13.1. Thus,

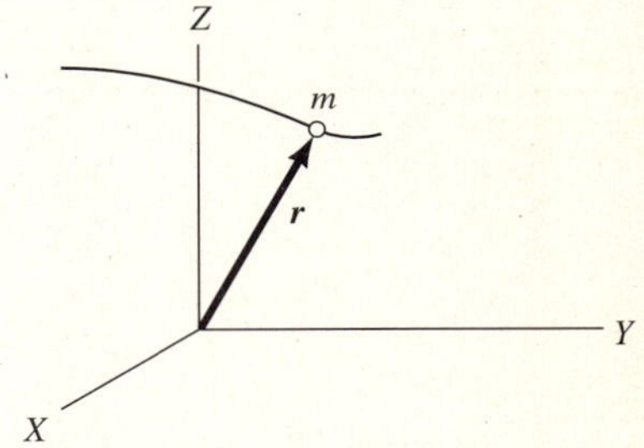

Figure 13.1. Particle moving relative to an inertial reference.

$$\boldsymbol{F} = m\frac{d^2\boldsymbol{r}}{dt^2} = m\frac{d\boldsymbol{V}}{dt} \tag{13.1}$$

Multiply each side of this equation by $d\boldsymbol{r}$ as a dot product and integrate from r_1 to r_2 along the path of motion:

$$\int_{r_1}^{r_2} \boldsymbol{F} \cdot d\boldsymbol{r} = m\int_{r_1}^{r_2} \frac{d\boldsymbol{V}}{dt} \cdot d\boldsymbol{r} = m\int_{t_1}^{t_2} \frac{d\boldsymbol{V}}{dt} \cdot \frac{d\boldsymbol{r}}{dt}\,dt$$

In the last integral, we multiplied and divided by dt, thus changing the variable of integration to t. Since $d\boldsymbol{r}/dt = \boldsymbol{V}$, we then have

$$\int_{r_1}^{r_2} F \cdot d\,r = m\int_{t_1}^{t_2}\left(\frac{dV}{dt}\cdot V\right)dt = \frac{1}{2}m\int_{t_1}^{t_2}\frac{d}{dt}(V\cdot V)\,dt$$

$$= \frac{1}{2}m\int_{t_1}^{t_2}\frac{d}{dt}V^2\,dt = \frac{1}{2}m\int_{V_1}^{V_2} d(V^2)$$

On carrying out the integration, we arrive at the familiar equation

$$\int_{r_1}^{r_2} \boldsymbol{F} \bullet d\boldsymbol{r} = \tfrac{1}{2}m(V_2^2 - V_1^2) \tag{13.2}$$

where the left side is the well-known expression for *work* (to be denoted at times as $\mathcal{W}_{1-2}$)[1] and the right side is clearly the change in *kinetic energy* as the mass moves from position $\boldsymbol{r}_1$ to position $\boldsymbol{r}_2$.

We shall see in Section 13.7 that for any system of particles, including, of course, rigid bodies, we get a work–energy equation of the form 13.2, where the velocity is that of the mass center, the force is the resultant external force on the system, and the path of integration is that of the mass center. Clearly, then, we can use a single particle model (and consequently Eq. 13.2) for:

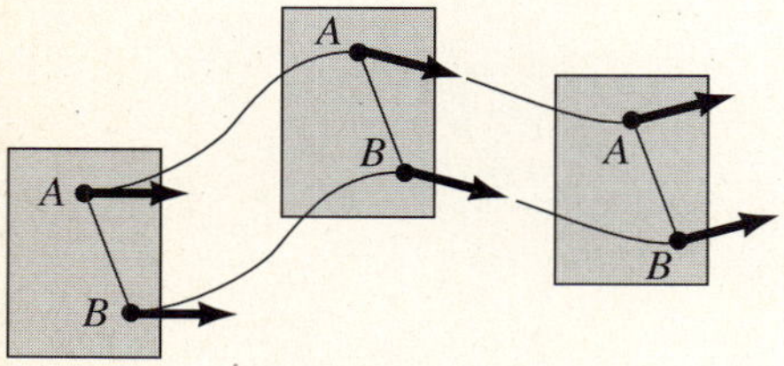

Figure 13.2. Translating body.

1. *A rigid body moving without rotation.* Such a motion was discussed in Chapter 11 and is called translation. Note that lines in a translating body remain parallel to their original directions, and points in the body move over a path which has identically the same form for all points. This condition is illustrated in Fig. 13.2 for two points *A* and *B*. Furthermore, each point in the body has at any instant of time *t* the same velocity as any other point. Clearly the motion of the center of mass fully characterizes the motion of the body and Eq. 13.2 will be used often for this situation.
2. Sometimes for *a body whose size is small compared to its trajectory*. Here the paths of points in the body differ very little from that of the mass center and knowing where the center of mass is tells us with sufficient accuracy all we need to know about the position of the body. However, keep in mind that the *velocity* and *acceleration* relative to the center of mass of a part of the body may be *very large*, irrespective of how small the body may be when compared to the trajectory of its center of mass. Then, information about the velocity and acceleration of this part of the body relative to the center of mass would require a more detailed consideration beyond a simple one-particle model centered around the center of mass.

Thus, as in our considerations of Newton's law in Chapter 12, when the motion of the mass center characterizes with sufficient accuracy what we want to know about the motion of a body, we use a particle at the mass center for energy considerations.

Next, suppose that we have a component of Newton's law in one direction, say the *x* direction:

$$F_x \boldsymbol{i} = m\frac{dV_x}{dt}\boldsymbol{i} \tag{13.3}$$

[1]It is important to note that the work done by a force system depends on the path over which the forces move, except in the case of conservative forces to be considered in Section 13.3. Thus, $\mathcal{W}_{1-2}$ is called a *path function* in thermodynamics. However, kinetic energy depends only on the instantaneous state of motion of the particle and is independent of the path. Kinetic energy is called, accordingly, a *point function* in thermodynamics.

In the next section, we shall present a more general definition of work.

Taking the dot product of each side of this equation with $dx\boldsymbol{i} + dy\boldsymbol{j} + dz\boldsymbol{k}$ (= $d\boldsymbol{r}$), we get, after integrating in the manner set forth at the outset:

$$\int_{x_1}^{x_2} F_x\, dx = \frac{m}{2}\left[(V_x)_2^2 - (V_x)_1^2\right] \tag{13.3a}$$

Similarly,

$$\int_{y_1}^{y_2} F_y\, dy = \frac{m}{2}\left[(V_y)_2^2 - (V_y)_1^2\right] \tag{13.3b}$$

$$\int_{z_1}^{z_2} F_z\, dz = \frac{m}{2}\left[(V_z)_2^2 - (V_z)_1^2\right] \tag{13.3c}$$

Thus, the foregoing equations demonstrate that the work done on a particle in any direction equals the change in kinetic energy associated with the component of velocity in that direction.

Instead of employing Newton's law, we can now use the energy equations developed in this section for solving certain classes of problems. This energy approach is particularly handy when velocities are desired and forces are functions of position. However, please understand that any problem solvable with the energy equation can be solved from Newton's law; the choice between the two is mainly a question of convenience and the manner in which the information is given.

Example 13.1

An automobile is moving at 30 m/s (see Fig. 13.3) when the driver jams on his brakes and goes into a skid in the direction of motion. The car has a mass of 2 Mg, and the dynamic coefficient of friction between the rubber tires and the concrete road is 0.60. How far, l, will the car move before stopping?

A constant friction force acts, which from Coulomb's law is $\mu_d N = (0.60)(2 \times 10^3 g) = 11.77$ kN. This force is the only force performing work, and clearly it is changing the kinetic energy of the vehicle from that corresponding to the speed of 30 m/s to zero. (You will learn in thermodynamics that this work facilitates a transfer of kinetic energy of the vehicle to an increase of internal energy of the vehicle, the road, and the air, as well as the wear of brake parts.) From the **work–energy equation** 13.2, we get[2]

$$-11.77 \times 10^3 = \frac{1}{2} 2 \times 10^3 (0 - 30^2)$$

Hence,

$$l = 76.5 \text{ m}$$

(Perhaps every driver should solve this problem periodically.)

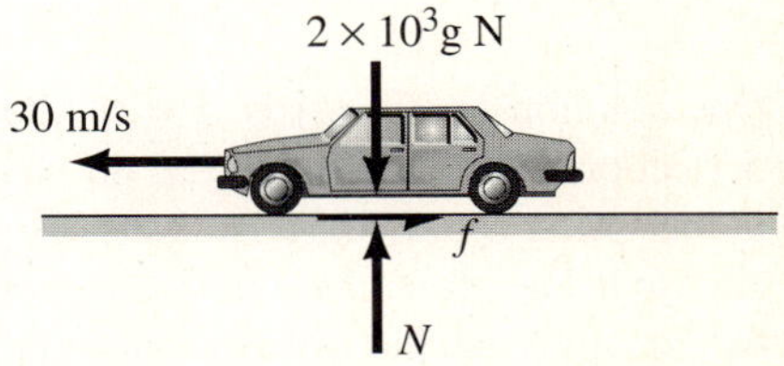

Figure 13.3. Car moving with brakes locked.

[2]Note that the sign of the work done is negative since the friction force is opposite in sense to the motion.

Example 13.2

Shown in Fig. 13.4 is a light platform B guided by vertical rods. The platform is positioned so that the spring has been compressed 10 mm. In this configuration a body A weighing 100 N is placed on the platform and released suddenly. If the guide rods give a total constant resistance force f to downward movement of the platform of 5 N, what is the largest distance that the weight falls? The spring used here is a *nonlinear* spring requiring $0.5x^2$ N of force for a deflection of x mm.

We take as the position of interest for the body the location δ below the initial configuration at which location the body A reaches zero velocity for the first time after having been released. The change in kinetic energy over the interval is accordingly zero. Thus, zero net work is done by the forces acting on the body A during displacement δ. These forces comprise the force of gravity, the friction force from the guides, and finally the force from the spring. Using as the origin for our measurements the *undeformed* top end position of the spring,[3] we can say:

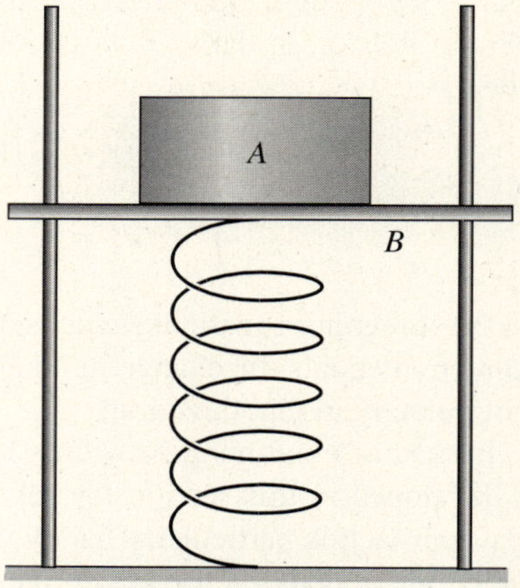

Figure 13.4. Preloaded nonlinear spring.

$$\int_{10}^{(10+\delta)} \boldsymbol{F} \bullet d\boldsymbol{r} = \int_{10}^{(10+\delta)} \left(W_A - f - 0.5x^2\right) dx$$
$$= \int_{10}^{(10+\delta)} \left(100 - 5 - 0.5x^2\right) dx = 0 \tag{a}$$

Integrating, we get

$$95\delta - \frac{0.5}{3}\left[(10+\delta)^3 - 10^3\right] = 0 \tag{b}$$

Therefore,

$$\delta^3 + 30\delta^2 - 270\delta = 0 \tag{c}$$

One solution to Eq. (c) is $\delta = 0$. Clearly, no work is done if there is no deflection. But this solution has no meaning for this problem since the force in the spring is only $0.5x^2 = 0.5(10)^2 = 50$ N, when the weight of 100 N is released. Therefore, there must be a nonzero positive value of δ that satisfies the equation and has physical meaning. Factoring out one δ from the equation, we then set the resulting quadratic expression equal to zero. Two roots result and the positive root $\delta = 7.25$ mm is the one with physical meaning.

[3]Since the force in the spring is a function of the elongation of the spring from its *undeformed* geometry, we must put the origin of our reference at a position corresponding to the undeformed geometry. At this position, both x and the spring force are zero simultaneously.

In the following example we deal with two bodies which can be considered as particles, rather than with one body as has been the case in the previous examples. We shall deal with these bodies separately in this example. Later in the chapter, we shall consider *systems* of particles, and in that context we will be able to consider this problem as a system of particles with less work needed to reach a solution.

Example 13.3

In Fig. 13.5, we have shown bodies A and B interconnected through a block and pulley system. Body B has a mass of 100 kg, whereas body A has a mass of 900 kg. Initially the system is stationary with B held at rest. What speed will B have when it reaches the ground at a distance $h = 3$ m below after being released? What will be the corresponding speed of A? Neglect the masses of the pulleys and the rope. Consider the rope to be inextensible.

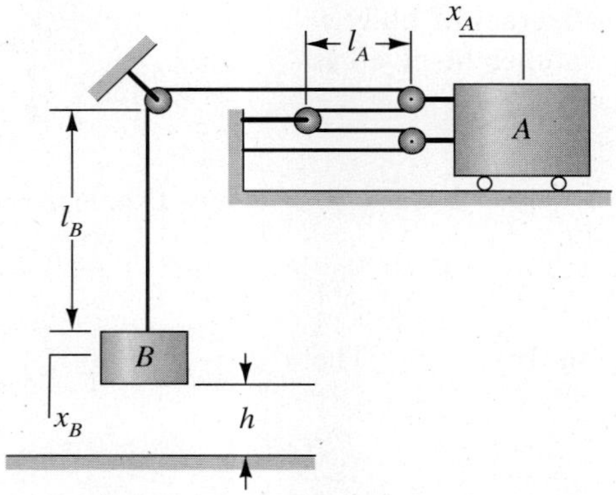

Figure 13.5. System of blocks and pulleys.

You will note from Fig. 13.5 that, as the bodies move, only the distances l_B and l_A change; the other distances involving the ropes do not change. And because the rope is taken as inextensible, we conclude that at all times

$$l_B + 4l_A = \text{constant} \quad \text{(a)}$$

Differentiating with respect to time, we can find that

$$\dot{l}_B + 4\dot{l}_A = 0$$

Therefore,

$$\dot{l}_B = -4\dot{l}_A \quad \text{(b)}$$

On inspecting Fig. 13.5, you should have no trouble in concluding that $\dot{l}_A = -V_A$ and that $\dot{l}_B = V_B$. Hence, from Eq. (b), we can conclude that

$$V_B = 4V_A \quad \text{(c)}$$

Next take the differential of Eq. (a):

$$dl_B + 4dl_A = 0$$

Therefore,

$$dl_B = -4dl_A \quad \text{(d)}$$

Example 13.3 (Continued)

Note that $dx_B = dl_B$ and that $dx_A = -dl_A$. Hence we see from Eq. (d) that a movement magnitude Δ_A of body A results in a movement magnitude, $4\Delta_A$, of body B:

$$\Delta_B = 4\Delta_A \tag{e}$$

With these kinematical conclusions as a background, we are now ready to proceed with the work–energy considerations.

For this purpose, we have shown a free-body diagram of body B in Fig. 13.6. The **work–energy equation** for body B can then be given as follows:

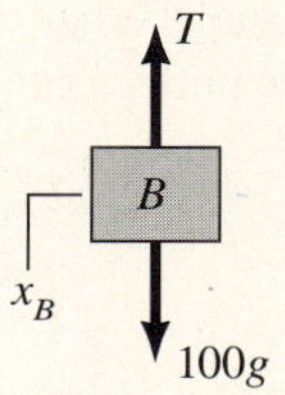

Figure 13.6. Free-body diagram of B.

$$(100g - T)h = \tfrac{1}{2}100V_B^2$$

Therefore,

$$(981 - T)(3) = \tfrac{1}{2}100V_B^2 \tag{f}$$

Now consider the free-body diagram of body A in Fig. 13.7. The **work–energy equation** for body A is then

$$(4T)(\Delta_A) = \tfrac{1}{2}900V_A^2 \tag{g}$$

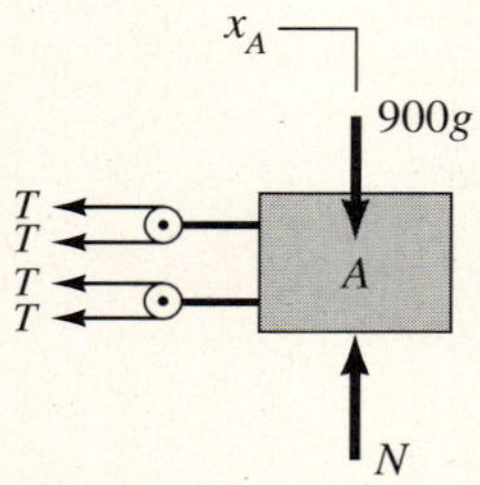

Figure 13.7. Free-body diagram of A.

But according to Eq. (e),

$$\Delta_A = \frac{\Delta_B}{4} = \frac{h}{4} = 0.75 \text{ m} \tag{h}$$

And according to Eq. (c),

$$V_A = \tfrac{1}{4}V_B \tag{i}$$

Substituting the results from Eqs. (h) and (i) into (g), we get

$$(4T)\left(\frac{3}{4}\right) = \frac{1}{2}900\left(\frac{V_B^2}{16}\right) \tag{j}$$

Example 13.3 (Continued)

Adding Eqs. (f) and (j), we can eliminate T to form the following equation with V_B as the only unknown:

$$(981)(3) = \frac{1}{2}(V_B^2)(100 + \frac{900}{16})$$

Therefore,

$$V_B = 6.14 \text{ m/s downward}$$

Hence,

$$V_A = 1.534 \text{ m/s to the left}$$

13.2 Power Considerations

The rate at which work is performed is called *power* and is a very useful concept for engineering purposes. Employing the notation $\mathcal{W}_k$ to represent work, we have

$$\text{power} = \frac{d\mathcal{W}_k}{dt} \tag{13.4}$$

Since $d\mathcal{W}_k$ for any given force $\boldsymbol{F}_i$ is $\boldsymbol{F}_i \cdot d\boldsymbol{r}_i$, we can say that the power being developed by a system of n forces at time t is, for a reference xyz,

$$\text{power} = \frac{\sum_{i=1}^{n} \boldsymbol{F}_i \cdot d\boldsymbol{r}_i}{dt} = \sum_{i=1}^{n} \boldsymbol{F}_i \cdot \boldsymbol{V}_i \tag{13.5}$$

where $\boldsymbol{V}_i$ is the velocity of the point of application of the ith force at time t as seen from reference xyz.[4]

In the following example we shall illustrate the use of the power concept. Note, however, that we shall find use of Newton's law advantageous in certain phases of the computation.

[4]We could have defined work $\mathcal{W}_k$ in terms of power as follows:

$$\mathcal{W}_k = \int_{t_1}^{t_2} (\boldsymbol{F} \cdot \boldsymbol{V})\, dt$$

When the force acts on a particular particle, the result above becomes the familiar $\int_{r_1}^{r_2} \boldsymbol{F} \cdot d\boldsymbol{r}$, where $\boldsymbol{r}$ is the position vector of the particle since $\boldsymbol{V}\, dt = d\boldsymbol{r}$. There are times when the force acts on *continuously changing* particles as time passes (see Section 13.8). The more general formulation above can then be used effectively.

Example 13.4

In hilly terrain, motors of an electric train are sometimes advantageously employed as brakes, particularly on downhill runs. This is accomplished by switching devices that change the electrical connections of the motors so as to correspond to connections for generators. This allows power developed during braking to be returned to the power source. In this way, we save much of the energy lost when employing conventional brakes—a considerable saving in every round trip. Such a train consisting of a single car is shown in Fig. 13.8 moving down a 15° incline at an initial speed of 3 m/s. This car has a mass of 20 Mg and has a cogwheel drive. If the conductor maintains an adjustment of the fields in his generators so as to develop a constant power *output* of 50 kW, how long does it take before the car moves at the rate of 5 m/s? Neglect the wind resistance and rotational effects of the wheels. The efficiency of the generators is 90%.

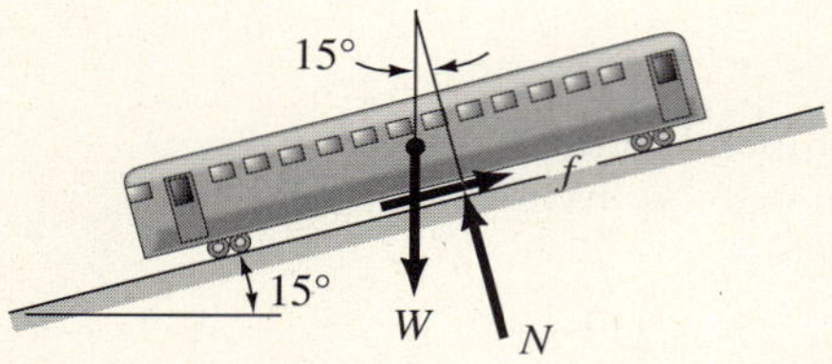

Figure 13.8. Train moving downhill with generators acting as brakes.

We have shown all the forces acting on the car in the diagram. **Newton's law** along the direction of the incline can be given as

$$W \sin 15° - f = M \frac{dV}{dt} \tag{a}$$

where f is the traction force from the rails developed by the generator action. Multiplying by V to get power, we get

$$W \sin 15°\, V - fV = MV \frac{dV}{dt} \tag{b}$$

If the efficiency of the generators (i.e., the power output divided by the power input) is 0.90, we can compute fV, which is the power input to the generators from the wheels, in the following manner:

$$\frac{\text{generator output}}{0.90} = fV \tag{c}$$

Example 13.4 (Continued)

Hence,

$$fV = \frac{(50 \times 10^3)}{0.90} = 55.56 \text{ kW} \qquad \text{(d)}$$

Equation (b) can now be given as

$$(20 \times 10^3)(9.81)(0.259)V - 55.56 \times 10^3 = 20 \times 10^3\, V\frac{dV}{dt}$$

Therefore,

$$2.54V - 2.78 = V\frac{dV}{dt} \qquad \text{(e)}$$

We can separate the variables as follows:

$$dt = \frac{V\,dV}{2.54V - 2.78} \qquad \text{(f)}$$

Integrating, using formula 1 in Appendix I, we get

$$t = \frac{1}{2.54^2}\left[2.54V - 2.78 + 2.78\ln(2.54V - 2.78)\right] + C \qquad \text{(g)}$$

To get the constant of integration C, note that when $t = 0$, $V = 3$ m/s. Hence,

$$0 = \frac{1}{2.54^2}\left\{(2.54)(3) - 2.78 + 2.78\ln\left[(2.54)(3) - 2.78\right]\right\} + C$$

Therefore,

$$C = -1.430$$

We thus have for Eq. (g):

$$t = \frac{1}{2.54^2}\left[2.54V - 2.78 + 2.78\ln(2.54V - 2.78)\right] - 1.430$$

When $V = 5$ m/s, we get for the desired value of t:

$$t = \frac{1}{2.54^2}\left\{(2.54)(5) - 2.78 + 2.78\ln\left[(2.54)(5) - 2.78\right]\right\} - 1.430$$

$t = 1.096$ s

PROBLEMS

13.1. What value of constant force P is required to bring the 50 kg body, which starts from rest, to a velocity of 10 m/s in 6 m? Neglect friction.

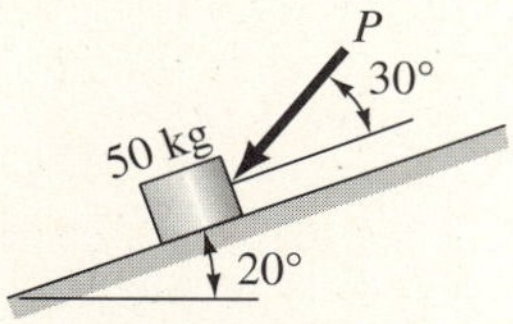

Figure P.13.1.

13.2. A light cable passes over a frictionless pulley. Determine the velocity of the 50 kg block after it has moved 10 m from rest. Neglect the inertia of the pulley. The dynamic coefficient of friction between block and incline is 0.2

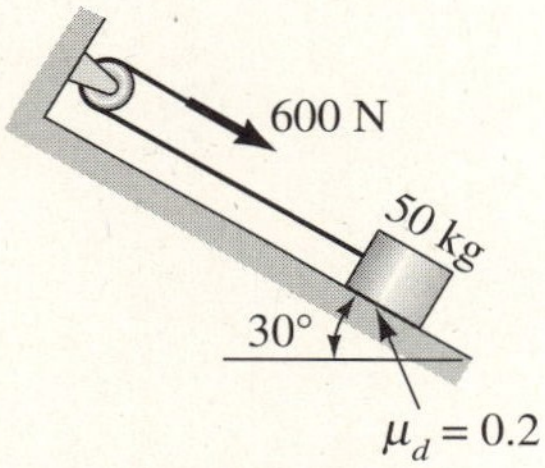

Figure P.13.2.

13.3. In Problem 13.2, the pulley has a radius of 0.3 m and has a resisting torque at the bearing of 15 N m. Neglect the inertia of the pulley and the mass of the cable. Compute the kinetic energy of the 50 kg block after it has moved 10 m from rest.

13.4. A light cable is wrapped around two drums fixed between a pair of blocks. The system has a mass of 50 kg. If a 250 N tension is exerted on the free end of the cable, what is the velocity change of the system after 3 m of travel down the incline? The body starts from rest. Take μ_d for all surfaces as 0.05.

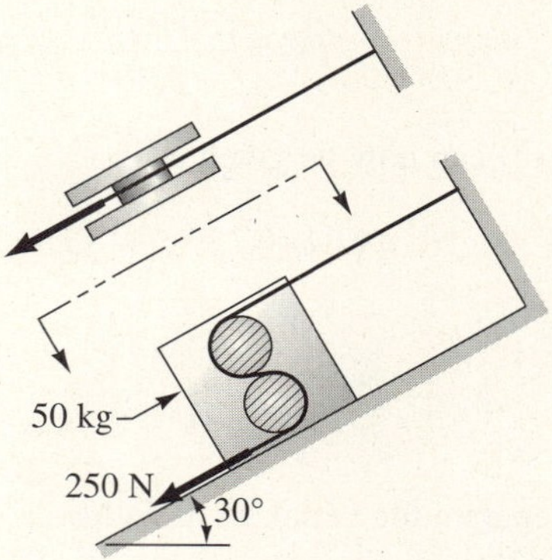

Figure P.13.4.

13.5. A 50 kg mass on a spring is moved so that it extends the spring 50 mm from its unextended position. If the dynamic coefficient of friction between the mass and the supporting surface is 0.3,

(a) What is the velocity of the mass as it returns to the undeformed configuration of the spring?

(b) How far will the spring be compressed when the mass stops instantaneously before starting to the left?

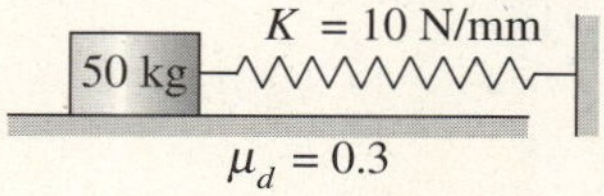

Figure P.13.5.

13.6. A truck–trailer is shown carrying three crushed junk automobile cubes each having a mass of 1.2 Mg. An electromagnet is used to pick up the cubes as the truck moves by. Suppose the truck starts at position 1 by applying a constant 70 N m total torque on the drive wheels. The magnet picks up only one cube C during the process. What will the velocity of the truck be when it has moved a total of 30 m? The truck unloaded has a mass of 2.4 Mg and has a tire diameter of 0.45 m. Neglect the rotational effects of the tires and wind friction.

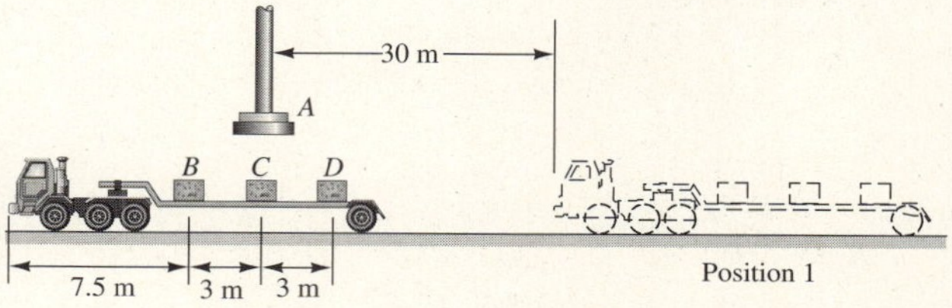

Figure P.13.6.

13.7. Do Problem 13.6 if the first cube *B* and the last cube *D* are removed as they go by the magnet.

13.8. A passenger ferry is shown moving into its dock to unload passengers. As it approaches the dock, it has a speed of 1.5 m/s. If the pilot reverses his engines just as the front of the ferry comes abreast of the first pilings at *A*, what constant reverse thrust will stop the ferry just as it reaches the ramp *B*? The ferry weighs 4.45 MN. Assume that the ferry does not hit the side pilings and undergoes no resistance from them. Neglect the drag of the water.

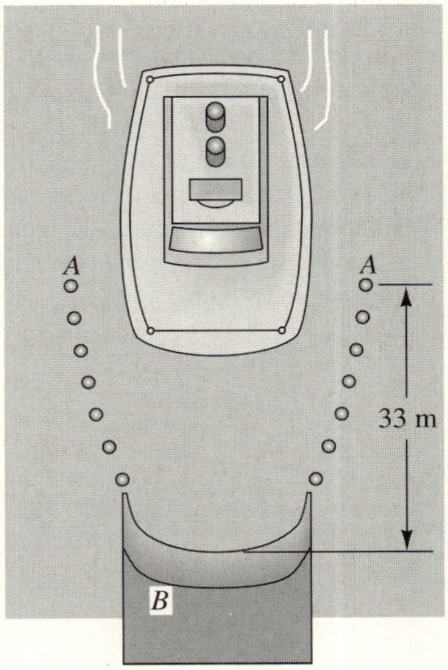

Figure P.13.8.

13.9. Do Problem 13.8 assuming that the ferry rubs against the pilings as a result of a poor entrance and undergoes a resistance against its forward motion given as

$$f = 9(x + 50)\text{ N}$$

where *x* is measured in metres from the first pilings at *A* to the front of the ferry.

13.10. A freight car weighing 90 kN is rolling at a speed of 1.7 m/s toward a spring-stop system. If the spring is nonlinear such that it develops a $0.045x^2$ kN force for a deflection of *x* mm, what is the maximum deceleration that the car *A* undergoes?

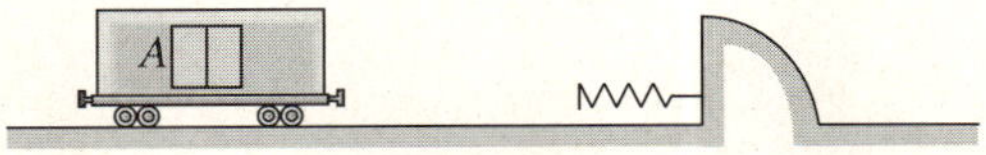

Figure P.13.10.

13.11. A 1 kN force is applied to a 3 kN block at the position shown. What is the speed of the block after it moves 2 m? There is Coulomb friction present. Assume at all times that the pressure at the bottom of the block is uniform. Neglect the height of the block in your calculations. Roller at right end moves with the block.

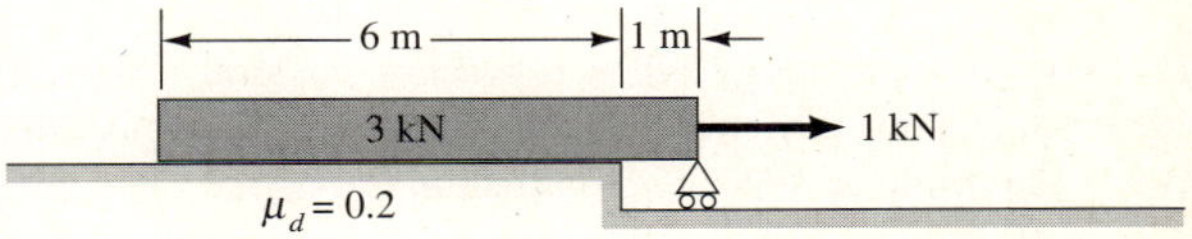

Figure P.13.11.

13.12. Two blocks *A* and *B* are connected by an inextensible cord running over a frictionless and massless pulley at E. The system starts from rest. What is the velocity of the system after it has moved 1 m? The coefficient of dynamic friction μ_d equals 0.22 for bodies *A* and *B*.

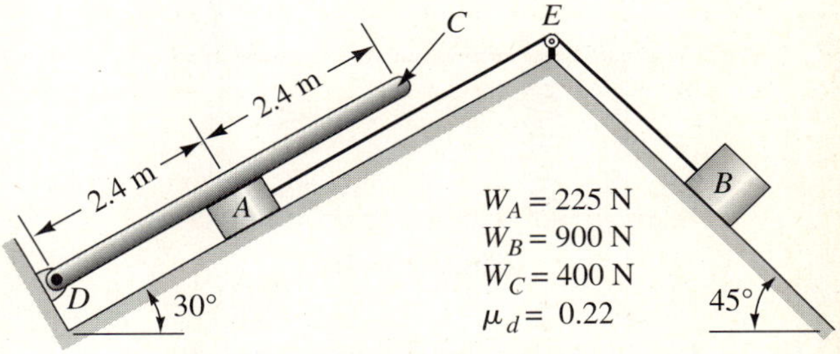

Figure P.13.12.

13.13. What are the velocities of blocks *A* and *B* when, after starting from rest, block *B* moves a distance of 100 mm? The dynamic coefficient of friction is 0.2 at all surfaces.

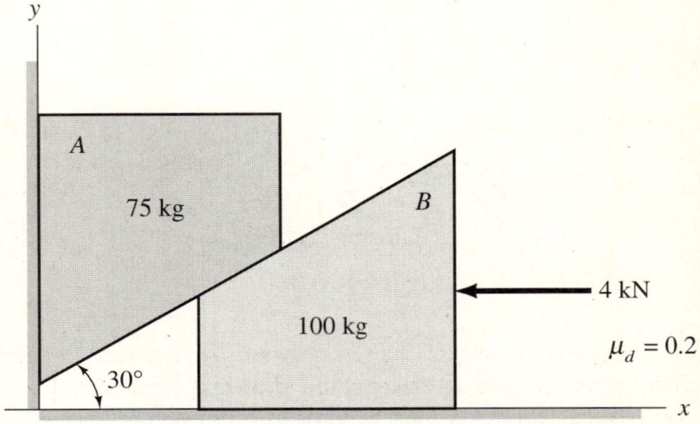

Figure P.13.13.

13.14. A particle of mass 5 kg is acted on by the following force field:

$$\boldsymbol{F} = 5x\boldsymbol{i} + (16 + 2y)\boldsymbol{j} + 20\boldsymbol{k} \text{ N}$$

When it is at the origin, the particle has a velocity $\boldsymbol{V}_0$ given as

$$\boldsymbol{V}_0 = 5\boldsymbol{i} + 10\boldsymbol{j} + 8\boldsymbol{k} \text{ m/s}$$

What is its kinetic energy when it reaches position (20, 5, 10) while moving along a frictionless path? Does the shape of the path between the origin and (20, 5, 10) affect the result?

13.15. A plate *AA* is held down by screws *C* and *D* so that a force of 245 N is developed in each spring. Mass *M* of 100 kg is placed on plate *AA* and released suddenly. What is the maximum distance that plate *AA* descends if the plate can slide freely down the vertical guide rods? Take $K = 3.6$ kN/m.

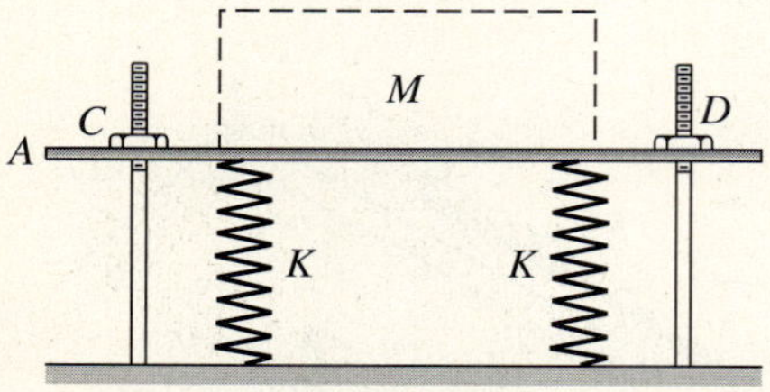

Figure P.13.15.

13.16. A 100 kg block is dropped on the system of springs. If $K_1 = 9$ kN/m and $K_2 = 3$ kN/m, what is the maximum force developed on the body?

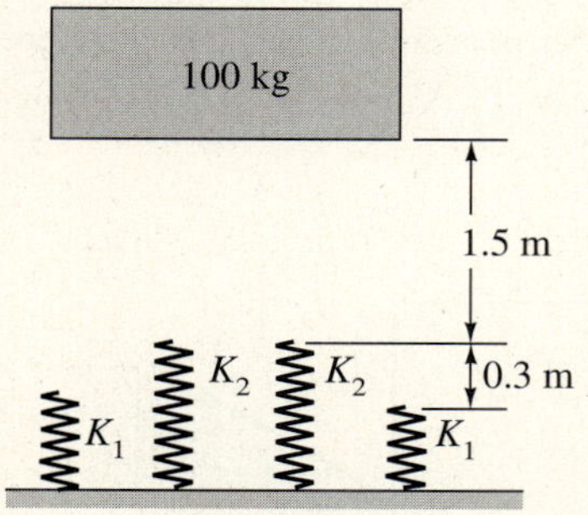

Figure P.13.16.

13.17. A block of mass 25 kg is shown on an inclined surface. The block is released at the position shown at a rest condition. What is the maximum compression of the spring? The spring has a spring constant *K* of 1.8 N/mm, and the dynamic coefficient of friction between the block and the incline is 0.3.

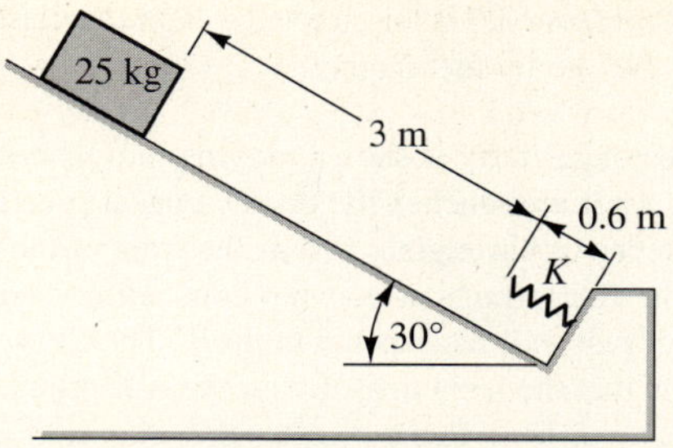

Figure P.13.17.

13.18. A classroom demonstration unit is used to illustrate vibrations and interactions of bodies. Body *A* has a mass of 0.5 kg and is moving to the left at a speed of 1.6 m/s at the position indicated. The body rides on a cushion of air supplied from the tube *B* through small openings in the tube. If there is a constant friction force of 0.1 N, what speed will *A* have when it returns to the position shown in the diagram? There are two springs at *C*, each having a spring constant of 15 N/m.

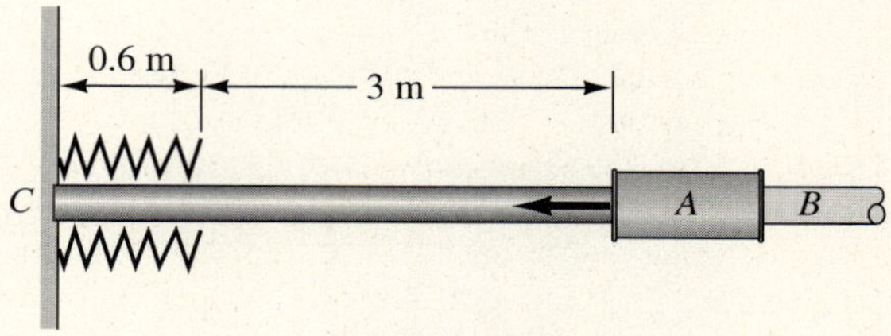

Figure P.13.18.

13.19. An electron moves in a circular orbit in a plane at right angles to the direction of a uniform magnetic field *B*. If the strength of *B* is slowly changed so that the radius of the orbit is halved, what is the ratio of the final to the initial angular speed of the electron? Explain the steps you take. The force $\boldsymbol{F}$ on a charged particle is $q\boldsymbol{V} \times \boldsymbol{B}$, where q is the charge and $\boldsymbol{V}$ is the velocity of the particle.

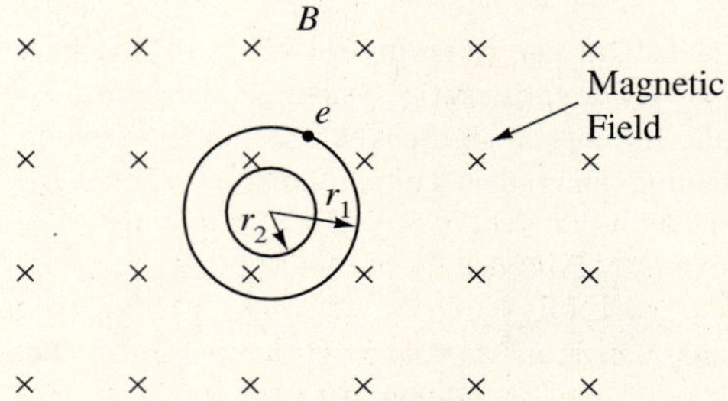

Figure P.13.19.

13.20. A light rod CD rotates about pin C under the action of constant torque T of 1 kN m. Body A having a mass of 100 kg slides on the horizontal surface for which the dynamic coefficient of friction is 0.4. If rod CD starts from rest, what angular speed is attained in one complete revolution? The entire weight of A is borne by the horizontal surface.

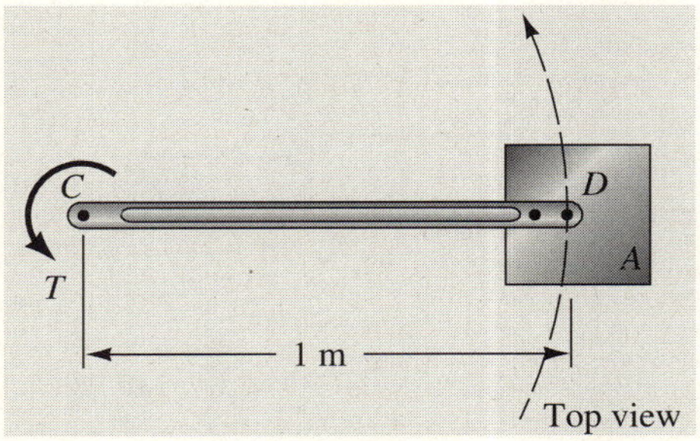

Figure P.13.20.

13.21. An astronaut is attached to his orbiting space laboratory by a light wire. The astronaut is propelled by a small attached compressed air device. The propulsive force is in the direction of the man's height from foot to head. When the wire is extended its full length of 6 m, the propulsion system is started, giving the astronaut a steady push of 22 N. If this push is at right angles to the wire at all times, what speed will the astronaut have in one revolution about A? The weight on earth of the astronaut plus equipment is 1.1 kN. The mass of the laboratory is large compared to that of the man and his equipment.

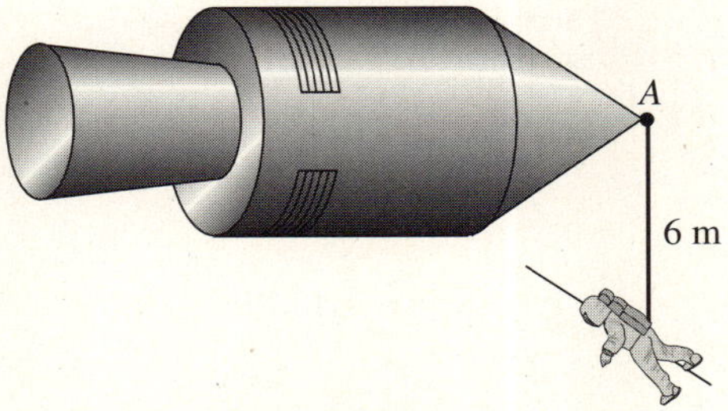

Figure P.13.21.

13.22. Body A, having a mass of 100 kg, is connected to body B by an inextensible light cable. Body B has a mass of 80 kg and is on small wheels. The dynamic coefficient of friction between A and the horizontal surface is 0.2. If the system is released from rest, how far d must B move along the incline before reaching a speed of 2 m/s?

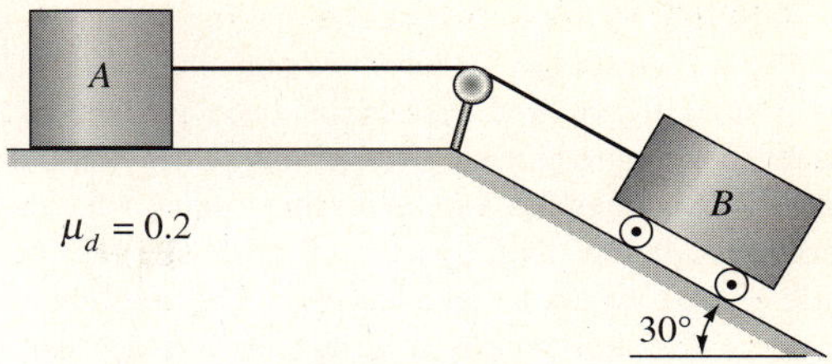

Figure P.13.22.

13.23. A conveyor has drum D driven by a torque of 70 N m. Bodies A and B on the conveyor each have a mass of 14 kg. The dynamic coefficient of friction between the conveyor belt and the conveyor bed is 0.2. If the conveyor starts from rest, how fast along the conveyor do A and B move after traveling 0.6 m? Drum C rotates freely, and the tension in the belt on the underside of the conveyor is 90 N. The diameter of both drums is 0.3 m. Neglect the mass of drums and belt. A and B do not slip on belt.

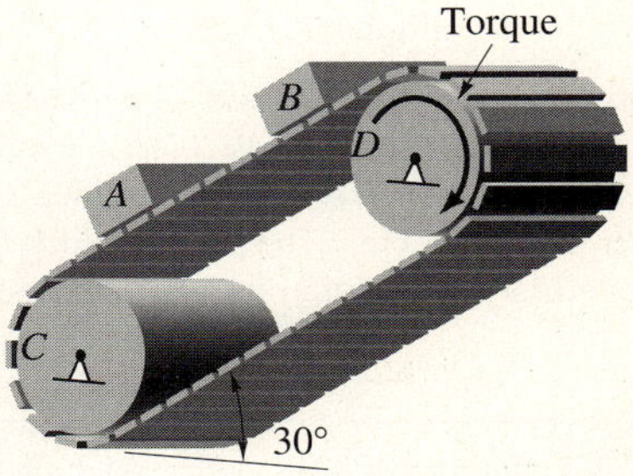

Figure P.13.23.

13.24. Bodies A and B are connected to each other through two light pulleys. Body A has a mass of 500 kg, whereas body B has a mass of 200 kg. A constant force F of value 10 kN is applied to body A whose surface of contact has a dynamic coefficient of friction equal to 0.4. If the system starts from rest, what distance d does B ascend before it has a speed of 2 m/s? [*Hint:* Considering pulley E, we have instantaneous rotation about point e. Hence, $V_b = \frac{1}{2} V_c$.]

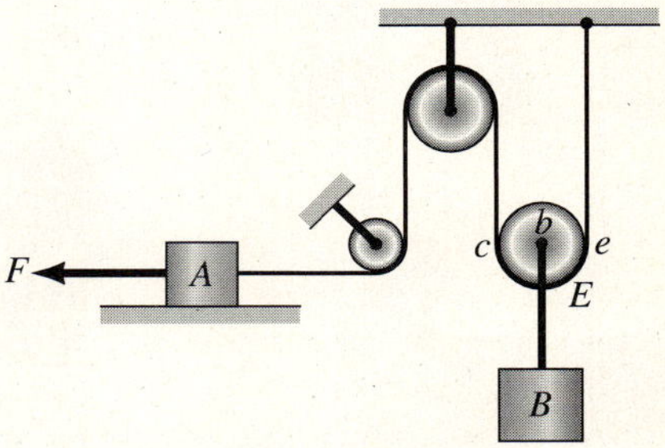

Figure P.13.24.

13.25. A rope tow for skiers is shown pulling 20 skiers up a 20° incline. The driving pulley A has a diameter of 1.5 m. The idler pulley B rotates freely. The system has been stopped to allow a fallen skier to untangle himself. The driving pulley starts from rest and is given a torque of 7 kN m. With this torque, what distance d do skiers move before their speed is 5 m/s? The tension on the slack side of the tow can be taken as zero. The coefficient of friction between skis and slope is 0.15 and the average mass of the skiers is 70 kg. Neglect the mass of the rope and the pulleys.

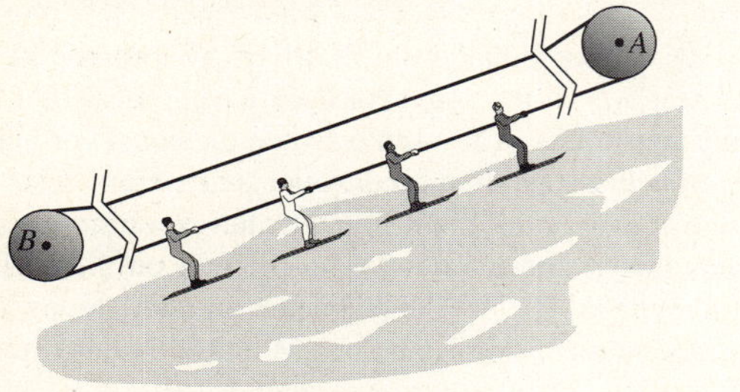

Figure P.13.25.

13.26. A uniform block A has a mass of 25 kg. The block is hinged at C and is supported by a small block B as shown in the diagram. A constant force F of 400 N is applied to block B. What is the speed of B after it moves 1.6 m? The mass of block B is 2.5 kg and the dynamic coefficient of friction for all contact surfaces is 0.3.

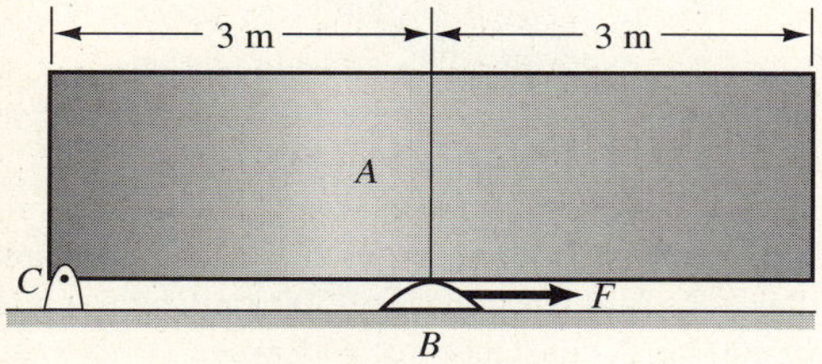

Figure P.13.26.

13.27. Block A weighs 900 N and block B weighs 680 N. If the system starts from rest, what is the speed of block B after it moves 0.3 m? Neglect the weight of the pulleys.

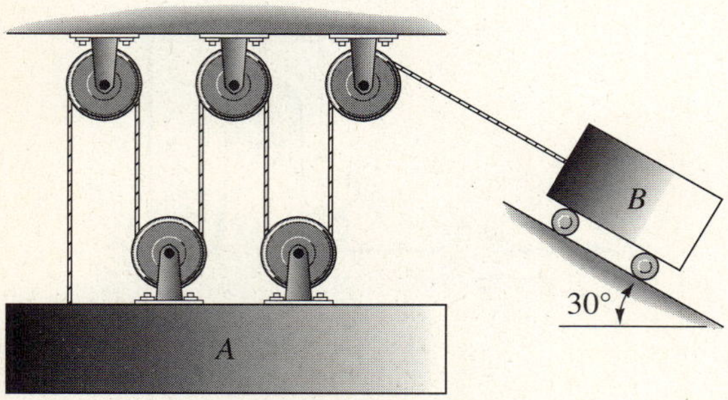

Figure P.13.27.

13.28. A weight W is to be lowered by a man. He lets the rope slip through his hands while maintaining a tension of 130 N on the rope. What is the maximum weight W that he can handle if the weight is not to exceed a speed of 5 m/s starting from rest and dropping 3 m? Use the coefficients of friction shown in the diagram. Neglect the mass of the rope. There are three wraps of rope around the post.

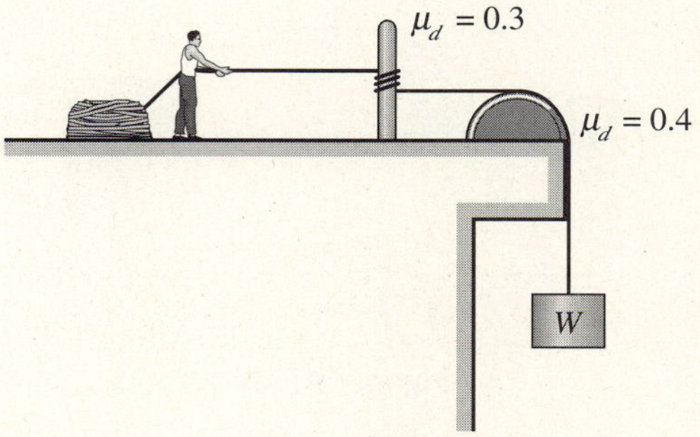

Figure P.13.28.

13.29. A vehicle B is being let down a 30° incline. The vehicle is attached to a weight A that restrains the motion. Vehicle B weighs 9 kN. What should the minimum restraining weight A be if, after starting from rest, the system does not exceed 2.4 m/s after moving 3 m? There are two wraps around the post.

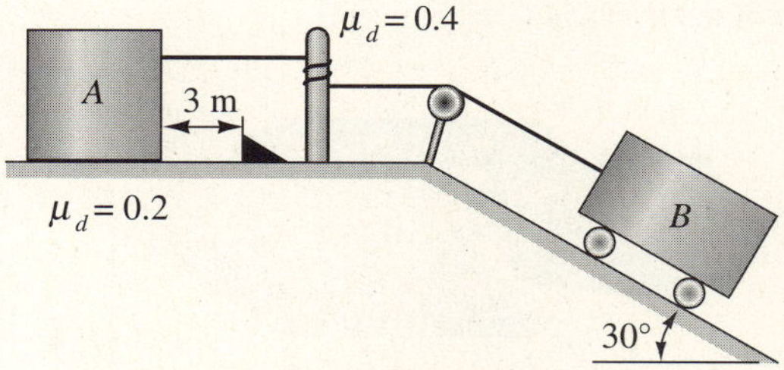

Figure P.13.29.

***13.30.** A spiral path is given parametrically in terms of the parameter τ as follows:

$$x_p = A \sin \eta\tau \text{ m}$$
$$y_p = A \cos \eta\tau \text{ m}$$
$$z_p = C\tau \text{ m}$$

where A, η, and C are known constants. A particle P of mass 0.5 kg is released from a position of rest 0.3 m above the xy plane. The particle is constrained by a spring (K = 30 N/m) coiled around the path. The spring is unstretched when P is released. Neglect friction and find how far P drops. Take $\eta = \pi/2$, $A = C = 1$.

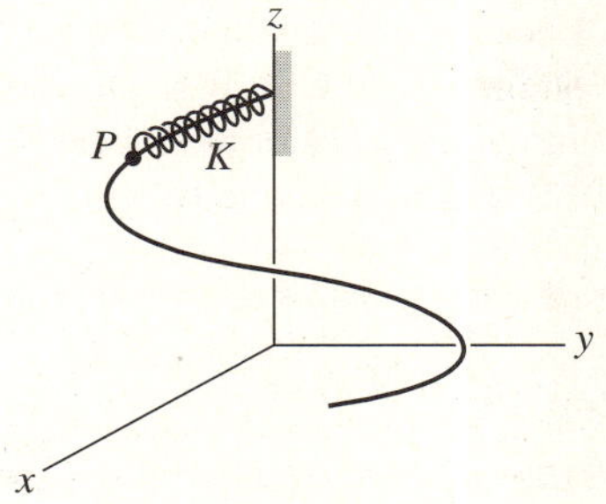

Figure P.13.30.

13.31. A body A of mass 0.5 kg is moving at time $t = 0$ with a speed V of 0.3 m/s on a smooth cylinder as shown. What is the speed of the body when it arrives at B? Take $r = 0.6$ m.

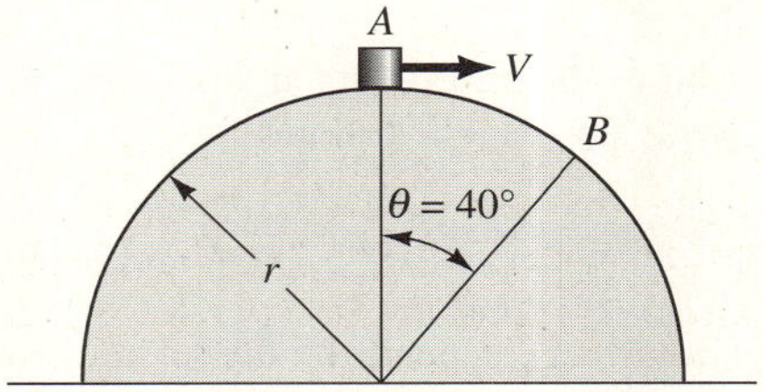

Figure P.13.31.

13.32. An automobile engine under test is rotating at 4.4×10^3 r/min and develops a torque of 40 Nm. What is the power developed by the engine? If the system has a mechanical efficiency of 0.90, what is the output of the generator? [*Hint:* The work of a torque equals the torque times the angle of rotation in radians.]

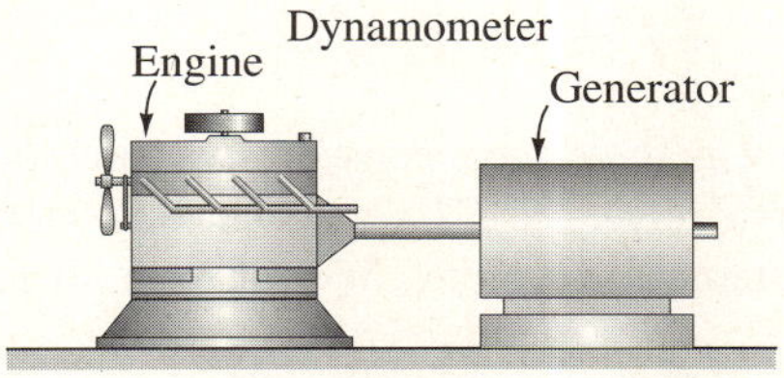

Figure P.13.32.

13.33. A rocket is undergoing static thrust tests in a test stand. A thrust of 1.35 MN is developed while 1.36 m^3 of fuel (density 800 kg/m^3) is burned per second. The exhaust products of combustion have a speed of 1.5 m/s relative to the rocket. What power is being developed on the rocket? What is the power developed on the exhaust gases?

13.34. A 14 Mg streetcar accelerates from rest at a constant rate a_0 until it reaches a speed V_1, at which time there is zero acceleration. The wind resistance is given as κV^2. Formulate expressions for power developed for the stated ranges of operation.

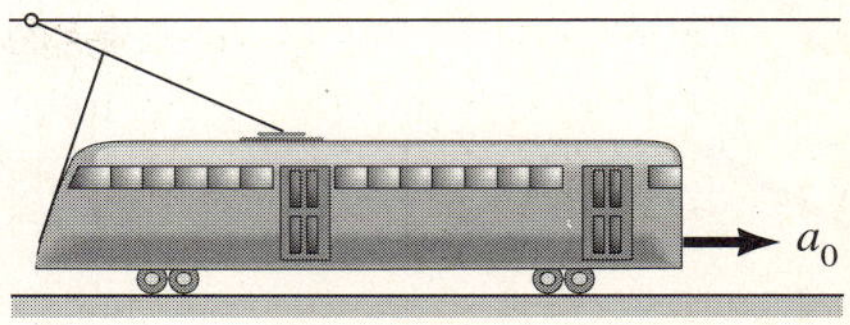

Figure P.13.34.

13.35. What is the maximum horsepower that can be developed on a streetcar of mass 14 Mg? The car has a coefficient of static friction of 0.20 between wheels and rail and a drag given as $32V^2$ N, where V is in m/s. All wheels are drive wheels.

13.36. A 7.5 Mg streetcar starts from rest when the conductor draws 5 kW of power from the line. If this input is maintained constant and if the mechanical efficiency of the motors is 90%, how long does the streetcar take to reach a speed of 3 m/s? Neglect wind resistance.

13.37. A children's boat ride can be found in many amusement parks. Small boats each weighing 50 kg are rotated in a tank of water. If the system is rotating with a speed $\dot{\theta}$ of 10 r/min, what is the kinetic energy of the system? Assume that each boat has two 30 kg children on board and that the kinetic energy of the supporting structure can be accounted for by "lumping" an additional 15 kg into each boat. If a wattmeter indicates that 4 kW of power is being absorbed by the motor turning the system, what is the drag for each boat? Take the mechanical efficiency of the motor to be 80%.

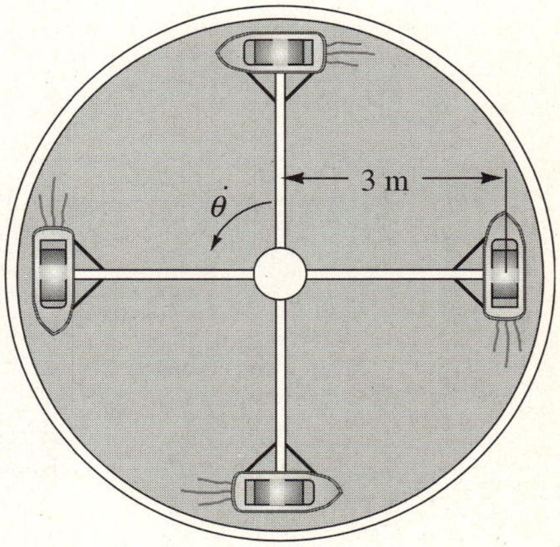

Figure P.13.37.

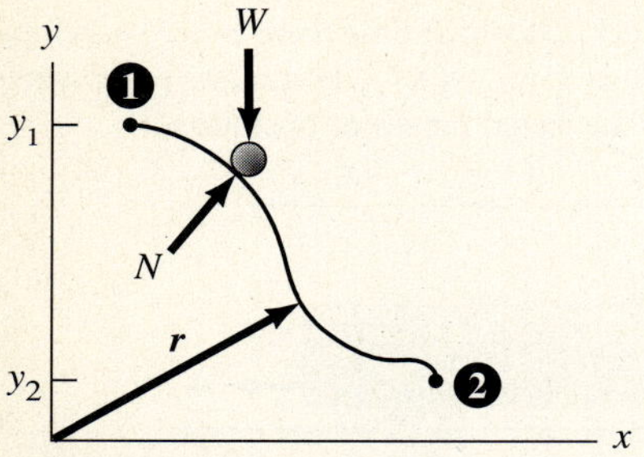

Figure 13.9. Particle moving along frictionless path.

13.3 Conservative Force Fields

In Section 10.6 we discussed an important class of forces called conservative forces. For convenience, we shall now repeat this discussion.

Consider first a body acted on only by gravity W as an active force (i.e., a force that can do work) and moving along a frictionless path from position 1 to position 2, as shown in Fig. 13.9. The work done by gravity $\mathcal{W}_{1-2}$ is then

$$\begin{aligned}\mathcal{W}_{1-2} &= \int_1^2 \boldsymbol{F} \cdot d\boldsymbol{r} = \int_1^2 (-W\boldsymbol{j}) \cdot d\boldsymbol{r} = -W\int_1^2 dy \\ &= -W(y_2 - y_1) = W(y_1 - y_2)\end{aligned} \tag{13.6}$$

Note that the work done *does not depend* on the path, but depends only on the positions of the end points of the path. *Force fields whose work like gravity is independent of the path are called conservative force fields.* In general, we can say for conservative force field $\boldsymbol{F}(x, y, z)$ that, along any path between positions 1 and 2, the work is

$$\mathcal{W}_{1-2} = \int_1^2 \boldsymbol{F} \cdot d\boldsymbol{r} = \mathcal{V}_1(x, y, z) - \mathcal{V}_2(x, y, z) \tag{13.7}$$

where $\mathcal{V}$ is a function of position of the end points and is called the *potential energy function*.[5] We may rewrite Eq. 13.7 as follows:

$$-\int_1^2 \boldsymbol{F} \cdot d\boldsymbol{r} = \mathcal{V}_2(x, y, z) - \mathcal{V}_1(x, y, z) = \Delta\mathcal{V} \tag{13.8}$$

Note that the potential energy, $\mathcal{V}(x, y, z)$, depends on the reference xyz used or, as we shall often say, the *datum* used. However, the *change* in potential energy, $\Delta\mathcal{V}$, is *independent* of the datum used.[6] Since we shall be using the change in potential energy, the datum is arbitrary and is chosen for convenience. From Eq. 13.8, we can say that *the change in potential energy*, $\Delta\mathcal{V}$ (= $\mathcal{V}_2 - \mathcal{V}_1$), of a conservative force field is *the negative of the work done by this conservative force field on a particle in going from position 1 to position 2 along any path.* For any *closed* path, clearly the work done by a conservative force field $\boldsymbol{F}$ is then

$$\oint \boldsymbol{F} \cdot d\boldsymbol{r} = 0 \tag{13.9}$$

Hence, this is a second way to define a conservative force field. How is the potential energy function $\mathcal{V}$ related to $\boldsymbol{F}$? To answer this query, consider that an infinitesimal path $d\boldsymbol{r}$ starts from point 1. We can then give Eq. 13.8 as

$$\boldsymbol{F} \cdot d\boldsymbol{r} = -d\mathcal{V} \tag{13.10}$$

[5]We shall also use the notation P.E. or simply PE for $\mathcal{V}$. Note that we used V for potential energy in Statics as is common practice. Here in Dynamics we have switched to $\mathcal{V}$ to avoid confusion with V the velocity.

[6]Thus, considering Eq. 13.6, the value of y itself for a particle at any time depends on the position of the origin O of the xyz reference. However, changing the position of O but keeping the same direction of the xyz axes (i.e., *changing the datum*) does not affect the value of $y_2 - y_1$.

Expressing the dot product on the left side in terms of components, and expressing $d\mathcal{V}$ as a total differential, we get

$$F_x\,dx + F_y\,dy + F_z\,dz = -\left(\frac{\partial \mathcal{V}}{\partial x}dx + \frac{\partial \mathcal{V}}{\partial y}dy + \frac{\partial \mathcal{V}}{\partial z}dz\right) \qquad (13.11)$$

We can conclude from this equation that

$$\begin{aligned} F_x &= -\frac{\partial \mathcal{V}}{\partial x} \\ F_y &= -\frac{\partial \mathcal{V}}{\partial y} \\ F_z &= -\frac{\partial \mathcal{V}}{\partial z} \end{aligned} \qquad (13.12)$$

In other words,

$$\begin{aligned} \boldsymbol{F} &= -\left(\frac{\partial \mathcal{V}}{\partial x}\boldsymbol{i} + \frac{\partial \mathcal{V}}{\partial y}\boldsymbol{j} + \frac{\partial \mathcal{V}}{\partial z}\boldsymbol{k}\right) \\ &= -\left(\frac{\partial}{\partial x}\boldsymbol{i} + \frac{\partial}{\partial y}\boldsymbol{j} + \frac{\partial}{\partial z}\boldsymbol{k}\right)\mathcal{V} \\ &= -\textbf{grad}\ \mathcal{V} = -\boldsymbol{\nabla}\ \mathcal{V} \end{aligned} \qquad (13.13)$$

The operator **grad** or $\boldsymbol{\nabla}$ that we have introduced is called the *gradient* operator[7] and is given as follows for rectangular coordinates:

$$\textbf{grad} \equiv \boldsymbol{\nabla} \equiv \left(\frac{\partial}{\partial x}\boldsymbol{i} + \frac{\partial}{\partial y}\boldsymbol{j} + \frac{\partial}{\partial z}\boldsymbol{k}\right) \qquad (13.14)$$

We can now say as a third definition that a *conservative force field must be a function of position and expressible as the gradient of a scalar function.* The *inverse* to this statement is also valid. That is, *if a force field is a function of position and the gradient of a scalar field, it must then be a conservative force field.*

Two examples of conservative force fields will now be presented and discussed.

Constant Force Field. If the force field is constant at all positions, it can always be expressed as the gradient of a scalar function of the form $\mathcal{V} = -(ax + by + cz)$, where a, b, and c are constants. The constant force field, then, is $\boldsymbol{F} = a\boldsymbol{i} + b\boldsymbol{j} + c\boldsymbol{k}$.

In limited changes of position near the earth's surface (a common situation), we can consider the gravitational force on a particle of mass, m, as a

[7]The gradient operator comes up in many situations in engineering and physics. In short, the gradient represents a *driving action.* Thus, in the present case, the gradient is a driving action to cause mass to move. And, the gradient of temperature causes heat to flow. Finally, the gradient of electric potential causes electric charge to flow.

constant force field given by $-mg\boldsymbol{k}$ (or $-W\boldsymbol{k}$). Thus, the constants for the general force field given above are $a = b = 0$ and $c = -mg$. Clearly, $\text{PE} = mgz$ for this case.

Force Proportional to Linear Displacements. Consider a body limited by constraints to move along a straight line. Along this line is developed a force directly proportional to the displacement of the body from some position O at $x = 0$ along the line. Furthermore, this force is always directed toward point O; it is then termed a *restoring* force. We can give this force as

$$\boldsymbol{F} = -Kx\boldsymbol{i} \tag{13.15}$$

where x is the displacement from point O. An example of this force is that of the linear spring (Fig.13.10) discussed in Section 12.3. The potential energy of this force field is given as follows wherein x is measured from the **undeformed** geometry (don't forget this important factor) of the spring:

$$\text{PE} = \frac{Kx^2}{2} \tag{13.16}$$

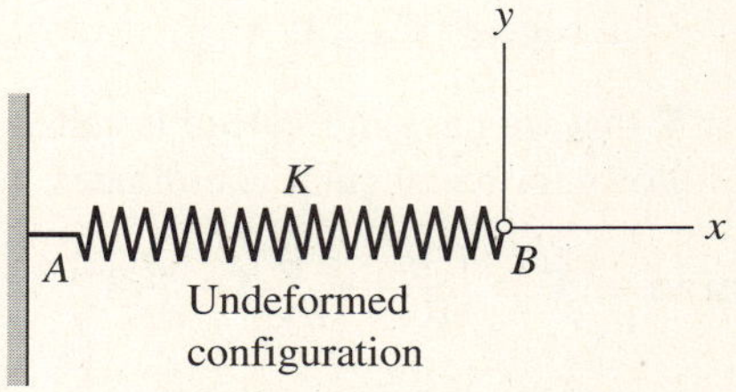

Figure 13.10. Linear spring.

What is the physical meaning of the term PE? Note that the change in potential energy has been defined (see Eq. 13.8) as the *negative* of the work done by a conservative force as the particle on which it acts goes from one position to another. Clearly, the change in the potential energy is then *directly equal* to the work done by the *reaction* to the conservative force during this displacement. In the case of the *spring*, the reaction force would be the force *from* the surroundings acting *on* the spring at point B (Fig. 13.10). During extension or compression of the spring from the undeformed position, this force (from the surroundings) does a *positive* amount of work. This work can be considered as a measure of the energy *stored* in the spring. Why? Because when allowed to return to its original position, the spring will do this amount of positive work on the surroundings at B, provided that the return motion is slow enough to prevent oscillations; and so on. Clearly then, since PE equals work of the surroundings on the spring, then PE is in effect the stored energy in the spring. In a general case, PE is the energy stored in the force field as measured from a given datum.

In previous chapters, several additional force fields were introduced: the gravitational central force field, the electrostatic field, and the magnetic field. Let us see which we can add to our list of conservative force fields.

Consider first the central gravitational force field where particle m, shown in Fig. 13.11, experiences a force given by the equation

$$\boldsymbol{F} = -G\frac{Mm}{r^2}\hat{\boldsymbol{r}} \tag{13.17}$$

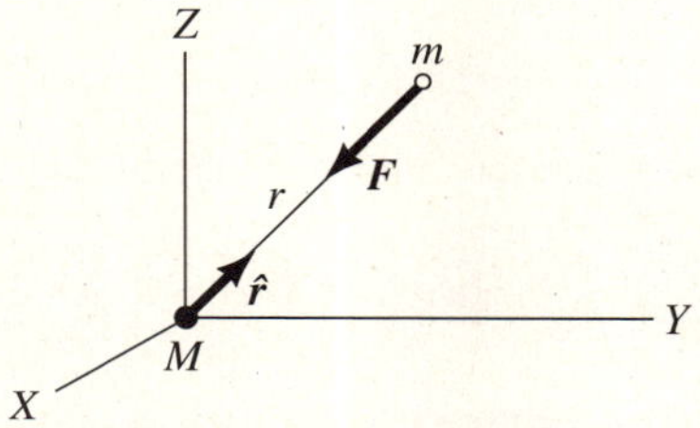

Figure 13.11. Central force on m.

Clearly, this force field is a function of spatial coordinates and can easily be expressed as the gradient of a scalar function in the following manner:

$$\boldsymbol{F} = -\mathbf{grad}\left(-\frac{GMm}{r}\right) \tag{13.18}$$

Hence, this is a conservative force field. The potential energy is then

$$\text{PE} = -\frac{GMm}{r} \tag{13.19}$$

Next, the force on a particle of unit positive charge from a particle of charge q_1 is given by Coulomb's law as

$$\boldsymbol{E} = \frac{q_1}{4\pi\epsilon_0 r^2}\hat{\boldsymbol{r}} \tag{13.20}$$

Since this equation has the same form as Eq. 13.17 (i.e., is also a function of $1/r^2$), we see immediately that the force field from q_1 is conservative. The potential energy per unit charge is then

$$\text{PE} = \frac{q_1}{4\pi\epsilon_0 r} \tag{13.21}$$

The remaining field introduced was the magnetic field where $\boldsymbol{F} = q\boldsymbol{V} \times \boldsymbol{B}$. For this field, the force on a charged particle depends on the velocity of the particle. The condition that the force be a function of position is not satisfied, therefore, and the magnetic field does *not* form a conservative force field.

13.4 Conservation of Mechanical Energy

Let us now consider the motion of a particle upon which only a conservative force field does work. We start with Eq. 13.2:

$$\int_{r_1}^{r_2} \boldsymbol{F} \cdot d\boldsymbol{r} = \tfrac{1}{2}mV_2^2 - \tfrac{1}{2}mV_1^2 \qquad (13.22)$$

Using the definition of potential energy, we replace the left side of the equation in the following manner:

$$(\text{PE})_1 - (\text{PE})_2 = \tfrac{1}{2}mV_2^2 - \tfrac{1}{2}mV_1^2 \qquad (13.23)$$

Rearranging terms, we reach the following useful relation:

$$(\text{PE})_1 + \tfrac{1}{2}mV_1^2 = (\text{PE})_2 + \tfrac{1}{2}mV_2^2 \qquad (13.24)$$

Since positions 1 and 2 are arbitrary, obviously *the sum of the potential energy and the kinetic energy for a particle remains constant at all times during the motion of the particle*. This statement is sometimes called the *law of conservation of mechanical energy for conservative systems*. The usefulness of this relation can be demonstrated by the following examples.

Example 13.5

A particle is dropped with zero initial velocity down a frictionless chute (Fig. 13.12). What is the magnitude of its velocity if the vertical drop during the motion is h ft?

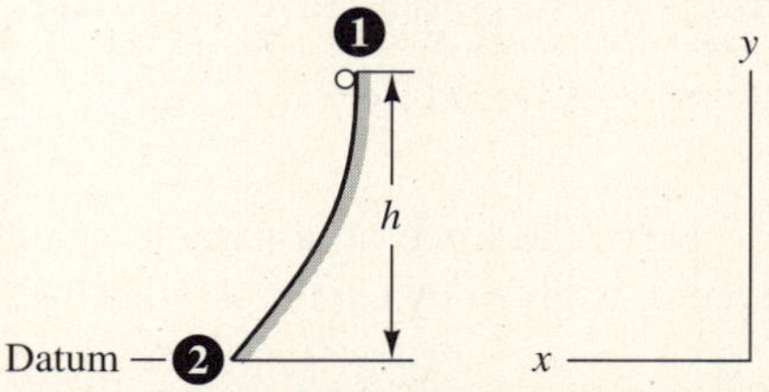

Figure 13.12. Particle on frictionless chute.

For small trajectories, we can assume a uniform force field $-mg\boldsymbol{j}$. Since this is the only force that can perform work on the particle (the normal force from the chute does no work), we can employ the **conservation-of-mechanical-energy** equation. If we take position 2 as a datum, we then have from Eq. 13.24:

$$mgh + 0 = 0 + \frac{1}{2}mV_2^2$$

Solving for V_2, we get

$$V_2 = \sqrt{2gh}$$

The advantages of the energy approach for conservative fields become apparent from this problem. That is, not all the forces need be considered in computing velocities, and the path, however complicated, is of no concern. If friction were present, a nonconservative force would perform work, and we would have to go back to the general relation given by Eq. 13.2 for the analysis.

Example 13.6

A mass is dropped onto a spring that has a spring constant K and a negligible mass (see Fig. 13.13). What is the maximum deflection δ? Neglect the effects of permanent deformation of the mass and any vibration that may occur.

In this problem, only conservative forces act on the body as it falls. Using the lowest position of the body as a datum, we see that the body falls a distance $h + \delta$. We shall equate the *mechanical energies* at the uppermost and lowest positions of the body. Thus,

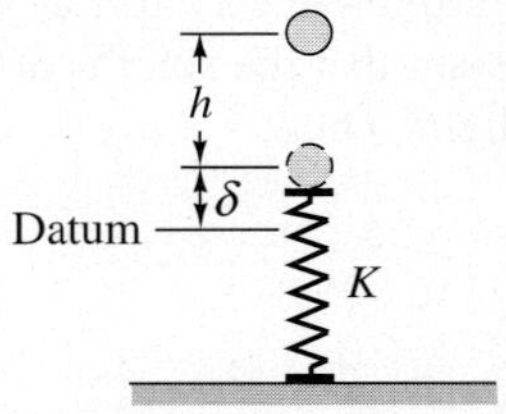

Figure 13.13. Mass dropped on spring.

$$\underbrace{mg(h+\delta)}_{\text{PE gravity}} + \underbrace{0}_{\text{PE spring}} + \underbrace{0}_{\text{KE}} = \underbrace{0}_{\text{PE gravity}} + \underbrace{\tfrac{1}{2}K\delta^2}_{\text{PE spring}} + \underbrace{0}_{\text{KE}} \quad \text{(a)}$$

Rearranging the terms,

$$\delta^2 - \frac{2mg}{K}\delta - \frac{2mgh}{K} = 0 \quad \text{(b)}$$

We may solve for a physically meaningful δ from this equation by using the quadratic formula.

Example 13.7

A ski jumper moves down the ramp aided only by gravity (Fig. 13.14). If the skier moves 33 m in the horizontal direction and is to land very smoothly at B, what must be the angle θ for the landing incline? Neglect friction. Also determine h.

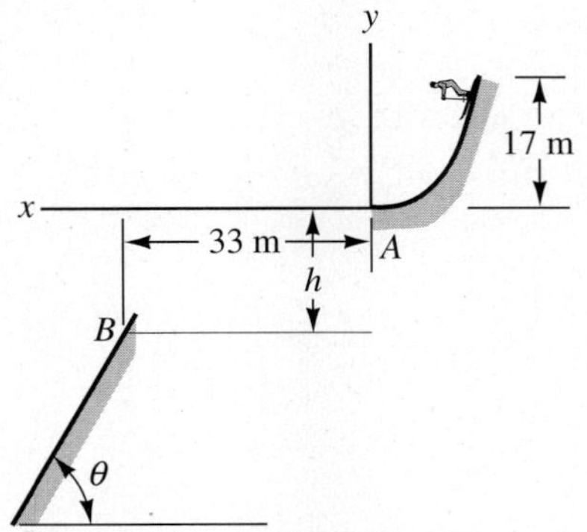

Figure 13.14. A ski jump with a landing ramp at an angle θ to be determined.

We first use **conservation of mechanical energy** along the ramp. Thus

$$(mg)(17) = \tfrac{1}{2}mV^2$$
$$\therefore V = \sqrt{(2g)(17)} = 18.26 \text{ m/s}$$

Example 13.7 (Continued)

Using a reference xy at A as shown in Fig. 13.14 and measuring time from the instant that the skier is at the origin, we now use **Newton's law** for the free flight. Thus

$$\ddot{y} = -9.81$$
$$\dot{y} = -9.81t + C_1$$
$$y = -9.81\frac{t^2}{2} + C_1 t + C_2$$

When $t = 0, \dot{y} = 0$, and we take $y = 0$. Hence,

$$C_1 = C_2 = 0$$

Also,

$$\ddot{x} = 0$$
$$\dot{x} = C_3$$
$$x = C_3 t + C_4$$

When $t = 0, \dot{x} = 18.26$, and $x = 0$,

$$\therefore C_3 = 18.26 \qquad C_4 = 0$$

Thus we have

$$\dot{y} = -9.81t \quad \text{(a)} \qquad \dot{x} = 18.26 \quad \text{(c)}$$
$$y = -9.81\frac{t^2}{2} \quad \text{(b)} \qquad x = 18.26t \quad \text{(d)}$$

To get h, set $x = 33$ in Eq. (d) and solve for the time t.

$$\therefore 33 = 18.26t \qquad t = 1.807 \text{ sec}$$

Hence, going to Eq. (b) we get

$$h = \left| -9.81\left(\frac{1.807^2}{2}\right) \right| = \boxed{16.01 \text{ m}}$$

Now get $\dot{y}$ at landing. Using Eq. (a) we have

$$\dot{y} = -(9.81)(1.807) = -17.73 \text{ m/s}$$

Also, we have at all times

$$\dot{x} = 18.26 \text{ m/s}$$

For best landing, $\boldsymbol{V}$ is parallel to incline

$$\therefore \tan\theta = \frac{-\dot{y}}{\dot{x}} = \frac{17.73}{18.26}$$

$$\boxed{\theta = 44.15^\circ}$$

Example 13.8

A block A of mass 0.2 kg slides on a frictionless surface as shown in Fig. 13.15. The spring constant K_1 is 25 N/m and initially, at the position shown, it is stretched 0.40 m. An elastic cord connects the top support to point C on A. It has a spring constant K_2 of 10.26 N/m. Furthermore, the cord disconnects from C at the instant that C reaches point G at the end of the straight portion of the incline. If A is released from rest at the indicated position, what value of θ corresponds to the end position B where A just loses contact with the surface? The elastic cord (at the top) is initially unstretched.

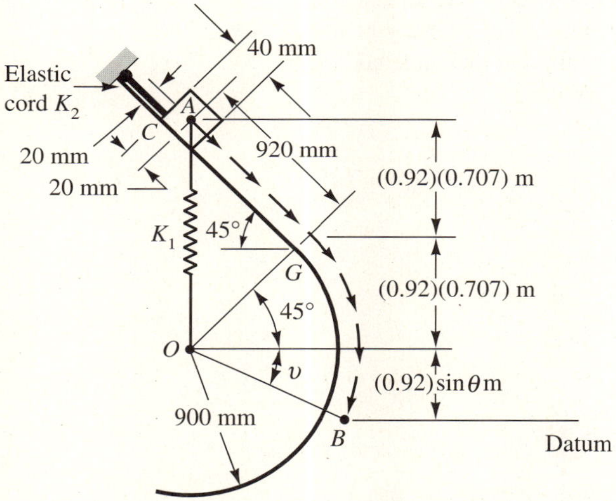

Figure 13.15. Mass A slides along frictionless surface.

We have conservative forces performing work on A so we have **conservation of mechanical energy.** Using the datum at B and using l_0 as the unstretched length of the spring with δ as the elongation of the spring, we then say that

$$mgz_1 + \frac{mV_1^2}{2} + \frac{1}{2}K_1\delta_1^2 = mgz_2 + \frac{mV_2^2}{2} + \frac{1}{2}K_1\delta_2^2 + \frac{1}{2}K_2(\overline{CG})^2$$

where the last term is the energy in the elastic cord when it disconnects at G. Therefore, noting that $\overline{CG} =$ 0.94 m and that $\overline{OB} =$ 0.92 m, we have on observing vertical distances in Fig. 13.15:

$$(0.200)(9.81)[(0.92)(0.707) + (0.92)(0.707) + (0.92)\sin\theta] + 0 + \tfrac{1}{2}(25)(0.40)^2$$
$$= 0 + \tfrac{1}{2}(0.20)V_2^2 + \tfrac{1}{2}(25)(0.92 - l_0)^2 + \tfrac{1}{2}(10.26)(0.94)^2 \qquad \text{(a)}$$

Example 13.8 (Continued)

To get l_0, examine the initial configuration of the system. With an initial stretch of 0.40 m for the spring, we can say observing again vertical distances in Fig. 13.15:

$$l_0 = [(0.92)(0.707) + (0.92)(0.707)] - 0.40$$
$$= 0.901 \text{ m}$$

Equation (a) can then be written as

$$V_2^2 = 0.1490 + 18.05 \sin\theta \qquad \text{(b)}$$

We now use **Newton's law** at the point of interest B where A just loses contact. This condition is shown in detail in Fig. 13.16, where you will notice that the contact force N has been taken as zero and thus deleted from the free-body diagram. In the radial direction, we have

$$-F_{sp} + (0.2)(g)\sin\theta = -0.2\left(\frac{V_2^2}{0.92}\right)$$

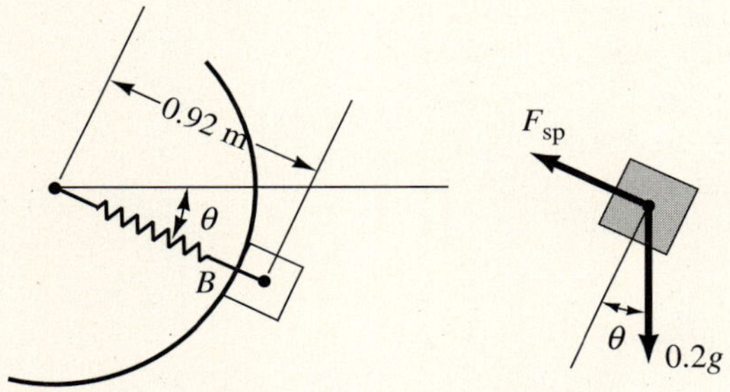

Figure 13.16. Contact is first lost at θ.

Therefore,

$$-(25)(0.92 - 0.901) + (0.2)(9.81)\sin\theta = -\left(\frac{V_2^2}{4.60}\right)$$

This equation can be written as

$$V_2^2 = 2.20 - 9.03 \sin\theta \qquad \text{(c)}$$

Solving Eqs. (b) and (c) simultaneously for θ, we get

$$\theta = 4.34°$$

13.5 Alternative Form of Work–Energy Equation

With the aid of the material in Section 13.4, we shall now set forth an alternative energy equation which has much physical appeal and which resembles the *first law of thermodynamics* as used in other courses. Let us take the case where certain of the forces acting on a particle are conservative while others are not. Remember that for conservative forces the negative of the change in potential energy between positions 1 and 2 equals the work done by these forces as the particle goes from position 1 to position 2 along any path. Thus, we can restate Eq. 13.2 in the following way:

$$\int_1^2 \boldsymbol{F} \cdot d\boldsymbol{r} - \Delta(\text{PE})_{1,2} = \Delta(\text{KE})_{1,2} \qquad (13.25)$$

where the integral represents the work of *nonconservative* forces and the Δ represents the final state minus the initial state. Calling the integral $\mathcal{W}_{1-2}$, we then have, on rearranging the equation:

$$\Delta(\text{KE} + \text{PE}) = \mathcal{W}_{1-2} \qquad (13.26)$$

In this form, we say that the work of *nonconservative* forces goes into changing the kinetic energy plus the potential energy for the particle. Since potential energies of such common forces as linear restoring forces, coulombic forces, and gravitational forces are so well known, the formulation above is useful in solving problems if it is understood thoroughly and applied properly.[8]

[8]Equation 13.26, you may notice, is actually a form of the first law of thermodynamics for the case of no heat transfer.

Example 13.9

Three coupled streetcars (Fig. 13.17) are moving at a speed of 9 m/s down a 7° incline. Each car has a weight of 198 kN. Specifications from

Figure 13.17 Coupled streetcars.

Example 13.9 (Continued)

the buyer requires that the cars must stop within 50 m beyond the position where the brakes are fully applied so as to cause the wheels to lock. What is the maximum number of brake failures that can be tolerated and still satisfy this specification? We will assume for simplicity that the weight is loaded equally among all the wheels of the system. There are 24 brake systems, one for each wheel. Take $\mu_d = 0.45$.

The friction force f on any one wheel where the brake has operated is ascertained from **Coulomb's law** as

$$f = \frac{198 \times 10^3 \cos 7°}{8}(0.45) = 11.05 \text{ kN}$$

We now consider the **work–energy relation** 13.26 for the case where a minimum number of good brakes, n, just causes the trains to stop in 50 m. We shall neglect the kinetic energy due to rotation of the rather small wheels. This assumption permits us to use a single particle to represent the three cars, wherein this particle moves a distance of 50 m. Using the end configuration of the train as the datum for potential energy of gravity, we have for Eq. 13.26:

$$\Delta \text{KE} + \Delta \text{PE} = \mathcal{W}_{1-2}$$

$$\left(0 - 3\left\{\frac{1}{2}\frac{198 \times 10^3}{g}(9)^2\right\}\right) + \left[0 - (3)(198 \times 10^3)(50)\sin 7°\right] = -(n)(11.05 \times 10^3)(50)$$

$$n = 10.99$$

The number of brake failures that can accordingly be tolerated is 24 − 11 = 13.

Number of brake failures to be tolerated = 13

Another example of conservation of mechanical energy will be in the next section (Example 13.11) for the case of a system of particles.

PROBLEMS

13.38. A railroad car traveling 1.4 m/s runs into a stop at a railroad terminal. A vehicle having a mass of 1.8 Mg is held by a linear restoring force system that has an equivalent spring constant of 20 kN/m. If the railroad car is assumed to stop suddenly and if the wheels in the vehicle are free to turn, what is the maximum force developed by the spring system? Neglect rotational inertia of the wheels of the vehicle.

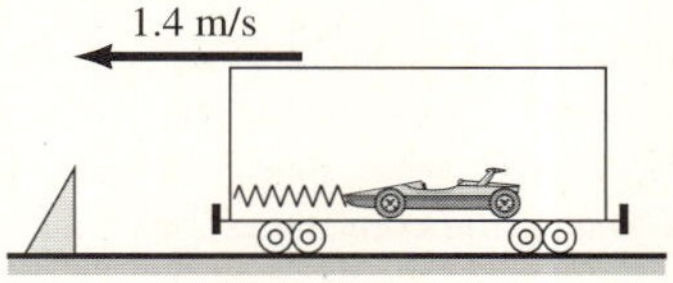

Figure P.13.38.

13.39. A mass of 15 kg is moving at a speed of 15 m/s along a horizontal frictionless surface, which later inclines upward at an angle 45°. A spring of constant $K = 1$ N/mm is present along the incline. How high does the mass move?

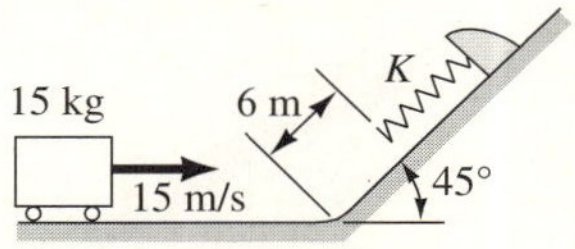

Figure P.13.39.

13.40. A block weighing 5 kg is released from rest where the springs acting on the body are horizontal and have a tension of 50 N each. What is the velocity of the block after it has descended 100 mm if each spring has a spring constant $K = 1$ N/mm?

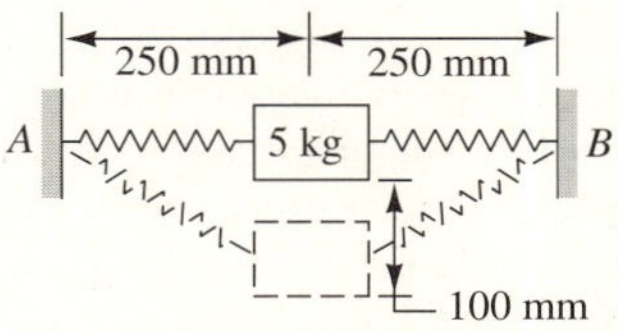

Figure P.13.40.

13.41. A nonlinear spring develops a force given as $0.06x^2$ N, where x is the amount of compression of the spring in millimetres. Does such a spring develop a conservative force? If so, what is the potential energy stored in the spring for a deflection of 60 mm?

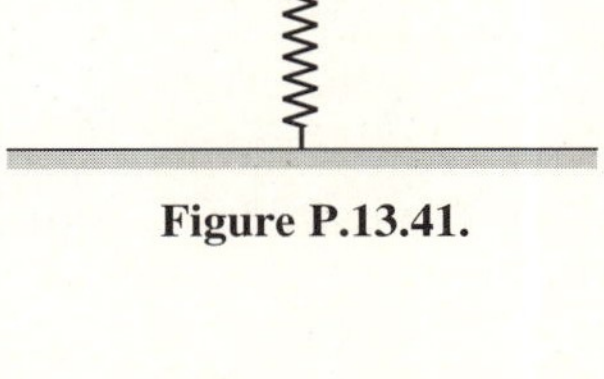

Figure P.13.41.

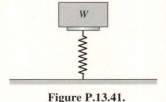

13.42. In Problem 13.41, a weight W of 225 N is released suddenly from rest on the nonlinear spring. What is the maximum deflection of the spring?

13.43. A vector operator that you will learn more about in fluid mechanics and electromagnetic theory is the *curl* vector operator. This operator is defined for rectangular coordinates in terms of its action on $\boldsymbol{V}$ as follows:

$$\textbf{curl}\, \boldsymbol{V}(x, y, z) = \left(\frac{\partial V_z}{\partial y} - \frac{\partial V_y}{\partial z}\right)\boldsymbol{i} + \left(\frac{\partial V_x}{\partial z} - \frac{\partial V_z}{\partial x}\right)\boldsymbol{j} + \left(\frac{\partial V_y}{\partial x} - \frac{\partial V_x}{\partial y}\right)\boldsymbol{k}$$

(When the curl is applied to a fluid velocity field $\boldsymbol{V}$ as above, the resulting vector field is twice the angular velocity field of infinitesimal elements in the flow.) Show that if $\boldsymbol{F}$ is expressible as $\boldsymbol{\nabla}\phi(x, y, z)$, then it must follow that **curl** $\boldsymbol{F} = \boldsymbol{0}$. The converse is also true, namely that *if* **curl** $\boldsymbol{F} = \boldsymbol{0}$, *then* $\boldsymbol{F} = \boldsymbol{\nabla}\phi(x, y, z)$ *and is thus a conservative force field.*

13.44. Determine whether the following force fields are conservative or not.

(a) $\boldsymbol{F} = (10z + y)\boldsymbol{i} + (15yz + x)\boldsymbol{j} + \left(10x + \dfrac{15y^2}{2}\right)\boldsymbol{k}$

(b) $\boldsymbol{F} = (z \sin x + y)\boldsymbol{i} + (4yz + x)\boldsymbol{j} + \left(2y^2 - 5 \cos x\right)\boldsymbol{k}$

See Problem 13.43 before doing this problem.

13.45. Given the following conservative force field:

$$\boldsymbol{F} = (10z + y)\boldsymbol{i} + (15yz + x)\boldsymbol{j} + \left(10x + \frac{15y^2}{2}\right)\boldsymbol{k}\ N$$

find the force potential to within an arbitrary constant. What work is done by the force field on a particle going from $\boldsymbol{r}_1 = 10\boldsymbol{i} + 2\boldsymbol{j} + 3\boldsymbol{k}$ m to $\boldsymbol{r}_2 = -2\boldsymbol{i} + 4\boldsymbol{j} - 3\boldsymbol{k}$ m? [*Hint:* Note that if $\partial\phi/\partial x$ equals some function $(xy^2 + z)$, then we can say on integrating that

$$\phi = \frac{x^2y^2}{2} + zx + g(y, z)$$

where $g(y, z)$ is an arbitrary function of y and z. Note we have held y and z constant during the integration.]

13.46. If the following force field is conservative,

$$\boldsymbol{F} = (5z \sin x + y)\boldsymbol{i} + (4yz + x)\boldsymbol{j} + (2y^2 - 5 \cos x)\boldsymbol{k}\ \text{N}$$

(where x, y, and z are in m), find the force potential up to an arbitrary constant. What is the work done on a particle starting at the origin and moving in a circular path of radius 0.6 m to form a semicircle along the positive x axis? (See the hint in Problem 13.45.)

13.47. A body A can slide in a frictionless manner along a stiff rod CD. At the position shown, the spring along CD has been compressed 150 mm and A is at a distance of 1.2 m from D. The spring connecting A to E has been elongated 25 mm. What is the speed of A after it moves 0.3 m? The spring constants are K_1 = 0.2 N/mm and K_2 = 0.1 N/m. The mass of A is 15 kg.

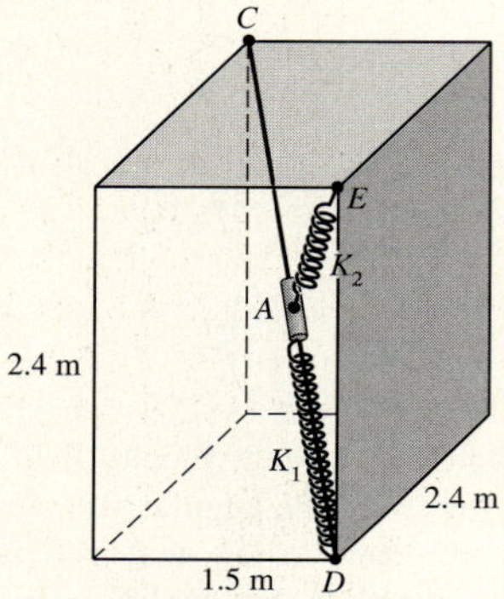

Figure P.13.47.

13.48. A collar A of mass 15 kg slides on a frictionless tube. The collar is connected to a linear spring whose spring constant K is 1 N/mm. If the collar is released from rest at the position shown, what is its speed when the spring is at elevation EF? The spring is stretched 75 mm at the initial position of the collar.

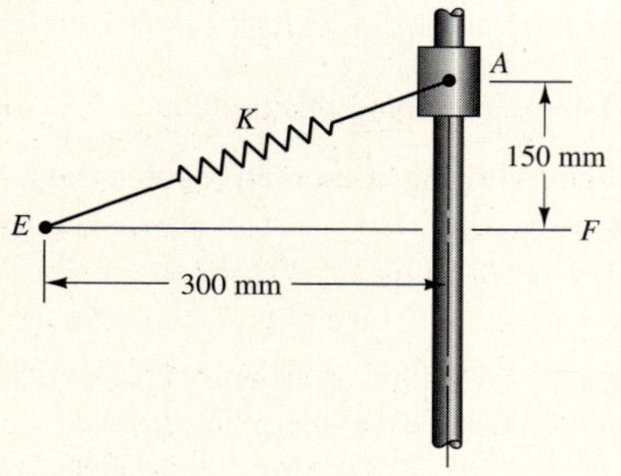

Figure P.13.48.

13.49. A mass M of 20 kg slides with no friction along a vertical rod. Two springs each of spring constant K_1 = 2 N/mm and a third spring having a spring constant K_2 = 3 N/mm are attached to the mass M. At the starting position when θ = 30°, the springs are unstretched. What is the velocity of M after it descends a distance d of 20 mm?

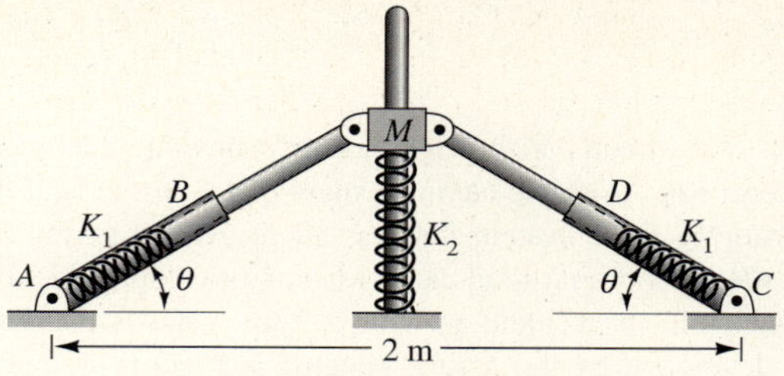

Figure P.13.49.

13.50. A collar A having a mass of 5 kg can slide without friction on a pipe. If released from rest at the position shown, where the spring is unstretched, what speed will the collar have after moving 50 mm? The spring constant is 2 N/mm.

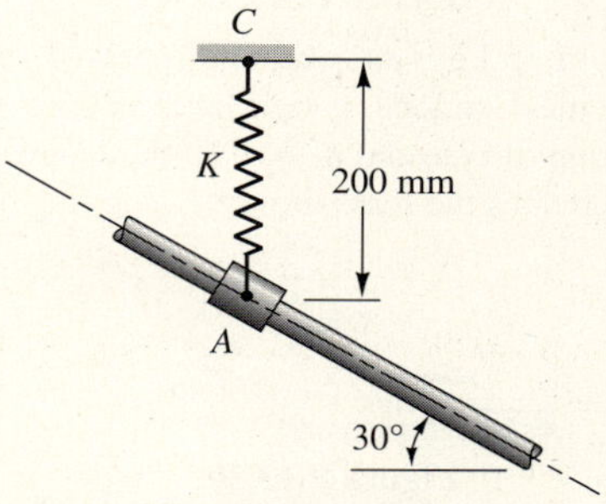

Figure P.13.50.

13.51. A slotted rod A is moving to the left at a speed of 2 m/s. Pins are moved to the left by this rod. These pins must slide in a slot under the rod as shown in the diagram. The pins are connected by a spring having a spring constant K of 1.5 N/mm. The spring is unstretched in the configuration shown. What distance d do the pins reach before stopping instantaneously? The mass of the slotted rod is 10 kg. The spring is held in the slotted rod so as not to buckle outward. Neglect the mass of the pins.

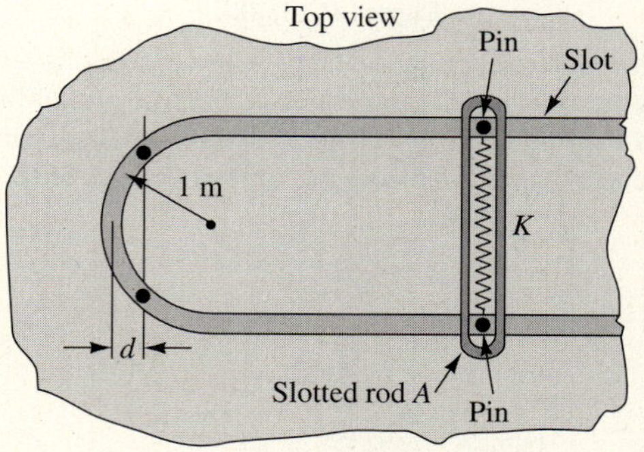

Figure P.13.51.

13.52. The top view of a slotted bar of mass 15 kg is shown. Two pins guided by the slotted bar ride in slots which have the equation of a hyperbola $xy = 1.5$, where x and y are in metres. The pins are connected by a linear spring having a spring constant K of 1 N/mm. When the pins are 0.6 m from the y axis, the spring is stretched 200 mm and the slotted bar is moving to the right at a speed of 0.6 m/s. What is $\dot{V}$ of the bar? [*Hint:* Differentiate energy equation.]

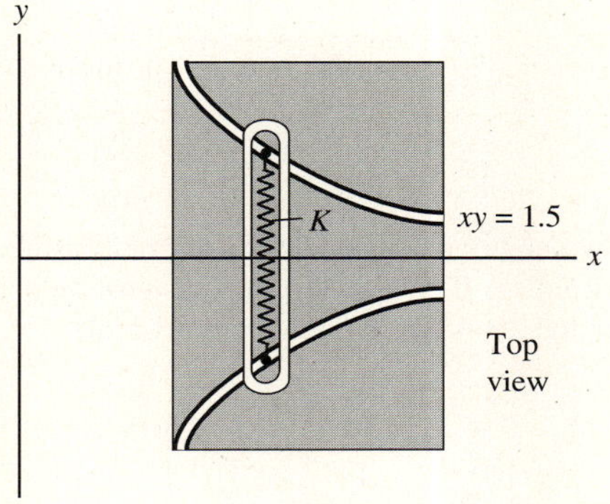

Figure P.13.52.

13.53. In Problem 13.52, what is the speed of the slotted bar when $x = 675$ mm?

13.54. Perhaps many of you as children constructed toy guns from half a clothespin, a wooden block, and bands of rubber cut from the inner tube of an automobile tire [see diagram (a)]. Rubber band A holds the half-clothespin to the wooden "gun stock." The "ammunition" is a rubber band B held by the clothespin at C by friction and stretched to go around the block at the other end. The rubber band B when laid flat as in (b) has a length of 180 mm. To "load the ammunition" takes a force of 90 N at C. If the gun is pointed upward, estimate how high the fired rubber band will go when "fired" if it has a mass of 0.1 kg. To "fire" the gun you push lowest part of clothespin toward the nail (see diagram) to release at C.

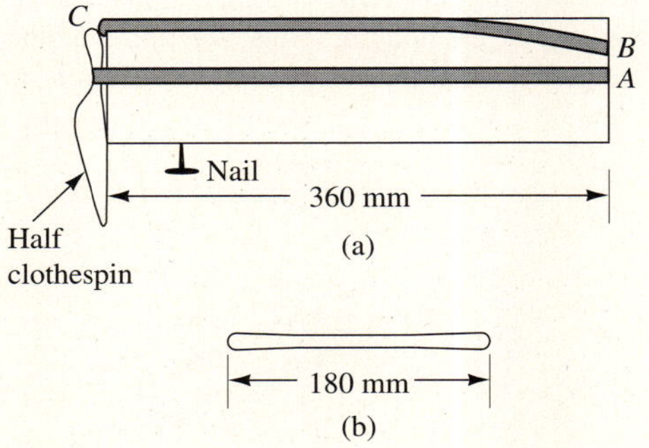

Figure P.13.54.

13.55. A meteor has a speed of 15 km/s when it is 320 Mm from the center of the earth. What will be its speed when it is 160 km from the earth's surface?

13.56 Do Problem 13.2 using the energy equation in the usual form of the first law of thermodynamics.

13.57. Do Problem 13.5 using the energy equation in the usual form of the first law of thermodynamics.

13.58. Do Problem 13.17 using the energy equation in the usual form of the first law of thermodynamics.

13.59. Do Problem 13.18 using the energy equation in the usual form of the first law of thermodynamics.

13.60. A constant-torque electric motor A is hoisting a weight W of 140 N. An inextensible cable connects the weight W to the motor over a stationary drum of diameter $D = 0.3$ m. The diameter d of the motor drive is 150 mm, and the delivered torque is 200 N m. The dynamic coefficient of friction between the drum and cable is 0.2. If the system is started from rest, what is the speed of the weight W after it has been raised 1.5 m?

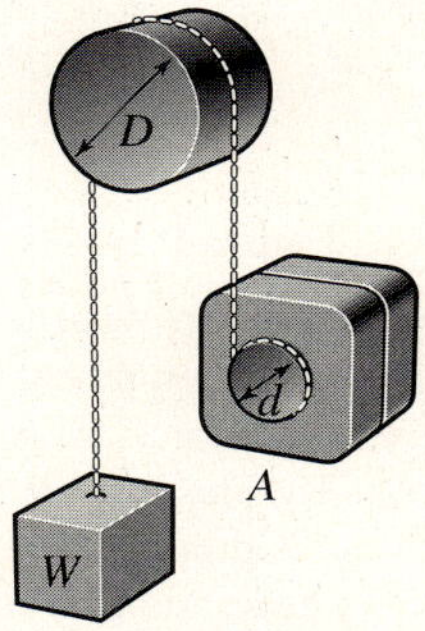

Figure P.13.60.

13.61. A body A of mass 5 kg, can slide along a fixed rod B–B. A spring is connected between fixed point C and the mass. AC is 0.6 m in length when the spring is unextended. If the body is released from rest at the configuration shown, what is its speed when it reaches the y axis? Assume that a constant friction force of 1.6 N acts on the body A. The spring constant K is 0.2 N/mm.

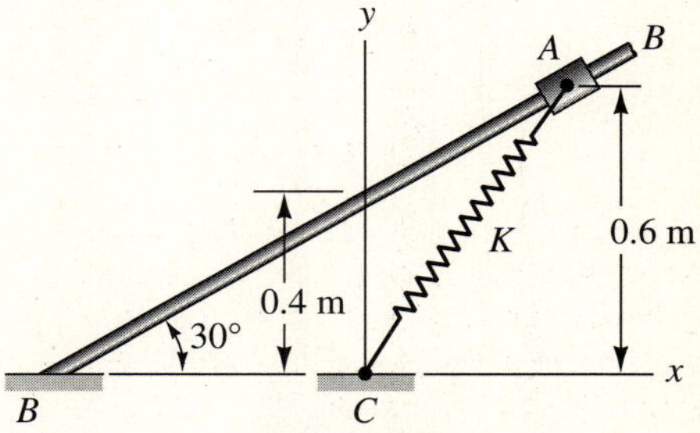

Figure P.13.61.

13.62. A body A is released from rest on a vertical circular path as shown. If a constant resistance force of 1 N acts along the path, what is the speed of the body when it reaches B? The mass of the body is 0.5 kg and the radius r of the path is 1.6 m.

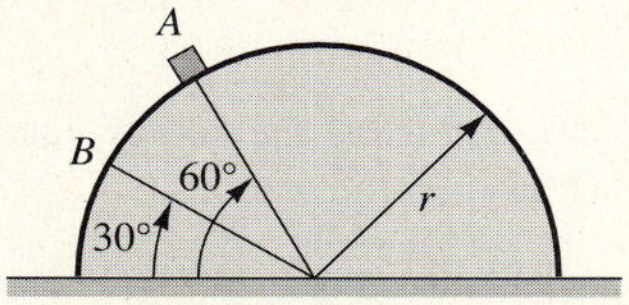

Figure P.13.62.

13.63. A cylinder slides down a rod. What is the distance δ that the spring is deflected at the instant that the disk stops instantaneously? Take $\mu_d = 0.3$.

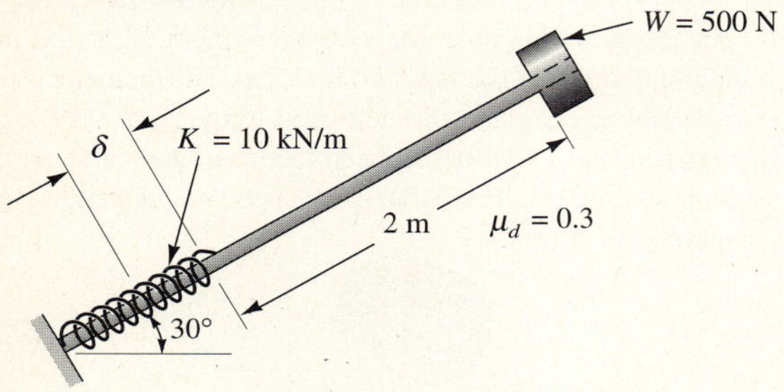

Figure P.13.63.

13.64. In ordnance work a very vital test for equipment is the *shock* test, in which a piece of equipment is subjected to a certain level of acceleration of short duration. A common technique for this test is the *drop test*. The specimen is mounted on a rigid carriage, which upon release is dropped along guide rods onto a set of lead pads resting on a heavy rigid anvil. The pads deform and absorb the energy of the carriage and specimen. We estimate through other tests that the energy E absorbed by a pad versus compression distance δ is given as shown, where the curve can be taken as a parabola. For four such pads, each placed directly on the anvil, and a height h of 3 m, what is the compression of the pads? The carriage and specimen together weigh 50g N. Neglect the friction of the guides.

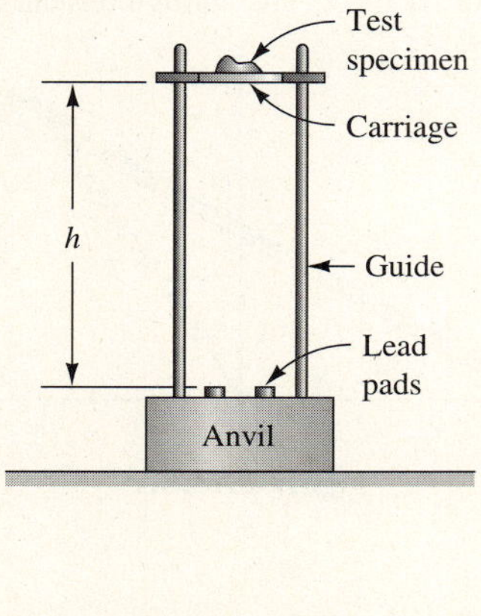

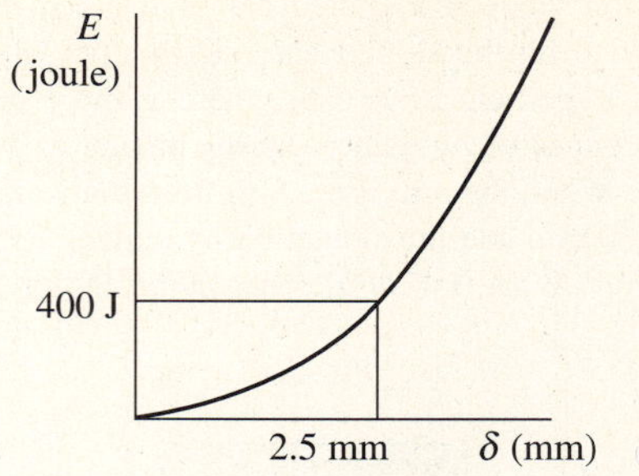

Figure P.13.64.

13.65. Two bodies are connected by an inextensible cord over a frictionless pulley. If released from rest, what velocity will they reach when the 250 kg body has dropped 1.5 m?

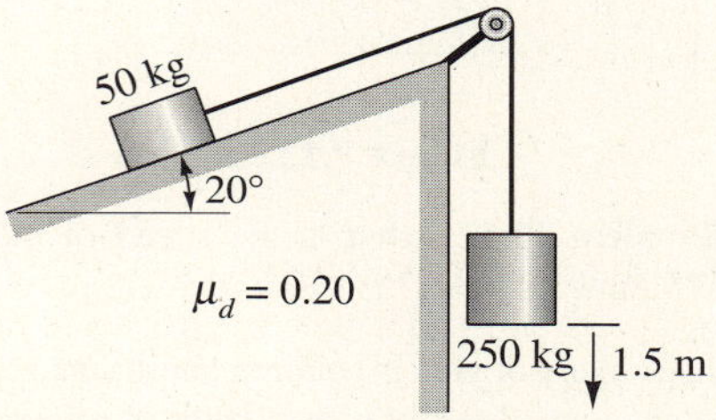

Figure P.13.65.

13.66. Suppose in Example 13.9 that only the brakes on train A operate and lock. What is the distance d before stopping? Also, determine the force in each coupling of the system.

13.67. A large constant force F is applied to a body of weight W resting on an inclined surface for which the coefficient of dynamic friction is μ_d. The body is acted on by a spring having a spring constant K. If initially the spring is compressed a distance δ, compute the velocity of the body in terms of F and the other parameters that are given, when the body has moved from rest a distance up the incline of $\frac{3}{2}\,\delta$.

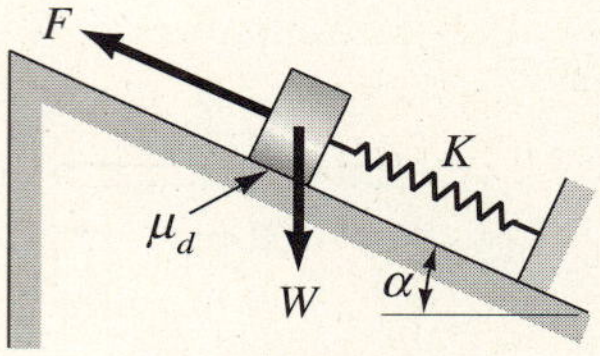

Figure P.13.67.

Part B: Systems of Particles

13.6 Work–Energy Equations

We shall now examine a system of particles from an energy viewpoint. A general aggregate of n particles is shown in Fig. 13.18. Considering the ith particle, we can say, by employing Eq. 13.2:

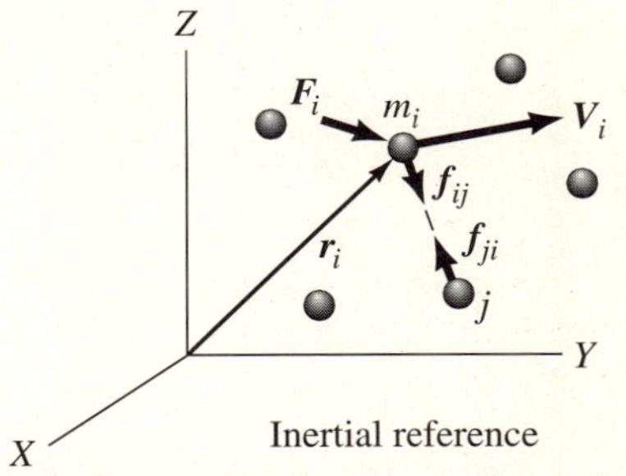

Figure 13.18. System of particles.

$$\int_1^2 \boldsymbol{F}_i \cdot d\boldsymbol{r}_i + \int_1^2 \left(\sum_{\substack{j=1 \\ j\neq i}}^{n} \boldsymbol{f}_{ij} \right) \cdot d\boldsymbol{r}_i = \left(\tfrac{1}{2}m_i V_i^2\right)_2 - \left(\tfrac{1}{2}m_i V_i^2\right)_1 \tag{13.27}$$

where, as in Chapter 12, $\boldsymbol{f}_{ij}$ is the force from the jth particle onto the ith particle, as illustrated in the diagram, and is thus an internal force. In contrast, $\boldsymbol{F}_i$ represents the total *external* force on the ith particle. In words, Eq. 13.27 says that for a displacement between $\boldsymbol{r}_1$ and $\boldsymbol{r}_2$ along some path, the energy relations for the ith particle are:

$$\text{external work} + \text{internal work} = (\text{change in kinetic energy relative to } XYZ) \tag{13.28}$$

Furthermore, we can adopt the point of view set forth in Section 13.5 and identify conservative forces, both external and internal, so as to utilize potential energies for these forces in the energy equation. To qualify as a conservative force, an internal force would have to be a function of only the spatial configuration of the system and expressible as the gradient of a scalar function. Clearly, forces arising from the gravitational attraction between the particles, electrostatic forces from electric charges on the particles, and forces from elastic connectors between the particles (such as springs) are all conservative internal forces.

We now sum Eqs. 13.27 for all the particles in the system to get the energy equation for a *system of particles.* We do *not* necessarily get a cancellation of contributions of the internal forces as we did for Newton's law in Chapter 12 because we are now adding the *work* done by each internal force on each particle. And even though we have pairs of internal forces that are equal and opposite, the *movements* of the corresponding particles in general are *not* equal. The result is that the work done by a pair of equal and opposite internal forces is not always zero. However, in the case of a *rigid body*, the contact forces between pairs of particles making up the body have the same motion, and so in this case the internal work is *zero* from such forces.[9] Also, if there is a system of rigid bodies interconnected by pin or ball joint connections, and if there is no friction at these movable connections, then again there will be no internal work. (Why?) We can then say for the system of particles that

$$\Delta(\text{KE} + \text{PE}) = \mathcal{W}_{1-2} \tag{13.29}$$

[9]We shall show this more directly in Chapter 17.

where $\mathcal{W}_{1-2}$ represents the net work done by *internal and external* nonconservative forces, and PE represents the total potential energy of the conservative *internal and external* forces. Clearly, if there are no nonconservative forces present then Eq. (13.29) degenerates to the conservation-of-mechanical-energy principle. As pointed out earlier, since we are employing the *change* in potential energy, the datums chosen for measuring PE are of little significance here.[10] For instance, any convenient datum for measuring the potential energy due to gravity of the earth yields the same result for the term ΔPE.

Looking back on Eq. (13.27), which on summation over all the particles gave rise to Eq. (13.29), namely the equation to be used for a system of particles, we wish to make the following point. It is the fact that the work contribution of each force stems from the *movement of each force with its specific point of application.* This should be clear from the use of $d\mathbf{r}_i$, with i identifying each particle. This will be an important consideration later.

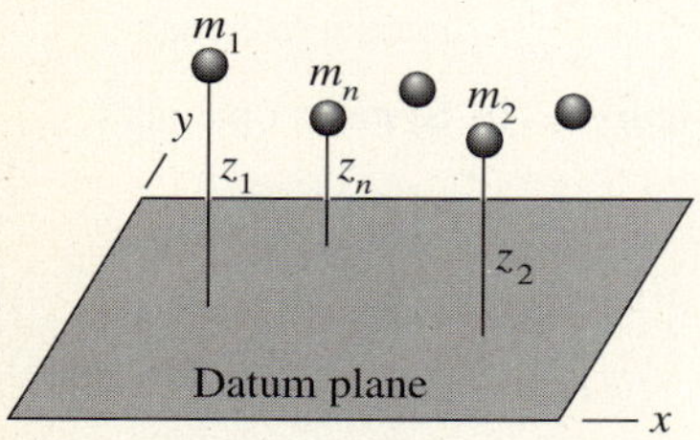

Figure 13.19. Particles above reference plane.

Let us now consider the action of gravity on a system of particles. The potential energy relative to a datum plane, xy, for such a system (see Fig. 13.19) is simply

$$\text{PE} = \sum_i m_i g z_i$$

Note that the right side of this equation represents the first moment of the weight of the system about the xy plane. This quantity can be given in terms of the center of gravity and the entire weight of W as follows:

$$\text{PE} = W z_c$$

where z_c is the vertical distance from the datum plane to the center of gravity. Note that if g is constant, the center of gravity corresponds to the center of mass. And so for any system of particles, the change in potential energy is readily found by concentrating the entire weight at the center of gravity or, as is almost always the case, at the center of mass.

Before proceeding with the problems we wish to emphasize certain salient features governing the work–energy principle for a system of particles.

1. In computing work, we must remember to have the forces move **with their points of application** (see Eq. 13.27).
2. Both **internal** and **external** forces may be present as conservative and as nonconservative forces and must be accounted for.
3. The kinetic energy must be the **total** kinetic energy and not just that of the mass center.

[10]One precaution in this regard must again be brought to your attention. You will remember that in the spring-force formula, $-Kx$, the term x represents the elongation or contraction of the spring from the *undeformed* condition. This condition must not be violated in the potential-energy expression $\frac{1}{2}Kx^2$.

Example 13.10

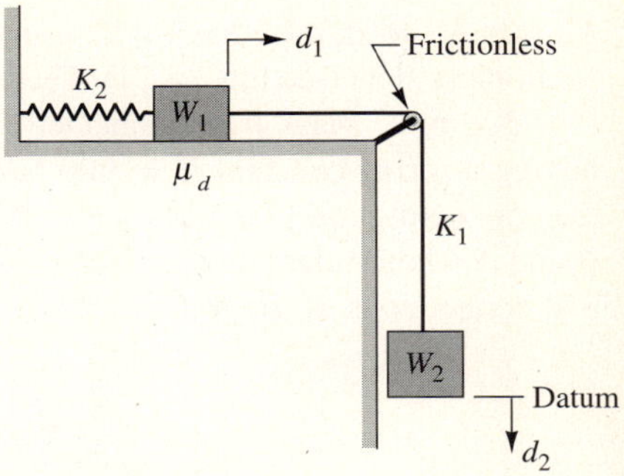

Figure 13.20. Elastically connected bodies.

In Fig. 13.20, two blocks have weights W_1 and W_2, respectively. They are connected by a flexible, *elastic* cable of negligible mass which has an equivalent spring constant of K_1. Body 1 is connected to the wall by a spring having a spring constant K_2 and slides along a horizontal surface for which the dynamic coefficient of friction with the body is μ_d. Body 2 is supported initially by some external agent so that, at the outset of the problem, the spring and cable are unstretched. What is the total kinetic energy of the system when, after release, body 2 has moved a distance d_2 and body 1 has moved a smaller distance d_1?

Use Eq. 13.29. Only one nonconservative force exists in the system, the external friction force on body 1. Therefore, the work term of the equation becomes

$$\mathcal{W}_{1-2} = -W_1\mu_d d_1 \qquad \text{(a)}$$

Three conservative forces are present; the spring force and the gravitational force are *external* and the force from the elastic cable is *internal*. (We neglect mutual gravitational forces between the bodies.) Using the initial position of W_2 as the datum for gravitational potential energy, we have, for the total change in potential energy:

$$\Delta\text{PE} = \left[\tfrac{1}{2}K_2d_1^2 - 0\right] + \left[\tfrac{1}{2}K_1\left(d_2 - d_1\right)^2 - 0\right] + \left[0 - W_2d_2\right]$$

We can compute the desired change in kinetic energy from Eq. 13.29 as

$$\Delta\text{KE} = -W_1\mu_d d_1 - \tfrac{1}{2}K_2d_1^2 - \tfrac{1}{2}K_1\left(d_2 - d_1\right)^2 + W_2d_2 \qquad \text{(b)}$$

As an additional exercise, you should arrive at this result by using the basic Eq. 13.28, where you cannot rely on familiar formulas for potential energies.

In the following example, we will see how using a system of particles approach can make for great simplification in a problem over the procedure of dealing with particles individually. Also we will have a case where there is no nonconservative work, which then results in a conservation of mechanical energy.

Example 13.11

Masses A and B, each having a mass of 75 kg, are constrained to move in frictionless slots (see Fig. 13.21). They are connected by a light rod of length l = 300 mm. Mass B is connected to two massless linear springs, each having a spring constant K = 900 N/m. The springs are unstretched when the connecting rod to masses A and B is vertical. What are the velocities of B and A when A descends a distance of 25 mm? There is no friction in the end connections of the rod.[11]

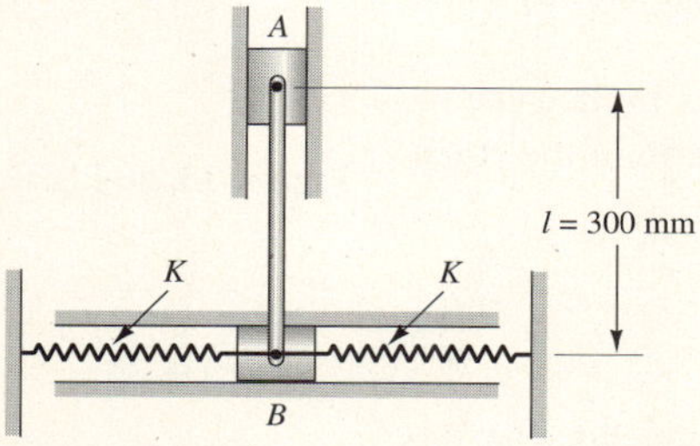

Figure 13.21. Two interconnected masses constrained by linear springs.

We shall use a **system of particles** approach for this case. This will eliminate the need to calculate the work of the rod on each mass that we would need had we elected to deal with each mass separately. For a system of particles approach, this work is internal between rigid bodies, having zero value as a result of **Newton's third law** and having ideal pin-connected joints.

We next note that only conservative forces are present (gravitational force and spring forces) so the first law form of our energy equation degenerates to conservation of mechanical energy. In Fig. 13.22, we show

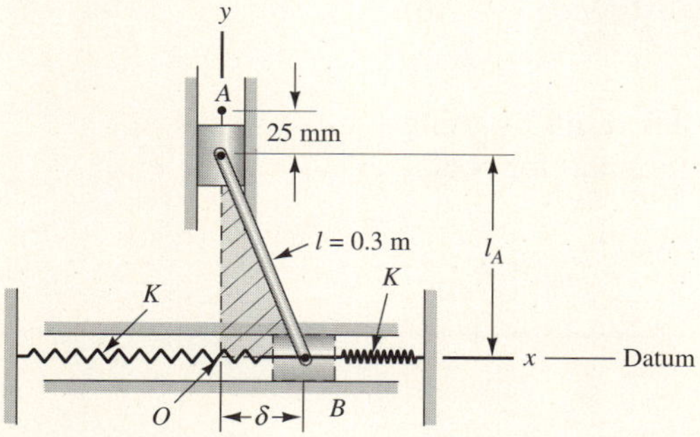

Figure 13.22. System in a configuration wherein mass A has dropped 25 mm.

[11]This is an example of an **unstable equilibrium** (a topic considered in Chapter 10) in that any slight movement of block B from its central position to the right or to the left causes A to start to accelerate downward.

Example 13.11 (Continued)

the system in a configuration where mass A has dropped a distance of 25 mm. We can then say for the beginning and end configurations using the datum shown in Fig. 13.22,

$$\Delta[\text{KE} + \text{PE}] = \mathcal{W}_{1-2} = 0$$

$$\left(\tfrac{1}{2}M_A V_A^2 - 0\right) + \left(\tfrac{1}{2}M_B V_B^2 - 0\right) + \\ \left[M_A g(0.3 - 0.025) - M_A g(0.3)\right] + \left[2\tfrac{1}{2}(900)\left(\delta^2\right) - 0\right] = 0 \quad \text{(a)}$$

We have three unknowns here. They are δ, V_A, and V_B. Observing the shaded triangle in Fig. 13.22 and using the Pythagorean theorem, we have

$$l_A^2 + \delta^2 = 0.3^2 \quad \text{(b)}$$

Taking the time derivative we get

$$2l_A \dot{l}_A + 2\delta\dot{\delta} = 0$$

We note that $\dot{l}_A = V_A$ and that $\dot{\delta} = V_B$. We then see from the preceding equation that

$$V_B = -\frac{l_A}{\delta} V_A \quad \text{(c)}$$

Now, returning to Fig. 13.22, we can compute δ for the case at hand. Noting that A has descended a distance of 0.025 m, thus making $l_A = 0.3 - 0.025 = 0.275$ m, we next go to Eq. (b) to get δ. Thus

$$(0.275)^2 + \delta^2 = 0.3^2$$
$$\therefore \delta = 0.1199 \text{ m} \quad \text{(d)}$$

Substituting data from Eqs. (c) and (d) into Eq. (a) (conserving mechanical energy) we get

$$V_A = -0.1524 \text{ m/s} \quad \therefore \text{from (c)}, \; V_B = -\frac{0.275}{0.1199} V_A = 0.3496 \text{ m/s}$$

Thus,

$V_A = 0.1524$ m/s downward $\quad V_B = 0.3496$ m/s to the right or the left

13.7 Kinetic Energy Expression Based on Center of Mass

In this and the next section, we shall introduce the center of mass into our discussion in order to develop useful expressions for the kinetic energy of an aggregate of particles. Also, we shall develop the work–energy equation for the center of mass set forth at the outset of this chapter.

Consider a system of n particles, shown in Fig. 13.23. The total kinetic energy relative to xyz of the system of particles can be given as

$$\text{KE} = \sum_{i=1}^{n} \tfrac{1}{2} m_i V_i^2 \tag{13.30}$$

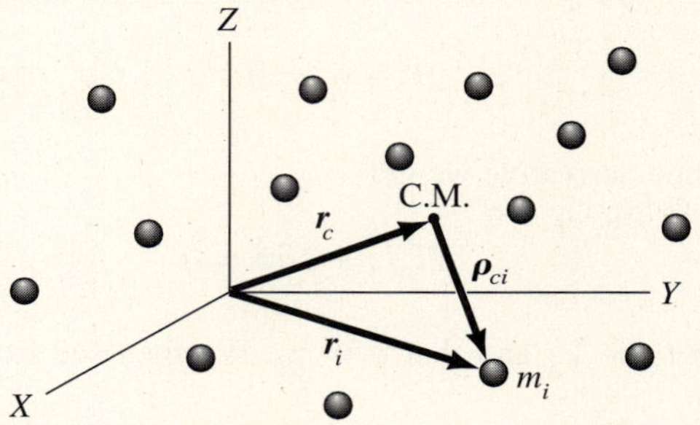

Figure 13.23. System of particles with center of mass.

We shall now express Eq. 13.30 in another way by introducing the mass center. Note in the diagram we have employed the vector $\boldsymbol{\rho}_{ci}$ as the displacement vector from the center of mass to the ith particle. We can accordingly say:

$$\boldsymbol{r}_i = \boldsymbol{r}_c + (\boldsymbol{r}_i - \boldsymbol{r}_c) = \boldsymbol{r}_c + \boldsymbol{\rho}_{ci} \tag{13.31}$$

Differentiating with respect to time, we get

$$\dot{\boldsymbol{r}}_i = \dot{\boldsymbol{r}}_c + (\dot{\boldsymbol{r}}_i - \dot{\boldsymbol{r}}_c) = \dot{\boldsymbol{r}}_c + \dot{\boldsymbol{\rho}}_{ci}$$

Therefore,

$$\boldsymbol{V}_i = \boldsymbol{V}_c + \dot{\boldsymbol{\rho}}_{ci} \tag{13.32}$$

From our earlier discussions on simple relative motion we can say that $\dot{\boldsymbol{\rho}}_{ci}$ is the motion of the ith particle *relative to the mass center*.[12] Substituting

[12]Note that

$$\dot{\boldsymbol{\rho}}_{ci} = \dot{\boldsymbol{r}}_i - \dot{\boldsymbol{r}}_c$$

That is, $\dot{\boldsymbol{\rho}}_{ci}$ is the *difference* between the velocity of the ith particle and that of the mass center. This is then the velocity of the particle *relative* to the center of the mass (i.e., relative to a reference translating with c or to a nonrotating observer moving with c.)

the relation above into the expression for kinetic energy, Eq. 13.30, we get

$$\text{KE} = \sum_{i=1}^{n} \tfrac{1}{2} m_i (\boldsymbol{V}_c + \dot{\boldsymbol{\rho}}_{ci})^2 = \sum_{i=1}^{n} \tfrac{1}{2} m_i (\boldsymbol{V}_c + \dot{\boldsymbol{\rho}}_{ci}) \bullet (\boldsymbol{V}_c + \dot{\boldsymbol{\rho}}_{ci})$$

Carrying out the dot product, we have

$$\text{KE} = \tfrac{1}{2} \sum_{i=1}^{n} m_i V_c^2 + \sum_{i=1}^{n} m_i \, \boldsymbol{V}_c \bullet \dot{\boldsymbol{\rho}}_{ci} + \tfrac{1}{2} \sum_{i=1}^{n} m_i \dot{\boldsymbol{\rho}}_{ci}^2 \qquad (13.33)$$

Since $\boldsymbol{V}_c$ is common for all values of the summation index, we can extract it from the summation operation, and this leaves

$$\text{KE} = \tfrac{1}{2} \left(\sum_{i=1}^{n} m_i \right) V_c^2 + \boldsymbol{V}_c \bullet \left(\sum_{i=1}^{n} m_i \dot{\boldsymbol{\rho}}_{ci} \right) + \tfrac{1}{2} \sum_{i=1}^{n} m_i \dot{\boldsymbol{\rho}}_{ci}^2 \qquad (13.34)$$

Perform the following replacements:

$$\sum_{i=1}^{n} m_i \text{ by } M, \quad \text{and} \quad \sum_{i=1}^{n} m_i \dot{\boldsymbol{\rho}}_{ci} \text{ by } \frac{d}{dt} \sum_{i=1}^{n} m_i \boldsymbol{\rho}_{ci}$$

We then have

$$\text{KE} = \tfrac{1}{2} M V_c^2 + \boldsymbol{V}_c \bullet \frac{d}{dt} \sum_{i=1}^{n} m_i \boldsymbol{\rho}_{ci} + \tfrac{1}{2} \sum_{i=1}^{n} m_i \dot{\boldsymbol{\rho}}_{ci}^2 \qquad (13.34)$$

But the expression

$$\sum_{i=1}^{n} m_i \boldsymbol{\rho}_{ci}$$

represents the first moment of mass of the system about the center of mass for the system. Clearly by definition, this quantity must always be zero. The expression for kinetic energy becomes

$$\text{KE} = \tfrac{1}{2} M V_c^2 + \tfrac{1}{2} \sum_{i=1}^{n} m_i \dot{\boldsymbol{\rho}}_{ci}^2 \qquad (13.35)$$

Thus, we see that the *kinetic energy* for some reference *can be considered to be composed of two parts: (1) the kinetic energy of the total mass moving relative to that reference with the velocity of the mass center, plus (2) the kinetic energy of the motion of the particles relative to the mass center.*

Example 13.12

A hypothetical vehicle is moving at speed V_0 in Fig. 13.24. On this vehicle are two bodies each of mass m sliding along a horizontal rod at a speed v relative to the rod. This rod is rotating at an angular speed ω rad/sec relative to the vehicle. What is the kinetic energy of the two bodies relative to the ground (XYZ) when they are at a distance r from point A?

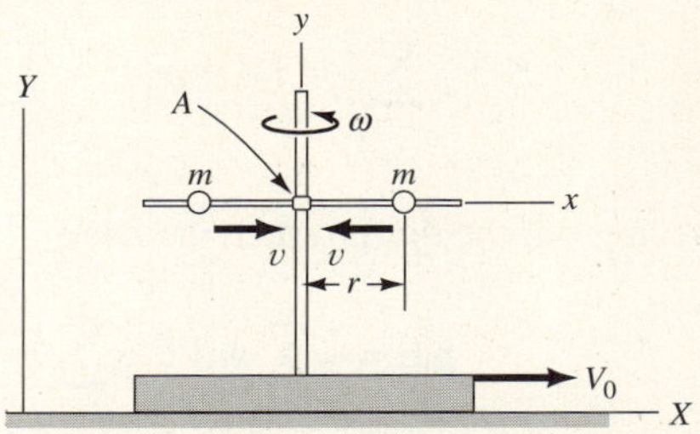

Figure 13.24. Moving device.

Clearly, the center of mass corresponds to point A and is thus moving at a speed V_0 relative to the ground. Hence, we have as part of the kinetic energy the term

$$\frac{1}{2}MV_c^2 = mV_0^2 \tag{a}$$

The velocity of each ball relative to the center of mass is easily formed using cylindrical components. Thus, imagining a reference xyz at A translating with the vehicle relative to XYZ, we have for the velocity of each ball relative to xyz:

$$\dot{\rho}^2 = \dot{r}^2 + (\omega r)^2 = v^2 + (\omega r)^2 \tag{b}$$

The total kinetic energy of the two masses relative to the ground is then

$$\boxed{\text{KE} = mV_0^2 + m\left[v^2 + (\omega r)^2\right]} \tag{c}$$

In Example 13.12, we considered a case where the bodies involved constituted a finite number of *discrete* particles. In the next example, we consider a case where we have a *continuum* of particles forming a rigid body. The formulation given by Eq. 13.35 can still be used but now, instead of summing for a finite number of discrete particles, we must integrate to account for the infinite number of infinitesimal particles comprising the system. We are thus taking a glimpse, for simple cases, of rigid-body dynamics to be studied later in the text. Those that do not have time for studying such energy problems in detail will be able to solve simple but useful rigid-body dynamics problems on the basis of these examples as well as later examples in this chapter.

Example 13.13

A thin uniform hoop of radius R is rolling without slipping such that O, the mass center, moves at a speed V (Fig. 13.25). If the hoop weighs W lb, what is the kinetic energy of the hoop relative to the ground?

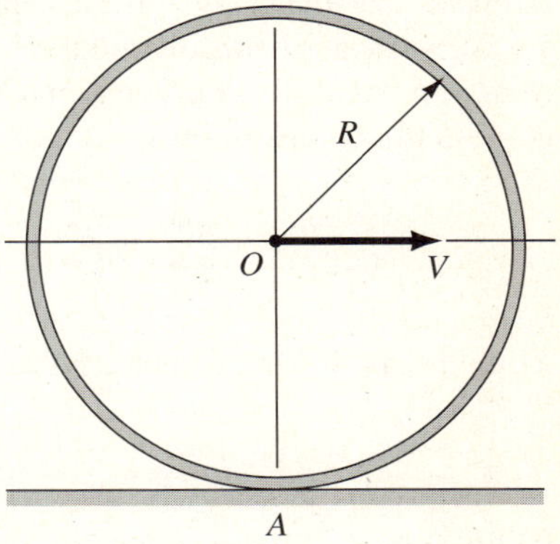

Figure 13.25. Rolling hoop.

Clearly, the hoop cannot be considered as a finite number of discrete finite particles as in the previous example, and so we must consider an infinity of infinitesimal contiguous particles. It is simplest to employ here the center-of-mass approach. The main problem then is to find the kinetic energy of the hoop relative to the mass center O, that is, relative to a reference xy translating with the mass center as seen from the ground reference XY (see Fig. 13.26). The motion relative to xy is clearly simple rotation; accordingly, we must find the angular velocity of the hoop for this reference. The no slipping condition means that the point of contact of the hoop with the ground has instantaneously a zero velocity. Observe the motion from a stationary reference XY. As you may have learned in physics, and as will later be shown (Chapter 15), the body has a *pure instantaneous rotational* motion about the point of contact. The angular velocity ω for this motion is then easily evaluated by considering point O rotating about the instantaneous center of rotation A. Thus,

$$\omega = \frac{V}{R} \tag{a}$$

Since reference xy *translates* relative to reference XY, an observer on xy sees the *same angular velocity* ω for the hoop as the observer on XY. Accordingly, we can now readily evaluate the second term on the right side of Eq. 13.35. As particles, use elements of the hoop which are $R\,d\theta$ in length, as shown in Fig. 13.26, and which have a mass per unit length of $W/(2\pi Rg)$. We then have, on replacing summation by integration, the result

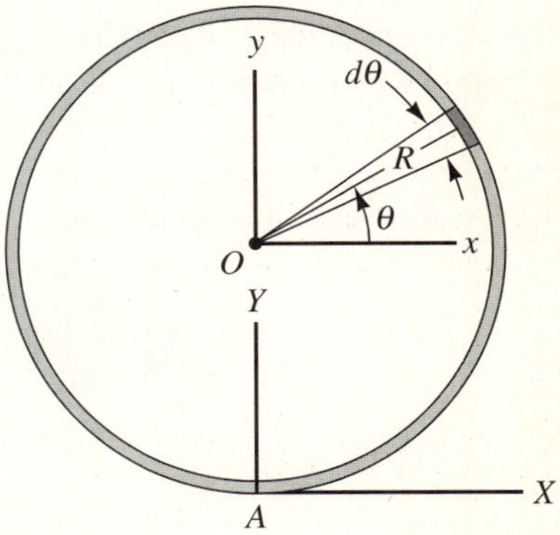

Figure 13.26. xy translates with O relative to XY.

$$\begin{aligned}\frac{1}{2}\sum_{i=1}^{n} m_i \dot{\rho}_{ci}^2 &= \frac{1}{2}\int_0^{2\pi}\left[\left(\frac{W}{2\pi Rg}\right)(R\,d\theta)\right](\omega R)^2 \\ &= \frac{1}{2}\int_0^{2\pi}\left[\frac{W}{2\pi Rg}(R\,d\theta)\right]\left(\frac{V}{R}R\right)^2 \\ &= \frac{1}{2}\frac{W}{g}V^2\end{aligned} \tag{b}$$

The kinetic energy of the hoop is then in accordance with Eq. 13.35:

$$\text{KE} = \underbrace{\frac{1}{2}\frac{W}{g}V^2}_{(\text{KE})_{\text{C.M.}}} + \underbrace{\frac{1}{2}\frac{W}{g}V^2}_{(\text{KE})_{\text{Rel. to C.M.}}} = \boxed{\frac{W}{g}V^2} \tag{c}$$

Example 13.13 (Continued)

Suppose that the body were a generalized cylinder of mass M (see Fig. 13.27) such as a tire of radius R having O as the center of mass with axisymmetrical distribution of mass about the axis at O. Then, we would express Eq. (b) as follows:

$$\frac{1}{2}\sum_{i}^{n} m_i \dot{\rho}_{ci}^2 = \frac{1}{2}\iiint_M (dm)(r\omega)^2 \tag{d}$$

You will recall from Chapter 9 that

$$\iiint_M r^2\, dm$$

is the *second moment of inertia* of the body taken about the z axis at O. That is,

$$I_{zz} = \iiint_M r^2\, dm$$

Thus, we have for the kinetic energy of such a body:

$$\text{KE} = \frac{1}{2} M V^2 + \frac{1}{2} I_{zz}\omega^2 \tag{e}$$

You may also recall from Chapter 9 that we could employ the *radius of gyration* k to express I_{zz} as follows:

$$I_{zz} = k^2 M \tag{f}$$

Hence, Eq. (e) can be given as

$$\text{KE} = \frac{1}{2} M V^2 + \frac{1}{2} k^2 M \omega^2$$

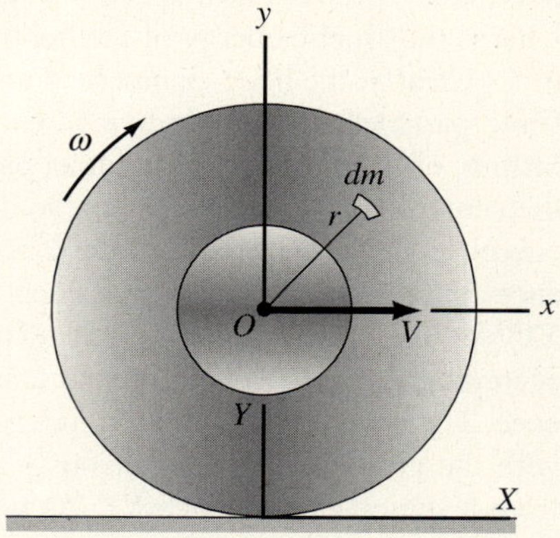

Figure 13.27. Rolling generalized cylinder of mass M.

We shall examine the kinetic energy formulations of rigid bodies carefully in Chapter 17. Here, we have used certain familiar results from physics pertaining to kinematics of plane motion of a nonslipping rolling rigid body. For a more general undertaking, we shall have to carefully consider more general aspects of kinematics of rigid-body motion. This will be done in Chapter 15. Also, in the last example we see one term of the inertia tensor I_{ij} showing up. The vital role of the inertia tensor in the dynamics of rigid bodies will soon be seen.

13.8 Work–Kinetic Energy Expressions Based on Center of Mass

The work–kinetic energy expressions of Section 13.6 were developed for a system of particles without regard to the mass center. We shall now introduce this point into the work–kinetic energy formulations. You will recall from Chapter 12 that *Newton's law* for the mass center of any system of particles is

$$\boldsymbol{F} = M\ddot{\boldsymbol{r}}_c \tag{13.36}$$

where $\boldsymbol{F}$ is the total *external* force on the system of particles. By the same development as presented in Section 13.1, we can readily arrive at the following equation:

$$\int_1^2 \boldsymbol{F} \cdot d\boldsymbol{r}_c = \left(\tfrac{1}{2}MV_c^2\right)_2 - \left(\tfrac{1}{2}MV_c^2\right)_1 \tag{13.37}$$

It is *vital* to understand from the left side of Eq. 13.37, where we note the term $d\boldsymbol{r}_c$, that the *external forces must all move with the center of mass* for the computation of the proper work term in this equation.[13] We wish next to point out that the single particle model represents a special case of the use of Eq. 13.37. Specifically, the single particle model represents the case where the motion of the center of mass of a body sufficiently describes the motion of the body and where the external forces on the body essentially move with the center of mass of the body. Such cases were set forth in Section 13.1.

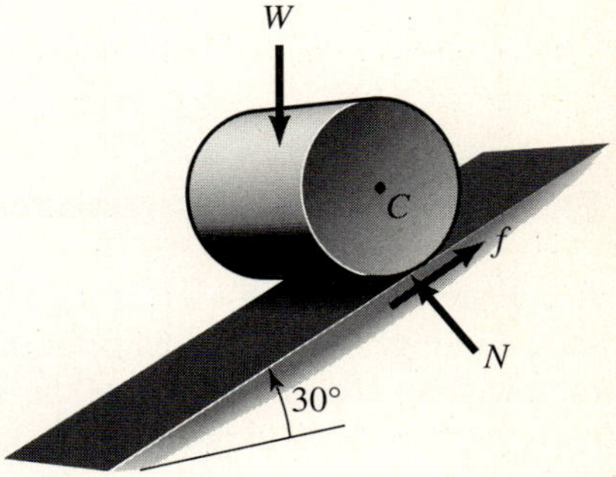

Figure 13.28. Cylinder on incline.

Before proceeding to the examples, let us consider for a moment the case of the cylinder rolling without slipping down an incline (see Fig. 13.28). We shall consider the cylinder as an *aggregate of particles* which form a rigid body—namely a cylinder. When using such an approach, we require that *all the forces both external and internal must move with their respective points of application*. Let us then consider the external work done on the particles making up the cylinder other than the work done by gravity. Clearly, only particles on the *rim* of the cylinder are acted on by external forces other than gravity. Consider one such particle during one rotation of the cylinder. This particle will have acting on it a friction force f and a normal force N at the *instant* when the particle is in *contact* with the inclined surface. The particle will have *zero* external force (except for gravity) at all other positions during the cycle. Now, at the instant of this contact, the normal force N has zero velocity in its direction because of the rigidity of the bodies. Therefore, N transmits no power and does no work on the particle during the cycle under consideration. Also, the friction

[13]This is in direct contrast to the work–energy equation for a system of particles wherein as was pointed out emphatically earlier, each external force moves with its *actual* point of application. Also, only external forces are involved for the center-of-mass approach, in contrast to the system of particles where internal forces may also be involved. Note that in Examples 13.1, 13.2, and 13.9 we were using a particle approach and thus were really considering the motion of the center of mass. The friction forces then moved with the center of mass.

force f acts on a particle having zero velocity at the instant of contact because of the *no slipping* condition. Accordingly, f transmits no power and does no work on the particle during this cycle.[14] This result must be true for each and every particle on the rim of the cylinder. Thus, clearly, f and N do no work when the cylinder rolls down the incline. Also, because of the rigidity of the body the internal forces do no work as pointed out earlier. Thus, only gravity does work.

However, in considering the motion of the *center of mass C* of the cylinder in Fig. 13.28, we note that force f now *moves* with C and hence *does* work.

At the risk of being repetitive, we now summarize the key features for properly using the center-of-mass approach.

1. Only **external** forces are involved.
2. Forces all move with the **center of mass** when computing work (see Eq. 13.37).
3. Only the kinetic energy of the center of mass is used.

[14]Another way of understanding this is to express work differently as was done in footnote 4. Once again, we multiply and divide the usual expression for work by dt as follows:

$$\int_1^2 \boldsymbol{F} \cdot d\boldsymbol{r} = \int_1^2 \boldsymbol{F} \cdot \frac{d\boldsymbol{r}}{dt}\, dt = \int_1^2 \boldsymbol{F} \cdot \boldsymbol{V}\, dt$$

We can now clearly see with $\boldsymbol{V} = \boldsymbol{0}$ at the point of contact that there will be zero work from the friction force f there.

Example 13.14

A cylinder with a mass of 25 kg is released from rest on an incline, as shown in Fig. 13.29. The diameter of the cylinder is 0.6 m. If the cylinder rolls without slipping, compute the speed of the centerline C after it has moved 1.6 m along the incline. Also, ascertain the friction force acting on the cylinder. Use the result from Problem 13.76 that the kinetic energy of a cylinder rotating about its own stationary axis is $\frac{1}{4}MR^2\omega^2$, where ω is the angular speed in rad/s.

In Fig. 13.29 we have shown the free body of the cylinder. We proceed to use the **work–energy equation** for a **system of particles**. Recall that we can concentrate the weight at the center of mass (Section 13.6). Accordingly, using the lowest position as a datum and noting from our earlier discussion that the friction force f does no work we have

$$\Delta(\text{PE} + \text{KE}) = \mathcal{W}_{1-2}$$

$$[0 - (25)(9.81)(1.6)\sin 30°] + \left\{\left[\tfrac{1}{2}(25)V_c^2 + \tfrac{1}{4}(25)(0.3)^2(\omega^2)\right] - 0\right\} = 0 \qquad \text{(a)}$$

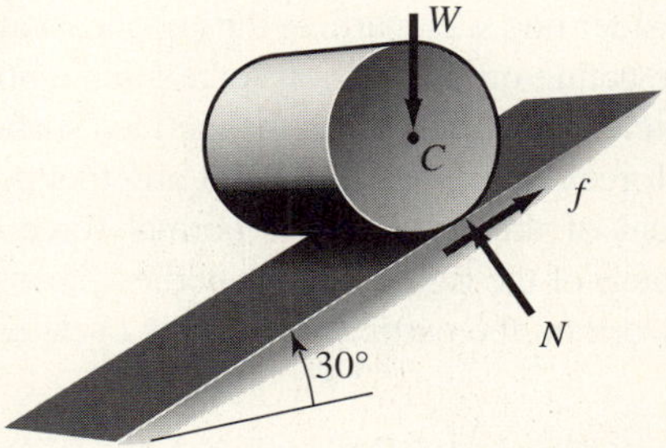

Figure 13.29. Free-body diagram of cylinder.

Example 13.14 (Continued)

where the kinetic energy of the cylinder is given as the kinetic energy of the mass taken at the mass center (straight line motion of C) plus kinetic energy of the cylinder relative to the center of mass (pure rotation about C). Noting from Example 13.13 that

$$\omega = \frac{V}{R} = \frac{V}{0.30}$$

we substitute into Eq. (a) and solve for V_c. We get

$$V_c = 3.23 \text{ m/s} \tag{b}$$

Now to find f, we *consider the motion of the mass center of the cylinder*. This means that we use Eq. 13.37 for the **center of mass**. Now *all external forces must move with the center of mass*; thus, f does work. Since the center of mass moves along a path always at right angles to N, this force still does no work. Accordingly, we can say:

$$-f(1.6) + W(1.6 \sin 30°) = \tfrac{1}{2}MV_c^2$$

$$-f(1.6) + (25)(9.81)(1.6) \sin 30° = \tfrac{1}{2}(25)(3.23^2)$$

$$f = 41.1 \text{ N}$$

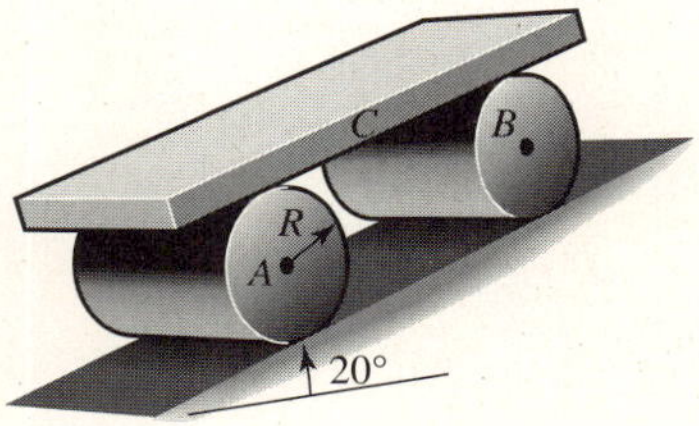

Figure 13.30. Three rigid bodies moving without slipping at any of the contact points.

Before going further, let us consider the two cylinders and the block in Fig. 13.30 as simply a **system of particles**. If there is no slipping between the block and the cylinders, the velocities of the particles on the block and the cylinders at the points of contact between these bodies have the *same* velocity at any time t. Furthermore, the friction force on the cylinder from the block is *equal* and *opposite* to the friction force on the block from the cylinder at the point of contact. We can then conclude that there is zero net work done by the friction forces between block and cylinders when considering them as an aggregate of particles.

Also, in the next problem, we will consider as a system of particles, rigid bodies which are joined by rigid connectors with frictionless interconnections.

Example 13.15

An external torque T of 50 N m is applied to a solid cylinder B (see Fig. 13.31), which has a mass of 30 kg and a radius of 0.2 m. The cylinder rolls without slipping. Block A, having a mass of 20 kg, is dragged up the 15° incline. The dynamic coefficient of friction μ_d between block A and the incline is 0.25. The connections at C and D are frictionless.

(a) What is the velocity of the system after moving a distance d of 2 m?

(b) What is the friction force on the cylinder?

Neglect the mass of the connecting rod.

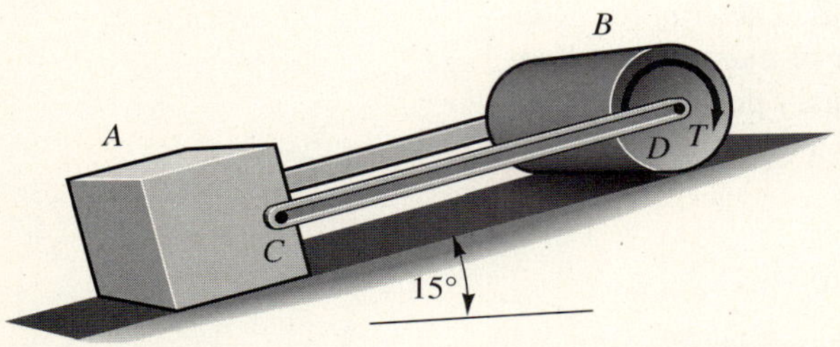

Figure 13.31. Torque-driven cylinder moves without slipping.

We show a free-body diagram of the system in Fig. 13.32. We begin by employing a **system of particles** point of view. Note there are pairs of internal forces present between the rod CD and body A at the contact point

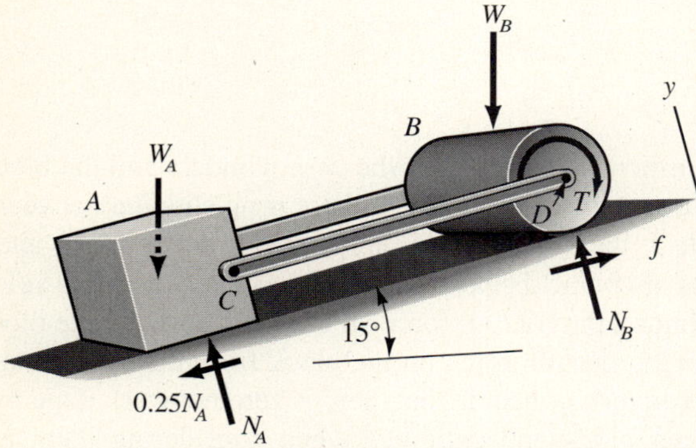

Data

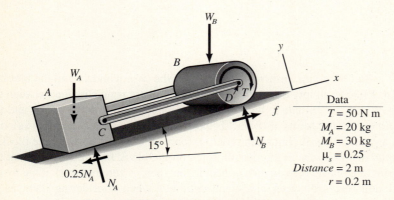

$T = 50$ N m
$M_A = 20$ kg
$M_B = 30$ kg
$\mu_s = 0.25$
Distance = 2 m
$r = 0.2$ m

Figure 13.32. Free-body diagram of the system.

Example 13.15 (Continued)

and similarly between CD and cylinder B. These force pairs are equal and opposite because of Newton's third law. And because the forces in each pair move exactly the same distance at the respective points of contact, there will be zero internal work from these force pairs. Hence, using the uppermost configuration as the datum,

$$\mathcal{W}_{1\to 2} = \Delta\text{PE} + \Delta\text{KE}$$

$$T\theta - (\mu_d N_A)(d) = \left[(W_A + W_B)(d)(\sin 15°) - 0\right] \qquad \text{(a)}$$

$$+\left[(M_A + M_B)\frac{V^2}{2} + \frac{1}{4}M_B r_B^2 \omega_B^2 - 0\right]$$

Note that as the cylinder moves without slipping a distance d along the incline, the circumference of the cylinder must come into contact with the incline along the very same distance d. Hence, by dividing d by the radius r, we get the rotation of the cylinder in radians associated with the movement of its center.

$$\theta = \frac{d}{r}$$

We then get for Eq. (a), on substituting data for the problem

$$(50)\left(\frac{2}{0.2}\right) - (0.25)[(20g)(\cos 15°)](2) = [(50g)(2)(0.2588)]$$

$$+\left[(50)\frac{V^2}{2} + \frac{1}{4}(30)(0.2^2)\frac{V^2}{(0.2)^2}\right]$$

$$V = 2.158 \text{ m/s}$$

Next, use the **center-of-mass** approach. We have on noting that a couple which is translating does no work. (Why?)

$$\int_1^2 \boldsymbol{F} \bullet d\boldsymbol{r}_C = \tfrac{1}{2} M_{\text{Total}}(V_2^2 - V_1^2)$$

$$-(0.25)(N_A)(d) - (W_A + W_B)(\sin 15°)(d) + fd = \tfrac{1}{2}(50)(2.158^2)$$

$$\therefore -(0.25)[(20g)(\cos 15°)](2) - (50g)(0.2588)(2) + f(2) = 116.4$$

$$f = 232.5 \text{ N}$$

Example 13.16

In Example 13.15, suppose that cylinder B is *slipping*. What is the dynamic coefficient of friction $(\mu_d)_B$ between the cylinder and the incline so that the system reaches a speed of 1.5 m/s after moving a distance $d = 2$ m starting from rest?

Using the **center-of-mass** approach, we have

$$\int_1^2 \boldsymbol{F} \cdot d\boldsymbol{r}_C = \left(\tfrac{1}{2}MV_C^2\right)_2 - \left(\tfrac{1}{2}MV_C^2\right)_1$$

$$-(W_A \cos 15°)(0.25)(d) + (W_B \cos 15°)(\mu_d)_B(d) - (W_A + W_B)(\sin 15°)(d) = \frac{1}{2}\frac{W_A + W_B}{g}V^2 - 0$$

$$-(20g)(\cos 15°)(0.25)(2) + (\mu_d)_B(30g)(\cos 15°)(2) - (50g)(\sin 15°)(2) = \frac{1}{2}(50)(1.5^2) - 0$$

$$(\mu_d)_B = 0.7122$$

In the next example, we shall consider a case where *internal forces* do work.

Example 13.17

A diesel powered electric train moves up a 7° grade in Fig. 13.33. If a torque of 750 N m is developed at each of its six pairs of drive wheels, what is the increase of speed of the train after it moves 100 m? Initially, the train has a speed of 4.4 m/s. The train weighs 90 kN. The drive wheels have a diameter of 600 mm. Neglect the rotational energy of the drive wheels.

Figure 13.33. Diesel–electric train.

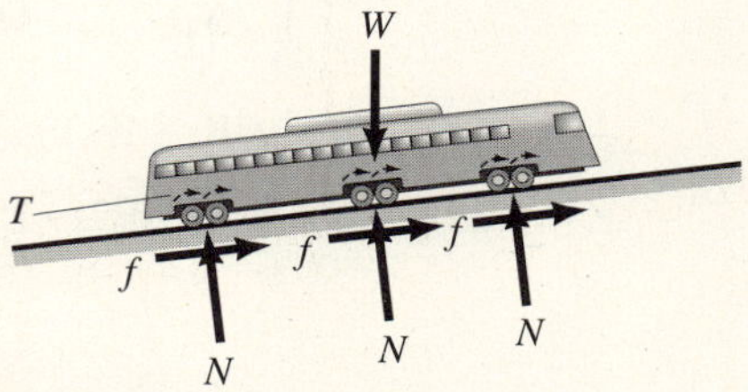

Figure 13.34. External and internal forces and torques.

We shall consider the train as a **system of particles** including the 6 pairs of wheels and the body. We have shown the train in Fig. 13.34 with the external forces, W, N, and f. In addition, we have shown certain internal

Example 13.17 (Continued)

torques T.[15] The torques shown act on the *rotors* of the motors, and, as the train moves, these torques rotate and accordingly do work. The *reactions* to these torques are equal and opposite to T according to Newton's third law and act on the *stators* of the motors (i.e., the field coils). The stators are stationary, and so the reactions to T do *no* work as the train moves. Thus we have an example wherein, using a **system of particles approach**, internal forces perform a nonzero amount of work. We now employ Eq. 13.29. Thus

$$\Delta \text{PE} + \text{KE} = \mathcal{W}_{1\to 2} \tag{a}$$

For the rolling without slipping condition, the friction forces f do no work. We then have

$$\left[(90\times 10^3)(100\sin 7^\circ - 0)\right] + \left\{\frac{1}{2}\frac{90\times 10^3}{g}V^2 - \frac{1}{2}\frac{90\times 10^3}{g}(4.4)^2\right\} = (6)(750)(\theta) \tag{b}$$

where θ is the clockwise rotation of the rotor in radians. Assuming direct drive from rotor to wheel, we can compute θ as follows for the 100-m distance d over which the train moves:

$$\theta = \frac{d}{r} = \frac{100}{0.3} = 333.3 \text{ rad} \tag{c}$$

Substituting into Eq. (b) and solving for V, we get

$$V = 10.36 \text{ m/s}$$

Hence, the increase of speed of the train is

$$\Delta V = 10.36 - 4.4 = \boxed{5.96 \text{ m/s}} \tag{d}$$

To determine the friction forces f, we now adopt a **center-of-mass** approach. Thus all forces now move with the center of mass. And they must be *external forces*. Accordingly we have

$$\left[6f - (90\times 10^3)(\sin 7^\circ)\right](100) = \frac{1}{2}\frac{90\times 10^3}{9.81}(10.36)^2 - \frac{1}{2}\frac{90\times 10^3}{9.81}(4.4)^2$$

$$\boxed{f = 2.50 \text{ kN}}$$

[15]Figure 13.34. accordingly, is *not* a free-body diagram.

PROBLEMS

13.68. A chain of total length L is released from rest on a smooth support as shown. Determine the velocity of the chain when the last link moves off the horizontal surface. In this problem, neglect the friction. Also, do not attempt to account for centrifugal effects stemming from the chain links rounding the corner.

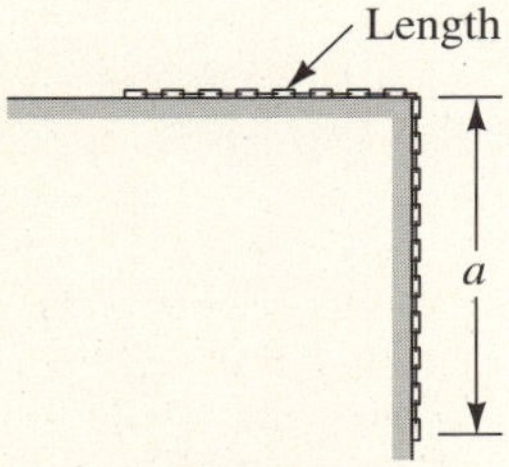

Figure P.13.68.

13.69. A chain is 15 m long and has a mass of 50 kg. A force P of 360 N has been applied at the configuration shown. What is the speed of the chain after force P has moved 3 m? The dynamic coefficient of friction between the chain and the supporting surface is 0.3. Utilize an approximate analysis.

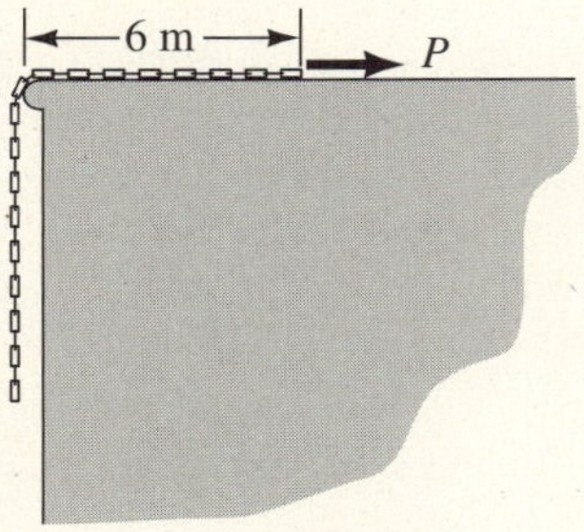

Figure P.13.69.

13.70. A bullet of weight W_1 N is fired into a block of wood weighing W_2 N. The bullet lodges in the wood, and both bodies then move to the dashed position indicated in the diagram before falling back. Compute the amount of internal work done during the action. Discuss the effects of this work. The bullet has a speed V_0 before hitting the block. Neglect the mass of the supporting rod and friction at A.

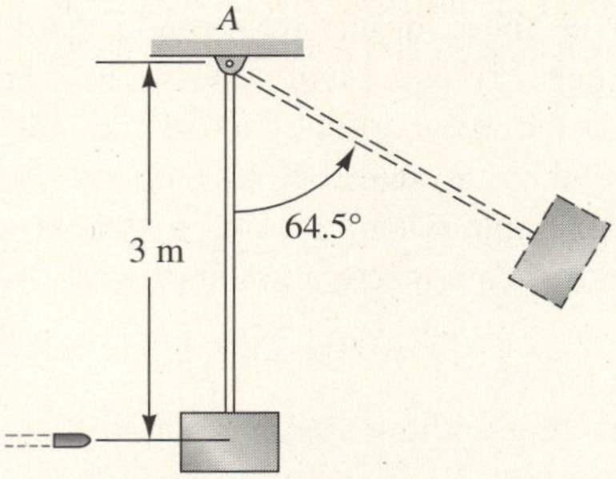

Figure P.13.70.

13.71. A device is mounted on a platform that is rotating with an angular speed of 10 rad/s. The device consists of two masses (each is 1.5 kg) rotating on a spindle with an angular speed of 5 rad/s relative to the platform. The masses are moving radially outward with a speed of 3 m/s, and the entire platform is being raised at a speed of 1.5 m/s. Compute the kinetic energy of the system of two particles when they are 1 ft from the spindle.

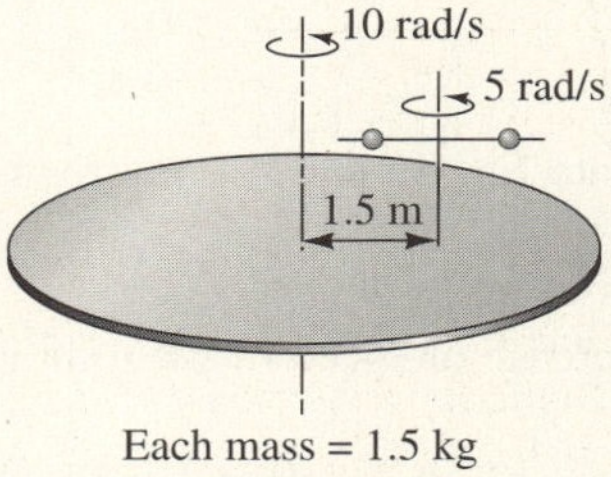

Figure P.13.71.

13.72. A hoop, with four spokes, rolls without slipping such that the center C moves at a speed V of 1.7 m/s. The diameter of the hoop is 3.3 m and the weight per unit length of the rim is 14 N/m. The spokes are uniform rods also having a weight per unit length of 14 N/m. Assume that rim and spokes are thin. What is the kinetic energy of the body?

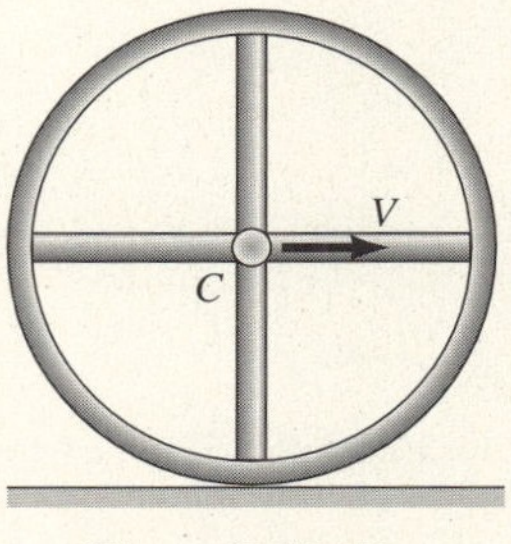

Figure P.13.72.

13.73. Three weights A, B, and C slide frictionlessly along the system of connected rods. The bodies are connected by a light, flexible, inextensible wire that is directed by frictionless small pulleys at E and F. If the system is released from rest, what is its speed after it has moved 300 mm? Employ the following data for the body masses:

Body A: 5 kg

Body B: 4 kg

Body C: 7.5 kg

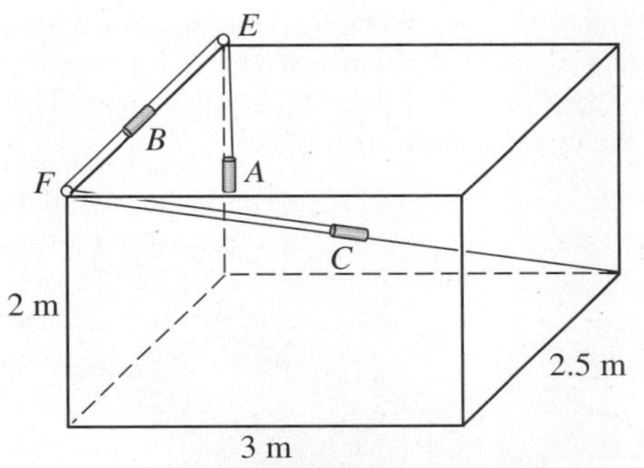

Figure P.13.73.

13.74. Bodies E and F slide in frictionless grooves. They are interconnected by a light, flexible, inextensible cable (not shown). What is the speed of the system after it has moved 0.6 m? The masses of bodies E and F are 5 kg and 10 kg, respectively. B is equidistant from A and C. E remains in top groove.

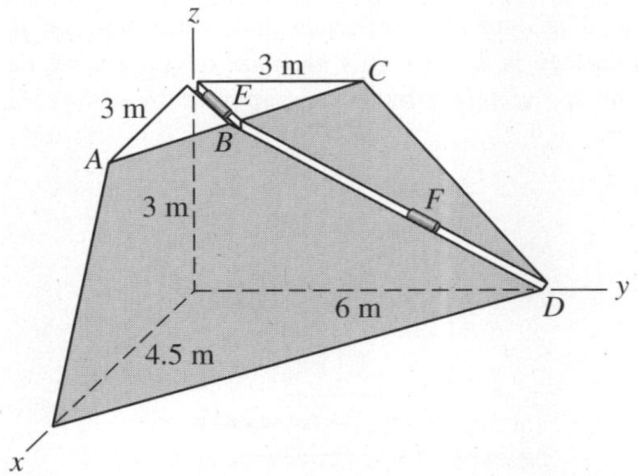

Figure P.13.74.

13.75. A tank is moving at the speed V of 4.4 m/s. What is the kinetic energy of each of the treads for this tank if they each have a mass per unit length of 300 kg/m?

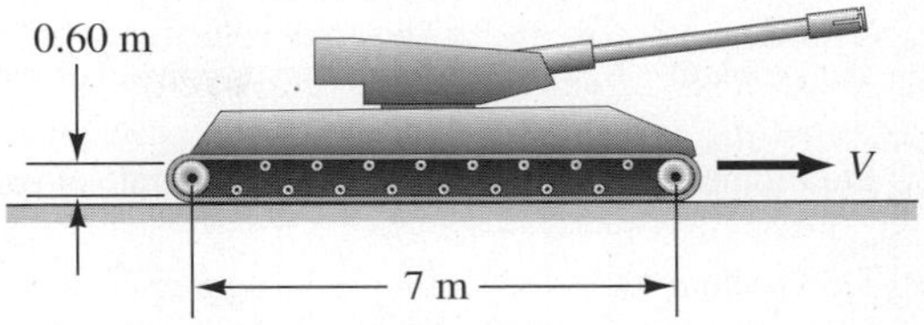

Figure P.13.75.

13.76. A cylinder of radius R rotates about its own axis with an angular speed of ω. If the total mass is M, show that the kinetic energy is $\frac{1}{4}MR^2\omega^2$.

13.77. Cylinders B and C each of mass 50 kg and have a diameter of 2 ft. Body A, mass 150 kg, rides on these cylinders. If there is no slipping anywhere, what is the kinetic energy of the system when the body A is moving at a speed V of 3 m/s? Use result of Problem 13.76.

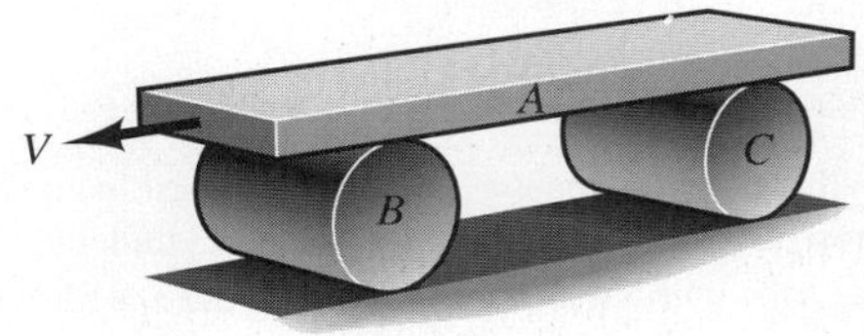

Figure P.13.77.

13.78. A pendulum has a bob with a comparatively large uniform disc of diameter 0.6 m and mass M of 1.5 kg. At the instant shown, the system has an angular speed θ of 0.3 rad/s. If we neglect the mass of the rod, what is the kinetic energy of the pendulum at this instant? What error is incurred if one considers the bob to be a particle as we have done earlier for smaller bobs? Use the result of Problem 13.76.

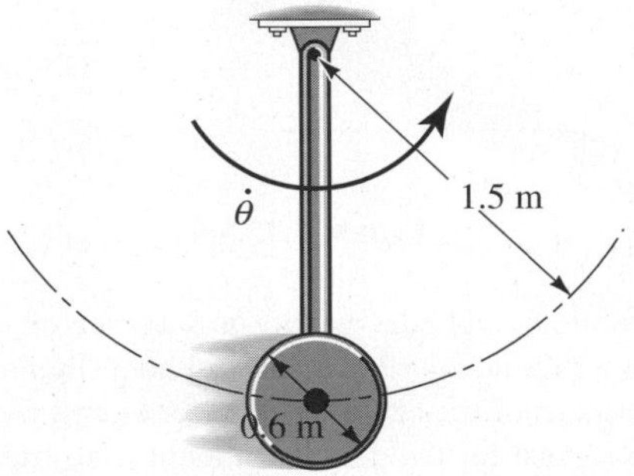

Figure P.13.78.

13.79. In Problem 13.78 compute the maximum angle that the pendulum rises.

13.80. Do Example 13.3 by treating as an aggregate of particles.

13.81. Do Problem 13.27 by treating as an aggregate of particles.

13.82. Do Problem 13.22 by treating as an aggregate of particles.

13.83. Do Problem 13.24 by treating as an aggregate of particles.

13.84. A constant force F is applied to the axis of a cylinder, as shown, causing the axis to increase its speed from 0.3 m/s to 0.9 m/s in 3 m without slipping. What is the friction force acting on the cylinder? The cylinder has a mass of 50 kg.

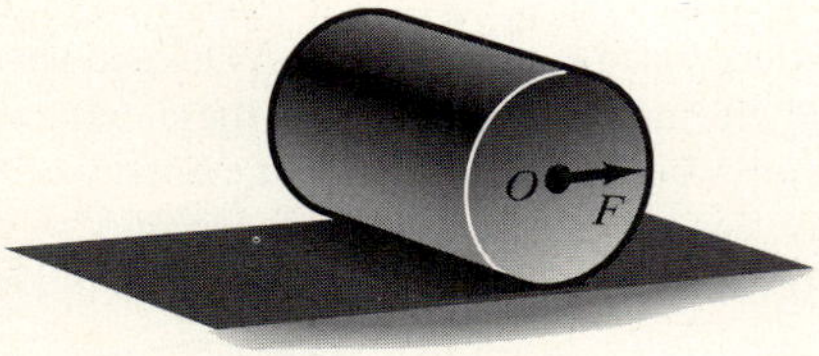

Figure P.13.84.

13.85. A cylinder with a mass of 25 kg is released from rest on an incline, as shown. The inner diameter D of the cylinder is 300 mm. If the cylinder rolls without slipping, compute the speed of the centerline O after the cylinder has moved 1.6 m along the incline. Ascertain the friction force acting on the cylinder. The radius of gyration k at O is $0.30/\sqrt{2}$ m.

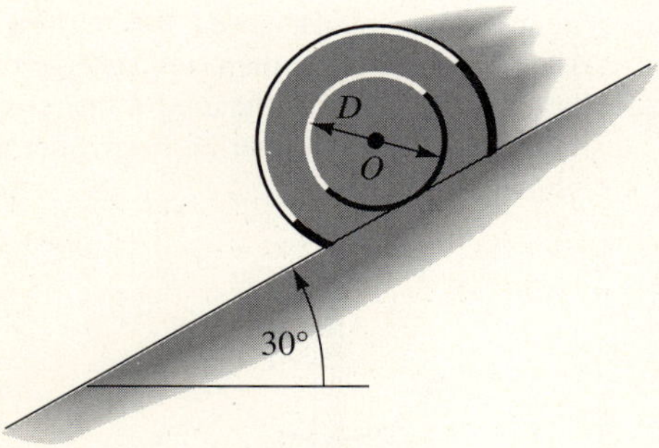

Figure P.13.85.

13.86. A uniform cylinder having a diameter of 0.6 m and a mass of 50 kg rolls down a 30° incline without slipping, as shown. What is the speed of the center after it has moved 6 m? Compare this result with that for the case when there is no friction present. [*Hint:* Use the result of Problem 13.76.]

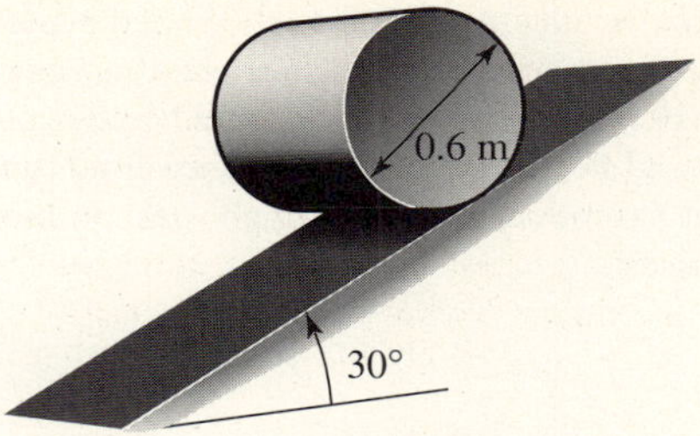

Figure P.13.86.

13.87. Cylinders A and B each have a mass of 25 kg and a diameter of 300 mm. Block C, riding on A and B, has a mass of 100 kg. If the system is released from rest at the configuration shown, what is the speed of C after the cylinders have made half a revolution? Use the result of Problem 13.76.

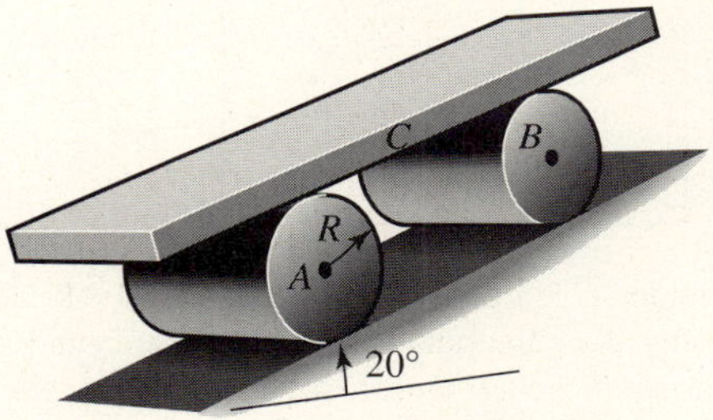

Figure P.13.87.

13.88. Shown are two identical blocks A and B, each of mass 25 kg. A force F of 450 N is applied to the lower block, causing it to move to the right. Block A, however, is restrained by the wall C. If block B reaches a speed of 3 m/s in 0.6 m starting from rest at the position shown in the diagram, what is the restraining force from the wall? The dynamic coefficient of friction between B and the ground surface is 0.3. Do this problem first by using Eq. 13.28. Then check the result by using separate free-body diagrams, and so on.

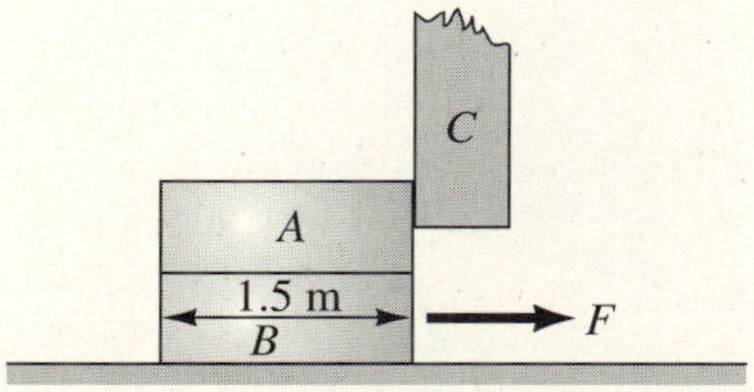

Figure P.13.88.

13.89. What is the tension T to accelerate the end of the cable downward at the rate of 1.5 m/s^2? From body C, weighing $50g$ N, is lowered a body D weighing $12.5g$ N at the rate of 1.5 m/s^2 relative to body C. Neglect the inertia of pulleys A and B and the cable. [*Hint:* From earlier courses in physics, recall that pulley B is rotating instantaneously about point e, and hence point b has an acceleration half that of point f. We will consider such relations carefully at a later time.]

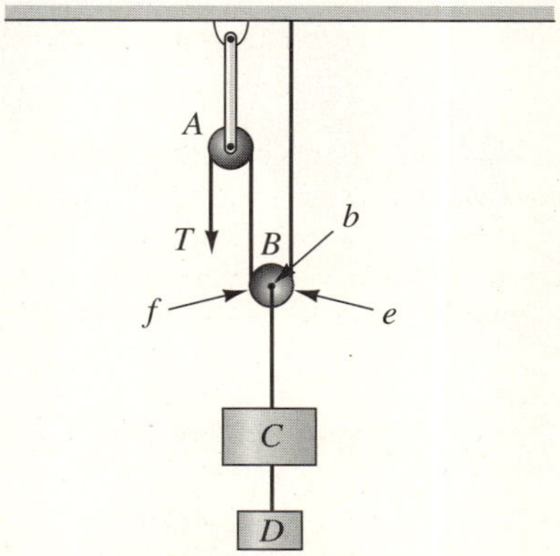

Figure P.13.89.

13.90. Cylinder C is connected by a light rod AB and can roll without slipping along the stationary cylinder D. Cylinder C weighs 30 N. A constant torque T = 20 N m is applied to AB when it is vertical and stationary. What is the angular speed of AB when it has rotated 90°°? The system of bodies is in the vertical plane. Recall from physics that a body which is rolling without slipping has instantaneous rotation about the point of contact.

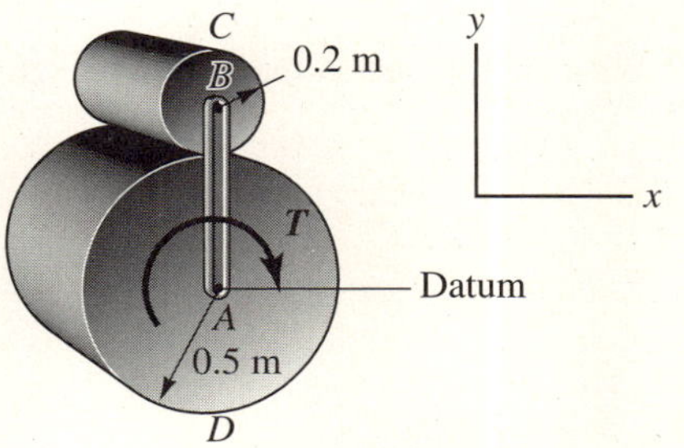

Figure P.13.90.

13.91. An 800 N force F is pulling the vehicle. The cylinders A and B each weigh 1 kN and roll without slipping. The indicated spring has a spring constant K equal to 50 N/mm and is compressed a distance δ of 20 mm. The pad slides on the upper guide with a dynamic coefficient of friction μ_d equal to 0.3. Neglect all masses except the cylinders', whose diameter D is 0.2 m.

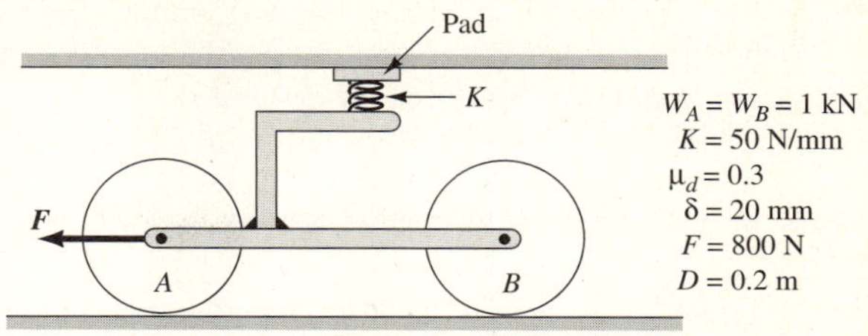

Figure P.13.91.

(a) What is the velocity of the vehicle after it moves a distance of 1.7 m starting from rest?

(b) What is the total friction force f on the cylinders from the ground?

13.92. A cylinder weighing 500 N rolls without slipping, first on a horizontal surface and then along a 30° incline.

(a) How far up the incline does it move?

(b) What are the friction forces on the cylinder along the horizontal surface and along the incline?

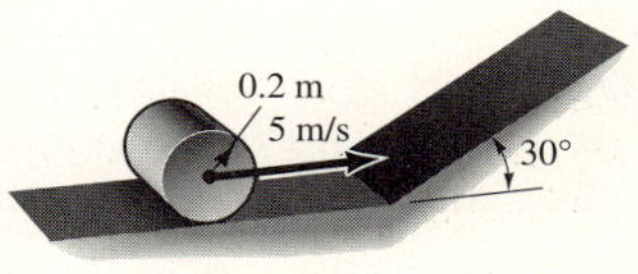

Figure P.13.92.

13.93. A hoop with four spokes is released from rest from a vertical position.

(a) What is the velocity of point C after it moves 1.3 m?

(b) What is the tension in the wire?

The rim and the spokes each have a weight per unit length of 15 N/m and are to be considered as thin. The wire is wrapped around the hoop and is the sole support.

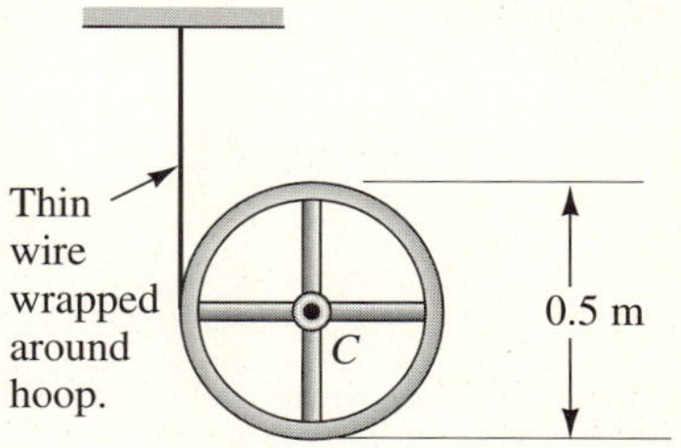

Figure P.13.93.

13.94. Three cylinders roll without slipping starting from rest. What is the speed of the system after moving 0.3 m? What is the *total* friction from the two walls on the system?

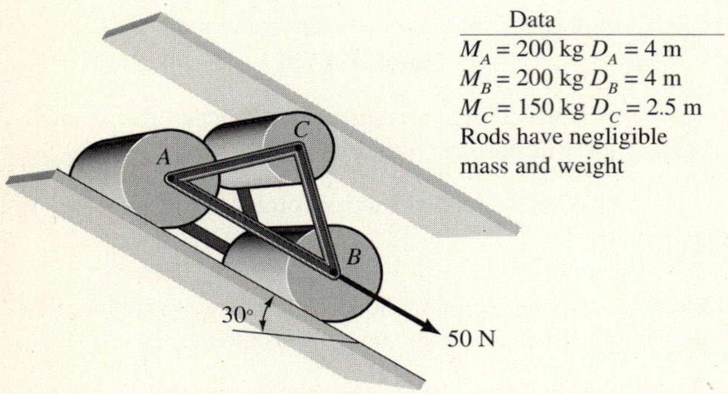

Figure P.13.94.

13.95. Find the velocity V_A after starting from rest and moving 5 m. What is the friction force from the incline on cylinder *A*? There is only rolling without slipping for the cylinders.

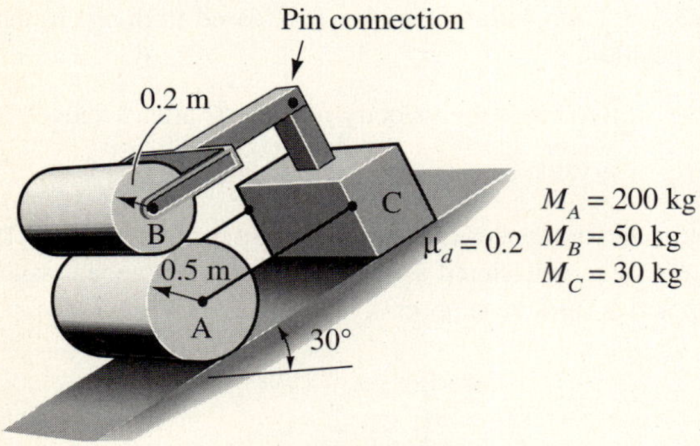

Figure P.13.95.

13.96. A cylinder is about to roll down an incline without slipping. It is connected to a linear spring. What is the angular speed of the cylinder after it rotates 20° starting from rest? The spring is originally unstretched.

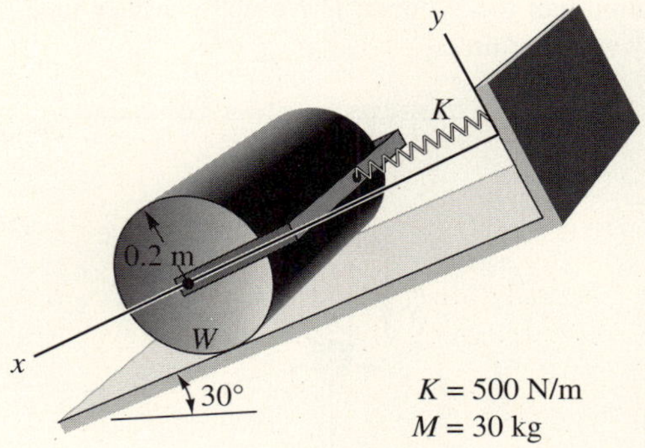

Figure P.13.96.

13.97. Three cylinders are connected together by light rods. Cylinders *A* have a mass of 5 kg each and cylinder *B* has a mass of 3 kg. If there is no slipping anywhere,

(a) What is the speed of the system after moving 0.8 m? The system starts from rest.

(b) What are the friction forces from the ground on each cylinder *A*?

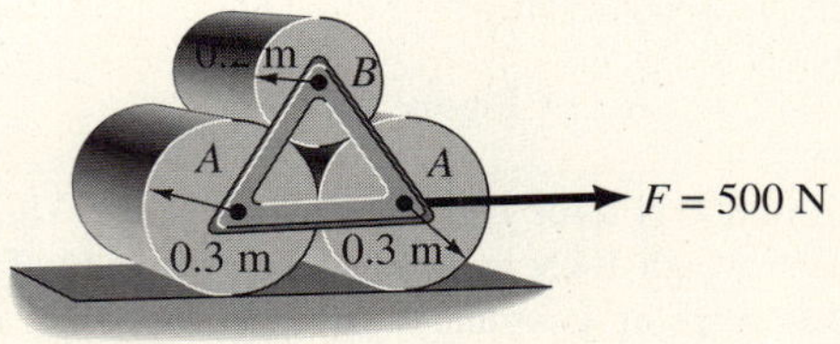

Figure P.13.97.

13.9 Closure

In this chapter, we presented the energy method as applied to particles. In Part A, we presented three forms of the energy equation applied to a *single* particle. The basic equation was

$$\int_1^2 \boldsymbol{F} \cdot d\boldsymbol{r} = \tfrac{1}{2}\left(MV^2\right)_2 - \tfrac{1}{2}\left(MV^2\right)_1 \qquad (13.38)$$

For the case of only conservative forces acting, we presented the equation for the *conservation of mechanical energy*:

$$(\text{PE})_1 + (\text{KE})_1 = (\text{PE})_2 + (\text{KE})_2 \qquad (13.39)$$

Finally, for both conservative and nonconservative forces, we presented an equation resembling the *first law of thermodynamics* as it is usually employed:

$$\Delta(\text{PE} + \text{KE}) = \mathcal{W}_{1-2} \qquad (13.40)$$

In Part B, we considered a *system of particles* and presented the above equation again, but this time the work and potential-energy terms are from both *internal* and *external* force systems.[16] Furthermore, all work and potential-energy terms are evaluated by using the *actual movement* of the points of application of internal and external forces.

Next, we presented the work–energy equation for the *center of mass* of any system of particles:

$$\int_1^2 \boldsymbol{F} \cdot d\boldsymbol{r}_c = \tfrac{1}{2}\left(MV_c^2\right)_2 - \tfrac{1}{2}\left(MV_c^2\right)_1 \qquad (13.41)$$

where $\boldsymbol{F}$, the resultant *external force*, *moves with the center of mass* in the computation of the work expression. We pointed out that the *single particle* model is a *special case* of the use of Eq. 13.41 applicable when the motion of the center of mass of a body sufficiently describes the motion of a body and where the external forces on the body move with the center of mass of the body.

To illustrate the use of the work–energy equation for a system of particles, we considered various elementary plane motions of simple rigid bodies. A more extensive treatment of the energy method applied to rigid bodies is found in Chapter 17.

We now turn to yet another useful set of relations derived from Newton's law, namely the methods of linear impulse-momentum and angular impulse-momentum for a particle and systems of particles.

[16]As will be seen in Chapter 14, this equation for a system of particles is the *only one* that involves internal forces. Note, however, that for a *rigid body* the internal forces *do no work*.

PROBLEMS

13.98. A tractor exerts a force of 3.6 kN on a block A, which has a dynamic coefficient of friction with block B of 0.7. Block B has a dynamic coefficient of friction of 0.2 with the ground. If block A has a mass of 200 kg and block B 300 kg, what is the speed of block A when, after starting from rest, the tractor has moved 0.6 m? What is the acceleration of block B?

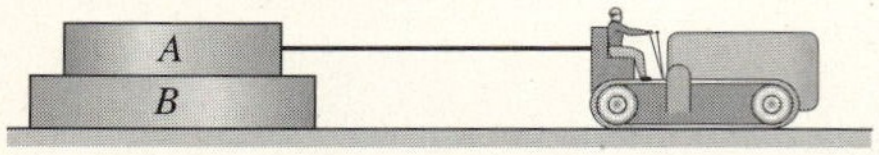

Figure P.13.98.

13.99. A body A is released from a condition of rest on a frictionless circular surface. The body then moves on a horizontal surface CD whose dynamic coefficient of friction with the body is 0.2. A spring having a spring constant $K = 900$ N/m is positioned at C as shown in the diagram. How much will the spring be compressed? The body has a mass of 5 kg.

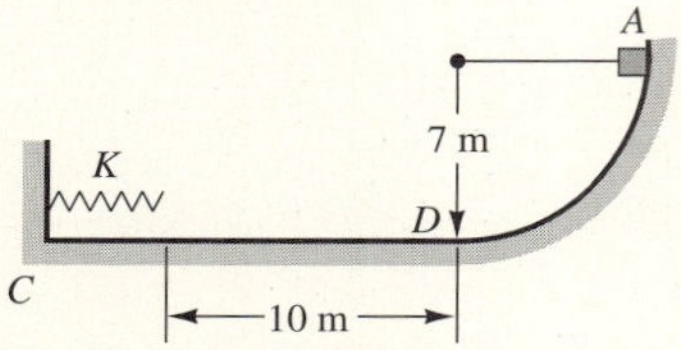

Figure P.13.99.

13.100. A cylinder is about to roll down an incline dragging block B. After starting from rest, what is the angular speed of the cylinder when it has moved 0.5 m? Use the following data:

$$(R_A)_{OUTSIDE} = 2.5 \text{ m} \qquad (R_A)_{INSIDE} = 1 \text{ m}$$
$$M_A = 100 \text{ kg} \qquad M_B = 30 \text{ kg}$$

The wire is thin and wraps around the inner cylinder of A. The kinetic energy of the compound cylinder due to rotation about its centerline is given as 0.8 times that of a solid cylinder of outside radius $r = 2.5$ m.

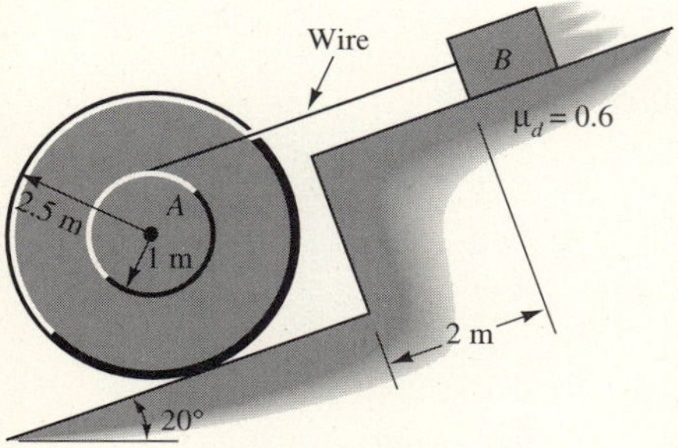

Figure P.13.100.

13.101. The cylinders in the system roll without slipping.

(a) What is the velocity of the system after it moves 1 m starting from rest?

(b) What is the total friction force f_{TOT} for the two cylinders?

(c) What is the acceleration of the system?

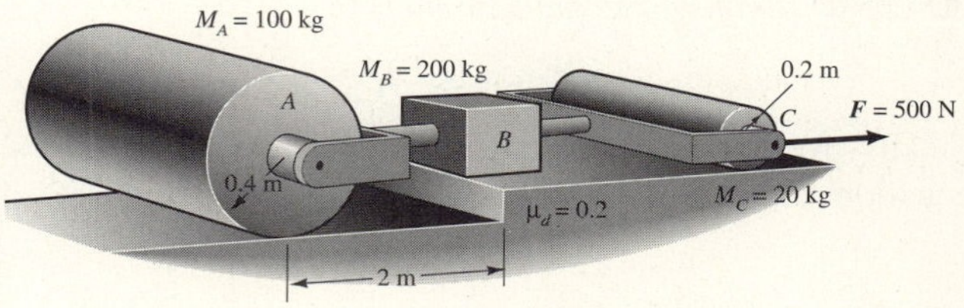

Figure P.13.101.

13.102. A 80 kg man runs up an escalator while it is not in operation in 10 s. What is the power developed by the man? If the escalator is moving at a speed of 0.6 m/s and carrying, on the average, 2000 people per hour, what is the power requirement on the driving motor assuming that the average mass of a passenger is 68 kg? Take the mechanical efficiency of the drive system to be 80%. Assume that passengers enter and leave at the same speed of 0.6 m/s and that there are equal numbers of passengers on the escalator at any one time.

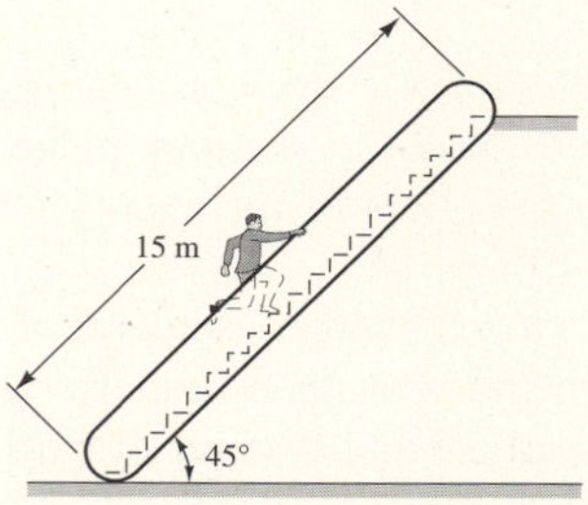

Figure P.13.102.

13.103. Grain is coming out of a hopper at the rate of 2 kg/s and falls onto a conveyor system that takes the grain into a bin. The conveyor belt moves at a steady speed of 2 m/s. What power is needed to operate the system for an efficiency of 0.6? What power is needed if we double the belt speed?

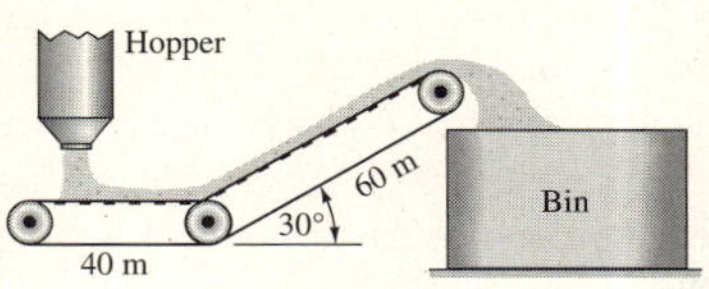

Figure P.13.103.

13.104. A self-propelled vehicle A has a mass of 220 kg. A gasoline engine develops torque on the drive wheels to help move A up the incline. A counterweight B of 1.3 kN is also shown in the diagram. What power is needed when A is moving up at a speed of 0.6 m/s and has an acceleration of 1 m/s^2? Neglect the weight of the pulley. [*Hint:* The pulley rolls along cord dg without slipping. It therefore has an instantaneous center of rotation at d. What does this mean about the relative value of velocity of point b on the pulley and point a?]

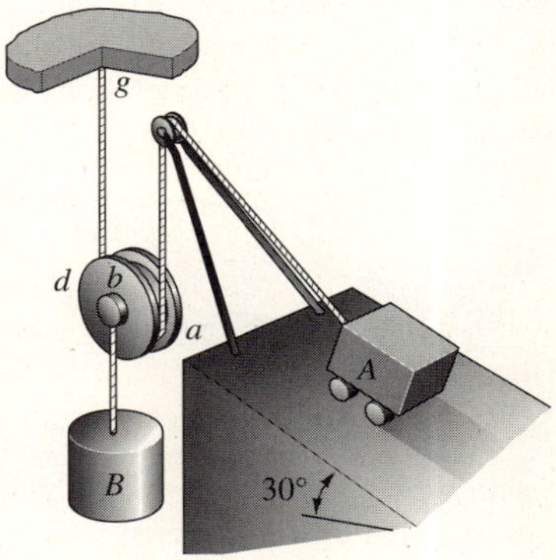

Figure P.13.104.

***13.105.** Set up an integro-differential equation (involving derivatives and integrals) for θ in Problem 13.31 if there is Coulombic friction with $\mu_d = 0.2$.

13.106. At what angle θ does body A of Problem 13.31 leave the circular surface?

***13.107.** Show that the work–energy equation for a particle can be expressed in the following way:

$$\int_0^x F\,dx = \int_0^V V\,d(mV)$$

Integrating the right side by parts,[17] and using relativistic mass $m_0/\sqrt{1 - V^2/c^2}$, where m_0 is the *rest mass* and c is the speed of light, show that a relativistic form of this equation can be given as

$$\int_0^x F\,dx = \frac{m_0 c^2}{\sqrt{1 - V^2/c^2}} - m_0 c^2 = mc^2 - m_0 c^2$$

so that the *relativistic kinetic energy* is

$$\text{KE} = mc^2 - m_0 c^2$$

***13.108.** By combining the kinetic energy as given in Problem 13.107 and $m_0 c^2$ to form E, the total energy, we get the famous formula of Einstein:

$$E = mc^2$$

in which energy is equated with mass. How much energy is equivalent to 27 μg of matter? How high could a weight of 500 N be lifted with such energy?

13.109. A 45 kg boy climbs up a rope in gym in 10 s and slides down in 4 s after he reaches uniform speed downward. What is the power developed by the boy going up? What is the average power dissipated on the rope by the boy going down after reaching uniform speed? The distance moved before reaching uniform speed downward is 0.6 m.

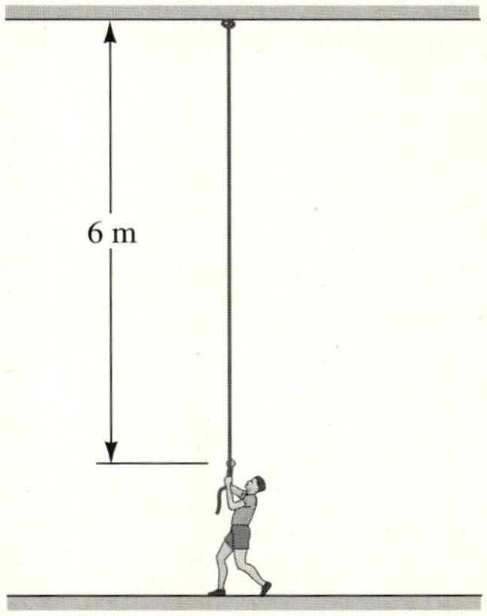

Figure P.13.109.

[17]To integrate by parts, note that

$$d(uv) = u\,dv + v\,du$$

Now integrate these terms:

$$\int_1^2 d(uv) = \int_1^2 u\,dv + \int_1^2 v\,du$$

Therefore,

$$\int_1^2 u\,dv = (uv)\Big|_1^2 - \int_1^2 v\,du$$

The last formulation is called integration by parts.

13.110. An aircraft carrier is shown in the process of launching an airplane via a catapult mechanism. Before leaving the catapult, the plane has a speed of 52 m/s relative to the ship. If the plane is accelerating at the rate of $1g$ and if it has a mass of 18 Mg, what power is being developed by the catapult system at the end of launch on the plane if we neglect drag? The thrust from the jet engines of the plane is 100 kN.

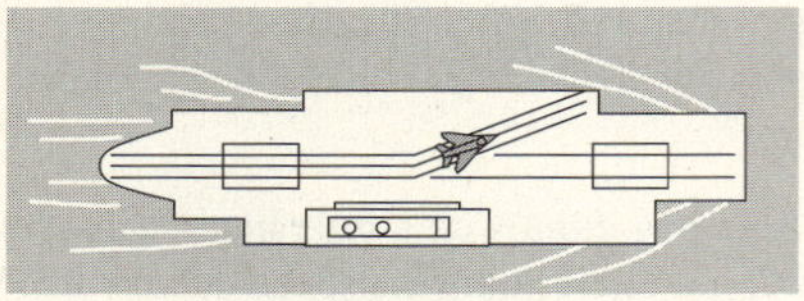

Figure P.13.110.

13.111. Vehicle B, weighing 25 kN, is to go down a 30° incline. The vehicle is connected to body A through light pulleys and a capstan. What should body A weigh if starting from rest it restricts body B to a speed of 5 m/s when B moves 3 m? There are two wraps of rope around the capstan.

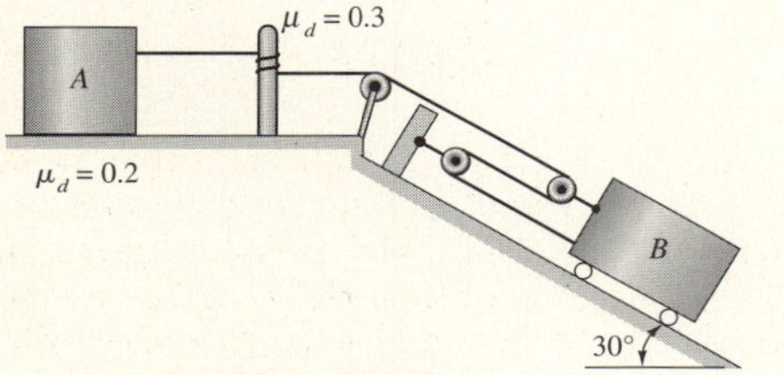

Figure P.13.111.

13.112. A jet passenger plane is moving along the runway for a takeoff. If each of its four engines is developing 45 kN of thrust, what is the power developed when the plane is moving at a speed of 65 m/s?

13.113. Block B, with a mass of 200 kg, is being pulled up an incline. A motor C pulls on one cable, developing 3 kN. The other cable is connected to a counterweight A having a mass of 150 kg. If B is moving at a speed of 2 m/s, what is its acceleration? [*Hint:* Start with Newton's law for A and B.]

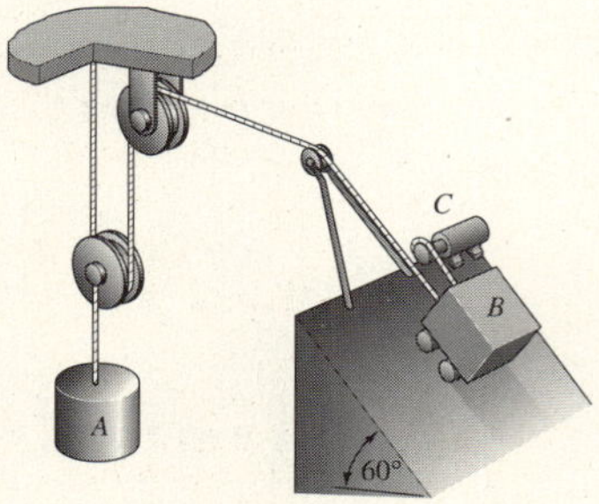

Figure P.13.113.

13.114. A block G slides along a frictionless path as shown. What is the minimum initial speed that G should have along the path if it is to remain in contact when it gets to A, the uppermost position of the path? The block weighs 9 N. What is the normal force on the path when for the condition described the block is at position B?

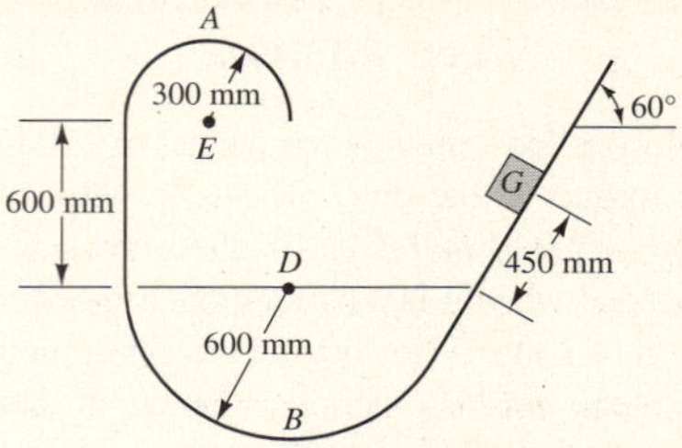

Figure P.13.114.

13.115. Cylinders A and B have masses of 50 kg each. Cylinder A can only rotate about a stationary axis while cylinder B rolls without slipping. Block C has a mass of 100 kg. Starting from rest, what is the speed of C after moving 0.1 m? Force P is 500 N and the diameter of the cylinders is 0.2 m.

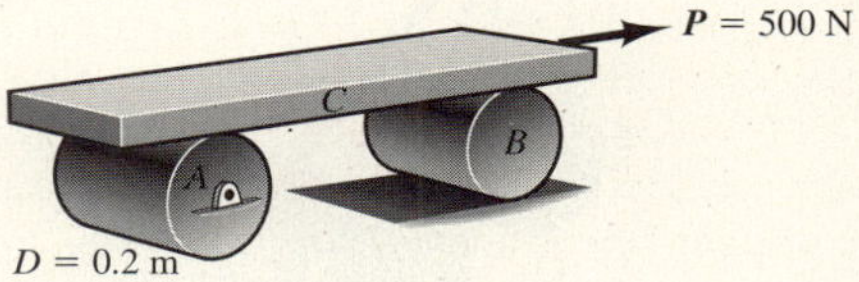

Figure P.13.115.

13.116. A system of 4 solid cylinders and a heavy block move vertically downward aided by a 1 kN force F. What is the angular speed of the wheels after the system descends 0.5 m after starting from rest? What is the friction force from the walls on each wheel? The wheels roll without slipping.

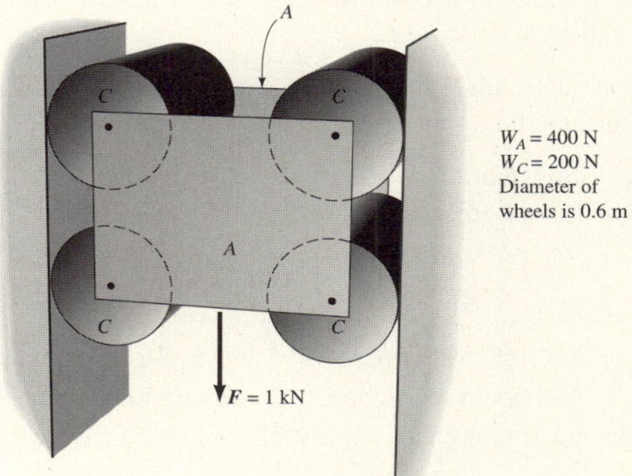

Figure P.13.116.

13.117. A collar B having a mass of 100 g moves along a frictionless curved rod in a vertical plane. A light rubber band connects B to a fixed point A. The rubber band is 250 mm in length when unstretched. A force of 30 N is required to extend the band 50 mm. If the collar is released from rest, what distance must d be so that the downward normal force on the rod at C is 20 N?

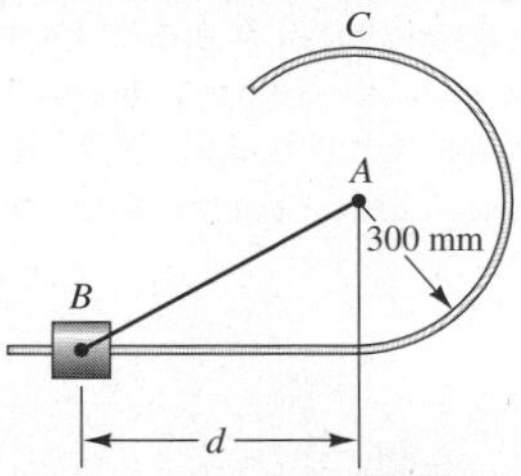

Figure P.13.117.

13.118. When your author was a graduate student he built a system for examining the effects of high-speed moving loads over elastically supported beams (see the diagram). A "vehicle" slides along a slightly lubricated square tube guide. At the base of the vehicle is a spring-loaded light wheel which will run over the beam (not shown). The vehicle is catapulted to a high speed by a stretched elastic cord (shock cord) which is pulled back from position A–A to the position B shown prior to "firing." At A–A the shock cord is elongated 250 mm, while at the firing position it is elongated 750 mm. A force of 1.8 N is required for each mm of elongation of the cord. If the cord has a total mass of 0.7 kg and the vehicle mass is 0.3 kg, what is the speed of the vehicle when the cord reaches A–A after firing? Take into account in some reasonable way the kinetic energy of the cord, but neglect friction.

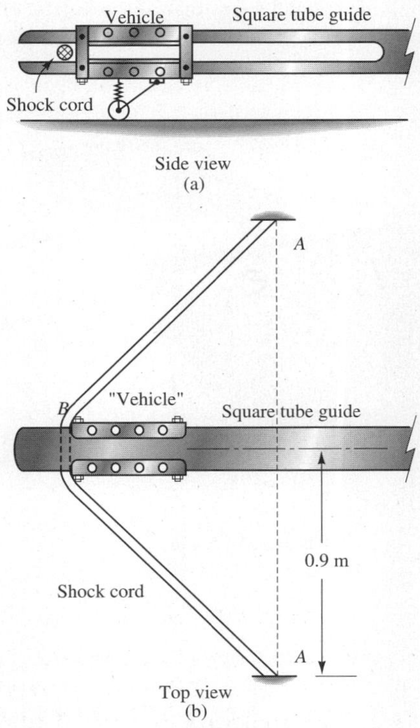

Figure P.13.118.

13.119. A body B of mass 60 kg slides in a frictionless slot on an inclined surface as shown. An elastic cord connects B to A. The cord has a "spring constant" of 360 N/m. If the body B is released from rest from a position where the elastic cord is unstretched, what is body B's speed after it moves 0.3 m?

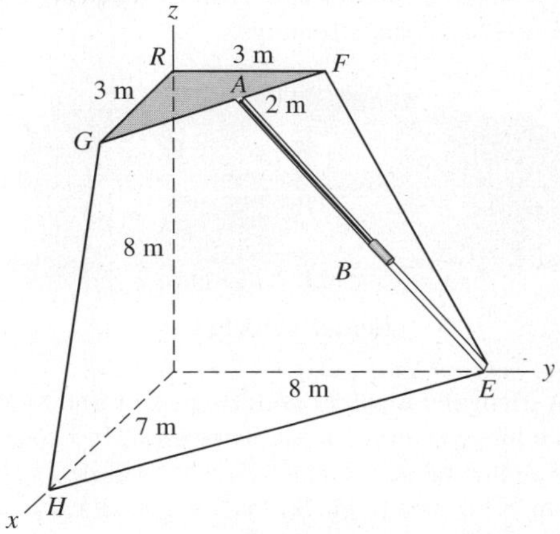

Figure P.13.119.

13.120. A collar slides on a frictionless tube as shown. The spring is unstretched when in the horizontal position and has a spring constant of 180 N/m. What is the minimum mass of A to just reach A' when released from rest from the position shown in the diagram? What is the force on the tube when A has traveled half the distance to A'?

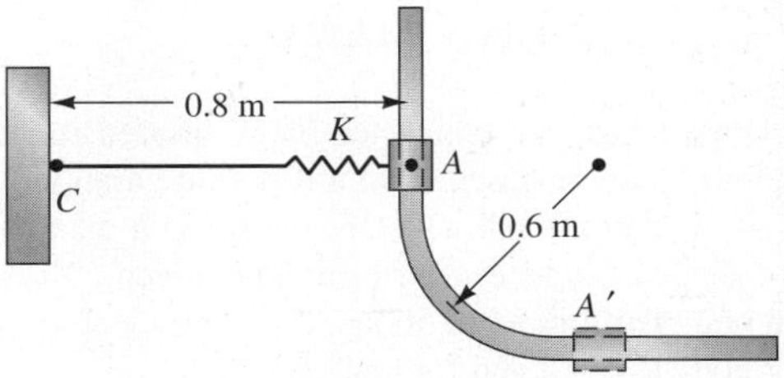

Figure P.13.120.

13.121. A 15 kg vehicle has two bodies (each with mass 1 kg) mounted on it, and these bodies rotate at an angular speed of 50 rad/s relative to the vehicle. If a 500 N force acts on the vehicle for a distance of 17 m, what is the kinetic energy of the system, assuming that the vehicle starts from rest and the bodies in the vehicle have constant rotational speed? Neglect friction and the inertia of the wheels.

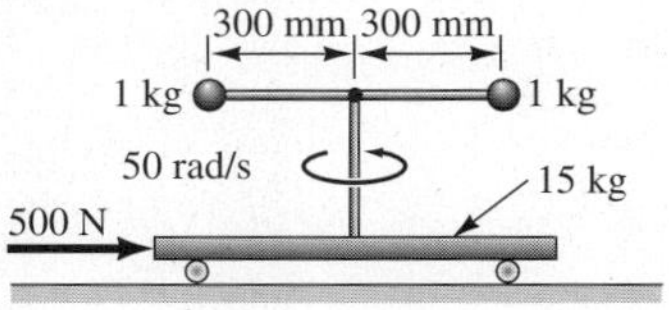

Figure P.13.121.

13.122. Two identical solid cylinders each weighing 100 N support a load *A* weighing 50 N. If a force *F* of 300 N acts as shown, what is the speed of the vehicle after moving 5 m? Also, what is the total friction force on each wheel? Neglect the mass of the supporting system connecting the cylinders. Note that the kinetic energy of the angular motion of a cylinder about its own axis is $\frac{1}{4}MR^2\omega^2$. The system starts from rest.

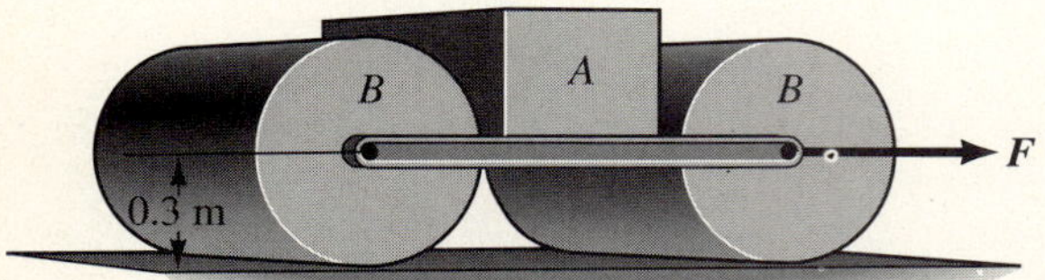

Figure P.13.122.

13.123. A triangular block of uniform density and total mass 50 kg rests on a hinge and on a movable block *B*. If a constant force *F* of 650 N is exerted on the block *B*, what will be its speed after it moves 3 m? The mass of block *B* is 5 kg, and the dynamic coefficient of friction for all contact surfaces is 0.3.

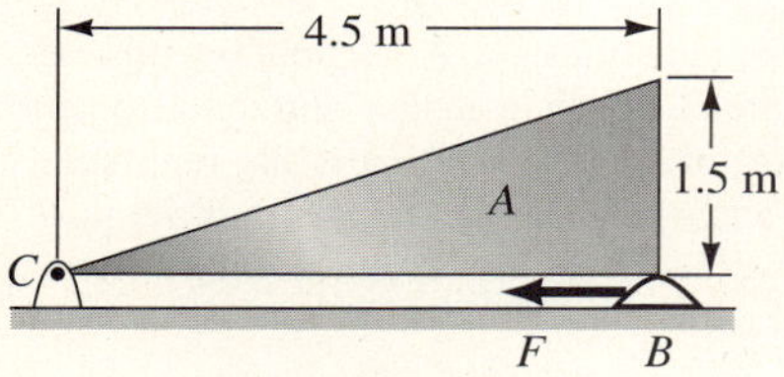

Figure P.13.123.

13.124. Three blocks are connected by an inextensible flexible cable. The blocks are released from a rest configuration with the cable taut. If A can only fall a distance equal to *h* m, what is the velocity of bodies C and B after each has moved a distance of 1 m? Each body has a mass of 50 kg. The coefficient of dynamic friction for body C is 0.3 and for body B is 0.2.

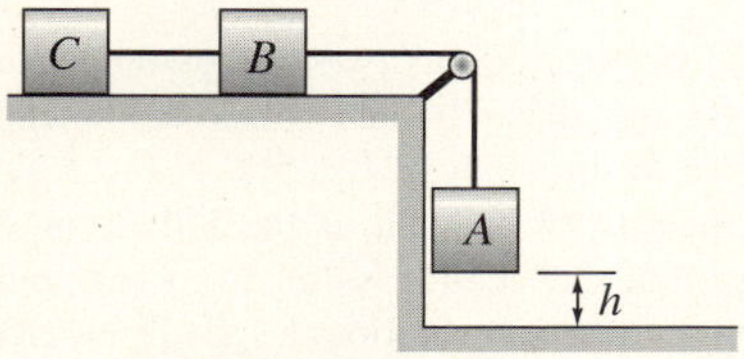

Figure P.13.124.

13.125. Two discs move on a horizontal frictionless surface shown looking down from above. Each disc weighs 20 N. A rectangular member B weighing 50 N is pulled by a force F having a value of 200 N. If there is no slipping anywhere except on the horizontal support surface, what is the speed of B after it moves 180 mm? Determine the friction forces from the walls onto the cylinders.

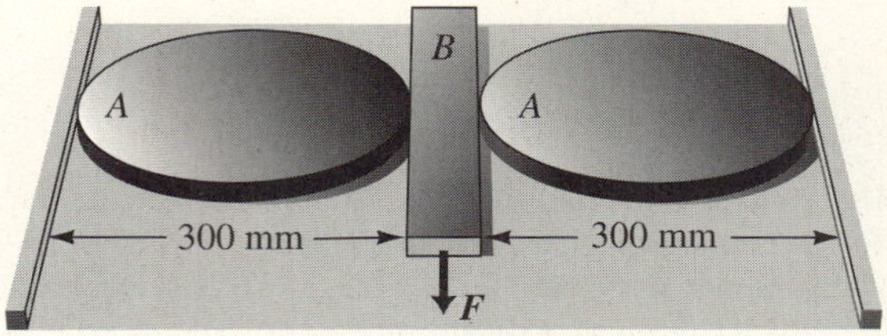

Figure P.13.125.

13.126. Rod *AB* is pinned to block *C* and is welded to cylinder *D*. Cylinder *E* rolls without slipping along the incline and rotates around cylinder *D*, which does not rotate at all. There is a constant friction torque between *D* and *E* of 25 N m. Starting from rest, what is the speed of the system after moving 0.1 m along the incline? What is the frictional force between the ground and cylinder *E*? Neglect the mass of rod *AB*. Take the kinetic energy of rotation of *E* as 0.8 times that of the kinetic energy of rotation of a solid cylinder of diameter D = 0.3 m. The length of *AB* is 0.5 m.

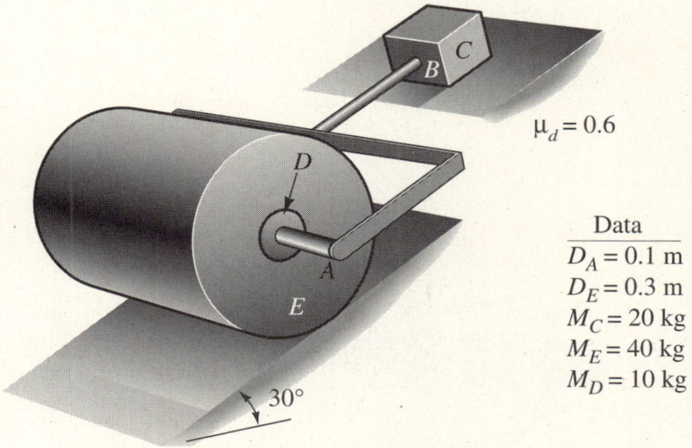

Figure P.13.126.

CHAPTER 14

Methods of Momentum for Particles

Part A: Linear Momentum

14.1 Impulse and Momentum Relations for a Particle

In Section 12.3, we integrated differential equations of motion for particles that are acted upon by forces that are functions of time. In this chapter, we shall again consider such problems and shall present alternative formulations, called *methods of momentum*, for handling certain of these problems in a convenient and straightforward manner. We start by considering Newton's law for a particle:

$$\boldsymbol{F} = m\frac{d\boldsymbol{V}}{dt} \tag{14.1}$$

Multiply both sides by dt and integrate from some initial time t_i to some final time t_f:

$$\int_{t_i}^{t_f} \boldsymbol{F}\,dt = \int_{t_i}^{t_f} m\frac{d\boldsymbol{V}}{dt}\,dt = m\boldsymbol{V}_f - m\boldsymbol{V}_i \tag{14.2}$$

Note first that this is a vector equation, in contrast to the work–kinetic energy equation 13.2. The integral

$$\int_{t_i}^{t_f} \boldsymbol{F}\,dt$$

which we shall denote as $\boldsymbol{I}$, is called the *impulse* of the force $\boldsymbol{F}$ during the time interval $t_f - t_i$, whereas $m\boldsymbol{V}$ is the *linear-momentum vector* of the particle.

Equation 14.2, then, states that *the impulse **I** over a time interval equals the change in linear momentum of a particle during that time interval.* As we shall demonstrate later, the impulse of a force may be known even though the force itself is not known.

Finally, you must remember that to produce an impulse, a force need only exist for a time interval. Sometimes we use the work integral so much that we tend to think—erroneously—that a force acting on a stationary body does not produce an impulse.

We now illustrate the use of the impulse-momentum equation.

Example 14.1

A particle initially at rest is acted on by a force whose variation with time is shown graphically in Fig. 14.1. If the particle has a mass of 15 kg and is constrained to move rectilinearly in the direction of the force, what is the speed after 15 s?

From the definition of the impulse, the area under the force–time curve will, in the one-dimensional example, equal the impulse magnitude. Thus, we simply compute this area between the times $t = 0$ and $t = 15$ s:

$$\text{impulse} = \underbrace{\tfrac{1}{2}(10)(50)}_{\text{area 1}} + \underbrace{(5)(75)}_{\text{area 2}} = 625 \text{ N s}$$

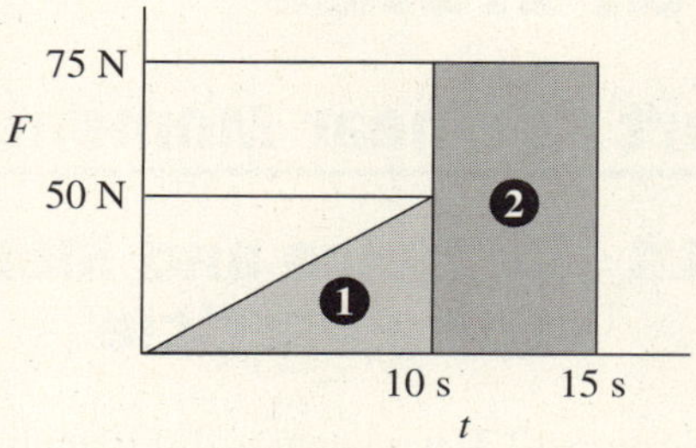

Figure 14.1. Force-versus-time plot.

The final velocity, then, is given as

$$625 = (15)(V_f) - 0$$

Therefore,

$$V_f = 41.7 \text{ m/s}$$

Note that the impulse-momentum equation is useful when the force variation during a time interval is a curve that cannot be conveniently expressed mathematically. The impulse, which is the area under an F versus t curve, can then be found with the help of a *planimeter*, thus permitting a quick solution of the velocity change during the time interval.[1]

[1]A planimeter is a mechanical device for measuring the area of a plane region bounded by an arbitrary curve.

Example 14.2

A particle A with a mass of 1 kg has an initial velocity $\boldsymbol{V}_0 = 10\boldsymbol{i} + 6\boldsymbol{j}$ m/s. After particle A strikes particle B, the velocity becomes $\boldsymbol{V} = 16\boldsymbol{i} - 3\boldsymbol{j} + 4\boldsymbol{k}$ m/s. If the time of encounter is 10 ms, what average force was exerted on the particle A? What is the change of linear momentum of particle B?

The impulse $\boldsymbol{I}$ acting on A is immediately determined by computing the change in linear momentum during the encounter:

$$\begin{aligned} \boldsymbol{I}_A &= (1)(16\boldsymbol{i} - 3\boldsymbol{j} + 4\boldsymbol{k}) - (1)(10\boldsymbol{i} + 6\boldsymbol{j}) \\ &= 6\boldsymbol{i} - 9\boldsymbol{j} + 4\boldsymbol{k} \text{ N s} \end{aligned}$$

Since

$$\int_{t_i}^{t_f} \boldsymbol{F}_A \, dt = (\boldsymbol{F}_{\text{av}})_A \Delta t$$

the average force $(\boldsymbol{F}_{\text{av}})_A$ becomes

$$(\boldsymbol{F}_{\text{av}})_A(0.010) = 6\boldsymbol{i} - 9\boldsymbol{j} + 4\boldsymbol{k}$$

Therefore,

$$(\boldsymbol{F}_{\text{av}})_A = 600\boldsymbol{i} - 900\boldsymbol{j} + 400\boldsymbol{k} \text{ N}$$

On the basis of the principle that action equals reaction, an equal but opposite average force must act on the object B during the 10 ms time interval. Thus, the impulse on particle B is $-\boldsymbol{I}_A$. Equating this impulse to the change in linear momentum, we get

$$\Delta(m\boldsymbol{V})_B = -\boldsymbol{I}_A = -6\boldsymbol{i} + 9\boldsymbol{j} - 4\boldsymbol{k} \text{ N s}$$

During impacts where the exact force variation is unknown, the impulse momentum principle is very useful. We shall examine impacts in more detail in a later section.

Example 14.3

Two bodies, 1 and 2, are connected by an inextensible and weightless cord (Fig. 14.2). Initially, the bodies are at rest. If the dynamic coefficient of friction is μ_d for body 1 on the surface inclined at angle α, compute the velocity of the bodies at any time t before body 1 has reached the end of the incline.

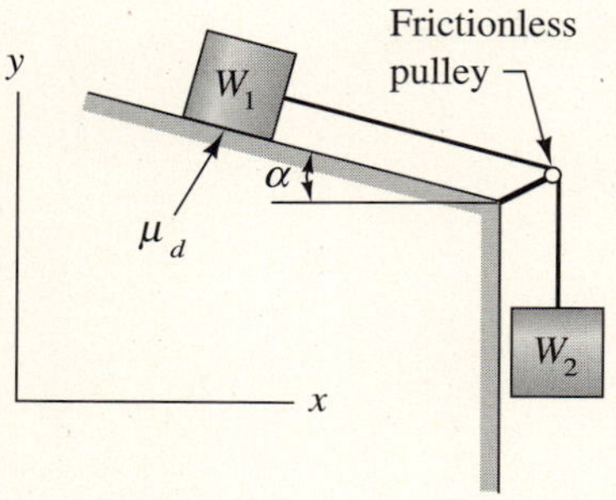

Figure 14.2. Two bodies connected by a cord.

Since only constant forces exist and since a time interval has been specified, we can use momentum considerations advantageously. The free-body diagrams of bodies 1 and 2 are shown in Fig. 14.3. Equilibrium considerations lead to the conclusion that $N_1 = W_1 \cos \alpha$, so the friction force f_1 is

$$f_1 = \mu_d N_1 = \mu_d W_1 \cos \alpha$$

For body 1, take the component of the **linear impulse-momentum** equation along the incline:

$$\int_0^t (-\mu_d W_1 \cos \alpha + W_1 \sin \alpha + T)dt = \frac{W_1}{g}(V - 0)$$

Carrying out the integration, we have

$$(-\mu_d W_1 \cos \alpha + W_1 \sin \alpha + T)t = \frac{W_1}{g} V \qquad \text{(a)}$$

For body 2, we have for the **momentum equation** in the vertical direction:

$$\int_0^t (W_2 - T)dt = \frac{W_2}{g}(V - 0)$$

where, because of the inextensible property of the cable and the frictionless condition of the pulley, the magnitudes of the velocity V and the force T are the same for bodies 1 and 2. Integrating the equation above, we write:

$$(W_2 - T)t = \frac{W_2}{g} V \qquad \text{(b)}$$

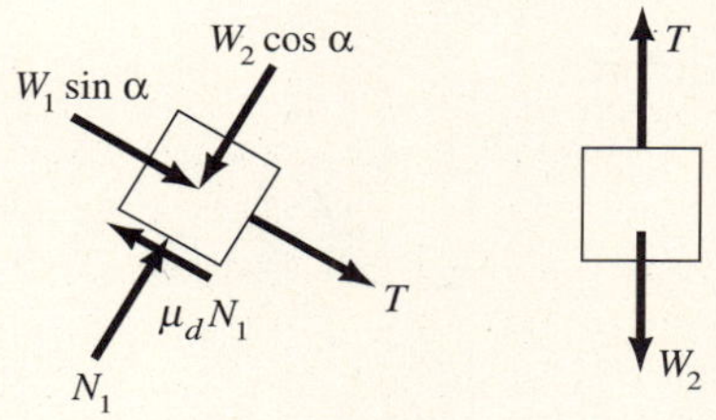

Figure 14.3. Free-body diagrams of W_1 and W_2.

Example 14.3 (Continued)

By adding Eqs. (a) and (b), we can eliminate T and solve for the desired unknown V. Thus,

$$(-\mu_d W_1 \cos\alpha + W_1 \sin\alpha + W_2)t = \frac{V}{g}(W_1 + W_2)$$

Therefore,

$$V = \frac{gt}{W_1 + W_2}(W_2 + W_1 \sin\alpha - \mu_d W_1 \cos\alpha) \qquad \text{(c)}$$

Note that we have used considerations of linear momentum for a *single* particle each time in solving this problem.

Example 14.4

A conveyor belt is moving from left to right at a constant speed V of 0.3 m/s in Fig. 14.4. Two hoppers drop objects onto the belt at the total rate n of 4 per second. The objects each have a weight W of 9 N and fall a height h of 0.3 m before landing on the conveyor belt. Farther along the belt (not shown) the objects are removed by personnel so that, for steady-state operation, the number N of objects on the belt at any time is 10. If the dynamic coefficient of friction between belt and conveyor bed is 0.2, estimate the average difference in tension $T_2 - T_1$ of the belt to maintain this operation. The weight of the belt on the conveyor bed is 45 N.

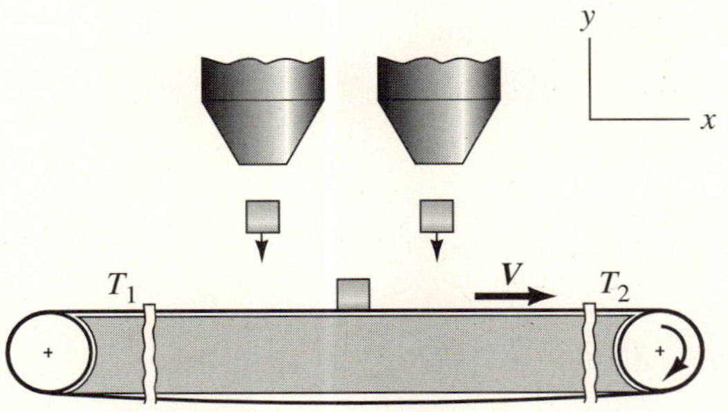

Figure 14.4. Objects falling on moving conveyor.

Example 14.4 (Continued)

We shall superimpose the following effects to get the desired result.

1. A friction force from the bed onto the belt results from the static weight of the ten objects riding on the belt and the weight of the portion of belt on the bed.
2. A friction force from the bed onto the belt results from the force in the y direction needed to change the *vertical* linear momentum of the falling objects from a value corresponding to the free-fall velocity just before impact ($\sqrt{2gh}$) to a value of zero after impact.
3. Finally, the belt must supply a force in the x direction to change the *horizontal* linear momentum of the falling objects from a value of zero to a value corresponding to the speed of the belt.

Thus, we have for the first contribution, which we donate as ΔT_1, the following result:

$$\Delta T_1 = (NW + 45)\mu_d = [(10)(9) + 45](0.2) = 27 \text{ N} \qquad \text{(a)}$$

As for the second contribution, we can only compute an average value $(\Delta T_2)_{av}$ by noting that each impacting object is given a vertical change in linear momentum equal to

$$\begin{aligned}\text{vertical change in linear momentum per object} &= \frac{W}{g}(\sqrt{2gh}) \\ &= \frac{9}{g}\sqrt{(2g)(0.3)} \\ &= 2.226 \text{ kg m/s}\end{aligned}$$

where we have assumed a free fall starting with zero velocity at the hopper. For four impacts per second, we have as the total vertical change in linear momentum per second the value $(4)(2.226) = 8.904$ kg m/s. The average vertical force during the 1 s interval to give the impulse needed for this change in linear momentum is clearly 8.904 N. Since this result is correct for every second, 8.904 N is the average normal force that the bed of the conveyor must transmit to the belt for arresting the vertical motion of the falling objects. The desired $[(\Delta T_2]_{av}$ for the belt arising from friction is accordingly given as

$$[(\Delta T)_2]_{av} = (\mu_d)(8.904) = 1.781 \text{ N} \qquad \text{(b)}$$

Finally, for the last contribution $[(\Delta T)_3]_{av}$, we note that the belt must give in the horizontal direction for each impacting object a change in linear momentum having the value

$$\begin{aligned}\text{horizontal change in linear momentum per object} &= \frac{W}{g}(0.3) \\ &= 0.275 \text{ kg m/s}\end{aligned}$$

Example 14.4 (Continued)

For four impacts per second we have as the total horizontal change in linear momentum developed by the belt during 1 s the value (4)(0.275) = 1.10 kg m/s. The average horizontal force during 1 s needed for this change in linear momentum is clearly 1.10 N. Thus, we have

$$[(\Delta T)_3]_{av} = 1.10 \text{ N} \tag{c}$$

The total average difference in tension is then

$$(\Delta T)_{av} = 27 + 1.781 + 1.10 = 29.88 \text{ N} \tag{d}$$

14.2 Linear-Momentum Considerations for a System of Particles

In Section 14.1, we considered impulse-momentum relations for a single particle. Although Examples 14.3 and 14.4 involved more than one particle, nevertheless the impulse-momentum considerations were made on one particle at a time. We now wish to set forth impulse-momentum relations for a *system* of particles.

Let us accordingly consider a system of n particles. We may start with Newton's law as developed previously for a system of particles:

$$\boldsymbol{F} = \sum_{j=1}^{n} m_j \frac{d\boldsymbol{V}_j}{dt} \tag{14.3}$$

Since we know that the internal forces cancel, $\boldsymbol{F}$ must be the *total external* force on the system of n particles. Multiplying by dt, as before, and integrating between t_i and t_f, we write:

$$\int_{t_i}^{t_f} \boldsymbol{F}\, dt = \boldsymbol{I}_{\text{ext}} = \left(\sum_{j=1}^{n} m_j \boldsymbol{V}_j\right)_f - \left(\sum_{j=1}^{n} m_j \boldsymbol{V}_j\right)_i \tag{14.4}$$

Thus, we see that *the impulse of the total external force on the system of particles during a time interval equals the sum of the changes of the linear-momentum vectors of the particles during the time interval.*

We now consider an example.

Example 14.5

A 27 kN truck is moving at a speed of 30 m/s. [See Fig. 14.5(a).] The driver suddenly applies his brakes at time $t = 0$ so as to lock his wheels in a panic stop. Load A weighing 9 kN breaks loose from its ropes and at time $t = 4$ s is sliding *relative to the truck* at a speed of 0.9 m/s. What is the speed of the truck at that time? Take μ_d between the tires and pavement to be 0.4.

Since we *do not know* the nature of the forces between the truck and load A while the latter is breaking loose, it is easiest to consider the *system* of two particles comprising the truck and the load simultaneously whereby the aforementioned forces become *internal* and are *not* considered. Accordingly, we have shown the system with all the external loads in Fig. 14.5(b). Clearly, $N = (27 + 9) = 36$ kN and the friction force is $(0.4)(36) = 14.4$ kN. We now employ Eq. 14.4 in the x direction as follows:

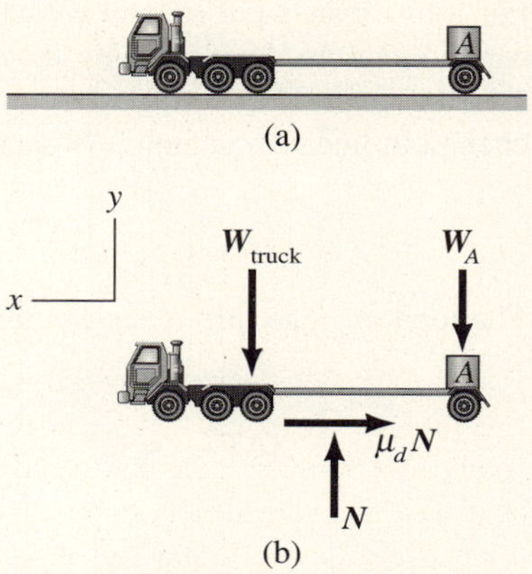

Figure 14.5. Truck undergoing panic stop.

$$\int_{t_1}^{t_2} F_x\,dt = \left(\sum_j m_j \boldsymbol{V}_j\right)_2 - \left(\sum_j m_j \boldsymbol{V}_j\right)_1$$

$$\int_0^4 (-14.4\times 10^3)\,dt = \left[\frac{(27\times 10^3)}{g}V_2 + \frac{(9\times 10^3)}{g}(V_2 + 0.9)\right] - \left[\frac{(36\times 10^3)}{g}(30)\right] \qquad \text{(a)}$$

Note that the first quantity inside the first brackets on the right side of Eq. (a) is the momentum of the truck at $t = 4$ s, and the second quantity inside the same brackets is the momentum of the load at this instant. We may readily solve for V_2:

$$V_2 = 14.1 \text{ m/s}$$

Introducing *mass-center* quantities into Eq. 14.4 is easy and sometimes advantageous. You will remember that:

$$M\boldsymbol{r}_c = \sum_{j=1}^{n} m_j \boldsymbol{r}_j \qquad (14.5)$$

Differentiating with respect to time, we get

$$M\boldsymbol{V}_c = \sum_{j=1}^{n} m_j \boldsymbol{V}_j \qquad (14.6)$$

Thus, we see from this equation that *the total linear momentum of a system of particles equals the linear momentum of a particle that has the total mass of the system and that moves with the velocity of the mass center.* Using Eq. 14.6 to replace the right side of Eq. 14.4, we can say:

$$\int_{t_i}^{t_f} \mathbf{F}\, dt = \mathbf{I}_{\text{ext}} = M(\mathbf{V}_c)_f - M(\mathbf{V}_c)_i \qquad (14.7)$$

Thus, *the total external impulse on a system of particles equals the change in linear momentum of a hypothetical particle having the mass of the entire aggregate and moving with the mass center.*

When the separate motions of the individual particles are reasonably simple, as a result of constraints, and the motion of the mass center is not easily available, then Eq. 14.4 can be employed for linear-momentum considerations as was the case for Example 14.5. On the other hand, when the motions of the particles individually are very complex and the motion of the mass center of the system is reasonably simple, then clearly Eq. 14.7 can be of great value for linear-momentum considerations. Also, as in the case of energy considerations, we note that the single-particle model is really a *special case* of the center-of-mass formulation above, wherein the motion of the center of mass of a body describes sufficiently the motion of the body in question.

Example 14.6

A truck in Fig. 14.6 has two rectangular compartments of identical size for the purpose of transporting water. Each compartment has the dimensions 6 m × 3 m × 2.4 m. Initially, tank A is full and tank B is empty. A pump in tank A begins to pump water from A to B at the rate Q_1 of 0.3 m^3/s and 10 s later is delivering water at the rate Q_2 of 0.9 m^3/s. If the level of the water in the tanks remains horizontal, what is the average horizontal force needed to restrain the truck from moving during this interval?

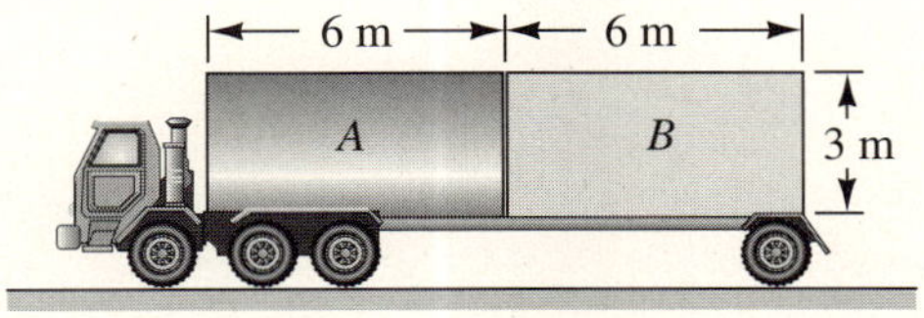

Figure 14.6. Truck with tank compartments.

Example 14.6 (Continued)

In this setup, the mass center of the water in the tanks is moving from left to right and moving *nonuniformly* during the time interval of interest. We show the water in Fig. 14.7 at some time t where the level in

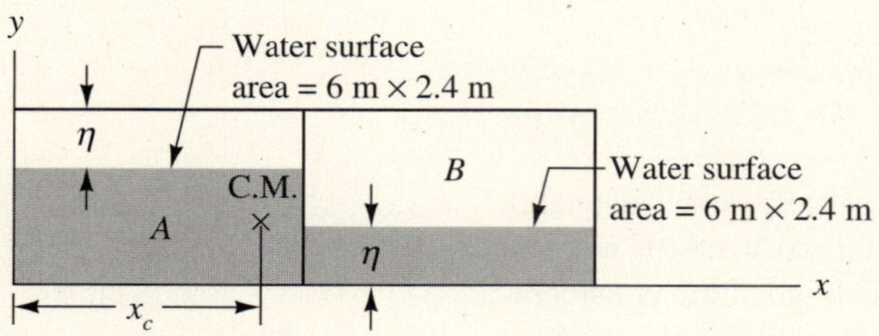

Figure 14.7. Compartments showing flow of water.

tank A has dropped an amount η while, by conservation of mass, the level in tank B has risen exactly the same amount η. The position x_c of the center of mass at this instant can be readily calculated in terms of η. Thus, using the basic definition of the center of mass, we can say:

$$Mx_c = (M_A)(x_A) + (M_B)(x_B)$$

$$[(6)(2.4)(3)](\rho)(x_c) = [(6)(2.4)(3 - \eta)](\rho)(3) + [(6)(2.4)(\eta)](\rho)(9) \quad \text{(a)}$$

Since we are interested in the time rate of change of x_c so that we can profitably employ Eq. 14.7, we next differentiate with respect to time as follows:

$$[(6)(2.4)(3)](\rho)(\dot{x}_c) = -[(6)(2.4)\dot{\eta}](\rho)(3) + [(6)(2.4)(\dot{\eta})](\rho)(9) \quad \text{(b)}$$

But $(6)(2.4)\dot{\eta}$ is the volume of flow[2] from tank A to tank B at time t. Using Q to represent this volume flow, we get for the equation above:

$$[(6)(2.4)(3)](\rho)\dot{x}_c = -(\rho)(3)Q + (\rho)(9)Q = (6)(\rho)Q$$

Solving for $\dot{x}_c$, we have

$$\dot{x}_c = \frac{1}{7.2}Q \quad \text{(c)}$$

[2]Remember that 6 m × 2.4 m is the area of the top water surface in each tank, as shown in Fig. 14.7.

Example 14.6 (Continued)

Now consider the momentum equation in the x direction for the water using the center of mass. We can say from Eq. 14.7:

$$\int_0^{10} F\,dt = \left[(M\dot{x}_c)_2 - (M\dot{x}_c)_1\right]$$

Therefore,

$$(F_{av})(10) = [(6)(2.4)(3)(\rho)][(\dot{x}_c)_2 - (\dot{x}_c)_1]$$
$$= [(6)(2.4)(3)(\rho)]\left[\frac{1}{7.2}(Q_2 - Q_1)\right] \quad \text{(d)}$$

where we have used Eq. (c) in the last step. Putting in $Q_2 = 0.9\ \text{m}^3/\text{s}$ and $Q_1 = 0.3\ \text{m}^3/\text{s}$, we then get for the average force during the 10 s interval of interest on using $\rho = 1\ \text{Mg/m}^3$:

$$F_{av} = 360\ \text{N} \quad \text{(e)}$$

This is the average horizontal force that the truck exerts on the water. Clearly, this force is also what the ground must exert on the truck in the horizontal direction to prevent motion of the truck during the water transfer operation.

From another viewpoint, this system is not unlike a propulsion system like a jet engine to be studied with the aid of a control volume (see Section 5.4) in your fluids course.

If the total external force on a system of particles is zero, it is clear from the previous discussion that there can be no change in the linear momentum of the system. This is the principle of *conservation of linear momentum*, which means, furthermore, that *with a zero total impulse on an aggregate of particles, there can be no change in the velocity of the mass center*. If at some time t_0 the velocity of the mass center of such a system of particles is zero, then this velocity must remain zero if the impulse on the system of particles is zero. That is, no matter what movements and gyrations the elements of the system may have, they must be such that the center of mass must remain stationary. We reached the same conclusion in Chapter 12, where we found from Newton's law that if the total external force on a system of particles is zero, then the acceleration of the center of mass is zero.[3]

[3]Problems 12.103 and 12.104 are examples of this condition.

14.3 Impulsive Forces

Let us now examine the action involved in the explosion of a bomb that is initially suspended from a wire, as shown in Fig. 14.8. First, consider the situation *directly after* the explosion has been set off. Since very large forces are present from expanding gases, a *fragment* of the bomb receives an appreciable impulse during this short time interval. Also, directly after the explosion, the gravitational forces are no longer counteracted by the supporting wire, so

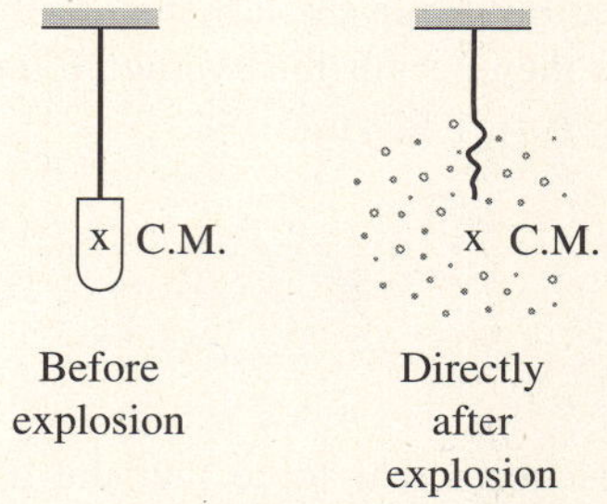

Figure 14.8. Exploding bomb.

there is an additional impulse acting on the fragment. But since the gravitational force is small compared to forces from the explosion, the gravitational impulse on a fragment can be considered negligibly small for the short period of time under discussion compared to that of the expanding gases acting on the fragment. A plot of the impulsive force (from the explosion) and the force of gravity on a fragment is shown in Fig. 14.9. It is clear from this diagram

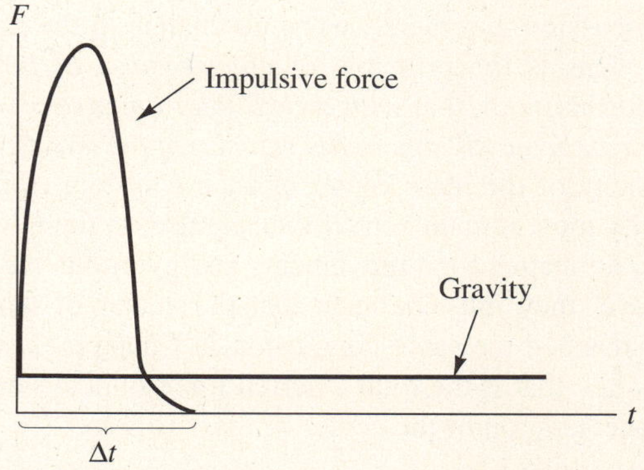

Figure 14.9. Plot of impulsive force and gravity force.

that the impulse from the explosion lasts for a very short time Δt and can be significant, whereas the impulse from gravity during the same short time is by comparison negligible. Forces that act over a very short time but have nevertheless appreciable impulse are called *impulsive forces*. In actions involving very small time intervals, we need only consider impulsive forces. Furthermore, during a very short time Δt an impulsive force acting on a particle can change the velocity of the particle in accordance with the impulse-momentum equation an appreciable amount while the particle undergoes very little change in position during the time Δt.[4] It is simplest in many cases to consider the *change in velocity of a particle from an impulsive force to occur over zero distance*.

Up to now, we have only considered a fragment of the bomb. Now let us consider all the fragments of the bomb taken as a system of particles. Since the explosive action is *internal* to the bomb, the action causes impulses that for any direction have equal and opposite counterparts, and thus *the total impulse on the bomb due to the explosion is zero*. We can thus conclude that *directly after* the explosion *the center of mass of the bomb has not moved appreciably* despite the high velocity of the fragments in all directions, as illustrated in Fig. 14.8. As time progresses beyond the short time interval described above, the gravitational impulse increases and has significant effect. If there were no friction, the center of mass would descend from the position of support as a freely falling particle under this action of gravity.

The following problems will illustrate these ideas.

[4]This idealization can be explained more precisely as follows. For an impulsive force F acting on a body of mass M, we can say from the linear momentum equation

$$\int_0^{\Delta t} F\,dt = F_{AV}\,\Delta t = MV_{\max} \qquad \therefore V_{\max} = \left(\frac{F_{AV}}{M}\right)(\Delta t)$$

The maximum movement of the body M during this time interval according to Newton's law on using the above result for V is then

$$x_{\max} = \int_0^{\Delta t} V_{\max}\,dt = \int_0^{\Delta t}\left(\frac{F_{AV}}{M}\Delta t\right)dt = \left(\frac{F_{AV}}{M}\right)(\Delta t)^2$$

Note that $V_{\max}$ is proportional to Δt while x is proportional to $(\Delta t)^2$. Clearly for a *very small* interval Δt the value of the movement x of the mass M can be considered *second order* compared to the value of the velocity V. For simplicity, with minimal error, we can say that the *mass M does not move while undergoing a change of velocity in response to an impulsive force*.

Example 14.7

Some top-flight tennis players hit the ball on a service at the instant that the ball is at the top of its trajectory after being released by the free hand. The ball is often given a speed V of 54 m/s by the racquet directly after the impact is complete. If the time of duration of the impact process is 5 ms, what is the magnitude of the average force from the racquet on the ball during this time interval? Take the mass of the ball as 42 g.

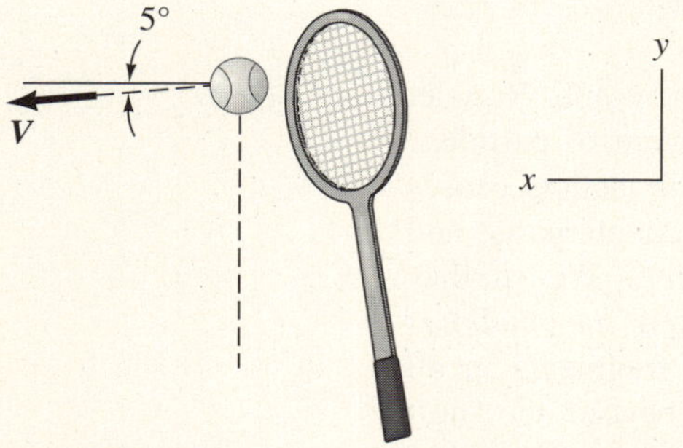

Figure 14.10. Impact of a tennis ball at service.

We have here acting on the ball during a very small time interval an **impulsive** force and the force of gravity. We will ignore the gravity force during the time of impact and we will consider that the ball achieves a post impact velocity while not moving, as explained earlier in the model for impulsive force behavior. As shown in Fig. 14.10, the impulse $\boldsymbol{I}$ generated on the ball by the racquet accordingly is

$$\boldsymbol{I} = \left(42 \times 10^{3}\right)[(54)][\cos 5° \boldsymbol{i} - \sin 5° \boldsymbol{j}]$$
$$= 2.268[0.996\boldsymbol{i} - 0.0872\boldsymbol{j}] \text{ N s (kg m/s)}$$

Next, we go to the **impulse momentum** equation. Thus

$$(\boldsymbol{F}_{av})(5 \times 10^{-3}) = 2.268(0.996\boldsymbol{i} - 0.0872\boldsymbol{j})$$

The magnitude of the average force is finally given as follows:

$$|\boldsymbol{F}_{av}| = 453.5 \text{ N}$$

After the impact, the ball will have a trajectory determined by gravity, wind forces, and the initial post-impact conditions.

Example 14.8

A 9 kN idealized cannon with a recoil spring (K = 4 kN/m) fires a 45 N projectile with a muzzle velocity of 625 m/s at an angle of 50° (Fig. 14.11). Determine the maximum compression of the spring.

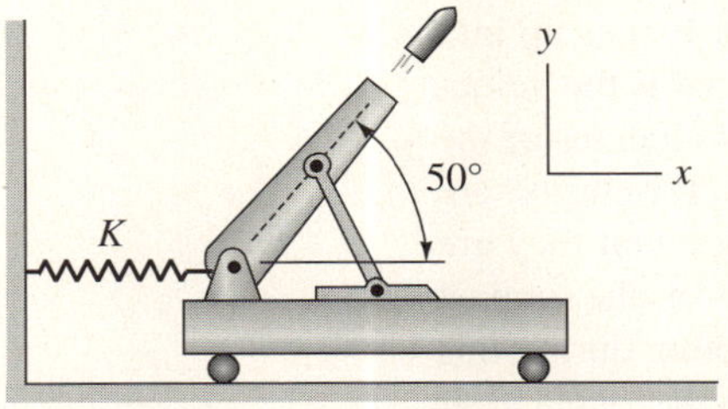

Figure 14.11. Idealized cannon..

The firing of the cannon takes place in a very short time interval. The force on the projectile and the force on the cannon from the explosion are impulsive forces. As a result, the cannon can be considered to achieve a recoil velocity instantaneously without having moved appreciably. Like the exploding bomb, the impulse on the cannon *plus* projectile is zero, as a result of the firing process. Since the linear momentum of the cannon plus projectile is zero just before firing, this linear momentum must be zero directly after firing. Thus, just after firing, we can say for the x direction:

$$(MV_x)_{\text{cannon}} + (MV_x)_{\text{projectile}} = 0 \tag{a}$$

Using V_c for the cannon velocity along the x axis and $V_p = V_c + 625 \cos 50°$ for the projectile velocity along the x axis we get

$$\frac{9\times 10^3}{g}V_c + \frac{45}{g}\left[(625)(\cos 50°) + V_c\right] = 0$$

Solving for V_c, we get

$$V_c = -2.00 \text{ m/s} \tag{b}$$

After this initial impulsive action, which results in an instantaneous velocity being imparted to the cannon, the motion of the cannon is then impeded by the spring. We may now use **conservation of mechanical energy** for a particle in this phase of motion of the cannon. Denoting δ as the maximum deflection of the spring, we can say:

$$\frac{1}{2}\frac{9\times 10^3}{g}(2.00^2) = \frac{1}{2}(4\times 10^3)(\delta^2)$$

Therefore,

$$\delta = 0.958 \text{ m}$$

Example 14.9

For target practice, a 9 N rock is thrown into the air and fired on by a pistol. The pistol bullet, of mass 57 g and moving with a speed of 312 m/s, strikes the rock as it is descending vertically at a speed of 6.25 m/s. [See Fig. 14.12(a).] Both the velocity of the bullet and the rock are parallel to the *xy* plane. Directly after the bullet hits the rock, the rock breaks up into two pieces, *A* weighing 5.78 N and *B* weighing 3.22 N. What is the velocity of *B* after collision for the given coplanar postcollision velocities of the bullet and the piece *A* shown in Fig. 14.12(b)? The bodies, for clarity, are shown separated in the diagram. Keep in mind, nevertheless, that they are very close to each other at post-impact. In our model of the impact process, they would not even have moved relative to each other during this process. The indicated 219 m/s and 25 m/s velocities are in the *xy* plane. If we neglect wind resistance, how high up does the center of mass of the rock and bullet system rise after collision?

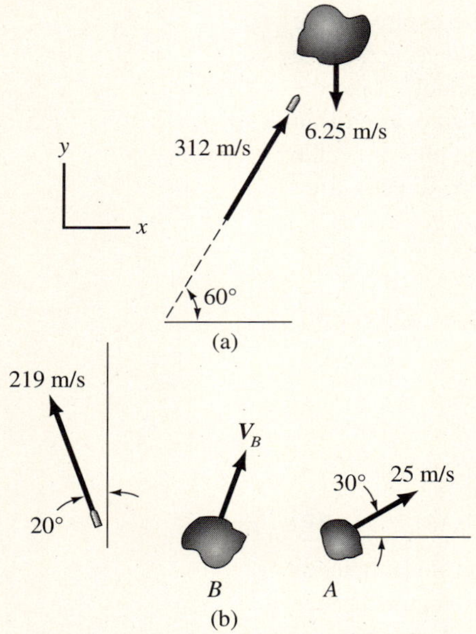

Figure 14.12. Bullet striking a rock.

Linear momentum is conserved during the collision, so we can equate linear momenta directly before and directly after collision. Thus,

$$(0.057)(312)(0.5\boldsymbol{i} + 0.866\boldsymbol{j}) + \frac{9}{g}(-6.25\boldsymbol{j})$$
$$= (0.057)(219)(-\sin 20°\boldsymbol{i} + \cos 20°\boldsymbol{j})$$
$$+ \frac{5.78}{g}25(0.866\boldsymbol{i} + 0.5\boldsymbol{j}) + \frac{3.22}{g}\left[(V_B)_x\boldsymbol{i} + (V_B)_y\boldsymbol{j}\right]$$

Example 14.9 (Continued)

We may solve for the desired quantities $(V_B)_x$ and $(V_B)_y$ to get

$$(V_B)_x = 1.235 \text{ m/s}$$
$$(V_B)_y = -28.7 \text{ m/s}$$

We now compute the velocity of the center of mass just before collision. Thus,

$$M\boldsymbol{V}_c = \left(\frac{9}{g} + 0.057\right)\boldsymbol{V}_c = \frac{9}{g}(-6.25)\boldsymbol{j} + (0.057)(312)(0.5\boldsymbol{i} + 0.866\boldsymbol{j})$$

Therefore,

$$\boldsymbol{V}_c = 9.125\boldsymbol{i} + 9.92\boldsymbol{j} \text{ m/s}$$

Hence, for the center of mass there is an initial velocity upward of 9.92 m/s just before collision. Directly after collision, since there has been no appreciable external impulse on the system during collision, the center of mass *still has* this upward speed. But now considering larger time intervals, we must take into account the action of gravity, which gives the center of mass a downward acceleration of 9.81 m/s^2. Thus,

$$\ddot{y}_c = -9.81$$
$$\dot{y}_c = -9.81t + C_1$$
$$y_c = -9.81\frac{t^2}{2} + C_1 t + C_2$$

When $t = 0$, $\dot{y}_c = 9.92$ and we take $y_c = 0$ for convenience. Hence we have

$$\dot{y}_c = -9.81t + 9.92 \quad \text{(a)}$$
$$y_c = -9.81\frac{t^2}{2} + 9.92t \quad \text{(b)}$$

Set $\dot{y}_c$ in (a) equal to zero and solve for t. We get

$$t = 1.011 \text{ s}$$

Substitute this value of t in Eq. (b) and solve for y_c, which now gives the desired maximum elevation of the center of mass after collision. Thus,

$$(y_c)_{max} = 5.01 \text{ m} \quad \text{(c)}$$

PROBLEMS

14.1. A body of mass 50 kg reaches an incline of 30° while it is moving at 15 m/s. If the dynamic coefficient of friction is 0.3, how long before the body stops?

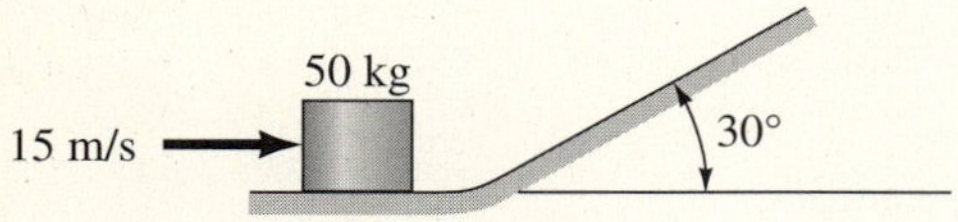

Figure P.14.1.

14.2. A particle of mass 1 kg is initially stationary at the origin of a reference. A force having a known variation with time acts on the particle. That is,

$$\boldsymbol{F}(t) = t^2\boldsymbol{i} + (6t + 10)\boldsymbol{j} + 1.6t^3\boldsymbol{k} \text{ N}$$

where t is in seconds. After 10 s, what is the velocity of the body?

14.3. A unidirectional force acting on a particle of mass 16 kg is plotted. What is the velocity of the particle at 40 s? Initially, the particle is at rest.

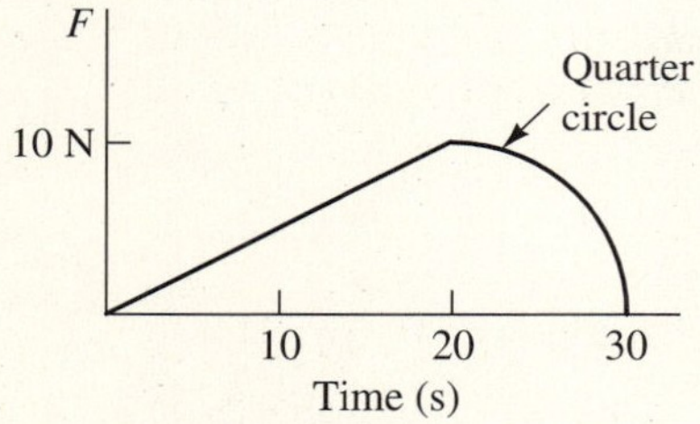

Figure P.14.3.

14.4. A 50 kg block is acted on by a force P, which varies with time as shown. What is the speed of the block after 80 s? Assume that the block starts from rest and neglect friction. The time axis gives time intervals.

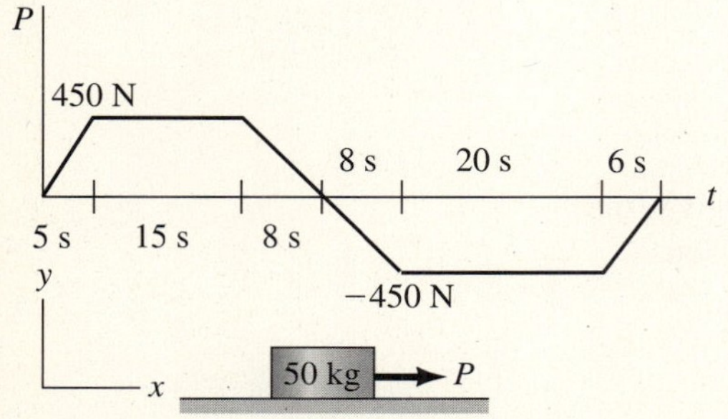

Figure P.14.4.

14.5. If the coefficient of static friction is 0.5 in Problem 14.4 and the coefficient of dynamic friction is 0.3, what is the speed of the block after 28 s?

14.6. A body is dropped from rest. (a) Determine the time required for it to acquire a velocity of 16 m/s. (b) Determine the time needed to increase its velocity from 16 m/s to 23 m/s.

14.7. A body having a mass of 2.5 kg is acted on by the following force:

$$\boldsymbol{F} = 36t\boldsymbol{i} + (27 + 13\sqrt{t})\boldsymbol{j} + (70 + 13t^2)\boldsymbol{k} \text{ N}$$

where t is in seconds. What is the velocity of the body after 5 s if the initial velocity is

$$V_1 = 2\boldsymbol{i} + \boldsymbol{j} - 3.3\boldsymbol{k} \text{ m/s?}$$

14.8. A body with a mass of 16 kg is required to change its velocity from $\boldsymbol{V}_1 = 2\boldsymbol{i} + 4\boldsymbol{j} - 10\boldsymbol{k}$ m/s to a velocity $\boldsymbol{V}_2 = 10\boldsymbol{i} - 5\boldsymbol{j} + 20\boldsymbol{k}$ m/s in 10 s. What average force $\boldsymbol{F}_{av}$ over this time interval will do the job?

14.9. In Problem 14.8, determine the force as a function of time for the case where force varies linearly with time starting with a zero value.

14.10. A hockey puck moves at 9 m/s from left to right. The puck is intercepted by a player who whisks it at 24 m/s toward goal A, as shown. The puck is also rising from the ice at a rate of 3 m/s. What is the impulse on the puck, whose mass is 140 g?

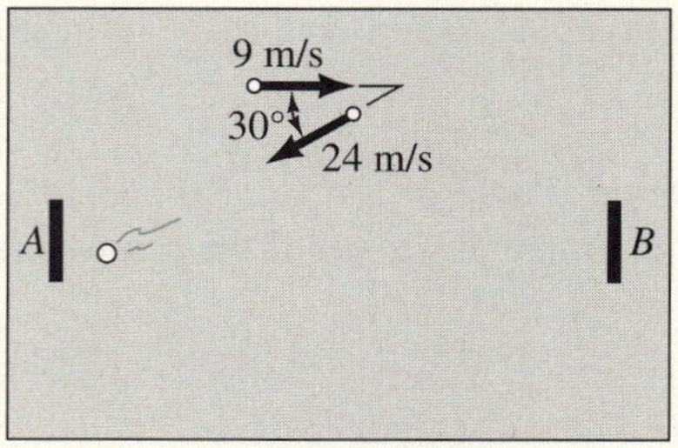

Figure P.14.10.

14.11. Gravel is released from a hopper at the rate of 1 kg/s. At the exit of the hopper it has a speed of 0.15 m/s. The belt is moving at a constant speed of 3 m/s. If there is 20 kg of gravel on the conveyor belt at all times and if the belt on the conveyor bed has a weight of 50 N, what is the difference in tension $T_2 - T_1$ for the belt to maintain operation? The dynamic coefficient of friction between bed and belt is 0.4. Assume that the gravel drops 0.2 m from the hopper outlet.

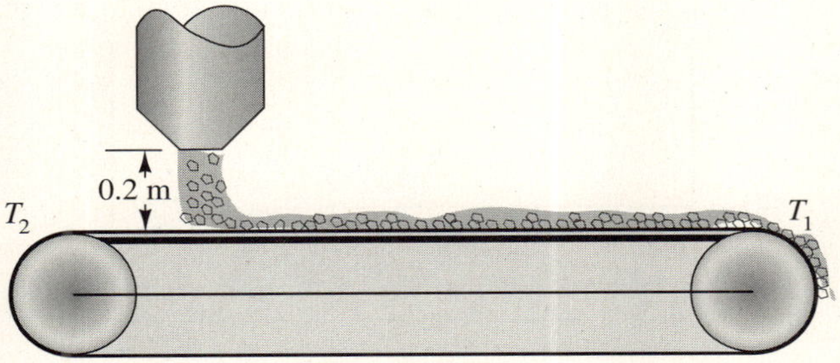

Figure P.14.11.

14.12. Do Problem 12.5 by methods of momentum.

14.13. Do Problem 12.6 by methods of momentum.

14.14. A commuter train made up of two cars is moving at a speed of 20 m/s. The first car has a mass of 20 Mg and the second 15 Mg.

(a) If the brakes are applied simultaneously to both cars, determine the minimum time the cars travel before stopping. The coefficient of static friction between the wheels and rail is 0.3.

(b) If the brakes on the first car only are applied, determine the time the cars travel before stopping and the force F transmitted between the cars.

14.15. Compute the velocity of the bodies after 10 s if they start from rest. The cable is inextensible, and the pulleys are frictionless. For the contact surfaces, $\mu_d = 0.2$.

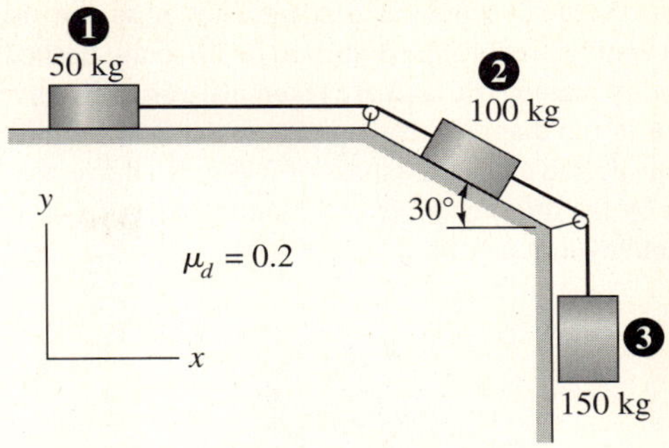

Figure P.14.15.

14.16. Two boxes per second each weighing 450 N land on a circular conveyor at a speed of 1.5 m/s in the direction of the chute. If there are 6 boxes on the circular belt at any one time, determine the average torque needed to rotate the belt at an angular speed of 0.2 rad/s. The dynamic coefficient of friction between the belt and the conveyer bed is 0.3. What power is needed for operating this belt? Neglect the rotational effect on the boxes themselves as they drop from the chute onto the conveyor. Also, does the radial change in velocity of the boxes affect the torque needed by the conveyor? Neglect any radial slipping of the boxes as they land.

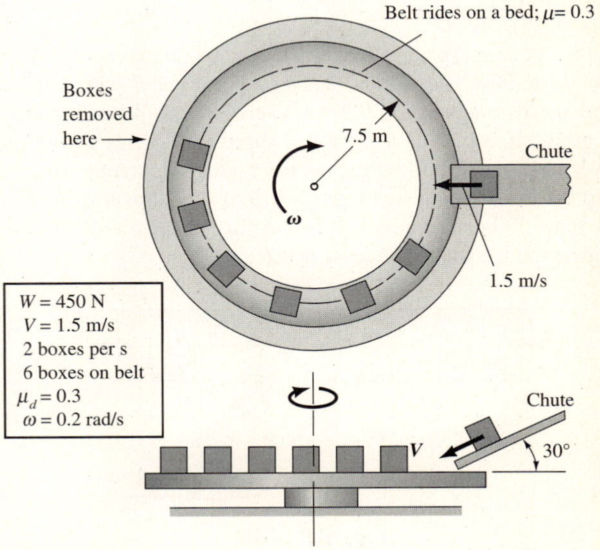

Figure P.14.16.

14.17. A vertical conveyor has sprocket A as the driver, and sprocket B turns freely. The bodies to be lifted are pushed onto the conveyor by a plunger C and are taken off from the conveyor at D as shown in the diagram. If the belt runs at 2 m/s and the bodies being transported each has a mass of 250 g, what average torque is required by the driving sprocket A? On the average, 40 bodies are on the conveyor at any time.

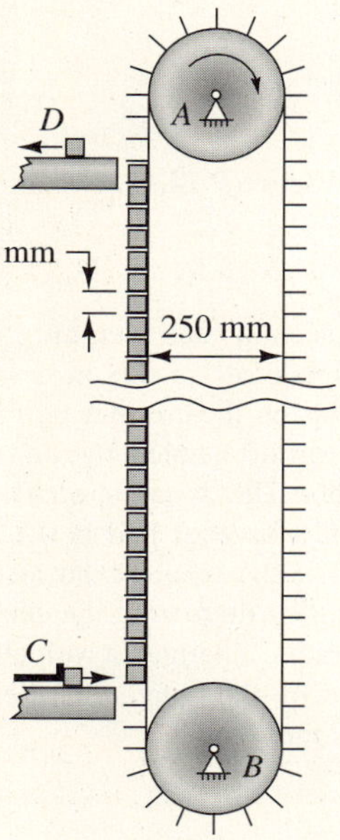

Figure P.14.17.

14.18. A conveyor A is feeding boxes onto a conveyor B. Each box has a mass of 1 kg and lands on conveyor B with a downward-speed component of 0.9 m/s. Conveyor belt A has a speed of 60 mm/s. If conveyor B runs at a speed of 1.5 m/s and if five boxes land per second on the average, what net average force T_2 must be exerted on the conveyor belt B to slide it over its bed? At any time, 50 boxes are on belt B. Take $\mu_d = 0.2$ for all surfaces. Neglect the weight of conveyor belt B.

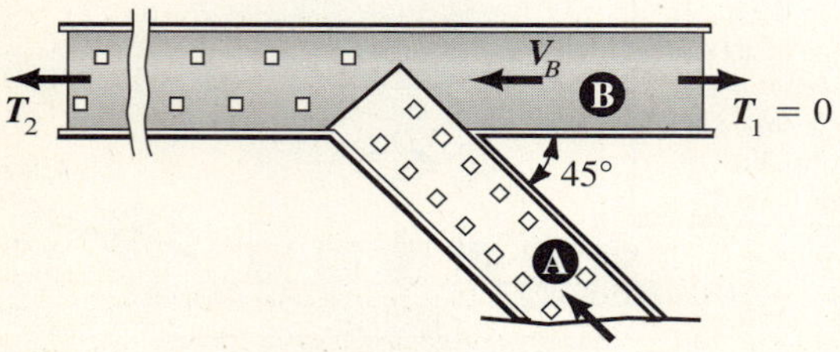

Figure P.14.18.

14.19. An idealized one-dimensional pressure wave (i.e., pressure is a function of one coordinate and time) generated by an explosion travels at a speed V of 360 m/s, as shown at time $t = 0$. The peak pressure of this wave is 35 kPa. What impulse per square metre is delivered to a wall oriented at right angles to the x axis? The wave is reflected from the wall, and the pressure at the wall is double the incoming pressure at all times. Do the problem for two time intervals corresponding to the interval (a) from when the wave front first touches the wall to when the peak reaches the wall and (b) from when the peak hits the wall to when the end of the wave reaches the wall.

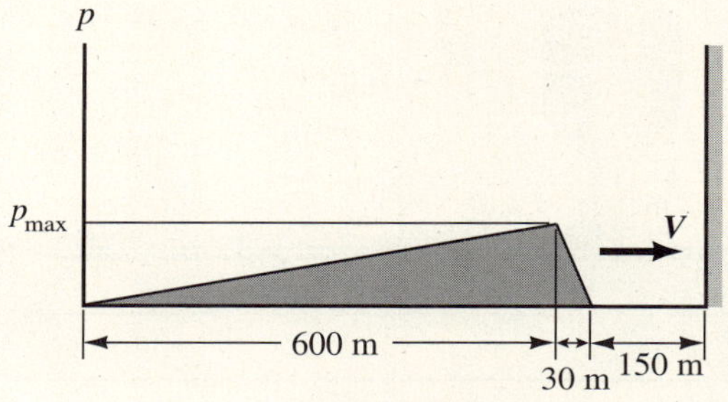

Figure P.14.19.

14.20. Blocks A and B move on frictionless surfaces. The blocks are interconnected with a light bar. Body A weighs 140 N; the weight of body B is not known. A constant force F of 450 N is applied at the configuration shown. If a speed of 7.6 m/s is reached by A after 1 s, what impulse is developed on the vertical wall?

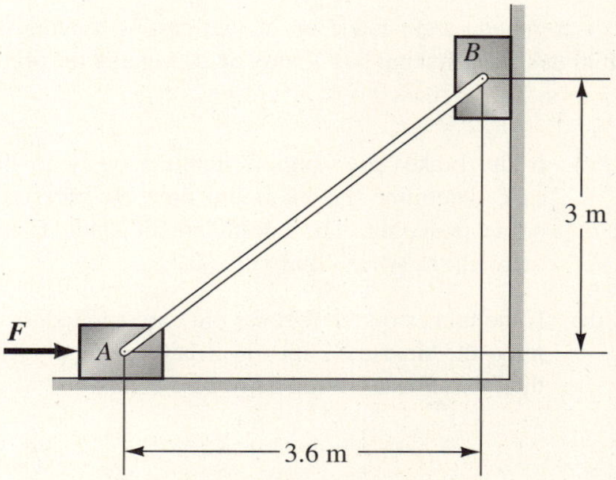

Figure P.14.20.

14.21. In Problem 14.20, compute the impulse on the horizontal surface. A moves 1.2 m in 1 s and $W_B = 90$ N.

14.22. An antitank airplane fires two 90 N projectiles at a tank at the same time. The muzzle velocity of the guns is 1 km/s relative to the plane. If the plane before firing weighs 65 kN and is moving with a velocity of 90 m/s, compute the change in its speed when it fires the two projectiles.

14.23. A toboggan has just entered the horizontal part of its run. It carries three people of mass 54 kg, 82 kg and 68 kg, respectively. Suddenly, a pedestrian of mass 90 kg strays onto the course and is turned end for end by the toboggan, landing safely among the riders. Since the toboggan path is icy, we can neglect friction with the toboggan path for all actions described here. If the toboggan is traveling at a speed of 16 m/s just before collision occurs, what is the speed after the collision when the pedestrian has become a rider? The toboggan has a mass of 14 kg.

Figure P.14.23.

14.24. An 890 N rowboat containing a 668 N man is pushed off the dock by an 800 N man. The speed that is imparted to the boat is 0.30 m/s by this push. The man then leaps into the boat from the dock with a speed of 0.60 m/s relative to the dock in the direction of motion of the boat. When the two men have settled down in the boat and before rowing commences, what is the speed of the boat? Neglect water resistance.

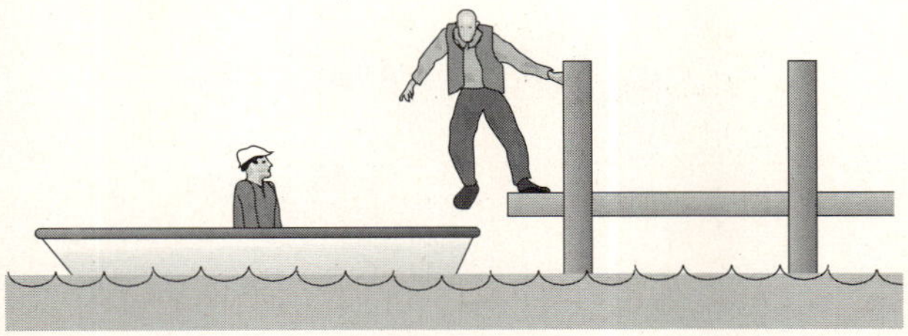

Figure P.14.24.

14.25. Two vehicles connected with an inextensible cable are rolling along a road. Vehicle B, using a winch, draws A toward it so that the relative speed is 1.5 m/s at $t = 0$ and 3 m/s at $t = 20$ s. Vehicle A weighs 9 kN and vehicle B weighs 13.5 kN. Each vehicle has a rolling resistance that is 0.01 times the vehicle's weight. What is the speed of A relative to the ground at $t = 20$ s if A is initially moving to the right at a speed of 9 m/s?

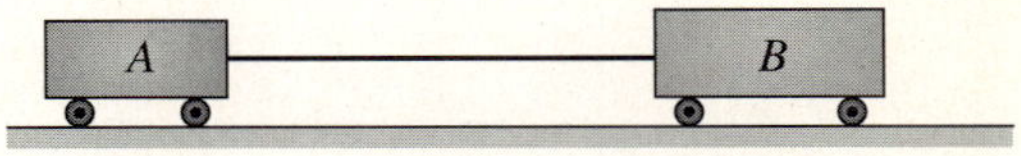

Figure P.14.25.

14.26. Treat Example 14.3 as a two-particle system in the impulse-momentum considerations. Verify the results of Example 14.3 for V. (Be sure to include *all* external forces for the system.)

14.27. Determine the velocity of body A and body B after 3 s if the system is released from rest. Neglect friction and the inertia of the pulleys.

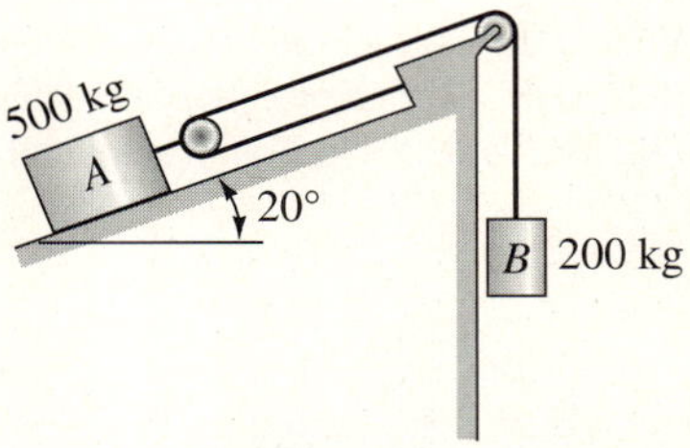

Figure P.14.27.

14.28. Do Problem 14.27 by considering a system of particles. (Be sure to include *all* external forces for the system of bodies A and B.)

14.29. A 40 kN truck is moving at the speed of 12 m/s carrying a 15 kN load A. The load is restrained only by friction with the floor of the truck where there is a dynamic coefficient of friction of 0.2 and the static coefficient of friction is 0.3. The driver suddenly jams his brakes on so as to lock all wheels for 1.5 s. At the end of this interval, the brakes are released. What is the final speed V of the truck neglecting wind resistance and rotational inertia of the wheels after load A stops slipping? The dynamic coefficient of friction between the tires and the road is 0.4.

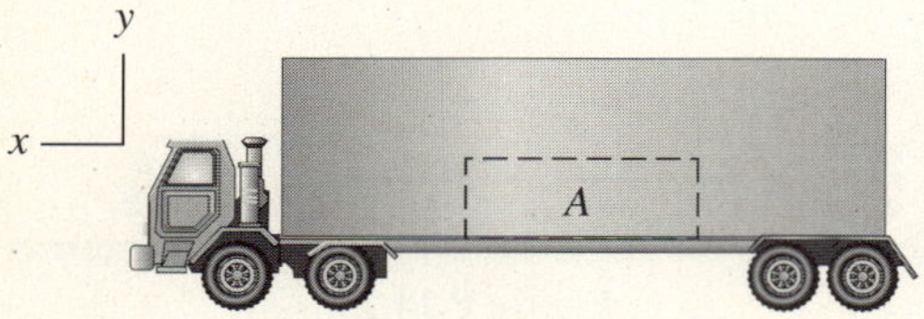

Figure P.14.29.

14.30. A 1.3 Mg Jeep is carrying three 100 kg passengers. The Jeep is in four-wheel drive and is under test to see what maximum speed is possible in 5 s from a start on an icy road surface for which $\mu_s = 0.1$. Compute V_{max} at $t = 5$ s.

Figure P.14.30.

14.31. Two adjacent tanks A and B are shown. Both tanks are rectangular with a width of 4 m. Gasoline from tank A is being pumped into tank B. When the level of tank A is 0.7 m from the top, the rate of flow Q from A to B is 300 L/s, and 10 s later it is 500 L/s. What is the average horizontal force from the fluids onto the tank during this 10 s time interval? The density of the gasoline is 800 kg/m^3. Tank A is originally full and tank B is originally empty.

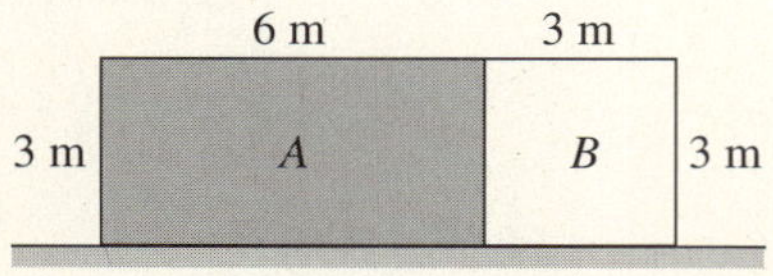

Figure P.14.31.

14.32. Two tanks A and B are shown. Tank A is originally full of water ($\rho = 1$ Mg/m^3), while tank B is empty. Water is pumped from A to B. If initially 3 m^3/s of water is being pumped and if this flow increases at the rate of 0.3 m^3/s for 30 s thereafter, what is the average vertical force onto the tanks from the water during this time period, aside from the static dead weight of the water?

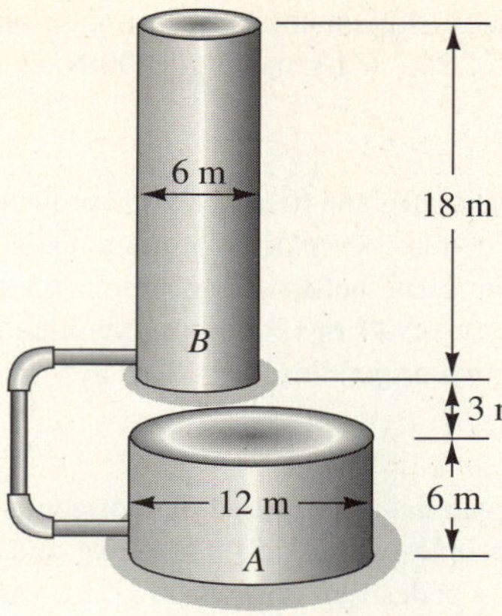

Figure P.14.32

14.33. A device to be detonated is shown in (a) suspended above the ground. Ten seconds after detonation, there are four fragments having the following masses and position vectors relative to reference XYZ:

$$m_1 = 5 \text{ kg}$$
$$\boldsymbol{r}_1 = 1000\boldsymbol{i} + 2000\boldsymbol{j} + 900\boldsymbol{k} \text{ m}$$
$$m_2 = 3 \text{ kg}$$
$$\boldsymbol{r}_2 = 800\boldsymbol{i} + 1800\boldsymbol{j} + 2500\boldsymbol{k} \text{ m}$$
$$m_3 = 4 \text{ kg}$$
$$\boldsymbol{r}_3 = 400\boldsymbol{i} + 1000\boldsymbol{j} + 2000\boldsymbol{k} \text{ m}$$
$$m_4 = 6 \text{ kg}$$
$$\boldsymbol{r}_4 = X_4\boldsymbol{i} + Y_4\boldsymbol{j} + Z_4\boldsymbol{k}$$

Find the position $\boldsymbol{r}_4$ if the center of mass of the device is initially at position $\boldsymbol{r}_0$, where

$$\boldsymbol{r}_0 = 600\boldsymbol{i} + 1200\boldsymbol{j} + 2300\boldsymbol{k} \text{ m}$$

Neglect wind resistance.

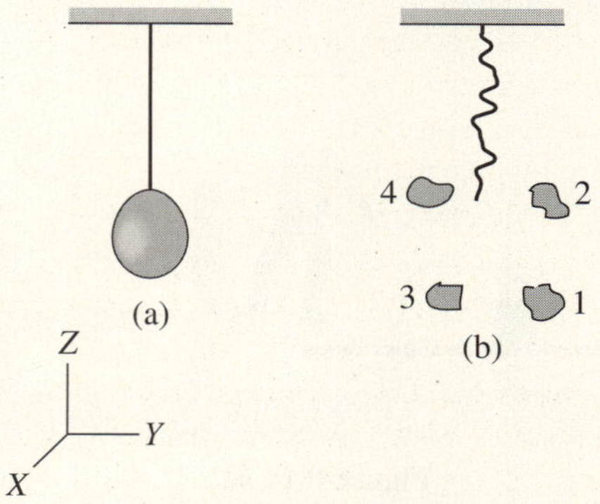

Figure P.14.33.

14.4 Impact

In Section 14.3, we discussed impulsive forces. We shall in this section discuss in detail an action in which impulsive forces are present. This situation occurs when two bodies collide but do not break. The time interval during collision is very small, and comparatively large forces are developed on the bodies during the small time interval. This action is called *impact*. For such actions with such short time intervals, the force of gravity generally causes a negligible impulse. The impact forces on the colliding bodies are always equal and opposite to each other, so the net impulse on the *pair* of bodies during collision is *zero*. This means that the total linear momentum directly after impact (postimpact) equals the total linear momentum directly before impact (preimpact).

We shall consider at this time two types of impact for which certain definitions are needed. We shall call the normal to the *plane of contact* during the collision of two bodies the *line of impact*. If the centers of mass of the two colliding bodies lie along the line of impact, the action is called *central impact* and is shown for the case of two spheres in Fig. 14.13.[5] If, in addition, the velocity vectors of the mass centers approaching the collision are collinear with the line of impact, the action is called *direct central impact*. This action is illustrated by $\mathbf{V}_1$ and $\mathbf{V}_2$ in Fig 14.13. Should one (or both) of the velocities have a line of action not collinear with the line impact—for example, $\mathbf{V}'_1$ and/or $\mathbf{V}'_2$—the action is termed *oblique central impact*.

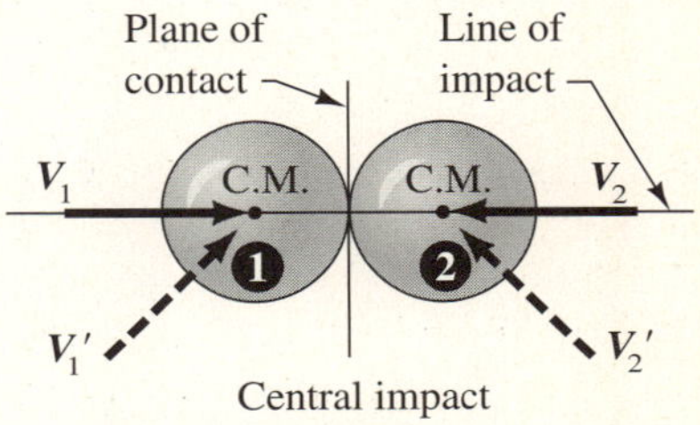

Figure 14.13. Central impact of two spheres.

In either case, linear momentum is conserved during the short time interval from directly before the collision (indicated with the subscript i) to directly after the collision (indicated with subscript f). That is,

$$(m_1\mathbf{V}_1)_i + (m_2\mathbf{V}_2)_i = (m_1\mathbf{V}_1)_f + (m_2\mathbf{V}_2)_f \tag{14.8}$$

In the *direct-central-impact* case for *smooth* bodies (i.e., bodies with no friction), this equation becomes a single scalar equation since $(\mathbf{V}_1)_f$ and $(\mathbf{V}_2)_f$ are collinear with the line of impact. Usually, the initial velocities are known and the final values are desired, which means that we have for this case one scalar equation involving two unknowns. Clearly, we must know more about the manner of interaction of the bodies, since Eq. 14.8 as it stands is valid for materials of any deformability (e.g., putty or hardened steel) and takes no account of such important considerations. Thus, we cannot consider the bodies undergoing impact only as particles as has been the case thus far, but must, in addition, consider them as deformable bodies of finite size in order to generate enough information to solve the problem at hand.

For the *oblique-impact case*, we can write components of the linear-momentum equation along the line of impact and for smooth (frictionless) bodies, along two other directions at right angles to the line of impact. If we know the initial velocities, then we have six unknown final velocity components and only three equations. Thus, we need even more information to establish fully the final velocities after this more general type of impact. We now consider each of these cases in more detail in order to establish these additional relations.

[5]Noncentral or *eccentric impact* is examined in Chapter 17 for the case of plane motion.

Case 1. Direct Central Impact. Let us first examine the direct-central-impact case. We shall consider the period of collision to be made up of two subintervals of time. The *period of deformation* refers to the duration of the collision, starting from initial contact of the bodies and ending at the instant of maximum deformation. During this period, we shall consider that impulse $\int D\,dt$ acts oppositely on each of the bodies. The second period, covering the time from the maximum deformation condition to the instant at which the bodies just separate,[6] we shall term the *period of restitution*. The impulse acting oppositely on each body during this period we shall indicate as $\int R\,dt$. If the bodies are *perfectly elastic*, they will reestablish their initial shapes during the period of restitution (if we neglect the internal vibrations of the bodies), as shown in Fig. 14.14(a). When the bodies do not reestablish their initial shapes [Fig. 14.14(b)], we say that *plastic deformation* has taken place.

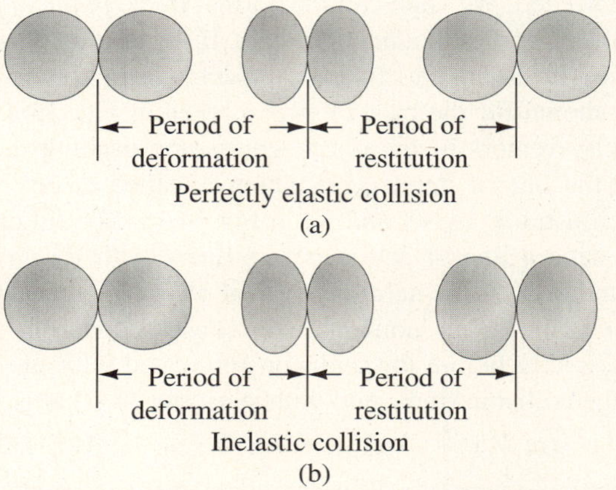

Figure 14.14. Collision process.

The ratio of the impulse during the restitution period $\int R\,dt$ to the impulse during the deformation period $\int D\,dt$ is a number ϵ, which depends mainly on the physical properties of the bodies in collision. We call this number the *coefficient of restitution*. Thus,

$$\epsilon = \frac{\text{impulse during restitution}}{\text{impulse during deformation}} = \frac{\int R\,dt}{\int D\,dt} \tag{14.9}$$

We must strongly point out that the coefficient of restitution depends also on the size, shape, and approach velocities of the bodies before impact. These dependencies result from the fact that plastic deformation is related to the magnitude and nature of the force distributions in the bodies and also to the rate of loading. However, values of ϵ have been established for different materials and can be used for approximate results in the kind of computations

[6]If they don't separate, the end of the second period occurs when the bodies cease to deform. We call such a process a **plastic** impact.

to follow. We shall now formulate the relation between the coefficient of restitution and the initial and final velocities of the bodies undergoing impact.

Let us consider *one* of the bodies during the two phases of the collision. If we call the velocity at the maximum deformation condition $(V)_D$, we can say for mass 1:

$$\int D\,dt = \left[(m_1V_1)_D - (m_1V_1)_i\right] = -m_1\left[(V_1)_i - (V_1)_D\right] \qquad (14.10)$$

During the period of restitution, we find that

$$\int R\,dt = -m_1\left[(V_1)_D - (V_1)_f\right] \qquad (14.11)$$

Dividing Eq. 14.11 by Eq. 14.10, canceling out m_1, and noting the definition in Eq. 14.9, we can say:

$$\epsilon = \frac{(V_1)_D - (V_1)_f}{(V_1)_i - (V_1)_D} \qquad (14.12)$$

A similar analysis for the other mass (2) gives

$$\epsilon = \frac{(V_2)_D - (V_2)_f}{(V_2)_i - (V_2)_D} = \frac{(V_2)_f - (V_2)_D}{(V_2)_D - (V_2)_i} \qquad (14.13)$$

In this last expression, we have changed the sign of numerator and denominator. At the intermediate position at the end of deformation and the beginning of restitution the masses have essentially the same velocity. Thus, $(V_1)_D = (V_2)_D$. Since the quotients in Eqs. 14.12 and 14.13 are equal to each other, we can add numerators and denominators to form another equal quotient, as you can demonstrate yourself. Noting the abovementioned equality of the V_D terms, we have the desired result:

$$\epsilon = -\frac{(V_2)_f - (V_1)_f}{(V_2)_i - (V_1)_i} = -\frac{\text{relative velocity of separation}}{\text{relative velocity of approach}} \qquad (14.14)$$

This equation involves the coefficient ϵ, which is presumably known or estimated, and the initial and final velocities of the bodies undergoing impact. Thus, with this equation we can solve for the final velocities of the bodies after collision when we use the linear-momentum equation 14.8 for the case of direct central impact.

During a *perfectly elastic* collision, the impulse for the period of restitution equals the impulse for the period of deformation,[7] so the coefficient of restitution is *unity* for this case. For inelastic collisions, the coefficient of restitution is less than unity since the impulse is diminished on restitution as a

[7]The impulses are equal because during the period of restitution the body can be considered to undergo identically the reverse of the process corresponding to the deformation period. Thus, from a thermodynamics point of view, we are considering the elastic impact to be a *reversible* process.

result of the failure of the bodies to resume their original geometries. For a *perfectly plastic* impact, $\epsilon = 0$ [i.e., $(V_2)_f = (V_1)_f$] and the bodies remain in contact. Thus ϵ ranges from 0 to 1.

Case 2. Oblique Central Impact. Let us now consider the case of oblique central impact. The velocity components along the line of impact can be related by the scalar component of the linear-momentum equation 14.8 in this direction and also by Eq. 14.14, where velocity components along the line of impact are used and where the coefficient of restitution may be considered (for smooth bodies) to be the same as for the direct-central-impact case. If we know the initial conditions, we can accordingly solve for those velocity components after impact in the direction of the line of impact. As for the other rectangular components of velocity, we can say that for smooth bodies, these velocity components are unaffected by the collision, since no impulses act in these directions on either body. That is, the velocity components normal to the line of impact for each body are the same immediately after impact as before. Thus, the final velocity components of both bodies can be established, and the motions of the bodies can be determined within the limits of the discussion. The following examples are used to illustrate the use of the preceding formulations.

Note that the mass and materials of the colliding bodies for both direct or oblique central impact can be different from each other.

Example 14.10

Two billiard balls (of the same size and mass) collide with the velocities of approach shown in Fig. 14.15. For a coefficient of restitution of 0.90, what are the final velocities of the balls directly after they part? What is the loss in kinetic energy?

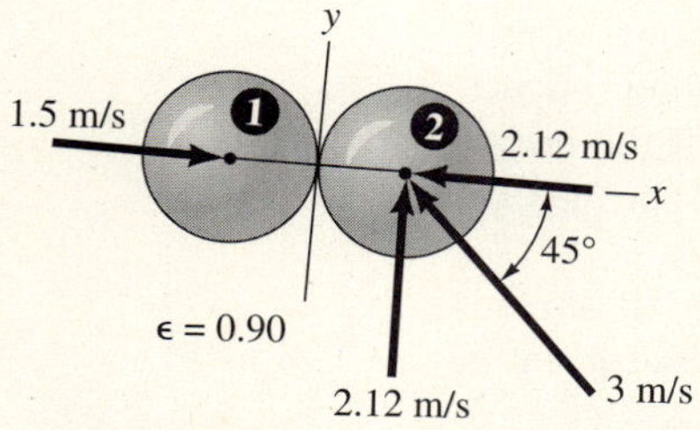

Figure 14.15. Oblique central impact.

A reference is established so that the x axis is along line of impact and the y axis is in the plane of contact such that the reference plane is par-

Example 14.10 (Continued)

allel to the billiard table. The approach velocities have been decomposed into components along these axes. The velocity components $(V_1)_y$ and $(V_2)_y$ are unchanged during the action. Along the line of impact, **linear-momentum** considerations lead to

$$1.5m - 2.12m = m[(V_1)_x]_f + m[(V_2)_x]_f \qquad \text{(a)}$$

Using the **coefficient-of-restitution** relation (Eq. 14.14), we have

$$\epsilon = 0.90 = -\frac{[(V_2)_x]_f - [(V_1)_x]_f}{-2.12 - 1.5} \qquad \text{(b)}$$

We thus have two equations, (a) and (b), for the unknown components in the x direction. Simplifying these equations, we have

$$[(V_1)_x]_f + [(V_2)_x]_f = -0.62 \qquad \text{(c)}$$

$$[(V_1)_x]_f + [(V_2)_x]_f = -3.26 \qquad \text{(d)}$$

Adding, we get

$$[(V_1)_x]_f = -1.94 \text{ m/s}$$

Solving for $[(V_2)_x]_f$ in Eq. (c), we write

$$[(V_2)_x]_f - 1.94 = -0.62$$

Therefore,

$$[(V_2)_x]_f = 1.32 \text{ m/s}$$

The final velocities after collision are then

$$(\mathbf{V}_1)_f = -1.94\boldsymbol{i} \text{ m/s}$$

$$(\mathbf{V}_2)_f = 1.32\boldsymbol{i} + 2.12\boldsymbol{j} \text{ m/s}$$

The loss in kinetic energy is given as

$$(\text{KE})_i - (\text{KE})_f = (\tfrac{1}{2}m1.5^2 + \tfrac{1}{2}m3^2) - \left[\tfrac{1}{2}m1.94^2 + \tfrac{1}{2}m(2.12^2 + 1.32^2)\right]$$

$$\Delta\text{KE} = \tfrac{1}{2}m[2.25 + 9 - (3.764 + 4.494 + 1.742)]$$

$$= 0.625m \text{ N m}$$

Please note that mechanical energy is conserved *only* if ϵ is unity (i.e., a perfectly elastic impact). For all other cases, there is always dissipation of mechanical energy into heat and permanent deformation. However, *all* impacts involve conservation of linear momentum for the system.

Example 14.11

A pile driver is used to force a pile A into the ground (Fig. 14.16) as part of a program to properly prepare the foundation for a tall building. The device consists of a piston C on which a pressure p is developed from steam or air.

The piston is connected to a 4.5 kN hammer B. The assembly is suddenly released and accelerates downward a distance h of 0.6 m to impact on pile A weighing 1.8 kN. If the earth develops a constant resisting force to movement of 110 kN, what distance d will the pile move for a drop involving no contribution from p (which is then zero gauge pressure). Take the impact as *plastic*. The weight of the piston and the connecting rod is 450 N.

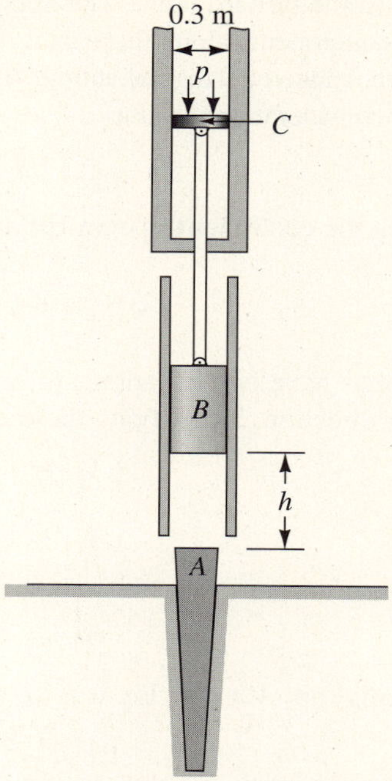

Figure 14.16. Steam-driven pile driver.

We begin by using **conservation of mechanical energy** for the freely falling system to a position just before impact (preimpact). Using the initial configuration as the datum we have

$$0 + (4.5 \times 10^3 + 450)(0.6) = \frac{1}{2}\left(\frac{4950}{g}\right)V^2 + 0$$
$$\therefore V = \sqrt{2gh} = \sqrt{(2)(9.81)(0.6)} = 3.43 \text{ m/s}$$

Now we get to the impact process. We have **conservation of linear momentum** while the bodies remain hypothetically at the position at which contact is first made. Thus, for plastic impact we can say

$$\frac{4950}{g}(3.43) + 0 = \frac{6750}{g}V \qquad \therefore V = 2.52 \text{ m/s}$$

Finally, we come to the post-impact process where we shall use the **work energy equation** for the pile driver and the pile.

$$(6750 - 110 \times 10^3)(d) = 0 - \frac{1}{2}\left(\frac{6750}{g}\right)(2.52^2)$$

where the term on the left side must be negative because the net force on the system (103.25 kN) is in the opposite direction to the motion (see Eq. 13.2). Solving for d we get

$$d = 21.2 \text{ m}$$

*14.5 Collision of a Particle with a Massive Rigid Body

In Section 14.4, we employed conservation-of-momentum considerations and the concept of the coefficient of restitution to examine the impact of two smooth bodies of comparable size. Now we shall extend this approach to include the impact of a spherical body with a much larger and more massive *rigid* body, as shown in Fig. 14.17.

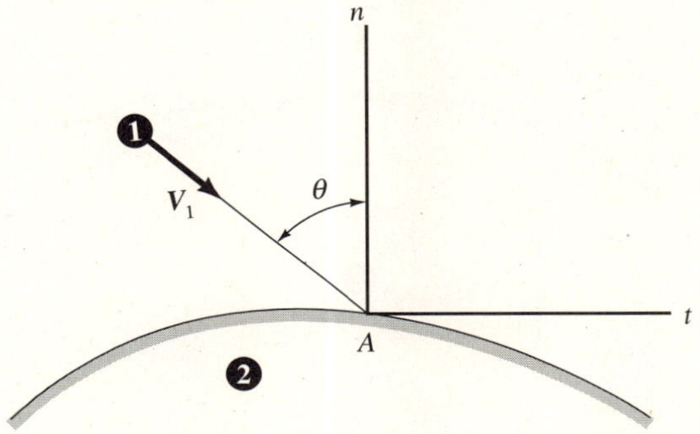

Figure 14.17. Small body collides with large body.

The procedure we shall follow is to consider the massive body to be a spherical body of *infinite* mass with a radius equal to the local radius of curvature of the surface of the massive body at the point of contact A. This condition is shown in Fig. 14.18. The line of impact then becomes identical with the normal n to the surface of the massive body at the point of impact. Note that the case we show in the diagram corresponds to oblique central impact. With no friction, clearly only the components along the line of impact n can change as a result of impact. But in this case, the velocity of the sphere representing the massive body must undergo no change in value after impact because of its infinite mass.[8] We cannot make good use here of the conservation of the linear-momentum equation in the n direction because the infinite mass of the hypothetical body (2) will render the equation indeterminate. However, we can use Eq. 14.14, assuming we have a coefficient of restitution ϵ for the action. Noting that the velocity of the massive body *does not change*, we accordingly get

$$\epsilon = -\frac{\left[(V_1)_n\right]_f - \left[(V_2)_n\right]}{\left[(V_1)_n\right]_i - \left[(V_2)_n\right]} \qquad (14.15)$$

Thus, knowing the velocities of the bodies before impact, as well as the quantity ϵ, we are able to compute the velocity of the particle after impact. If the

[8]Otherwise, there would be an infinite change in momentum for this sphere.

collision is perfectly elastic, $\epsilon = 1$, and we see from Eq. 14.15 that for a stationary massive body

$$[(V_1)_n]_i = -[(V_1)_n]_f$$

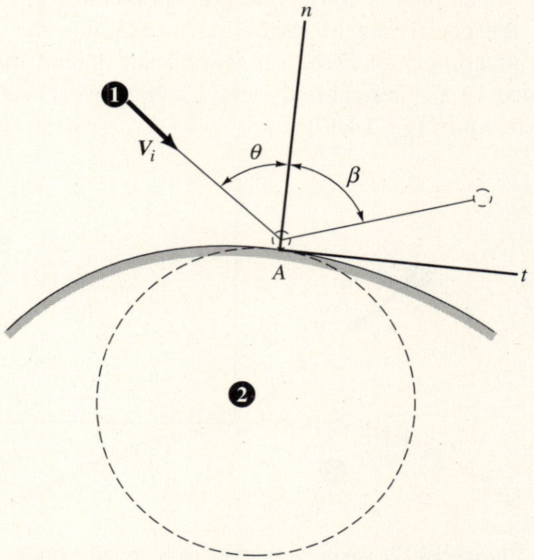

Figure 14.18. Angle of incidence and angle of reflection.

This means that the angle of incidence θ equals the angle of reflection β. For $\epsilon < 1$ (i.e., for an inelastic collision), the angle of reflection β will clearly exceed θ as shown in Fig. 14.18.

We now illustrate the use of these formulations.

Example 14.12

A ball is dropped onto a concrete floor from height h (Fig. 14.19). If the coefficient of restitution is 0.90 for the action, to what height h' will the ball rise on the rebound?

Here the massive body has an infinite radius at the surface. Furthermore, we have a direct central impact. Accordingly, from Eq. 14.15 we have

$$\epsilon = -\frac{(V)_f - 0}{(V)_i - 0} = \frac{\sqrt{2gh'}}{\sqrt{2gh}}$$

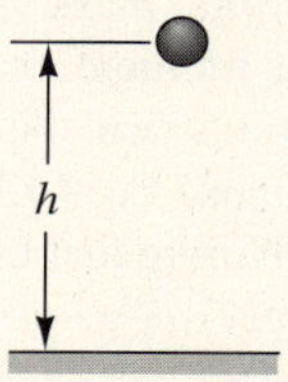

Figure 14.19. Ball dropped on concrete floor.

Solving for h', we get

$$h' = \epsilon^2 h = 0.81h$$

In the following interesting example as well as in some homework problems, we will have to determine, for a given uniform distribution of stationary particles in space, how many of these particles collide per unit time with a rigid body translating through this cloud of particles at constant speed V_0. To illustrate how this may be accomplished easily, we have shown a cone-cylinder moving through such a cloud of particles at constant speed V_0 in Fig. 14.20.

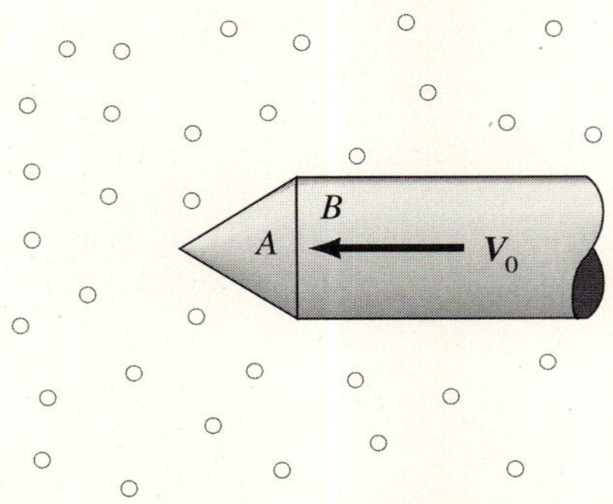

Figure 14.20. A cone–cylinder moving through a cloud of particles.

During a time interval Δt, the cone A moves a distance $V_0\,\Delta t$, colliding with all the particles in the volume swept out by the conical surface during this time interval as shown in Fig. 14.21, where this region is outlined with dashed lines. This volume can easily be calculated. It is that of a right circular cylinder shown in Fig. 14.22 having a cross section corresponding to the *projected* area of the cone taken along the axis parallel to the direction of motion of the moving body. Clearly, by adding the volume of cone A to the right circular cylinder along its axis at the forward end and then deleting the same volume at the rear end, we reproduce the dashed volume in Fig 14.21 during the time interval Δt. In general, the volume swept out by a body during a time interval can readily be found by using the *projected* area of the body in the direction of motion. We then use this area to sweep out a volume during this time interval. This negates having to deal with the actual more complicated three-dimensional end surface itself. We shall make use of this procedure in the following example.

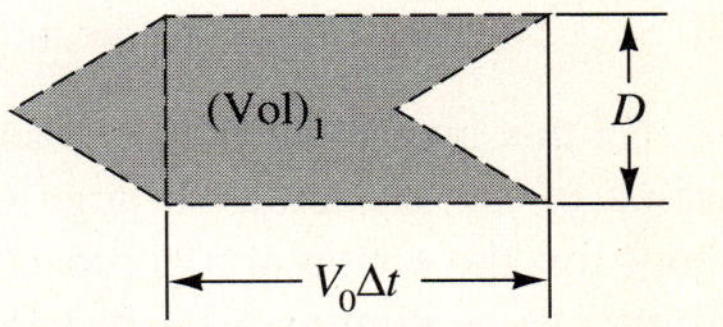

Figure 14.21. Dashed region is volume swept out by the cone A during Δt.

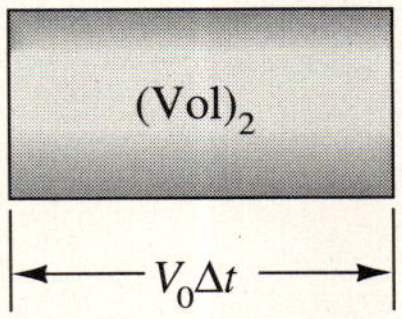

Figure 14.22. Volume swept by cone A.

Example 14.13

A satellite in the form of a sphere with radius R [Fig. 14.23(a)] is moving above the earth's surface in a region of highly rarefied atmosphere. We wish to estimate the drag on the satellite. Neglect the contribution from the antennas.

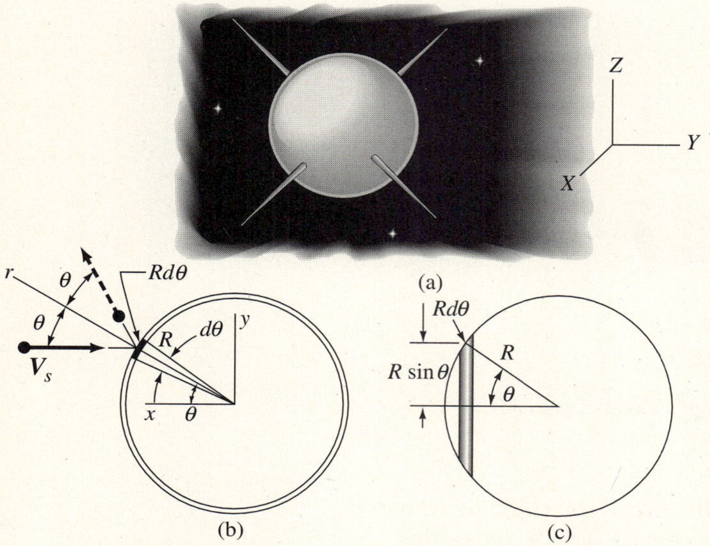

Figure 14.23. Satellite moving at high speed V_s in space.

In this highly rarefied atmosphere, we shall assume that the average spacing of the molecules is large enough relative to the satellite that we cannot use the continuum approach of fluid dynamics, wherein matter is assumed to be continuously distributed. Instead, we must consider collisions of the individual molecules with the satellite, which is a noncontinuum approach, as discussed in Section 1.7. The mass per molecule is m kg and the number density of the molecules is uniformly n molecules/m^3. Since the satellite is moving with a speed V_s much greater than the speed of the molecules (the molecules move at about the speed of sound), we can assume that the molecules are stationary relative to inertial space reference XYZ and that only the satellite is moving. Furthermore, we assume that when the satellite hits a molecule there is an elastic, frictionless collision.

To study this problem, we have shown a section of the satellite in Fig. 14.23(b). A reference xyz is fixed to the satellite at its center. We shall consider this reference also to be an inertial reference—a step that for small drag will introduce little error for the ensuing calculations. Relative to this reference, the molecules approach the satellite with a horizontal velocity V_s, as shown for one molecule. They then collide with the surface

Example 14.13 (Continued)

with an angle of incidence measured by the polar coordinate θ. Finally, they deflect with an equal angle of reflection of θ. The component of the impulse given to the molecule in the x direction $(I_{mol})_x$ is

$$(I_{mol})_x = (mV_s \cos 2\theta) - (-mV_s) = mV_s(1 + \cos 2\theta) \quad \text{(a)}$$

This is the impulse component that would be given to any molecule hitting a strip that is $R\,d\theta$ in width and which is revolved around the x axis as shown in Fig. 14.23(c). The number of such collisions per second for this strip can readily be calculated as follows:[9]

collisions for strip per second

$$= \begin{bmatrix}\text{projected area}\\ \text{of the strip}\\ \text{in } x \text{ direction}\end{bmatrix}\begin{bmatrix}\text{distance the}\\ \text{strip moves}\\ \text{in 1 s}\end{bmatrix}\begin{bmatrix}\text{number of}\\ \text{molecules per}\\ \text{unit volume}\end{bmatrix} \quad \text{(b)}$$
$$= [(R\,d\theta\cos\theta)(2\pi R\sin\theta)][V_s][n]$$
$$= 2\pi R^2 n V_s \sin\theta\cos\theta\,d\theta$$

The impulse component dI_x provided by the strip in 1 s is the product of the right sides of Eqs. (a) and (b). Thus,

$$dI_x = 2\pi mnR^2V_s^2(\sin\theta\cos\theta)(1 + \cos 2\theta)\,d\theta \quad \text{(c)}$$

Noting that $2\sin\theta\cos\theta = \sin 2\theta$, we have

$$dI_x = \pi mnR^2V_s^2(\sin 2\theta + \sin 2\theta\cos 2\theta)\,d\theta$$
$$= \pi mnR^2V_s^2\left(\sin 2\theta + \frac{\sin 4\theta}{2}\right)d\theta$$

Integrating from $\theta = 0$ to $\theta = \pi/2$,[10] we get the total impulse for 1 sec by the sphere:

$$I_x = \pi mnR^2V_s^2\left(\int_0^{\pi/2}\sin 2\theta\,d\theta + \tfrac{1}{2}\int_0^{\pi/2}\sin 4\theta\,d\theta\right)$$
$$= \pi mnR^2V_s^2\left(-\tfrac{1}{2}\cos 2\theta\Big|_0^{\pi/2} - \tfrac{1}{8}\cos 4\theta\Big|_0^{\pi/2}\right) \quad \text{(d)}$$
$$= \pi mnR^2V_s^2(1 + 0) = \pi mnR^2V_s^2$$

The average force needed to give this impulse by the satellite is clearly $\pi mnR^2V_s^2$, and so the reaction to this force is the desired drag.

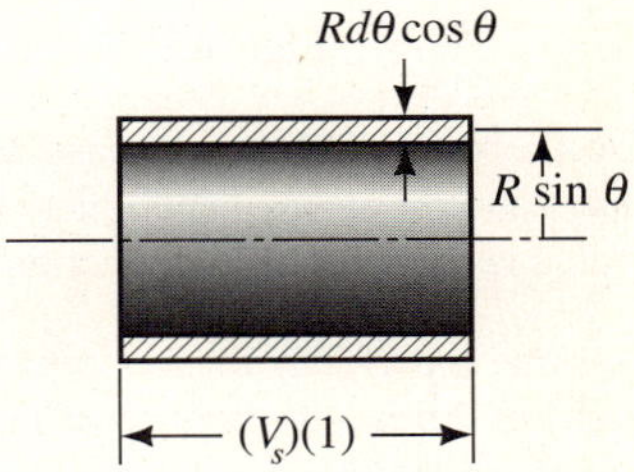

[9]As is shown here the volume swept out by the strip in one second will be a right circular tube of length $V_s\,\Delta t = (V_s)(1)$ and thickness $R\,d\theta\cos\theta$ and having a radius equal to $R\sin\theta$.

[10]We integrate only up to $\pi/2$ because collisions take place only on the *front* part of the sphere. (Note also, we are already rotating for any θ completely around the axis of the sphere.) This is so since, in our model, the molecules are moving only from left to right toward the sphere with no collisions possible beyond $\theta = \pi/2$.

PROBLEMS

14.34. Two cylinders move along a rod in a frictionless manner. Cylinder A has a mass of 10 kg and moves to the right at a speed of 3 m/s, while cylinder B has a mass of 5 kg and moves to the left at a speed of 2.5 m/s. What is the speed of cylinder B after impact for a coefficient of restitution ϵ of 0.8? What is the loss in kinetic energy?

Figure P.14.34.

14.35. In Problem 14.34, what coefficient of restitution is needed for body A to be stationary after impact?

14.36. Two smooth cylinders of identical radius roll toward each other such that their centerlines are perfectly parallel. Cylinder A has a mass of 10 kg, and cylinder B has a mass of 7.5 kg. What is the speed at which cylinder A moves directly after collision for a coefficient of restitution $\epsilon = 0.75$?

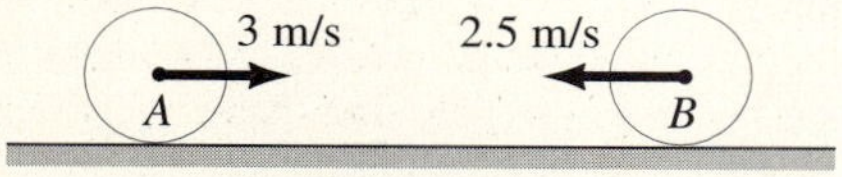

Figure P.14.36.

14.37. Cylinder A, of mass 5 kg, moves toward stationary cylinder B, of mass 20 kg, at the speed of 6 m/s. Mass B is attached to a spring having a spring constant K equal to 2 N/mm. If the collision has a coefficient of restitution $\epsilon = 0.9$, what is the maximum deflection δ of the spring? Assume that there is no friction along the rod and that the spring has negligible mass.

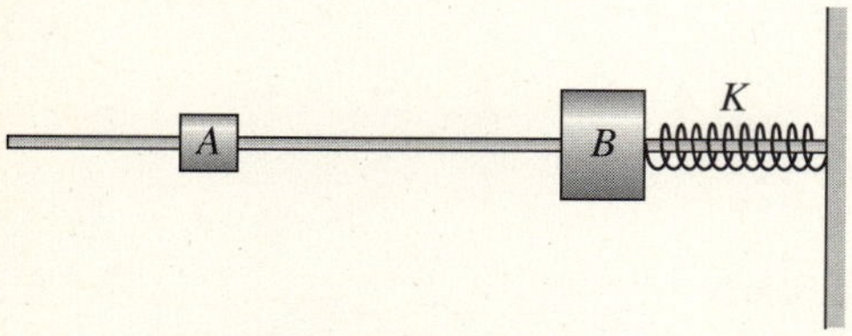

Figure P.14.37.

14.38. Do Problem 14.37 for the case where there is a perfectly plastic impact and the spring is nonlinear such that a force of 0.18 $x^{3/2}$ N of force is required for a deflection of x mm.

14.39. Assume a perfectly plastic impact as the 5 kg body falls from a height of 2.6 m onto a plate of mass 2.5 kg. This plate is mounted on a spring having a spring constant of 1.77 kN/m. Neglect the mass of the spring as well as friction, and compute the maximum deflection of the spring after impact.

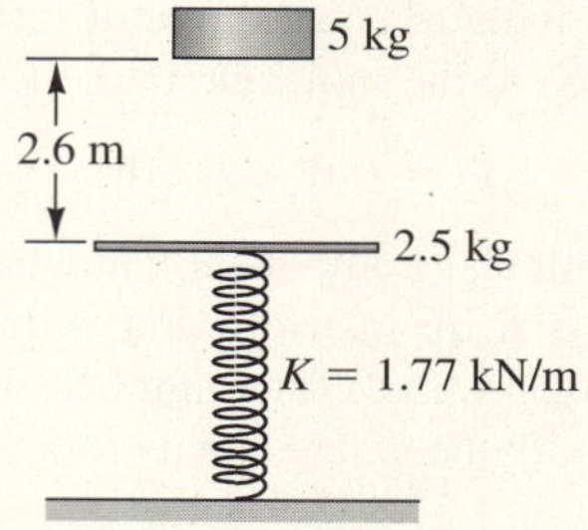

Figure P.14.39.

14.40. Identical spheres B, C, and D lie along a straight line on a frictionless surface. Sphere A, which is identical to the others, moves toward the other spheres at a speed V_A in a direction collinear with the centers of the spheres. For perfectly elastic collisions, what are the final velocities of the bodies?

14.41. In Problem 14.40, (a) What is the final velocity of sphere D if $\epsilon = 0.80$ for all spheres and $V_A = 15$ m/s? (b) Set up a relation for the speed of the $(n + 1)$th sphere in terms of the speed of the nth sphere, again for $\epsilon = 0.80$ and $V_A = 15$ m/s.

14.42. A spherical mass M_1 of 10 kg is held at an angle θ_1 of 60° before being released. It strikes mass M_2 of 5 kg with an impact having a coefficient of restitution equal to 0.75. Mass M_2 is held by a light rod of length 0.6 m at the end of which is a torsional spring requiring 700 N m per radian of rotation. The spring has no torque when l_2 is vertical. What is the maximum rotation of l_2 after impact? The length of $l_1 = 450$ mm. [*Hint:* The work of a couple C rotating on angle $d\theta$ is $C\ d\theta$. A trial-and-error solution for θ_2 will be necessary.]

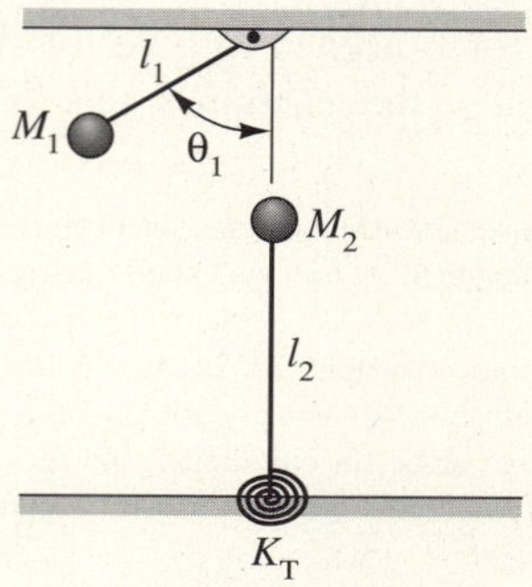

Figure P.14.42.

14.43. Cylinder A, of mass 10 kg, is moving at a speed of 6 m/s when it is at a distance 3 m from cylinder B, which is stationary. Cylinder B, of mass 7.5 kg, has a dynamic coefficient of friction with the rod on which it rides of 0.3. Cylinder A has a dynamic coefficient of friction of 0.1 with the rod. What is the coefficient of restitution if cylinder B comes to rest after collision at a distance 3.5 m to the right of the initial position?

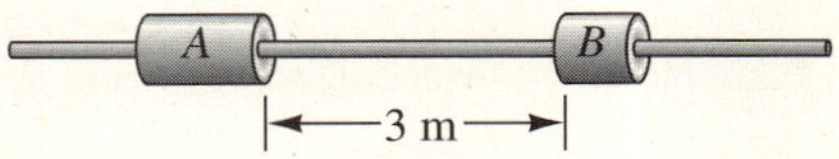

Figure P.14.43.

14.44. A load is being lowered at a speed of 2 m/s into a barge. The barge weighs 1 MN, and the load weighs 100 kN. If the load hits the barge at 2 m/s and the collision is plastic, what is the maximum depth that the barge is lowered into the water, assuming that the position of loading is such as to maintain the barge in a horizontal position? The width of the barge is 10 m. What are the weaknesses (if any) of your analysis? The density of water is 1 Mg/m^3. [Hint: Recall the Archimedes Principle]

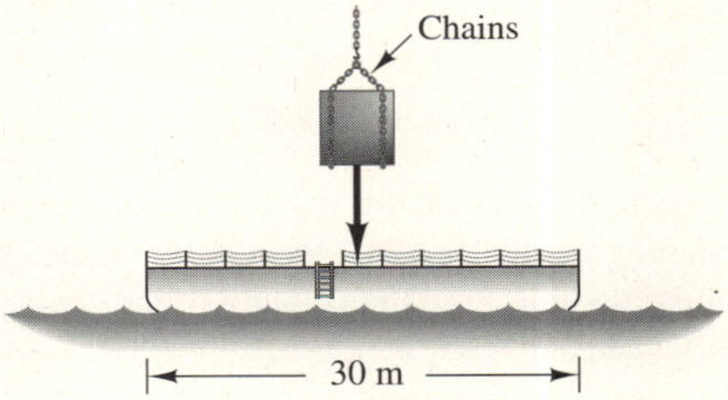

Figure P.14.44.

14.45. A tractor-trailer weighing 50 kN without a load carries a 10 kN load A as shown. The driver jams on his brakes until they lock for a panic stop. The load A breaks loose from its ropes. When the truck has stopped the load is 3 m from the left end of the trailer wall (see diagram) and is moving at a speed of 4 m/s relative to the truck. The coefficient of dynamic friction between the load A and the trailer is 0.2 and between the tires and road is 0.5. If there is a plastic impact between A and the trailer and the driver keeps his brakes locked, how far d does the truck then move?

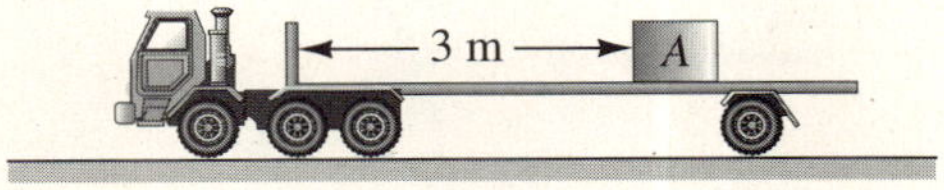

Figure P.14.45.

14.46. Two identical cylinders, each of mass 5 kg, slide on a frictionless rod. Each is fastened to a linear spring (K = 5 kN/m) whose unstretched length is 0.65 m. The spring mass is negligible. If the cylinders are released from rest by raising the restraints,

(a) What is their speed just after colliding with a coefficient of restitution of 0.6?

(b) How close do they come to the walls?

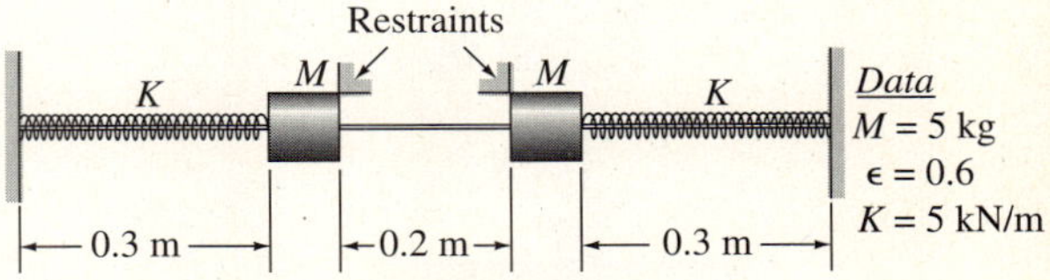

Figure P.14.46.

14.47. A light arm, connected to a mass A, is released from rest at a horizontal orientation. Determine the maximum deflection of the linear spring (K = 3 kN/m) after A impacts with body B with a coefficient of restitution equal to 0.8. If body B does not reach the spring, indicate this fact. Note that there is Coulomb friction between the body B and the floor with μ_d = 0.6. Consider bodies A and B to be small.

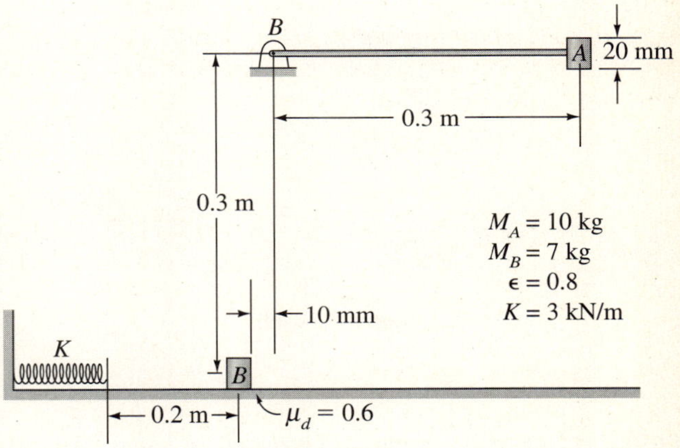

Figure P.14.47.

14.48. Mass M_A slides down the frictionless rod and hits mass M_B, which rests on a linear spring. The coefficient of restitution ϵ for the impact is 0.8. What is the total maximum deflection δ of the spring?

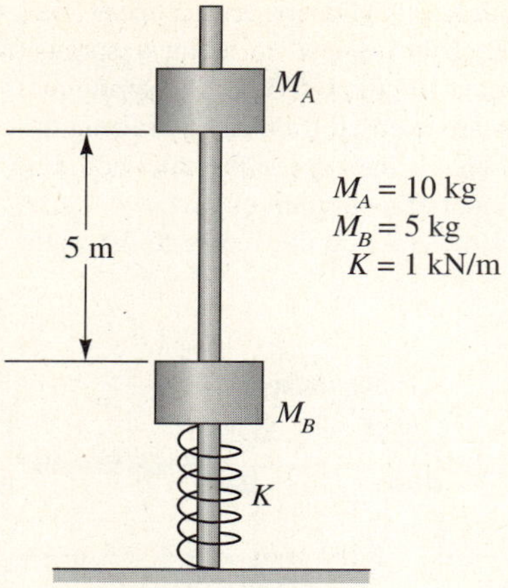

Figure P.14.48.

14.49. A cart A having a mass of 5 kg is released from rest at the position shown. As it rolls along, a constant resisting force of 4 N acts between the wheels and the surface. The cart collides with a block B having a mass of 3 kg. The coefficient of restitution is 0.5. Determine the maximum deflection δ_{max} of the spring having a spring constant of 15 N/m. Also, determine the maximum angle of rotation θ_{max} of the light rod supporting block B. The blocks are small.

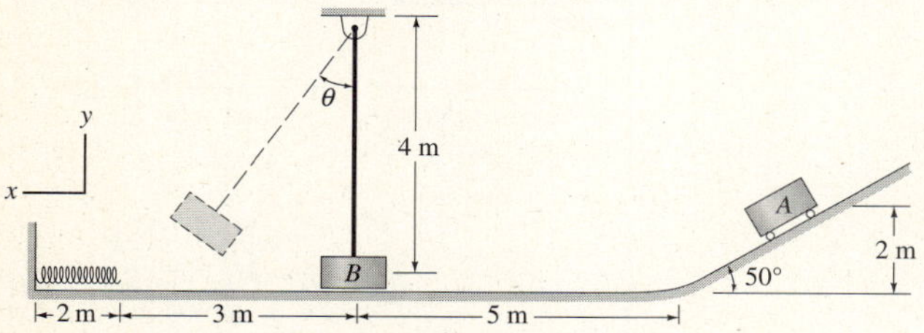

Figure P.14.49.

14.50. A three-seater racing scull is poised for a start. The mass of the scull is 140 kg, and each occupant is about 70 kg. We want to know the speed of the scull after 2 s. At the sound of the starting gun, each man exerts a 140 N constant push on the water from each oar in the direction of the axis of the boat. At the 2 s mark, each man is moving to the left relative to the hull with a speed of 0.3 m/s. Neglect the inertia of the oars as well as water and air friction.

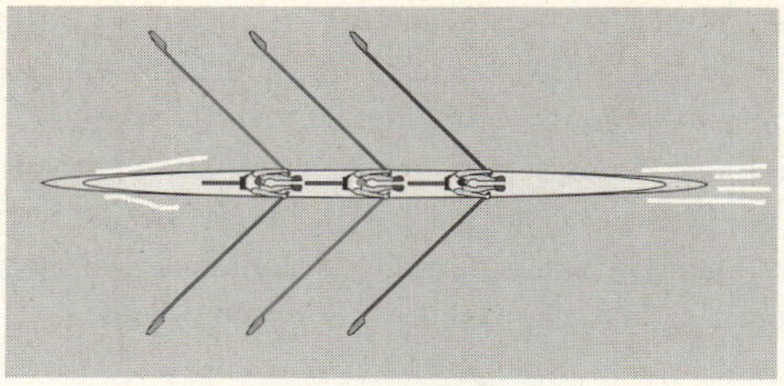

Figure P.14.50.

14.51. Do Example 14.11 for a constant pressure $p = 350$ kPa.

14.52. A thin disc A of mass 2.5 kg translates along a frictionless surface at a speed of 6 m/s. The disc strikes a square stationary plate B, mass 5 kg, at the center of a side. What are the velocity and direction of motion of the plate and the disc after collision? Assume that the surfaces of the plate and disc are smooth. Take $\epsilon = 0.7$.

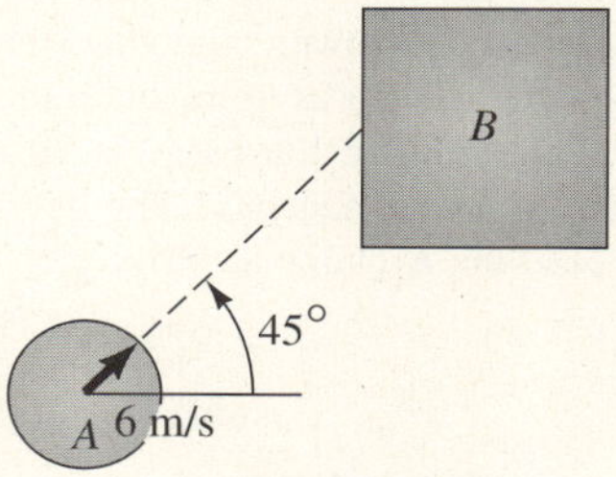

Figure P.14.52.

14.53. In Problem 14.52 at the instant of contact between the bodies a clamping device firmly connects the bodies together so as to form one rigid unit. Find the velocity of the center of mass of the system after impact.

14.54. The theory of collisions of *subatomic* particles is called the theory of *scattering*. The coefficient-of-restitution concept presented for macroscopic bodies in this chapter cannot be used. However, conservation of momentum can be used.

A neutron N shown moving with a speed V_0 strikes a stationary proton P. After collision the velocity of the neutron is V_N and that of the proton is V_P, as shown in the diagram. For an *elastic* collision, prove that $\phi + \theta = \pi/2$. [*Hint:* Use the vector polygon concept and the Pythagorean theorem. Also, take the mass of proton and neutron to be equal.]

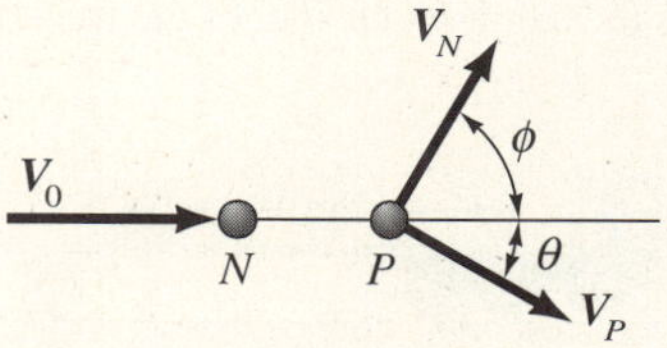

Figure P.14.54.

***14.55.** A neutron N is moving toward a stationary helium nucleus He (atomic number 2) with kinetic energy 10 MeV. If the collision is inelastic, causing a loss of 20% of the kinetic energy, what is the angle θ after collision? See the first paragraph (only) of Problem 14.54. [*Hint:* There is no need (if one is clever) to have to convert the atomic number to kilograms.]

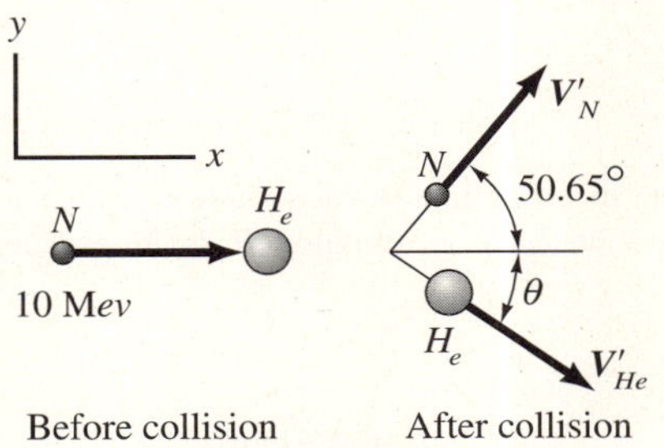

Figure P.14.55.

14.56. Cylinders A and B are free to slide without friction along a rod. Cylinder A is released from rest with spring K_1 to which it is connected initially unstretched. The impact with cylinder B has a coefficient of restitution ϵ equal to 0.8. Cylinder B is at rest before the impact supported in the position shown by spring K_2. Assume springs are massless.

(a) How much is the lower spring compressed initially?

(b) How much does cylinder B descend after impact before reaching its lowest position?

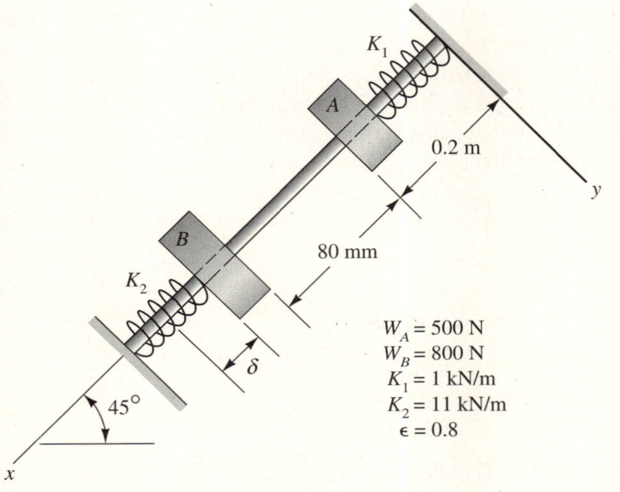

Figure P.14.56.

14.57. Masses A and B slide on a rod which is frictionless. The spring is initially *compressed* from 0.8 m to the position shown. The system is released from rest. A and B undergo a *plastic* impact. The spring is massless.

(a) What is the speed of the masses after B moves 0.2 m?

(b) What is the loss in *mechanical energy* for the system?

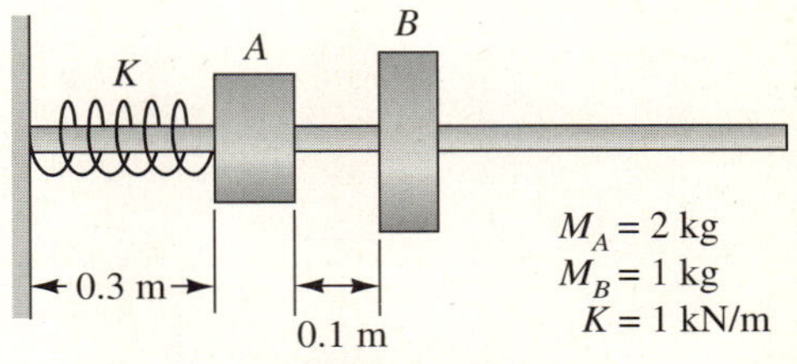

Figure P.14.57.

14.58. A ball is thrown against a floor at an angle of 60° with a speed at impact of 16 m/s. What is the angle of rebound α if $\epsilon = 0.7$? Neglect friction.

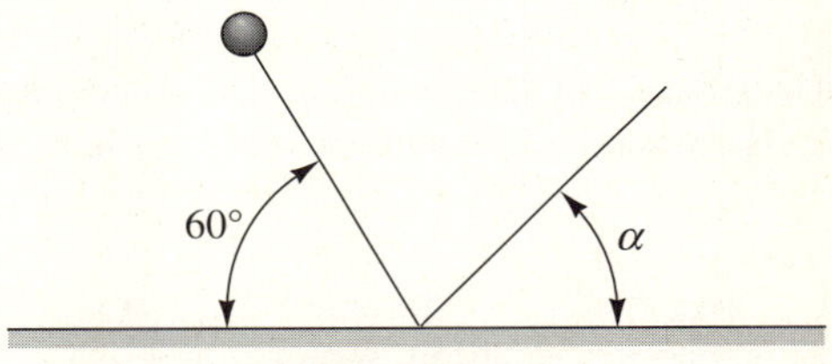

Figure P.14.58.

14.59. A ball strikes the xy plane of a handball court at $\boldsymbol{r} = 0.9\boldsymbol{i} + 2.1\boldsymbol{j}$ m. The ball has initially a velocity $\boldsymbol{V}_1 = -3\boldsymbol{i} - 3\boldsymbol{j} - 4.5\boldsymbol{k}$ m/s. The coefficient of restitution is 0.8. Determine the final velocity $\boldsymbol{V}_2$ after it bounces off the xy, yz, and xz planes once. Neglect gravity and friction.

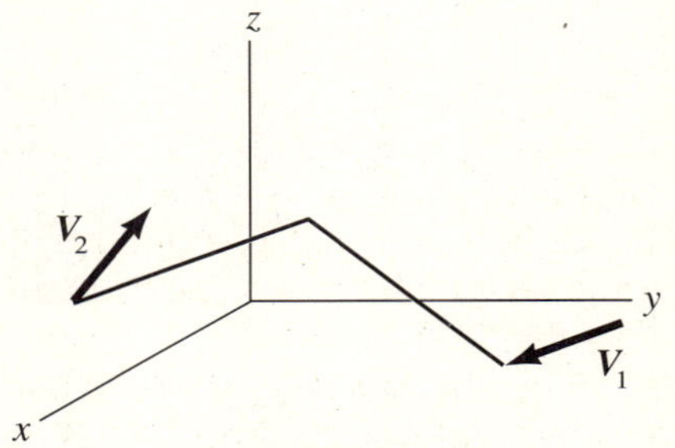

Figure P.14.59.

14.60. A space vehicle in the shape of a cone–cylinder is moving at a speed V m/s, many times the speed of sound through highly rarefied atmosphere. If each molecule of the gas has a mass m kg and if there are, on the average, n molecules per cubic metre, compute the drag on the cone–cylinder. The cone half-angle is 30°. Take the collision to be perfectly elastic.

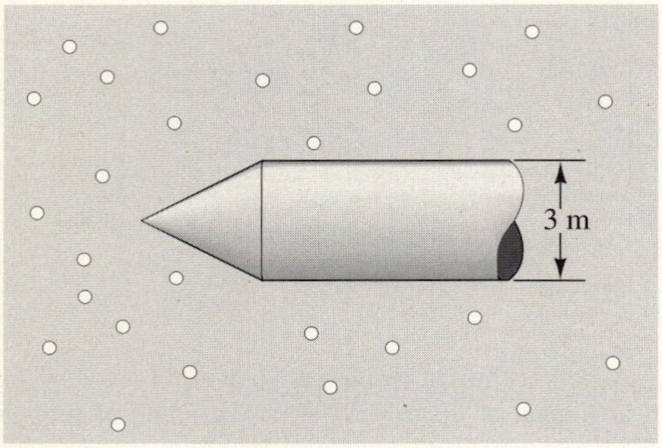

Figure P.14.60.

14.61. Do Problem 14.60 for a case where the collisions are assumed to be inelastic. Assume the coefficient of restitution to be 0.8.

14.62. A double-wedge airfoil section for a space glider is shown. If the glider moves in highly rarefield atmosphere at a speed V many times greater than the speed of sound, what is the drag per unit length of this airfoil? Assume the collision to be perfectly elastic. There are n molecules per m^3, each having a mass m in kg.

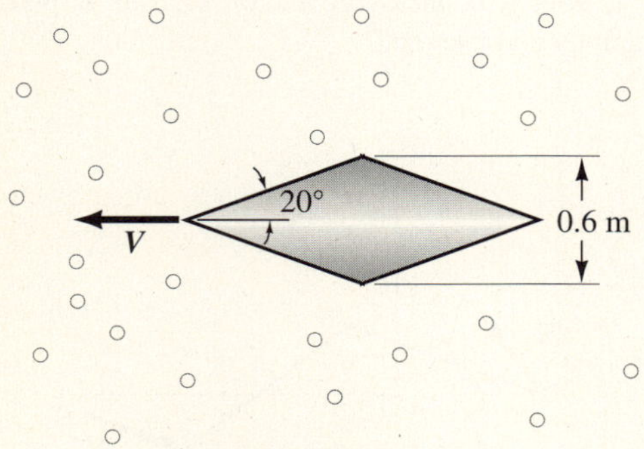

Figure P.14.62.

14.63. Consider a parallel beam of light having an energy flux of S W/m^2, shining normal to a flat surface that completely absorbs the energy. You learned in physics that an impulse dI is developed on the surface during time dt given by the formula

$$dI = \frac{S}{c}\, dt\, dA$$

where c is the speed of light in vacuo in m/s. If the surface reflects the light, then we have an impulse dI developed on the surface given as

$$dI = 2\frac{S}{c}\, dt\, dA$$

Compute the force stemming from the reflection of light shining normal to a perfectly reflecting mirror having an area of 1 m^2. The light has an energy flux S of 20 W/m^2. Take the speed c = 300 Mm/s. What is the radiation pressure p_{rad} on the mirror?

***14.64.** The Echo satellite when put into orbit is inflated to a 45-m-diameter sphere having a skin made up of a laminate of aluminum over mylar over aluminum. This skin is highly reflectant of light. Because of the small mass of this satellite, it may be affected by small forces such as that stemming from the reflection of light. If a parallel beam of light having an energy density S of 0.5 W/mm^2 impinges on the Echo satellite, what total force is developed on the satellite from this source? From physics (see Problem 14.63), the radiation pressure, p_{rad}, on a reflecting surface from a beam of light inclined by θ° from the normal to the surface is

$$p_{\text{rad}} = 2\frac{S}{c}\cos^2\theta$$

The pressure is in the direction of the incident radiation.

Part B: Moment of Momentum

14.6 Moment-of-Momentum Equation for a Single Particle

At this time, we shall introduce another auxiliary statement that follows from Newton's law and that will have great value when extended to the case of a rigid body. We start with Newton's law for a particle in the following form:

$$\boldsymbol{F} = \frac{d}{dt}(m\boldsymbol{V}) = \dot{\boldsymbol{P}}$$

where the symbol $\boldsymbol{P}$ represents the linear momentum of the particle. We next take the moment of each side of the equation about a point a in space (see Fig. 14.24):

$$\boldsymbol{\rho}_a \times \boldsymbol{F} = \boldsymbol{\rho}_a \times \dot{\boldsymbol{P}} \tag{14.16}$$

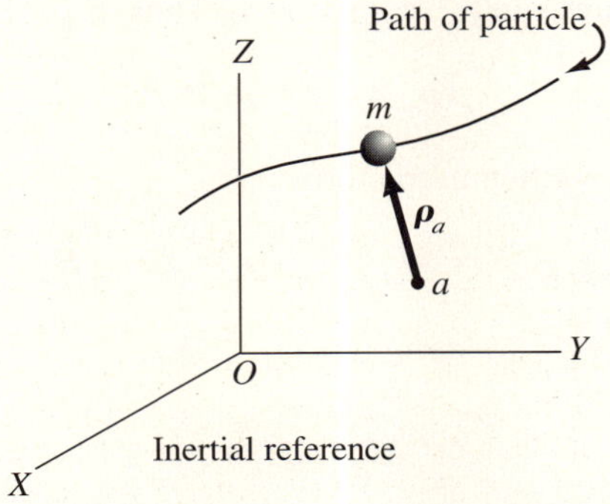

Figure 14.24. Point a fixed in inertial space.

If this point a is positioned at a fixed location in XYZ, we can simplify the right side of Eq. 14.16. Accordingly, examine the expression $(d/dt)(\boldsymbol{\rho}_a \times \boldsymbol{P})$:

$$\frac{d}{dt}(\boldsymbol{\rho}_a \times \boldsymbol{P}) = \boldsymbol{\rho}_a \times \dot{\boldsymbol{P}} + \dot{\boldsymbol{\rho}}_a \times \boldsymbol{P} \tag{14.17}$$

But the expression $\dot{\boldsymbol{\rho}}_a \times \boldsymbol{P}$ can be written as $\dot{\boldsymbol{\rho}}_a \times m\dot{\boldsymbol{r}}$. The vectors $\boldsymbol{\rho}_a$ and $\boldsymbol{r}$ are measured in the same reference from a fixed point a to the particle and from the origin to the particle, respectively (see Fig. 14.25). They are thus different at all times to the extent of a constant vector $\overrightarrow{Oa}$. Note that

$$\boldsymbol{r} = \overrightarrow{Oa} + \boldsymbol{\rho}_a$$

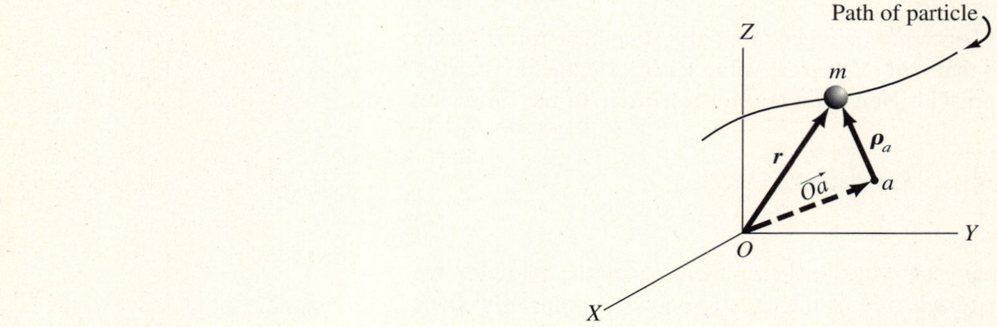

Figure 14.25. Position vectors to m and a.

Therefore,

$$\dot{\boldsymbol{r}} = \dot{\boldsymbol{\rho}}_a$$

Accordingly, the expression $\dot{\boldsymbol{\rho}}_a \times m\dot{\boldsymbol{r}}$ is zero. Thus, Eq. 14.17 becomes

$$\frac{d}{dt}(\boldsymbol{\rho}_a \times \boldsymbol{P}) = \boldsymbol{\rho}_a \times \dot{\boldsymbol{P}} \tag{14.18}$$

and Eq. 14.16 can be written in the form

$$\boldsymbol{\rho}_a \times \boldsymbol{F} = \boldsymbol{M}_a = \frac{d}{dt}(\boldsymbol{\rho}_a \times \boldsymbol{P}) = \dot{\boldsymbol{H}}_a$$

Therefore,

$$\boldsymbol{M}_a = \dot{\boldsymbol{H}}_a \tag{14.19}$$

where $\boldsymbol{H}_a$ *is the moment about point a of the linear momentum vector.* Also, $\boldsymbol{H}$ is termed the *angular momentum vector.* Equation 14.19, then, states that *the moment $\boldsymbol{M}_a$ of the resultant force on a particle about a point a, fixed in an inertial reference, equals the time rate of change of the moment about point a of the linear momentum of the particle relative to the inertial reference.* This is the desired alternative form of Newton's law.

The scalar component of Eq. 14.19 along some axis, say the z axis, can be useful. Thus,

$$M_z = \dot{H}_z$$

where M_z is the torque of the total external force about the z axis and H_z is the moment of the momentum (or angular momentum) about the z axis.

Example 14.14

A boat containing a man is moving near a dock (see Fig. 14.26). He throws out a light line and lassos a piling on the dock at A. He starts drawing in on the line so that when he is in the position shown in the diagram, the line is taut and has a length of 7.5 m. His speed V_1 is 1.5 m/s in a direction normal to the line. If the net horizontal force F on the boat from tension in the line and from water resistance is maintained at 220 N essentially in the direction of the line, what is the component of his velocity toward piling A (i.e., V_A) after the man has pulled in 1 m of line? The boat and the man have a combined mass of 160 kg.

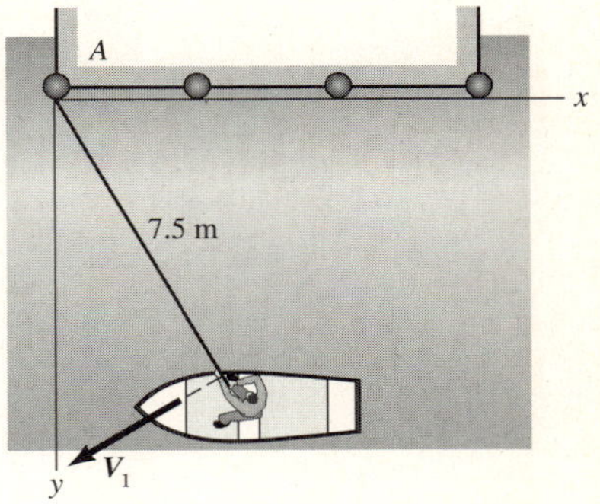

Figure 14.26. Man pulls toward piling.

We may consider the boat and man as a particle for which we can apply the **moment of momentum** equation. Thus,

$$\boldsymbol{M}_A = \dot{\boldsymbol{H}}_z \qquad \text{(a)}$$

Clearly, here $\boldsymbol{M}_A = \boldsymbol{0}$ since $\boldsymbol{F}$ goes through A at all times. Thus, $\boldsymbol{H}_A$ is a constant—that is, the angular momentum about A must be constant. Observing Fig. 14.27, we can say accordingly

$$\boldsymbol{r}_1 \times m\boldsymbol{V}_1 = \boldsymbol{r}_2 \times m\boldsymbol{V}_2$$

Since $\boldsymbol{r}_1$ is perpendicular to $\boldsymbol{V}_1$ and $\boldsymbol{r}_2$ is perpendicular to $(\boldsymbol{V}_2)_t$, we get a simple scalar product from above. Thus

$$(7.5)(m)(1.5) = (6.5)(m)(V_2)_t$$

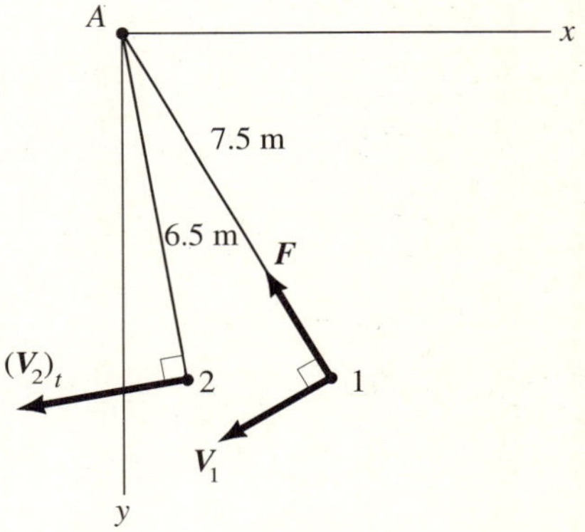

Figure 14.27. Boat at positions 1 and 2.

Therefore,

$$(V_2)_t = 1.73 \text{ m/s} \qquad \text{(b)}$$

We need more information to get the desired result V_A toward the piling. We have not yet used the fact that $F = 220$ N. Accordingly, we now employ the **work–kinetic energy** equation from Chapter 13. Thus,

$$\int_1^2 \boldsymbol{F} \bullet d\boldsymbol{s} = \left(\frac{1}{2}MV^2\right)_2 - \left(\frac{1}{2}MV^2\right)_1$$

$$(220)(1) = \frac{1}{2}160(V_2)^2 - \frac{1}{2}160(2.25)$$

Therefore,

$$V_2 = 2.34 \text{ m/s} \qquad \text{(c)}$$

Now V_2 is the *total* velocity of the boat at position 2. To get the desired component V_A toward the piling, we can say, using Eqs. (b) and (c):

$$V_2^2 = (V_2)_t^2 + V_A^2$$

$$(2.34)^2 = (1.73)^2 + V_A^2$$

Therefore,

$$V_A = 1.58 \text{ m/s}$$

14.7 More on Space Mechanics

Many problems of space mechanics can be solved by using energy and angular-momentum methods of this and the preceding chapter without considering the detailed trajectory equations of Chapter 12. Let us therefore set forth some salient factors concerning the motion of a space vehicle moving in the vicinity of a planet or star with the engine shut off and with negligible friction from the outside.[11]

After the space vehicle has been propelled at great speed by its rocket engines to a position outside the planet's atmosphere (the final powered velocity is called the *burnout* velocity), the vehicle then undergoes plane, gravitational, *central-force motion* (Section 12.6). If it continues to go around the planet, the vehicle is said to go into *orbit* and the trajectory is that of a circle or that of an ellipse. If, on the other hand, the vehicle escapes from the influence of the planet, then the trajectory will either be a parabola or a hyperbola. In the case of an elliptic orbit, the position closest to the surface of the planet is called *perigee* (see Fig. 14.28) and the position farthest from the surface of the planet is called *apogee*. Notice that at apogee and perigee the velocity vectors $\boldsymbol{V}_a$ and $\boldsymbol{V}_p$ of the vehicle are parallel to the surface of the planet and so at these points (and only at these points)

$$V = V_\theta; \qquad V_r = 0$$

In the case of a *circular* orbit of radius r and velocity V_c, we can use Newton's law and the gravitational law to state

$$\frac{GMm}{r^2} = m(r\omega^2)$$

where M is the mass of the planet and ω is the angular speed of the radius vector to the vehicle. Replacing the acceleration term $r\omega^2$ by V_c^2/r and solving for V_c, we get

$$V_c = \sqrt{\frac{GM}{r}} \tag{14.20}$$

Knowing GM and r, we can readily compute the speed V_c for a particular circular orbit. In Section 12.6 we showed that GM can be easily computed using the relation

$$GM = gR^2 \tag{14.21}$$

[11]Those readers who have studied Sections 12.7 and 12.8 have already gone into these factors in considerable depth.

where g is the acceleration of gravity at the surface of the planet and R is the radius of the planet.

In gravitational central-force motion, only the conservative force of gravity is involved, and so we must have *conservation of mechanical energy*. Furthermore, since this force is directed to O, the center of the planet, at all times (see Fig. 14.28), then the moment about O of the gravitational force must be zero. As a consequence, we must have *conservation of angular momentum* about O.[12]

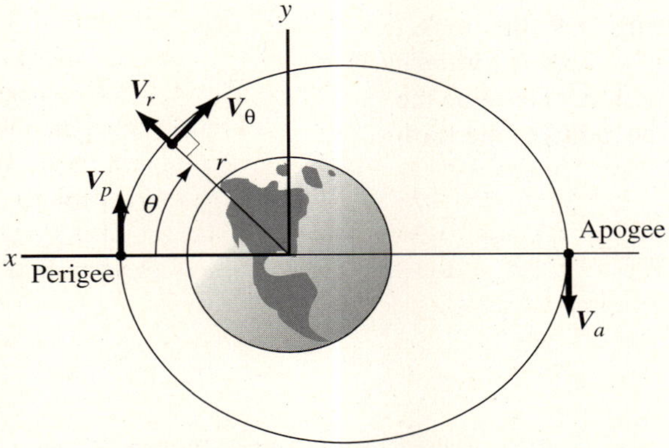

Figure 14.28. Elliptic orbit with perigee and apogee.

We shall illustrate in the next example the dual use of the conservation-of-angular-momentum principle and the conservation-of-mechanical-energy principle for space mechanics problems. In the homework problems you will be asked to solve again some of the space problems of Chapter 12 using the principles above without getting involved with the trajectory equations. Such problems, you will then realize, are sometimes more easily solved by using the two principles discussed above rather than by using the trajectory equations.

[12]Those who have studied the trajectory equations of Chapter 12 might realize that

$$C = rV_\theta = \text{constant}$$

is actually a statement of the conservation of angular momentum since mrV_θ is the moment about O of the linear momentum relative to O.

Example 14.15

A space-shuttle vehicle on a rescue mission (see Fig. 14.29) is sent into a circular orbit at a distance of 1.2 Mm above the earth's surface. This orbit is inserted so as to be in the same plane as that of a spacecraft whose rocket engines will not start, thus preventing it from initiating a procedure for returning to earth. The goal of the shuttle is to enter a trajectory that will permit docking with the disabled spacecraft and then to rescue the occupants. The timing of insertion of the circular orbit of the space shuttle has so been chosen that the space shuttle by firing its rockets at the position shown can, by the proper change of speed, reach apogee at the same time and same location as does the crippled space vehicle. At this time, docking procedures can be carried out. Considering that the rocket engines of the space-shuttle vehicle operate during a *very short distance* of travel[13] to achieve the proper velocity V_0 for the mission, determine the change in speed that the space shuttle must achieve. The radius of the earth is 6.373 Mm.

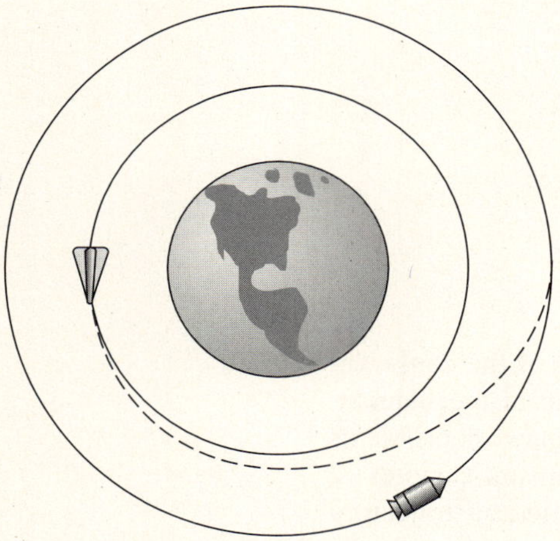

Figure 14.29. Rescue mission for space-shuttle vehicle.

[13]During this part of the flight we do *not* have central-force motion.

Example 14.15 (Continued)

We shall first compute GM. Thus, working with metres and seconds,

$$GM = gR^2 = (9.81)(6.373 \times 10^6)^2 = 398.4 \times 10^{12} \text{ m}^3/\text{s}^2$$

The velocity for the circular orbit for the space shuttle is then

$$V_c = \sqrt{\frac{GM}{r}} = \sqrt{\frac{398.4 \times 10^{12}}{7.573 \times 10^6}} = 7.253 \text{ km/s} \qquad \text{(a)}$$

From **conservation of angular momentum** for the space-shuttle rescue orbit we can say

$$\begin{aligned} m r_0 V_0 &= m(rV)_{\text{apogee}} \\ (7.573 \times 10^6)(V_0) &= (10 \times 10^6)(V)_{\text{apogee}} \end{aligned}$$

Therefore,

$$V_0 = 1.320 V_{\text{apogee}} \qquad \text{(b)}$$

where V_0 is the speed of the space shuttle just *after* firing rockets. Next, we use the principle of **conservation of mechanical energy** for the rescue orbit. Thus,

$$\begin{aligned} -\frac{GMm}{r_0} + \frac{mV_0^2}{2} &= -\frac{GMm}{r} + \frac{mV_{\text{apogee}}^2}{2} \\ -\frac{398.4 \times 10^{12}}{7.573 \times 10^6} + \frac{V_0^2}{2} &= -\frac{398.4 \times 10^{12}}{10 \times 10^6} + \frac{V_{\text{apogee}}^2}{2} \end{aligned} \qquad \text{(c)}$$

Substitute for V_{apogee} using Eq. (b) and solve for V_0. We get

$$V_0 = 7.742 \text{ km/s} \qquad \text{(d)}$$

Hence, using Eq. (a), we can say:

$$\Delta V = 7.742 - 7.253 = \boxed{489 \text{ m/s}}$$

PROBLEMS

14.65. A particle rotates at 30 rad/s along a frictionless surface at a distance 0.6 m from the center. A flexible cord restrains the particle. If this cord is pulled so that the particle moves inward at a velocity of 1.5 m/s, what is the magnitude of the total velocity when the particle is 0.3 m from the center?

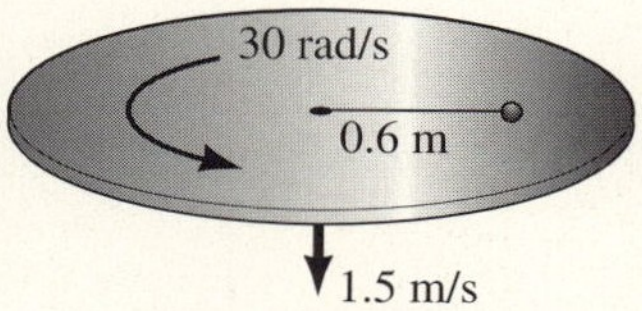

Figure P.14.65.

14.66. A satellite has an apogee of 7.128 Mm. It is moving at a speed of 10.13 km/s. What is the transverse velocity of the satellite when $r = 6.97$ Mm?

14.67. A system is shown rotating freely with an angular speed ω of 2 rad/s. A mass A of 1.5 kg is held against a spring such that the spring is compressed 100 mm. If the device a holding the mass in position is suddenly removed, determine how far toward the vertical axis of the system the mass will move. The spring constant K is 0.531 N/mm. Neglect all friction and inertia of the bar. The spring is not connected to the mass.

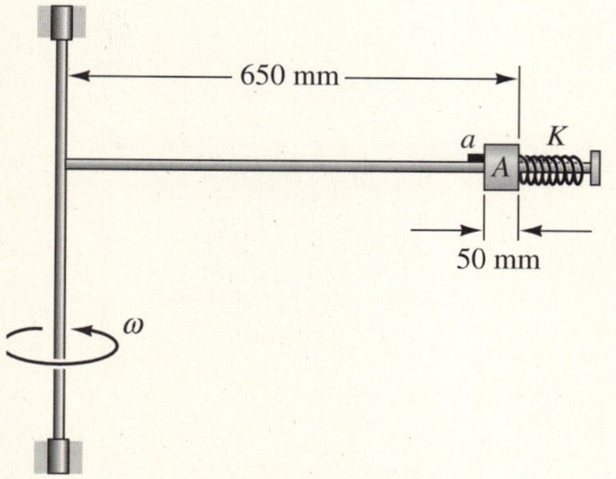

Figure P.14.67.

14.68. Do Problem 14.67 for the case where there is Coulombic friction between the mass A and the horizontal rod with a constant μ_d equal to 0.4.

14.69. A body A of mass 5 kg is moving initially at a speed of V_1 of 6 m/s on a frictionless surface. An elastic cord AO, which has a length l of 6 m, becomes taut but not stretched at the position shown in the diagram. What is the radial speed toward O of the body when the cord is stretched 0.6 m? The cord has an equivalent spring constant of 50 N/m

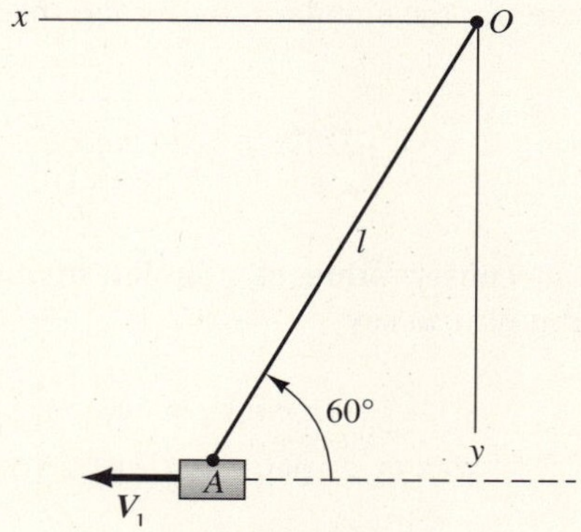

Figure P.14.69.

14.70. A small ball B of mass 1 kg is rotating about a vertical axis at a speed ω_1 of 15 rad/s. The ball is connected to bearings on the shaft by light inextensible strings having a length l of 0.6 m. The angle θ_1 is 30°. What is the angular speed ω_2 of the ball if bearing A is moved up 150 mm?

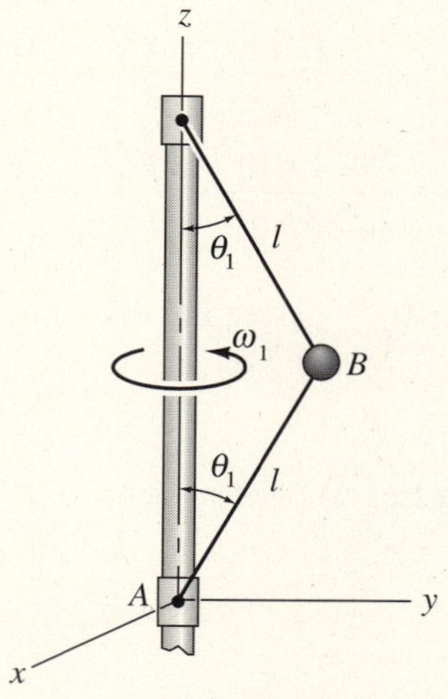

Figure P.14.70.

14.71. A mass m of 1 kg is swinging freely about the z axis at a speed ω_1 of 10 rad/s. The length l_1 of the string is 250 mm. If the tube A through which the connecting string passes is moved down a distance d of 90 mm, what is ω_2 of the mass? You should get a fourth-order equation for ω_2 which has as the desired root $\omega_2 = 21.05$ rad/s.

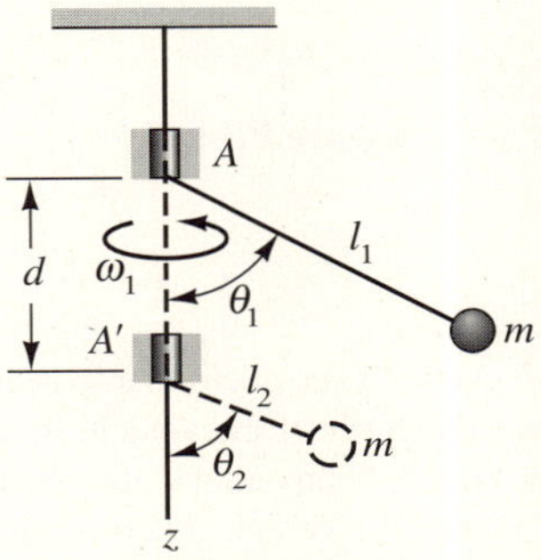

Figure P.14.71.

14.72. A small 1 kg ball B is rotating at angular speed ω_1 of 10 rad/s about a horizontal shaft. The ball is connected to the bearings with light elastic cords which when unstretched are each 300 mm in length. A force of 65 N is required to stretch the cord 25 mm. The distance d_1 between the bearings is originally 500 mm. If bearing A is moved to shorten d by 150 mm, what is the angular velocity ω_2 of the ball? Neglect the effects of gravity and the mass of the elastic cords. [*Hint:* You should arrive at a transcendental equation for θ_2 whose solution is 54.49°.]

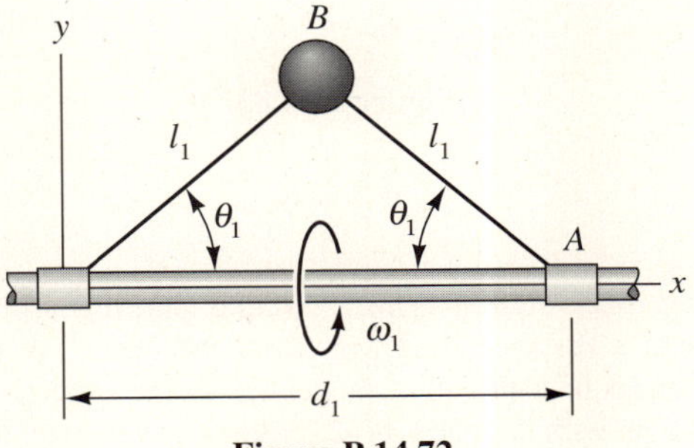

Figure P.14.72.

14.73. A space vehicle is moving at a speed of 10.3 km/s at position A, which is perigee at a distance of 250 km from the earth's surface. What are the radial and transverse velocity components as well as the distance from the earth's surface at B? The trajectory is in the xy plane.

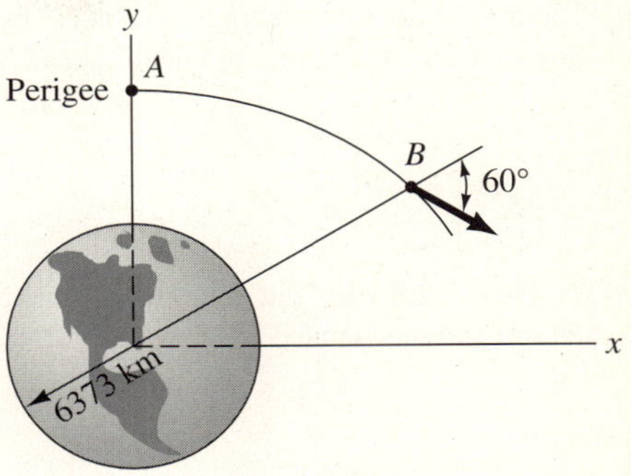

Figure P.14.73.

14.74. A space vehicle is in orbit A around the earth. At position (1) it is 8 Mm from the center of the earth and has a velocity of 9 km/s. The transverse velocity at (1) is 6.75 km/s. At apogee, it is desired to continue in the circular orbit shown dashed. What change in speed is needed to change orbits when firing at apogee?

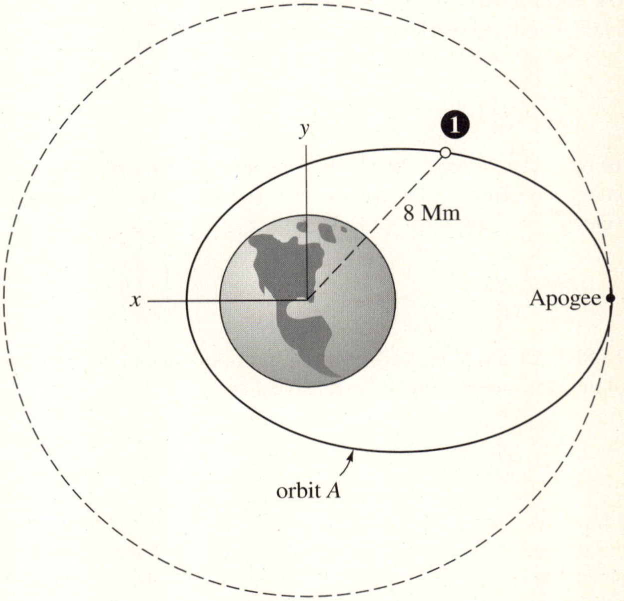

Figure P.14.74.

14.75. Do Problem 12.75 using the principles of conservation of momentum and conservation of mechanical energy.

14.76. In Problem 12.86 find the radial velocity by using the method of conservation of angular momentum and mechanical energy.

14.77. Do Problem 12.82 by the method of conservation of angular momentum and mechanical energy.

14.78. In Problem 12.87, find the height of the bullet above the surface of the planet by the methods of conservation of angular momentum and mechanical energy.

14.79. In Problem 12.119, find the maximum elevation above the earth's surface by the methods of conservation of angular momentum and mechanical energy.

14.80. Do Problem 12.114 by methods of conservation of angular momentum and mechanical energy. [*Hint:* The escape velocity $= \sqrt{2GM/r} = \sqrt{2}\,V_c$.]

14.81. Do Problem 12.113 using the principles of conservation of angular momentum and mechanical energy.

14.82. A space vehicle is in a circular orbit 1.2 Mm above the surface of the earth. A projectile is shot from this space vehicle at a speed relative to the vehicle of 1.4 km/s in a radial direction as seen from the vehicle. What are the *apogee* and the *perigee* distances from the center of the earth for the trajectory of the projectile?

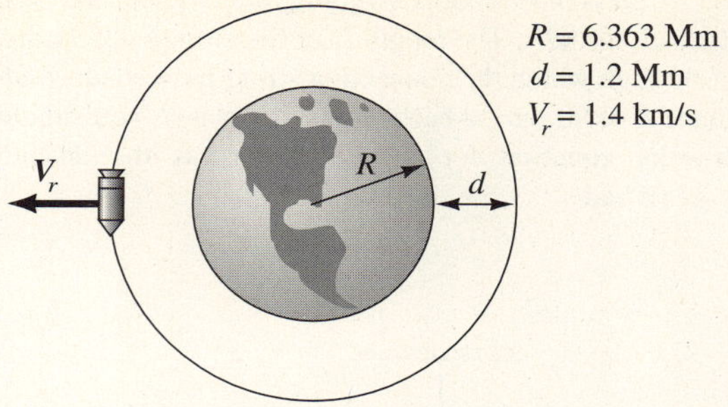

Figure P.14.82.

14.83. A space vehicle is in a circular parking orbit 500 km above the surface of the earth. If the vehicle is to reach an apogee at location 2 which is 850 km above the earth's surface, what increase in velocity must the vehicle attain by firing its rockets for a short time at location 1? The radius of the earth is 6.37 Mm.

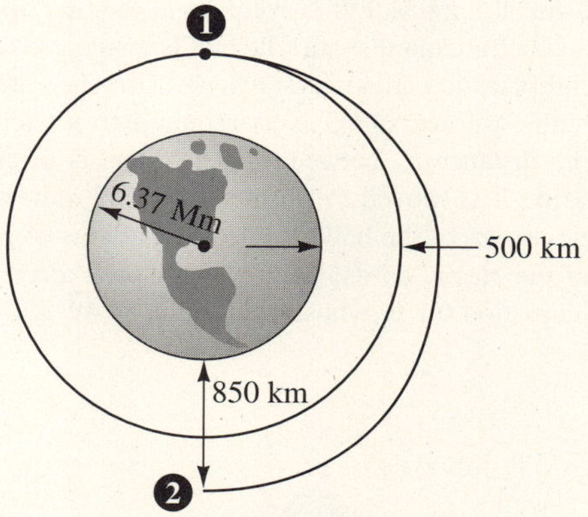

Figure P.14.83.

14.84. A space station is in a circular parking orbit around the earth at a distance of 8 Mm from the center. A projectile is fired ahead in a direction tangential to the trajectory of the space station with a speed of 2.25 km/s relative to the space station. What is the maximum distance from earth reached by the projectile?

14.85. A skylab is in a circular orbit about the earth 500 km above the earth's surface. A space-shuttle vehicle has rendezvoused with the skylab and now, after disengaging from the skylab, its rocket engines are fired so as to move the vehicle with a speed of 800 m/s relative to the skylab in the opposite direction to that of the skylab. Assume that the firing of the rocket takes place over a short distance and does not affect the skylab. What speed would the space-shuttle vehicle have when it encounters appreciable atmosphere at about 50 km above the earth's surface? What is the radial velocity at this position?

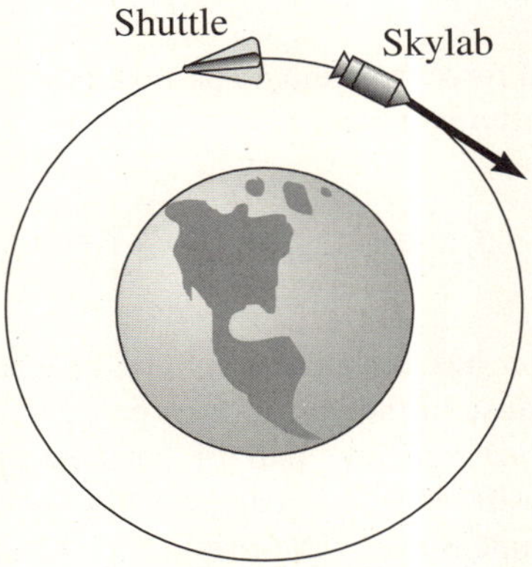

Figure P.14.85.

14.86. A space probe is approaching Mars. When the probe is 80 Mm from the center of Mars it has a speed V_0 of 4.47 km/s with a component $(V_r)_0$ toward the center of Mars of 4.34 km/s. How close does the probe come to the surface of Mars? If retro-rockets are fired at this lowest position A, what change in speed is needed to alter the trajectory into a circular orbit as shown? The acceleration of gravity at the surface of Mars is 3.78 m/s^2, and the radius R of the planet is 3.39 Mm.

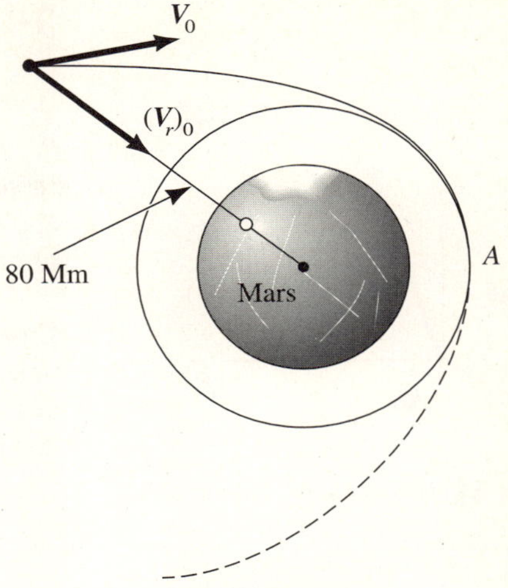

Figure P.14.86.

14.87. In Problem 14.86, a midcourse correction is to be made to get the probe within 1.6 Mm from the surface of Mars. If V_0 at r_0 = 80 Mm is still to be 4.47 km/s, what should be the radial velocity component $(V_r)_0$?

14.88. The Apollo command module is in a circular parking orbit about the moon at a distance of 161.0 km above the surface of the moon. The lunar exploratory module is to detach from the command module. The lunar-module rockets are fired briefly to give a velocity V_0 relative to the command module in the opposite direction. If the lunar module is to have a transverse velocity of 1.5 km/s when it is 80 km from the surface of the moon before rockets are fired again, what must V_0 be? What is the radial velocity at this position? The radius of the moon is 1.733 Mm, and the acceleration of gravity is 1.70 m/s^2 at the surface.

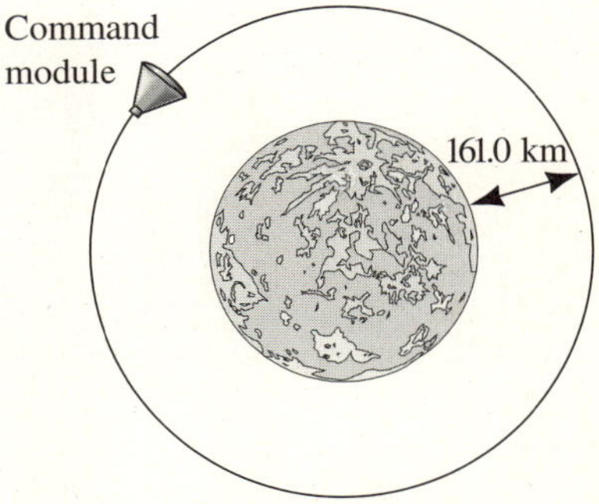

Figure P.14.88.

14.8 Moment-of-Momentum Equations for a System of Particles

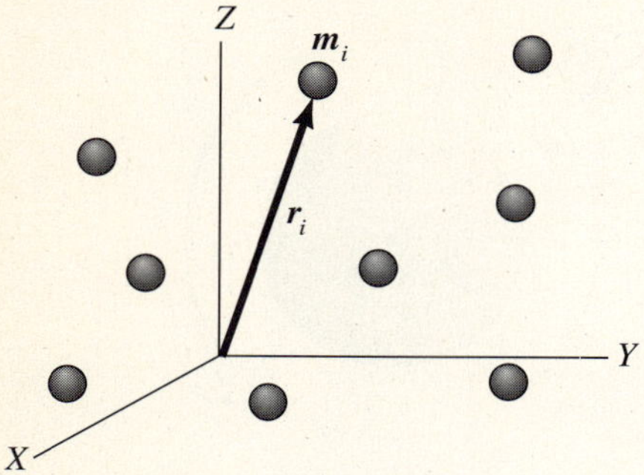

Figure 14.30. System of n particles.

We shall now develop the moment-of-momentum equations for an aggregate of particles. The resulting equations will be of vital importance when we apply them to rigid bodies in later chapters. We shall consider a number of cases.

Case 1. Fixed Reference Point in Inertial Space. An aggregate of n particles and an inertial reference are shown in Fig. 14.30. The moment of momentum equation for the ith particle is now written about the origin of this reference:

$$\boldsymbol{r}_i \times \boldsymbol{F}_i + \boldsymbol{r}_i \times \left(\sum_{\substack{j=1 \\ i \neq j}}^{n} \boldsymbol{f}_{ij} \right) = \frac{d}{dt}(\boldsymbol{r}_i \times \boldsymbol{P}_i) \qquad (14.22)$$

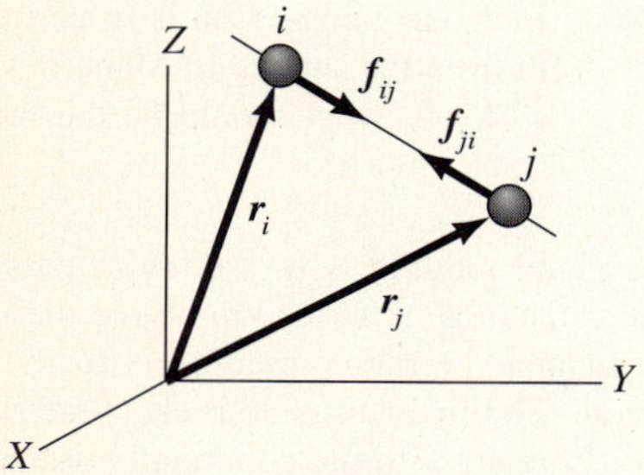

Figure 14.31. Internal equal and opposite forces.

where, as usual, $\boldsymbol{f}_{ij}$ is the internal force from the jth particle on the ith particle. We now sum this equation for all n particles:

$$\sum_{i=1}^{n} \boldsymbol{r}_i \times \boldsymbol{F}_i + \sum_{i=1}^{n}\sum_{j=1}^{n} (\boldsymbol{r}_i \times \boldsymbol{f}_{ij}) = \frac{d}{dt}\left[\sum_{i=1}^{n} (\boldsymbol{r}_i \times \boldsymbol{P}_i) \right] = \dot{\boldsymbol{H}}_{\text{total}} \qquad (14.23)$$

where the summation operation has been put after the differentiation on the right side (permissible because of the distributive property of differentiation with respect to addition). For any pair of particles, the internal forces will be equal and opposite and collinear (see Fig. 14.31). Hence, the forces will have a zero moment about the origin. (This result is most easily understood by remembering that, for purposes of taking moments about a point, forces are transmissible.) We can then conclude that the expression

$$\sum_{i=1}^{n}\sum_{j=1}^{n} (\boldsymbol{r}_i \times \boldsymbol{f}_{ij})$$

in this equation is zero. Realizing that $\sum_i \boldsymbol{r}_i \times \boldsymbol{F}_i$ is the total moment of the external forces about the origin, we have as a result for Eq. 14.23:

$$\boldsymbol{M}_a = \dot{\boldsymbol{H}}_a \qquad (14.24)$$

Thus, *the total moment* $\boldsymbol{M}$ *of external forces acting on an aggregate of particles about a point a fixed in an inertial reference* (the point in the development was picked as the origin merely for convenience) *equals the time rate of change of the total moment of the linear momentum relative to the inertial reference, where this moment is taken about the aforementioned point a.*[14]

We may express Eq. 14.24 in a different form by considering the *center of mass*. In a manner analogous to kinetic energy of an aggregate of particles, we can first show that the angular momentum of an aggregate of particles about a fixed point can be given as the angular momentum of the center of mass about the fixed point plus the angular momentum of the particles relative to the center of mass.

Accordingly, consider the center of mass c of an aggregate of particles as shown in Fig. 14.32. For the ith particle, we can say:

$$\boldsymbol{r}_i = \boldsymbol{r}_c + \boldsymbol{\rho}_{ci} \tag{14.25}$$

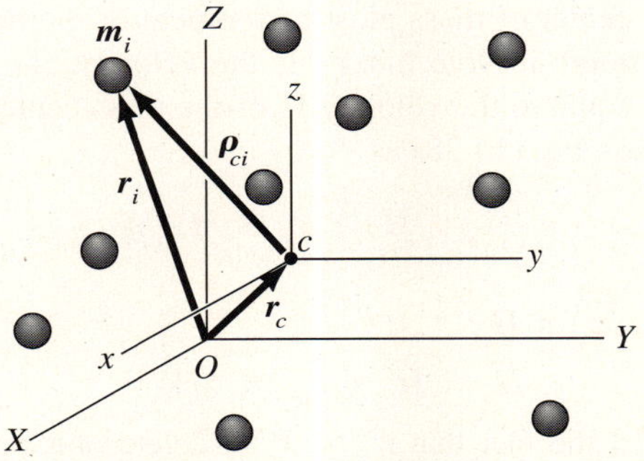

Figure 14.32. c is center of mass of aggregate.

The angular momentum for the aggregate of particles about O is then

$$\begin{aligned} \boldsymbol{H}_0 &= \sum_i (\boldsymbol{r}_c + \boldsymbol{\rho}_{ci}) \times \boldsymbol{P}_i \\ &= \sum_i (\boldsymbol{r}_c + \boldsymbol{\rho}_{ci}) \times \left[(m_i)(\dot{\boldsymbol{r}}_c + \dot{\boldsymbol{\rho}}_{ci})\right] \end{aligned} \tag{14.26}$$

Carry out the cross product and extract $\boldsymbol{r}_c$ from the summations:

$$\boldsymbol{H}_0 = \boldsymbol{r}_c \times M\dot{\boldsymbol{r}}_c + \boldsymbol{r}_c \times \sum_i m_i \dot{\boldsymbol{\rho}}_{ci} + \left(\sum_i m_i \boldsymbol{\rho}_{ci}\right) \times \dot{\boldsymbol{r}}_c + \sum_i \boldsymbol{\rho}_{ci} \times m_i \dot{\boldsymbol{\rho}}_{ci} \tag{14.27}$$

But since c is the center of mass, it follows that

$$\sum_i m_i \boldsymbol{\rho}_{ci} = \boldsymbol{0}$$

$$\sum_i m_i \dot{\boldsymbol{\rho}}_{ci} = \boldsymbol{0}$$

[14]Point a could also be moving with a constant velocity V_0 relative to inertial reference XYZ. However, a would then be fixed in another inertial reference $X'Y'Z'$, which is translating with respect to XYZ at a speed V_0.

Going back to Eq. 14.27, we see that the second and third expressions on the right side are to be deleted and we get then the desired result for $\boldsymbol{H}_0$:

$$\boldsymbol{H}_0 = \boldsymbol{r}_c \times M\dot{\boldsymbol{r}}_c + \sum_i \boldsymbol{\rho}_{ci} \times m_i\dot{\boldsymbol{\rho}}_{ci} = \boldsymbol{r}_c \times M\dot{\boldsymbol{r}}_c + \boldsymbol{H}_c$$

where $\boldsymbol{H}_c$ is the moment about the center of mass of the linear momentum as seen from the center of mass for the aggregate.[15] This may be rewritten and expressed for *any* fixed point a where, using $\boldsymbol{r}_{ac}$ as the position vector from fixed point a to the center of mass c, we have

$$\boldsymbol{H}_a = \boldsymbol{H}_c + \boldsymbol{r}_{ac} \times M\dot{\boldsymbol{r}}_{ac} \tag{14.28}$$

Thus, in a manner analogous to the case of kinetic energy (see Section 13.7), the moment of momentum about point a is the sum of the moment of momentum relative to the center of mass plus the moment of momentum of the center of mass about point a. Note that $\dot{\boldsymbol{r}}_{ac}$ is the velocity of c relative to fixed point a and is thus equal to the velocity $\boldsymbol{V}_c$ of the mass center relative to XYZ. Thus, we can express Eq. (14.28) as

$$\boldsymbol{H}_a = \boldsymbol{H}_c + \boldsymbol{r}_{ac} \times M\boldsymbol{V}_c \tag{14.29}$$

Furthermore, we have for $\dot{\boldsymbol{H}}_a$:

$$\dot{\boldsymbol{H}}_a = \dot{\boldsymbol{H}}_c + \boldsymbol{r}_{ac} \times M\dot{\boldsymbol{V}}_c$$

where we have used the fact that $\dot{\boldsymbol{r}}_{ac} = \boldsymbol{V}_c$ to delete one expression. Note in effect we have put dots over $\boldsymbol{H}_a$, $\boldsymbol{H}_c$, and $\boldsymbol{V}_c$ in Eq (14.29) to reach to above equation. We may now restate Eq. 14.24 for *a fixed point a* as follows on replacing $\dot{\boldsymbol{H}}_a$ using the above equation. Then, using $\boldsymbol{a}_c$ for $\dot{\boldsymbol{V}}_c$ we have the desired result:

$$\boldsymbol{M}_a = \dot{\boldsymbol{H}}_c + \boldsymbol{r}_{ac} \times M\boldsymbol{a}_c \tag{14.30}$$

Case 2. Reference Point at the Center of Mass. We can use Eq. 14.30 for this purpose. First, we will replace $\boldsymbol{M}_a$ using the left side of Eq. 14.23. But in so doing, we will replace $\boldsymbol{r}_i$ in the first expression by $(\boldsymbol{r}_c + \boldsymbol{\rho}_{ci})$. Note next that Eq. 14.30 calls for stationary point a. We will want a to be the origin O of XYZ and so $\boldsymbol{r}_{ac}$ becomes simply $\boldsymbol{r}_c$. Finally, we replace $\boldsymbol{a}_c$ by $\ddot{\boldsymbol{r}}_c$ in Eq. 14.30 and we have after these steps

$$\sum_i (\boldsymbol{r}_c + \boldsymbol{\rho}_{ci}) \times \boldsymbol{F}_i + \sum_j \sum_i \boldsymbol{r}_i \times \boldsymbol{f}_{ij} = \dot{\boldsymbol{H}}_c + \boldsymbol{r}_c \times M\ddot{\boldsymbol{r}}_c$$

The internal forces $\boldsymbol{f}_{ij}$ give zero contribution in this equation as explained earlier and we have on rearranging the remaining terms in the equation

$$\boldsymbol{r}_c \times \sum_i \boldsymbol{F}_i + \sum_i \boldsymbol{\rho}_{ci} \times \boldsymbol{F}_i = \boldsymbol{r}_c \times M\ddot{\boldsymbol{r}}_c + \dot{\boldsymbol{H}}_c$$

[15]That is, as seen from a reference *xyz* translating with *c* relative to *XYZ*—in other words, as seen by a nonrotating observer moving with *c*.

From Newton's law for the center of mass, we know that $\sum \boldsymbol{F}_i = M\ddot{\boldsymbol{r}}_c$ and so the first terms on the left and right sides of the equation above cancel. The remaining expression on the left side of the equation is the moment about the center of mass of the external forces. We then get

$$\boldsymbol{M}_c = \dot{\boldsymbol{H}}_c \tag{14.31}$$

We thus get the same formulation for the center of mass as for a fixed point in inertial space. Please note that $\boldsymbol{H}_c$ is the moment about the center of mass of the linear momentum as seen from the center of mass but that the time derivative is as seen from inertial reference *XYZ*.

Case 3. Point Accelerating Toward or Away from the Mass Center. There is yet a third point of interest to be considered and that is a point *a* accelerating toward or away from the mass center of the aggregate (Fig. 14.33).

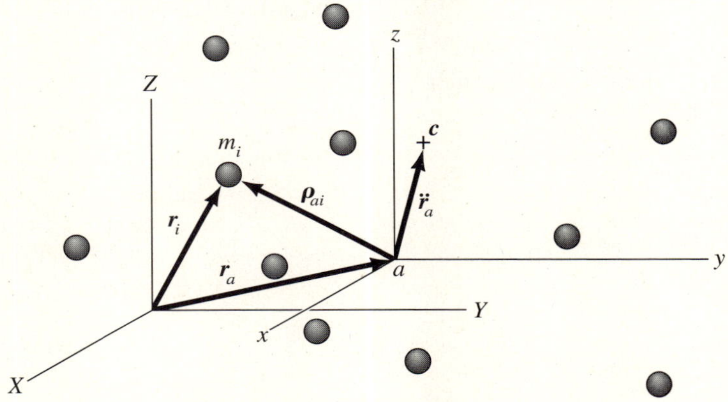

Figure 14.33. Point *a* accelerates toward or away from *c*.

For such a point, we can again give the same simple equation presented for cases 1 and 2. Thus,

$$\boldsymbol{M}_a = \dot{\boldsymbol{H}}_a \tag{14.32}$$

where $\boldsymbol{H}$ is taken relative to point *a* (i.e., relative to axes *xyz* translating with point *a*). We have asked for the derivation of this equation in Problem 14.96.

The component of the equation $\boldsymbol{M}_a = \dot{\boldsymbol{H}}_a$ for any one of the three cases in say the *x* direction,

$$M_x = \dot{H}_x$$

can be very useful. Here, M_x is the torque about the *x* axis, and H_x is the moment of momentum (angular momentum) about the *x* axis. We now examine such a problem in the following example.

Example 14.16

A heavy chain of length 6 m lies on a light plate A which is freely rotating at an angular speed of 1 rad/s (see Fig. 14.34). A channel C acts as a guide for the chain on the plate, and a stationary pipe acts as a guide for the chain below the plate. What is the speed of the chain after it moves 1.5 m starting from rest relative to the platform? Neglect friction, the angular momentum of the plate, and the angular momentum of the vertical section of the chain about its own axis. The chain mass per unit length, m, is 15 kg/m.

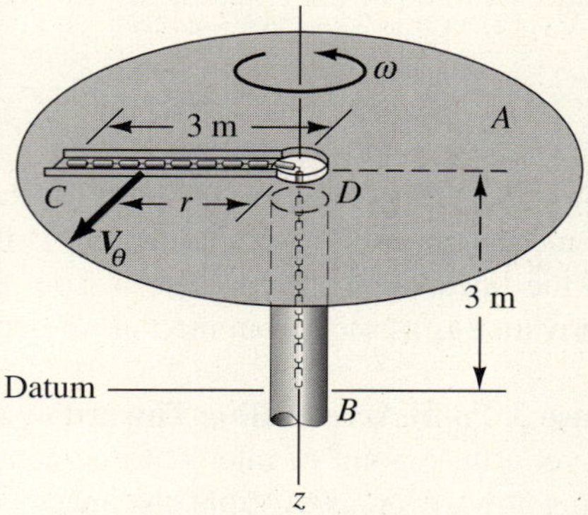

Figure 14.34. Sliding chain.

We shall first apply the **moment-of-momentum** equation about point D for the chain and plate. Taking the component of this equation along the z axis, we can say:

$$M_z = (\dot{H})_z \tag{a}$$

Clearly $M_z = 0$, and so we have conservation of angular momentum. That is,

$$H_z = \text{constant}$$
$$(H_z)_1 = (H_z)_2 \tag{b}$$

where 1 and 2 refer to the initial condition and the condition after the chain moves 1.5 m. We can then say:

$$\int_0^3 r(V_\theta)_1(mdr) = \int_0^{1.5} r(V_\theta)_2(mdr)$$
$$\int_0^3 r(\omega_1 r)(mdr) = \int_0^{1.5} r(\omega_2 r)(mdr)$$
$$(1)(m)\int_0^3 r^2\, dr = (\omega_2)(m)\int_0^{1.5} r^2\, dr$$

Therefore,

$$\omega_2 = 8 \text{ rad/s} \tag{c}$$

To find the speed of movement of the chain, we must next go to *energy* considerations. Because only conservative forces are acting here, we may employ the **conservation-of-mechanical-energy** principle. In so doing, we shall use as a datum the end of the chain B at the initial condition (see Fig. 14.34). We can then say:

$$(\text{PE})_1 = (3)(\text{m}g)(3) + (3)(\text{m}g)(1.5) = 1.987 \text{ kN m}$$

Observing Fig. 14.35, we can say for condition 2:

$$(\text{PE})_2 = (1.5)(\text{m}g)(3) + (3)(\text{m}g)(1.5) - (1.5)(\text{m}g)(0.75)$$
$$= 1.159 \text{ kN m}$$

As for kinetic energy, we have

$$(\text{KE})_1 = \frac{1}{2}(3\text{m})\left(V^2_{\text{channel}}\right)_1 + \frac{1}{2}(3\text{m})\left(V^2_{\text{pipe}}\right)_1 + \frac{1}{2}\int_0^3 (r\omega_1)^2\, mdr$$

Example 14.16 (Continued)

where the first two expressions on the right side give the kinetic energy from the motion relative to the channel and pipe, respectively. The last expression is the kinetic energy due to rotation of that part of the chain that is in the channel. Clearly, $V_{\text{channel}} = V_{\text{pipe}} = 0$ initially, and so we have

$$(\text{KE})_1 = \frac{1}{2}(1)^2(15)\int_0^3 r^2\,dr = 67.5 \text{ N m}$$

Furthermore, at condition 2, we have (see Fig. 14.35)

$$(\text{KE})_2 = \frac{1}{2}(1.5m)\left(V^2_{\text{channel}}\right)_2 + \frac{1}{2}(4.5m)\left(V^2_{\text{pipe}}\right)_2 + \frac{1}{2}\int_0^{1.5}(r\omega_2)^2\, m dr$$

Note that $(V_{\text{channel}})_2 = (V_{\text{pipe}})_2$. Simply calling this quantity V_2, we have

$$(\text{KE})_2 = \frac{1}{2}(6)(m)\left(V_2^2\right) + \frac{1}{2}(8^2)(m)\int_0^{1.5} r^2\,dr$$
$$= 45\,V_2^2 + 540$$

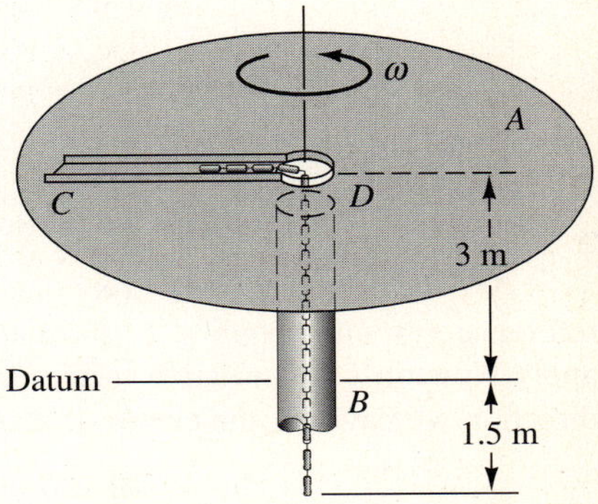

Figure 14.35. Chain after motion of 1.5 m.

We can now state

$$(\text{PE})_1 + (\text{KE})_1 = (\text{PE})_2 + (\text{KE})_2$$
$$1.987\times 10^3 + 67.5 = 1.159\times 10^3 + (45\,V_2^2 + 540)$$

Therefore,

$$V_2 = 2.81 \text{ m/s}$$

We can conclude that the chain is moving at a speed of 2.81 m/s along the channel and down the stationary pipe and that the plate *A* is rotating at an angular speed of 8 rad/s.

Much time will be spent later in the text in applying $\boldsymbol{M}_a = \dot{\boldsymbol{H}}_a$ to a rigid body. There, the rigid body is considered to be made up of an infinite number of contiguous elements. Summations then give way to integration, and so on. The final equations of this section accordingly are among the most important in mechanics.

In the homework assignments, we have included, as in Chapter 13, several very simple rigid-body problems to illustrate the use of the equation $\boldsymbol{M}_a = \dot{\boldsymbol{H}}_a$ and to give an early introduction to rigid-body mechanics.[16] We now illustrate such a problem.

[16]The instructor may wish not to get into rigid-body dynamics at this time. This approach presents no loss in continuity.

Example 14.17

A uniform cylinder of radius 400 mm and mass 100 kg is acted on at its center by a force of 500 N (see Fig. 14.36). What is the friction force f? Take $\mu_s = 0.2$.

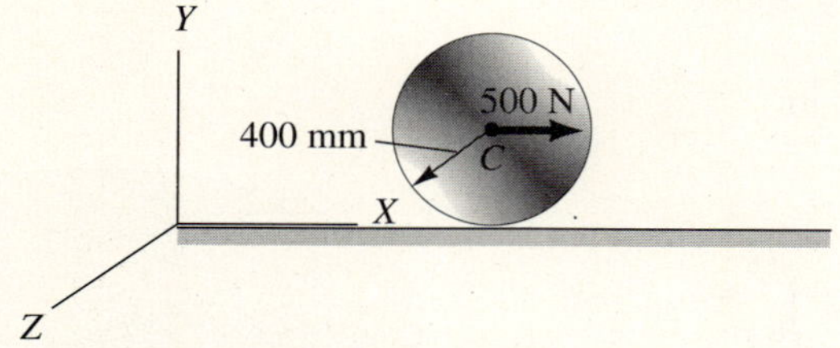

Figure 14.36. Rolling cylinder.

We have shown a free-body diagram of the cylinder in Fig. 14.37. A reference xyz with origin at C translates with the center of mass. We first apply **Newton's law** relative to inertial reference XYZ.[17] Thus, for the X direction we have for the center of mass C:

$$500 - f = 100\ddot{X}_c \qquad \text{(a)}$$

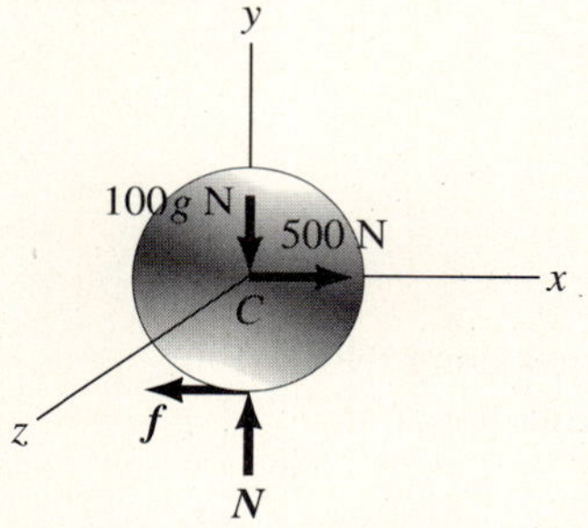

Figure 14.37. Free body.

Next, we write the **moment-of-momentum** equation about the z axis, which goes through the *center of mass*. Thus, noting that we have simple circular motion relative to the z axis for all particles of the cylinder and observing Fig. 14.38:

$$M_z = \frac{d}{dt}(H_z)$$

$$-(f)(0.40) = \frac{d}{dt}\left[\int_0^{2\pi}\int_0^{0.40} \underbrace{(\rho\, dr\, d\theta\, l)}_{dm}(r)\underbrace{(r\omega)}_{V_\theta}\right]$$

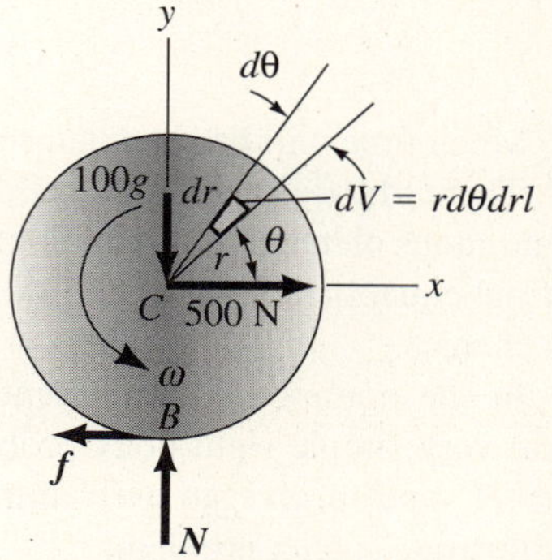

Figure 14.38. Element dV in cylinder. Unknown ω shown as positive.

[17]The reader is cautioned that although the diagram of this problem looks like those solved by energy methods in Section 13.8 of the previous chapter, this is not such a problem since distance is not involved. Instead, we deal with the linear acceleration and the angular acceleration.

Example 14.17 (Continued)

where $\boldsymbol{\rho}$ is the mass density of the cylinder and l is the thickness of the cylinder. Evaluating the integral and differentiating with respect to time as seen from XYZ, we get

$$0.40f = -(\rho l)(2\pi)(\dot{\omega})\left(\frac{0.4^4}{4}\right) \qquad \text{(b)}$$
$$f = -0.1005(\rho l)\dot{\omega}$$

We can determine ρl as follows from geometry:

$$M = 100 = (\rho l)[\pi(0.4)^2] \qquad \text{(c)}$$
$$\rho l = 198.9 \text{ kg/m}^2$$

We have two equations (a) and (b) with three unknowns, f, $\ddot{X}_c$, and $\dot{\omega}$. We now need another independent equation. This equation can be found from **kinematics**. Thus, assuming a *no-slipping* condition, we have pure instantaneous rotation about point B. From your work in physics (we shall later prove this) we can say for point C of the cylinder:

$$(0.4)\omega = -\dot{X}$$

Therefore,

$$(0.4)\dot{\omega} = -\ddot{X} \qquad \text{(d)}$$

Substituting for ρl and $\dot{\omega}$ in Eq. (b) using Eqs. (c) and (d), we get

$$f = (-0.1005)(198.9)\left(-\frac{\ddot{X}}{0.4}\right) \qquad \text{(e)}$$

Now solve for $\ddot{X}$ from Eq. (a) and substitute into Eq. (e):

$$f = (0.1005)(198.9)\left(\frac{5-0.01f}{0.4}\right)$$

Solving for f, we get

$$f = 166.6 \text{ N}$$

We must now check to see whether our no-slipping assumption is valid. The maximum possible friction force clearly is

$$f_{\max} = (100)(9.81)(0.2) = 196.2 \text{ N}$$

which is greater than the actual friction force, so that the no-slip assumption is consistent with our results.

*14.9 Looking Ahead—Basic Laws of Continua

In the preceding three chapters, we have presented three alternate approaches. They were, broadly speaking:

1. Direct application of Newton's law.
2. Energy methods.
3. Linear-momentum methods and moment-of-momentum methods

These all come from a *common* source (i.e., Newton's law) and so we can use any one for particle and rigid-body problems.

Later, when you study more complex continua such as a flowing fluid with heat transfer and compression you will have to satisfy four basic laws. These basic laws are:

1. Conservation of mass.
2. Linear momentum and moment of momentum (these are now Newton's law).
3. First law of thermodynamics.
4. Second law of thermodynamics.

For more general continua, the above mentioned four basic laws[18] are *independent* of each other (i.e., they must be separately satisfied) whereas in particle and rigid-body mechanics that we have been studying, 2 and 3 directly above, are equivalent to each other; 1 is satisfied by simply keeping the mass M constant; and 4 is satisfied by making sure that friction impedes the relative motion between two bodies in contact.

Furthermore, we applied the approaches of the preceding three chapters to free bodies. For more general continuum studies, such as fluid flow, we can apply the four basic laws to systems (i.e., free bodies) and also to so-called control volumes (fixed volumes in space) as discussed in the Looking Ahead Section 5.4.

Thus, in this book, we are considering a very simple phase of continuum mechanics whereby, in effect, we need only consider explicitly one of the basic laws. Your view will broaden as you move through the curriculum.

In some mechanics books there is presented an elementary presentation for determining the force developed by a stream of water or other fluid on a

[18]Electrical engineering students will, in addition, spend considerable time studying four other basic laws which are the famous Maxwell equations.

vane or some other object that deflects the stream of fluid. Your author has refrained from including this material. It is felt that the procedure for this computation should be undertaken in a more mature and thorough manner in a fluid mechanics course. There, the **linear momentum equation** is developed around the concept of the control volume and must be executed with care. In essence, the total external force at the control surface from fluids as well as those from solids such as vanes plus external forces such as gravity acting on what is inside the control volume—all equals the net efflux rate of linear momentum of flow through the control surface plus the rate of change of linear momentum inside the control volume. The execution of this principle should be done with care and precision and should not be undertaken lightly in a very limited simple-minded way.

Using a similar approach, the other basic laws are developed in the fluid mechanics course. Thus, for **moment of momentum**, we use the moments about any convenient point in inertial space of external forces acting at the control surface and also of external forces acting on anything inside the control volume and we equate the sum of these moments with the rate of flow of angular momentum through the control surface plus the rate of change of angular momentum inside the control volume.

Similarly, for the **first law of thermodynamics**, we add the rate of heat flow through the control surface, plus the rate of work passing through the control surface from fluids, plus the equivalent rate of work from electric currents in wires passing through the control surface, and finally plus the rate of work transmitted by shafts or any other devices passing through the control surface all at any time t. This is then equated to the flow of mechanical and internal energy through the control surface plus the rate of change of mechanical and internal energy inside the control volume at time t.

For the **conservation of mass**, we equate the net efflux rate of mass through the control surface with the rate of decrease of mass inside the control volume.

Although a similar approach can be made for the **second law of thermodynamics**, it is not as useful in this form and we shall defer this to your thermodynamics course.

Your author believes these laws in the above form should be taken up in *concert* and in *generality* for the greatest understanding and benefit for the student. Elementary treatments found in many mechanics books do not contribute to the understanding of the basic laws needed in fluid mechanics[19] and in later courses and may even present a hindrance for deeper understanding needed later.

[19]See I. H. Shames, *Mechanics of Fluids*, McGraw-Hill, 3rd ed., 1992, Chapters 5 and 6.

PROBLEMS

14.89. A system of particles is shown at time t moving in the xy plane. The following data apply:

$$
\begin{aligned}
m_1 &= 1 \text{ kg}, & V_1 &= 5\boldsymbol{i} + 5\boldsymbol{j} \text{ m/s} \\
m_2 &= 0.7 \text{ kg}, & V_2 &= -4\boldsymbol{i} + 3\boldsymbol{j} \text{ m/s} \\
m_3 &= 2 \text{ kg}, & V_3 &= -4\boldsymbol{j} \text{ m/s} \\
m_4 &= 1.5 \text{ kg}, & V_4 &= 3\boldsymbol{i} - 4\boldsymbol{j} \text{ m/s}
\end{aligned}
$$

(a) What is the total linear momentum of the system?
(b) What is the linear momentum of the center of mass?
(c) What is the total moment of momentum of the system about the origin and about point (2, 6)?

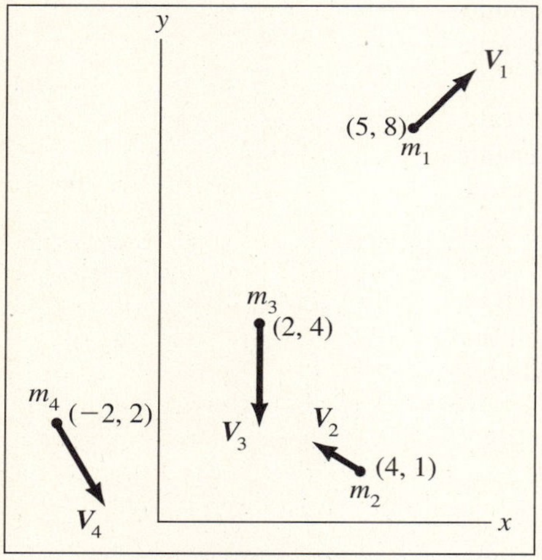

Figure P.14.89.

14.90. A system of particles at time t has the following velocities and masses:

$$
\begin{aligned}
V_1 &= 6 \text{ m/s}, & m_1 &= 0.5 \text{ kg} \\
V_2 &= 5.5 \text{ m/s}, & m_2 &= 1.5 \text{ kg} \\
V_3 &= 4.5 \text{ m/s}, & m_3 &= 1 \text{ kg} \\
V_4 &= 1.5 \text{ m/s}, & m_4 &= 0.5 \text{ kg}
\end{aligned}
$$

Determine (a) the total linear momentum of the system, (b) the angular momentum of the system about the origin, and (c) the angular momentum of the system about point a.

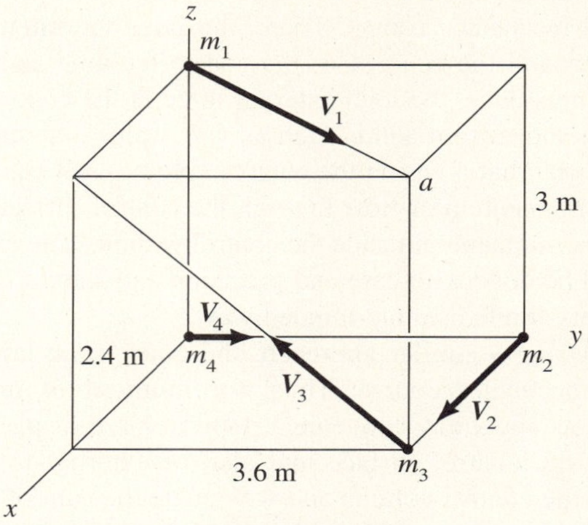

Figure P.14.90.

14.91. A system of particles at time t_1 has masses $m_1 = 1$ kg, $m_2 = 0.5$ kg, $m_3 = 1.5$ kg and locations and velocities as shown in (a). The same system of masses is shown in (b) at time t_2. What is the total linear impulse on the system during this time interval? What is the total angular impulse $\int \boldsymbol{M}\, dt$ during this time interval about the origin?

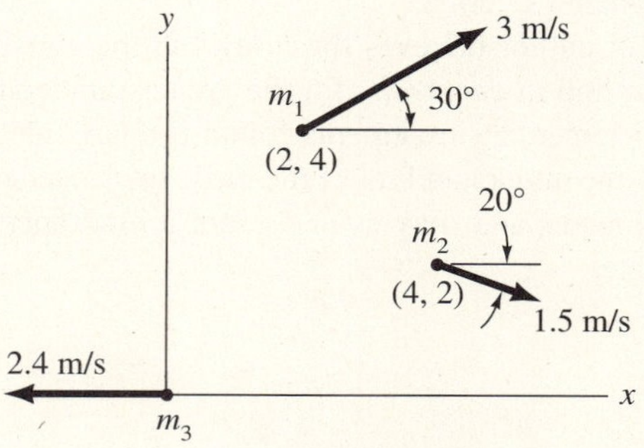

(a) Time t_1

Figure P.14.91-a.

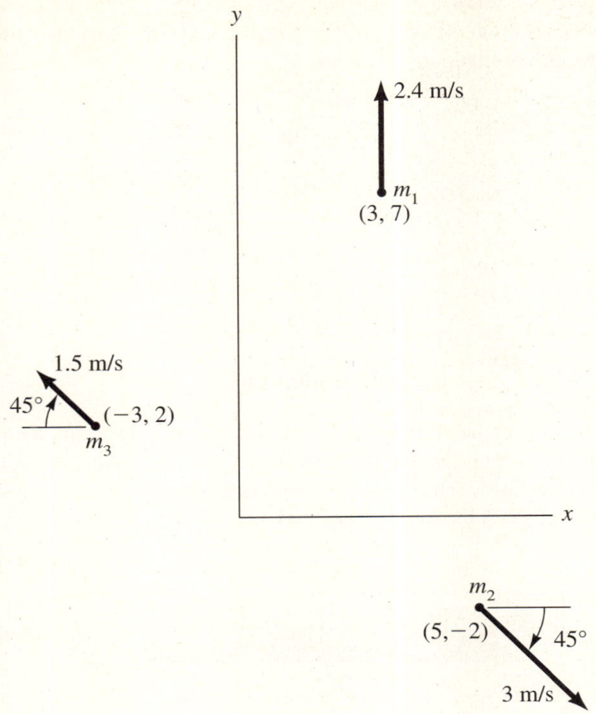

(b) Time t_2

Figure P.14.91-b.

14.92. Two masses slide along bar *AB* at a constant speed of 1.5 m/s. Bar *AB* rotates freely about axis *CD*. Consider only the mass of the sliding bodies to determine the angular acceleration of *AB* when the bodies are 1.5 m from *CD* if the angular velocity at that instant is 10 rad/s.

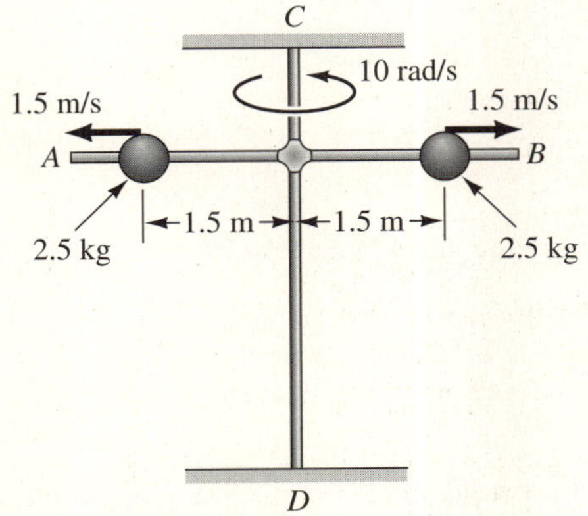

Figure P.14.92.

14.93. A mechanical system is composed of three identical bodies *A*, *B*, and *C* each of mass 1.5 kg moving along frictionless rods 120° apart on a wheel. Each of these bodies is connected with an inextensible cord to the freely hanging weight *D*. The connection of the cords to *D* is such that no torque can be transmitted to *D*. Initially, the three masses *A*, *B*, and *C* are held at a distance of 0.6 m from the centerline while the wheel rotates at 3 rad/s. What is the angular speed of the wheel and the velocity of descent of *D* if, after release of the radial bodies, body *D* moves 0.3 m? Assume that body *D* is initially stationary (i.e., is not rotating). Body *D* has a mass of 50 kg.

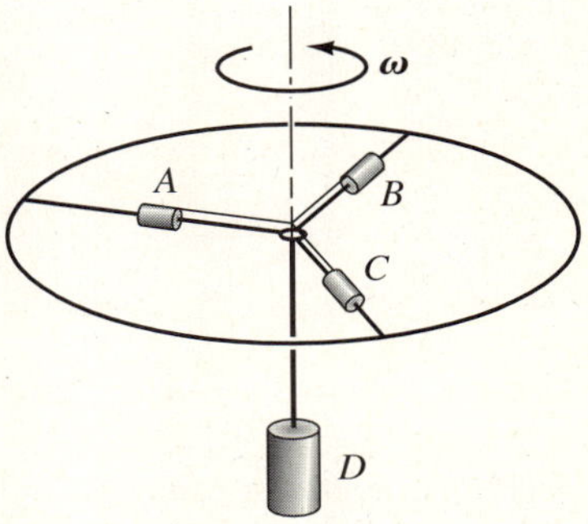

Figure P.14.93.

14.94. Two sets of particles *a*, *b*, and *c*, *d* (each particle of mass *m*) are moving along two shafts *AB* and *CD*, which are, in turn, rigidly attached to a crossbar *EF*. All particles are moving at a constant speed V_1 away from *EF*, and their positions at the moment of interest are as shown. The system is rotating about *G*, and a constant torque of magnitude *T* is acting in the plane of the system. Assume that all masses other than the concentrated masses are negligible and that the angular velocity of the system at the instant of discussion is ω. Determine the instantaneous angular acceleration in terms of m, T, ω, s_1, and s_2.

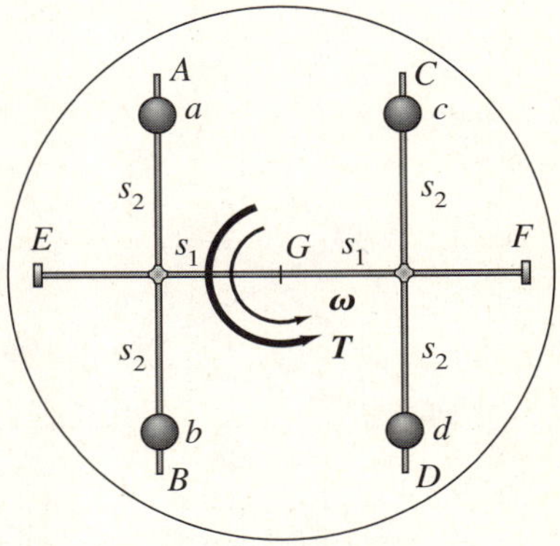

Figure P.14.94.

14.95. A uniform rod with a mass of 7 kg/m lies flat on a frictionless surface. A force of 250 N acts on the rod as shown in the diagram. What is the angular acceleration of the rod? What is the acceleration of the mass center?

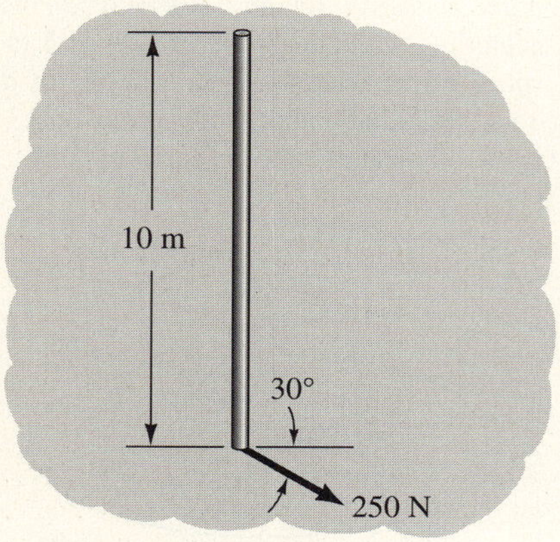

Figure P.14.95

***14.96.** Consider an aggregate of particles with C as the mass center and point A accelerating toward or away from C. Start with the expression for $\dot{\boldsymbol{H}}$ about O given as

$$\boldsymbol{M}_0 = \dot{\boldsymbol{H}}_0 = \frac{d}{dt}\left[\sum_i \boldsymbol{r}_i \times m_i \dot{\boldsymbol{r}}_i\right] = \sum_i \boldsymbol{r}_i \times \dot{\boldsymbol{P}}_i = \sum_i (\boldsymbol{r}_A \times \boldsymbol{\rho}_{Ai}) \times \dot{\boldsymbol{P}}$$

Formulate $\boldsymbol{M}_0$ in terms of $\boldsymbol{F}_i$ and use Newton's law to eliminate terms. Next show from the resulting equation that

$$\boldsymbol{M}_A = \left(\sum_i m_i \boldsymbol{\rho}_{Ai}\right) \times \ddot{\boldsymbol{r}}_A + \sum_i \boldsymbol{\rho}_{Ai} \times m_i \ddot{\boldsymbol{\rho}}_{Ai}$$

Replace $\sum_i m_i \boldsymbol{\rho}_{Ai}$ by $M\boldsymbol{\rho}_{Ac}$ Explain why it follows then that

$$\boldsymbol{M}_A = \dot{\boldsymbol{H}}_A$$

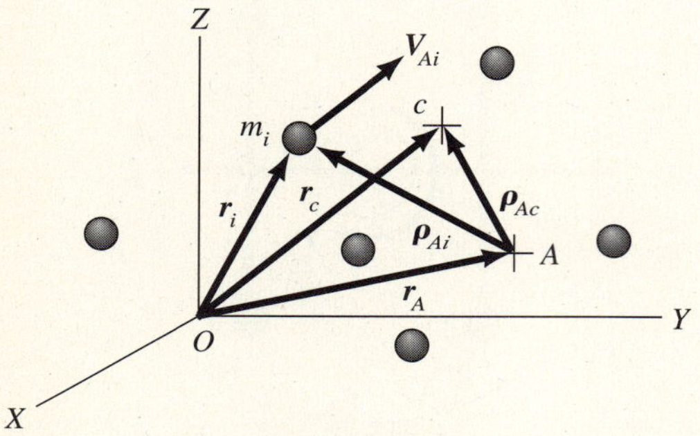

Figure P.14.96.

14.97. A uniform cylinder of radius 1 m rolls without slipping down a 30° incline. What is the angular acceleration of the cylinder if it has a mass of 50 kg?

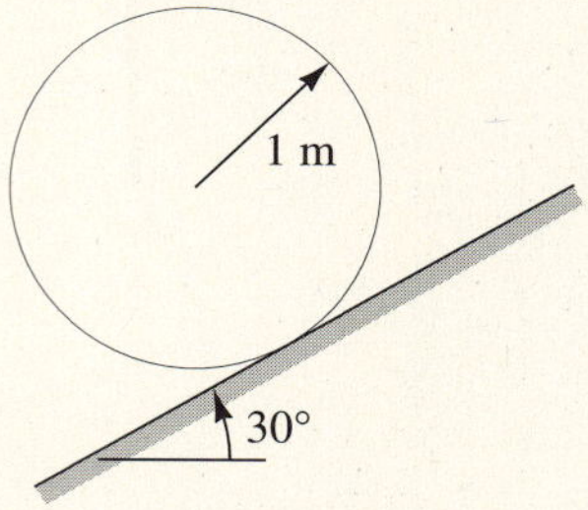

Figure P.14.97.

14.98. A cylinder of length 3 m and mass 45 kg is acted on by a torque $T = (11.25t + 21t^2)$ N m (where t is in seconds) about its geometric axis. What is the angular speed after 10 s? The cylinder is at rest when the torque is applied.

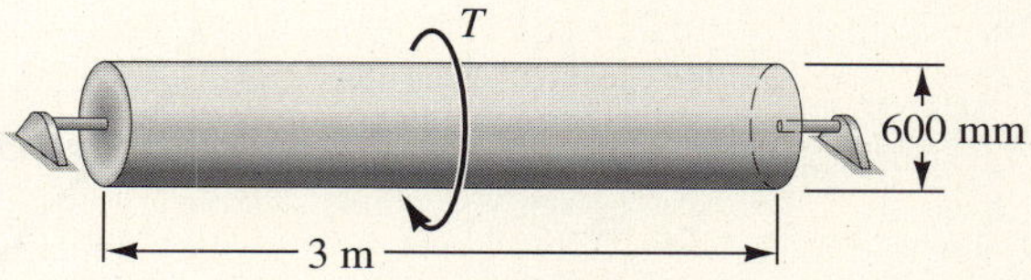

Figure P.14.98.

14.99. A constant torque T of 800 N m is applied to a uniform cylinder of radius 400 mm and mass 50 kg. A 1.5 kN weight is attached to the cylinder with a light cable. What is the acceleration of W?

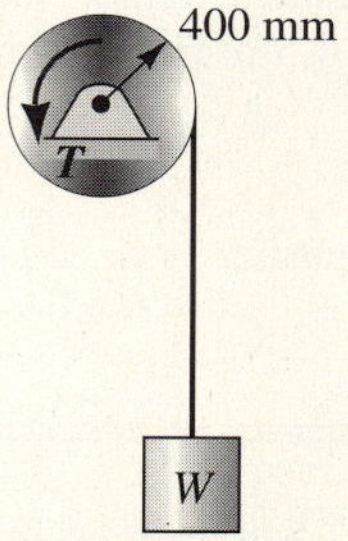

Figure P.14.99.

***14.100.** In Problem 14.99, the torque T is $T = (300 + 0.2t^2)$ N m, where t is in seconds. When $t = 0$, the system is at rest. Determine the acceleration of W at the instants when it has zero velocity for $t > 0$.

14.101. A constant torque T of 8 kN m is applied to a uniform cylinder of radius 0.3 m. A light inextensible cable is wrapped partly around an identical cylinder and is then connected to a block M of mass 50 kg. What is the acceleration of W if the cable does not slip on the cylinders? Take $\mu_d = 0.3$ for the block. For the cylinders, $M_A = M_B = 50$ kg.

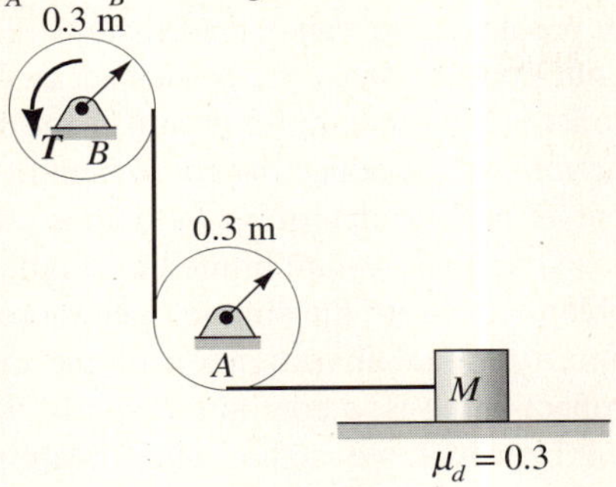

Figure P.14.101.

14.102. A canal with a rectangular cross section is shown having a width of 9 m and a depth of 1.5 m. The velocity of the water is assumed to be zero at the banks and to vary parabolically over the section as shown in the diagram. If δ is the radial distance from the centerline of the channel, the transverse velocity V_θ is given as

$$V_\theta = \tfrac{1}{6}(20 - \delta^2) \text{ m/s}$$

What is the angular momentum H_0 about O at any time t of the water in the circular portion of the canal (i.e., between the x and y axes)? The radial component V_r is zero.

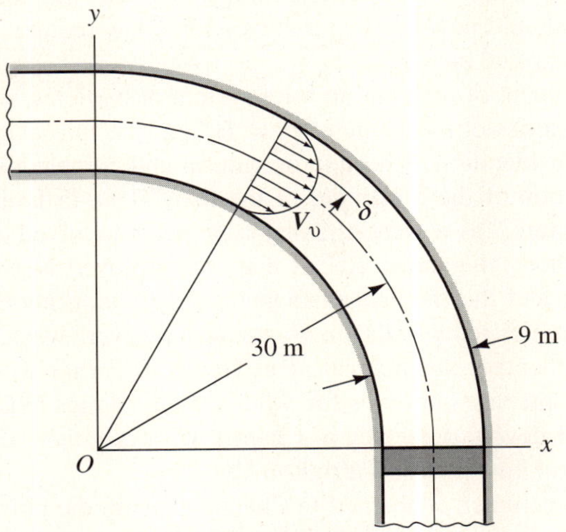

Figure P.14.102.

14.103. A hoop with mass per unit length 6.5 kg/m lies flat on a frictionless surface. A 500 N force is suddenly applied. What is the angular acceleration of the hoop? What is the acceleration of the mass center?

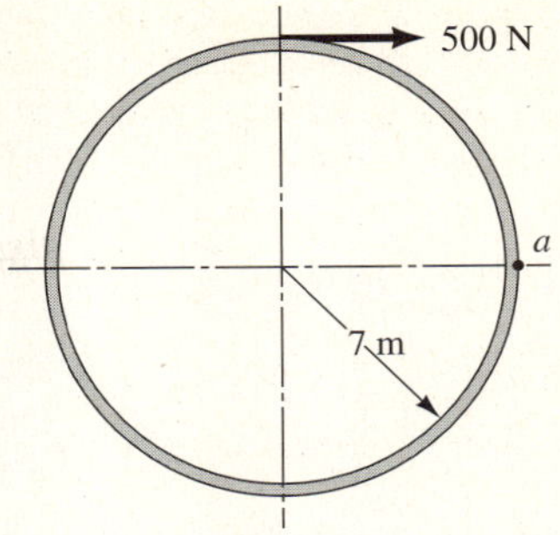

Figure P.14.103.

14.104. Do Problem 14.103 for the case where a force given as:

$$\boldsymbol{F} = 50\boldsymbol{i} + 75\boldsymbol{j} \text{ N}$$

is applied at point a instead of the 500 N force.

14.105. A cylinder of mass 25 kg lies on a frictionless surface. Two forces are applied simultaneously as shown in the diagram. What is the angular acceleration of the cylinder? What is the acceleration of the mass center?

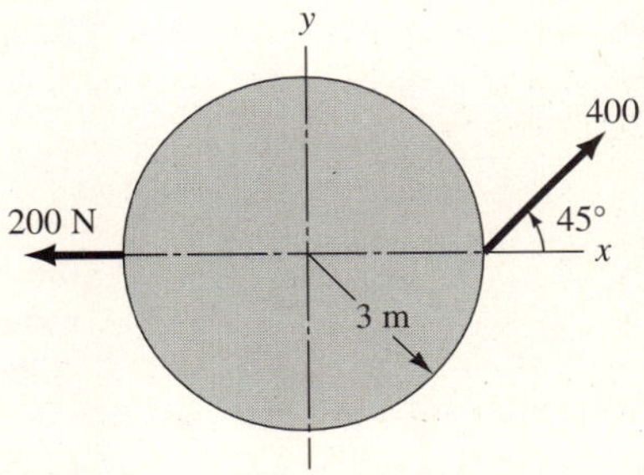

Figure P.14.105.

14.106. A thin uniform hoop rolls without slipping down a 30° incline. The hoop material has a mass of 7.5 kg/m and has a radius R of 1.2 m. What is the angular acceleration of the hoop?

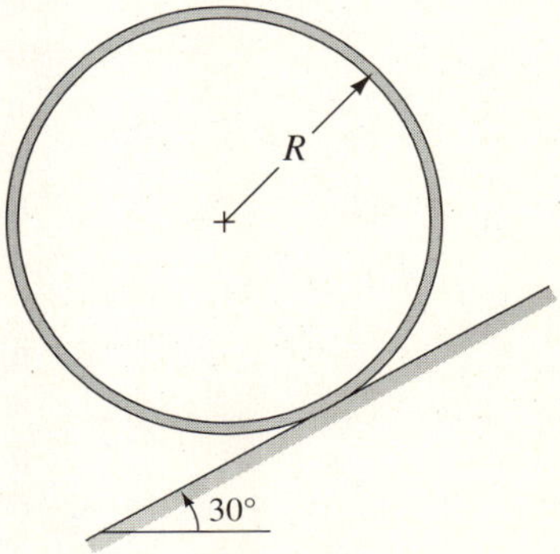

Figure P.14.106.

14.10 Closure

One of the topics studied in this chapter is the impact of bodies under certain restricted conditions. For such problems, we can consider the bodies as particles before and after impact, but during impact the bodies act as deformable media for which a particle model is not meaningful or sufficient. By making an elementary picture of the action, we introduce the coefficient of restitution to yield additional information we need to determine velocities after impact. This is an empirical approach, so our analyses are limited to simple problems. To handle more complex problems or to do the simple ones more precisely, we would have to make a more rational investigation of the deformation actions taking place during impact—that is, a continuum approach to part of the problem would be required. However, we cannot make a careful study of the deformation aspects in this text since the subject of high-speed deformation of solids is a difficult one that is still under careful study by engineers and physicists.

Note in the last two chapters we started with Newton's law $\boldsymbol{F} = M\boldsymbol{a}$ and performed the following operations:

1. Took the dot product of both sides using position vector $\boldsymbol{r}$.
2. Multiplied both sides by dt and integrated.
3. Took the cross product of both sides using position vector $\boldsymbol{r}$.

These steps permitted a surprisingly large number of very useful formulations and concepts that have occupied us for some considerable time. These were the energy methods, the linear-momentum methods, and the moment-of-momentum methods. It should now be clear that Newton's law requires considerable study to fully explore its use.

In our study of moment of momentum for a system of particles, we set forth one of the key equations of mechanics, $\boldsymbol{M}_A = \dot{\boldsymbol{H}}_A$, and we introduced in the examples several considerations whose more careful and complete study will occupy a good portion of the remainder of the text. Thus, in Example 14.17 we have "in miniature," as it were, the major elements involved in the study of much of rigid-body dynamics. Recall that we employed Newton's law for the mass center and the moment-of-momentum equation about the mass center to reach the desired results. In so doing, however, we had to make use of certain elementary kinematical ideas from our earlier work in physics. Accordingly, to prepare ourselves for rigid-body dynamics in Chapters 16 and 17, we shall devote ourselves in Chapter 15 to a rather careful examination of the general kinematics of a rigid body.

Although we shall be much concerned in Chapter 15 with the kinematics of rigid bodies, we shall not cease to consider particles. You will see that an understanding of rigid-body kinematics will permit us to formulate very powerful relations for the general relative motions of a particle involving references that move in any arbitrary manner with respect to each other.

14.107. A disc is rotated in the horizontal plane with a constant angular speed ω of 30 rad/s. A body A with a mass of 0.2 kg is moved in a frictionless slot at a uniform speed of 0.3 m/s relative to the platform by a force F as shown. What is the linear momentum of the body relative to the ground reference XY when r = 0.6 m and θ = 45°? What is the impulse developed on the body as it goes from r = 0.6 m to r = 0.3 m? Neglect the mass of the disc.

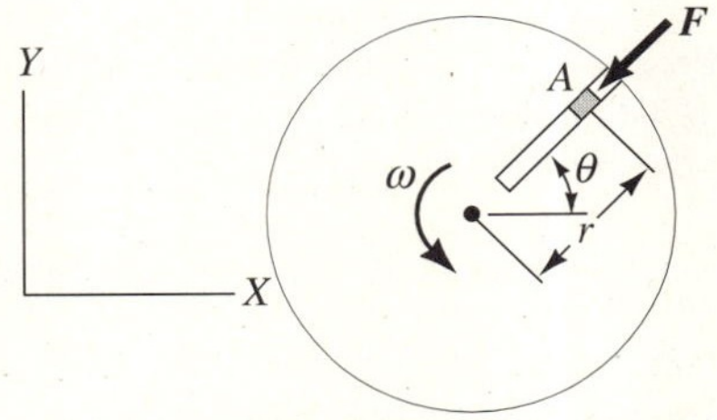

Figure P.14.107.

14.108. Three bodies are towed by a force $F = (400 + 200e^{-t})$ N as shown. If M_1 = 15 kg, M_2 = 30 kg, and M_3 = 25 kg, what is the speed 5 s after the application of the given force? The dynamic coefficient of friction is 0.3 for all surfaces.

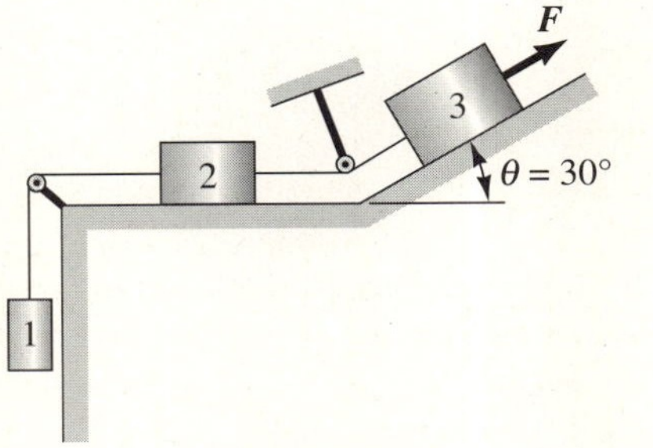

Figure P.14.108.

14.109. A space vehicle is in a circular parking orbit around the earth at a distance of 100 km from the earth's surface. What increase in speed must be given to the vehicle by firing its rockets so as to attain a radial velocity of 550 m/s at an elevation of 200 km?

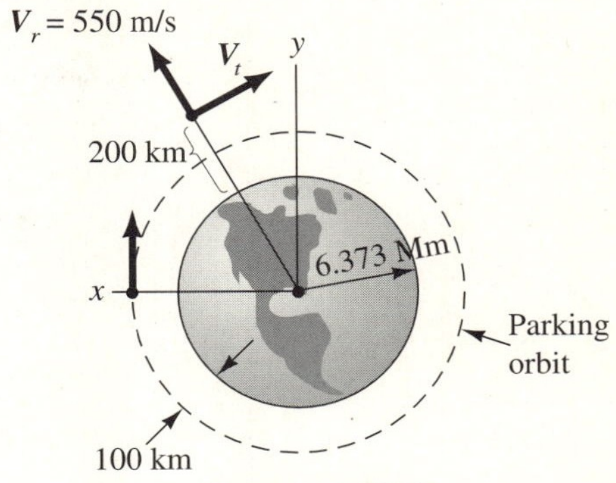

Figure P.14.109.

14.110. A space ship is in a circular parking orbit around the earth at 320 km above the earth's surface. At space headquarters, they wish to get the vehicle to a position 16 Mm from the center of the earth with a velocity at this position of 11 km/s. The command is given to fire rockets directly to the rear for a specified short time interval. What is the change in speed needed for this maneuver? What are the radial and tangential velocity components of the 11 km/s velocity vector?

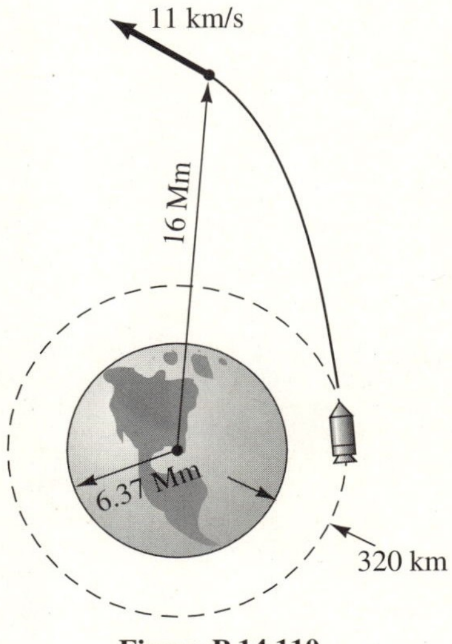

Figure P.14.110.

14.111. A small elastic ball is dropped from a height of 5 m onto a rigid cylindrical body having a radius of 1.5 m. At what position on the x axis does the ball land after the collision with the cylinder?

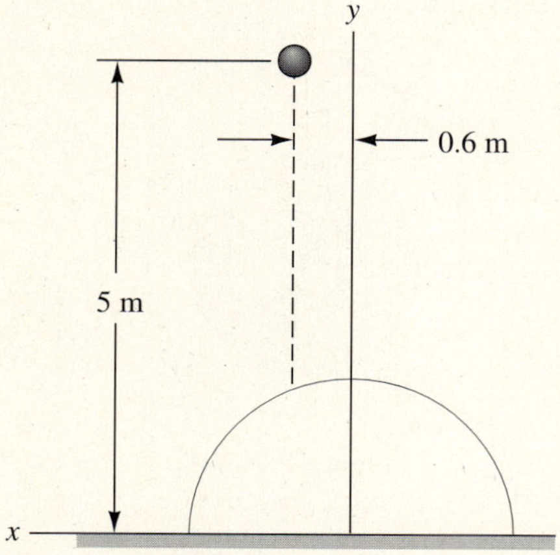

Figure P.14.111.

14.112. Do Problem 14.111 for an inelastic impact with $\epsilon = 0.6$.

14.113. A small elastic sphere is dropped from position (1, 1.5, 15) m onto a hard spherical body having a radius of 2.5 m positioned so that the z axis of the reference shown is along a diameter. For a perfectly elastic collision, give the speed of the small sphere directly after impact.

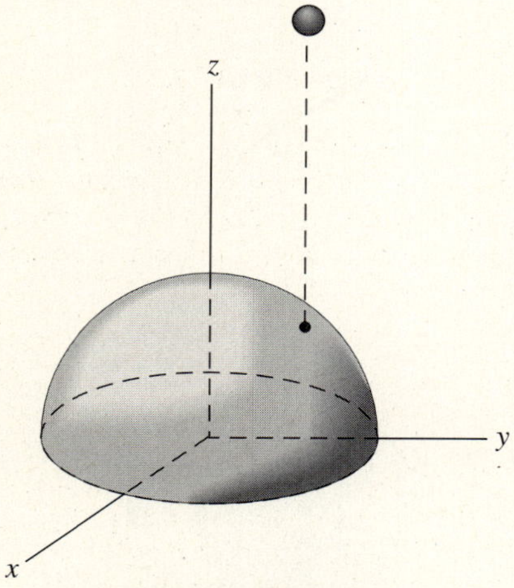

Figure P.14.113.

14.114. Do Problem 14.113 for an inelastic impact with $\epsilon = 0.6$.

14.115. A bullet hits a smooth, hard, massive two-dimensional body whose boundary has been shown as a parabola. If the bullet strikes 1.5 m above the x axis and if the collision is perfectly elastic, what is the maximum height reached by the bullet as it ricochets? Neglect air resistance and take the velocity of the bullet on impact as 700 m/s with a direction that is parallel to the x axis.

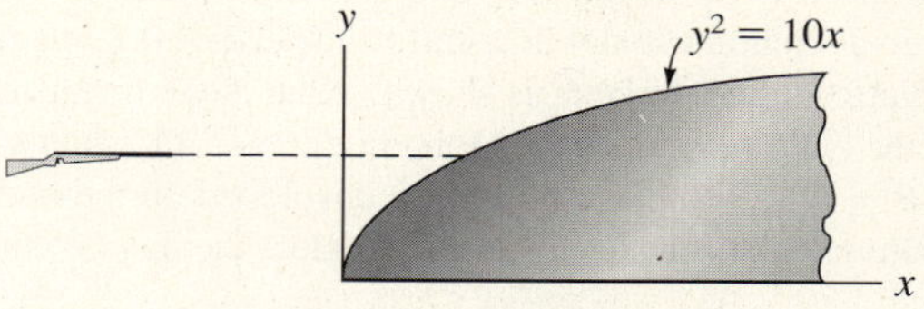

Figure P.14.115.

14.116. In Problem 14.115, assume an inelastic impact with $\epsilon = 0.6$. At what position along x does the bullet strike the parabola after the impact?

14.117. A space vehicle is in a circular "parking" orbit (1) around the earth 200 km above the earth's surface. It is to transfer to another circular orbit (2) 500 km above the earth's surface. The transfer to the second orbit is done in two stages.

1. Fire rockets so the vehicle has an *apogee* equal to the radius of the second circular orbit. What change of speed is required for this maneuver?

2. At *apogee* rockets are fired again to get into the second circular orbit. What is this second change of speed?

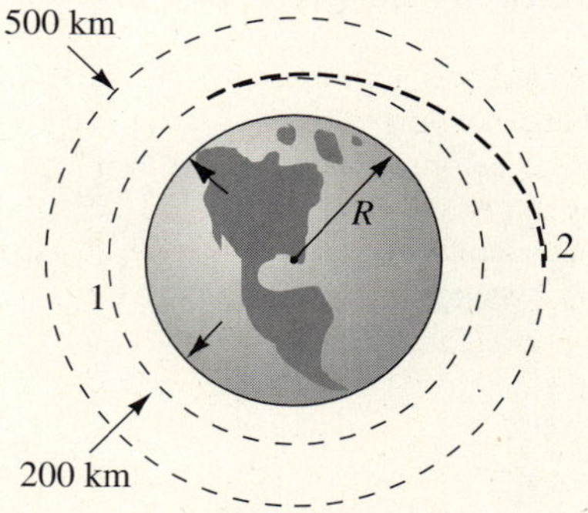

Figure P.14.117.

***14.118.** A tugboat of mass 90 Mg is moving toward a stationary barge of mass 180 Mg and carrying a mass C of 45 Mg. The tug is moving at 2.5 m/s and its propellers are developing a thrust of 22 kN when it contacts the barge. As a result of the soft padding at the nose of the tug, consider that there is plastic impact. If the load C is not tied in any way to the barge and has a dynamic coefficient of friction of 0.1 with the slippery deck of the barge, what is the speed V of the barge 2 s after the tug first contacts the barge? The load C slips during a 1 s interval starting at the beginning of the contact.

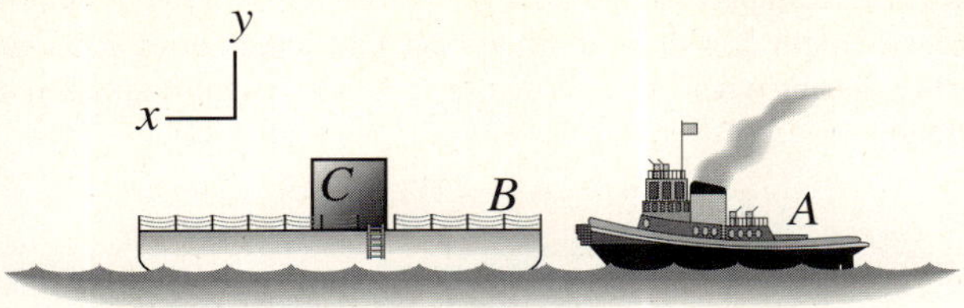

Figure P.14.118.

14.119. A hopper drops small cylinders each weighing 10 N onto a conveyor belt which is moving at a speed of 3 m/s. At the top, the cylinders are dropped off as shown. If at any time t there are 14 cylinders on the belt and if 10 cylinders are dropped per second from a hopper from a height of 300 mm above the belt, what average torque is needed to operate the conveyor? The weight of the belt that is on the conveyor bed is 100 N. The coefficient of friction μ_d between the belt and the bed is 0.3. The radius of the driving cylinder is 300 mm. Neglect bearing friction.

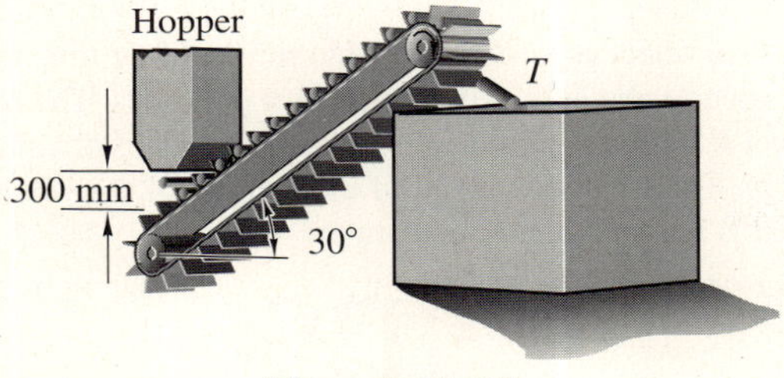

Figure P.14.119.

14.120. A body A weighing 18 kN is allowed to slide down an incline on a barge as shown. Body A moves a distance of 7.5 m along the incline before it is stopped at B. If we neglect water resistance, how far does the barge shift in the horizontal direction? If the maximum speed of body A relative to the incline of the barge is 0.6 m/s, what is the maximum speed of the barge relative to the water? The weight of the barge is 180 kN.

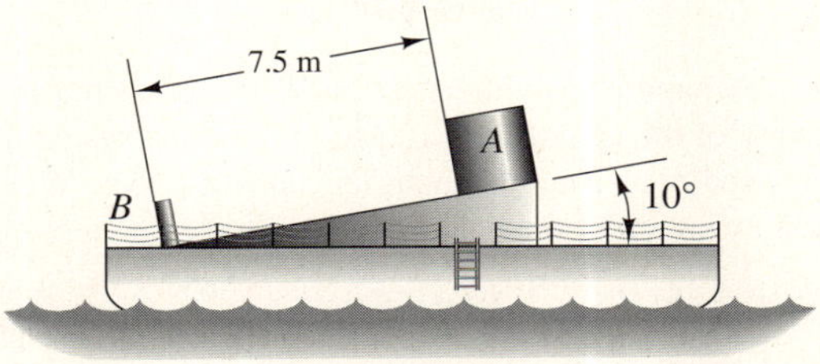

Figure P.14.120.

14.121. A water droplet of diameter 2 mm is falling in the atmosphere at the rate of 2 m/s. As a result of an updraft, a second water droplet of diameter 1 mm impinges on the aforementioned droplet. The velocity of the second droplet just prior to impingement is $3\boldsymbol{i} + 1\boldsymbol{j}$ m/s. After impingement three droplets are formed moving parallel to the xy plane. We have the following information:

$$D_1 = 0.6 \text{ mm}, \quad V_1 = 2 \text{ m/s}, \quad \theta_1 = 45^\circ$$
$$D_2 = 1.2 \text{ mm}, \quad V_2 = 1 \text{ m/s}, \quad \theta_2 = 30^\circ$$

Find D_3, V_3, and θ_3.

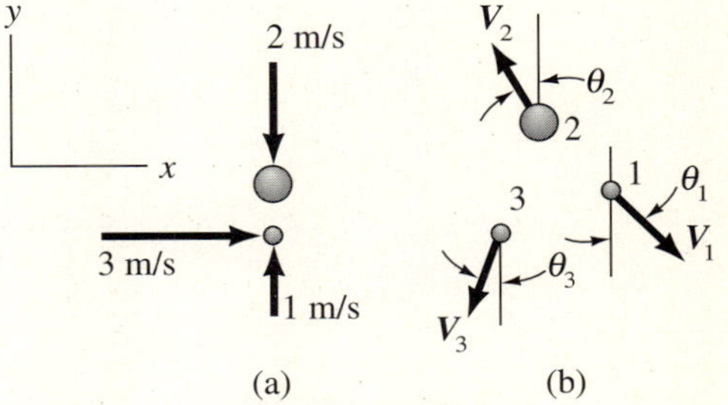

Figure P.14.121.

14.122. If the coefficient of restitution is 0.8 for the two spheres, what are the maximum angles from the vertical that the spheres will reach after the first impact? Neglect the mass of the cables.

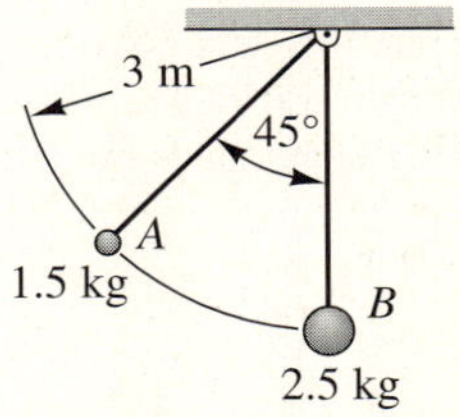

Figure P.14.122.

14.123. Thin discs A and B slide along a frictionless surface. Each disc has a radius of 25 mm. Disc A has a mass of 85 g, whereas disc B has a mass of 227 g. What are the speeds of the discs after collision for $\epsilon = 0.7$? Assume that the discs slide on a frictionless surface.

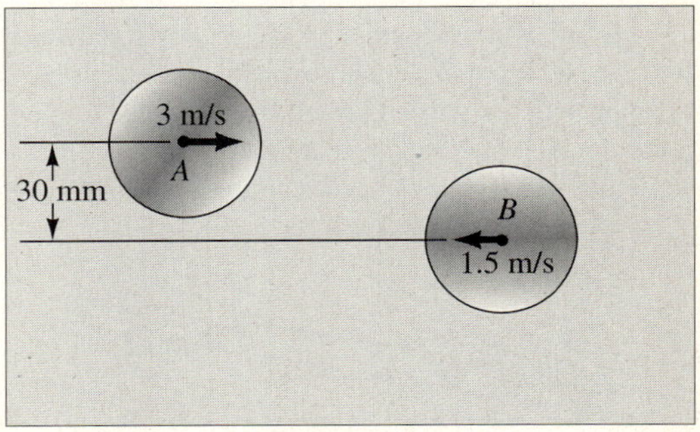

Figure P.14.123.

14.124. A BB is shot at the hard, rigid surface. The speed of the pellet is 100 m/s as it strikes the surface. If the direction of the velocity for the pellet is given by the following unit vector:

$$\boldsymbol{\epsilon} = -0.6\boldsymbol{i} - 0.8\boldsymbol{k}$$

what is the final velocity vector of the pellet for a collision having $\epsilon = 0.7$?

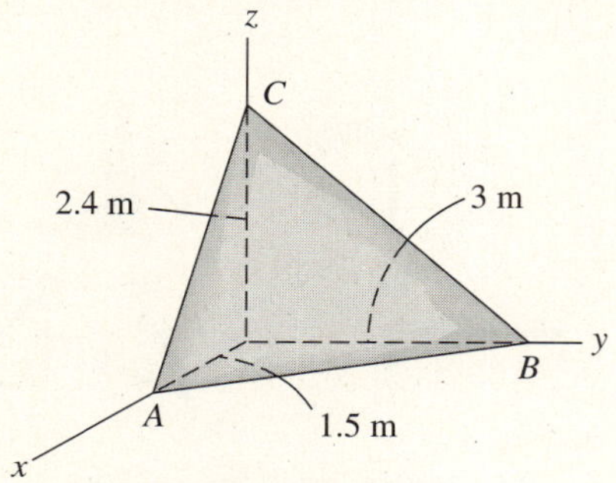

Figure P.14.124.

***14.125.** A chain of wrought iron, with length of 7 m and a mass of 100 kg, is held so that it just touches the support *AB*. If the chain is released, determine the total impulse during 2 s in the vertical direction experienced by the support if the impact is plastic (i.e., the chain does not bounce up) and if we move the support so that the links land on the platform and not on each other? [*Hint:* Note that any chain *resting* on *AB* delivers a vertical impulse. Also check to see if the entire chain lands on *AB* before 2 s.]

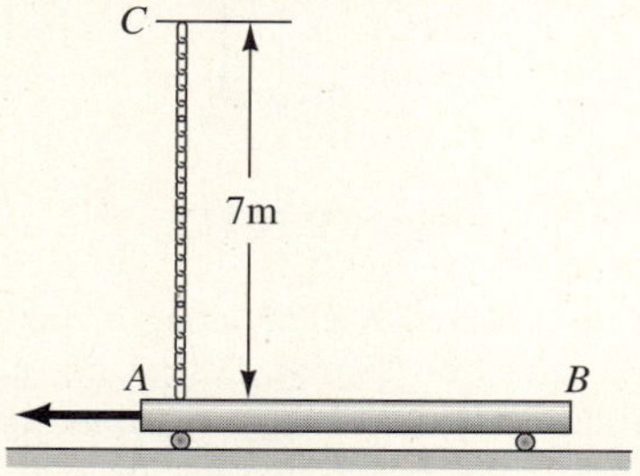

Figure P.14.125.

14.126. Two trucks are shown moving up a 10° incline. Truck *A* weighs 30 kN and is developing a 15 kN driving force on the road. Truck *B* weighs 20 kN and is connected with an inextensible cable to truck *A*. By operating a winch *b*, truck *B* approaches truck *A* with a constant acceleration of 0.3 m/s^2. If at time $t = 0$ both trucks have a speed of 10 m/s, what are their speeds at time $t = 15$ s?

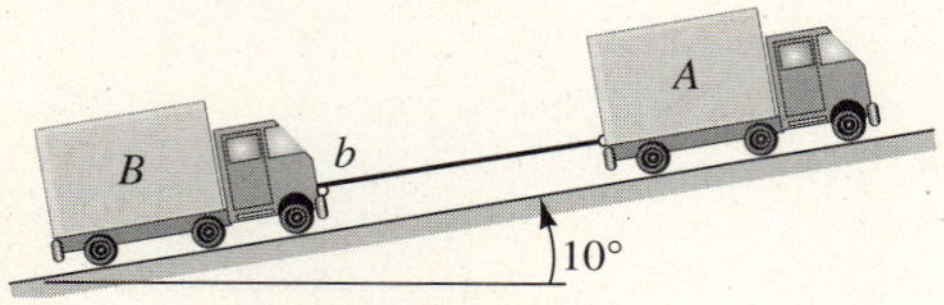

Figure P.14.126.

14.127. Compute the angular momentum about *O* of a uniform rod, of length $L = 3$ m and mass per unit length m of 7.5 kg/m, at the instant when it is vertical and has an angular speed ω of 3 rad/s.

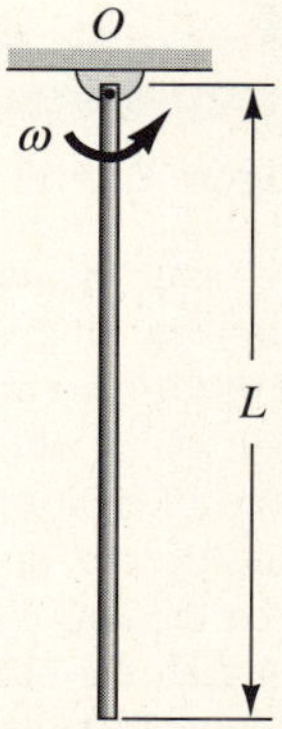

Figure P.14.127.

14.128. A wheel consisting of a thin rim and four thin spokes is shown rotating about its axis at a speed ω of 2 rad/s. The radius of the wheel R is 0.6 m and the mass per unit length of rim and spoke is 3 kg/m. What is the moment of momentum of the wheel about *O*? What is the total linear momentum?

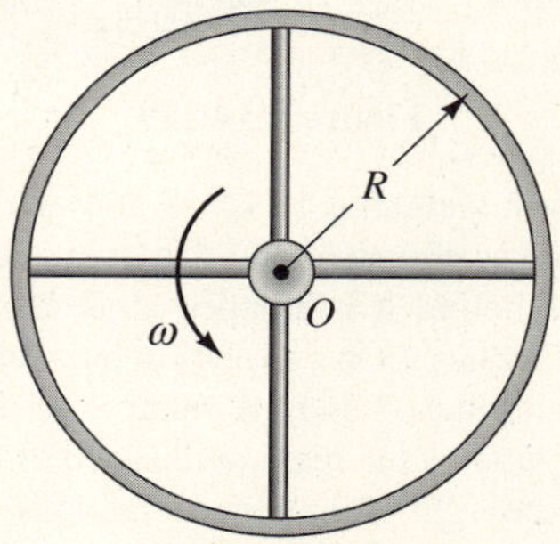

Figure P.14.128.

14.129. Two uniform cylinders *A* and *B* are connected as shown. The density of the cylinders is 10 Mg/m^3, and the system is rotating at a speed ω of 10 rad/s about its geometric axis. What is the angular momentum of the body?

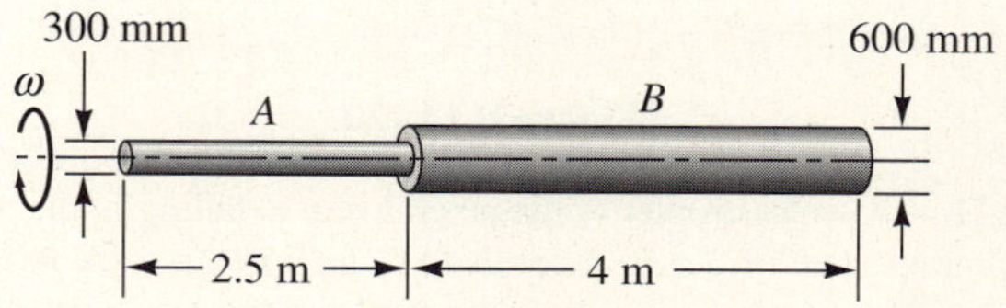

Figure P.14.129.

14.130. A closed container is full of water. By rotating the container for some time and then suddenly holding the container stationary, we develop a rotational motion of the water, which, you will learn in fluid mechanics, resembles a vortex. If the velocity of the fluid elements is zero in the radial direction and is given as $0.9/r$ m/s in the transverse direction, what is the angular momentum of the water?

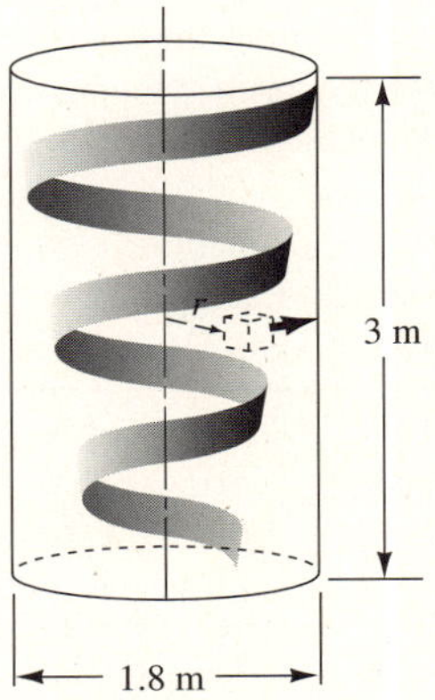

Figure P.14.130.

14.131. Identical thin masses A and B slide on a light horizontal rod that is attached to a freely turning light vertical shaft. When the masses are in the position shown in the diagram, the system rotates at a speed ω of 5 rad/s. The masses are released suddenly from this position and move out toward the identical springs, which have a spring constant K = 150 N/mm. Set up the equation for the compression δ of the spring once all motions of the bodies relative to the rod have damped out. The mass of each body is 5 kg. Neglect the mass of the rods and coulombic friction. Show that δ = 2.081 mm satisfies your equation.

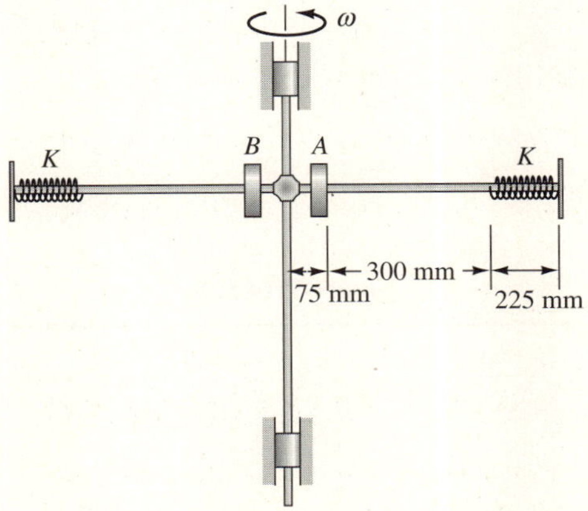

Figure P.14.131.

14.132. A spacecraft has a burnout velocity V_0 of 8.3 km/s at an elevation of 80 km above the earth's surface. The launch angle α is 15°. What is the maximum elevation h from the earth's surface for the spacecraft?

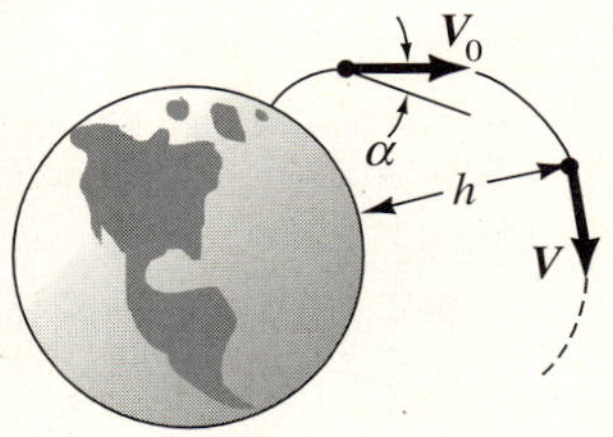

Figure P.14.132.

14.133. A set of particles, each having a mass of 8 kg, rotates about axis A–A. The masses are moving out radially at a constant speed of 1.5 m/s at the same time that they are rotating about the A–A, axis. When they are 0.3 m from A–A, the angular velocity is 5 rad/s and at that instant a torque is applied in the direction of motion which varies with time t in seconds as

$$\text{torque} = (7.5t^2 + 15t) \text{ N m}$$

What is the angular velocity when the masses have moved out radially at constant speed to 0.6 m?

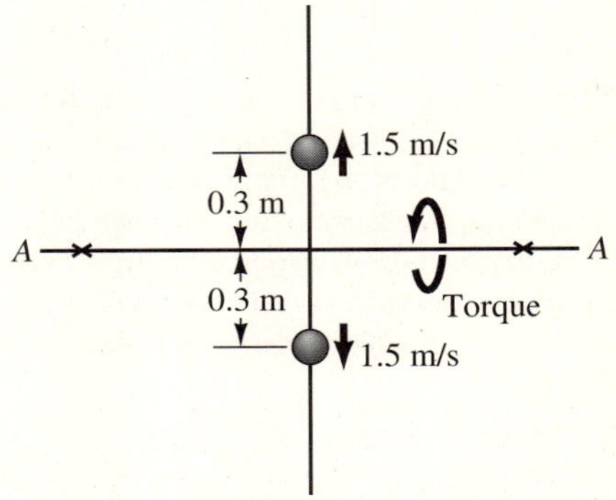

Figure P.14.133.

14.134. A torpedo boat of mass 45 Mg moves at 20 m/s away from an engagement. To go even faster, all four 50-caliber machine guns are ordered to fire simultaneously toward the rear. Each weapon fires at a muzzle velocity of 900 m/s and fires 500 rounds per minute. Each slug has a mass of 60 g. How much is the average force on the boat increased by this action? Neglect the rate of change of the total mass of boat.

14.135. A device to be detonated with a small charge is suspended in space [see (a)]. Directly after detonation, four fragments are formed moving away from the point of suspension. The following information is known about these fragments:

$$m_1 = 1 \text{ kg}$$
$$\boldsymbol{V}_1 = 200\boldsymbol{i} - 100\boldsymbol{j} \text{ m/s}$$
$$m_2 = 2 \text{ kg}$$
$$\boldsymbol{V}_2 = 125\boldsymbol{i} + 180\boldsymbol{j} - 100\boldsymbol{k} \text{ m/s}$$
$$m_3 = 1.6 \text{ kg}$$
$$\boldsymbol{V}_3 = -200\boldsymbol{i} + 150\boldsymbol{j} + 180\boldsymbol{k} \text{ m/s}$$
$$m_4 = 3.2 \text{ kg}$$

What is the velocity $\boldsymbol{V}_4$?

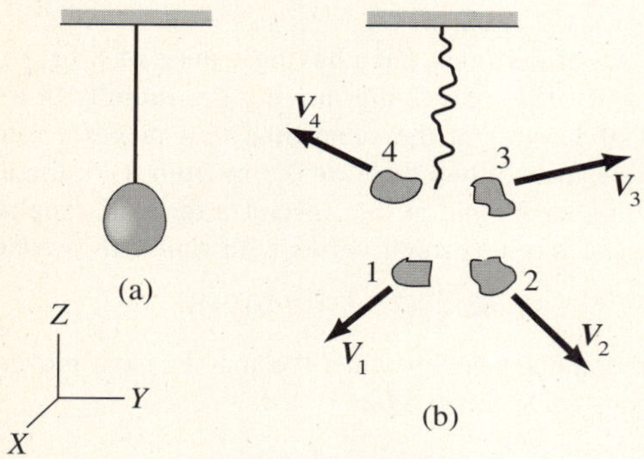

Figure P.14.135.

14.136. A hawk is a predatory bird which often attacks smaller birds in flight. A hawk having a mass of 1.3 kg is swooping down on a sparrow having a mass of 150 g. Just before seizing the sparrow with its claws, the hawk is moving downward with a speed V_H of 5.5 m/s. The sparrow is moving horizontally at a speed V_S of 4.2 m/s. Directly after seizure, what is the speed of the hawk and its prey? What is the loss in kinetic energy in joules?

14.137. The principal mode of propulsion of an octopus is to take in water through the mouth and then after closing the inlet to eject the water to the rear. If a 2.5 kg octopus after taking in 0.5 kg of water is moving at a speed of 1 m/s, what is its speed directly after ejecting the water? The water is ejected at an average speed to the rear of 3 m/s relative to the initial speed of the octopus. What power is being developed by the octopus in the above action if it occurs in 1 s?

***14.138.** In the *fission* process in a nuclear reactor, a ^{235}U nucleus first absorbs or captures a neutron [see (a)]. A short time later, the ^{235}U nucleus breaks up into fission products plus neutrons, which may subsequently be captured by other ^{235}U nuclei and maintain a *chain reaction*. Energy is released in each fission. In (b) we have shown the results of a possible fission. The following information is known for this fission:

	Mass No.	Kinetic Energy (MeV)	Direction of $\boldsymbol{V}$
Product *A*	138	*E*	$\boldsymbol{\epsilon}_A = 0.3\boldsymbol{i} - 0.2\boldsymbol{j} + 0.98\boldsymbol{k}$
Product *B*	96	90	$\boldsymbol{\epsilon}_B = l_B\boldsymbol{i} + m_B\boldsymbol{j} + n_B\boldsymbol{k}$
Neutron 1	1	10	$\boldsymbol{\epsilon}_1 = 0.6\boldsymbol{i} + 0.8\boldsymbol{j}$
Neutron 2	1	10	$\boldsymbol{\epsilon}_2 = 0.4\boldsymbol{i} - 0.6\boldsymbol{j} - 0.693\boldsymbol{k}$

What is the energy *E* of product *A* in MeV and what is the vector $\boldsymbol{\epsilon}_B$ for the velocity of product *B*? Assume that before fission the nucleus of ^{235}U plus captured neutrons is stationary: [*Hint:* You do not have to actually convert MeV to joules or atomic number to kilograms to carry out the problem.]

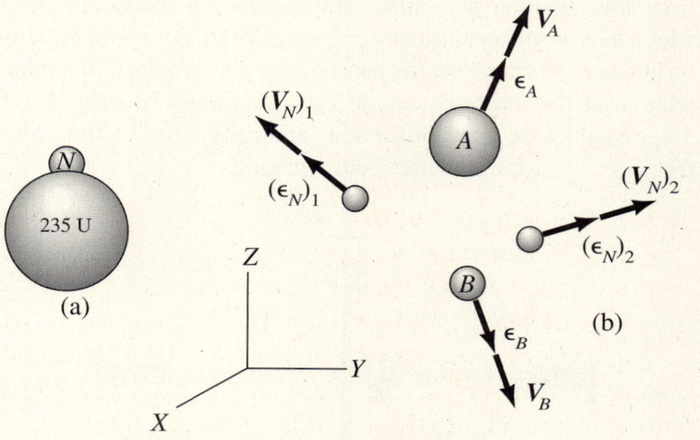

Figure P.14.138.

CHAPTER 15

Kinematics of Rigid Bodies: Relative Motion

15.1 Introduction

In Chapter 11, we studied the kinematics of a particle. During virtually all of this study only a single reference was used. However, at the end of that chapter, we briefly introduced the use of two references—for the case of *simple relative motion* involving two references *translating* with respect to each other.

One of the things we shall do in this chapter is to generalize the formulations for multireference analysis. There are two reasons for doing this. First, we shall be able to analyze complicated motions in a more simple systematic way by using several references. Second, the motion of a particle is often known relative to a moving body (such as an airplane), to which we can fix a reference *xyz*, while the motion of the plane (and hence *xyz*) is known relative to an *inertial reference XYZ* (such as the ground). Now **Newton's law** in the form $\boldsymbol{F} = m\boldsymbol{a}$, is valid *only* for an inertial reference. Hence, to use **Newton's law** for the particle we must express the acceleration of the particle relative to the inertial reference directly. Accordingly, for practical reasons we must become involved in multireference systems.

A reference is a rigid body, and, before we can set forth multireference considerations, we must first study the kinematics of a rigid body. In so doing, we will also set the stage for our main effort in the remaining portion of the text involving the dynamics of rigid bodies.

15.2 Translation and Rotation of Rigid Bodies

For purposes of dynamics, a rigid body is considered to be composed of a continuous distribution of particles having fixed distances between each other. We shall profitably define once again two simple types of motion of a rigid body:

Translation. As pointed out in Chapter 11, if a body moves so that all the particles have at time t the same velocity relative to some reference, the body is said to be in *translation* relative to this reference at this time. The velocity of a translating body can vary with time and so can be represented as $\mathbf{V}(t)$. Accordingly, translational motion does not necessarily mean motion along a straight line. For example, the body shown in Fig. 15.1 is in translation over the interval indicated because at each instant, each particle in the body has a common velocity. A characteristic of translational motion is that a straight line between two points of the body such as *ab* in Fig. 15.1 always retains an orientation parallel to its *original* direction during the motion.

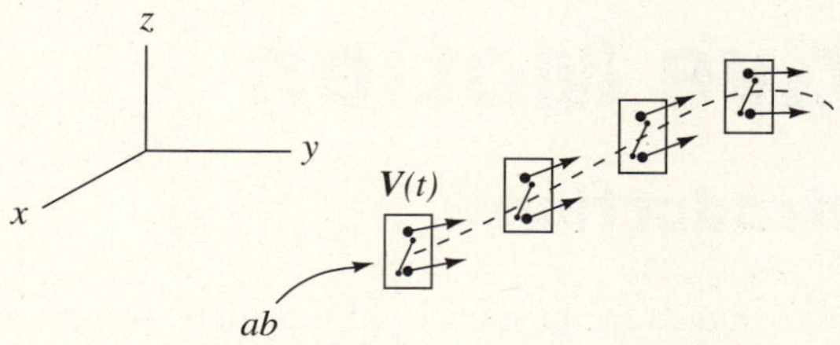

Figure 15.1. Translation of a body.

Rotation. If a rigid body moves so that along some straight line all the particles of the body, or a hypothetical extension of the body, have zero velocity relative to some reference, the body is said to be in *rotation* relative to this reference. The line of stationary particles is called the *axis of rotation.*

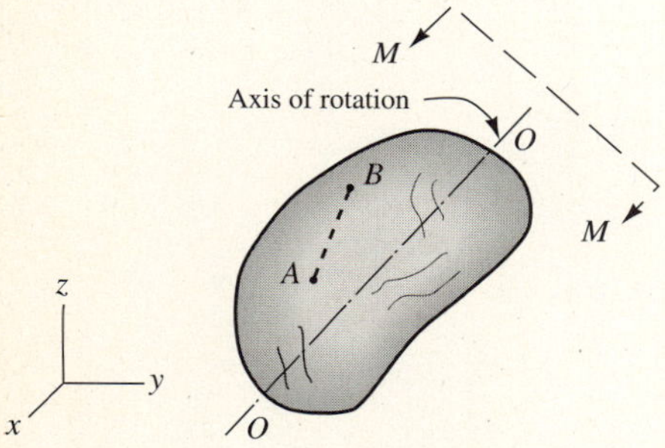

Figure 15.2. Rotation of a body.

We shall now consider how we measure the rotation of a body. A single revolution is defined as the amount of rotation in either a clockwise or a counterclockwise direction about the axis of rotation that brings the body back to its original position. Partial revolutions can conveniently be measured by observing *any* line segment such as *AB* in the body (Fig. 15.2) from a viewpoint *M-M* directed along the axis of rotation. In Fig. 15.3, we have shown this view of *AB* at the beginning of the partial rotation as seen along the axis of rotation, as well as the view *A′B′* at the end of the partial rotation. The angle β that these lines form will be the same for the initial and final

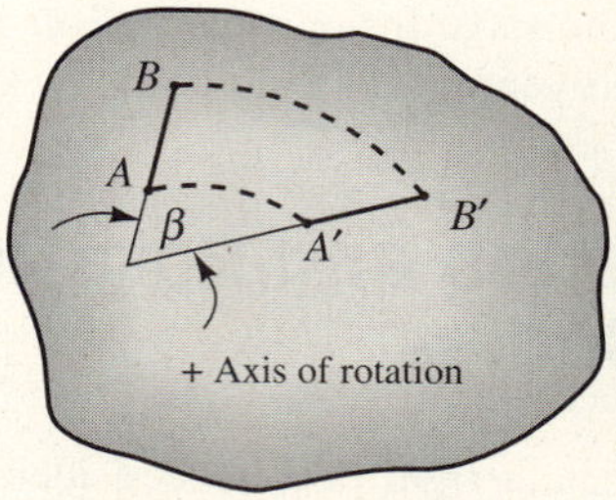

Figure 15.3. Measure of a partial rotation.

projections viewed along the axis of rotation of *any* line segment so examined in the partial rotation of the rigid body. Accordingly, the angle β so formed during a partial rotation is the measure of rotation.

In Chapter 1, we pointed out that finite rotations, although they have a magnitude and a direction along the axis of rotation, are not vectors. The superposition of rotations is not commutative, and therefore rotations do not add according to the parallelogram law, which, you will recall, is a requirement of all vector quantities. However, we can show (see Appendix IV) that as rotations become *infinitesimal*, they satisfy in the limit the commutative law of addition, so that infinitesimal rotations $d\beta$ are vector quantities. Therefore, the *angular velocity* is a vector quantity having a magnitude $d\beta/dt$ with an orientation parallel to the axis of rotation and a sense in accordance with the right-hand-screw rule. We shall employ $\boldsymbol{\omega}$ to represent the angular velocity vector. Note that this definition does not prescribe the line of action of this vector, for the line of action may be considered at positions other than the axis of rotation. The line of action depends on the situation at hand (as will be discussed in later sections).

15.3 Chasles' Theorem

We have just considered two simple motions of a body, translation and rotation. We shall now demonstrate that at each instant, the motion of any rigid body can be thought of as the superposition of both a translational motion and a rotational motion.

Consider for simplicity a body moving in a plane. Positions of the body are shown tinted at times t and $(t + \Delta t)$ in Fig. 15.4. Let us select any point B of the body. Imagine that the body is displaced without rotation from its position at time t to the position at time $(t + \Delta t)$ so that point B reaches its correct final position B'. The displacement vector for this translation is shown at $\Delta \boldsymbol{R}_B$. To reach the correct orientation for $(t + \Delta t)$, we must now rotate the body an angle $\Delta\phi$ about an axis of rotation which is normal to the plane and which passes through point B'.

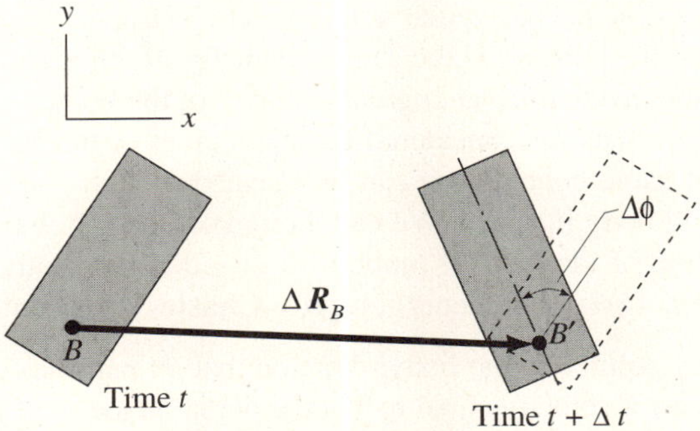

Figure 15.4. Translation and rotation of a rigid body.

What changes would occur had we chosen some other point C for such a procedure? Consider Fig. 15.5, where we have included an alternative procedure by translating the body so that point C reaches the correct final position C'. Next, we must rotate the body an amount $\Delta\phi$ about an axis of rotation which is normal to the plane and which passes through C' in order to get to the final orientation of the body. Thus, we have indicated two routes. We conclude from the diagram that the displacement $\Delta\boldsymbol{R}_C$ differs from $\Delta\boldsymbol{R}_B$, but there is no difference in the amount of rotation $\Delta\phi$. Thus, in general, $\Delta\boldsymbol{R}$ *and the axis of rotation will depend on the point chosen, while the amount of rotation* $\Delta\phi$ *will be the same for all such points.*

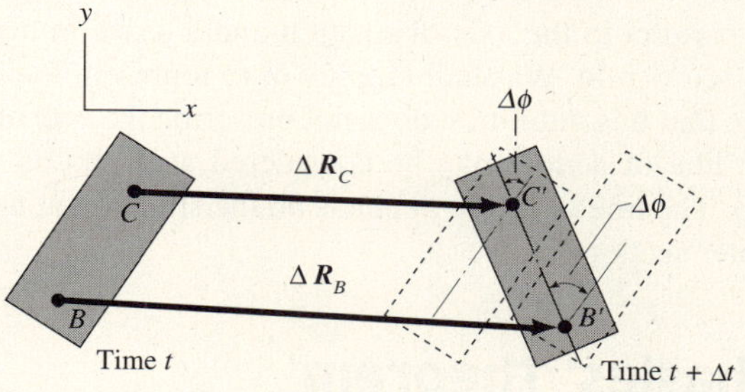

Figure 15.5. Translation and rotation of a rigid body using points B and C.

Consider now the ratios $\Delta\boldsymbol{R}/\Delta t$ and $\Delta\phi/\Delta t$. These quantities can be regarded as an average translational velocity and an average rotational speed, respectively, of the body, which we could superpose to get from the initial position to the final position in the time Δt. Thus, $\Delta\boldsymbol{R}/\Delta t$ and $\Delta\phi/\Delta t$ represent an average measure of the motion during the time interval Δt. *If we go to the limit by letting* $\Delta t \to 0$, *we have instantaneous translational and angular velocities which, when superposed, give the instantaneous motion of the body.* The displacement vector of the chosen point B in the previous discussion represents the translation of the body during the time Δt. Furthermore, the chosen point B undergoes no other motion during Δt other than that occurring during translation. Thus, we can conclude that, in the limit, the *translational velocity* used for the body corresponds to the *actual instantaneous* velocity of the chosen point B at time t. The angular velocity $\boldsymbol{\omega}$ to be used in the movement of the body, as described above, is the same vector for *all* points B chosen. Accordingly, $\boldsymbol{\omega}$ is the *instantaneous angular* velocity of the body.

We have thus far considered the movement of the body along a plane surface. The same conclusions can be reached for the general motion of an arbitrary rigid body in space. We can then make the following statements for the description of the general motion of a rigid body relative to some reference at time t. These statements comprise **Chasles' theorem**.

1. Select any point B in the body. Assume that all particles of the body have at the time t a velocity equal to $\boldsymbol{V}_B$, the actual velocity of the point B.
2. Superpose a pure rotational velocity $\boldsymbol{\omega}$ about an axis of rotation going through point B.

With V_B and $\boldsymbol{\omega}$, the actual instantaneous motion of the body is determined, and $\boldsymbol{\omega}$ will be the same for all points B which might be chosen. Thus, only the translational velocity and the axis of rotation change when different points B are chosen. However, clearly understand that the *actual instantaneous axis of rotation* at time t is the one going through those points of the body having zero velocity at time t.

15.4 Derivative of a Vector Fixed in a Moving Reference

Two references XYZ and xyz move arbitrarily relative to each other in Fig. 15.6. Assume we are observing xyz from XYZ. Since a reference is a rigid system, we can apply **Chasles' theorem** to reference xyz. Thus, to fully describe the motion of xyz relative to XYZ, we choose the origin O, and we superpose a translation velocity $\dot{\boldsymbol{R}}$, equal to the velocity of O, onto a rotational velocity $\boldsymbol{\omega}$ with an axis of rotation through O.

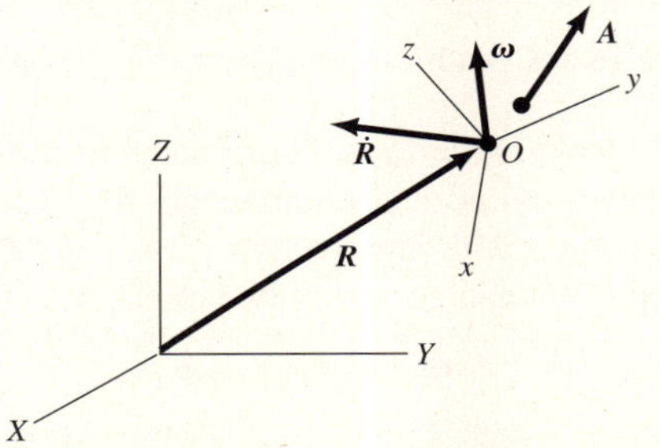

Figure 15.6. Vector $\boldsymbol{A}$ fixed in xyz moving relative to XYZ.

Now suppose that we have a vector $\boldsymbol{A}$ of *fixed length* and of *fixed orientation* as seen from reference xyz. We say that such a vector is "fixed" in reference xyz. Clearly, the time rate of change of $\boldsymbol{A}$ as seen from reference xyz must be zero. We can express this statement mathematically as

$$\left(\frac{d\boldsymbol{A}}{dt}\right)_{xyz} = \boldsymbol{0}$$

However, as seen from XYZ, the time rate of change $\boldsymbol{A}$ will *not* necessarily be zero. To evaluate $(d\boldsymbol{A}/dt)_{XYZ}$, we make use of **Chasles' theorem** in the following manner:

1. Consider the *translational* motion $\dot{\boldsymbol{R}}$. This motion does not alter the direction of $\boldsymbol{A}$ as seen from XYZ. Also, the magnitude of $\boldsymbol{A}$ is fixed; thus, vector $\boldsymbol{A}$ cannot change as a result of this motion.[1]
2. We next consider *solely* a pure rotation about a stationary axis collinear with $\boldsymbol{\omega}$ and passing through point O.

[1]The *line of action* of A, however, will change as seen from XYZ. But a change of line of action does not signify a change in the vector, as pointed out in Chapter 1 on the discussion of equality of vectors.

To best observe this rotation, we shall employ at O a *stationary* reference $X'Y'Z'$ positioned so that Z' coincides with the axis of rotation. This reference is shown in Fig. 15.7. Now the vector $\mathbf{A}$ is rotating at this instant about

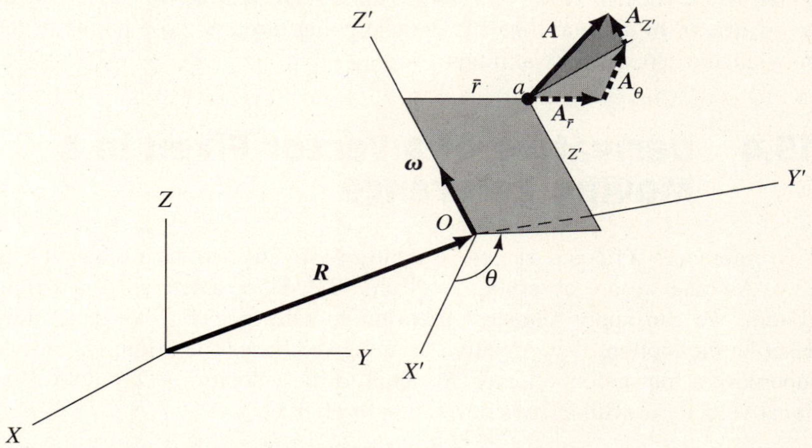

Figure 15.7. Cylindrical components for vector $\mathbf{A}$.

the Z' axis. We have shown cylindrical coordinates to the end of $\mathbf{A}$ (i.e., at point a); and have shown cylindrical components $A_{\bar{r}}$, A_θ, and $A_{Z'}$. In Fig. 15.8, we have shown point a with unit vectors $\boldsymbol{\epsilon}_{\bar{r}}$, $\boldsymbol{\epsilon}_\theta$, and $\boldsymbol{\epsilon}_{Z'}$, for cylindrical coordinates at this point. We can accordingly express $\mathbf{A}$ as

$$\mathbf{A} = A_{\bar{r}}\boldsymbol{\epsilon}_{\bar{r}} + A_\theta\boldsymbol{\epsilon}_\theta + A_{Z'}\boldsymbol{\epsilon}_{Z'}$$

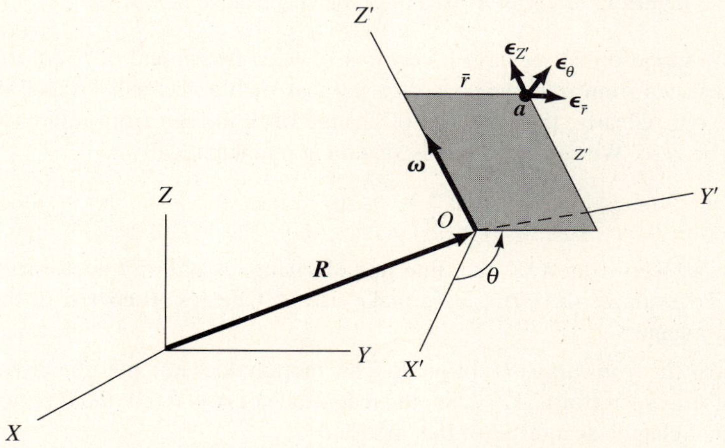

Figure 15.8. Unit vectors for cylindrical coordinates.

Clearly, as $\mathbf{A}$ rotates about Z', the values of the cylindrical scalar components of $\mathbf{A}$ for $X'Y'Z'$, namely $A_{\bar{r}}$, A_θ, and $A_{Z'}$, do not change. Hence, as seen from

$X'Y'Z'$, $\dot{A}_{\bar{r}} = \dot{A}_\theta = \dot{A}_Z = 0$. Also, noting that $\dot{\boldsymbol{\epsilon}}_{Z'} = \mathbf{0}$, we can say that

$$\left(\frac{d\boldsymbol{A}}{dt}\right)_{X'Y'Z'} = A_{\bar{r}}\left(\frac{d\boldsymbol{\epsilon}_{\bar{r}}}{dt}\right)_{X'Y'Z'} + A_\theta\left(\frac{d\boldsymbol{\epsilon}_\theta}{dt}\right)_{X'Y'Z'}$$

We have already evaluated the time derivatives of the unit vectors for cylindrical coordinates. Hence, using Eqs. 11.28 and 11.32 and noting that $\dot{\theta}$ corresponds to ω, we have

$$\left(\frac{d\boldsymbol{A}}{dt}\right)_{X'Y'Z'} = A_{\bar{r}}\omega\boldsymbol{\epsilon}_\theta - A_\theta\omega\boldsymbol{\epsilon}_{\bar{r}}$$

But the right hand side is simply the cross product of $\boldsymbol{\omega}$ and $\boldsymbol{A}$ as you can see by carrying out the cross product with cylindrical components. Thus,

$$\begin{aligned}\boldsymbol{\omega} \times \boldsymbol{A} &= \omega\boldsymbol{\epsilon}_{Z'} \times \left(A_{\bar{r}}\boldsymbol{\epsilon}_{\bar{r}} + A_\theta\boldsymbol{\epsilon}_\theta + A_{Z'}\boldsymbol{\epsilon}_{Z'}\right)\\ &= \omega A_{\bar{r}}\boldsymbol{\epsilon}_\theta - \omega A_\theta\boldsymbol{\epsilon}_{\bar{r}}\end{aligned}$$

We conclude that

$$\left(\frac{d\boldsymbol{A}}{dt}\right)_{X'Y'Z'} = \boldsymbol{\omega} \times \boldsymbol{A}$$

Since $X'Y'Z'$ is stationary relative to XYZ, we would observe the same time derivative from the latter reference as from the former reference. That is, $(d/dt)_{XYZ} = (d/dt)_{X'Y'Z'}$ and we can conclude that

$$\left(\frac{d\boldsymbol{A}}{dt}\right)_{XYZ} = \boldsymbol{\omega} \times \boldsymbol{A} \tag{15.1}$$

The foregoing result gives the time rate of change of a vector $\boldsymbol{A}$ fixed in reference xyz moving arbitrarily relative to reference XYZ. From this result, we see that $(d\boldsymbol{A}/dt)_{XYZ}$ depends only on the vectors $\boldsymbol{\omega}$ and $\boldsymbol{A}$ and not on their lines of action. Thus, we can conclude that the time rate of change of $\boldsymbol{A}$ fixed in xyz is not altered when:

1. The vector $\boldsymbol{A}$ is fixed at some other location in xyz provided the vector itself is not changed.
2. The actual axis of rotation of the xyz system is shifted to a new parallel position.

We can differentiate the terms in Eq. 15.1 a second time. We thus get

$$\left(\frac{d^2\boldsymbol{A}}{dt^2}\right)_{XYZ} = \left(\frac{d\boldsymbol{\omega}}{dt}\right)_{XYZ} \times \boldsymbol{A} + \boldsymbol{\omega} \times \left(\frac{d\boldsymbol{A}}{dt}\right)_{XYZ} \tag{15.2}$$

Using Eq. 15.1 to replace $(d\mathbf{A}/dt)_{XYZ}$ and using $\dot{\boldsymbol{\omega}}$ to replace $(d\boldsymbol{\omega}/dt)_{XYZ}$, since the reference being used for this derivative is clear,[2] we get

$$\left(\frac{d^2\mathbf{A}}{dt^2}\right)_{XYZ} = \dot{\boldsymbol{\omega}} \times \boldsymbol{A} + \boldsymbol{\omega} \times (\boldsymbol{\omega} \times \boldsymbol{A}) \tag{15.3}$$

You can compute higher-order derivatives by continuing the process. We suggest that only Eq. 15.1 be remembered and that all subsequent higher-order derivatives be evaluated when needed.

In this discussion thus far, we have considered a vector $\boldsymbol{A}$ fixed in a reference xyz. But a reference xyz is a rigid system and can be considered a *rigid body*. Thus, the words "*fixed in a reference xyz*" in the previous discussion can be replaced by the words "*fixed in a rigid body*." The angular velocity $\boldsymbol{\omega}$ used in Eq. 15.1 is then the angular velocity of the rigid body in which $\boldsymbol{A}$ is fixed. We shall illustrate this condition in the following examples, which you are urged to study very carefully. An understanding of these examples is vital for attaining a good working grasp of rigid-body kinematics.

As an aid in carrying out computations involving the triple cross product, we wish to point out that the product

$$\omega_1\boldsymbol{k} \times (\omega_1\boldsymbol{k} \times C\boldsymbol{j}) = -\omega_1^2 C\boldsymbol{j}$$

That is, the product is minus the product of the scalars and has a direction corresponding to the last unit vector, $\boldsymbol{j}$. Remembering this will greatly facilitate our computations.[3]

Additionally, consider a situation where the angular velocity of body A relative to body B is given as $\boldsymbol{\omega}_1$, while the angular velocity of body B relative to the ground is $\boldsymbol{\omega}_2$. What is the *total* angular velocity $\boldsymbol{\omega}_T$ of body A relative to the ground? In such a case, we must remember that the angular velocity $\boldsymbol{\omega}_1$ of body A *relative* to body B is actually the *difference* between the total angular velocity $\boldsymbol{\omega}_T$ of body A as seen from the ground and the angular velocity $\boldsymbol{\omega}_2$ of body B as seen from the ground. Thus,

$$\boldsymbol{\omega}_1 = \boldsymbol{\omega}_T - \boldsymbol{\omega}_2$$

Solving for $\boldsymbol{\omega}_T$, we get

$$\boldsymbol{\omega}_T = \boldsymbol{\omega}_1 + \boldsymbol{\omega}_2$$

We see from above that to get the total angular velocity $\boldsymbol{\omega}_T$, we simply add the various relative angular velocities just as we would with any pair of vectors.

[2]When it is clear from the discussion what reference is involved for a time derivative, we shall use the dot to indicate a time derivative.

[3]Of course, if the $\boldsymbol{j}$ vector were a $\boldsymbol{k}$ vector, then clearly we would arrive at a null value for the triple vector product.

Example 15.1

A disc C is mounted on a shaft AB in Fig. 15.9. The shaft and disc rotate with a constant angular speed ω_2 of 10 rad/s relative to the platform to which bearings A and B are attached. Meanwhile, the platform rotates at a constant angular speed ω_1 of 5 rad/s relative to the ground in a direction parallel to the Z axis of the ground reference XYZ. What is the angular velocity vector $\boldsymbol{\omega}$ for the disc C relative to XYZ? What are $(d\boldsymbol{\omega}/dt)_{XYZ}$ and $(d^2\boldsymbol{\omega}/dt^2)_{XYZ}$?

The total angular velocity $\boldsymbol{\omega}$ of the disc relative to the ground is easily given at all times as follows:

$$\boldsymbol{\omega} = \boldsymbol{\omega}_1 + \boldsymbol{\omega}_2 \text{ rad/s} \qquad \text{(a)}$$

At the instant of interest as depicted by Fig. 15.9, we have for $\boldsymbol{\omega}$:

$$\boldsymbol{\omega} = 5\boldsymbol{k} + 10\boldsymbol{j} \text{ rad/s}$$

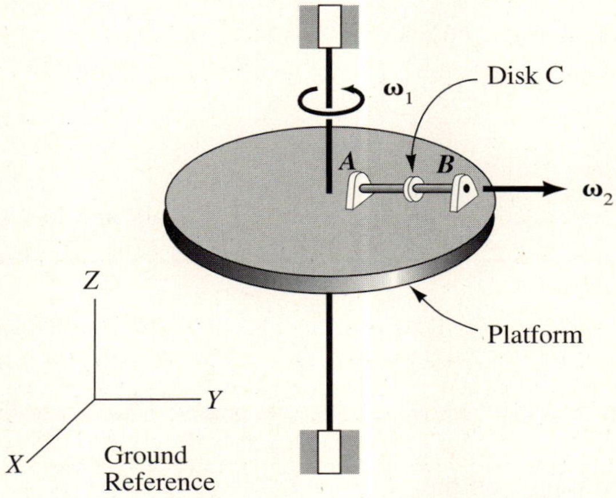

Figure 15.9. Rotating disc on rotating platform.

To get the first time derivative of $\boldsymbol{\omega}$, we go back to Eq. (a), which is always valid and hence can be differentiated with respect to time. Using a dot to represent the time derivative as seen from XYZ, we have

$$\dot{\boldsymbol{\omega}} = \dot{\boldsymbol{\omega}}_1 + \dot{\boldsymbol{\omega}}_2 \qquad \text{(b)}$$

Consider now the vector $\boldsymbol{\omega}_2$. Note that this vector is constrained in direction to be always collinear with the axis AB of the bearings of the shaft. This clearly is a physical requirement. Also, since ω_2 is of constant value, we may think of the vector $\boldsymbol{\omega}_2$ as *fixed* to the platform along AB. Therefore, since the platform has an angular velocity of $\boldsymbol{\omega}_1$ relative to XYZ, we can say:

$$\dot{\boldsymbol{\omega}}_2 = \boldsymbol{\omega}_1 \times \boldsymbol{\omega}_2 \qquad \text{(c)}$$

Example 15.1 (Continued)

As for $\dot{\boldsymbol{\omega}}_1$, namely the other vector in Eq. (b), we note that as seen from *XYZ*, $\boldsymbol{\omega}_1$ is a constant vector and so at all times $\dot{\boldsymbol{\omega}}_1 = \mathbf{0}$. Hence Eq. (b) can be written as follows:

$$\dot{\boldsymbol{\omega}} = \boldsymbol{\omega}_1 \times \boldsymbol{\omega}_2 \qquad \text{(d)}$$

This equation is valid at all times and so can be differentiated again. At the instant of interest as depicted by Fig. 15.9, we have for $\dot{\boldsymbol{\omega}}$:

$$\dot{\boldsymbol{\omega}} = 5\boldsymbol{k} \times 10\boldsymbol{j} = -50\boldsymbol{i} \text{ rad/s}^2 \qquad \text{(e)}$$

To get $\ddot{\boldsymbol{\omega}}$, we now differentiate (d) with respect to time. We have

$$\begin{aligned}\ddot{\boldsymbol{\omega}} &= \dot{\boldsymbol{\omega}}_1 \times \boldsymbol{\omega}_2 + \boldsymbol{\omega}_1 \times \dot{\boldsymbol{\omega}}_2 \\ &= \mathbf{0} + \boldsymbol{\omega}_1 \times (\boldsymbol{\omega}_1 \times \boldsymbol{\omega}_2)\end{aligned} \qquad \text{(f)}$$

where we have used the fact that $\dot{\boldsymbol{\omega}}_1 = \mathbf{0}$ at all times as well as Eq. (c) for $\dot{\boldsymbol{\omega}}_2$. At the instant of interest, we have

$$\ddot{\boldsymbol{\omega}} = 5\boldsymbol{k} \times (5\boldsymbol{k} \times 10\boldsymbol{j}) = -250\boldsymbol{j} \text{ rad/s}^3$$

Example 15.2

In Example 15.1, consider a position vector $\boldsymbol{\rho}$ between two points on the rotating disc (see Fig. 15.10). The length of $\boldsymbol{\rho}$ is 100 mm and, at the instant of interest, is in the vertical direction. What are the first and second time derivatives of $\boldsymbol{\rho}$ at this instant as seen from the ground reference?

It should be obvious that the vector $\boldsymbol{\rho}$ is fixed to the disc which has at all times an angular velocity relative to *XYZ* equal to $\boldsymbol{\omega}_1 + \boldsymbol{\omega}_2$. Hence, at all times we can say:

$$\dot{\boldsymbol{\rho}} = (\boldsymbol{\omega}_1 + \boldsymbol{\omega}_2) \times \boldsymbol{\rho} \qquad \text{(a)}$$

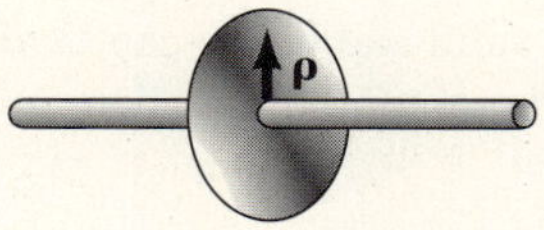

Figure 15.10. Displacement vector $\boldsymbol{\rho}$ in disc.

At the instant of interest, we have noting that $\boldsymbol{\rho} = 100\boldsymbol{k}$

$$\dot{\boldsymbol{\rho}} = (5\boldsymbol{k} + 10\boldsymbol{j}) \times 100\boldsymbol{k} = 1000\boldsymbol{i} \text{ mm/s} \qquad \text{(b)}$$

To get the second derivative of $\boldsymbol{\rho}$, go back to Eq. (a) and differentiate:

$$\ddot{\boldsymbol{\rho}} = (\dot{\boldsymbol{\omega}}_1 + \dot{\boldsymbol{\omega}}_2) \times \boldsymbol{\rho} + (\boldsymbol{\omega}_1 + \boldsymbol{\omega}_2) \times \dot{\boldsymbol{\rho}}$$

Example 15.2 (Continued)

Noting that $\dot{\boldsymbol{\omega}}_1 = \mathbf{0}$ at all times and, as discussed in Example 15.1, that $\boldsymbol{\omega}_2$ is fixed in the platform, we can say:

$$\ddot{\boldsymbol{\rho}} = (\mathbf{0} + \boldsymbol{\omega}_1 \times \boldsymbol{\omega}_2) \times \boldsymbol{\rho} + (\boldsymbol{\omega}_1 + \boldsymbol{\omega}_2) \times \dot{\boldsymbol{\rho}} \qquad \text{(c)}$$

At the instant of interest we have, on noting Eq. (b):

$$\ddot{\boldsymbol{\rho}} = (5\boldsymbol{k} \times 10\boldsymbol{j}) \times 100\boldsymbol{k} + (5\boldsymbol{k} + 10\boldsymbol{j}) \times 1000\boldsymbol{i} \text{ mm/s}^2$$

$$\ddot{\boldsymbol{\rho}} = 10\boldsymbol{j} - 10\boldsymbol{k} \text{ m/s}^2$$

Although we shall later formally examine the case of the time derivative of vector $\boldsymbol{A}$ as seen from XYZ when $\boldsymbol{A}$ is *not fixed* in a body or a reference xyz, we can handle such cases less formally with what we already know. We illustrate this in the following example.

Example 15.3

For the disc in Fig. 15.9, $\omega_2 = 6$ rad/s and $\dot{\omega}_2 = 2$ rad/s^2, both relative to the platform at the instant of interest. At this instant, $\omega_1 = 2$ rad/s and $\dot{\omega}_1 = -3$ rad/s^2 for the platform relative to the ground. Find the angular acceleration vector $\dot{\boldsymbol{\omega}}$ for the disc relative to the ground at the instant of interest.

The angular velocity of the disc relative to the ground at all times is

$$\boldsymbol{\omega} = \boldsymbol{\omega}_1 + \boldsymbol{\omega}_2 \qquad \text{(a)}$$

For $\dot{\boldsymbol{\omega}}$, we can then say

$$\dot{\boldsymbol{\omega}} = \dot{\boldsymbol{\omega}}_1 + \dot{\boldsymbol{\omega}}_2 \qquad \text{(b)}$$

It is apparent on inspecting Fig. 15.11 that at all times $\boldsymbol{\omega}_1$ is vertical, and so we can say:

$$\dot{\boldsymbol{\omega}}_1 = \frac{d}{dt}_{XYZ}(\omega_1 \boldsymbol{k}) = \dot{\omega}_1 \boldsymbol{k} \qquad \text{(c)}$$

Example 15.3 (Continued)

However, $\boldsymbol{\omega}_2$ is changing direction and, most importantly, is changing magnitude. Because of the latter, $\boldsymbol{\omega}_2$ cannot be considered fixed in a reference or a rigid body for purposes of computing $\dot{\boldsymbol{\omega}}_2$. To get around this difficulty, we fix a unit vector $\boldsymbol{j}'$ *onto the platform* to be collinear with the centerline of the shaft *AB* as shown in Fig. 15.11. We know the angular velocity of this unit vector; it is $\boldsymbol{\omega}_1$ at all times. We can then express $\boldsymbol{\omega}_2$ in the following manner, which is valid at all times:

$$\boldsymbol{\omega}_2 = \omega_2 \boldsymbol{j}' \tag{d}$$

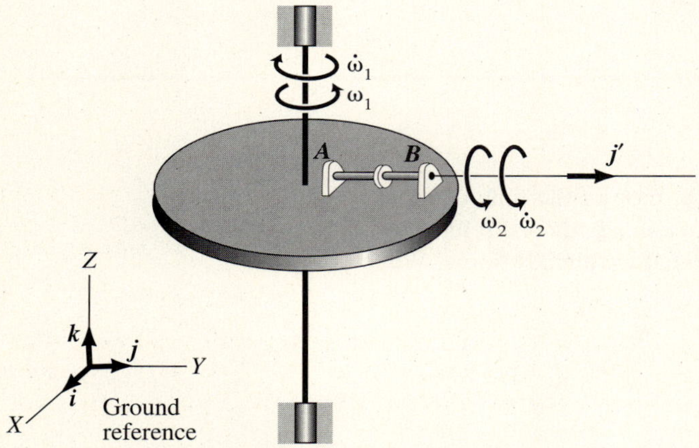

Figure 15.11. Unit vector $\boldsymbol{j}'$ fixed to platform.

We can differentiate the above with respect to time as follows:

$$\dot{\boldsymbol{\omega}}_2 = \dot{\omega}_2 \boldsymbol{j}' + \omega_2 \dot{\boldsymbol{j}}'$$

But $\boldsymbol{j}'$ is *fixed* to the platform which has angular velocity $\boldsymbol{\omega}_1$ relative to *XYZ* at all times. Hence, we have for the above,

$$\dot{\boldsymbol{\omega}}_2 = \dot{\omega}_2 \boldsymbol{j}' + \omega_2(\boldsymbol{\omega}_1 \times \boldsymbol{j}') \tag{e}$$

Thus, Eq. (b) then can be given as

$$\dot{\boldsymbol{\omega}} = \dot{\omega}_1 \boldsymbol{k} + \dot{\omega}_2 \boldsymbol{j}' + \omega_2(\boldsymbol{\omega}_1 \times \boldsymbol{j}')$$

This expression is valid at all times and could be differentiated again. At the instant of interest, we can say, noting that $\boldsymbol{j}' = \boldsymbol{j}$ at this instant,

$$\dot{\boldsymbol{\omega}} = -3\boldsymbol{k} + 2\boldsymbol{j} + 6(2\boldsymbol{k} \times \boldsymbol{j})$$

$$\dot{\boldsymbol{\omega}} = -12\boldsymbol{i} + 2\boldsymbol{j} - 3\boldsymbol{k} \text{ rad/s}^2$$

PROBLEMS

15.1. Is the motion of the cabin of a ferris wheel rotational or translational if the wheel moves at uniform speed and the occupants cause no disturbances? Why?

15.2. A cylinder rolls without slipping down an inclined surface. What is the actual axis of rotation at any instant? Why? How is this axis moving?

15.3. A reference xyz is moving such that the origin O has at time t a velocity relative to reference XYZ given as

$$V_0 = 6\boldsymbol{i} + 12\boldsymbol{j} + 13\boldsymbol{k} \text{ m/s}$$

The xyz reference has an angular velocity $\boldsymbol{\omega}$ relative to XYZ at time t given as

$$\boldsymbol{\omega} = 10\boldsymbol{i} + 12\boldsymbol{j} + 2\boldsymbol{k} \text{ rad/s}$$

What is the time rate of change relative to XYZ of a directed line segment $\boldsymbol{\rho}$ going from position (3,2,–5) to (–2,4,6) in xyz? What is the time rate of change relative to XYZ of position vectors $\boldsymbol{i}'$ and $\boldsymbol{k}'$?

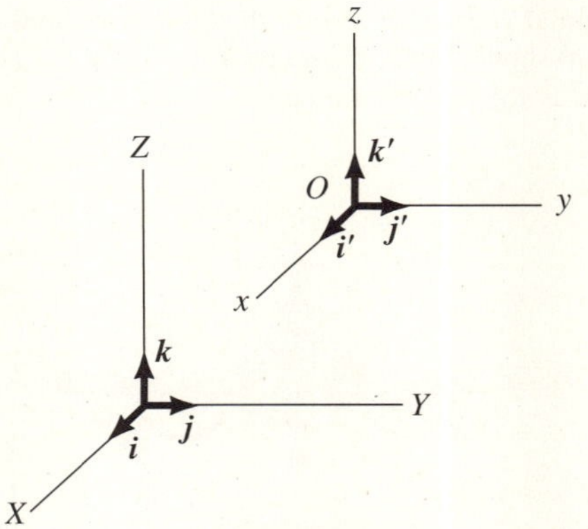

Figure P.15.3.

15.4. A reference xyz is moving relative to XYZ with a velocity of the origin given at time t as

$$V_0 = 6\boldsymbol{i} + 4\boldsymbol{j} + 6\boldsymbol{k} \text{ m/s}$$

The angular velocity of reference xyz relative to XYZ is

$$\boldsymbol{\omega} = 3\boldsymbol{i} + 14\boldsymbol{j} + 2\boldsymbol{k} \text{ rad/s}$$

What is the time rate of change as seen from XYZ of a directed line segment $\boldsymbol{\rho}_{1,2}$ in xyz going from position 1 to position 2 where the position vectors in xyz for these points are, respectively,

$$\boldsymbol{\rho}_1 = 2\boldsymbol{i}' + 3\boldsymbol{j}' \text{ m}$$
$$\boldsymbol{\rho}_2 = 3\boldsymbol{i}' - 4\boldsymbol{j}' + 2\boldsymbol{k}' \text{ m}$$

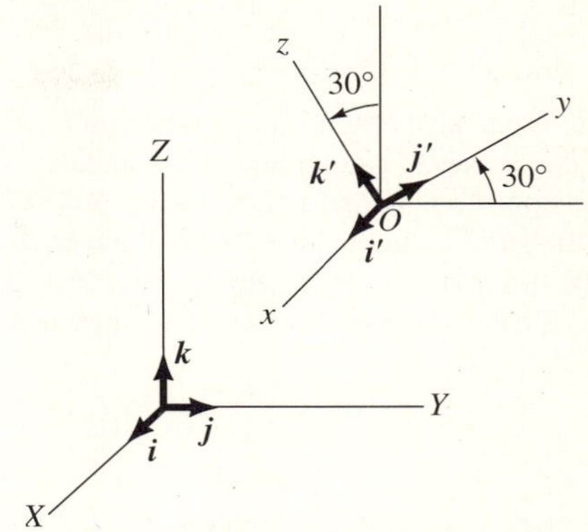

Figure P.15.4.

15.5. Find the second derivatives as seen from XYZ of the vector $\boldsymbol{\rho}$ and the unit vector $\boldsymbol{i}'$ specified in Problem 15.3. The angular acceleration of xyz relative to XYZ at the instant of interest is

$$\dot{\boldsymbol{\omega}} = 5\boldsymbol{i} + 2\boldsymbol{j} + 3\boldsymbol{k} \text{ rad/s}^2$$

15.6. Find the second derivative as seen from XYZ of the vector $\boldsymbol{\rho}_{1,2}$ specified in Problem 15.4. Take the angular acceleration of xyz relative to XYZ at the instant of interest as

$$\dot{\boldsymbol{\omega}} = 15\boldsymbol{i} + 2\boldsymbol{k} \text{ rad/s}^2$$

15.7. A platform is rotating with a constant speed ω_1 of 10 rad/s relative to the ground. A shaft is mounted on the platform and rotates relative to the platform at a speed ω_2 of 5 rad/s. What is the angular velocity of the shaft relative to the ground? What are the first and second time derivatives of the angular velocity of the shaft relative to the ground?

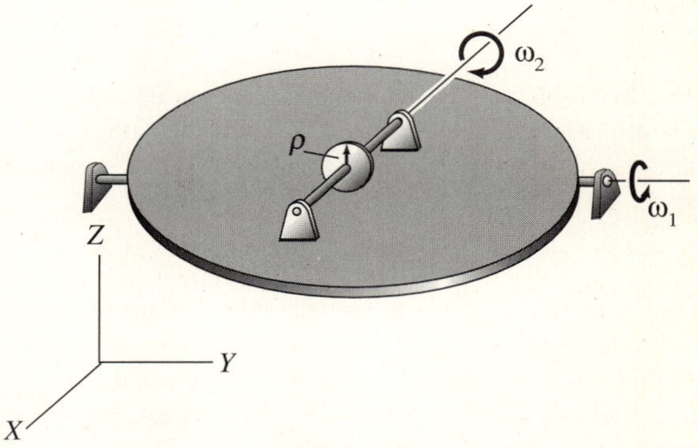

Figure P.15.7.

15.8. In Problem 15.7, what are the first and second time derivatives of a directed line segment $\boldsymbol{\rho}$ in the disc at the instant that the system has the geometry shown? The vector $\boldsymbol{\rho}$ is of length 10 mm.

15.9. A tank is maneuvering its gun into position. At the instant of interest, the turret A is rotating at an angular speed $\dot{\theta}$ of 2 rad/s relative to the tank and is in position $\theta = 20°$. Also, at this instant, the gun is rotating at an angular speed $\dot{\phi}$ of 1 rad/s relative to the turret and forms an angle $\phi = 30°$ with the horizontal plane. What are $\boldsymbol{\omega}$, $\dot{\boldsymbol{\omega}}$, and $\ddot{\boldsymbol{\omega}}$ of the gun relative to the ground?

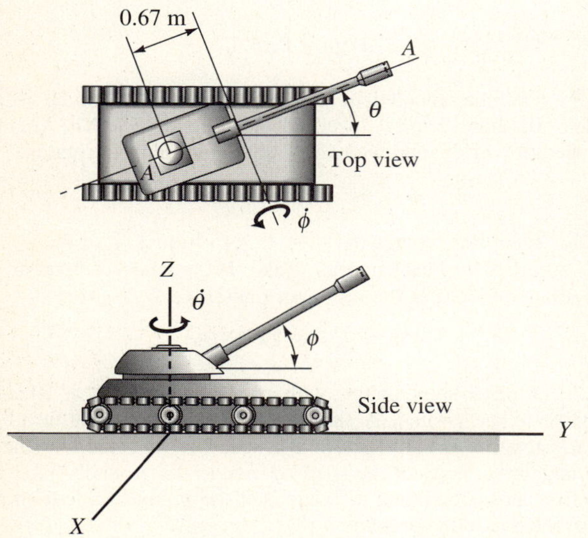

Figure P.15.9.

15.10. In Problem 15.9, determine $\boldsymbol{\omega}$ and $\dot{\boldsymbol{\omega}}$ assuming that the tank is also rotating about the vertical axis at a rate of 0.2 rad/s relative to the ground in a clockwise direction as viewed from above.

15.11. A particle is made to move at constant speed V equal to 10 m/s along a straight groove on a plate B. The plate rotates at a constant angular speed ω_2 equal to 3 rad/s relative to a platform C while the platform rotates with a constant angular speed ω_1 of 5 rad/s relative to the ground reference XYZ. Find the first and second derivatives of $\boldsymbol{V}$ as seen from the ground reference.

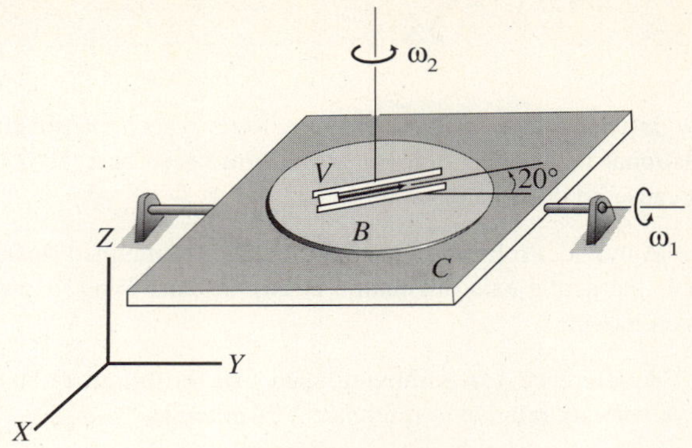

Figure P.15.11.

15.12. A jet fighter plane has just taken off and is retracting its landing gear. At the end of its run on the ground, the plane is moving at a speed of 55 m/s. If the diameter of the tires is 460 mm and if we neglect the loss of angular speed of the wheels due to wind friction after the plane is in the air, what is the angular speed $\boldsymbol{\omega}$ and the angular acceleration $\dot{\boldsymbol{\omega}}$ of the left wheel (under the wing) at the instant shown in the diagram? Take $\omega_2 = 0.4$ rad/s and $\dot{\omega}_2$ is 0.2 rad/s^2 at the instant of interest.

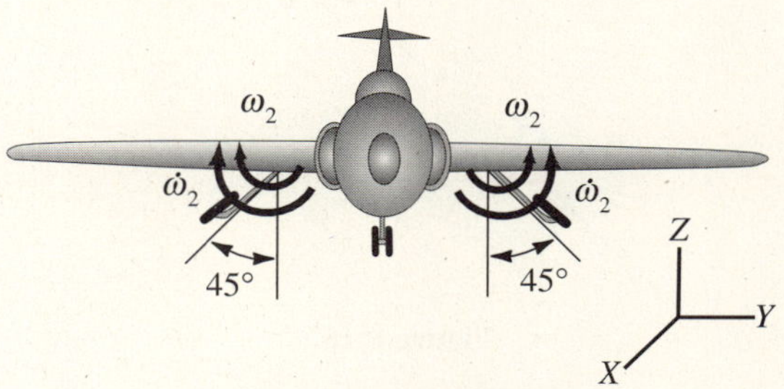

Figure P.15.12.

15.13. A truck is carrying a cockpit for a worker who repairs overhead road fixtures. At the instant shown in the diagram, the base D is rotating with constant speed ω_2 of 1 rad/s relative to the truck. Arm AB is rotating at constant angular speed ω_1 of 2 rad/s relative to DA. Cockpit C is rotating relative to AB so as to always keep the man upright. What are $\boldsymbol{\omega}$, $\dot{\boldsymbol{\omega}}$, and $\ddot{\boldsymbol{\omega}}$ of arm AB relative to the ground at the instant of interest? The truck is stationary.

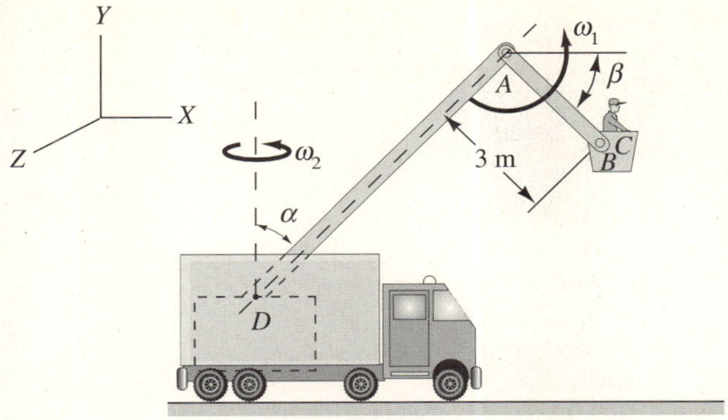

Figure P.15.13.

15.14. An electric motor M is mounted on a plate A which is welded to a shaft D. The motor has a constant angular speed ω_2 relative to plate A of 1750 r/min. Plate A at the instant of interest is in a vertical position as shown and is rotating with an angular speed ω_1 equal to 100 r/min and a rate of change of angular speed $\dot{\omega}_1$ equal to 0.5 r/s²—all relative to the ground. The normal projection of the centerline of the motor shaft onto the plate A is at an angle of 45° with the edge of the plate FE. Compute the first and second time derivatives of $\boldsymbol{\omega}$, the angular velocity of the motor, as seen from the ground.

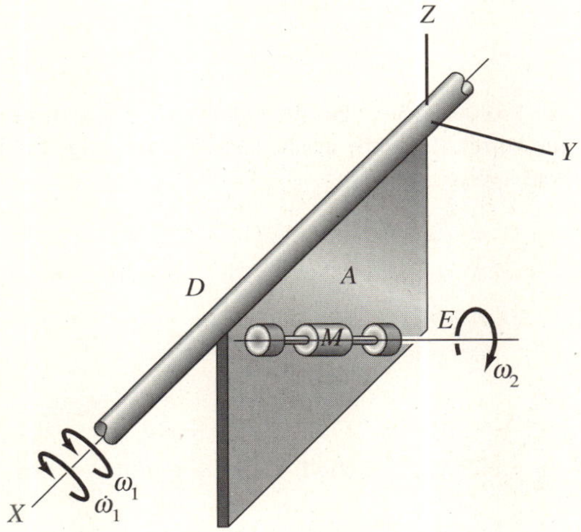

Figure P.15.14.

15.15. A racing car is moving at a constant speed of 320 m/s when the driver turns his front wheels at an increasing rate, $\dot{\omega}_1$, of 20 mrad/s². If $\omega_1 = 16.8$ mrad/s at the instant of interest, what are $\boldsymbol{\omega}$ and $\dot{\boldsymbol{\omega}}$ of the front wheels at this instant? The diameter of the tires is 0.75 m.

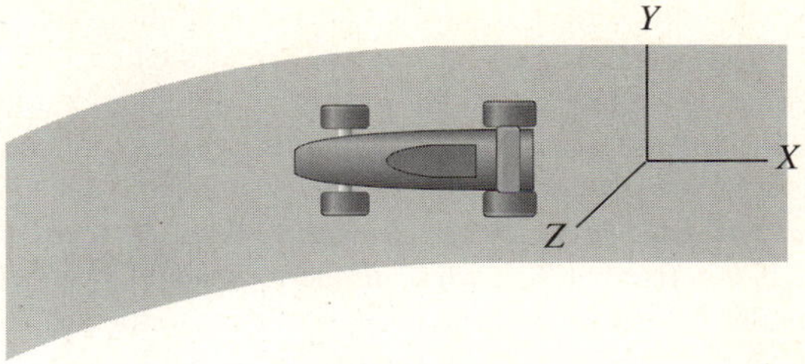

Figure P.15.15.

15.16. A cone is rolling without slipping such that its centerline rotates at the rate ω_1 of 5 revolutions per second about the **Z** axis. What is the angular velocity $\boldsymbol{\omega}$ of the body relative to the ground? What is the angular acceleration vector for the body?

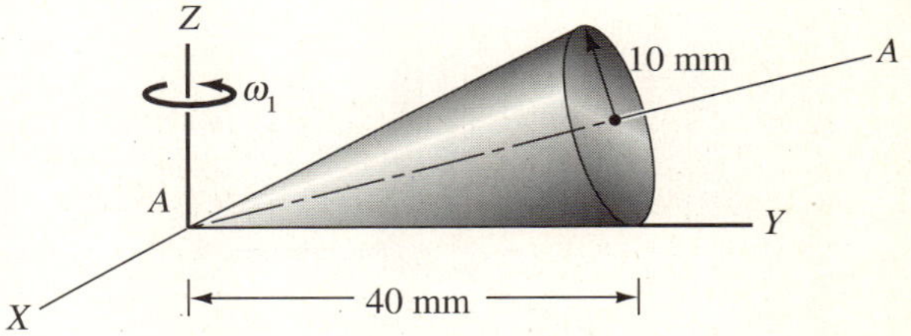

Figure P.15.16.

15.17. A small cone A is rolling without slipping inside a large conical cavity B. What is the angular velocity $\boldsymbol{\omega}$ of cone A relative to the large cone cavity B if the centerline of A undergoes an angular speed ω_1 of 5 rotations per second about the Z axis?

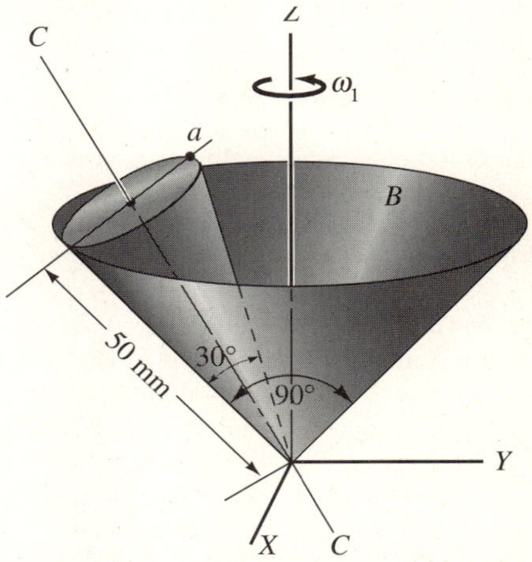

Figure P.15.17.

15.18. An amusement park ride consists of a stationary vertical tower with arms that can swing outward from the tower and at the same time can rotate about the tower. At the ends of the arms, cockpits containing passengers can rotate relative to the arms. Consider the case where cockpit A rotates at angular speed ω_2 relative to arm BC, which rotates at angular speed ω_1 relative to the tower. If θ is fixed at 90°, what are the total angular velocity and the angular acceleration of the cockpit relative to the ground? Use $\omega_1 = 0.2$ rad/s and $\omega_2 = 0.6$ rad/s.

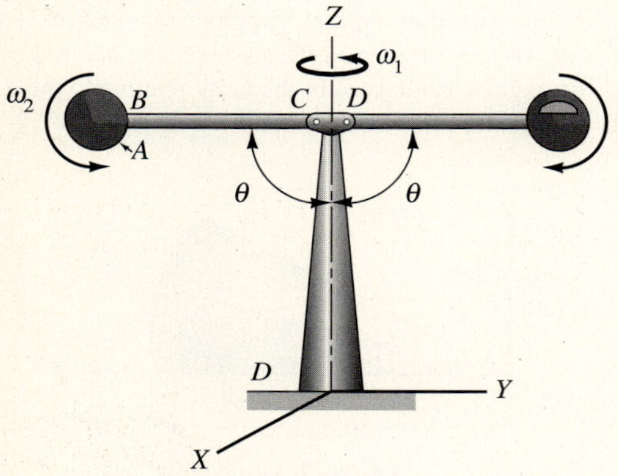

Figure P.15.18.

15.19. In Problem 15.18, find $\dot{\boldsymbol{\omega}}$ of the cockpit for the case where $\dot{\theta} = \omega_3 = 0.8$ rad/s at the instant that $\theta = 90°$.

15.20. Mass A is connected to an inextensible wire. Supports C and D are moving as shown.

(a) What is the velocity vector of mass A?

(b) If cylinder G is free to rotate and there is no slipping, what is its angular velocity?

The following data apply:

$$h = 2 \text{ m} \qquad (V_x)_2 = 0.24 \text{ m/s}$$
$$L = 3 \text{ m} \qquad (V_y)_2 = 0.21 \text{ m/s}$$
$$l = 2 \text{ m} \qquad R = 1 \text{ m}$$
$$(V_x)_1 = 0.5 \text{ m/s} \qquad \alpha = 45°$$
$$(V_y)_1 = 0.6 \text{ m/s}$$

The last four problems of this set are designed for those students who have studied Example 15.3.

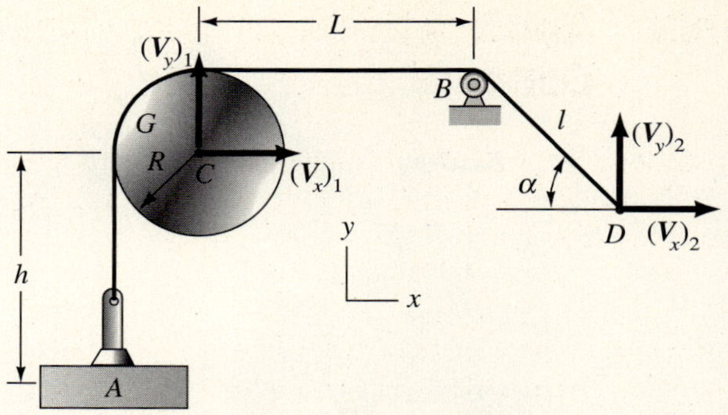

Figure P.15.20.

15.21. In Problem 15.18, find $\dot{\boldsymbol{\omega}}$ of the cockpit A for the case where $\dot{\omega}_1 = 0.2$ rad/s² and $\dot{\omega}_2 = 0.3$ rad/s².

15.22. In Problem 15.13, find $\dot{\boldsymbol{\omega}}$ of beam AB relative to the ground if at the instant shown the following data apply:

$$\omega_1 = 0.3 \text{ rad/s}$$
$$\dot{\omega}_1 = 0.2 \text{ rad/s}^2$$
$$\omega_2 = 0.6 \text{ rad/s}$$
$$\dot{\omega}_2 = -0.1 \text{ rad/s}^2$$

15.23. In Problem 15.9, find the angular acceleration vector $\dot{\boldsymbol{\omega}}$ for the gun barrel, if, for the instant shown in the diagram, the following data apply:

$$\dot{\phi} = 0.30 \text{ rad/s}, \qquad \theta = 20°$$
$$\ddot{\phi} = 0.26 \text{ rad/s}^2, \qquad \phi = 30°$$
$$\dot{\theta} = 0.17 \text{ rad/s}$$
$$\ddot{\theta} = -0.34 \text{ rad/s}^2$$

15.24. In Problem 15.11, find $\dot{\mathbf{V}}$ if at the instant shown in the diagram:

$$\omega_1 = 5 \text{ rad/s}$$
$$\dot{\omega}_1 = 10 \text{ rad/s}^2$$
$$\omega_2 = 2 \text{ rad/s}$$
$$\dot{\omega}_2 = 3 \text{ rad/s}^2$$
$$V = 10 \text{ m/s}$$
$$\dot{V} = 5 \text{ m/s}^2$$

15.5 Applications of the Fixed-Vector Concept

In Section 15.4, we considered the time derivative, as seen from a reference XYZ, of a vector $\boldsymbol{A}$ fixed in a rigid body or fixed in reference xyz. The result was a simple formula:

$$\dot{\boldsymbol{A}} = \boldsymbol{\omega} \times \boldsymbol{A}$$

where $\boldsymbol{\omega}$ is the angular velocity relative to XYZ of the body or the reference in which $\boldsymbol{A}$ is fixed. In this section, we shall use the preceding formula for a *vector connecting two points a and b in a rigid body* (see Fig. 15.12). This vector, which we denote as $\boldsymbol{\rho}_{ab}$, clearly is fixed in the rigid body. The body in accordance with **Chasles' theorem** has a velocity $\dot{\boldsymbol{R}}$ relative to XYZ corresponding to some point O in the body plus an angular velocity $\boldsymbol{\omega}$ relative to XYZ with the axis of rotation going through O. We can then say on observing from XYZ:

$$\dot{\boldsymbol{\rho}}_{ab} = \boldsymbol{\omega} \times \boldsymbol{\rho}_{ab} \tag{15.4}$$

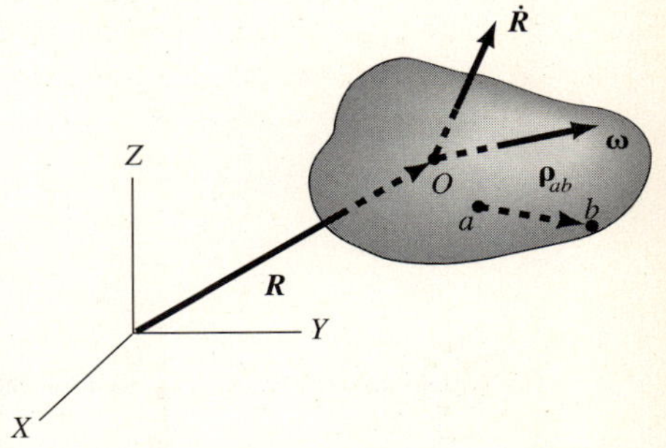

Figure 15.12. $\boldsymbol{\rho}_{ab}$ fixed in rigid body.

Now consider position vectors at a and b as shown in Fig. 15.13. We can say:

$$\boldsymbol{r}_a + \boldsymbol{\rho}_{ab} = \boldsymbol{r}_b$$

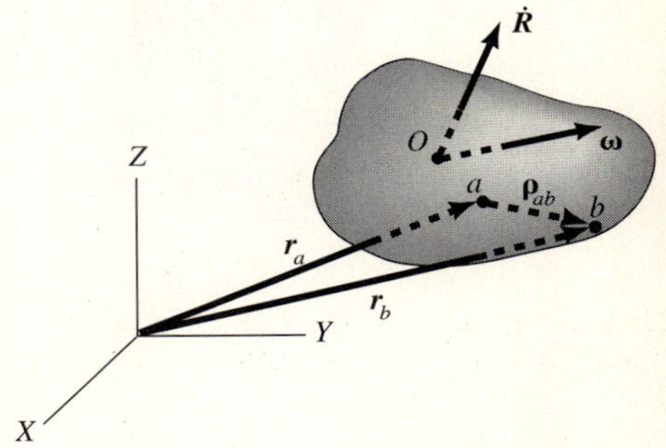

Figure 15.13. Insert position vectors.

Taking the time derivative as seen from XYZ, we have

$$\left(\frac{d\boldsymbol{r}_a}{dt}\right)_{XYZ} + \left(\frac{d\boldsymbol{\rho}_{ab}}{dt}\right)_{XYZ} = \left(\frac{d\boldsymbol{r}_b}{dt}\right)_{XYZ}$$

This equation can be written as

$$\left(\frac{d\boldsymbol{\rho}_{ab}}{dt}\right)_{XYZ} = \boldsymbol{V}_b - \boldsymbol{V}_a \tag{15.5}$$

Since $(d\boldsymbol{\rho}_{ab}/dt)_{XYZ}$ is the *difference* between the velocity of point b and that at point a as noted above, we can say that $(d\boldsymbol{\rho}_{ab}/dt)_{XYZ}$ is the velocity of point b *relative* to point a.[4] Next, using Eq. 15.4 to replace $(d\boldsymbol{\rho}_{ab}/dt)_{XYZ}$, we have, on rearranging terms, a very useful equation:

$$\boldsymbol{V}_b = \boldsymbol{V}_a + \boldsymbol{\omega} \times \boldsymbol{\rho}_{ab} \tag{15.6}$$

In using the foregoing equation, we must be sure that we get the sequence of subscripts correct on $\boldsymbol{\rho}$ since a change in ordering brings about a change in sign (i.e., $\boldsymbol{\rho}_{ab} = -\boldsymbol{\rho}_{ba}$). This equation is a statement of the physically obvious result that *the velocity of particle b of a rigid body as seen from XYZ equals the velocity of any other particle a of this body as seen from XYZ plus the velocity of particle b relative to particle a.*

[4]That is, $(d\boldsymbol{\rho}_{ab}/dt)_{XYZ}$ is the velocity of b as seen by an observer translating relative to XYZ with point a, i.e., as seen by a nonrotating observer moving with a.

Differentiating Eq. 15.6 again, we can get a relation involving the acceleration vectors of two points on a rigid body:

$$\boldsymbol{a}_b = \boldsymbol{a}_a + \left(\frac{d\boldsymbol{\omega}}{dt}\right)_{XYZ} \times \boldsymbol{\rho}_{ab} + \boldsymbol{\omega} \times \left(\frac{d\boldsymbol{\rho}_{ab}}{dt}\right)_{XYZ}$$

Hence, we have on using Eq. 15.4 in the last expression

$$\boldsymbol{a}_b = \boldsymbol{a}_a + \dot{\boldsymbol{\omega}} \times \boldsymbol{\rho}_{ab} + \boldsymbol{\omega} \times (\boldsymbol{\omega} \times \boldsymbol{\rho}_{ab}) \tag{15.7}$$

We have thus formulated relations between *the motions of two points of a rigid body as seen from a single reference*. Such relations can be very useful in the study of machine elements.

Before going to the examples, let us now consider the case of a circular cylinder rolling *without slipping* (see Fig. 15.14). The point of contact A of the cylinder with the ground has instantaneously *zero velocity* and hence we have pure instantaneous rotation at any time t about an instantaneous axis of rotation at the line of contact. The velocity of any point B of the cylinder can then easily be found by using Eq. 15.6 for points B and A. Thus:

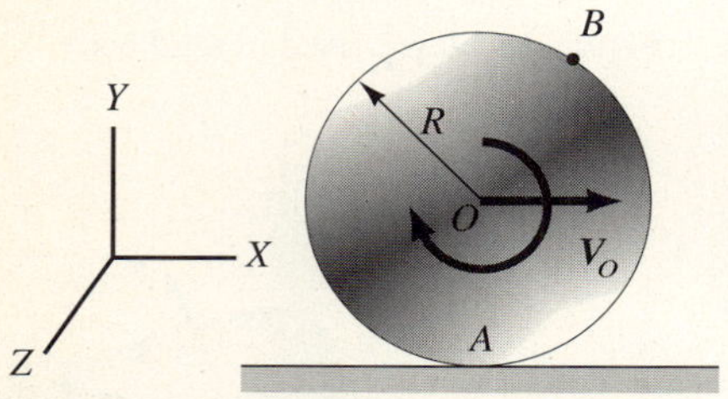

Figure 15.14. Cylinder rolling without slipping on a flat surface.

$$\boldsymbol{V}_B = \boldsymbol{V}_A + \boldsymbol{\omega} \times \boldsymbol{\rho}_{AB}$$

Therefore,

$$\boldsymbol{V}_B = \boldsymbol{0} + \boldsymbol{\omega} \times \boldsymbol{\rho}_{AB} = \boldsymbol{\omega} \times \boldsymbol{\rho}_{AB}$$

From the above equation it is clear that for computing the velocity of any point on the cylinder we can think of the cylinder as *hinged* at the point of contact. In particular for point O, the center of the cylinder, we get from above:

$$V_0 = -\omega R\boldsymbol{i}$$

If the velocity V_0 is known, clearly the angular velocity has a magnitude of V_0/R.

Another way of relating V and ω is to realize that the distance s that O moves must equal the length of circumference coming into contact with the ground. That is, measuring θ from the X axis to the Y axis:

$$s = -R\theta$$

Differentiating we get:

$$V_0 = -R\dot{\theta} = -R\omega$$

thus reproducing the previous result. Differentiating again, we get

$$a_0 = -R\ddot{\theta} = -R\alpha \tag{15.8}$$

relating now the acceleration of O and the angular acceleration α. Clearly, the acceleration vector for O must be parallel to the ground. Again, for computing $\boldsymbol{a}_0$, we have a simple situation.

Next, let us determine the acceleration vector for the *point of contact A* of the cylinder. Thus, we can say for points A and O:

$$\boldsymbol{a}_O = \boldsymbol{a}_A + \dot{\boldsymbol{\omega}} \times \boldsymbol{\rho}_{AO} + \boldsymbol{\omega} \times (\boldsymbol{\omega} \times \boldsymbol{\rho}_{AO})$$

Therefore,

$$-R\ddot{\theta}\boldsymbol{i} = \boldsymbol{a}_A + \ddot{\theta}\boldsymbol{k} \times R\boldsymbol{j} + \dot{\theta}\boldsymbol{k} \times (\dot{\theta}\boldsymbol{k} \times R\boldsymbol{j}) \tag{15.9}$$

Carrying out the products:

$$-R\ddot{\theta}\boldsymbol{i} = \boldsymbol{a}_A - R\ddot{\theta}\boldsymbol{i} - R\dot{\theta}^2\boldsymbol{j}$$

Therefore, cancelling terms, we get

$$\boldsymbol{a}_A = R\dot{\theta}^2\boldsymbol{j} \tag{15.10}$$

We see that *point A is accelerating upward toward the center of the cylinder.*[5] This information will be valuable for us in Chapter 16 when we study rigid-body dynamics.

[5]This conclusion must apply also to a sphere rolling without slipping on a flat surface. As for acceleration of other points of the cylinder, we do not have a simple formula but must insert data for these points into the acceleration formula valid for two points of a rigid body.

Example 15.4

Wheel D rotates at an angular speed ω_1 of 2 rad/s counterclockwise in Fig. 15.15. Find the angular speed ω_E of gear E relative to the ground at the instant shown in the diagram.

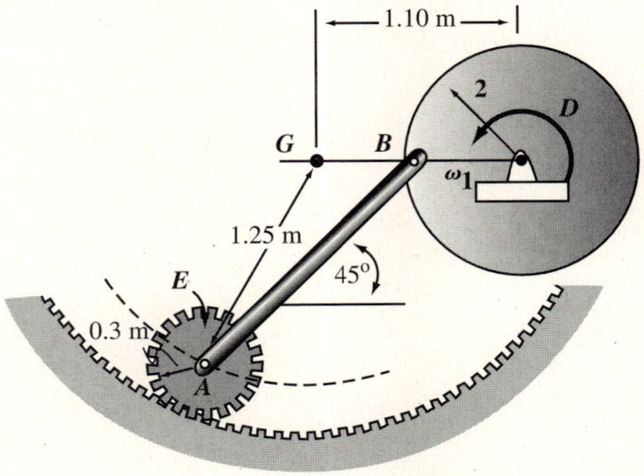

Figure 15.15. Two-dimensional device.

We have information about two points of one of the rigid bodies, namely AB, of the device. At B, the velocity must be downward with the

Example 15.4 (Continued)

value of $(\omega_1)(r_D) = 1.2$ m/s as shown in Fig. 15.16. Furthermore, since point A must travel a circular path of radius GA we know that A has velocity V_A with a direction at right angles to GA. Accordingly, since the angle between GA and the vertical is $(90° - 45° - \alpha) = (45° - \alpha)$ as can readily be seen on inspecting Fig. 15.16, then the angle between V_A and the horizontal must also be $(45° - \alpha)$ because of the *mutual perpendicularity* of the sides of these angles. If we can determine velocity V_A, we can get the desired angular speed of gear A immediately.

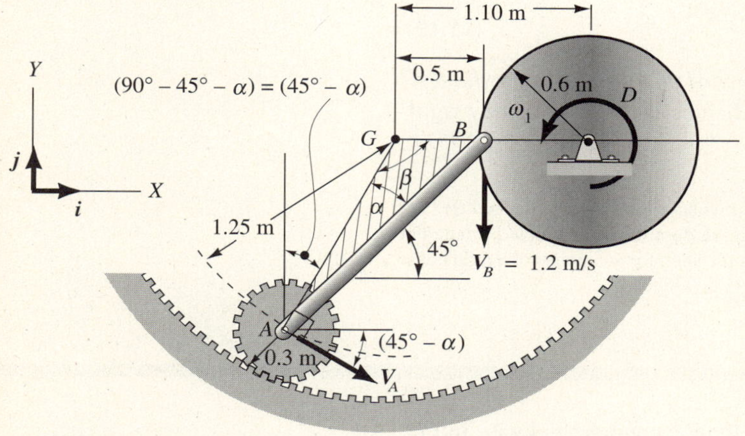

Figure 15.16. Velocity vectors for two points of a rigid body shown.

Before examining rigid body AB, we have some geometrical steps to take. Considering triangle GAB in Fig. 15.16, we can first solve for α using the law of sines as follows:

$$\frac{GA}{\sin(\measuredangle GBA)} = \frac{GB}{\sin\alpha}$$

Therefore, since $\measuredangle GBA = 45°$

$$\frac{1.25}{\sin 45°} = \frac{0.5}{\sin\alpha} \tag{a}$$

Solving for α, we get

$$\alpha = 16.43° \tag{b}$$

The angle β is then easily evaluated considering the angles in the triangle GBA. Thus,

$$\begin{aligned}\beta &= 180° - \alpha - \measuredangle GBA \\ &= 180° - 16.43° - 45° = 118.57°\end{aligned} \tag{c}$$

Example 15.4 (Continued)

Finally, we can determine AB of the triangle, again using the law of sines. Thus,

$$\frac{AB}{\sin\beta} = \frac{GA}{\sin 45°}$$
$$\frac{AB}{\sin 118.57°} = \frac{1.25}{0.707}$$

Solving for AB, we get

$$AB = 1.553 \text{ m} \tag{d}$$

We now can consider bar AB as our rigid body. For the points A and B on this body, we can say:

$$\boldsymbol{V}_A = \boldsymbol{V}_B + \boldsymbol{\omega}_{AB} \times \boldsymbol{\rho}_{BA}$$

Noting that the motion is coplanar and that $\boldsymbol{\omega}_{AB}$ must then be normal to the plane of motion, we have[6]

$$V_A[\cos(45° - \alpha)\boldsymbol{i} - \sin(45° - \alpha)\boldsymbol{j}]$$
$$= -1.25\boldsymbol{j} + \omega_{AB}\boldsymbol{k} \times 1.553(-\cos 45°\boldsymbol{i} - \sin 45°\boldsymbol{j})$$

Inserting the value $\alpha = 16.43°$, we then get the following vector equation:

$$V_A(0.878)\boldsymbol{i} - V_A(0.478)\boldsymbol{j} = -1.25\boldsymbol{j} - 1.098\omega_{AB}\boldsymbol{j} + 1.098\omega_{AB}\boldsymbol{i} \tag{e}$$

The scalar equations are

$$\begin{aligned} 0.878V_A &= 1.098\omega_{AB} \\ -0.478V_A &= -1.25 - 1.098\omega_{AB} \end{aligned} \tag{f}$$

Solving, we get[7]

$$\begin{aligned} V_A &= -3.125 \text{ m/s} \\ \omega_{AB} &= -2.499 \text{ rad/s} \end{aligned} \tag{g}$$

Thus, point A moves in a direction *opposite* to that shown in Fig. 15.16. We now can readily evaluate ω_E, which clearly must have a value of

$$\omega_E = \frac{V_A}{r_E} = \frac{3.125}{0.3} = \boxed{10.42 \text{ rad/s}}$$

in the counterclockwise direction.

[6]Our practice will be to consider unknown angular velocities as *positive*. The sign for the unknown angular velocity coming out of the computations will then correspond to the *actual convention* sign for the angular velocity.

[7]By having assumed ω_{AB} as positive and thus *counterclockwise* for the reference xy employed, we conclude from the presence of the minus sign that the assumption is wrong and that ω_{AB} must be *clockwise* for the reference used. It is significant to note that as a result of the initial positive assumption, the result $\omega_{AB} = -2.499$ rad/s gives at the same time the *correct convention sign* for the actual angular velocity for the reference used.

Example 15.5

In the device in Fig. 15.17, find the angular velocities and angular accelerations of both bars.

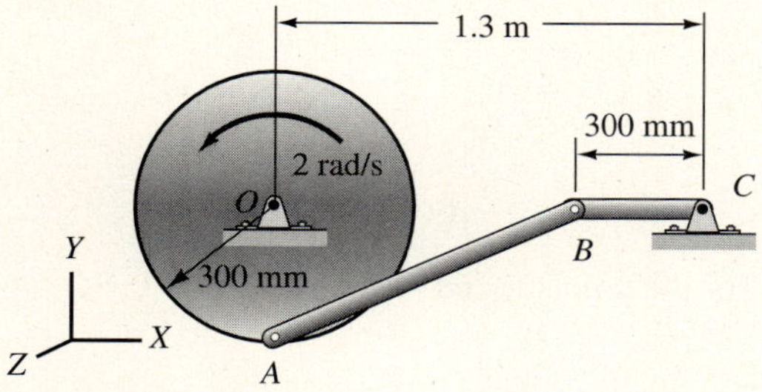

Figure 15.17. Two-dimensional device.

We shall consider points A and B of bar AB. Note first that at the instant shown:

$$\boldsymbol{V}_B = -(0.3)(\omega_{BC})\boldsymbol{j}\ \text{m/s} \tag{a}$$

$$\begin{aligned} \boldsymbol{V}_A &= (2)(0.3)\boldsymbol{i} \\ &= 0.6\boldsymbol{i}\ \text{m/s} \end{aligned} \tag{b}$$

Noting that $\boldsymbol{\omega}_{AB}$ must be oriented in the Z direction because we have plane motion in the XY plane, we have for Eq. 15.6:

$$\begin{aligned} \boldsymbol{V}_B &= \boldsymbol{V}_A + \boldsymbol{\omega}_{AB} \times \boldsymbol{\rho}_{AB} \\ -0.3\omega_{BC}\,\boldsymbol{j} &= 0.6\boldsymbol{i} + (\omega_{AB}\,\boldsymbol{k}) \times (\boldsymbol{i} + 0.3\,\boldsymbol{j}) \\ \therefore -0.3\omega_{BC}\,\boldsymbol{j} &= 0.6\boldsymbol{i} + \omega_{AB}\,\boldsymbol{j} - 0.3\omega_{AB}\,\boldsymbol{i} \end{aligned} \tag{c}$$

Note we have assumed ω_{BC} and ω_{AB} as positive and thus counterclockwise. The scalar equations are:

$$\begin{aligned} 0.6 &= 0.3\omega_{AB} \\ -0.3\omega_{BC} &= \omega_{AB} \end{aligned} \tag{d}$$

We then get

$$\begin{aligned} \omega_{AB} &= 2\ \text{rad/s} \\ \omega_{BC} &= -6.67\ \text{rad/s} \end{aligned} \tag{e}$$

Therefore, ω_{AB} is counterclockwise while ω_{BC} must be clockwise.

Let us now turn to the angular acceleration considerations for the bars. We consider separately now points A and B of bar AB. Thus,

$$\begin{aligned} \boldsymbol{a}_A &= (r\omega^2)\boldsymbol{j} = (0.3)(2^2)\boldsymbol{j} = 1.2\boldsymbol{j}\ \text{m/s}^2 \\ \boldsymbol{a}_B &= \rho_{BC}\omega_{BC}^2\boldsymbol{i} + \rho_{BC}\dot{\omega}_{BC}(-\boldsymbol{j}) \\ &= (0.3)(-6.67^2)\boldsymbol{i} - 0.3\dot{\omega}_{BC}\boldsymbol{j} \\ &= 13.33\boldsymbol{i} - 0.3\dot{\omega}_{BC}\boldsymbol{j} \end{aligned}$$

Example 15.5 (Continued)

Again, we have assumed $\dot{\boldsymbol{\omega}}_{BC}$ positive and thus counterclockwise. Considering bar AB, we can say for Eq. 15.7:

$$\boldsymbol{a}_B = \boldsymbol{a}_A + \dot{\boldsymbol{\omega}}_{AB} \times \boldsymbol{\rho}_{AB} + \boldsymbol{\omega}_{AB} \times (\boldsymbol{\omega}_{AB} \times \boldsymbol{\rho}_{AB}) \qquad \text{(f)}$$

Noting that $\dot{\boldsymbol{\omega}}_{AB}$ must be in the Z direction, we have for the foregoing equation:

$$\begin{aligned} 13.33\boldsymbol{i} - 0.3\dot{\omega}_{BC}\boldsymbol{j} & \\ &= 1.2\boldsymbol{j} + \dot{\omega}_{AB}\boldsymbol{k} \times (\boldsymbol{i} + 0.3\boldsymbol{j}) + (2\boldsymbol{k}) \times [2\boldsymbol{k} \times (\boldsymbol{i} + 0.3\boldsymbol{j})] \end{aligned} \qquad \text{(g)}$$

The scalar equations are

$$\begin{aligned} 17.33 &= -0.3\dot{\omega}_{AB} \\ -0.3\dot{\omega}_{BC} &= \dot{\omega}_{AB} \end{aligned}$$

We get

$$\begin{aligned} \dot{\omega}_{AB} &= -57.8 \text{ rad/s}^2 \\ \dot{\omega}_{BC} &= 192.6 \text{ rad/s}^2 \end{aligned}$$

Clearly, for the reference used, $\dot{\omega}_{AB}$ must be clockwise and $\dot{\omega}_{BC}$ must be counterclockwise.

Example 15.6

(a) In Example 15.5, find the *instantaneous axis of rotation* for the rod AB.

The intersection of the instantaneous axis of rotation with the xy plane will be a point E in a hypothetical rigid-body extension of bar AB having zero velocity at the instant of interest. We can accordingly say:

$$\boldsymbol{V}_E = \boldsymbol{V}_A + \boldsymbol{\omega}_{AB} \times \boldsymbol{\rho}_{AE}$$

Therefore,

$$\boldsymbol{0} = 0.6\boldsymbol{i} + (2\boldsymbol{k}) \times (\Delta x\boldsymbol{i} + \Delta y\boldsymbol{j}) \qquad \text{(a)}$$

where Δx and Δy are the components of the directed line segment from point A to the center of rotation E. The scalar equations are:

$$\begin{aligned} 0 &= 0.6 - 2\Delta y \\ 0 &= 2\Delta x \end{aligned}$$

Clearly, $\Delta y = 0.3$ and $\Delta x = 0$. Thus, the center of rotation is point O.

Example 15.6 (Continued)

We could have easily deduced this result by inspection in this case. The velocity of each point of bar *AB* must be at *right angles* to a line from the center of rotation to the point. The velocity of point *A* is in the horizontal direction and the velocity of point *B* is in the vertical direction. Clearly, as seen from Fig. 15.18, point *O* is the only point from which lines to points *A* and *B* are normal to the velocities at these points.

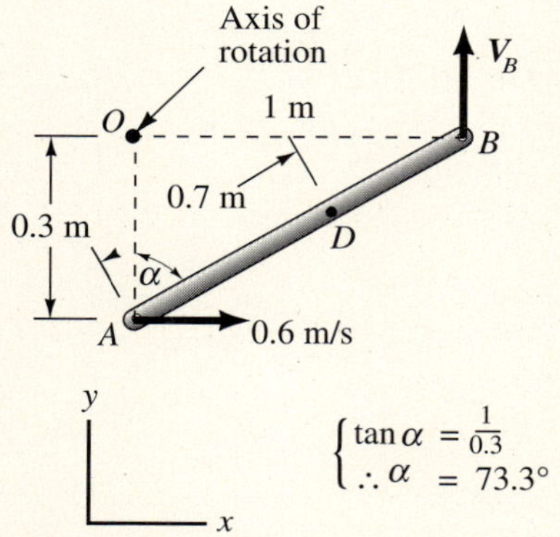

Figure 15.18. Instantaneous axis of rotation of *AB*.

(b) Now using the instantaneous axis of rotation, find the magnitudes of the velocity and acceleration of point *D* (Fig. 15.18) using data from the previous example.

In Fig. 15.19, we show the velocity vector normal to line *OD*. Using the law of cosines for triangle *AOD*, we can find *OD* which is a key distance for this example. Thus noting from Fig. 15.18 that $\alpha = 73.3°$, we have

$$\overline{OD} = \left[0.7^2 + 0.3^2 - (2)(0.7)(0.3)(\cos 73.3°)\right]^{1/2} = 0.6777 \text{ m}$$

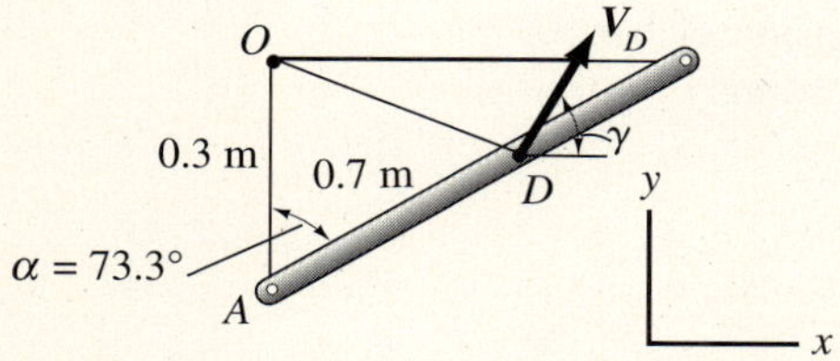

Figure 15.19. Velocity vector for point *D*.

We then say from rotational motion about the instantaneous center of rotation *O*,

$$V_D = (0.6777)(\omega_{AB}) = (0.6777)(2) = 1.355 \text{ m/s}$$

Example 15.6 (Continued)

For the acceleration, we have (see Fig. 15.20)

$$a_D = \left[(a_D)_c^2 + (a_D)_t^2\right]^{1/2}$$

where $(a_D)_c$ and $(a_D)_t$, respectively, are the centripetal and tangential components of acceleration at point D. Noting that r for point D is 0.6777 m, we get for the above

$$\therefore a_D = \left\{\left(\frac{V_D^2}{r}\right)^2 + \left[(r)(\dot{\omega}_{AB})\right]^2\right\}^{1/2}$$

$$= \left\{\left(\frac{1.355^2}{0.6777}\right)^2 + \left[(0.6777)(57.8)\right]^2\right\}^{1/2} = 39.26 \text{ m/s}^2 \quad \text{(b)}$$

We now get the vectors $\mathbf{V}_D$ and $\boldsymbol{a}_D$. For this purpose we determine the angle β of the tinted triangle in Fig. 15.20 by first using the law of sines for triangle AOD

$$\frac{0.7}{\sin(90° - \beta)} = \frac{0.6777}{\sin 73.3°}$$

$$\therefore \beta = 8.373°$$

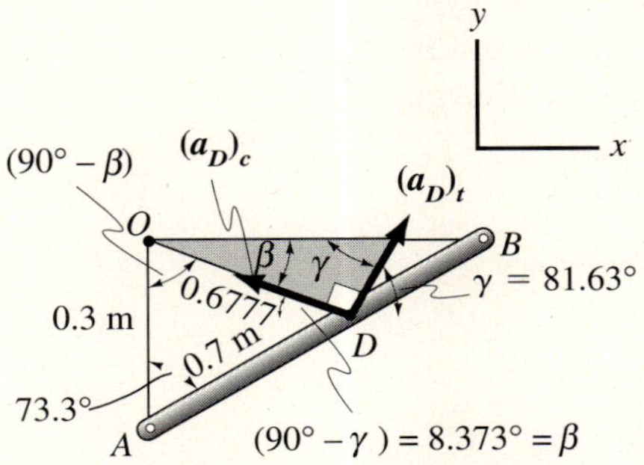

Figure 15.20. Acceleration components of point D.

Hence, looking at the tinted triangle it is clear that $\gamma = 90° - 8.373° = 81.63°$. We can now give V_D (see Fig. 15.19).

$$V_D = V_D(\cos\gamma i + \sin\gamma j) = 1.355(\cos 81.63 i + \sin 81.63 j)$$

$$\mathbf{V}_D = 0.1972\boldsymbol{i} + 1.341\boldsymbol{j} \text{ m/s}$$

Example 15.6 (Continued)

For the acceleration vector, we refer back to Eq. (b) for components of $\boldsymbol{a}_D$. Noting Fig. 15.20, we have

$$a_D = r(\dot{\omega}_{AB})[\cos\gamma \boldsymbol{i} + \sin\gamma \boldsymbol{j}] + \frac{V_D^2}{r}[-\cos\beta \boldsymbol{i} + \sin\beta \boldsymbol{j}]$$
$$= (0.6777)(-57.8)[\cos 81.63° \boldsymbol{i} + \sin 81.63° \boldsymbol{j}] + \frac{1.355^2}{0.6777}[-\cos 8.373° \boldsymbol{i} + \sin 8.373° \boldsymbol{j}]$$

$$\therefore \quad \boldsymbol{a}_D = -8.38\boldsymbol{i} - 38.36\boldsymbol{j} \text{ m/s}^2$$

*Example 15.7

A disk E is rotating about a fixed axis HG at a constant angular speed ω_1 of 5 rad/s in Fig. 15.21. A bar CD is held by the wheel at D by a ball-joint connection and is guided along a rod AB cantilevered at A and B by a collar at C having a second ball-joint connection with CD, as shown in the diagram. Compute the velocity of C.

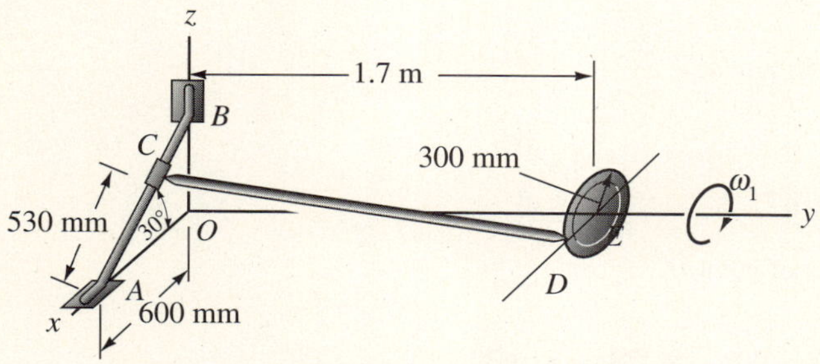

Figure 15.21. Three-dimensional device.

We shall need the vector $\boldsymbol{\rho}_{DC}$. Thus,

$$\boldsymbol{\rho}_{DC} = \boldsymbol{r}_C - \boldsymbol{r}_D$$
$$= [(0.6 - 0.53\cos 30°)\boldsymbol{i} + 0.53\sin 30°\boldsymbol{k}] - (1.7\boldsymbol{j} + 0.3\boldsymbol{i})$$
$$= -0.159\boldsymbol{i} - 1.7\boldsymbol{j} + 0.265\boldsymbol{k} \text{ m}$$

Example 15.7 (Continued)

Now employ Eq. 15.6 for rod CD. Thus,

$$\boldsymbol{V}_C = \boldsymbol{V}_D + \boldsymbol{\omega}_{CD} \times \boldsymbol{\rho}_{DC}$$

Therefore, assuming C is going from B to A

$$V_C(\cos 30°\boldsymbol{i} - \sin 30°\boldsymbol{k})$$
$$= (5)(0.3)\boldsymbol{k} + (\omega_x \boldsymbol{i} + \omega_y \boldsymbol{j} + \omega_z \boldsymbol{k}) \times (-0.159\boldsymbol{i} - 1.7\boldsymbol{j} + 0.265\boldsymbol{k})$$
$$V_C(0.866\boldsymbol{i} - 0.5\boldsymbol{k}) = 1.50\boldsymbol{k} - 1.7\omega_x \boldsymbol{k} - 0.265\omega_x \boldsymbol{j} + 0.159\omega_y \boldsymbol{k}$$
$$+0.265\omega_y \boldsymbol{i} - 0.159\omega_z \boldsymbol{j} + 1.7\omega_z \boldsymbol{i}$$

The scalar equations are:

$$0.866V_C = 0.265\omega_y + 1.7\omega_z \quad \text{(a)}$$
$$0 = -0.265\omega_x - 0.159\omega_z \quad \text{(b)}$$
$$-0.5V_C = 1.50 - 1.7\omega_x + 0.159\omega_y \quad \text{(c)}$$

From these equations, we cannot solve for ω_x, ω_y, and ω_z because the spin of CD about its own axis (allowed by the ball joints) can have *any value* without affecting the velocity of slider C. However, we can determine V_C, as we shall now demonstrate.

In Eq. (b), solve for ω_x in terms of ω_z.

$$\omega_x = -0.6\omega_z \quad \text{(d)}$$

In Eq. (a), solve for ω_y in terms of ω_z:

$$\omega_y = 3.27V_C - 6.415\omega_z \quad \text{(e)}$$

Substitute for ω_x and ω_y in Eq. (c) using the foregoing results:

$$-0.5V_C = 1.50 - (1.7)(-0.6\omega_z) + (0.159)(3.27V_C - 6.415\omega_z)$$

Therefore,

$$-1.020V_C = 1.5 + 1.020\omega_z - 1.020\omega_z$$
$$V_C = -1.471 \text{ m/s}$$

Hence,

$$\boldsymbol{V}_c = -1.471(\cos 30°\boldsymbol{i} - \sin 30°\boldsymbol{k})$$

$$\boldsymbol{V}_c = -1.274\boldsymbol{i} + 0.7355\boldsymbol{k} \text{ m/s}$$

Clearly, contrary to our assumption C is going from A to B.

Before going on to the next section, we wish to point out a simple relation that will be of use in the remainder of the chapter. Suppose that you have a moving particle whose position vector $\boldsymbol{r}$ has a magnitude that is constant (see Fig. 15.22). This position vector, however, has an angular velocity $\boldsymbol{\omega}$ relative to xyz. We wish to know the velocity of the particle P relative to xyz.

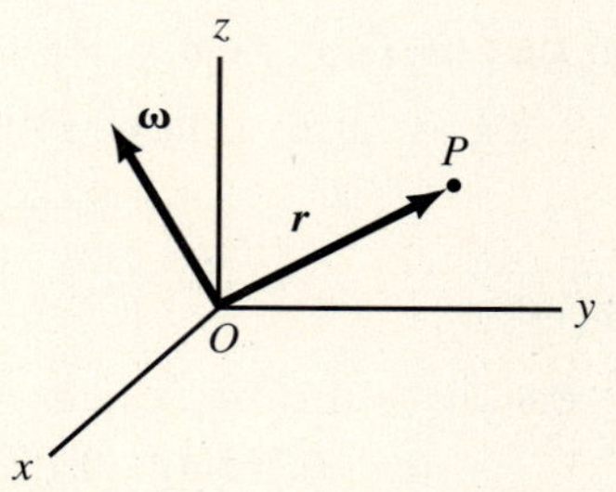

Figure 15.22. Position vector $\boldsymbol{r}$ has constant magnitude but rotates relative to xyz with angular velocity $\boldsymbol{\omega}$.

We could imagine for this purpose that particle P is part of a rigid body attached to xyz at O and rotating with angular velocity $\boldsymbol{\omega}$. This situation is shown in Fig. 15.23. Using Eq. 15.6, we can then say:

$$\boldsymbol{V}_P = \boldsymbol{V}_O + \boldsymbol{\omega} \times \boldsymbol{\rho}_{OP}$$

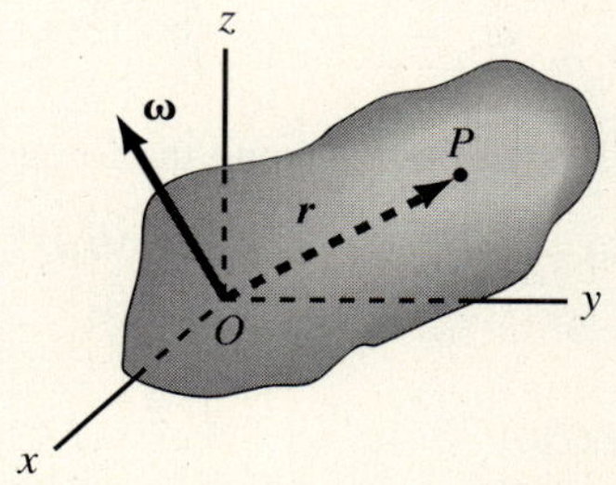

Figure 15.23. P now considered as a point in a rigid body attached at O and having angular velocity $\boldsymbol{\omega}$.

But $\boldsymbol{V}_O = \boldsymbol{0}$ and $\boldsymbol{\rho}_{OP}$ is simply $\boldsymbol{r}$. Hence, we have

$$\boldsymbol{V}_P = \boldsymbol{\omega} \times \boldsymbol{r}$$

We thus have a simple formula for the velocity of a particle moving at a fixed distance from the origin of xyz. This velocity is simply the cross product of the angular velocity $\boldsymbol{\omega}$ of the position vector about xyz and the position vector $\boldsymbol{r}$.

PROBLEMS

15.25. A body is spinning about an axis having direction cosines $l = 0.5$, $m = 0.5$, and $n = 0.707$. The angular speed is 50 rad/s. What is the velocity of a point in the body having a position vector $\boldsymbol{r} = 6\boldsymbol{i} + 4\boldsymbol{j}$ m?

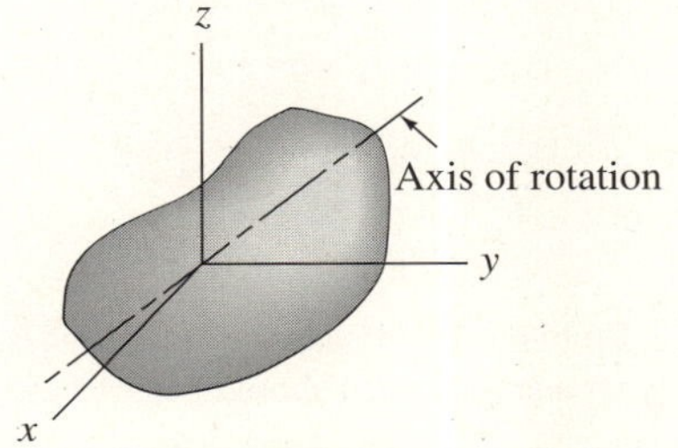

Figure P.15.25.

15.26. In Problem 15.25, what is the relative velocity between a point in the body at position $x = 10$ m, $y = 6$ m, $z = 3$ m and a point in the body at position $x = 2$ m, $y = -3$ m, $z = 0$ m?

15.27. If the body in Problem 15.25 is given an additional angular velocity $\boldsymbol{\omega}_2 = 6\boldsymbol{j} + 10\boldsymbol{k}$ rad/s, what is the direction of the axis of rotation? Compute the velocity at $\boldsymbol{r} = 10\boldsymbol{j} + 3\boldsymbol{k}$ m if the actual axis of rotation goes through the origin.

15.28. A wheel is rolling along at 17 m/s without slipping. What is the angular speed? What is the velocity of point B on the rim of the wheel at the instant shown?

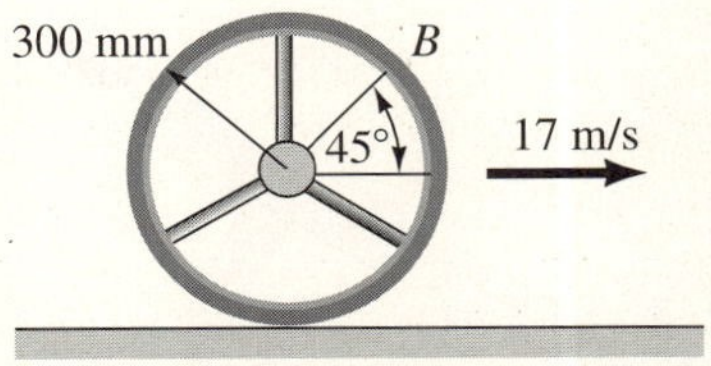

Figure P.15.28.

15.29. A flexible cord is wrapped around a spool and is pulled at a velocity of 3 m/s relative to the ground. If there is no slipping at C, what is the velocity of points O and D at the instant shown?

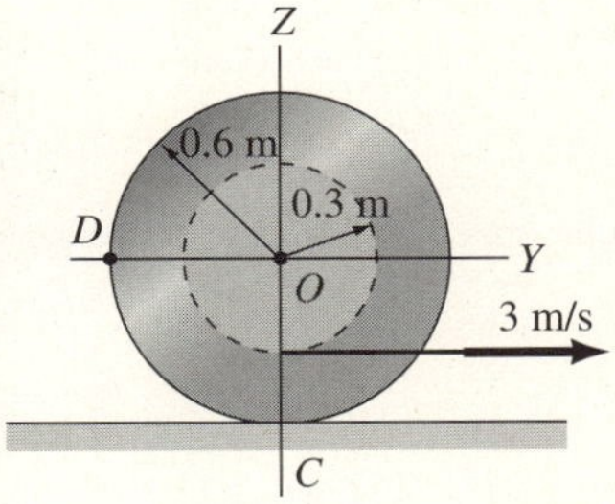

Figure P.15.29.

15.30. A piston P is shown moving downward at the constant speed of 0.3 m/s. What is the speed of slider A at the instant of interest?

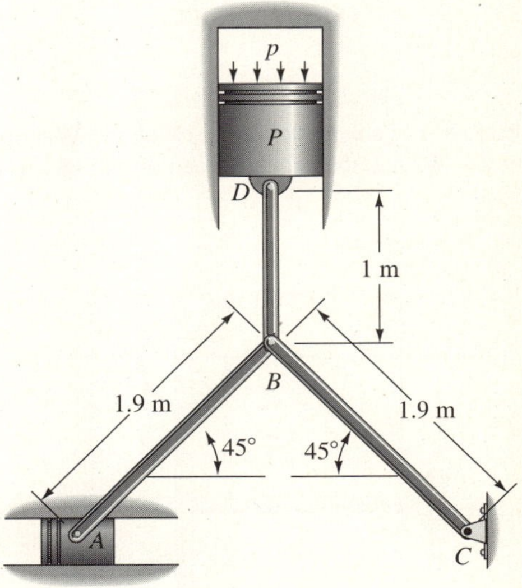

Figure P.15.30.

15.31. A rod AB is 1 m in length. If the end A slides down the surface at a speed V_A of 3 m/s, what is the angular speed of AB at the instant shown?

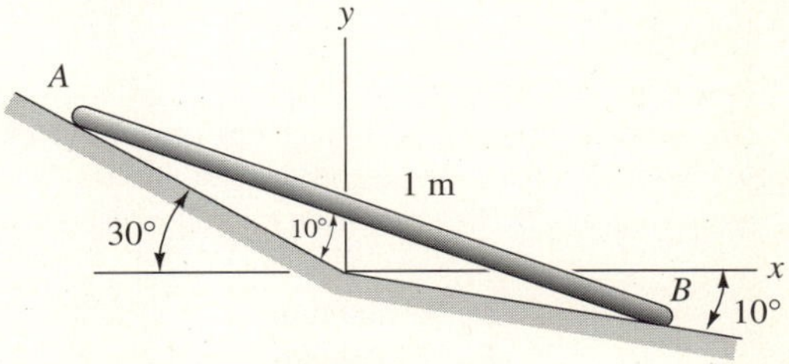

Figure P.15.31.

15.32. A plate moves along a horizontal surface. Components of the velocity for three corners are:

$$(V_A)_x = 2 \text{ m/s}$$
$$(V_B)_y = -3 \text{ m/s}$$
$$(V_C)_y = 5 \text{ m/s}$$

What is the angular speed of the plate and what is the velocity of corner D?

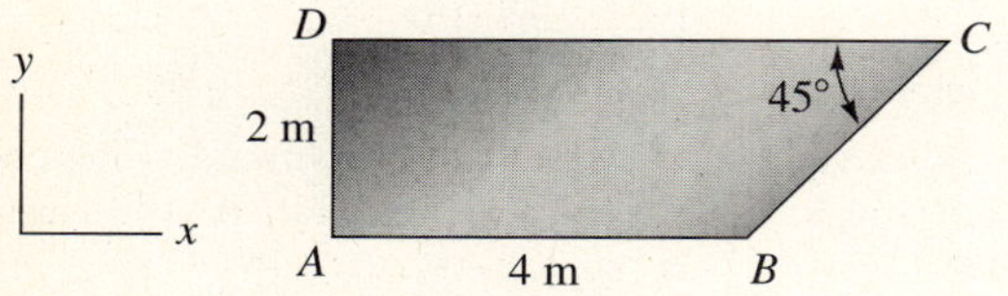

Figure P.15.32.

15.33. Rod DC has an angular speed ω_1 of 5 rad/s at the configuration shown. What is the angular speed of bar AB?

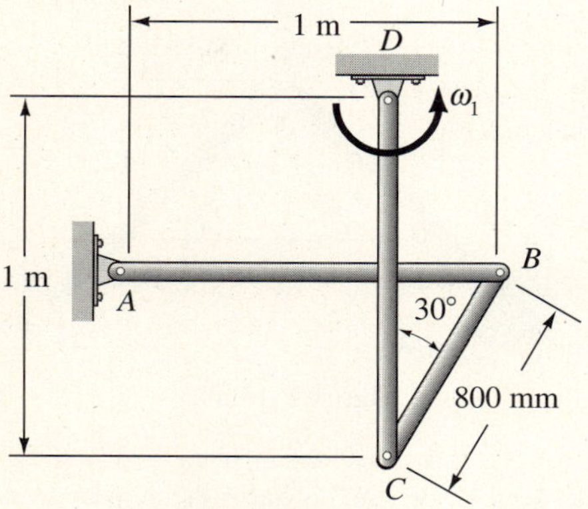

Figure P.15.33.

15.34. A system of meshing gears includes gear A, which is held stationary. Rod $A'C'$ rotates with a speed ω_1 of 5 rad/s. What is the angular speed of gear C? The gears have the following diameters:

$$D_A = 600 \text{ mm}$$
$$D_B = 350 \text{ mm}$$
$$D_C = 200 \text{ mm}$$

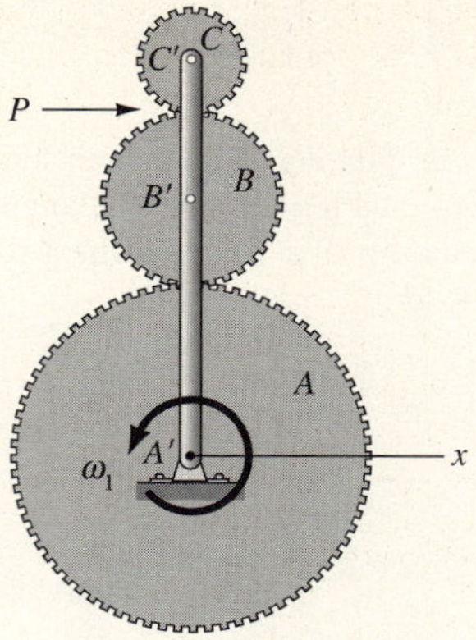

Figure P.15.34.

15.35. In Problem 15.34 take $\omega_1 = 10$ rad/s. If gear C is to translate, what angular speed should gear A have?

15.36. A bar moves in the plane of the page so that end A has a velocity of 7 m/s and decelerates at a rate of 3.3 m/s^2. What are the velocity and acceleration of point C when BA is at 30° to the horizontal?

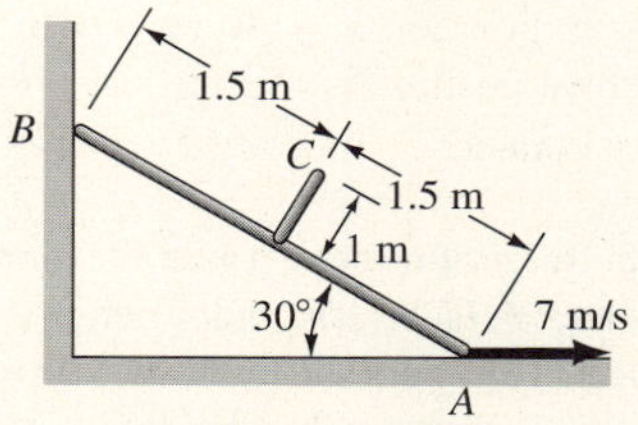

Figure P.15.36.

15.37. Bar AB is rotating at a constant speed of 5 rad/s clockwise in a device. What is the angular velocity of bar BD and body EFC? Determine the velocity of point D [*Hint:* What is the direction of the velocity of point G?]

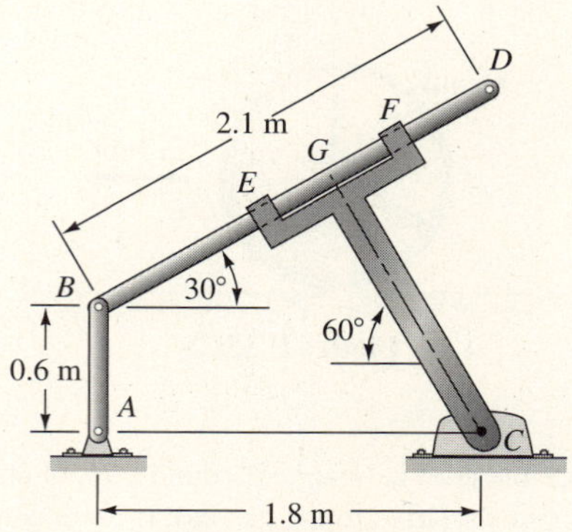

Figure P.15.37.

15.38. A wheel rotates with an angular speed of 20 rad/s. A connecting rod connects points *A* on the wheel with a slider at *B*. Compute the angular velocity of the connecting rod and the velocity of the slider when the apparatus is in the position shown in the diagram.

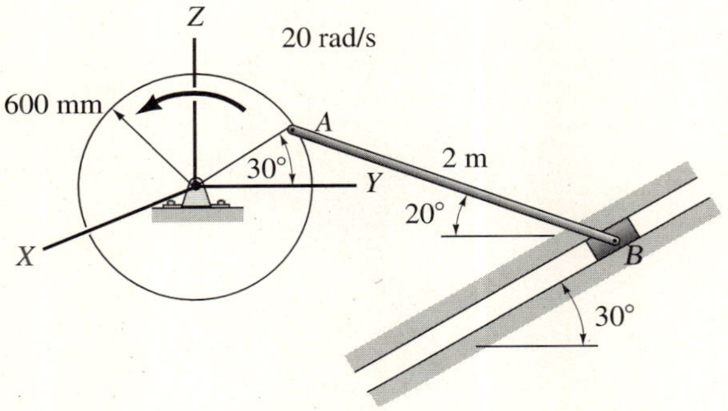

Figure P.15.38.

15.39. In Problem 15.38, if $V_B = 14.30$ m/s and $\omega_{AB} = -9.33$ rad/s, where is the instantaneous axis of rotation of connecting rod *AB*?

15.40. A piston, connecting rod, and crankshaft of an engine are represented schematically. The engine is rotating at 3000 r/min. At the position shown, what is the velocity of pin *A* relative to the engine block and what is the angular velocity of the connecting rod *AB*?

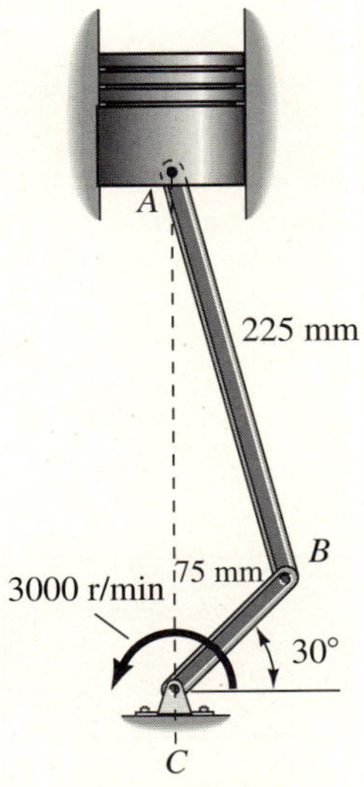

Figure P.15.40.

15.41. Member *AB* is rotating at a constant speed of 4 rad/s in a counterclockwise direction. What is the angular velocity of bar *BC* for the position shown in the diagram? What is the velocity of point *D* at the center of bar *BC*? Bar *BC* is 0.9 m in length.

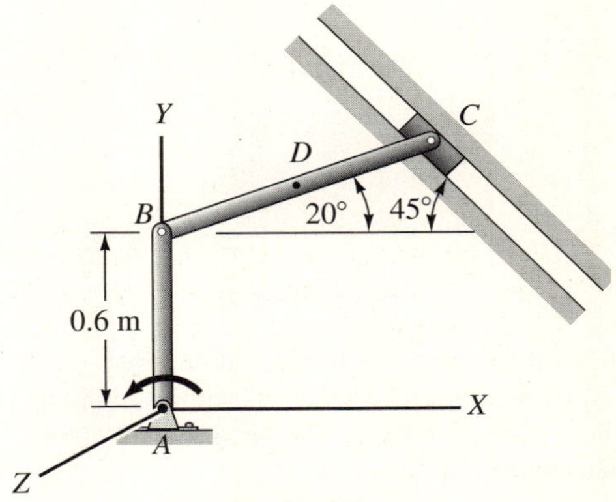

Figure P.15.41.

15.42. In Problem 15.41, determine in the simplest manner the instantaneous axis of rotation for bar *BC*.

15.43. Suppose that bar *AB* of Problem 15.41 has an angular velocity of 3 rad/s counterclockwise and a counterclockwise angular acceleration of 5 rad/s^2. What is the angular acceleration of bar *BC*, which is 75 mm in length?

15.44. A rod is moving on a horizontal surface and is shown at time *t*. What is V_y of end *A* and ω of the rod at the instant shown? [*Hint:* Use the fact that the rod is inextensible.]

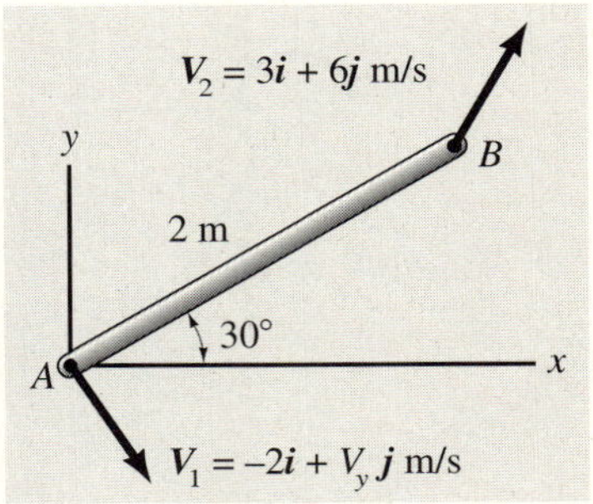

Figure P.15.44.

15.45. A plate *ABCD* moves on a horizontal surface. At time *t* corners *A* and *B* have the following velocities:

$$V_A = 3\boldsymbol{i} + 2\boldsymbol{j} \text{ m/s}$$
$$V_B = (V_B)_x\boldsymbol{i} + 5\boldsymbol{j} \text{ m/s}$$

Find the location of the instantaneous axis of rotation.

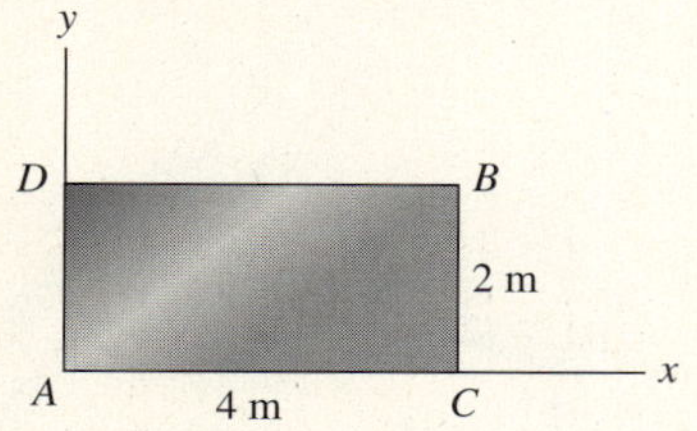

Figure P.15.45.

15.46. Find the velocity and acceleration relative to the ground of pin *B* on the wheel. The wheel rolls without slipping. Also, find the angular velocity of the slotted bar in which the pin *B* of the wheel slides when θ of the bar is 30°.

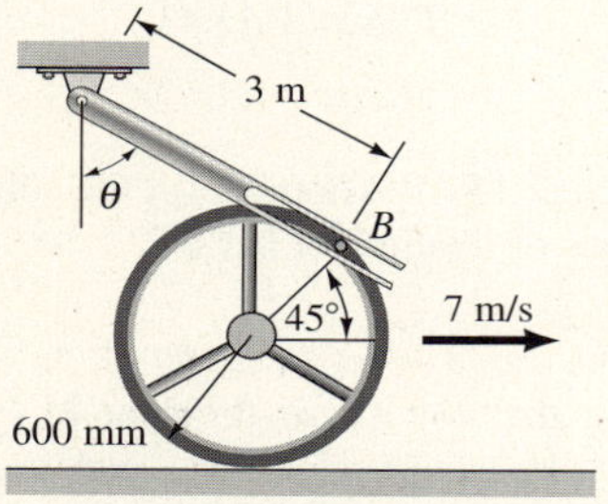

Figure P.15.46.

15.47. If $\omega_1 = 5$ rad/s and $\dot{\omega}_1 = 3$ rad/s² for bar *CD*, compute the angular velocity and angular acceleration of the gear *D* relative to the ground. Solve the problem using Eqs. 15.6 and 15.7, and then check the result by considering simple circular motion of point *D*.

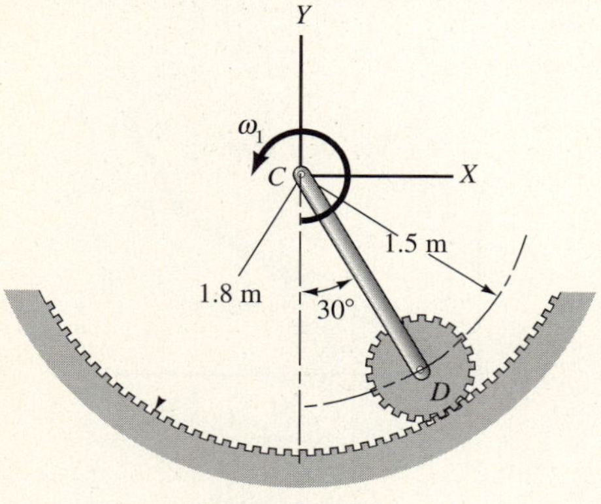

Figure P.15.47.

15.48. A mechanism with two sliders is shown. Slider *A* at the instant of interest has a speed of 3 m/s and is accelerating at the rate of 1.7 m/s². If member *AB* is 2.5 m in length, what are the angular velocity and angular acceleration for this member?

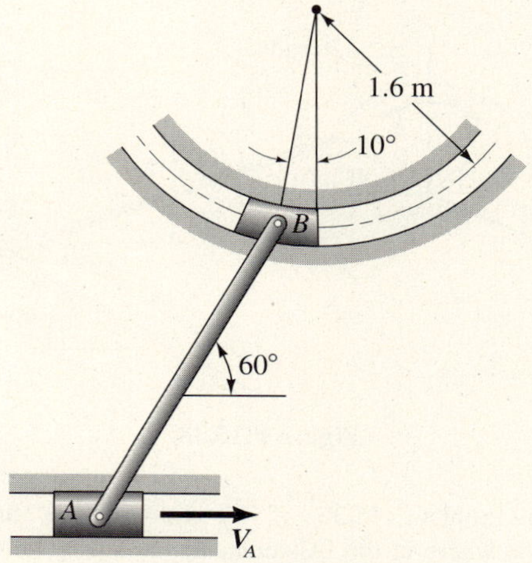

Figure P.15.48.

15.49. In Problem 15.48 find the instantaneous center of rotation of bar *AB* if V_A is 2.7 m/s.

15.50. The velocity of corner *A* of the block is known to be at time *t*:

$$V_A = 10\boldsymbol{i} + 4\boldsymbol{j} - 3\boldsymbol{k} \text{ m/s}$$

The angular speed about edge $\overline{AD}$ is 2 rad/s, and the angular speeds about the diagonals $\overline{AF}$ and $\overline{HE}$ are known to be 3 rad/s and 6 rad/s, respectively. What is the velocity of corner *B* at this instant?

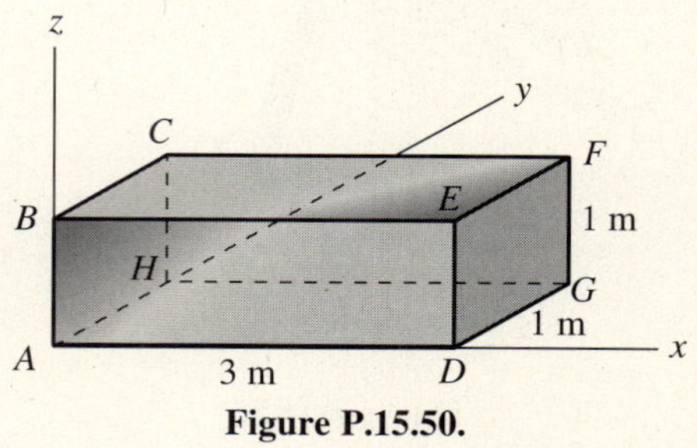

Figure P.15.50.

15.51. A rigid sphere is moving in space. The velocities for two points A and B on the surface have the values at time t:

$$V_A = 6\boldsymbol{i} + 3\boldsymbol{j} + 2\boldsymbol{k} \text{ m/s}$$
$$V_B = (V_B)_x\boldsymbol{i} + 6\boldsymbol{j} - 4\boldsymbol{k} \text{ m/s}$$

The position vectors for points A and B are at time t:

$$\boldsymbol{r}_A = 10\boldsymbol{i} + 15\boldsymbol{j} + 12\boldsymbol{k} \text{ m}$$
$$\boldsymbol{r}_B = 7\boldsymbol{i} + 20\boldsymbol{j} + 18\boldsymbol{k} \text{ m}$$

What is the angular velocity of the sphere? At the instant of interest the sphere has zero spin about axis AB.

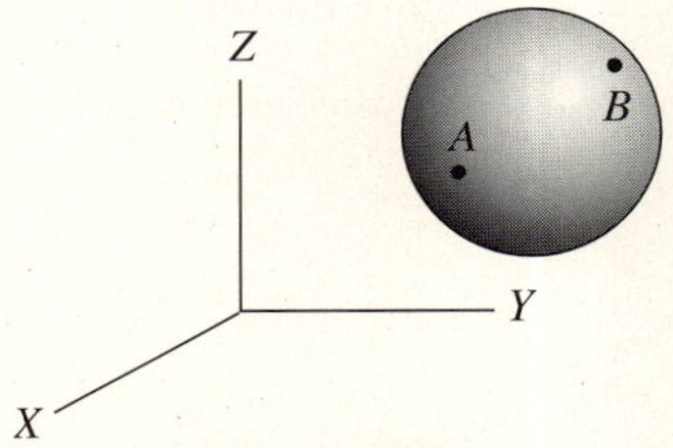

Figure P.15.51.

15.52. A conveyor element moves down the incline at a speed of 1.5 m/s. A shaft and platform move with the conveyor element but have a spin of 0.5 rad/s about the centerline AB. Also, the shaft swings in the YZ plane at a speed ω_1 of 1 rad/s. What is the velocity and acceleration of point D on the platform at the instant it is in the YZ plane, as shown in the diagram? Note that at the instant of interest AB is vertical.

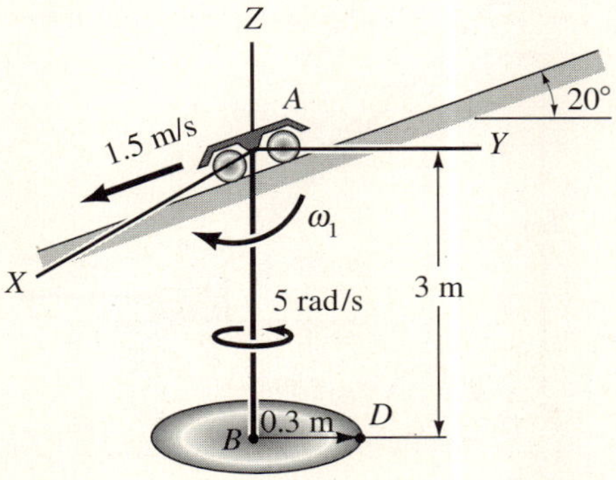

Figure P.15.52

15.53. A conveyor element moves down an incline at a speed of 15 m/s. A plate hangs down from the conveyor element and, at the instant of interest shown in the diagram, is spinning about AB at the rate of 5 rad/s. Also, the axis AB swings in the YZ plane at the rate ω_1 of 10 rad/s and $\dot{\omega}_1 = 3$ rad/s² at the instant of interest. DB is parallel to the X axis at this instant. Find the velocity and acceleration of point D at the instant shown.

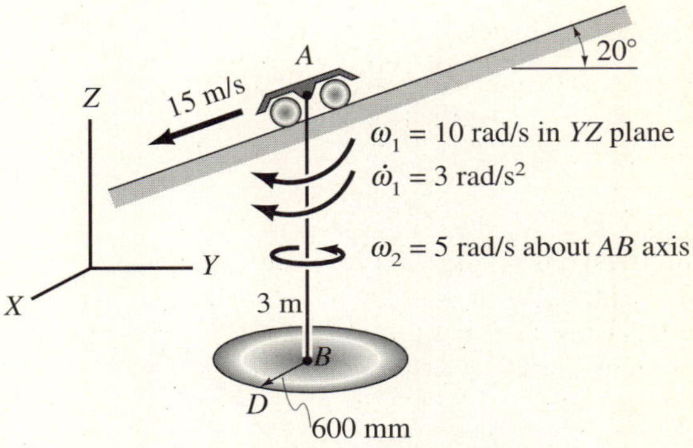

Figure P.15.53.

15.54. A cylinder rolls without slipping. It has an angular velocity $\omega = 0.3$ rad/s and an angular acceleration $\dot{\omega} = 14$ mrad/s². What are the angular velocity and angular acceleration of member AB?

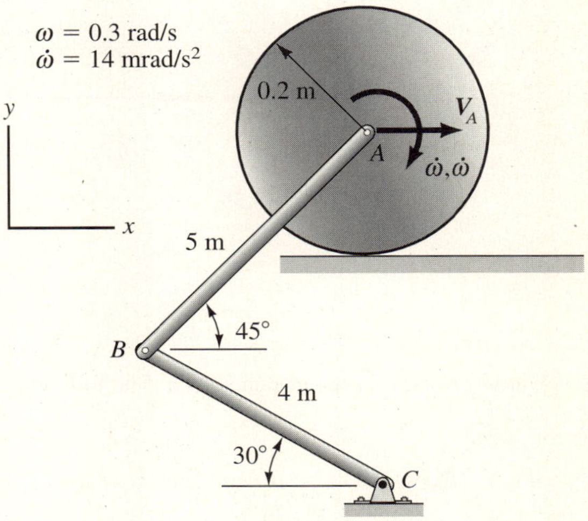

Figure P.15.54.

15.55. Slider A moves in a parabolic slot with speed $\dot{s} = 3$ m/s and $\ddot{s} = 1$ m/s^2 at the instant shown in the diagram. Cylinder E is connected to A by rod AB.

(a) Find the angular velocity of cylinder E at the time of interest.

(b) Also, find the angular acceleration of cylinder E and rod AB at this instant.

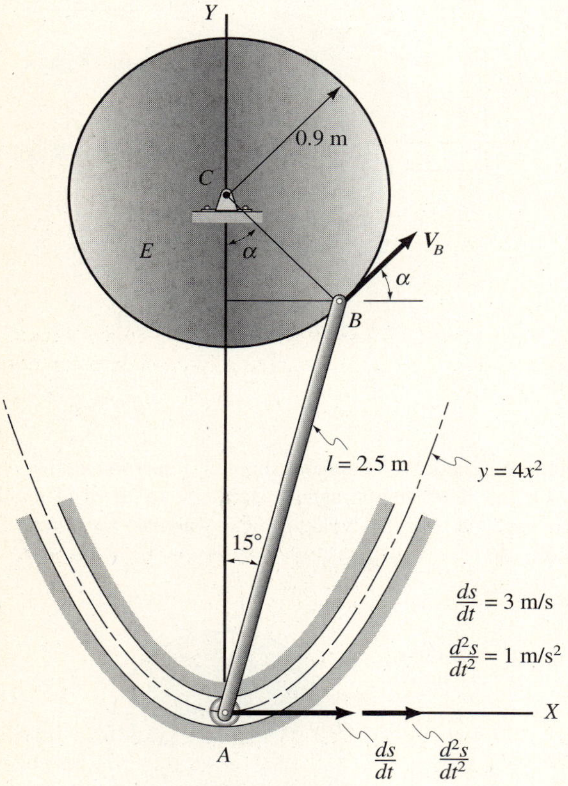

Figure P.15.55.

15.56. Find ω_A and $\dot{\omega}_A$ at the instant shown. The following data apply:

$R_A = 0.3$ m $\quad R_B = 0.2$ m $\quad CD = 5$ m $\quad V_B = 0.2$ m/s

Disc A rolls without slipping.

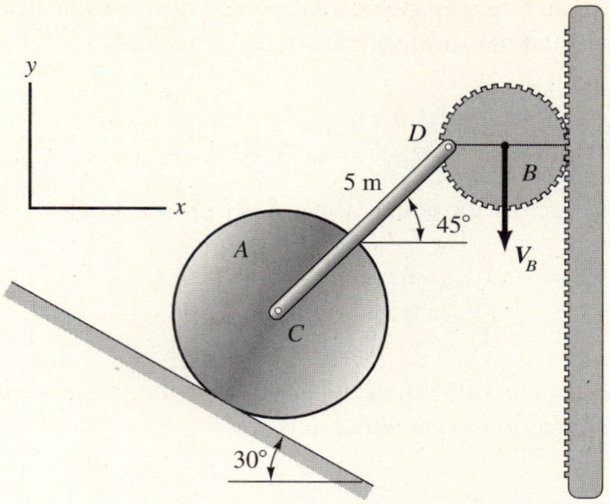

Figure P.15.56.

15.57. Find the velocity and acceleration of the center of A.

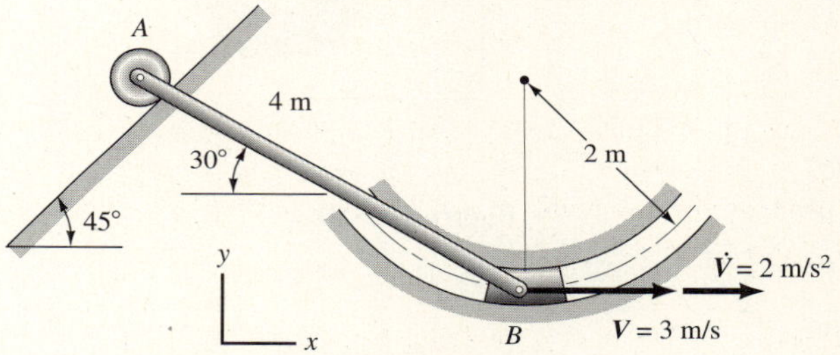

Figure P.15.57.

15.58. What is the angular velocity of rod AD? What is the magnitude of the velocity of point C of rod AD? Rod BC is vertical at the instant of interest.

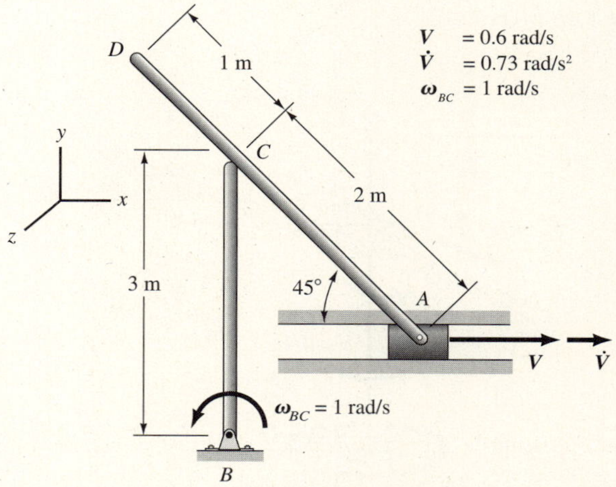

Figure P.15.58.

15.59. What are the angular velocities of the two rods? Slider A has a speed of 0.4 m/s, whereas slider C has a speed of 1.2 m/s.

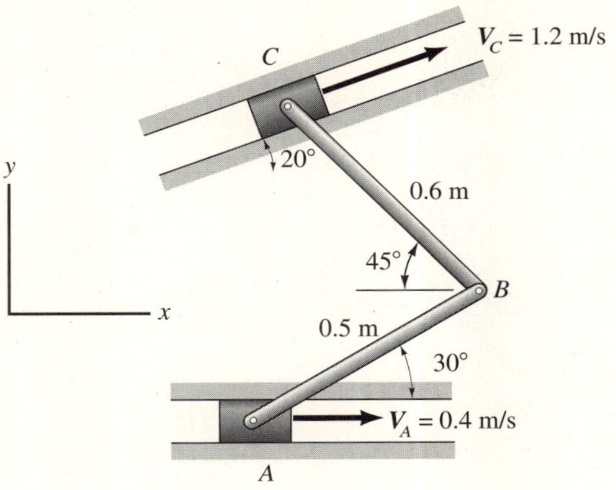

Figure P.15.59.

15.60. In Problem 15.33, find the instantaneous center O of rotation for rod CB in the simplest possible manner. What is the speed of the midpoint of CB found using O? From Problem 15.33, ω_{BC} = 2.89 rad/s.

15.61. Find ω_E and $\dot{\omega}_E$ at the instant shown.

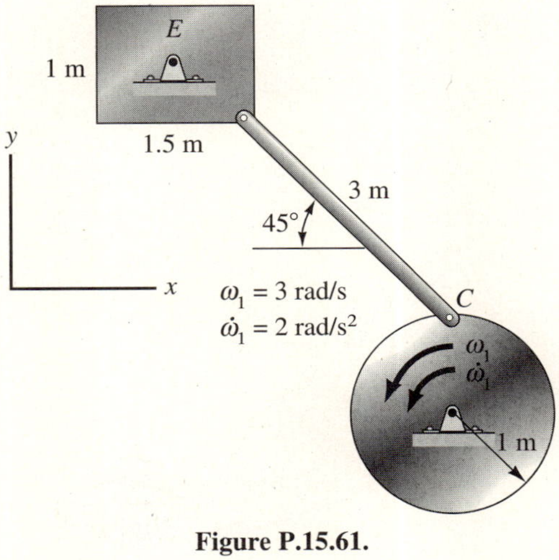

Figure P.15.61.

15.62. A bent rod is pinned to a slider at A and a cylinder at B. Find the velocity and acceleration of the slider at the instant depicted in the diagram.

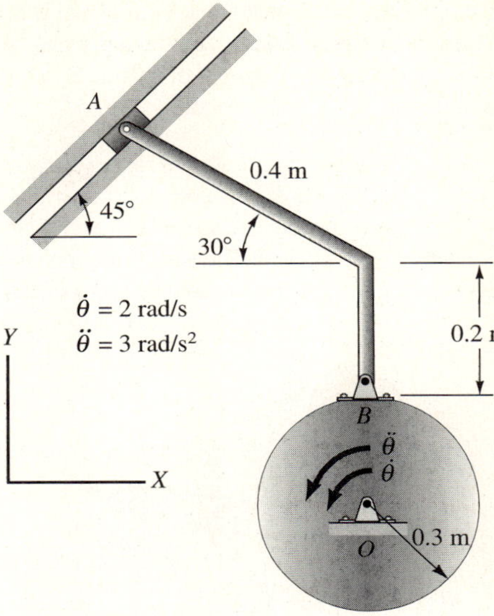

Figure P.15.62.

15.63. A cylinder rolls without slipping. Develop a formula for $\boldsymbol{a}_B$ in terms of V_0, $\dot{V}_0$, and R. Then get a formula for $\boldsymbol{a}_C$ in terms of V_0, $\dot{V}_0$, R, and d.

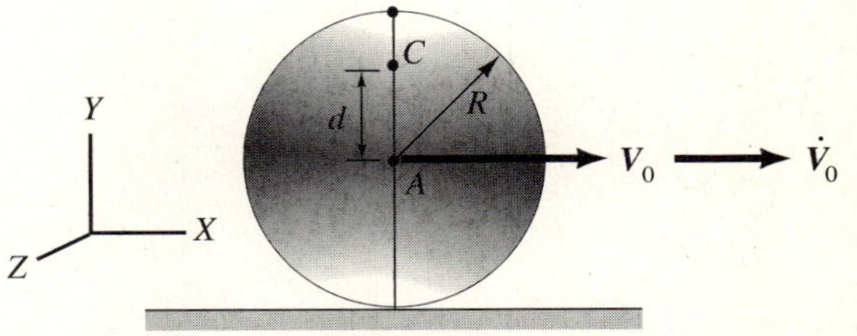

Figure P.15.63.

15.64. Two stationary half-cylinders F and I are shown, on which roll cylinders G and H. If the motion is such that line BA has an angular speed of 2 rad/s clockwise, what is the angular speed and the angular acceleration of cylinder H relative to the ground? The cylinders roll without slipping.

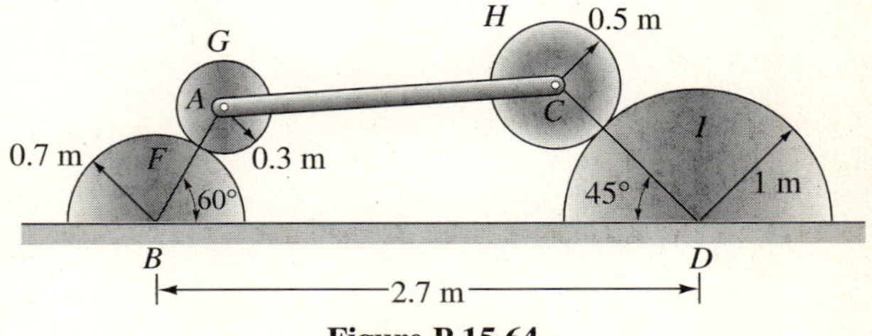

Figure P.15.64.

15.65. In Problem 15.64, assume that cylinder G is rotating at a speed of 5 rad/s clockwise as seen from the ground. What is the speed and rate of change of speed of point C relative to the ground? Assume that no slipping occurs.

15.66. A wheel D of radius $R_1 = 150$ mm rotates at a speed $\omega_1 = 5$ rad/s as shown. A second wheel C is connected to wheel D by connecting rod AB. What is the angular speed of wheel C at the instant shown? The radius $R_2 = 300$ mm. The wheels are separated by a distance $d = 600$ mm. At A and at B there are ball-and-socket connections.

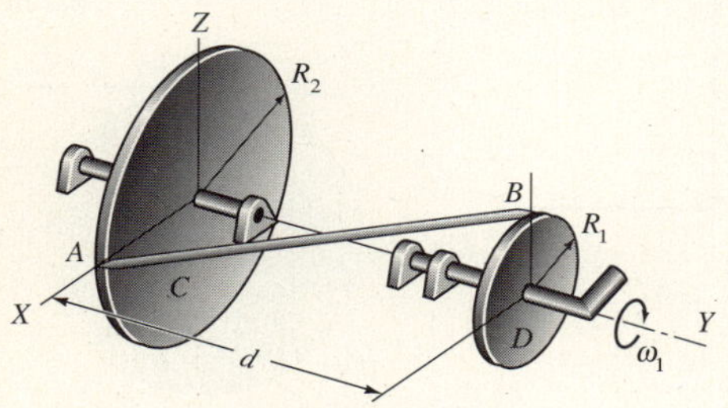

Figure P.15.66.

***15.67.** A bar AB can slide along members CD and FG of a rigid structure. If A is moving at a speed of 300 mm/s along CD toward D and is at this instant a distance of 300 mm from C, what is the speed of B along FG? At A and B there are ball-and-socket-joint connections.

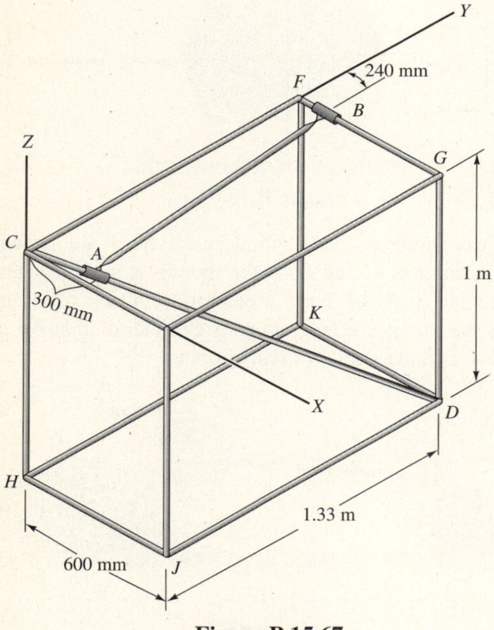

Figure P.15.67.

15.68. Member AB connects two sliders A and B. If $V_B = 5$ m/s and $\dot{V}_B = 3$ m/s², what are ω_{AB} and $\dot{\omega}_{AB}$ at the configuration shown?

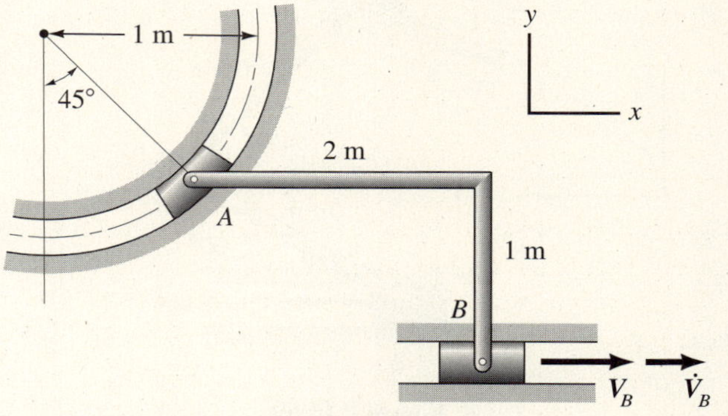

Figure P.15.68.

15.69. Find ω_{AB} and $\dot{\omega}_{AB}$. Cylinder D rolls without slipping with angular motion given as

$$\omega_D = 0.2 \text{ rad/s and } \dot{\omega}_D = 0.3 \text{ rad/s}^2$$

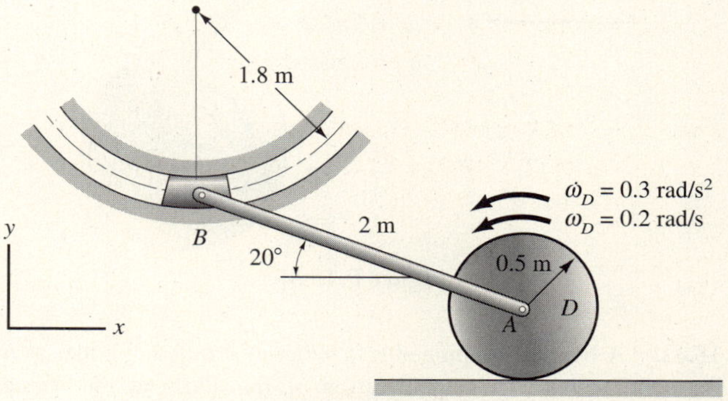

Figure P.15.69.

15.6 General Relationship Between Time Derivatives of a Vector for Different References

In Section 15.4, we considered the time derivatives of a vector $\boldsymbol{A}$ "fixed" in a reference *xyz* moving arbitrarily relative to *XYZ*. Our conclusions were:

$$\left(\frac{d\boldsymbol{A}}{dt}\right)_{xyz} = \boldsymbol{0}$$

$$\left(\frac{d\boldsymbol{A}}{dt}\right)_{XYZ} = \boldsymbol{\omega} \times \boldsymbol{A}$$

We now wish to extend these considerations to include time derivatives of a vector $\boldsymbol{A}$ which is not necessarily fixed in reference *xyz*. Primarily, our intention in this section is to relate time derivatives of such vectors $\boldsymbol{A}$ as seen both from reference *xyz* and from *XYZ*, two references moving arbitrarily relative to each other.

For this purpose, consider Fig. 15.24, where we show a moving particle P with a position vector $\boldsymbol{\rho}$ in reference *xyz*. Reference *xyz* moves arbitrarily relative to reference *XYZ* with translational velocity $\dot{\boldsymbol{R}}$ and angular velocity $\boldsymbol{\omega}$ in accordance with **Chasles' theorem**. We shall now form a relation between $(d\boldsymbol{\rho}/dt)_{xyz}$ and $(d\boldsymbol{\rho}/dt)_{XYZ}$. We shall then extend this result so as to relate the time derivative of any vector $\boldsymbol{A}$ as seen from any two references.

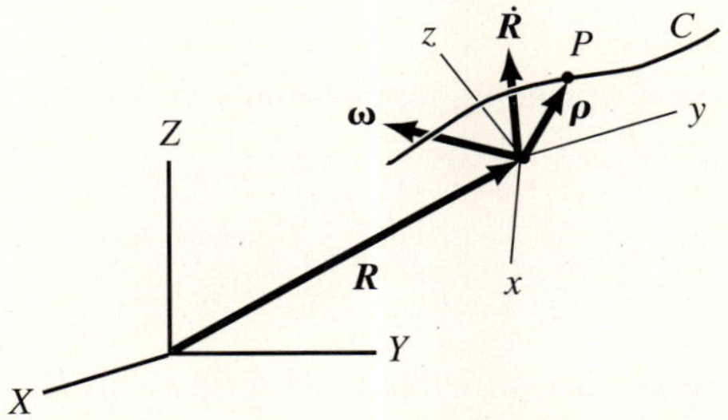

Figure 15.24. *xyz* moves relative to *XYZ*.

To reach the desired results effectively, we shall express the vector $\boldsymbol{\rho}$ in terms of components parallel to the *xyz* reference:

$$\boldsymbol{\rho} = x\boldsymbol{i} + y\boldsymbol{j} + z\boldsymbol{k} \tag{15.11}$$

where $\boldsymbol{i}, \boldsymbol{j},$ and $\boldsymbol{k}$ are unit vectors for reference *xyz*. Differentiating this equation with respect to time for the *xyz* reference, we have:[8]

$$\left(\frac{d\boldsymbol{\rho}}{dt}\right)_{xyz} = \dot{x}\boldsymbol{i} + \dot{y}\boldsymbol{j} + \dot{z}\boldsymbol{k} \tag{15.12}$$

[8]Note that $\dot{x}$, $\dot{y}$, and $\dot{z}$ are time derivatives of scalars and accordingly there is no identification with any reference as far as the time derivative operation is concerned.

If we next take the derivative of $\boldsymbol{\rho}$ with respect to time for the XYZ reference, we must remember that $\boldsymbol{i}, \boldsymbol{j}$, and $\boldsymbol{k}$ of Eq. 15.11 generally will each be a function of time, since these vectors will generally have some rotational motion relative to the XYZ reference. Thus, if dots are used for the time derivatives:

$$\left(\frac{d\boldsymbol{\rho}}{dt}\right)_{XYZ} = (\dot{x}\boldsymbol{i} + \dot{y}\boldsymbol{j} + \dot{z}\boldsymbol{k}) + (x\dot{\boldsymbol{i}} + y\dot{\boldsymbol{j}} + z\dot{\boldsymbol{k}}) \tag{15.13}$$

The unit vector $\boldsymbol{i}$ is a vector *fixed* in reference xyz, and accordingly $\dot{\boldsymbol{i}}$ equals $\boldsymbol{\omega} \times \boldsymbol{i}$. The same conclusions apply to $\boldsymbol{j}$ and $\boldsymbol{k}$. The last expression in parentheses can then be stated as

$$\begin{aligned}(x\dot{\boldsymbol{i}} + y\dot{\boldsymbol{j}} + z\dot{\boldsymbol{k}}) &= x(\boldsymbol{\omega} \times \boldsymbol{i}) + y(\boldsymbol{\omega} \times \boldsymbol{j}) + z(\boldsymbol{\omega} \times \boldsymbol{k}) \\ &= \boldsymbol{\omega} \times (x\boldsymbol{i}) + \boldsymbol{\omega} \times (y\boldsymbol{j}) + \boldsymbol{\omega} \times (z\boldsymbol{k}) \\ &= \boldsymbol{\omega} \times (x\boldsymbol{i} + y\boldsymbol{j} + z\boldsymbol{k}) = \boldsymbol{\omega} \times \boldsymbol{\rho}\end{aligned} \tag{15.14}$$

In Eq. 15.13 we can replace $(\dot{x}\boldsymbol{i} + \dot{y}\boldsymbol{j} + \dot{z}\boldsymbol{k})$ by $(d\boldsymbol{\rho}/dt)_{xyz}$, in accordance with Eq. 15.12, and $(x\dot{\boldsymbol{i}} + y\dot{\boldsymbol{j}} + z\dot{\boldsymbol{k}})$ by $\boldsymbol{\omega} \times \boldsymbol{\rho}$, in accordance with Eq. 15.14. Hence,

$$\left(\frac{d\boldsymbol{\rho}}{dt}\right)_{XYZ} = \left(\frac{d\boldsymbol{\rho}}{dt}\right)_{xyz} + \boldsymbol{\omega} \times \boldsymbol{\rho} \tag{15.15}$$

We can generalize the preceding result for any vector $\boldsymbol{A}$:

$$\left(\frac{d\boldsymbol{A}}{dt}\right)_{XYZ} = \left(\frac{d\boldsymbol{A}}{dt}\right)_{xyz} + \boldsymbol{\omega} \times \boldsymbol{A} \tag{15.16}$$

where, you must remember, $\boldsymbol{\omega}$ without subscripts will always be the *angular velocity of the xyz reference relative to the XYZ reference.* Note that Eq. 15.1 is a special case of Eq. 15.16 since for $\boldsymbol{A}$ fixed in xyz, $(d\boldsymbol{A}/dt)_{xyz} = \boldsymbol{0}$. We shall have much use for this relationship in succeeding sections.

15.7 Relationship Between Velocities of a Particle for Different References

We shall now define the velocity of a particle again in the presence of several references:

The velocity of a particle relative to a reference is the derivative as seen from this reference of the position vector of the particle in the reference.

In Fig. 15.24, the velocities of the particle *P* relative to the *XYZ* and the *xyz* references are, respectively,[9]

$$\boldsymbol{V}_{XYZ} = \left(\frac{d\boldsymbol{r}}{dt}\right)_{XYZ}, \qquad \boldsymbol{V}_{xyz} = \left(\frac{d\boldsymbol{\rho}}{dt}\right)_{xyz} \tag{15.17}$$

Since a vector can always be decomposed into *any* set of orthogonal components, $\boldsymbol{V}_{XYZ}$ can be expressed in components parallel to the *xyz* reference at any time *t* while $\boldsymbol{V}_{xyz}$ may be expressed in components parallel to the *XYZ* reference at any time *t*.

Now, we shall relate these velocities by first noting that

$$\boldsymbol{r} = \boldsymbol{R} + \boldsymbol{\rho} \tag{15.18}$$

Differentiating with respect to time for the *XYZ* reference, we have

$$\left(\frac{d\boldsymbol{r}}{dt}\right)_{XYZ} \equiv \boldsymbol{V}_{XYZ} = \left(\frac{d\boldsymbol{R}}{dt}\right)_{XYZ} + \left(\frac{d\boldsymbol{\rho}}{dt}\right)_{XYZ} \tag{15.19}$$

The term $(d\boldsymbol{R}/dt)_{XYZ}$ is clearly the velocity of the origin of the *xyz* reference relative to the *XYZ* reference, according to our definitions, and we denote this velocity as $\dot{\boldsymbol{R}}$. The term $(d\boldsymbol{\rho}/dt)_{XYZ}$ can be replaced, by use of Eq. 15.15, in which $(d\boldsymbol{\rho}/dt)_{xyz}$ is the velocity of the particle relative to the *xyz* reference. Denoting $(d\boldsymbol{\rho}/dt)_{xyz}$ simply as $\boldsymbol{V}_{xyz}$, we find that the foregoing equation then becomes the desired relation:

$$\boldsymbol{V}_{XYZ} = \boldsymbol{V}_{xyz} + \dot{\boldsymbol{R}} + \boldsymbol{\omega} \times \boldsymbol{\rho} \tag{15.20}$$

We reiterate the understanding that $\boldsymbol{\omega}$ *without* subscripts represents the angular velocity of *xyz* relative to *XYZ*. This $\boldsymbol{\omega}$ always goes into the last expression of Eq. 15.20.

Note that in Sections 15.4 and 15.5 we considered the motion of *two* particles in a rigid body as seen from a *single* reference. Now we are considering the motion of a *single* particle as seen from *two* references.

The multireference approach can be very useful. For instance, we could know the motion of a particle relative to some device, such as a rocket, to which we attach a reference *xyz*. Furthermore, from telemetering devices, we know the translational and rotational motion (**Chasles' theorem**) of the rocket (and hence *xyz*) relative to an inertial reference *XYZ*. It is often important to know the motion of the aforementioned particle relative directly to the inertial reference. The multireference approach clearly is invaluable for such problems.

We now illustrate the use of Eq. 15.20. We shall proceed in a particular methodical way which we encourage you to follow in your homework problems. In these problems, we remind you the dot over a vector generally represents a time derivative as seen from *XYZ*.

[9]Generally, we have employed $\boldsymbol{r}$ as a position vector and $\boldsymbol{\rho}$ as a displacement vector. With two references, we shall often use $\boldsymbol{\rho}$ to denote a position vector for one of the references.

Example 15.8

An airplane moving at 60 m/s is undergoing a roll of 30 mrad/s (Fig. 15.25). When the plane is horizontal, an antenna is moving out at a speed of 2.4 m/s relative to the plane and is at a position of 3 m from the centerline of the plane. If we assume that the axis of roll corresponds to the centerline, what is the velocity of the antenna end relative to the ground when the plane is horizontal?

A *stationary reference XYZ on the ground* is shown in the diagram. A moving reference *xyz is fixed to the plane* with the x axis along the axis of roll and the y axis collinear with the antenna. We announce this formally as follows:

Fix *xyz* to plane.
Fix *XYZ* to ground.

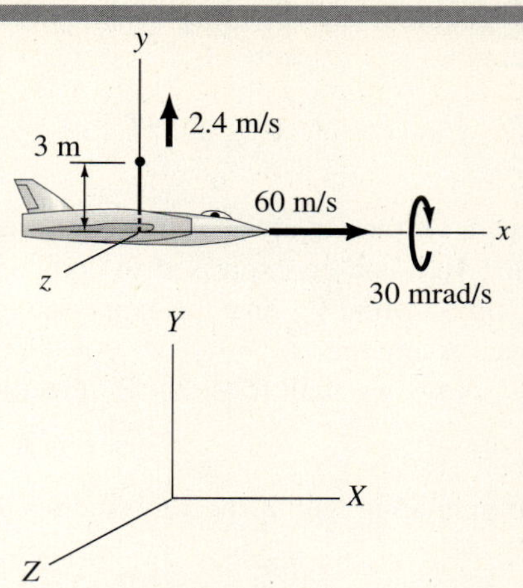

Figure 15.25. *xyz* fixed to plane; *XYZ* fixed to ground.

We then proceed in the following manner:

A. Motion of particle (antenna end) relative to *xyz*[10]

$$\boldsymbol{\rho} = 3\boldsymbol{j}\ \text{m}$$
$$\boldsymbol{V}_{xyz} = 2.4\boldsymbol{j}\ \text{m/s}$$

B. Motion of *xyz* (moving reference) relative to *XYZ* (fixed reference)

$$\dot{\boldsymbol{R}} = 60\boldsymbol{i}\ \text{m/s}$$
$$\boldsymbol{\omega} = -0.03\boldsymbol{i}\ \text{rad/s}$$

We now employ Eq. 15.20 to get

$$\boldsymbol{V}_{XYZ} = \boldsymbol{V}_{xyz} = \dot{\boldsymbol{R}} + \boldsymbol{\omega} \times \boldsymbol{\rho}$$
$$= 2.4\boldsymbol{j} + 60\boldsymbol{i} + (-0.03\boldsymbol{i}) \times (3\boldsymbol{j})$$

$$\boldsymbol{V}_{XYZ} = 60\boldsymbol{i} + 2.4\boldsymbol{j} - 0.09\boldsymbol{k}\ \text{m/s}$$

[10]Note that since the corresponding axes of the references are parallel to each other at the instant of interest, the unit vectors ***i***, ***j***, and ***k*** apply to either reference at the instant of interest. We will arrange *xyz* and *XYZ* this way whenever possible.

Note from the preceding example that in Part *A*, we are using the dynamics of a particle as presented in Chapters 11–14, while in Part *B* we are implementing **Chasles' theorem** as presented in this chapter. Your author based on long experience urges the student to work in this methodical manner.

Example 15.9

A tank is moving up an incline with a speed of 2.8 m/s in Fig. 15.26. The turret is rotating at a speed ω_1 of 2 rad/s relative to the tank, and the gun barrel is being lowered (rotating) at a speed ω_2 of 0.3 rad/s relative to the turret. What is the velocity of point *A* of the gun barrel relative to the tank and relative to the ground? The gun barrel is 3 m in length. We proceed as follows (see Fig. 15.27).

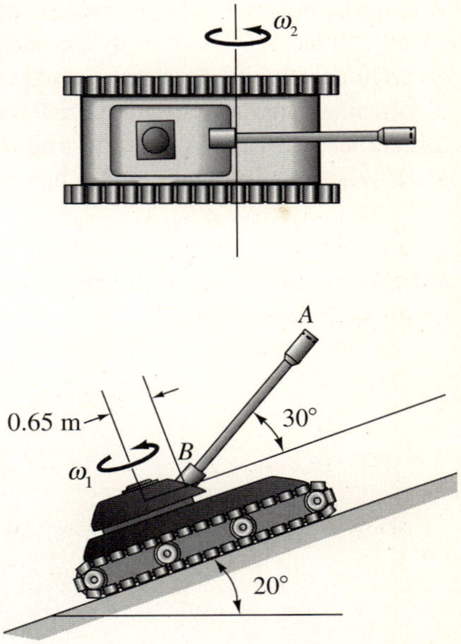

Figure 15.26. Tank with turret and gun barrel in motion.

Fix *xyz* to turret.
Fix *XYZ* to tank.

A. Motion of particle relative to *xyz*

$$\boldsymbol{\rho} = 3(\cos 30°\boldsymbol{j} + \sin 30°\boldsymbol{k}) = 2.60\boldsymbol{j} + 1.5\boldsymbol{k} \text{ m}$$

Since $\boldsymbol{\rho}$ is fixed in the gun barrel, which has an angular velocity $\boldsymbol{\omega}_2$ relative to *xyz*, we have

$$\boldsymbol{V}_{xyz} = \left(\frac{d\boldsymbol{\rho}}{dt}\right)_{xyz} = \boldsymbol{\omega}_2 \times \boldsymbol{\rho} = (-3\boldsymbol{i}) \times (2.60\boldsymbol{j} + 1.5\boldsymbol{k})$$
$$= -0.780\boldsymbol{k} + 0.45\boldsymbol{j} \text{ m/s}$$

B. Motion of *xyz* relative to *XYZ*

$$\boldsymbol{R} = 0.65\boldsymbol{j}$$

Since $\boldsymbol{R}$ is fixed in the turret, which is rotating with angular speed $\boldsymbol{\omega}_1$ relative to *XYZ*, we have

$$\dot{\boldsymbol{R}} = \omega_1 \times \boldsymbol{R} = 2\boldsymbol{k} \times 0.65\boldsymbol{j} = -1.3\boldsymbol{i} \text{ m/s}$$
$$\boldsymbol{\omega} = \boldsymbol{\omega}_1 = 2\boldsymbol{k} \text{ rad/s}$$

We can now substitute into the basic equation relating $\boldsymbol{V}_{xyz}$ to $\boldsymbol{V}_{XYZ}$. That is,

$$\boldsymbol{V}_{XYZ} = \boldsymbol{V}_{xyz} + \dot{\boldsymbol{R}} + \boldsymbol{\omega} \times \boldsymbol{\rho}$$
$$= (-0.780\boldsymbol{k} + 0.45\boldsymbol{j}) - 1.3\boldsymbol{i} + (2\boldsymbol{k}) \times (2.60\boldsymbol{j} + 1.5\boldsymbol{k})$$

$$\boldsymbol{V}_{XYZ} = -6.5\boldsymbol{i} + 0.45\boldsymbol{j} - 0.780\boldsymbol{k} \text{ m/s}$$

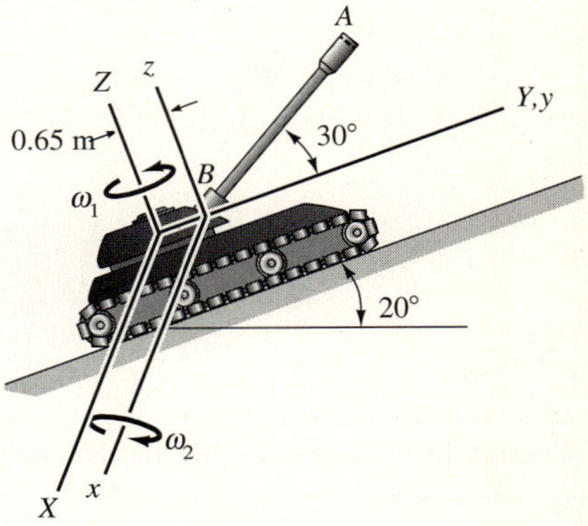

Figure 15.27. *xyz* fixed to turret; *XYZ* fixed to tank.

This result is the desired velocity of *A* relative to the tank. Since the tank is moving with a speed of 2.8 m/s relative to the ground, we can say that *A* has a velocity relative to the ground given as

$$\boldsymbol{V}_{\text{ground}} = \boldsymbol{V}_{XYZ} + 2.8\boldsymbol{j}$$

$$\boldsymbol{V}_{\text{ground}} = -6.5\boldsymbol{i} + 3.25\boldsymbol{j} - 0.780\boldsymbol{k} \text{ m/s}$$

Example 15.10

A gunboat in heavy seas is firing its main battery (see Fig. 15.28). The gun barrel has an angular velocity ω_1 relative to the turret, while the turret has an angular velocity ω_2 relative to the ship. If we wish to have the velocity components of the emerging shell to be zero in the stationary X and Z directions at a certain specific time t, what should ω_1 and ω_2 be at this instant? At this instant, the ship has a translational velocity given as

$$\boldsymbol{V}_{ship} = 0.02\boldsymbol{i} + 0.016\boldsymbol{k} \text{ m/s}$$

Take the inclination of the barrel to be $\theta = 30°$. Determine also the velocity of the gun barrel tip A.

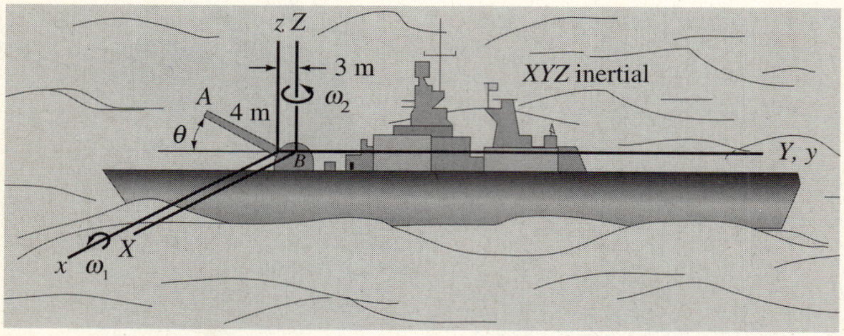

Figure 15.28. A gunboat in heavy seas firing its main battery.

We proceed to solve this problem by the following positioning of axes shown on Fig. 15.28.

Fix xyz to turret.
Fix XYZ to the ground (inertial reference).

We can now proceed with the detailed analysis of the problem.

A. Motion of *A* relative to *xyz*

$$\boldsymbol{\rho} = -(4)(0.866)\,\mathrm{j} + (4)(0.5)\,\mathrm{k} = -3.464\,\mathrm{j} + 2\,\mathrm{k} \text{ m}$$

$$\mathrm{V}_{xyz} = \omega_1 \times \boldsymbol{\rho} = \omega_1\,\mathrm{i} \times (-3.464\,\mathrm{j} + 2\,\mathrm{k}) = -3.464\omega_1\boldsymbol{k} - 2\omega_1\boldsymbol{j} \text{ m/s}$$

Example 15.10 (Continued)

B. Motion of *xyz* relative to *XYZ*

$$\mathrm{R} = -3\mathrm{j}\ \mathrm{m}$$
$$\dot{\mathrm{R}} = \omega_2 \mathrm{k} \times (-3\mathrm{j}) + (0.02\mathrm{i} + 0.016\mathrm{k}) = (3\omega_2 + 0.02)\mathrm{i} + 0.016\mathrm{k}\ \mathrm{m/s}$$
$$\omega = \omega_2 \mathrm{k}\ \mathrm{rad/s}$$

We can now proceed with the calculations.

$$V_{XYZ} = V_{xyz} + \dot{R} + \mathrm{v} \times \rho$$
$$= (-3.464\omega_1 k - 2\omega_1 j) + (3\omega_2 + 0.02)i + 0.016k + (\omega_2 k) \times (-3.464 j + 2k)$$
$$= -3.464\omega_1 k - 2\omega_1 i + 3\omega_2 i + 0.02 i + 0.016 k + 3.464\omega_2 i$$
$$\therefore V_{XYZ} = (3\omega_2 + 3.464\omega_2 + 0.02)i + (-2\omega_1)j + (-3.464\omega_1 + 0.016)k\ \mathrm{m/s}$$

Let $(V_{XYZ})_X = \mathbf{0}$

$$\therefore\ 6.464\omega_2 = -0.02 \qquad \omega_2 = -3.09\ \mathrm{mrad/s}$$

Let $(V_{XYZ})_Z = \mathbf{0}$

$$\therefore\ -3.464\omega_1 = -0.016 \qquad \omega_1 = 4.62\ \mathrm{mrad/s}$$

Finally, we can give V_{XYZ} as $\quad V_{XYZ} = -2\omega_1 j = 9.238 \times 10^{-3} j\ \mathrm{m/s}$

In some of the homework problem diagrams, in the remainder of the chapter, a set of axes *xyz* has been shown as a suggestion for use by the student. This has been done to help clarify the geometry of the diagram. Also, if the student chooses to use these axes he/she will be able to compare more easily his/her solution with that of the author as presented in the instructor's manual. However, (and note this carefully) the student must decide independently as to how to *fix this reference* and to state clearly as we have done in the examples *how this reference has been fixed.*

Also, we strongly urge the student to make careful clear diagrams and to follow the orderly progression of steps (**A. Motion of particle etc., etc.** followed by **B. Motion of *xyz* etc., etc.**).

PROBLEMS

15.70. A space laboratory, in order to simulate gravity, rotates relative to inertial reference *XYZ* at a rate ω_1. For occupant *A* to feel comfortable, what should ω_1 be? Clearly, at the center room *B*, there is close to zero gravity for zero-*g* experiments. A conveyor along one of the spokes transports items from the living quarters at the periphery to the zero-gravity laboratory at the center. In particular, a particle *D* has a velocity toward *B* of 5 m/s relative to the space station. What is its velocity relative to the inertial reference *XYZ*?

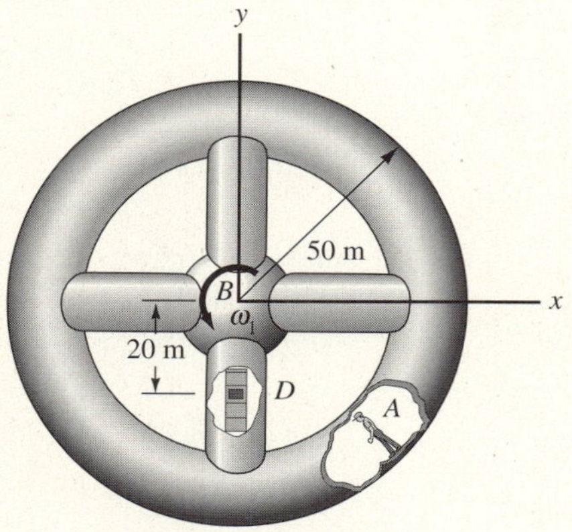

Figure P.15.70.

15.71. Bodies *a* and *b* slide away from each other each with a constant velocity of 1.5 m/s along the axis *C–C* mounted on a platform. The platform rotates relative to the ground reference *XYZ* at an angular velocity of 10 rad/s about axis *E–E* and has an angular acceleration of 5 rad/s^2 relative to the ground reference *XYZ* at the time when the bodies are at a distance r = 1 m from *E–E*. Determine the velocity of particle *b* relative to the ground reference.

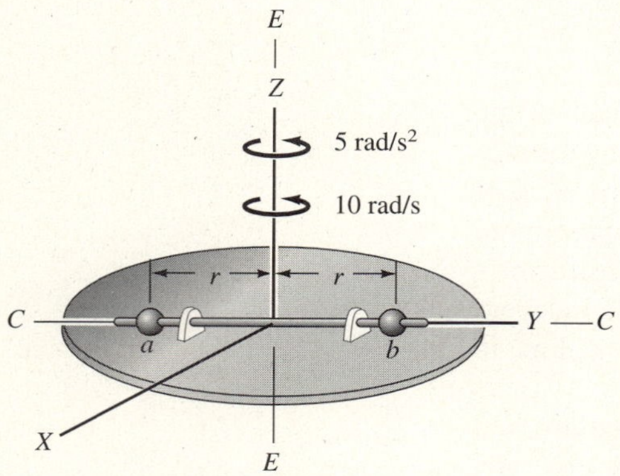

Figure P.15.71.

15.72. A particle rotates at a constant angular speed of 10 rad/s on a platform, while the platform rotates with a constant angular speed of 50 rad/s about axis *A–A*. What is the velocity of the particle *P* at the instant the platform is in the *XY* plane and the radius vector to the particle forms an angle of 30° with the *Y* axis as shown?

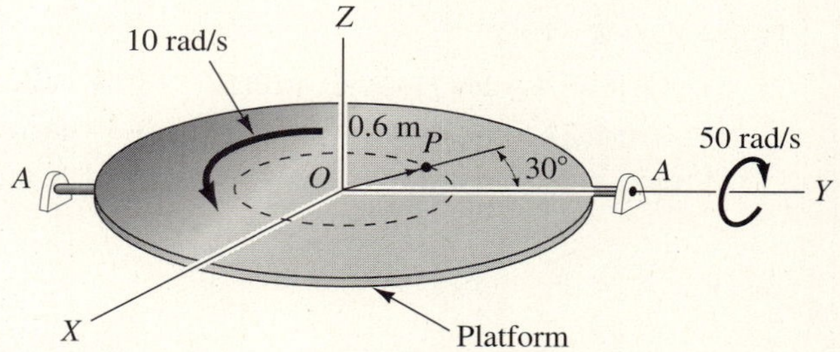

Figure P.15.72.

15.73. A platform *A* is rotating with constant angular speed ω_1 of 1 rad/s. A second platform *B* rides on *A*, contains a row of test tubes, and has a constant angular speed ω_2 of 0.2 rad/s relative to the platform *A*. A third platform *C* is in no way connected with platforms *A* and *B*. *E* on platform *C* is positioned above *A* and *B* and carries dispensers of chemicals which are electrically operated at proper times to dispense drops into the test tubes held by *B* below. What should the angular speed ω_3 be for platform *E* at instant shown if it is to dispense a drop of chemical having a zero tangential velocity relative to the test tube below?

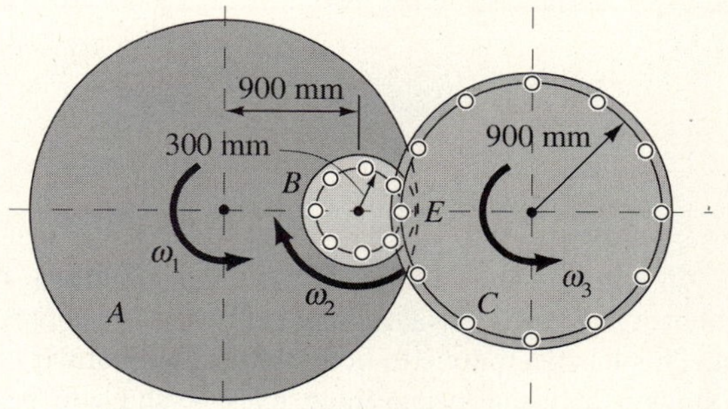

Figure P.15.73.

15.74. In an amusement park ride, the cockpit containing two occupants can rotate at an angular speed ω_1 relative to the main arm *OB*. The arm can rotate with angular speed ω_2 relative to the ground. For the position shown in the diagram and for ω_1 = 2 rad/s and ω_2 = 0.2 rad/s, find the velocity of point *A* (corresponding to the position of the eyes of an occupant) relative to the ground.

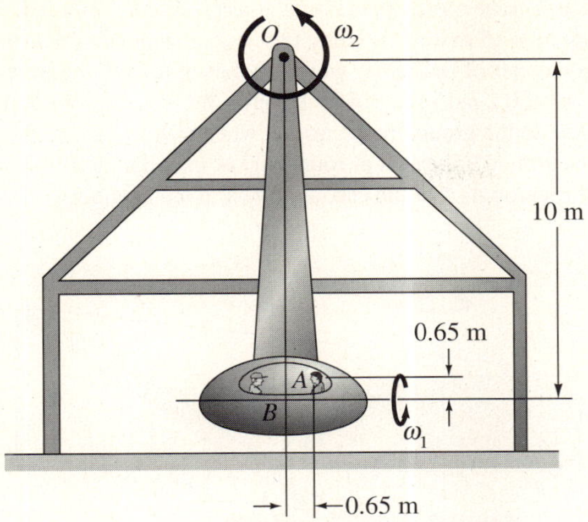

Figure P.15.74.

15.75. A water sprinkler has 11×10^{-3} m^3/s of water fed into the base. The sprinkler turns at the rate ω_1 of 1 rad/s. What is the speed of the jet of water relative to the ground at the exits? The outlet area of the nozzle cross section is 480 mm^2. [*Hint:* The volume of flow through a cross section is VA, where V is the velocity and A is the area of the cross section.]

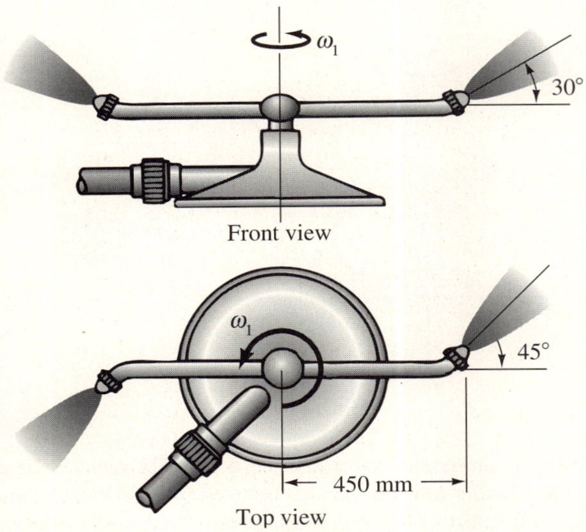

Figure P.15.75.

15.76. A tank is moving over rough terrain while firing its main gun at a fixed target. The barrel and turret of the gun partly compensate for the motion of the tank proper by giving the barrel an angular velocity $\boldsymbol{\omega}_1$ relative to the turret and, simultaneously, by giving the turret an angular velocity $\boldsymbol{\omega}_2$ relative to the tank proper such that any instant the velocity of end A of the barrel has zero velocity in the X and Z directions relative to the ground reference. What should these angular velocities be for the following translational motion of the tank:

$$\boldsymbol{V}_{\text{TANK}} = 10\boldsymbol{i} + 4\boldsymbol{k} \text{ m/s}$$

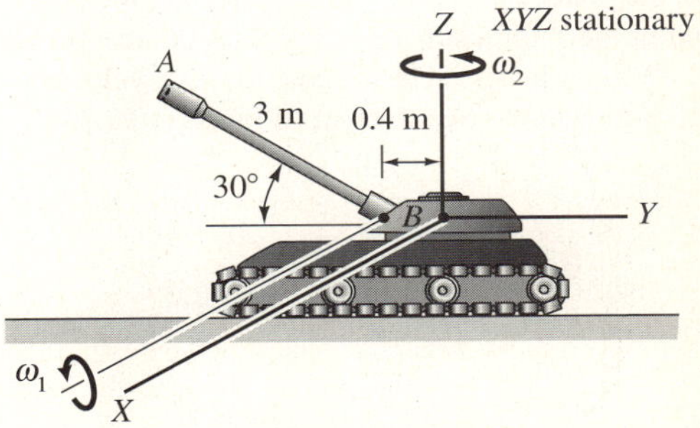

Figure P.15.76.

15.77. We can show that Eq. 15.6 is actually a special case of Eq. 15.20. For this purpose, consider a rigid body moving relative to XYZ. Choose two points a and b in the body. The body has a translational velocity corresponding to the velocity of point a and a rotational velocity $\boldsymbol{\omega}$ as shown in the diagram. Now embed a reference xyz into the body with origin at point a. Next, use this diagram and consider point b to show that Eq. 15.20 can be reformulated to be identical to Eq. 15.6.

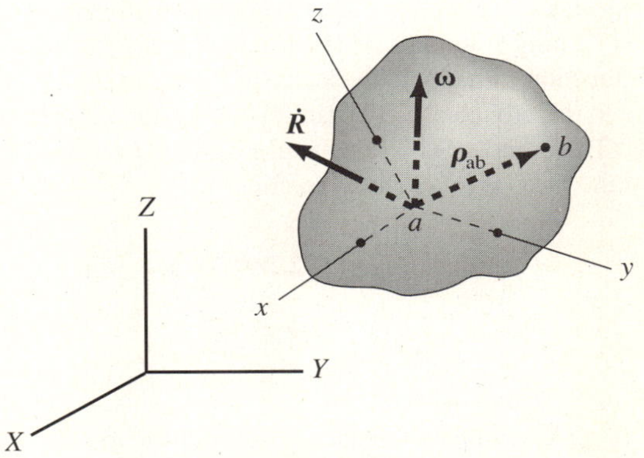

Figure P.15.77.

15.78. A simple-impulse type of turbomachine called a *Pelton* water wheel has a single jet of water issuing out of a nozzle and impinging on the system of buckets attached to a wheel. The runner, which is the assembly of buckets and wheel, has a radius of r to the center of the buckets. The shape of the bucket is also shown where a horizontal midsection of the bucket has been taken. Note that the jet is split in two parts by the bucket and is rotated relative to the bucket in the horizontal plane as measured by β. If we neglect gravity and friction, the speed of the water relative to the bucket is unchanged during the action. Suppose that 8 liters of water per second flow through the nozzle, whose cross-sectional area at the exit is 2000 mm^2. If $r = 1$ m, what should ω_1 be (in r/min) for the water on average to have zero velocity relative to the ground in the Y direction when it comes off the bucket? Take $\beta = 10°$. (Why is it desirable to have the exit velocity equal to zero in the Y direction?) See the hint of Problem 15.75.

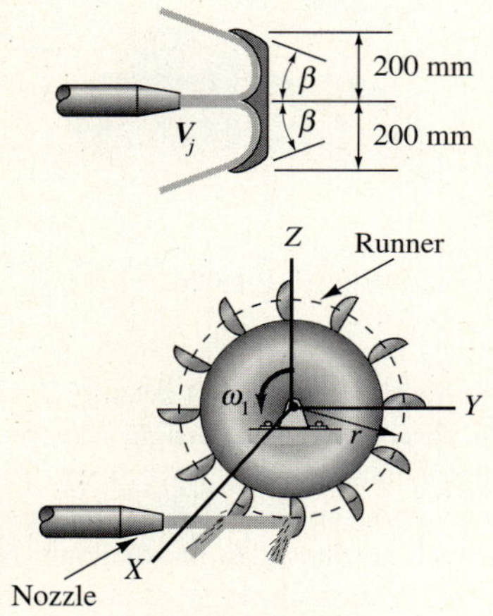

Figure P.15.78.

15.79. A propeller-driven airplane is moving at a speed of 36 m/s. Also, it is undergoing a yaw rotation of 0.25 rad/s and is simultaneously undergoing a loop rotation of 0.25 rad/s. The propeller is rotating at the rate of 100 r/min with a sense in the positive Y direction. What is the velocity of the tip of the propeller a relative to the ground at the instant that the plane is horizontal as shown? The propeller is 3 m in total length and at the instant of interest the blade is in a vertical position.

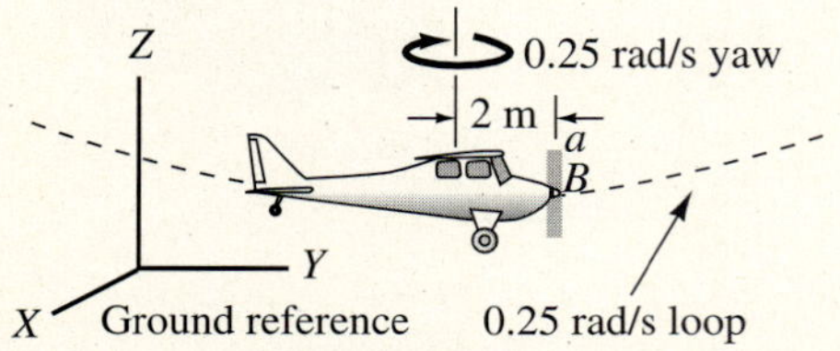

Figure P.15.79.

15.80. A crane moves to the right at a speed of 1.4 m/s. The boom OB, which is 15 m long, is being raised at an angular speed ω_2 relative to the cab of 0.4 rad/s, while the cab is rotating at an angular speed ω_1 of 0.2 rad/s relative to the base. What is the velocity of pin B relative to the ground at the instant when OB is at an angle of 35° with the ground? The axis of rotation O of the boom is 1 m from the axis of rotation A–A of the cab, as shown in the diagram.

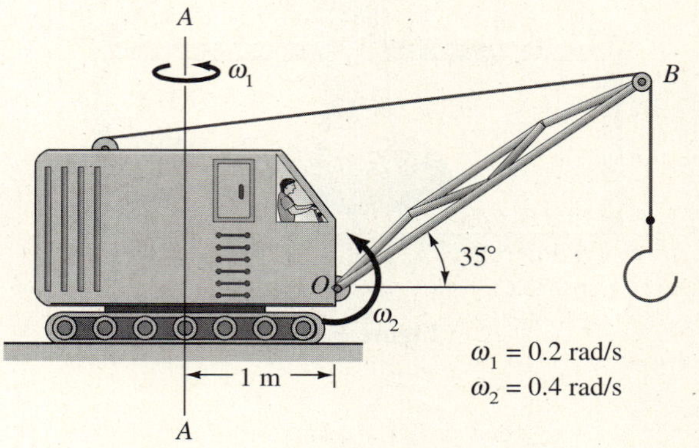

Figure P.15.80.

15.81. A power shovel main arm AC rotates with angular speed ω_1 of 0.3 rad/s relative to the cab. Arm ED rotates at a speed ω_2 of 0.4 rad/s relative to the main arm AC. The cab rotates about axis A–A at a speed ω_3 of 0.15 rad/s relative to the tracks which are stationary. What is the velocity of point D, the center of the shovel, at the instant of interest shown in the diagram? AB has a length of 5 m and BD has a length of 4 m.

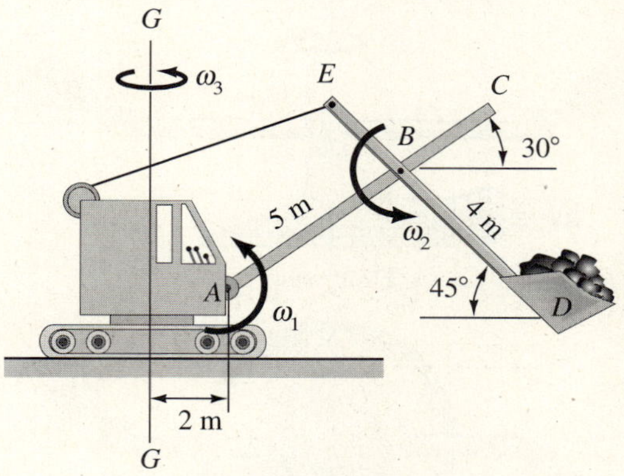

Figure P.15.81.

15.82. An antiaircraft gun is shown in action. The values of ω_1 and ω_2 are 0.3 rad/s and 0.6 rad/s, respectively. At the instant shown, what is the velocity of a projectile *normal* to the direction of the gun barrel when it just leaves the gun barrel as seen from the ground?

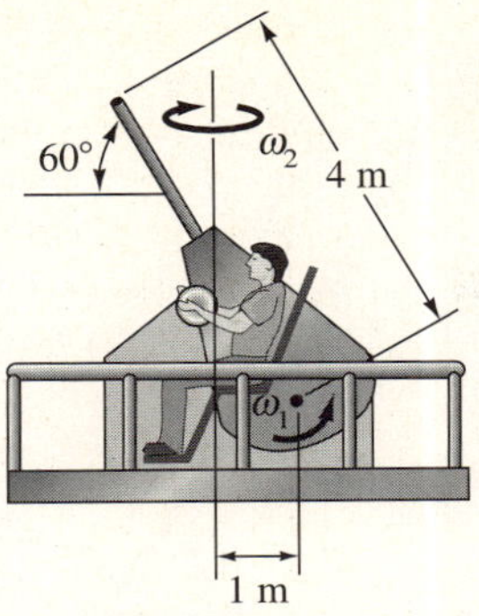

Figure P.15.82.

15.83. A cone is rolling without slipping about the Z axis such that its centerline rotates at the rate ω_1 of 5 rad/s. Use a multireference approach to determine the total angular velocity of the body relative to the ground.

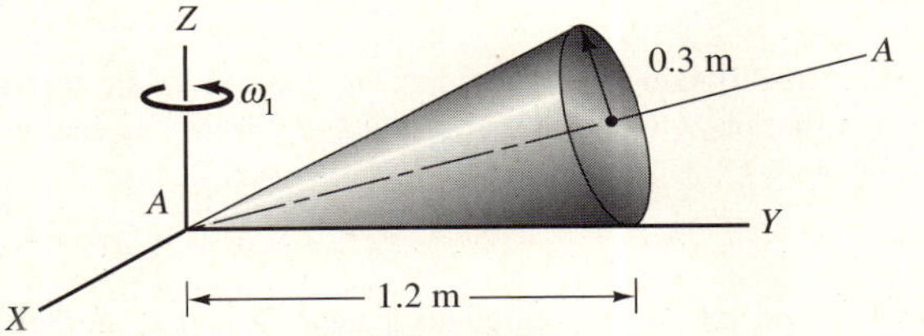

Figure P.15.83.

15.84. Find the velocity of gear tooth A relative to the ground reference XYZ. Note that ω_1 and ω_2 are both relative to the ground. Bevel gear A is free to rotate in the collar at C. Take ω_1 = 2 rad/s and ω_2 = 4 rad/s.

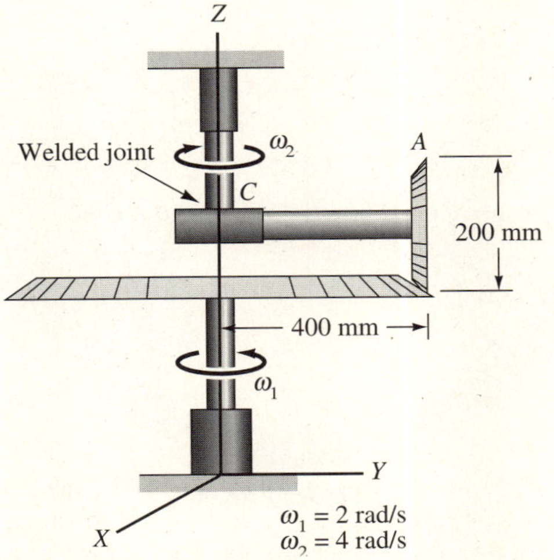

Figure P.15.84.

15.85. In a merry-go-round, the main platform rotates at the rate ω_1 of 10 revolutions per minute. A set of 45° bevel gears causes B to rotate at an angular speed $\dot{\theta}$ relative to the platform. The horse is mounted on AB, which slides in a slot at C and is moved at A by shaft B, as indicated in the diagram, where part of the merry-go-round is shown. If AB = 0.3 m and AC = 4.5 m, compute the velocity of point C relative to the platform. Then, compute the velocity of point C relative to the ground. Take θ = 45° at the instant of interest. What is the angular velocity of the horse relative to the platform and relative to the ground at the instant of interest?

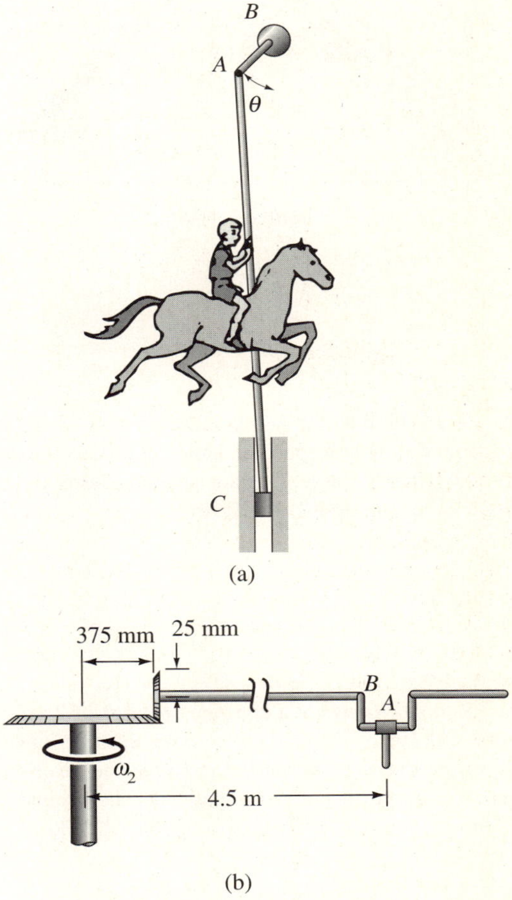

Figure P.15.85.

15.86. Rod BO rotates at a constant angular speed $\dot{\theta}$ of 5 rad/s clockwise. A collar A on the rod is pinned to a slider C, which moves in the groove shown in the diagram. When $\theta = 60°$, compute the speed of the collar A relative to the ground. What is the speed of collar A relative to the rod?

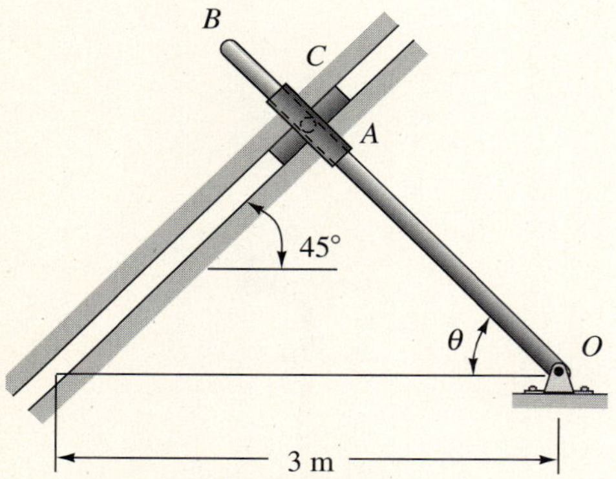

Figure P.15.86.

15.87. Work Problem 15.86 assuming that pin O is on rollers moving to the right at a speed of 1 m/s relative to the ground. In addition, OB rotates at a constant angular speed $\dot{\theta}$ of 5 rad/s clockwise all at the instant of interest.

15.88. Rod AD rotates at a constant speed $\dot{\theta}$ of 2 rad/s. Collar C on the rod DA is constrained to move in the circular groove shown in the diagram. When the rod is at the position shown, compute the speed of collar C relative to the ground. What is the speed of collar C relative to the rod AD? Point A is stationary.

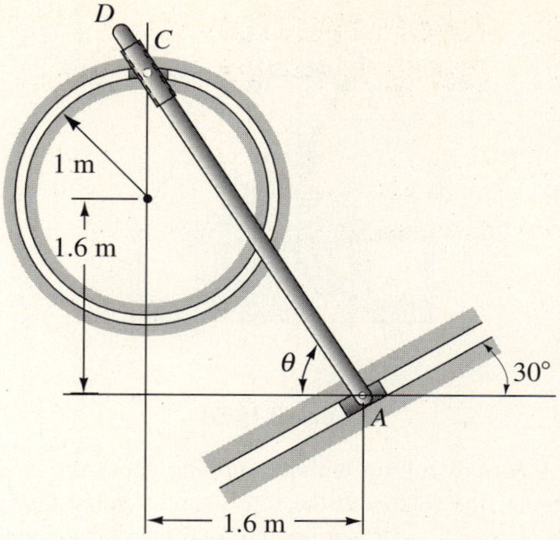

Figure P.15.88.

15.89. In Problem 15.88, assume, in addition to the rotation of bar AD, that pin A is moving at a speed of 1.6 m/s up the grooved incline.

15.90. Rod AC is connected to a gear D and is guided by a bearing B. Bearing B can rotate only in the plane of the gears. If the angular speed of AC is 5 rad/s clockwise, what is the angular speed of gear D relative to the ground? The diameter of gear D is 0.6 m.

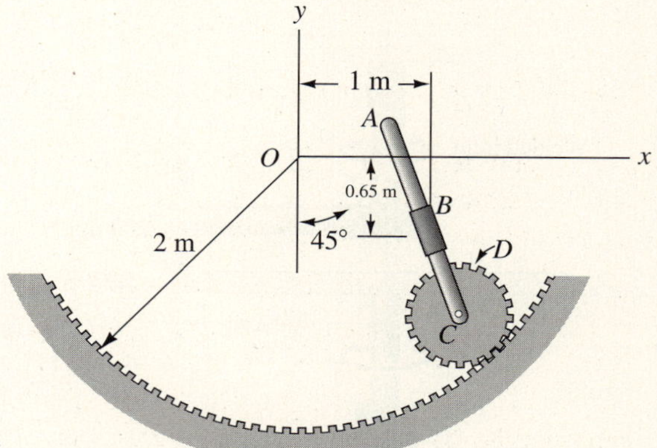

Figure P.15.90.

*15.8 Acceleration of a Particle for Different References

The acceleration of a particle relative to a coordinate system is simply the time derivative, as seen from the coordinate system, of the velocity relative to the coordinate system. Thus, observing Fig. 15.29, we can say:

$$\begin{aligned} \boldsymbol{a}_{XYZ} &= \left(\frac{d}{dt}\mathbf{V}_{XYZ}\right)_{XYZ} = \left(\frac{d^2\boldsymbol{r}}{dt^2}\right)_{XYZ} \\ \boldsymbol{a}_{xyz} &= \left(\frac{d}{dt}\mathbf{V}_{xyz}\right)_{xyz} = \left(\frac{d^2\boldsymbol{\rho}}{dt^2}\right)_{xyz} \end{aligned} \tag{15.21}$$

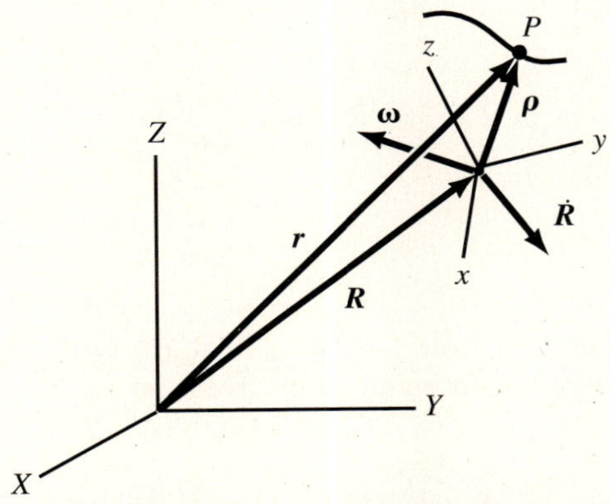

Figure 15.29. *xyz* moves arbitrarily relative to *XYZ*.

This notation may at first seem cumbersome to you, but it will soon be simplified.

Let us now relate the acceleration vectors of a particle for two references moving arbitrarily relative to each other. We do this by differentiating with respect to time the terms in Eq. 15.20 for the *XYZ* reference. Thus,

$$\left(\frac{d\,\mathbf{V}_{XYZ}}{dt}\right)_{XYZ} \equiv \boldsymbol{a}_{XYZ} = \left(\frac{d\,\mathbf{V}_{xyz}}{dt}\right)_{XYZ} + \ddot{\boldsymbol{R}} + \left[\frac{d}{dt}(\boldsymbol{\omega}\times\boldsymbol{\rho})\right]_{XYZ} \tag{15.22}$$

Now carry out the derivative of the cross product using the product rule.

$$\boldsymbol{a}_{XYZ} = \left(\frac{d\mathbf{V}_{xyz}}{dt}\right)_{XYZ} + \ddot{\boldsymbol{R}} + \boldsymbol{\omega}\times\left(\frac{d\boldsymbol{\rho}}{dt}\right)_{XYZ} + \left(\frac{d\boldsymbol{\omega}}{dt}\right)_{XYZ}\times\boldsymbol{\rho} \tag{15.23}$$

To introduce more physically meaningful terms, we can replace

$$\left(\frac{d\boldsymbol{V}_{xyz}}{dt}\right)_{XYZ} \quad \text{and} \quad \left(\frac{d\boldsymbol{\rho}}{dt}\right)_{XYZ}$$

using Eq. 15.16 in the following way:

$$\left(\frac{d\boldsymbol{V}_{xyz}}{dt}\right)_{XYZ} = \left(\frac{d\boldsymbol{V}_{xyz}}{dt}\right)_{xyz} + \boldsymbol{\omega} \times \boldsymbol{V}_{xyz}$$

$$\left(\frac{d\boldsymbol{\rho}}{dt}\right)_{XYZ} = \left(\frac{d\boldsymbol{\rho}}{dt}\right)_{xyz} + \boldsymbol{\omega} \times \boldsymbol{\rho}$$

Substituting into Eq. 15.23, we get

$$\boldsymbol{a}_{XYZ} = \left(\frac{d\boldsymbol{V}_{xyz}}{dt}\right)_{xyz} + \boldsymbol{\omega} \times \boldsymbol{V}_{xyz} + \ddot{\boldsymbol{R}} + \boldsymbol{\omega} \times \left(\frac{d\boldsymbol{\rho}}{dt}\right)_{xyz} + \boldsymbol{\omega} \times (\boldsymbol{\omega} \times \boldsymbol{\rho}) + \left(\frac{d\boldsymbol{\omega}}{dt}\right)_{XYZ} \times \boldsymbol{\rho}$$

You will note that $(d\boldsymbol{V}_{xyz}/dt)_{xyz}$ is $\boldsymbol{a}_{xyz}$; that $(d\boldsymbol{\rho}/dt)_{xyz}$ is $\boldsymbol{V}_{xyz}$; and that $(d\boldsymbol{\omega}/dt)_{XYZ}$ is $\dot{\boldsymbol{\omega}}$. Hence, rearranging terms, we have

$$\boldsymbol{a}_{XYZ} = \boldsymbol{a}_{xyz} + \ddot{\boldsymbol{R}} + 2\boldsymbol{\omega} \times \boldsymbol{V}_{xyz} + \dot{\boldsymbol{\omega}} \times \boldsymbol{\rho} + \boldsymbol{\omega} \times (\boldsymbol{\omega} \times \boldsymbol{\rho}) \qquad (15.24)$$

where $\boldsymbol{\omega}$ and $\dot{\boldsymbol{\omega}}$ are the angular velocity and acceleration, respectively, of the *xyz* reference relative to the *XYZ* reference. The vector $2(\boldsymbol{\omega} \times \boldsymbol{V}_{xyz})$ is called the *Coriolis acceleration vector*; we shall examine its interesting effects in Section 15.10.

Although Eq. 15.24 may seem somewhat terrifying at first, you will find that, by using it, problems that would otherwise be tremendously difficult can readily be carried out in a systematic manner. *You should keep in mind when solving problems that any of the methods developed in Chapter 11 can be used for determining the motion of the particle relative to the xyz reference or for determining the motion of the origin of xyz relative to the XYZ reference.* We shall now examine several problems, in which we shall use the notation, $\boldsymbol{\omega}_1$, $\boldsymbol{\omega}_2$, etc., to denote the various angular velocities involved. The notation, $\boldsymbol{\omega}$ (i.e., without subscripts), however, will we repeat be reserved to represent the angular velocity of the *xyz* reference relative to the *XYZ* reference.

Example 15.11

A stationary truck is carrying a cockpit for a worker who repairs overhead fixtures. At the instant shown in Fig. 15.30, the base D is rotating at angular speed ω_2 of 0.1 rad/s with $\dot{\omega}_2 = 0.2$ rad/s^2 relative to the truck. Arm AB is rotating at angular speed ω_1 of 0.2 rad/s with $\dot{\omega}_1 = 0.8$ rad/s^2 relative to DA. Cockpit C is rotating relative to AB so as to always keep the man upright. What are the velocity and acceleration vectors of the man relative to the ground if $\alpha = 45°$ and $\beta = 30°$ at the instant of interest? Take $DA = 13$ m.

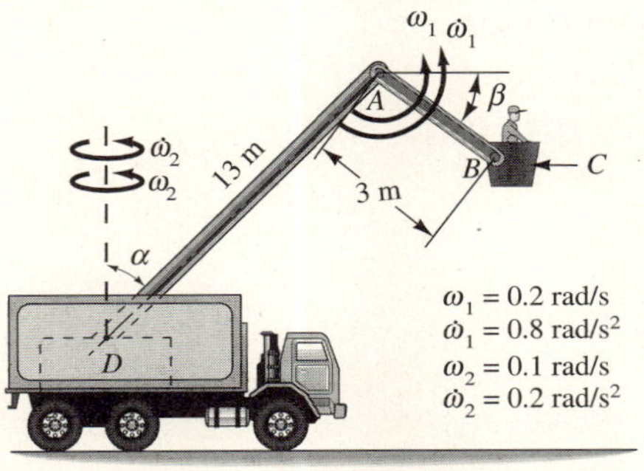

Figure 15.30. Truck with moving cockpit.

Because of the rotation of the cockpit C relative to arm AB to keep the man vertical, clearly, each particle in that body including the man has the same motion as point B of arm AB. Therefore, we shall concentrate our attention on this point.

Fix xyz to arm DA.
Fix XYZ to truck.

This situation is shown in Fig. 15.31.

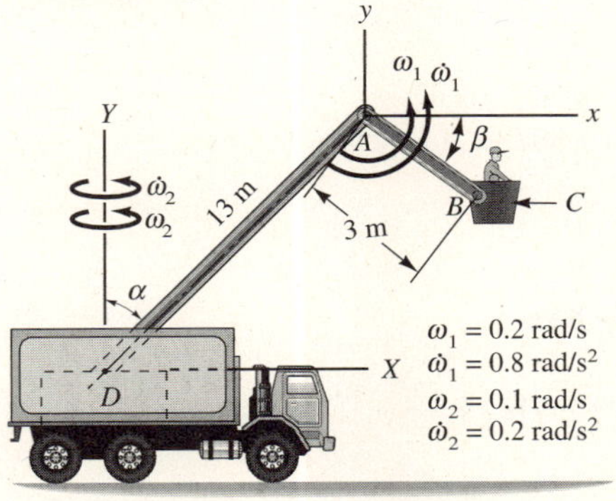

Figure 15.31. xyz fixed to DA; XYZ fixed to truck.

A. Motion of B relative to xyz

$$\boldsymbol{\rho} = 3(\cos\beta\boldsymbol{i} - \sin\beta\boldsymbol{j}) = 2.60\boldsymbol{i} - 1.5\boldsymbol{j} \text{ m}$$

Example 15.11 (Continued)

Since $\boldsymbol{\rho}$ is fixed in AB, which has angular velocity $\boldsymbol{\omega}_1$ relative to xyz, we have

$$\begin{aligned} \boldsymbol{V}_{xyz} &= \boldsymbol{\omega}_1 \times \boldsymbol{\rho} = (0.2\boldsymbol{k}) \times (2.60\boldsymbol{i} - 1.5\boldsymbol{j}) \\ &= 0.520\boldsymbol{j} + 0.3\boldsymbol{i} \text{ m/s} \\ \boldsymbol{a}_{xyz} &= \left(\frac{d\boldsymbol{\omega}_1}{dt}\right)_{xyz} \times \boldsymbol{\rho} + \boldsymbol{\omega}_1 \times \left(\frac{d\boldsymbol{\rho}}{dt}\right)_{xyz} \end{aligned}$$

As seen from xyz, only the value of $\boldsymbol{\omega}_1$ and not its direction is changing. Also note that $(d\boldsymbol{\rho}/dt)_{xyz} = \boldsymbol{V}_{xyz}$. Hence,

$$\begin{aligned} \boldsymbol{a}_{xyz} &= (0.8\boldsymbol{k}) \times (2.60\boldsymbol{i} - 1.5\boldsymbol{j}) + (0.2\boldsymbol{k}) \times (0.520\boldsymbol{j} + 0.3\boldsymbol{i}) \\ &= 1.09\boldsymbol{i} + 2.14\boldsymbol{j} \text{ m/s}^2 \end{aligned}$$

B. Motion of *xyz* relative to *XYZ*

$$\boldsymbol{R} = 13(0.707\boldsymbol{i} + 0.707\boldsymbol{j}) = 9.19\boldsymbol{i} + 9.19\boldsymbol{j} \text{ m}$$

Since $\boldsymbol{R}$ is fixed in DA, and since DA rotates with angular velocity $\boldsymbol{\omega}_2$ relative to XYZ, we have

$$\begin{aligned} \dot{\boldsymbol{R}} &= \boldsymbol{\omega}_2 \times \boldsymbol{R} = (0.1\boldsymbol{j}) \times (9.19\boldsymbol{i} + 9.19\boldsymbol{j}) \\ &= -0.919\boldsymbol{k} \text{ m/s} \\ \ddot{\boldsymbol{R}} &= \dot{\boldsymbol{\omega}}_2 \times \boldsymbol{R} + \boldsymbol{\omega}_2 \times \dot{\boldsymbol{R}} \\ &= (0.2\boldsymbol{j}) \times (9.19\boldsymbol{i} + 9.19\boldsymbol{j}) + (0.1\boldsymbol{j}) \times (-0.919\boldsymbol{k}) \\ &= -1.838\boldsymbol{k} - 0.0919\boldsymbol{i} \text{ m/s}^2 \\ \boldsymbol{\omega} &= \boldsymbol{\omega}_2 = 0.1\boldsymbol{j} \text{ rad/s} \\ \dot{\boldsymbol{\omega}} &= \dot{\boldsymbol{\omega}}_2 = 0.2\boldsymbol{j} \text{ rad/s}^2 \end{aligned}$$

Hence,

$$\begin{aligned} \boldsymbol{V}_{XYZ} &= \boldsymbol{V}_{xyz} + \dot{\boldsymbol{R}} + \boldsymbol{\omega} \times \boldsymbol{\rho} \\ &= 0.520\boldsymbol{j} + 0.3\boldsymbol{i} - 0.919\boldsymbol{k} + (0.1\boldsymbol{j}) \times (2.60\boldsymbol{i} - 1.5\boldsymbol{j}) \end{aligned}$$

$$\boldsymbol{V}_{XYZ} = 0.3\boldsymbol{i} + 0.520\boldsymbol{j} - 1.179\boldsymbol{k} \text{ m/s}$$

$$\begin{aligned} \boldsymbol{a}_{XYZ} &= \boldsymbol{a}_{xyz} + \ddot{\boldsymbol{R}} + 2\boldsymbol{\omega} \times \boldsymbol{V}_{xyz} + \dot{\boldsymbol{\omega}} \times \boldsymbol{\rho} + \boldsymbol{\omega} \times (\boldsymbol{\omega} \times \boldsymbol{\rho}) \\ &= 1.09\boldsymbol{i} + 2.14\boldsymbol{j} - 1.838\boldsymbol{k} - 0.0919\boldsymbol{i} \\ &\quad + 2(0.1\boldsymbol{j}) \times (0.520\boldsymbol{j} + 0.3\boldsymbol{i}) + (0.2\boldsymbol{j}) \times (2.60\boldsymbol{i} - 1.5\boldsymbol{j}) \\ &\quad + (0.1\boldsymbol{j}) \times [(0.1\boldsymbol{j}) \times (2.60\boldsymbol{i} - 1.5\boldsymbol{j})] \end{aligned}$$

$$\boldsymbol{a}_{XYZ} = 0.978\boldsymbol{i} + 2.14\boldsymbol{j} - 2.42\boldsymbol{k} \text{ m/s}$$

Notice that the essential aspects of the analysis come in the consideration of parts A and B of the problem, while the remaining portion involves direct substitution and vector algebraic operations.

Example 15.12

A wheel rotates with an angular speed ω_2 of 5 rad/s on a platform which rotates with a speed ω_1 of 10 rad/s relative to the ground as shown in Fig. 15.32. A valve gate *A* moves down the spoke of the wheel, and when the spoke is vertical the valve gate has a speed of 6 m/s, an acceleration of 3 m/s^2 along the spoke, and is 0.3 m from the shaft centerline of the wheel. Compute the velocity and acceleration of the valve gate relative to the ground at this instant.

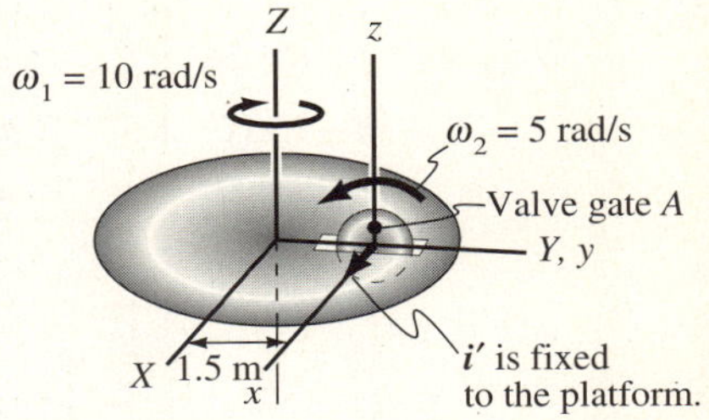

Figure 15.32. *xyz* fixed to wheel; *XYZ* fixed to ground.

Fix xyz to wheel.
Fix XYZ to ground.

A. Motion of particle relative to *xyz*

$$\boldsymbol{\rho} = 0.3\boldsymbol{k} \text{ m}$$
$$\boldsymbol{V}_{xyz} = -6\boldsymbol{k} \text{ m/s}$$
$$\boldsymbol{a}_{xyz} = -3\boldsymbol{k} \text{ m/s}^2$$

B. Motion of *xyz* relative to *XYZ*

$$\boldsymbol{R} = 1.5\boldsymbol{j} \text{ m}$$

Since $\boldsymbol{R}$ is fixed to the platform:

$$\dot{\boldsymbol{R}} = \boldsymbol{\omega}_1 \times \boldsymbol{R} = (-10\boldsymbol{k}) \times (1.5\boldsymbol{j}) = 15\boldsymbol{i} \text{ m/s}$$
$$\ddot{\boldsymbol{R}} = \dot{\boldsymbol{\omega}}_1 \times \boldsymbol{R} + \boldsymbol{\omega}_1 \times \dot{\boldsymbol{R}}$$
$$= \boldsymbol{0} + (-10\boldsymbol{k}) \times (15\boldsymbol{i}) = -150\boldsymbol{j} \text{ m/s}^2$$
$$\boldsymbol{\omega} = \boldsymbol{\omega}_2 + \boldsymbol{\omega}_1 = 5\boldsymbol{i} - 10\boldsymbol{k} \text{ rad/s}$$
$$\dot{\boldsymbol{\omega}} = \dot{\boldsymbol{\omega}}_2 + \dot{\boldsymbol{\omega}}_1$$

Note that $\boldsymbol{\omega}_2$ is of constant magnitude but, because of the bearings of the wheel, $\boldsymbol{\omega}_2$ must rotate with the platform. In short, we can say that $\boldsymbol{\omega}_2$ is *fixed* to the platform and so $\dot{\boldsymbol{\omega}}_2 = \boldsymbol{\omega}_1 \times \boldsymbol{\omega}_2$. Hence,

$$\dot{\boldsymbol{\omega}} = \boldsymbol{\omega}_1 \times \boldsymbol{\omega}_2 + \boldsymbol{0}$$
$$= (-10\boldsymbol{k}) \times (5\boldsymbol{i}) = -50\boldsymbol{j} \text{ rad/s}^2$$

We then have

$$\boldsymbol{V}_{XYZ} = \boldsymbol{V}_{xyz} + \dot{\boldsymbol{R}} + \boldsymbol{\omega} \times \boldsymbol{\rho}$$
$$= -6\boldsymbol{k} + 15\boldsymbol{i} + (5\boldsymbol{i} - 10\boldsymbol{k}) \times 0.3\boldsymbol{k}$$

$$\boldsymbol{V}_{XYZ} = 15\boldsymbol{i} - 1.5\boldsymbol{j} - 6\boldsymbol{k} \text{ m/s}$$

Example 15.12 (Continued)

Also,

$$\begin{aligned} \boldsymbol{a}_{XYZ} &= \boldsymbol{a}_{xyz} + \ddot{\boldsymbol{R}} + 2\boldsymbol{\omega} \times \boldsymbol{V}_{xyz} + \dot{\boldsymbol{\omega}} \times \boldsymbol{\rho} + \boldsymbol{\omega} \times (\boldsymbol{\omega} \times \boldsymbol{\rho}) \\ &= -3\boldsymbol{k} - 150\boldsymbol{j} + 2(5\boldsymbol{i} - 10\boldsymbol{k}) \times (-6\boldsymbol{k}) + (-50\boldsymbol{j}) \times 0.3\boldsymbol{k} \\ &\qquad + (5\boldsymbol{i} - 10\boldsymbol{k}) \times [(5\boldsymbol{i} - 10\boldsymbol{k}) \times 0.3\boldsymbol{k}] \end{aligned}$$

$$\boldsymbol{a}_{XYZ} = -30\boldsymbol{i} - 90\boldsymbol{j} - 10.5\boldsymbol{k} \text{ m/s}^2$$

Example 15.13

In Example 15.12, the wheel accelerates at the instant under discussion with $\dot{\omega}_2 = 5$ rad/s^2, and the platform accelerates with $\dot{\omega}_1 = 10$ rad/s^2 (see Fig. 15.32). Find the velocity and acceleration of the valve gate A.

If we review the contents of parts A and B of Example 15.12, it will be clear that only $\ddot{\boldsymbol{R}}$ and $\dot{\boldsymbol{\omega}}$ are affected by the fact that $\dot{\omega}_1 = 10$ rad/s^2 and $\dot{\omega}_2 = 5$ rad/s^2. In this regard, consider $\boldsymbol{\omega}_2$. It is no longer of constant value and cannot be considered as *fixed* in the platform. However we can express $\boldsymbol{\omega}_2$ as $\omega_2\boldsymbol{i}'$ *at all times*, wherein $\boldsymbol{i}'$ is *fixed* in the platform as shown in Fig. 15.32. Thus, we can say for $\boldsymbol{\omega}$:

$$\boldsymbol{\omega} = \omega_2\boldsymbol{i}' + \boldsymbol{\omega}_1$$

Therefore,

$$\begin{aligned} \dot{\boldsymbol{\omega}} &= \dot{\omega}_2\boldsymbol{i}' + \omega_2\dot{\boldsymbol{i}}' + \dot{\boldsymbol{\omega}}_1 \\ &= 5\boldsymbol{i}' + 5(\boldsymbol{\omega}_1 \times \boldsymbol{i}') - 10\boldsymbol{k} \\ &= 5\boldsymbol{i}' + 5(-10\boldsymbol{k}) \times \boldsymbol{i}' - 10\boldsymbol{k} \end{aligned}$$

At the instant of interest, $\boldsymbol{i}' = \boldsymbol{i}$. Hence,

$$\dot{\boldsymbol{\omega}} = 5\boldsymbol{i} - 50\boldsymbol{j} - 10\boldsymbol{k} \text{ rad/s}^2$$

Hence, we use the above $\dot{\boldsymbol{\omega}}$ in part B of Example 15.12 to compute $\boldsymbol{V}_{XYZ}$ and $\boldsymbol{a}_{XYZ}$. The computation of $\ddot{\boldsymbol{R}}$ is straightforward and so we can compute $\boldsymbol{a}_{XYZ}$ accordingly. We leave the details to the reader.

An understanding of Examples 15.11, 15.12, and 15.13, involving two angular velocities of component parts is sufficient for most of the homework problems of this section covering a wide range of applications. In the next example, we have three angular velocities to deal with. We urge you to examine it carefully if time allows. It is an interesting problem, and comprehension of the three different analyses given will ensure a strong grasp of multireference kinematics.

Example 15.14[11]

To simulate the flight conditions of a space vehicle, engineers have developed the *centrifuge*, shown diagrammatically in Fig. 15.33. A main *arm*, 12 m long, rotates about the *A–A* axis. The pilot sits in a *cockpit*, which can rotate about axis *C–C*. The *seat* for the pilot can rotate inside the cockpit about an axis shown as *B–B*. These rotations are controlled by a computer that is set to simulate certain maneuvers corresponding to the entry and exit from the earth's atmosphere, malfunctions of the control system, and so on. When a pilot sits in the cockpit, his/her head has a position which is 0.9 m from the seat as shown in Fig. 15.33. At the instant of interest the main arm is rotating at 10 r/min and accelerating at 5 r/min^2. The cockpit is rotating at a constant speed about *C–C* relative to the main arm at 10 r/min. Finally, the seat is rotating at a constant speed of 5 r/min relative to the cockpit about axis *B–B*. How many g's of acceleration relative to the ground is the pilot's head subject to?[12] Note that the three axes, *A–A, C–C*, and *B–B*, are orthogonal to each other at time t.

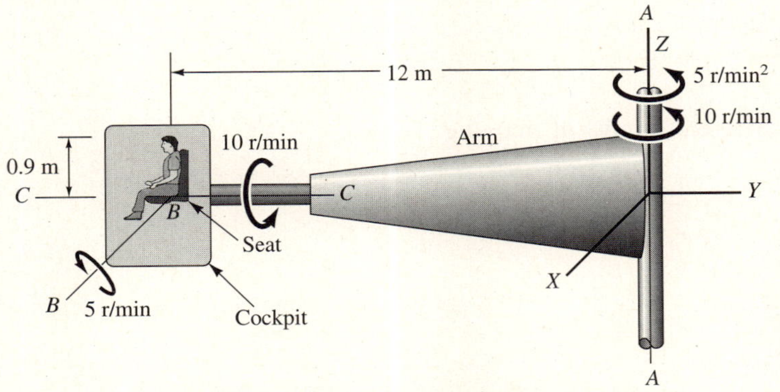

Figure 15.33. Centrifuge for simulating flight conditions.

[11]Example 15.14 was given as two homework problems in both the first and second editions of this text. They were so instructive that for subsequent editions the author decided to move the problems into the main text.

[12]A g of acceleration is an amount of acceleration equal to that of gravity (9.81 m/s^2). Thus, a $4g$ acceleration is equivalent to an acceleration of 39.24 m/s^2.

Example 15.14 (Continued)

In Fig. 15.34 the arm of the centrifuge rotates relative to the ground at an angular velocity of ω_1. The cockpit meanwhile rotates relative to the arm with angular speed ω_2. Finally, the seat rotates relative to the cockpit at an angular speed ω_3. For constant ω_2, we see that, because of bearings in the arm, the vector $\boldsymbol{\omega}_2$ is "fixed" in the arm. Also, for constant ω_3, because of bearings in the cockpit, the vector $\boldsymbol{\omega}_3$ is "fixed" in the cockpit. Before we examine the acceleration of the pilot's head, note that at the instant of interest:

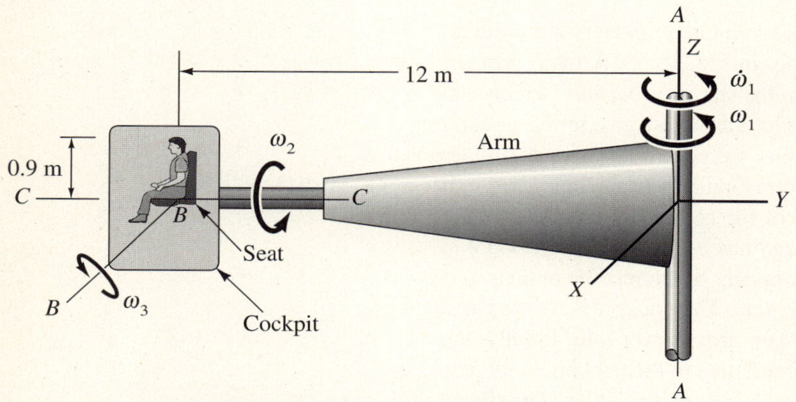

Figure 15.34. Centrifuge listing $\boldsymbol{\omega}$'s.

$$\omega_1 = \omega_2 = 10 \text{ r/min} = 1.048 \text{ rad/s}$$
$$\dot{\omega}_1 = 5 \text{ r/min}^2 = 8.73 \text{ mrad/s}^2$$
$$\omega_3 = 5 \text{ r/min} = 0.524 \text{ rad/s}$$

We shall do this problem using three different kinds of moving references *xyz*.

ANALYSIS I

Fix *xyz* to arm.
Fix *XYZ* to ground.

Note in Fig. 15.35 that *xyz* and the arm to which it is fixed are shown dark. Note also that the axes *xyz* and *XYZ* are parallel to each other at the instant of interest.

Example 15.14 (Continued)

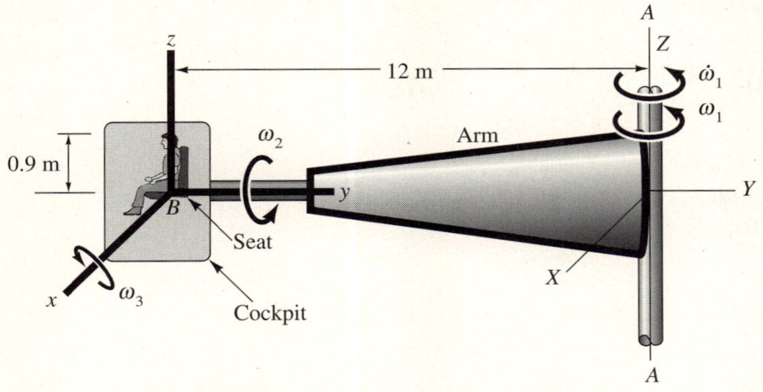

Figure 15.35. Centrifuge with *xyz* fixed to arm.

A. Motion of particle relative to *xyz*

$$\boldsymbol{\rho} = 0.9\boldsymbol{k} \text{ m}$$

Note that $\boldsymbol{\rho}$ is "fixed" to the seat and that the seat has an angular velocity of $(\boldsymbol{\omega}_2 + \boldsymbol{\omega}_3)$ relative to the arm and thus to *xyz*. Hence,

$$\begin{aligned} \boldsymbol{V}_{xyz} &= (\boldsymbol{\omega}_2 + \boldsymbol{\omega}_3) \times \boldsymbol{\rho} \\ &= (1.048\boldsymbol{j} + 0.524\boldsymbol{i}) \times 0.9\boldsymbol{k} = 0.943\boldsymbol{i} - 0.472\boldsymbol{j} \text{ m/s} \end{aligned}$$

$$\boldsymbol{a}_{xyz} = \left(\frac{d\boldsymbol{V}_{xyz}}{dt}\right)_{xyz} = \left[\frac{d}{dt_{xyz}}(\boldsymbol{\omega}_2 + \boldsymbol{\omega}_3)\right] \times \boldsymbol{\rho} + (\boldsymbol{\omega}_2 + \boldsymbol{\omega}_3) \times \left(\frac{d\boldsymbol{\rho}}{dt}\right)_{xyz}$$

Clearly, relative to the arm, and thus to *xyz*, $\boldsymbol{\omega}_2$ is constant. And $\boldsymbol{\omega}_3$ is fixed in the cockpit that has an angular velocity of $\boldsymbol{\omega}_2$ relative to *xyz*. Thus, we have

$$\begin{aligned} \boldsymbol{a}_{xyz} &= (\boldsymbol{0} + \boldsymbol{\omega}_2 \times \boldsymbol{\omega}_3) \times \boldsymbol{\rho} + (\boldsymbol{\omega}_2 + \boldsymbol{\omega}_3) \times \boldsymbol{V}_{xyz} \\ &= (1.048\boldsymbol{j} \times 0.524\boldsymbol{i}) \times 0.9\boldsymbol{k} + (1.048\boldsymbol{j} + 0.524\boldsymbol{i}) \times (0.943\boldsymbol{i} - 0.472\boldsymbol{j}) \\ &= -1.236\boldsymbol{k} \text{ m/s}^2 \end{aligned}$$

B. Motion of *xyz* relative to *XYZ*

$$\boldsymbol{R} = -12\boldsymbol{j} \text{ m}$$

Note that $\boldsymbol{R}$ is fixed in the arm, which has an angular velocity $\boldsymbol{\omega}_1$ relative to *XYZ*. Hence,

$$\begin{aligned} \dot{\boldsymbol{R}} &= \boldsymbol{\omega}_1 \times \boldsymbol{R} = 1.048\boldsymbol{k} \times (-12\boldsymbol{j}) = 12.58\boldsymbol{i} \text{ m/s} \\ \ddot{\boldsymbol{R}} &= \boldsymbol{\omega}_1 \times \dot{\boldsymbol{R}} + \dot{\boldsymbol{\omega}}_1 \times \boldsymbol{R} \\ &= 1.048\boldsymbol{k} \times 12.58\boldsymbol{i} + 8.73 \times 10^{-3}\boldsymbol{k} \times (-12\boldsymbol{j}) \\ &= 13.18\boldsymbol{j} + 0.105\boldsymbol{i} \text{ m/s}^2 \\ \boldsymbol{\omega} &= \boldsymbol{\omega}_1 = 1.048\boldsymbol{k} \text{ rad/s} \\ \dot{\boldsymbol{\omega}} &= \dot{\boldsymbol{\omega}}_1 = 8.73 \times 10^{-3}\boldsymbol{k} \text{ rad/s}^2 \end{aligned}$$

Example 15.14 (Continued)

We can now substitute into the following equation:

$$\boldsymbol{a}_{XYZ} = \boldsymbol{a}_{xyz} + \ddot{\boldsymbol{R}} + 2\boldsymbol{\omega} \times \boldsymbol{V}_{xyz} + \dot{\boldsymbol{\omega}} \times \boldsymbol{\rho} + \boldsymbol{\omega} \times (\boldsymbol{\omega} \times \boldsymbol{\rho})$$

Therefore,

$$\boldsymbol{a}_{XYZ} = 1.094\boldsymbol{i} + 15.16\boldsymbol{j} - 1.236\boldsymbol{k} \text{ m/s}^2$$

$$|\boldsymbol{a}_{XYZ}| = \frac{\sqrt{1.094^2 + 15.16^2 + 1.236^2}}{9.81} = \boxed{1.554g}$$

ANALYSIS II

Fix xyz to cockpit.
Fix XYZ to ground.

This situation is shown in Fig. 15.36.

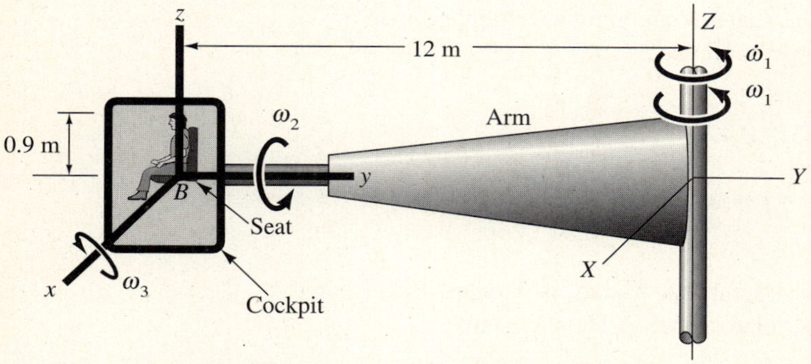

Figure 15.36. Centrifuge with xyz fixed to cockpit.

A. Motion of particle relative to xyz

$$\boldsymbol{\rho} = 0.9\boldsymbol{k} \text{ m}$$

Note that $\boldsymbol{\rho}$ is fixed to the seat, which has an angular velocity of $\boldsymbol{\omega}_3$ relative to the cockpit and thus relative to xyz. Hence,

$$\boldsymbol{V}_{xyz} = \boldsymbol{\omega}_3 \times \boldsymbol{\rho} = 0.524\boldsymbol{i} \times 0.9\boldsymbol{k} = -0.4716\boldsymbol{j} \text{ m/s}$$

$$\boldsymbol{a}_{xyz} = \left(\frac{d\boldsymbol{\omega}_3}{dt}\right)_{xyz} \times \boldsymbol{\rho} + \boldsymbol{\omega}_3 \times \left(\frac{d\boldsymbol{\rho}}{dt}\right)_{xyz}$$

But $\boldsymbol{\omega}_3$ is constant as seen from the cockpit and thus from xyz. Hence,

$$\boldsymbol{a}_{xyz} = \boldsymbol{0} \times \boldsymbol{\rho} + \boldsymbol{\omega}_3 \times \boldsymbol{V}_{xyz} = 0.524\boldsymbol{i} \times (-0.4716\boldsymbol{j})$$

$$= -0.247\boldsymbol{k} \text{ m/s}^2$$

Example 15.14 (Continued)

B. Motion of *xyz* relative to *XYZ*. The origin of *xyz* in this analysis has the same motion as the origin of *xyz* in the previous analysis. Thus, we use the results of analysis I for $\boldsymbol{R}$ and its time derivatives.

$$\boldsymbol{R} = -12\boldsymbol{j} \text{ m}$$
$$\dot{\boldsymbol{R}} = 12.58\boldsymbol{i} \text{ m/s}$$
$$\ddot{\boldsymbol{R}} = 13.18\boldsymbol{j} + 0.105\boldsymbol{i} \text{ m/s}^2$$
$$\boldsymbol{\omega} = \boldsymbol{\omega}_1 + \boldsymbol{\omega}_2 = 1.048\boldsymbol{j} + 1.048\boldsymbol{k} \text{ rad/s}$$
$$\dot{\boldsymbol{\omega}} = \dot{\boldsymbol{\omega}}_1 + \dot{\boldsymbol{\omega}}_2$$

We are given $\boldsymbol{\omega}_1$ about the Z axis and $\boldsymbol{\omega}_2$ is fixed in the arm, which is rotating with angular velocity $\boldsymbol{\omega}_1$ relative to the *XYZ* reference. Hence,

$$\dot{\boldsymbol{\omega}} = \dot{\boldsymbol{\omega}}_1 + \boldsymbol{\omega}_1 \times \boldsymbol{\omega}_2 = 8.73 \times 10^{-3}\boldsymbol{k} + (1.048\boldsymbol{k} \times 1.048\boldsymbol{j})$$
$$= -1.098\boldsymbol{i} + 8.73 \times 10^{-3}\boldsymbol{k} \text{ rad/s}^2$$

We can now substitute into the key equation, 15.24:

$$\boldsymbol{a}_{XYZ} = \boldsymbol{a}_{xyz} + \ddot{\boldsymbol{R}} + 2\boldsymbol{\omega} \times \boldsymbol{V}_{xyz} + \dot{\boldsymbol{\omega}} \times \boldsymbol{\rho} + \boldsymbol{\omega} \times (\boldsymbol{\omega} \times \boldsymbol{\rho})$$
$$= 1.094\boldsymbol{i} + 15.16\boldsymbol{j} - 1.236\boldsymbol{k} \text{ m/s}^2$$

$$|\boldsymbol{a}_{XYZ}| = 1.554g$$

ANALYSIS III

Fix *xyz* to seat.
Fix *XYZ* to ground.

This situation is shown in Fig. 15.37.

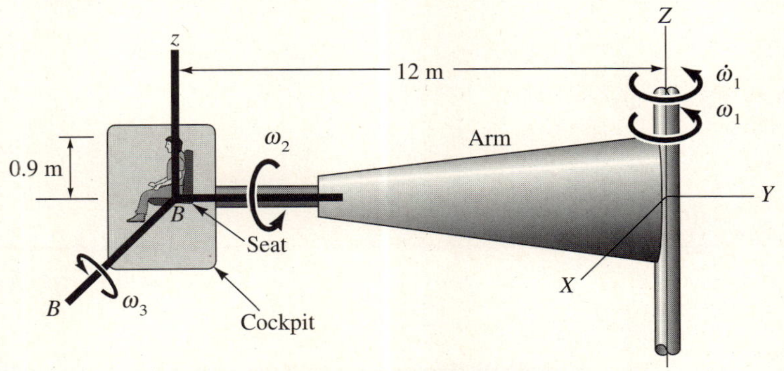

Figure 15.37. Centrifuge with *xyz* fixed to seat.

Example 15.14 (Continued)

A. Motion of particle relative to *xyz*

$$\boldsymbol{\rho} = 0.9\boldsymbol{k} \text{ m}$$

Since the particle is fixed to the seat and is thus fixed in *xyz*, we can say:

$$\boldsymbol{V}_{xyz} = \boldsymbol{0}$$
$$\boldsymbol{a}_{xyz} = \boldsymbol{0}$$

B. Motion of *xyz* relative to *XYZ*. Again, the origin of *xyz* has identically the same motion as in the previous analyses. Thus, we have the same results as before for $\boldsymbol{R}$ and its derivatives.

$$\boldsymbol{R} = -12\boldsymbol{j} \text{ m}$$
$$\dot{\boldsymbol{R}} = 12.58\boldsymbol{i} \text{ m/s}$$
$$\ddot{\boldsymbol{R}} = 13.18\boldsymbol{j} + 0.105\boldsymbol{i} \text{ m/s}^2$$
$$\boldsymbol{\omega} = \boldsymbol{\omega}_1 + \boldsymbol{\omega}_2 + \boldsymbol{\omega}_3 = 1.048\boldsymbol{k} + 1.048\boldsymbol{j} + 0.524\boldsymbol{i}$$
$$\dot{\boldsymbol{\omega}} = \dot{\boldsymbol{\omega}}_1 + \dot{\boldsymbol{\omega}}_2 + \dot{\boldsymbol{\omega}}_3$$

Note that $\dot{\boldsymbol{\omega}}_1$ is given. Also, $\boldsymbol{\omega}_2$ is fixed in the arm, which rotates with angular speed $\boldsymbol{\omega}_1$ relative to *XYZ*. Finally, $\boldsymbol{\omega}_3$ is fixed in the cockpit, which has an angular velocity $\boldsymbol{\omega}_2 + \boldsymbol{\omega}_1$ relative to *XYZ*. Thus,

$$\begin{aligned}\dot{\boldsymbol{\omega}} &= \dot{\boldsymbol{\omega}}_2 + \boldsymbol{\omega}_1 \times \boldsymbol{\omega}_2 + (\boldsymbol{\omega}_1 + \boldsymbol{\omega}_2) \times \boldsymbol{\omega}_3 \\ &= 8.73 \times 10^{-3}\boldsymbol{k} + (1.048\boldsymbol{k} \times 1.048\boldsymbol{j}) + (1.048\boldsymbol{k} + 1.048\boldsymbol{j}) \times (0.524\boldsymbol{i}) \\ &= -1.098\boldsymbol{i} + 0.549\boldsymbol{j} - 0.540\boldsymbol{k}\end{aligned}$$

We now go to the basic equation, 15.24.

$$\boldsymbol{a}_{XYZ} = \boldsymbol{a}_{xyz} + \ddot{\boldsymbol{R}} + 2\boldsymbol{\omega} \times \boldsymbol{V}_{xyz} + \dot{\boldsymbol{\omega}} \times \boldsymbol{\rho} + \boldsymbol{\omega} \times (\boldsymbol{\omega} \times \boldsymbol{\rho})$$

Substituting, we get

$$\boldsymbol{a}_{XYZ} = 1.094\boldsymbol{i} + 15.16\boldsymbol{j} - 1.236\boldsymbol{k} \text{ m/s}^2$$

$$|\boldsymbol{a}_{XYZ}| = 1.554g$$

In the final example of this series, we have a case where it is advantageous to use cylindrical coordinates in parts of the problem and then later to convert to rectangular coordinates.

Example 15.15

A submersible (see Fig. 15.38) is moving relative to the ground reference XYZ so as to have the following motion at the instant of interest for point A fixed to shaft $\overline{CD}$ which in turn is fixed to the submersible:

$$\boldsymbol{V} = 3\boldsymbol{i} + 0.6\boldsymbol{j} \text{ m/s}$$
$$\boldsymbol{a} = 2\boldsymbol{i} + 3\boldsymbol{j} - 0.5\boldsymbol{k} \text{ m/s}^2$$

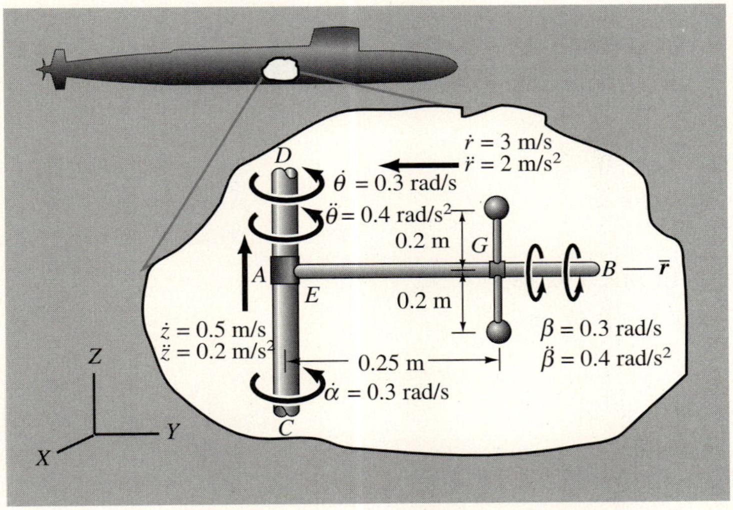

Figure 15.38. Rotating device inside a moving submersible. CD is fixed to the submersible.

At the instant of interest, the vessel has an angular speed of rotation $\dot{\alpha}$ = 0.3 rad/s about the centerline of $\overline{CD}$ as seen from the ground reference. A horizontal rod $\overline{EB}$ has the following angular motion about $\overline{CD}$:

$$\dot{\theta} = 0.3 \text{ rad/s} \qquad \ddot{\theta} = 0.4 \text{ rad/s}^2$$

Two spheres, each of mass 1 kg, are mounted on a rod turning about $\overline{EB}$ with the following angular motion

$$\dot{\beta} = 0.2 \text{ rad/s} \qquad \ddot{\beta} = 0.5 \text{ rad/s}^2$$

Also, the rod and the attached spherical masses advance toward $\overline{CD}$ at the following rate:

$$\dot{r} = -3 \text{ m/s} \qquad \ddot{r} = -2 \text{ m/s}^2$$

at a time when r = 0.25 m. Finally, at the instant of interest, the horizontal rod $\overline{EB}$ moves up along vertical rod $\overline{CD}$ with a speed of 0.5 m/s and a rate of change of speed of 0.2 m/s². What force must rod $\overline{EB}$ exert at point G at this instant *due only to the motion of the two spheres*?

Example 15.15 (Continued)

We will first consider the motion of the *center of mass* of the rotating spheres which clearly must be G. We now proceed to get the acceleration of G relative to XYZ (see Fig. 15.39).

Fix xyz to the vessel at A.
Fix XYZ to the ground.

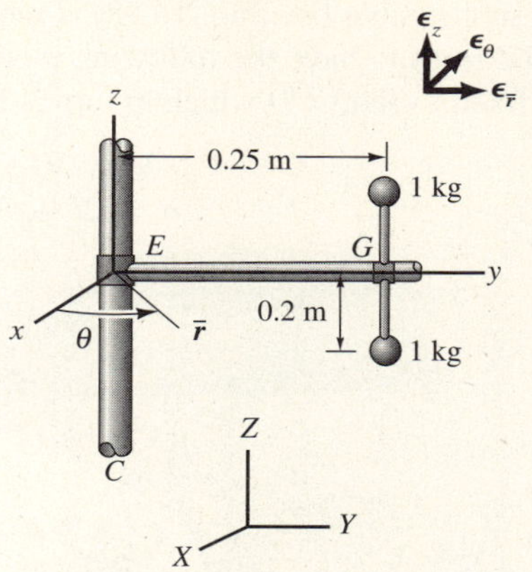

Figure 15.39. Reference xyz fixed to the vessel.

A. Motion of G relative to xyz (using cylindrical coordinates). Use Figs. 15.38 and 15.39.

$$\boldsymbol{\rho} = 0.25\boldsymbol{\epsilon}_{\bar{r}} = 0.25\boldsymbol{j} \text{ m}$$

$$\boldsymbol{V}_{xyz} = \dot{\bar{r}}\boldsymbol{\epsilon}_{\bar{r}} + \bar{r}\dot{\theta}\boldsymbol{\epsilon}_\theta + \dot{z}\boldsymbol{\epsilon}_z = -3\boldsymbol{\epsilon}_{\bar{r}} + (0.25)(0.3)\boldsymbol{\epsilon}_\theta + 0.5\boldsymbol{\epsilon}_z$$
$$= -3\boldsymbol{j} - 0.075\boldsymbol{i} + 0.5\boldsymbol{k} = -0.075\boldsymbol{i} - 3\boldsymbol{j} + 0.5\boldsymbol{k} \text{ m/s}$$

$$\boldsymbol{a}_{xyz} = (\ddot{\bar{r}} - \bar{r}\dot{\theta}^2)\boldsymbol{\epsilon}_{\bar{r}} + (\bar{r}\ddot{\theta} + 2\dot{\bar{r}}\dot{\theta})\boldsymbol{\epsilon}_\theta + \ddot{z}\boldsymbol{\epsilon}_z$$
$$= \left[-2 - (0.25)(0.3)^2\right]\boldsymbol{\epsilon}_{\bar{r}} + [(0.25)(0.4) + (2)(-3)(0.3)]\boldsymbol{\epsilon}_\theta + 0.2\boldsymbol{\epsilon}_z$$
$$= -2.023\boldsymbol{\epsilon}_{\bar{r}} - 1.7\boldsymbol{\epsilon}_\theta + 0.2\boldsymbol{\epsilon}_z = 1.7\boldsymbol{i} - 2.023\boldsymbol{j} + 0.2\boldsymbol{k} \text{ m/s}^2$$

B. Motion of xyz relative to XYZ

$$\dot{\boldsymbol{R}} = 3\boldsymbol{i} + 0.6\boldsymbol{j} \text{ m/s}$$
$$\ddot{\boldsymbol{R}} = 2\boldsymbol{i} + 3\boldsymbol{j} - 0.5\boldsymbol{k} \text{ m/s}^2$$
$$\boldsymbol{\omega} = 0.3\boldsymbol{k} \text{ rad/s}$$
$$\dot{\boldsymbol{\omega}} = \boldsymbol{0} \text{ rad/s}^2$$

We may now express $\boldsymbol{a}_{XYZ}$ for point G. Thus

$$\boldsymbol{a}_{XYZ} = \boldsymbol{a}_{xyz} + \ddot{\boldsymbol{R}} + \dot{\boldsymbol{\omega}} \times \boldsymbol{\rho} + 2\boldsymbol{\omega} \times \boldsymbol{V}_{xyz} + \boldsymbol{\omega} \times (\boldsymbol{\omega} \times \boldsymbol{\rho})$$
$$= (1.7\boldsymbol{i} - 2.023\boldsymbol{j} + 0.2\boldsymbol{k}) + (2\boldsymbol{i} + 3\boldsymbol{j} - 0.5\boldsymbol{k})\boldsymbol{0} \times \boldsymbol{\rho} + 2(0.3\boldsymbol{k})$$
$$\times (-0.075\boldsymbol{i} - 3\boldsymbol{j} + 0.5\boldsymbol{k}) + (0.3\boldsymbol{k}) \times (0.3\boldsymbol{k} \times 0.25\boldsymbol{j})$$
$$= 5.5\boldsymbol{i} + 0.9095\boldsymbol{j} - 0.3\boldsymbol{k} \text{ m/s}^2$$

Now we apply **Newton's law** to the mass center G at the instant of interest. Denoting the force from the rod AB onto G as $\boldsymbol{F}_{\text{ROD}}$, we get

$$\boldsymbol{F}_{\text{ROD}} - 2mg\boldsymbol{k} = 5.5\boldsymbol{i} + 0.9095\boldsymbol{j} - 0.3\boldsymbol{k}$$
$$\therefore \boldsymbol{F}_{\text{ROD}} = (2)(1)(9.81)\boldsymbol{k} + 5.5\boldsymbol{i} + 0.9095\boldsymbol{j} - 0.3\boldsymbol{k}$$

$$\boldsymbol{F}_{\text{ROD}} = 5.5\boldsymbol{i} + 0.9095\boldsymbol{j} + 19.32\boldsymbol{k} \text{ N}$$

This is our desired result.

PROBLEMS

15.91. A truck has a speed V of 9 m/s and an acceleration $\dot{V}$ of 1.34 m/s^2 at time t. A cylinder of radius equal to 0.6 m is rolling without slipping at time t such that relative to the truck it has an angular speed ω_1 and angular acceleration $\dot{\omega}_1$ of 2 rad/s and 1 rad/s^2, respectively. Determine the velocity and acceleration of the center of the cylinder relative to the ground.

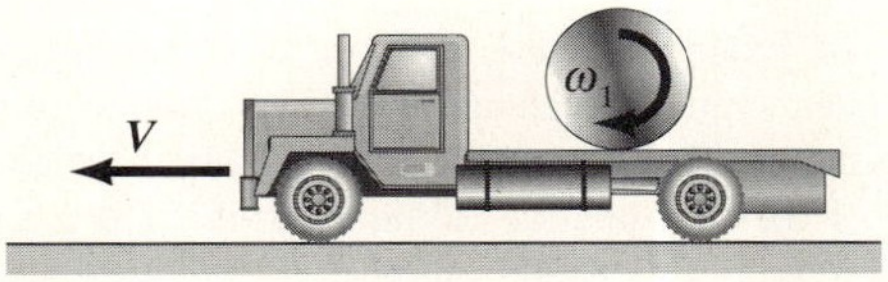

Figure P.15.91.

15.92. A wheel rotates with an angular speed ω_2 of 5 rad/s relative to a platform, which rotates with a speed ω_1 of 10 rad/s relative to the ground as shown. A collar moves down the spoke of the wheel, and, when the spoke is vertical, the collar has a speed of 6 m/s, an acceleration of 3 m/s^2 along the spoke, and is positioned 0.3 m from the shaft centerline of the wheel. Compute the velocity and acceleration of the collar relative to the ground at this instant. Fix xyz to platform and use cylindrical coordinates.

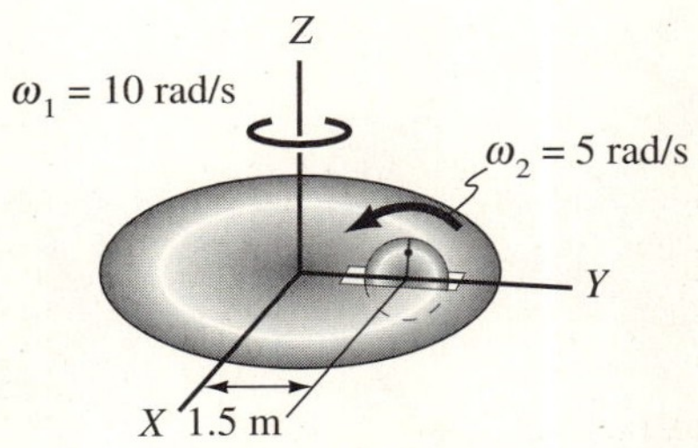

Figure P.15.92.

15.93. In Problem 15.71, determine the acceleration of the particle at the instant of interest.

15.94. In Problem 15.72, find the acceleration of the particle P relative to the ground reference.

15.95. In Problem 15.74, find the acceleration of point A relative to the ground.

15.96. In Problem 15.79, find the acceleration of the tip of the propeller relative to the ground reference. Take the yaw rotation to be zero and the loop rotation radius r to be 500 m.

15.97. In Problem 15.80, find the acceleration of point B relative to the ground.

15.98. In Problem 15.80, find the acceleration of point B relative to the ground for the following data at the instant of interest shown in the diagram.

$$\omega_1 = 0.2 \text{ rad/s}$$
$$\dot{\omega}_1 = -0.1 \text{ rad/s}^2$$
$$\omega_2 = 0.4 \text{ rad/s}$$
$$\dot{\omega}_2 = 0.3 \text{ rad/s}^2$$

15.99. In Problem 15.82, determine the acceleration of the top tip of the gun relative to the ground.

15.100. In Problem 15.74, find the acceleration of point A relative to the ground for the configuration shown. Take ω_1 = 2 rad/s, $\dot{\omega}_1$ = 3 rad/s^2, ω_2 = 0.1 rad/s, and $\dot{\omega}_2$ = 2 rad/s^2. How many g's of acceleration is this point subject to?

15.101. In Problem 15.81, find the acceleration of D relative to the ground. [*Hint:* Use two position vectors to get $\boldsymbol{\rho}$.]

15.102. Find the acceleration of gear tooth A relative to the ground in Problem 15.84.

15.103. In Problem 15.92, the wheel accelerates at the instant under discussion with 5 rad/s^2 relative to the platform, and the platform increases its angular speed at 10 rad/s^2 relative to the ground. Find the velocity and acceleration of the collar relative to the ground.

15.104. As with the velocity equation 15.20, we can easily show that Eq. 15.7, relating accelerations between two points on a rigid body, is actually a special case of Eq. 15.24. Thus, consider the diagram showing a rigid body moving arbitrarily relative to XYZ. Choose two points a and b in the body and embed a reference xyz in the body with the origin at a. Now express the acceleration of point b as seen from the two references. Show how this equation can be reformulated as Eq. 15.7.

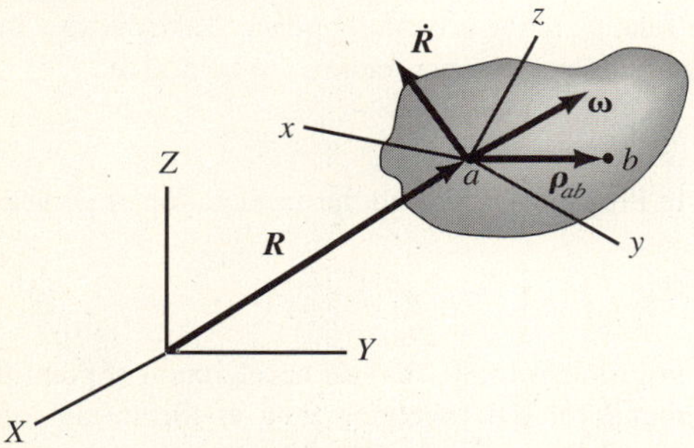

Figure P.15.104.

15.105. Solve Problem 15.81 for the following data:

$$\begin{aligned} \omega_1 &= 0.3 \text{ rad/s} \\ \dot{\omega}_1 &= 0.2 \text{ rad/s}^2 \\ \omega_2 &= 0.40 \text{ rad/s} \\ \dot{\omega}_2 &= 0.10 \text{ rad/s}^2 \\ \omega_3 &= 0.15 \text{ rad/s} \\ \dot{\omega}_3 &= -0.2 \text{ rad/s}^2 \end{aligned}$$

15.106. In Problem 15.82, find the component of acceleration of the projectile relative to the ground which is normal to the gun barrel at the instant that the projectile just leaves the barrel. Use the following data:

$$\begin{aligned} \omega_1 &= 0.3 \text{ rad/s} \\ \dot{\omega}_1 &= 0.2 \text{ rad/s}^2 \\ \omega_2 &= -0.6 \text{ rad/s} \\ \dot{\omega}_2 &= -0.4 \text{ rad/s}^2 \end{aligned}$$

15.107. In Problem 15.88, find the magnitude of the acceleration of collar C relative to the ground for the following data at the instant shown:

$$\begin{aligned} \dot{\theta} &= 2 \text{ rad/s} \\ \ddot{\theta} &= 4 \text{ rad/s}^2 \end{aligned}$$

15.108. A truck is moving at a constant speed $V = 1.7$ m/s at time t. The truck loading compartment has at this instant a constant angular speed $\dot{\theta}$ of 0.1 rad/s at an angle $\theta = 45°$. A cylinder of radius 300 mm rolls relative to the compartment at a speed ω_1 of 1 rad/s, accelerating at a rate $\dot{\omega}_1$ of 0.5 rad/s^2 at time t. What are the velocity and acceleration of the center of the cylinder relative to the ground at time t? The distance d at time t is 5 m.

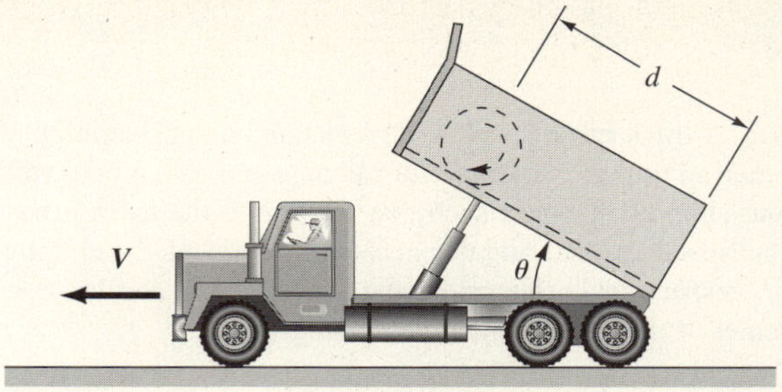

Figure P.15.108.

***15.109.** In Example 15.14, suppose at the instant of interest that there is an angular acceleration $\dot{\omega}_2 = 0.3$ rad/s^2 of the cockpit relative to the arm and that there is an angular acceleration $\dot{\omega}_3 = 0.2$ rad/s^2 of the seat relative to the cockpit. Find the number of g's to which the pilot's head is subjected. Follow analysis I in the example. [*Hint:* The angular velocity $\boldsymbol{\omega}_3$ can always be expressed as $\omega_3\hat{\mathbf{c}}$, where $\hat{\mathbf{c}}$ is a unit vector *fixed to the cockpit* having a direction along the x axis at the instant of interest.]

15.110. A ferris wheel is out of control. At the instant shown, it has an angular speed ω_1 equal to 0.2 rad/s and a rate of change of angular speed $\dot{\omega}_1$ of 40 mrad/s^2 relative to the ground. At this instant a "chair" shown in the diagram has an angular speed ω_2 relative to the ferris wheel equal to 0.25 rad/s and a rate change of speed $\dot{\omega}_2$, again relative to the ferris wheel, equal to 30 mrad/s^2. In the figure, we have shown details of the passenger at this instant. Note that the hinge of the seat is at A. How many g's of acceleration is the passenger's head subject to?

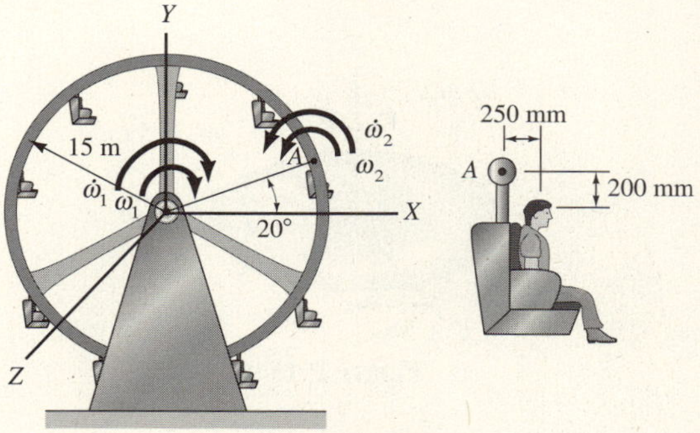

Figure P.15.110.

15.111. A shaft BC rotates relative to platform A at a speed $\omega_1 = 0.34$ rad/s. A rod is welded to BC and is vertical at the instant of interest. A tube is fixed to the vertical rod in which a small piston head is moving relative to the tube at a speed V of 3 m/s with a rate of change of speed $\dot{V}$ of 0.4 m/s^2. The platform A has an angular velocity relative to the ground given as $\omega_2 = 0.8$ rad/s with a rate of change of speed of 0.5 rad/s^2. Find the acceleration vector of the piston head relative to the ground.

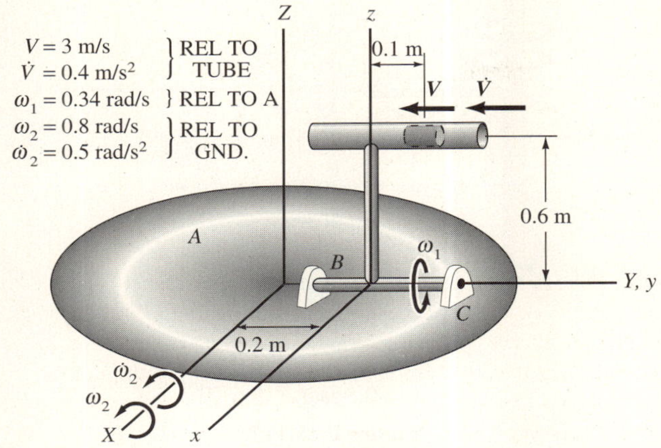

Figure P.15.111.

15.112. A communications satellite has the following motion relative to an inertial reference XYZ.

$$\boldsymbol{\omega}_1 = 3\boldsymbol{i} + 4\boldsymbol{j} + 10\boldsymbol{k} \text{ rad/s} \qquad \dot{\boldsymbol{\omega}}_1 = 2\boldsymbol{i} + 3\boldsymbol{k} \text{ rad/s}^2$$
$$\boldsymbol{V}_0 = \boldsymbol{a}_0 = \boldsymbol{0}$$

A wheel at A is rotating relative to the satellite at a constant speed $\omega_2 = 5$ rad/s. What is the acceleration of point D on the wheel relative to XYZ at the instant shown? The following additional data apply:

$$OA = 1 \text{ m} \quad AD = 0.2 \text{ m}$$

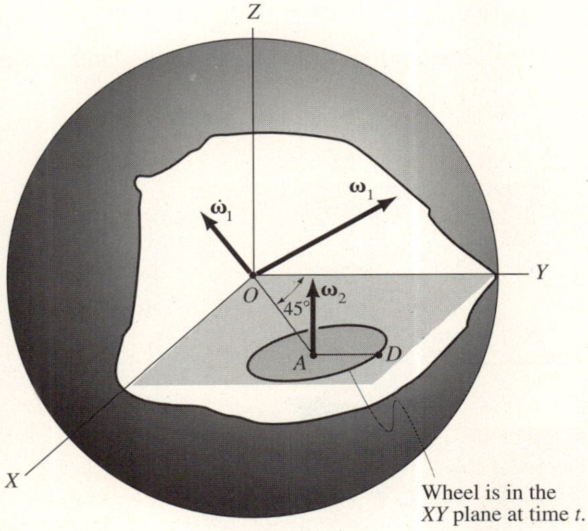

Figure P.15.112.

15.113. A particle moves in a slot of a gear with speed $V = 2$ m/s and a rate of change of speed $\dot{V} = 1.2$ m/s^2 both relative to the gear. Find the acceleration vector for the particle at the configuration shown relative to the ground reference XYZ.

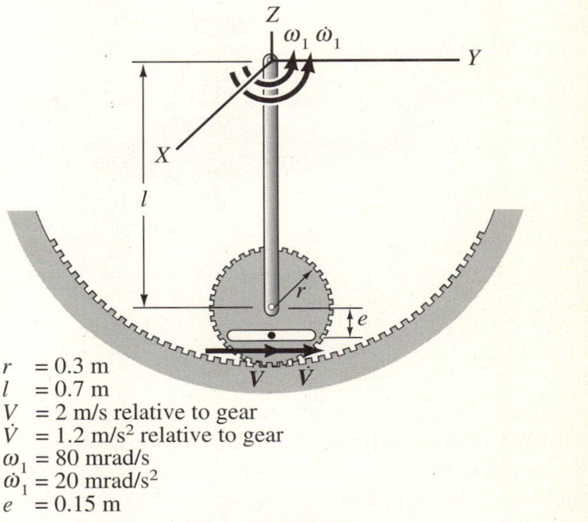

Figure P.15.113.

15.114. A submarine is undergoing an evasive maneuver. At the instant of interest, it has a speed $V_s = 10$ m/s and an acceleration $a_s = 15$ m/s^2 at its center of mass. It also has an angular velocity about its center of mass C of $\omega_1 = 0.5$ rad/s and an angular acceleration $\dot{\omega}_1 = 20$ mrad/s^2. Inside is a part of an inertial guidance system that consists of a wheel spinning with speed $\omega_2 = 20$ rad/s about a vertical axis of the ship. Along a spoke shown at the instant of interest, a particle is moving toward the center with r, $\dot{r}$ and $\ddot{r}$ as given in the diagram. What is the acceleration of the particle relative to inertial reference XYZ? Just write out the formulations for $\boldsymbol{a}_{XYZ}$ but do not carry out the cross products.

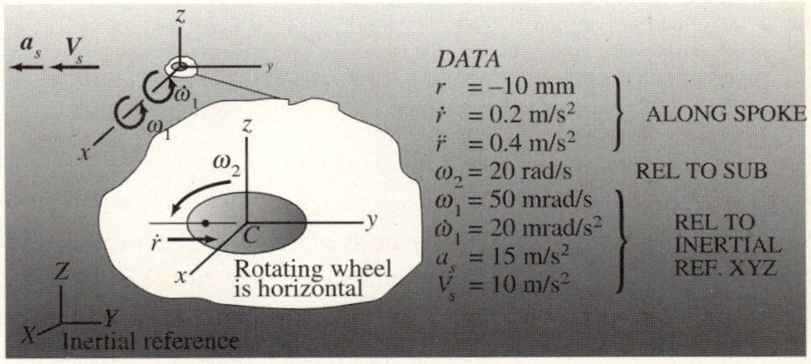

Figure P.15.114.

15.115. Find $\boldsymbol{V}$ and $\boldsymbol{a}$ of B relative to XYZ. The single blade portion AB is rotating as shown relative to the helicopter with speed ω_2 about an axis parallel to the X axis at the instant of interest while the entire blade system is rotating about the vertical axis with speed ω_1.

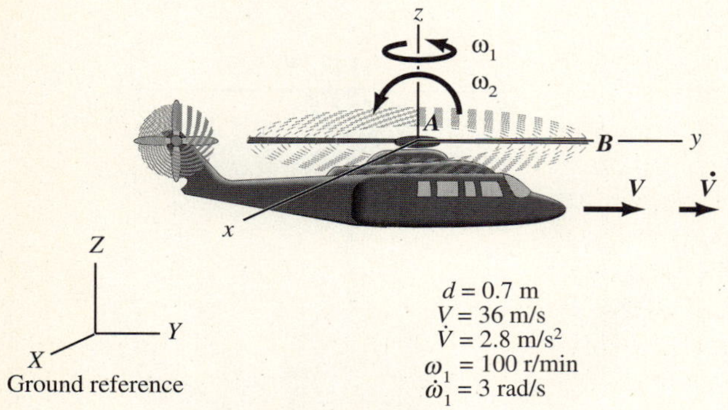

Figure P.15.115.

15.116. An F-16 fighter plane is moving at a constant speed of 220 m/s while undergoing a loop as shown in the diagram. At the instant of interest, it has an angular roll velocity ω_2 of 80 mrad/s relative to the ground. A solenoid is activated at this instant and moves the moving portion of a gate valve downward at a speed of 3 m/s with an acceleration of 5 m/s² all relative to the plane. What is the total external force on the moving portion of the gate valve if it is 1 m above the axis of roll at the instant of interest? The moving portion of the valve has a weight of 10 N.

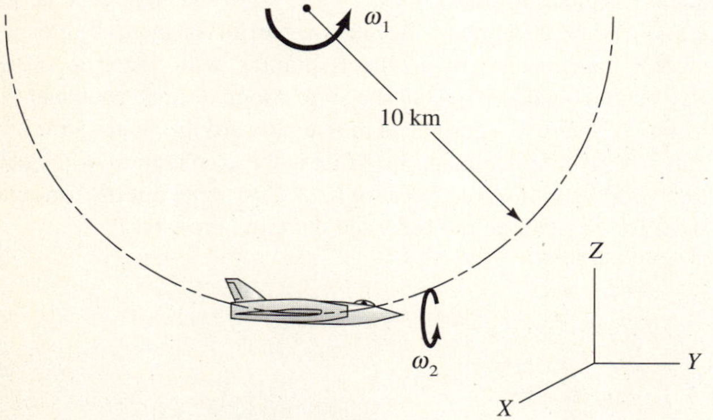

Figure P.15.116.

15.117. A robot moves a body held by its "jaws" G as shown in the diagram. What is the velocity and acceleration of point A at the instant shown relative to the ground? Arm EH is welded to the vertical shaft MN. Arm HKG is one rigid member which rotates about EH. How do you want to fix xyz?

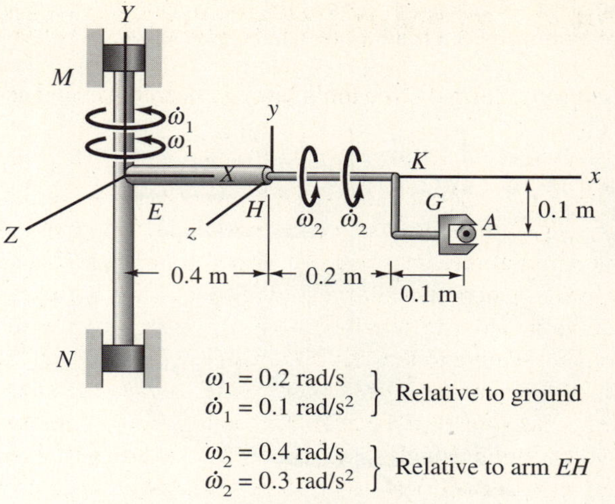

Figure P.15.117.

15.118. The turret of the main gun of a destroyer has at time t an angular velocity $\omega_1 = 2$ rad/s and a rate of change of angular velocity $\dot{\omega}_1 = 3$ rad/s² both relative to the ship. The gun barrel has $\omega_2 = 0.5$ rad/s and $\dot{\omega}_2 = 0.3$ rad/s² relative to the turret.

(a) Find the acceleration of the tip A of the gun at time t relative to the destroyer.

(b) If the destroyer has a translational acceleration relative to land equal to

$$\boldsymbol{a}_{\text{Destroyer}} = 0.05\boldsymbol{i} + 0.26\boldsymbol{j} - 2.2\boldsymbol{k} \text{ m/s}^2$$

what is the acceleration of A relative to the land at time t?

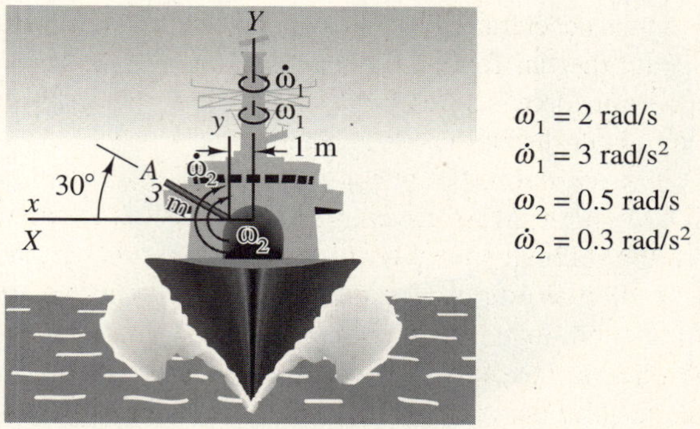

Figure P.15.118.

*15.9 A New Look at Newton's Law

The proper form of Newton's law has been presented as

$$\boldsymbol{F} = m\boldsymbol{a}_{XYZ} \tag{15.25}$$

where the acceleration is measured relative to an inertial reference. There are times when the motion of a particle is known and makes sense only relative to a noninertial reference. Such a case would arise, for example, in an airplane or rocket, where machine elements must move in a certain way relative to the vehicle in order to function properly. Therefore, the motion of the machine element relative to the vehicle is known. If, however, the vehicle is undergoing a severe maneuver relative to inertial space, we cannot use Eq. 15.25 with the acceleration of the machine element measured relative to the vehicle. This is so since the vehicle is not at that instant an inertial reference, and to disregard this fact will lead to erroneous results. In such problems, the motion of the vehicle may be known relative to inertial space, and we can employ to good advantage the multireference analysis of the previous section. Attaching the reference xyz to the vehicle and XYZ to inertial space, we can then use Newton's law in the following way:

$$\boldsymbol{F} = m\left[\boldsymbol{a}_{xyz} + \ddot{\boldsymbol{R}} + 2\boldsymbol{\omega} \times \boldsymbol{V}_{xyz} + \dot{\boldsymbol{\omega}} \times \boldsymbol{\rho} + \boldsymbol{\omega} \times (\boldsymbol{\omega} \times \boldsymbol{\rho})\right] \tag{15.26}$$

Clearly, the bracketed expression is the required quantity $\boldsymbol{a}_{XYZ}$ needed for Newton's law. It is the usual practice to write Eq. 15.26 in the following form:

$$\boldsymbol{F} - m\left[\ddot{\boldsymbol{R}} + 2\boldsymbol{\omega} \times \boldsymbol{V}_{xyz} + \dot{\boldsymbol{\omega}} \times \boldsymbol{\rho} + \boldsymbol{\omega} \times (\boldsymbol{\omega} \times \boldsymbol{\rho})\right] = m\boldsymbol{a}_{xyz} \tag{15.27}$$

This equation may now be considered as Newton's law written for a *noninertial* reference xyz. The terms $-m\ddot{\boldsymbol{R}}$, $-m(2\boldsymbol{\omega} \times \boldsymbol{V}_{xyz})$, and so on, are then considered as forces and are termed *inertial forces*. Thus, we can take the viewpoint that for a noninertial reference, xyz, we can still say force $\boldsymbol{F}$ equals mass times acceleration, $\boldsymbol{a}_{xyz}$, provided that we include with the applied force $\boldsymbol{F}$, all the inertial forces. Indeed, we shall adopt this viewpoint in this text. The inertial force $-2m\boldsymbol{\omega} \times \boldsymbol{V}_{xyz}$ is the very interesting *Coriolis force*, which we shall later discuss in some detail.

The inertial forces result in baffling actions that are sometimes contrary to our intuition. Most of us during our lives have been involved in actions where the reference used (knowingly or not) has been with sufficient accuracy an inertial reference, usually the earth's surface. We have, accordingly, become conditioned to associating an acceleration proportional to, and in the same direction as, the applied force. Occasions do arise when we find ourselves relating our motions to a reference that is highly noninertial. For example, fighter pilots and stunt pilots carry out actions in a cockpit of a plane while the plane is undergoing severe maneuvers. Unexpected results frequently occur for flyers if they use the cockpit interior as a reference for their actions. Thus, to move their hands from one position to another relative to the cockpit sometimes requires an exertion that is not the one anticipated, causing considerable confusion. The next example will illustrate this, and the sections that follow will explore further some of these interesting effects.[13]

[13]At this juncture we should remind ourselves that "physical feel" or "intuition" is really a direct consequence of past experiences. For this reason, many of the things you will later formally learn will initially be at variance with your physical feel and intuition. Thus, certain phenomena occurring in supersonic fluid flow will seem very strange, since our direct experience with fluid flow (faucets, swimming, and so on) has been entirely subsonic. Because you have not moved with speeds approaching the speed of light and because you have not been prowling around the nucleus of an atom, you will find the tenets of relativity theory and quantum mechanics absolutely bizarre. Should you have little feel for the Coriolis force at this time, do not be unduly concerned (unless you have spent a lot of time moving about high-speed merry-go-rounds). We must in such instances rely on the theory. Working with the theory, we can often build up a strong "physical feel" in the new areas.

Example 15.16

The plan view of a rotating platform is shown in Fig. 15.40. A man is seated at the position labeled A and is facing point O of the platform. He is carrying a mass of 0.3 kg at the rate of 3 m/s in a direction straight ahead of him (i.e., toward the center of the platform). If this platform has an angular speed of 10 rad/s and an angular acceleration of 5 rad/s^2 relative to the ground at this instant, what force $\boldsymbol{F}$ must he exert to cause the mass to accelerate at 1.5 m/s^2 toward the center?

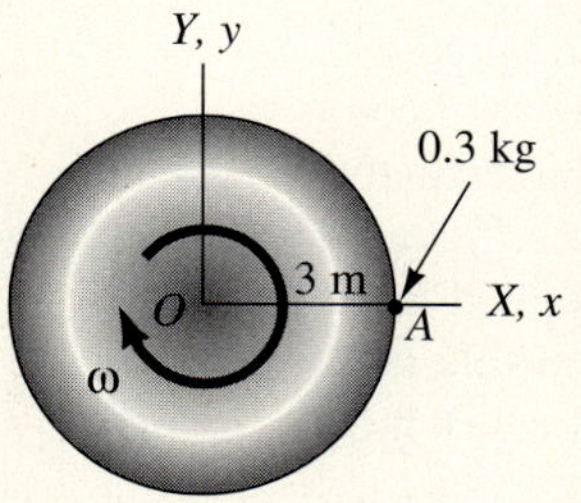

Figure 15.40. Rotating platform.

For purposes of determining inertial forces, we proceed as follows:

Fix xyz to platform.
Fix XYZ to ground.

Example 15.16 (Continued)

A. Motion of mass relative to *xyz* reference

$$\boldsymbol{\rho} = 3\boldsymbol{i}\ \text{m},\quad \boldsymbol{V}_{xyz} = -3\boldsymbol{i}\ \text{m/s},\quad \boldsymbol{a}_{xyz} = -\ 1.5\boldsymbol{i}\ \text{m/s}^2$$

B. Motion of xyz relative to *XYZ*

$$\dot{\boldsymbol{R}} = \boldsymbol{0},\quad \ddot{\boldsymbol{R}} = \boldsymbol{0},\quad \omega = -10\boldsymbol{k}\ \text{rad/s},\quad \dot{\boldsymbol{\omega}} = -5\boldsymbol{k}\ \text{rad/s}^2$$

Hence,

$$\boldsymbol{a}_{XYZ} = -\ 1.5\boldsymbol{i} + 2(-10\boldsymbol{k})\times(-3\boldsymbol{i}) + (-5\boldsymbol{k})\times 3\boldsymbol{i} + (-10\boldsymbol{k})\times(-10\boldsymbol{k}\times 3\boldsymbol{i})$$

Therefore,

$$\boldsymbol{a}_{XYZ} = -1.5\boldsymbol{i} + (60\boldsymbol{j} - 30\boldsymbol{j} - 300\boldsymbol{i})$$

Employing **Newton's law** (Eq. 15.27) for the mass, we get

$$\boldsymbol{F} - 0.3(60\boldsymbol{j} - 15\boldsymbol{j} - 300\boldsymbol{i}) = 0.3(-1.5\boldsymbol{i})$$

Solving for **F**, we get

$$\boldsymbol{F} = 13.5\boldsymbol{j}\ -90.45\boldsymbol{i}\ \text{N}$$

This force $\boldsymbol{F}$ is the *total* external force on the mass. Since the man must exert this force and also withstand the pull of gravity (the weight) in the $-\boldsymbol{k}$ direction, the force exerted by the man on the mass is

$$\boldsymbol{F}_{\text{man}} = 13.5\boldsymbol{j} - 90.45\boldsymbol{i} + 0.3g\boldsymbol{k}\ \text{N} \tag{a}$$

If the platform were *not* rotating at all, it could serve as an inertial reference. Then, we would have for the total external force $\boldsymbol{F}'$:

$$\boldsymbol{F}' = 0.3(-1.5\boldsymbol{i}) = 0.45\boldsymbol{i}\ \text{N}$$

The force exerted by the man, $\mathbf{F}'_{\text{man}}$, is then

$$\boldsymbol{F}'_{\text{man}} = -0.45\boldsymbol{i} + 0.3g\boldsymbol{k}\ \text{N} \tag{b}$$

This force is considerably different from that given in Eq. (a).

As a matter of interest, we note that aviators of World War I were required to carry out such maneuvers on a rapidly rotating and accelerating platform so as to introduce them safely to these "peculiar" effects.

*15.10 The Coriolis Force

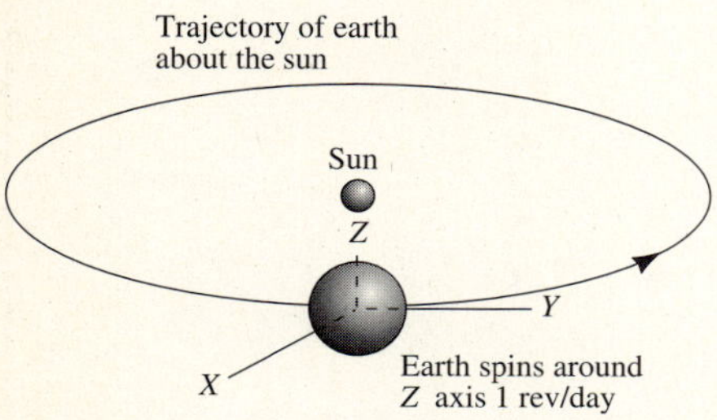

Figure 15.41. Proposed inertial reference.

Of great interest is the Coriolis force, defined in Section 15.9, particularly as it relates to certain terrestrial actions. For many of our problems, the earth's surface serves with sufficient accuracy as an inertial reference. However, where the time interval of interest is large (such as in the flight of rockets, or the flow of rivers, or the movement of winds and ocean currents), we must consider such a reference as noninertial in certain instances and accordingly, when using **Newton's law**, we must include some or all of the inertial forces given in Eq. 15.27. For such problems (as you will recall from Chapter 11, we often use an inertial reference that has an origin at the center of the earth (see Fig. 15.41) with the Z axis collinear with the N-S axis of the earth and moving such that the earth rotates one revolution per 24 hr relative to the reference. Thus, the reference approaches a translatory motion about the sun. To a high degree of accuracy, it is an inertial reference.

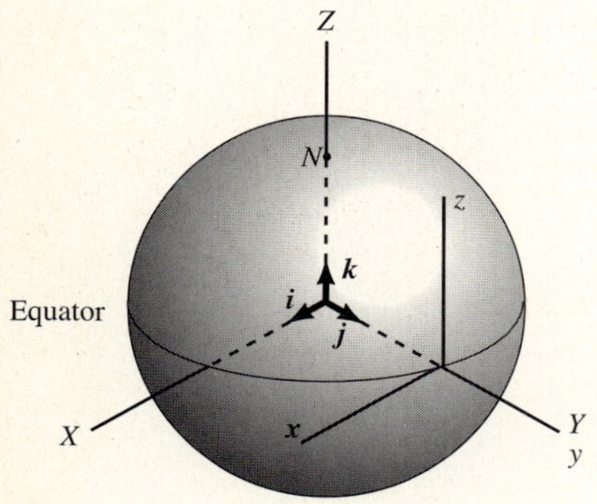

Figure 15.42. xyz fixed to earth.

We start by considering particles that are stationary relative to the earth. We choose a reference xyz fixed to the earth at the equator as shown in Fig. 15.42. The angular velocity of xyz fixed anywhere on the earth's surface can readily be evaluated as follows:

$$\boldsymbol{\omega} = \frac{2\pi}{(24)(3600)}\boldsymbol{k} = 72.7\,\boldsymbol{k}\ \ \mu\text{rad/s}$$

Newton's law, in the form of Eq. 15.27, for a "stationary" particle positioned at the origin of xyz simplifies to

$$\boldsymbol{F} - m\ddot{\boldsymbol{R}} = \boldsymbol{0} \tag{15.28}$$

since $\boldsymbol{\rho}$, $\boldsymbol{V}_{xyz}$, and $\boldsymbol{a}_{xyz}$ are zero vectors. Let us next evaluate the inertial force, $-m\ddot{\boldsymbol{R}}$, for the particle, using $R = 6.37$ Mm:

$$\begin{aligned}-m\ddot{\boldsymbol{R}} &= -m(-|\boldsymbol{R}|\omega^2\boldsymbol{j}) = m(6.37\times10^6)(72.7\times10^{-6})^2\boldsymbol{j}\\ &= m(33.67)\boldsymbol{j}\text{ mN}\end{aligned}$$

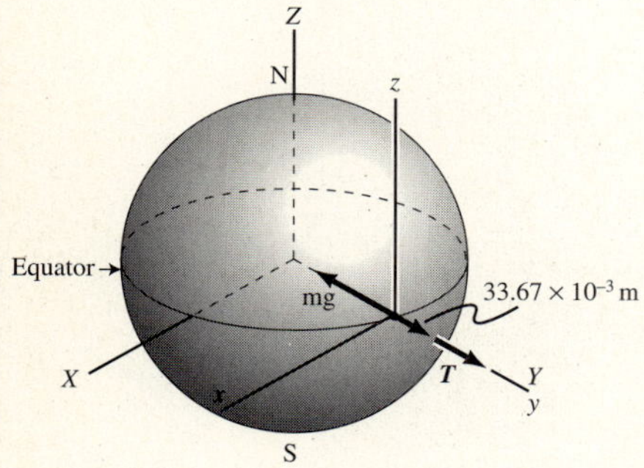

Figure 15.43. Plumb bob at equator.

Clearly, this is a "centrifugal force," as we learned in physics. Note in Fig. 15.43 that the direction of this force is collinear with the gravitational force on a particle, but with opposite sense. Note further that the centrifugal force has a magnitude that is $(33.67 \times 10^{-3}) \times 100 = 0.34\%$ of the gravitational force at the indicated location. Thus, clearly, in the usual engineering problems, such effects are neglected.

Assume that the particle is restrained from resting on the surface of the earth by a flexible cord. In accordance with Eq. 15.28, the external force $\boldsymbol{F}$ (which includes gravitational attraction and the force from the cord) and the centrifugal force $-m\ddot{\boldsymbol{R}}$ add up to zero, and hence these forces are in equilibrium. They are shown in Fig. 15.43 in which T represents the contribution of the cord. Clearly, a force T radially out from the center of the earth will restrain the particle, and so the direction of the flexible cord will point toward the center of the earth. On the other hand, at a nonequatorial location this will not be true. The gravity force points toward the center of the earth (see Fig. 15.44), but the centrifugal force—now having the value $m[R(\sin\theta)\omega^2]$—points radially out from the Z axis, and thus T, the restraining force, must be inclined somewhat from a direction toward the center of the earth. Therefore, except at the equator or at the poles (where the centrifugal force is zero), a *plumb bob* does not point directly toward the center of the earth. This deviation is very small and is negligible for most but not all engineering work.

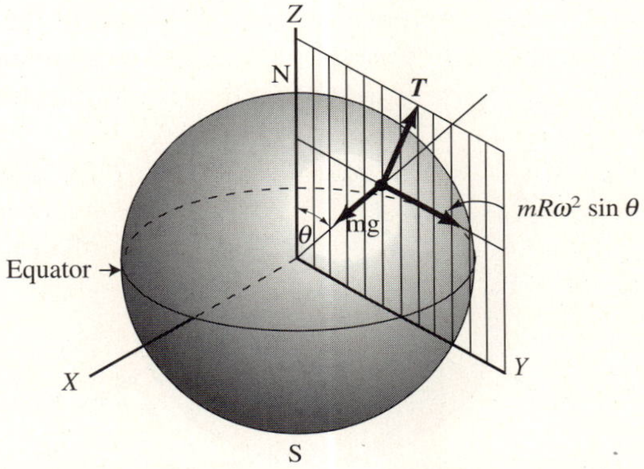

Figure 15.44. Plumb bob does not point to center of earth.

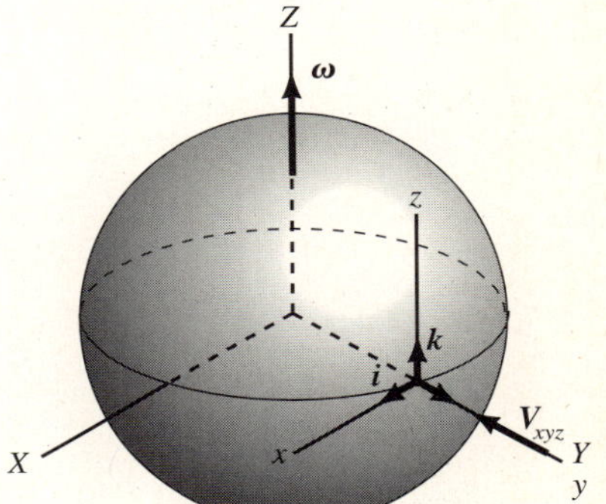

Figure 15.45. Free fall at the equator.

Consider now a body that is held above the earth's surface so as to always be above the same point on the earth's surface. (The body thus moves with the earth with the same angular motion.) If the body is released we have what is called a *free fall*. The body will attain initially a downward velocity $\boldsymbol{V}_{xyz}$ relative to the earth's surface (see Fig. 15.45). Now in addition to a centrifugal force described earlier, we have a *Coriolis force* given as

$$\boldsymbol{F}_{\text{Coriolis}} = -2m\boldsymbol{\omega} \times \boldsymbol{V}_{xyz}$$

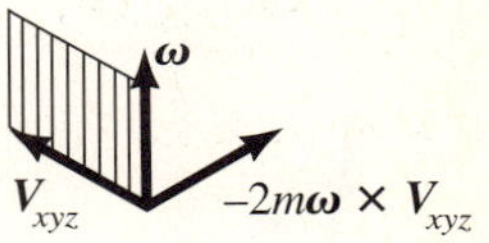

Figure 15.46. Direction of Coriolis force.

In Fig. 15.46 we have shown the $\boldsymbol{\omega}$ and $\boldsymbol{V}_{xyz}$ vectors. The Coriolis force must point to the right as you should verify (do not forget the minus sign). If we dropped a mass from a position in xyz above a target, therefore, the mass as a result of the Coriolis force would curve slightly away from the target (see Fig. 15.47) even if there were no friction, wind, etc., to complicate matters. Furthermore, the induced motion in the x direction itself induces Coriolis-force components of a smaller order in the y direction, and so forth. You will surely begin to appreciate how difficult a "free fall" can really become when great precision is attempted.

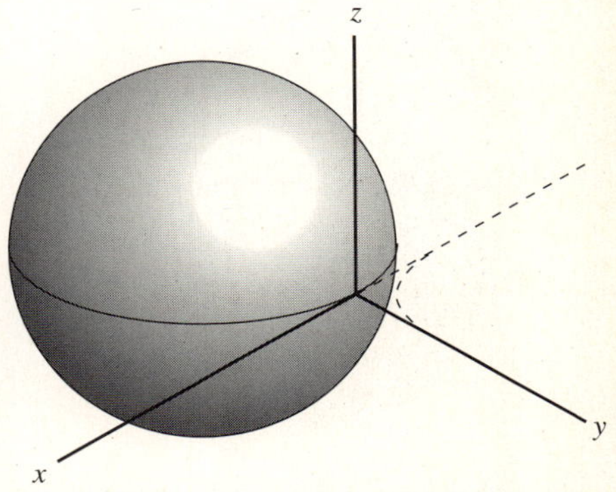

Figure 15.47. Primary Coriolis effect on free fall.

Finally, consider a current of air or a current of water moving in the Northern Hemisphere. In the absence of a Coriolis force, the fluid would move in the direction of the pressure drop. In Fig. 15.48 the pressure drop has been shown for simplicity along a meridian line pointing toward the equator. For fluid motion in this direction, a Coriolis force will be present in the negative y direction and so the fluid will follow the dashed-line path BA. The prime induced motion is to the right of the direction of flow developed by the pressure alone. By similar argument, you can demonstrate that, in the Southern Hemisphere, the Coriolis force induces a motion to the left of the flow that would be present under the action of the pressure drop alone. Such effects are of significance in meteorology and oceanography.[14]

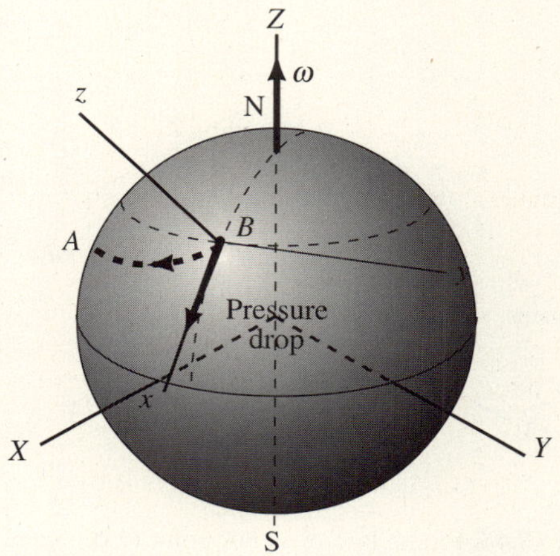

Figure 15.48. Coriolis effect on wind.

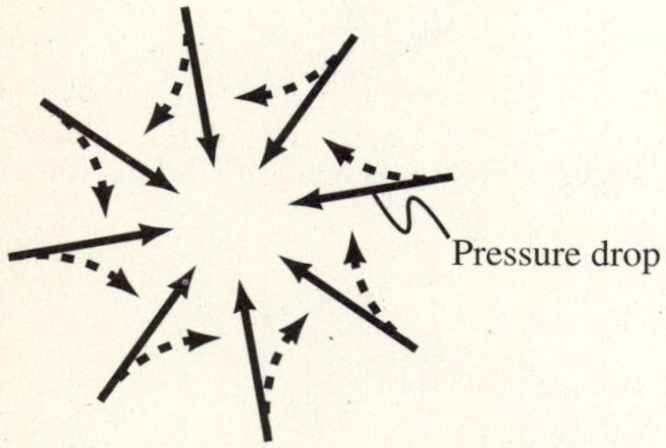

Figure 15.49. Beginning of whirlpool.

The conclusions in the preceding paragraph explain why cyclones and whirlpools rotate in a counterclockwise direction in the Northern Hemisphere and a clockwise direction in the Southern Hemisphere. In order to start, a whirlpool or cyclone needs a low-pressure region with pressure increasing radially outward. The pressure drops are shown as full lines in Fig. 15.49. For such a pressure distribution, the air will begin to move radially inward. As this happens, the Coriolis force causes the fluid in the Northern Hemisphere to swerve to the right of its motion, as indicated by the dashed lines. This result is the beginning of a counterclockwise motion. You can readily demonstrate that in the Southern Hemisphere a clockwise rotation will be induced.

[14]Keep in mind that the Coriolis force in these situations is small but, because it persists during long time intervals and because the resultant of other forces is often also small, this force usually must carefully be taken into account in studies of meteorology and oceanography.

Note that the famous Gulf Stream going north from the Caribbean swings eastward toward the British Isles as a result of the Coriolis force. With the prevailing winds, this results in a more moderate climate for these lands.

PROBLEMS

15.119. A reference *xyz* is attached to a space probe, which has the following motion relative to an inertial reference *XYZ* at a time *t* when the corresponding axes of the references are parallel:

$$\ddot{\boldsymbol{R}} = 100\boldsymbol{j} \text{ m/s}^2$$
$$\boldsymbol{\omega} = 10\boldsymbol{i} \text{ rad/s}$$
$$\dot{\boldsymbol{\omega}} = -8\boldsymbol{k} \text{ rad/s}^2$$

If a force $\boldsymbol{F}$ given as

$$\boldsymbol{F} = 500\boldsymbol{i} + 200\boldsymbol{j} - 300\boldsymbol{k} \text{ N}$$

acts on a particle of mass 1 kg at position

$$\boldsymbol{\rho} = 0.5\boldsymbol{i} + 0.6\boldsymbol{j} \text{ m}$$

what is the acceleration vector relative to the probe? The particle has a velocity $\boldsymbol{V}$ relative to *xyz* of

$$\boldsymbol{V} = 10\boldsymbol{i} + 20\boldsymbol{j} \text{ m/s}$$

15.120. In the space probe of Problem 15.119, what must the velocity vector $\boldsymbol{V}_{xyz}$ of the particle be to have the acceleration

$$\boldsymbol{a}_{xyz} = 495.2\boldsymbol{i} + 100\boldsymbol{j} \text{ m/s}^2$$

if all other conditions are the same? Is there a component of $\boldsymbol{V}_{xyz}$ that can have any value for this problem?

15.121. A mass *A* of 0.1 kg is made to rotate at a constant angular speed of $\omega_2 = 15$ rad/s relative to a platform. This motion is in the plane of the platform, which, at the instant of interest, is rotating at an angular speed $\omega_1 = 10$ rad/s and decelerating at a rate of 5 rad/s^2 relative to the ground. If we neglect the mass of the rod supporting the mass *A*, what are the axial force and shear force at the base of the rod (i.e., at 0)? The rod at the instant of interest is shown in the diagram. The shear force is the total force acting on a cross-section of the member in a direction *tangent* to the section.

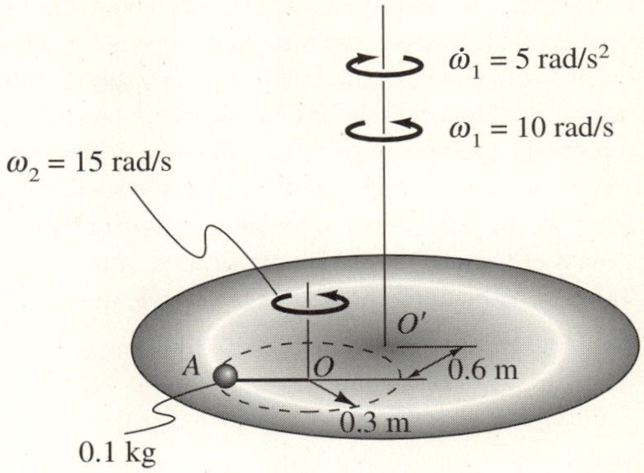

Figure P.15.121.

15.122. In Problem 15.91, what is the total external force acting on the cylinder for the case when

$$V = 1.5 \text{ m/s}$$
$$\dot{V} = -0.6 \text{ m/s}^2$$
$$\omega_1 = 2 \text{ rad/s}$$
$$\dot{\omega}_1 = 1 \text{ rad/s}^2?$$

The mass of the cylinder is 50 kg.

15.123. A truck is moving at constant speed *V* of 5 m/s. A crane *AB* is at time *t* at $\theta = 45°$ with $\dot{\theta} = 1$ rad/s and $\ddot{\theta} = 0.2$ rad/s^2. Also at time *t*, the base of *AB* rotates with speed $\omega_1 = 1$ rad/s relative to the truck. If *AB* is 10 m in length, what is the axial force along *AB* as a result of mass *M* of 50 kg at *B*?

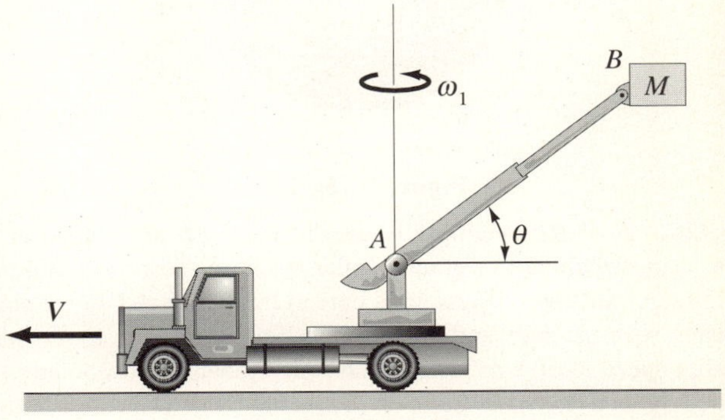

Figure P.15.123.

15.124. An exploratory probe shot from the earth is returning to the earth. On entering the earth's atmosphere, it has a constant angular velocity component ω_1 of 10 rad/s about an axis normal to the page and a constant component ω_2 of 50 rad/s about the vertical axis. The velocity of the probe at the time of interest is 1.3 km/s vertically downward with a deceleration of 160 m/s². A small sphere is rotating at $\omega_3 = 5$ rad/s inside the probe, as shown. At the time of interest, the probe is oriented so that the trajectory of the sphere in the probe is in the plane of the page and the arm is vertical. What are the axial force in the arm and the bending moment at its base (neglect the mass of the arm) at this instant of time, if the sphere has a mass of 300 g? (The bending moment is the couple moment acting on the cross section of the beam.)

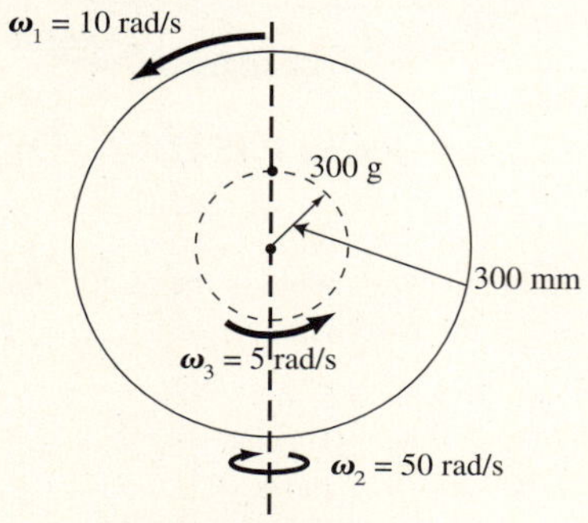

Figure P.15.124.

15.125. A river flows at 0.6 m/s average velocity in the Northern Hemisphere at a latitude of 40° in the north–south direction. What is the Coriolis acceleration of the water relative to the center of the earth? What is the Coriolis force on 1 kg of water?

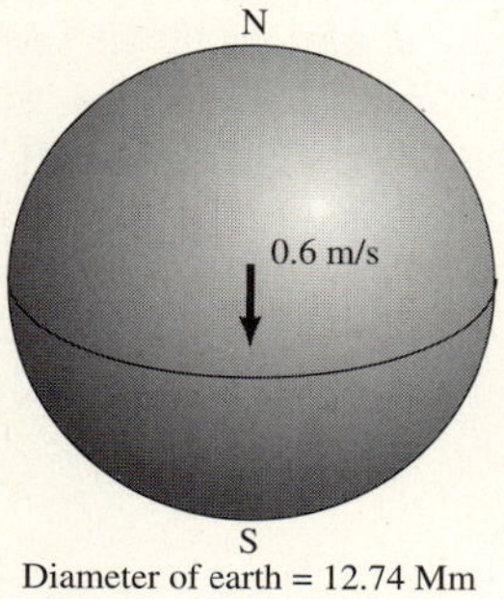

Figure P.15.125.

15.126. A clutch assembly is shown. Rods AB are pinned to a disc at B, which rotates at an angular speed $\omega_1 = 1$ rad/s and $\dot{\omega}_1 = 2$ rad/s² at time t. These rods extend through a rod EF, which rotates with the rods and at the same time is moving to the left with a speed V of 1 m/s. At the instant shown, corresponding to time t, what is the axial force on the member AB as a result of the motion of particle A having a mass of 0.6 kg?

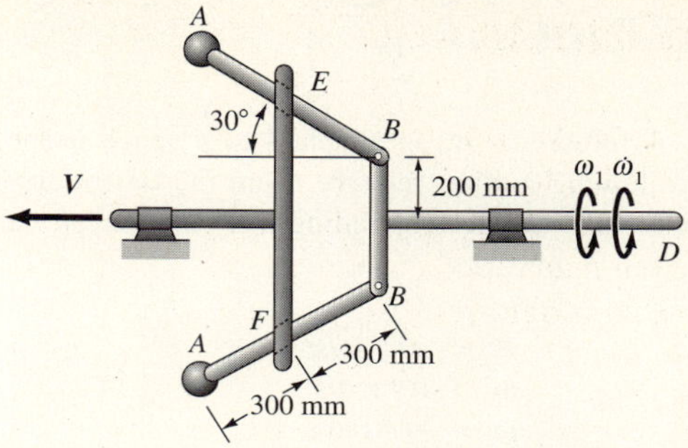

Figure P.15.126.

15.127. A flyball governor is shown. The weights C and D each have a mass of 200 g. At the instant of interest, $\theta = 45°$ and the system is rotating about axis AB at a speed ω_1 of 2 rad/s. At this instant, collar B is moving upward at a speed of 0.5 m/s. If we neglect the mass of the members, find the axial forces in the members at the instant of interest. What is the total shear force F_s on the members? (The shear force is the force component tangent to the cross section of the member.)

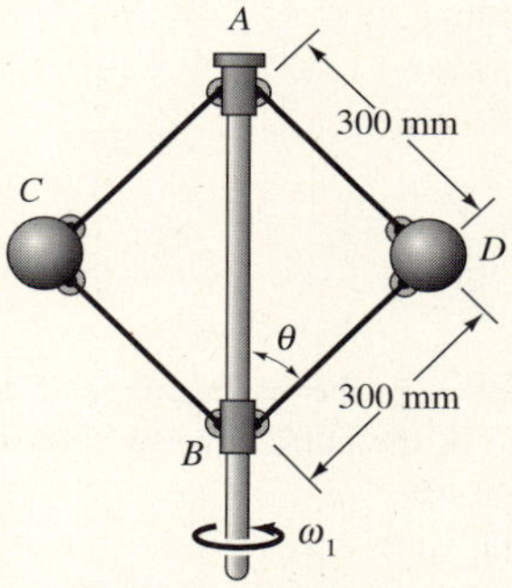

Figure P.15.127.

15.128. A man throws a ball of mass 85 g from one side of a rotating platform to a man diametrically opposite, as shown. What is the Coriolis acceleration and force on the ball? Relative to the platform, in what direction does the ball tend to go as a result of the Coriolis force?

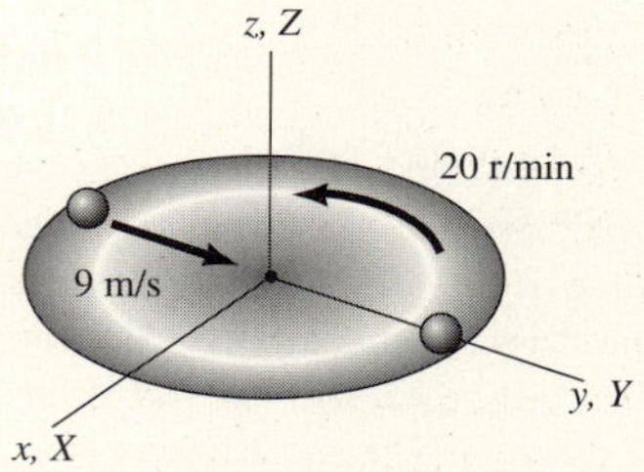

Figure P.15.128.

15.11 Closure

In this chapter, we first presented Chasles' theorem for describing the motion of a rigid body. Making use of Chasles' theorem for describing the motion of a reference *xyz* moving relative to a second reference *XYZ*, we presented next a simple but much used differentiation formula for vectors $\boldsymbol{A}$ *fixed* in the reference *xyz* or a rigid body. Thus,

$$\left(\frac{d\boldsymbol{A}}{dt}\right)_{XYZ} = \boldsymbol{\omega} \times \boldsymbol{A}$$

where $\boldsymbol{\omega}$ is the angular velocity of *xyz* or the rigid body relative to *XYZ*.

We next considered *two points fixed in a rigid body in the presence of a single reference*. We can relate velocities and accelerations of the points relative to the aforementioned reference as follows:

$$\begin{aligned} \boldsymbol{V}_b &= \boldsymbol{V}_a + \boldsymbol{\omega} \times \boldsymbol{\rho}_{ab} \\ \boldsymbol{a}_b &= \boldsymbol{a}_a + \dot{\boldsymbol{\omega}} \times \boldsymbol{\rho}_{ab} + \boldsymbol{\omega} \times (\boldsymbol{\omega} \times \boldsymbol{\rho}_{ab}) \end{aligned}$$

where $\boldsymbol{\omega}$ is the angular velocity of the body relative to the single reference. These relations can be valuable in studies of kinematics of machine elements.

We then considered *one particle in the presence of two references xyz and XYZ*. We expressed the velocity and acceleration as seen from the two references as follows:

$$\begin{aligned} \boldsymbol{V}_{XYZ} &= \boldsymbol{V}_{xyz} + \dot{\boldsymbol{R}} + \boldsymbol{\omega} \times \boldsymbol{\rho} \\ \boldsymbol{a}_{XYZ} &= \boldsymbol{a}_{xyz} + \ddot{\boldsymbol{R}} + 2\boldsymbol{\omega} \times \boldsymbol{V}_{xyz} + \dot{\boldsymbol{\omega}} \times \boldsymbol{\rho} + \boldsymbol{\omega} \times (\boldsymbol{\omega} \times \boldsymbol{\rho}) \end{aligned}$$

In computing $\boldsymbol{V}_{xyz}$, $\boldsymbol{a}_{xyz}$, $\dot{\boldsymbol{R}}$, and $\ddot{\boldsymbol{R}}$, we use the various techniques presented in Chapter 11 for computing the velocity and acceleration of a particle relative to a given reference. Thus, use can be made of Cartesian components, path components, and cylindrical components as presented in that chapter. We then explored some interesting and often unexpected effects that occur when we use a noninertial reference.

You will have occasion to use these two important formulations in your basic studies of solid and fluid mechanics as well as in your courses in kinematics of machines and machine design.

Now that we can express the motion of a rigid body in terms of a velocity vector $\dot{\boldsymbol{R}}$ and an angular velocity vector $\boldsymbol{\omega}$, our next job will be to relate these quantities with the forces acting on the body. You may recall from your physics course and from the end of Chapter 14 that for a body rotating about a fixed axis in an inertial reference, we could relate the torque T and the angular acceleration α as

$$T = I\alpha$$

where I is the mass moment of inertia of the body about the axis of rotation. In Chapter 16, we shall see that this motion is a special case of plane motion, which itself is a special case of general motion.

PROBLEMS

15.129. A light plane is circling an airport at constant elevation. The radius R of the path = 3 km and the speed of the plane is 35 m/s. The propeller of the plane is rotating at 100 r/min relative to the plane in a clockwise sense as seen by the pilot. What are $\boldsymbol{\omega}$, $\dot{\boldsymbol{\omega}}$, and $\ddot{\boldsymbol{\omega}}$ of the propeller as seen from the ground at the instant shown in the diagram?

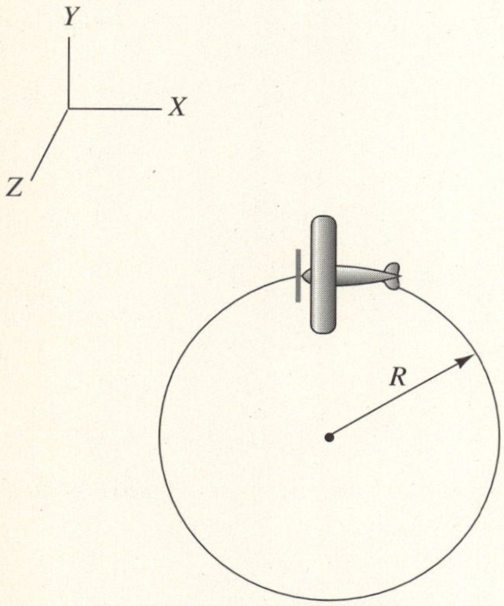

Figure P.15.129.

***15.130.** In Problem 15.7, find the angular acceleration $\dot{\boldsymbol{\omega}}$ of the disc for the configuration shown in the diagram, if, at the instant shown, the following data apply:

$$\omega_1 = 3 \text{ rad/s}$$
$$\dot{\omega}_1 = 2 \text{ rad/s}^2$$
$$\omega_2 = -10 \text{ rad/s}$$
$$\dot{\omega}_2 = -4 \text{ rad/s}^2$$

15.131. A slider A has at the instant of interest a speed V_A of 3 m/s with a deceleration of 2 m/s². Compute the angular velocity and angular acceleration of bar AB at the instant of interest. What is the position of the instantaneous axis of rotation of bar AB?

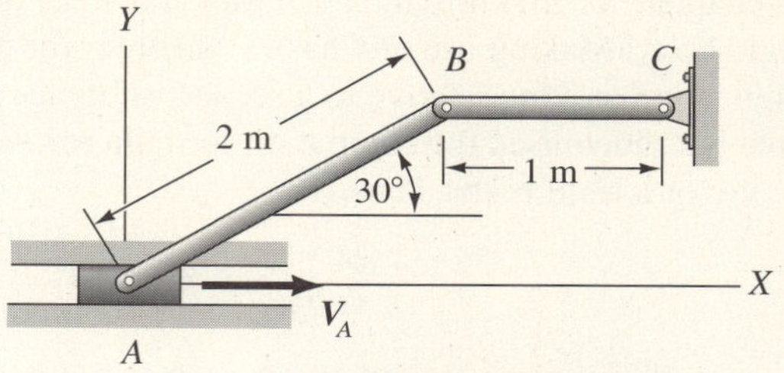

Figure P.15.131.

15.132. A cylinder C rolls without slipping on a half-cylinder D. Rod BA is 7 m long and is connected at A to a slider which at the instant of interest is moving in a groove at the speed V of 3 m/s and increasing its speed at the rate of 2 m/s². What is the angular speed and the angular acceleration of cylinder C relative to the ground?

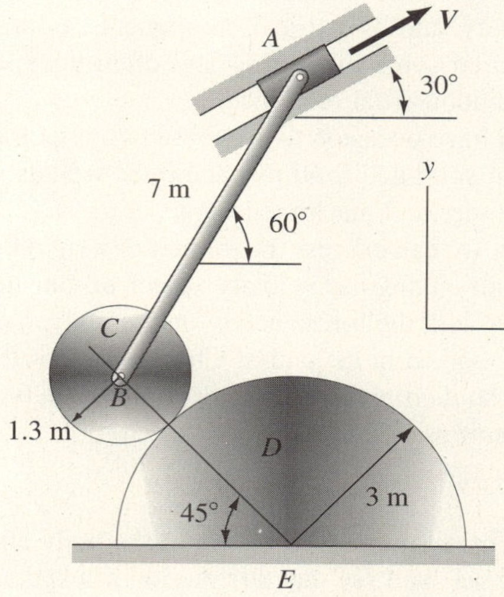

Figure P.15.132.

15.133. A wheel is rotating with a constant angular speed ω_1 of 10 rad/s relative to a platform, which in turn is rotating with a constant angular speed ω_2 of 5 rad/s relative to the ground. Find the velocity and acceleration relative to the ground at a point b on the wheel at the instant when it is directly vertically above point a.

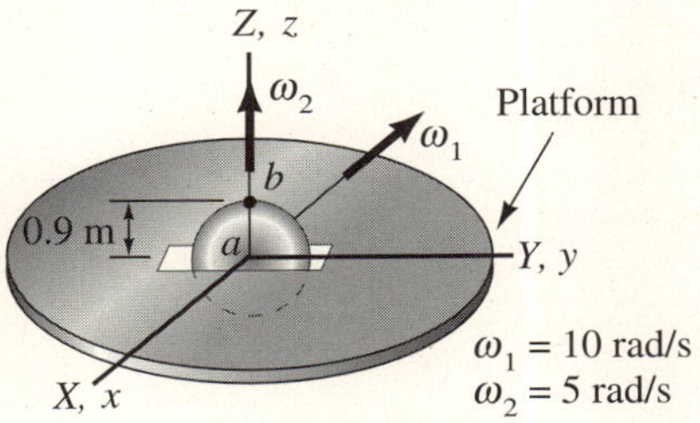

Figure P.15.133.

***15.134.** Solve Problem 15.133 for the case where ω_1 is increasing in value at the rate of 5 rad/s^2 and where ω_2 is increasing in value at the rate of 10 rad/s^2.

***15.135.** A barge is shown with a derrick arrangement. The main beam AB is 40 ft in length. The whole system at the instant of interest is rotating with a speed ω_1 of 1 rad/s and an acceleration $\dot{\omega}_1$ of 2 rad/s^2 relative to the barge. Also, at this instant $\theta = 45°$, $\dot{\theta} = 2$ rad/s, and $\ddot{\theta} = 1$ rad/s^2. What are the velocity and acceleration of point B relative to the barge?

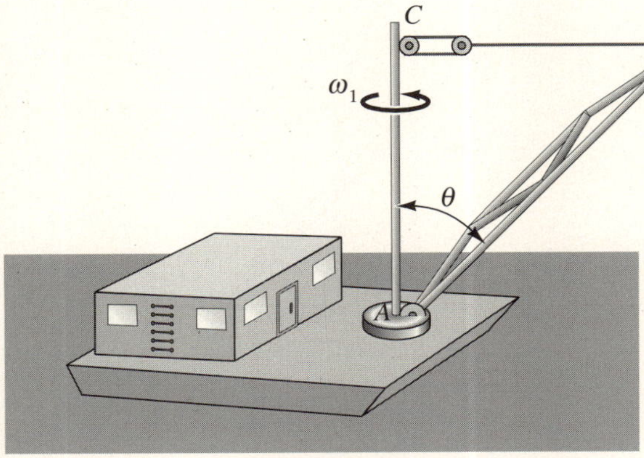

Figure P.15.135.

***15.136.** In Problem 15.86, find the acceleration of the collar C relative to the ground if for the configuration shown:

$$\dot{\theta} = 5 \text{ rad/s}$$
$$\ddot{\theta} = 8 \text{ rad/s}^2$$

15.137. In Example 15.11, find axial force for the beam AB at A resulting from cockpit C, which weighs (with occupant) $136g$ N. The following data apply:

$$\beta = 20°$$
$$\alpha = 60°$$
$$\omega_1 = 0.2 \text{ rad/s}$$
$$\omega_2 = 0.1 \text{ rad/s}$$

15.138. Find ω_B and $\dot{\omega}_B$ if ω_A and $\dot{\omega}_A$ are, respectively, 2 rad/s and 3 rad/s^2 counterclockwise.

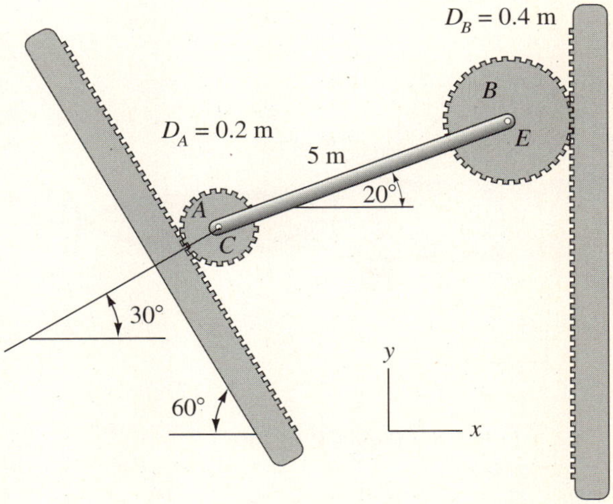

Figure P.15.138.

15.139. Find ω_{AC} and $\dot{\omega}_{AC}$ at the instant shown.

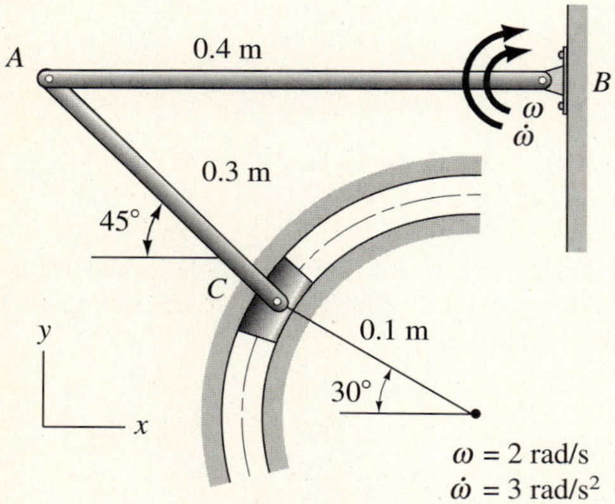

Figure P.15.139.

15.140. A missile travels in a straight line with respect to inertial reference XYZ at speed $V_1 = 3$ m/s and change of speed $a_1 = 1.5$ m/s^2. At the same instant the missile rolls about its direction of flight at an angular speed $\omega_1 = 5$ rad/s and change in angular speed $\dot{\omega}_1 = 5$ rad/s^2. Inside the missile a rod is rotating at a constant angular speed (relative to the missile) $\omega_2 = 10$ rad/s about an axis perpendicular to the page. Find the velocity and acceleration of the tip of the rod relative to XYZ at the instant shown.

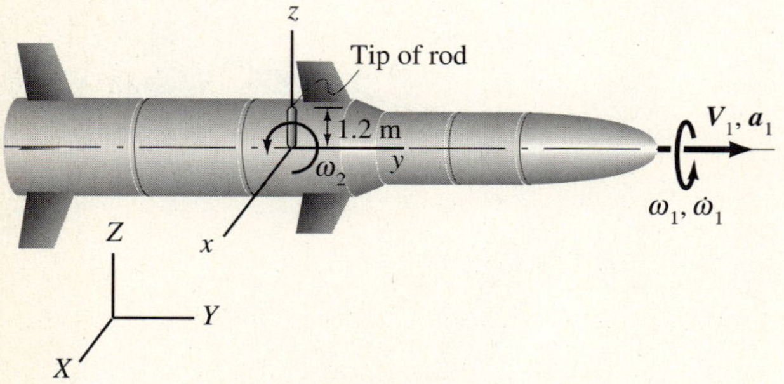

Figure P.15.140.

15.141. A transport plane is undergoing a severe maneuver. As shown, it is moving at a constant speed of 110 m/s, is rolling at the rate of $\omega_1 = 0.2$ rad/s, and is in a loop of radius 1000 m. A solenoid is causing a small machine element to move relative to the plane at a velocity of 10 m/s and an acceleration of 3 m/s^2 both directed downward. If the mass of the machine element is 10 kg, what is the force on it from the plane at this instant? The machine element is at position A at the time of interest.

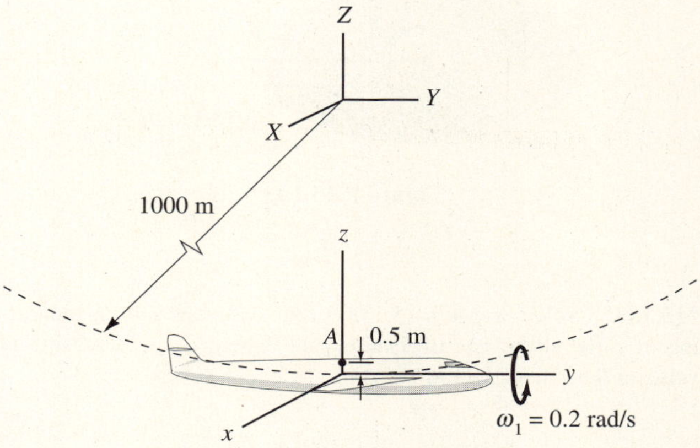

Figure P.15.141.

15.142. A rod moves in the plane of the paper in such a way that end A has a speed of 3 m/s. What is the velocity of point B of the rod when the rod is inclined at 45° to the horizontal? B is at the upper support.

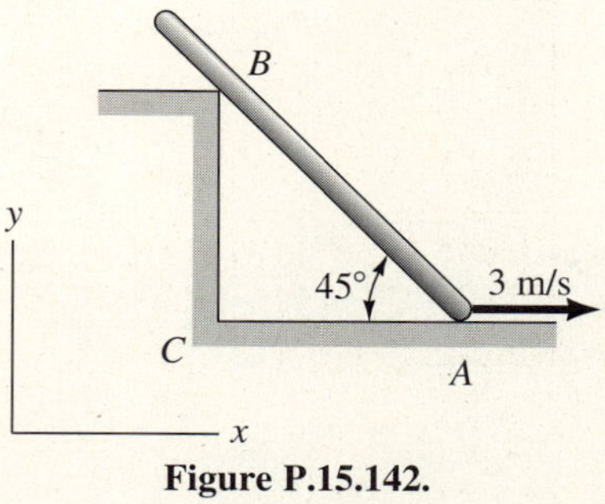

Figure P.15.142.

15.143. A vehicle on a monorail has a speed $\dot{S} = 10$ m/s and an acceleration $\ddot{S} = 4$ m/s^2 relative to the ground reference XYZ when it reaches point A. Inside the vehicle, a 3 kg mass slides along a rod which at the time of interest is parallel to the X axis. This rod rotates about a vertical axis with $\omega = 1$ rad/s and $\dot{\omega} = 2$ rad/s^2 relative to the vehicle at the time of interest. Also at this time, the radial distance d of the mass is 0.2 m and its radial velocity $\dot{r} = 0.4$ m/s inward. What is the dynamic force on the mass at this instant?

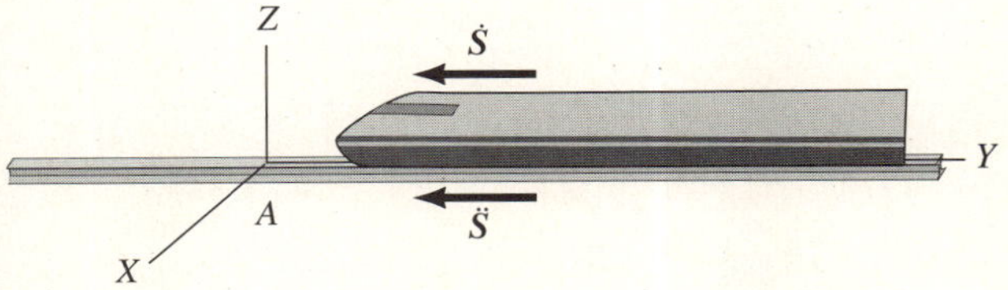

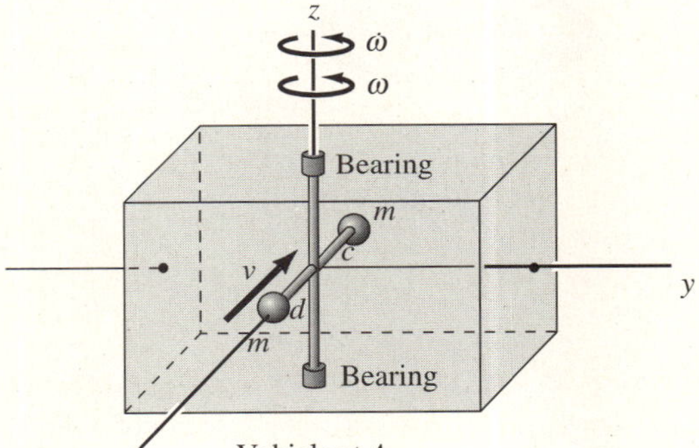

Vehicle at A

When vehicle is at A:
$\omega = 1$ rad/s $\dot{\omega} = 2$ rad/s^2 $v = 0.4$ m/s $d = 0.2$ m Relative to vehicle
$\dot{S} = 10$ m/s $\ddot{S} = 4$ m/s^2 Relative to ground XYZ

Figure P.15.143.

15.144. Cylinder A rolls without slipping. What are ω_{BC}, $\dot{\omega}_{BC}$, and $\dot{\omega}_{CD}$. At the instant shown, $V_A = 5$ m/s and $\dot{V}_A = 3$ m/s^2. Use an intuitive approach only as a check over a formal approach.

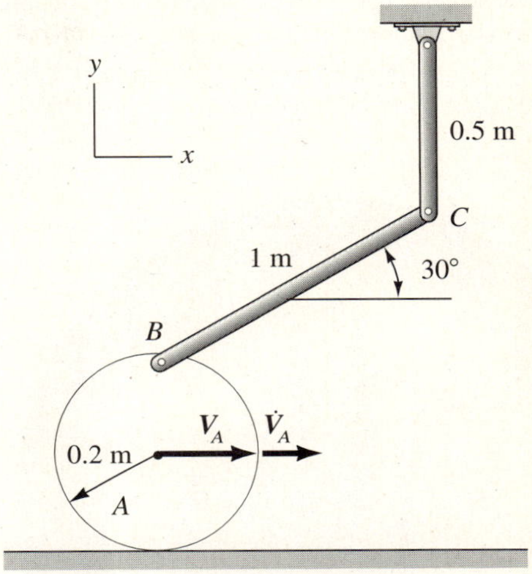

Figure P.15.144.

15.145. What is the magnitude of the velocity of the slider C at the instant shown in the diagram?

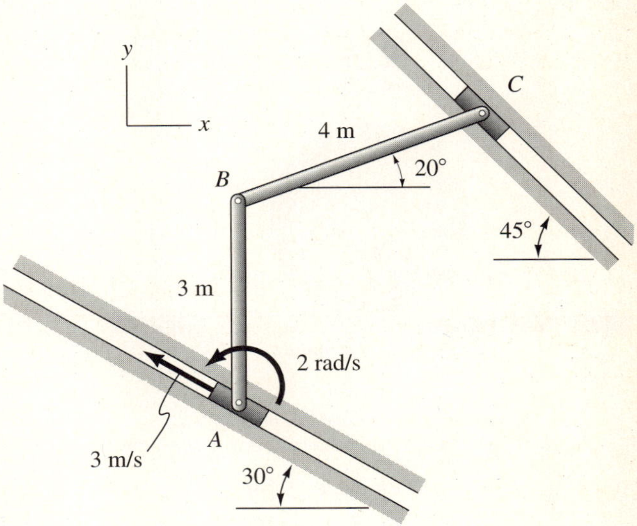

Figure P.15.145.

15.146. Rod AB is mounted on a vertical rod AK *fixed* to a horizontal platform D which rotates with angular speed ω_1 relative to the ground reference XYZ. The shaft AB meanwhile rotates relative to the platform with speed ω_2. Disc G rotates relative to rod AB with speed ω_3 whose value can be determined by a no slipping condition for the disc and platform contact surface. Find the velocity and acceleration vectors for point E on the disc G relative to XYZ. [Suggestion: Fix a second reference to the shaft AB as shown.]

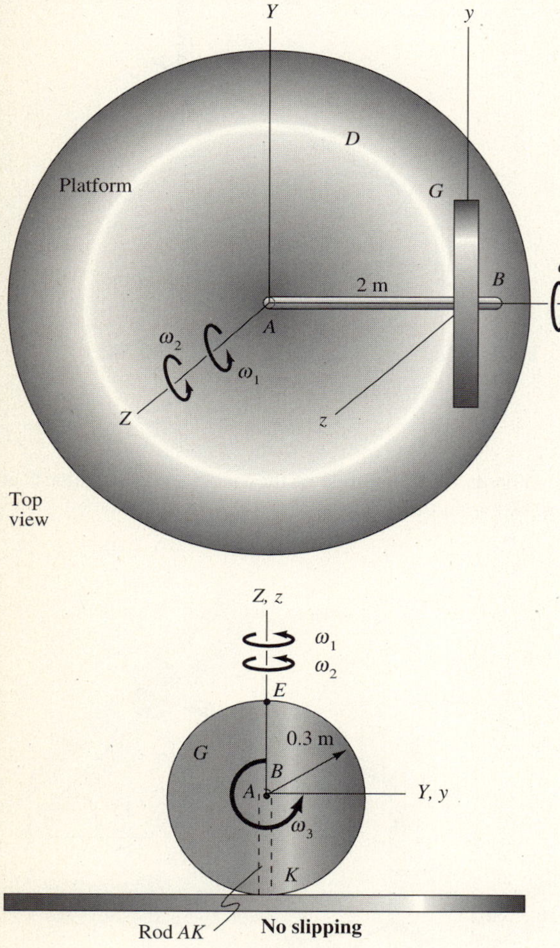

ω_1 = 3 rad/s relative to ground (for platform D)
ω_2 = 2 rad/s 1 rad/s relative to platform (for centerline AB)

Figure P.15.146.

15.147. A conveyor system is shown. Rod AB is welded to plate D. End A is connected to a vehicle which moves with velocity V_2 along a rail. System AB and connected plate D have angular velocities ω_1 and ω_2 relative to the vehicle A as shown in the diagram. On plate D, a particle G has motion given as V_1 and $\dot{V}_1$ along spoke BE which at the instant of interest is parallel to the X axis of the stationary reference XYZ. Determine the acceleration of particle G at the time of interest. Use as a second reference xyz fixed to the plate with the origin at B.

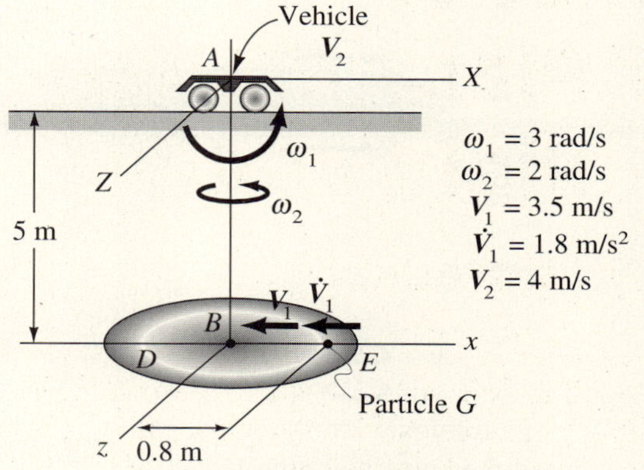

Figure P.15.147.

CHAPTER 16

Kinetics of Plane Motion of Rigid Bodies

16.1 Introduction

In **kinematics** we learned that the motion of a rigid body at any time t can be considered to be a superposition of a translational motion and a rotational motion. The translational motion may have the actual instantaneous velocity of any point of the body, and the angular velocity of the rotation, $\boldsymbol{\omega}$, then has its axis of rotation through the chosen point. A convenient point is, of course, the center of mass of the rigid body. The translatory motion can then be found from particle dynamics. You will recall that the motion of the center of mass of any aggregate of particles (this includes a rigid body) is related to the total external force by the equation

$$\boldsymbol{F} = M\dot{\boldsymbol{V}}_c \tag{16.1}$$

where M is the total mass of the aggregate. Integrating this equation, we get the motion of the center of mass. To ascertain fully the motion of the body, we must next find $\boldsymbol{\omega}$. As we saw in Chapter 14,

$$\boldsymbol{M}_A = \dot{\boldsymbol{H}}_A \tag{16.2}$$

for any system of particles where the point A about which moments of force and linear momentum are to be taken can be (1) the mass center, (2) a point fixed in an inertial reference, or (3) a point accelerating toward or away from the mass center. For these points, we shall later show that the angular velocity vector $\boldsymbol{\omega}$ is involved in the equation above when it is applied to rigid bodies. Also, the inertia tensor will be involved. After we find the motion of the mass center from Eq. 16.1 and the angular velocity $\boldsymbol{\omega}$ from Eq. 16.2, we get the instantaneous motion by letting the entire body have the velocity $\boldsymbol{V}_c$ plus the angular velocity $\boldsymbol{\omega}$, with the axis of rotation going through the center of mass.[1]

[1]Those readers and/or instructors who wish to go to the three-dimensional approach first so as to have plane motion emerge as a special case, should now go to Chapter 18. After the general development and after looking at the solution of three-dimensional problems, one may wish to come back to Chapter 16 to study plane motion dynamics in detail. This approach is entirely optional.

16.2 Moment-of-Momentum Equations

Consider now a rigid body wherein each particle of the body moves parallel to a plane. Such a body is said to be in *plane motion* relative to this plane. We shall consider that axes XY are in the aforementioned plane in the ensuing discussion. The Z axis is then normal to the velocity vector of each point in the body. Furthermore, we consider only the situation where XYZ is an *inertial reference*. A body undergoing plane motion relative to XYZ as described above is shown in Fig. 16.1.

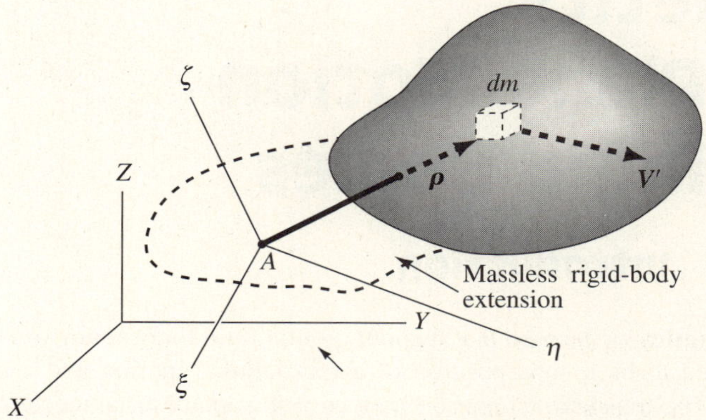

Figure 16.1. Body undergoing plane motion parallel to XY plane. $\xi\eta\zeta$ translates with A.

Choose some point A which is part of this body or a hypothetical massless rigid body extension of this body. An element dm of the body is shown at a position $\boldsymbol{\rho}$ from A. The velocity $\mathbf{V}'$ of dm relative to A is simply the velocity of dm relative to any reference $\xi\eta\zeta$, which *translates* with A relative to XYZ. Similarly, the linear momentum of dm relative to A (i.e., $\mathbf{V}'\,dm$) is the linear momentum of dm relative to $\xi\eta\zeta$ translating with A. We can now give the moment of this momentum (i.e., the angular momentum) $d\mathbf{H}_A$ about A as

$$d\mathbf{H}_A = \boldsymbol{\rho} \times \mathbf{V}'\,dm = \boldsymbol{\rho} \times \left(\frac{d\boldsymbol{\rho}}{dt}\right)_{\xi\eta\zeta} dm$$

But since A is fixed in the body (or in a hypothetical massless extension of the body) and dm is a part of the body having mass, the vector $\boldsymbol{\rho}$ must be *fixed* in the body and, accordingly,

$$\left(\frac{d\boldsymbol{\rho}}{dt}\right)_{\xi\eta\zeta} = \boldsymbol{\omega} \times \boldsymbol{\rho}$$

where $\boldsymbol{\omega}$ is the angular velocity of the body relative to $\xi\eta\zeta$. However, since $\xi\eta\zeta$ translates relative to XYZ, $\boldsymbol{\omega}$ is the angular velocity of the body relative to XYZ as well. Hence, we can say:

$$d\mathbf{H}_A = \boldsymbol{\rho} \times (\boldsymbol{\omega} \times \boldsymbol{\rho})dm \tag{16.3}$$

Note that the angular velocity $\boldsymbol{\omega}$ for the plane motion relative to the XY plane must have a direction *normal* to the XY plane.

Having helped us reach Eq. (16.3), we no longer need reference $\xi\eta\zeta$ and so we now dispense with it. Instead we *fix* reference xyz to the body at point A such that the z axis is *normal* to the plane of motion while the other two axes have arbitrary orientations normal to z (see Fig. 16.2).

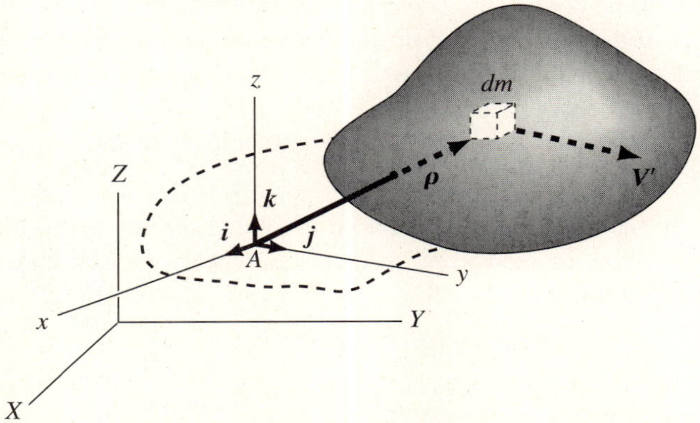

Figure 16.2. Reference xyz fixed to body at A; reference XYZ is inertial.

Note that the z axis will remain normal to XY as the body moves because of the plane motion restriction. Next, we evaluate Eq. 16.3 in terms of components relative to xyz as follows:

$$(dH_A)_x\boldsymbol{i} + (dH_A)_y\boldsymbol{j} + (dH_A)_z\boldsymbol{k}$$
$$= (x\boldsymbol{i} + y\boldsymbol{j} + z\boldsymbol{k}) \times \left[(\omega\boldsymbol{k}) \times (x\boldsymbol{i} + y\boldsymbol{j} + z\boldsymbol{k})\right]dm$$

The scalar equations resulting from the foregoing vector equations are

$$(dH_A)_x = -\omega xz\,dm$$
$$(dH_A)_y = -\omega yz\,dm$$
$$(dH_A)_z = \omega(x^2 + y^2)\,dm$$

Integrating over the entire body, we get[2]

$$\begin{aligned}
(H_A)_x &= -\iiint_M \omega xz\,dm = -\omega\iiint_M xz\,dm = -\omega I_{xz} \\
(H_A)_y &= -\iiint_M \omega yz\,dm = -\omega\iiint_M yz\,dm = -\omega I_{yz} \\
(H_A)_z &= \iiint_M \omega(x^2 + y^2)\,dm = \omega\iiint_M (x^2 + y^2)\,dm = \omega I_{zz}
\end{aligned} \tag{16.4}$$

[2]Note that the massless extension of the rigid body has zero density and hence does not contribute to the integration.

We now have the angular momentum components for reference xyz at A.[3] Note that, because xyz is *fixed* to the body, the inertia terms I_{xz}, I_{yz}, and I_{zz} must be *constants*.

In order to employ the moment-of-momentum equation, $\boldsymbol{M}_A = \dot{\boldsymbol{H}}_A$, at time t, we next restrict point A of the body to be any one of the following three cases.[4]

1. Point A of the body is the *center of mass* of the body.
2. Point A of the body is *fixed* or moving with *constant velocity* at time t in inertial reference XYZ (i.e., point A has zero acceleration at time t relative to XYZ).
3. Point A of the body is *accelerating* toward or away from the mass center at time t.

Using the results of Section 15.6 involving the relation between derivatives of vectors as seen from different references, we next can say, using references XYZ and xyz:

$$\left(\frac{d\boldsymbol{H}_A}{dt}\right)_{XYZ} = \left(\frac{d\boldsymbol{H}_A}{dt}\right)_{xyz} + \boldsymbol{\omega} \times \boldsymbol{H}_A$$

where $\boldsymbol{\omega}$ is the angular velocity of xyz and thus of the body relative to XYZ. Hence, the *moment-of-momentum* equation can be stated as follows:

$$\boldsymbol{M}_A = \left(\frac{d\boldsymbol{H}_A}{dt}\right)_{xyz} + \boldsymbol{\omega} \times \boldsymbol{H}_A \qquad (16.5)$$

Using Eqs. 16.4 for the components of $\boldsymbol{H}_A$, we get for this equation:

$$\boldsymbol{M}_A = \frac{d}{dt_{xyz}}(-\omega I_{xz}\boldsymbol{i} - \omega I_{yz}\boldsymbol{j} + \omega I_{zz}\boldsymbol{k}) + \omega\boldsymbol{k} \times (-\omega I_{xz}\boldsymbol{i} - \omega I_{yz}\boldsymbol{j} + \omega I_{zz}\boldsymbol{k})$$

Noting that $\boldsymbol{i}, \boldsymbol{j}$, and $\boldsymbol{k}$ are constant vectors as seen from xyz, as are the inertia terms, we get

$$\boldsymbol{M}_A = -\dot{\omega} I_{xz}\boldsymbol{i} - \dot{\omega} I_{yz}\boldsymbol{j} + \dot{\omega} I_{zz}\boldsymbol{k} - \omega^2 I_{xz}\boldsymbol{j} + \omega^2 I_{yz}\boldsymbol{i}$$

The scalar forms of the equation above are then

$$(M_A)_x = -I_{xz}\dot{\omega} + I_{yz}\omega^2 \qquad (16.6a)$$

$$(M_A)_y = -I_{yz}\dot{\omega} - I_{xz}\omega^2 \qquad (16.6b)$$

$$(M_A)_z = I_{zz}\dot{\omega} \qquad (16.6c)$$

[3]We now see the motivation for presenting earlier the definitions of mass moments and products of inertia. Clearly the inertia tensor enters prominently in the evaluation of the angular motion of a rigid body.

[4]The "body" here includes the hypothetical, massless, rigid-body extension as well as the actual body.

It is important to note emphatically that the angular velocity as given by ω (and later by $\dot{\theta}$) is always taken *relative to the inertial reference XYZ*, whereas the moments of forces (as given by $(M_A)_x$, $(M_A)_y$, and $(M_A)_z$ as well as the inertia tensor components are always taken about the axes *xyz fixed to the body at A* (Eqs. (16.6)). Eqs. (16.6) are the *general angular momentum equations for plane motion*. The last equation is probably familiar to you from your work in physics. There you expressed it as

$$T = I\alpha \tag{16.7}$$

or as

$$T = I\ddot{\theta} \tag{16.8}$$

We shall now consider special cases of plane motion, starting with the most simple case and going toward the most general case. However, please remember that for *all* plane motions relative to an inertial reference, the moment of the forces about the z axis at A *always* equals $I_{zz}\dot{\omega}$. The other two equations of 16.6 may get simplified for various special plane motions.

Also, to use Eqs. 16.6, we must remember that point A is *part of the body* (because we took $\boldsymbol{\rho}$ to be fixed in the body so we could use $d\boldsymbol{\rho}/dt = \boldsymbol{\omega} \times \boldsymbol{\rho}$) and is also one of the *three acceptable points* presented in Chapter 14. Furthermore, the xyz axes are *fixed to the body* to render the inertia tensor components constant.

Finally, we note from the derivation that ω, θ, and their derivatives are measured from the inertial reference XYZ.

16.3 Pure Rotation of a Body of Revolution About Its Axis of Revolution

A uniform body of revolution is shown in Fig. 16.3. If the body undergoes pure rotation about the axis of revolution fixed in inertial space reference XYZ, we then have plane motion parallel to any plane for which the axis of revolution is a normal. A reference xyz is *fixed* to the body such that the z axis is collinear with the axis of revolution. Since all points along the axis of revolution are fixed in inertial space XYZ, we can choose for the origin of reference xyz any point A along this axis. The x and y axes forming a right handed triad then have arbitrary orientation. For simplicity, we choose xyz collinear with axes XYZ at time t. Clearly, the plane zy is a plane of symmetry for this body, and the x axis is normal to this plane of symmetry. From our work in Chapter 9, recall[5] that, as a consequence, $I_{xy} = I_{xz} = 0$.

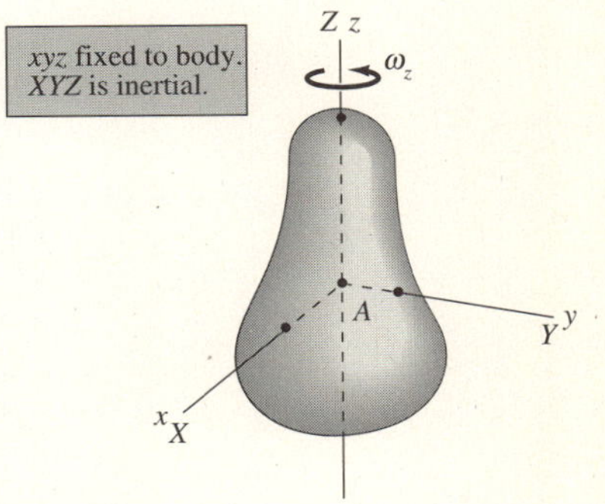

Figure 16.3. Rigid uniform body of revolution at time t.

[5]We pointed out in Chapter 9 that if an axis, such as the x axis, is normal to a plane of symmetry, then the products of inertia with x as a subscript must be zero. This is similarly true for other axes normal to a plane of symmetry.

Similarly, with y normal to a plane of symmetry, xz, we conclude that $I_{yx} = I_{yz} = 0$. Hence, xyz are principal axes. Returning to Eq. 16.6, we find that only one equation of the set has nonzero moment, and that is the familiar equation

$$M_z = I_{zz}\dot{\omega}_z \tag{16.9}$$

The other pair of equations from 16.6 yield

$$\begin{aligned} M_x &= 0 \\ M_y &= 0 \end{aligned} \tag{16.10}$$

Now we turn to *Newton's law*. For this we must use the inertial reference XYZ. Note that at the instant t shown in Fig. 16.3, axes xyz and axes XYZ have been taken as collinear. This means that at this instant the forces on the body needed in Newton's law such as F_X can be denoted as F_x since the directions of X and x are the same at this instant and it is only the direction that is significant here. Since the center of mass of the body is stationary at all times (it is on the axis of rotation), we can accordingly say from *Newton's law*:

$$\begin{aligned} \sum F_x &= 0 \\ \sum F_y &= 0 \\ \sum F_z &= 0 \end{aligned} \tag{16.11}$$

Thus, the applied forces at any time t, the supporting forces, and the weight of the body of revolution satisfy *all* the equations of equilibrium *except* for motion about the axis of revolution where Eq. 16.9 applies.

Notice that the key equation (16.9) has the *same form* as Newton's law for *rectilinear translation* of a particle along an axis, say the x axis. We write both equations together as follows:

$$M_z = I_{zz}\ddot{\theta} \tag{16.12a}$$

$$F_X = M\ddot{X} \tag{16.12b}$$

In Chapter 12 we integrated Eq. 16.12b for various kinds of force functions: time functions, velocity functions, and position functions. The same techniques used then to integrate Eq. 16.12b can now be used to integrate Eq. 16.12a, where the moment functions can also be time functions, angular velocity functions, and angular position functions.

We illustrate these possibilities in the following examples. The first example involves a torque which in part is a function of angular position θ.

Example 16.1

A stepped cylinder having a radius of gyration $k = 0.40$ m and a mass of 200 kg is shown in Fig. 16.4. The cylinder supports a weight W of mass 100 kg with an inextensible cord and is restrained by a linear spring whose constant K is 2 N/mm. What is the angular acceleration of the stepped cylinder when it has rotated 10° after it is released from a state of rest? The spring is initially unstretched. What are the supporting forces at this time?

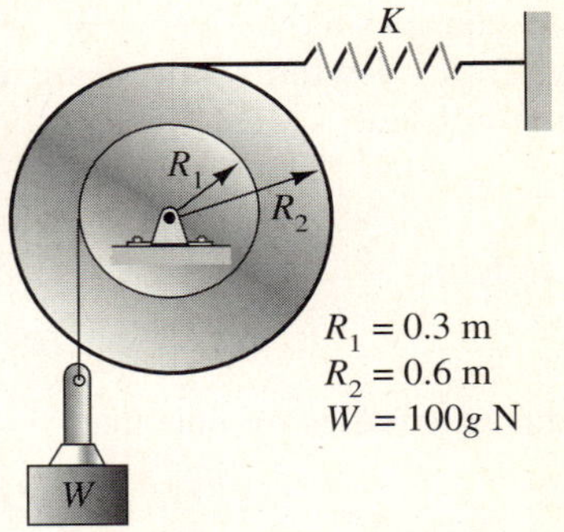

Figure 16.4. Stepped cylinder.

We have shown free-body diagrams of the stepped cylinder and the weight W in Fig. 16.5. A tension T from the cord is shown acting both on the weight W and the stepped cylinder. We have here for the stepped cylinder a body of revolution rotating about its axis of symmetry along which we have chosen point A. Axes xyz are fixed to the body at A with z along the axis of rotation. Furthermore we have shown inertial axes XYZ at A collinear with xyz at the time t.

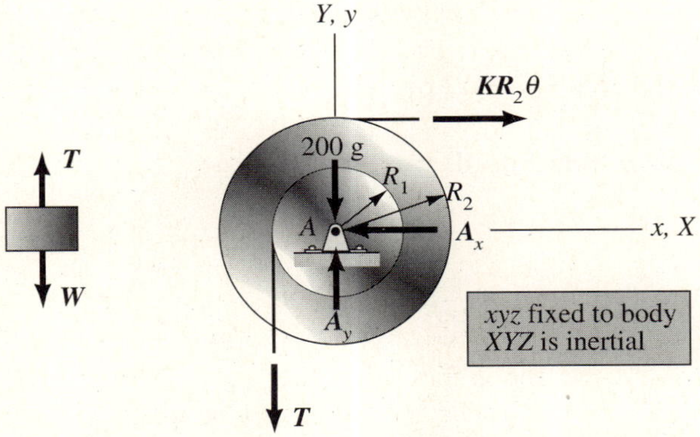

Figure 16.5. Free-body diagrams of components at time t.

We can accordingly apply the **moment-of-momentum** equation about the centerline of the cylinder:

$$TR_1 - KR_2^2\theta = I\ddot{\theta} = (Mk^2)\ddot{\theta}$$

$$T(0.3) - \left[(2\times10^3)\right](0.6)^2\theta = (200)(0.4)^2\ddot{\theta} \qquad \text{(a)}$$

where as indicated earlier $\dot{\theta}$ is the angular velocity of the body relative to stationary axes XYZ and where θ is the rotation of the cylinder in radians from a position corresponding to the unstretched condition of the spring.

Example 16.1 (Continued)

Now considering the weight W, which is in a translatory motion, we can say, from **Newton's law** using inertial reference XYZ as required by Newton's law

$$T - W = M\ddot{Y}$$

Therefore,

$$T - (100)(9.81) = 100\ddot{Y} \qquad \text{(b)}$$

From **kinematics** we note that

$$R_1\ddot{\theta} = -\ddot{Y}$$

Therefore,

$$0.3\ddot{\theta} = -\ddot{Y} \qquad \text{(c)}$$

Substituting for T in Eq. (a) using Eq. (b) and for $\ddot{Y}$ using Eq. (c), we then have

$$\left[(100)(9.81) + (100)(-0.3\ddot{\theta})\right](0.3) - \left[(2\times10^3)\right](0.6)^2\theta = (200)(0.4)^2\ddot{\theta} \qquad \text{(d)}$$

When $\theta = (10°)(2\pi/360°) = 0.1745$ rad, we get for $\ddot{\theta}$ from Eq. (d) the desired result:

$$\ddot{\theta} = 4.11 \text{ rad/s}^2$$

From Eqs. (b) and (c), we then have for T:

$$T = 981 - 123.3 = 858 \text{ N}$$

We next use **Newton's law** for the center of mass A of the stepped cylinder. Thus, considering Fig. 16.5, realizing that the center of mass is in equilibrium and assuming collinear orientation of the two sets of axes at time t, we can say $A_X \equiv A_x$ and $A_Y \equiv A_y$ at time t since only direction is involved here. Thus summing forces,

$$-858 - 200g + A_y = 0$$

$$A_y = 2.82 \text{ kN}$$

$$-A_x + [(2\times10^3)](0.6)(0.1745) = 0$$

$$A_x = 209 \text{ N}$$

It should be clear on examining Fig. 16.4 that the motion of the cylinder, after W is released from rest, will be rotational oscillation. This motion ensues because the spring develops a restoring torque much as the spring in the classic spring–mass system (Fig. 16.6) supplies a restoring force. We shall study torsional oscillation or vibration in Chapter 19 when we consider vibrations. The key concepts and mathematical techniques for both motions you will find to be identical.

The torque in the next example is, in part, a function of time.

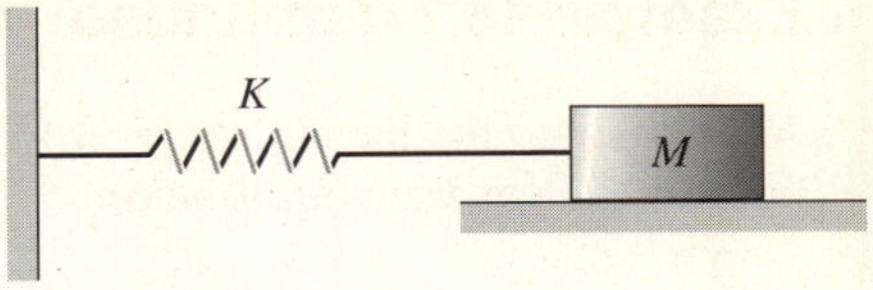

Figure 16.6. Classic spring–mass system.

Example 16.2

A cylinder A is rotating at a speed ω of 1750 r/min (see Fig. 16.7) when the light handbrake system is applied using force F = $(10t + 300)$ N with t in seconds. If the cylinder has a radius of gyration of 200 mm and a mass of 500 kg, how long a time does it take to halve the speed of the cylinder? The dynamic coefficient of friction between the belt and the cylinder is 0.3.

We start by showing the free body of the cylinder and of the brake lever in Fig. 16.8. From the belt formula of Chapter 7, we can say for the belt tensions on the cylinder

$$\frac{T_1}{T_2} = e^{\mu_d \beta} = e^{(0.3)(\frac{3}{4})(2\pi)} = 4.11$$

$$\therefore T_1 = 4.11 T_2 \qquad \text{(a)}$$

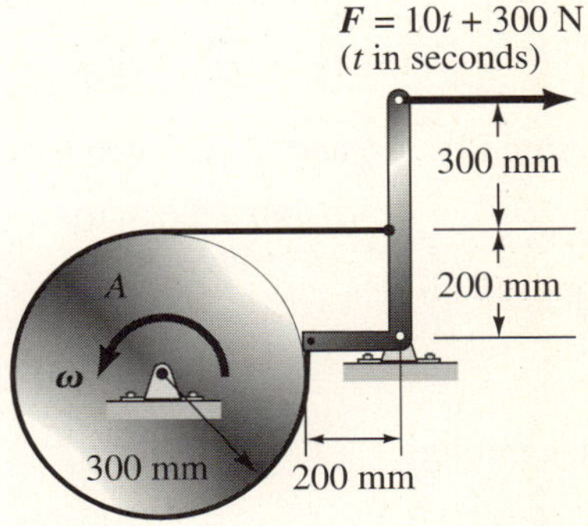

Figure 16.7. Cylinder and handbrake system.

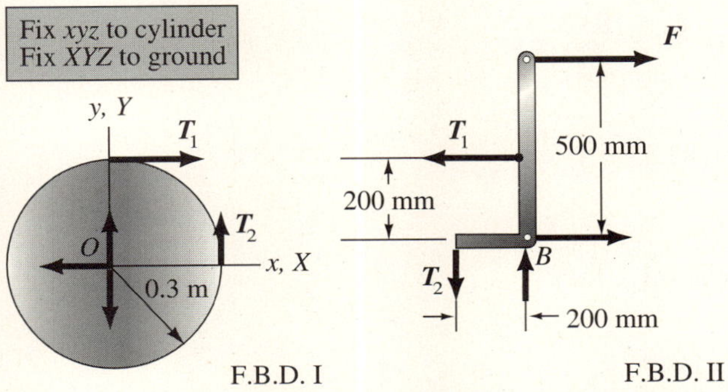

Figure 16.8. Free-body diagrams of cylinder and handbrake lever.

Example 16.2 (Continued)

Now going to the handbrake lever in F.B.D. II and taking moments about point B, we get from **equilibrium**

$$-F(0.5) + T_1(0.2) + T_2(0.2) = 0$$

Inserting $F = 10t + 300$ N, we get for $(T_1 + T_2)$

$$T_1 + T_2 = 2.5(10t + 300) \qquad \text{(b)}$$

Substitute for T_1 using Eq. (a).

$$4.11T_2 + T_2 = 2.5(10t + 300)$$

$$\therefore T_2 = 0.489(10t + 300) \qquad \text{(c)}$$

Also from (a)

$$T_1 = (4.11)(0.489)(10t + 300) = 2.01(10t + 300) \qquad \text{(d)}$$

Now going to F.B.D. I in Fig. 16.8 and using axes xyz fixed to the cylinder at O, we next write the **moment-of-momentum** equation.

$$-T_1(0.3) + T_2(0.3) = (500)(0.2)^2\,(\ddot{\theta})$$

Using Eqs. (c) and (d) we then have

$$(0.489 - 2.01)(10t + 300)(0.3) = (500)(0.2)^2\ddot{\theta}$$

Hence

$$\ddot{\theta} = -0.02283(10t + 300)$$

Integrating

$$\dot{\theta} = -0.02283\left(10\frac{t^2}{2} + 300t\right) + C_1 \qquad \text{(e)}$$

When $t = 0$,

$$\dot{\theta} = (1750)\left(\frac{2\pi}{60}\right) = 183.3 \text{ rad/s} \qquad \therefore C_1 = 183.3$$

Thus

$$\dot{\theta} = -0.02283(5t^2 + 300t) + 183.3$$

Set

$$\dot{\theta} = \left(\frac{1}{2}\right)(183.3) = 91.63 \text{ rad/s} \quad \text{and solve for } t$$

$$91.63 = -0.02283(5t^2 + 300t) + 183.3$$

Example 16.2 (Continued)

Rearranging, we get

$$t^2 + 60t - 803 = 0$$

Using the quadratic formula, we have

$$t = \frac{-60 \pm \sqrt{60^2 + (4)(803)}}{2}$$

$$t = 11.27 \text{ s}$$

16.4 Pure Rotation of a Body with Two Orthogonal Planes of Symmetry

Consider next a uniform body having *two* orthogonal planes of symmetry. Such a body is shown in Fig. 16.9, where in (a) we have shown the aforementioned planes of symmetry and in (b) we have shown a view along the intersection of the planes of symmetry. We shall consider pure rotation of such a body about a stationary axis collinear with the intersection of the planes of symmetry, which we take as the z axis. The origin A can be taken anywhere along the axis of rotation, and the x and y axes are fixed in the planes of symmetry, as shown in the diagram. XYZ is taken collinear with xyz at time t. We leave it to the reader to show that the identical equations apply to this case as to the previous case of a body of revolution.

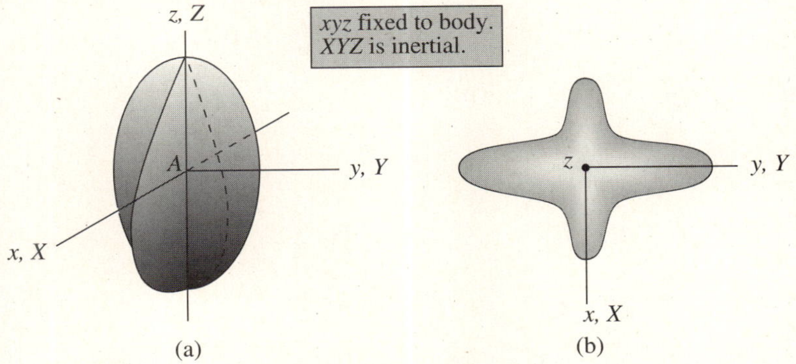

Figure 16.9. Body with two orthogonal planes of symmetry at time t.

The torque in the next example is a function of angular speed.[6]

[6]From now on it will be understood that in the diagrams xyz will be fixed to the body and that XYZ will be fixed to the ground and hence to be an inertial reference. This will avoid cluttering the diagrams unnecessarily.

Example 16.3

A thin-walled shaft is shown in Fig. 16.10. On it are welded identical plates A and B, each having a mass of 10 kg. Also welded onto the shaft at right angles to A and B are two identical plates C and D, each having a mass of 6 kg. The thin-walled shaft is of diameter 100 mm and has a mass of 15 kg. The wind resistance to rotation of this system is given for small angular velocities as $0.2\dot{\theta}$ N m, with $\dot{\theta}$ in rad/s. Starting from rest, what is the time required for the system to reach 100 r/min if a torque T of 5 N m is applied? What are the forces on the bearings G and E when this speed is reached?

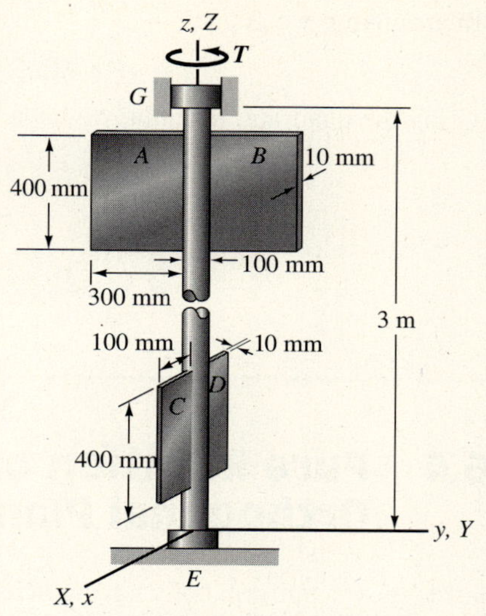

Figure 16.10. Device with rotational resistance.

We have here a body with two orthogonal planes of symmetry. The body is rotating about the axis of symmetry which we take as the z axis for axes xyz fixed to the body at E. As usual we position xyz at time t to be collinear with inertial reference XYZ.

The **moment-of-momentum** equation about the z axis is given as follows using moment of inertia formulas for plates (see inside covers) along with the parallel-axis formula and noting for the thin-walled shaft that we use $I_{zz} = Mr^2$ where, as an approximation, we take the outside radius for r.

$$5 - 0.2\dot{\theta} = \Big\{(15)(0.05)^2 + 2\Big[\tfrac{1}{12}(10)(0.3^2 + 0.01^2) + 10(0.2)^2\Big] + 2\Big[\tfrac{1}{12}(6)(0.1^2 + 0.01^2) + (6)(0.1)^2\Big]\Big\}\ddot{\theta} \quad \text{(a)}$$

This becomes

$$5 - 0.2\dot{\theta} = 1.118\frac{d\dot{\theta}}{dt}$$

We can separate the variables as follows:

$$\frac{1.118\,d\dot{\theta}}{5 - 0.2\dot{\theta}} = dt \qquad \text{(b)}$$

Now make the following substitution:

$$5 - 0.2\dot{\theta} = \eta$$

Therefore, taking the differential

$$-0.2d\dot{\theta} = d\eta$$

Hence, we have for Eq. (b):

$$-5.59\frac{d\eta}{\eta} = dt$$

Example 16.3 (Continued)

Integrate to get

$$-5.59 \ln \eta = t + C_1$$

Hence, replacing η

$$-5.59 \ln(5 - 0.2\dot{\theta}) = t + C_1 \tag{c}$$

When $t = 0$, $\dot{\theta} = 0$, and we have for C_1:

$$-5.59 \ln(5) = C_1$$

Therefore,

$$C_1 = -8.99$$

Hence, Eq. (c) becomes

$$-5.59 \ln(5 - 0.2\dot{\theta}) = t - 8.99$$

Let $\dot{\theta} = (100/60)(2\pi) = 10.47$ rad/s in the above equation. The desired time t for this speed to be reached then is

$$t = 3.03 \text{ s}$$

We now consider the supporting forces for the system. For reasons set forth in Section 16.3 we know that

$$M_x = 0$$
$$M_y = 0$$

for the other **moment of momentum** equations. Also from **Newton's law**, while noting that xyz and XYZ are collinear at time t,

$$\sum F_x = 0$$
$$\sum F_y = 0$$
$$\sum F_z = 0$$

for the center of mass. Clearly, the dead weights of bodies (in the z direction) give rise to a constant supporting force of $[2(10 + 6) + 15]g = 461$ N at bearing E. All other forces are zero.

$E_z = 461$ N
All other support forces are zero

16.5 Pure Rotation of Slablike Bodies

We now consider bodies that have a *single* plane of symmetry, such as is shown in Fig. 16.11. Such bodies we shall call *slablike* bodies. We have oriented the body in Fig. 16.8 so that the plane of symmetry is parallel to the XY plane. We shall now consider the pure rotation of such a body about a fixed axis normal to the XY plane and going through a point A in the plane of symmetry of the body. We fix a reference xyz to the body at point A with xy in the plane of symmetry and z along the axis of rotation.

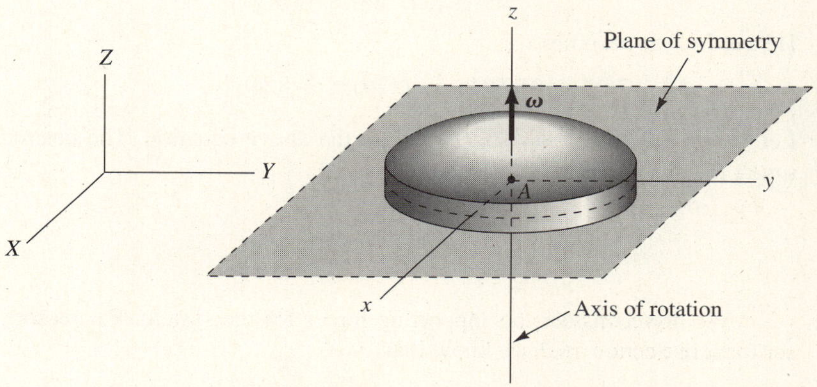

Figure 16.11. Slablike body undergoing pure rotation.

The angular velocity $\boldsymbol{\omega}$ is then along the z axis. Since z is normal to the plane of symmetry, it is clear immediately that $I_{zx} = I_{zy} = 0$. And so the *moment of momentum* equations become for this case:

$$\begin{aligned} M_x &= 0 \\ M_y &= 0 \\ M_z &= I_{zz}\dot{\omega}_z \end{aligned} \tag{16.13}$$

If the center of mass is not at a position along the axis of rotation, then we no longer have equilibrium conditions for the center of mass. It will be undergoing circular motion. However, through *Newton's law* we can relate the external forces on the body to the acceleration of the mass center. We may then have to use the *kinematics* of rigid-body motion to yield enough equations to solve the problem. We now illustrate this case.

Example 16.4

A uniform rod of weight W and length L supported by a pin connection at A and a wire at B is shown in Fig. 16.12. What is the force on pin A at the instant that the wire is released? What is the force at A when the rod has rotated 45°?

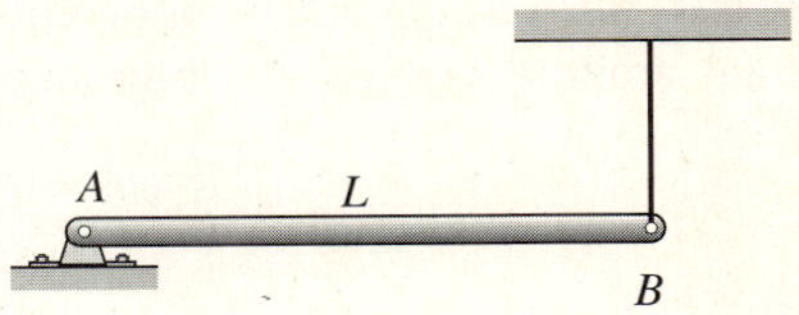

Figure 16.12. Rod supported by wire.

Part A. A free-body diagram of the rod is shown in Fig. 16.13 at the instant that the wire is released at B. We fix xyz to the body at A. XYZ is stationary. The **moment-of-momentum** equation about the axis of rotation at A, on using the formula for I of a rod about a transverse axis at the end, yields

$$\frac{WL}{2} = I\ddot{\theta} = \frac{1}{3}\left(\frac{W}{g}\right)L^2\ddot{\theta}$$

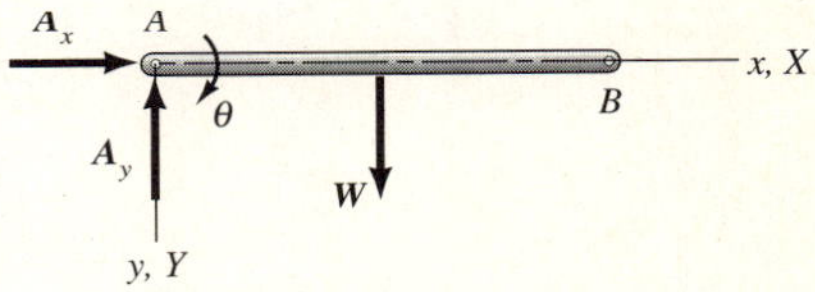

Figure 16.13. Wire suddenly cut.

Therefore,

$$\ddot{\theta} = \frac{3}{2}\frac{g}{L} \quad \text{at time } t = 0 \tag{a}$$

Using simple **kinematics** of plane circular motion, we can determine the acceleration of the mass center at $t = 0$. Thus, using the inertial reference

$$\ddot{X} = 0, \qquad \ddot{Y} = \frac{L}{2}\ddot{\theta} = \frac{3}{4}g \tag{b}$$

where we have used Eq. (a) in the last step. Next express **Newton's law** for the mass center using $A_y \equiv A_Y, A_x \equiv A_X$:[7]

$$\frac{W}{g}\ddot{X} = A_x, \qquad \frac{W}{g}\ddot{Y} = W - A_y$$

Accordingly, at time $t = 0$ we have, on noting Eqs. (b):

$$A_x = 0, \qquad A_y = \frac{1}{4}W \tag{c}$$

Thus, we see that at the instant of releasing the wire there is an upward force of $\frac{1}{4}W$ on the left support.

[7]Note that we can replace A_X by A_x, etc., because of the common direction of X and x as well as the other corresponding axes at time t. However, we cannot replace $\ddot{X}$ by $\ddot{x}$ and $\ddot{Y}$ by $\ddot{y}$. The reason for this is that while the orientations of the respective axes are the same at time t, the fact remains that because of the rotational motion of xyz relative to XYZ, the velocities and accelerations of a particle relative to xyz and XYZ will be different requiring us to use only the XYZ axes when dealing with derivatives of the particle coordinates in Newton's law.

Example 16.4 (Continued)

Part B. We next express the **moment-of-momentum** equation for the rod at any arbitrary position θ. Observing Fig. 16.14, we get

$$\frac{WL}{2}\cos\theta = \frac{1}{3}\frac{W}{g}L^2\ddot{\theta}$$

Therefore,

$$\ddot{\theta} = \frac{3}{2}\frac{g}{L}\cos\theta \qquad \text{(d)}$$

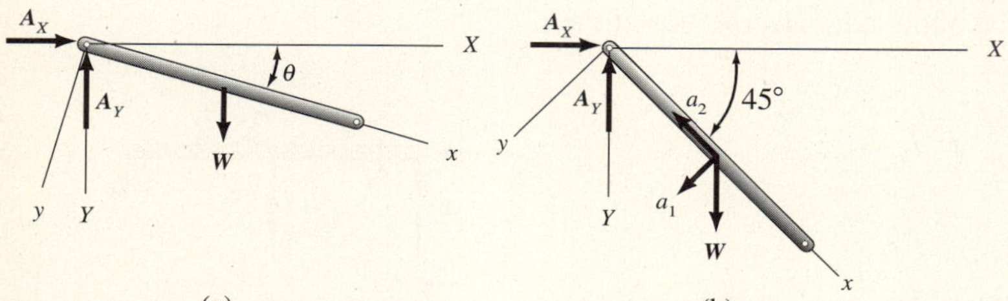

(a) (b)

Figure 16.14. (a) Rod at position θ; (b) rod at $\theta = 45°$.

Consequently, at $\theta = 45°$ we have

$$\ddot{\theta} = (1.5)(0.707)\frac{g}{L} = 1.060\frac{g}{L} \qquad \text{(e)}$$

We shall also need $\dot{\theta}$, and accordingly we now rewrite Eq. (d) as follows:

$$\ddot{\theta} \equiv \left(\frac{d\dot{\theta}}{d\theta}\right)\left(\frac{d\theta}{dt}\right) \equiv \left(\frac{d\dot{\theta}}{d\theta}\right)(\dot{\theta}) = \frac{3}{2}\frac{g}{L}\cos\theta \qquad \text{(f)}$$

Separating variables, we get

$$\dot{\theta}\,d\dot{\theta} = \frac{3}{2}\frac{g}{L}\cos\theta\,d\theta$$

which we integrate to get

$$\frac{\dot{\theta}^2}{2} = \frac{3}{2}\frac{g}{L}\sin\theta + C$$

When $\theta = 0$, $\dot{\theta} = 0$; accordingly, $C = 0$. We then have

$$\dot{\theta}^2 = 3\frac{g}{L}\sin\theta \qquad \text{(g)}$$

At the instant of interest, $\theta = 45°$ and we get for $\dot{\theta}^2$:

$$\dot{\theta}^2 = 3\frac{g}{L}(0.707) = 2.12\frac{g}{L} \qquad \text{(h)}$$

For $\theta = 45°$, we can now give the acceleration component a_1 of the center of mass directed normal to the rod and component a_2 directed along

Example 16.4 (Continued)

the rod [see Fig. 16.14(b)]. From **kinematics** we can say, using Eqs. (e) and (h):

$$a_1 = \frac{L}{2}\ddot{\theta} = \frac{L}{2}\left(1.060\frac{g}{L}\right) = 0.530g$$

$$a_2 = \frac{L}{2}(\dot{\theta})^2 = \frac{L}{2}\left(2.12\frac{g}{L}\right) = 1.060g \qquad \text{(i)}$$

Now, employing **Newton's law** for the mass center, we have on noting that xy is *no longer collinear* with XY.

$$A_X = \frac{W}{g}(-a_1 \sin 45° - a_2 \cos 45°)$$

Therefore, using Eqs. (i)

$$A_X = -1.124W$$

Also, from **Newton's law**

$$-A_Y + W = \frac{W}{g}(-a_2 \sin 45° + a_1 \cos 45°)$$

Therefore, again using Eq. (i)

$$A_Y = 1.375W \qquad \text{(j)}$$

Consider next the case of a body undergoing pure rotation about an axis which, for some point A in the body (or massless hypothetical extension of the body), is a *principal* axis (see Fig. 16.15). For a reference xyz fixed at A with z collinear with the axis of rotation, it is clear that $I_{zx} = I_{zy} = 0$, and hence the moment of momentum equations simplify to the exact same forms as presented here for the rotation of slablike bodies.

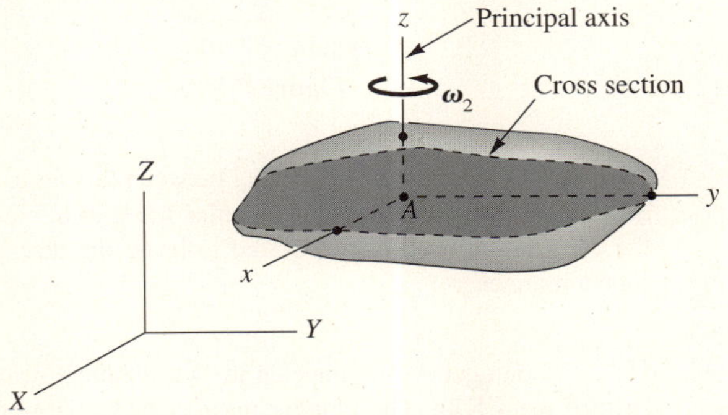

Figure 16.15. Axis z is a principal axis for point A.

PROBLEMS

16.1. A shaft and disc of steel having a density of 7.63 Mg/m^3 are subjected to a constant torque T of 67.5 N m, as shown. After 1 min, what is the angular velocity of the system? How many revolutions have occurred during this interval? Neglect friction of the bearings. Use $\frac{1}{2}Mr^2 = I$ for disc.

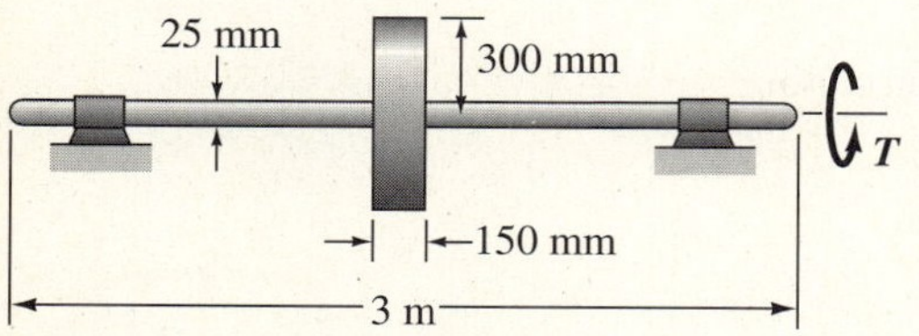

Figure P.16.1.

16.2. In Problem 16.1 include wind and bearing friction losses by assuming that they are proportional to angular speed $\dot{\theta}$. If the disc will halve its speed after 5 min from a speed of 300 r/min when there is no external applied torque, what is the resisting torque at 600 r/min?

16.3. A 180 kg flywheel is shown. A 45 kg block is held by a light cable wrapped around the hub of the flywheel, the diameter of which is 0.3 m. If the initially stationary weight descends 0.9 m in 5 s, what is the radius of gyration of the flywheel?

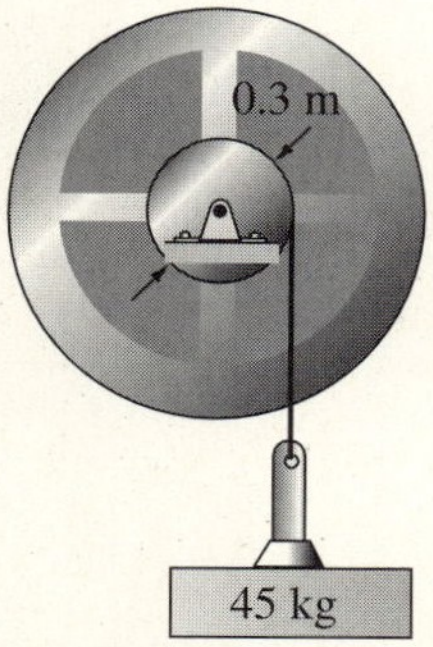

Figure P.16.3.

16.4 A stepped cylinder has the dimensions R_1 = 0.3 m, R_2 = 0.65 m, and the radius of gyration, k, is 0.35 m. The mass of the stepped cylinder is 100 kg. Weights A and B are connected to the cylinder. If weight B is of mass 80 kg and weight A is of mass 50 kg, how far does A move in 5 s? In which direction does it move?

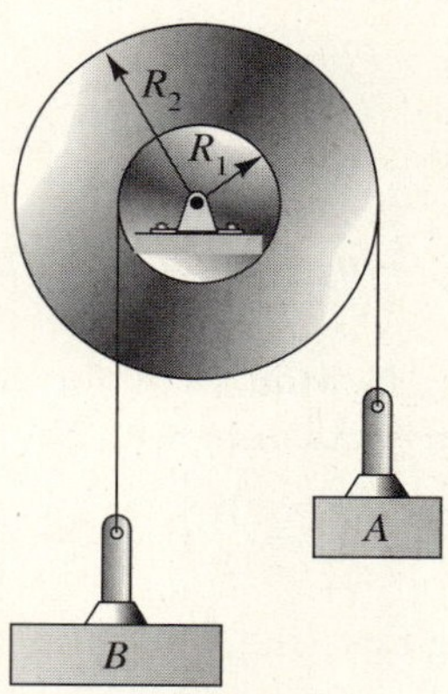

Figure P.16.4.

16.5. Two discs E and F of diameter 0.3 m rotate in frictionless bearings. Disc F, mass 50 kg and rotates with angular speed ω_1 of 10 rad/s, whereas disc E, mass 15 kg, rotates with angular speed ω_2 of 5 rad/s. Neglecting the angular momentum of the shafts, what is the total angular momentum of the system relative to the ground? Use Eq. 16.4 to compute H from first principles. Consider that the discs are forced together along the axis of rotation. What is the common angular velocity when friction has reduced relative motion between the discs to zero?

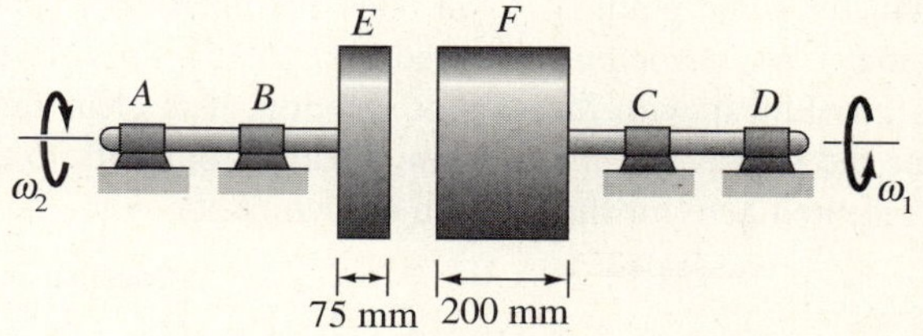

Figure P.16.5.

16.6 In Problem 16.5, if μ_s = 0.2 between the discs, and it takes 30 s for them to reach the same angular speed of 6.5 rad/s, what is the constant normal force required to bring the discs together in such a manner?

16.7. A plunger A is connected to two identical gears B and C, each of mass 5 kg. The plunger mass is 20 kg. How far does the plunger drop in 1 s if released from rest?

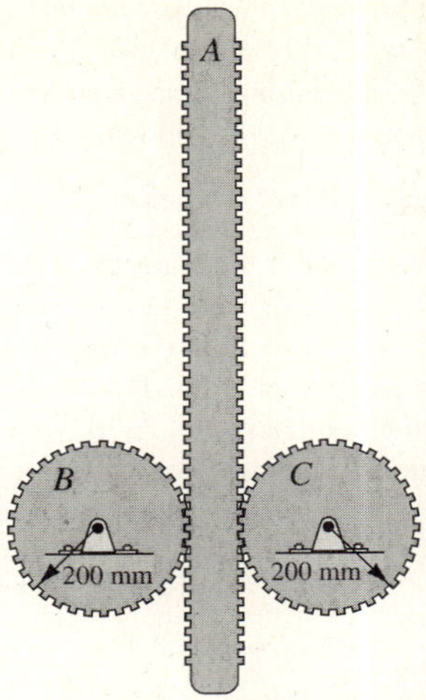

Figure P.16.7.

16.8. A pulley A and its rotating accessories have a mass of 1000 kg and a radius of gyration of 0.25 m. A simple hand brake is applied as shown using a force P. If the dynamic coefficient of friction between belt and pulley is 0.2, what must force P be to change ω from 1750 r/min to 300 r/min in 60 s?

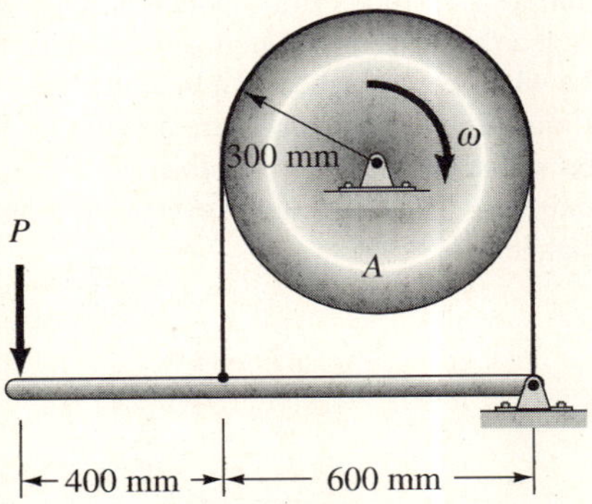

Figure P.16.8.

16.9. A flywheel is shown. There is a viscous damping torque due to wind and bearing friction which is known to be -0.04ω N m, where ω is in rad/s. If a torque $T = 100$ N m is applied, what is the speed in 5 min after starting from rest? The mass of the wheel is 500 kg, and the radius of gyration is 0.5 m.

Figure P.16.9.

16.10. Two cylinders and a rod are oriented in the vertical plane. The rod is guided by bearings (not shown) to move vertically. The following are the weights of the three bodies:

$$W_A = 1000 \text{ N} \quad W_B = 300 \text{ N} \quad W_C = 200 \text{ N}$$

What is the acceleration of the rod? There is rolling without slipping.

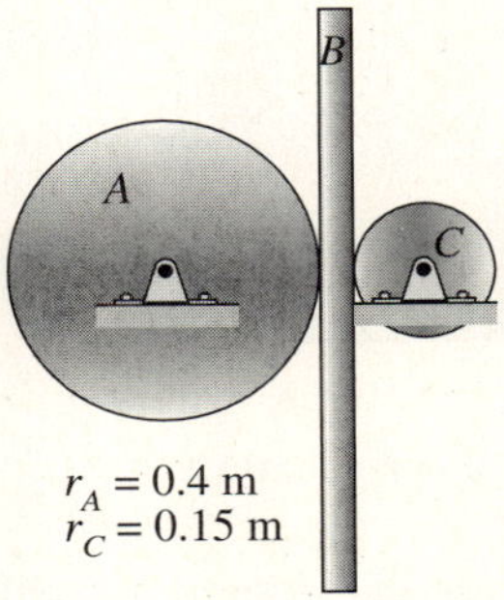

Figure P.16.10.

16.11. A bar A of mass 36 kg is supported at one end by rollers that move with the bar and a stationary cylinder B of mass 31 kg that rotates freely. A 450 N force is applied to the end of the bar. What is the friction force between the bar and the cylinder as a function of x? Indicate the ranges for non-slipping and slipping conditions. Take $\mu_s = 0.5$ and $\mu_d = 0.3$.

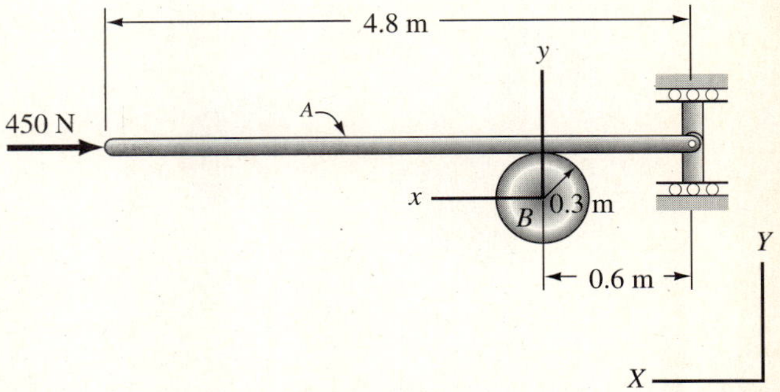

Figure P.16.11.

16.12. Cylinder A has an angular speed ω_1 of 3 rad/s when it is lowered onto cylinder B, which has an angular speed ω_2 of 5 rad/s before contact is made. What are the final angular velocities of the cylinders resulting from friction at the surfaces of contact? The mass of A is 250 kg and of B is 200 kg. If $\mu_d = 0.3$ for the contact surface of the cylinders and if the normal force transmitted from A to B is 45 N, how long does it take for the cylinders to reach a constant speed?

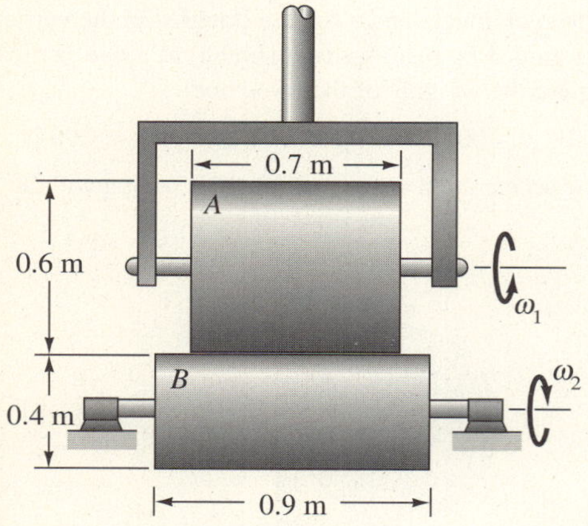

Figure P.16.12.

16.13. A driving cylinder A has a torque T_A of 30 N m applied to it while the driven cylinder B has a resisting torque T_B of 10 N m. Cylinder B has a mass of 15 kg, a radius of gyration of 100 mm, and a diameter of 400 mm. Cylinder A has a mass of 50 kg, a radius of gyration of 200 mm, and a diameter of 800 mm. Rod CD is a light rod connecting the cylinders. What is the angular acceleration of cylinder A at the instant shown if the system is stationary at this instant? Rod CD is 1 m long.

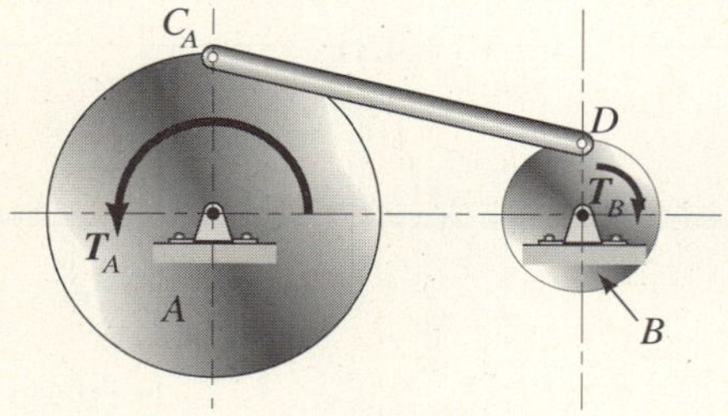

Figure P.16.13.

16.14. Do Problem 16.13 for the case where the angular velocity of B is 20 rad/s clockwise at the instant of interest.

16.15. A loading vehicle is on a platform that can rotate in order to unload potash into the hold of a ship. The wheels of the vehicle are free to rotate. The vehicle is constrained from moving to the right by a stop at A. If the platform has the following rotational data at time t

$$\theta = 40 \text{ mrad} \qquad \dot{\theta} = 50 \text{ mrad/s} \qquad \ddot{\theta} = 12 \text{ mrad/s}^2$$

what are the forces normal to the platform acting on the wheels of the vehicle and what is the force at the stop in a direction parallel to the platform? The total weight of the vehicle with load is 12 kN. The radius of each wheel is 0.15 m. The radius of gyration at the center of mass of the vehicle is $k = 2.76$ m. Note that the center of mass of the vehicle lies along an axis which passes through the axis of rotation and is perpendicular to the platform. Also, note that θ is measured clockwise here.

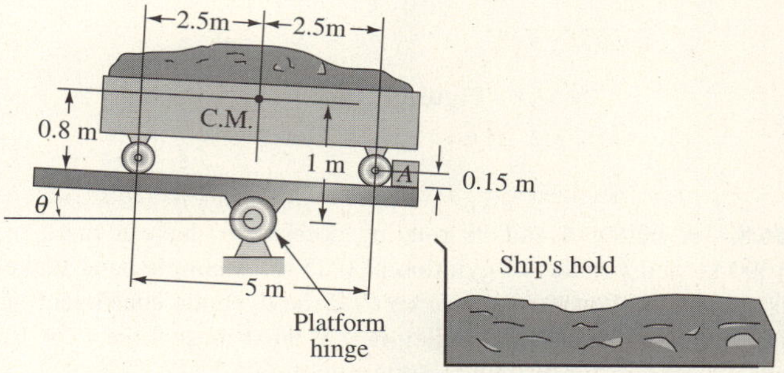

Figure P.16.15.

16.16. A torque T of 100 N m is applied to a wheel D having a mass of 50 kg, a diameter of 600 mm, and a radius of gyration of 280 mm. The wheel D is attached by a light member AB to a slider C having a mass of 30 kg. If the system is at rest at the instant shown, what is the acceleration of slider C? What is the axial force in member AB? Neglect friction everywhere, and neglect the inertia of member AB.

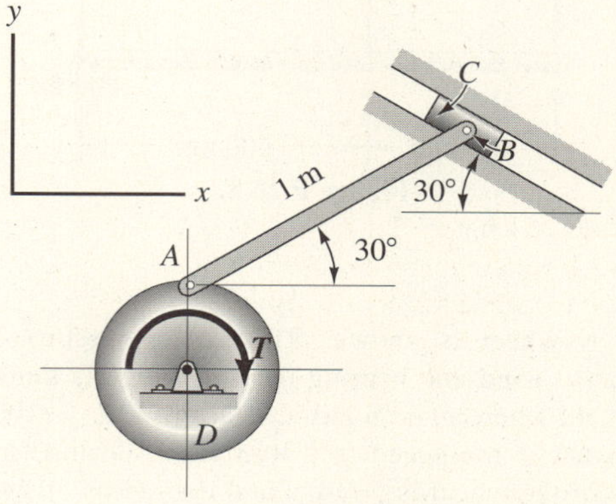

Figure P.16.16.

16.17. Do Problem 16.16 for the case where at the instant of interest $\omega_D = 2$ rad/s counterclockwise.

16.18. A torque T of 50 N m is applied to the device shown. The bent rods are of mass per unit length 5 kg/m. Neglecting the inertia of the shaft, how many rotations does the system make in 10 s? Are there forces coming onto the bearings other than from the dead weights of the system?

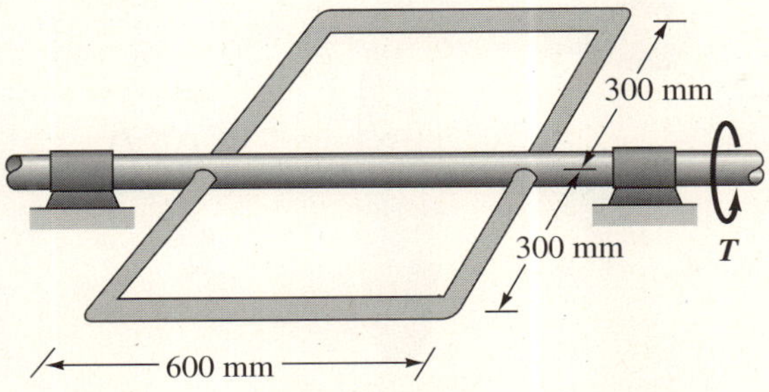

Figure P.16.18.

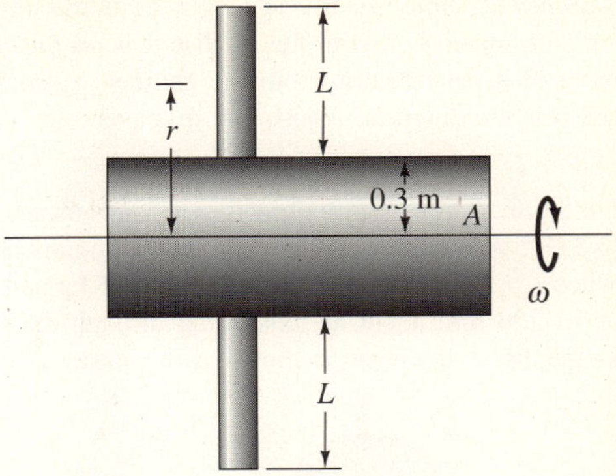

Figure P.16.20.

16.19. An idealized torque-versus-angular-speed curve for a shunt, direct-current motor is shown as curve A. The motor drives a pump which has a resisting torque-versus-speed curve shown in the diagram as curve B. Find the angular speed of the system as a function of time, after starting, over the range of speeds given in the diagram. Take the moment of inertia of motor, connecting shaft, and pump to be I.

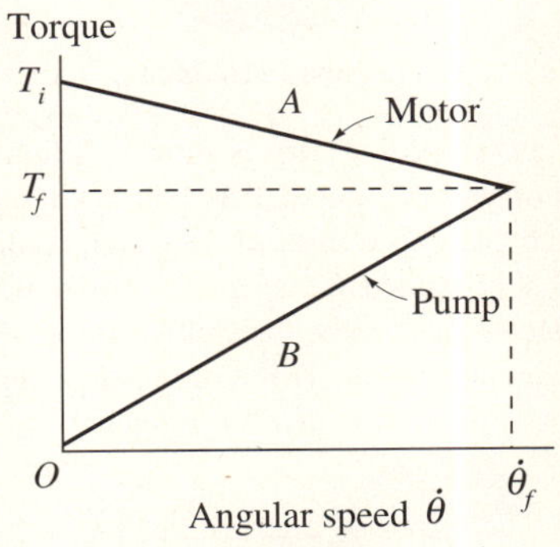

Figure P.16.19.

16.20. Rods of length L have been welded onto a rigid drum A. The system is rotating at a speed ω of 5000 r/min. By this time, you may have studied stress in a rod in your strength of materials class. In any case, the stress is the normal force per unit area of cross-section of the rod. If the cross-sectional area of the rod is 1.3×10^3 mm^2 and the mass per unit length is 7.5 kg/m, what is the normal stress τ_{rr} on a section at any position r? The length L of the rods is 0.6 m. What is τ_{rr} at $r = 0.45$ m? Consider the upper rod when it is vertical.

***16.21.** In Problem 16.20, consider that the mass per unit length varies linearly from 7.5 kg/m at $r = 0.3$ m (at the bottom of the rod) to 9 kg/m at $r = 0.9$ m (at the top of the rod). Find τ_{rr} at any position r and then compute τ_{rr} for $r = 0.45$.

16.22. A plate of mass 15 kg/m^2 is supported at A and B. What are the force components at B at the instant support A is removed?

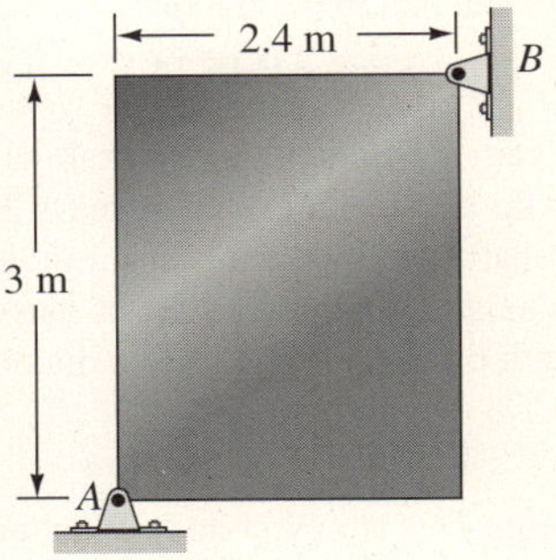

Figure P.16.22.

16.23. When the uniform rigid bar is horizontal, the spring at C is compressed 75 mm. If the bar has a mass of 25 kg, what is the force at B when support A is removed suddenly? The spring constant is 9 N/mm.

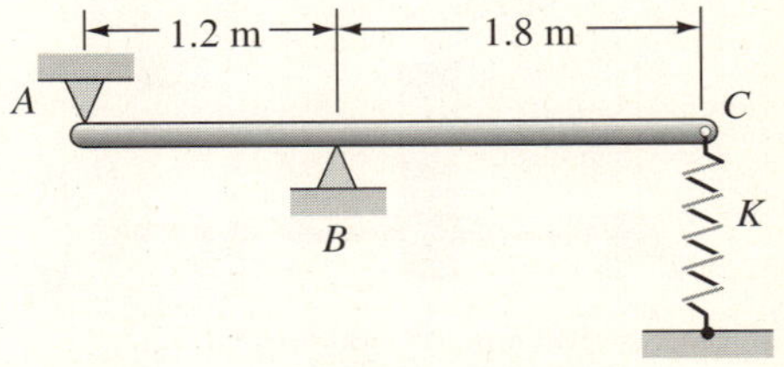

Figure P.16.23.

16.24. An electric motor E drives a light shaft through a coupling D which transmits only torque. A disc A is on the shaft and has its center of mass 200 mm from the geometric center of the disc as shown in the diagram. A torque T given as

$$T = 5 \times 10^{-3}t^2 + 30 \times 10^{-3}t \text{ N m}$$

is applied to the shaft from the motor (t is in seconds measured from when the system is at rest). What are the force components on the bearings when $t = 30$ s? The disc has a mass of 15 kg and a radius of gyration of 250 mm about an axis going through the center of mass. Take the disc at the position shown as the instant of interest.

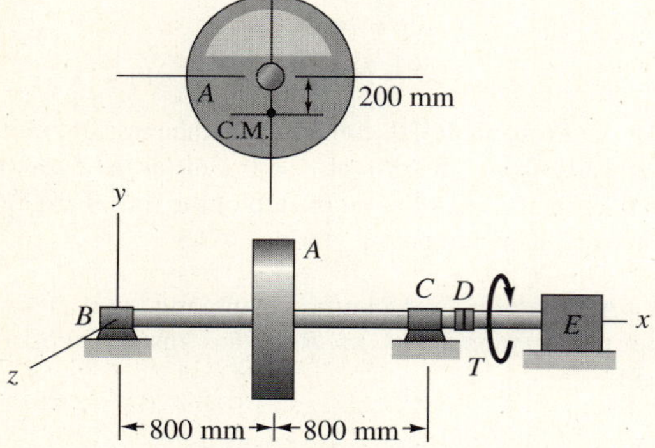

Figure P.16.24.

16.25. A single cam A is mounted on a shaft as shown. The cam has a mass of 10 kg and has a center of mass 300 mm from the centerline of the shaft. Also, the cam has a radius of gyration of 180 mm about an axis through the center of mass. The shaft has a mass per unit length of 10 kg/m and has a diameter of 30 mm. A torque T given as

$$T = 10^{-3}t^2 + 10 \text{ N m}$$

is applied at coupling D (t is in seconds). What are the force components in the bearings after 25 s if the cam has the position shown in the diagram at this instant?

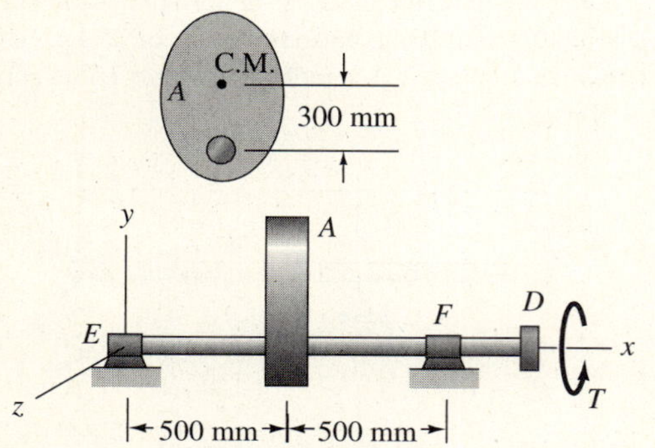

Figure P.16.25.

16.26. A circular plate is rotating at a speed of 10×10^3 r/min. A hole has been cut out of the plate so that the beam of light is allowed through for very short intervals of time. Such a device is called a *chopper*. If the plate was 9 kg originally and 85 g of material was removed for the hole, what are the forces in the bearings A and B from the circular plate for the instant shown?

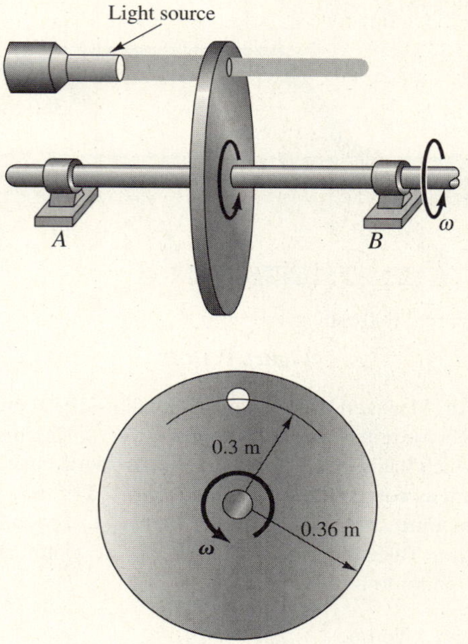

Figure P.16.26.

16.27. Part of a conveyor system is shown. A link belt is meshed around portions of gears A and B. The belt has a mass of 5 kg/m. Furthermore, each gear has a mass of 3 kg and a radius of 300 mm. If a force of 100 N is applied at one end as shown, what is the maximum possible force T that can be transmitted at the other end if $\ddot{\theta}_A = 5$ rad/s^2? The gears turn freely. The system is in a vertical plane.

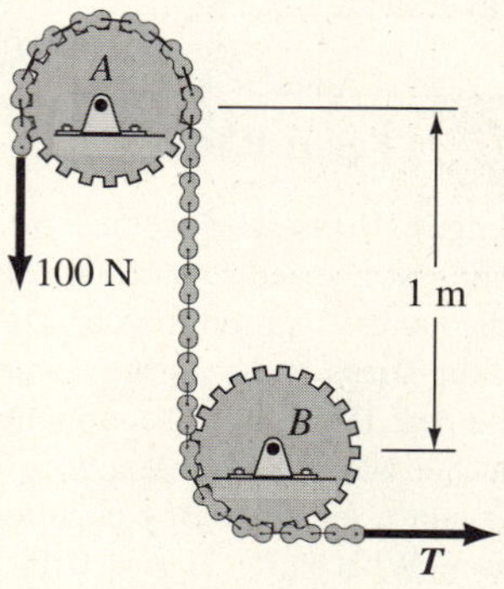

Figure P.16.27.

16.28. A hollow cylinder A of mass 100 kg can rotate over a stationary solid cylinder B having a mass of 70 kg. The surface of contact is lubricated so that there is a resisting torque between the bodies given as $0.2\,\dot{\theta}_A$ N m with $\dot{\theta}_A$ in radians per second. The outer cylinder is connected to a device at C which supplies a force equal to $-50\ddot{Y}$ N. Starting from a counter-clockwise angular speed of 1 rad/s, what is the angular speed of cylinder A after the force at C starts to move downward and moves 0.7 m?

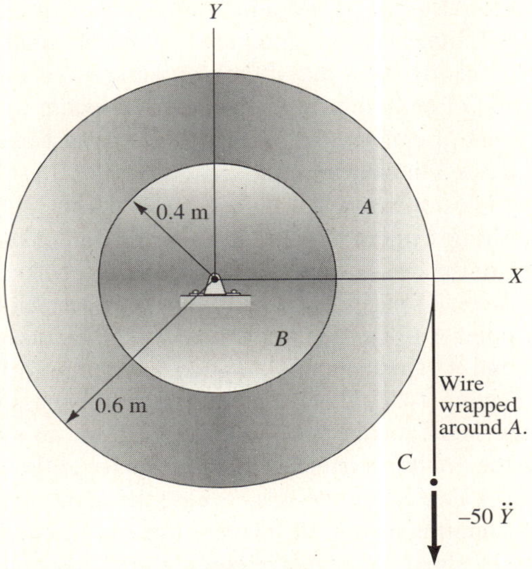

Figure P.16.28.

16.29. A box C of mass 70 kg rests on a conveyor belt. The driving drum B has a mass of 45 kg and a radius of gyration of 100 mm. The driven drum A has a mass of 32 kg and a radius of gyration of 75 mm. The belt has a mass of 4.5 kg/m. Supporting the belt on the top side is a set of 20 rollers each with a mass of 1.4 kg, a diameter of 50 mm, and a radius of gyration of 20 mm. If a torque T of 70 N m is developed on the driving drum, what distance does C travel in 1 s starting from rest? Assume that no slipping occurs.

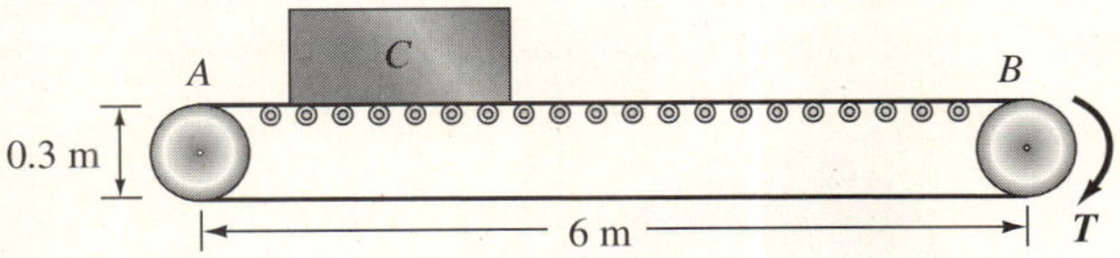

Figure P.16.29.

16.30. A uniform slender member is supported by a hinge at A. A force P is suddenly applied at an angle α with the horizontal. What value should P have and at what distance d should it be applied to result in zero reactive forces at A at the configuration shown if $\alpha = 45°$? The weight of the member AB is W. What is the angular acceleration of the bar for these conditions at the instant of interest?

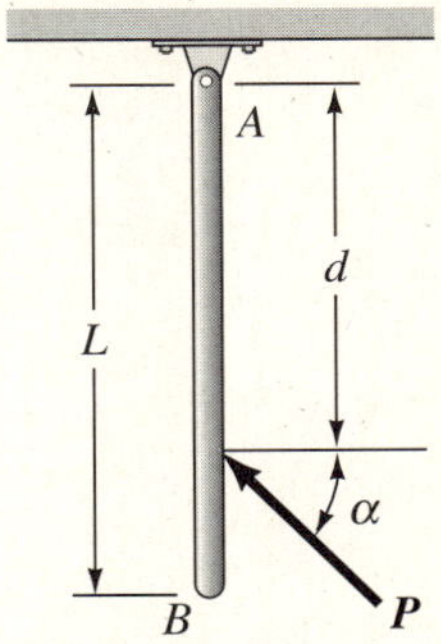

Figure P.16.30.

***16.31.** A rod AB is welded to a rod CD, which in turn is welded to a shaft as shown. The shaft has the following angular motion at time t:

$$\omega = 10 \text{ rad/s}$$
$$\dot{\omega} = 40 \text{ rad/s}^2$$

What are the shear force, axial force, and bending moment along CD at time t as a function of r? The rods have a mass per unit length of 5 kg/m. Neglect gravity.

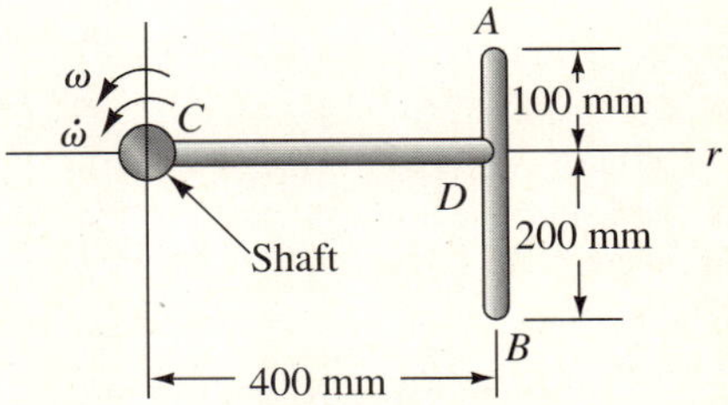

Figure P.16.31.

16.32. A four-bar linkage is shown (the ground is the fourth linkage). Each member is 300 mm long and has a mass per unit length of 10 kg/m. A torque T of 5 N m is applied to each of bars AB and DC. What is the angular acceleration of bars AB and CD?

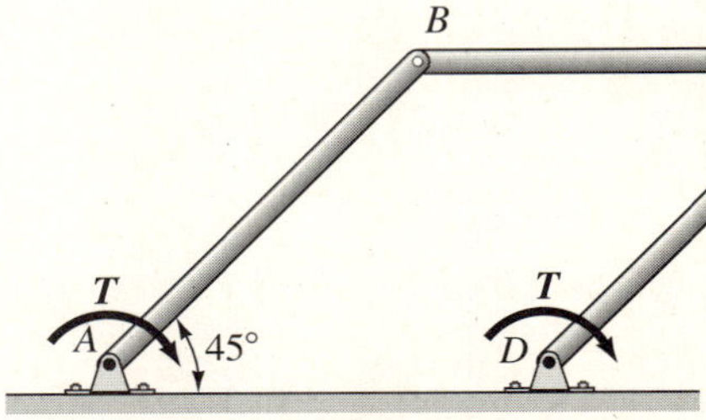

Figure P.16.32.

16.6 Rolling Slablike Bodies

We now consider the rolling without slipping of slablike bodies such as cylinders, spheres, or plane gears. As we have indicated in Chapter 15, the point of contact of the body has instantaneously *zero velocity*, and we have *pure instantaneous rotation about this contact point*. We pointed out that for getting velocities of points on such a rolling body, we could imagine that there is a *hinge* at the point of contact. Also, the acceleration of the *center* of a rolling without slipping sphere or cylinder can be computed using the simple formula—$R\ddot{\theta}$. Finally, you can readily show that if the angular speed is zero, we can compute the acceleration of any point in the cylinder or sphere by again imagining a *hinge* at the point of contact. For other cases, we must use more detailed kinematics, as discussed in Chapter 15.

A very important conclusion we reached in Chapter 15 for cylinders and spheres was that for rolling without slipping the acceleration of the contact point on the cylinder or sphere is *toward the geometric center of the cylinder or sphere*. If the center of mass of the body lies anywhere along the line AO from the contact point A to the geometric center O, then clearly we can use Eq. 16.6 for the point A. This action is justified since point A is then an example of case 3 in Section 16.2 (A accelerates toward the mass center and is part of the cylinder). Thus, for the body in Fig. 16.16 for no slipping we can use $T = I\alpha$ about the point of contact A of the cylinder at the instant shown. However, in Fig. 16.17 we cannot do this because the point of contact A of the cylinder is not accelerating toward the center of mass as in the previous case. We can use $T = I\alpha$ about the *center of mass* in the latter case.

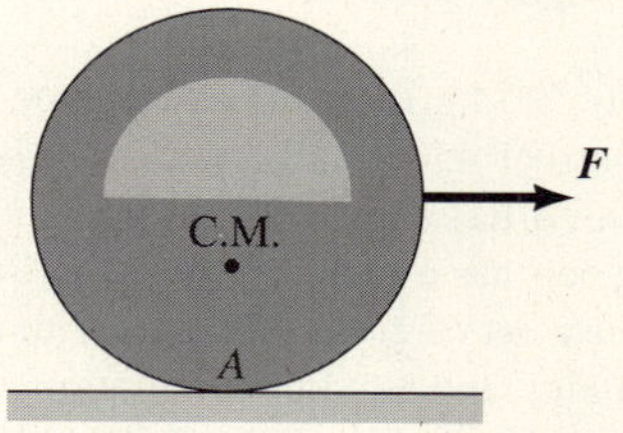

Figure 16.16. Point A accelerates toward center of mass.

We shall now examine a problem involving rolling without slipping. The equations of motion, you can readily deduce, are the same as in the previous section.

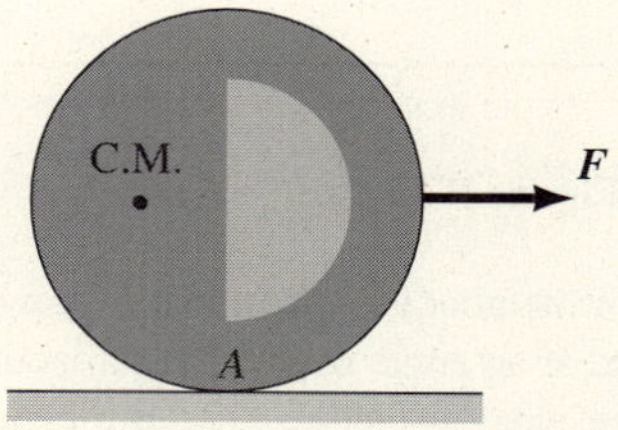

Figure 16.17. Point A does not accelerate toward center of mass.

Example 16.5

A steam roller is shown going up a 5° incline in Fig. 16.18. Wheels A have a radius of gyration of 0.45 m and a mass each of 220 kg, whereas roller B has a radius of gyration of 0.3 m and a mass of 2.2 Mg. The vehicle, minus the wheels and roller but including the operator, has a mass of 3.2 Mg with a center of mass positioned as shown in the diagram. The steam roller is to accelerate at the rate of 0.3 m/s^2. In part A of the problem, we are to determine the torque T_{eng} from the engine onto the drive wheels.

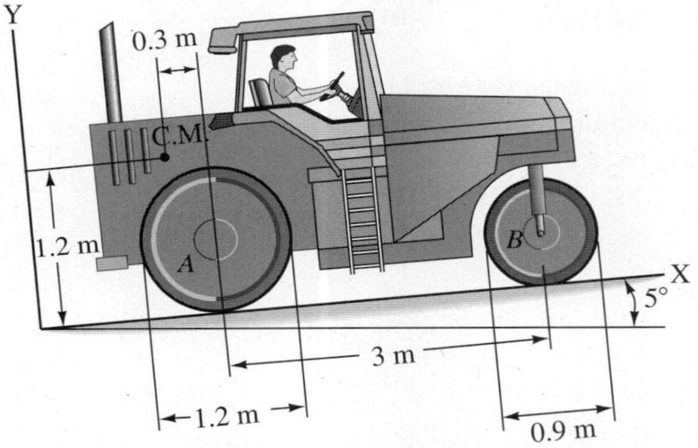

Figure 16.18. Steam roller moving up incline.

Part A. In Fig. 16.19, we have shown free-body diagrams of the drive wheels and the roller. Note we have combined the two drive wheels into a single 440 kg wheel. In each case, the point of contact on the wheel accelerates toward the mass center of the wheel and we can put to good

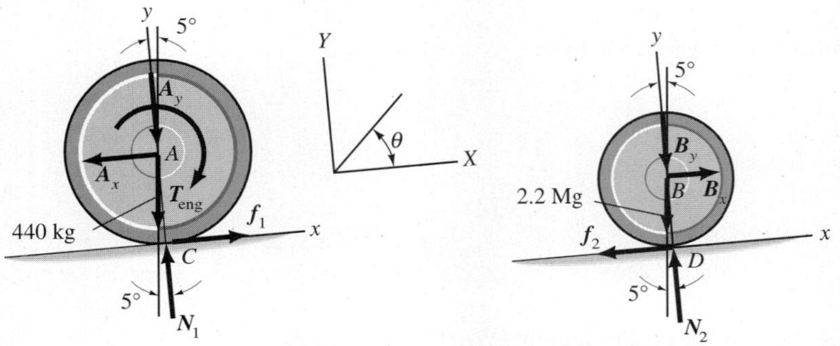

Figure 16.19. Free-body diagrams of driving wheels and roller.

use the **moment-of-momentum** equation (16.9) for the points of contact on the cylinders. Accordingly, we fix xyz to cylinder A and we fix another

Example 16.5 (Continued)

reference xyz to cylinder B at their respective points of contact as has been shown.[8] Hence, for cylinder A we have

$$(A_x + 440g \sin 5°)(0.6) - T_{eng} = 440(0.45^2 + 0.3^2)\ddot{\theta}_A \qquad \text{(a)}$$

where we have employed the parallel-axis theorem in computing the moment of inertia about the line of contact at C. Similarly, for the roller, we have

$$(2200g \sin 5° - B_x)(0.45) = 2200(0.3^2 + 0.45^2)\ddot{\theta}_B \qquad \text{(b)}$$

We have here two equations and no fewer than five unknowns. By considering the free body of the vehicle minus wheels shown diagrammatically in Fig. 16.20, we can say from **Newton's law** noting once again that $A_X \equiv A_x$, etc., because of the parallel orientation of axes XYZ with axes xyz of Fig. 16.19(a) and Fig. 16.19(b) at time t

$$A_x - B_x - 3200g \sin 5° = 3200(0.3) \qquad \text{(c)}$$

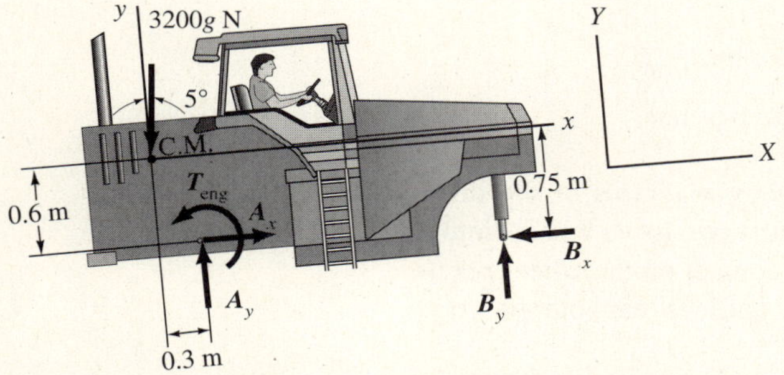

Figure 16.20. Free-body diagram of vehicle without wheels and roller.

Finally, from **kinematics** we can say:

$$\ddot{\theta}_A = -\frac{\ddot{X}}{-r_A} = -\frac{0.3}{0.6} = -0.5 \text{ rad/s}^2$$

$$\ddot{\theta}_B = -\frac{\ddot{X}}{-r_B} = -\frac{0.3}{0.45} = -0.667 \text{ rad/s}^2 \qquad \text{(d)}$$

where $\ddot{X} = 0.3$ m/s^2 is the acceleration of the vehicle up the incline. We can now readily solve the equations. We get B_x directly from Eq. (b) on

[8]We shall use for simplicity axes xyz fixed to different bodies in this and other problems. The directions will thus be the same and this will make for simplicity in the formulations. However, keep in mind that the computation of moments of inertia for any body applies to *the particular reference xyz* fixed to that body. Keeping this in mind, we should not encounter any difficulties and we shall benefit from the simpler notation.

Example 16.5 (Continued)

replacing $\ddot{\theta}_B$ by –0.667. Next, we get A_x from Eq. (c). Finally, going back to Eq. (a), we can solve for T_{eng}. The results are:

$$T_{eng} = 4.209 \text{ kN m}$$
$$A_x = 6.531 \text{ kN}$$
$$B_x = 2.835 \text{ kN}$$

Part B. Determine next the normal forces N_1 and N_2 at the wheels and roller, respectively.

We can express **Newton's law** for the wheels and roller in the direction normal to the incline by using the free-body diagrams of Fig. 16.19. Thus,

$$N_1 - A_y - 440g\cos 5° = 0 \qquad \text{(e)}$$

$$N_2 - B_y - 2200g\cos 5° = 0 \qquad \text{(f)}$$

Next, we consider the free body of the vehicle without the wheels and roller (Fig. 16.20). **Newton's law** in the y direction for the center of mass then becomes

$$A_y + B_y - 3200g\cos 5° = 0 \qquad \text{(g)}$$

The **moment-of-momentum** equation about the center of mass of the vehicle without wheels and roller is

$$A_x(0.6) - B_x(0.75) + A_y(0.3) + B_y(3.3) + T_{eng} = 0 \qquad \text{(h)}$$

We have four equations in four unknowns. Solve for A_y in Eq. (g), and substitute into Eq. (h). Inserting known values for A_x, B_x, and T_{eng}, we have

$$(6530)(0.6) - (2835)(0.75) + (3200g\cos 5° - B_y)(0.3) + (3.3)(B_y) + 4209 = 0$$

Therefore,

$$B_y = 5.128 \text{ kN}$$

Now from Eq. (g) we get A_y:

$$A_y = 3200g\cos 5° + 5128 = 36.401 \text{ kN}$$

Finally, from Eqs. (e) and (f) we get N_1 and N_2.

$$N_1 = 36.401 + 440g\cos 5° = 40.70 \text{ kN}$$

$$N_2 = -5.128 + 2200g\cos 5° = 16.37 \text{ kN}$$

Hence, on each wheel we have a normal force of 20.35 kN, and for the roller we have a normal force of 16.37 kN.

*Example 16.6

A gear A weighing 100 N is connected to a stepped cylinder B (see Fig. 16.21) by a light rod DC. The stepped cylinder weighs 1 kN and has a radius of gyration of 250 mm along its centerline. The gear A has a radius of gyration of 120 mm along its centerline. A force $F = 1.5$ kN is applied to the gear at D. What is the compressive force in member DC if, at the instant that F is applied, the system is stationary?

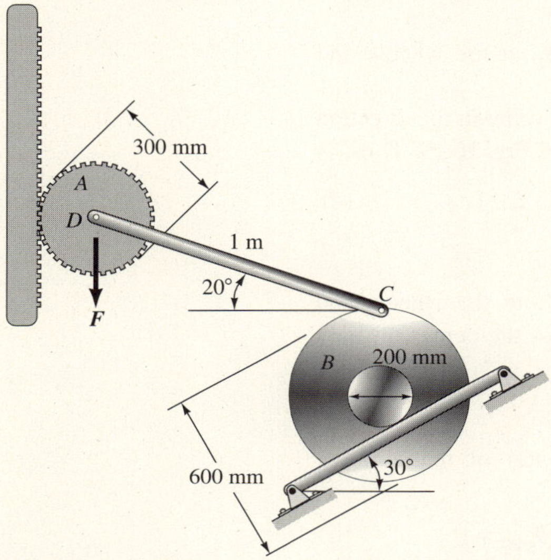

Figure 16.21. Stepped cylinder connected to a gear by a light rod.

Noting that DC is a two-force compressive member, we draw the free-body diagrams for the gear and the stepped cylinder in Fig. 16.22.

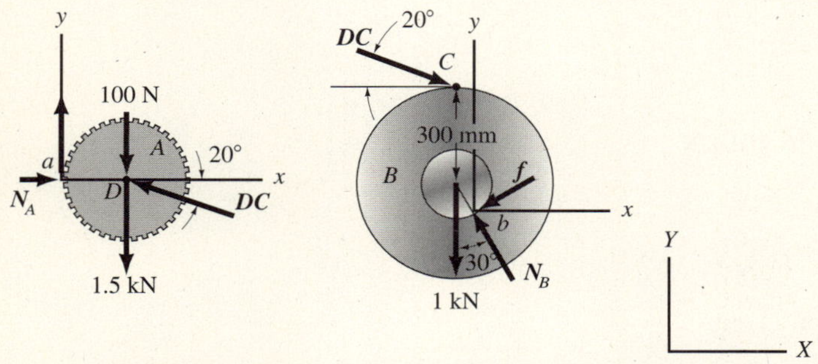

Figure 16.22. Free-body diagram of gear and cylinder.

The **moment-of-momentum** equations about the contact points for both bodies (points a and b, respectively) are

Example 16.6 (Continued)

$$-(100+1500)(0.15)+DC(\sin 20°)(0.15)=\left(\frac{100}{9.81}\right)\left[(0.12)^2+(0.15)^2\right]\ddot{\theta}_A$$

$$(DC\sin 20°+1000)(\sin 30°)(0.10)-(DC\cos 20°)(0.30+0.10\cos 30°)$$
$$=\frac{1000}{9.81}\left[(0.25)^2+(0.10)^2\right]\ddot{\theta}_B$$

These equations simplify to the following pair:

$$0.0513DC-240=0.376\ddot{\theta}_A \qquad \text{(a)}$$

$$-0.3462DC+50=7.39\ddot{\theta}_B \qquad \text{(b)}$$

Clearly, we need an equation from **kinematics** at this time. Considering rod DC, we can say:[9]

$$\boldsymbol{a}_c=\boldsymbol{a}_D+\dot{\boldsymbol{\omega}}_{DC}\times\boldsymbol{\rho}_{DC}+\boldsymbol{\omega}_{DC}\times(\boldsymbol{\omega}_{DC}\times\boldsymbol{\rho}_{DC})$$

$$\ddot{\theta}_B\,\boldsymbol{k}\times[(0.30+0.10\cos 30°)\,\boldsymbol{j}-(0.10\sin 30°)\,\boldsymbol{i}]$$
$$=\ddot{Y}_D\,\boldsymbol{j}+\dot{\omega}_{DC}\,\boldsymbol{k}\times(\cos 20°\,\boldsymbol{i}-\sin 20°\,\boldsymbol{j})+\boldsymbol{0}$$

The scalar equations are

$$-0.3866\ddot{\theta}_B=0.342\dot{\omega}_{DC} \qquad \text{(c)}$$

$$-0.05\ddot{\theta}_B=\ddot{Y}_D+0.940\dot{\omega}_{DC} \qquad \text{(d)}$$

Also, from **kinematics** we can say, considering gear A:

$$\ddot{Y}_D = 0.15\ddot{\theta}_A \qquad \text{(e)}$$

Multiply Eq. (c) by 0.940/0.342 and rewrite Eq. (d) below it with $\ddot{Y}_D$ replaced by using Eq. (e):

$$-1.063\ddot{\theta}_B=0.940\dot{\omega}_{DC}$$

$$-0.05\ddot{\theta}_B-0.15\ddot{\theta}_A=0.940\dot{\omega}_{DC}$$

Subtracting, we get

$$-1.013\ddot{\theta}_B+0.15\ddot{\theta}_A=0$$

Therefore,

$$\ddot{\theta}_A = 6.75\ddot{\theta}_B \qquad \text{(f)}$$

Solving Eq. (a), (b), and (f) simultaneously gives us for DC the result

$$DC = 1.51 \text{ kN (compression)}$$

Also, note that $\ddot{Y}$ comes out negative indicating that D accelerates downward.

[9]Note that because cylinder B has zero angular velocity, we can imagine it to be hinged at b for computing $\boldsymbol{a}_C$. Also, we do not know the sign of $\ddot{Y}_D$ and so we leave it as positive and thus let the mechanics yield the correct sign at the end of the calculations.

16.7 General Plane Motion of a Slablike Body

We now consider *general plane motion* of slablike bodies. The motion to be studied will be parallel to the plane of symmetry. Accordingly, we use the center of mass. The angular velocity vector $\boldsymbol{\omega}$ will be normal to the plane of symmetry, and, in accordance with Chasles' theorem, will be taken to pass through the center of mass. The translational velocity vector V_c will be parallel to the plane of symmetry. We fix a reference xyz at the center of mass of the body such that the xy plane coincides with the plane of symmetry as shown in Fig. 16.23. As usual, take the inertial reference parallel to xyz at time t. Note that the actual instantaneous axis of rotation is also shown. For the same

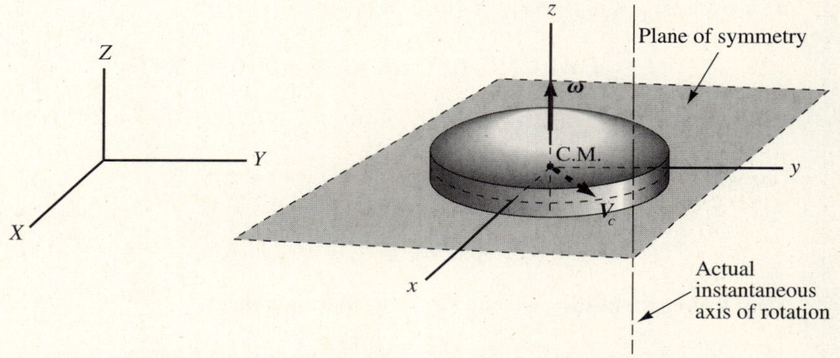

Figure 16.23. Slablike body undergoing general plane motion parallel to XY.

reasons used in Section 16.5 for slablike bodies, the *moment-of-momentum* equations become

$$M_x = 0 \tag{16.14a}$$

$$M_y = 0 \tag{16.14b}$$

$$M_z = I_{zz}\dot{\omega} \tag{16.14c}$$

Furthermore, considering the center of mass, we must have *equilibrium* in the z direction.

$$\sum F_z = 0 \tag{16.15}$$

while the full form of *Newton's law* holds in the x and y directions.

As in Section 16.5, we can generalize the results of this section to include the plane motion of a body having at the center of mass a *principal* axis z normal to the plane of motion XY. Clearly, for principal axes xyz at A, we get the same equations of motion as for the slablike body.

Example 16.7

Find the acceleration of block B shown in Fig. 16.24. The system is in a vertical plane and is released from rest. The cylinders roll without slipping along the vertical walls and along body B. Neglect friction along the guide rod. The 150 N m torque M_A is applied to cylinder A.

We first draw free-body diagrams of the three bodies comprising the system as shown in Fig. 16.25 where it will be noticed that we have deleted the horizontal forces since they play no role in this problem. As usual, XYZ is our inertial reference. Points a, b, d, and e in F.B.D. I and F.B.D. III, respectively, are contact points in the respective bodies where, we repeat, there is rolling without slipping. Furthermore, it should be clear that points a and b are accelerating toward respective mass centers. Accordingly we fix xyz to the cylinders at these points.

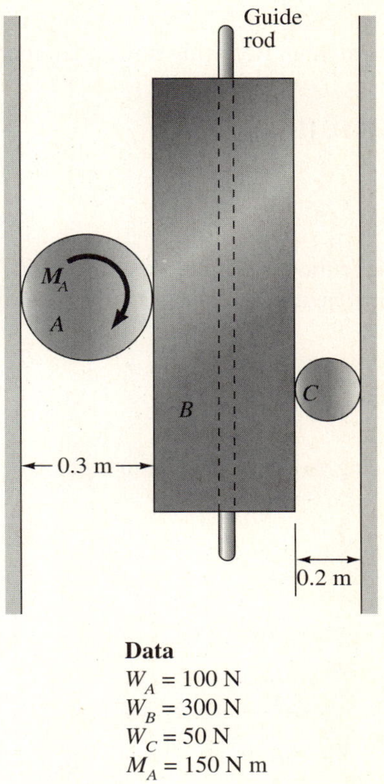

Figure 16.24. A block and two cylinders in a vertical plane.

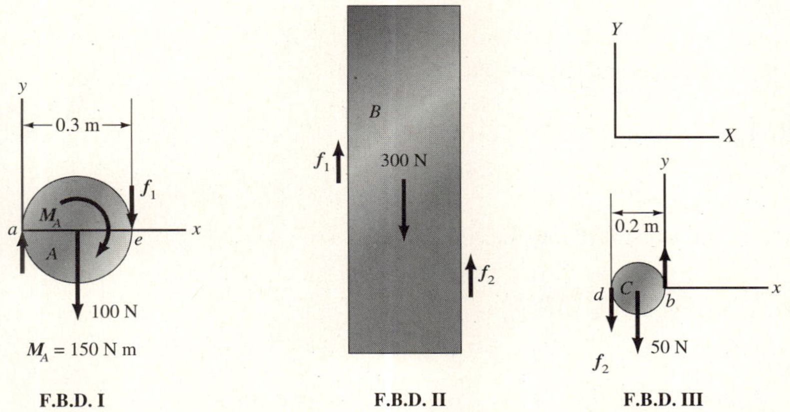

Figure 16.25. Free-body diagrams of the system elements with horizontal forces deleted.

We may immediately write the **moment-of-momentum** equations for the two cylinders about their respective points of contact a and b. Thus

F.B.D. I

$$-(f_1)(0.3)-(100)(0.15)-150=\left[\frac{1}{2}\frac{100}{g}(0.15)^2+\frac{100}{g}(0.15)^2\right]\ddot{\theta}_A$$

$$\therefore -0.3f_1-165=0.344\ddot{\theta}_A \qquad \text{(a)}$$

F.B.D. III

$$(f_2)(0.2)+(50)(0.1)=\left[\frac{1}{2}\frac{50}{g}(0.1)^2+\frac{50}{g}(0.1)^2\right]\ddot{\theta}_C$$

$$0.2f_2+5=0.07645\ddot{\theta}_C \qquad \text{(b)}$$

Example 16.7 (Continued)

Next going to F.B.D. II we employ **Newton's law** since we have simple translation. Referring now to the inertial reference *XYZ* we have

F.B.D. II

$$-300 + f_1 + f_2 = \frac{300}{g}\ddot{Y}_B \qquad \text{(c)}$$

Since the three bodies are interconnected by nonslip rolling conditions we must next consider the **kinematics** of the system. Thus

$$0.3\ddot{\theta}_A = \ddot{Y}_B$$
$$0.2\ddot{\theta}_C = -\ddot{Y}_B$$

Using the above results to replace $\ddot{\theta}_A$ and $\ddot{\theta}_C$ in Eqs. (a) and (b) we get

$$-0.3f_1 - 165 = 1.1467\ddot{Y}_B \qquad \text{(d)}$$
$$0.2f_2 + 5 = -0.3823\ddot{Y}_B \qquad \text{(e)}$$

Now solve for f_1 and f_2 in the above equations and substitute into Eq. (c). We get

$$-300 + \frac{1}{0.3}(-1.1467\ddot{Y}_B - 165) + \frac{1}{0.2}(-0.3823\ddot{Y}_B - 5) = \frac{300}{g}\ddot{Y}_B$$
$$36.31\ddot{Y}_B = -875$$

$$\therefore \quad \ddot{Y}_B = -24.10 \text{ m/s}^2$$

Now from Eqs. (d) and (e) we can determine f_1 and f_2.

$$f_1 = -457.9 \text{ N} \qquad f_2 = 21.07 \text{ N}$$

Thus, cylinder *A* forces body *B* downward while cylinder *C* resists this motion.

Notice, unless we want to determine the friction forces at the walls there is no need to use **Newton's law** for the cylinders. Also note that we could not use the **moment-of-momentum** equations for points *c* and *d* of the cylinders even though there is rolling without slipping there. The reason for this, as you must know, is that these points *do not accelerate toward or away from the mass centers of the cylinders.*

Example 16.8

A stepped cylinder having a weight of 450 N and a radius of gyration k of 300 mm is shown in Fig. 16.26(a). The radii R_1 and R_2 are, respectively, 300 mm and 600 mm. A total pull T equal to 180 N is exerted on the ropes attached to the inner cylinder. What is the ensuing motion? The coefficients of static and dynamic friction between the cylinder and the ground are, respectively, 0.1 and 0.08.

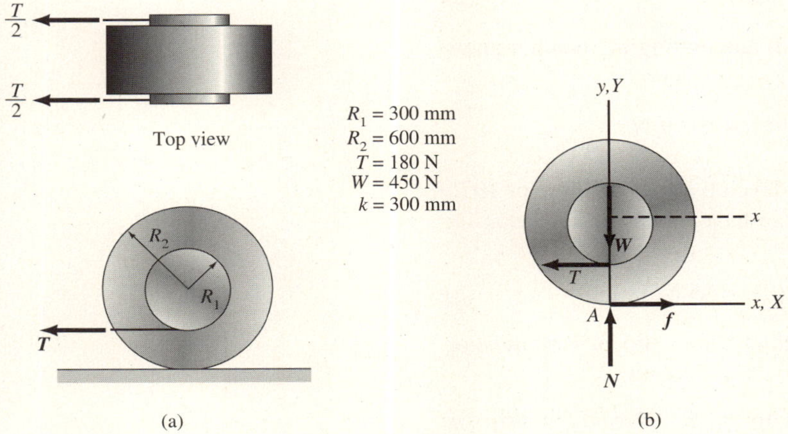

Figure 16.26. (a) Stepped cylinder; (b) free-body diagram of cylinder. *XY* is stationary.

A free-body diagram of the cylinder is shown in Fig. 16.26(b). Let us assume first that there is *no slipping* at the contact surface. Of course, we will have to later check this supposition. We have then pure instantaneous rotation about contact point A. Fix xyz to the body at A. XYZ as usual is stationary. We can then say for the **moment-of-momentum** equation about the axis of contact:

$$T(R_2 - R_1) = \left(\frac{W}{g}k^2 + \frac{W}{g}R_2^2\right)\ddot{\theta} \qquad \text{(a)}$$

wherein we have used the parallel-axis theorem for moment of inertia. Inserting numerical values, we can solve directly for $\ddot{\theta}$ at the instant that the force T is applied. Thus,

$$(180)(0.3) = \left[\frac{450}{g}(0.3)^2 + \frac{450}{g}(0.6)^2\right]\ddot{\theta}$$

Therefore,

$$\ddot{\theta} = 2.62 \text{ rad/s}^2 \qquad \text{(b)}$$

Example 16.8 (Continued)

We must now check our assumption of no slipping. Employ **Newton's law** for the mass center. In the X direction we get

$$-180 + f = \frac{W}{g}\ddot{X} \tag{c}$$

Using **kinematics** we note that

$$\ddot{X} = -R_2\ddot{\theta} = -0.6\ddot{\theta} \tag{d}$$

Substituting into Eq. (c) for $\ddot{X}$ using Eq. (d) and putting in known numerical values, we can solve for f:

$$f = 180 - \frac{450}{9.81}[(0.6)(2.62)] = 107.9 \text{ N}$$

Thus, for no slipping, we must be able to develop a friction force of 107.9 N. The maximum friction force that we can have, however, is, according to **Coulomb's law**,

$$f_{\text{max}} = W\mu_s = (450)(0.1) = 45 \text{ N} \tag{e}$$

Accordingly, we must conclude that the cylinder *does* slip, and we must reexamine the problem as a *general plane-motion* problem.

Using μ_d = 0.08, we now take f to be 36 N and employ the **moment-of-momentum** equation for the **center of mass** with xyz now fixed at the **center of mass**. We then have (Fig. 16.26(b))

$$fR_2 - TR_1 = \frac{W}{g}k^2\ddot{\theta} \tag{f}$$

Inserting numerical values, we get for $\ddot{\theta}$:

$$\ddot{\theta} = -7.85 \text{ rad/s}^2 \tag{g}$$

Now, using **Newton's law** in the X direction for the mass center, we get

$$-T + f = \frac{W}{g}\ddot{X} \tag{h}$$

Inserting numerical values, we get for $\ddot{X}$:

$$\ddot{X} = -3.14 \text{ m/s}^2 \tag{i}$$

Thus, the cylinder has a linear acceleration of 3.14 m/s^2 to the left and an angular acceleration of 7.85 rad/s^2 in the clockwise direction. Equations (g) and (i) are valid at all times, so we can integrate them if we like to get θ and X at any time t.

Example 16.9

A 4.905 kN flywheel rotating at a speed ω of 200 r/min (see Fig. 16.27) breaks away from the steam engine that drives it and falls on the floor. If the coefficient of dynamic friction between the floor and the flywheel surface is 0.4, at what speed will the flywheel axis move after 2 s? At what speed will it hit the wall *A*? The radius of gyration of the flywheel is 1 m and its diameter is 2.30 m. Do not consider effects of bouncing in your analysis. Neglect rolling resistance (Section 7.7) and wind friction losses.

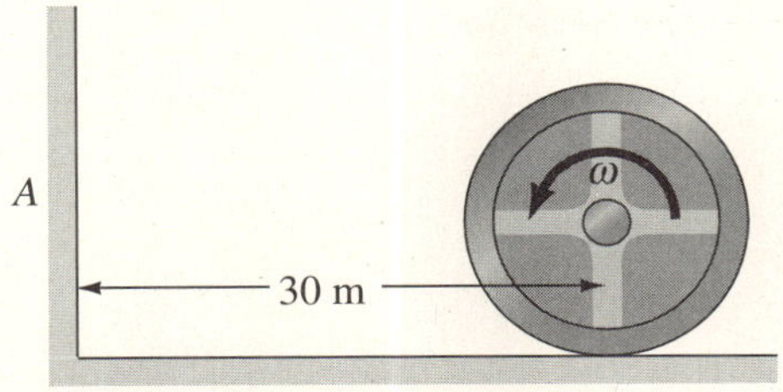

Figure 16.27. Runaway flywheel at initial position.

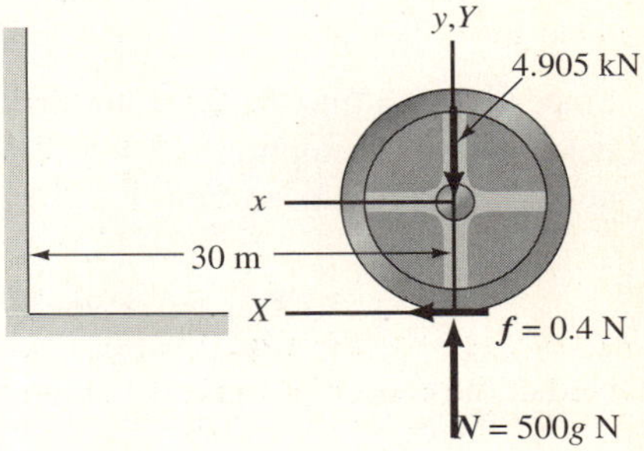

Figure 16.28. *xy* fixed at initial position.

We assume slipping occurs when the flywheel first touches the floor (see Fig. 16.28). **Newton's law** for the center of mass of the flywheel is

$$(0.4)N = \left(\frac{4905}{9.81}\right)\ddot{X}$$

Therefore,

$$\ddot{X} = 3.92 \text{ m/s}^2$$

Integrate twice:

$$\dot{X} = 3.92t + C_1 \qquad \text{(a)}$$

$$X = 1.962t^2 + C_1 t + C_2 \qquad \text{(b)}$$

At $t = 0$, $\dot{X} = 0$ and $X = 0$. Hence, $C_1 = 0$ and $C_2 = 0$. The **moment-of-momentum** equation for axes fixed to the body at the center of mass is next given.

$$(0.4)(N)\left(\frac{2.30}{2}\right) = \left(\frac{4905}{9.81}\right)(1)^2\ddot{\theta}$$

Therefore,

$$\ddot{\theta} = 4.51 \text{ rad/s}^2$$

Integrate twice:

$$\dot{\theta} = 4.51t + C_3 \qquad \text{(c)}$$

$$\theta = 2.26t^2 + C_3 t + C_4 \qquad \text{(d)}$$

Example 16.9 (Continued)

When $t = 0$, $\theta = 0$, and $\dot{\theta} = -(200)(2\pi/60) = -20.94$ rad/s. Hence, $C_3 = -20.94$ and $C_4 = 0$.

We now ask when does the slipping stop? Clearly, it stops when there is *zero velocity* of the point of contact of the cylinder.[10] From **kinematics** we have for this condition:

$$\dot{X} + \left(\frac{2.30}{2}\right)\dot{\theta} = 0 \qquad \text{(e)}$$

Substituting from Eq. (a) and (c) for $\dot{X}$ and $\dot{\theta}$, respectively, we have for Eq. (e):

$$3.92t + \left(\frac{2.30}{2}\right)(4.51t - 20.94) = 0$$

Therefore,

$$t = 2.64 \text{ s}$$

Since we get a time here greater than zero, we can be assured that the initial slipping assumption is valid. The position $X_{\text{N.S.}}$ at the time of initial no-slipping is deduced from Eq. (b). Thus,

$$X_{\text{N.S.}} = (1.962)(2.64)^2 = 13.67 \text{ m}$$

Accordingly, the flywheel hits the wall *after* it starts rolling without slipping. At $t = 2$ s, there is still slipping, and we can use Eq. (a) to find $\dot{X}$ at this instant. Thus,

$$(\dot{X})_{2\text{sec}} = (3.92)(2) = 7.85 \text{ m/s}$$

The speed, once there is no further slipping, is constant, and so the speed at the wall is found by using $t = 2.64$ s in Eq. (a). Thus,

$$(\dot{X})_{\text{wall}} = (3.92)(2.64) = 10.35 \text{ m/s}$$

[10]Note that the friction will accelerate the center of mass of the flywheel (which starts out with a zero velocity) while at the same time friction will decrease the angular speed of the flywheel (which starts out at its maximum angular speed). Thus the contact point of the flywheel will be subject to two opposing speeds, one of which is increasing and the other of which is decreasing. When there occurs a cancellation of these speed, we have rolling without slipping and there ceases to be coulombic friction present. And so, neglecting the other resistances to motion there ceases to be any change of speed of the flywheel.

PROBLEMS

16.33. A stepped cylinder is released from a rest configuration where the spring is stretched 200 mm. A constant force F of 360 N acts on the cylinder, as shown. The cylinder has a mass of 146 kg and has a radius of gyration of 1 m. What is the friction force at the instant the stepped cylinder is released? Take $\mu_s = 0.3$ for the coefficient of friction. The spring constant K is 270 N/m.

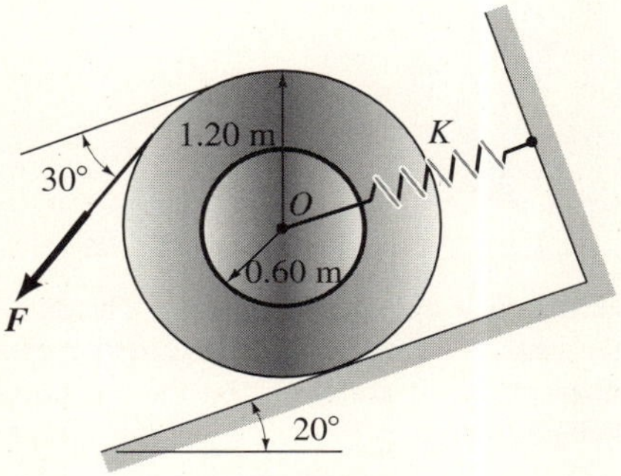

Figure P.16.33.

16.34. A stepped cylinder is held on an incline with an inextensible cord wrapped around the inner cylinder and an outside agent (not shown). If the tension T on the cord at the instant that the cylinder is released by the outside agent from the position shown is 450 N, what is the initial angular acceleration? What is the acceleration of the mass center? Use the following data:

$$W = 1.35 \text{ kN}$$
$$k = 0.9 \text{ m}$$
$$R_1 = 0.6 \text{ m}$$
$$R_2 = 1.2 \text{ m}$$
$$\mu_s = 0.1$$

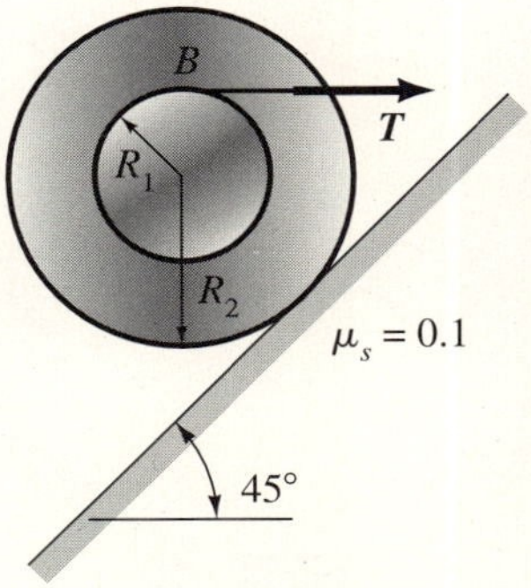

Figure P.16.34.

16.35. The cylinder shown is acted on by a 500 N force. At the contact point A, there is viscous friction such that the friction force is given as

$$f = 0.015V_A$$

where V_A is the velocity of the cylinder at the contact point in m/s. The mass of the cylinder is 14 kg, and the radius of gyration k is 0.3 m. Set up a third-order differential equation for finding the position of O as a function of time.

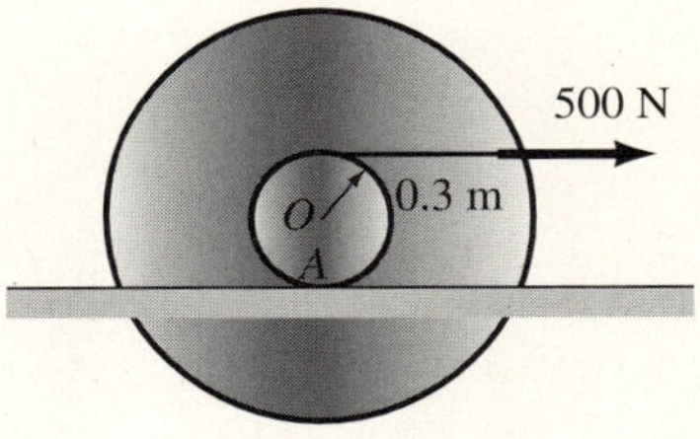

Figure P.16.35.

16.36. The cylinder shown weighs 445 N and has a radius of gyration of 0.27 m. What is the minimum coefficient of friction at A that will prevent the body from moving? Using half of this coefficient of friction, how far d does point O move in 1.2 s if the cylinder is released from rest?

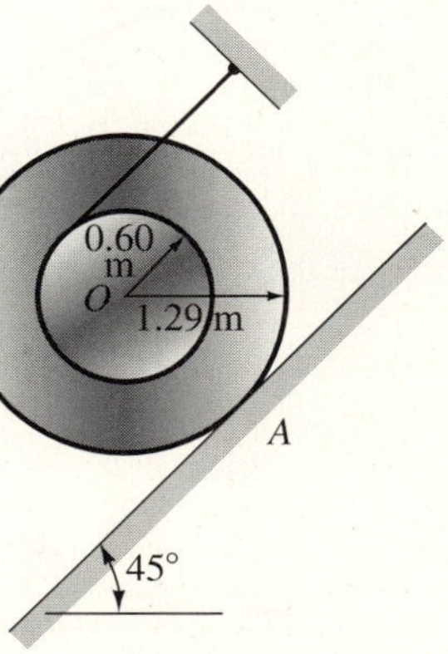

Figure P.16.36.

16.37. The velocities of two points of a cylinder, namely A and B, are

$$V_A = 6 \text{ m/s} \quad V_B = 2 \text{ m/s}$$

What is the velocity of point D? If the cylinder has a mass of 4.2 kg, what is the angular acceleration for a dynamic coefficient of friction $\mu_d = 0.35$?

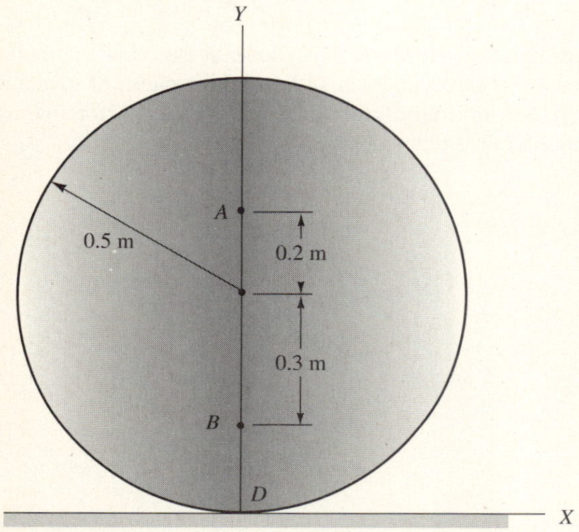

Figure P.16.37.

16.38. A thin ring having a mass of 0.4 kg is released from rest and rolls without slipping under the action of a 10 N force. Two identical metal sectors each of mass 0.83 kg are attached to the ring. What is the angular acceleration of the ring? Each sector has a radius of gyration at its centroid equal to $k = 0.18$ m.

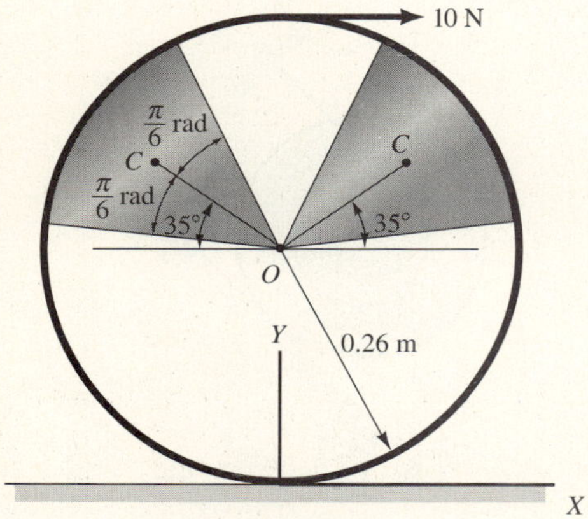

Figure P.16.38.

16.39. A light rod AB connects a plate C with a cylinder D which may roll without slipping. A torque T of value 70 N m is applied to plate C. What is the angular acceleration of cylinder D when the torque is applied? The plate has a mass of 50 kg and the cylinder 100 kg.

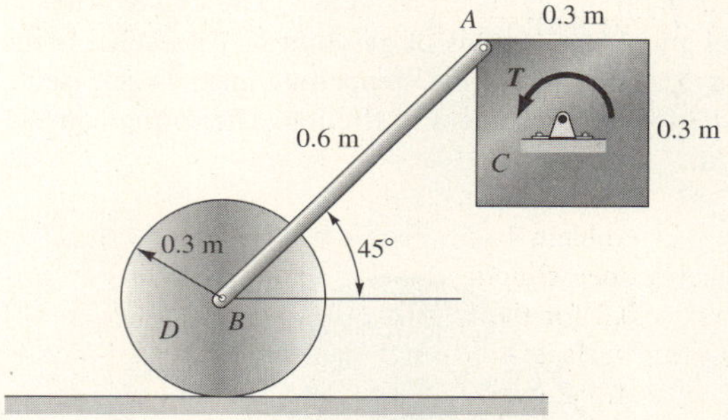

Figure P.16.39.

16.40. A semicircular cylinder A is shown. The diameter of A is 0.3 m, and the mass is 50 kg. What is the angular acceleration of A at the position shown if at this instant there is no slipping and the semicylinder is stationary?

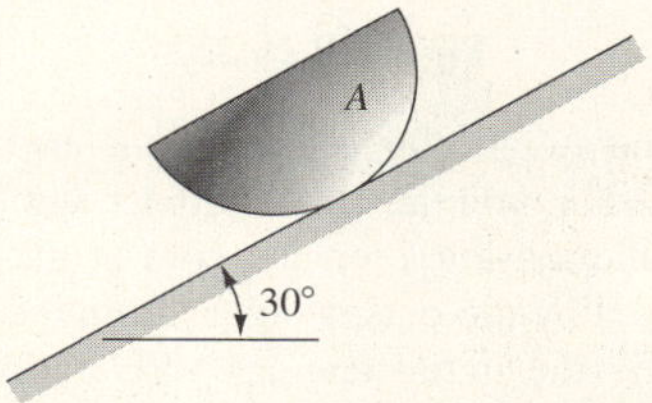

Figure P.16.40.

16.41. A 20 kg bar AB connects two gears G and H. These gears each have a mass of 5 kg and a diameter of 300 mm. A torque T of 5 N m is applied to gear H. What is the angular speed of H after 20 s if the system starts from rest? The system is in a horizontal plane. Bar AB is 2 m long.

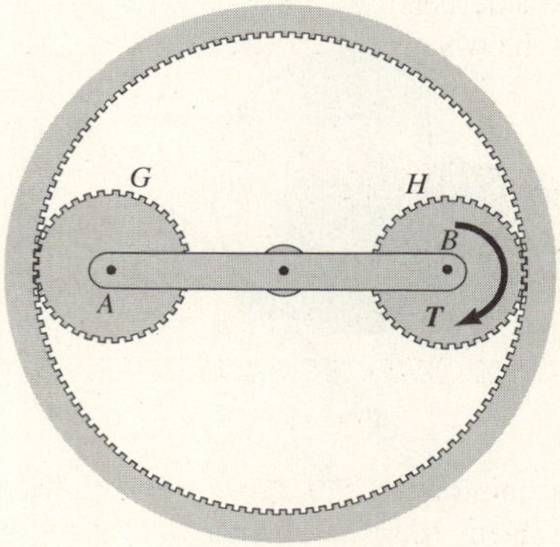

Figure P.16.41.

16.42. A bar C weighing 445 N rolls on cylinders A and B, each weighing 223 N. What is the acceleration of bar C when the 90 N force is applied as shown? There is no slipping.

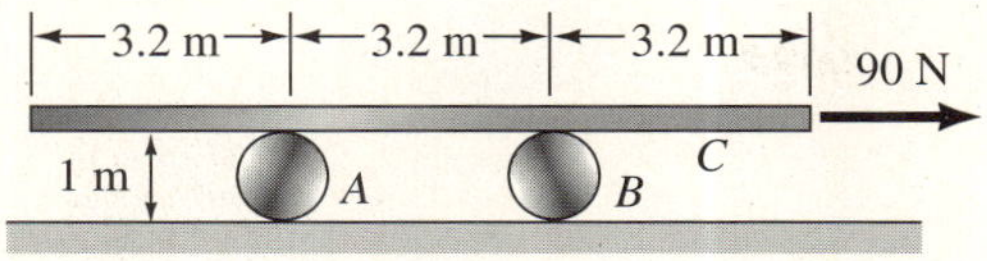

Figure P.16.42.

16.43. In Problem 16.42, at what position of the bar relative to the wheels does slipping first occur after the force is applied? Take $\mu_s = 0.2$ for the bottom contact surface and $\mu_s = 0.1$ for the contact surface between bar and cylinders. From Problem 16.42, $\ddot{x}_c = 1.442$ m/s^2.

16.44. A platform B, of mass 15 kg and carrying block A of mass 50 kg, rides on gears D and E as shown. If each gear has a mass of 15 kg, what distance will platform B move in 0.1 s after the application of a 500 N force as shown?

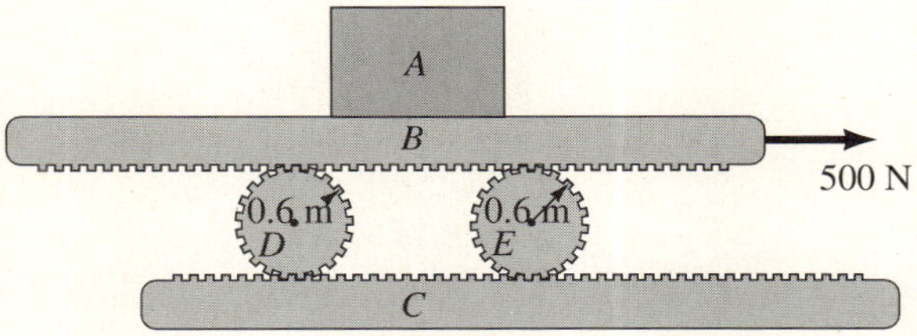

Figure P.16.44.

16.45. A crude cart is shown. A horizontal force P of 500 N is applied to the cart. The coefficient of static friction between wheels and ground is 0.6. If $D = 0.9$ m, what is the acceleration of the cart to the right? The mass of each wheel is 25 kg. Neglect friction in the axle bearings. The total mass of cart with load is 146 kg. Treat the wheels as simple solid cylinders.

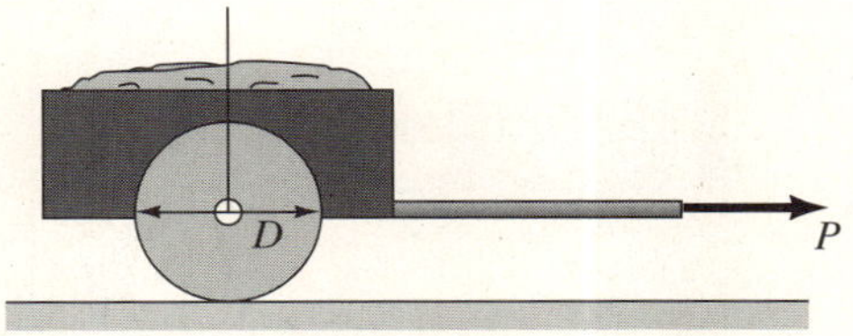

Figure P.16.45.

16.46. What minimum force component P is required to cause the cart in Problem 16.45 to move so that the wheels slip rather than roll without slipping?

16.47. A pulley system is shown. Sheave A has a mass of 25 kg and has a radius of gyration of 250 mm. Sheave B has a mass of 15 kg and has a radius of gyration of 150 mm. If released from rest, what is the acceleration of the 50 kg block? There is no slipping.

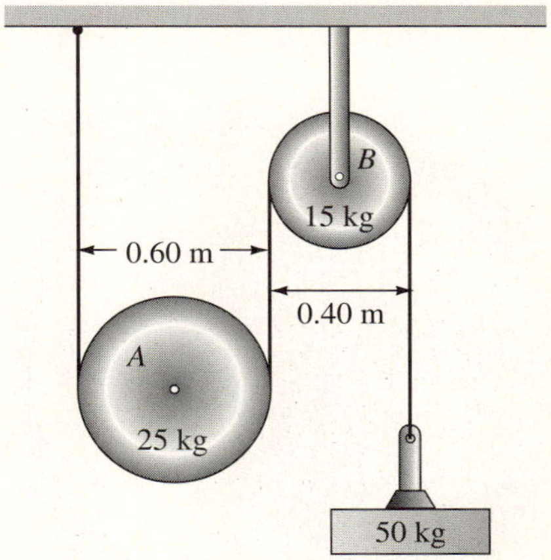

Figure P.16.47.

16.48. A steam locomotive drive system is shown. Each drive wheel weighs 5 kN and has a radius of gyration of 400 mm. At the instant shown, a pressure $p = 0.5$ N/mm^2 above atmospheric acts on the piston to drive the train backward. If the train is moving at 1 m/s backward at the instant shown, what is its acceleration? Members AB and BC are to be considered stiff but light in comparison to other parts of the engine. Also, the piston assembly can be considered light. Only the driving car is in action in this problem. It has one driving system, on each side of the locomotive as shown below. It has two additional wheels of the size and mass described above on each side of the locomotive plus additional small wheels whose rotational inertia we shall neglect. The drive train minus its eight large wheels has a weight of 150 kN. Assume no slipping, and neglect friction in the piston assembly.

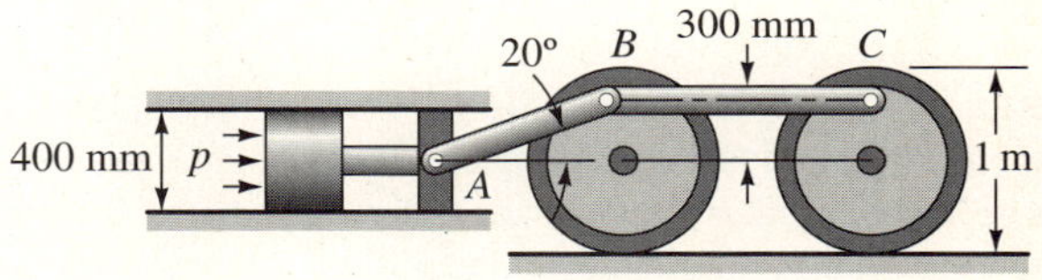

Figure P.16.48.

16.49. A system of interconnected gears is shown. Gear B rotates about a fixed axis, and gear D is stationary. If a torque T of 2.5 N m is applied to gear B at the configuration shown, what is the angular acceleration of gear A? Gear A has a mass of 1.36 kg while gear B has a mass of 4.55 kg. The system is in a vertical orientation relative to the ground. What vertical force is transmitted to stationary gear D?

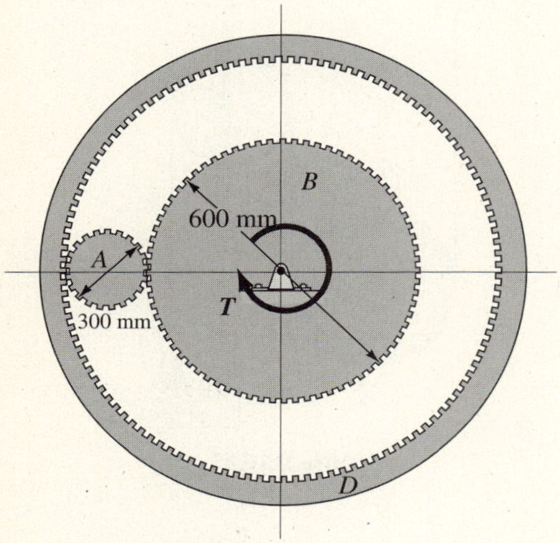

Figure P.16.49.

16.50. A solid semicircular cylinder of weight W and radius R is released from rest from the position shown. What is the friction force at that instant?

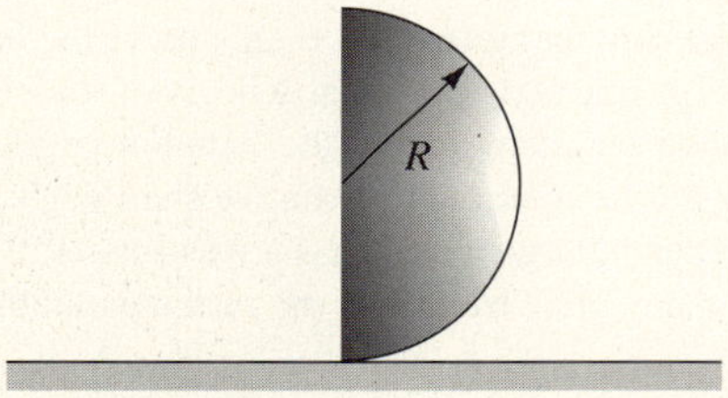

Figure P.16.50.

16.51. A cylinder is shown made up of two semicylinders A and B of mass 7 kg and 14 kg, respectively. If the cylinder has a diameter of 75 mm, what is the angular acceleration when it is released from a stationary configuration at the position shown? Assume no slipping.

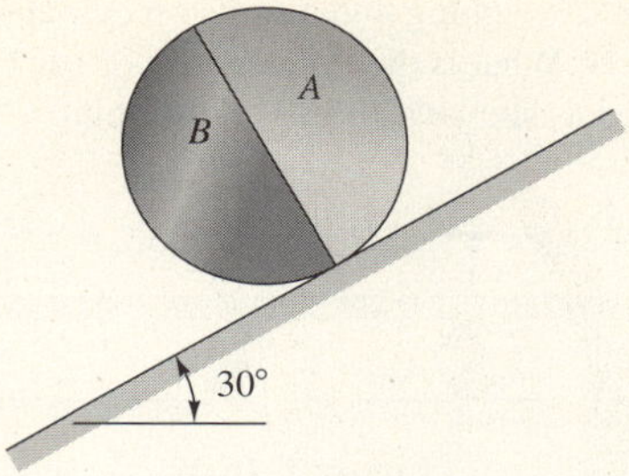

Figure P.16.51.

16.52. A thin-walled cylinder is shown held in position by a cord AB. The cylinder has a mass of 10 kg and has an outside diameter of 600 mm. What are the normal and friction forces at the contact point C at the instant that cord AB is cut? Assume that no slipping occurs.

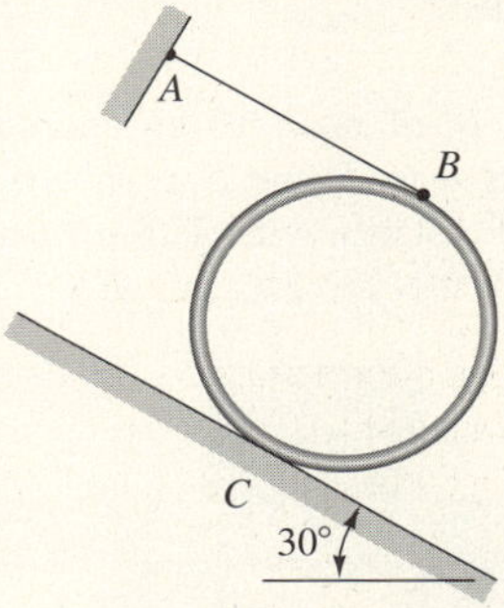

Figure P.16.52.

16.53. A bent rod $CBEF$ is welded to a shaft. At the ends C and F are identical gears G and H, each of mass 3 kg and radius of gyration 70 mm. The gears mesh with a large stationary gear D. A torque T of 50 N m is applied to the shaft. What is the angular speed of the shaft after 10 s if the system is initially at rest? The bent rod has a mass per unit length of 5 kg/m. Will there be forces on the bearings of the shaft other than those from gravity? Why?

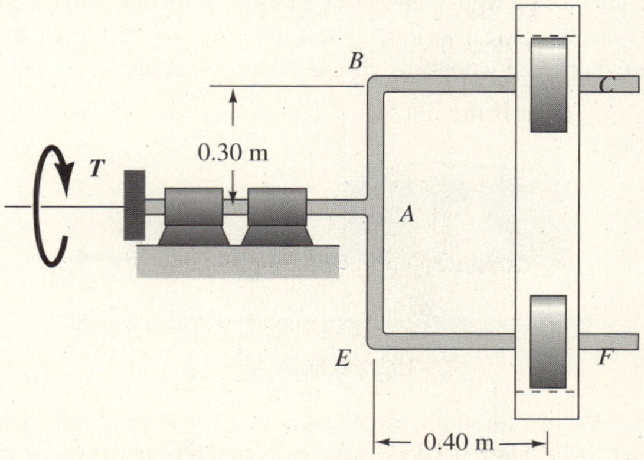

Figure P.16.53. *(Continued on next page)*

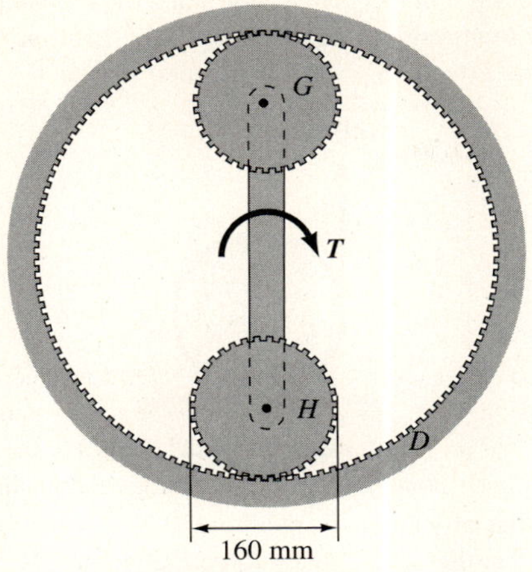

Figure P.16.53.

16.54. A tractor and driver has a mass of 1.35 Mg. If a total torque T of 300 N m is developed on the two drive wheels by the motor, what is the acceleration of the tractor? The large drive wheels each have a mass of 90 kg, a diameter of 1 m, and a radius of gyration of 400 mm. The small wheels each have a mass 20 kg and have a diameter of 300 mm with a radius of gyration of 100 mm.

Figure P.16.54.

16.55. A block B of mass 50 kg rides on two identical cylinders C and D, mass 25 kg each, as shown. On top of block B is a block A of mass 50 kg. Block A is prevented from moving to the left by a wall. If we neglect friction between A and B and between A and the wall and we consider no slipping at the contact surfaces of the cylinders, what is the angular speed of the cylinders after 2 s for $P = 350$ N?

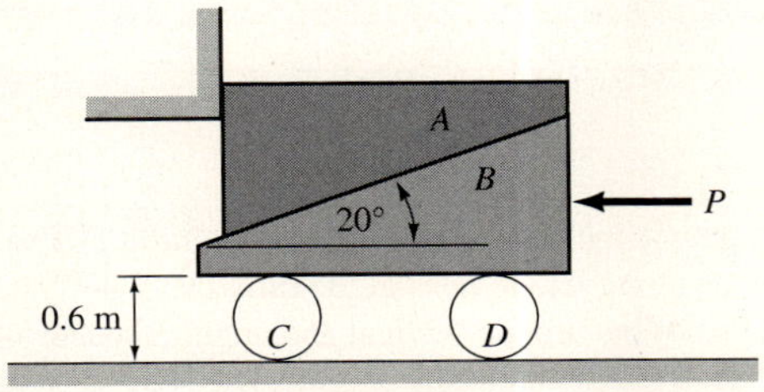

Figure P.16.55.

16.56. A cable is wrapped around two pulleys A and B. A force T is applied to the end of the cables at G. Each pulley has a mass of 2.5 kg and has a radius of gyration of 100 mm. The diameter of the pulleys is 300 mm. A body C of mass 50 kg is supported by pulley B. Suspended from body C is a body D of mass 12.5 kg. Body D is lowered from body C so as to accelerate at the rate of 1.5 m/s^2 relative to body C. What force T is then needed to pull the cable downward at G at the increasing rate of 1.5 m/s^2?

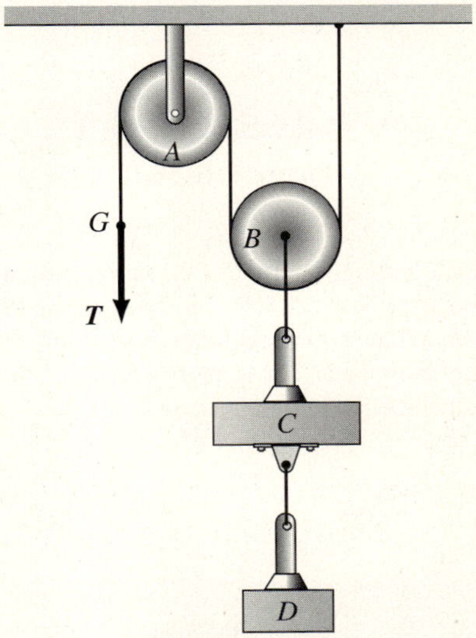

Figure P.16.56.

16.57. A cylinder A is acted on by a torque T of 1 kN m. The cylinder has a mass of 75 kg and a radius of gyration of 400 mm. A light rod CD connects cylinder A with a second cylinder B having a mass of 50 kg and a radius of gyration of 200 mm. What is the force in member CD when torque T is applied? The system is stationary at the instant the torque is applied. Assume no slipping of cylinder C along the incline.

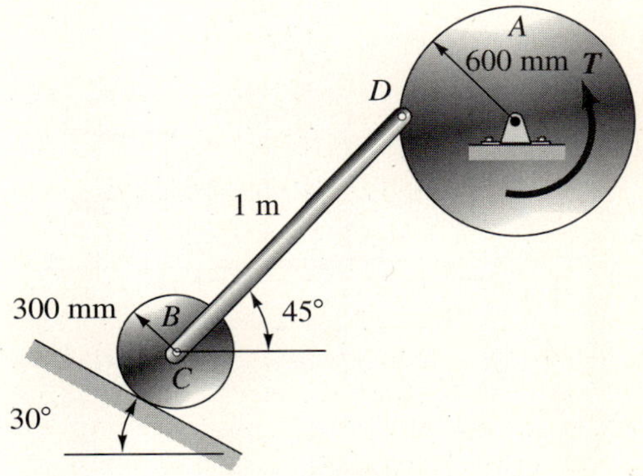

Figure P.16.57.

16.58. A ring rests on a smooth surface as shown from above. The ring has a mean radius of 2 m and a mass of 30 kg. A force of 400 N is applied to the ring at B. What is the acceleration of point A on the ring?

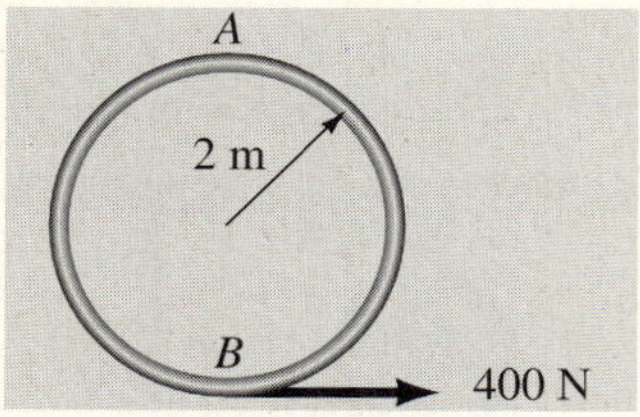

Figure P.16.58.

16.59. A block having a mass of 300 kg rests on a smooth surface as shown from above. A 500 N force is applied at an angle of 45° to the side. What is the acceleration of point B of the block when the force is applied? The center of mass of the block coincides with the geometric center.

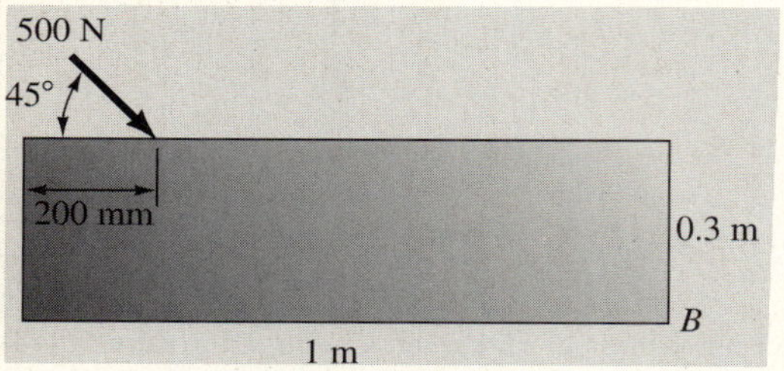

Figure P.16.59.

16.60. A bent rod rests on a smooth surface. The rod has a mass of 20 kg. What is the acceleration of point A when a force P = 100 N is applied?

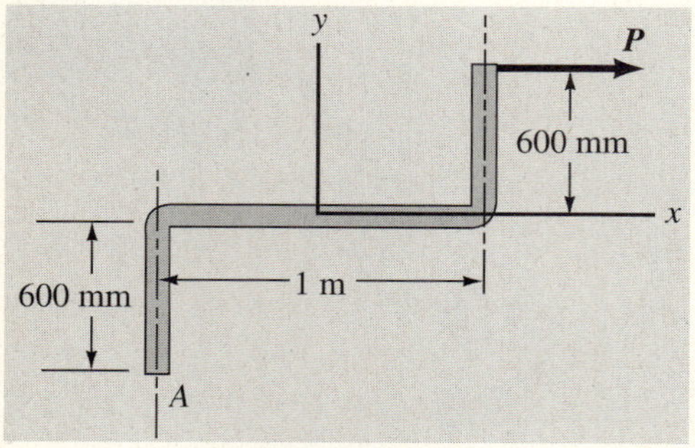

Figure P.16.60.

16.61. A constant force of 500 N is exerted on a rope wrapped around a 25 kg cylinder. How high does the cylinder rise in 2 s? How many rotations has the cylinder had at that time? Neglect initial frictional effects by the ground support.

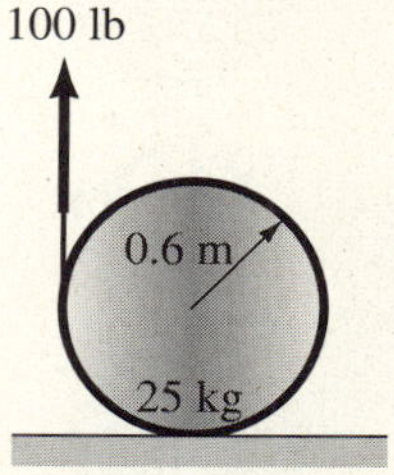

Figure P.16.61.

16.62. A bar of weight W equal to 450 N is at rest on a horizontal surface S at the instant that a force P equal to 250 N is applied. Show that the center of rotation at the instant that the force P is applied is 0.98 m from the left end. The coefficient of friction μ_s equals 0.2. The length of the bar is 3 m. Assume that the normal force on the surface is uniform.

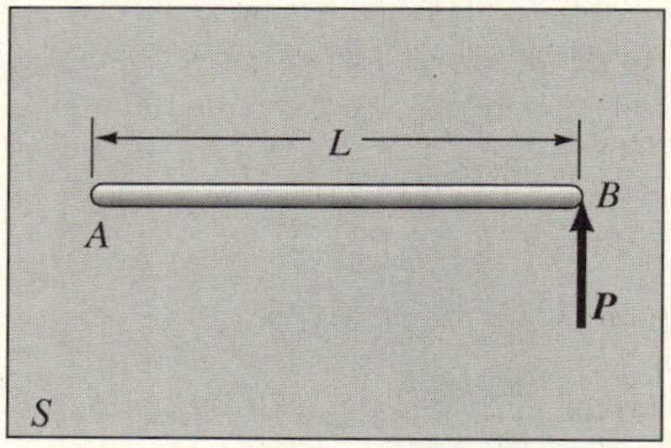

Figure P.16.62.

16.63. A cart B is given a constant acceleration of 5 m/s². On the cart is a cylinder A having a mass of 5 kg and a diameter of 600 mm. If the system is initially stationary, how far does the cylinder move relative to the cart in 1.5 s? Assume no slipping.

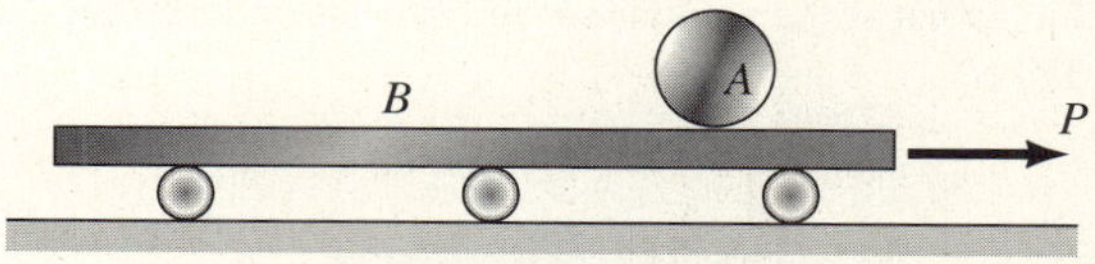

Figure P.16.63.

16.64. A gear A meshes with a stationary pinion B. The hub H of gear A moves along frictionless guide rails DE. The system is in a vertical plane. What are the vertical and angular speeds of the gear after 2 s if the system is initially at rest? The gear has a mass of 20 kg.

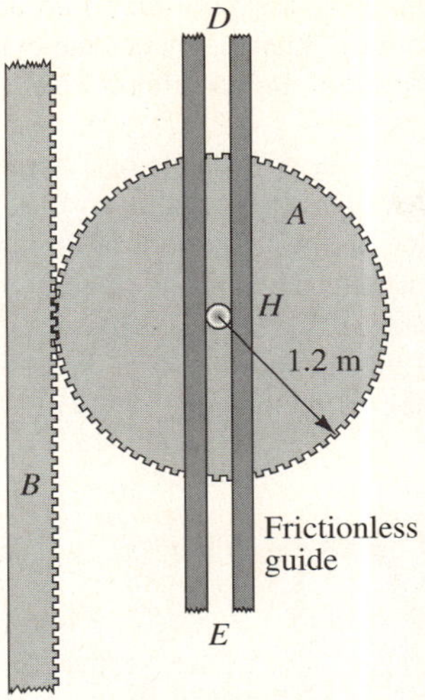

Figure P.16.64.

16.65. In Problem 15.43, compute the forces at pin B and pin C. The mass of rod BC is 11.5 kg and the mass of rod AB is 8.2 kg. The following data stem from the solution of Problem 15.43:

$$\omega_{BC} = 3.35 \text{ rad/s} \quad \dot{\omega}_{BC} = 39.7 \text{ rad/s}^2 \quad a_c = 35.0 \text{ m/s}^2$$

Neglect the mass of the slider at C, but do not consider it to be frictionless. A strain gauge informs us that there is a torque of 27 N m acting on rod AB at A.

16.66. A circular disc is shown with a circular hole. It rests on a frictionless surface and the view shown is from above. A force F = 0.2 N acts on the disc. The thickness of the disc is 50 mm and the density is 5.6 Mg/m^3. What is the initial linear acceleration of the center of mass and the angular acceleration of the disc?

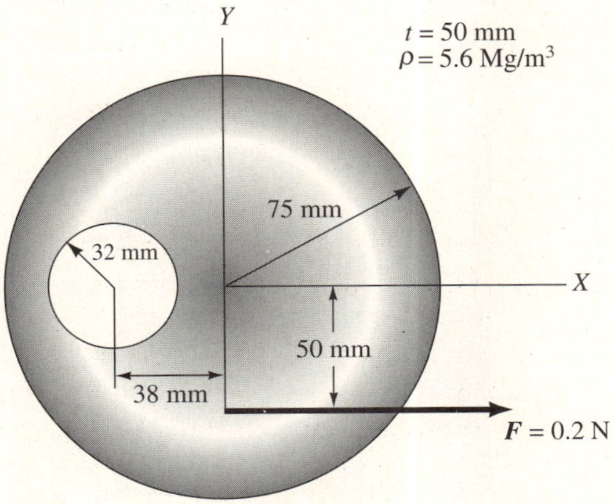

Figure P.16.66.

16.67. In the preceding problem the disc is in the vertical plane and is held up by a horizontal surface where the coefficients of friction are μ_s = 0.005 and μ_d = 0.003. What are the initial linear acceleration of the center of mass and the initial angular acceleration of the disc? Start by assuming no slipping.

16.68. A cylinder A slides off the flatbed of a truck (Fig. P.16.68) onto the road with zero angular velocity. The mass of the cylinder is 100 kg; its radius is 1 m; and its radius of gyration about the axis through the center of mass is 0.75 m. The coefficient of friction μ_d between the cylinder and the pavement is 0.6. If marks on the pavement from the cylinder while it is sliding extend over a distance along the road of 3 m, and if the axis of the cylinder remains perpendicular to the sides of the road during the action, what was the approximate speed of the truck when the cylinder slid off. Neglect the speed of the cylinder relative to the truck when it slides off. [*Hint:* What do the pavement marks signify?]

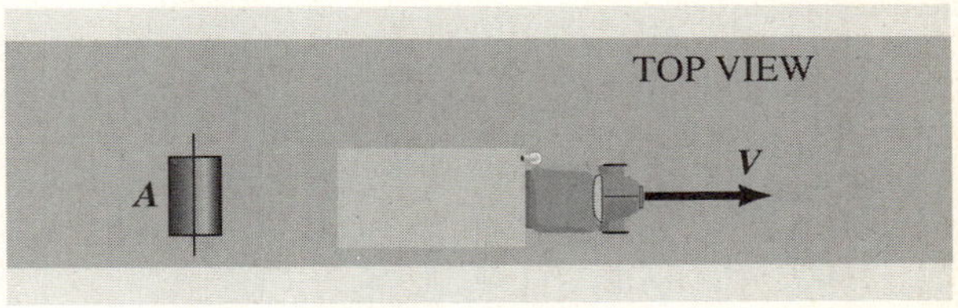

Figure P.16.68.

16.69. Three forces act on a plate resting on a frictionless surface. They are F_1 = 450 N, F_2 = 900 N, and a force F_3 = 1600 N whose direction θ is to be determined so that the plate has a counterclockwise angular acceleration of 0.2 rad/s^2. Determine also the acceleration vector for the center of mass of the plate. The mass of the plate is 80 kg.

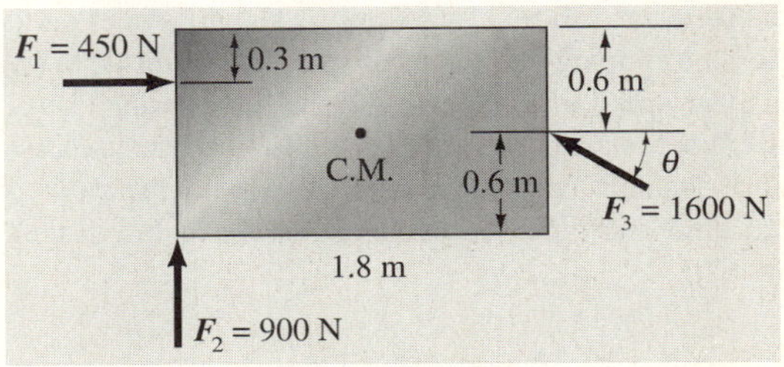

Figure P.16.69.

16.70. In Problem 16.63, what force P is needed to uniformly accelerate the cart so that the cylinder A moves 1 m in 2 s relative to the cart. Cart B has a mass of 10 kg. Neglect the inertia of the small rollers supporting the cart, and assume there is no slipping.

16.71. A wedge B is shown with a cylinder A of mass 20 kg and diameter 500 mm on the incline. The wedge is given a constant acceleration of 20 m/s^2 to the right. How far d does the cylinder move in 0.5 s relative to the incline if there is no slipping? The system starts from rest.

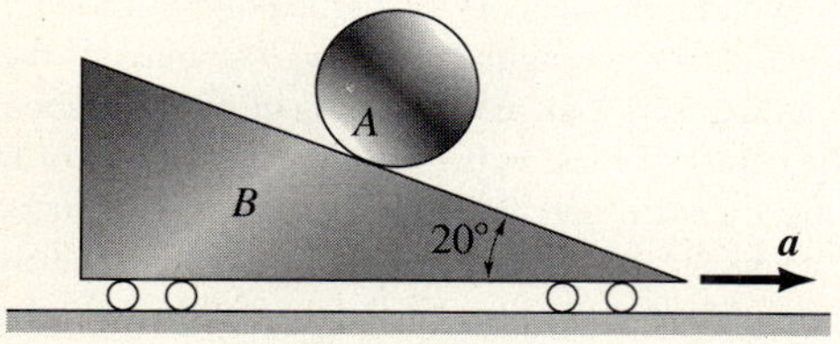

Figure P.16.71.

16.72. A block weighing 100 N is held by three inextensible guy wires. What are the forces in wires AC and BD at the instant that wire EC is cut?

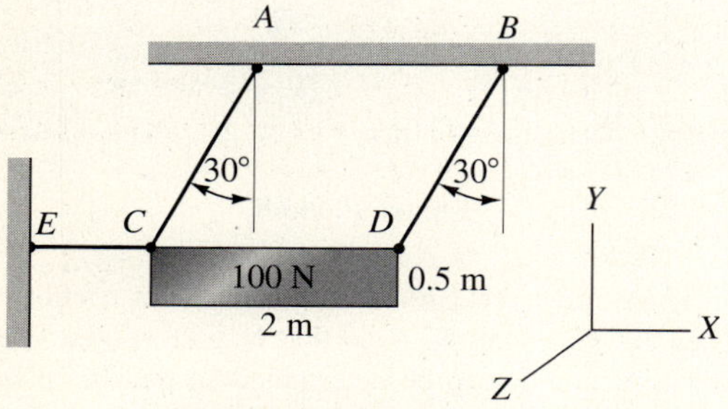

Figure P.16.72.

16.73. A bowler releases his ball at a speed of 3 m/s. If the ball has a diameter of 250 mm, what spin ω should be put on the ball so there is no slipping? If he puts on only half of this spin ω, keeping the same speed of 3 m/s, what is the final (terminal) speed of the ball? What is the speed after 0.3 s? The ball has a mass of 1.8 kg and a dynamic coefficient of friction with the floor of 0.1. Neglect wind resistance and rolling resistance (as discussed in Section 7.7).

Figure P.16.73.

16.74. In Problem 16.73, suppose ω is 1.4 times the value ω = 24 rad/s needed for no slipping. What is the terminal speed of the ball? What is the speed of the ball after 0.3 s?

16.75. A tug is pushing on the side of a barge which is loaded with sand and which has a total mass of 4.5 Mg. The tug generates a 3 kN force which is always normal to the barge. If the barge rotates 5° in 20 s, what is the moment of inertia of the barge at the center of mass which we take at the geometric center of the barge? The distance d for this maneuver is 5 m. What is the acceleration of the center of mass of the barge at t = 35 s? Neglect the resistance of the water.

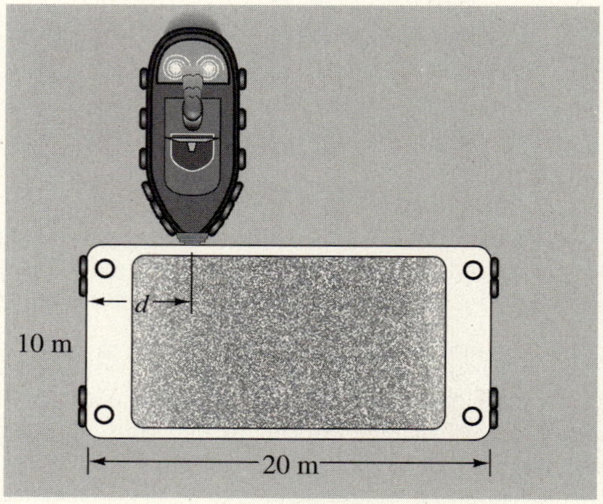

Figure P.16.75.

16.76. A plate A having a mass of 50 kg is supported by a rod at one end and by a linear spring having a spring constant K = 35 N/mm at the other. If the rod BC is suddenly released at B, what is the angular acceleration of the plate? Also, what is the acceleration of the plate mass center? The plate is oriented as shown at the instant of cutting the left support.

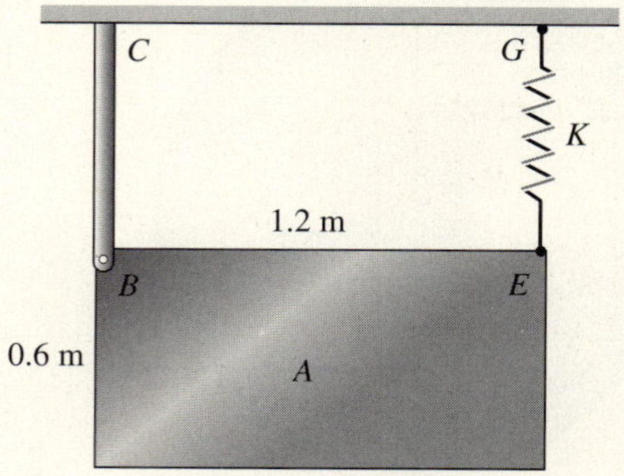

Figure P.16.76.

16.77. The great English liner, the *Queen Elizabeth* (QE II), is the last transatlantic luxury ship left. All the others have been either scrapped or made into pleasure excursion ships. The QE II is 213 m long and has a mass of 54.5 Gg.

(a) At a top speed of 20 m/s, with the engines producing 82 MW, what is the thrust coming from the propellers?
(b) In a harbor, two tugboats are turning an initially stationary QE II as shown in the diagram. Determine an approximate value for the angular acceleration of the ship. Consider the ship to be a long uniform rod. Include an additional one fourth of the mass of the ship to account for the water which must be moved to accommodate the movement of the ship. Each tug develops a force of 22 kN.
(c) If the tugs remain perpendicular to the QE II, and if we assume constant angular acceleration, how many minutes are required to turn the ship 10 degrees?

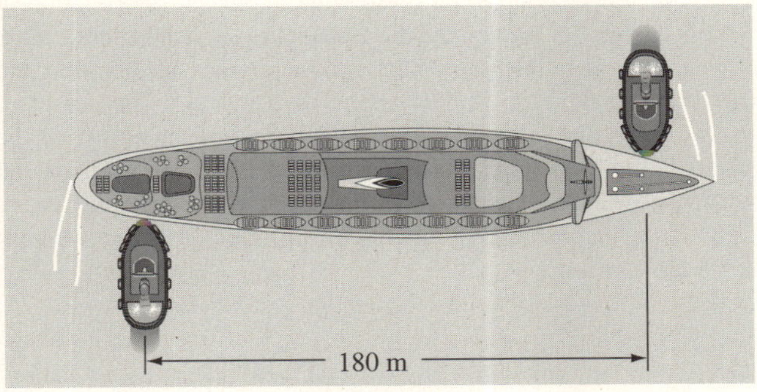

Figure P.16.77.

16.78. A cable supports cylinder *A* of mass 40 kg and then wraps around a light cylinder *B* and finally supports cylinder *C* of mass 20 kg. If the system is released from rest, what are the accelerations of the centers of *A* and *C*? There is no slipping.

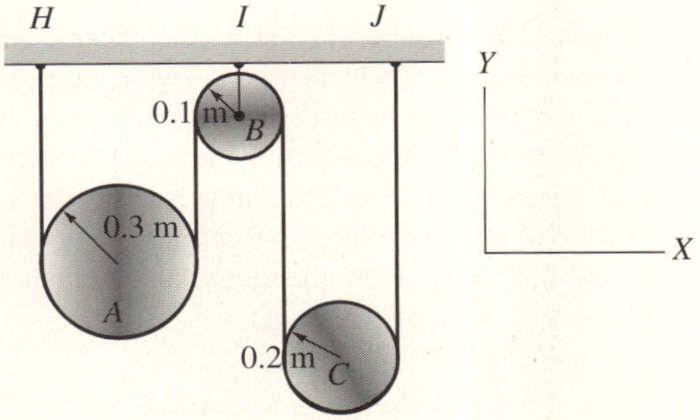

Figure P.16.78.

16.79. A cylinder of mass 20 kg can rotate inside a block *B* whose mass is 35 kg. There is a constant resisting torque for this rotation given as 200 Nm. A horizontal 1000 N force is applied to a cable firmly wrapped around the cylinder. What is the velocity of the block after moving 0.5 m? What is the angular velocity of the cylinder when the block reaches this position? The coefficient of dynamic friction between block *B* and the floor is 0.3. The system is shown in a vertical orientation.

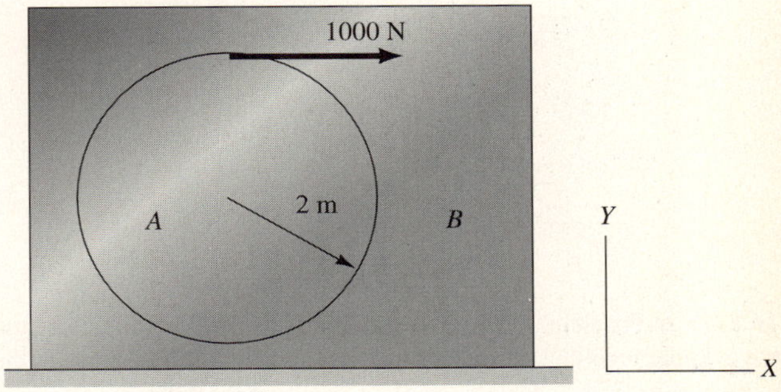

Figure P.16.79.

16.80. A rod *AB* of length 3 m and weight 445 N is shown immediately after it has been released from rest. Compute the tension in wires *EA* and *DB* at this instant.

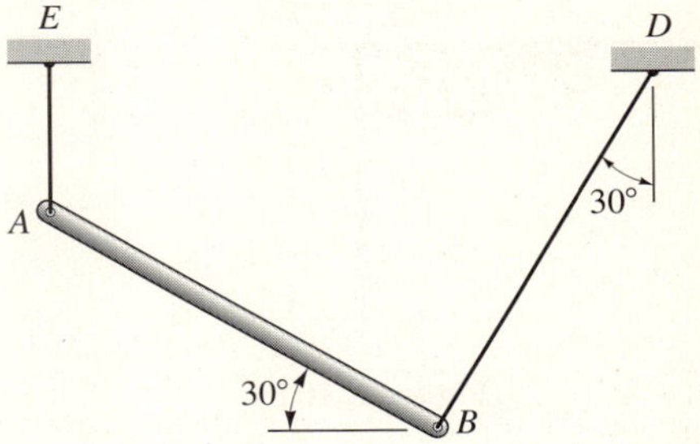

Figure P.16.80.

16.81. Rod *AB* is released from the configuration shown. What are the supporting forces at this instant if we neglect friction? The rod weighs 900 N and is 6 m in length.

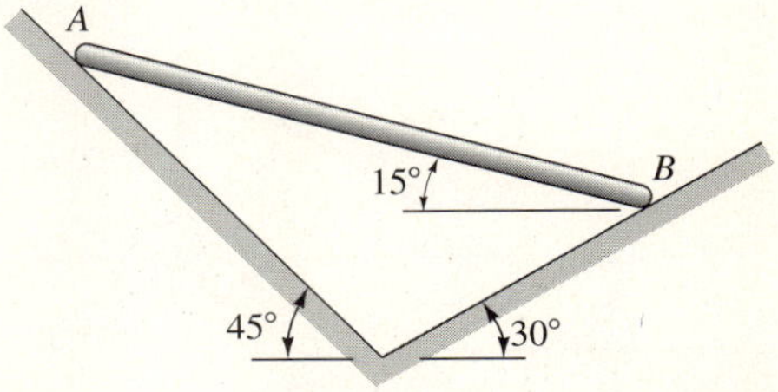

Figure P.16.81.

16.82. A cylinder of mass 50 kg with a radius of 0.3 m is held fixed on an incline that is rotating at 0.5 rad/s. The cylinder is released when the incline is at position θ equal to 30°. If the cylinder is 6 m from the bottom A at the instant of release, what is the initial acceleration of the center of the cylinder relative to the incline? There is no slipping.

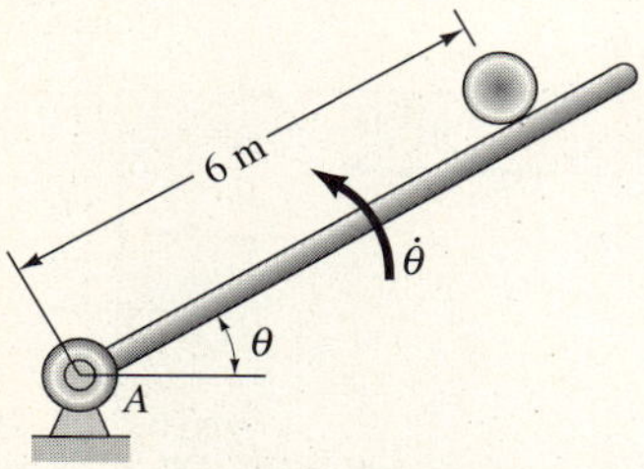

Figure P.16.82.

16.83. In Problem 16.82, $\ddot{\theta} = 2$ rad/s^2 at the instant of interest and the cylinder is rolling downward at a speed ω_1 of 3 rad/s relative to the incline. What is the acceleration of the center of the cylinder relative to the incline at the instant of interest? There is no slipping.

16.84. Two identical bars, each having a mass of 9 kg, hang freely from the vertical. A force of 45 N is applied at the center of the upper bar AB. What are the angular accelerations of the bars?

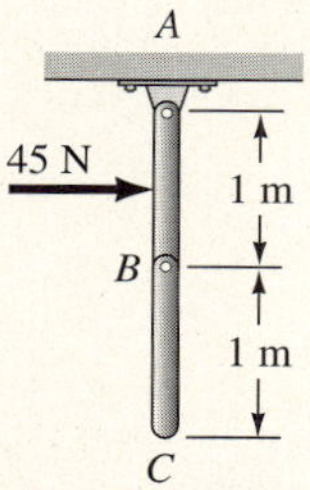

Figure P.16.84.

16.85. A rod BA is made to rotate at a constant speed ω of 100 r/min by a torque not shown. It drives a rod BC, having a mass of 5 kg, which in turn moves a gear D having a mass of 3 kg and a radius of gyration of 200 mm. The diameter of the gear is 450 mm. At the instant shown, what are the forces transmitted by the pins at C and at B?

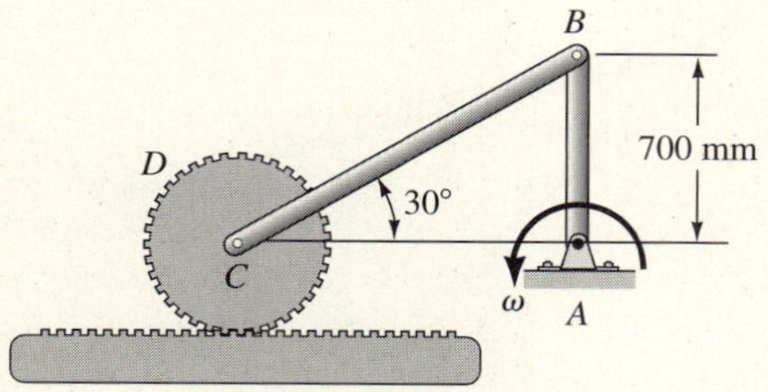

Figure P.16.85.

16.86. A 1 m rod AB of mass 10 kg is suspended at one end by a cord BC and at the other end rides on an inclined surface on small wheels. Initially the wheels are being held at the position shown. At the instant the wheels are released, what is the angular acceleration of rod AB? Neglect friction on the inclined surface.

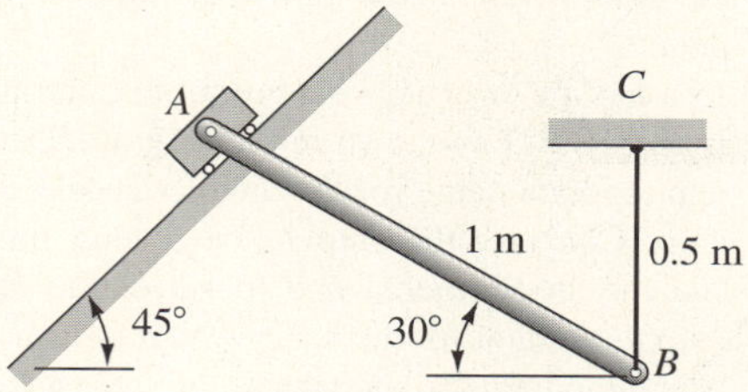

Figure P.16.86.

16.87. A truck is carrying a 10 kN crate A held by two steel cables each under a tensile force of 800 N. At corner B there is a stop. The truck undergoes a crash with a constant deceleration of 4.6 g's. Cable FG breaks and the crate starts to rotate about corner B where we have the stop. What are the angular acceleration and the forces at the stop at the instant of the break? The center of mass of the crate is at the geometrical center of the crate and the radius of gyration about the center of mass is 2.3 m. Consider that when the rotational acceleration first occurs, the crate has yet to move and that cable DE still has the original tension, whereas cable FG has snapped.

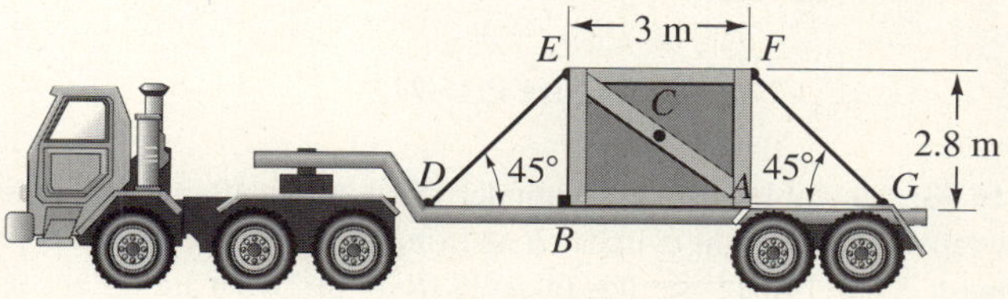

Figure P.16.87.

16.88. An object with a cylindrical hub, thin spokes, and a thin rim rests on a frictionless horizontal surface. The hub weighs 100 N and the spokes and rim each weigh 15 N/m. What is the acceleration of center A when the indicated force is applied? What is the angular acceleration of the system? Finally, what is the acceleration of point B? Use calculus when considering the spokes and the rim. For the hub, the angular momentum about an axis through the center of mass is given as

$$\int_V \mathbf{r} \times dmV = \frac{1}{2} MR^2 \omega \mathbf{k}$$

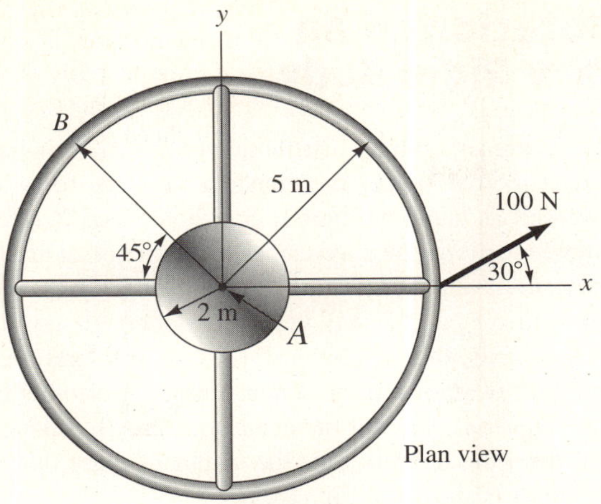

Figure P.16.88.

16.89. We are looking down from above at a baseball player swinging his bat in a horizontal plane in the process of hitting a baseball at point B. The player is holding the bat such that the resultant of the force from his hands is at point A. At what distance d should the ball hit the bat to render as zero the normal force component from the batter's hands onto the centerline of the bat? This point B is called the *center of percussion*. Because this kind of hit feels effortless to the batter it is often called the "sweet spot" by athletes in both baseball and tennis. The radius of gyration about the center of mass is k_z.

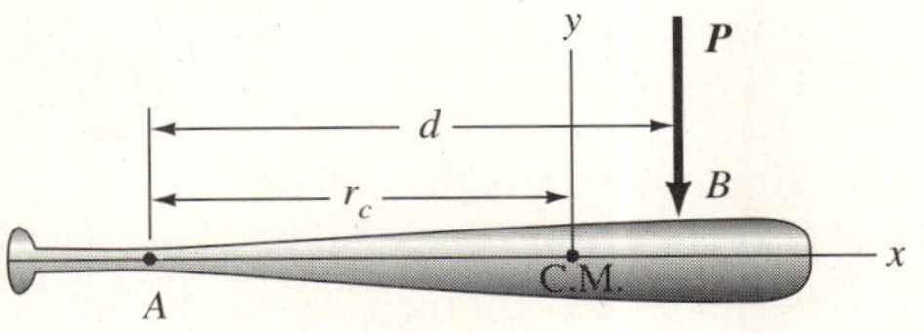

Figure P.16.89.

16.90. A torque T = 10 N m is applied to body C. If there is no slipping, how many rotations does cylinder B make in 1 s if the system starts from rest? A is stationary at all times. The system starts from rest. We are observing the system from above. The following data apply.

$$M_C = 50 \text{ kg} \qquad M_B = 30 \text{ kg}$$
$$k_C = 0.2 \text{ m} \qquad k_B = 0.1 \text{ m}$$

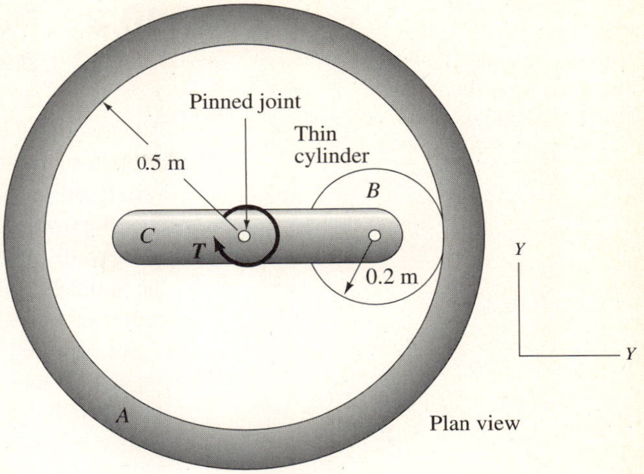

Figure P.16.90.

16.91. In Problem 15.48, we determined the following results from kinematics:

$$\omega_{AB} = -0.609 \text{ rad/s} \qquad V_B = 4.38 \text{ m/s}$$
$$\dot{\omega}_{AB} = 14.71 \text{ rad/s}^2 \quad \dot{V}_B = -33.28 \text{ m/s}^2$$

The mass of rod AB is 10 kg. If we neglect the masses of the sliders, what are the forces coming onto the end pins of the rod? The horizontal slot in which slider A is moving is frictionless.

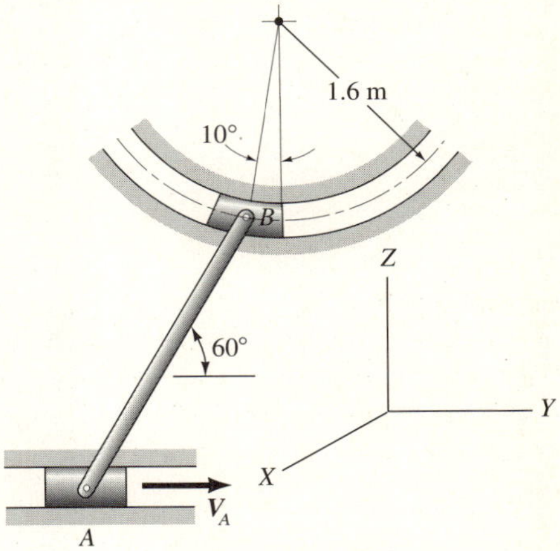

Figure P.16.91.

16.92. In the preceding problem, the slider at A no longer moves in the slot without friction and we do not know the friction force there. However, we have a strain gage mounted on rod AB giving data indicating a 200 N compressive axial force at A. Using the data of the previous problem, compute the force components at the ends of the rod.

16.8 Pure Rotation of an Arbitrary Rigid Body

We now consider a body having an arbitrary distribution of mass rotating about an axis of rotation fixed in inertial space. We consider this axis to be the z axis fixed in the body as well as being an inertial coordinate axis Z. We can take the origin of xyz anywhere along the z axis since all such points are fixed in inertial space. The *moment-of-momentum* equations to be used will now be the general equations 16.6 since I_{zx} and I_{zy} will generally not equal zero. If the center of mass is along the z axis, then it obviously has no acceleration, and so we can then apply the rules of statics to the center of mass. For other cases we shall often need to use *Newton's law* for the center of mass. In this regard it will be helpful to note from the definition of the center of mass that for a system of rigid bodies such as is shown in Fig. 16.29

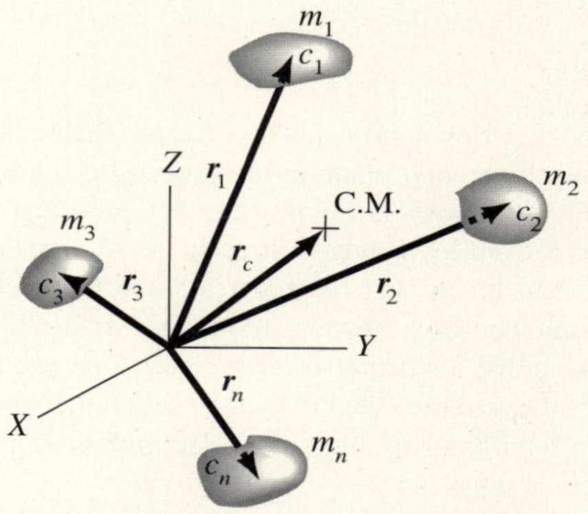

Figure 16.29. n rigid bodies having a total mass M.

$$M\boldsymbol{r}_c = \sum_{i=1}^{n} m_i \boldsymbol{r}_i \tag{16.16}$$

where m_i is the mass of the ith rigid body, $\boldsymbol{r}_i$ is the position vector to the center of mass of the ith rigid body, M is the total mass, and $\boldsymbol{r}_c$ is the position vector to the center of mass of the system. We can then say on differentiating:

$$M\dot{\boldsymbol{r}}_c = \sum_{i=1}^{n} m_i \dot{\boldsymbol{r}}_i \tag{16.17}$$

$$M\ddot{\boldsymbol{r}}_c = \sum_{i=1}^{n} m_i \ddot{\boldsymbol{r}}_i \tag{16.18}$$

In *Newton's law* for the mass center of a system of rigid bodies, we conclude that we can use the centers of mass of the component parts of the system as given on the right side of Eq. 16.18 rather than the center of mass of the total mass.

Example 16.10

A shaft has protruding arms each of which weighs 40 N/m (see Fig. 16.30). A torque T gives the shaft an angular acceleration $\dot{\omega}$ of 2 rad/s^2. At the instant shown in the diagram, ω is 5 rad/s. If the shaft without arms

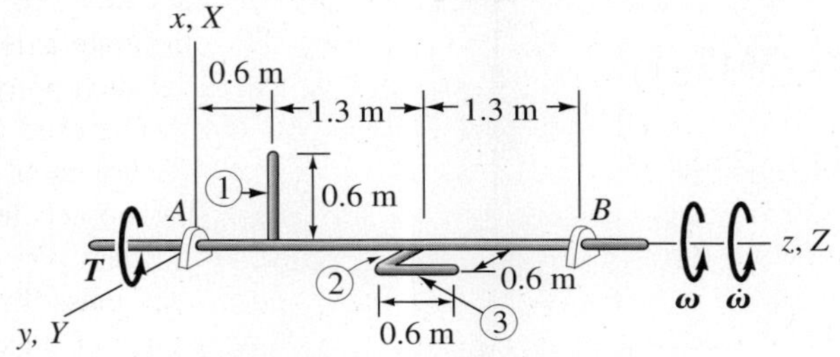

Figure 16.30. Rotating shaft with arms.

weighs 180 N, compute the vertical and horizontal forces at bearings *A* and *B* (see Fig. 16.31). Note that we have numbered the various arms for convenient identification.

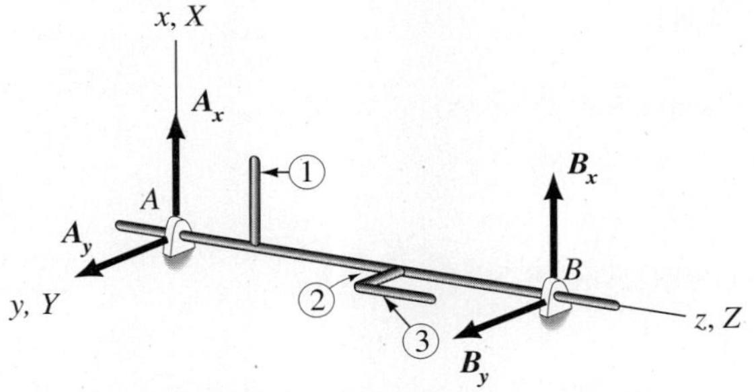

Figure 16.31. Supporting forces.

We first fix a reference *xyz* to the shaft at *A*. Also at *A* we fix an inertial reference *XYZ* to the ground. We can directly use Eqs. 16.6a and 16.6b about point *A*. For this reason, we shall compute the required products of inertia of the shaft system for reference *xyz*. Accordingly, using the parallel-axis theorem, we have:

$$(I_{xz})_{\text{arm}(1)} = 0 + \frac{(40)(0.6)}{g}[(0.6)(0.3)] = 0.440 \text{ kg m}^2$$

$$(I_{xz})_{\text{arm}(2)} = 0 + \frac{(40)(0.6)}{g}[(1.9)(0)] = 0$$

$$(I_{xz})_{\text{arm}(3)} = 0 + \frac{(40)(0.6)}{g}[(2.20)(0)] = 0$$

Hence, for the system, $I_{xz} = 0.440$ kg m^2.

Example 16.10 (Continued)

We next consider I_{zy}. Accordingly, we have

$$(I_{zy})_{\text{arm}(1)} = 0 + \frac{(40)(0.6)}{g}[(0.6)(0)] = 0$$
$$(I_{zy})_{\text{arm}(2)} = 0 + \frac{(40)(0.6)}{g}[(1.9)(0.3)] = 1.394 \text{ kg m}^2$$
$$(I_{zy})_{\text{arm}(3)} = 0 + \frac{(40)(0.6)}{g}[(2.20)(0.6)] = 3.23 \text{ kg m}^2$$
$$(I_{zy})_{\text{shaft}} = 0 + 0 = 0$$

Hence, for the system, $I_{zy} = 4.62 \text{ kg m}^2$.

We can now employ the **moment of momentum** equations (Eqs. 16.6) to get the required moments M_x and M_y about point A needed for the motion we are considering. Thus, we have

$$M_x = -(2)(0.440) + (5^2)(4.62) = 114.7 \text{ N m} \qquad \text{(a)}$$
$$M_y = -(2)(4.62) - (5^2)(0.440) = -20.2 \text{ N m} \qquad \text{(b)}$$

Summing moments of all the forces acting on the system about the y axis at A, we can say (see Fig. 16.31):

$$M_y = -20.2 = -(40)(0.6)(0.6) - (40)(0.6)(1.9) - (40)(0.6)(2.2) - (180)(1.6) + (B_x)(3.2)$$

Therefore, we require

$$\boxed{B_x = 118.9 \text{ N}} \qquad \text{(c)}$$

Summing moments about the x axis at A, we can say

$$M_x = 114.7 = -B_y(3.2)$$

Therefore, we require

$$\boxed{B_y = -35.8 \text{ N}} \qquad \text{(d)}$$

We next use **Newton's law** considering the three arms to be three particles at their mass centers as has been shown in Fig. 16.32. In the x

Example 16.10 (Continued)

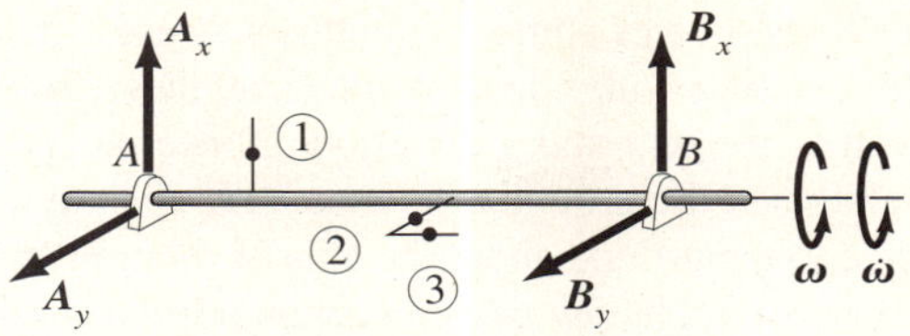

Figure 16.32. Arms replaced by mass centers.

direction at time t we have, using Eq. 16.18 and noting that each of the aforementioned particles has circular motion:

$$118.9 + A_x - 180 - (3)[(40)(0.6)] = -\frac{(40)(0.6)}{g}(0.3)(\omega^2)$$
$$-\frac{(40)(0.6)}{g}(0.3)(\dot{\omega}) - \frac{(40)(0.6)}{g}(0.6)(\dot{\omega})$$

Setting $\omega = 5$ rad/s and $\dot{\omega} = 2$ rad/s^2, we get

$$A_x = 110.3 \text{ N} \tag{e}$$

In the y direction, we can say similarly at time t

$$A_y - 35.8 =$$
$$\frac{(40)(0.6)}{g}(0.3)(\dot{\omega}) - \frac{(40)(0.6)}{g}(0.3)(\omega^2) - \frac{(40)(0.6)}{g}(0.6)(\omega^2)$$

Therefore,

$$A_y = -17.78 \text{ N} \tag{f}$$

The forces acting on the shaft are shown in Fig. 16.33. The reactions to these forces are then the desired forces on the bearings. In the z direction it should be clear that there is no force on the bearings.

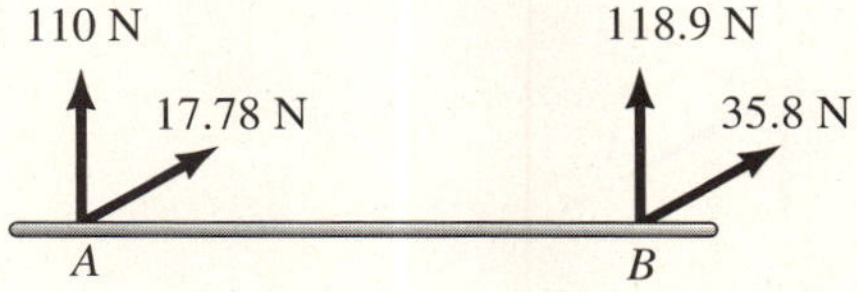

Figure 16.33. Forces on shaft.

If, in the last example, we had ignored the constant forces of gravity, we would have determined forces at bearings A and B that are due entirely to the motion of the body. Forces computed in this way are called *dynamic forces*. If the body were rotating with constant speed ω, these forces would clearly have constant values in the x and y directions. Since the xy axes are rotating with the body relative to the ground reference XYZ, such dynamic forces must also rotate relative to the ground about the axis of rotation with the speed ω of the body. This means that, in any *fixed* direction normal to the shaft at a bearing, there will be a *sinusoidal force variation* with a frequency corresponding to the angular rotation of the shaft. Such forces can induce vibrations of large amplitude in the structure or support if a natural frequency or multiple of a natural frequency is reached in these bodies.[11] When a shaft creates rotating forces on the bearings by virtue of its own rotation, the shaft is said to be unbalanced. We shall set up criteria for balancing a rotating body in the next section.

*16.9 Balancing

We shall now set forth the criteria for the condition of dynamic balance in a rotating body. Then, we shall set forth the requirements needed to achieve balance in a rotating body. Consider then some arbitrary rigid body rotating with angular speed ω and a rate of change of angular speed $\dot{\omega}$ about axis AB (Fig. 16.34). We shall set up general equations for determining the supporting forces at the bearings. Consider point a on the axis of rotation at the bearing A and establish a set of axes xyz fixed to the rotating body with the z axis corresponding to the axis of rotation. The x and y axes are chosen for convenience. Axes XYZ are, as usual, inertial axes. Using the *moment-of-momentum* equations (a) and (b) in Eq. 16.6 and including only *dynamic* forces, we get for point a:

$$B_y l = -I_{xz}\dot{\omega} + I_{yz}\omega^2 \qquad (16.19a)$$

$$B_x l = -I_{yz}\dot{\omega} - I_{xz}\omega^2 \qquad (16.19b)$$

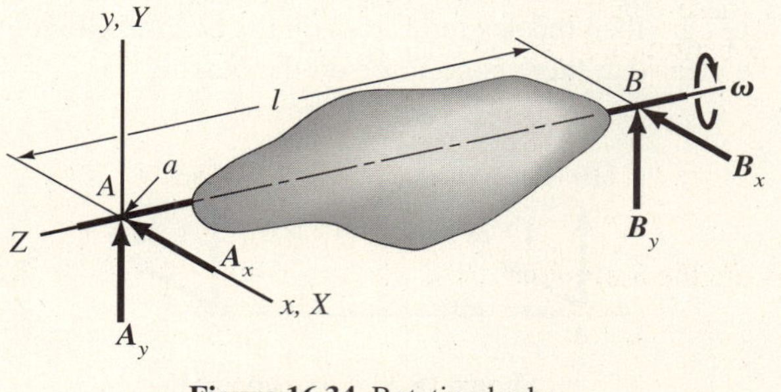

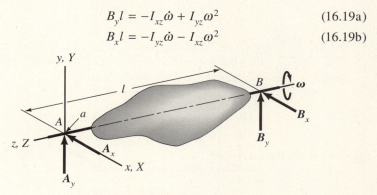

Figure 16.34. Rotating body.

[11]Natural frequencies will be discussed in Chapter 19.

If the axis of rotation is a *principal axis* at bearing A, then $I_{yz} = I_{xz} = 0$ in accordance with the results of Chapter 9. The dynamic forces at bearing B are then zero.

Next, we shall show that if in addition the *center of mass lies along the axis of rotation*, this axis is a principal axis for *all points* on it. In Fig. 16.35 a set of axes $x'y'z'$ fixed to the body and parallel to the xyz axes has been set up at an arbitrary point E along the axis of rotation. We can see from the arrangement of the axes that for any element of the body dm:

$$y' = y, \qquad x' = x, \qquad z' = D + z \tag{16.20}$$

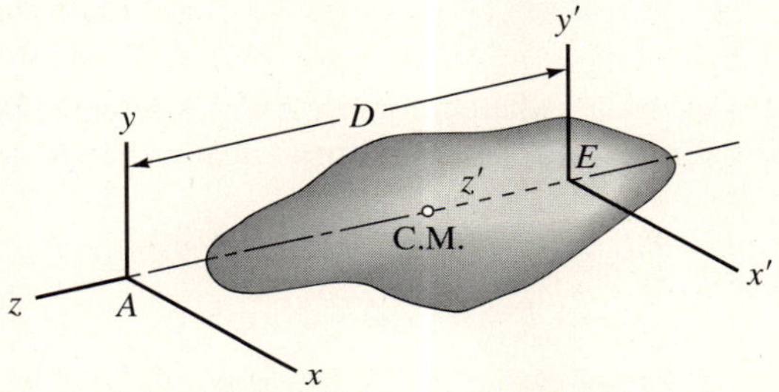

Figure 16.35. Reference $x'y'z'$ fixed at E..

Also, we know for the xyz reference that

$$I_{xz} = \int_M xz\,dm = 0, \qquad I_{yz} = \int_M yz\,dm = 0 \tag{16.21}$$

And if the center of mass is along the centerline, we can say:

$$\int_M y\,dm = \int_M y'\,dm = My_c = 0 \tag{16.22}$$
$$\int_M x\,dm = \int_M x'\,dm = Mx_c = 0$$

We shall now show that all products of inertia involving the z' axis at E are zero under these conditions and, consequently, that the z' axis is a principal axis at E. Substituting from Eqs. 16.20 into 16.21, we get

$$\int_M x'(z' - D)\,dm = 0 \tag{a}$$

$$\int_M y'(z' - D)\,dm = 0 \tag{b}$$

If we carry out the multiplication in the integrand of the above Eqs. (a) and (b), we get

$$\int_M x'z'\,dm - D\int_M x'\,dm = 0 \tag{c}$$

$$\int_M y'z'\,dm - D\int_M y'\,dm = 0 \tag{d}$$

As a result of Eq. 16.22, the second integrals of Eqs. (c) and (d) are zero, and we conclude that the products of inertia $I_{x'z'}$ and $I_{y'z'}$ are zero. Now the xy axes and hence the $x'y'$ axes can have any orientation as long as they are normal to the axis of rotation. This means that at E the z' axis yields a zero product of inertia for all axes normal to it. As a result of our deliberations of Chapter 9, we can conclude that z' is a principal axis for point E. And since E is any point on the axis of rotation, we can say the following:

If the axis of rotation is a principal axis at any point along this axis and if the center of mass is on the axis of rotation, then the axis of rotation is a principal axis at all points on it.

We now consider Fig. 16.36, where reference $x'y'z'$ is set up at bearing B. We can next employ the **moment-of-momentum** equation (16.6) for these axes at B. We get

$$-A_y l = -I_{x'z'}\dot{\omega} + I_{y'z'}\omega^2$$

$$-A_x l = -I_{y'z'}\dot{\omega} - I_{x'z'}\omega^2$$

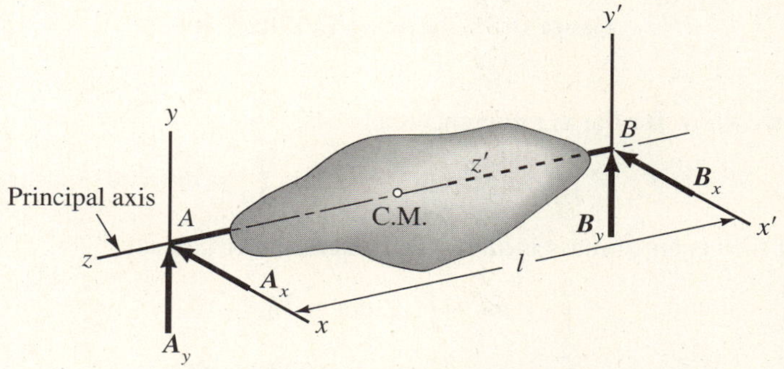

Figure 16.36. Reference $x'y'z'$ fixed at B.

With the z axis a principal axis at A and the center of mass along the axis of rotation, the z' axis at B must be a principal axis, and hence $I_{y'z'} = I_{x'z'} = 0$. The dynamic forces at bearing A, therefore, are zero. The rotating system is thus balanced.

We can now conclude that:

For a rotating system to be dynamically balanced, it is necessary and sufficient (1) that at any point along the axis of rotation this axis is a principal axis and (2) that the center of mass is along the axis of rotation.

We next illustrate how we can make use of these results to balance a rotating body.

Example 16.11

A rotating member carries two particles having masses $M_1 = 2.5$ kg and $M_2 = 4$ kg at radial distances $r_1 = 0.3$ m and $r_2 = 0.45$ m, respectively. The masses and a reference *xyz* fixed to the shaft are shown in Fig. 16.37. They are to be balanced by two other particles having masses M_3 and M_4 (shown dashed) which are to be placed in the balancing planes *A* and *B*, respectively. If the masses are placed in these planes at a distance of 0.3 m from the axis of rotation, determine the value of these weights and their position relative to the *xyz* reference.

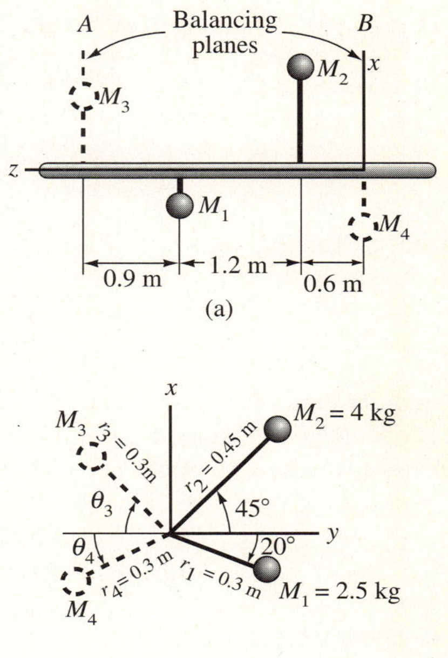

Figure 16.37. Rotating system to be balanced.

We have two unknown masses and two unknown angles, that is, four unknowns [see Fig. 16.37(b)], to evaluate in this problem. The condition that the mass center be on the centerline yields the following relations:[12]

$$\int_M y\,dm = 0:$$

$$M_1 r_1 \cos 20° + M_2 r_2 \cos 45° - M_3 r_3 \cos\theta_3 - M_4 r_4 \cos\theta_4 = 0$$

$$\int_M x\,dm = 0:$$

$$-M_1 r_1 \sin 20° + M_2 r_2 \sin 45° + M_3 r_3 \sin\theta_3 - M_4 r_4 \sin\theta_4 = 0$$

When the numerical values of r_1, r_2, etc., are inserted, these equations become

$$W_3 \cos\theta_3 + W_4 \cos\theta_4 = 6.592 \quad \text{(a)}$$

$$W_3 \sin\theta_3 - W_4 \sin\theta_4 = -3.388 \quad \text{(b)}$$

Now we require that the products of inertia I_{yz} and I_{xz} be zero for the *xyz* reference positioned so that *xy* is in the balancing plane *B*.

$$I_{xz} = 0:$$

$$M_1(1.8)(-r_1 \sin 20°) + M_2(0.6)(r_2 \sin 45°) + M_3(2.7)(r_3 \sin\theta_3) = 0 \quad \text{(c)}$$

$$I_{yz} = 0:$$

$$M_1(1.8)(r_1 \cos 20°) + M_2(0.6)(r_2 \cos 45°) + M_3(2.7)(-r_3 \cos\theta_3) = 0 \quad \text{(d)}$$

Equations (c) and (d) can be put in the form

$$2.7 M_3 \sin\theta_3 = -1.007 \quad \text{(e)}$$

$$2.7 M_3 \cos\theta_3 = 6.774 \quad \text{(f)}$$

[12]We are considering the weights to be particles in this discussion. In some homework problems you will be asked to balance rotating systems for which the particle model will not be proper. You will then have to carry out integrations and/or employ the formulas and transfer theorems for first moments of mass and products of inertia.

Example 16.11 (Continued)

Dividing Eq. (f) into Eq. (e), we get

$$\tan \theta_3 = -0.1487$$

$$\theta_3 = 171.5° \text{ or } 351.5°$$

and so, from Eq. (e),

$$M_3 = -\frac{1.007}{2.7}\frac{1}{\sin\theta_3} = 2.52 \text{ kg}$$

In order to have a positive mass M_3, we chose θ_3 to be 351.5° rather than 171.5°. Now we return to Eqs. (a) and (b). We can then say, on substituting known values of M_3 and θ_3:

$$M_4 \cos\theta_4 = 6.592 - 2.492 = 4.10 \qquad \text{(g)}$$
$$M_4 \sin\theta_4 = 3.388 - 0.372 = 3.016 \qquad \text{(h)}$$

Dividing Eq. (g) into Eq. (h), we get

$$\tan \theta_4 = 0.736$$

$$\theta_4 = 36.4° \text{ or } 216.4°$$

Hence, from Eq. (g), we have

$$M_4 = \frac{4.10}{\cos\theta_4} = 5.09 \text{ kg}$$

where we use $\theta_4 = 36.4°$ rather than 216.4° to prevent a negative W_4. The final orientation of the balanced system is shown in Fig. 16.38.

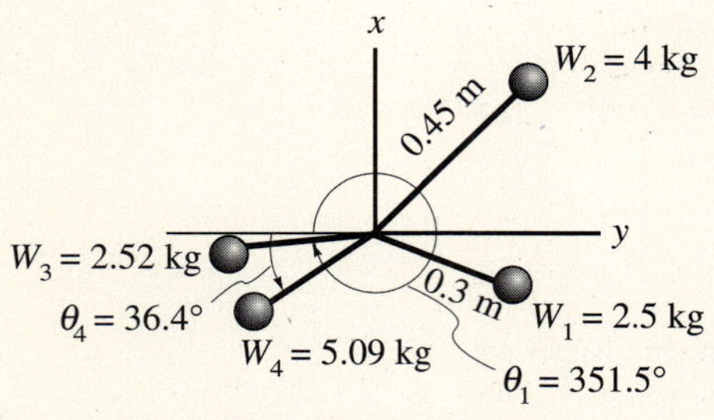

Figure 16.38. Balanced system.

PROBLEMS

16.93. A shaft shown supported by bearings A and B is rotating at a speed ω of 3 rad/s. Identical blocks C and D of mass 15 kg each are attached to the shaft by light structural members. What are the bearing reactions in the x and y directions if we neglect the weight of the shaft?

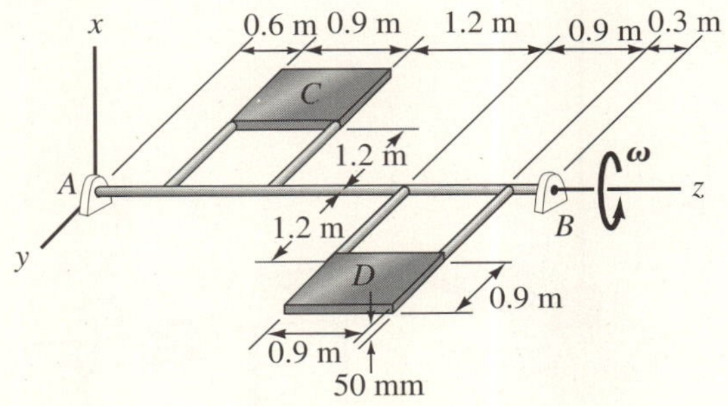

Figure P.16.93.

16.94. Shaft AB is rotating at a constant speed ω of 20 rad/s. Two rods having a weight of 10 N each are welded to the shaft and support a disc D weighing 30 N. What are the supporting forces at the instant shown?

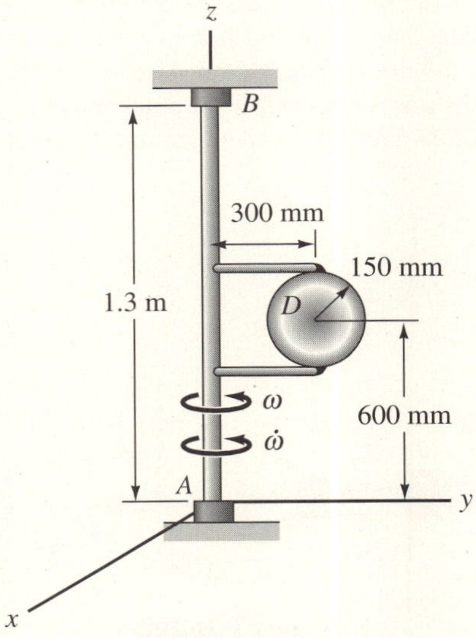

Figure P.16.94.

16.95. Do Problem 16.94 for the case in which $\omega = 20$ rad/s and $\dot{\omega} = 38$ rad/s^2 at the instant of interest as shown.

16.96. A uniform wooden panel is shown supported by bearings A and B. A 50 kg mass is connected with an inextensible cable to the panel at point G over a light pulley D. If the system is released from rest at the configuration shown, what is the angular acceleration of the panel, and what are the forces at the bearings? The panel has a mass of 30 kg.

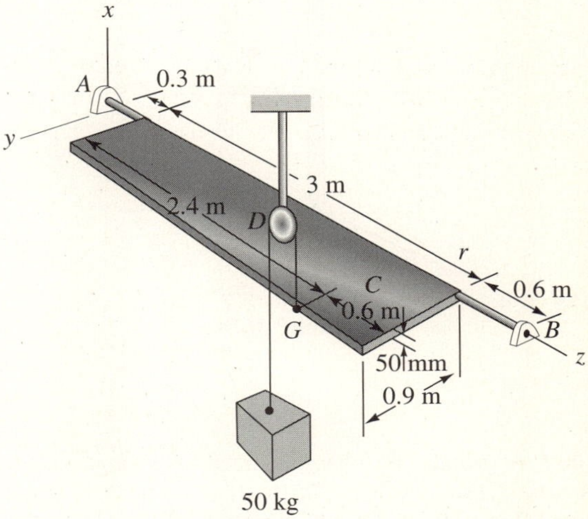

Figure P.16.96.

16.97. Do Problem 16.96 when there is a frictional torque at the bearings of 14 N m and the pulley has a radius of 0.9 m and a moment of inertia of 0.4 kg m^2.

16.98. A thin rectangular plate weighing 50 N is rotating about its diameter at a speed ω of 25 rad/s. What are the supporting forces in the x and y directions at the instant shown when the plate is parallel to the yz plane?

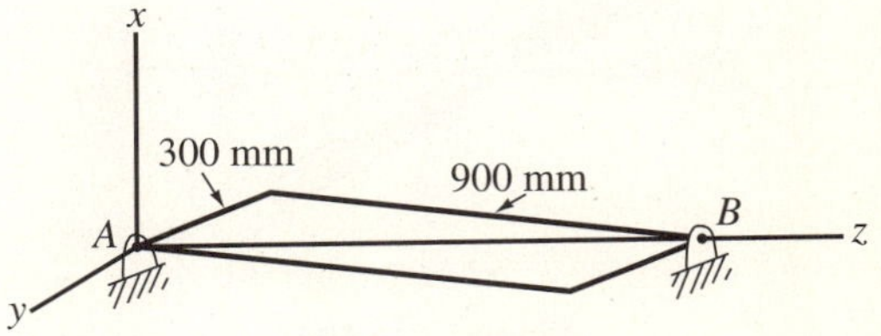

Figure P.16.98.

16.99. A shaft is shown rotating at a speed of 20 rad/s. What are the supporting forces at the bearings? The rods welded to the shaft weight 40 N/m. The shaft weighs 80 N.

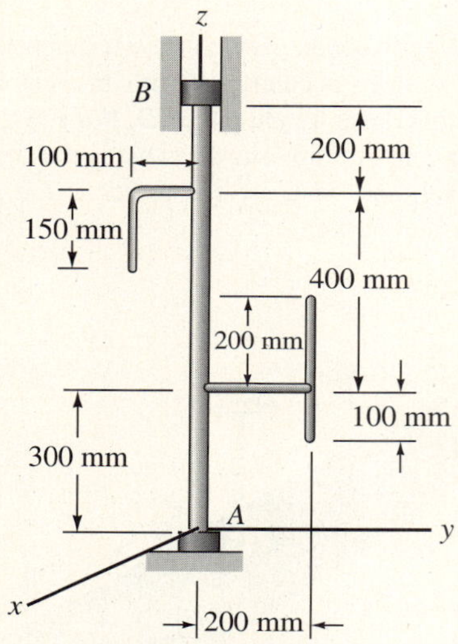

Figure P.16.99.

16.100. A cylinder is shown mounted at an angle of 30° to a shaft. The cylinder weighs 400 N. If a torque T of 20 N m is applied, what is the angular acceleration of the system? What are the supporting forces in the x and y directions at the configuration shown wherein the system is stationary? Neglect the mass of the shaft. The centerline of the cylinder is in the xz plane at the instant shown.

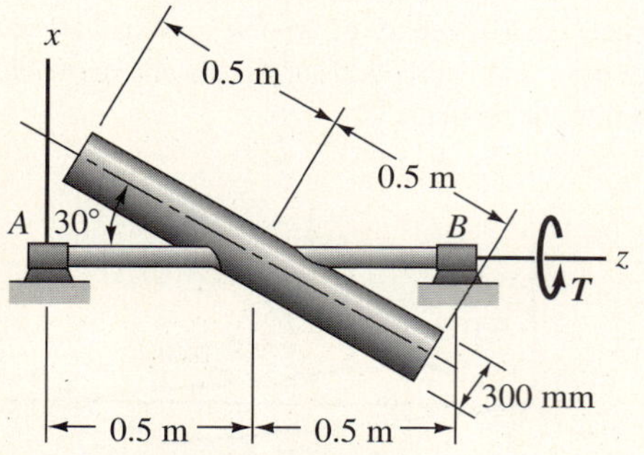

Figure P.16.100.

16.101. A bent shaft has applied to it a torque T including gravity given as

$$T = 10 + 5t \text{ N m}$$

where t is in seconds. What are the supporting forces at the bearings in the x and y directions when $t = 3$ s? The shaft is made from a rod 20 mm in diameter and weighing 70 N/m. At $t = 3$ s, the position of the shaft is as shown.

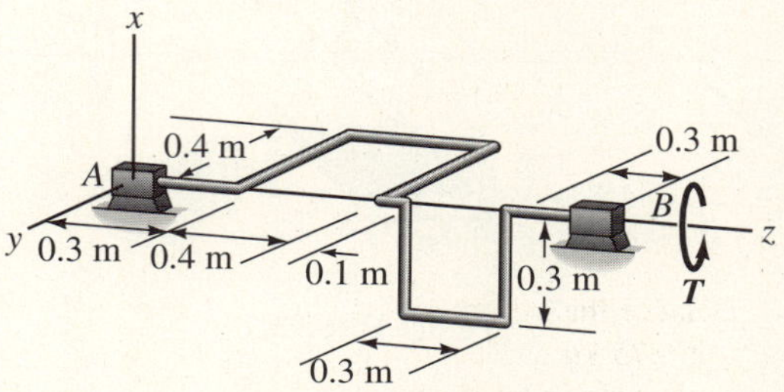

Figure P.16.101.

***16.102.** A shaft has an angular velocity ω of 10 rad/s and an angular acceleration $\dot{\omega}$ of 5 rad/s^2 at the instant of interest. What are the bending moments at this instant about the x and y axes just to the right of the bearing at A? Also, what are the twisting moment and shear forces there? The shaft and attached rods have a diameter of 20 mm and a weight per unit length of 50 N/m.

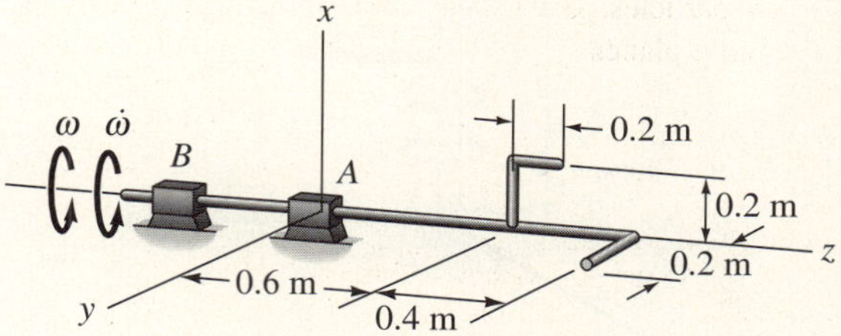

Figure P.16.102.

16.103. Balance the system in planes A and B at a distance 0.3 m from centerline. Use two masses.

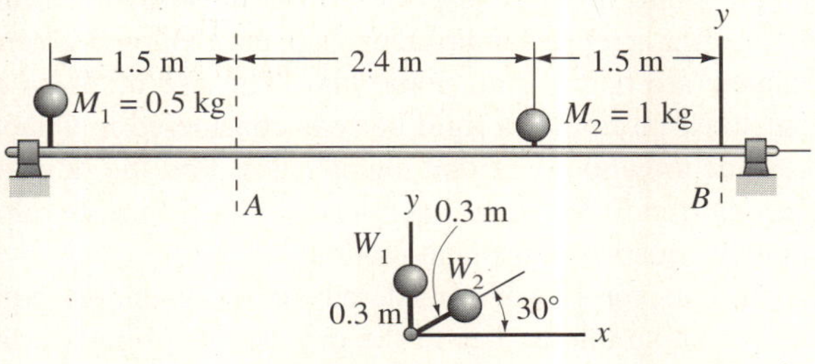

Figure P.16.103.

16.104. Balance the system in Problem 16.103 by using a mass in plane A of 0.75 kg and a mass in plane B of 0.5 kg. You may choose suitable radii in these planes.

16.105. Balance the shaft of Problem 16.94 using rods of mass 5 kg/m and welded to the shaft normal to the centerline just next to bearings A and B. Determine the lengths of these rods and their orientations relative to the xy axes.

16.106. A disc and a cylinder are mounted on a shaft. The disc has been mounted eccentrically so that the center of mass is 12 mm from the centerline of the shaft. Balance the shaft using balancing planes 1.5 m in from bearing A and 0.9 m in from bearing B, respectively. The balancing masses are each 1.5 kg and can be regarded as particles. Give the proper position of the balancing masses in these planes.

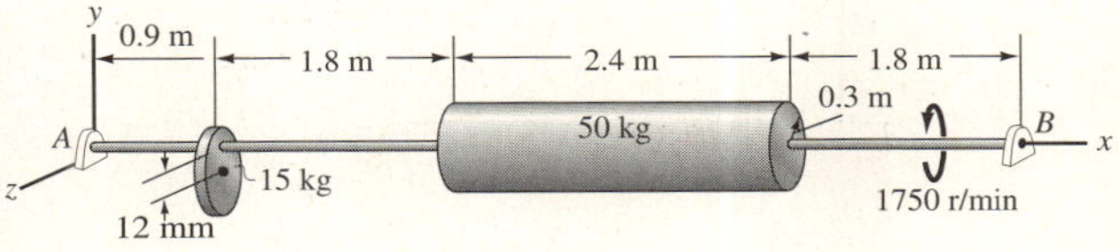

Figure P.16.106.

16.107. Balance the shaft described in Problem 16.106 by removing a small chunk of metal from each of the end faces of the 50 kg cylinder at a position 250 mm from the shaft centerline. What are the masses of these chunks and what are their orientations?

16.108. Balance the shaft in Problem 16.99 using balancing planes just next to bearings A and B. At bearing B use a small balancing sphere of mass 3 kg and at bearing A use a rod having a mass per unit length of 3.5 kg/m.

16.109. A disc is shown mounted off-center at B on a shaft CD that rotates with angular speed ω. The diameter of the shaft is 50 mm. The disc has a mass of 25 kg and has a diameter of 1.8 m. Balance the system using two rods, each of mass 15 kg/m and having a diameter of 50 mm. The rods are to be attached normal to the shaft at position 0.3 m in from bearing C and 0.6 m in from bearing D. Determine the lengths of these rods and their inclination.

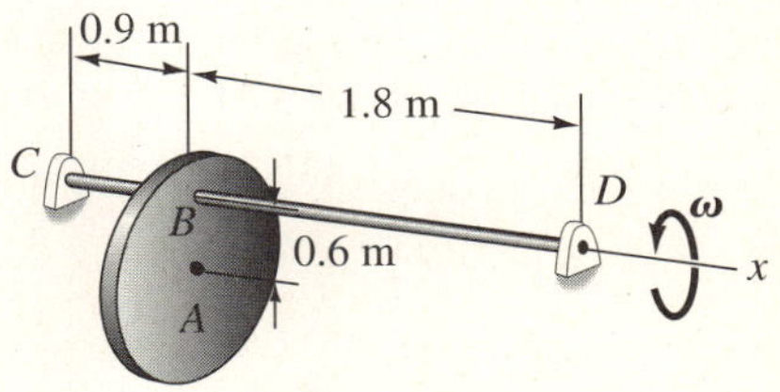

Figure P.16.109.

16.110. Balance the shaft shown for Problem 16.101. Use a balancing plane just next to bearing A and one just next to bearing B. Attach a circular plate normal to the shaft at each bearing, and cut a hole with diameter 60 mm at the proper position in the plate to balance the system. The plates are 30 mm thick and have a density of 8.2 Mg/m^3.

16.10 Closure

In this chapter, we have developed the moment-of-momentum equations for plane motion of a rigid body. We applied this equation to various cases of plane motion starting from the simplest case and going to the most difficult case. Many problems of engineering interest can be taken as plane-motion problems; the results of this chapter are hence quite important. The use of $\boldsymbol{M}_A = \dot{\boldsymbol{H}}_A$ applied to three-dimensional motion of a rigid body is considered in Chapter 18 (starred chapter). Students who cover that chapter will find the development of the key equations (the Euler equations) very similar to the development of Eqs. 16.6, the key equations for plane motion.

Recall next that in Chapter 13 we considered the work–energy equations for the plane motion of simple bodies in the process of rolling without slipping. We did this to help illustrate the use of the work–energy equations for an aggregate of particles. Also, this undertaking served to motivate a more detailed study of kinematics of rigid bodies and to set forth in miniature the more general procedures to come later. We are therefore now ready to examine energy methods as applied to rigid bodies in a more general way. This will be done in Part A of Chapter 17. We shall develop the work–energy equations for three-dimensional motion and apply them to all kinds of motions, including plane motions. The student should not have difficulty in going directly to the general case; indeed, a better understanding of the subject should result.

In Part B of Chapter 17, we shall consider the impulse-momentum equations for rigid bodies. This will be an extension of the useful impulse-momentum methods discussed in Chapter 14 for particles and aggregates of particles. Again, we shall be able to go to the general case and then apply the results to three- and two-dimensional motions (plane motions).

PROBLEMS

16.111. A circular plate A is used in electric meters to damp out rotations of a shaft by rotating in a bath of oil. The plate and its shaft have a mass of 300 g and a radius of gyration of 100 mm. If the shaft and plate very thin down from 30 r/min to 20 r/min in 5 s, what angular acceleration can be developed by a 5 mN m torque when ω of the shaft is 10 r/min? Assume that the damping torque is proportional to the angular speed.

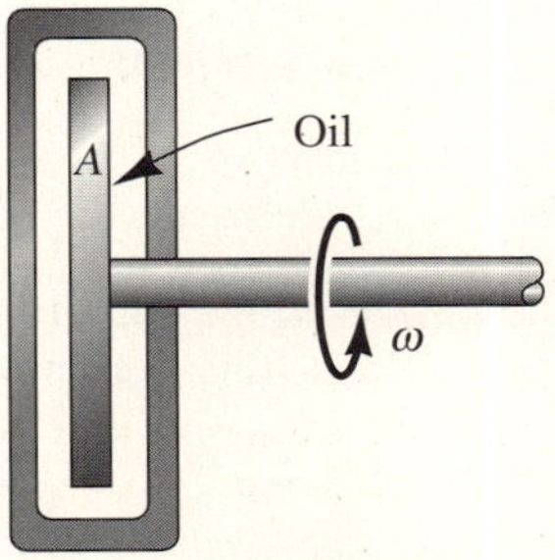

Figure P.16.111.

16.112. The dynamic coefficient of friction for contact surfaces E and G is 0.2 and for A is 0.3. If a force P of 11 kN is applied, what will be the tension in the cord HB? Start by assuming no slipping at A. Check your assumption at the end of the calculation.

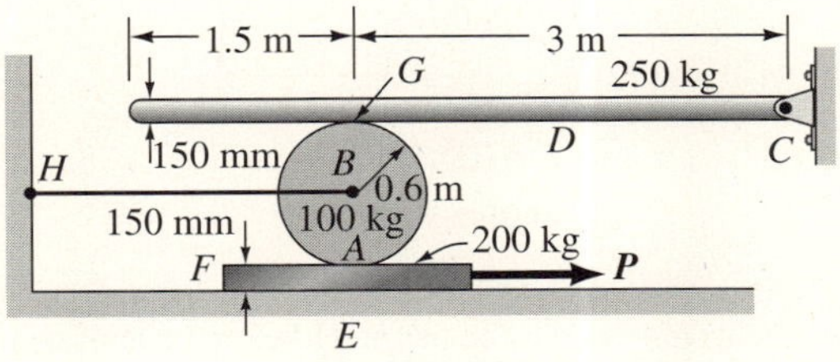

Figure P.16.112.

16.113. A torque T of 15 N m is applied to a rod AB as shown. At B there is a pin which slides in a frictionless slot in a disc E whose mass is 10 kg and whose radius of gyration is 300 mm. If the system is at rest at the instant shown, what is the angular acceleration of the rod and the disc? The rod has a mass of 18 kg.

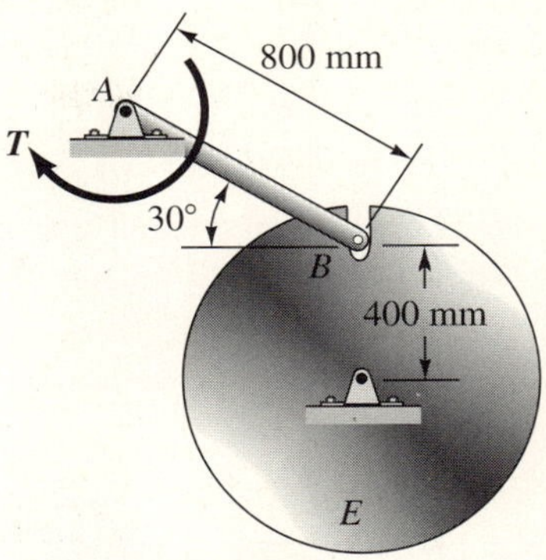

Figure P.16.113.

16.114. A platform A has a torque T applied about its axis of rotation. The platform has a mass of 1000 kg and a radius of gyration of 2 m. A block B rests on the platform but is prevented from sliding off by very thin stops C and D. The block has a mass of 1 kg and has dimensions 200 mm × 200 mm × 200 mm. The center of mass of the block is at its geometric center. If a torque $T = 20t^2 + 50t$ N m is applied with t in seconds, when and how does the block first tip? (Because of the small size and mass of B, consider the system to be a slablike body.)

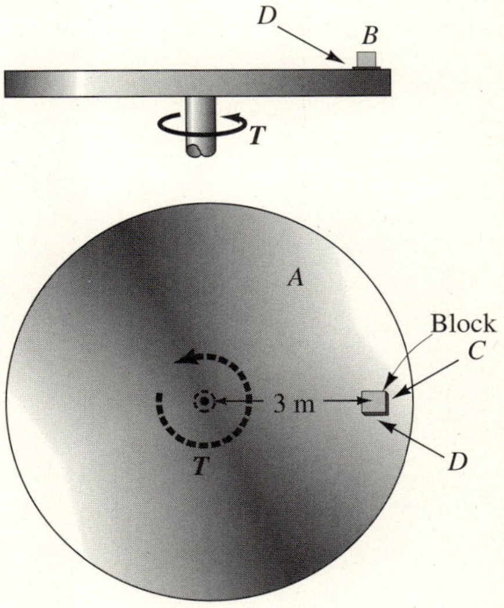

Figure P.16.114.

16.115. A motor B drives a gear C which connects with gear D to driven device A. The top system of shaft and gear has a moment of inertia I_1 about the axis of rotation of 120 g m^2, whereas the bottom system of device A and gear D has a moment of inertia about its axis of rotation of I_2 equal to 40 g m^2. If motor B develops a torque given as

$$T = 3.4 - 0.02t^2 \text{ N m}$$

with t in seconds, what is the angular speed of gear D 6 s after starting from a stationary configuration? How many revolutions has it undergone during this time interval?

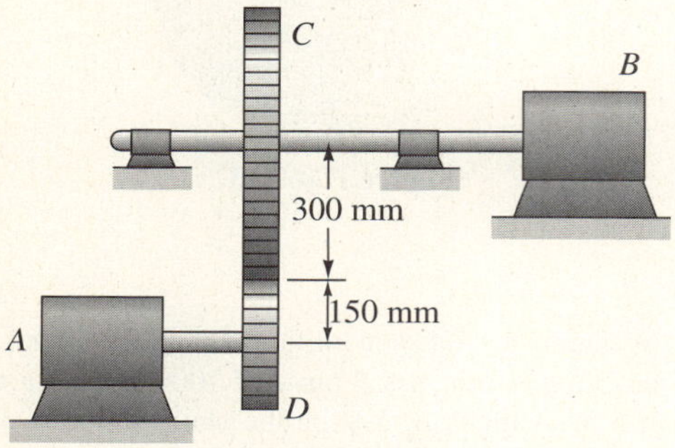

Figure P.16.115.

16.116. A 50 kg container A is being transported by a conveyor as shown. A torque T of 200 N m is applied to the driving drum. Both driving and driven drums each have a mass of 10 kg and a radius of gyration of 130 mm. The belt has a mass per unit length of 3 kg/m and a dynamic coefficient of friction of 0.3 with the conveyor bed. What is the acceleration of container A?

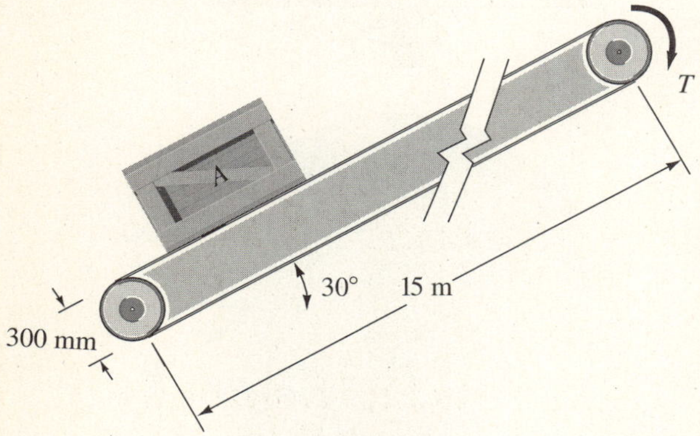

Figure P.16.116.

16.117. A four-bar linkage is shown. A torque T of 10 N m is applied to member AB, which at the instant shown is rotating clockwise at an angular of speed of 3 rad/s. Bars AB and CD are 300 mm long. Bar BC is a circular arc of radius 400 mm and length 450 mm. All bars have a mass of 10 kg/m. What is the angular acceleration of bar AB at the instant shown?

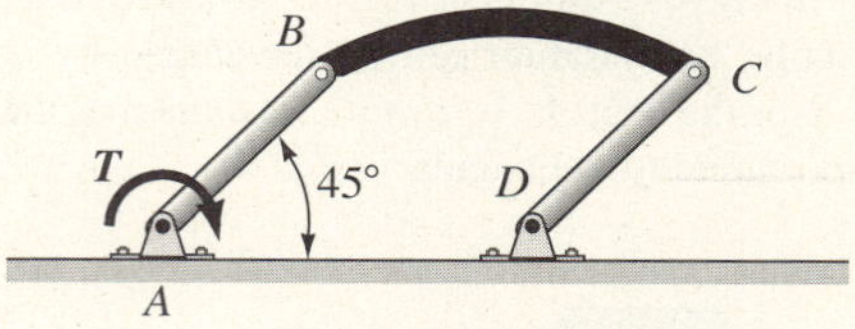

Figure P.16.117.

16.118. Bar AB having a mass of 20 kg is connected to two gears at its ends. Each such gear has a diameter of 300 mm and a mass of 5 kg. The two aforementioned gears mesh with a stationary gear E. If a torque of 100 N m is applied to the bar AB, what is the angular acceleration of the small gears?

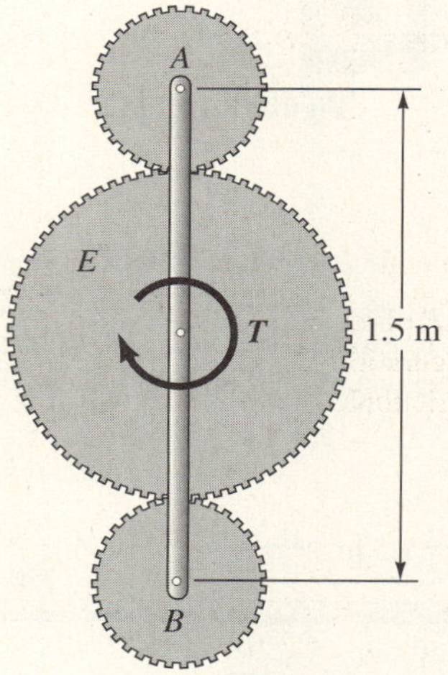

Figure P.16.118.

16.119. An astronaut is assembling a space laboratory in outer space. At the instant shown, he has an H-beam made of composite materials[13] coplanar with two other similar beams. The members are stationary relative to each other. Using a small rocket system that he is wearing, he develops a force F of 200 N always normal to the beam A and in the plane of the beams. At what position d should he exert this force to have the beam A parallel to the tops of beams C and D in 2 s? What is the acceleration of the center of mass of A at time $t = 1$ s? Beam A has a mass 200 kg and is 10 m long. Consider beam A to be a long slender rod.

[13]Made up of plastics and various kinds of fibers.

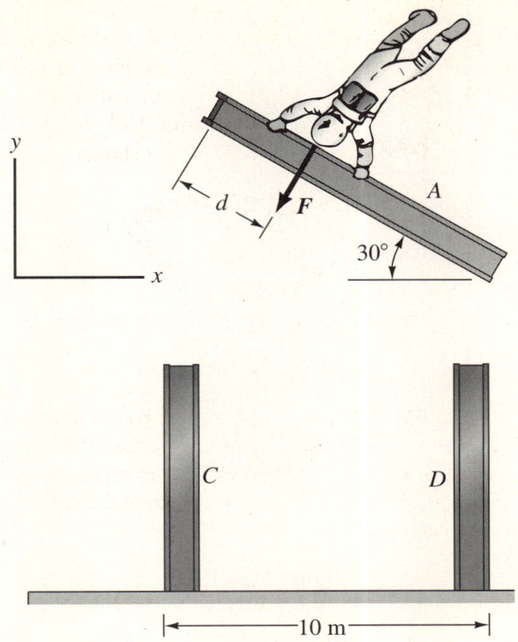

Figure P.16.119.

16.120. Rod AB of length 6 m and mass 100 kg is released from rest at the configuration shown. AC is a weightless wire, and the incline at B is frictionless. What is the tension in the wire AC at this instant?

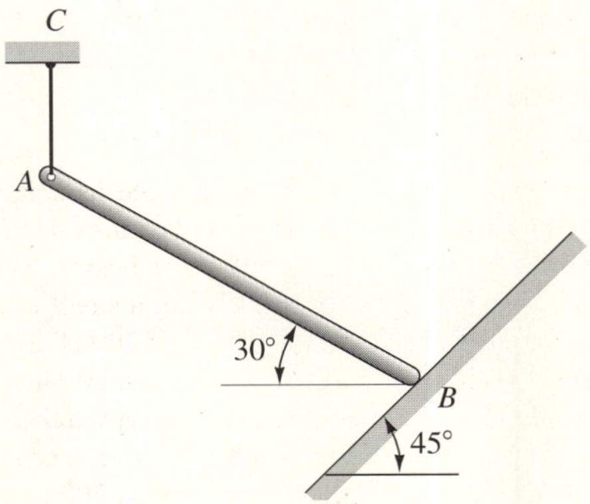

Figure P.16.120.

16.121. Rod AB of length 1 m and mass 10 kg is pinned to a disc D having a mass of 20 kg and a diameter of 0.5 m. A torque T = 15 N m is applied to the disc. What are the angular accelerations of the rod AB and disc at this instant?

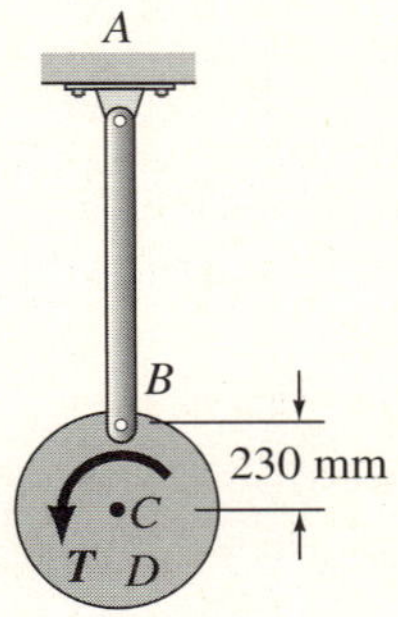

Figure P.16.121.

16.122. A four-bar linkage is shown. Bar AB has the following angular motion at the instant shown.

$$\omega_1 = 2 \text{ rad/s}$$
$$\dot{\omega}_1 = 3 \text{ rad/s}^2$$

What are the supporting forces at D at this instant for the following data?

mass of AB = 4 kg
mass of BC = 3 kg
mass of CD = 6 kg

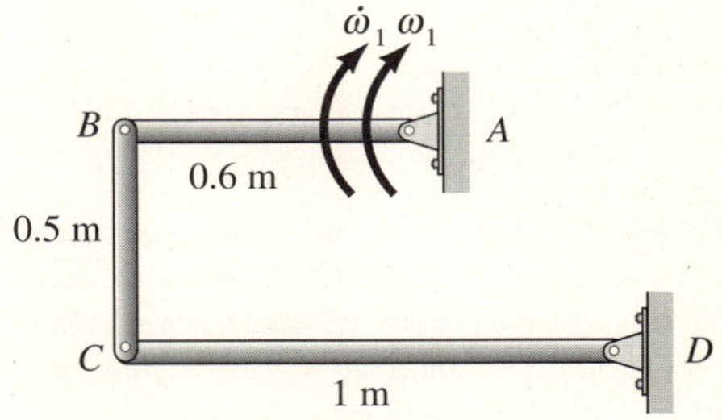

Figure P.16.122.

16.123. Starting from rest, find the acceleration of device A when the 10 kN force is applied. Proceed as follows:

(a) Draw and label four free-body diagrams.
(b) Write a system of equations whose number equals the number of unknowns.
(c) Do not solve the equations but do put in any numerical data.

Data:

M_J (for 2 wheels) = 50 kg
M_E (for 2 wheels) = 100 kg
M_A = 220 kg
M_C = 40 kg
M_E (for 2 wheels) = 100 kg
k_J = 80 mm
k_E = 0.12 m
D_J = 0.2 m
D_E = 0.3 m

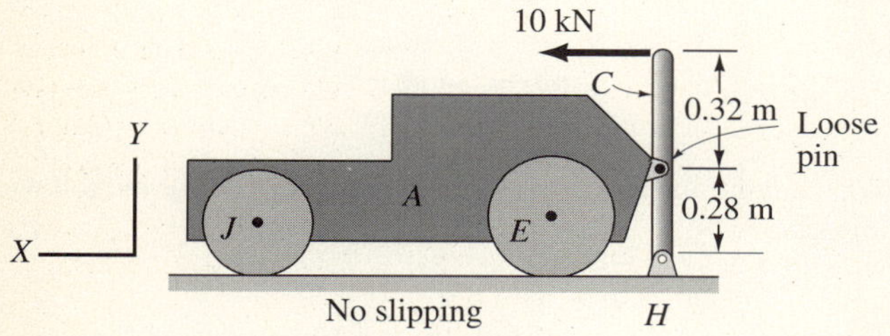

Figure P.16.123.

16.124. A cylinder A can rotate freely about its fixed centerline. Two smaller identical cylinders B have axes of rotation on cylinder A and initially will start to roll without slipping along the indicated walls. What is the initial angular acceleration of cylinder A starting from rest under the action of a torque T = 5 N m? The system is in a vertical plane. The following data apply:

M_A = 1.8 kg M_B = 1.4 kg r_A = 1.3 m r_B = 40 mm

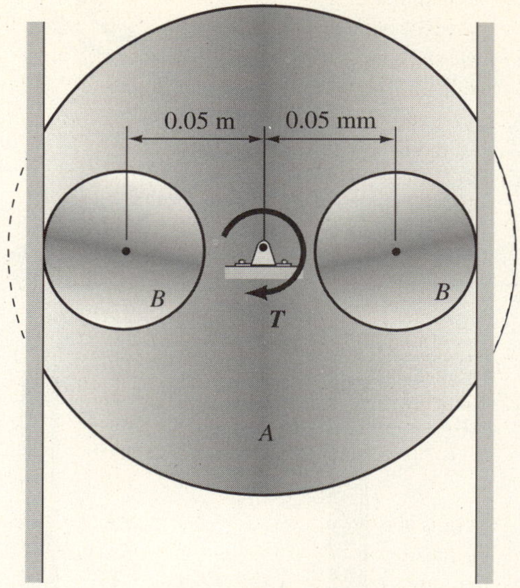

Figure P.16.124.

16.125. In Example 15.5 determine the force components at pins A and B. The mass of rod AB is 8 kg and the mass of BC is 5 kg. Use the kinematic results of the example.

16.126. A 10 m I-beam having a mass of 400 kg is being pulled by an astronaut with his space propulsion rig as shown. The force is 40 N and is always in the same direction. Initially, the beam is stationary relative to the astronaut, and the connecting cord is at right angles to the beam. Consider the beam to be a long slender rod. What is the angular acceleration when the beam has rotated 15°°? What is the position of the center of mass after 10 s? (Can you integrate the differential equation for θ to get familiar functions? Explain.)

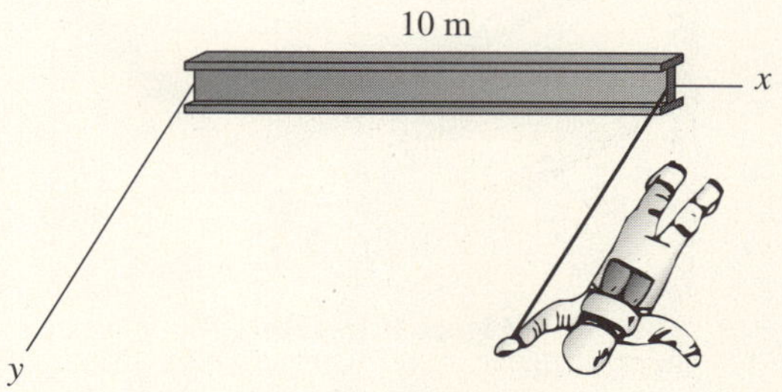

Figure P.16.126.

16.127. A rectangular box having a mass of 20 kg is being transported on a conveyor belt. The center of mass of the box is 150 mm above the conveyor belt, as shown. What is the maximum starting torque T for which the box will not tip? The belt has a mass per unit length of 2 kg/m, and the driving and driven drums have a mass of 5 kg each and a radius of gyration of 130 mm. The dynamic coefficient of friction between the belt and conveyor bed is 0.2.

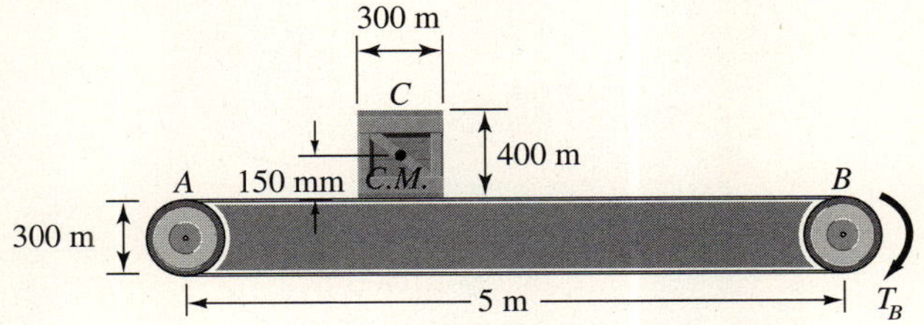

Figure P.16.127.

16.128. A ring is shown supported by wire AB and a smooth surface. The ring has a mass of 10 kg and a mean radius of 2 m. A body D having a mass of 3 kg is fixed to the ring as shown. If the wire is severed, what is the acceleration of body D?

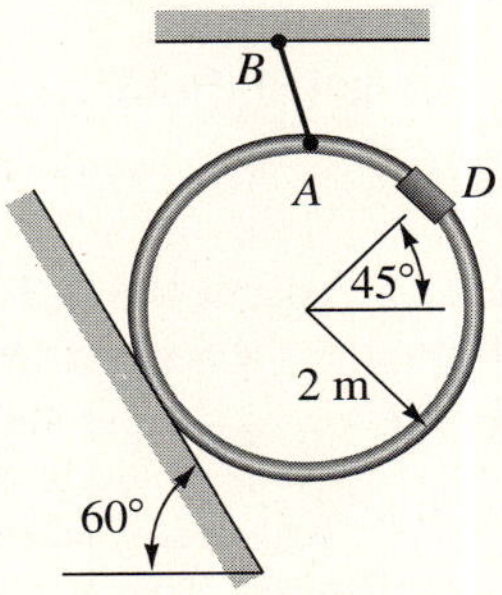

Figure P.16.128.

16.129. If the rod shown is released from rest at the configuration shown, what are the supporting forces at A and B at that instant? The rod has a mass of 30 kg and is 3 m long. The static coefficient of friction is 0.2 for all surface contacts.

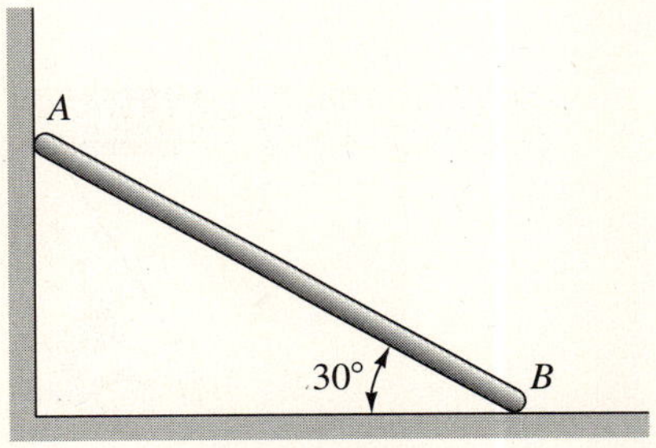

Figure P.16.129.

16.130. Do Problem 16.129 for the case where end A is moving downward at a speed of 3 m/s at the instant shown and where $\mu_d = 0.2$.

16.131. In Problem 16.129, find by inspection the instantaneous axis of rotation for the rod. What are the magnitude and direction of the acceleration vector for the axis of rotation at the instant the rod is released? We know from Problem 16.129 that $\dot{\omega} = 3.107$ rad/s^2 and $\boldsymbol{a}_c = 2.37\boldsymbol{i} - 4.10\boldsymbol{j}$ m/s^2 for the center of mass.

16.132. Identical bars AB and BC are pinned as shown with frictionless pins. Each bar is 2.3 m in length and has a mass of 9 kg. A force of 450 N is exerted at C when the bars are inclined at 60°. What is the angular acceleration of the bars?

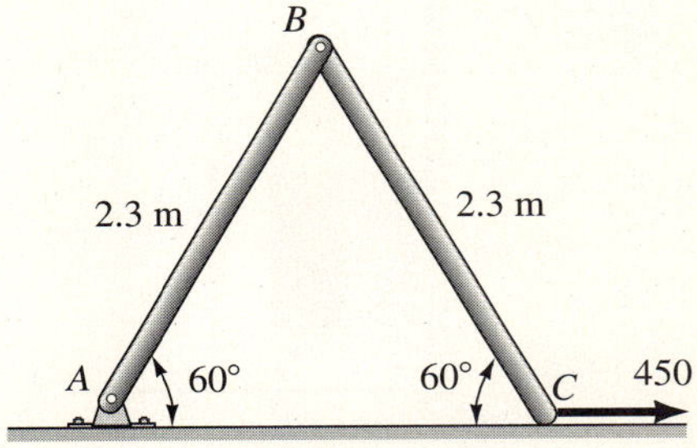

Figure P.16.132.

16.133. A compressor is shown. Member AB is rotating at a constant speed ω_1 of 100 r/min. Member BC has a mass of 2 kg and piston C has a mass of 1 kg. The pressure p on the piston is 10 kPa. At the instant shown, what are the forces transmitted by pins B and C?

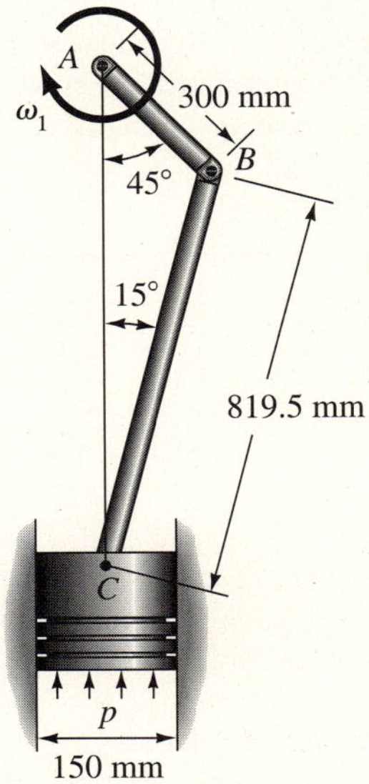

Figure P.16.133.

***16.134.** A thin vertical shaft rotates with angular speed ω of 5 rad/s in bearings *A* and *D* as shown in the diagram. A uniform plate *B*, mass 25 kg, is attached to the shaft as is a disc *C*, mass 15 kg. What are the bearing reactions at the configuration shown? The shaft has a mass of 10 kg and the thickness of disc and plate is 50 mm.

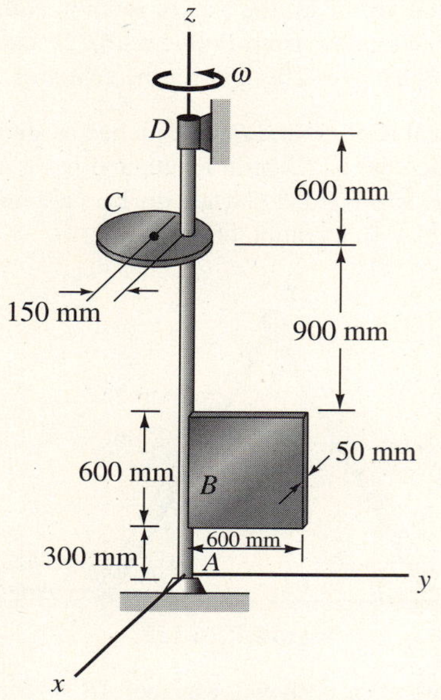

Figure P.16.134.

***16.135.** Do Problem 16.111 for the case where $\omega = 5$ rad/s at the instant of interest.

***16.136.** In Example 16.10, balance the rotating system by properly placing 3.6 kg spherical masses in balancing planes 300 mm inside from the bearings *A* and *B*. Assume that the balancing masses are particles.

16.137. A 12 kg cylinder which is spinning at a rate ω_0 of 500 r/min is placed on an 8 degree incline. The coefficients of friction are $\mu_s = 0.4$ and $\mu_d = 0.3$. How far does the cylinder move before there is rolling without slipping? How much time elapses before the cylinder stops moving instantaneously?

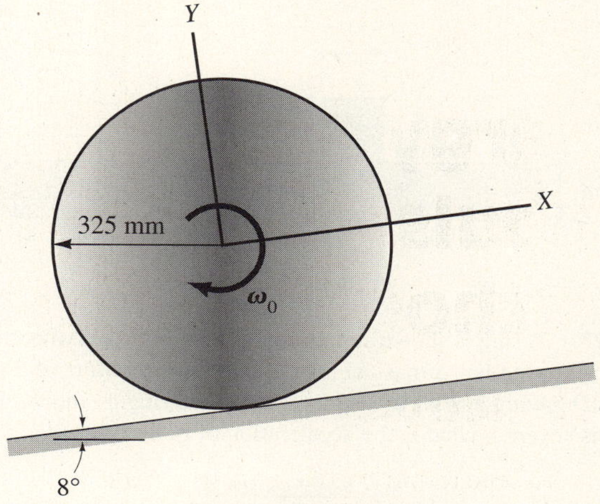

Figure P.16.137.

16.138. In Example 15.11, take ω_2 and $\dot{\omega}_2$ both equal to zero but keep all other data. Determine the force system at *A* knowing that the following data apply:

$$W_{AB} = 800 \text{ N} \qquad M_{man\text{-}cockpit} = 135 \text{ kg}$$

Note that the man and cockpit are translating and can be considered as a particle of mass 135 kg.

CHAPTER 17

Energy and Impulse–Momentum Methods for Rigid Bodies

17.1 Introduction

Let us pause to reflect on where we have been thus far in dynamics and where we are about to go. In Chapter 12, you will recall, we worked directly with *Newton's law* and integrated it several times to consider the motion of a *particle*. Then, in Chapter 13 and 14, we formulated certain useful integrated forms from *Newton's law* and thereby presented the *energy methods* and the *linear impulse-momentum* methods also for a *particle*. At the end of Chapter 14, we derived the important *angular momentum* equation, $\boldsymbol{M}_A = \dot{\boldsymbol{H}}_A$. In Chapter 16, we returned to Newton's law and along with the angular momentum equation, $\boldsymbol{M}_A = \dot{\boldsymbol{H}}_A$, carried out integrations to solve *plane motion* problems of rigid bodies. In the present chapter, we shall come back to *energy methods* and *linear impulse-momentum methods*—this time for the *general motion* of rigid bodies. In addition, we shall use a certain integrated form of the angular momentum equation $\boldsymbol{M}_A = \dot{\boldsymbol{H}}_A$, namely the *angular impulse-momentum equation*. These equations at times will be applied to a single rigid body. At other times, we shall apply them to several interconnected rigid bodies considered as a whole. When we do the latter, we say we are dealing with a *system* of rigid bodies. We shall consider energy methods first.

Part A: Energy Methods

17.2 Kinetic Energy of a Rigid Body

First, we shall derive a convenient expression for the kinetic energy of a rigid body. We have already found (Section 13.7) that the kinetic energy of an aggregate of particles relative to any reference is the sum of two parts, which we list again as:

1. The kinetic energy of a hypothetical particle that has a mass equal to the total mass of the system and a motion corresponding to that of the mass center of the system, plus
2. The kinetic energy of the particles relative to the mass center.

Mathematically,

$$\text{KE} = \tfrac{1}{2} M |\dot{\boldsymbol{r}}_c|^2 + \tfrac{1}{2} \sum_{i=1}^{n} m_i |\dot{\boldsymbol{\rho}}_i|^2 \tag{17.1}$$

where $\boldsymbol{\rho}_i$ is the position vector from the mass center to the ith particle.

Let us now consider the foregoing equation as applied to a rigid body which is a special "aggregate of particles" (Fig. 17.1). In such a case, the velocity of any particle relative to the mass center becomes

$$\dot{\boldsymbol{\rho}}_i = \boldsymbol{\omega} \times \boldsymbol{\rho}_i \tag{17.2}$$

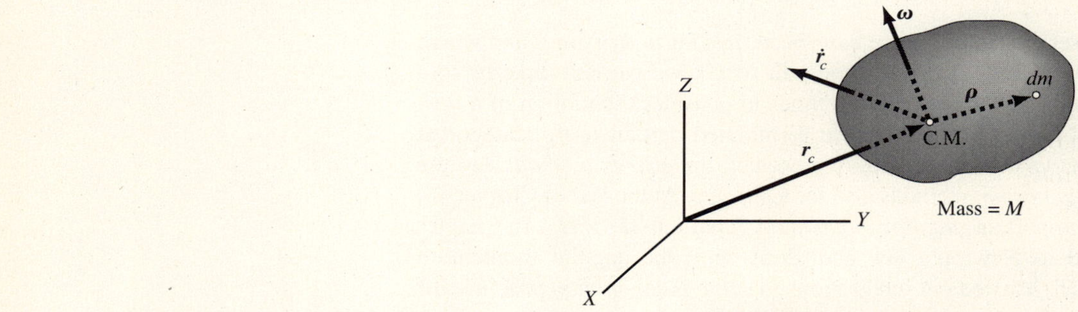

Figure 17.1. Rigid body.

where $\boldsymbol{\omega}$ is the angular velocity of the body relative to reference XYZ in which we are computing the kinetic energy. For the rigid body, the discrete particles of mass m_i become a continuum of infinitesimal particles each of mass dm, and the summation in Eq. 17.1 then becomes an integration. Thus, we can say for the rigid body, replacing $|\dot{\boldsymbol{r}}_c|^2$ by V_c^2.

$$\text{KE} = \tfrac{1}{2} M V_c^2 + \tfrac{1}{2} \iiint_M |\boldsymbol{\omega} \times \boldsymbol{\rho}|^2 \, dm \tag{17.3}$$

where $\boldsymbol{\rho}$ represents the position vector from the center of mass to any element of mass dm. Let us now choose a set of orthogonal directions xyz at the center of mass, so we can carry out the preceding integration in terms of the scalar components of $\boldsymbol{\omega}$ and $\boldsymbol{\rho}$. This step is illustrated in Fig. 17.2. We first express the integral in Eq. 17.3 in the following manner:

$$\iiint_M |\boldsymbol{\omega} \times \boldsymbol{\rho}|^2 \, dm = \iiint_M (\boldsymbol{\omega} \times \boldsymbol{\rho}) \cdot (\boldsymbol{\omega} \times \boldsymbol{\rho}) \, dm \tag{17.4}$$

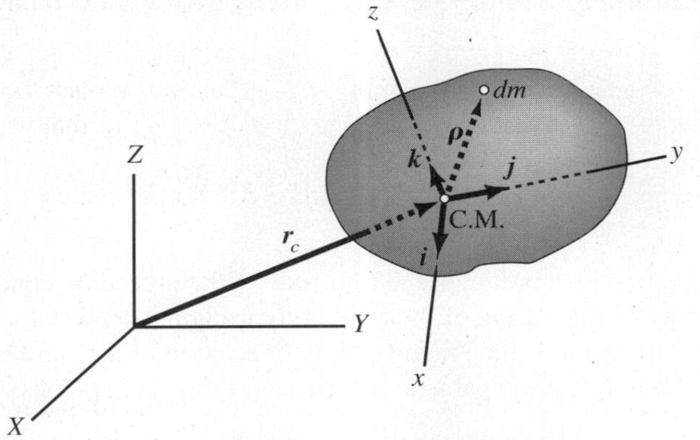

Figure 17.2. Fix *xyz* at center of mass.

Inserting the scalar components, we get:

$$\iiint_M |\boldsymbol{\omega}\times\boldsymbol{\rho}|^2\,dm = \iiint_M \Big\{\Big[\big(\omega_x\boldsymbol{i}+\omega_y\boldsymbol{j}+\omega_z\boldsymbol{k}\big)\times\big(x\boldsymbol{i}+y\boldsymbol{j}+z\boldsymbol{k}\big)\Big] \cdot\Big[\big(\omega_x\boldsymbol{i}+\omega_y\boldsymbol{j}+\omega_z\boldsymbol{k}\big)\times\big(x\boldsymbol{i}+y\boldsymbol{j}+z\boldsymbol{k}\big)\Big]\Big\}\,dm$$

Carrying out first the cross products and then the dot product in the integrand and collecting terms, we form the following relation on extracting the ω's from the integrals.

$$\begin{aligned}\iiint_M |\boldsymbol{\omega}\times\boldsymbol{\rho}|^2\,dm = {} & \left[\iiint_M (z^2+y^2)\,dm\right]\omega_x^2 - \left[\iiint_M xy\,dm\right]\omega_x\omega_y - \left[\iiint_M xz\,dm\right]\omega_x\omega_z - \\ & \left[\iiint_M yx\,dm\right]\omega_y\omega_x + \left[\iiint_M (x^2+z^2)\,dm\right]\omega_y^2 - \left[\iiint_M yz\,dm\right]\omega_y\omega_z - \\ & \left[\iiint_M zx\,dm\right]\omega_z\omega_x - \left[\iiint_M zy\,dm\right]\omega_z\omega_y + \left[\iiint_M (x^2+y^2)\,dm\right]\omega_z^2\end{aligned}$$

You will recognize that the integrals are the moments and products of inertia for the *xyz* reference. Thus,[1]

$$\begin{aligned}\iiint_M |\boldsymbol{\omega}\times\boldsymbol{\rho}|^2\,dm = {} & I_{xx}\omega_x^2 - I_{xy}\omega_x\omega_y - I_{xz}\omega_x\omega_z \\ & -I_{yx}\omega_y\omega_x + I_{yy}\omega_y^2 - I_{yz}\omega_y\omega_z \\ & -I_{zx}\omega_z\omega_x - I_{zy}\omega_z\omega_y + I_{zz}\omega_z^2\end{aligned}$$

[1]Note that we have deliberately used a matrixlike array for ease in remembering the formulation.

We can now give the kinetic energy of a rigid body in the following form:

$$\begin{aligned} KE = \tfrac{1}{2} MV_c^2 + \tfrac{1}{2}(&I_{xx}\omega_x^2 - I_{xy}\omega_x\omega_y - I_{xz}\omega_x\omega_z \\ &-I_{yx}\omega_y\omega_x + I_{yy}\omega_y^2 - I_{yz}\omega_y\omega_z \\ &-I_{zx}\omega_z\omega_x - I_{zy}\omega_z\omega_y + I_{zz}\omega_z^2) \end{aligned} \tag{17.5}$$

Note that the first expression on the right side of the preceding equation is the kinetic energy of translation of the rigid body using the center of mass, while the second expression is the kinetic energy of rotation of the rigid body about its center of mass. If principal axes are chosen, Eq. 17.5 becomes

$$KE = \tfrac{1}{2} MV_c^2 + \tfrac{1}{2}(I_{xx}\omega_x^2 + I_{yy}\omega_y^2 + I_{zz}\omega_z^2) \tag{17.6}$$

Note that for this condition the kinetic energy terms for rotation have the same form as the kinetic energy term that is due to translation, with the moment of inertia corresponding to mass and angular velocity corresponding to linear velocity.

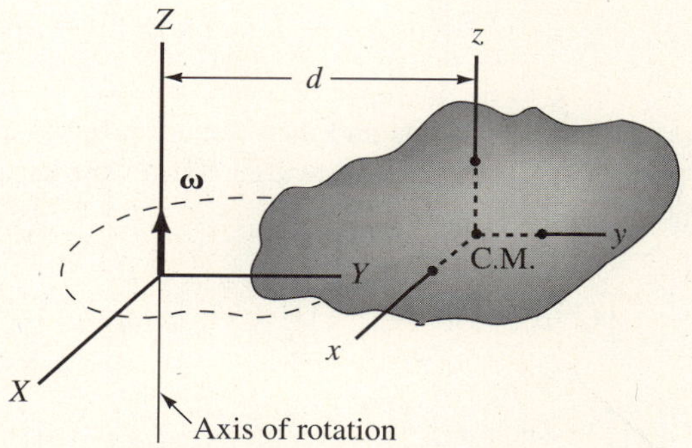

Figure 17.3. Rigid body undergoing pure rotation about Z axis.

As a special case, we shall consider the calculation of kinetic energy of any rigid body undergoing *pure rotation* relative to *XYZ* with angular velocity $\boldsymbol{\omega}$ about an actual axis of rotation[2] going through part of the body or through a rigid hypothetical massless extension of the body. We have shown this situation in Fig. 17.3, where the Z axis is chosen to be collinear with the $\boldsymbol{\omega}$ vector and the axis of rotation. The reference *xyz* at the center of mass is chosen parallel to *XYZ*. Clearly, $\omega_x = \omega_y = 0$ and $\omega_z = \omega$ so Eq. 17.5 becomes

$$KE = \tfrac{1}{2} MV_c^2 + \tfrac{1}{2} I_{zz}\omega^2$$

[2]An actual axis of rotation in *XYZ* is a line along which the velocity relative to *XYZ* is zero.

where I_{zz} is about an axis which goes through the mass center parallel to Z. Note that $V_c = \omega d$, where d is the distance between the axis of rotation Z and the z axis at the center of mass. We then have

$$\begin{aligned} KE &= \tfrac{1}{2}(Md^2)\omega^2 + \tfrac{1}{2}I_{zz}\omega^2 \\ &= \tfrac{1}{2}(I_{zz} + Md^2)\omega^2 \end{aligned}$$

But the bracketed expression is the moment of inertia of the body about the axis of rotation Z. Denoting this moment of inertia simply as I, we get for the kinetic energy:

$$KE = \tfrac{1}{2}I\omega^2 \tag{17.7}$$

This simple expression for pure rotation is completely analogous to the kinetic energy of a body in pure translation.

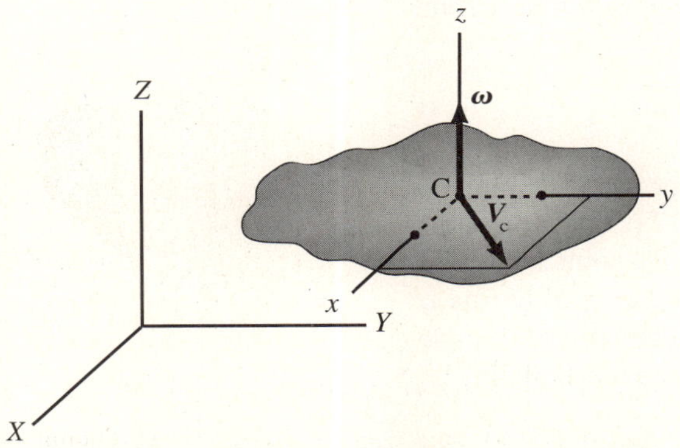

Figure 17.4. Plane motion relative to the *XY* plane.

For a body undergoing *general plane motion* (see Fig. 17.4) parallel to the *XY* plane, where *xyz* are taken at the center of mass and oriented parallel to *XYZ*, we get from Eq. 17.5:

$$KE = \tfrac{1}{2}MV_c^2 + \tfrac{1}{2}I_{zz}\omega_z^2 \tag{17.8}$$

We now illustrate the calculation of the kinetic energy in the following example.

Example 17.1

Compute the kinetic energy of the crank system in the configuration shown in Fig. 17.5. Piston A has a mass of 1 kg, rod AB is 0.6 m long and has a mass of 2.5 kg, and flywheel D has a mass of 50 kg with a radius of gyration of 0.36 m. The radius r is 0.3 m. At the instant of interest, piston A is moving to the right at a speed V of 3 m/s.

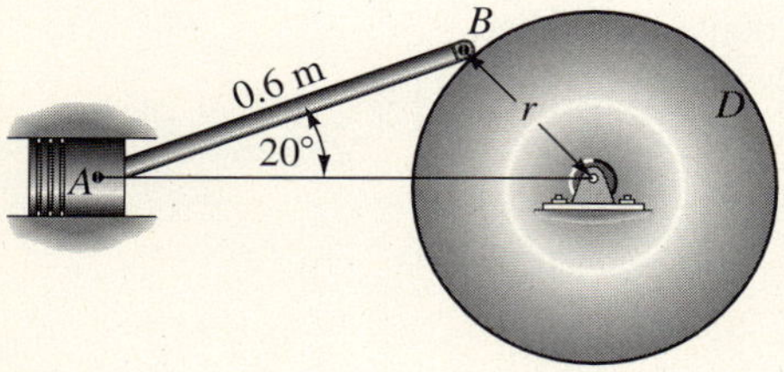

Figure 17.5. Crank system.

We have here a translatory motion (piston A), a plane motion (rod AB), and a pure rotation (flywheel D). Thus, for piston A we have for the kinetic energy:

$$(\text{KE})_A = \frac{1}{2}MV^2 = \frac{1}{2}(1)(3^2) = 4.5 \text{ N m} \qquad \text{(a)}$$

For the rod AB, we must first consider *kinematical* aspects of the motion. For this purpose we have shown rod AB again in Fig. 17.6, where V_A is the known velocity of point A and V_B is the velocity vector for point B oriented at an angle α such that V_B is perpendicular to OB. We can readily find α for the configuration of interest by trigonometric considerations of triangle ABO. To do this, we use the law of sines to first compute the angle β:

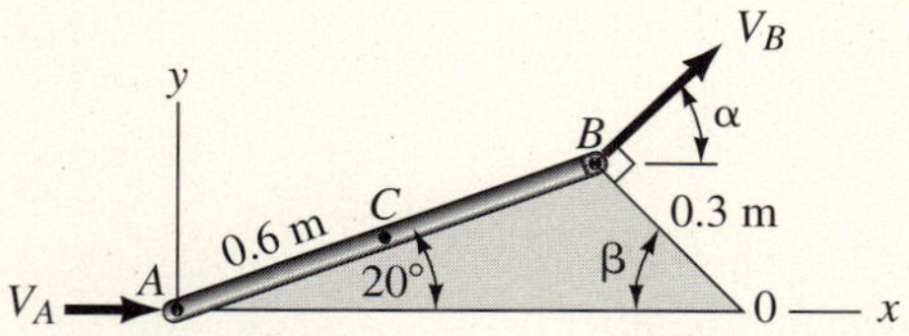

Figure 17.6. Kinematics of rod AB.

$$\frac{0.6}{\sin\beta} = \frac{0.3}{\sin 20°}$$

Therefore,

$$\beta = 43.2° \qquad \text{(b)}$$

Example 17.1 (Continued)

Because V_B is at right angles to OB, we have for the angle α:

$$\alpha = 90° - \beta = 46.8° \qquad \text{(c)}$$

From **kinematics** of a rigid body we can now say:

$$\boldsymbol{V}_B = \boldsymbol{V}_A + (\omega_{AB}\boldsymbol{k}) \times \boldsymbol{\rho}_{AB}$$

Hence,

$$V_B(\cos\alpha\boldsymbol{i} + \sin\alpha\boldsymbol{j} = 3\boldsymbol{i} + \omega_{AB}\boldsymbol{k} \times (0.6\cos 20°\boldsymbol{i} + 0.6\sin 20°\boldsymbol{j})$$
$$\therefore V_B(0.684\boldsymbol{i} + 0.729\boldsymbol{j}) = 3\boldsymbol{i} + 0.564\omega_{AB}\boldsymbol{j} - 0.205\omega_{AB}\boldsymbol{i} \qquad \text{(d)}$$

From this we solve for ω_{AB} and V_B. Thus,

$$V_B = 3.16 \text{ m/s}$$
$$\omega_{AB} = 4.09 \text{ rad/s} \qquad \text{(e)}$$

To get the velocity of the mass center C of AB, we proceed as follows:

$$\begin{aligned} V_C &= V_A + (\omega_{AB}\boldsymbol{k}) \times \boldsymbol{\rho}_{AC} \\ &= 10\boldsymbol{i} + 4.09\boldsymbol{k} \times (0.940\boldsymbol{i} + 0.342\boldsymbol{j})(0.3) \qquad \text{(f)} \\ &= 3\boldsymbol{i} + 1.158\boldsymbol{j} - 0.420\boldsymbol{i} = 2.580\boldsymbol{i} + 1.158\boldsymbol{j} \text{ m/s} \end{aligned}$$

We can now calculate $(\text{KE})_{AB}$, the kinetic energy of the rod:

$$\begin{aligned} (\text{KE})_{AB} &= \frac{1}{2}M_{AB}V_c^2 + \frac{1}{2}I_{zz}\omega_{AB}^2 \\ &= \frac{1}{2}(2.5)(2.58^2 + 1.153^2) + \frac{1}{2}\left(\frac{1}{12} \times 2.5 \times 0.6^2\right)(4.09^2) \qquad \text{(g)} \\ &= 10.61 \text{ N m} \end{aligned}$$

Finally, we consider the flywheel D. The angular speed ω_D can easily be computed using V_B of Eq. (e). Thus,

$$\omega_D = \frac{V_B}{r} = \frac{3.16}{0.3} = 10.53 \text{ rad/s} \qquad \text{(h)}$$

Accordingly, we get for $(\text{KE})_D$:

$$(\text{KE})_D = \left[\frac{1}{2}(50)(0.36^2)\right](10.53^2) = 359.3 \text{ N m} \qquad \text{(i)}$$

The total kinetic energy of the system can now be given as

$$\text{KE} = (\text{KE})_A + (\text{KE})_{AB} + (\text{KE})_D$$

$$= 4.5 + 10.61 + 359.3 = \boxed{374.4 \text{ N m}} \qquad \text{(j)}$$

17.3 Work–Energy Relations

We presented in Chapter 13 the work–energy relation for a *single* particle m_i in a *system* of n particles (see Fig. 17.7) to reach the following equation:

$$\int_1^2 \boldsymbol{F}_1 \bullet d\boldsymbol{r}_i + \int_1^2 \sum_{\substack{j=1 \\ j\neq i}}^{n} \boldsymbol{f}_{ij} \bullet d\boldsymbol{r}_i = \tfrac{1}{2}(m_i V_i^2)_2 - \tfrac{1}{2}(m_i V_i^2)_1 = (\Delta \text{KE})_i \quad (17.9)$$

where $\boldsymbol{f}_{ij}$ is the force from particle j onto particle i and is an internal force. (Note that since a particle cannot exert a force on itself, $\boldsymbol{f}_{ii} = \boldsymbol{0}$.) Now consider that the particle m_i is part of a rigid body, as shown in Fig. 17.8. From **Newton's third law** we can say that

$$\boldsymbol{f}_{ij} = -\boldsymbol{f}_{ji} \quad (17.10)$$

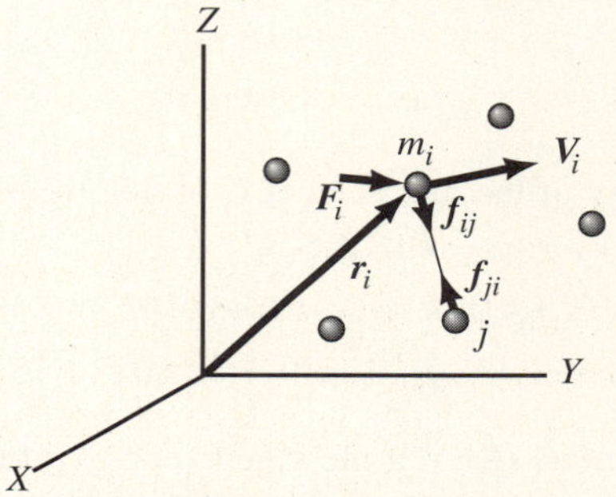

Figure 17.7. System of particles.

It might be intuitively obvious to the reader that for any motion of a rigid body the totality of internal forces $\boldsymbol{f}_{ij}$ can do no work. If not, read the following proof to verify this claim.

Suppose the rigid body moves an infinitesimal amount. We employ *Chasles' theorem*, whereby we give the entire body a displacement $d\boldsymbol{r}$ corresponding to the actual displacement of particle m_i (see Fig. 17.8). The total work done by $\boldsymbol{f}_{ij}$ and $\boldsymbol{f}_{ji}$ is clearly zero for this displacement as a result of

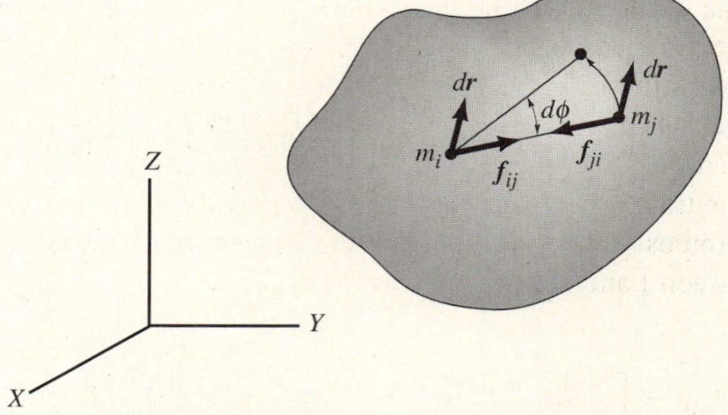

Figure 17.8. Particles of a rigid body.

Eq. 17.10. In addition, we will have a rotation $d\boldsymbol{\phi}$ about an axis of rotation going through m_i. We can decompose $d\boldsymbol{\phi}$ into orthogonal components, such that one component $d\phi_1$ is along the line between m_i and m_j (and thus collinear with $\boldsymbol{f}_{ij}$) and two components are at right angles to this line (see Fig. 17.9). Clearly, the work done by the forces $\boldsymbol{f}_{ij}$ and $\boldsymbol{f}_{ji}$ for $d\boldsymbol{\phi}_1$ is zero. Also, the movement of m_j for the other components of $d\boldsymbol{\phi}$ is at right angles to $\boldsymbol{f}_{ji}$, and again there is no work done. Consequently, the work done by $\boldsymbol{f}_{ij}$ and $\boldsymbol{f}_{ji}$ is zero during the total infinitesimal movement. And since a finite movement is a sum of such infinitesimal movements, the work done for a finite movement of m_i and m_j is zero. But a rigid body consists of *pairs* of interacting particles such as m_i and m_j. Hence, on summing Eq. 17.9 for all particles of a rigid body, we can conclude that the *work done by forces internal to a rigid body for any rigid-body movement is always zero.*

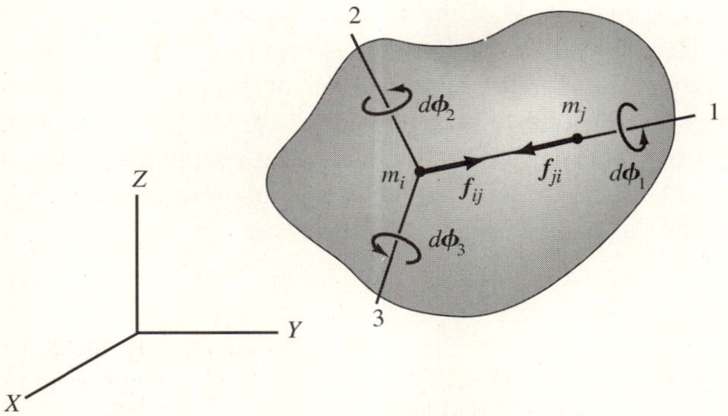

Figure 17.9. Rectangular components of $d\boldsymbol{\phi}$.

We must clearly point out here that although the internal forces *in* a rigid body can do no work, forces *between rigid bodies* of a *system* of rigid bodies *can* do a net amount of work even though Newton's third law applies and even though these forces are *internal to the system.* We shall say more about this later when we discuss systems of rigid bodies.

We accordingly compute the work done on a rigid body in moving from configuration I to configuration II by summing the work terms for all the *external* forces. Thus, for the body shown in Fig. 17.10, we can express the work between I and II in the following manner:

$$(\text{work})_{\text{I,II}} = \int_{\substack{\text{I} \\ \text{path 1}}}^{\text{II}} \boldsymbol{F}_1 \cdot d\boldsymbol{s}_1 + \int_{\substack{\text{I} \\ \text{path 2}}}^{\text{II}} \boldsymbol{F}_2 \cdot d\boldsymbol{s}_2 + \cdots + \int_{\substack{\text{I} \\ \text{path } n}}^{\text{II}} \boldsymbol{F}_n \cdot d\boldsymbol{s}_n \quad (17.11)$$

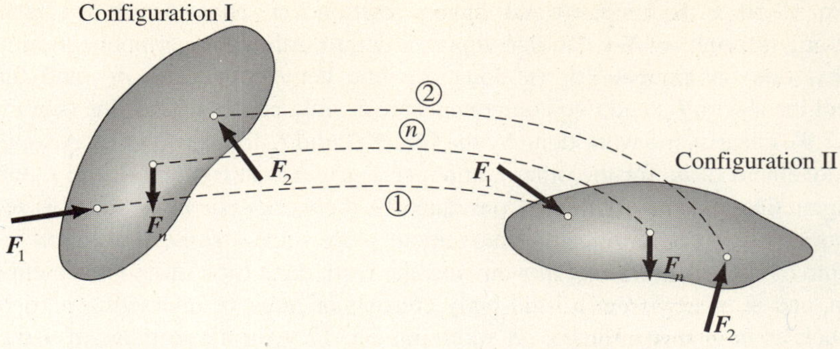

Figure 17.10. Rigid body moves from configuration I to configuration II.

In this equation, we must remember, the dot products of *nonconservative* forces are to be integrated over the *actual paths* along which the *points of application of the forces on the rigid body move*. We must take into account the variations of direction and magnitude of these nonconservative forces along their paths. For conservative forces we can use the concept of potential energy.

Although we can treat *couples* as sets of discrete forces in the foregoing manner, it is often useful to take advantage of the special properties of couples and to treat them separately. It should be a simple matter for you to show (see Fig. 17.11) that a torque T about an axis upon rotating an angle $d\theta$ about the axis does an amount of work $d\mathcal{W}_K$ given as

$$d\mathcal{W}_K = T\,d\theta$$

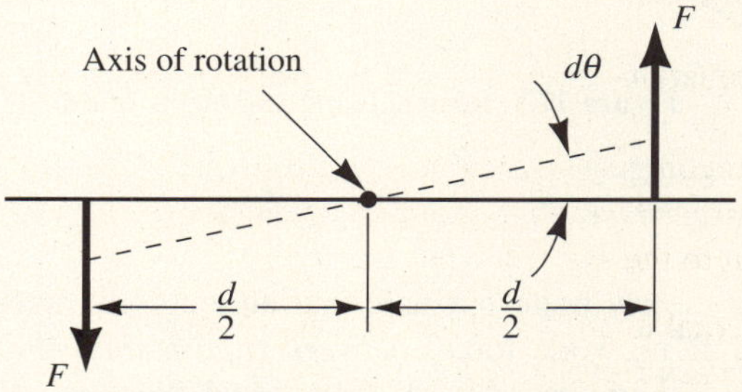

Figure 17.11. Work of torque T about an axis rotating an angle $d\theta$ about the axis.

Dividing and multiplying by dt, we can say further that

$$d\mathcal{W}_K = T\frac{d\theta}{dt}\,dt = T\dot{\theta}\,dt$$

Integrating, we get

$$\mathcal{W}_K = \int_{t_1}^{t_2} T\dot{\theta}\,dt \tag{17.12}$$

In this case the torque T and angular speed $\dot{\theta}$ are about the same axis. The generalization of Eq. 17.12 for any moment $\boldsymbol{M}$ and any angular velocity $\boldsymbol{\omega}$ (see Fig. 17.12) then is

$$\mathcal{W}_K = \int_{t_1}^{t_2} \boldsymbol{M} \cdot \boldsymbol{\omega}\, dt \tag{17.13}$$

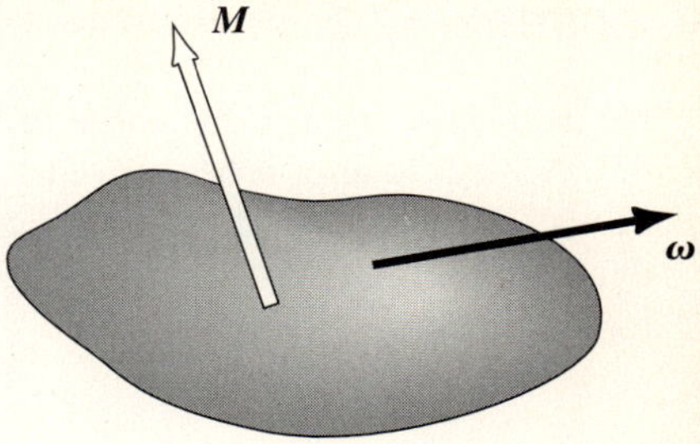

Figure 17.12. $d\mathcal{W}_K = \boldsymbol{M} \cdot \boldsymbol{\omega}\, dt$.

We thus have formulations for finding the work done by external forces and couples on a rigid body. For the conservative forces, we know from Chapter 13 that we can use for work a quantity that is minus the change in potential energy from I to II without having to specify the path taken.

Using this information for computing work, we can then say for any rigid body:

$$\mathcal{W}_K \text{ from I to II} = \Delta\text{KE} \tag{17.14}$$

where $\mathcal{W}_K$ is the work by *external* forces.

If there are only *conservative* external forces present, we can also say for the rigid body:

$$(\text{PE})_\text{I} + (\text{KE})_\text{I} = (\text{PE})_\text{II} + (\text{KE})_\text{II} \tag{17.15}$$

If both conservative *and* nonconservative external forces are present, we can say:

$$\text{nonconservative } \mathcal{W}_K \text{ from I to II} = \Delta\text{PE} + \Delta\text{KE} \tag{17.16}$$

These three equations parallel the three we developed for a particle in Chapter 13.

The foregoing equations are expressed for a *single* rigid body. For a *system* of *interconnected rigid* bodies, we distinguish between two types of forces internal to the system. They are

1. Forces internal to any rigid body of the system.
2. Forces *between* rigid bodies of the system.

For a system of bodies, as in the case of a single rigid body, forces of category 1 can do no work. However, if the forces *between two bodies* of 2 system do not move the same distance over the same path, then there may be a net amount of work done on the system by these internal forces. We must include such work contributions when employing Eqs. 17.14–17.16 *for a system of interconnected rigid bodies.*[3] Example 17.4 is an example of this situation.

[3]Recall from Chapter 13, that Eq. 17.16 and hence Eqs. 17.14 and 17.15 are valid for *any* aggregate of particles provided we include the work of internal forces both conservative and nonconservative.

Example 17.2

Neglect the weight of the cable in Fig. 17.13, and find the speed of the 450 N block A after it has moved 1.7 m along the incline from a position of rest. The static coefficient of friction along the incline is 0.32, and the dynamic coefficient of friction is 0.30. Consider the pulley B to be a uniform cylinder.

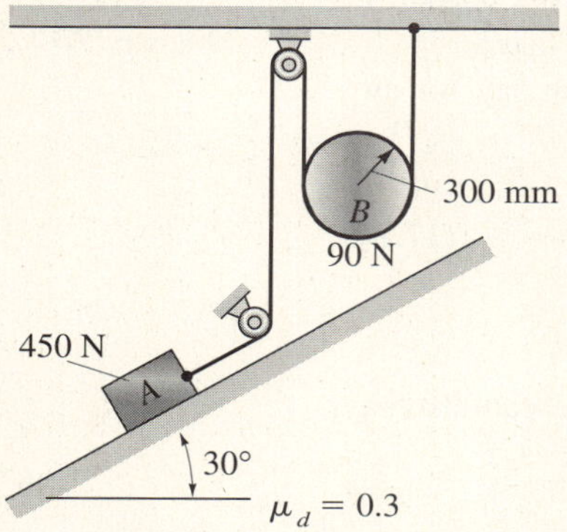

Figure 17.13. Pulley system.

We must first decide which way the block moves along the incline. To overcome friction and move down the incline, the block must create a force in the downward direction of the cable exceeding 90/2 = 45 N. Considering the block A alone (see Fig. 17.14), we can readily decide that the maximum force T_1 to allow A to start sliding downward is

$$(T_1)_{max} = -(0.32)\,N + 450\sin 30°$$
$$= -(0.32)(450)\cos 30° + 450\sin 30° = 100.3\text{ N}$$

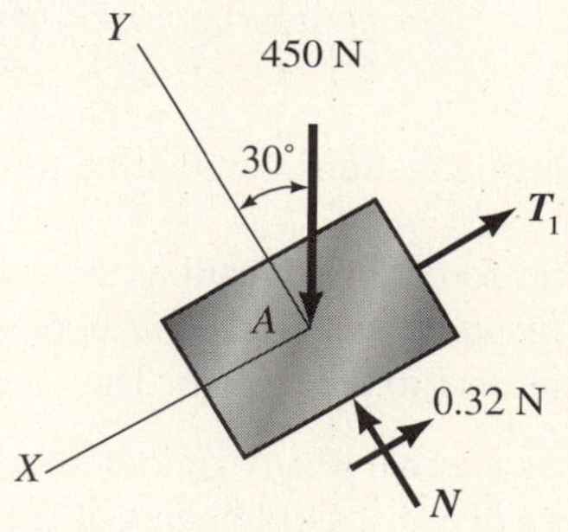

Figure 17.14. Free body of block.

Clearly, the block goes down the incline.

We now use the **work–kinetic energy** equation separately for each body. Thus, for the block we have, using now the dynamic coefficient of friction

$$(450\sin 30°)(1.7) - (450)(\cos 30°)(0.30)(1.7) - T_1(1.7) = \frac{1}{2}\frac{450}{g}V_A^2$$

Therefore,

$$T_1 = 108.1 - 13.49V_A^2 \qquad \text{(a)}$$

Example 17.2 (Continued)

We now consider the cylinder for which the free-body diagram is displayed in Fig. 17.15. Note that the cylinder is in effect rolling without slipping along the right supporting cable. Hence, T_2 does no work as explained in Chapter 13.[4] Thus, the **work–kinetic energy** equation is as follows using the formula $\frac{1}{2}Mr^2$ for I of the cylinder:

$$T_1(1.7) - 90\left(\frac{1.7}{2}\right) = \frac{1}{2}\frac{90}{g}V_C^2 + \frac{1}{2}\left[\frac{1}{2}\left(\frac{90}{g}\right)(0.30)^2\right]\omega^2$$

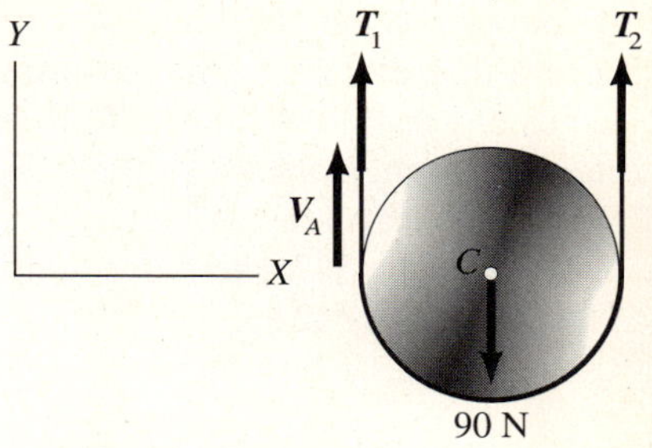

Figure 17.15. Free body of cylinder.

Therefore,

$$T_1 = 45 + 2.70V_C^2 + 0.1214\omega^2 \qquad \text{(b)}$$

From **kinematics** we can conclude on inspection that

$$V_C = \frac{1}{2}V_A \qquad \text{(c)}$$

$$\omega = -\frac{V_A}{0.60} \qquad \text{(d)}$$

Subtracting Eq. (b) from Eq. (a) to eliminate T_1, and then substituting from Eq. (c) and Eq. (d) for V_C and ω, we then get for V_A:

$$V_A = 2.09 \text{ m/s} \qquad \text{(e)}$$

This problem can readily be solved as a system, i.e, without disconnecting the bodies. You are asked to do this in Problem 17.25.

[4]Recall that the point of contact of the cylinder has zero velocity, and hence the friction force (in this case, T_2) transmits no power to the cylinder.

In the previous example, we considered the problem to be composed of two *discrete* bodies. We proceeded by expressing equations for each body separately. In the following example, we will consider a *system* of bodies expressing equations for the whole system directly. In this problem, the forces between any two bodies of the system have the *same velocity*; consequently, from Newton's third law, we can conclude that these forces contribute zero net work. We shall illustrate a case where this condition is not so in Example 17.5.

Example 17.3

A conveyor is moving a mass M of 30 kg in Fig. 17.16. Cylinders A and B have a diameter of 0.3 m and mass 15 kg each. Also, they each have a radius of gyration of 0.12 m. Rollers C, D, E, F, and G each have a diameter of 75 mm, mass 4.5 kg each, and have a radius of gyration of 25 mm. What constant torque will increase the speed of M from 0.3 m/s to 0.9 m/s in 1.5 m of travel? There is no slipping at any of the rollers and drums. The belt mass is 12 kg.

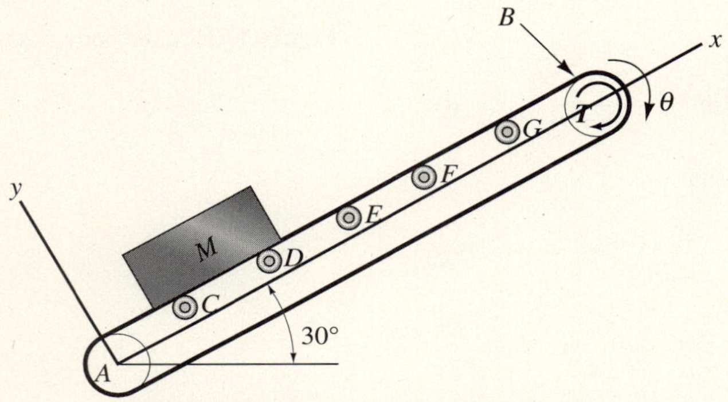

Figure 17.16. Conveyor moving mass M.

We shall use the **work–kinetic energy** relation specified in Eq. 17.14 for this problem. Only external forces and torques do work for the system; the interactive forces between bodies do work in amounts that clearly cancel each other because of the condition of no slipping. Hence,

$$T\theta - \text{Mg}(1.5)(\sin 30°) = (\text{KE})_2 - (\text{KE})_1 \qquad \text{(a)}$$

where T is the applied torque. The general expression for the kinetic energy is

$$\text{KE} = 5\left(\frac{1}{2}I_{\text{roll}}\omega_{\text{roll}}^2\right) + 2\left(\frac{1}{2}I_{\text{cyl}}\omega_{\text{cyl}}^2\right) + \frac{1}{2}M_{\text{belt}}V_{\text{belt}}^2 + \frac{1}{2}\frac{W}{g}V_{\text{mass}}^2 \qquad \text{(b)}$$

From **kinematics** we can say:

$$\omega_{\text{roll}} = -\frac{V_{\text{belt}}}{\left(\frac{1}{2}\right)(0.075)} = -26.67V_{\text{belt}}$$

$$\omega_{\text{cyl}} = -\frac{V_{\text{belt}}}{\left(\frac{1}{2}\right)(0.3)} = -6.67V_{\text{belt}}$$

$$V_{\text{mass}} = V_{\text{belt}}$$

Example 17.3 (Continued)

Using these results, we can give the kinetic energy at the end and at the beginning of the interval of interest as

$$\begin{aligned}(KE)_2 &= 5\left\{\left(\tfrac{1}{2}\right)(4.5)(0.025)^2[(26.67)(0.9)]^2\right\} + 2\left\{\left(\tfrac{1}{2}\right)(15)(0.12)^2[(6.67)(0.9)]^2\right\}\\ &\quad + \frac{1}{2}(12)(0.9)^2 + \frac{1}{2}(30)(0.9)^2\\ &= 28.84 \text{ N m}\end{aligned}$$

$$\begin{aligned}(KE)_2 &= 5\left\{\left(\tfrac{1}{2}\right)(4.5)(1)^2[(26.67)(0.3)]^2\right\} + 2\left\{\left(\tfrac{1}{2}\right)(15)(0.12)^2[(6.67)(0.3)]^2\right\}\\ &\quad + \frac{1}{2}(12)(0.3^2) + \frac{1}{2}(30)(0.3^2)\\ &= 3.21 \text{ N m}\end{aligned}$$

Substituting these results into Eq. (a), we get

$$T\theta - (30g)(1.5)(\sin 30°) = 28.84 - 3.21 \qquad \text{(c)}$$

From **kinematics** again we can say for θ, on considering the rotation of a cylinder and the distance traveled by the belt:

$$(r_{\text{cyl}})(\theta) = 1.5$$

Therefore,

$$\theta = \frac{1.5}{0.15} = 10 \text{ rad}$$

Substituting back into Eq. (c), we can then solve for the desired torque T:

$$T = \tfrac{1}{10}\left[28.84 - 3.21 + (30g)(1.5)(\sin 30°)\right]$$

$$T = 24.64 \text{ N m}$$

In the next example, we have a case of internal forces between bodies that satisfy Newton's third law but do not have identical velocities.

Example 17.4[5]

A diesel-powered electric train moves up a 7° grade in Fig. 17.17. If a torque of 750 N m is developed at each of its six pairs of drive wheels, what is the increase of speed of the train after it moves 100 m? Initially, the train has a speed of 5 m/s. The train weighs 90 kN. The drive wheels have a diameter of 600 mm. Neglect the rotational energy of the drive wheels.

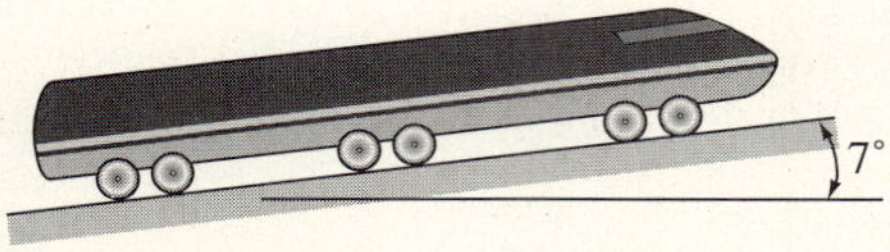

Figure 17.17. Diesel–electric train.

We shall consider the train as a *system of rigid bodies* including the 6 pairs of wheels and the body. We have shown the train in Fig. 17.18 with the external forces, W, N, and f. In addition, we have shown certain internal torques M.[6] The torques shown act on the *rotors* of the motors, and, as the train moves, these torques rotate and accordingly do work. The *reactions* to these torques are equal and opposite to M according to **Newton's third law** and act on the *stators* or the motors (i.e., the field coils). The stators are stationary, and so the reactions to M do *no* work as the train moves. Thus, we have an example of equal and opposite internal forces between bodies of a system performing a nonzero net amount of work. We now employ Eq. 17.16. Thus,

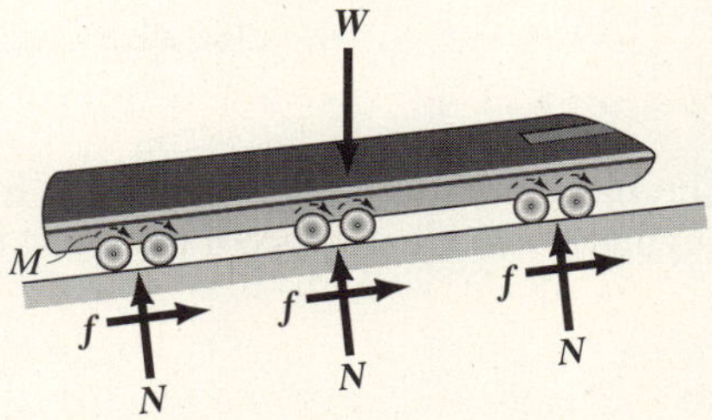

Figure 17.18. External and internal forces and torques.

$$\Delta PE + \Delta KE = \mathcal{W}_K \quad \text{(a)}$$

Using the initial configuration as the datum, we have[7]

$$\left[(90\times 10^3)(100\sin 7^\circ - 0)\right] + \left\{\frac{1}{2}\frac{90\times 10^3}{g}V^2 - \frac{1}{2}\frac{90\times 10^3}{g}(5)^2\right\} = (6)(750)(\theta) \quad \text{(b)}$$

[5]This problem was undertaken in Chapter 13 as a system of particles. Here, we consider it from the viewpoint of a system of interconnected rigid bodies.

[6]Figure 17.18, accordingly, is *not* a free-body diagram.

[7]Recall that for a *rolling* body with *no slipping*, the friction force does *no work*. If we were considering the *center of mass* of the train, then f *would move* with the center of mass and then do work as we shall see in Example 17.5.

Example 17.4 (Continued)

where θ is the clockwise rotation of the rotor in radians. Assuming direct drive from rotor to wheel, we can compute θ as follows for the 100 m distance over which the train moves:

$$\theta = \underbrace{\left(\frac{100}{2\pi r}\right)}_{\text{rev}} \underbrace{(2\pi)}_{\text{rad/rev}} = \frac{100}{0.3}\ \text{rad} \qquad \text{(c)}$$

Substituting into Eq. (b) and solving for V, we get

$$V = 10.63\ \text{m/s}$$

Hence,

$$\Delta V = 10.63 - 5 = \boxed{5.63\ \text{m/s}}$$

In Chapter 13, we also developed a work–energy equation involving the *mass center* of any system of particles. You will recall that

$$\int_1^2 \boldsymbol{F} \bullet d\boldsymbol{r}_c = \tfrac{1}{2}(MV_c^2)_2 - \tfrac{1}{2}(MV_c^2)_1 \qquad \text{(17.17a)}$$

where $\boldsymbol{F}$ is the *total external force* (only!) which hypothetically *moves with the center of mass*. This equation applies to a rigid body. Note that an external torque makes no work contribution here since equal and opposite forces each having identical motion (that of the mass center) can do no net amount of work. For a system of *interconnected rigid bodies*, we can say:

$$\int_1^2 \boldsymbol{F} \bullet d\boldsymbol{r}_c = \left[\sum_i \tfrac{1}{2} M_i (V_c)_i^2\right]_2 - \left[\sum_i \tfrac{1}{2} M_i (V_c)_i^2\right]_1 \qquad \text{(17.17b)}$$

The force $\boldsymbol{F}$ includes *only external forces* (internal forces between interconnecting bodies are equal and opposite and must move with the mass center of the system; hence they contribute no work to the left side of the foregoing equation). On the other hand, external friction forces on wheels rolling without slipping must move with the mass center of the system in this formulation and thus can *do work*, in contrast to the previous approach, in which the mass center is not used. On the right side, we have summed the kinetic energies of each of the mass centers of the constituent bodies of the system.[8] We now illustrate the use of Eq. 17.17b.

[8]We discussed this topic in Section 16.8.

Example 17.5

A vehicle for traversing swamplands is shown in Fig. 17.19. The vehicle has four-wheel drive and weighs 22.5 kN. Each wheel weighs 2 kN has a diameter of 2.5 m and has a radius of gyration of 180 mm. If each wheel gets a torque of 100 N m, what is the speed after 20 m of travel starting from rest? Also, determine the friction force from the ground on each wheel. The weight of the vehicle includes that of passengers and baggage. Neglect rolling resistance since the vehicle in this problem is moving on a hard surface. Consider rolling without slipping.

Figure 17.19. A vehicle used in swampland.

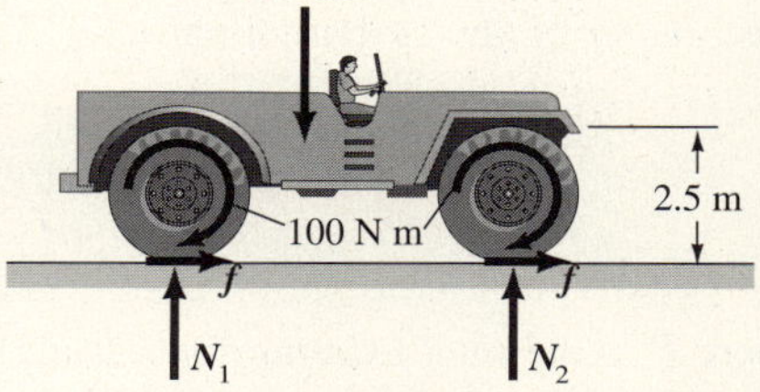

Figure 17.20. External forces and internal torques on swampland vehicle. Query: Is this a freebody diagram?

We have shown the free-body diagram of the system in Fig. 17.20. We will first use the **system of particles** approach. This includes internal torques from the vehicle frame onto the wheels and the reaction torques from the wheels onto the frame. The former will do internal work because these torques rotate with the wheels. The reaction torques on the frame do not rotate and obviously do no work. Also, because of the no slipping condition, the friction forces from the ground onto the tires do no work as has been explained at length in Chapter 13. Hence, we can say

$$(4)(T)(\theta) = (\text{KE})_2 - (\text{KE})_1$$

Recalling that $\theta =$ distance/radius, we get

$$\therefore (4)(100)\left(\frac{20}{1.25}\right) = \left[\left(\frac{4}{2}\right)\left(\frac{2000}{9.81}\right)(0.18)^2\left(\frac{V}{1.25}\right)^2\right] + \frac{1}{2}\left(\frac{22.5\times 10^3}{9.81}\right)\left(V^2\right)$$

$$\therefore \quad V = 2.354 \text{ m/s}$$

To get the friction forces f (why are they the same for each wheel?) we use the **center of mass approach**. Thus

$$W_k = (\Delta \text{KE})_{CM}$$

$$\therefore (4)(f)(20) = \frac{1}{2}\left[\frac{22.5\times 10^3}{9.81}\right](2.354)^2$$

$$\therefore \quad f = 79.4 \text{ N}$$

We thus have the desired information.

PROBLEMS

In the following problems, neglect friction unless otherwise instructed.

17.1. A uniform solid cylinder of radius 0.6 m and mass 100 kg rolls without slipping down a 45° incline and drags the 50 kg block *B* with it. What is the kinetic energy of the system if block *B* is moving at a speed of 3 m/s? Neglect the mass of connecting agents between the bodies.

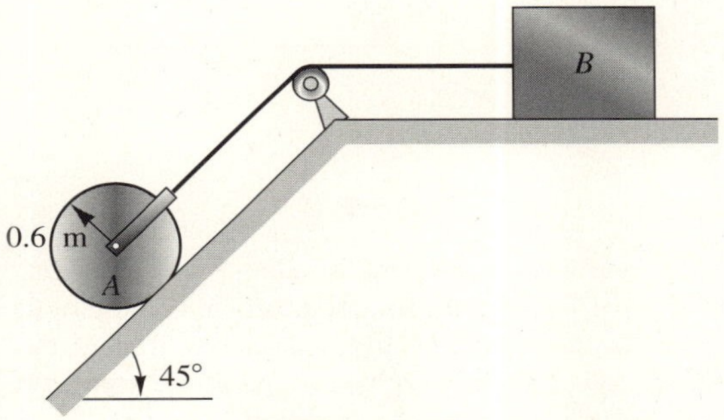

Figure P.17.1.

17.2. A steam roller with driver has a mass of 4.5 Mg. Wheel A, mass 900 kg, and has a radius of gyration of 0.25 m. Drive wheels B have a total mass of 450 kg and a radius of gyration of 0.5 m. If the steam roller is coasting at a speed of 1.5 m/s with motor disconnected, what is the total kinetic energy of the system?

Figure P.17.2.

17.3. A thin disc weighing 450 N is suspended from an overhead conveyor moving at a speed of 10 m/s. If the disc rotates at a speed of 5 rad/s in the plane of the page (i.e., *ZY* plane), compute the kinetic energy of the disc relative to the ground.

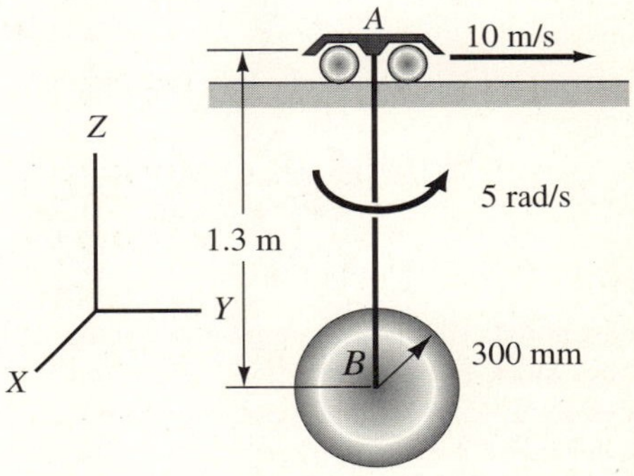

Figure P.17.3.

17.4. Two slender rods *CD* and *EA* are pinned together at *B*. Rod *EA* is rotating at a speed ω equal to 2 rad/s. Rod *CD* rides in a vertical slot at *D*. For the configuration shown in the diagram, compute the kinetic energy of the rods. Rod *CD* weighs 50 N and rod *EA* weighs 80 N.

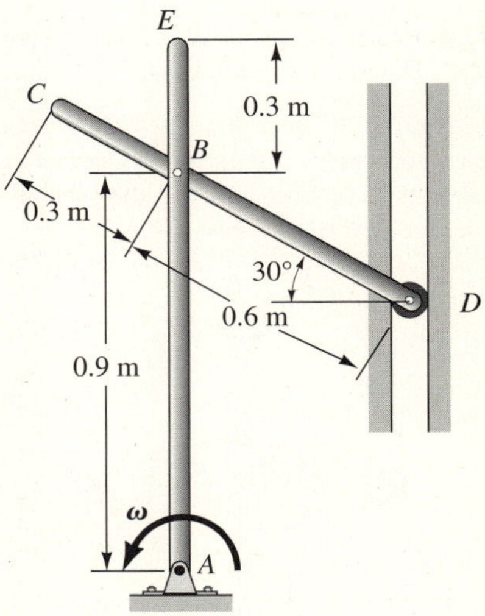

Figure P.17.4.

17.5. Consider the connecting rod AB to be a slender rod of mass 1 kg, and compute its kinetic energy for the data given.

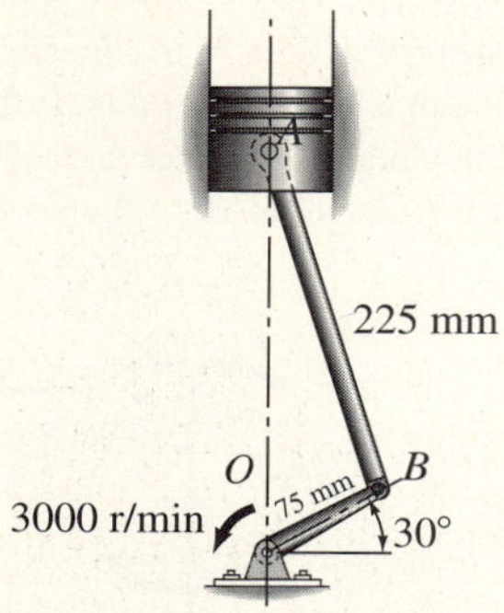

Figure P.17.5.

17.6. Identical rods CB and AB are pinned together at B. Rod BC is pinned to a block D weighing 225 N. Each rod is 600 mm in length and weighs 45 N. Rod BA rotates counterclockwise at a constant speed ω of 3 rad/s. Compute the kinetic energy of the system when BA is oriented (a) at an angle of 60° with the vertical and (b) at an angle of 90° with the vertical (the latter position is shown dashed in the diagram).

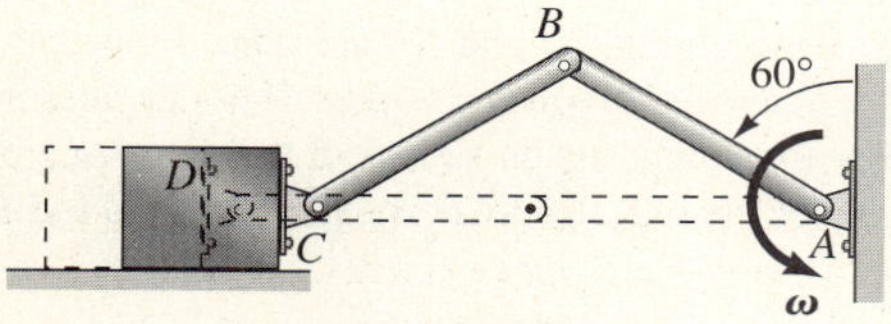

Figure P.17.6.

17.7. Find the kinetic energy of the rotating system described in Example 16.10. The diameter of the shaft is 50 mm.

17.8. The centerline of gear A rotates about axis M–M at an angular speed ω_1 of 3 rad/s. The mean diameter of gear A is 0.3 m. If gear A has a mass of 5 kg, what is the kinetic energy of the gear? Consider the gear to be a disc.

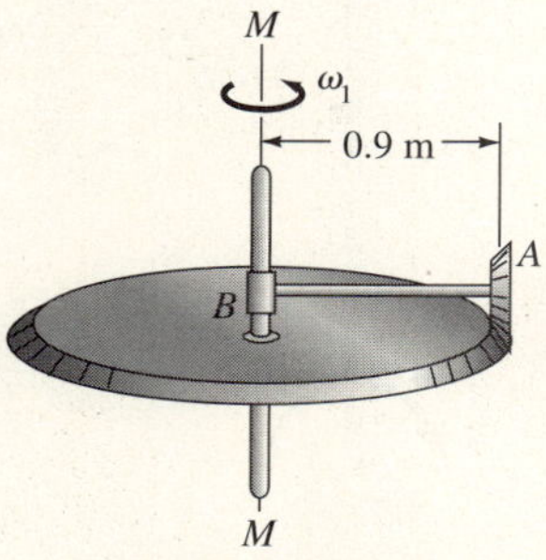

Figure P.17.8.

17.9. A cone B of mass 10 kg rolls without slipping inside a conical cavity C. The cone has a length of 3 m. The centerline of the cone rotates with an angular speed ω_1 of 5 rad/s about the Y axis. Compute the kinetic energy of the cone.

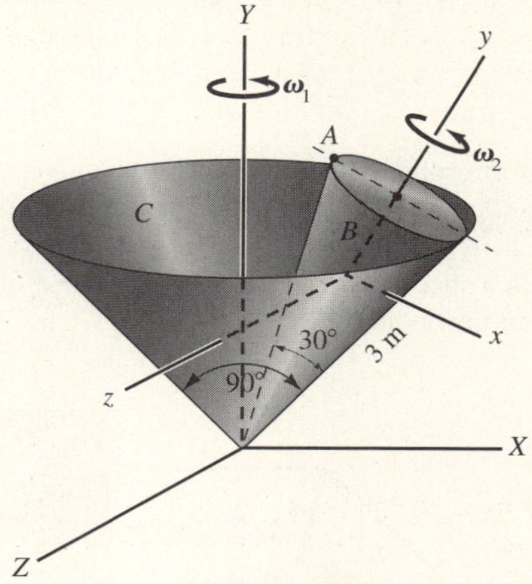

Figure P.17.9.

17.10. A uniform cylinder has a radius r and a weight W_1. A weight W_2, which we shall consider a particle because of its small physical dimensions, is placed at G a distance a from O, the center of the disc, such that OG is vertical. What is the angular velocity of the cylinder when, after it is released from rest, the point G reaches its lowest elevation, as shown at the right? The cylinder rolls without slipping.

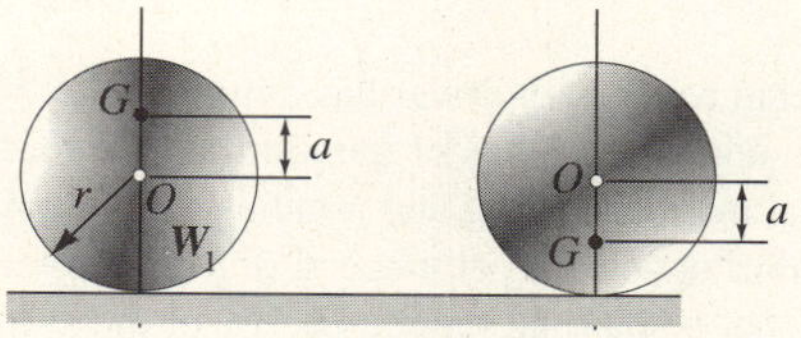

Figure P.17.10.

17.11. A homogeneous solid cylinder of radius 300 mm is shown with a fine wire held fixed at A and wrapped around the cylinder. If the cylinder is released from rest, what will its velocity be when it has dropped 3 m?

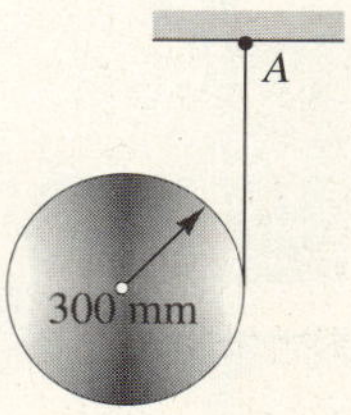

Figure P.17.11.

17.12. Three identical bars, each of length l and weight W, are connected to each other and to a wall with smooth pins at A, B, C, and D. A spring having spring constant K is connected to the center of bar BC at E and to a pin at F, which is free to slide in the slot. Compute the angular speed $\dot{\theta}$ as a function of time if the system is released from rest when AB and DC are at right angles to the wall. The spring is unstretched at the outset of the motion. Neglect friction.

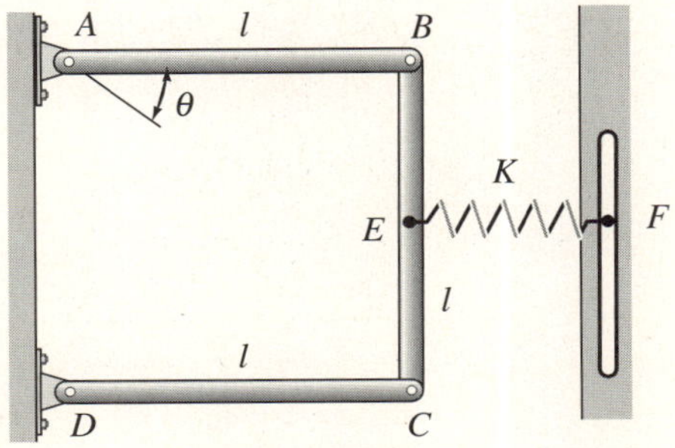

Figure P.17.12.

17.13. A 3-m rod AB weighing 225 N is guided at A by a slot and at B by a smooth horizontal surface. Neglect the mass of the slider at A, and find the speed of B when A has moved 1 m along the slot after starting from a rest configuration shown in the diagram.

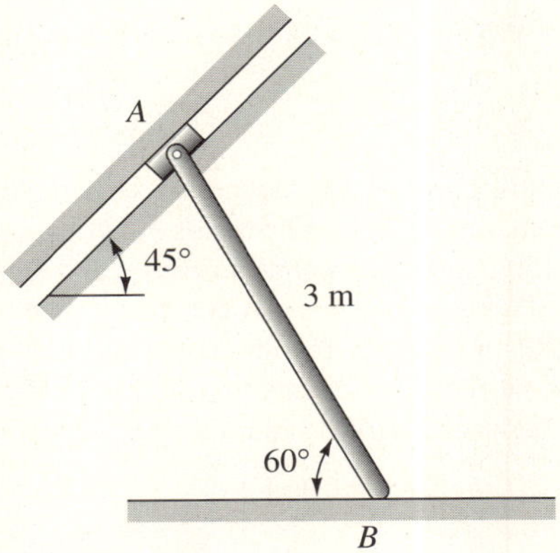

Figure P.17.13.

17.14. A stepped cylinder has radii of 600 mm for the smaller radius and 1.3 m for the larger radius. A rectangular block A weighing 225 N is welded to the cylinder at B. The spring constant K is 0.18 N/mm. If the system is released from a configuration of rest, what is the angular speed of the cylinder after it has rotated 90°? The radius of gyration for the stepped cylinder is 1 m and its mass is 36 kg. The spring is unstretched in the position shown.

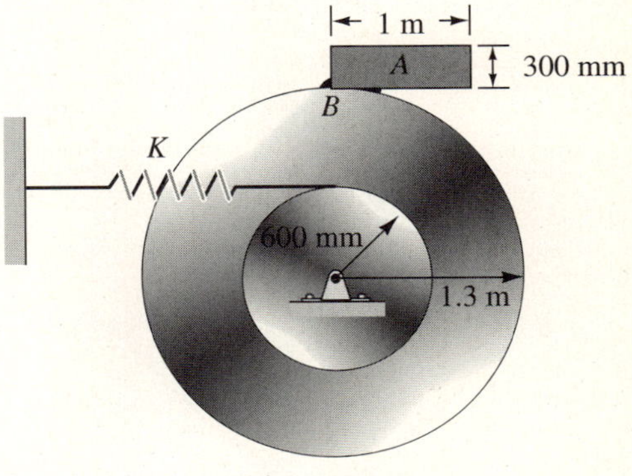

Figure P.17.14.

17.15. A cylinder of diameter 0.6 m is composed of two semi-cylinders C and D weighing 25 kg and 40 kg, respectively. Bodies A and B, weighing 10 kg and 25 kg, respectively, are connected by a light, flexible cable that runs over the cylinder. If the system is released from rest for the configuration shown, what is the speed of B when the cylinder has rotated 90°? Assume no slipping.

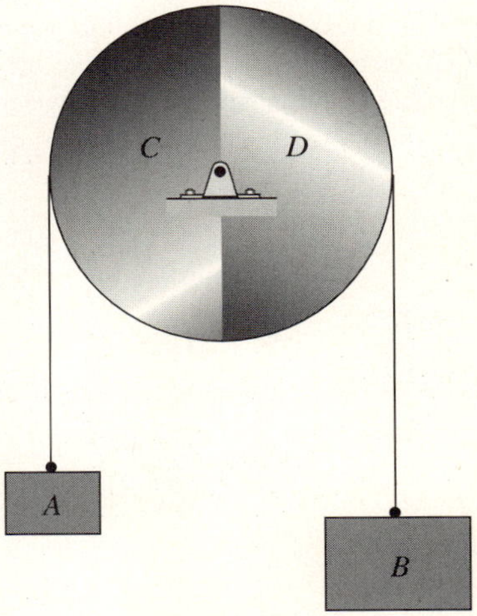

Figure P.17.15.

17.16. Four identical rods, each of length $l = 1.3$ m and weight 90 N, are connected at the frictionless pins *A*, *B*, *C*, and *D*. A compression spring of spring constant $K = 5.3$ N/mm connects pins *B* and *C*, and a weight W_2 of 450 N is supported at pin D. The system is released from a configuration where $\theta = 45°$. If the spring is not compressed at that configuration, show that the maximum deflection of the weight W_2 is 0.1966 m.

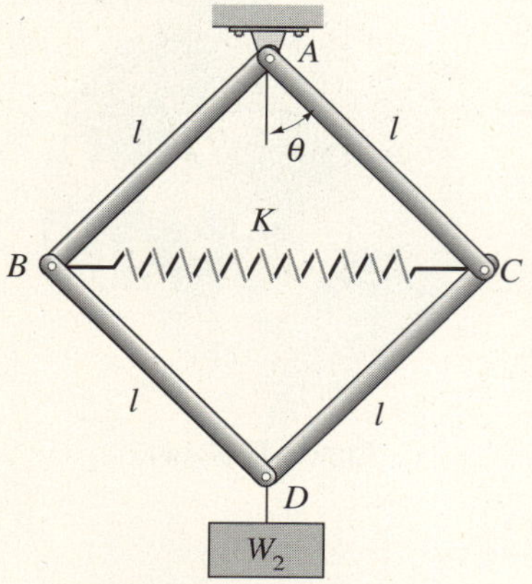

Figure P.17.16.

17.17. A stepped cylinder of mass 15 kg with a radius of gyration of 0.3 m is connected to a 15 m chain of mass 50 kg. The chain hangs down from the horizontal surface a distance of 3 m when the system is released. Determine the speed of the chain when 9 additional metres of chain have come off the horizontal surface. The coefficient of dynamic friction between the chain and the horizontal surface is 0.2, and the smaller diameter of the stepped cylinder is 100 mm.

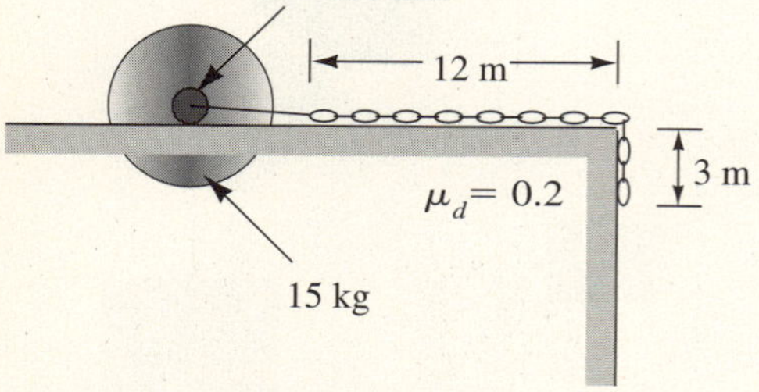

Figure P.17.17.

17.18. The linkage system rests on a frictionless plane. The lengths *AB*, *BF*, etc., are each 300 mm, and the bars, all of the same stock, weigh 67.5 N/m. A force *F* of 450 N is applied at *D*. What is the speed of *D* after it moves 300 mm? The system is stationary in the configuration shown. The view of the linkage system is from above.

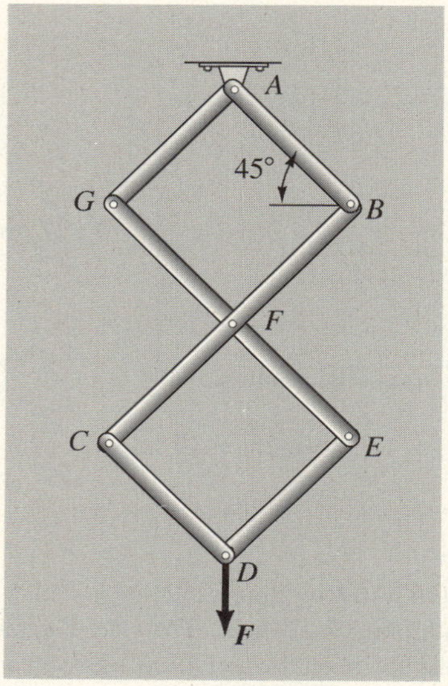

Figure P.17.18.

17.19. Work Problem 17.18 for the case where the system is in a vertical plane.

17.20. A belt of mass 5 kg is mounted over two pulleys of radii 0.3 m and 0.6 m, respectively. The radius of gyration and mass for pulley *A* are 150 mm and 25 kg, respectively, and for pulley *B* are 225 mm and 100 kg, respectively. A constant torque of 2.2 N m is applied to pulley *A*. After 30 revolutions of pulley *A*, what will its angular speed be if the system starts from rest? There is no slipping between belts and pulleys, and pulley *B* turns freely.

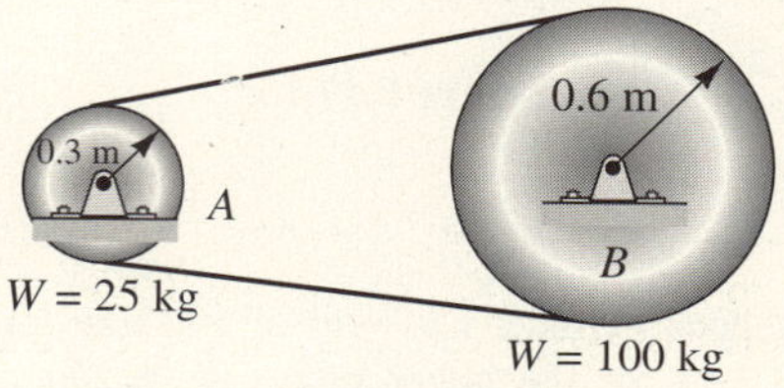

Figure P.17.20.

17.21. Two identical members, AB and BC, are pinned together at B. Also member BC is pinned to the wall at C. Each member has a mass of 15 kg and is 6 m long. A spring having a spring constant K = 300 N/m is connected to the centers of the members. A force P = 450 N is applied to member AB at A. If initially the members are inclined 45° to the ground and the spring is unstretched, what is $\dot{\beta}$ after A has moved 0.6 m? System is in a vertical plane.

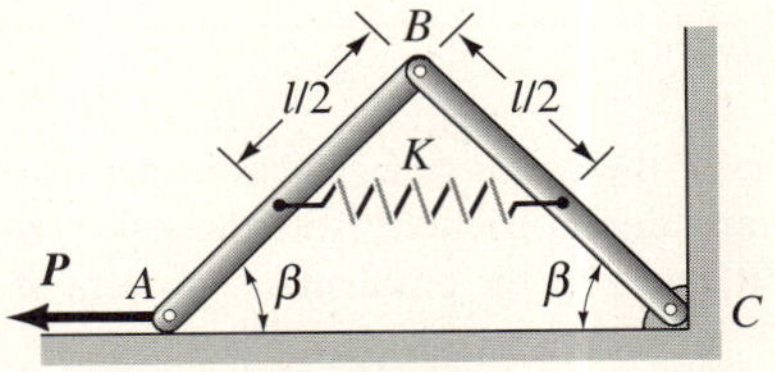

Figure P.17.21.

17.22. A flexible cord of total length 15 m and mass 25 kg is pinned to a wall at A and is wrapped around a cylinder having a radius of 1.2 m and mass 15 kg. A 250 N force is applied to the end of the cord. What is the speed of the cylinder after the end of the cord has moved 3 m? The system starts from rest in the configuration shown in the diagram. Neglect potential energy considerations arising from the sag of the upper cord.

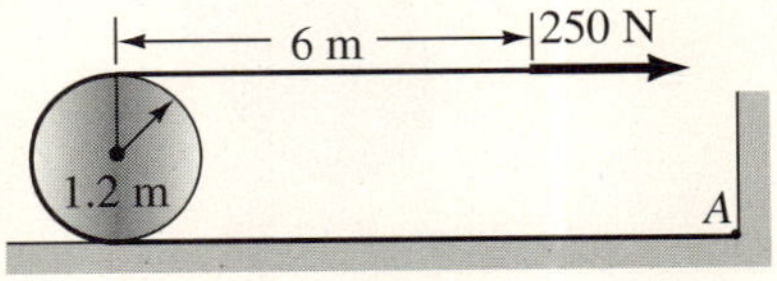

Figure P.17.22.

17.23. In Problem 17.5, suppose that an average pressure of 140 kPa (above atmosphere) exists in the cylinder. What is the r/min after the crankshaft has rotated 60° from position shown? The crank rod OB has a mass of 0.5 kg and has a radius of gyration about O of 50 mm. The diameter of the piston is 100 mm, and its mass is 220 g. The crankshaft is rotating at 3000 r/min at the position shown. Take the center of mass of the crank rod at the midpoint of OB.

17.24. A right circular cone of mass 15 kg, height 1.2 m, and cone angle 20° is allowed to roll without slipping on a plane surface inclined at an angle of 30° to the horizontal. The cone is started from rest when the line of contact is parallel to the X axis. What is the angular speed of the centerline of the cone when it has its maximum kinetic energy?

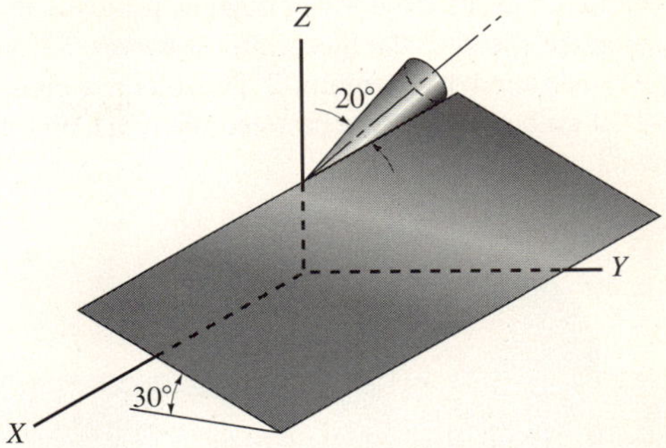

Figure P.17.24.

17.25. Work Example 17.2 by considering the system to be the block, pulley, and cable.

17.26. A weight W_1 is held with a light flexible wire. The wire runs over a stationary semicylinder of radius R equal to 0.3 m. A pulley weighing 150 N and having a radius of gyration of unity rides on the wire and supports a weight W_2 of 75 N. If W_1 weighs 600 N and the dynamic coefficient of friction for the semicylinder and wire is 0.2 what is the drop in the weight W_1 for an increase in speed of 1.5 m/s of weight W_1 starting from rest? The diameter d of the small pulley is 0.3 m.

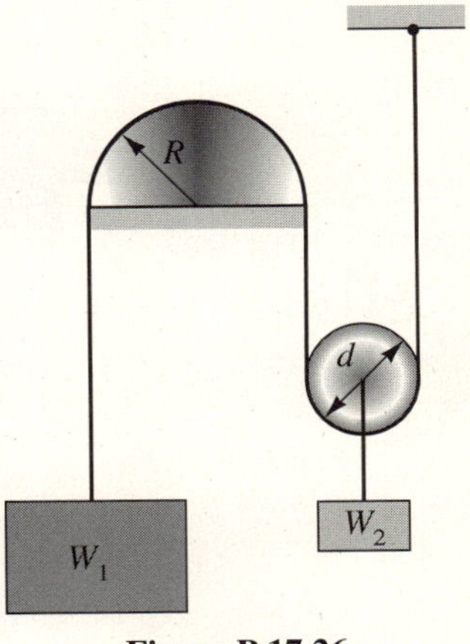

Figure P.17.26.

17.27. A solid uniform block A moves along two frictionless angle-iron supports at a speed of 6 m/s. One of the supports is inclined at an angle of 20° from the horizontal at B and causes the block to rotate about its front lower edge as it moves to the right of B. What is the speed of the block after it moves 300 mm to the right of B (measured horizontally)? The block weighs 450 N. Consider that no binding occurs between the block and the angle-iron supports.

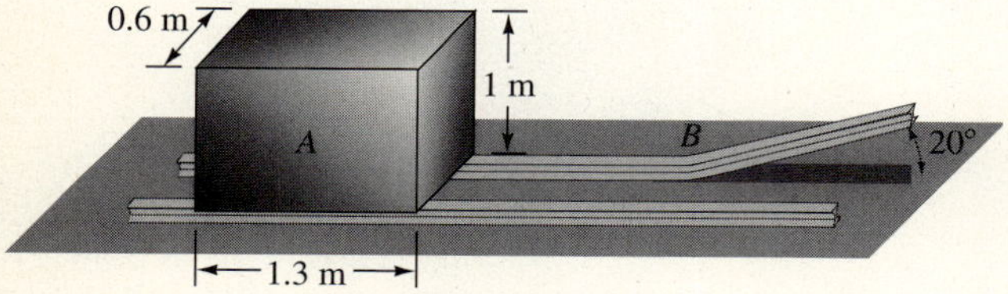

Figure P.17.27.

17.28. In Problem 17.27, will the block reach an instantaneous zero velocity and then slide back or will it tip over onto face A?

17.29. A torque $T = 0.30$ N m is applied to a bevel gear B. Bevel gear D meshes with gear B and drives a pump A. Gear B has a radius of gyration of 150 mm and a weight of 50 N, whereas gear D and the impeller of pump A have a combined radius of gyration of 50 mm and a weight of 100 N. Show that the *work of the contact forces between gears B and D is zero*. Next, find how many revolutions of gear B are needed to get the pump up to 200 r/min from rest. Treat the problem as a system of bodies.

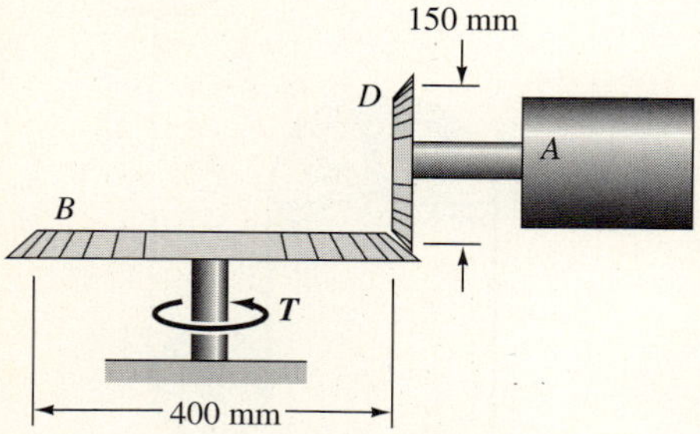

Figure P.17.29.

17.30. A torque T of 0.5 N m acts on worm gear E, which meshes with gear A, which drives a gear train. After five revolutions of gear H, what is its angular speed? One revolution of worm gear E corresponds to 0.2 revolution of gear A. Use the following data:

	k (mm)	W (N)	D (mm)
A	30	10	100
B	30	10	100
C	100	40	300
H	200	100	500

Neglect inertia of the worm gear, and consider an energy loss of 10% of the input due to friction. Note the italicized statement of Problem 17.30, and treat the problem as a system of bodies.

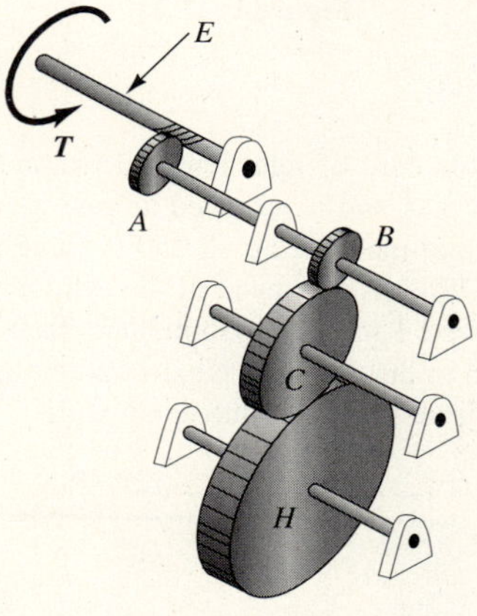

Figure P.17.30.

17.31. A force F of 450 N acts on block A weighing 435 N. Block A rides on identical uniform cylinders B and C, each weighing 290 N and having a radius of 300 mm. If there is no slipping, what is the speed of A after it moves 1 m?

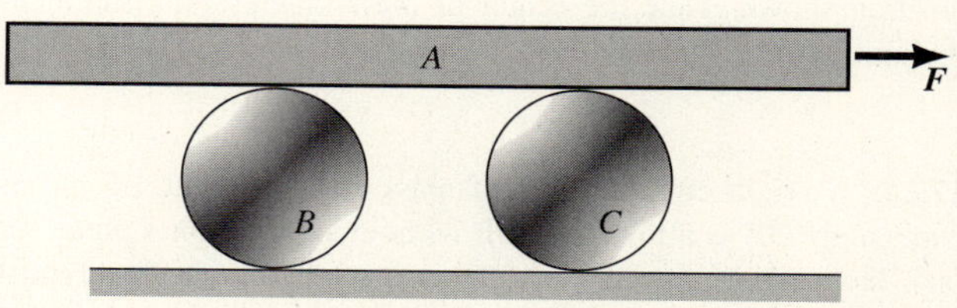

Figure P.17.31.

17.32. Work Example 17.5 using the center-of-mass approach for the whole system. What are the friction forces on the wheels from the ground?

17.33. An electric train (one car) uses its motors as electric generators for braking action. Suppose that this train is moving down a 15° incline at a speed initially of 10 m/s and, during the next 100 m, the generators develop 1.5 kW h of energy. What is the speed of the train at the end of this interval? The train with passengers weighs 200 kN. Each of the eight wheels weighs 900 N and has a radius of gyration of 250 mm and a diameter of 600 mm. Neglect wind resistance, and consider that there is no slipping. The efficiency of the generators for developing power is 90%. Do not use center of mass approach.

17.34. Work Problem 17.33 using the center of mass approach for the whole train. Also, find the average friction force from the rail onto the wheels. Consider that each wheel is attached to a generator.

17.35. A windlass has a rotating part which has a mass of 35 kg and has a radius of gyration of 0.3 m. When the suspended weight of 90 N is dropping at a speed of 6 m/s, a 450 N force is applied to the lever at *A*. This action applies the brake shoe at *B*, where there is a coefficient of friction of 0.5. How far will the 90 N weight drop before stopping?

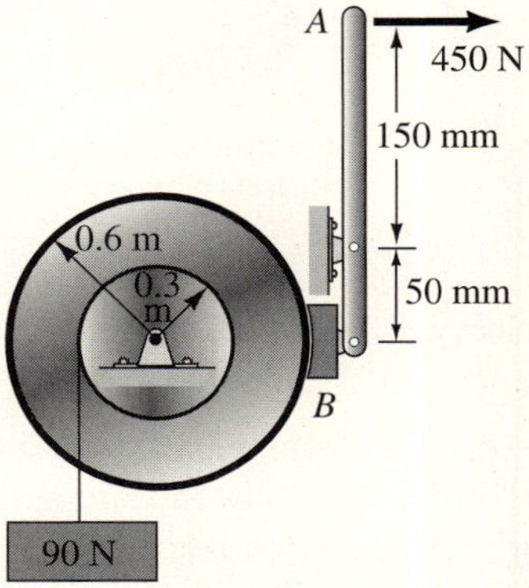

Figure P.17.35.

17.36. A square-threaded screw has a diameter of 50 mm and is inclined 45° to the horizontal. The pitch of the thread is 5 mm, and it is single-threaded. A body *A* weighing 290 N and having a radius of gyration of 300 mm screws onto the shaft. A torque *T* of 45 N m is applied to *A* as shown. What is the angular speed of *A* after three revolutions starting from a rest configuration? Neglect friction.

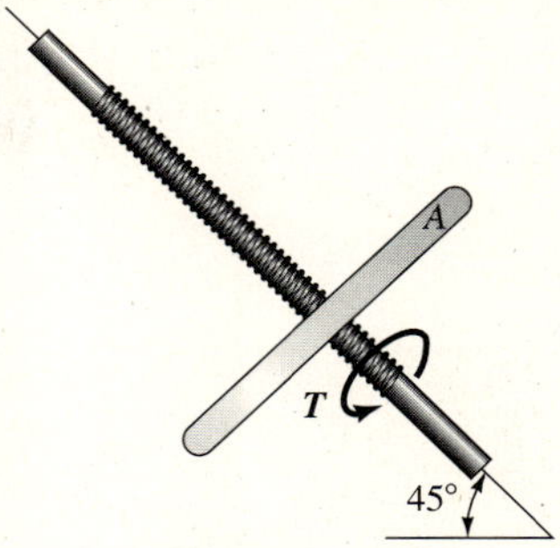

Figure P.17.36.

17.37. A uniform block *A* of mass 30 kg is pulled by a force *P* of 220 N as shown. The block moves along the rails on small, light wheels. One rail descends at an angle of 15° at point *B*. If the force *P* always remains horizontal, what is the speed of the block after it has moved 1.5 m in the horizontal direction? The block is stationary at the position shown. Assume that the block does not tilt forward.

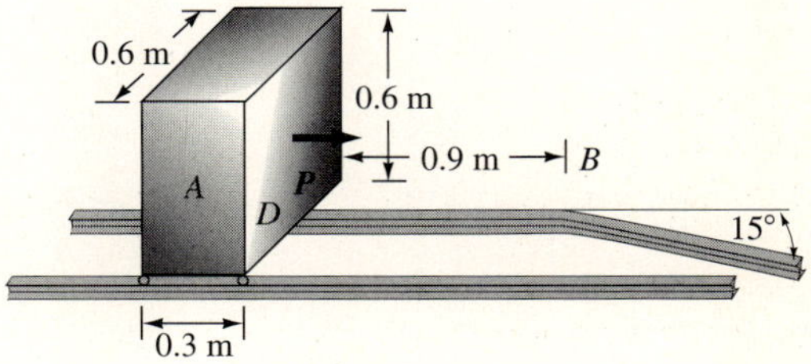

Figure P.17.37.

17.38. A solid uniform rod *AB* connects two light slider bearings *A* and *B*, which move in a frictionless manner along the indicated guide rods. The rod *AB* has a mass of 70 kg and a diameter of 50 mm. Smooth ball-joint connections exist between the rod and the bearings. If the rod is released from rest at the configuration shown, what is the speed of the bearing *A* when it has dropped 0.6 m?

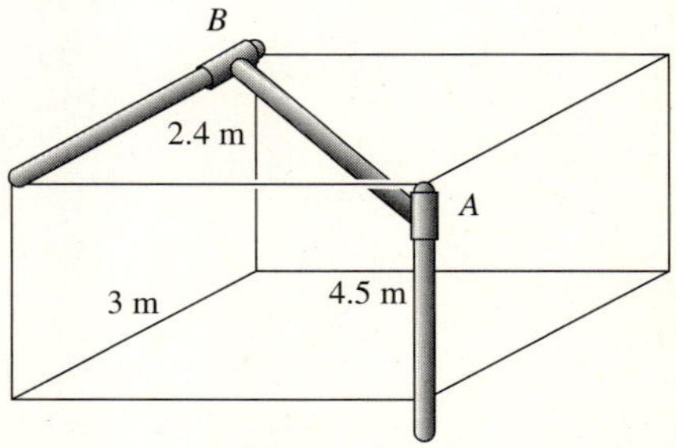

Figure P.17.38.

Part B: Impulse-Momentum Methods

17.4 Angular Momentum of a Rigid Body About Any Point in the Body

As we go to three dimensions, we will need formulations for linear momentum and angular momentum of rigid bodies. The linear momentum is simply $\iiint \boldsymbol{V}\, dm = M\boldsymbol{V}_c$. We shall now formulate an expression for the more complicated angular momentum $\boldsymbol{H}$ of a rigid body about a point. For this purpose, we choose a point A in a rigid body or hypothetical massless extension of the rigid body as shown in Fig. 17.21. An element of mass dm at a position $\boldsymbol{\rho}$ from A is shown. The velocity $\boldsymbol{V}'$ of dm relative to A is simply the velocity of dm relative to reference $\xi\eta\zeta$ which *translates* with A relative to XYZ. Similarly, the linear momentum of dm relative to A is the linear momentum of dm relative to a reference $\xi\eta\zeta$ translating with A. We can now give the angular momentum $d\boldsymbol{H}_A$ for element dm about A as

$$d\boldsymbol{H}_A = \boldsymbol{\rho} \times \boldsymbol{V}'\, dm = \boldsymbol{\rho} \times \left(\frac{d\boldsymbol{\rho}}{dt}\right)_{\xi\eta\zeta} dm \qquad (17.18a)$$

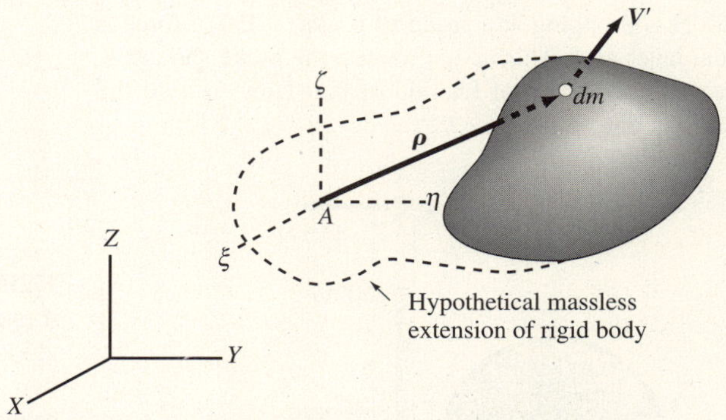

Figure 17.21. Velocity of *dm* relative to point *A*.

But since A is fixed in the body (or in the hypothetical massless extension of the body), the vector $\boldsymbol{\rho}$ must be fixed in the body. Accordingly, $(d\boldsymbol{\rho}/dt)_{\xi\eta\zeta} = \boldsymbol{\omega} \times \boldsymbol{\rho}$, where $\boldsymbol{\omega}$ is the angular velocity of the body relative to $\xi\eta\zeta$. However, since $\xi\eta\zeta$ translates with respect to XYZ, $\boldsymbol{\omega}$ *is also the angular velocity of the body relative to XYZ as well*. Hence, we can say:

$$d\boldsymbol{H}_A = \boldsymbol{\rho} \times (\boldsymbol{\omega} \times \boldsymbol{\rho})\, dm \qquad (17.18b)$$

We shall find it convenient to express Eq. 17.18 in terms of orthogonal components. For this purpose, imagine an arbitrary reference xyz fixed to the body or rigid-body extension of the body having the origin at A and any arbitrary

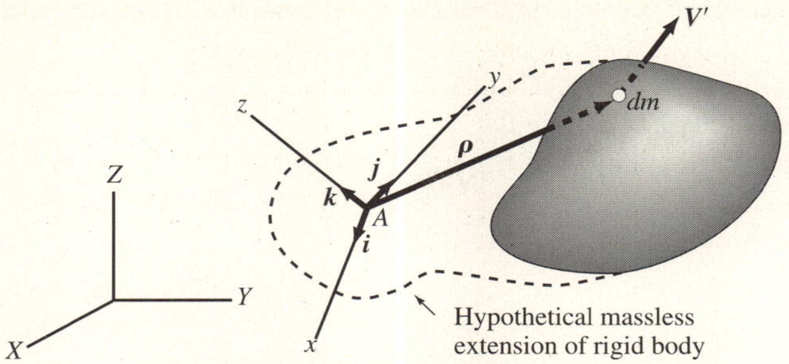

Figure 17.22. Reference *xyz* at *A*.

orientation relative to *XYZ*,[9] as shown in Fig. 17.22. We next decompose each of the vectors in Eq. 17.18b into rectangular components in the $\boldsymbol{i}$, $\boldsymbol{j}$, and $\boldsymbol{k}$ directions associated with the *x*, *y*, and *z* axes, respectively. Thus,

$$d\boldsymbol{H}_A = (dH_A)_x \boldsymbol{i} + (dH_A)_y \boldsymbol{j} + (dH_A)_z \boldsymbol{k} \tag{17.19a}$$

$$\boldsymbol{\rho} = x\boldsymbol{i} + y\boldsymbol{j} + z\boldsymbol{k} \tag{17.19b}$$

$$\boldsymbol{\omega} = \omega_x \boldsymbol{i} + \omega_y \boldsymbol{j} + \omega_z \boldsymbol{k} \tag{17.19c}$$

We then have for Eq. 17.18b:

$$(dH_A)_x \boldsymbol{i} + (dH_A)_y \boldsymbol{j} + (dH_A)_z \boldsymbol{k} = (x\boldsymbol{i} + y\boldsymbol{j} + z\boldsymbol{k}) \times [(\omega_x \boldsymbol{i} + \omega_y \boldsymbol{j} + \omega_z \boldsymbol{k}) \times (x\boldsymbol{i} + y\boldsymbol{j} + z\boldsymbol{k})]\, dm \tag{17.20}$$

Carrying out the cross products and collecting terms, we have

$$(dH_A)_x = \omega_x(y^2 + z^2)\, dm - \omega_y xy\, dm - \omega_z xz\, dm \tag{17.21a}$$

$$(dH_A)_y = -\omega_x yx\, dm + \omega_y(x^2 + z^2)\, dm - \omega_z yz\, dm \tag{17.21b}$$

$$(dH_A)_z = -\omega_x zx\, dm - \omega_y zy\, dm + \omega_z(x^2 + y^2)\, dm \tag{17.21c}$$

If we integrate these relations for all the mass elements *dm* of the rigid body, we see that the components of the inertia tensor for point *A* appear:

$$(H_A)_x = I_{xx}\omega_x - I_{xy}\omega_y - I_{xz}\omega_z \tag{17.22a}$$

$$(H_A)_y = -I_{yx}\omega_x + I_{yy}\omega_y - I_{yz}\omega_z \tag{17.22b}$$

$$(H_A)_z = -I_{zx}\omega_x - I_{zy}\omega_y + I_{zz}\omega_z \tag{17.22c}$$

We thus have components of the angular momentum vector $\boldsymbol{H}_A$ for a rigid body about point *A* in terms of an arbitrary set of directions *x*, *y*, and *z* at point *A*.

We now illustrate the calculation of $\boldsymbol{H}_A$ in the following example.

[9]At this time we can forget about the axes ξ, η, and ζ. They only become necessary when we ask the question: What is the velocity or linear momentum of a particle relative to point *A*. To repeat, the velocity or linear momentum of a particle relative to point *A* is the velocity or momentum relative to a reference $\xi\eta\zeta$ translating with point *A* as seen from *XYZ* or, in other words, relative to a nonrotating observer moving with *A*.

Example 17.6

A disc B has a mass M and is rotating around centerline E–E in Fig. 17.23 at a speed ω_1 relative to E–E. Centerline E–E, meanwhile, has an angular speed ω_2 about the vertical axis. Compute the angular momentum of the disc about point A as seen from ground reference XYZ.

The angular velocity of the disc relative to the ground is

$$\boldsymbol{\omega} = \omega_1 \boldsymbol{i} + \omega_2 \boldsymbol{j} \tag{a}$$

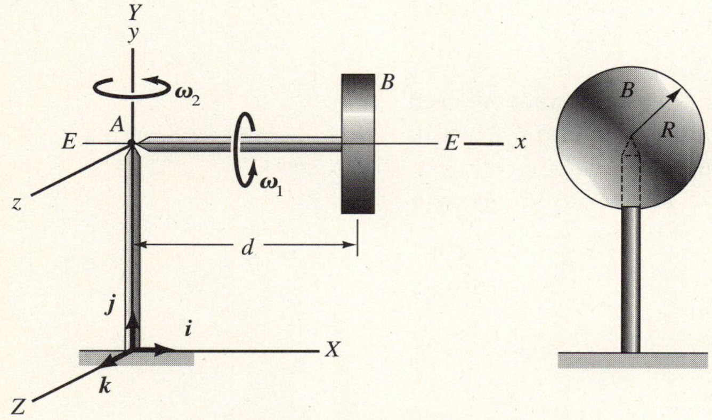

Figure 17.23. Rotating disc.

Consider a set of axes xyz with the origin at A and fixed to the body having cylinder B. At the instant of interest, the xyz axes are parallel to the inertial reference XYZ and we can say (see Fig. 17.24):

$$\omega_x = \omega_1, \qquad \omega_y = \omega_2, \qquad \omega_z = 0 \tag{b}$$

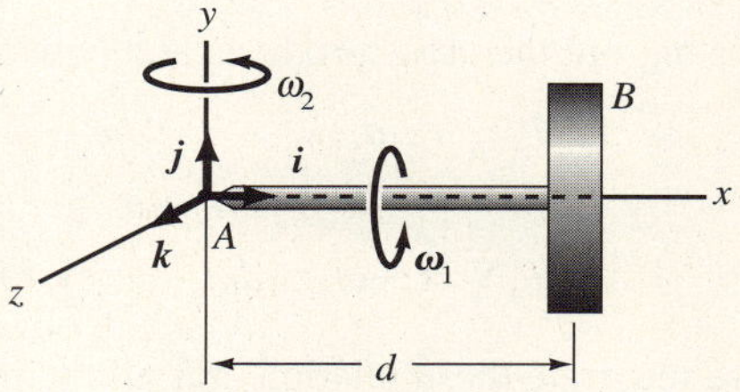

Figure 17.24. xyz axes at A.

The inertia tensor for the disc taken at A is next presented.

$$\begin{array}{lll} I_{xx} = \frac{1}{2}MR^2 & I_{xy} = 0 & I_{xz} = 0 \\ I_{yx} = 0 & I_{yy} = \frac{1}{4}MR^2 + Md^2 & I_{yz} = 0 \\ I_{zx} = 0 & I_{zy} = 0 & I_{zz} = \frac{1}{4}MR^2 + Md^2 \end{array} \tag{c}$$

Note that the product-of-inertia terms are zero because the xy and the xz planes are planes of symmetry. Clearly, moments of inertia for the xyz axes are principal moments of inertia. Now going to Eq. 17.22, we have

$$\begin{aligned} (H_A)_x &= \frac{MR^2}{2}\omega_1 \\ (H_A)_y &= \left(\tfrac{1}{4}MR^2 + Md^2\right)\omega_2 \\ (H_A)_z &= 0 \end{aligned} \tag{d}$$

As seen in Example 17.5, when xyz are *principal* axes, $\boldsymbol{H}_A$ simplifies to

$$\boldsymbol{H}_A = I_{xx}\omega_x \boldsymbol{i} + I_{yy}\omega_y \boldsymbol{j} + I_{zz}\omega_z \boldsymbol{k} \qquad (17.23)$$

which is analogous to the linear momentum vector $\boldsymbol{P}$. That is,

$$\boldsymbol{P} = MV_x \boldsymbol{i} + MV_y \boldsymbol{j} + MV_z \boldsymbol{k}$$

Note that mass plays the same role as does I, and $\boldsymbol{V}$ plays the same role as does $\boldsymbol{\omega}$.

In Chapter 16, we considered with some care the plane motion of a slablike body, the motion being parallel to the plane of symmetry of the body. We have shown such a case in Fig. 17.25. A reference xyz is shown fixed to the body at A with xy at the midplane of the body and with the z axis oriented normal to the plane of motion. Recall now that for a slablike body the plane xy must be a plane of symmetry or be a principal plane, and consequently that $I_{zx} = I_{zy} = 0$. Also, the only nonzero component of $\boldsymbol{\omega}$ is ω_z. Going back to Eq. 17.22, we see that only $(H_A)_z$ is nonzero, with the result

$$(H_A)_z = (I_{zz})_A \omega_z \qquad (17.24)$$

Because of the importance of plane motion of slablike bodies, we shall often use the foregoing simple formula.

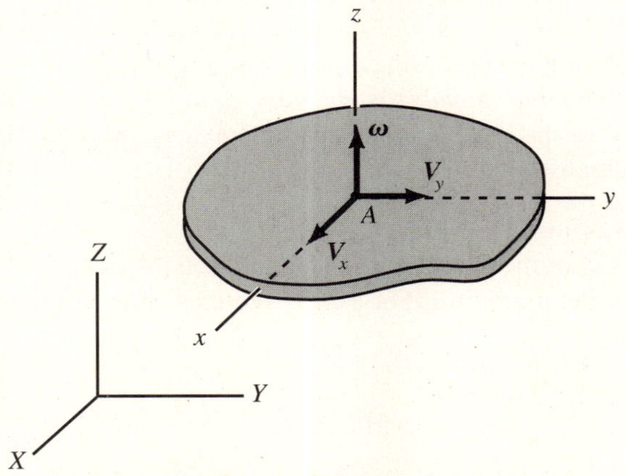

Figure 17.25. Slablike body in plane motion.

We leave it for you to show that the Eq. 17.24 also applies to a body of revolution rotating about its axis of symmetry in inertial space, where z is taken along this axis. Also, Eq. 17.24 is valid for a body having two orthogonal planes of symmetry rotating about an axis corresponding to the intersection of these planes of symmetry in inertial space, where z is taken along this axis. We are now ready to relate linear and angular momenta with force system causing the motion.

17.5 Impulse-Momentum Equations

You will recall that **Newton's law** for the center of mass of any body is

$$\boldsymbol{F} = M\frac{d\boldsymbol{V}_c}{dt} = \frac{d}{dt}(M\boldsymbol{V}_c)$$

where $\boldsymbol{F}$ is the total *external* force on the body. The corresponding *linear impulse-momentum* equation can then be given as

$$\int_{t_1}^{t_2} \boldsymbol{F}\,dt = \boldsymbol{I}_{\text{lin}} = (M\boldsymbol{V}_C)_2 - (M\boldsymbol{V}_C)_1 \tag{17.25}$$

where $\boldsymbol{I}_{\text{lin}}$ is the *linear impulse*. For a system of n rigid bodies we have, for the foregoing equation:

$$\int_{t_1}^{t_2} \boldsymbol{F}\,dt = \boldsymbol{I}_{\text{lin}} = \left[\sum_{i=1}^{n} M_i(\boldsymbol{V}_C)_i\right]_2 - \left[\sum_{i=1}^{n} M_i(\boldsymbol{V}_C)_i\right]_1 \tag{17.26}$$

where $\boldsymbol{F}$ is the total *external* force on the system, M_i is the mass of the ith body, and $(\boldsymbol{V}_C)_i$ is the velocity of the center of mass of the ith body. We are justified in forming the preceding equation as a result of Eq. 16.17.

For the *angular impulse-momentum equation*, we consider points A which are part of the rigid body or massless extension of the rigid body and which, in addition, are either:

1. The center of mass.
2. A point fixed or moving at constant $\boldsymbol{V}$ in inertial space.
3. A point accelerating toward or away from the center of mass.

In such cases we can say:

$$\boldsymbol{M}_A = \dot{\boldsymbol{H}}_A$$

where $\boldsymbol{H}_A$ is given by Eq. 17.22. Integrating with respect to time, we then get the desired *angular impulse-momentum equation*:

$$\int_{t_1}^{t_2} \boldsymbol{M}_A\,dt = \boldsymbol{I}_{\text{ang}} = (\boldsymbol{H}_2)_A - (\boldsymbol{H}_1)_A \tag{17.27}$$

where $\boldsymbol{I}_{\text{ang}}$ is the *angular impulse*. We now illustrate the use of the *linear impulse-* and *angular impulse-momentum equations*.

Example 17.7

A thin bent rod is sliding along a smooth surface (Fig. 17.26). The center of mass has the velocity

$$V_C = 10\boldsymbol{i} + 15\boldsymbol{j} \text{ m/s}$$

and the angular speed ω is 5 rad/s counterclockwise. At the configuration shown, the rod is given two simultaneous impacts as a result of a collision. These impacts have the following impulse values:

$$\int_{t_1}^{t_2} F_1\, dt = 5 \text{ N s}$$

$$\int_{t_1}^{t_2} F_2\, dt = 3 \text{ N s}$$

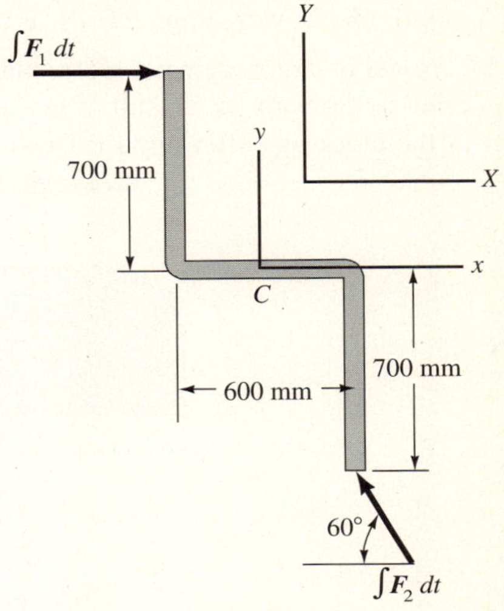

Figure 17.26. Bent rod slides on smooth horizontal surface.

What is the angular speed of the rod and the linear velocity of the mass center, directly after the impact? The rod weighs 35 N/m.

The velocity of the mass center after the impact can easily be determined using the **linear impulse-momentum equation** (Eq. 17.25). Thus, we have

$$5\boldsymbol{i} + 3\sin 60°\boldsymbol{j} - 3\cos 60°\boldsymbol{i} = (0.7+0.7+0.6)\left(\frac{35}{g}\right)(\boldsymbol{V}_2 - 10\boldsymbol{i} - 15\boldsymbol{j})$$

Solving for $\boldsymbol{V}_2$:

$$\boldsymbol{V}_2 = 10.49\boldsymbol{i} + 15.36\boldsymbol{j} \text{ m/s} \tag{a}$$

For the angular velocity, we use the **angular impulse momentum equation** (Eq. 17.27) simplified for the case of plane motion of a slablike body. Again using the center of mass at which we fix xyz, we have for Eq. (17.27):

$$\int_1^2 \boldsymbol{M}\, dt = (I_{zz}\omega_2 - I_{zz}\omega_1)\boldsymbol{k} \tag{b}$$

Putting in numerical data and canceling $\boldsymbol{k}$, we get

$$-(5)(0.7) + (3)(\sin 60°)(0.3) - (3)(\cos 60°)(0.7) = I_{zz}(\omega_2 - 5) \tag{c}$$

We next compute I_{zz} at C:

$$I_{zz} = \frac{1}{12}\left[\frac{35}{g}(0.6)\right](0.6)^2$$

$$+2\left[\frac{1}{12}\left(\frac{35}{g}\right)(0.7)(0.7)^2 + \left(\frac{35}{g}\right)(0.7)(0.3^2 + 0.35^2)\right] \tag{d}$$

$$= 1.330 \text{ kg m}^2$$

Going back to Eq. (c), we can now give ω_2:

$$\omega_2 = 2.16 \text{ rad/s}$$

Example 17.8

A solid block weighing 300 N is suspended from a wire (see Fig. 17.27) and is stationary when a horizontal impulse $\int F\,dt$ equal to 100 N s is applied to the body as a result of an impact. What is the velocity of corner A of the block just after impact: Does the wire remain taut?

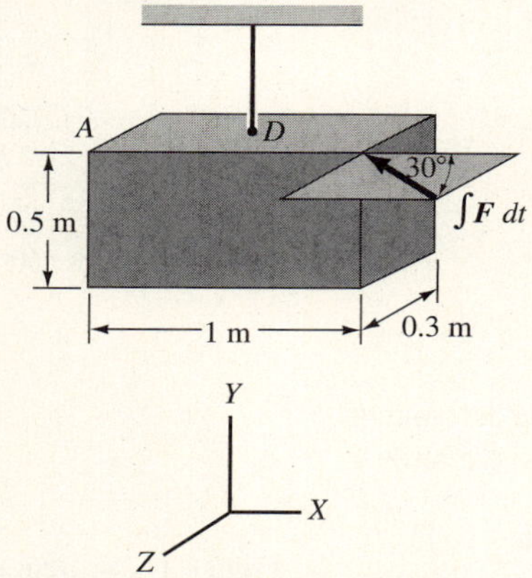

Figure 17.27. Stationary block under impact.

For the **linear momentum equation** we can say for the center of mass (see Fig. 17.28) in the z, x, and y directions:

$$-100 \sin 30° = \frac{300}{g}\left[(V_c)_z - 0\right]$$

$$-100 \cos 30° = \frac{300}{g}\left[(V_c)_x - 0\right] \qquad \text{(a)}$$

$$0 = \frac{300}{g}\left[(V_c)_y - 0\right]$$

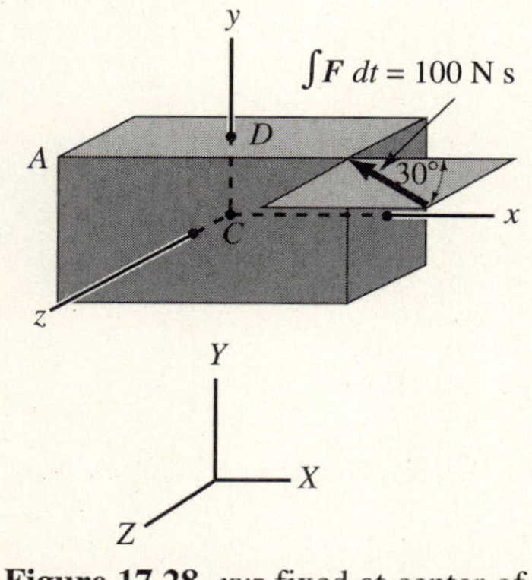

Figure 17.28. xyz fixed at center of mass.

Example 17.8 (Continued)

From these equations, we get

$$V_C = -2.83\boldsymbol{i} - 1.635\boldsymbol{k} \text{ m/s} \tag{b}$$

For the **angular momentum equation** about C, we can say, on noting that xyz are principal axes:

$$\begin{aligned}
-(100)(\sin 30°)(0.25) &= I_{xx}\omega_x - 0 \\
(100)(\sin 30°)(0.5) - 100(\cos 30°)\left(\frac{0.3}{2}\right) &= I_{yy}\omega_y - 0 \\
(100)(\cos 30°)(0.25) &= I_{zz}\omega_z - 0
\end{aligned} \tag{c}$$

Note that

$$\begin{aligned}
I_{xx} &= \frac{1}{12}\left(\frac{300}{g}\right)(0.3^2 + 0.5^2) = 0.866 \text{ kg m}^2 \\
I_{yy} &= \frac{1}{12}\left(\frac{300}{g}\right)(1^2 + 0.3^2) = 2.78 \text{ kg m}^2 \\
I_{zz} &= \frac{1}{12}\left(\frac{300}{g}\right)(1^2 + 0.5^2) = 3.19 \text{ kg m}^2
\end{aligned} \tag{d}$$

We then get for $\boldsymbol{\omega}$ as seen from XYZ from Eq. (c):

$$\boldsymbol{\omega} = -14.43\boldsymbol{i} + 4.32\boldsymbol{j} + 6.79\boldsymbol{k} \text{ rad/s}$$

Hence, for the velocity of point A, we have

$$\begin{aligned}
\boldsymbol{V}_A &= \boldsymbol{V}_C + \boldsymbol{\omega} \times \boldsymbol{\rho}_{CA} \\
&= -2.83\boldsymbol{i} - 1.63\boldsymbol{k} + (-14.43\boldsymbol{i} + 4.32\boldsymbol{j} + 6.79\boldsymbol{k}) \\
&\quad \times (-0.5\boldsymbol{i} + 0.25\boldsymbol{j} + 0.15\boldsymbol{k})
\end{aligned}$$

$$V_A = -3.88\boldsymbol{i} - 1.230\boldsymbol{j} - 3.08\boldsymbol{k} \text{ m/s}$$

Finally, to decide if wire remains taut, find the velocity of point D after impact.

$$\begin{aligned}
\mathbf{V}_D &= \mathbf{V}_C + \boldsymbol{\omega} \times \boldsymbol{\rho}_{CD} \\
&= -2.83\boldsymbol{i} - 1.635\boldsymbol{k} + (-14.43\boldsymbol{i} + 4.32\boldsymbol{j} + 6.79\boldsymbol{k}) \times (0.25\boldsymbol{j}) \\
&= -4.53\boldsymbol{i} - 5.24\boldsymbol{k} \text{ m/s}
\end{aligned}$$

Since there is zero velocity component in the y direction stemming from the given impulse, we can conclude that the wire remains taut.

In the following example we consider a problem involving a system of interconnected bodies.

Example 17.9

A tractor has a mass of 900 kg, including the driver (Fig. 17.29). The large driver wheels each have a mass of 90 kg with a radius of 0.6 m and a radius of gyration of 0.55 m. The small wheels have a mass of 18 kg each, with a radius of 0.3 m and a radius of gyration of 0.25 m. The tractor is pulling a bale of cotton weighing 135 kg. The coefficient of friction between the bale and the ground is 0.2. What torque is needed on the drive wheels from the motor for the tractor to go from 1.5 m/s to 3 m/s in 25 s? Assume that the tires do not slip.

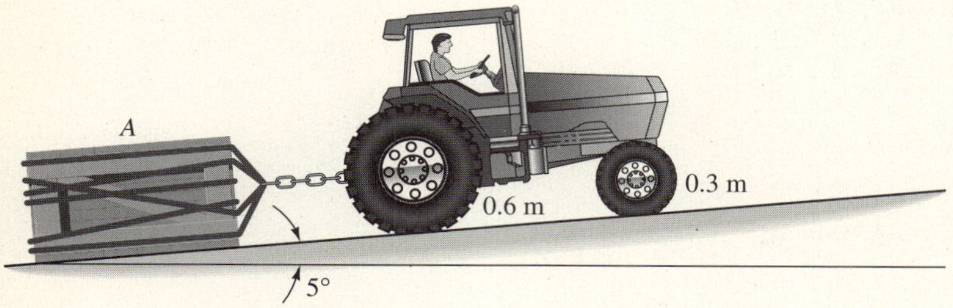

Figure 17.29. Tractor pulling bale of cotton.

We have shown a free-body diagram of the system in Fig. 17.30. Noting on inspection that $N_1 = 300 \cos 5°$, we can give the **linear momentum equation** for the system in the X direction as

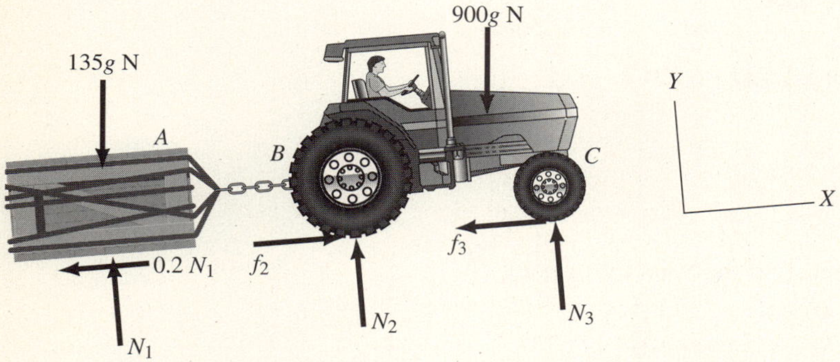

Figure 17.30. Free-body diagram of system.

$$[f_2 - f_3 - (0.2)(135g)\cos 5° - 135g \sin 5° - 900g \sin 5°](25) = 1035(3 - 1.5)$$

Therefore,

$$f_2 - f_3 = 1210.6 \qquad \text{(a)}$$

Example 17.9 (Continued)

We next consider free-body diagrams of the wheels in Fig. 17.31. The **impulse-angular momentum equation** for the drive wheels then can be given about the center of mass as

$$\left[-T+f_2(0.6)\right](25)=180(0.55)^2\left[(\omega_B)_2-(\omega_B)_1\right]$$

Noting from **kinematics** that $\omega_B = -V/0.6$, we have

$$-T+0.6f_2=-5.445 \qquad \text{(b)}$$

The **impulse-angular momentum equation** about the center of mass for the front wheels is then

$$-(f_3)(0.3)(25)=36(0.25)^2\left[(\omega_C)_2-(\omega_C)_1\right]$$

Noting from **kinematics** again that $\omega_C = -V/0.3$, we get from the equation above:

$$f_3 = 1.5 \text{ N} \qquad \text{(c)}$$

From Eq. (a) we may now solve for f_2. Thus,

$$f_2=1.5+1210.6=1.212 \text{ N}$$

Finally, from Eq. (b) we get the desired torque T:

$$T=5.445+(0.6)(1212)$$

$$T = 733 \text{ N m}$$

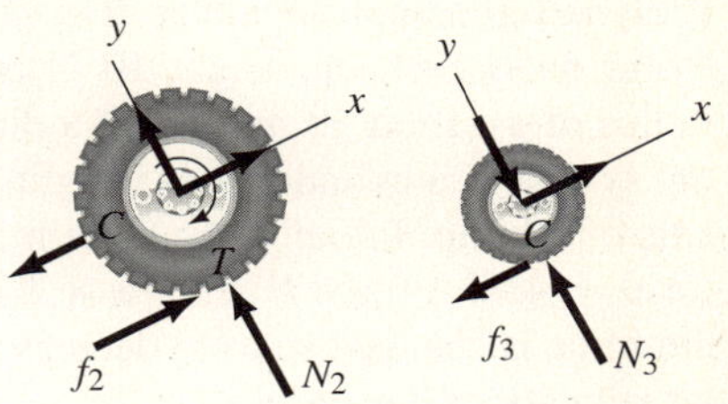

Figure 17.31. Free-body diagrams of wheels. xy axes are fixed to wheels.

In Example 17.8, there was no obvious convenient stationary point or stationary axis which could be considered as part of a rigid-body extension of *all* the bodies at any time. Therefore, in order to use the formulas for $\boldsymbol{H}$ given by Eq. 17.22, we considered rigid bodies *separately*. In the following example, we have a case where there is a stationary axis present which can be considered as part of (or a hypothetical rigid-body extension of) all bodies in the system at the instants of interest. And for this reason, we shall consider the angular momentum equation for the entire system using this stationary axis. Also, if the torque about such a common axis for a system of bodies is zero, then the angular momentum of the system about the aforestated axis must be *conserved*. In the example to follow, we shall also illustrate conservation of angular momentum about an axis for such a case.

Example 17.10

A flyball-governor apparatus (Fig. 17.32) consists of four identical arms (solid rods) each of weight 10 N and two spheres of weight 18 N and radius of gyration 30 mm about a diameter. At the base and rotating with the system is a cylinder B of weight 20 N and radius of gyration along its axis of 50 mm. Initially, the system is rotating at a speed ω_1 of 500 r/min for $\theta = 45°$. A force F at the base B maintains the configuration shown. If the force is changed so as to decrease θ from 45° to 30°, what is the angular velocity of the system?

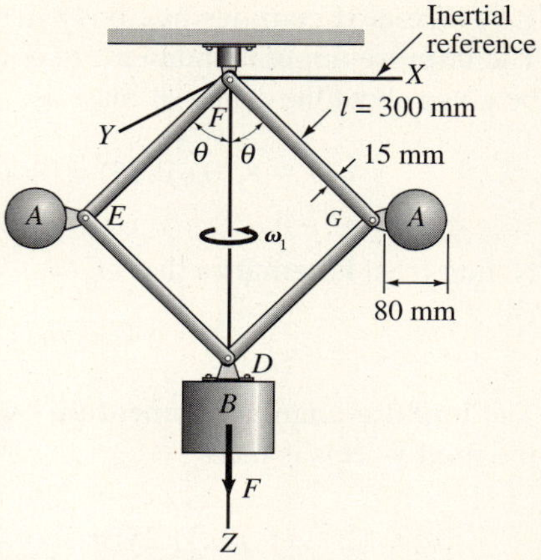

Figure 17.32. Flyball governor apparatus; Y axis is normal to page.

Clearly, there is zero torque from external forces about the stationary axis FD which we take as a Z axis at all times. Hence, we have conservation of angular momentum about this axis at all times. And, since the axis is an axis of rotation for all bodies of the system,[10] we can use Eq. 17.22 for computing H about FD for all bodies in the system. As a first step we shall need I_{ZZ} for the members of the system.

Consider first member FG, which is shown in Fig. 17.33. The axes $\xi\eta\zeta$ are principal axes of inertia for the rod at F. The η axis is collinear with the Y axis, and these are normal to the page. The axes XYZ are reached by $\xi\eta\zeta$ by rotating $\xi\eta\zeta$ about the η axis an angle θ. Using the transformation equations for I_{ZZ} (see Eq. 9.13), we can say:

$$(I_{ZZ})_{FG} = I_{\xi\xi}\left[\cos\left(\frac{\pi}{2}+\theta\right)\right]^2 + I_{\eta\eta}\left(\cos\frac{\pi}{2}\right)^2 + I_{\zeta\zeta}(\cos\theta)^2$$

$$= \left[\frac{1}{3}\frac{10}{g}(0.3)^2\right]\sin^2\theta + 0 + \left[\frac{1}{2}\frac{10}{g}(0.0075)^2\right]\cos^2\theta \qquad \text{(a)}$$

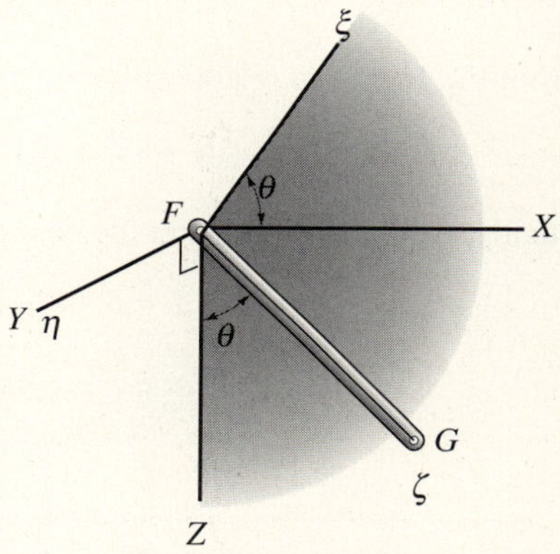

Figure 17.33. $\xi\eta\zeta$ are principal axes of FG at F.

For the sphere we have, using the parallel axis theorem

$$(I_{ZZ})_{\text{sphere}} = \frac{18}{g}(0.03)^2 + \frac{18}{g}[(0.3)\sin\theta + 0.04]^2 \qquad \text{(b)}$$

Finally, for cylinder B we have

$$(I_{ZZ})_{\text{cyl}} = \frac{20}{g}(0.05)^2 \qquad \text{(c)}$$

Conservation of angular momentum about the Z axis then prescribes the following:

$$\left[4(I_{ZZ})_{FG} + 2(I_{ZZ})_{\text{sphere}} + (I_{ZZ})_{\text{cyl}}\right]_{\theta=45°} \frac{(500)(2\pi)}{60}$$

$$= \left[4(I_{ZZ})_{FG} + 2(I_{ZZ})_{\text{sphere}} + (I_{ZZ})_{\text{cyl}}\right]_{\theta=30°} \omega_2$$

[10]That is, at any time t the stationary axis FD either is part of a rigid body directly or is part of a hypothetical extension of a rigid body for all bodies of the system.

Example 17.10 (Continued)

Substituting from Eqs. (a), (b), and (c), we have

$$\left(4\left[\frac{1}{3}\frac{10}{g}(0.3)^2(0.707)^2+\frac{1}{2}\frac{10}{g}(0.0075)^2(0.707)^2\right]\right.$$
$$\left.+2\left\{\frac{18}{g}(0.03)^2+\frac{18}{g}[(0.3)(0.707)+0.04]^2\right\}+\frac{20}{g}(0.05)^2\right)\frac{(500)(2\pi)}{60}$$
$$=\left(4\left[\frac{1}{3}\frac{10}{g}(0.3)^2(0.5)^2+\frac{1}{2}\frac{10}{g}(0.0075)^2(0.866)^2\right]\right.$$
$$\left.+2\left\{\frac{18}{g}(0.03)^2+\frac{18}{g}[(0.3)(0.5)+0.04]^2\right\}+\frac{20}{g}(0.05)^2\right)\omega_2$$

Therefore,

$$\omega_2 = 92.4 \text{ rad/s} = 883 \text{ r/min}$$

Before closing the section, we note that we have worked with *fixed points* or axes in inertial space and with the *mass center*. What about a point *accelerating toward the mass center*? A common example of such a point is the point of contact A of a cylinder rolling without slipping on a circular arc with the center of mass of the cylinder coinciding with the geometric center of the cylinder. We can then say:

$$\boldsymbol{M}_A = \dot{\boldsymbol{H}}_A$$

The question then arises: Can we form the familiar angular momentum equation from above about point A? In other words, is the following equation valid for plane motion?

$$\int_{t_1}^{t_2} M_A\,dt = I_A\omega_2 - I_A\omega_1 \qquad (17.28)$$

The reason we might hesitate to do this is that the point of contact *continually changes* during a time interval when the cylinder is rolling. However, we have asked you to prove in Problem 17.60 that for rolling without slipping along a circular or straight path, the equation above is still valid.[11] At times it can be very useful.

[11]The statement is actually valid for point A when there is rolling of the cylinder without slipping on a *general* path.

PROBLEMS

17.39. A uniform cylinder C of radius of 0.3 m and thickness 75 mm rolls without slipping at its center plane on the stationary platform B such that the centerline of CD makes 2 revolutions per second relative to the platform. What is the angular momentum vector for the cylinder about the center of mass of the cylinder? The cylinder has a mass of 30 kg.

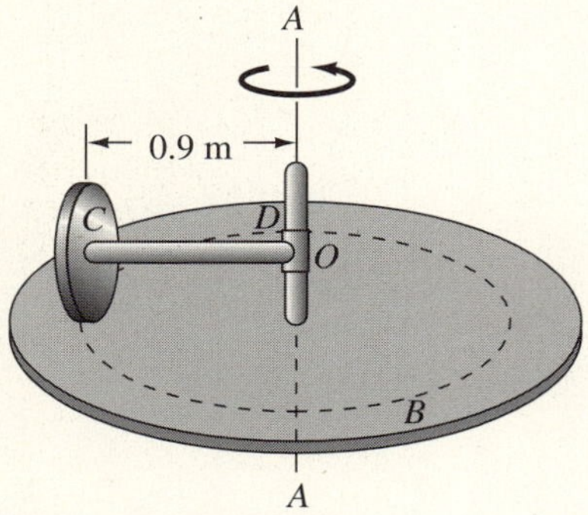

Figure P.17.39.

17.40. In Problem 17.39, find the angular momentum of the disc about the stationary point O along the vertical axis A–A.

17.41. A platform rotates at an angular speed of ω_1, while a cylinder or radius r and length a mounted on the platform rotates relative to the platform at an angular speed of ω_2. When the axis of the cylinder is collinear with the stationary Y axis, what is that angular momentum vector of the cylinder about the center of mass of the cylinder? The mass of the cylinder is M.

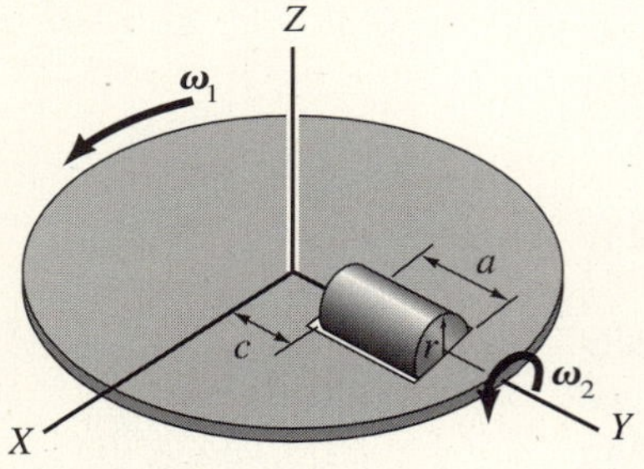

Figure P.17.41.

17.42. A disc A rotates relative to an inclined shaft CD at the ratio ω_2 of 3 rad/s while shaft CD rotates about vertical axis FE at the rate ω_1 of 4 rad/s relative to the ground. What is the angular momentum of the disc about its mass center as seen from the shaft CD? What is the angular momentum of the disc about its mass center as seen from the ground? The disc weighs 290 N.

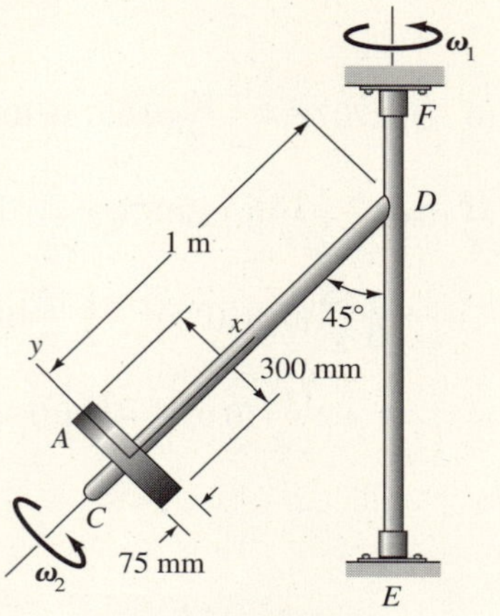

Figure P.17.42.

17.43. Work Problem 16.5 by methods of momentum.

17.44. A flywheel having a mass of 1000 kg is brought to speed of 200 r/min in 360 s by an electric motor developing a torque of 60 N m. What is the radius of gyration of the wheel?

Figure P.17.44.

17.45. Work Problem Example 16.2 by method of angular momentum.

17.46. Work Problem 16.8 by method of angular momentum.

17.47. A light plane is coming in for a landing at a speed of 30 m/s. The wheels have zero rotation just before touching the runway. If $\frac{1}{10}$ the weight of the plane is maintained by the upward force of the runway for the first second, what is the approximate length of the skid mark left by the wheel on the runway? The wheels each weigh 100 N and have a radius of gyration of 180 mm and a diameter of 450 mm. The plane weighs with load 8 kN. The coefficient of friction between the tire and runway is 0.3.

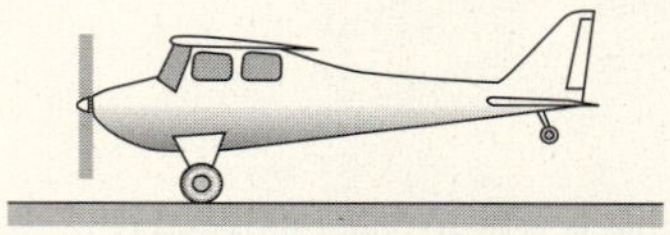

Figure P.17.47.

17.48. Work Problem 17.47 for the case where the upward force from the ground on the plane during the first second after touchdown is

$$N = 8 \times 10^3 t^2 \text{ N}$$

where t is in seconds after touchdown.

17.49. A circular conveyor carries cylinders a from position A through a heat treatment furnace. The cylinders are dropped onto the conveyor at A from a stationary position above and picked up at B. The conveyor is to turn at an average speed ω of 2 r/min. The cylinders are dropped onto the conveyor at the rate of 9 per minute. If the resisting torque due to friction is 1 N m, what average torque T is needed to maintain the prescribed angular motion? Each cylinder weighs 300 N and has a radius of gyration of 150 mm about its axis.

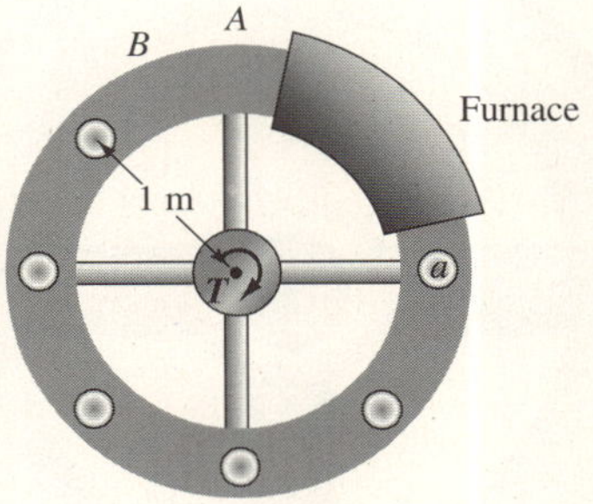

Figure P.17.49.

17.50. A circular *towing tank* has a main arm A which has a mass of 1000 kg and a radius of gyration of 1 m. On the arm rides the model support B, having a mass of 200 kg and having a radius of gyration about the vertical axis at its mass center of 600 mm. If a torque T of 50 N m is developed on A when B is at position $r = 1.8$ m, what will be the angular speed 5 s later if B moves out at a constant radial speed of 0.1 m/s. The initial angular speed of the arm A is 2 r/min. Neglect the drag of the model.

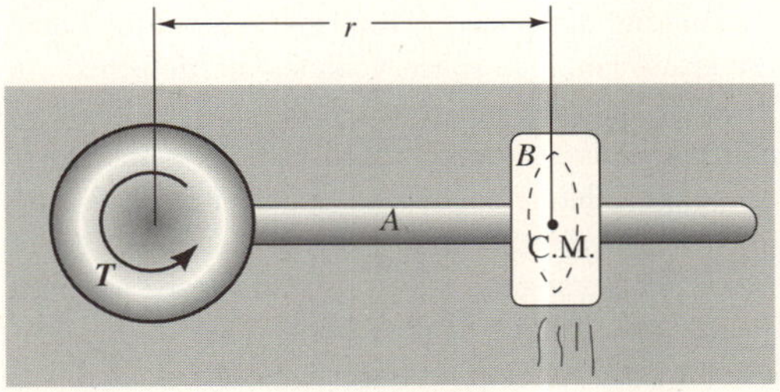

Figure P.17.50.

17.51. A steam roller with driver has a mass of 4.5 Mg. Wheel A weighs 18 Mg and has a radius of gyration of 0.24 m. Drive wheels B have a total mass of 9 Mg and a radius of gyration of 0.46 m. If a total torque of 540 N m is developed by the engine on the drive wheels, what is the speed of the steam roller after 10 s starting from rest? There is no slipping.

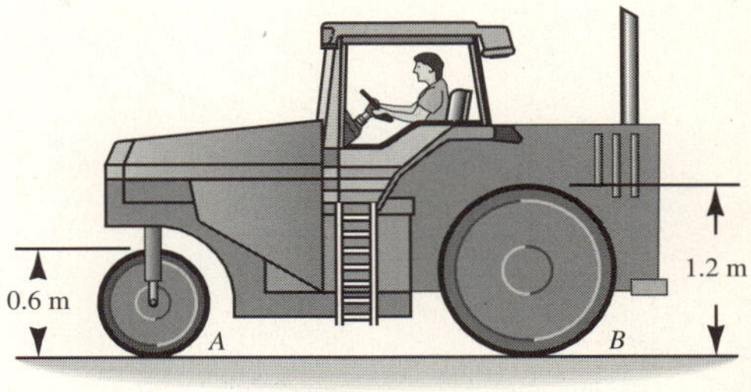

Figure P.17.51.

17.52. An electric motor D drives gears C, B, and device A. The diameters of gears C and B are 150 mm and 400 mm, respectively. The mass of A is 90 kg. The combined mass of the motor armature and gear C is 22.5 kg, while the radius of gyration of this combination is 200 mm. Also, the mass of B is 9 kg. If a constant counterclockwise torque of 80 N m is developed on the armature of the motor, what is the speed of A in 2 s after starting from rest? Neglect the inertia of the small wheels under A.

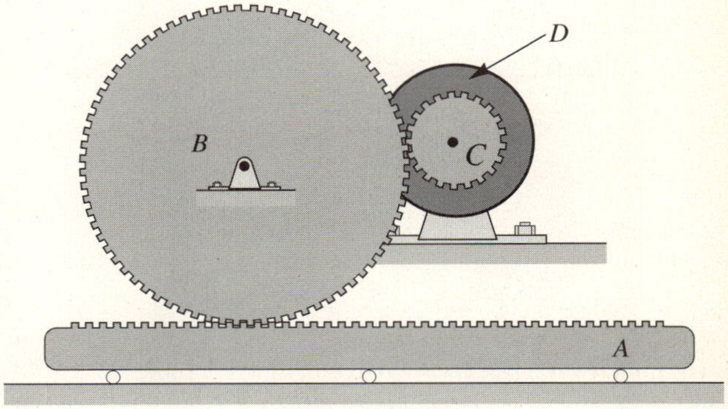

Figure P.17.52.

17.53. A conveyor is moving a mass M of 30 kg. Cylinders A and B have a diameter of 0.3 m and mass of 15 kg each. Also, they each have a radius of gyration of 0.24 m. Rollers C, D, E, F and G each have a diameter of 75 mm, mass of 5 kg each, and have a radius of gyration of 50 mm. What constant torque T will increase the speed of W from 0.9 m/s to 1.5 m/s in 3 s? The belt mass is 25 kg.

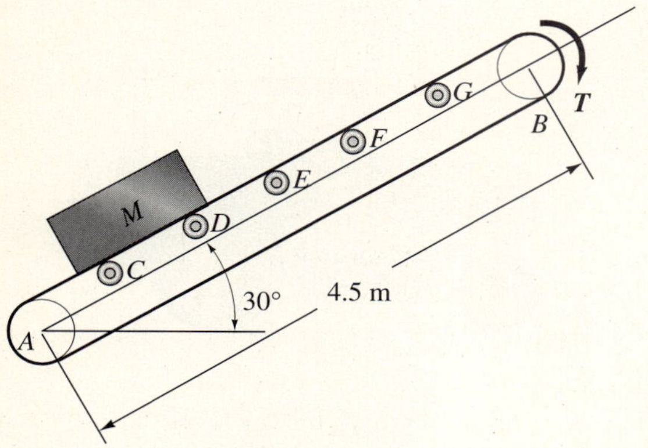

Figure P.17.53.

17.54. A rectangular block A is rotating freely at a speed ω of 200 r/min about a light hollow shaft. Attached to A is a circular rod B which can rotate out from the block A about a hinge at C. This rod weighs 20 N. When the system is rotating at the speed ω of 200 r/min, the rod B is vertical as shown. If the catch at the upper end of B releases so that B falls to a horizontal orientation (shown as dashed in the diagram), what is the new angular velocity? The block A weighs 60 N. Neglect the inertia of the shaft.

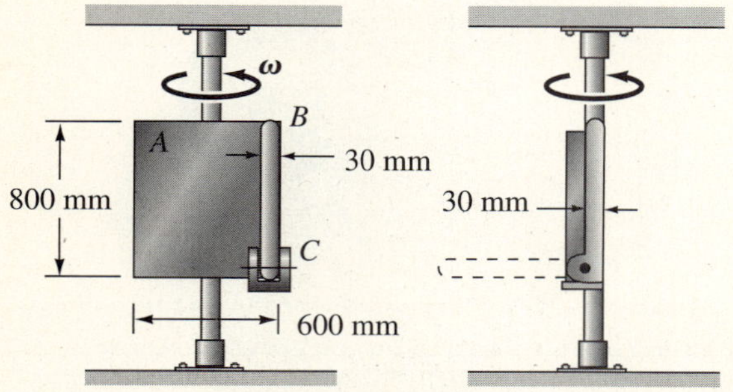

Figure P.17.54.

17.55. A space laboratory is in orbit and has an angular velocity ω of 0.5 rad/s relative to inertial space so as to have a partial "gravitational" force for the living quarters on the outside ring. The space lab has a mass of 4.5 Mg and a radius of gyration about its axis equal to 20 m. Two space ships C and D, each of mass 1000 kg, are shown docked so as also to get the benefit of "gravity." The center of mass of each vehicle is 0.7 m from the wall of the space lab. The radius of gyration of each vehicle about an axis at the mass center parallel to the axis of the space lab is 0.5 m. Small rocket engines A and B are turned on to develop a thrust each of 250 N. How long should they be on to increase the angular speed so as to have $1g$ of gravity at the outer radius of the space lab?

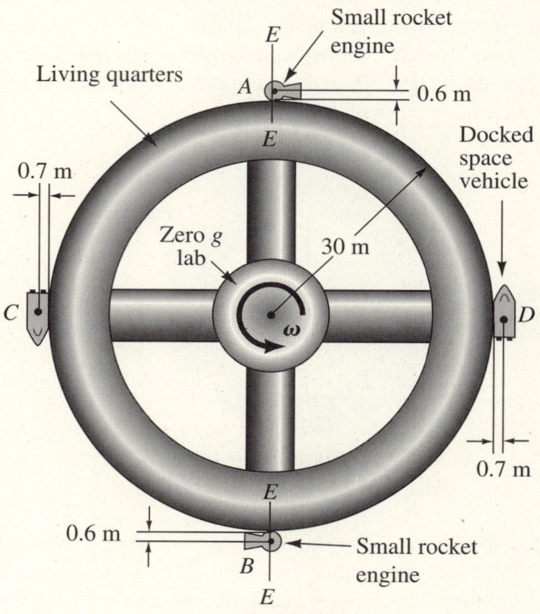

Figure P.17.55.

17.56. In Problem 17.55 the space lab and docked vehicles C and D are rotating at a speed ω for $1g$ gravity at the outer periphery of the space lab. The space vehicles are detached simultaneously. In this detachment process devices on the space lab produce an impulsive torque on each vehicle so that each vehicle ceases to rotate relative to inertial space. For how long and in what directions should the small rockets A and B of the space lab be fired to bring the angular speed ω of the space lab back to normal? The rockets can be rotated about axes E–E and, for this maneuver, they are developing a thrust of 100 N each.

17.57. A turbine is rotating freely with a speed ω of 6000 r/min. A blade breaks off at its base at the position shown. What is the velocity of the center of mass of the blade just after the fracture? Does the blade have an angular velocity just after fracture assuming that no impulsive torques or forces occur at the fracture? Explain.

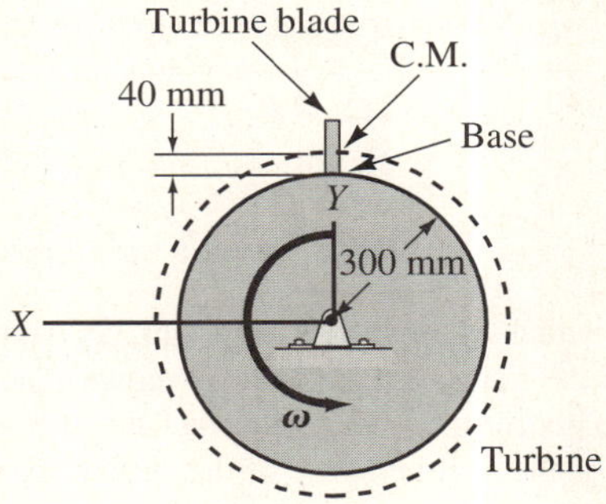

Figure P.17.57.

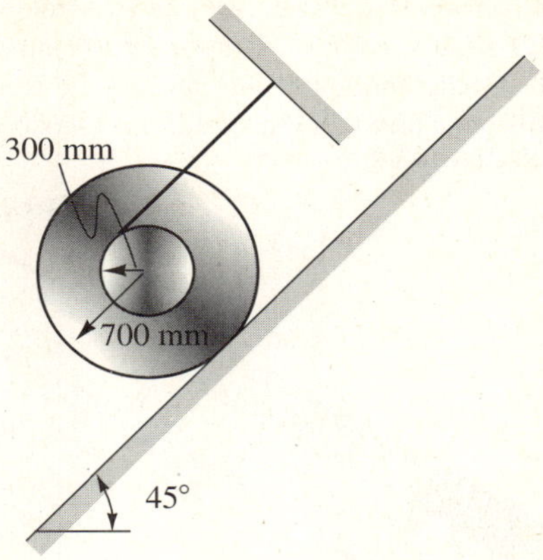

Figure P.17.59.

17.58. Two rods are welded to a drum which has an angular velocity of exactly 2000 r/min. The rod A breaks at the base at the position shown. If we neglect wind friction, how high up does the center of mass of the rod go? What is the angular orientation of the rod at the instant that the center of mass reaches its apex? Assume that there are no impulsive torques or forces at fracture. The Y axis is vertical. Neglect friction. Use exact value of 2000 r/min throughout.

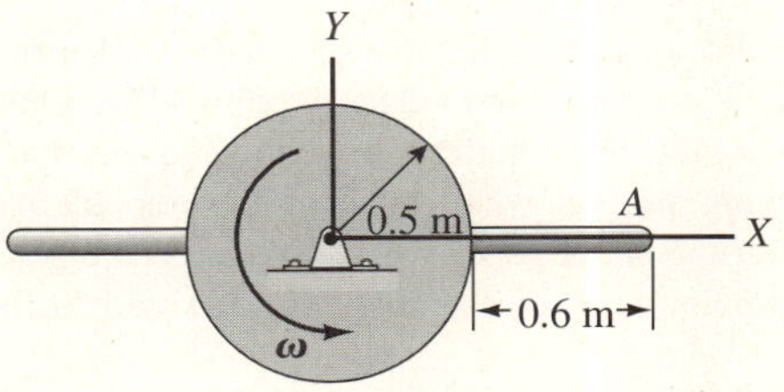

Figure P.17.58.

17.59. A stepped cylinder is released on a 45° incline where the dynamic coefficient of friction at the contact is 0.2 and the static coefficient of friction is 0.22. What is the angular speed of the stepped cylinder after 4 s? The stepped cylinder has a weight of 500 N and a radius of gyration about its axis of 250 mm. Be sure to check to see if the cylinder moves at all!

17.60. Prove that you can apply the angular momentum equation about the contact point A on a cylinder rolling without slipping on a circular (and hence including a straight) path. A force P always normal to OA acts on the cylinder as do a couple moment T and weight W. Specifically, prove that

$$\int_{t_1}^{t_2} (2Pr + W \sin\theta\, r + T)\,dt = M(k^2 + r^2)(\omega_2 - \omega_1) \quad \text{(a)}$$

where ω is the total angular speed of the cylinder. [*Hint:* Express the *angular momentum* equation about C and then, from *Newton's law* using the cylindrical component in the transverse direction, show on integrating that

$$\int_{t_1}^{t_2} (f + P + W\sin\theta)\,dt = -M(R - r)(\dot{\theta}_2 - \dot{\theta}_1)$$

where f is the friction force at the point of contact. Now let xy rotate with line OC. From *kinematics* first show that $R\theta = -r\phi$, where ϕ is the rotation of the cylinder relative to xy. Then, show that $\omega = -[(R - r)/r]\dot{\theta}$. From these three considerations, you should readily be able to derive Eq. (a).]

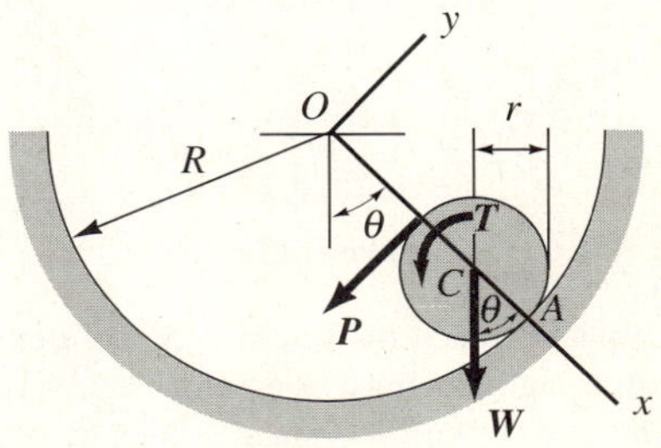

Figure P.17.60.

17.61 A slab A weighing 2 kN rides on two rollers each having a mass of 50 kg and a radius of gyration of 200 mm. If the system starts from rest, what minimum constant force T is needed to prevent the slab from exceeding the speed of 3 m/s down the incline in 4 s? There is no slipping.

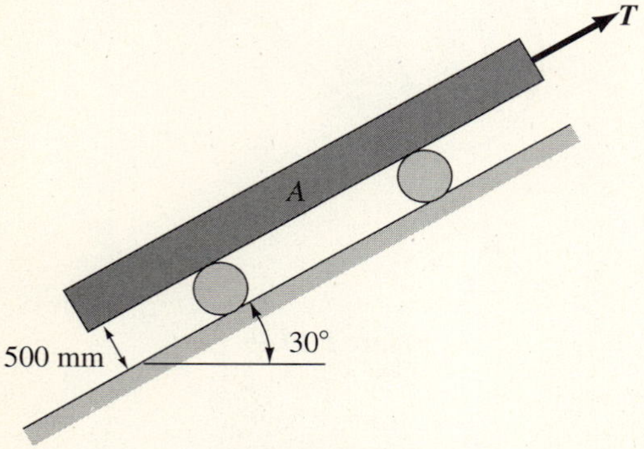

Figure P.17.61.

17.62. A main gear A rotates about a fixed axis and meshes with four identical floating gears B. The floating gears, in turn, mesh with the stationary gear F. If a torque T of 200 N m is applied to the main gear A, what is its angular speed in 5 s? The following data apply:

$$M_A = 100 \text{ kg}, \qquad k_A = 250 \text{ mm}$$
$$M_B = 20 \text{ kg}, \qquad k_B = 40 \text{ mm}$$

The system is horizontal. Read the first sentence of Problem 17.60.

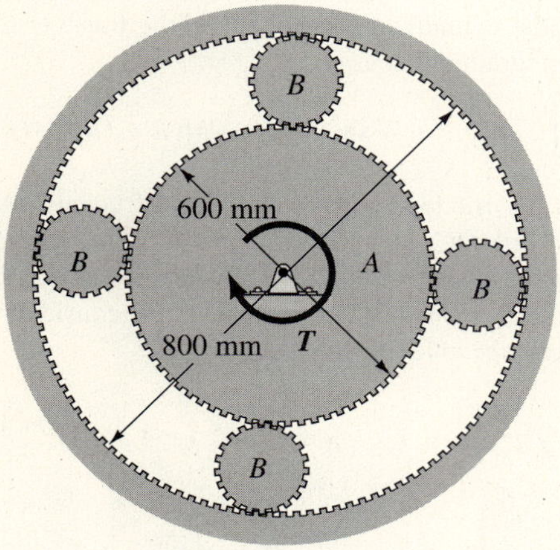

Figure P.17.62.

17.63. Two spheres, each of weight 2 N and diameter 30 mm, slide in smooth troughs inclined at an angle of 30°. A torque T of 2 N m is applied as shown for 3 s and is then zero. How far d up the inclines do the spheres move? The support system exclusive of the spheres has a weight of 10 N and a radius of gyration for the axis of rotation of 100 mm. Neglect friction and wind-resistance losses. Treat the spheres as particles.

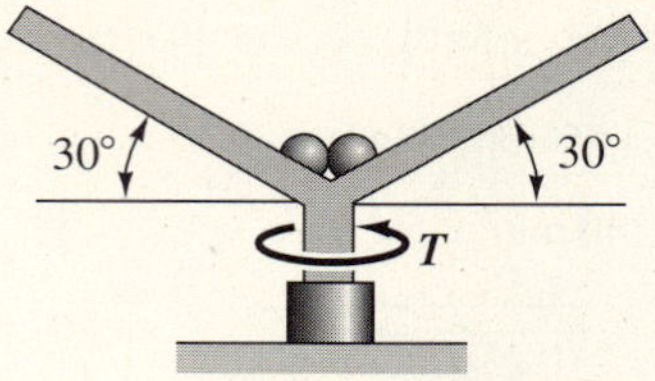

Figure P.17.63.

17.64. Two small spheres, each of weight 3 N, slide on a smooth rod curved in the shape of a parabola as shown. The curved rod is mounted on a platform on which a torque T of 1.5 N m is applied for 2 s. What is the angular speed of the system after 2 s? Neglect friction. The curved rod and the platform have a mass of 5 kg and a radius of gyration of 200 mm.

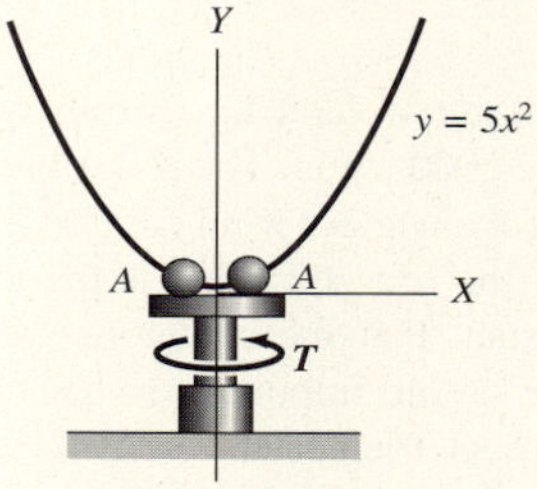

Figure P.17.64.

17.65. A disc of mass 3 kg is suspended between two wires. A bullet is fired at the disc and lodges at point A, for which r = 100 mm and θ = 45°. If the bullet is traveling at a speed of 600 m/sec before striking the disc, what is the angular velocity of the system directly after the bullet gets lodged? The bullet weighs 1 N (*Caution:* What point in the disc should you work with? It is not 0!)

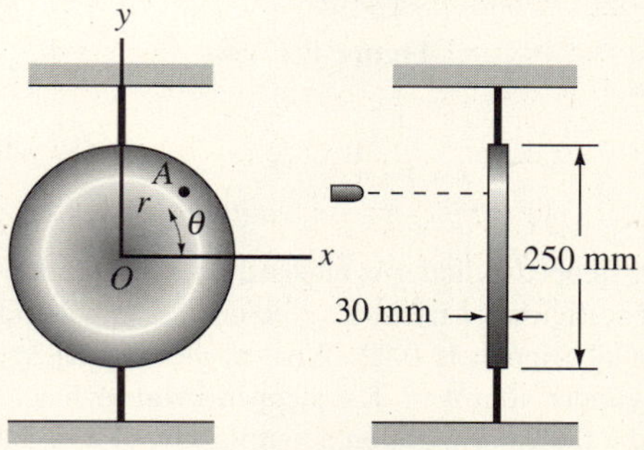

Figure P.17.65.

17.6 Impulsive Forces and Torques: Eccentric Impact

In Chapter 14, we introduced the concept of an *impulsive force*. Recall that an impulsive force F acts over a very short time interval Δt but has a very high value during this interval such that the impulse $\int_0^{\Delta t} F\,dt$ is significant. The impulse of other ordinary forces (not having very high peaks during Δt) is usually neglected for the short interval Δt. The same concept applies to torques, so that we have *impulsive torques*. The impulsive force and impulsive torque concepts are most valuable for the consideration of impact of bodies. Here, the collision forces and torques are impulsive while the other forces, such as gravity forces, have negligible impulse during collision.

In Chapter 14, we considered the case of *central impact* between bodies. Recall that for such problems the mass centers of the colliding bodies lie along the line of impact.[12] At this time, we shall consider the *eccentric impact of slab-like* bodies undergoing plane motion such as shown in Fig. 17.34. For eccentric impact, at *least one of the mass centers does not lie along line of impact*. The bodies in Fig. 17.32 have just begun contact whereby point A of one body has just touched point B of the other body. The velocity of point A just before contact (preimpact) is given as $(V_A)_i$, while the velocity just before contact for point B is $(V_B)_i$. (The i stands for "initial," as in earlier work.) We shall consider only *smooth bodies*, so that the impulsive forces acting on the body at the point of contact are *collinear* with the line of impact. As a result of the impulsive forces, there is a *period of deformation*, as in our earlier studies, and a *period of restitution*. In the period of deformation, the bodies are deforming, while in the period of restitution there is a complete or partial recovery of the original geometries. At the end of the period of deformation, the points A and B have the *same velocity* and we denote this velocity as V_D. Directly after impact (post impact), the velocities of points A and B are denoted as $(V_A)_f$ and $(V_B)_f$, respectively, where the subscript f is used to connote the final velocity resulting solely from the impact process. We shall be able to use the *linear impulse-momentum equation* and the *angular impulse-momentum* equation to relate the velocities, both linear and angular, for preimpact and postimpact states. These equations do not take into account the nature of the material of the colliding bodies, and so additional information is needed for solving these problems.

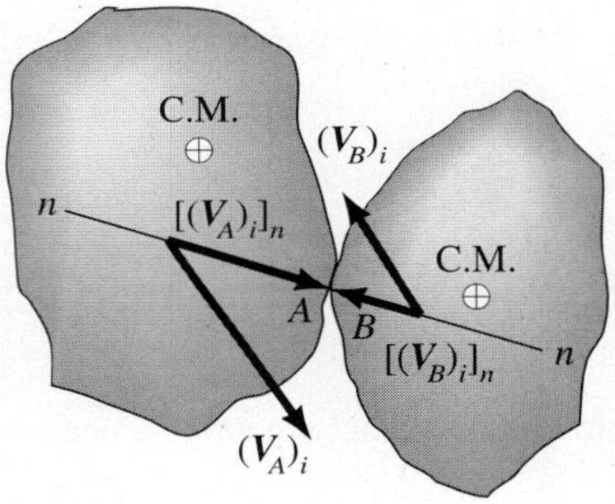

Figure 17.34. Eccentric impact between two bodies.

For this reason, we use the ratio between the impulse on each body during the period of restitution, $\int R\,dt$, and the impulse on each body during the period of deformation, $\int D\,dt$. As in central impact, the ratio is a number ϵ, called the *coefficient of restitution*, which depends primarily on the materials of the bodies in collision. Thus,

$$\epsilon = \frac{\int R\,dt}{\int D\,dt} \tag{17.29}$$

[12]The line of impact is normal to the plane of contact between the bodies.

We shall now show that the components along the line of impact *n–n* of $\mathbf{V}_A$ and $\mathbf{V}_B$, taken at pre- and postimpact, are related to ϵ by the very same relation that we had for central impact. That is,

$$\epsilon = -\frac{\left[(V_B)_f\right]_n - \left[(V_A)_f\right]_n}{\left[(V_B)_i\right]_n - \left[(V_A)_i\right]_n} \tag{17.30}$$

Recall that the numerator represents the relative velocity of separation along *n* of the points of contact, whereas the denominator represents the relative velocity of approach along *n* of these points.

We shall first consider the case where the bodies are in *no way constrained* in their plane of motion; we can then neglect all impulses except that coming from the impact. We now consider the body having contact point *A*. In Fig. 17.35 we have shown this body with impulse $\int D\,dt$ acting. Using the component of the *linear momentum equation* along line of impact *n–n*, we can say for the center of mass;

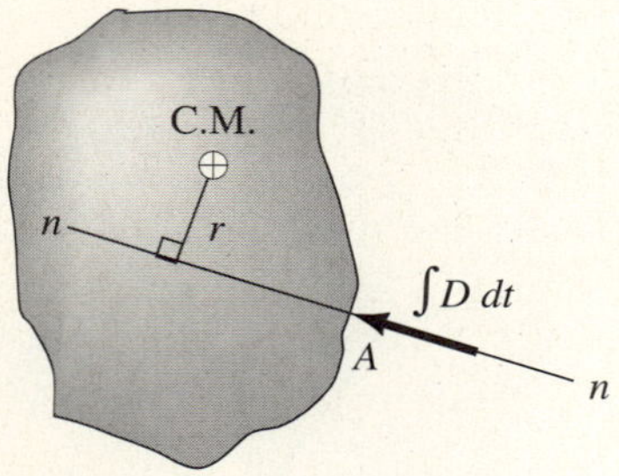

Figure 17.35. Impulse acting on one of the bodies.

$$\int D\,dt = M\left[(V_C)_D\right]_n - M\left[(V_C)_i\right]_n \tag{17.31}$$

where $(V_C)_i$ refers to the preimpact velocity of the center of mass and where $(V_C)_D$ is the velocity of the center of mass at the end of the deformation period. Similarly, for the period of restitution, we have

$$\int R\,dt = M\left[(V_C)_f\right]_n - M\left[(V_C)_D\right]_n \tag{17.32}$$

For angular momentum, we can say for the deformation period using *r* as the distance from the center of mass to *n–n*:

$$r\int D\,dt = I\omega_D - I\omega_i \tag{17.33}$$

Similarly, for the period of restitution:

$$r\int R\,dt = I\omega_f - I\omega_D \tag{17.34}$$

Now, substitute the right sides of Eqs. 17.31 and 17.32 into Eq. 17.29. We get on cancellation of *M*:

$$\epsilon = \frac{\left[(V_C)_f\right]_n - \left[(V_C)_D\right]_n}{\left[(V_C)_D\right]_n - \left[(V_C)_i\right]_n} = \frac{\left[(V_C)_D\right]_n - \left[(V_C)_f\right]_n}{\left[(V_C)_i\right]_n - \left[(V_C)_D\right]_n} \tag{17.35}$$

Next, substitute for the impulses in Eq. 17.29 using Eqs. 17.33 and 17.34. We get on canceling only *I*:

$$\epsilon = \frac{r\omega_f - r\omega_D}{r\omega_D - r\omega_i} = \frac{r\omega_D - r\omega_f}{r\omega_i - r\omega_D} \tag{17.36}$$

Adding the numerators and denominators of Eqs. 17.35 and 17.36, we can then say on rearranging the terms:

$$\epsilon = \frac{\{[(V_C)_D]_n + r\omega_D\} - \{[(V_C)_f]_n + r\omega_f\}}{\{[(V_C)_i]_n + r\omega_i\} - \{[(V_C)_D]_n + r\omega_D\}} \tag{17.37}$$

We pause now to consider the *kinematics* of the motion. We can relate the velocities of points A and C on the body (see Fig. 17.36) as follows:

$$\mathbf{V}_A = \mathbf{V}_C + \boldsymbol{\omega} \times \boldsymbol{\rho}_{CA} \tag{17.38}$$

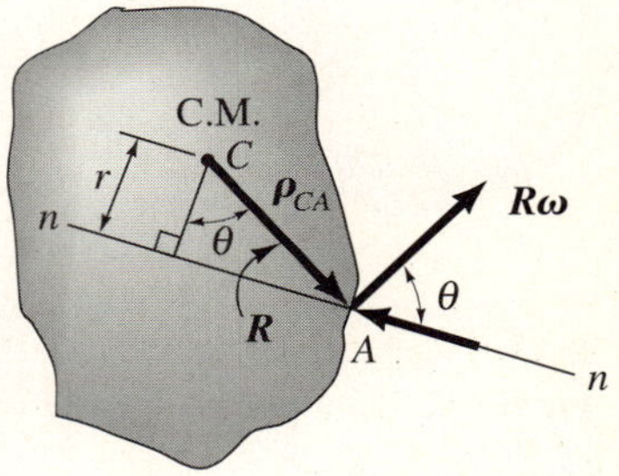

Figure 17.36. Slab with $|\boldsymbol{\rho}_{CA}|$ as R.

Note that the magnitude of $\boldsymbol{\rho}_{CA}$ is R as shown in the diagram. Since $\boldsymbol{\omega}$ is normal to the plane of symmetry of the body and thus to $\boldsymbol{\rho}_{CA}$, the value of the last term in Eq. 17.38 is $R\omega$ with a direction normal to R as has been shown in Fig. 17.36. The components of the vectors in Eq. 17.38 in direction n can then be given as follows:

$$(V_A)_n = (V_C)_n + \omega R \cos\theta \tag{17.39}$$

Since $R \cos\theta = r$ (see Fig. 17.36), we conclude that

$$(V_A)_n = (V_C)_n + r\omega \tag{17.40}$$

With the preceding result applied to the initial condition (i), the final condition (f), and the intermediate condition (D), we can now go back to Eq. 17.37 and replace the expressions inside the braces ({ }) by the left side of Eq. 17.40 as follows:

$$\epsilon = \frac{[(V_A)_D]_n - [(V_A)_f]_n}{[(V_A)_i]_n - [(V_A)_D]_n} \tag{17.41}$$

A similar process for the body having contact point B will yield the preceding equation with subscript B replacing subscript A:

$$\epsilon = \frac{[(V_B)_D]_n - [(V_B)_f]_n}{[(V_B)_i]_n - [(V_B)_D]_n} = \frac{[(V_B)_f]_n - [(V_B)_D]_n}{[(V_B)_D]_n - [(V_B)_i]_n} \tag{17.42}$$

Now add the numerators and denominators of the right side of Eq. 17.41 and the extreme right side of Eq. 17.42. Noting that

$$[(V_A)_D]_n = [(V_B)_D]_n \tag{17.43}$$

we get Eq. 17.30, thus demonstrating the validity of that equation.

Let us next consider the case where one or both bodies undergoing impact is constrained to *rotate about a fixed axis*. We have shown such a body in Fig. 17.37 where O is the axis of rotation and point A is the contact point. If an impulse is developed at the point of contact A (we have shown the impulse during the period of deformation), then clearly there will be an

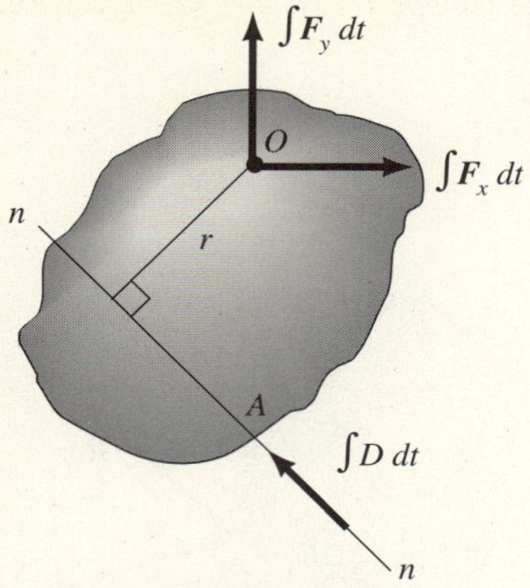

Figure 17.37. Impact for a body under constraint at O.

impulsive force at O, as shown in the diagram. We shall employ the *angular impulse-momentum* equation about the fixed point O. Thus, we have for the period of deformation and the period of restitution:

$$r\int D\,dt = I_0\omega_D - I_0\omega_i$$
$$r\int R\,dt = I_0\omega_f - I_0\omega_D$$

Now solve for the impulses in the equation above, and substitute into Eq. 17.29. Canceling I_0, the moment of inertia about the axis of rotation at O, we get

$$\epsilon = \frac{r\omega_f - r\omega_D}{r\omega_D - r\omega_i} = \frac{r\omega_D - r\omega_f}{r\omega_i - r\omega_D} \tag{17.44}$$

In Fig. 17.38 we see that

$$V_A = R\omega$$

Therefore,

$$(V_A)_n = R\omega\cos\theta = r\omega$$

Using the above result in Eq. 17.44, we get

$$\epsilon = \frac{\left[(V_A)_D\right]_n - \left[(V_A)_f\right]_n}{\left[(V_A)_i\right]_n - \left[(V_A)_D\right]_n} \tag{17.45}$$

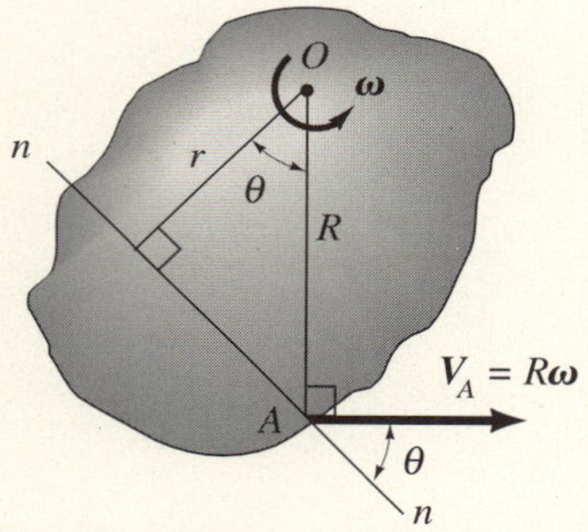

Figure 17.38. Velocity of point A is $R\omega$.

But this expression is identical to Eq. 17.41. And by considering the second body, which is either free or constrained to rotate, we get an equation corresponding to Eq. 17.42. We can then conclude that Eq. 17.30 is valid for impact where one or both bodies are constrained to rotate about a fixed axis.

In a typical impact problem, the motion of the bodies preimpact is given and the motion of the bodies postimpact is desired. Thus, there could be four unknowns—two velocities of the mass centers of the bodies plus two angular velocities. The required equations for solving the problem are formed from linear and angular momentum considerations of the bodies taken separately or taken as a system. Only impulsive forces are taken into consideration during the time interval spanning the impact. If the bodies are considered separately, we simply use the formulations of Section 17.5, remembering to observe Newton's third law at the point of impact between the two bodies. Furthermore, we must use the coefficient of restitution equation 17.30. Generally, kinematic considerations are also needed to solve the problem. When there are no other impulsive forces other than those occurring at the point of impact, it might be profitable to consider the bodies as one system. Then, clearly, as a result of Newton's third law, we must have *conservation of linear momentum* relative to an inertial reference, and also we must have *conservation of angular momentum* about *any one axis* fixed in inertial space.

We now illustrate these remarks in the following series of examples.

Example 17.11

A rectangular plate A weighing 20 N has two identical rods weighing 10 N each attached to it (see Fig. 17.39). The plate moves on a plane smooth surface at a speed of 5 m/s. Moving oppositely at 10 m/s is disc B, weighing 10 N. A perfectly elastic collision ($\epsilon = 1$) takes place at G. What is the speed of the center of mass of the plate just after collision (postimpact)? Solve the problem two ways: consider the bodies separately and consider the bodies as a system.

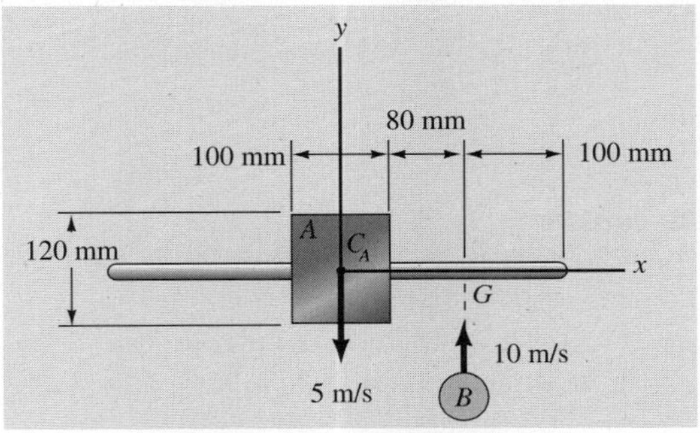

Figure 17.39. Colliding bodies.

Solution 1. We can consider the disc to be a particle since this body will only translate. Let point G be the point of contact on the rod. We then have, for Eq. 17.30:

$$\epsilon = 1 = -\frac{(V_G)_f - (V_B)_f}{-5 - 10}$$

Therefore,

$$(V_G)_f - (V_B)_f = 15 \qquad \text{(a)}$$

Now, consider linear and angular momentum for each of the bodies. For this purpose we have shown the bodies in Fig. 17.40 with only impulsive forces acting. We might call such a diagram an "impulsive free-body diagram." We then see that C_A must move in the plus or minus y direction after impact. We can then say for body A, using **linear impulse-momentum** and **angular impulse-momentum equations** (the latter about the center of mass):

$$\int F\,dt = \frac{40}{g}\left[\left(V_{C_A}\right)_f - (-5)\right] \qquad \text{(b)}$$

$$(0.13)\int F\,dt = \left[\frac{1}{12}\left(\frac{20}{g}\right)(0.10^2 + 0.12^2) + 2\left(\frac{1}{12}\right)\left(\frac{10}{g}\right)(0.18)^2 + 2\left(\frac{10}{g}\right)(0.14)^2\right](\omega_A - 0) \qquad \text{(c)}$$

$$= 0.0496\omega_A$$

Example 17.11 (Continued)

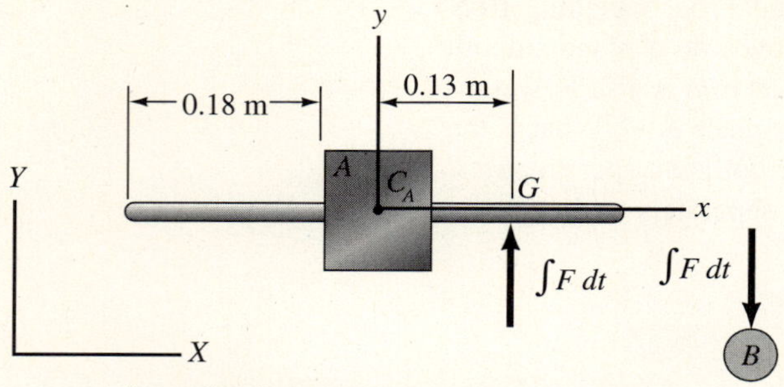

Figure 17.40. Impulsive free-body diagrams.

For body B we have for the **linear impulse-momentum equation**:

$$-\int F\,dt = \frac{10}{g}\left[(V_B)_f - 10\right] \qquad \text{(d)}$$

From **kinematics** we have for the plate with arms:

$$(V_G)_f = (V_{C_A})_f + (\omega_A)(0.13) \qquad \text{(e)}$$

We have now a complete set of equations which we can proceed to solve. We get:

$$(V_{C_A})_f = -0.305 \text{ m/s}$$

Solution 2. Equation (a) above from the coefficient of restitution and Eq. (e) above from kinematics also apply to solution 2. Substituting for $(V_G)_f$ in Eq. (a) of solution 1 using Eq. (e) of solution 1, we get for solution 2:

$$(V_{C_A})_f + 0.13\omega_A - (V_B)_f = 15 \qquad \text{(a)}$$

Conservation of linear momentum for the system leads to the requirement in the y direction that

$$\frac{40}{g}(-5) + \frac{10}{g}(10) = \frac{40}{g}(V_{C_A})_f + \frac{10}{g}(V_B)_f$$

Therefore,

$$4(V_{C_A})_f + (V_B)_f = -10 \qquad \text{(b)}$$

Also **angular momentum is conserved** about any fixed z axis. We choose the axis at the position corresponding to C_A at the time of impact. Noting that I for the plate and arms is 49.6 g m^2 from solution 1 (see Eq. (c)) and noting that B can be considered as a particle, we have

Example 17.11 (Continued)

$$\underbrace{\frac{10}{g}(10)(0.13)}_{\substack{\text{angular}\\\text{momentum}\\\text{preimpact}}} = \underbrace{(0.0496)\omega_A + \frac{10}{g}(V_B)_f(0.13)}_{\substack{\text{angular}\\\text{momentum}\\\text{postimpact}}}$$

Therefore,

$$\omega_A = -2.67(V_B)_f + 26.7 \qquad \text{(c)}$$

Solving Eqs. (a), (b), and (c) simultaneously we get:

$$(V_{C_A})_f = -0.305 \text{ m/s}$$

We leave it to you to demonstrate there has been *conservation of mechanical energy* during this perfectly elastic impact. We could have used this fact in place of Eq. (a) in solution 1.

Note that there was some saving of time and labor in using the system approach throughout for the preceding problem wherein a rigid body, namely the plate and its arms, collided with a body, the disc, which could be considered as a particle. In problems involving two colliding rigid bodies neither of which can be considered a particle, we must consider the bodies separately since the system approach does not yield a sufficient number of independent equations as you can yourself demonstrate.

In the preceding example, the bodies were not constrained except to move in a plane. If one or both colliding bodies is pinned, the procedure for solving the problem may be a little different than what was shown in Example 17.11. Note that there will be unknown supporting impulsive forces at the pin of any pinned body. If we are not interested in the supporting impulsive force for a pinned body, we only consider *angular momentum* about the pin for that pinned body; in this way the undesired unknown supporting impulsive forces at the pin do not enter the calculations. Other than this one factor, the calculations are the same as in the previous example.

You will recall from momentum considerations of particles that we considered the collision of a comparatively small body with a very massive one. We could not use the conservation of momentum equation for the collision between such bodies since the velocity change of the massive body went to zero as the mass (mathematically speaking) went to infinity, thus producing an indeterminacy in our idealized formulations. We shall next illustrate the procedure for the collision of a very massive body with a much smaller one. You will note that we cannot consider a system approach for linear or angular momentum conservation for the same reasons set forth in Chapter 14.

Example 17.12

A 10 kg rod AB is dropped onto a massive body (Fig. 17.41). What is the angular velocity of the rod postimpact for the following conditions:

A. Smooth floor; elastic impact.
B. Rough floor (no slipping); plastic impact.

In either case, the velocity of end B preimpact is

$$V_B = \sqrt{2gh} = \sqrt{(2)(9.81)(0.6)} = 3.43 \text{ m/s}$$

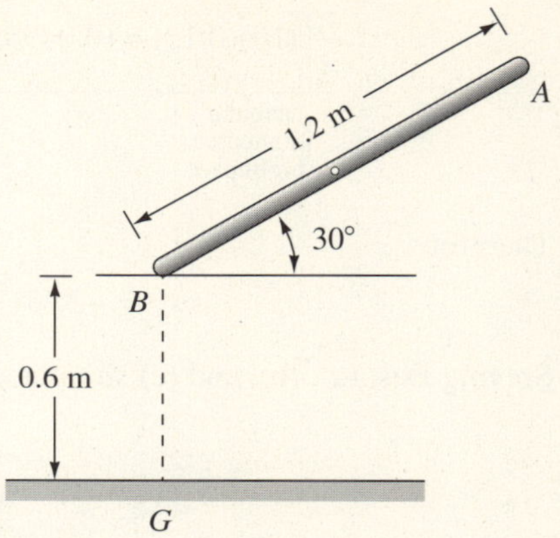

Figure 17.41. Falling rod on a massive body.

We now consider each case separately.

Case A. Equation 17.30 can be used here. Thus,

$$\epsilon = 1 = -\frac{\left[(V_B)_f\right]_n - 0}{-3.43 - 0}$$

Therefore,

$$\left[(V_B)_f\right]_n = 3.43 \text{ m/s} \qquad \text{(a)}$$

Next, considering rod AB in Fig. 17.42 we have for **linear impulse-** and **angular impulse-momentum** considerations (the latter about an axis at the center of mass).

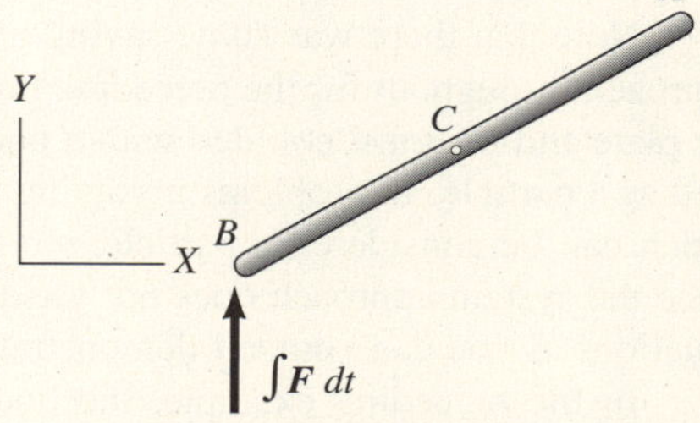

Figure 17.42. Impulsive free-body diagram.

$$\int F\,dt = 10\left[(V_C)_f - (-3.43)\right] \qquad \text{(b)}$$

$$-(0.6)(\cos 30°)\int F\,dt = \frac{1}{12}(10)\left(1.2^2\right)\omega_f \qquad \text{(c)}$$

From **kinematics**, we have[13]

$$(\boldsymbol{V}_B)_f = (\boldsymbol{V}_C)_f + \boldsymbol{\omega}_f \times \boldsymbol{\rho}_{CB}$$

$$\left[(V_B)_x\right]_f \boldsymbol{i} + 3.43\boldsymbol{j} = (V_C)_f \boldsymbol{j} + \omega_f \boldsymbol{k} \times (0.6)(-0.866\boldsymbol{i} - 0.5\boldsymbol{j})$$

$$\left[(V_B)_x\right]_f \boldsymbol{i} + 3.43\boldsymbol{j} = (V_C)_f \boldsymbol{j} - 0.5196\omega_f \boldsymbol{j} + 0.3\omega_f \boldsymbol{i}$$

Hence, the scalar equations are

$$3.43 = (V_C)_f - 0.5196\omega_f \qquad \text{(d)}$$

$$\left[(V_B)_x\right]_f = 0.3\omega_f \qquad \text{(e)}$$

[13]Note that since the initial velocity of C is vertical and since the impulsive force is vertical, the final velocity of C must be vertical. This fact is used in the kinematics.

Example 17.12 (Continued)

We now have a complete set of equations considering $\int F\,dt$ as an unknown.

Solving, we have for the desired unknowns:

$$\omega_f = -9.14 \text{ rad/s}$$
$$[(V_B)_x]_f = -2.74 \text{ m/}s$$

You can demonstrate that energy has been conserved in this action. We could have used this fact in lieu of Eq. (a) for this problem.

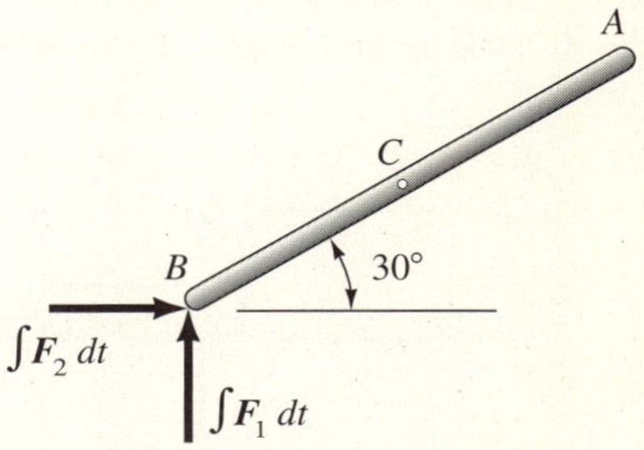

Figure 17.43. Impulsive free-body diagram.

Case B. Here we have no slipping on the rough surface and zero vertical movement of point *B*. Accordingly, we have shown rod *BA* with vertical and horizontal impulses in Fig. 17.43. The **linear momentum equations** for the center of mass then are

$$\int F_1\,dt = 10\left\{[(V_C)_y]_f - (-3.43)\right\} \quad \text{(f)}$$
$$\int F_2\,dt = 10\left\{[(V_C)_x]_f - 0\right\} \quad \text{(g)}$$

The **angular impulse-momentum equation** about the center of mass is

$$-0.6(\cos 30°)\int F_1\,dt + 0.6(\sin 30°)\int F_2\,dt = \frac{1}{12}(10)(1.2^2)\omega_f \quad \text{(h)}$$

From **kinematics**, noting that at postimpact there is *pure rotation* about point *B*, we have

$$[(V_C)_y]_f = 0.6(\cos 30°)(\omega_f) = 0.5196\omega_f \quad \text{(i)}$$
$$[(V_C)_x]_f = -0.6(\sin 30°)(\omega_f) = -0.3\omega_f \quad \text{(j)}$$

Substitute for $\int F_1\,dt$ and $\int F_2\,dt$ from Eqs. (f) and (g) into Eq. (h). We get on then employing Eqs. (i) and (j):

$$-(0.6)(0.866)(10)(0.5196\omega_f + 3.43) + (0.6)(0.5)(10)(-0.3\omega_f) = \frac{1}{12}(10)(1.2^2)\omega_f$$

Therefore,

$$\omega_f = -3.71 \text{ rad/s}$$

PROBLEMS

17.66. A rod AB slides on a smooth surface at a speed of 10 m/s and hits a disc D moving at a speed of 5 m/s. What is the postimpact velocity of point A for a coefficient of restitution $\epsilon = 0.8$? The rod weighs 30 N and the disc weighs 8 N. Solve by considering AB and D separately.

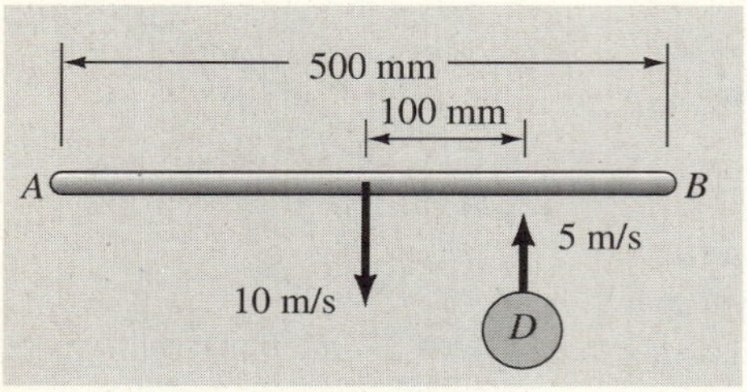

Figure P.17.66.

17.67. Solve Problem 17.66 by considering the two bodies as a system.

17.68. Rods AB and HD translate on a horizontal frictionless surface. When they collide at G we have a coefficient of restitution of 0.7. Rod HD weighs 70 N and rod AB weighs 40 N. What is the postimpact angular speed of AB?

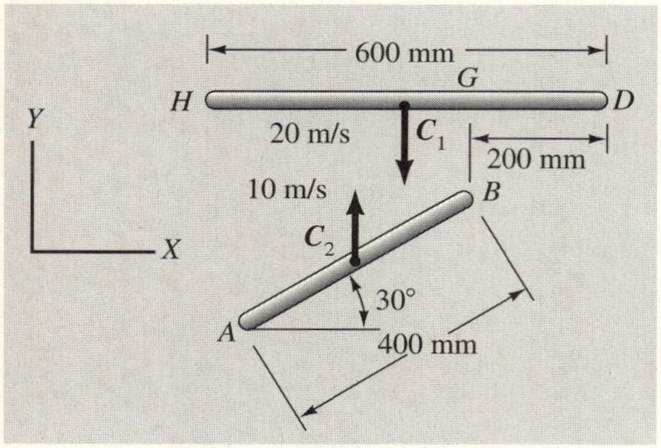

Figure P.17.68.

17.69. Two slender rods 10 kg each with small 45° protuberances are shown on a smooth horizontal surface. The rods are identical except for the position of the protuberances. Rod A moves to the left at a speed of 3 m/s while rod B is stationary. If the protuberance surfaces are smooth, and an impact having $\epsilon = 0.8$ occurs, what should d be in order for the postimpact angular speed of A to be 5 rad/s? Neglect the protuberance for determination of moments of inertia and centers of mass.

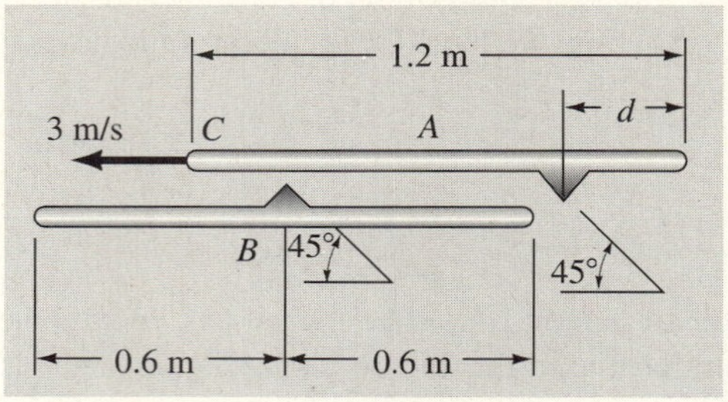

Figure P.17.69.

17.70. A 22 N sphere moving at a speed of 10 m/s hits the end of a 1 m rod having a mass of 10 kg. The coefficient of restitution for the impact is 0.9. What is the postimpact angular velocity of the rod if it is stationary just before impact? The rod is pinned at O.

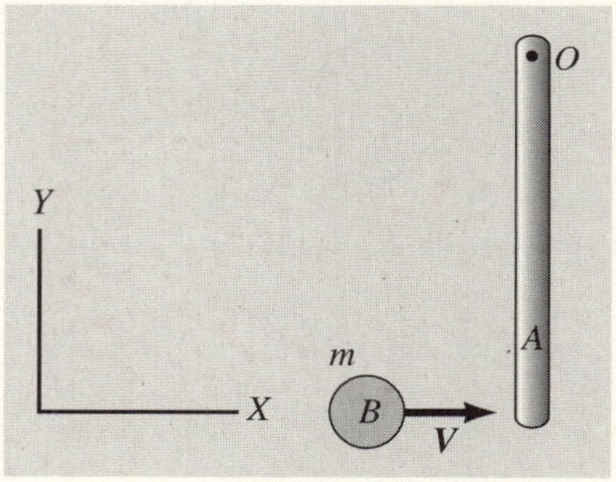

Figure P.17.70.

17.71. A rod A weighing 50 N rotates freely about a hinge at a speed of 10 rad/s on a frictionless surface just prior to hitting a disc B weighing 20 N and moving at a speed of 5 m/s. If the coefficient of restitution is 0.8, what is the postimpact angular speed of A? The contact surface between the rod and the disc is smooth.

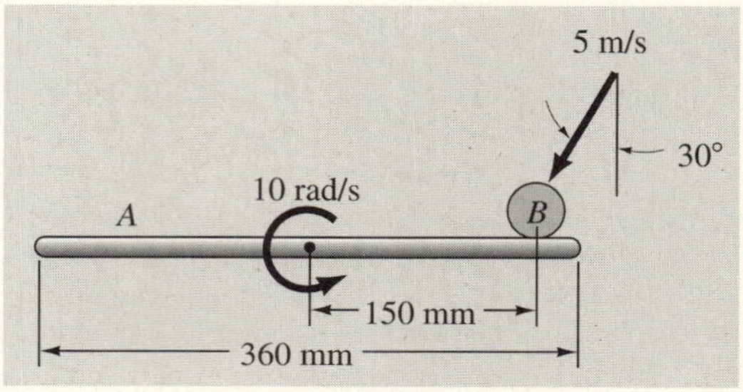

Figure P.17.71.

17.72. A stiff bent rod is dropped so that end A strikes a heavy table D. If the impact is plastic and there is no sliding at A, what is the postimpact speed of end B? The rod weighs per unit length 30 N/m.

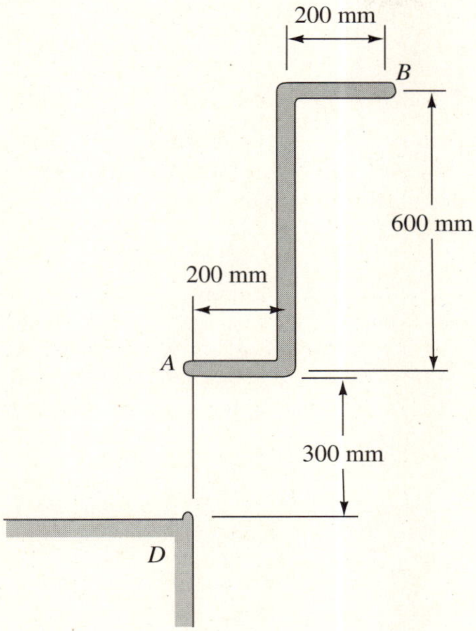

Figure P.17.72.

17.73. Solve Problem 17.72 for an elastic impact at A. Take the surface of contact to be smooth. Demonstrate that energy has been conserved.

17.74. A drumstick and a wooden learner's drum are shown in (a) of the diagram. At the beginning of a drum roll, the drumstick is horizontal. The action of the drummer's hand is simulated by force components F_x and F_y, which make A a stationary axis of rotation. Also, there is a constant couple M of 0.5 N m. If the action starts from a stationary position shown, what is the frequency of the drum roll for perfectly elastic collision between the drumstick and the learner's drum? The mass of the drumstick is 200 g. Idealize the drumstick as a uniform slender rod.

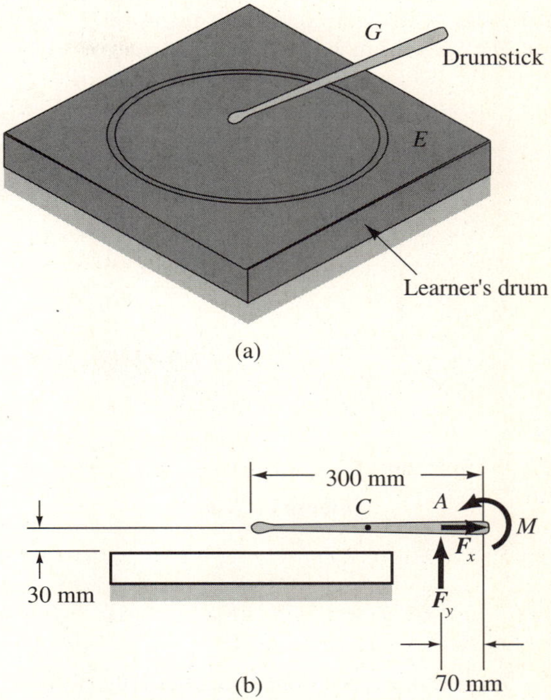

Figure P.17.74.

17.75. A block of ice 0.3 m × 0.3 m × 0.15 m slides along a surface at a speed of 3 m/s. The block strikes stop D in a plastic impact. Does the block turn over after the impact? What is the highest angular speed reached? Take $\rho = 1 \text{ Mg/m}^3$.

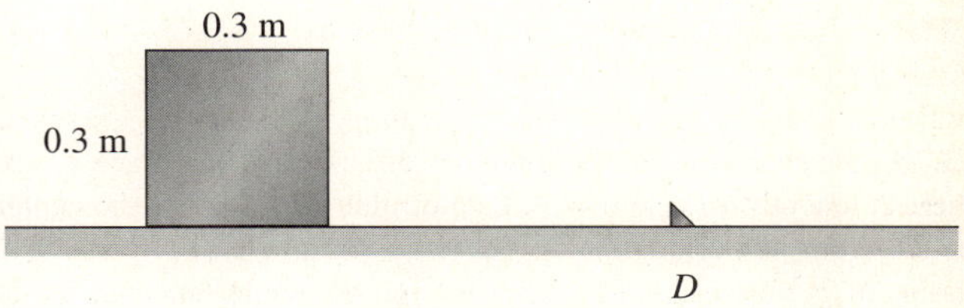

Figure P.17.75.

17.76. A horizontal rigid rod is dropped from a height 3 m above a heavy table. The end of the rod collides with the table. If the coefficient of impact ϵ between the end of the rod and the corner of the table is 0.6, what is the postimpact angular velocity of the rod? Also, what is the velocity of the center of mass post-impact: The rod is 0.3 m in length and has a mass of 0.7 kg.

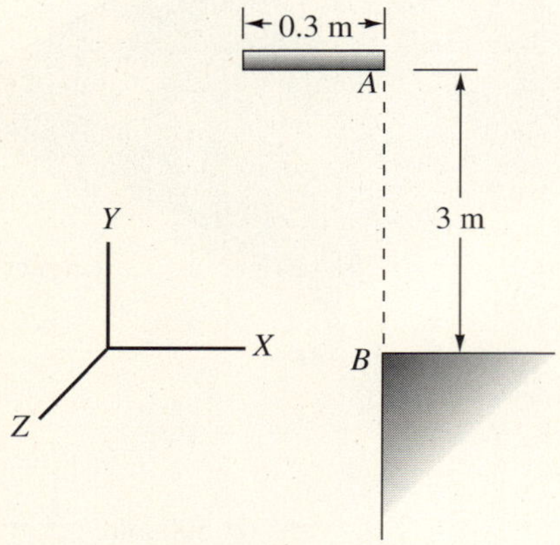

Figure P.17.76.

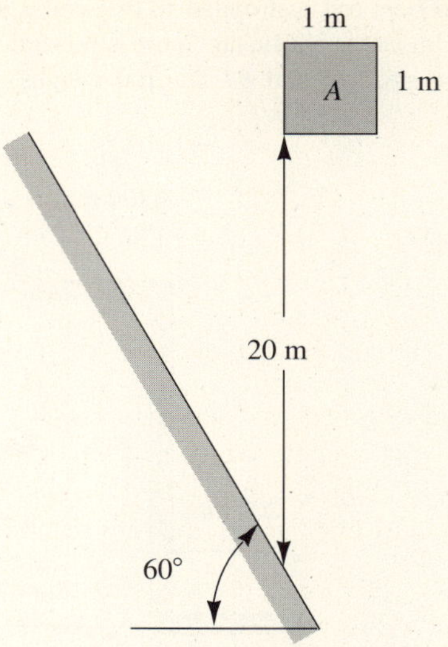

Figure P.17.78.

17.77. A rod A is dropped from a height of 120 mm above B. If an elastic collision takes place at B, at what time Δt later does a second collision with support D take place? The rod weighs 40 N.

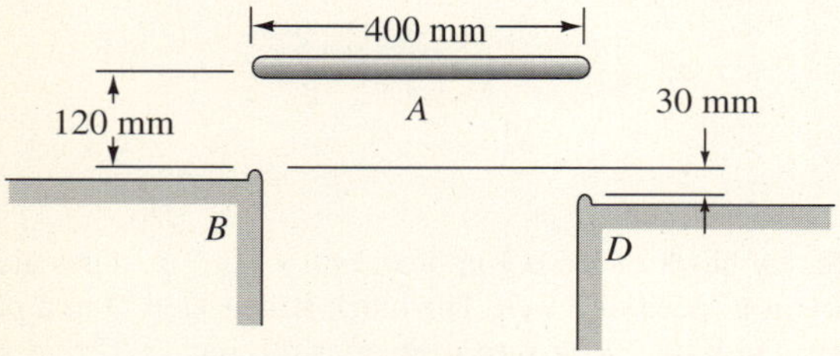

Figure P.17.77.

17.78. A packing crate falls on the side of a hill at a place where there is smooth rock. If $\epsilon = 0.8$, determine the speed of the center of the crate just after impact. The crate weighs 400 N and has a radius of gyration about an axis through the center and normal to the face in the diagram of $\sqrt{2}$ m.

17.79. An arrow moving essentially horizontally at a speed of 20 m/s impinges on a stationary wooden block which can rotate freely about light rods DE and GH. The arrowhead weighs 1.5 N, and the shaft, which is 250 mm long, weighs 0.7 N. If the block weighs 10 N, what is the angular velocity of the block just after the arrowhead becomes stuck in the wood? The arrowhead sinks into the wood at point A such that the shaft sticks out 250 mm from the surface of the wood.

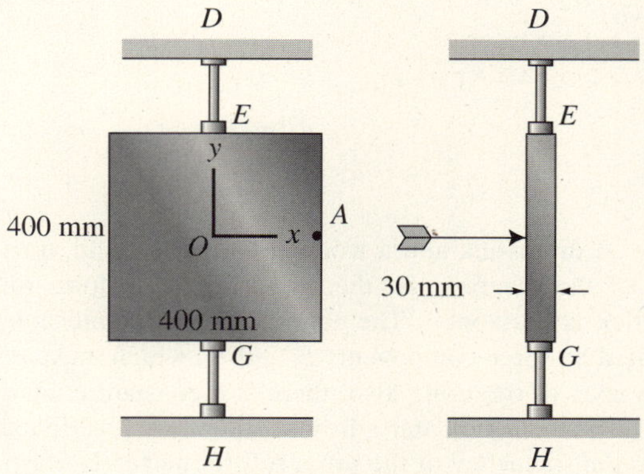

Figure P.17.79.

17.7 Closure

Let us now pause for an overview of the text up to this point. Recall the following:

(a) In Chapter 13, we studied energy methods for single particles and systems of particles with a short introduction to coplanar rigid body motions.
(b) Then in Chapter 14, we presented linear and angular momentum principles for a single particle and systems of particles with again a short introduction to plane motion of rigid bodies.

These methods are derived from Newton's law. In short, they are integrated forms of Newton's law which bring in useful concepts and greater ease in solving many problems. Next, after studying *Chasles' theorem* and kinematics of rigid bodies in Chapter 15, we went to Chapter 16.

(c) In Chapter 16, we studied the dynamics of plane motion for rigid bodies using Newton's law and the equation $\boldsymbol{M}_A = \dot{\boldsymbol{H}}_A$ directly rather than integrated of forms of these equations, namely the energy and momentum equations.

Now in Chapter 17, we went through a partial recycle of the above steps. That is, we went back to Chapters 13 and 14 this time for three-dimensional formulations and focused on rigid bodies and systems of rigid bodies. Naturally, there is an overlap with those earlier chapters plus an expansion of viewpoints to include a general formulation and use of $\boldsymbol{H}$ and a look at eccentric coplanar impact of more complex bodies. Again, with the aforementioned integrated forms of *Newton's law*, we were able to solve some interesting problems.

We are now at a stage comparable to what we were in just preceding Chapter 16. In Chapter 18, we go back again to Newton's law and $\boldsymbol{M}_A = \dot{\boldsymbol{H}}_A$, this time for the dynamics of general three-dimensional motion of rigid bodies. This is considered the province of a second course in Dynamics and accordingly is a starred chapter. Interested students who go ahead will unlock the mysteries of gyroscopic motion and the performance of gyros among other interesting applications and concepts. In the process, those students will meet the famous Euler equations of motion whose simplified, special form we have been using up to this point.

PROBLEMS

17.80. Gear E rotates at an angular speed ω of 5 rad/s and drives four smaller "floating" gears A, B, C, and D, which roll within stationary gear F. What is the kinetic energy of the system if gear E has a mass of 25 kg and each of the small gear has a mass of 5 kg?

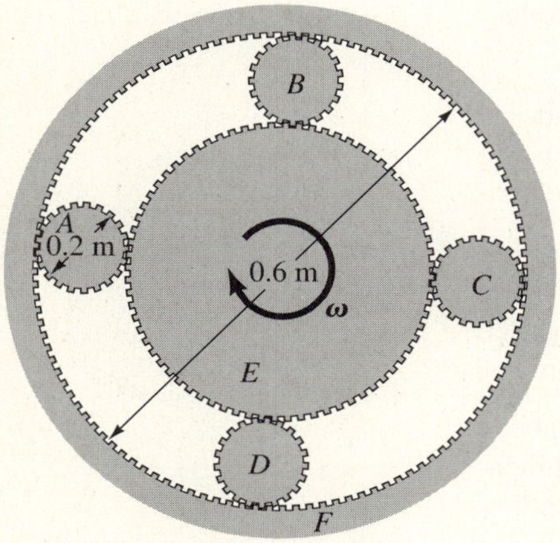

Figure P.17.80.

17.81. A homogeneous rectangular parallelepiped weighing 200 N rotates at 20 rad/s about a main diagonal held by bearings A and B, which are mounted on a vehicle moving at a speed 20 m/s. What is the kinetic energy of the rectangular parallelepiped relative to the ground?

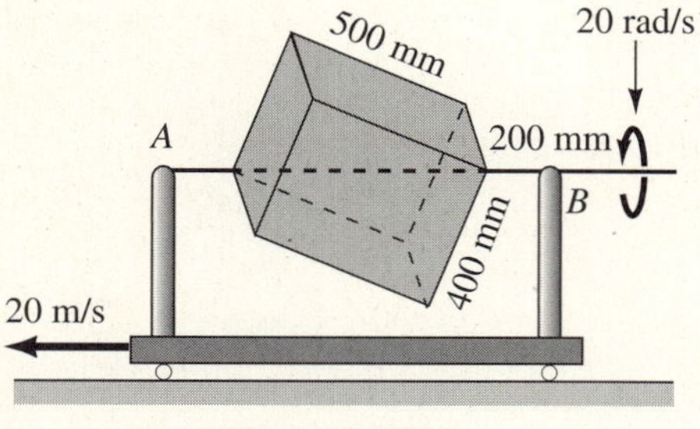

Figure P.17.81.

17.82. A rod AB is held at a position $\theta = 45°$ and released. HA is a light wire to which the rod is attached. When AB is vertical, it strikes a stop E and undergoes an elastic collision. What position d of the stop will result in a zero postimpact velocity of point A? The rod has a mass $M = 10$ kg.

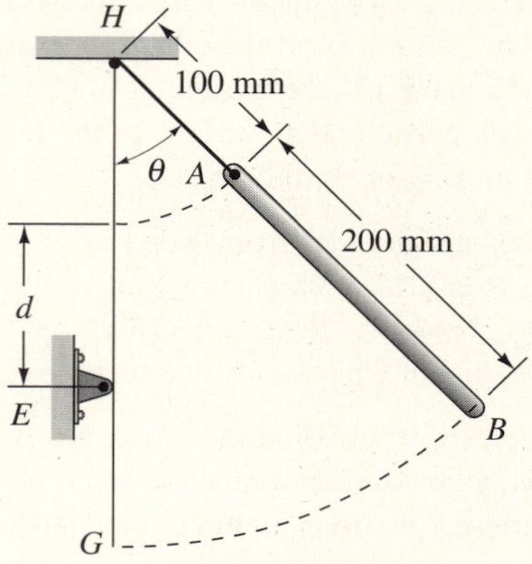

Figure P.17.82.

17.83. A stepped cylinder, mass 50 kg, has a radius of gyration of 1.2 m. A 25 kg block A is welded to the cylinder. If the spring is unstretched in the configuration shown and has a spring constant K of 0.1 N/mm, what is the angular speed of the cylinder after it rotates 90°? Assume that the cylinder rolls without slipping.

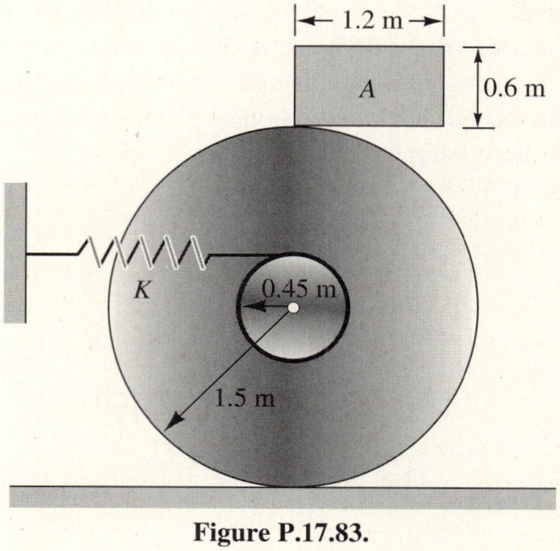

Figure P.17.83.

17.84. A tractor with driver has a mass of 1.4 Mg. If a torque of 280 N m is developed on each of the drive wheels by the motor, what is the speed of the tractor after it moves 3 m? The large drive wheels each have a mass of 90 kg, a diameter of 0.9 m and a radius of gyration of 0.3 m. The small wheels each have a mass of 18 kg and a diameter of 0.3 m with a radius of gyration of 0.12 m. Do not use the center of mass approach for the whole system.

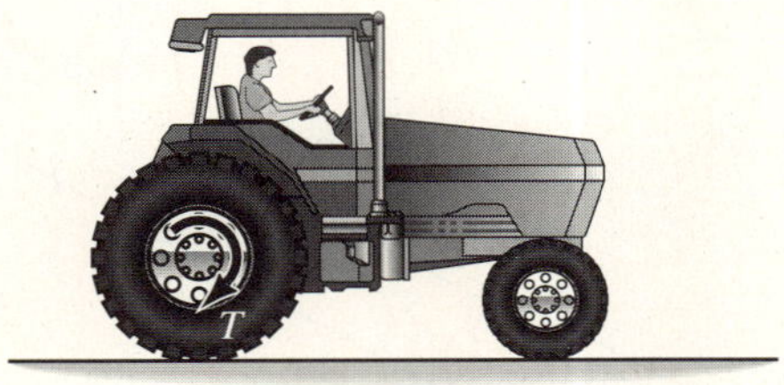

Figure P.17.84.

17.85. Work Problem 17.84 using the center of mass of the whole system. Find the friction forces on each wheel.

17.86. What is the angular momentum vector about the center of mass of a homogeneous rectangular parallelepiped rotating with an angular velocity of 10 rad/s about a main diagonal? The sides of the rectangular parallelepiped are 0.3 m, 0.6 m, and 1.2 m, as shown, and the mass is 1.8 Mg.

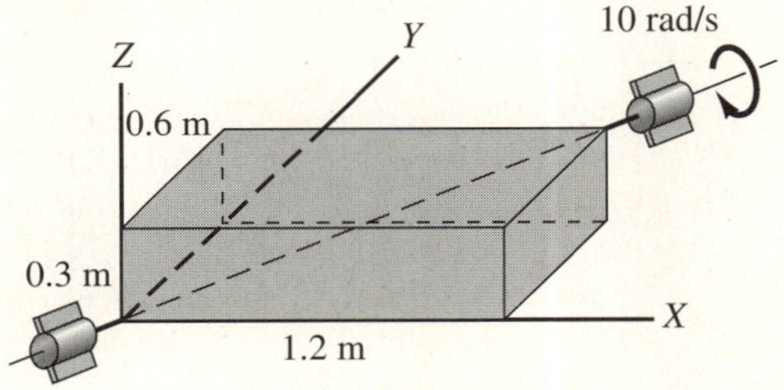

Figure P.17.86.

17.87. A force P of 200 N is applied to a cart. The cart minus the four wheels weighs 150 N. Each wheel has a weight of 50 N, a diameter of 400 mm, and a radius of gyration of 150 mm. The load A weighs 500 N. If the cart starts from rest, what is its speed in 20 s? The wheels roll without slipping.

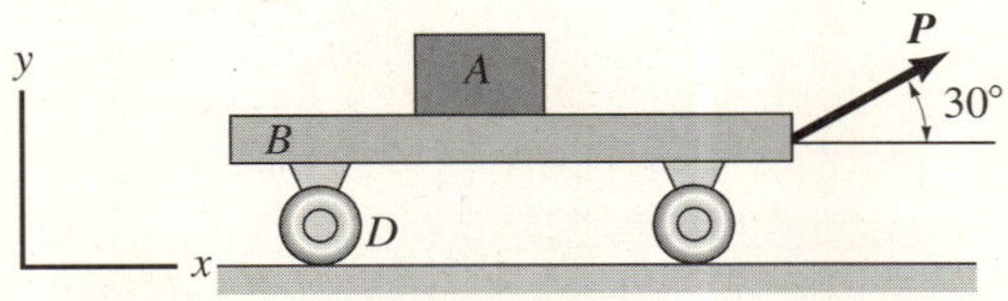

Figure P.17.87.

***17.88.** A rod weighing 90 N is guided by two slider bearings A and B. Smooth ball joints connect the rod to the bearings. A force F of 45 N acts on bearing A. What is the speed of A after it has moved 200 mm? The system is stationary for the configuration shown. Neglect friction.

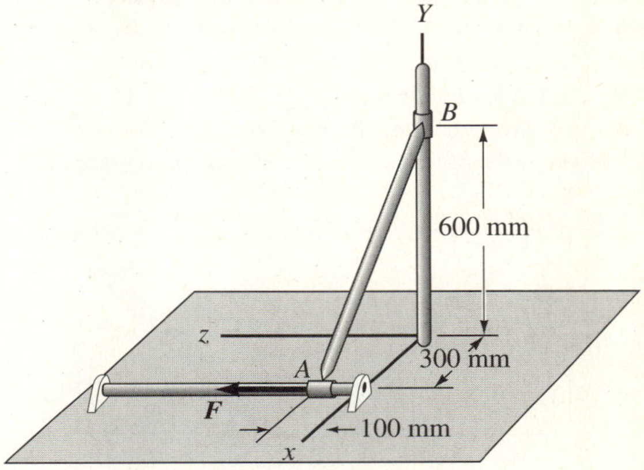

Figure P.17.88.

***17.89.** A device with thin walls contains water. The device is supported on a platform on which a torque T of 0.5 N m is applied for 2 s and is zero thereafter. What is the angular velocity ω of the system at the end of 2 s assuming that, as a result of low viscosity, the water has no rotation relative to the ground in the vertical tubes. The system of tubes and platform have a weight of 50 N and a radius of gyration of 200 mm. [*Hint:* The pressure in the liquid is equal at all times to the density times g times the distance d below the free surface of the liquid.] Note that for water $\rho = 1\ \text{Mg/m}^3$. Note: water heights in the tubes change with angular speed.

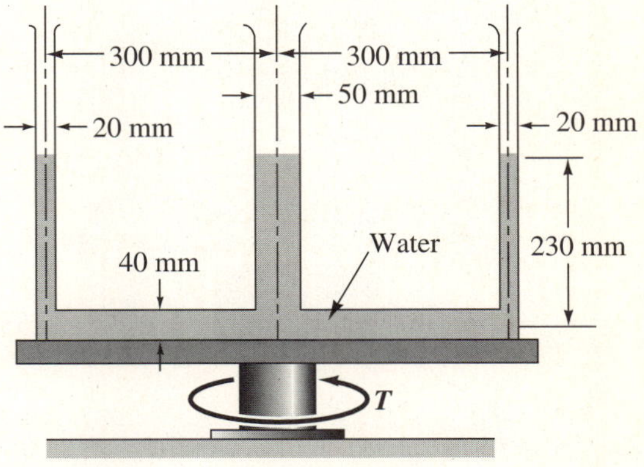

Figure P.17.89.

***17.90.** In Problem 17.89 what is the final angular speed of the system when viscosity has had its full effect and there is no movement of the water relative to the container walls? Disregard other frictional effects.

17.91. A bullet weighing 0.5 N is fired at a wooden block weighing 60 N. What is the angular speed of the block after the bullet has lodged in the block at the right end? Two light hollow rods support the block at the longitudinal midplane. The bullet has a velocity of 400 m/s in a direction along the longitudinal midplane of the block. [*Hint:* Where is the instantaneous axis of rotation postimpact?]

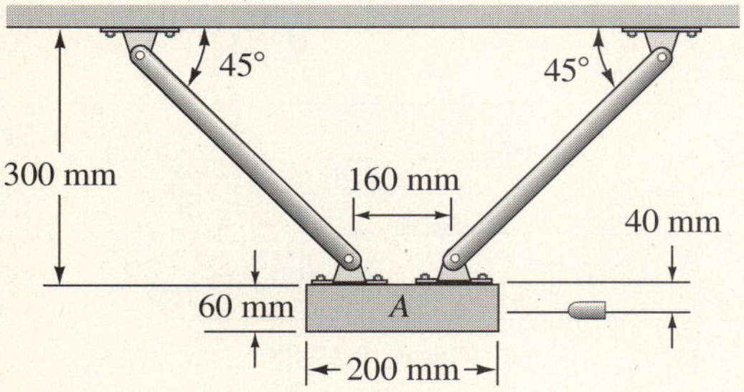

Figure P.17.91.

17.92. A device consists of two identical rectangular plates each weighing 100 N welded to a rod weighing 50 N. The device rests horizontally on a smooth surface. As a result of a collision, a horizontal impulse of 30 N s is delivered to point *D*. What is the velocity of corner *E* on plate *A* postimpact?

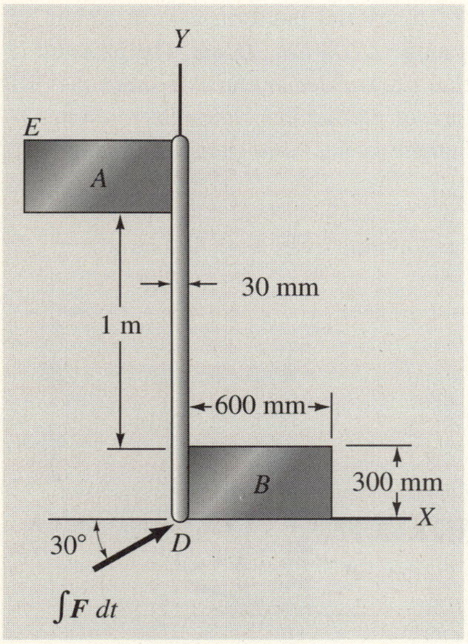

Figure P.17.92.

17.93 A cylinder *A* of weight 150 N and radius of gyration 100 mm is placed onto a conveyor belt which is moving at the constant speed of $V_B = 10$ m/s. Give the speed of the axis of the cylinder at time $t = 5$ s. The coefficient of friction between the cylinder and the belt is 0.5.

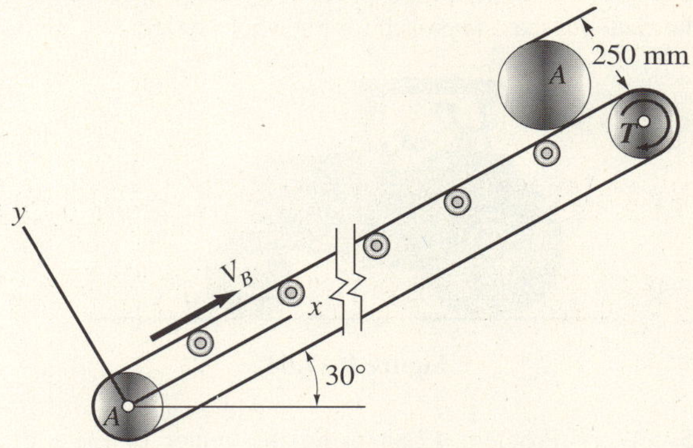

Figure P.17.93.

17.94 A rectangular rod *A* rests on a smooth horizontal surface. A disc *B* moves toward the rod at a speed of 10 m/s. What is the postimpact angular velocity of the rod for a coefficient of restitution of 0.6? The rod weighs 50 N, and the disc *B* weighs 8 N.

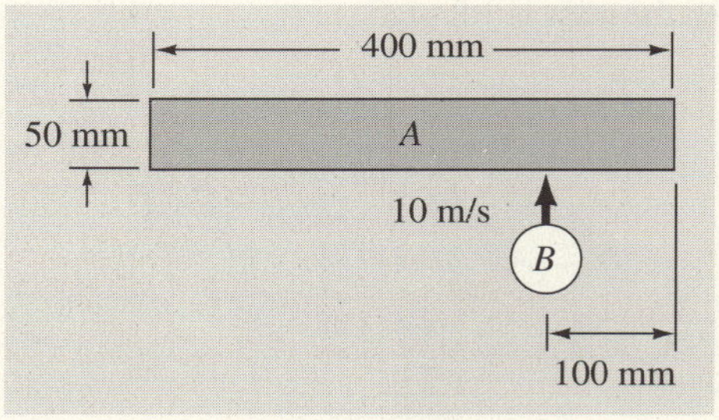

Figure P.17.94.

17.95. A block *B* of mass 5 kg is dropped onto the end of a rectangular rod *A* of mass 10 kg. What is the angular speed of rod *A* postimpact for $\epsilon = 0.7$? The rod is pinned at its center as shown.

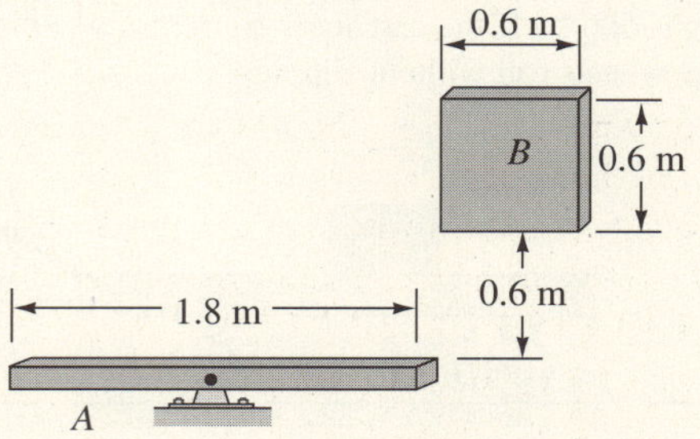

Figure P.17.95.

CHAPTER 18

Dynamics of General Rigid-Body Motion

18.1 Introduction[1]

Consider a rigid body moving arbitrarily relative to an inertial reference XYZ as shown in Fig. 18.1.

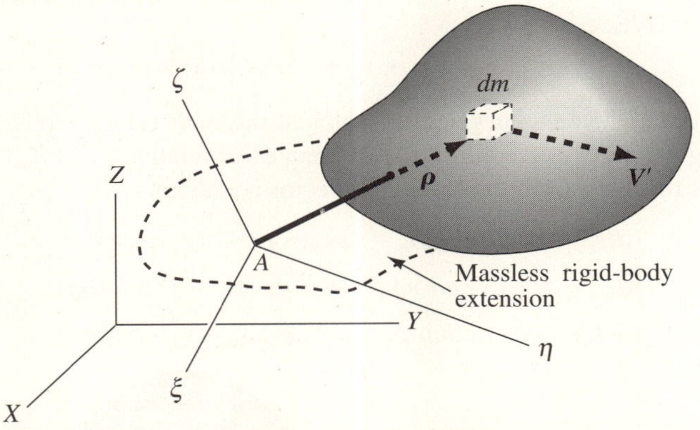

Figure 18.1. A rigid body undergoing arbitrary motion relative to inertial reference XYZ. Reference $\xi\eta\zeta$ translates with A.

Choose some point A in this body or in a hypothetical massless extension of this body. An element dm of the body is shown at a position $\boldsymbol{\rho}$ from A. The velocity $\mathbf{V}'$ of dm relative to A is simply the velocity of dm relative to a reference $\xi\eta\zeta$ which *translates* with A relative to XYZ. Similarly, the linear momentum of dm relative to A (i.e., $\mathbf{V}'\, dm$) is the linear momentum of dm relative to

[1]Until Eq. 18.1, we will be repeating the development of the formulations of the components of $\boldsymbol{H}_A$ first done in Section 16.2. This will eliminate the inconvenience of having to turn back at various times as we move along in Chapter 18.

$\xi\eta\zeta$ translating with A. We can now give the moment of this momentum (i.e., the angular momentum) $d\boldsymbol{H}_A$ about A as

$$d\boldsymbol{H}_A = \boldsymbol{\rho} \times \boldsymbol{V}'\,dm = \boldsymbol{\rho} \times \left(\frac{d\boldsymbol{\rho}}{dt}\right)_{\xi\eta\zeta} dm$$

but since A is fixed in the body (or in a hypothetical massless extension of the body) the vector $\boldsymbol{\rho}$ must be fixed in the body and, accordingly,

$$\left(\frac{d\boldsymbol{\rho}}{dt}\right)_{\xi\eta\zeta} = \boldsymbol{\omega} \times \boldsymbol{\rho}$$

where $\boldsymbol{\omega}$ is the angular velocity of the body relative to $\xi\eta\zeta$. However, since $\xi\eta\zeta$ translates relative to XYZ, $\boldsymbol{\omega}$ is the angular velocity of the body relative to XYZ as well. Hence, we can say:

$$d\boldsymbol{H}_A = \boldsymbol{\rho} \times (\boldsymbol{\omega} \times \boldsymbol{\rho})\,dm \tag{18.1}$$

Having helped us reach Eq. 18.1, we no longer need reference $\xi\eta\zeta$ translating with point A. Instead, we now insert reference xyz having its origin fixed at A but having at this time an *arbitrary* angular velocity $\boldsymbol{\Omega}$ relative to XYZ. We will have use of $\boldsymbol{\Omega}$ shortly.

Next, we evaluate $d\boldsymbol{H}_A$ in Eq. 18.1 in terms of components relative to xyz as follows:

$$(dH_A)_x\boldsymbol{i} + (dH_A)_y\boldsymbol{j} + (dH_A)_z\boldsymbol{k} = (x\boldsymbol{i} + y\boldsymbol{j} + z\boldsymbol{k}) \times \left[(\boldsymbol{\omega}) \times (x\boldsymbol{i} + y\boldsymbol{j} + z\boldsymbol{k})\right]dm$$

Note that $\boldsymbol{\omega}$ is the angular velocity vector of the *body* relative to XYZ and $\boldsymbol{\Omega}$ is the angular velocity vector of the reference xyz relative to XYZ. The scalar equations resulting from the foregoing vector equations are

$$\begin{aligned}(dH_A)_x &= (x^2\,dm)\omega_x - (xy\,dm)\omega_y - (xz\,dm)\omega_z\\(dH_A)_y &= -(yx\,dm)\omega_x + (y^2\,dm)\omega_y - (yz\,dm)\omega_z\\(dH_A)_z &= -(zx\,dm)\omega_x - (zy\,dm)\omega_y + (z^2\,dm)\omega_z\end{aligned}$$

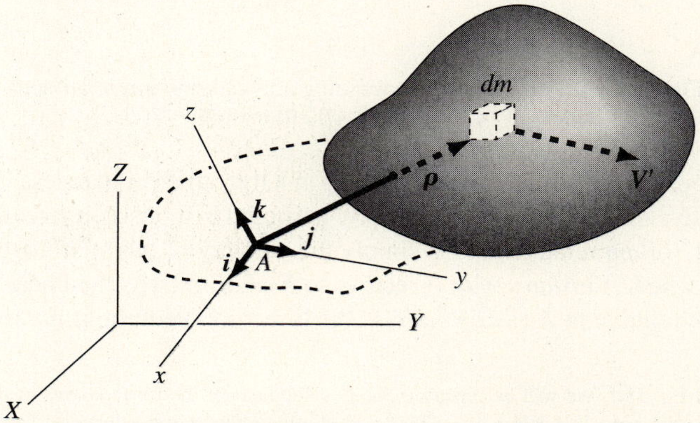

Figure 18.2. xyz fixed to body at A.

We next integrate the above equations over the domain of the rigid body deleting of course the massless hypothetical extension. We see on noting the definitions of the mass moments and products of inertia that we get the components of $\boldsymbol{H}_A$ along an arbitrary set of axes xyz at A. We thus have at any time t:

$$\begin{aligned}(H_A)_x &= I_{xx}\omega_x - I_{xy}\omega_y - I_{xz}\omega_z \\ (H_A)_y &= -I_{yx}\omega_x + I_{yy}\omega_y - I_{yz}\omega_z \\ (H_A)_z &= -I_{zx}\omega_x - I_{zy}\omega_y + I_{zz}\omega_z\end{aligned} \tag{18.2}$$

where we repeat $\boldsymbol{\omega}$ is the angular velocity of the body relative to XYZ at time t. Now the axes xyz served *only* to give a set of directions for $\boldsymbol{H}_A$ at time t. The reference xyz remember could have any angular velocity $\boldsymbol{\Omega}$ relative to XYZ.

Note that the angular velocity $\boldsymbol{\Omega}$ did not enter the formulations for $\boldsymbol{H}_A$ at time t. The reason for this result is that the values of the components of $\boldsymbol{H}_A$ along xyz at time t do not in any way depend on angular velocity $\boldsymbol{\Omega}$—they depend only on the instantaneous orientation of xyz at time t. To illustrate this point, suppose that we have two sets of axes xyz and $x'y'z'$ at point A in Fig. 18.3. At time t they coincide as has been shown in the diagram, but xyz has zero angular velocity relative to XYZ, whereas $x'y'z'$ has an angular velocity $\boldsymbol{\Omega}$ relative to XYZ. Clearly, one can say at time t:

$$\begin{aligned}(H_A)_x &= (H_A)_{x'} \\ (H_A)_y &= (H_A)_{y'} \\ (H_A)_z &= (H_A)_{z'}\end{aligned}$$

However, *the time derivatives* of the corresponding components will not be equal to each other at time t. Accordingly, in the next section, where $\dot{\boldsymbol{H}}_A$ is treated, we must properly account for $\boldsymbol{\Omega}$.

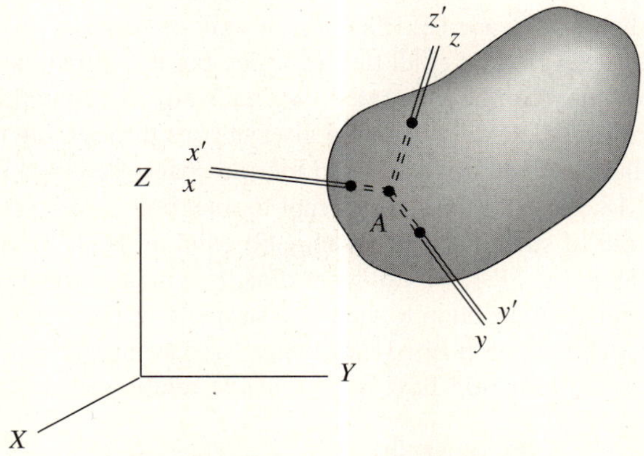

Figure 18.3. xyz and $x'y'z'$ coincide at time t.

18.2 Euler's Equations of Motion

We shall now restrict point A of Section 18.1 further, as we did in Chapter 17, by considering only those points for which the equation $\boldsymbol{M}_A = \dot{\boldsymbol{H}}_A$ is valid:

1. The mass center.
2. Points fixed or moving with constant $\boldsymbol{V}$ at time t in inertial space (i.e., points having zero acceleration at time t relative to inertial reference XYZ).
3. A point accelerating toward or away from the mass center.

We learned in Chapter 15 that derivatives of a vector as seen from different references could be related as follows:

$$\left(\frac{d\boldsymbol{A}}{dt}\right)_{XYZ} = \left(\frac{d\boldsymbol{A}}{dt}\right)_{xyz} + \boldsymbol{\omega} \times \boldsymbol{A}$$

where $\boldsymbol{\omega}$ is the angular velocity of xyz relative to XYZ. We shall employ this equation in the basic *moment-of-momentum equation* to shift the observation reference from XYZ to xyz in the following way where now $\boldsymbol{\Omega}$ represents the angular velocity of xyz. Thus

$$\boldsymbol{M}_A = \left(\frac{d\boldsymbol{H}_A}{dt}\right)_{XYZ} = \left(\frac{d\boldsymbol{H}_A}{dt}\right)_{xyz} + \boldsymbol{\Omega} \times \boldsymbol{H}_A \qquad (18.3)$$

The idea now is to choose the angular velocity $\boldsymbol{\Omega}$ of reference xyz at A in such a way that $(d\boldsymbol{H}_A/dt)_{XYZ}$ is most easily evaluated. With this accomplished, the next step is to attempt the integration of the resulting differential equation.

With regard to attempts at integration, we point out at this early stage that Eq. 18.3 is valid only as long as point A is one of the three qualified points just discussed. Clearly, if A is the mass center, then Eq. 18.3 is valid at all times and can be integrated with respect to time provided that the mathematics are not too difficult. However, if for cases (2) and (3) point A qualifies only at time t, then Eq. 18.3 is valid only at time t and accordingly cannot be integrated. If, on the other hand, for case (2), there is an axis of rotation fixed in inertial space, then Eq. 18.3 is valid at all times for any point A along the axis of rotation and accordingly can be integrated. We have already done this in Chapter 16. If, furthermore, the axis of rotation always goes through the fixed point A but does not have a fixed orientation in inertial space (see Fig. 18.4), we can again use Eq. 18.3 at all times and attempt to integrate it with respect to time. The carrying out of such integrations may be quite difficult, however.[2]

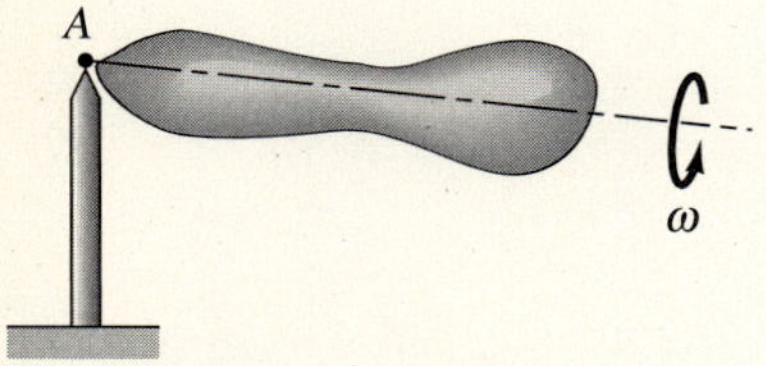

Figure 18.4. Axis of rotation goes through fixed point A.

Returning to Eq. 18.3, we can work directly with this equation selecting a reference xyz for each problem to yield the simplest working equation. On the other hand, we can develop Eq. 18.3 further for certain classes of references xyz. For example, we could have xyz translate relative to XYZ. This would

[2]In a later section we shall examine this case in some detail.

mean that $\boldsymbol{\Omega} = \mathbf{0}$ so that Eq. 18.3 would seem to be more simple for such cases.[3] However, the body will be rotating relative to xyz and the moments and products of inertia measured about xyz will then be time functions. Since the computation of these terms as time functions is generally difficult, such an approach has limited value. On the other hand, the procedure of *fixing xyz* in the body (as we did for the case of plane motion in Chapter 16) does lead to very useful forms of Eq. 18.3, and we shall accordingly examine these equations with great care. Note first that the moments and products of inertia will be constants for this case and that $\boldsymbol{\Omega} = \boldsymbol{\omega}$. Hence, we have

$$\boldsymbol{M}_A = \left(\frac{d\boldsymbol{H}_A}{dt}\right)_{xyz} + \boldsymbol{\omega} \times \boldsymbol{H}_A \tag{18.4}$$

Employing components for all vectors along axes xyz and utilizing Eq. 18.2, we get on dropping the subscript A:

$$\begin{aligned} M_x\boldsymbol{i} + M_y\boldsymbol{j} + M_z\boldsymbol{k} = \frac{d}{dt}_{xyz} &[(\omega_x I_{xx} - \omega_y I_{xy} - \omega_z I_{xz})\boldsymbol{i} \\ &+(-\omega_x I_{yx} + \omega_y I_{yy} - \omega_z I_{yz})\boldsymbol{j} + (-\omega_x I_{zx} - \omega_y I_{zy} + \omega_z I_{zz})\boldsymbol{k}] \\ &+\boldsymbol{\omega} \times [(\omega_x I_{xx} - \omega_y I_{xy} - \omega_z I_{xz})\boldsymbol{i} \\ &+(-\omega_x I_{yx} + \omega_y I_{yy} - \omega_z I_{yz})\boldsymbol{j} + (-\omega_x I_{zx} - \omega_y I_{zy} + \omega_z I_{zz})\boldsymbol{k}] \end{aligned}$$

In the first expression on the right side of the foregoing equation, the vectors $\boldsymbol{i}, \boldsymbol{j}$, and $\boldsymbol{k}$ are constant vectors as seen from xyz. Also, because xyz is fixed in the body, the moments and products of inertia are constant. Only ω_x, ω_y, and ω_z, the angular velocity components of the rigid body, are time functions. We then can say:

$$\begin{aligned} M_x\boldsymbol{i} + M_y\boldsymbol{j} + M_z\boldsymbol{k} &= (\dot{\omega}_x I_{xx} - \dot{\omega}_y I_{xy} - \dot{\omega}_z I_{xz})\boldsymbol{i} + (-\dot{\omega}_x I_{yx} + \dot{\omega}_y I_{yy} - \dot{\omega}_z I_{yz})\boldsymbol{j} \\ &+(-\dot{\omega}_x I_{zx} - \dot{\omega}_y I_{zy} + \dot{\omega}_z I_{zz})\boldsymbol{k} + (\omega_x I_{xx} - \omega_y I_{xy} - \omega_z I_{xz})(\boldsymbol{\omega} \times \boldsymbol{i}) \\ &+(-\omega_x I_{yx} + \omega_y I_{yy} - \omega_z I_{yz})(\boldsymbol{\omega} \times \boldsymbol{j}) + (-\omega_x I_{zx} - \omega_y I_{zy} + \omega_z I_{zz})(\boldsymbol{\omega} \times \boldsymbol{k}) \end{aligned}$$

Carrying out the cross products, collecting terms, and expressing as scalar equations, we get

$$\begin{aligned} M_x &= \dot{\omega}_x I_{xx} + \omega_y\omega_z(I_{zz} - I_{yy}) + I_{xy}(\omega_z\omega_x - \dot{\omega}_y) \\ &\quad - I_{xz}(\dot{\omega}_z + \omega_y\omega_x) - I_{yz}(\omega_y^2 - \omega_z^2) \end{aligned} \tag{18.5a}$$

$$\begin{aligned} M_y &= \dot{\omega}_y I_{yy} + \omega_z\omega_x(I_{xx} - I_{zz}) + I_{yz}(\omega_x\omega_y - \dot{\omega}_z) \\ &\quad - I_{yx}(\dot{\omega}_x + \omega_z\omega_y) - I_{zx}(\omega_z^2 - \omega_x^2) \end{aligned} \tag{18.5b}$$

$$\begin{aligned} M_z &= \dot{\omega}_z I_{zz} + \omega_x\omega_y(I_{yy} - I_{xx}) + I_{zx}(\omega_y\omega_z - \dot{\omega}_x) \\ &\quad - I_{zy}(\dot{\omega}_y + \omega_x\omega_z) - I_{xy}(\omega_x^2 - \omega_y^2) \end{aligned} \tag{18.5c}$$

[3]We shall later find it advantageous not to fix xyz to the body for certain problems.

This is indeed a formidable set of equations. For the important special case of *plane motion* the z axis is always normal to the XY plane which is the plane of motion. Hence $\boldsymbol{\omega}$ must be normal to plane XY and thus collinear with the z axis. Thus replacing ω_z by ω and setting the other components equal to zero, we get the *moment of momentum equations* for *general plane motion.* Thus we have

$$\begin{aligned}(M_A)_x &= -I_{xz}\dot{\omega} + I_{yz}\omega^2 \\ (M_A)_y &= -I_{yz}\dot{\omega} - I_{xz}\omega^2 \\ (M_A)_z &= I_{zz}\dot{\omega}\end{aligned}$$

The reader may now or at any time later go back to Chapter 16 for a rather careful study of the use of the above plane motion *moment of momentum* equations. We now continue with the three-dimensional approach.

Note now that if we choose reference xyz to be *principal* axes of the body at point A, then it is clear that the products of inertia are all zero in the system of equations 18.5, and this fact enables us to simplify the equations considerably. The resulting equations given below are the famous *Euler equations* of motion. Note that these equations relate the angular velocity and the angular acceleration to the moment of the external forces about the point A.

$$M_x = I_{xx}\dot{\omega}_x + \omega_y\omega_z(I_{zz} - I_{yy}) \qquad (18.6a)$$
$$M_y = I_{yy}\dot{\omega}_y + \omega_z\omega_x(I_{xx} - I_{zz}) \qquad (18.6b)$$
$$M_z = I_{zz}\dot{\omega}_z + \omega_x\omega_y(I_{yy} - I_{xx}) \qquad (18.6c)$$

In both sets of Eqs. 18.5 and 18.6, we have three simultaneous first-order differential equations. If the motion of the body about point A is known, we can easily compute the required moments about point A. On the other hand, if the moments are known functions of time and the angular velocity is desired, we have the difficult problem of solving simultaneous nonlinear differential equations for the unknowns ω_x, ω_y, and ω_z. However, in practical problems, we often know some of the angular velocity and acceleration components from constraints or given data, so, with the restrictions mentioned earlier, we can sometimes integrate the equations readily as we did in Chapter 16 for plane motion. At other times we use them to solve for certain desired *instantaneous values* of the unknowns.

We shall now discuss the use of Euler's equations.

18.3 Application of Euler's Equations

In this section, we shall apply the **Euler equations** to a number of problems. Before taking up these problems, let us first carefully consider how to express the components $\dot{\omega}_x$, $\dot{\omega}_y$, and $\dot{\omega}_z$ for use in **Euler's equations**. First note that $\dot{\omega}_x$, $\dot{\omega}_y$, and $\dot{\omega}_z$ are time derivatives as seen from XYZ of the components of $\boldsymbol{\omega}$ along reference xyz. A possible procedure then is to express ω_x, ω_y, and ω_z first in a way that ensures that these quantities are correctly stated over a time interval rather than at some instantaneous configuration. Once this is done, we can simply differentiate these scalar quantities with respect to time in this interval to get $\dot{\omega}_x, \dot{\omega}_y$, and $\dot{\omega}_z$.

To illustrate this procedure, consider in Fig. 18.5 the case of a block E rotating about rod AB, which in turn rotates about vertical axis CD. A reference xyz is fixed to the block at the center of mass so as to coincide with the principal axes of the block at the center of mass. When the block is vertical as shown, the angular speed and rate of change of angular speed relative to rod AB have the known values $(\omega_2)_0$ and $(\dot{\omega}_2)_0$, respectively. At that instant, AB has an angular speed and rate of change of angular speed about axis CD of known values $(\omega_1)_0$ and $(\dot{\omega}_1)_0$, respectively. We can immediately give the angular velocity components at the instant shown as follows:

$$\omega_x = 0$$
$$\omega_y = (\omega_2)_0$$
$$\omega_z = (\omega_1)_0$$

But to get the quantities $(\dot{\omega}_x)_0$, $(\dot{\omega}_y)_0$, and $(\dot{\omega}_z)_0$ for the instant of interest[4] we must first express ω_x, ω_y, and ω_z as *general functions of time* in order to permit differentiation with respect to time.

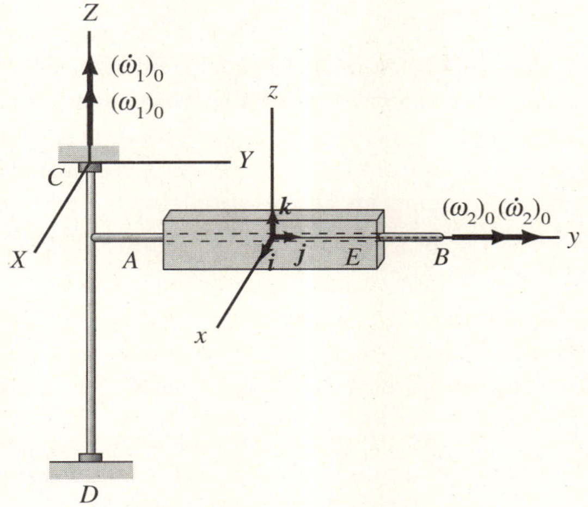

Figure 18.5. Block at time t_0.

To do this differentiating, we have shown the system at some arbitrary time in Fig. 18.6. Note that the x axis is at some angle β from the horizontal. When β becomes zero, we arrive back at the configuration of interest, and ω_1,

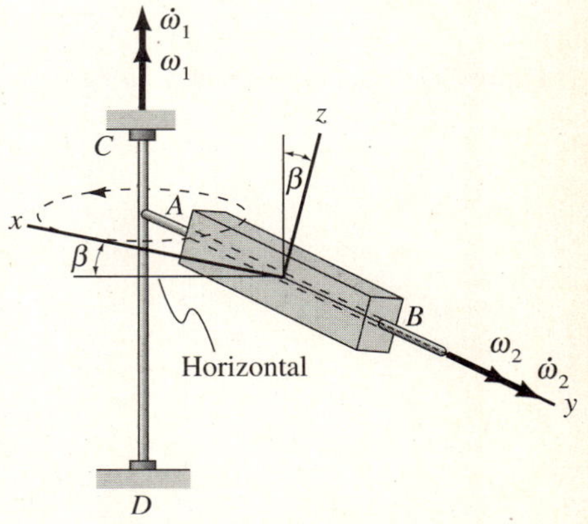

Figure 18.6. Block at time t.

[4]You may be tempted to say $(\dot{\omega}_x)_0 = 0$, $(\dot{\omega}_y)_0 = (\dot{\omega}_2)_0$, and $(\dot{\omega}_z)_0 = (\dot{\omega}_1)_0$ by just inspecting the diagram. You will now learn that this will not be correct.

$\dot{\omega}_1$, ω_2, and $\dot{\omega}_2$ become known values $(\omega_1)_0$, $(\dot{\omega}_1)_0$, $(\omega_2)_0$, and $(\dot{\omega}_2)_0$, respectively. The angular velocity components for this arbitrary situation are:

$$\omega_x = \omega_1 \sin\beta, \qquad \omega_y = \omega_2, \qquad \omega_z = \omega_1 \cos\beta \tag{18.7}$$

Since these relations are valid over the time interval of interest, we can differentiate them with respect to time and get

$$\begin{aligned} \dot{\omega}_x &= \dot{\omega}_1 \sin\beta + \omega_1 \cos\beta\dot{\beta} \\ \dot{\omega}_y &= \dot{\omega}_2 \\ \dot{\omega}_z &= \dot{\omega}_1 \cos\beta - \omega_1 \sin\beta\dot{\beta} \end{aligned} \tag{18.8}$$

It should be clear upon inspecting the diagram that $\dot{\beta} = -\omega_2$, and so the preceding terms become

$$\begin{aligned} \dot{\omega}_x &= \dot{\omega}_1 \sin\beta - \omega_1\omega_2 \cos\beta \\ \dot{\omega}_y &= \dot{\omega}_2 \\ \dot{\omega}_z &= \dot{\omega}_1 \cos\beta + \omega_1\omega_2 \sin\beta \end{aligned} \tag{18.9}$$

If we now let β become zero, we reach the configuration of interest and we get from Eqs. 18.7 and 18.8 the proper values of the angular velocity components and their time derivatives at this configuration:

$$\begin{aligned} \omega_x &= 0, & \dot{\omega}_x &= -(\omega_1)_0(\omega_2)_0 \\ \omega_y &= (\omega_2)_0, & \dot{\omega}_y &= (\dot{\omega}_2)_0 \\ \omega_z &= (\omega_1)_0, & \dot{\omega}_z &= (\dot{\omega}_1)_0 \end{aligned}$$

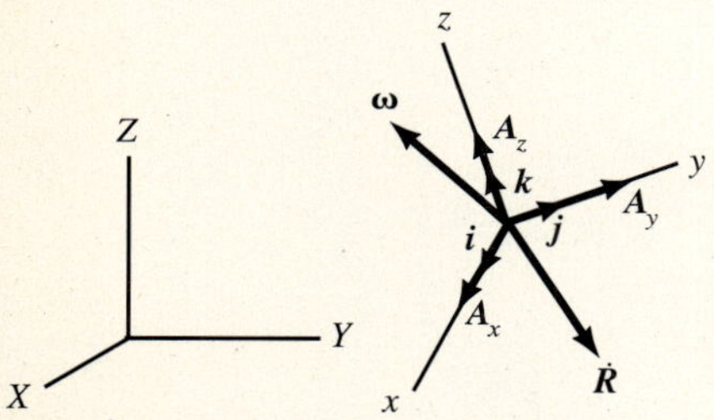

Figure 18.7. Components of $\boldsymbol{A}$ along xyz.

Actually, we do not have to employ such a procedure for the evaluation of these quantities. There is a simple direct approach that can be used, but we must preface the discussion of this method by some general remarks about the time derivative, as seen from the XYZ axes, of a vector $\boldsymbol{A}$. This vector $\boldsymbol{A}$ is expressed in terms of components always parallel to the xyz reference, which moves relative to XYZ (Fig. 18.7). We can then say, considering $\boldsymbol{i}, \boldsymbol{j}$, and $\boldsymbol{k}$ as unit vectors for reference xyz:

$$\begin{aligned} \left(\frac{d\boldsymbol{A}}{dt}\right)_{XYZ} &= \frac{d}{dt_{XYZ}}(A_x\boldsymbol{i} + A_y\boldsymbol{j} + A_z\boldsymbol{k}) \\ &= \dot{A}_x\boldsymbol{i} + \dot{A}_y\boldsymbol{j} + \dot{A}_z\boldsymbol{k} + A_x(\boldsymbol{\omega}\times\boldsymbol{i}) + A_y(\boldsymbol{\omega}\times\boldsymbol{j}) + A_z(\boldsymbol{\omega}\times\boldsymbol{k}) \end{aligned} \tag{18.10}$$

If we decompose the vector $(d\boldsymbol{A}/dt)_{XYZ}$ into components parallel to the xyz axes at time t and carry out the cross products on the right side in terms of xyz components, then we get, after collecting terms and expressing the results as scalar equations:

$$\left[\left(\frac{d\boldsymbol{A}}{dt}\right)_{XYZ}\right]_x = \dot{A}_x + A_z\omega_y - A_y\omega_z \tag{18.11a}$$

$$\left[\left(\frac{d\boldsymbol{A}}{dt}\right)_{XYZ}\right]_y = \dot{A}_y + A_x\omega_z - A_z\omega_x \tag{18.11b}$$

$$\left[\left(\frac{d\boldsymbol{A}}{dt}\right)_{XYZ}\right]_z = \dot{A}_z + A_y\omega_x - A_x\omega_y \tag{18.11c}$$

We can learn an important lesson from these equations. If you take the time derivative of a vector $\mathbf{A}$ with respect to a reference XYZ and express the *components* of this vector parallel to the axes of a reference xyz rotating relative to XYZ (these are the terms on the left side of the above equations), then the results are in general *not the same* as *first* taking the components of the vector $\mathbf{A}$ along the directions xyz and *then* taking time derivatives of these scalars. Thus,

$$\left[\left(\frac{d\mathbf{A}}{dt}\right)_{XYZ}\right]_x \neq \frac{d(A_x)}{dt} = \dot{A}_x \qquad \text{etc.} \tag{18.12}$$

How does this result relate to our problem where we are considering $\dot{\omega}_x$, $\dot{\omega}_y$, and $\dot{\omega}_z$? Clearly, these expressions are time derivatives of the components of the vector $\boldsymbol{\omega}$ along the moving xyz axes, and so in this respect they correspond to the terms on the right side of inequality 18.12. Let us then consider vector $\mathbf{A}$ to be $\boldsymbol{\omega}$ and examine Eq. 18.11:

$$\begin{aligned}
\left[\left(\frac{d\boldsymbol{\omega}}{dt}\right)_{XYZ}\right]_x &= \dot{\omega}_x + \omega_z\omega_y - \omega_y\omega_z \\
\left[\left(\frac{d\boldsymbol{\omega}}{dt}\right)_{XYZ}\right]_y &= \dot{\omega}_y + \omega_x\omega_z - \omega_z\omega_x \\
\left[\left(\frac{d\boldsymbol{\omega}}{dt}\right)_{XYZ}\right]_z &= \dot{\omega}_z + \omega_y\omega_x - \omega_x\omega_y
\end{aligned} \tag{18.13}$$

We see that the last two terms on the right side in each equation cancel for this case, leaving us

$$\begin{aligned}
\left[\left(\frac{d\boldsymbol{\omega}}{dt}\right)_{XYZ}\right]_x &= \dot{\omega}_x \\
\left[\left(\frac{d\boldsymbol{\omega}}{dt}\right)_{XYZ}\right]_y &= \dot{\omega}_y \\
\left[\left(\frac{d\boldsymbol{\omega}}{dt}\right)_{XYZ}\right]_z &= \dot{\omega}_z
\end{aligned} \tag{18.14}$$

We see that for the vector $\boldsymbol{\omega}$ (i.e., the angular velocity of the xyz reference relative to the XYZ reference), we have an exception to the rule stated earlier (Eq. 18.12). Here is the one case where the derivative of a vector as seen from one set of axes XYZ has components along the directions of another set of axes xyz rotating relative to XYZ, wherein these components are respectively equal to the simple time derivatives of the scalar components of the vector along the xyz directions. In other words, you can take the derivative of $\boldsymbol{\omega}$ first from the XYZ axes and then take scalar components along xyz, or you can take scalar components along xyz first and then take simple time derivatives of the scalar components, and the results are the same.

If we fully understand the exceptional nature of Eq. 18.14, we can compute $\dot{\omega}_x$, $\dot{\omega}_y$, and $\dot{\omega}_z$ in a straightforward manner by simply first determining

$(d\boldsymbol{\omega}/dt)_{XYZ}$ and then taking the components. This is a step that we have practiced a great deal in *kinematics*. For instance, for the problem introduced at the outset of this discussion, we see by inspecting Fig. 18.6 that at all times

$$\boldsymbol{\omega} = \omega_2 \boldsymbol{j} + \omega_1 \boldsymbol{k}_1 \tag{18.15}$$

where $\boldsymbol{k}_1$ is the unit vector in the fixed Z direction. Now differentiate with respect to time for the XYZ reference:

$$\dot{\boldsymbol{\omega}} = \dot{\omega}_2 \boldsymbol{j} + \omega_2 \dot{\boldsymbol{j}} + \dot{\omega}_1 \boldsymbol{k}_1 \tag{18.16}$$

But $\boldsymbol{j}$ is fixed in rod AB that is rotating with angular velocity $\omega_1 \boldsymbol{k}_1$ relative to XYZ. We then get

$$\dot{\boldsymbol{\omega}} = \dot{\omega}_2 \boldsymbol{j} + \omega_2 (\omega_1 \boldsymbol{k}_1) \times \boldsymbol{j} + \dot{\omega}_1 \boldsymbol{k}_1 \tag{18.17}$$

When the xyz axes are parallel to the XYZ axes, the unit vector $\boldsymbol{k}$ becomes the same as the unit vector $\boldsymbol{k}_1$, and ω_1, ω_2, etc., become known $(\omega_1)_0$, $(\omega_2)_0$, etc. We then get for that configuration:

$$\dot{\boldsymbol{\omega}}_0 = (\dot{\omega}_2)_0 \boldsymbol{j} - (\omega_1)_0 (\omega_2)_0 \boldsymbol{i} + (\dot{\omega}_1)_0 \boldsymbol{k} \tag{18.18}$$

The components of this equation give the desired values of $\dot{\omega}_x$, $\dot{\omega}_y$, and $\dot{\omega}_z$, at the instant of interest. Thus, we have

$$\begin{aligned} \dot{\omega}_x &= -(\omega_1)_0 (\omega_2)_0 \\ \dot{\omega}_y &= (\dot{\omega}_2)_0 \\ \dot{\omega}_z &= (\dot{\omega}_1)_0 \end{aligned} \tag{18.19}$$

which are the same results developed earlier.

In most of the following examples we shall proceed by the second method discussed here. That is, we shall get $\dot{\omega}_x$, $\dot{\omega}_y$, and $\dot{\omega}_z$, using the following formulations:

$$\dot{\omega}_x = \left[\left(\frac{d\boldsymbol{\omega}}{dt}\right)_{XYZ}\right]_x \tag{18.20a}$$

$$\dot{\omega}_y = \left[\left(\frac{d\boldsymbol{\omega}}{dt}\right)_{XYZ}\right]_y \tag{18.20b}$$

$$\dot{\omega}_z = \left[\left(\frac{d\boldsymbol{\omega}}{dt}\right)_{XYZ}\right]_z \tag{18.20c}$$

Example 18.1

In Fig. 18.8 a thin disc of radius $R = 1.2$ m and mass 150 kg rotates at an angular speed ω_2 of 100 rad/s relative to a platform. The platform rotates with an angular speed ω_1 of 20 rad/s relative to the ground. Compute the bearing reactions at A and B. Neglect the mass of the shaft and assume that bearing A restrains the system in the radial direction.

Clearly, we shall need to use **Euler's equations** as part of the solution to this problem, and so we *fix a reference xyz to the center of mass* of the disc as shown in Fig. 18.8. *XYZ is fixed to the ground*. In using **Euler's equations**, the key step is to get the angular velocity components and their time derivatives for the body as seen from *XYZ*. Accordingly, we have

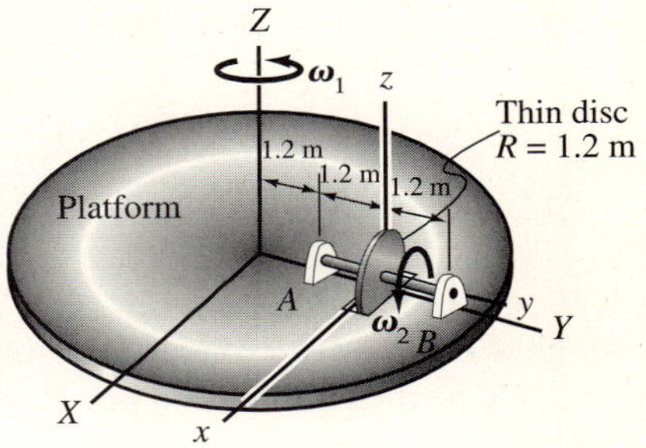

Figure 18.8. Rotating disc on platform.

$$\boldsymbol{\omega} = \boldsymbol{\omega}_1 + \boldsymbol{\omega}_2 = -20\boldsymbol{k} + 100\boldsymbol{j} \text{ rad/s}$$

Hence, we have for the *xyz* components:

$$\begin{aligned} \omega_x &= 0 \\ \omega_y &= 100 \text{ rad/s} \\ \omega_z &= -20 \text{ rad/s} \end{aligned}$$

Next, we have

$$\begin{aligned} \dot{\boldsymbol{\omega}} &= \dot{\boldsymbol{\omega}}_1 + \dot{\boldsymbol{\omega}}_2 = \boldsymbol{0} + \boldsymbol{\omega}_1 \times \boldsymbol{\omega}_2 \\ &= (-20\boldsymbol{k}) \times (100\boldsymbol{j}) = 2000\boldsymbol{i} \text{ rad/s}^2 \end{aligned}$$

We thus have for the *xyz* components:

$$\begin{aligned} \dot{\omega}_x &= 2000 \text{ rad/s}^2 \\ \dot{\omega}_y &= 0 \\ \dot{\omega}_z &= 0 \end{aligned}$$

Before going to the **Euler's equations**, we shall need the principal moments of inertia of the disc. Using formulas from the front inside covers, we get:

$$\begin{aligned} I_{yy} &= \frac{1}{2}MR^2 = \frac{1}{2}(150)(1.2^2) = 108 \text{ kg m}^2 \\ I_{xx} &= I_{zz} = \frac{1}{4}MR^2 = 54 \text{ kg m}^2 \end{aligned}$$

We can now substitute into the **Euler's equations**.

$$\begin{aligned} M_x &= (54)(2000) + (100)(-20)(54 - 108) = 216 \text{ kN m} \\ M_y &= (108)(0) + (0)(-20)(54 - 54) = 0 \\ M_z &= (54)(0) + (0)(100)(108 - 54) = 0 \end{aligned}$$

Example 18.1 (Continued)

Now the moment components above are generated by the bearing-force components (see Fig. 18.9). Hence, we can say:

$$M_x = 216\times 10^3 = 1.2B_z - 1.2A_z \quad \text{(a)}$$
$$M_y = 0 \quad \text{(b)}$$
$$M_z = 0 = 1.2A_x - 1.2B_x \quad \text{(c)}$$

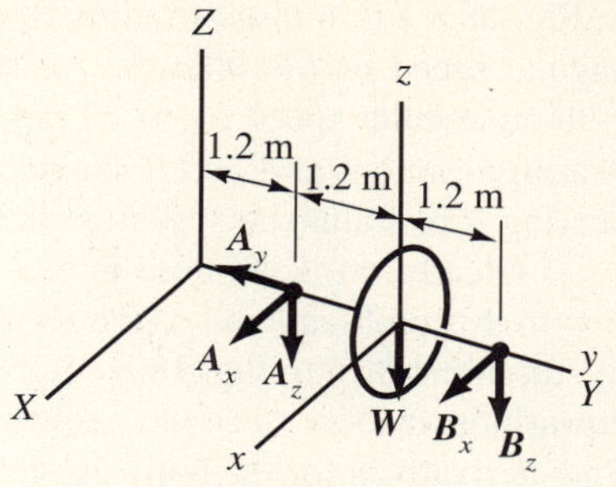

Figure 18.9. Bearing forces.

We have effectively two equations for four unknowns.

We next use **Newton's law** for the mass center:

$$\begin{aligned} A_z + B_z - 150g &= 0 \\ A_x + B_x &= 0 \\ -A_y &= -(150)(2.4)(20)^2 \end{aligned} \quad \text{(d)}$$

The first two equations are equilibrium equations. The third equation relates the radial force A_y, from the bearing A, and the radial acceleration of the center of mass of the disc which you will notice is in simple circular motion about the Z axis. We now have enough equations for all the unknowns. It is then a simple matter to evaluate the forces from the bearings. They are:

$$\begin{aligned} A_x &= B_x = 0 \\ A_y &= 144 \text{ kN} \\ B_z &= 90.74 \text{ kN} \\ A_z &= -89.27 \text{ kN} \end{aligned}$$

The *reactions* to these forces are the desired forces *onto* the bearings.

In Example 18.1, you may have been surprised at the large value of the bearing forces in the z direction. Actually, if we did not include the weight of the disc, then A_z and B_z would have formed a sizable couple. This couple stems from the fact that a body having a high angular momentum about one axis is made to move such that the aforementioned axis rotates about yet a second axis. Such a couple is called a *gyroscopic couple*. It occurs in no small measure for the front wheels of a car that is steered while moving at high speeds. It occurs in the jet engine of a plane that is changing its direction of flight. You will have opportunity to investigate these effects in the homework problems.

Example 18.2

A cylinder AB is rotating in bearings mounted on a platform (Fig. 18.10). The cylinder has an angular speed ω_2 and a rate of change of speed $\dot{\omega}_2$, both quantities being relative to the platform. The platform rotates with an angular speed ω_1 and has a rate of change of speed $\dot{\omega}_1$, both quantities being relative to the ground. Compute the moment of the supporting forces of the cylinder AB about the center of mass of the cylinder in terms of the aforementioned quantities and the moments of inertia of the cylinder.

We shall do this problem by two methods, one using axes fixed to the body and using **Euler's equations**, and the other using axes fixed to the platform and using Eq. 18.2.

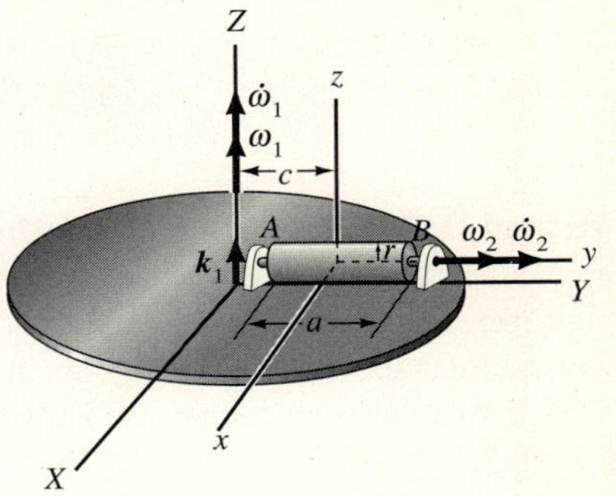

Figure 18.10. Rotating cylinder on platform.

Method 1: Reference fixed to cylinder. In Fig. 18.10 we have fixed axes xyz to the cylinder at the mass center. To get components of $\boldsymbol{M}$ parallel to the inertial reference, we consider the problem when the xyz reference is parallel to the XYZ reference. The angular velocity vector $\boldsymbol{\omega}$ for the body is then

$$\boldsymbol{\omega} = \boldsymbol{\omega}_1 + \boldsymbol{\omega}_2 = \omega_1\boldsymbol{k} + \omega_2\boldsymbol{j}$$

Hence,

$$\begin{aligned} \omega_x &= 0 \text{ rad/s} \\ \omega_y &= \omega_2 \text{ rad/s} \\ \omega_z &= \omega_1 \text{ rad/s} \end{aligned} \qquad \text{(a)}$$

Also, we can say noting that $\boldsymbol{j}$ is fixed to the centerline of the cylinder

$$\begin{aligned} \dot{\boldsymbol{\omega}} = \dot{\boldsymbol{\omega}}_1 + \dot{\boldsymbol{\omega}}_2 &= \dot{\omega}_1\boldsymbol{k} + \left(\frac{d}{dt}\right)_{XYZ}(\omega_2\boldsymbol{j}) \\ &= \dot{\omega}_1\boldsymbol{k} + \dot{\omega}_2\boldsymbol{j} + \omega_2(\boldsymbol{\omega}_1 \times \boldsymbol{j}) \\ &= \dot{\omega}_1\boldsymbol{k} + \dot{\omega}_2\boldsymbol{j} + \omega_2(\omega_1\boldsymbol{k} \times \boldsymbol{j}) \\ &= \dot{\omega}_1\boldsymbol{k} + \dot{\omega}_2\boldsymbol{j} - \omega_1\omega_2\boldsymbol{i} \end{aligned}$$

Accordingly, we have at the instant of interest

$$\begin{aligned} \dot{\omega}_x &= -\omega_1\omega_2 \text{ rad/s}^2 \\ \dot{\omega}_y &= \dot{\omega}_2 \text{ rad/s}^2 \\ \dot{\omega}_z &= \dot{\omega}_1 \text{ rad/s}^2 \end{aligned} \qquad \text{(b)}$$

The **Euler equations** then become

$$M_x = I_{xx}(-\omega_1\omega_2) + \omega_1\omega_2(I_{zz} - I_{yy}) \qquad \text{(c)}$$
$$M_y = I_{yy}\dot{\omega}_2 + 0 \qquad \text{(d)}$$
$$M_z = I_{zz}\dot{\omega}_1 + 0 \qquad \text{(e)}$$

Since $I_{zz} = I_{xx}$ we see that $I_{xx}(-\omega_1\omega_2)$ cancels $\omega_1\omega_2 I_{zz}$ in Eq. (c), and we then have the desired result:

Example 18.2 (Continued)

$$M = -I_{yy}\omega_1\omega_2\boldsymbol{i} + I_{yy}\dot{\omega}_2\boldsymbol{j} + I_{zz}\dot{\omega}_1\boldsymbol{k} \tag{f}$$

Method II: Reference fixed to platform. We shall now do this problem by having xyz at the mass center C of the cylinder again, but now fixed to the platform. In other words, the cylinder rotates relative to the xyz reference with angular speed ω_2. Keeping this in mind, we can still refer to Fig. 18.10.

Obviously, we cannot use **Euler's equations** here and must return to Eq. 18.2.

$$\boldsymbol{M}_c = \left(\frac{d\boldsymbol{H}_C}{dt}\right)_{xyz} + \boldsymbol{\Omega} \times \boldsymbol{H}_C \tag{a}$$

Because the cylinder is a body of revolution about the y axis, the products of inertia I_{xy}, I_{xz}, and I_{yz} are always zero, and I_{xx}, I_{yy}, and I_{zz} are *constants* at all times. Were these conditions not present, this method of approach would be very difficult, since we would have to ascertain the time derivatives of these inertia terms. Thus, using Eq. 18.2, remembering that $\boldsymbol{\omega}$, the angular velocity of the body, goes into $\boldsymbol{H}_C$ while $\boldsymbol{\Omega}$ is the angular velocity of xyz, we see that

$$\begin{aligned}\left(\frac{d\boldsymbol{H}_C}{dt}\right)_{xyz} &= \left(\frac{d}{dt}\right)_{xyz}(H_x\boldsymbol{i} + H_y\boldsymbol{j} + H_z\boldsymbol{k})_C \\ &= \left(\frac{d}{dt}\right)_{xyz}(I_{xx}\omega_x\boldsymbol{i} + I_{yy}\omega_y\boldsymbol{j} + I_{zz}\omega_z\boldsymbol{k}) \qquad \text{(b)} \\ &= \left(\frac{d}{dt}\right)_{xyz}(0\boldsymbol{i} + I_{yy}\omega_2\boldsymbol{j} + I_{zz}\omega_1\boldsymbol{k})\end{aligned}$$

Note that $\boldsymbol{i}$, $\boldsymbol{j}$, and $\boldsymbol{k}$ are constants as seen from xyz. Only ω_1 and ω_2 are time functions and undergo simple time differentiation of scalars. Thus,

$$\left(\frac{d\boldsymbol{H}_C}{dt}\right)_{xyz} = I_{yy}\dot{\omega}_2\boldsymbol{j} + I_{zz}\dot{\omega}_1\boldsymbol{k} \tag{c}$$

Noting further that $\boldsymbol{\Omega} = \omega_1\boldsymbol{k}$, we have on substituting Eq. (c) into Eq. (a):

$$\begin{aligned}M &= I_{yy}\dot{\omega}_2\boldsymbol{j} + I_{zz}\dot{\omega}_1\boldsymbol{k} + \omega_1\boldsymbol{k} \times (H_x\boldsymbol{i} + H_y\boldsymbol{j} + H_z\boldsymbol{k}) \\ &= I_{yy}\dot{\omega}_2\boldsymbol{j} + I_{zz}\dot{\omega}_1\boldsymbol{k} + \omega_1\boldsymbol{k} \times (0\boldsymbol{i} + I_{yy}\omega_2\boldsymbol{j} + I_{zz}\omega_1\boldsymbol{k}) \qquad \text{(d)}\end{aligned}$$

$$\boldsymbol{M} = -I_{yy}\omega_1\omega_2\boldsymbol{i} + I_{yy}\dot{\omega}_2\boldsymbol{j} + I_{yy}\dot{\omega}_1\boldsymbol{k}$$

This equation is identical to the one obtained using method I.

PROBLEMS

18.1. The moving parts of a jet engine consist of a *compressor* and a *turbine* connected to a common shaft. Suppose that this system is rotating at a speed ω_1 of 10×10^3 r/min and the plane it is in moves at a speed of 270 m/s in a circular loop of radius 3.2 km. What is the direction and magnitude of the gyroscopic torque transmitted to the plane from the engine through bearings A and B? The engine has a mass of 90 kg and a radius of gyration about its axis of rotation of 0.3 m. The radius of gyration at the center of mass for an axis normal to the centerline is 0.45 m. What advantage is achieved by using two oppositely turning jet engines instead of one large one?

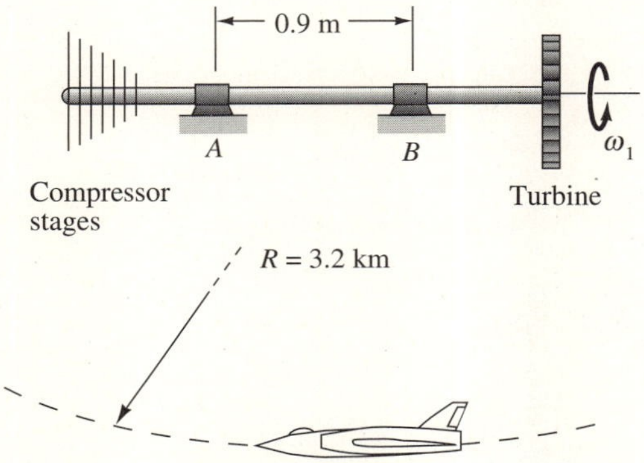

Figure P.18.1.

18.2. A space capsule (unmanned) is tumbling in space due to malfunctioning of its control system such that at time t, $\omega_1 = 3$ rad/s, $\omega_2 = 5$ rad/s, and $\omega_3 = 4$ rad/s. At this instant, small jets are creating a torque T of 30 N m. What are the angular acceleration components at this instant? The vehicle is a body of revolution with $k_z = 1$ m, $k_y = k_x = 1.6$ m, and has a mass of 1000 kg.

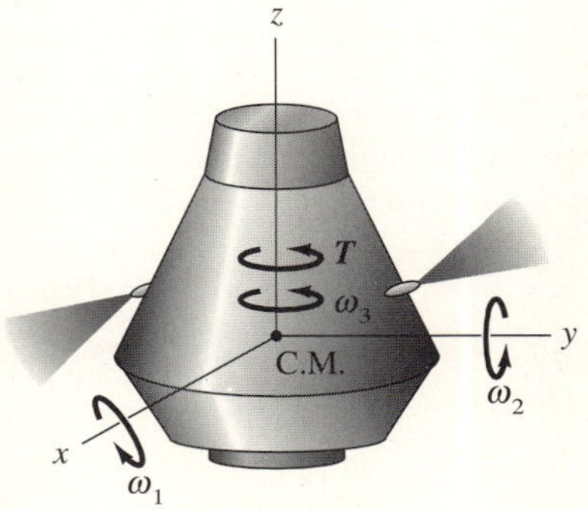

Figure P.18.2.

18.3. The left front tire of a car moving at 25 m/s along an unbanked road along a circular path having a mean radius of 140 m. The tire is 660 mm in diameter. The rim plus tire mass is 14 kg and has a combined radius of gyration of 230 mm about its axis. Normal to the axis, the radius of gyration is 190 mm. What is the gyroscopic torque on the bearings of this front wheel coming solely from the motion of the front wheel?

18.4. In Problem 18.3, suppose that the driver is turning the front wheel at a rate ω_3 of 0.2 rad/s at the instant where the radius of curvature of the path is 140 m. What is then the torque needed on the wheel solely from the motion of the wheel?

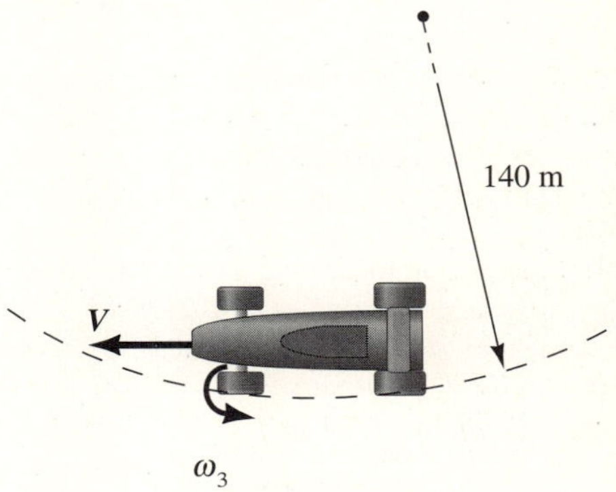

Figure P.18.4.

18.5. A student is holding a rapidly rotating wheel in front of him. He is standing on a platform that can turn freely. If the student exerts a torque M_1 as shown, what begins to happen initially? What happens a little later?

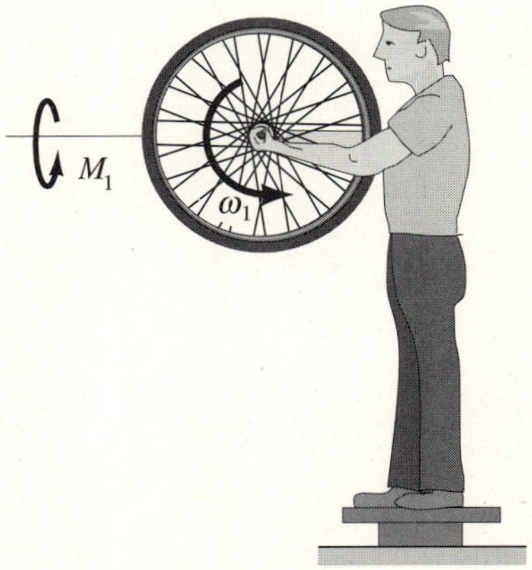

Figure P.18.5.

18.6. An electric motor is mounted on a rotating platform having an angular velocity ω_1 of 2 rad/s. The motor drives two fans at the rate ω_2 of 1750 r/min relative to the platform. The fans plus armature of the motor have a total weight of 100 N and a radius of gyration along the axis of 200 mm. About the Z axis, the radius of gyration is 200 mm. What is the torque coming onto the bearings of the motor as a result of the motion?

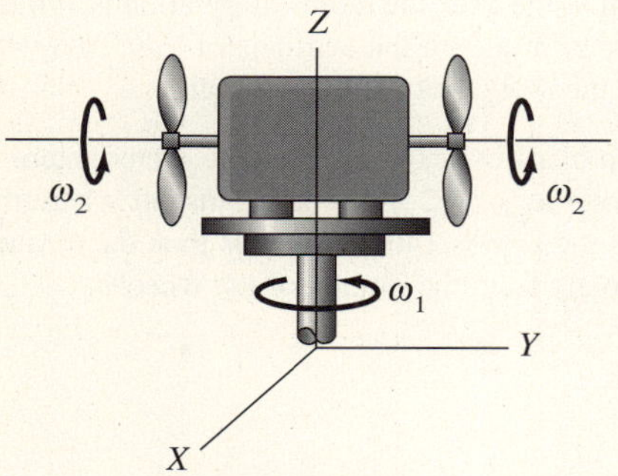

Figure P.18.6.

18.7. A lug wrench is translating in inertial space inside an orbiting space vehicle. A torque $\boldsymbol{T} = 2\boldsymbol{i} + 6\boldsymbol{j} + 5\boldsymbol{k}$ N m is exerted at the center of the wrench. What is the acceleration of end A at this instant? Approximate the wrench as two slender rods. The wrench weighs 12 N.

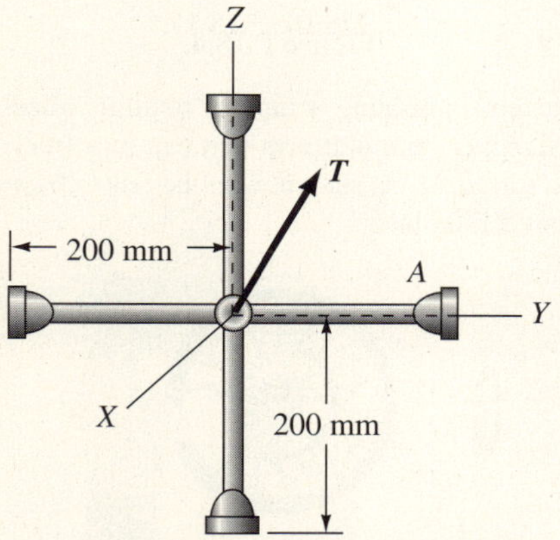

Figure P.18.7.

18.8. Work Problem 18.7 for the case where

$$\boldsymbol{\omega} = 3\boldsymbol{i} - 2\boldsymbol{j} + 4\boldsymbol{k} \text{ rad/s}$$

about an axis of rotation through the center of mass at the instant when $\boldsymbol{T}$ is applied.

18.9. A propeller-driven airplane is at the bottom of a loop of radius 600 m and traveling at 150 m/s. The propeller consists of two identical blades at right angles, mass 150 kg, has a radius of gyration of 0.6 m about its axis of rotation, and is rotating at 1200 r/min. If the propeller rotates counterclockwise as viewed from the rear of the plane, compute the torques coming onto the propeller at the bearings from the motion if one blade is vertical and the other is horizontal at the time of interest.

***18.10.** A thin disc, mass 15 kg, rotates on rod AB at a speed ω_2 of 100 rad/s in a clockwise direction looking from B to A. The radius of the disc is 0.3 m, and the disc is located 3 m from the centerline of the shaft CD, to which rod AB is fixed. Shaft CD rotates at $\omega_1 = 50$ rad/s in a counterclockwise direction as we look from C to D. Find the tensile force, bending moment, and shear force on rod AB at the end A due to the disc.

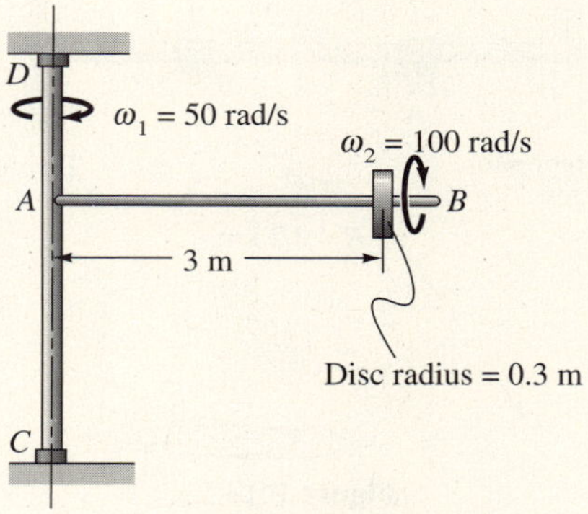

Figure P.18.10.

18.11. The turbine in a ship is parallel to the longitudinal axis of the ship and is rotating at a rate ω_1 of 800 r/min counterclockwise as viewed from stern to bow. The turbine weighs 445 kN and has radii of gyration at the mass center of 1 m about axes normal to the centerline and of 300 mm about the centerline. The ship has a pitching motion, which is approximately sinusoidal, given as

$$\theta = 0.3 \sin \frac{2\pi}{15} t \text{ rad}$$

The amplitude of the pitching is thus 0.3 rad and the period is 15 s. Determine the moment as a function of time about the mass center needed for the motion of the turbine.

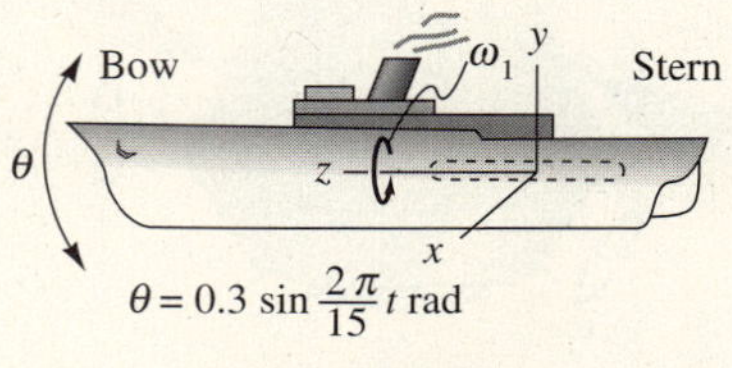

Figure P.18.11.

18.12. Explain how the roll of a ship can be stabilized by the action of a heavy rapidly spinning disc (gyroscope) rotating in a set of bearings in the ship as shown.

Figure P.18.12.

18.13. A 10-kg disc rotates with speed $\omega_1 = 10$ rad/s relative to rod AB. Rod AB rotates with speed $\omega_2 = 4$ rad/s relative to the vertical shaft, which rotates with speed $\omega_3 = 2$ rad/s relative to the ground. What is the torque coming onto the bearings at B due to the motion at a time when $\theta = 60°$? Take $\dot{\omega}_1 = \dot{\omega}_2 = \dot{\omega}_3 = 0$.

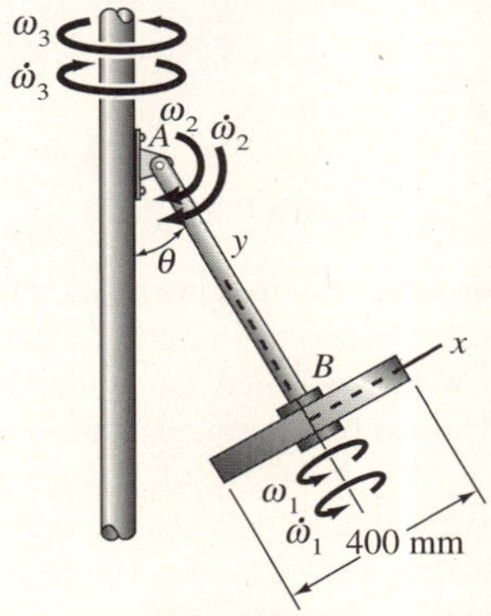

Figure P.18.13.

18.14. Solve Problem 18.13 for the case where $\dot{\omega}_1 = 2$ rad/s², $\dot{\omega}_2 = 3$ rad/s², and $\dot{\omega}_3 = 4$ rad/s² in the directions shown.

18.15. A man is seated in a centrifuge of the type described in Example 15.14. If $\omega_1 = 2$ rad/s, $\omega_2 = 3$ rad/s, and $\omega_3 = 4$ rad/s, what torque must the seat develop about the center of mass of the man as a result of the motion? The man weighs 700 N and has the following radii of gyration as determined by experiment while sitting in the seat:

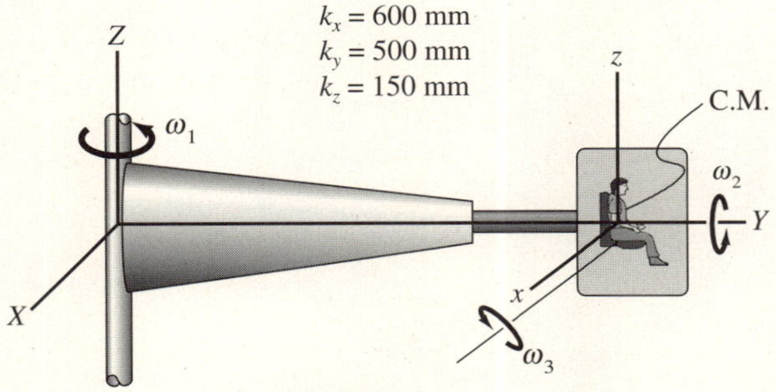

Figure P.18.15.

***18.16.** Work Problem 18.15 for the following data:

$$\dot{\omega}_1 = 3 \text{ rad/s}^2$$
$$\dot{\omega}_2 = -2 \text{ rad/s}^2$$
$$\dot{\omega}_3 = 4 \text{ rad/s}^2$$

The other data are the same.

18.17. A thin disc has its axis inclined to the vertical by an angle θ and rolls without slipping with an angular speed ω_1 about the supporting rod held at B with a ball-and-socket joint. If $l = 3$ m, $r = 0.6$ m, $\theta = 45°$, and $\omega_1 = 10$ rad/s, compute the angular velocity of the rod BC about axis O-O. If the disc mass is 18 kg, what is the total moment about point B from all forces acting on the system? Neglect the mass of the rod OC. [*Hint:* Use a reference xyz at B when two of the axes are in the plane of O-O and OC.]

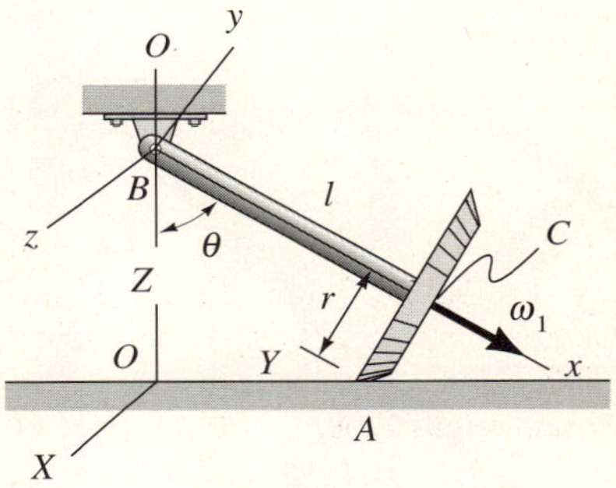

Figure P.18.17.

18.18. A gage indicator CD in an instrument is 20 mm long and is rotating relative to platform A at the rate $\omega_1 = 0.3$ rad/s. The platform A is fixed to a space vehicle which is rotating at speed $\omega_2 = 1$ rad/s about vertical axis LM. At the instant shown, what are the moment components from the bearings E and G about the center of mass of CD needed for the motion of CD and EG? The indicator weighs 0.25 N, and the shaft EG weighs 0.30 N, has a diameter of 1 mm, and a length of 10 mm.

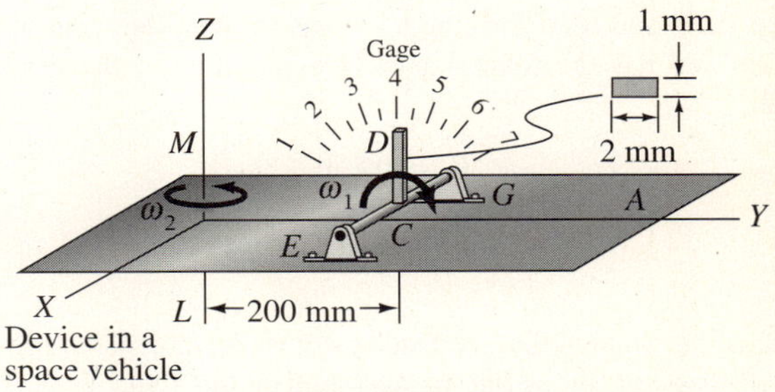

Figure P.18.18.

18.19. An eight-bladed fan is used in a wind tunnel to drive the air. The angular velocity ω_1 of the fan is 120 r/min. At the instant of interest, each blade is rotating about its own axis z (in order to change the angle of attack of the blade) such that $\omega_2 = 1$ rad/s and $\dot{\omega}_2$ is 0.5 rad/s^2. Each blade weighs 200 N and has the following radii of gyration at the mass center (C.M.):

$$k_x = 600 \text{ mm}$$
$$k_y = 580 \text{ mm}$$
$$k_z = 100 \text{ mm}$$

Consider only the dynamics of the system (and not the aerodynamics) to find the torque required about the z axis of the blade. If a couple moment of 4 N m is developed in the x direction at the base of the blade, what force component F_y is needed at the base of the blade for the motion described?

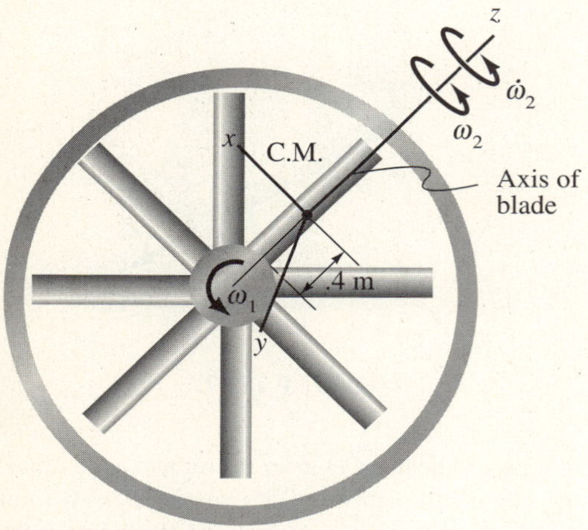

Figure P.18.19.

18.20. A swing-wing fighter plane is moving with a speed of Mach 1.3. (Note that Mach 1 corresponds to a speed of about 330 m/s) The pilot at the instant shown is swinging his wings back at the rate ω_1 of 0.3 rad/s. At the same time he is rolling at a rate ω_2 of 0.6 rad/s and is performing a loop as shown. The wing weighs 6.5 kN and has the following radii of gyration at the center of mass:

$$k_x = 0.8 \text{ m}$$
$$k_y = 4 \text{ m}$$
$$k_z = 6 \text{ m}$$

What is the moment that the fuselage must develop about the center of mass of the wing to accomplish the dynamics of the described motions? Do not consider aerodynamics.

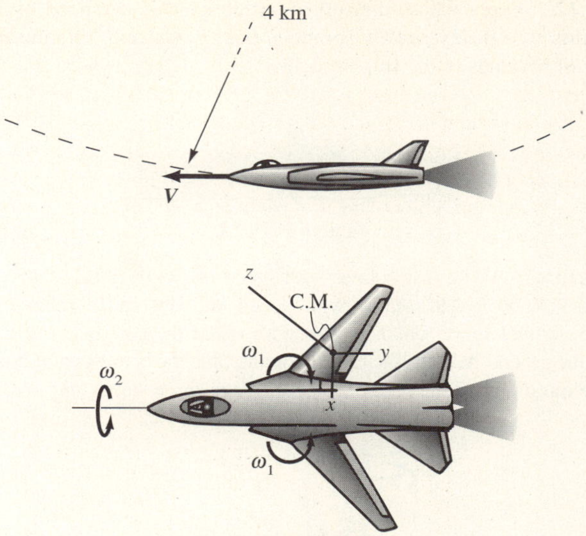

Figure P.18.20.

18.21. An orbiting skylab is rotating at an angular speed ω_1 of 0.4 rad/s to give a "gravitational" effect for the living quarters in the outer annulus. A many-bladed fan is mounted as shown on the outer "floor." The fan blades rotate at a speed ω_2 of 200 r/min and the base rotates at a speed ω_3 of 1 rad/s with $\dot{\omega}_3$ equal to 0.6 rad/s^2 all relative to the skylab. At the instant of interest $\boldsymbol{\omega}_1$, $\boldsymbol{\omega}_2$, and $\boldsymbol{\omega}_3$ are orthogonal to each other. If the blade has a mass of 100 g and if the radius of gyration about its axis is 200 mm, what is the torque coming onto the fan due solely to the motion? Take the transverse radii of gyration of the fan to be 120 mm.

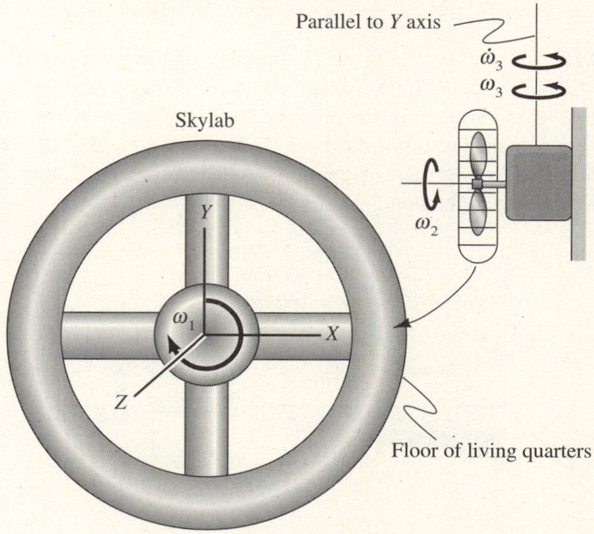

Figure P.18.21.

18.22. You will learn in fluid mechanics that when air moves across a rotating cylinder a force is developed normal to the axis of the cylinder (the *Magnus effect*). In 1926 Flettner used this principle to "sail" a vessel across the Atlantic. Two cylinders were kept at a constant rotational speed by a motor of $\omega_1 = 200$ r/min relative to the ship. Suppose that in rough seas the ship is rolling about the axis of the ship with a speed ω_2 of 0.8 rad/s. What couple moment components must the ship transmit to the *base* of the cylinder as a result only of the motion of the ship? Each cylinder has a mass of 320 kg and has a radius of gyration along the axis of 0.6 m and normal to the axis at the center of mass of 3 m.

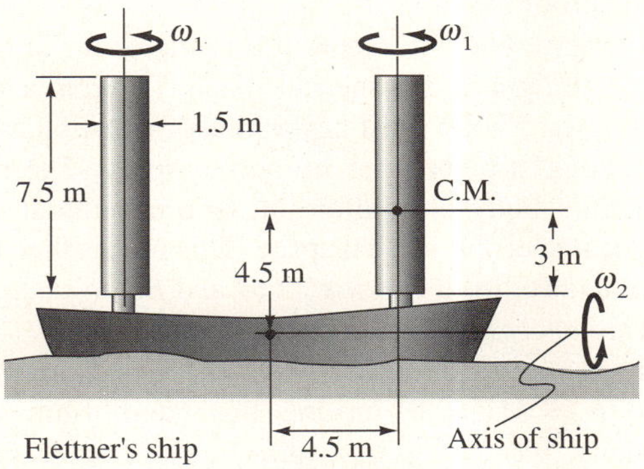

Figure P.18.22.

***18.23.** Part of a clutch system consists of identical rods *AB* and *AC* rotating relative to a shaft at the speed $\omega_2 = 3$ rad/s while the shaft rotates relative to the ground at a speed $\omega_1 = 40$ rad/s. As a result of the motion what are the bending moment components at the base of each rod if each rod weighs 10 N?

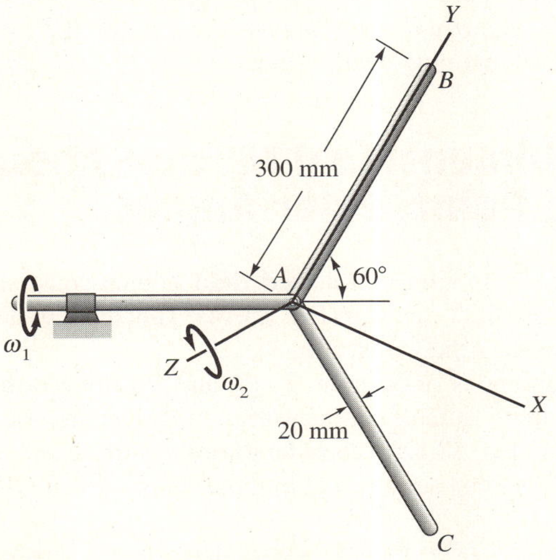

Figure P.18.23.

***18.24.** Solve Problem 18.23 for the case when $\dot{\omega}_1 = 2$ rad/s^2 and $\dot{\omega}_2 = 6$ rad/s^2 at the instant of interest.

18.25. A rod *D* weighing 2 N and having a length of 200 mm and a diameter of 10 mm rotates relative to platform *K* at a speed ω_1 of 120 rad/s. The platform is in a space vehicle and turns at a speed ω_2 of 50 rad/s relative to the vehicle. The vehicle rotates at a speed ω_3 of 30 rad/s relative to inertial space about the axis shown. What are the bearing forces at *A* and *B* normal to the axis *AB* resulting from the motion of rod *D*? What torque *T* is needed about *AB* to maintain a constant value of ω_1 at the instant of interest?

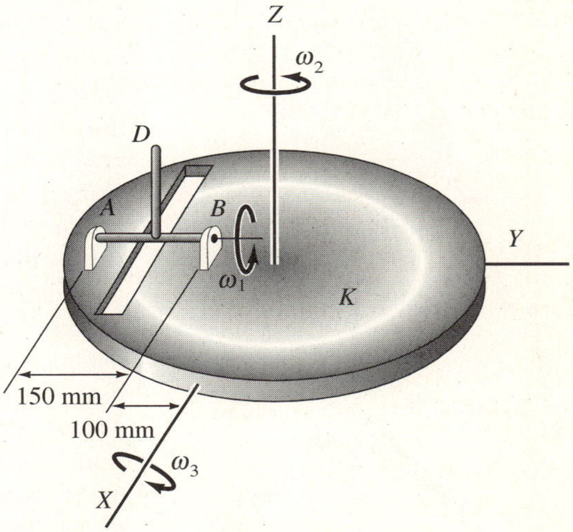

Figure P.18.25.

18.26. Solve Problem 18.6 using a reference at the center of mass of the armature but not fixed to the armature.

18.27. In Problem 18.10, find the moment about the C.M. of the disc using a reference *xyz* fixed to the rod *AB* and not to the disc. Take $\dot{\omega}_1 = 10$ rad/s^2 and $\dot{\omega}_2 = 30$ rad/s^2.

18.28. Work Problem 18.11 using reference *xyz* fixed to the ship with the origin at the center of mass of the turbine. Do not use **Euler's equations**.

18.29. Solve Problem 18.13 using a set of axes *xyz* at the center of mass of the disc but fixed to arm *AB*.

18.30. Work Problem 18.13 by using a reference *xyz* fixed to the arm at the center of mass of the disc for the case where $\dot{\omega}_1 = 2$ rad/s^2, $\dot{\omega}_2 = 3$ rad/s^2, and $\dot{\omega}_3 = 4$ rad/s^2 in directions shown.

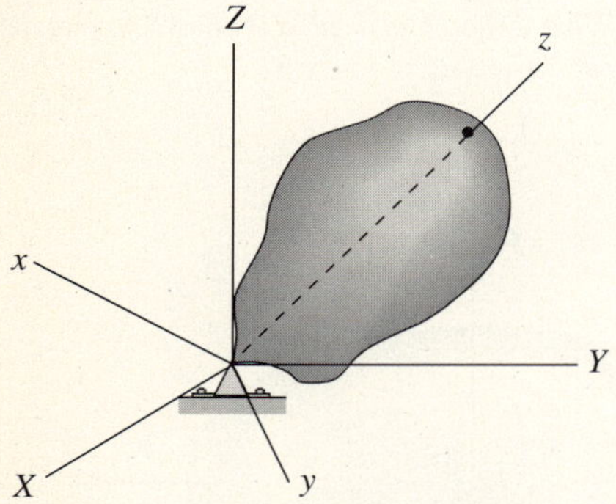

Figure 18.11. Body with a fixed point.

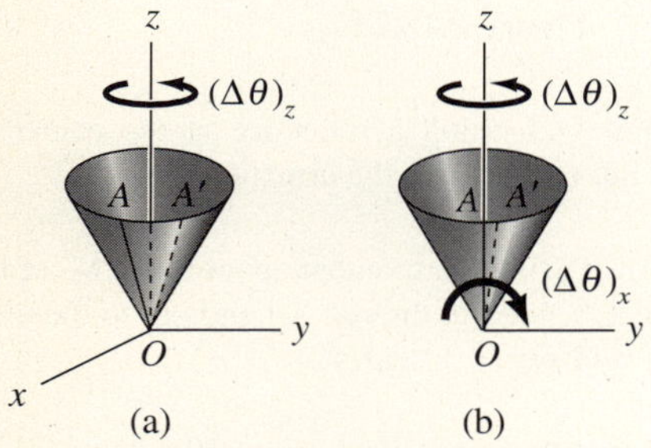

Figure 18.12. $\Delta\theta_z$ causes $\Delta\theta_x$.

18.4 Necessary and Sufficient Conditions for Equilibrium of a Rigid Body

In this chapter, we have employed **Newton's law** at the mass center as well as the equation $\boldsymbol{M}_A = \dot{\boldsymbol{H}}_A$ and from it we derived **Euler's equations** for rigid bodies. We can now go back to our work in Chapter 5 and put on firm ground the fact that $\boldsymbol{M} = \boldsymbol{0}$ and $\boldsymbol{F} = \boldsymbol{0}$ are necessary conditions for equilibrium of a rigid body. (You will recall we accepted these equations for statics at that time by intuition, pending a proof to come later.)

A particle is in equilibrium, you will recall, if it is stationary or moving with constant speed along a straight line in inertial space. To be in a state of equilibrium, every point in a rigid body must accordingly be stationary or be moving at uniform speed along straight lines in inertial space. The rigidity requirement thus limits a rigid body in equilibrium to translational motion along a straight line at constant speed in inertial space. This means that $\dot{\boldsymbol{V}}_C = \boldsymbol{0}$ and $\boldsymbol{\omega} = \boldsymbol{0}$ for equilibrium and so, from *Newton's law* and *Euler's equations*, we see that $\boldsymbol{F} = \boldsymbol{M}_A = \boldsymbol{0}$ are *necessary* conditions for equilibrium.

For a *sufficiency* proof, we go the other way. For a body initially in equilibrium, the condition $\boldsymbol{F} = \boldsymbol{M}_A = \boldsymbol{0}$[5] ensures that equilibrium will be maintained. More specifically, we shall start with a body in equilibrium at time t and apply a force system satisfying the preceding conditions. We address ourselves to the question: Does the body stay in a state of equilibrium? According to *Newton's law* there will be no change in the velocity of the mass center since $\boldsymbol{F} = \boldsymbol{0}$. And, with $\boldsymbol{\omega} = \boldsymbol{0}$ at time t, *Euler's equations* lead to the result for $\boldsymbol{M}_C = \boldsymbol{0}$, that $\dot{\omega}_x = \dot{\omega}_y = \dot{\omega}_z = 0$. Thus, the angular velocity must remain zero. With the velocity of the center of mass constant, and with $\dot{\boldsymbol{\omega}} = \boldsymbol{0}$ in inertial space we know that the body remains in equilibrium. Thus, if a body is *initially in equilibrium*, the condition $\boldsymbol{F} = \boldsymbol{0}$ and $\boldsymbol{M}_A = \boldsymbol{0}$ is *sufficient* for maintaining equilibrium.

18.5 Three-Dimensional Motion About a Fixed Point; Euler Angles

We shall now examine the motion of selected rigid bodies constrained to have a point fixed in an inertial reference (Fig. 18.11). This topic will lead to study of an important device—the gyroscope.

We now show that angles of rotation θ_x, θ_y, and θ_z, along orthogonal axes x, y, and z are not highly suitable for measuring the orientation of a body with a fixed point. Thus, in Fig. 18.12(a) consider a conical surface and observe straight line OA which is on this surface and in the xz plane. Rotate the cone

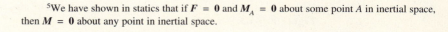

[5]We have shown in statics that if $\boldsymbol{F} = \boldsymbol{0}$ and $\boldsymbol{M}_A = \boldsymbol{0}$ about some point A in inertial space, then $\boldsymbol{M} = \boldsymbol{0}$ about any point in inertial space.

an angle $(\Delta\theta)_z$, thus causing OA to move to OA' shown as dashed. In Fig. 18.12(b), view line OA along a line of sight corresponding to the x axis. As a result of the rotation $(\Delta\theta)_z$ about the z axis, there will clearly be a rotation $(\Delta\theta)_x$ of this line about the x axis. Thus, we see that θ_x and θ_z are mutually dependent and thus not suitable for our use. This result stems from the fact that we are using directions that have a fixed mutual relative orientation. We now introduce a set of rotations that are *independent.* And, not unexpectedly, the axes for these rotations will not have a fixed relative orientation.

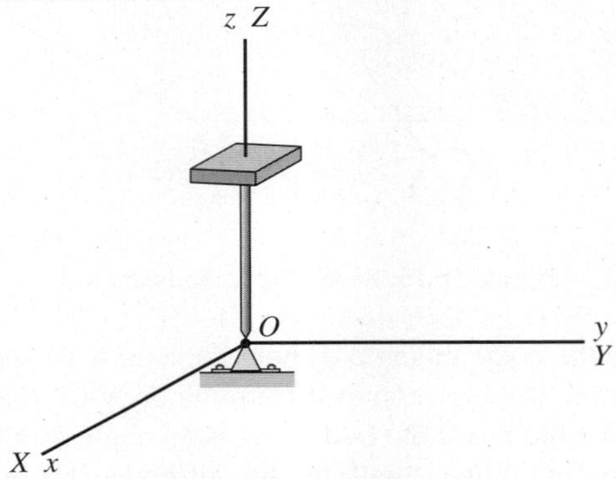

Figure 18.13. Rigid body.

Accordingly, consider the rigid body shown in Fig. 18.13. We shall specify a sequence of three independent rotations in the following manner:

1. Keep a reference xyz fixed in the body, and rotate the body about the Z axis through an angle ψ shown in Fig. 18.14.

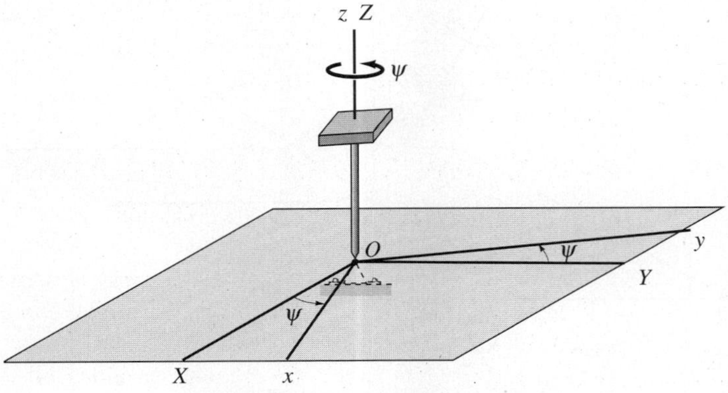

Figure 18.14. First rotation is about Z.

2. Now rotate the body about the x axis through an angle θ to reach the configuration in Fig. 18.15. Note that the z, Z, and y axes form a plane I normal to the XY plane and normal also to the x axis. The axis of rotation for this rotation (x axis) is called the *line of nodes.*

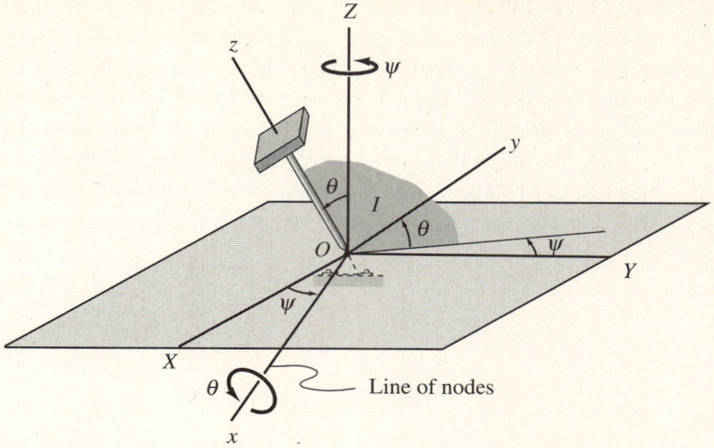

Figure 18.15. Second rotation is about x.

3. Finally, rotate the body an angle ϕ about the z axis. We provide the option here of *detaching* the xyz reference from the body for this movement (see Fig. 18.16), in which case the body rotates an angle ϕ relative to xyz, and the x axis remains collinear with the line of nodes. Or, we can permit the reference xyz to remain fixed in the body (see Fig. 18.17), in which case we can use components of $\boldsymbol{M}$ and $\boldsymbol{\omega}$ along these axes in employing *Euler's equations*. The line of nodes is now identified simply as the normal to plane I containing z and Z and is the same axis as shown in Fig. 18.16.

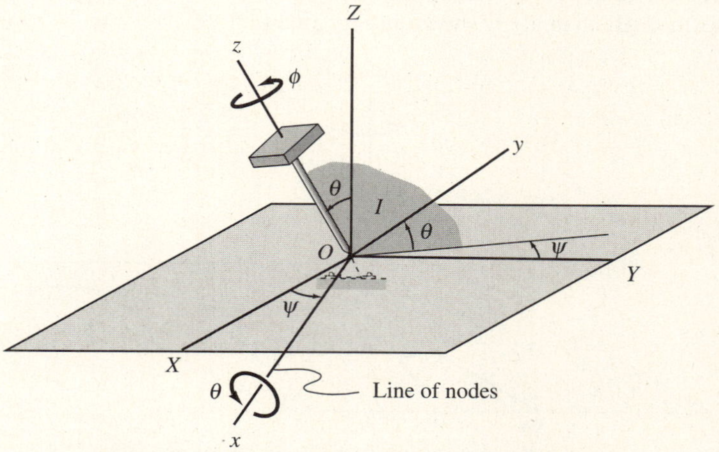

Figure 18.16. Third rotation is about z. xyz is detached.

We thus arrive at the desired orientation. Positive rotations in each case are those taken as counterclockwise as one looks to the origin O along the axis of rotation. (Thus, we have performed three positive rotations here.)

We call these angles the *Euler angles*, and we assign the following names.

$$\psi = \text{angle of precession}$$
$$\theta = \text{angle of nutation}$$
$$\phi = \text{angle of spin}$$

Furthermore, the z axis is usually called the *body axis*, and the Z axis is often called the *axis of precession*. The line of nodes then is normal to the body axis and the precession axis.

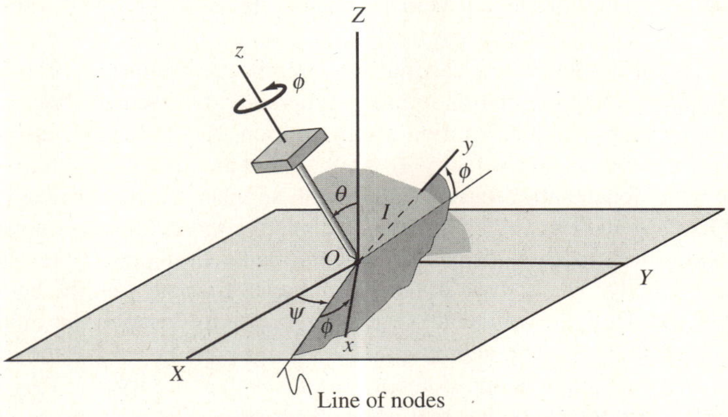

Figure 18.17. Axes xyz are fixed to body.

We have shown that the position of a body moving with one point fixed can be established by three independent rotations given in a certain sequence. For an infinitesimal change in position, this situation would be a rotation $d\psi$ about the Z axis, $d\theta$ about the line of nodes, and $d\phi$ about the body axis z. Because these rotations are infinitesimal, they can be construed as vectors, and the order mentioned above is no longer required. The limiting ratios of these changes in angles with respect to time give rise to three angular velocity vectors (Fig. 18.18), which we express in the following manner:

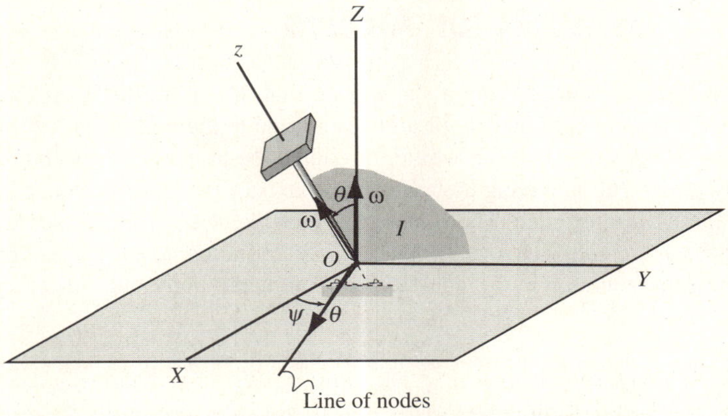

Figure 18.18. Precession, nutation, and spin velocity vectors.

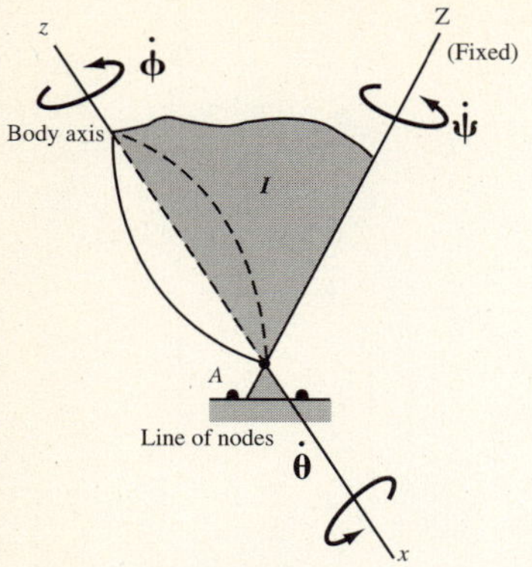

Figure 18.19. Line of nodes always normal to plane I.

$\dot{\boldsymbol{\psi}}$, directed along the Z axis
$\dot{\boldsymbol{\theta}}$, directed along the line of nodes
$\dot{\boldsymbol{\phi}}$, directed along the z body axis

Note that the nutation velocity vector $\dot{\boldsymbol{\theta}}$ is always normal to plane I, and consequently the nutation velocity vector is always normal to the spin velocity vector $\dot{\boldsymbol{\phi}}$ and the precession velocity vector $\dot{\boldsymbol{\psi}}$. However, the spin velocity vector $\dot{\boldsymbol{\phi}}$ will generally *not* be at right angles with the precession velocity $\dot{\boldsymbol{\psi}}$, and so this system of angular velocity vectors generally is *not* an orthogonal system.

Finally, it should be clear that the reference xyz moves with the body during precession and nutation motion of the body, but we can choose that it not move (relative to XYZ) during a spin rotation. Hence for this case, while the body has the angular velocity $\dot{\boldsymbol{\phi}} + \dot{\boldsymbol{\psi}} + \dot{\boldsymbol{\theta}}$ at any time t the reference xyz would have, for the aforestated condition, an angular velocity, denoted as $\boldsymbol{\Omega}$, equal to $\dot{\boldsymbol{\psi}} + \dot{\boldsymbol{\theta}}$. The velocity of the disc relative to xyz is, accordingly, $\dot{\phi}\boldsymbol{k}$.

Consider a body with two orthogonal planes of symmetry forming an axis in the body. The body is moving about a fixed point A on the body axis (Fig. 18.19). How do we decide what axes to use to describe the motion in terms of spin, precession, and nutation? First, we take the axis of the body to be the z axis; the angular speed about this axis is then the spin $\dot{\phi}$. This step is straightforward for the bodies that we shall consider. The next step is not. By inspection find a Z axis in a *fixed direction* so as to form with the aforementioned body axis z a plane I whose angular speed about the Z axis is either known or is sought. Such an axis Z is then the *precession* axis about which we have for the body axis z a precession speed $\dot{\psi}$. The *line of nodes* is then the axis which at all times remains *normal* to plane I containing the body axis z and the precession axis Z. The nutation speed $\dot{\theta}$ finally is the angular speed component of the body axis z about the line of nodes.

18.6 Equations of Motion Using Euler Angles

Consider next a body having a shape such that, at any point along the body axis z, the moments of inertia for all axes normal to the body axis at this point have the same value I'. Such would, of course, be true for the special case of a body of revolution having the z axis as the axis of symmetry.[6] We shall consider the motion of such a body about a fixed point O, which is somewhere along the axis z (see Fig. 18.20). Axes xyz are principal axes. This set of axes has the same nutation and precession motion as the body but will be chosen

[6] You will be asked to show in Problem 18.41 that, if I_{xx}, I_{yy}, and I_{zz} are principal axes and $I_{xx} = I_{yy} = I'$, then $I'_{xx} = I'_{yy} = I'$ for *any* axes x', y' formed by rotating xyz about the z axis. Thus, homogeneous cylinders having regular cross sections such as squares or octagons would meet the requirements of this section.

such that the body rotates with angular speed $\dot{\phi}$ relative to it. Since the reference is not fixed to the body, we cannot use **Euler's equations** but must go back to the equation $\boldsymbol{M}_0 = \dot{\boldsymbol{H}}_0$, which when carried out in terms of components parallel to the *xyz* reference becomes:

$$\boldsymbol{M}_0 = \left(\frac{d\boldsymbol{H}_0}{dt}\right)_{xyz} + \boldsymbol{\Omega} \times (H_x\boldsymbol{i} + H_y\boldsymbol{j} + H_z\boldsymbol{k}) \qquad (18.21)$$

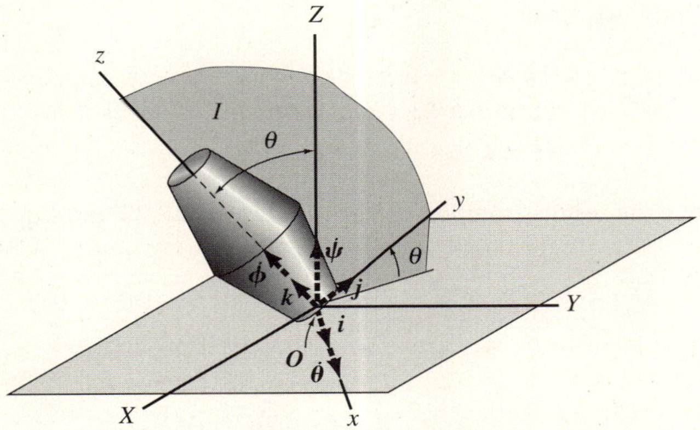

Figure 18.20. Body moving about fixed point *O*.

Since the *xyz* axes remain at *all times* principal axes, we have:

$$H_x = I'\omega_x, \qquad H_y = I'\omega_y, \qquad H_z = I\omega_z \qquad (18.22)$$

where I is the moment of inertia about the axis of symmetry and I' is the moment of inertia about an axis normal to the z axis at O. Considering Fig. 18.20, we can see by inspection that the angular velocity of the body relative to *XYZ* is at *all times* given by components parallel to *xyz* as follows:

$$\omega_x = \dot{\theta} \qquad (18.23a)$$
$$\omega_y = \dot{\psi}\sin\theta \qquad (18.23b)$$
$$\omega_z = \dot{\phi} + \dot{\psi}\cos\theta \qquad (18.23c)$$

Hence, the components of the angular momentum at all times are:

$$\begin{aligned}(H_x) &= I'\dot{\theta}\\ (H_y) &= I'\dot{\psi}\sin\theta\\ (H_z) &= I(\dot{\phi} + \dot{\psi}\cos\theta)\end{aligned} \qquad (18.24)$$

We then have for $\boldsymbol{H}_0$:

$$\boldsymbol{H}_0 = I'\dot{\theta}\boldsymbol{i} + I'\dot{\psi}\sin\theta\boldsymbol{j} + I(\dot{\phi} + \dot{\psi}\cos\theta)\boldsymbol{k}$$

Remembering that $\boldsymbol{i}, \boldsymbol{j}$ and $\boldsymbol{k}$ are constants as seen from xyz, we can say:

$$\left(\frac{d\boldsymbol{H}_0}{dt}\right)_{xyz} = \left(\frac{d}{dt}\right)_{xyz}\left[I'\dot{\theta}\boldsymbol{i} + I'\dot{\psi}\sin\theta\,\boldsymbol{j} + I(\dot{\phi} + \dot{\psi}\cos\theta)\boldsymbol{k}\right] \tag{18.25}$$
$$= I'\ddot{\theta}\boldsymbol{i} + I'(\ddot{\psi}\sin\theta + \dot{\psi}\dot{\theta}\cos\theta)\boldsymbol{j} + I(\ddot{\phi} + \ddot{\psi}\cos\theta - \dot{\psi}\dot{\theta}\sin\theta)\boldsymbol{k}$$

As for the angular velocity of reference xyz, we have on considering Eq. 18.24 with $\dot{\phi}$ deleted because xyz is not fixed to the body as far as spin is concerned:

$$\boldsymbol{\Omega} = \dot{\theta}\boldsymbol{i} + \dot{\psi}\sin\theta\boldsymbol{j} + \dot{\psi}\cos\theta\boldsymbol{k} \tag{18.26}$$

Consequently, we have

$$\begin{aligned} \boldsymbol{\Omega}\times\boldsymbol{i} &= -\dot{\psi}\sin\theta\boldsymbol{k} + \dot{\psi}\cos\theta\boldsymbol{j} \\ \boldsymbol{\Omega}\times\boldsymbol{j} &= \dot{\theta}\boldsymbol{k} - \dot{\psi}\cos\theta\boldsymbol{i} \\ \boldsymbol{\Omega}\times\boldsymbol{k} &= -\dot{\theta}\boldsymbol{j} + \dot{\psi}\sin\theta\boldsymbol{i} \end{aligned} \tag{18.27}$$

Substituting the results from Eqs. 18.25, 18.24, and 18.27 into Eq. 18.21, we get

$$\begin{aligned} M_x\boldsymbol{i} + M_y\boldsymbol{j} + M_z\boldsymbol{k} = {} & I'\ddot{\theta}\boldsymbol{i} + I'(\ddot{\psi}\sin\theta + \dot{\psi}\dot{\theta}\cos\theta)\boldsymbol{j} \\ & + I(\ddot{\phi} + \ddot{\psi}\cos\theta - \dot{\psi}\dot{\theta}\sin\theta)\boldsymbol{k} \\ & + I'\dot{\theta}(-\dot{\psi}\sin\theta\boldsymbol{k} + \dot{\psi}\cos\theta\boldsymbol{j}) \\ & + I'\dot{\psi}\sin\theta(\dot{\theta}\boldsymbol{k} - \dot{\psi}\cos\theta\boldsymbol{i}) \\ & + I(\dot{\phi} + \dot{\psi}\cos\theta)(-\dot{\theta}\boldsymbol{j} + \dot{\psi}\sin\theta\boldsymbol{i}) \end{aligned} \tag{18.28}$$

The corresponding scalar equations are:

$$M_x = I'\ddot{\theta} + (I - I')(\dot{\psi}^2\sin\theta\cos\theta) + I\dot{\phi}\dot{\psi}\sin\theta \tag{18.29a}$$
$$M_y = I'\ddot{\psi}\sin\theta + 2I'\dot{\theta}\dot{\psi}\cos\theta - I(\dot{\phi} + \dot{\psi}\cos\theta)\dot{\theta} \tag{18.29b}$$
$$M_z = I(\ddot{\phi} + \ddot{\psi}\cos\theta - \dot{\psi}\dot{\theta}\sin\theta) \tag{18.29c}$$

The foregoing equations are valid at all times for the motion of a homogeneous body having $I_{xx} = I_{yy} = I'$ moving about a fixed point on the axis of the body. Clearly, these equations are also applicable for motion about the center of mass for such bodies. Note that the equations are nonlinear and, except for certain special cases, are very difficult to integrate. They are, of course, very useful as they stand when computer methods are to be employed.

As a special case, we shall now consider a motion involving a constant nutation angle θ, a constant spin speed $\dot{\phi}$, and a constant precession speed $\dot{\psi}$. Such a motion is termed *steady precession*. To determine the torque $\boldsymbol{M}$ for a given steady precession, we set $\dot{\theta}$, $\ddot{\theta}$, $\ddot{\phi}$, and $\ddot{\psi}$ equal to zero in Eqs. 18.29. Accordingly, we get the following result:

$$M_x = \{I(\dot{\phi} + \dot{\psi}\cos\theta) - I'\dot{\psi}\cos\theta]\,\dot{\psi}\sin\theta \tag{18.30a}$$
$$M_y = 0 \tag{18.30b}$$
$$M_z = 0 \tag{18.30c}$$

We see that for such a motion, we require a *constant torque about the line of nodes* as given by Eq. 18.30a. Noting that $\dot{\phi} + \dot{\psi}\cos\theta = \omega_z$ from Eq. 18.23c, this torque may also be given as

$$M_x = (I\omega_z - I'\dot{\psi}\cos\theta)\dot{\psi}\sin\theta \tag{18.31}$$

Examining Fig. 18.20, we can conclude that for the body to maintain a constant spin speed $\dot{\phi}$ about its body axis (i.e., relative to xyz) while the body axis (and also xyz) is rotating at constant speed $\dot{\psi}$ about the Z axis at a fixed angle θ, we require a constant torque M_x having a value dependent on the motion of the body as well as the values of the moments of inertia of the body, and having a direction *always normal to the body and precession axes* (i.e., normal to plane I). Intuitively you may feel that such a torque should cause a rotation about its own axis (the *torque axis*) and should thereby change θ. Instead, the torque causes a rotation $\dot{\psi}$ of the body axis about an axis *normal* to the torque axis. As an example, consider the special case where θ has been chosen as 90° for motion of a disc about its center of mass (see Fig. 18.21). In accordance with Eq. 18.31, we have as a required torque for a steady precession the result

$$M_x = I\omega_z\dot{\psi} = I\dot{\phi}\dot{\psi} \tag{18.32}$$

Here for a given spin, the proper torque M_x about the line of nodes maintains a steady rotation $\dot{\psi}$ of the spin axis z about an axis (the Z axis), which is at *right angles* both to the torque and the spin axes and given by Eq. 18.32. Because of this unexpected phenomenon, toy manufacturers have developed various gyroscopic devices to surprise and delight children (as well as their parents). Here is yet another case where relying solely on intuition may lead to highly erroneous conclusions.

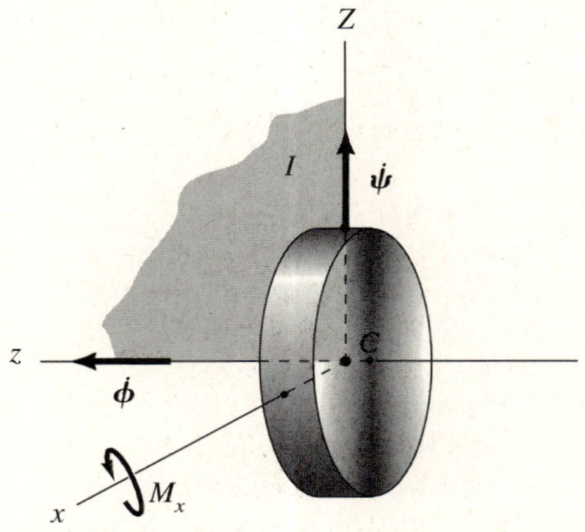

Figure 18.21. Steady precession; $\theta = 90°$.

We should strongly point out that steady precessions are not easily initiated. We must have, at the start, simultaneously the proper precession and spin speeds as well as the proper θ for the given applied torque. If these conditions are not properly met initially, a complicated motion ensues.

Example 18.3

A *single-degree-of-freedom gyro* is shown in Fig. 18.22. The spin axis of disc E is held by a gimbal A which can rotate about bearings C and D. These bearings are supported by the gyro case, which in turn is generally clamped to the vehicle to be guided. If the gyro case rotates about a vertical axis (i.e., normal to its base) while the rotor is spinning, the gimbal A will tend to rotate about CD in an attempt to align with the vertical. When gimbal A is resisted from rotation about CD by a set of torsional springs S with a combined torsional spring constant given as K_t, the gyro is called a *rate gyro*. If the rotation of the gyro case is constant (at speed ω_2), the gimbal A assumes a fixed orientation relative to the vertical as a result of the restraining springs and a damper (not shown). About the body axis z there is a constant angular rotation of the rotor of ω_1 rad/s maintained by a motor (not shown). This angular rotation is clearly is the spin speed $\dot{\phi}$. Next, note in Fig. 18.23 that the z axis and the fixed vertical axis z form a plane (plane I) which has a known angular speed ω_2 about this fixed vertical axis. Clearly, this fixed vertical axis will be our precession axis, and the precession speed $\dot{\psi}$ equals ω_2. The line of nodes has also been shown; it must at all times be normal to plane I and is thus collinear with axis C-D of gimbal A. With θ fixed, we have a case of steady precession.

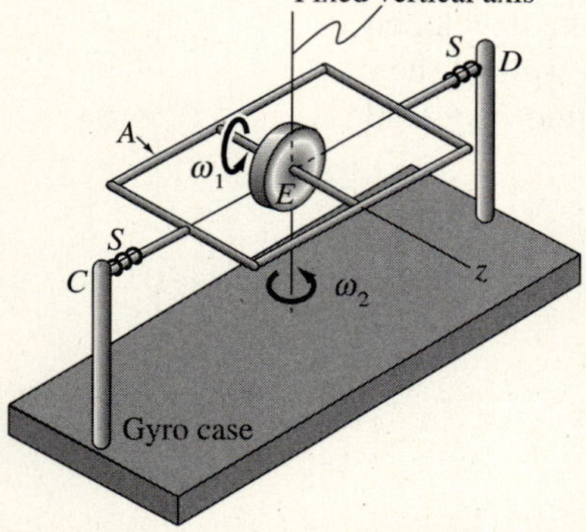

Figure 18.22. Single-degree-of-freedom gyro.

Given the following data:

$$I = 400 \text{ kg mm}^2$$
$$I' = 200 \text{ kg mm}^2$$
$$\dot{\phi} = 20 \text{ krad/s}$$
$$K_t = 6.7 \text{ N m/rad}$$
$$\dot{\psi} = 1 \text{ rad/s}$$

what is θ for the condition of steady precession? The torsional springs are unstretched when $\theta = 90°$.

We have for Eq. 18.30a:

Example 18.3 (Continued)

$$M_x = K_t\left(\frac{\pi}{2} - \theta\right) = \left[I(\dot{\phi} + \dot{\psi}\cos\theta) - I'\dot{\psi}\cos\theta\right]\dot{\psi}\sin\theta \qquad \text{(a)}$$

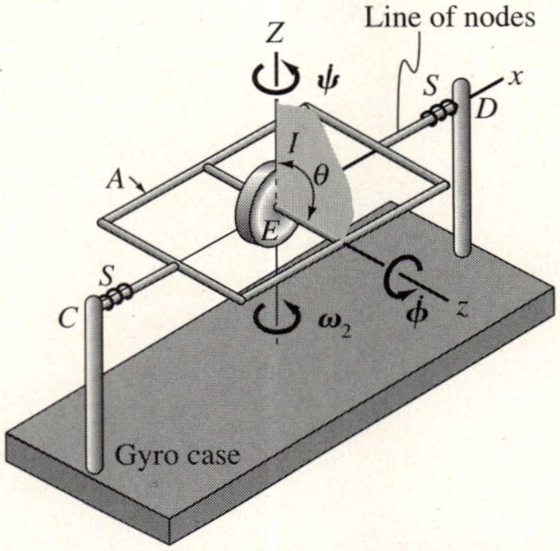

Figure 18.23. Gyro showing line of nodes and precession axis.

Putting in numerical values, we have

$$(6.7)\left(\frac{\pi}{2} - \theta\right)$$
$$= \left[400 \times 10^{-6}(20 \times 10^3 + (1)\cos\theta) - (200 \times 10^{-6})(1)\cos\theta\right]\sin\theta$$

Therefore,

$$6.7 \times 10^6\left(\frac{\pi}{2} - \theta\right) = (8 \times 10^6 + 200\cos\theta)\sin\theta$$

We can neglect the term (200 cos θ), and so we have

$$6.7\left(\frac{\pi}{2} - \theta\right) = 8\sin\theta$$

Therefore,

$$\frac{\pi}{2} - \theta = 1.194\sin\theta \qquad \text{(b)}$$

Solving by trial and error or by computer, we get

$$\theta = 43.2^\circ \qquad \text{(c)}$$

The way the *rate gyro* is used in practice is to maintain θ close to 90° by a small motor. The torque M_x developed to maintain this angle is measured, and from Eq. 18.30a we have available the proper $\dot{\psi}$, which tells us of the rate of rotation of the gyro case and hence the rate of rotation of the vehicle about an axis normal to the gyro case. Now $\dot{\psi}$ need not be constant as was the case in this problem. If it does not change very rapidly, the results from Eq. 18.30a can be taken as instantaneously valid even though the equation, strictly speaking, stems from steady precession where $\dot{\psi}$ should be constant.

Example 18.4

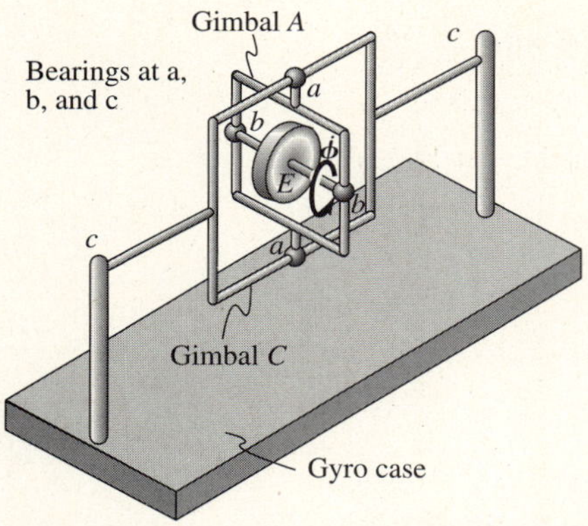

Figure 18.24. Two-degree-of-freedom gyro.

A *two-degree-of-freedom gyroscope* is shown in Fig. 18.24. The rotor E rotates in gimbal A, which in turn rotates in gimbal C. Note that the axes b–b of the rotor and a–a of the gimbal C are always at right angles to each other. Gimbal C is held by bearings c supported by the gyro case. Axes c–c and a–a must always be at right angles to each other, as can easily be seen from the diagram. This kind of suspension of the rotor is called a *Cardan suspension*. If the bearings at a, b, and c are frictionless, a torque cannot be transmitted from the gyro case to the rotor.[7] The rotor is said to be *torque-free* for this case.

If the rotor is given a rapid spin velocity in a given direction in inertial space (such as toward the North Star), then for the ideal case of frictionless bearings the rotor will maintain this direction even though the gyro case is given rapid and complicated motions in inertial space. This constancy of direction results since no torque can be transmitted to the rotor to alter the direction of its angular momentum. Thus, the two-dimensional gyro gives a fixed direction in inertial space for purposes of guidance of a vehicle such as a missile. In use, the gyro case is rigidly fixed to the frame of the missile, and measurements of the orientation of the missile are accomplished by having pickoffs mounted between the gyro case and the outer gimbal and between the outer and inner gimbals.

The presence of some friction in the gyro bearings is, of course, inevitable. The counteraction of this friction when possible and, when not, the accounting for its action is of much concern to the gyro engineer. Suppose that the gyro has been given a motion such that the spin axis b–b (see Fig. 18.25) has an angular speed ω_1 about axis c–c of 0.1 r/s while maintaining a fixed orientation of 85° with axis c–c. The gyro case is stationary,

[7]That is, except for the singular situation where the gimbal axes are coplanar.

Example 18.4 (Continued)

and the spin speed $\dot{\phi}$ of the disc relative to gimbal A is 10×10^3 r/min. What frictional torque must be developed on the rotor for this motion? From what bearings must such a torque arise? The radius of gyration for the disc is 50 mm for the axis of symmetry and 38 mm for the transverse axes at the center of mass. The weight of the disc is 4.5 N.

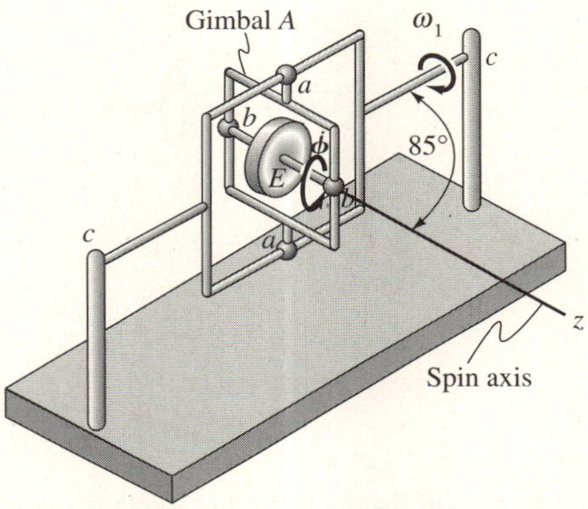

Figure 18.25. Two-degree-of-freedom gyro with body axis z.

Note in Fig. 18.26 that the spin axis z and fixed axis c–c form a plane that has a known angular speed ω_1 about axis c–c. Clearly, c–c then can be taken as the precession axis Z; the precession speed $\dot{\psi} = \omega_1$ is then 0.10 r/s. The line of nodes x is along axis a–a at all times. With $\theta = 85°$ at all times, we have steady precession. A constant torque M_x is required to maintain this motion. We can solve for M_x as follows (see Eq. 18.30(a)):

$$M_x = \left[I(\dot{\phi} + \dot{\psi}\cos\theta) - I'\dot{\psi}\cos\theta\right]\dot{\psi}\sin\theta$$

$$= \left\{\frac{4.5}{9.81}(0.05)^2\left[10\times10^3\,\frac{2\pi}{60} + (0.1)(2\pi)\cos 85°\right]\right.$$

$$\left. - \frac{4.5}{9.81}(0.038)^2(0.1)(2\pi)\cos 85°\right\}(0.1)(2\pi)\cos 85°$$

$$M_x = 0.752 \text{ N m}$$

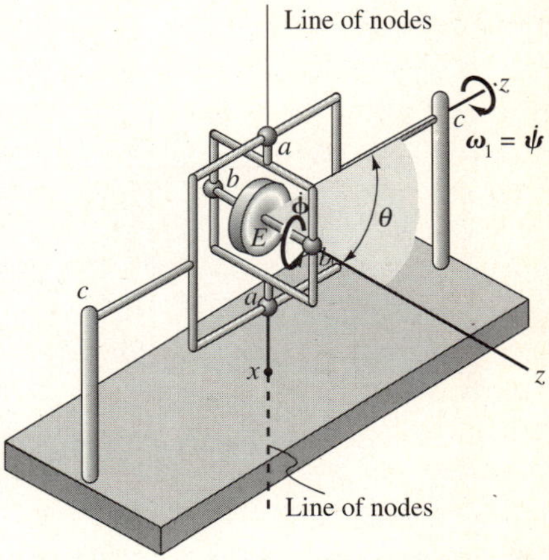

Figure 18.26. Axis c–c is precession axis.

Thus, bearings along the a–a axis interconnecting the two gimbals are developing the frictional torque.

PROBLEMS

18.31. The z axis coincides initially with the centerline of the block. The block is given the following rotations in the sequence listed: (a) $\psi = 30°$, (b) $\theta = 45°$, (c) $\phi = 20°$. What are the projections of the centerline OA along the XYZ axes in the final position?

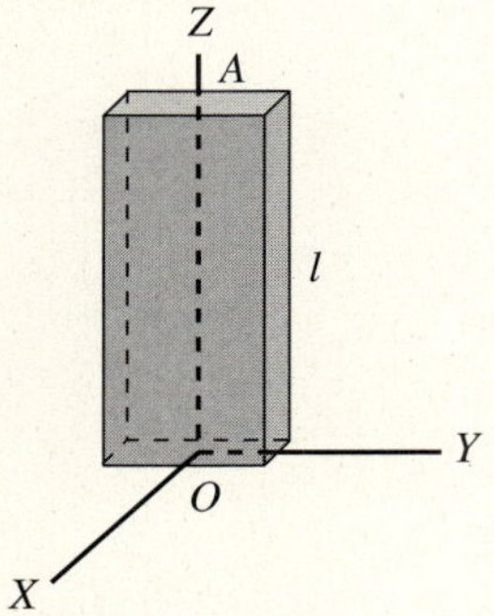

Figure P.18.31.

18.32. A disc A of mean diameter 300 mm rolls without slipping so that its centerline rotates an angular speed ω_1 of 2 rad/s about the Z axis. What are the precession, nutation, and spin angular velocity components?

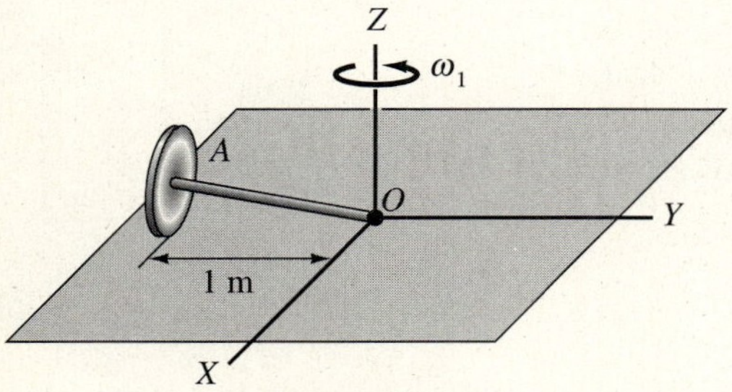

Figure P.18.32.

18.33. A body has the following components of angular velocity:

$$\dot{\phi} = 10 \text{ rad/s}$$
$$\dot{\theta} = 5 \text{ rad/s}$$
$$\dot{\psi} = 2 \text{ rad/s}$$

when the following Euler angles are known to be

$$\psi = 45°, \quad \theta = 30°$$

What is the magnitude of the total angular velocity?

18.34. If the disc shown were to be undergoing regular precession as shown at the rate of 0.3 rad/s, what would have to be the spin velocity $\dot{\phi}$? The disc weighs 90 N. Neglect the mass of the rod.

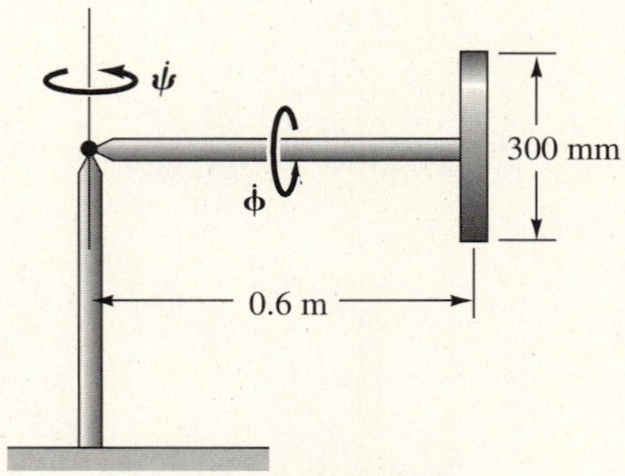

Figure P.18.34.

18.35. In Problem 18.34, explain how you institute such a motion. Would you get the steady precession if for the computed $\dot{\phi}$ you merely released the disc from a horizontal configuration of the disc centerline?

18.36. A 9 kg cylinder having a radius of 0.3 m and a length of 75 mm is connected by a 0.6 m rod to a fixed point O where there is a ball-joint connector. The cylinder spins about its own centerline at a speed of 50 rad/s. What external torque about O is required for the cylinder to precess uniformly at a rate of 0.5 rad/s about the Z axis at an inclination of 45° to the Z axis? (Compute the torque when the centerline of the cylinder is in the XZ plane.)

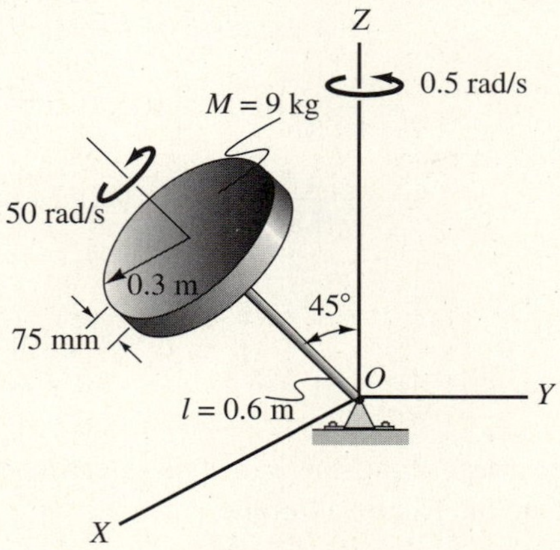

Figure P.18.36.

18.37. The centerline of the rod rotates uniformly in a horizontal plane with a constant torque of 2.29 N m applied about O. Each cylinder weighs 225 N and has a radius of 300 mm. The discs rotate on a bar AB with a speed ω_1 of 5000 r/min. Bar AB is held at O by a ball-joint connection. The applied torque is always perpendicular to AB and can only rotate about the vertical axis. What is the precession speed of the system?

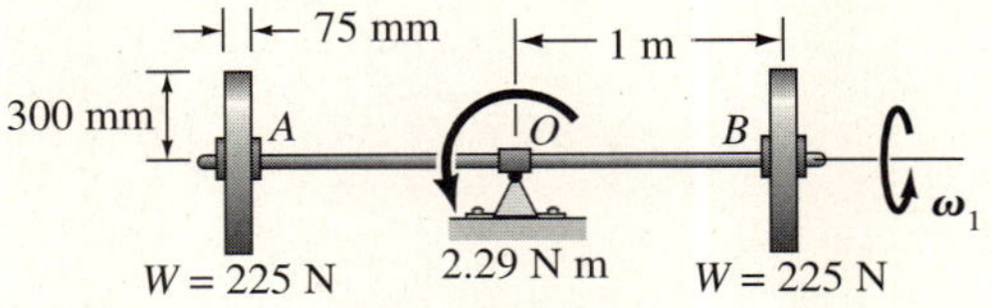

Figure P.18.37.

18.38. (a) In Problem 18.37, consider the disc at B to have an angular speed of 5000 r/min and the disc at A to have a speed of 2500 r/min. What is the precession speed for the condition of steady precession?

(b) If the disc at A and the disc at B have angular speeds of 5000 r/min in opposite directions, what is the initial motion of the system when a torque perpendicular to B is suddenly applied?

18.39. Two discs A and B roll without slipping at their mid-planes. Light shafts cd and ef connect the discs to a centerpost which rotates at an angular speed ω_1 of 2 rad/s. If each disc has a mass of 9 kg, what total force downward is developed by the discs on the ground support?

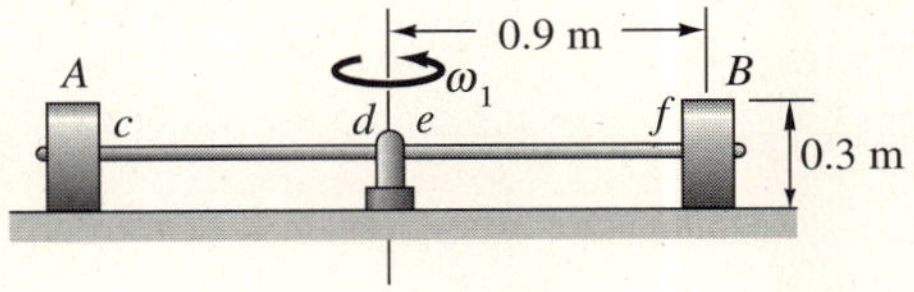

Figure P.18.39.

18.40. A disc is spinning about its centerline with speed $\dot{\phi}$ while the centerline is precessing uniformly at fixed angle θ about the vertical axis. The mass of the disc is M. Consider the evaluation of $\dot{\psi}$, and show that such a state of regular precession is possible if

$$\omega_z^2 > \frac{4I'Mgl}{I^2}$$

Also show that there are, for every θ, two possible precession speeds. In particular, show that as ω_z gets very large, the following precessional speeds are possible:

$$\dot{\psi}_1 = \frac{I\omega_z}{I'\cos\theta}, \qquad \dot{\psi}_2 = \frac{Mgl}{I\omega_z}$$

[*Hint:* Consider Eq. 18.31 for first part of proof. Then use a power expansion of the root when evaluating $\dot{\psi}$.]

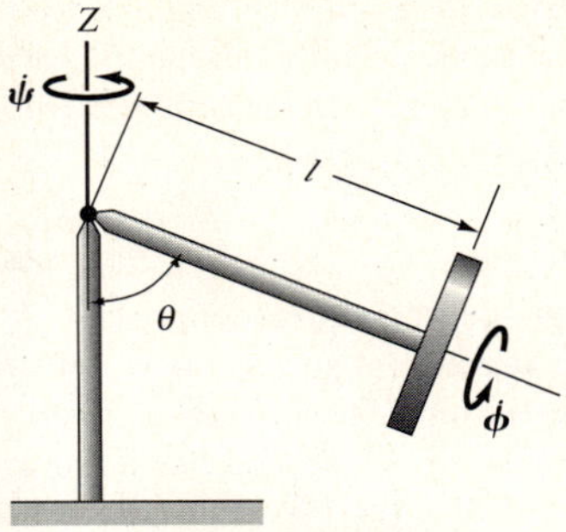

Figure P.18.40.

18.41. A uniform prism has a square cross section. Prove that $I_{\eta\eta}$ for any angle θ equals $I_{xx} = I_{yy}$. Thus, in order to use $I' = I_{xx} = I_{yy}$, as was done in the development in Section 18.6, the body need not be a body of revolution. Show in general that if $I_{xx} = I_{yy}$, and if xyz are principal axes, then $I_{\eta\eta} = I_{xx} = I_{yy}$.

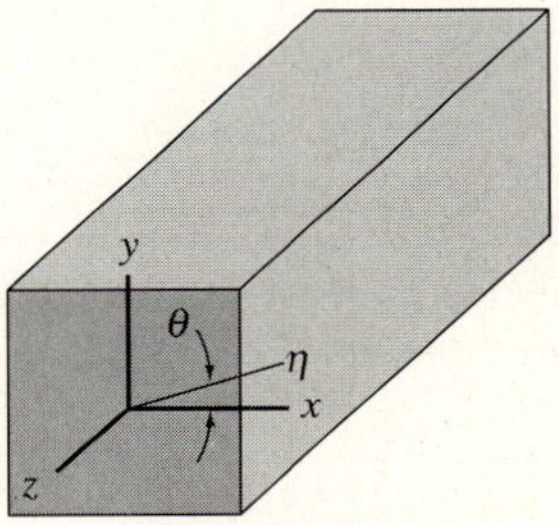

Figure P.18.41.

18.42. A block weighing 50 N and having a square cross-section is spinning about its body axis at a rate ω_1 of 100 rad/s. For an angle $\alpha = 30°$, what is the precession speed $\dot{\psi}$ of the axis of the block? Neglect the weight of rod AB. See Problem 18.41 before proceeding.

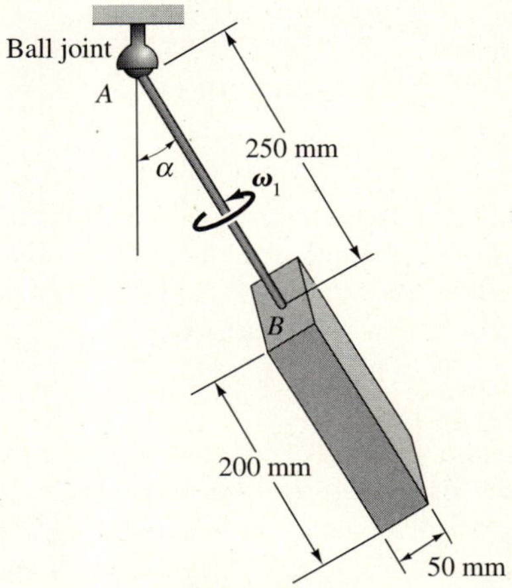

Figure P.18.42.

18.43. In Problem 18.42, at what angle α does a steady rotation of the axis of the block occur about the indicated vertical axis at the rate of 6 rad/s counterclockwise looking down from above? The angular speed ω_1 is 100 rad/s.

18.44. A single-degree-of-freedom gyro is mounted on a vehicle moving at constant speed V of 30 m/s on a track which is coplanar and is circular, having a mean radius of 60 m. The disc has a mass of 0.5 kg and has a radius of 50 mm. It is turning at a speed of 20×10^3 r/min relative to the gimbal. If the gimbal maintains a rotated position of 15° with the horizontal, what is the equivalent torsional spring constant about axis A–A for the gimbal suspension?

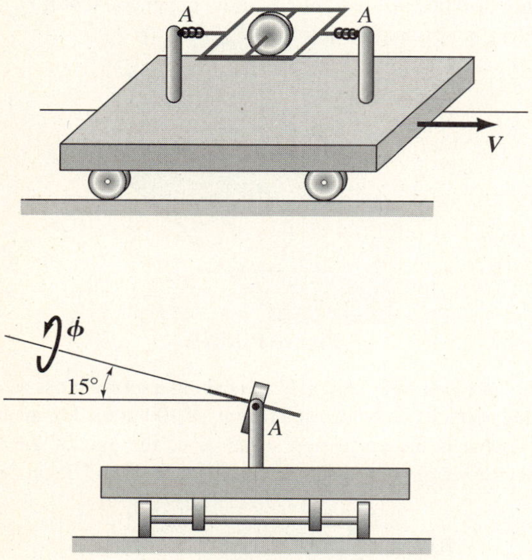

Figure P.18.44.

18.45. In Problem 18.44, where we found that K_t = 2.25 kN m/rad, suppose that the speed of the vehicle were adjusted to 15 m/s (i.e., half its given speed). What would then be the position of the gimbal for steady-state precession?

18.46. A plate is rotating about shaft CD at a constant speed ω_1 of 2 rad/s. A single-degree-of-freedom gyro is mounted on the plate. The gyro has mounted on it through a gimbal a disc of weight 3.4 N and radius 50 mm. The disc has a rotational speed ω_2 of 10×10^3 r/min relative to the gimbal. If the gimbal is to be maintained parallel to the plate, what torque is required? If the bearings are frictionless, explain how this torque is developed. What is the direction of the line of nodes for steady precession?

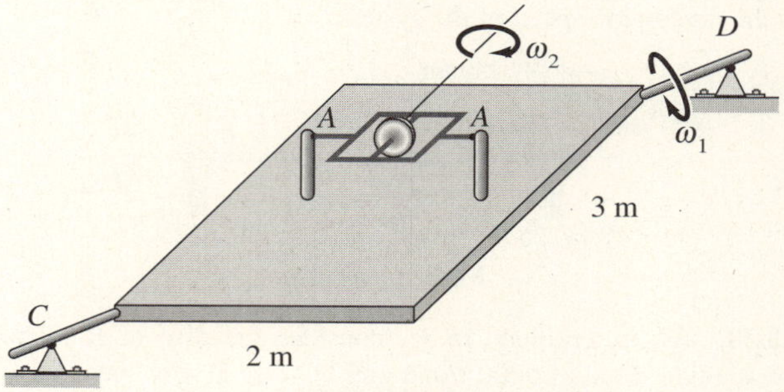

Figure P.18.46.

***18.47.** A solid sphere of diameter 200 mm and weight 50 N is spinning about its body axis at a speed $\dot{\phi}_0$ of 100 rad/s while its axis is held stationary at an angle θ_0 of 120°. The sphere is released suddenly. What is the precession speed $\dot{\psi}$ when θ has increased by 10°? Use the fact that

$$H_z = \text{const.} = I'\dot{\psi}\sin^2\theta + I\omega_z\cos\theta$$

(This equation can be reached by projecting H_x, H_y, and H_z in the Z direction.)

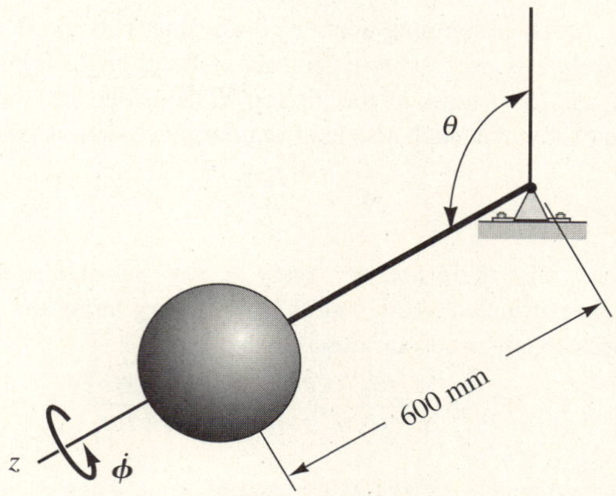

Figure P.18.47.

18.7 Torque-Free Motion

We shall now consider possible *torque-free motion* for a body having $I_{xx} = I_{yy} = I'$ for axes normal to the body axis. An example of a device that can approach such motion is the two-degree-of-freedom gyroscope described in Example 18.4.

Let us now examine the equations of torque-free motion. First, consider the general relation

$$\boldsymbol{M}_C = \dot{\boldsymbol{H}}_C \tag{18.33}$$

Since $\boldsymbol{M}_c$ is zero, $\boldsymbol{H}_c$ must be constant. Thus,

$$\boldsymbol{H}_C = \boldsymbol{H}_0 \tag{18.34}$$

where $\boldsymbol{H}_0$ is the *initial* angular momentum about the mass center. We shall first assume and later justify in this section that all torque-free motions will be *steady precessions* about an axis going through the center of mass and directed *parallel to the vector* $\boldsymbol{H}_0$. Accordingly, we choose Z to pass through the center of mass and to have a direction corresponding to that of $\boldsymbol{H}_0$, as shown in Fig. 18.27. The axis x' normal to axes z and Z forming plane I is then the line of nodes and y' is in plane I. The reference xyz is *fixed* to the body and hence spins about the z axis. Axes xy then rotate in the $x'y'$ plane as shown in the diagram. Using Fig. 18.27, we can then express $\boldsymbol{H}_0$ in terms of its x, y, and z components having unit vectors $\boldsymbol{i}, \boldsymbol{j}$, and $\boldsymbol{k}$ in the following way at all times:

$$\boldsymbol{H}_0 = H_0 \sin\theta \sin\phi \boldsymbol{i} + H_0 \sin\theta\cos\phi\boldsymbol{j} + H_0\cos\theta\boldsymbol{k} \tag{18.35}$$

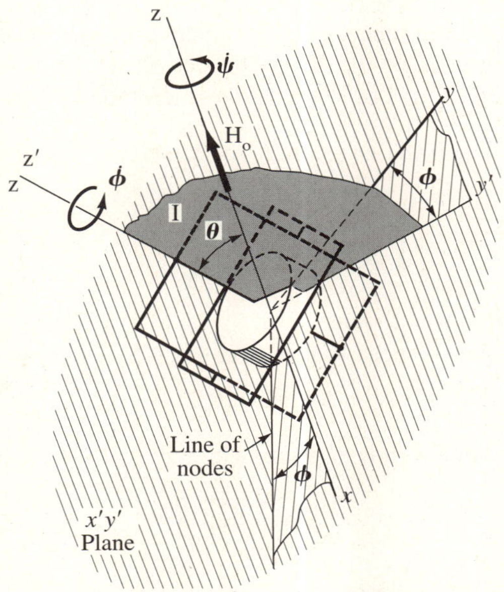

Figure 18.27. Two-degree-of-freedom gyroscope illustrating torque-free motion. Outer gimbal support is not shown.

Since xyz are principal axes, we can also state:

$$\boldsymbol{H}_0 = I'\omega_x\boldsymbol{i} + I'\omega_y\boldsymbol{j} + I\omega_z\boldsymbol{k} \tag{18.36}$$

Comparing Eqs. 18.35 and 18.36, we then have

$$\omega_x = \frac{H_0 \sin\theta \sin\phi}{I'} \tag{18.37a}$$

$$\omega_y = \frac{H_0 \sin\theta \cos\phi}{I'} \tag{18.37b}$$

$$\omega_z = \frac{H_0 \cos\theta}{I} \tag{18.37c}$$

By using the preceding formulations for ω_x, ω_y, and ω_z, we can write *Euler's equations* using axes xyz fixed to the body in a form that includes the constant H_0;

$$I'\frac{d}{dt}\left(\frac{H_0 \sin\theta \sin\phi}{I'}\right) + \frac{I - I'}{I'I} H_0^2 \sin\theta \cos\theta \cos\phi = 0 \tag{18.38a}$$

$$I'\frac{d}{dt}\left(\frac{H_0 \sin\theta \cos\phi}{I'}\right) + \frac{I' - I}{I'I} H_0^2 \sin\theta \cos\theta \sin\phi = 0 \tag{18.38b}$$

$$I\frac{d}{dt}\left(\frac{H_0 \cos\theta}{I}\right) = 0 \tag{18.38c}$$

From Eq. 18.38c, it is then clear that

$$\left(\frac{H_0 \cos\theta}{I}\right) = \text{constant} \tag{18.39}$$

Thus, since H_0 and I are constant, we can conclude from this equation that *the nutation angle* is a fixed angle θ_0.[8] Now consider Eq. 18.38b, using the fact that $\theta = \theta_0$. Canceling H_0 and carrying out the differentiation, we get

$$-(\sin\theta_0)(\sin\phi)(\dot{\phi}) + \frac{I' - I}{I'I} H_0 \sin\theta_0 \cos\theta_0 \sin\phi = 0$$

Therefore,

$$\dot{\phi} = \frac{I' - I}{I'I} H_0 \cos\theta_0 \tag{18.40}$$

Thus, the *spin speed*, $\dot{\phi}$, is constant.

To get the precession speed $\dot{\psi}$, note from Fig. 18.27 that

$$\omega_z = \dot{\phi} + \dot{\psi}\cos\theta_0$$

Now equate the right side of this equation with the right side of Eq. 18.37c:

$$\dot{\phi} + \dot{\psi}\cos\theta_0 = \frac{H_0 \cos\theta_0}{I} \tag{18.41}$$

[8] We now see that taking Z to be collinear with $\boldsymbol{H}_0$ at the outset is justified.

Substituting for $\dot{\phi}$ from Eq. 18.40 and solving for $\dot{\psi}$, we get

$$\dot{\psi} = \frac{H_0}{I} - H_0\frac{I' - I}{I'I}$$

Collecting terms, we have

$$\dot{\psi} = \frac{H_0}{I}\left(1 - \frac{I' - I}{I'}\right) = \frac{H_0}{I'} \tag{18.42}$$

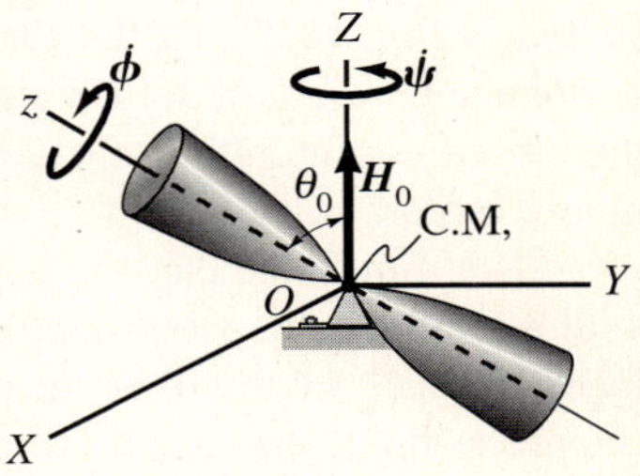

Figure 18.28. Body of revolution is also symmetric about C.M. where it is supported.

The results of the discussion for torque-free motion of the body of revolution can then be given as

$$\theta = \theta_0 \tag{18.43a}$$

$$\dot{\psi} = \frac{H_0}{I'} \tag{18.43b}$$

$$\dot{\phi} = \frac{I' - I}{I'I}H_0\cos\theta_0 \tag{18.43c}$$

Since $\dot{\theta}$, $\ddot{\theta}$, $\ddot{\psi}$, and $\ddot{\phi}$ are all zero, Eqs. 18.43 depict a case of steady precession, and so the assumption made earlier to this effect is completely consistent with the results emerging from *Euler's equation*. Thus, we can consider the assumption and the ensuing conclusions as correct.

Hence, if a body of revolution is torque-free—as, for example, in the case illustrated in Fig. 18.28, where the center of mass is fixed and where the body has initially any angular momentum vector $\boldsymbol{H}_0$, then at all times the angular momentum $\boldsymbol{H}$ is constant and equals $\boldsymbol{H}_0$. Furthermore, the body will have a regular precession that consists of a constant angular velocity $\dot{\psi}$ of the centerline about a Z axis collinear with $\boldsymbol{H}_0$ at a fixed inclination θ_0 from Z. Finally, there is a constant spin speed $\dot{\phi}$ about the centerline. Thus, two angular velocity vectors $\dot{\boldsymbol{\psi}}$ and $\dot{\boldsymbol{\phi}}$, are present, and the *total* angular velocity $\boldsymbol{\omega}$ is at an inclination of ϵ from the Z axis (see Fig. 18.29) and precesses with angular speed $\dot{\psi}$ about the Z axis. This must be true, since the direction of one component of $\boldsymbol{\omega}$, namely $\dot{\boldsymbol{\phi}}$, precesses in this manner while the other component, $\dot{\boldsymbol{\psi}}$, is fixed in the Z direction. The vector $\boldsymbol{\omega}$ then can be considered to continuously sweep out a cone, as illustrated in Fig. 18.29.

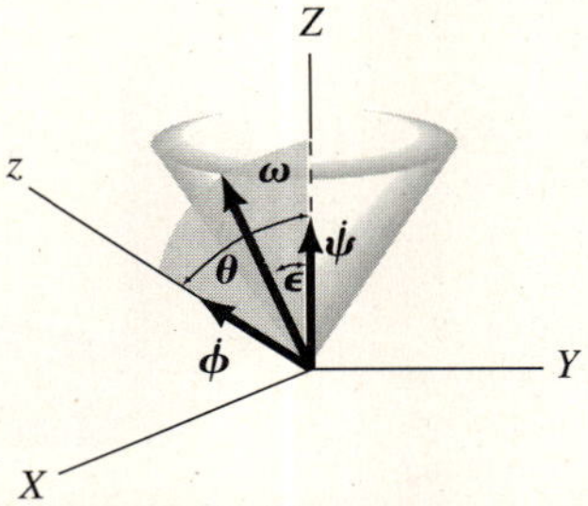

Figure 18.29. Vector $\boldsymbol{\omega}$ sweeps out a conical surface.

We now illustrate the use of the basic formulations for torque-free motion.

Example 18.5

A cylindrical space capsule in orbit is shown in Fig. 18.30. A quarter section of the cylinder can be opened about AB to a test configuration as shown in Fig. 18.31. At the earth's surface, the end plates of the capsule each have a weight of 1.4 kN and the cylindrical portion weighs 8.4 kN. In the closed configuration, the center of mass of the capsule is at the geometric center.

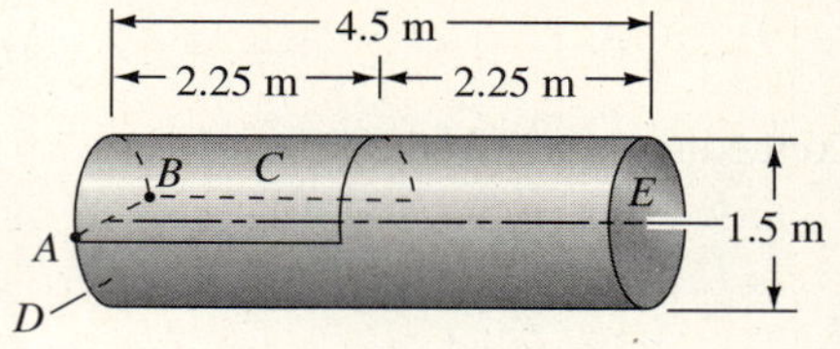

Figure 18.30. Space capsule.

If at time t the capsule is placed in a *test configuration* (Fig. 18.31) with a total instantaneous angular speed ω of 2 rad/s in the z direction in inertial space,[9] what will be the precession axis for the capsule and the rate of precession of the capsule when door C subsequently is closed by an internal mechanism?

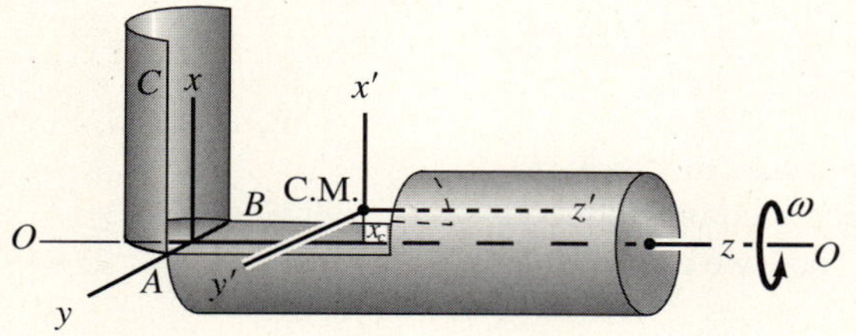

Figure 18.31. Space capsule in initial test configuration.

During a change of configuration, there is a zero net torque on the system so that $\boldsymbol{H}_C$, the angular momentum about the center of mass, is not changed. As a first step we shall compute $\boldsymbol{H}_C$ using data for the instantaneous test configuration. For this calculation we shall need the position of the center of mass.

A reference xyz has been fixed to the system, as shown in Fig. 18.31, and the system has been decomposed into simple portions in Fig. 18.32 for convenience in carrying out ensuing calculations. Employing formulas for positions of centers of mass as given in the front inside covers, we have for moments of mass about the origin of xyz:

$$\frac{11.2\times10^3}{g}\boldsymbol{r}_C = \sum_{i=1}^{6} M_i(\boldsymbol{r}_C)_i = \frac{2100}{g}\left(\frac{4.5}{4}\boldsymbol{i}-\frac{1.5}{\pi}\boldsymbol{k}\right)+\frac{700}{g}\left[-(0.424)\left(\frac{1.5}{2}\right)(\boldsymbol{k})\right]$$
$$+\frac{700}{g}\left[-(0.424)\left(\frac{1.5}{2}\right)(\boldsymbol{i})\right]+\frac{2100}{g}\left(\frac{4.5}{4}\boldsymbol{k}-\frac{1.5}{\pi}\boldsymbol{i}\right)$$
$$+\frac{4200}{g}\left(\frac{3}{4}\right)(4.5)\boldsymbol{k}+\frac{1400}{g}(4.5)\boldsymbol{k}$$

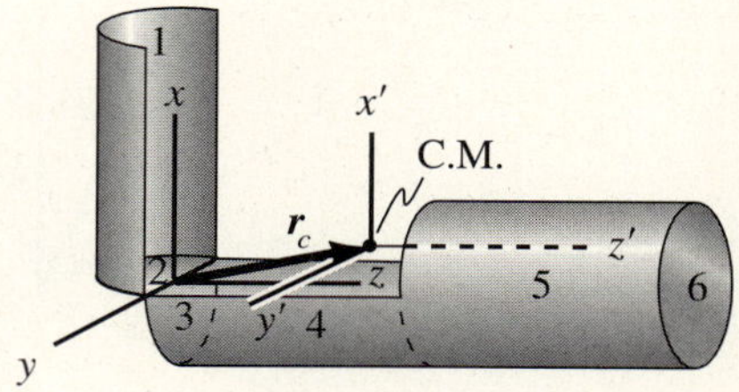

Figure 18.32. Space capsule subdivided into simple shapes.

[9]The initial motion with just rotation about the z axis can be only an *instantaneous* motion at the instant when the device is placed into the shown configuration. At this instant, $\boldsymbol{H}_C$ is at some angle relative to the z axis [as you will soon see in Eq. (d)] and is rotating about z at an angular speed $\boldsymbol{\omega}$. Subsequent motion, being torque-free, requires $\boldsymbol{H}_C$ to be *constant* and thus, in turn, means that the z axis will thereafter have to be rotating about $\boldsymbol{H}_C$.

Example 18.5 (Continued)

Therefore,

$$\boldsymbol{r}_c = 1.838\boldsymbol{k} + 0.102\boldsymbol{i} \text{ m} \tag{a}$$

To get $\boldsymbol{H}_C$, we next set up a second reference $x'y'z'$ at the center of mass as shown in Fig. 18.31. In accordance with Eq. 18.2, we have for $\boldsymbol{H}_C$, on noting that the only nonzero component of $\boldsymbol{\omega}$ is in the z' direction:

$$\boldsymbol{H}_C = -I_{x'z'}\omega\boldsymbol{i} - I_{y'z'}\omega\boldsymbol{j} + I_{z'z'}\omega\boldsymbol{k} \tag{b}$$

To compute $I_{z'z'}$ we proceed as follows, using the decomposed capsule parts of Fig. 18.32 and employing transfer theorems for moments of inertia:

$$(I_{z'z'})_1 = \frac{1}{2}\left(\frac{2100}{g}\right)\left(0.75^2 + \frac{2.25^2}{6}\right) + \frac{2100}{g}(1.125 - 0.102)^2 = 374.54 \text{ kg m}^2$$

$$(I_{z'z'})_2 = \frac{1}{4}\left(\frac{700}{g}\right)\left(0.75^2\right) + \frac{700}{g}(0.102)^2 = 10.78 \text{ kg m}^2$$

$$(I_{z'z'})_3 = \left\{\frac{1}{2}\left(\frac{700}{g}\right)(0.75)^2 - \frac{700}{g}[(0.424)(0.75)]^2\right\} + \frac{700}{g}[0.102 + (0.424)(0.75)]^2 = 25.48 \text{ kg m}^2$$

$$(I_{z'z'})_4 = \left[\frac{2100}{g}(0.75)^2 - \frac{2100}{g}\left(\frac{1.5}{\pi}\right)^2\right] + \frac{2100}{g}\left(0.102 + \frac{1.5}{\pi}\right)^2 = 143.49 \text{ kg m}^2$$

$$(I_{z'z'})_5 = \frac{4200}{g}(0.75)^2 + \frac{4200}{g}(0.102)^2 = 245.28 \text{ kg m}^2$$

$$(I_{z'z'})_6 = \frac{1}{2}\left(\frac{1400}{g}\right)(0.75)^2 + \frac{1400}{g}(0.102)^2 = 41.62 \text{ kg m}^2$$

Accordingly, we have for $I_{z'z'}$:

$$I_{z'z'} = \sum_{i=1}^{6}(I_{zz})_i = 841 \text{ kg m}^2 \tag{c}$$

We proceed in a similar manner to compute $I_{z'x'}$. We will illustrate this computation for portion 1 of the system and then give the total result. Thus, employing the parallel axis theorem, we have

$$(I_{z'x'})_1 = 0 + \left(\frac{2100}{g}\right)\left(-1.838 - \frac{1.5}{\pi}\right)\left(\frac{2.25}{2} - 0.102\right) = -528 \text{ kg m}^2$$

It is important to remember that the transfer distances (with proper signs) used for computing $(I_{z'x'})_1$ using the parallel-axis theorem are measured from $x'y'z'$, *about which you are making the calculation, to the center of mass of body 1.* For the entire system we have, by similar calculations:

$$I_{z'x'} = -454 \text{ kg m}^2$$

Example 18.5 (Continued)

Because of symmetry, furthermore:

$$I_{z'y'} = 0$$

From Eq. (b) we then have for $\boldsymbol{H}_C$:

$$\begin{aligned} \boldsymbol{H}_C &= 434\omega \boldsymbol{i} + 841\omega \boldsymbol{k} \\ &= 908\boldsymbol{i} + 1682\boldsymbol{k} \end{aligned} \qquad \text{(d)}$$

Since the precession axis is collinear with $\boldsymbol{H}_C$, we have thus established this axis (Z axis), as is shown in Fig. 18.33. A reference $\xi\eta\zeta$ has been set up at the center of mass for the closed configuration. The body axis ζ remains at a fixed angle θ with the Z axis and precesses about it with an angular speed $\dot{\psi}$ given in accordance with Eq. 18.43b as

$$\begin{aligned} \dot{\psi} &= \frac{H_C}{I_{\xi\xi}} \\ &= \frac{\sqrt{908^2 + 1682^2}}{\frac{1}{2}\left(\frac{8400}{g}\right)\left(0.75^2 + \frac{4.5^2}{6}\right) + 2\left(\frac{1}{4}\right)\left(\frac{1400}{g}\right)\left(0.75^2\right) + 2\left(\frac{1400}{g}\right)(2.25)^2} \qquad \text{(e)} \\ &= \frac{1911.4}{3170.9} = \boxed{0.603 \text{ rad/s}} \end{aligned}$$

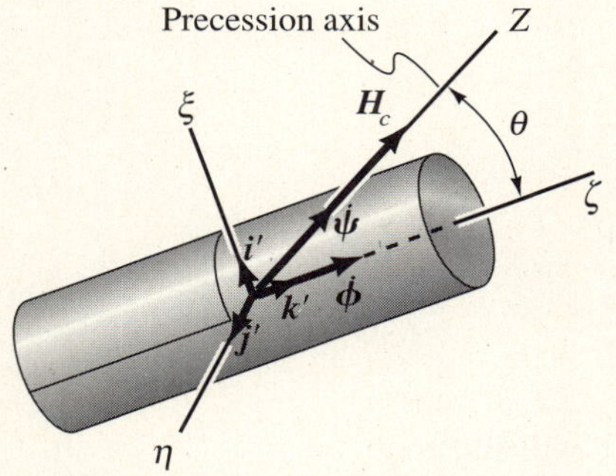

Figure 18.33. Space capsule in final configuration.

We can now state that the body axis will precess around an axis collinear with $\boldsymbol{H}_C$ as given by Eq. (d) with a speed of 0.603 rad/s. However, we are unable to determine θ and $\dot{\phi}$ with what information we now have available. We need more information as to the way the door was closed. Thus, knowing how much net work was done in the configuration change from the work energy relations, we can write another equation. And, with Eqs. 18.37 and Eq. 18.40, we can compute θ and $\dot{\phi}$. We shall present several problems with that information available in the homework exercises.

Let us now examine Eqs. 18.43 for the special case where $I = I'$. Here the spin velocity $\dot{\phi}$ must be zero, leaving only *one* angular motion, $\dot{\boldsymbol{\psi}}$, the precession velocity. This rotation is about the Z axis so the direction of angular velocity $\boldsymbol{\omega}$ of the body corresponds to $\boldsymbol{H}_0$. Since this could be a body of revolution, the moment-of-inertia condition for this case ($I = I'$) means that the moments of inertia for principal axes x, y, and z are mutually equal, and we can verify from Eq. 9.13 that all axes inclined to the xyz reference have the same moment of inertia I (and all therefore are principal axes at the point). Thus, the body, if homogeneous, could be a sphere, a cube, any regular polyhedron, or any body that possesses point symmetry. *No matter how we launch this body, the angular momentum* $\boldsymbol{H}$ *will be equal to* $I\boldsymbol{\omega}$ *and will thus always coincide with the direction of angular velocity* $\boldsymbol{\omega}$. This situation can also be shown analytically as follows:

$$\boldsymbol{H} = H_x\boldsymbol{i} + H_y\boldsymbol{j} + H_z\boldsymbol{k} \tag{18.44}$$

For principal axes, we have

$$\boldsymbol{H} = \omega_x I_{xx}\boldsymbol{i} + \omega_y I_{yy}\boldsymbol{j} + \omega_z I_{zz}\boldsymbol{k} \tag{18.45}$$

If $I_{xx} = I_{yy} = I_{zz} = I$, we have for the foregoing

$$\boldsymbol{H} = I(\omega_x\boldsymbol{i} + \omega_y\boldsymbol{j} + \omega_z\boldsymbol{k}) = I\boldsymbol{\omega} \tag{18.46}$$

indicating that $\boldsymbol{H}$ and $\boldsymbol{\omega}$ must be collinear. The situation just described represents the case of a thrown baseball or basketball.

There are *two* situations for the case $I_{xx} = I_{yy}$ in which $\boldsymbol{H}$ and $\boldsymbol{\omega}$ are collinear. Examining Eq. 18.43, we thus see that if θ_0 is 90°, then $\dot{\phi} = \dot{\theta} = 0$, leaving only precession $\dot{\psi}$ along the Z axis (see Fig. 18.27). The Z axis for the analysis corresponds to the direction of $\boldsymbol{H}$. We thus see that since $\boldsymbol{\omega} = \dot{\boldsymbol{\psi}}$ then $\boldsymbol{\omega}$ is collinear with $\boldsymbol{H}$. This case corresponds to a proper "drop kick" or "place kick" of a football [Fig. 18.34(a)] wherein the body axis z is at right angles to the Z axis.

The other case consists of $\theta_0 = 0$. This means that $\dot{\boldsymbol{\phi}}$ and $\dot{\boldsymbol{\psi}}$ have the same direction—that is, along the Z direction (see Fig. 18.27) which then means that $\boldsymbol{\omega}$ and $\boldsymbol{H}$ again are collinear. This case corresponds to a good football pass [Fig. 18.34(b)]. *For all other motions of bodies where* $I_{xx} = I_{yy}$, *the angular velocity vector* $\boldsymbol{\omega}$ *will not have the direction of angular momentum* $\boldsymbol{H}_0$.

Upon further consideration, we can make a simple model of torque-free motion. Start with the fixed cone described earlier (see Fig. 18.29) about the Z axis, where the cone surface is that swept out by the total angular velocity vector $\boldsymbol{\omega}$ of the torque-free body. Now consider a second cone about the spin axis z of the torque-free body (see Fig. 18.35) in direct contact with the initial stationary cone. Rotate the second cone about its axis with a speed and sense corresponding to $\dot{\phi}$ of the torque-free body and impose a no-slipping

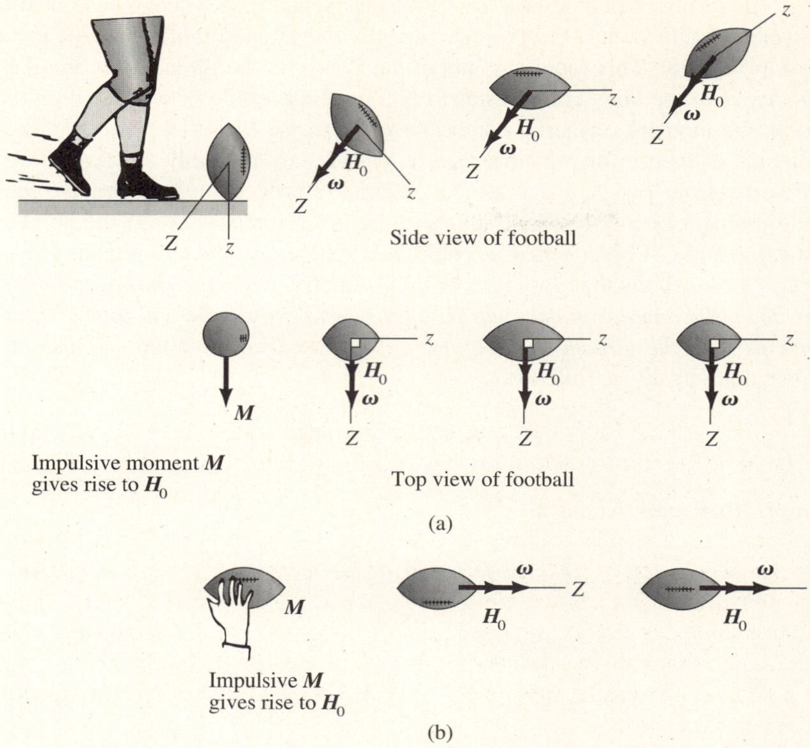

Figure 18.34. Two cases where $\boldsymbol{H}$ and $\boldsymbol{\omega}$ are collinear.

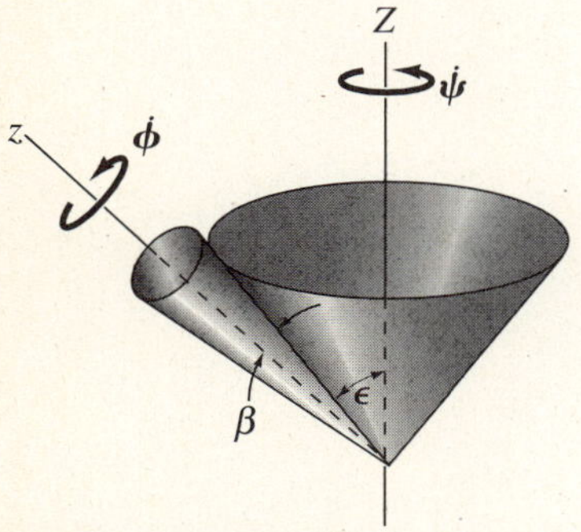

Figure 18.35. Rolling-cone model.

condition between the cones. Clearly, the second cone will precess about the Z axis at some speed $\dot{\psi}'$. Also, the total angular velocity $\boldsymbol{\omega}'$ of the moving cone will lie along the line of contact between cones and is thus collinear with $\boldsymbol{\omega}$. We shall now show that the speed $\dot{\boldsymbol{\psi}}'$ of the cone model equals $\dot{\boldsymbol{\psi}}$ of the torque-free body and, as a consequence, that $\boldsymbol{\omega}'$ for the moving cone equals $\boldsymbol{\omega}$ of the torque-free body. We know now that:

1. $\dot{\boldsymbol{\phi}}$ is the same for both the device and the physical case.
2. The direction of resultant angular velocity is the same for both cases.
3. The direction of the precession velocity must be the same for both cases (i.e., the Z direction).

This situation is shown in Fig. 18.36. Note that $\dot{\boldsymbol{\phi}}$ is the same in both the physical case and the mechanical model and that the directions of $\boldsymbol{\omega}$ and $\boldsymbol{\omega}'$ as well as $\dot{\boldsymbol{\psi}}$ and $\dot{\boldsymbol{\psi}}'$ are, respectively, the same for both diagrams. Accordingly, when we consider the construction of the parallelogram of vectors, we see that the vectors $\boldsymbol{\omega}$ and $\boldsymbol{\omega}'$ as well as $\dot{\boldsymbol{\psi}}$ and $\dot{\boldsymbol{\psi}}'$ must necessarily be equal for both the physical case and the model, respectively.

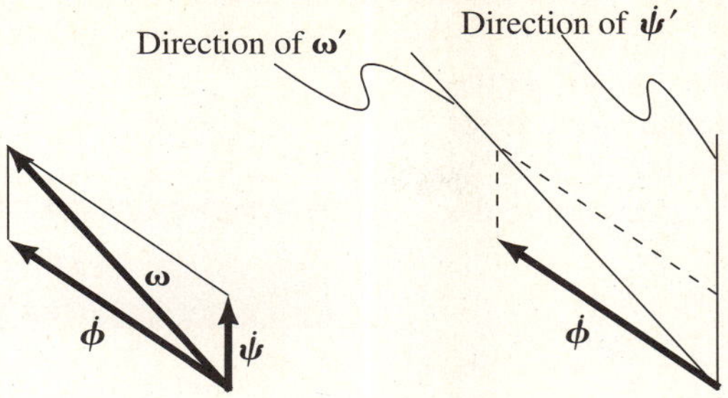

Figure 18.36. Angular vector diagrams.

We shall now investigate more carefully the relation between the sense of rotation for corresponding angular velocities between the model and the physical case for certain classes of geometries of the physical body.

1. $I' > I$. From Eq. 18.43c, we see that when θ_0 is less than $\pi/2$ rad, $\dot{\phi}$ is positive for this case.[10] Thus, the spin must be counterclockwise as we look along the z axis toward the origin. From Eq. 18.43b, we see that $\dot{\psi}$ is positive and thus counterclockwise as we look toward the origin along Z. Clearly, from these stipulations, the rolling-cone model shown in Fig. 18.35 has the proper motion for this case. The motion is termed *regular precession*.
2. $I' < I$. Here, the spin $\dot{\phi}$ will be negative for a nutation angle less than 90° as stipulated by Eq. 18.43c. However, the precession $\dot{\psi}$ must still be positive in accordance with Eq. 18.43b. The rolling-cone model thus far presented clearly cannot give these proper senses, but if the moving cone is *inside* the stationary cone (Fig. 18.37), we have motion that is consistent with the relations in Eq. 18.43. Such motion is called *retrograde precession*.

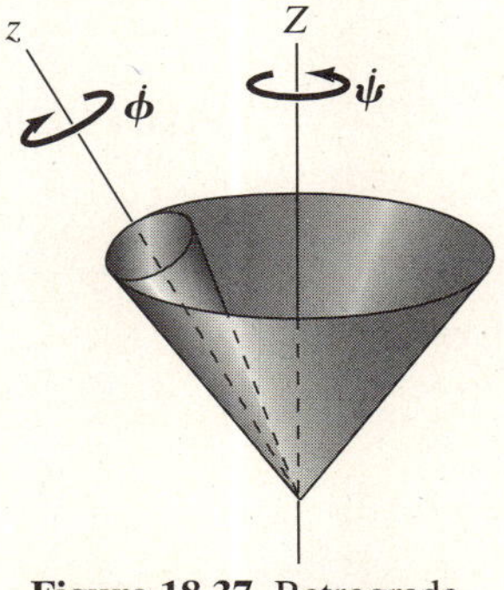

Figure 18.37. Retrograde precession.

[10]Recall from Eq. 18.35 that H_0 is just a magnitude. Also, note that this case corresponds to what has been shown in Fig. 18.29—namely, the case we have just discussed.

Example 18.6

The space capsule of Example 18.5 is shown again in Fig. 18.38 rotating about its axis of symmetry z in inertial space with an angular speed ω_z of 2 rad/s. As a result of an impact with a meteorite, the capsule is given an impulse $\boldsymbol{I}$ of 90 N s at position A as shown in the diagram. Ascertain the postimpact motion.

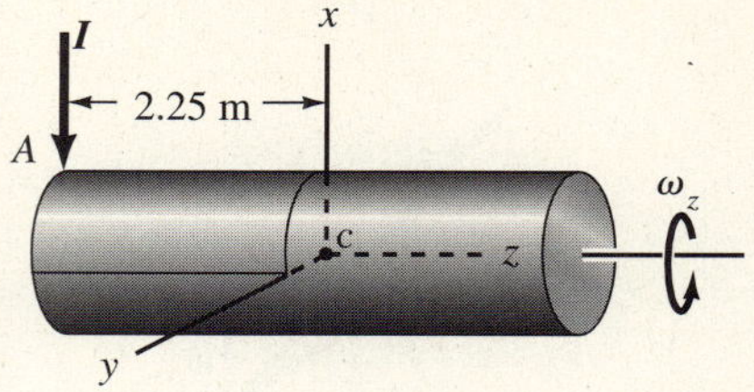

Figure 18.38. Space capsule with impulse.

The impact will give the cylinder an angular impulse:

$$\boldsymbol{I}_{\text{ang}} = (90)(2.25)\boldsymbol{j} = 202.5\boldsymbol{j} \text{ kg m}^2/\text{s} \tag{a}$$

From the **angular impulse-momentum** equation we can say for the impact:

$$\boldsymbol{I}_{\text{ang}} = \boldsymbol{H}_2 - \boldsymbol{H}_1$$

Therefore,

$$202.5\boldsymbol{j} = \boldsymbol{H}_2 - I_{zz}\omega_z\boldsymbol{k}$$

The postimpact angular momentum is then

$$\begin{aligned}\boldsymbol{H}_2 &= 202.5\boldsymbol{j} + I_{zz}\omega_z\boldsymbol{k}\\ &= 202.5\boldsymbol{j} + \left\{2\left[\frac{1}{2}\left(\frac{1400}{g}\right)(0.75)^2\right] + \frac{8400}{g}(0.75)^2\right\}2\boldsymbol{k} \qquad \text{(b)}\\ &= 202.5\boldsymbol{j} + 1123.9\boldsymbol{k} \text{ kg m}^2/\text{s}\end{aligned}$$

Since the ensuing motion is *torque-free*, we have thus established the direction of the precession axis (Z). This axis has been shown in Fig. 18.39 coinciding with $\boldsymbol{H}_2$ in the yz plane. Furthermore, we can give $\boldsymbol{H}_2$ postimpact as follows remembering that xyz are principal axes:[11]

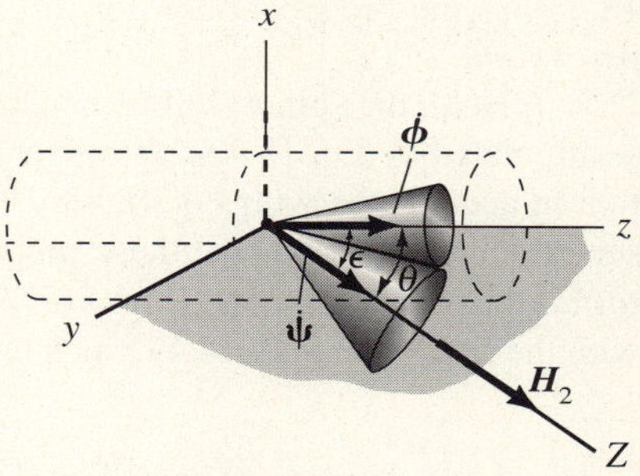

Figure 18.39. Cone model; Z is in yz plane.

$$\boldsymbol{H}_2 = 202.5\boldsymbol{j} + 1123.9\boldsymbol{k} = \omega_x I_{xx}\boldsymbol{i} + \omega_y I_{yy}\boldsymbol{j} + \omega_z I_{zz}\boldsymbol{k} \tag{c}$$

Hence, from Eq. (c) we have at postimpact:

$$\begin{aligned}\omega_x &= 0\\ \omega_y &= \frac{202.5}{I_{yy}} = \frac{202.5}{3170.9} = 63.9 \text{ mrad/s}\\ \omega_z &= \frac{1123.9}{I_{zz}} = \frac{1123.9}{561.9} = 2 \text{ rad/s}\end{aligned}$$

where for I_{yy} we used the value of $I_{\xi\xi}$ as computed in Eq. (e) of Example 18.5. Hence,

$$\boldsymbol{\omega} = 63.94 \times 10^{-3}\boldsymbol{j} + 2\boldsymbol{k} \text{ rad/s} \tag{d}$$

[11]Remember that the *position* of a body is considered *not* to change during the action of an impulsive force whereas its linear and angular *momenta do* change. Therefore, the xyz axes postimpact have not moved from the position corresponding to preimpact.

Example 18.6 (Continued)

We can now make good use of the cone model representing the motion. Accordingly, in Fig. 18.39 corresponding to postimpact we have shown two cones, one about the body axis z (this is the moving cone) and one about the Z axis (this is the stationary cone). The line of contact between the cones coincides with the total angular velocity vector $\boldsymbol{\omega}$. The capsule must subsequently have the same motion as the moving cone as it rolls with an angular speed $\dot{\phi} = 2$ rad/s about its own axis without slipping on the stationary cone having Z as an axis. We can easily compute the angles θ and ϵ using Eqs. (b) and (d). Thus:

$$\tan\theta = \frac{H_y}{H_z} = \frac{202.5}{1123.9} = 0.1802$$

Therefore,

$$\theta = 10.21°$$

$$\epsilon = \theta - \tan^{-1}\frac{\omega_y}{\omega_z} = 10.21° - \tan^{-1}\frac{0.0639}{2}$$

Therefore,

$$\epsilon = 8.36°$$

In Fig. 18.40 we have shown ω, $\dot{\phi}$, and $\dot{\psi}$ in the yz plane corresponding to postimpact. Knowing $\dot{\phi}$, ω, ϵ, and θ, we can easily compute $\dot{\psi}$. Thus, using the law of sines, we get

$$\frac{\dot{\psi}}{\sin(\theta - \epsilon)} = \frac{\omega}{\sin(\pi - \theta)}$$

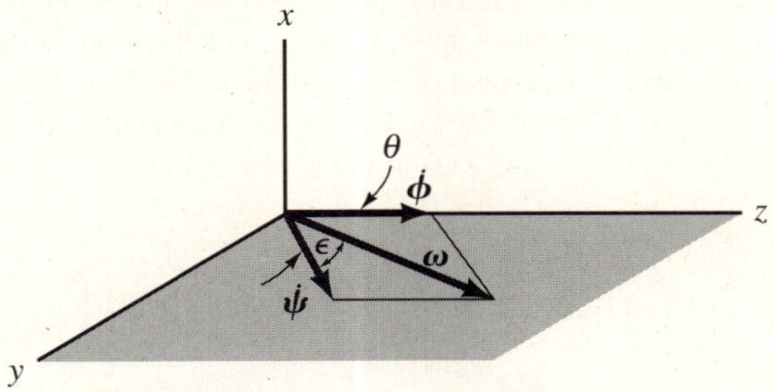

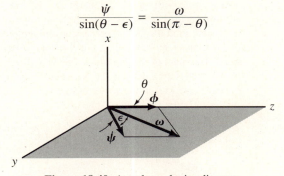

Figure 18.40. Angular-velocity diagram.

Therefore,

$$\dot{\psi} = \frac{\sin(\theta - \epsilon)\omega}{\sin\theta} = \frac{(0.0319)(2^2 + 0.0639^2)^{1/2}}{0.1773} = \boxed{0.360 \text{ rad/s}}$$

Thus, we can say that the body continues to spin at 2 rad/s about its axis, but now the axis precesses about the indicated Z direction at the rate of 0.360 rad/s.

PROBLEMS

18.48. A dynamical model of a device in orbit consists of 900 kg cylindrical shell *A* of uniform thickness and a disc *B* rotating relative to the shell at a speed ω_2 of 80 rad/s. The disc *B* is 0.3 m in diameter and has a mass of 45 kg. The shell is rotating at a speed ω_1 of 0.16 rad/s about axis *D–D* in inertial space. If the shaft *FF* about which *B* rotates is made to line up with *D–D* by an internal mechanism, what is the final angular momentum vector for the system? Neglect the mass of all bodies except the disc *B* and the shell.

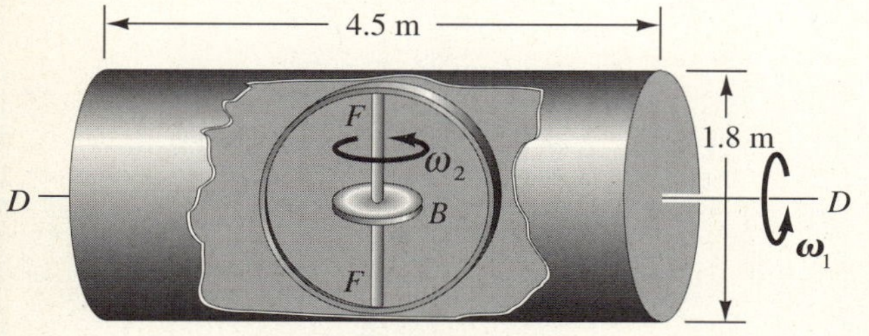

Figure P.18.48.

18.49. A solar energy power unit is in orbit having been given an angular speed ω_1 equal to 0.2 rad/s about the z axis at time t. Vane *B* is identical to vane *A* but is rotated 90° from vane *A*. By an internal mechanism, vane *B* is rotated about its axis to be parallel to vane *A*. What is the new angular speed of the system after this adjustment has taken place? What is the final direction of $\boldsymbol{\omega}$? The vanes on earth each weigh 900 N and can be considered as uniform blocks. The radii of gyration for the configuration corresponding to vane *A* are as follows at the mass center C.M.:

$$k_x = 0.15 \text{ m}$$
$$k_y = 1.5 \text{ m}$$
$$k_z = 1.05 \text{ m}$$

The unit *D* can be considered as equivalent to a uniform sphere of weight on earth of 1.35 kN and radius 0.3 m. (*Advice*: This is a simple problem despite seeming complexity.)

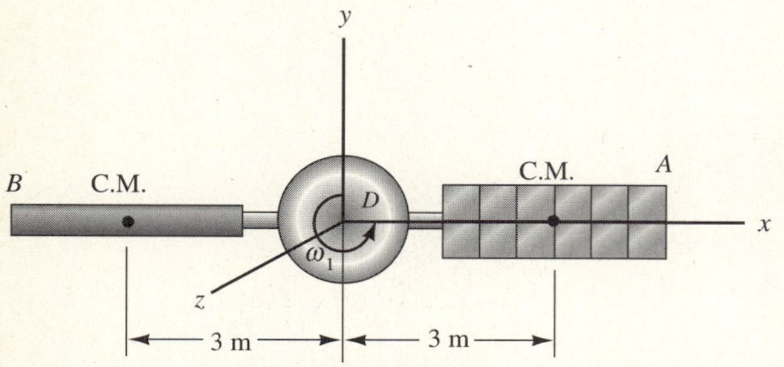

Figure P.18.49.

18.50. A projectile is shot out of a weapon in such a manner that it has an angular velocity $\boldsymbol{\omega}$ at a known angle α from the centerline as it leaves the weapon. Using the cone model, draw a picture depicting the ensuing motion. Denote θ on this diagram, and indicate the direction of $\boldsymbol{H}$. Assume that the spin $\dot{\phi}$ about the axis of symmetry is known, as are the moments of inertia at the C.M. Set up formulations leading to the valuation of the rate of precession of z about $\boldsymbol{H}$ and the angle θ between z and $\boldsymbol{H}$. [*Hint:* Use trigonometry as well as two of Eqs. 18.43.]

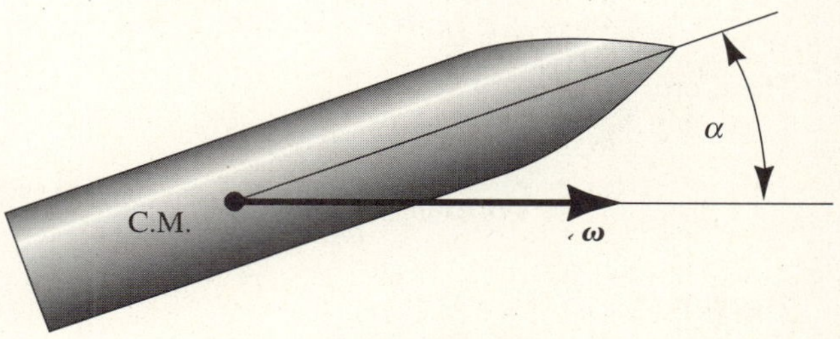

Figure P.18.50.

18.51. A space capsule is rotating about its axis of symmetry in inertial space with an angular speed ω_1 of 2 rad/s. As a result of an impact with a meteorite at point *A*, an impulse $\boldsymbol{I}$ of 180 N s is developed. Find the axis of precession and the precession velocity for postimpact motion. What are the spin velocity and nutation angle? The mass of the capsule is 1.36 Mg. The radius of gyration for the axis of symmetry is 0.6 m, whereas the radius of gyration for transverse axes at the center of mass is 0.75 m.

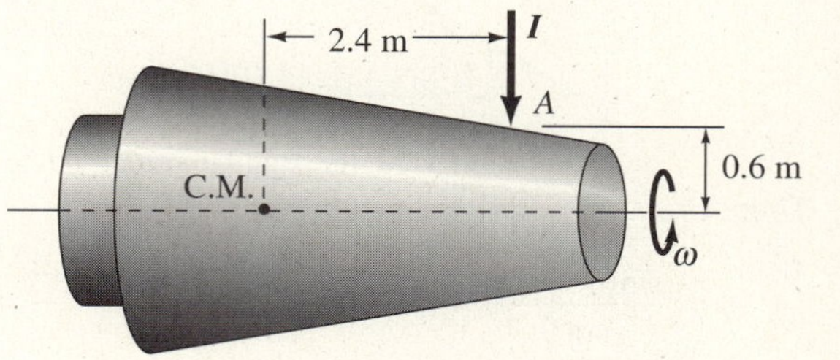

Figure P.18.51.

18.52. Work Problem 18.51 for the case where $\boldsymbol{I}$ is inclined to the right an angle of 45°.

18.53. A rocket casing is in orbit. The casing has a spin of 5 rad/s about its axis of symmetry. The axis of symmetry is oriented 30° from the precession axis as shown in the diagram. What is the precession speed and the angular momentum of the casing? The casing has a mass of 909 kg, a radius of gyration of 0.6 m about the axis of symmetry, and a radius of gyration about transverse axes at its center of mass of 1 m.

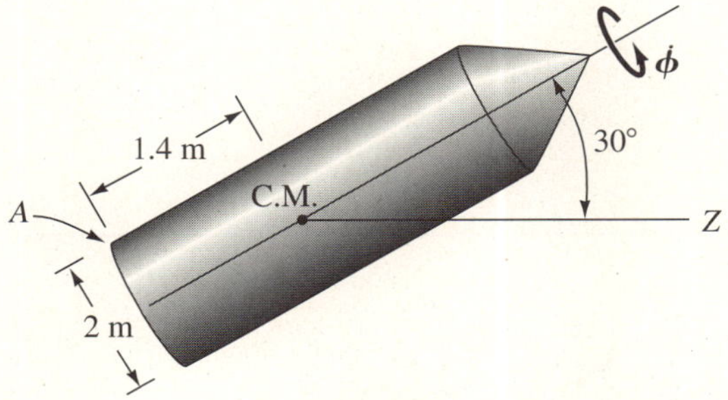

Figure P.18.53.

18.54. In Problem 18.53 assume that an impulse in the vertical direction is developed at point A as a result of an impact with a meteorite. If the impulse from the impact is 133 N s, what are the new precession axis and the rate of precession after impact? H_0 from Problem 18.53 is 2.952 Mg m² s.

18.55. An object representing dynamically a space device is made of three homogeneous blocks A, B, and C each of specific weight on earth of 6.075 kN/m³. Blocks A and C are identical and are hinged along aa and bb. At the configuration shown, the system is in orbit and is made to rotate instantaneously about an axis parallel to RR at a speed ω_1 of 3 rad/s. The block C is then closed by an internal mechanism. What are the subsequent precession axis and rate of precession?

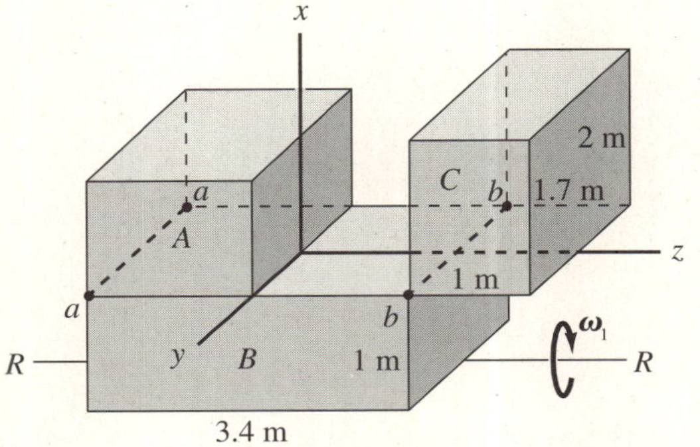

Figure P.18.55.

18.56. A space vehicle has zero rotation relative to inertial space. A jet at A is turned on to give a thrust of 50 N for 0.8 s. Identify the body axis and the subsequent line of nodes. Then, give the nutation angle, the spin speed $\dot{\phi}$, and the precession speed $\dot{\psi}$. The vehicle weighs 10 kN and has a radius of gyration $k_z = 1$ m and, transverse to the z axis at the center of mass, $k' = 0.8$ m. Consider the thrust to be impulsive.

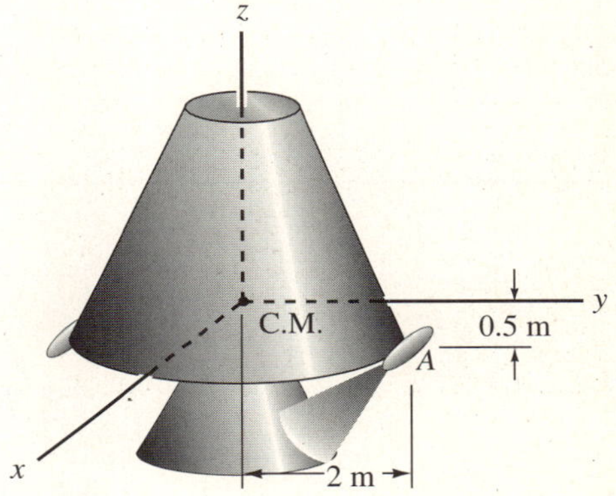

Figure P.18.56.

18.57. An intermediate-stage rocket engine is separated from the first stage by activating exploding bolts. The angular velocity of the spent engine is given as

$$\boldsymbol{\omega} = 2\boldsymbol{i} + 3\boldsymbol{j} + 0.2\boldsymbol{k} \text{ rad/s}$$

What is the spin speed $\dot{\phi}$, the precession speed $\dot{\psi}$, and the nutation speed $\dot{\theta}$? Give direction of Z axis. Identify the line of nodes. Note that $I_{zz} = 1.5$ Mg m² and $I_{xx} = I_{yy} = 2$ Mg m².

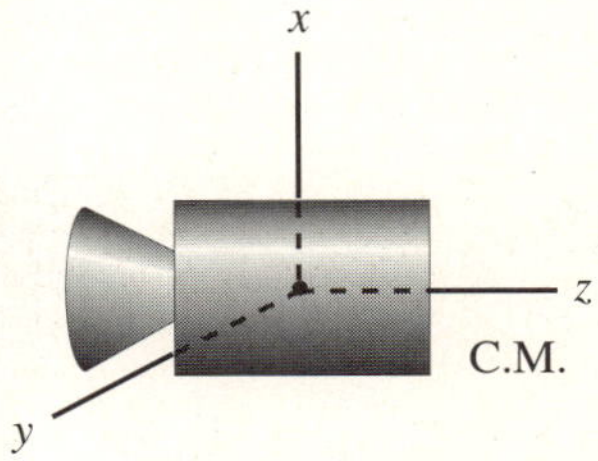

Figure P.18.57.

18.58. In Problem 18.55, if 15 J of mechanical energy is added to the bodies from a battery in the closure process, determine the nutation angle θ and spin speed $\dot{\phi}$. The following results are available from Problem 18.55:

$$\left.\begin{aligned} I &= 5.615 \text{ Mg m}^2 \\ I' &= 10.921 \text{ Mg m}^2 \end{aligned}\right\} \text{closed configuration}$$

$$H_0 = 22.27 \text{ Mg m}^2$$

$$I_{zz} = 6.877 \text{ Mg m}^2 \} \text{ open configuration}$$

[*Hint:* Make use of work-energy equation and Eq. 18.37.]

18.59. In Example 18.5, if 27 J of work is done on the system from an internal power source when going from a test configuration to a closed configuration, determine the nutation angle θ and the spin speed $\dot{\phi}$ for closed configuration. [*Hint:* Make use of work-energy equation and Eq. 18.37.]

18.8 Closure

This chapter brings to a close our study of the motion of rigid bodies. In the final chapter of this text we shall, for the most part, go back to particle mechanics to consider the dynamics of particles constrained to move about a fixed point in a small domain. This is the study of small vibrations (alluded to in Chapter 12) which we have held in abeyance so as to take full advantage of your course work in differential equations.

PROBLEMS

18.60. A plane just after takeoff is flying at a speed V of 56 m/s and is in the process of retracting its wheels. The back wheels (under wings) are being rotated at a speed ω_1 of 3 rad/s and at the instant of interest have rotated 30° as shown in the diagram. The plane is rising by following a circular trajectory of radius 1000 m. If at the instant shown, $\dot{V}$ is 14 m/s^2, what is the total moment coming onto the bearings of the wheel from the motion of the wheel? The diameter of the wheel is 600 mm and its weight is 900 N. The radius of gyration along its axis is 250 mm and transverse to its axis is 180 mm. Neglect wind and bearing friction.

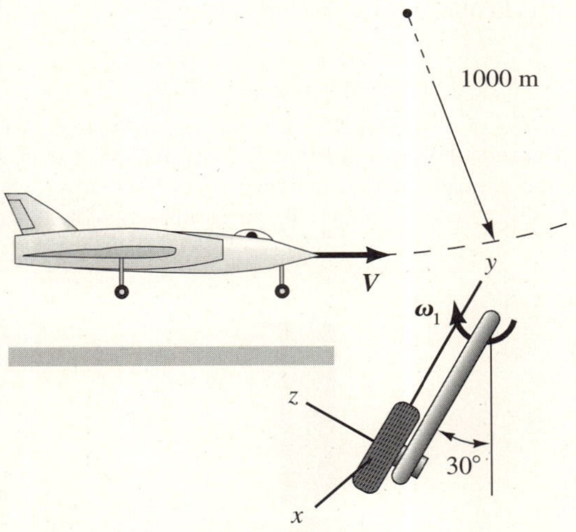

Figure P.18.60.

18.61. A disc A weighing 10 N and of diameter 100 mm rotates with constant speed ω_1 = 15 rad/s relative to G. (A motor on G, not shown, ensures this constant speed.) The shaft of motor B on C rotates with constant speed ω_2 = 8 rad/s relative to C and causes platform G to rotate relative to C. Finally, platform C rotates with angular speed ω_3 = 3 rad/s relative to the ground. What are the supporting forces on bearings H and K of the disc A on platform G? Bearing K alone supports disc A in the axial direction (i.e, it acts as a thrust bearing in addition to being a regular bearing).

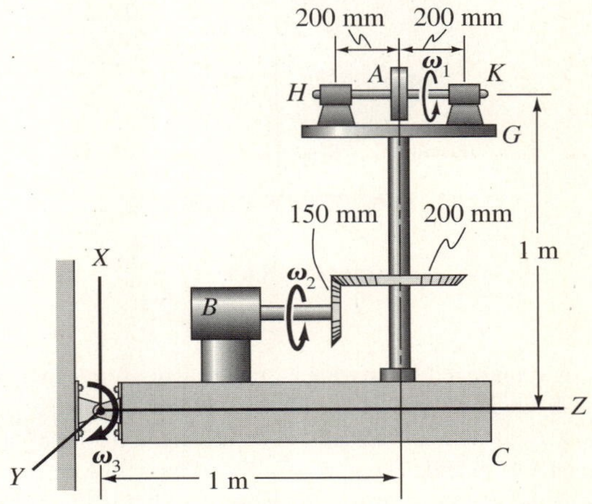

Figure P.18.61.

18.62. Discs A and B are rolling without slipping at their centerlines against an upper surface D. Each disc has a mass of 18 kg, and each spins about a shaft which connects to a centerpost E rotating at an angular speed ω_1. If a total of 90 N is developed upward on D, what is ω_1?

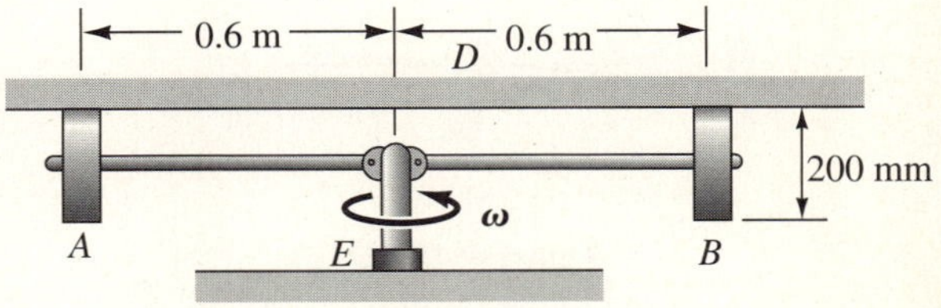

Figure P.18.62.

CHAPTER 19

Vibrations

19.1 Introduction

You will recall that in Chapter 12 we said we would defer a more general examination of particle motion about a fixed point until the very end of the text. We do this to take full advantage of any course in differential equations that you might be taking simultaneously with this course. Accordingly, we shall now continue the work begun in Chapter 12.

19.2 Free Vibration

Let us begin by reiterating what we have done earlier leading to the study of vibrations. Recall that we examined the case of a particle in rectilinear translation acted on either by a constant force, a force given as a function of time, a force that is a function of speed, or, finally, a force that is a function of position. In each case we could separate the variables and effect a quadrature to arrive at the desired algebraic equations, including constants of integration. In particular we considered, as a special case of a force given as a function of position, the linear restoring force resulting from the action (or equivalent action) of a linear spring. Thus, for the spring-mass system shown in Fig. 19.1, the differential equation of motion was shown to be

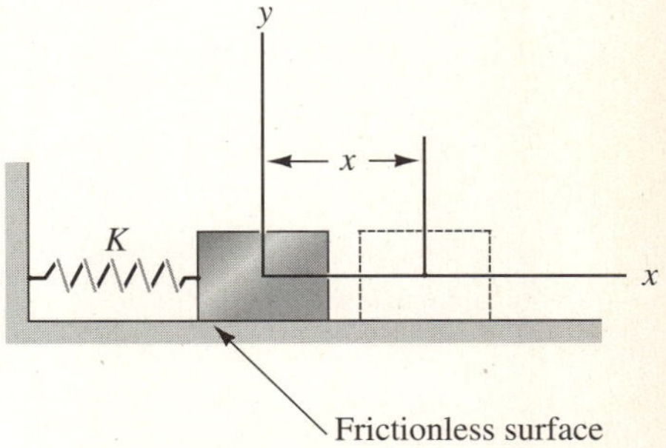

Figure 19.1. Spring–mass system.

$$\frac{d^2x}{dt^2} + \frac{K}{m}x = 0 \qquad (19.1)$$

where K is the spring constant and where x is measured from the static equilibrium position of the mass. You will now recognize this equation from your studies in mathematics as a second-order, linear differential equation with constant coefficients.

PROBLEMS

19.1. If a 5 kg mass causes an elongation of 50 mm when suspended from the end of a spring, determine the natural frequency of the spring–mass system.

19.2. (a) Show that the spring constant is doubled if the length of the spring is halved.

(b) Show that two springs having spring constants K_1 and K_2 have a combined spring constant of $K_1 + K_2$ when connected in parallel, and have a combined spring constant whose reciprocal is $1/K_1 + 1/K_2$ when combined in series.

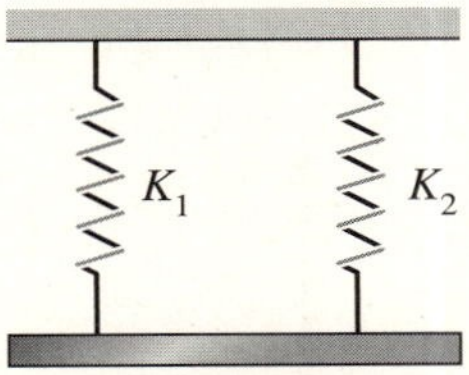

Parallel connection
(a)

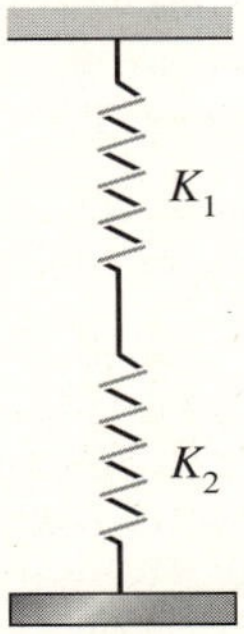

Series
connection
(b)

Figure P.19.2.

19.3. A mass M of 2 kg rides on a vertical frictionless guide rod. With only one spring K_1, the natural frequency of the system is 2 rad/s. If we want to increase the natural frequency threefold, what must the spring constant K_2 of a second spring be?

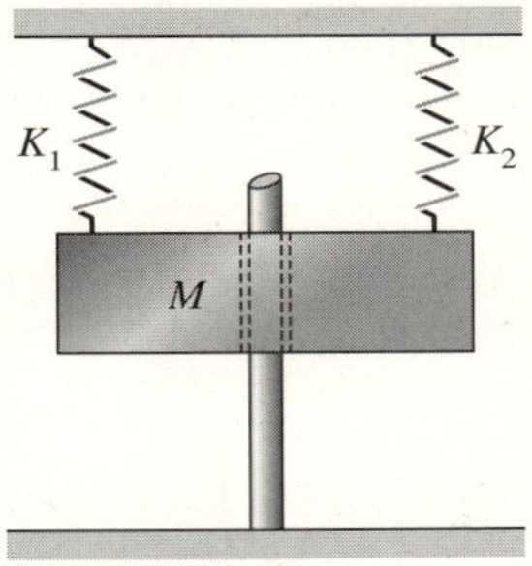

Figure P.19.3.

19.4. A mass M of 100 g rides on a frictionless guide rod. If the natural frequency with spring K_1 attached is 5 rad/s, what must K_2 be to increase the natural frequency to 8 rad/s?

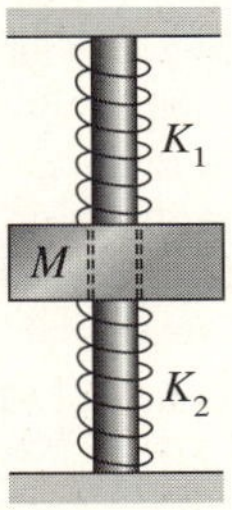

Figure P.19.4.

19.5. For small oscillations, what is the natural frequency of the system in terms of a, b, K, and W? (Neglect the mass of the rod.)

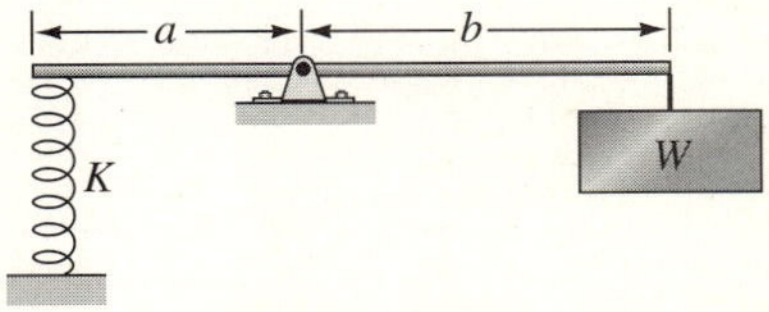

Figure P.19.5.

19.6. A rod is supported on two rotating grooved wheels. The contact surfaces have a coefficient of friction of μ_d. Explain how the rod will oscillate in the horizontal direction if it is disturbed in that direction. Compute the natural frequency of the system.

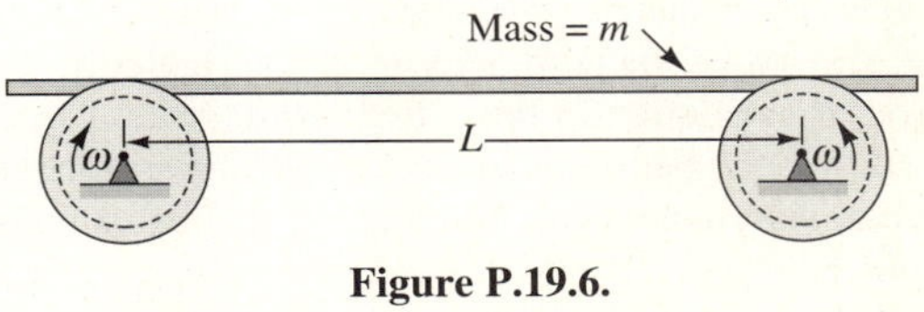

Figure P.19.6.

18.63. A right circular cone weighing 20 N is spinning like a top about a fixed point O at a speed $\dot{\phi}$ of 15 krad/s. What are two possible precession speeds for $\theta = 30°$?

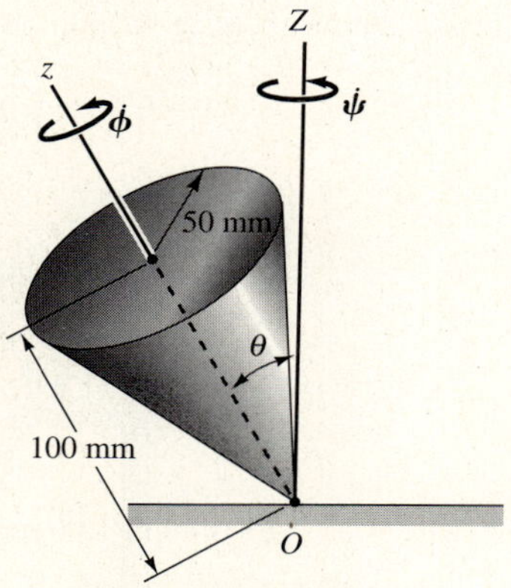

Figure P.18.63.

18.64. A submerged submarine is traveling at 10 m/s in a circular path. A single-degree-of-freedom gyroscope (rate gyro) turns 5° against two torsional springs each having a torsional spring constant of 1 N m/rad. The gyro shown in (b) is viewed from the rear (from stern to bow on the submarine). What is the radius of curvature R of the path of the sub from this reading? This disc weighs 4 N, has a radius of 50 mm, and rotates at 20×10^3 r/min.

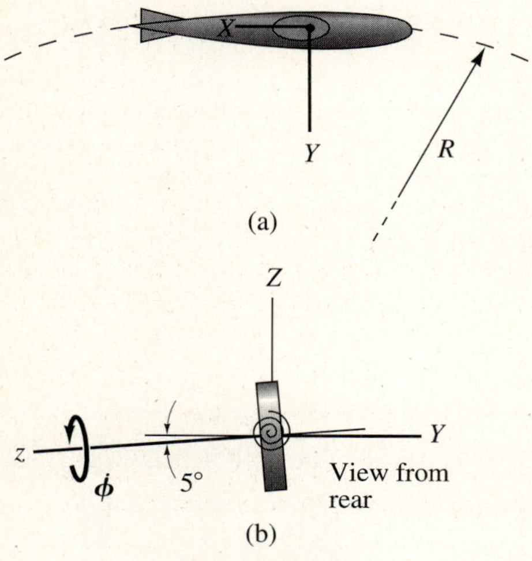

Figure P.18.64.

18.65. Work Problem 18.56 for the case where, before activation of the jet, the vehicle has the following angular velocity:

$$\boldsymbol{\omega} = 0.20\boldsymbol{i} + 0.3\boldsymbol{j} + 0.15\boldsymbol{k} \text{ rad/s}$$

18.66. In Problem 18.49, the initial angular velocity is:

$$\boldsymbol{\omega} = 0.3\boldsymbol{i} + 0.2\boldsymbol{j} + 0.4\boldsymbol{k} \text{ rad/s}$$

What are ω_x, ω_y, and ω_z after vane B has been rotated parallel to vane A by an internal mechanism?

18.67. A research space capsule is in orbit having imposed on it an instantaneous angular speed ω of 0.4 rad/s. The capsule consists of a spherical unit D which can be considered a uniform sphere of radius 0.6 m and weight 800 N. Arms A and B extend from the sphere and consist of cylinders which pick up and record dust particle collisions. Each such unit weighs 500 N and has the following radii of gyration at the mass center, C.M.

$$k \text{ (lateral axes)} = 0.8 \text{ m}$$
$$k \text{ (longitudinal axis)} = 0.2 \text{ m}$$

If arm A is rotated 90° to position E by an internal mechanism, what is the angular precession speed of the system? Give the direction of Z about which there is precession. All weights are as on earth.

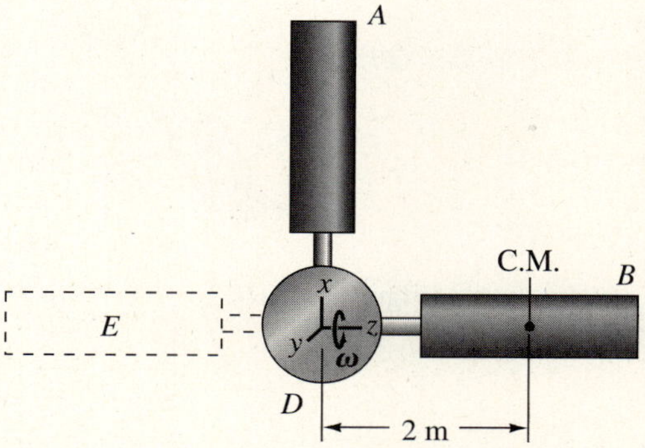

Figure P.18.67.

18.68. In Problem 18.67, 20 J of energy is used to cause the change in configuration. What is the spin speed and nutation angle? From a previous solution, $\dot{\psi} = 0.2064$ rad/s. [*Hint:* Make use of work-energy equation and Eq. 18.37.]

Instead of rearranging the equation to effect a quadrature, as we did in the previous case,[1] we shall take a more general viewpoint toward the solving of differential equations.

To solve a differential equation, we must find a function of time, $x(t)$, which when substituted into the equation satisfies the equation (i.e., reduces it to an identity $0 = 0$). We can either guess at $x(t)$ or use a formal procedure. You have learned in your differential equations course that the most general solution of the above equation will consist of a linear combination of two functions that cannot be written as multiples of each other (i.e, the functions are linearly independent). There will also be two arbitrary constants of integration. Thus, $C_1 \cos \sqrt{K/m}\, t$ and $C_2 \sin \sqrt{K/m}\, t$ will satisfy the equation, as we can readily demonstrate by substitution, and are independent in the manner described. We can therefore say

$$x = C_1 \cos \sqrt{\frac{K}{m}}t + C_2 \sin \sqrt{\frac{K}{m}}t \tag{19.2}$$

where C_1 and C_2 are the aforementioned constants of integration to be determined from the initial conditions.

We can conveniently represent each of the above functions by employing rotating vectors of magnitudes that correspond to the coefficients of the functions. This representation is shown in Fig. 19.2, where, if the vector C_1 rotates counterclockwise with an angular velocity of $\sqrt{K/m}$ radians per unit time and if C_1 lies along the x axis at time $t = 0$, then the projection of this vector along the x axis represents one of the functions of Eq. 19.2, namely $C_1 \cos \sqrt{K/m}\, t$. Vectors used in this manner are called *phasors*.

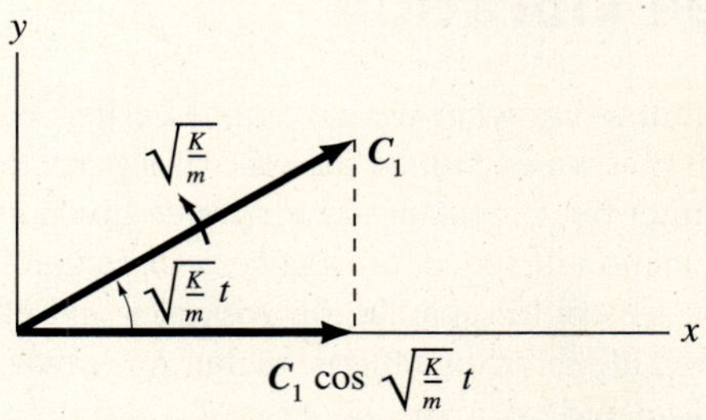

Figure 19.2. Phasor representation.

Consider now the function $C_2 \sin \sqrt{K/m}\, t$, which we can replace by $C_2 \cos (\sqrt{K/m}\, t - \pi/2)$, as we learned in elementary trigonometry. The phasor representation for this function, therefore, would be a vector of magnitude C_2 that rotates with angular velocity $\sqrt{K/m}$ and that is out of phase by $\pi/2$ with

[1]Recall that this can be done by replacing d^2x/dt^2 by $(dV/dx)(dx/dt)$, which is simply $V(dV/dx)$.

the phasor C_1 (Fig. 19.3). Thus, the projection of C_2 on the x axis is the other function of Eq. 19.2. Clearly, because vectors C_1 and C_2 rotate at the same angular speed, we can represent the combined contribution by simply summing the vectors and considering the projection of the resulting single vector

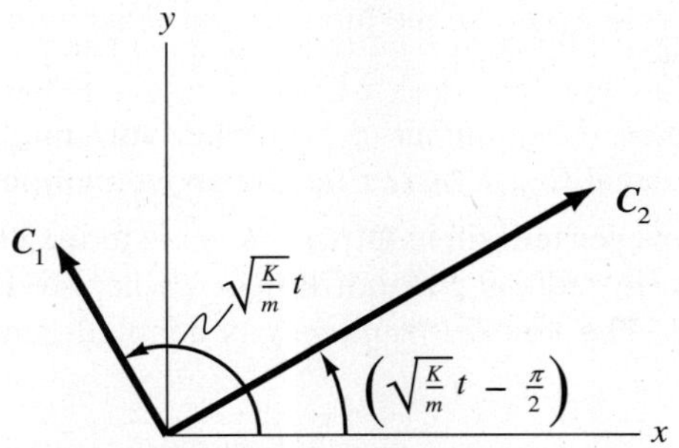

Figure 19.3. Phasors $\pi/2$ out of phase.

along the x axis. This summation is shown in Fig. 19.4 where vector C_3 replaces the vectors C_1 and C_2. Now we can say:

$$C_3 = \sqrt{C_1^2 + C_2^2}, \qquad \beta = \tan^{-1}\frac{C_2}{C_1} \tag{19.3}$$

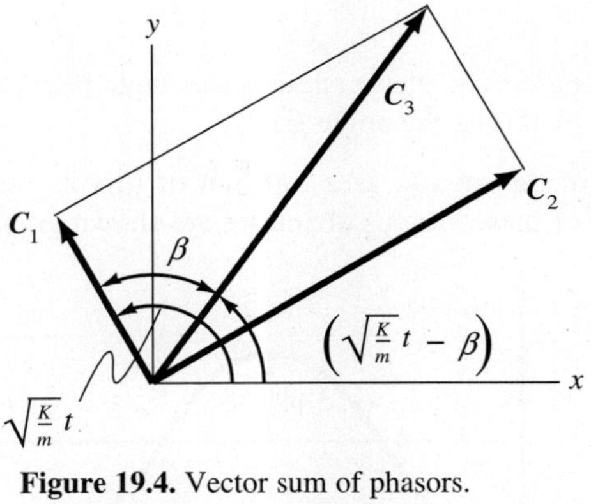

Figure 19.4. Vector sum of phasors.

Because C_1 and C_2 are arbitrary constants, C_3 and β are also arbitrary constants. Consequently, we can replace the solution given by Eq. 19.2 by another equivalent form:

$$x = C_3 \cos\left[\sqrt{\frac{K}{m}}t - \beta\right] \tag{19.4}$$

From this form, you probably recognize that the motion of the body is *harmonic motion*. In studying this type of motion, we shall use the following definitions:

Cycle. The cycle is that portion of a motion (or series of events in the more general usage) which, when repeated, forms the motion. On the phasor diagrams, a cycle would be the motion associated with one revolution of the rotating vector.

Frequency. The number of cycles per unit time is the frequency. The frequency is equal to $\sqrt{K/m}/2\pi$ for the above motion, because $\sqrt{K/m}$ has units of radians per unit time. Often $\sqrt{K/m}$ is termed the *natural frequency* of the system in radians per unit time or, when divided by 2π, in cycles per unit time. The *natural frequency* is denoted generally in the following ways:

$$\omega_n = \sqrt{K/m} \text{ rad/s}$$
$$f_n = \frac{1}{2\pi}\sqrt{K/m} \text{ Hz}$$

Period. The period, T, is the time of one cycle, and is therefore the reciprocal of frequency. That is,

$$T = \frac{2\pi}{\sqrt{K/m}} \tag{19.5}$$

Amplitude. The largest displacement attained by the body during a cycle is the amplitude. In this case, the amplitude corresponds to the coefficient C_3.

Phase angle. The phase angle is the angle between the phasor and the x axis when $t = 0$ (i.e., the angle β).

A plot of the motion as a function of time is presented in Fig. 19.5, where certain of these various quantities are shown graphically.

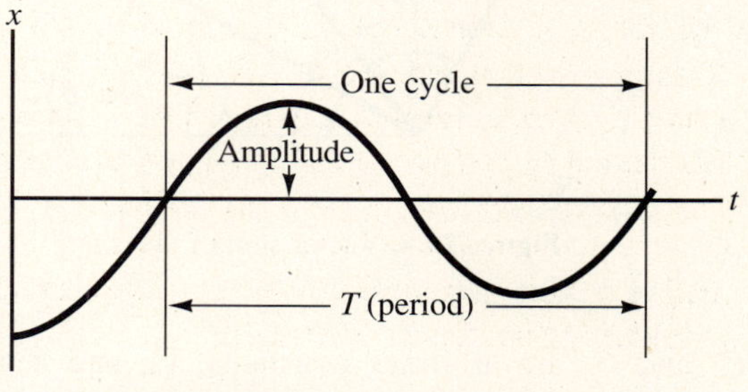

Figure 19.5. Plot of harmonic motion.

It is usually easier to use the earlier form of solution, Eq. 19.2, rather than Eq. 19.4 in satisfying initial conditions. Accordingly, the position and velocity can be given as

$$x = C_1 \cos\sqrt{\frac{K}{m}}t + C_2 \sin\sqrt{\frac{K}{m}}t$$

$$V = -C_1\sqrt{\frac{K}{m}} \sin\sqrt{\frac{K}{m}}t + C_2\sqrt{\frac{K}{m}} \cos\sqrt{\frac{K}{m}}t$$

The initial conditions to be applied to these equations are:

$$\text{when } t = 0 \qquad x = x_0,\ V = V_0$$

Substituting, we get

$$x_0 = C_1, \qquad V_0 = C_2\sqrt{\frac{K}{m}}$$

Therefore, the motion is given as

$$x = x_0 \cos\sqrt{\frac{K}{m}}t + \frac{V_0}{\sqrt{K/m}} \sin\sqrt{\frac{K}{m}}t \tag{19.6a}$$

$$V = -x_0\sqrt{\frac{K}{m}} \sin\sqrt{\frac{K}{m}}t + V_0 \cos\sqrt{\frac{K}{m}}t \tag{19.6b}$$

We can generalize from these results by noting that any agent supplying a linear restoring force for all rectilinear motions of a mass can take the place of the spring in the preceding computations. We must remember, however, that to behave this way the agent must have negligible mass. Thus, we can associate with such agents an *equivalent spring constant* K_e, which we can ascertain if we know the static deflection δ permitted by the agent on application of some known force F. We can then say:

$$K_e = \frac{F}{\delta}$$

Once we determine the equivalent spring constant, we immediately know that the natural frequency of the system is $(1/2\pi)\sqrt{K_e/m}$ cycles per unit time. This natural frequency is the number of cycles the system will repeat in a unit time if some initial disturbance is imposed on the mass. Note that this natural frequency depends only on the "stiffness" of the system and on the mass of the system and is not dependent on the amplitude of the motion.[2]

We shall now consider several problems in which we can apply what we have just learned about harmonic motion.

[2]Actually, when the amplitude gets comparatively large, the spring ceases to be linear, and the motion does depend on the amplitude. Our results do not apply for such a condition.

Example 19.1

A mass weighing 45 N is placed on the spring shown in Fig. 19.6 and is released very slowly, extending the spring a distance of 50 mm. What is the natural frequency of the system? If the mass is given a velocity instantaneously of 1.60 m/s down from the equilibrium position, what is the equation for displacement as a function of time?

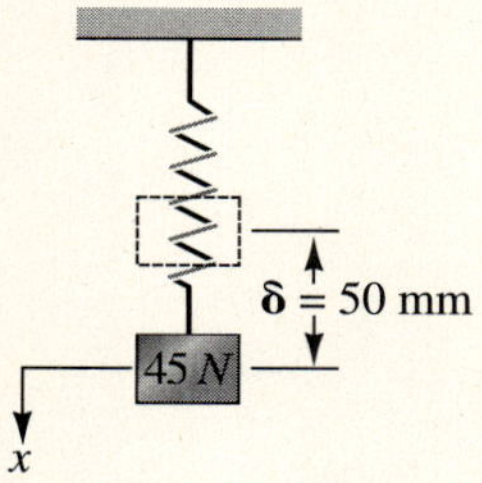

Figure 19.6. x measured from static deflection position.

The spring constant is immediately available by the equation

$$K = \frac{F}{\delta} = \frac{45 \text{ N}}{50 \text{ mm}} = 0.9 \text{ N/mm}$$

The equation of motion for the mass can be written for a reference whose origin is at the static equilibrium position shown in the diagram. Thus,

$$m\frac{d^2x}{dt^2} = W - K(x + \delta)$$

where δ is the distance from the unextended position of the spring to the origin of the reference. However, from our initial equation, $\delta = F/K = W/K$. Therefore, we have

$$m\frac{d^2x}{dt^2} = W - K(x + \frac{W}{K}) = -Kx$$

and the equation becomes identical to Eq. 19.1:

$$\frac{d^2x}{dt^2} + \frac{K}{m}x = 0 \qquad \text{(a)}$$

Thus, the motion will be an oscillation about the position of static equilibrium, which is an extended position of the spring. Measuring x from the *static equilibrium position* and considering the spring force as $-Kx$, we can thus disregard the weight on the body in writing Newton's law for the body to reach Eq. (a) most directly.

Example 19.1 (Continued)

Accordingly, we can use the results stemming from our main discussion. Employing the notation ω_n as the natural frequency in units of radians per unit time, we have

$$\omega_n = \sqrt{\frac{K}{m}} = \sqrt{\frac{\left(\frac{0.9\text{ N}}{\text{mm}}\right)\left(\frac{1000\text{ mm}}{\text{m}}\right)}{\frac{45}{g}\text{ kg}}} = \sqrt{\frac{\left(\frac{0.9\text{ N}}{\text{mm}}\right)\left(\frac{1000\text{ mm}}{\text{m}}\right)}{\frac{45}{g}\left(\frac{\text{N}}{\text{m/s}^2}\right)}} = 14.01\text{ rad/s}$$

The motion is now given by the equations

$$x = C_1 \sin 14.01t + C_2 \cos 14.01t$$
$$\dot{x} = 14.01C_1 \cos 14.01t - 14.01C_2 \sin 14.01t$$

From the specified initial conditions, we know that when $t = 0$, $x = 0$, and $\dot{x} = 1.6$ m/s. Therefore, the constants of integration are

$$C_2 = 0, \qquad C_1 = \frac{1.60}{14.01} = 0.1142$$

The desired equation, then, is

$$x = 0.1142 \sin 14.01t \text{ m}$$

Example 19.2

A body weighing 22 N is positioned in Fig. 19.7 on the end of a slender cantilever beam whose mass we can neglect in considering the motion of the body at its end.

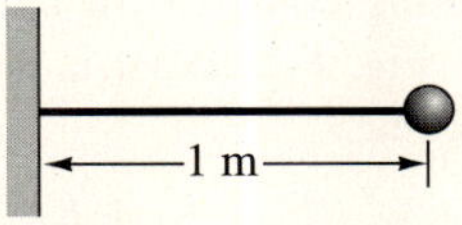

Figure 19.7. Slender cantilever beam with weight at end.

If we know the geometry and the composition of the cantilever beam, and if the deflection involved is small, we can compute from strength of materials the deflection of the end of the beam that results from a vertical load there. This deflection is directly proportional to the load. In

Example 19.2 (Continued)

this case, suppose that we have computed a deflection of 12.5 mm for a force of 4 N (see Fig. 19.8). What would be the natural frequency of the body weighing 22 N for small oscillations in the vertical direction?

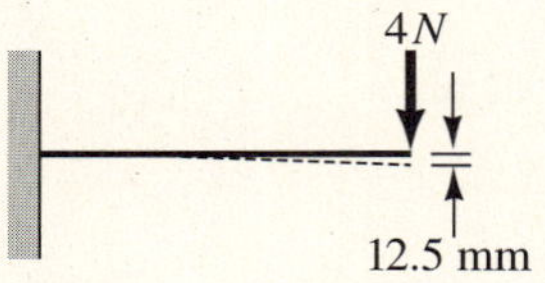

Figure 19.8. Beam acts as linear spring.

Because the motion is restricted to small amplitudes, we can consider the mass to be in rectilinear motion in the vertical direction in the same manner as the mass on the spring in the previous case. The beam now supplies the linear restoring force. The formulations of this section are once again applicable. The equivalent spring constant is found to be

$$K_e = \frac{F}{\delta} = \frac{4}{12.5} = 0.32 \text{ N/mm}$$

The natural frequency for vibration of the 22 N weight at the end of the cantilever is then

$$\omega_n = \sqrt{\frac{(0.32)(1000)}{22/9.81}} = \boxed{11.94 \text{ rad/s}}$$

Before starting out on the exercises, we wish to point out results of Problem 19.2 that will be of use to you. In that problem, you are asked to show that the equivalent spring constant for springs in *parallel* and subject to the same deflection [see Fig. P.19.2(a)] is simply the sum of the spring constants of the springs. That is,

$$K_e = K_1 + K_2 \tag{19.7}$$

For springs in series [Fig. P.19.2(b)] we have

$$\frac{1}{K_e} = \frac{1}{K_1} + \frac{1}{K_2} \tag{19.8}$$

Also, we wish to point out that, for small values of θ, we can approximate $\sin\theta$ by θ and $\cos\theta$ by unity. To justify this, expand $\sin\theta$ and $\cos\theta$ as power series and retain first terms.

19.7. A mass is held so it just makes contact with a spring. If the mass is released suddenly from this position, give the amplitude, frequency, and the center position of the motion. First use the *undeformed position* to measure x. Then, do the problem using x' from the static equilibrium position.

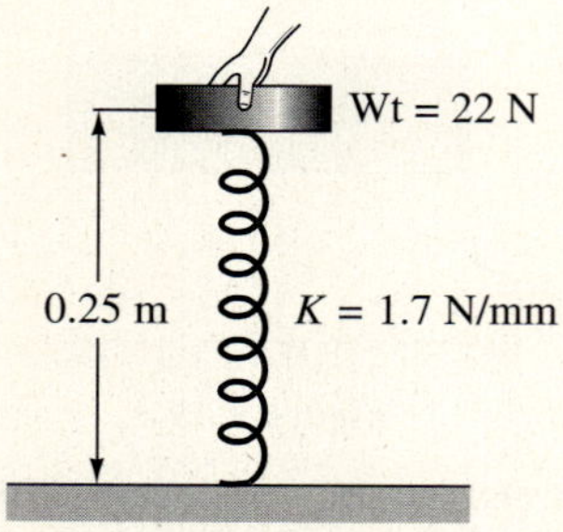

Figure P.19.7.

19.8. A *hydrometer* is a device to measure the *specific gravity* of liquids. The hydrometer weighs 0.36 N, and the diameter of the cylindrical portion above the base is 6 mm. If the hydrometer is disturbed in the vertical direction, what is the frequency of vibration in Hz as it bobs up and down? Recall from *Archimedes' principle* that the buoyant force equals the weight of the water displaced. Water has a density of 1000 kg/m^3.

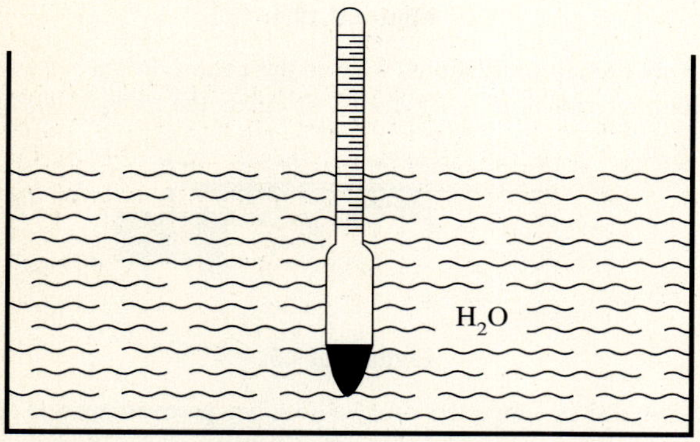

Figure P.19.8.

***19.9.** The hydrometer of Problem 19.8 is used to test the specific gravity of battery acid in a car battery. What is the period of oscillation in this case? [*Hint:* Note if hydrometer goes down, the battery acid surface will have to rise a certain amount simultaneously. The battery acid has a density of 1.1 Mg/m^3.]

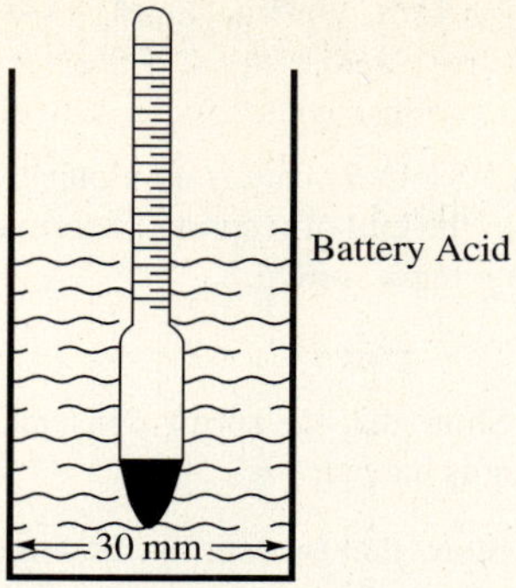

Figure P.19.9.

19.10. A 30 kg block is suspended using two light wires. What is the frequency in Hz at which the block will swing back and forth in the x direction if it is slightly disturbed in this direction? [*Hint:* For small θ, $\sin\theta \approx \theta$ and $\cos\theta \approx 1$.]

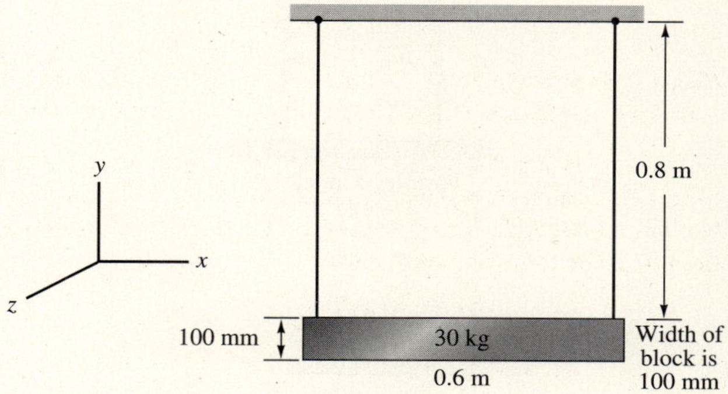

Figure P.19.10.

19.11. In Problem 19.10, what is the period of small oscillations for a small disturbance that causes the block to move in the z direction?

19.12. What is the natural frequency of motion for block A for small oscillation? Consider BC to have negligible mass and body A to be a particle. When body A is attached to the rod, the static deflection is 25 mm. The spring constant K_1 is 1.75 N/mm. Body A weighs 110 N. What is K_2?

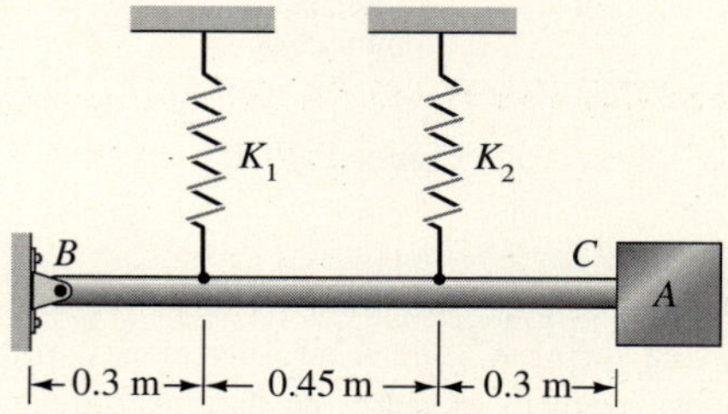

Figure P.19.12.

19.13. If bar ABC is of negligible mass, what is the natural frequency of free oscillation of the block for small amplitude of motion? The springs are identical, each having a spring constant K of 4.5 N/mm. The weight of the block is 45 N. The springs are unstretched when AB is oriented vertically as shown in the diagram.

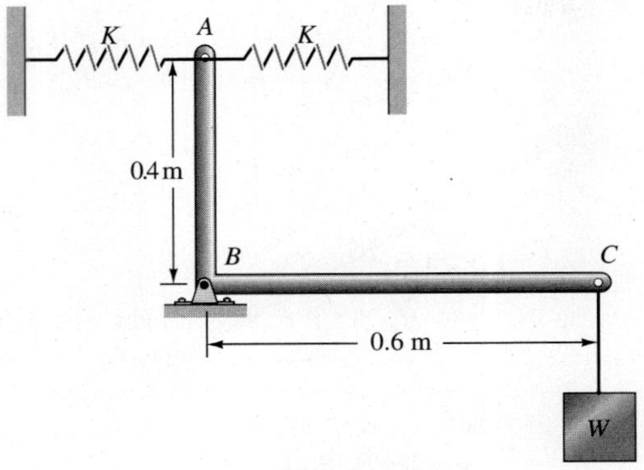

Figure P.19.13.

19.14. Work Problem 19.13 for the case where the springs are both stretched 25 mm when AB is vertical.

19.15. What are the differential equation of motion about the static-equilibrium configuration and the natural frequency of motion of body A for small motion of BC? Neglect inertial effects from BC. The following data apply:

$$K_1 = 2.7 \text{ N/m}$$
$$K_2 = 3.6 \text{ N/m}$$
$$K_3 = 5.4 \text{ N/m}$$
$$W_A = 135 \text{ N}$$

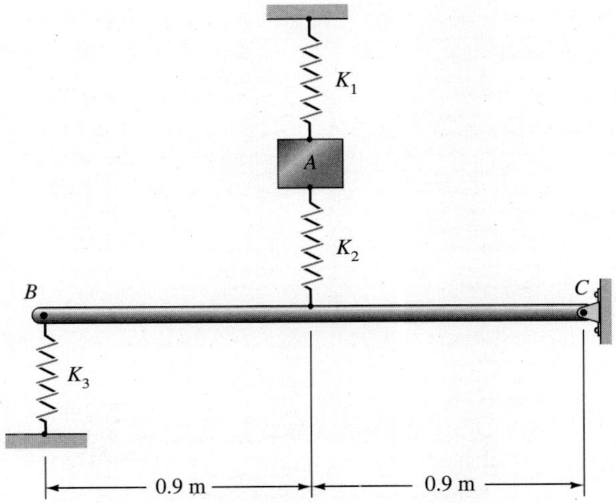

Figure P.19.15.

19.16. A horizontal platform is rotating with a uniform angular speed of ω rad/s. On the platform is a rod CD on which slides a cylinder A having weight W. The cylinder is connected to C through a linear spring having a spring constant K. What is the equation of motion for A relative to the platform after it has been disturbed? What is the natural frequency of oscillation? Take r_0 as the unstretched length of spring.

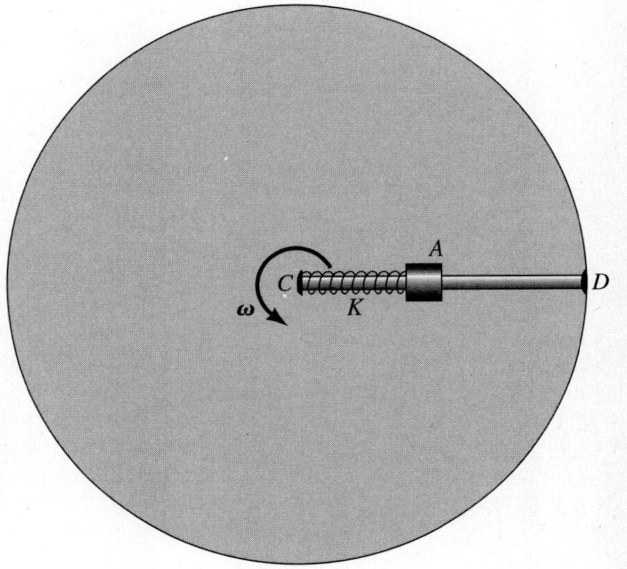

Figure P.19.16.

19.17. A rigid body A rests on a spring with stiffness K equal to 8.80 N/mm. A lead pad B falls onto the block A with a speed on impact of 7 m/s. If the impact is perfectly plastic, what are the frequency and amplitude of the motion of the system, provided that the lead pad sticks to A at all times? Take W_A = 134 N and W_B = 22 N. What is the distance moved by A in 20 ms? (*Caution:* Be careful about the initial conditions.)

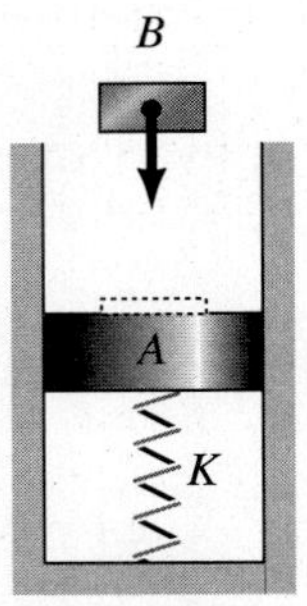

Figure P.19.17.

19.18. A small sphere of mass 2.5 kg is held by taut elastic cords on a frictionless plane. If a 225 N of force is needed to cause an elongation of 25 mm for each cord, what is the natural frequency of small oscillation of the weight in a transverse direction? Also, determine the natural frequency of the weight in a direction along the cord for small oscillations. Neglect the mass of the cord. The tension in the cord in the configuration shown is 450 N.

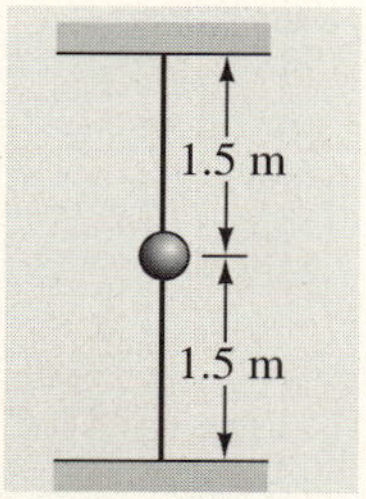

Figure P.19.18.

19.19. Body A weighs 445 N and is connected to a spring having a spring constant K_1 of 3.50 N/mm. At the right of A is a second spring having a spring constant K_2 of 8.80 N/mm. Body A is moved 150 mm to the left from the configuration of static equilibrium shown in the diagram, and it is released from rest. What is the period of oscillation for the body? [*Hint:* Work with half a cycle.]

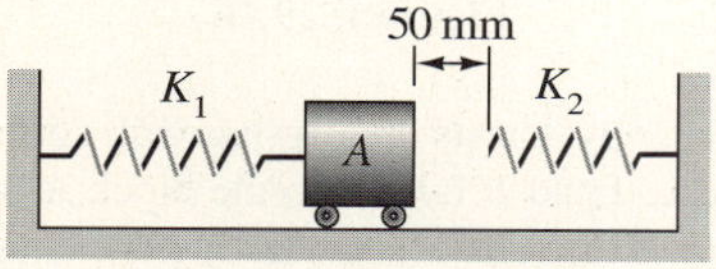

Figure P.19.19.

19.20. Body A, mass 25 kg, has a speed of 6 m/s to the left. If there is no friction, what is the period of oscillation of the body for the following data:

$$K_1 = 3.6 \text{ N/mm}$$
$$K_2 = 1.8 \text{ N/mm}$$

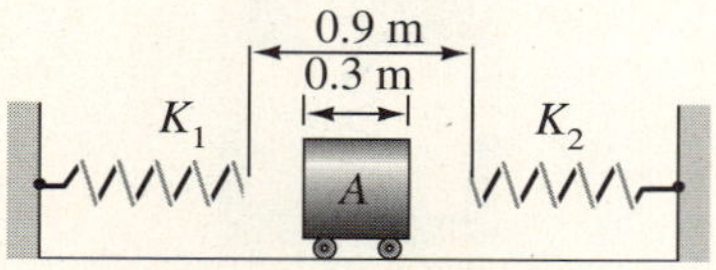

Figure P.19.20.

19.21. A spherical body A of mass 2 kg is attached by a light rod to a shaft BC which is inclined by an angle of 30°. For small, rotational oscillations about BC, what is the natural frequency of the system? [*Hint:* Recall that the moment about an axis n is $(\boldsymbol{r} \times \boldsymbol{F}) \cdot \hat{\boldsymbol{n}}$.]

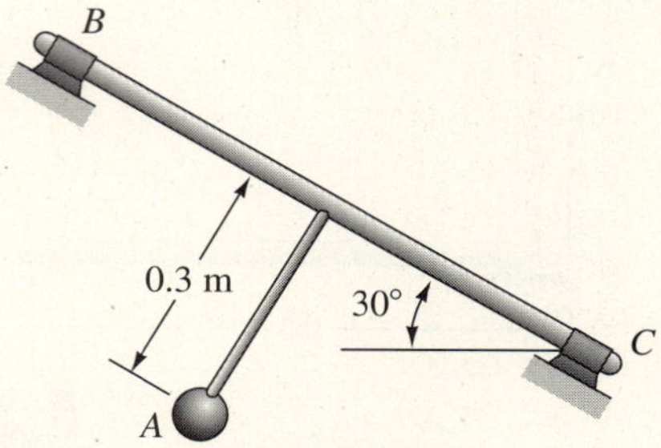

Figure P.19.21.

*__19.22.__ A cube A, 0.25 m on a side, has a specific gravity of 1.10 and is attached to a cone having a specific gravity of 0.8. What is the equation for up-and-down motion of the system? Neglect the mass and buoyant force for rod CD. For very small oscillations, what is the approximate natural frequency?

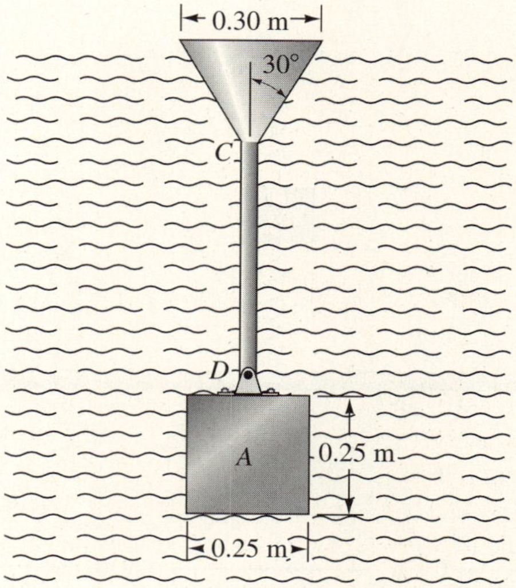

Figure P.19.22.

19.3 Torsional Vibration

We showed in Section 16.2 that, for a body constrained to rotate about an axis fixed in inertial space, the angular momentum equation about the fixed axis is

$$I_{zz}\dot{\omega}_z = I_{zz}\ddot{\theta} = M_z \tag{19.9}$$

Numerous homework problems involved the determination of $\dot{\theta}$ and θ for applied torques which either were constant, varied with time, varied with speed $\dot{\theta}$, or, finally varied with position θ. The analyses paralleled very closely the corresponding cases for rectilinear translation of Chapter 12. Primarily, the approach was that of separation of variables and then that of carrying out one or more quadratures.

Paralleling the case of the linear restoring force in rectilinear translation is the important case where M_z is a *linear restoring torque*. For example, consider a circular disc attached to the end of a light shaft as shown in Fig. 19.9.

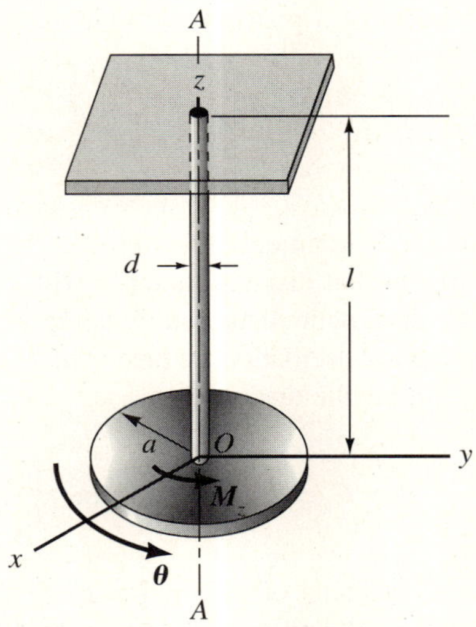

Figure 19.9. Shaft–disc analog of spring–mass system.

Note that the upper end of the shaft is fixed. If the disc is twisted by an external agent about the centerline A–A of the shaft, then the disc will rotate essentially as a rigid body, whereas the shaft, since it is so much thinner and longer, will twist and supply a restoring torque on the disc that tries to bring the disc back to its initial position. In considering the possible motions of such a system disturbed in the aforementioned manner, we idealize the problem by *lumping* all elastic action into the shaft and all inertial effects into the disc. We know from strength of materials that for a circular shaft of constant cross section the amount of twist θ induced by torque M_z is, in the elastic range of deformation,

$$\theta = \frac{M_z L}{GJ} \tag{19.10}$$

where G is the shear modulus of the shaft material, J is the polar moment of area of the shaft cross section, and L is the length of the shaft. We can set forth the concept of a torsional spring constant K_t given as

$$K_t = \frac{M_z}{\theta} \tag{19.11}$$

For the case at hand, we have

$$K_t = \frac{GJ}{L} \tag{19.12}$$

Thus, the thin shaft has the same role in this discussion as the light linear spring of Section 19.2. Employing Eq. 19.11 for M_z and using the proper sign to ensure that we have a restoring action, we can express Eq. 19.9 as follows:

$$\ddot{\theta} + \frac{K_t}{I_{zz}}\theta = 0 \tag{19.13}$$

Notice that this equation is identical in form to Eq. 19.1. Accordingly, all the conclusions developed in that discussion apply with the appropriate changes in notation. Thus, the disc, once disturbed by being given an angular motion, will have a *torsional* natural oscillation frequency of $(\omega_n)_t = \sqrt{K_t/I_{zz}}$ rad/sec. The equation of motion for the disc is

$$\theta = C_1 \cos\sqrt{\frac{K_t}{I_{zz}}}t + C_2 \sin\sqrt{\frac{K_t}{I_{zz}}}t \tag{19.14}$$

where C_1 and C_2 are constants of integration to be determined from initial conditions. Thus, for $\theta = \theta_0$ and $\dot{\theta} = \dot{\theta}_0$ at $t = 0$ we have

$$\theta = \theta_0 \cos\sqrt{\frac{K_t}{I_{zz}}}t + \frac{\dot{\theta}_0}{\sqrt{K_t/I_{zz}}} \sin\sqrt{\frac{K_t}{I_{zz}}}t \tag{19.15}$$

In the example just presented, the linear restoring torque stemmed from a long thin shaft. There could be other agents that can develop a linear restoring torque on a system otherwise free to rotate about an axis fixed in inertial space. We then talk about an *equivalent* torsional spring constant. We shall illustrate such cases in the following examples.

Example 19.3

What are the equation of motion and the natural frequency of oscillation for small amplitude of a simple plane pendulum shown in Fig. 19.10? The pendulum rod may be considered massless.

Because the pendulum bob is small compared to the radius of curvature of its possible trajectory of motion, we may consider it as a particle. The pendulum has one degree of freedom, and we can use θ as the independent coordinate.[3] Notice from the diagram that there is a restoring torque about point A developed by gravity given as

$$M_x = -WL\sin\theta \tag{a}$$

where W is the weight of the bob. If the amplitude of the motion θ is very small, we can replace $\sin\theta$ by θ and so for this case we have a linear restoring torque given as

$$M_x = -WL\theta \tag{b}$$

We then have an equivalent torsional spring constant for the system

$$K_t = WL \tag{c}$$

The equation of possible *small-amplitude* motions for the pendulum is given as

$$-WL\theta = (ML^2)\ddot{\theta} \tag{d}$$

where we have used the **moment-of-momentum equation** about the fixed point A. Rearranging terms, we get

$$\ddot{\theta} + \frac{WL}{ML^2}\theta = 0 \tag{e}$$

Noting that $W = Mg$, we have

$$\ddot{\theta} + \frac{g}{L}\theta = 0 \tag{f}$$

Accordingly, the natural frequency of oscillation is

$$\omega_n = \sqrt{\frac{g}{L}} \text{ rad/s} \tag{g}$$

The equation of motion for this system is

$$\theta = C_1 \cos\sqrt{\frac{g}{L}}t + C_2 \sin\sqrt{\frac{g}{L}}t \tag{h}$$

where C_1 and C_2 are computed from known conditions at some time t_0.

[3]One degree of freedom means that one independent coordinate locates the system.

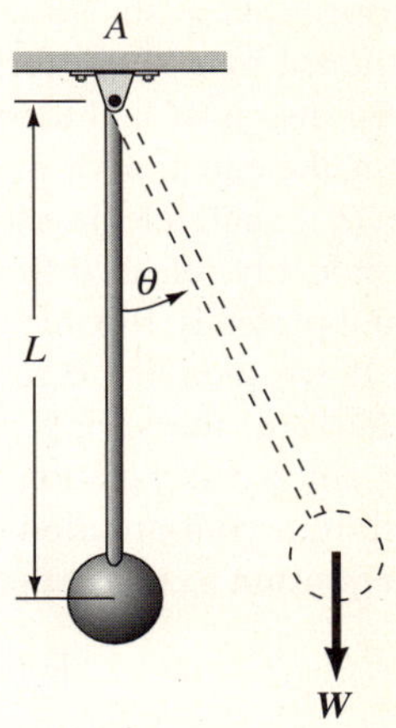

Figure 19.10. Pendulum.

Example 19.4

A stepped disc is shown in Fig. 19.11 supporting a weight W_1 while being constrained by a linear spring having a spring constant K. The mass of the stepped disc is M and the radius of gyration about its geometric axis is k. What is the equation of motion for the system if the disc is rotated a small angle θ_1 counterclockwise from its static-equilibrium configuration and then suddenly released from rest? Assume the cord holding W_1 is weightless and perfectly flexible.

If we measure θ from the static-equilibrium position as shown in Fig. 19.12(a) the spring is stretched an amount $R_2(\theta + \theta_0)$ wherein θ_0 is the amount of rotation induced by the weight W_1 to reach the static-equilibrium configuration. Consequently, applying the **angular momentum equation** to the stepped disc about the axis of rotation, we get

$$R_1 T - KR_2^2(\theta + \theta_0) = Mk^2\ddot{\theta} \tag{a}$$

Next consider the suspended weight W_1. Clearly we have only translation for this body, for which **Newton's law** gives us

$$T - W_1 = -\frac{W_1}{g} R_1\ddot{\theta} \tag{b}$$

where we have made the assumption that the **cord** is always taut and is inextensible and have considered the **kinematics** of the motion. We may replace T in Eq. (a) using Eq. (b) as follows:

$$R_1 W_1 - \frac{W_1}{g} R_1^2\ddot{\theta} - KR_2^2(\theta + \theta_0) = Mk^2\ddot{\theta} \tag{c}$$

Rearranging terms, we get

$$\left(Mk^2 + \frac{W_1}{g} R_1^2\right)\ddot{\theta} + KR_2^2\theta = R_1 W_1 - KR_2^2\theta_0 \tag{d}$$

Considering the **static-equilibrium** configuration of the system, we see on summing moments about the axis of rotation that the right side of the equation above is zero. Accordingly, we have for Eq. (d):

$$\ddot{\theta} + \frac{KR_2^2}{Mk^2 + (W_1/g)R_1^2}\theta = 0 \tag{e}$$

We can say immediately that the natural torsional frequency of the system is

$$\omega_n = \sqrt{\frac{KR_2^2}{Mk^2 + (W_1/g)R_1^2}} \text{ rad/s} \tag{f}$$

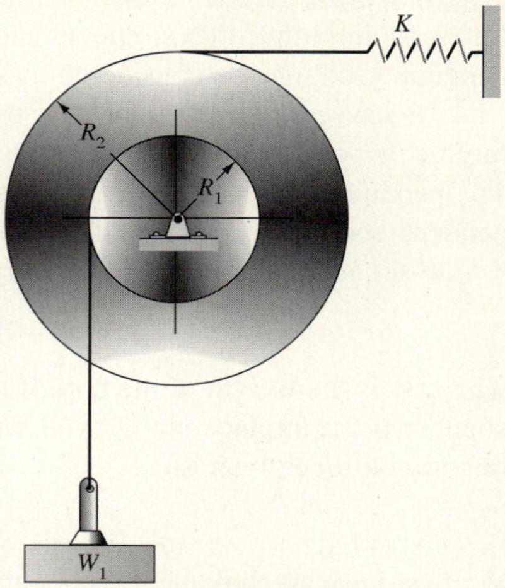

Figure 19.11. Stepped disc.

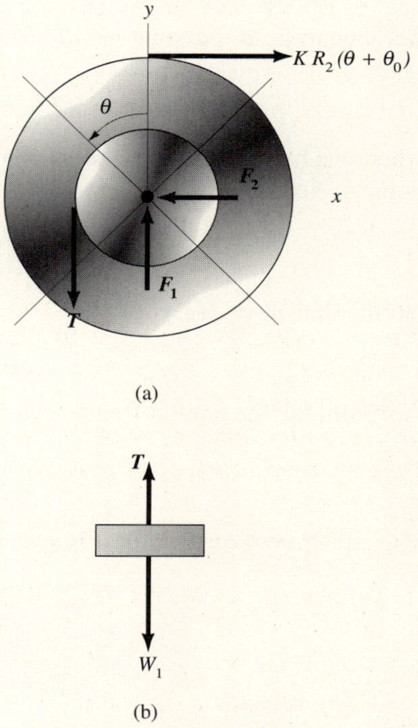

Figure 19.12. Free-body diagrams.

Example 19.4 (Continued)

The equation of motion is then

$$\theta = C_1 \cos\sqrt{\frac{KR_2^2}{Mk^2 + (W_1/g)R_1^2}}\,t + C_2 \sin\sqrt{\frac{KR_2^2}{Mk^2 + (W_1/g)R_1^2}}\,t \quad \text{(g)}$$

Submitting Eq. (g) to the initial conditions to determine C_1 and C_2, we get

$$\theta = \theta_1 \cos\sqrt{\frac{KR_2^2}{Mk^2 + (W_1/g)R_1^2}}\,t \quad \text{(h)}$$

It is important to note that we could have reached Eq. (e) more directly by using θ measured from the *equilibrium configuration*. That is, use the fact that the moment from the weight W_1 is counteracted by the static moment from the stretch $R_2\theta_0$ of the spring. Accordingly, only the moment from the force—$R_2K\theta$ from further stretch of the spring as well as the moment from the inertial force—$(W_1/g)R_1\ddot{\theta}$ of the hanging weight from Eq. (b) need be considered in the angular-momentum equation (a). Thus, we have from this viewpoint:

$$-\frac{W_1}{g} R_1^2\ddot{\theta} - KR_2^2\theta = Mk^2\ddot{\theta} \quad \text{(i)}$$

Rearranging, we have

$$\ddot{\theta} + \frac{KR_2^2}{Mk^2 + (W_1/g)R_1^2}\theta = 0 \quad \text{(j)}$$

Accordingly, we arrive at very same differential equation (e) in a more direct manner. We can again conclude as in Example 19.1 that *when the coordinate is measured from an equilibrium configuration we can forget about contributions of torques that are present for the equilibrium configuration and include only new torques developed when there is a departure from the equilibrium configuration.*

Before you start on the problems, we wish to point out that shafts directly connected to each other (see the shafts on the right side of the disc in Fig. P.19.23) are analogous to springs in *series* as far as the equivalent torsional spring constant is concerned. On the other hand, shafts on opposite sides of the disc are analogous to springs in *parallel* as far as the equivalent torsional spring constant is concerned. You should have no trouble justifying these observations.

PROBLEMS

19.23. Compute the equivalent torsional spring constant of the shaft on the disc.

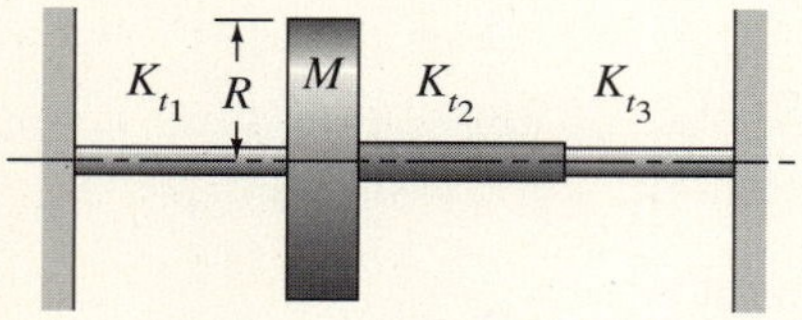

Figure P.19.23.

19.24. What is the equivalent torsional spring constant on the disc from the shafts? The modulus of elasticity G for the shafts is 100 GN/m^2. What is the natural frequency of the system? If the disc is twisted 10° and then released, what will its angular position be in 1 s? Neglect the mass of the shafts. The disc weighs 143 N.

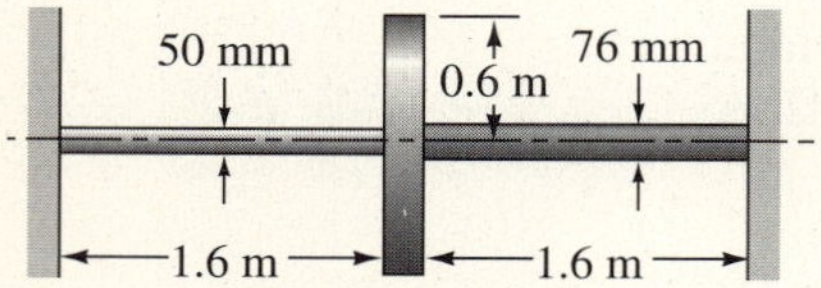

Figure P.19.24.

19.25. A small pendulum is mounted in a rocket that is accelerating upward at the rate of $3g$. What is the natural frequency of rotation of the pendulum if the bob has a mass of $50g$? Neglect the weight of the rod. Consider small oscillations.

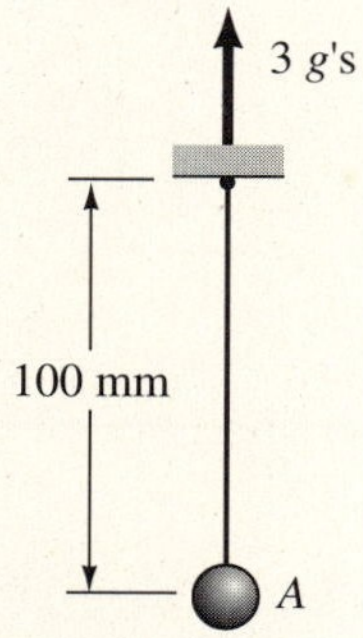

Figure P.19.25.

19.26. Work Problem 19.25 for the case where the rocket is decelerating at $0.6g$ in the vertical direction.

19.27. What is the natural frequency for small oscillations of the compound pendulum supported at A?

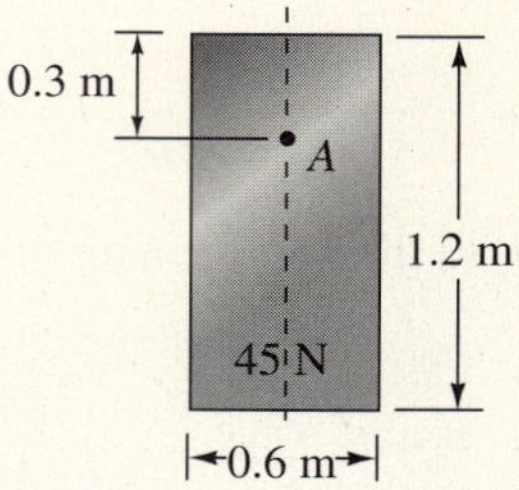

Figure P.19.27.

19.28. A slender rod weighing 140 N is held by a frictionless pin at A and by a spring having a spring constant of 8.80 N/mm at B

(a) What is the natural frequency of oscillation for small vibrations?

(b) If point B of the rod is depressed 25 mm at $t = 0$ from the static-equilibrium position, what will its position be when $t = 20$ ms?

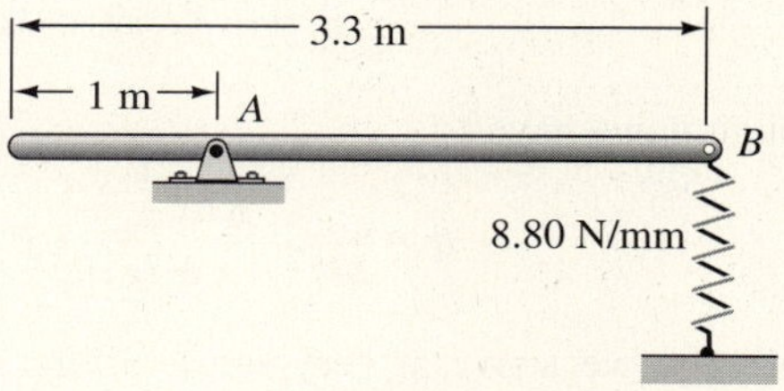

Figure P.19.28.

19.29. What is the natural frequency of the pendulum shown for small oscillations? Take into account the inertia of the rod whose mass is m. Also, consider the bob to be a sphere of diameter D and mass M, rather than a particle. The length of the rod is l.

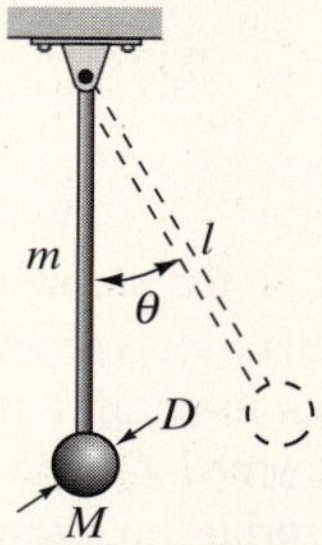

Figure P.19.29.

19.30. A cylinder of mass M and radius R is connected to identical springs and rotates without friction about O. For small oscillations, what is the natural frequency? The cord supporting W_1 is wrapped around the cylinder.

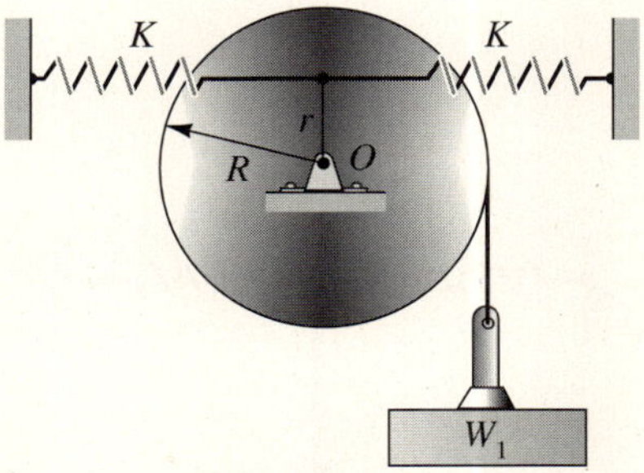

Figure P.19.30.

19.31. The author's 6.7 m Columbia sailboat is suspended by straps from a crane. The boat is made to swing freely about support A in the xy plane (the plane of the page). What is the radius of gyration about the z axis at the center of mass if the period of oscillation is 5 s? The boat has a mass of 1000 kg. Neglect the weight of supporting wires and belts.

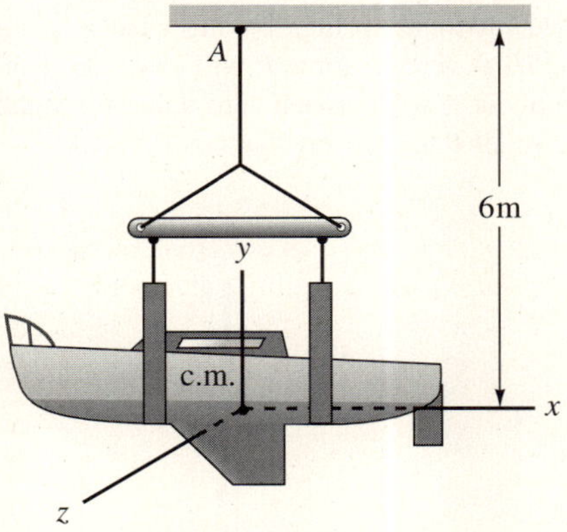

Figure P.19.31.

19.32. A uniform bar of length L and weight W is suspended by strings. What is the differential equation of motion for small torsional oscillation about a vertical axis at the center of mass at C? What is the natural frequency?

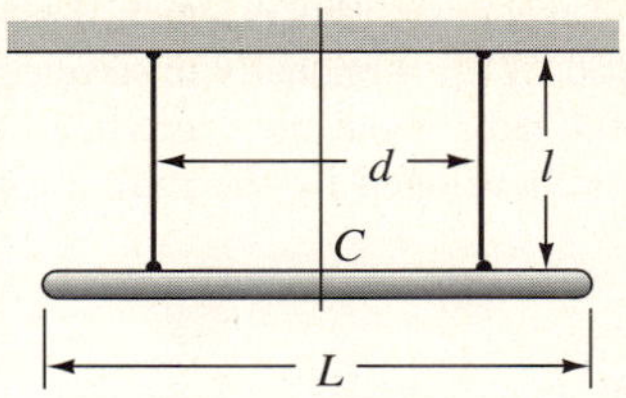

Figure P.19.32.

19.33. In Problem 19.12, do not consider body A to be a particle, and compute the natural frequency of the system for small vibrations. Take the dimension of A to be that of a 150 mm cube. If ω_n for the particle approach is 19.81 rad/s, what percentage error is incurred using the particle approach?

19.34. Gears A and B of mass 25 kg and 40 kg, respectively, are fixed to supports C and D as shown. If the shear modulus G for the shafts is 100 GN/m^2, what is the natural frequency of oscillation for the system?

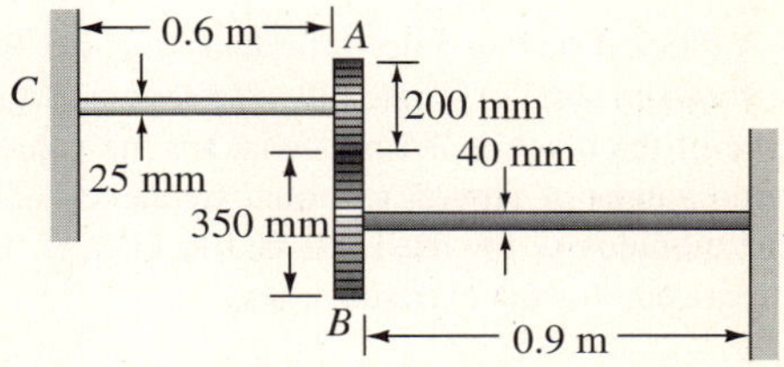

Figure P.19.34.

19.35. A four-bar linkage, $ABCD$, is disturbed slightly so as to oscillate in the xy plane. What is the frequency of oscillation if each bar has a mass of 5.0 g/mm?

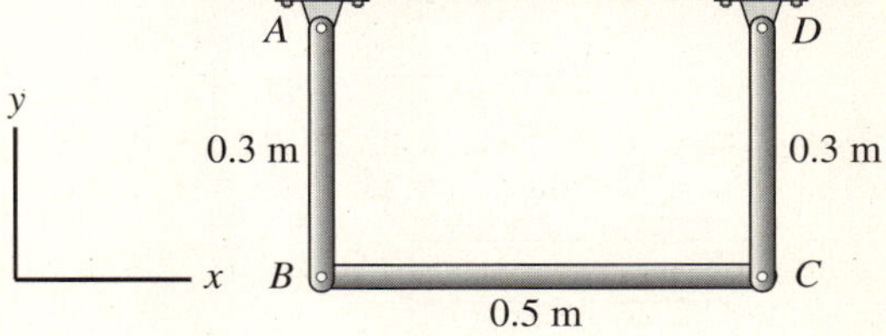

Figure P.19.35.

19.36. A plate A weighing 1 kN is attached to a rod CD. If at the instant that the rod CD is torsionally unstrained, the plate has an angular speed of 2 rad/s about the centerline of CD, what is the amplitude of twist developed by the rod? Take $G = 69$ GN/m^2 for the rod.

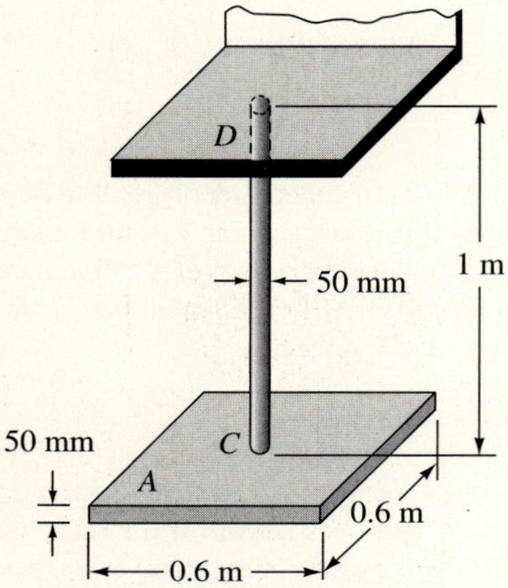

Figure P.19.36.

19.37. A block A having a uniform density of 4.8 Mg/m^3 is suspended by a fixed shaft of length 0.9 m as shown. If the area of the top surface of the block is 0.5 m^2, what are the values of a and b for extreme values of natural torsional frequency of the system? The shear modulus G for the shaft is 100 GN/m^2. Compute the natural frequency for the extreme cases.

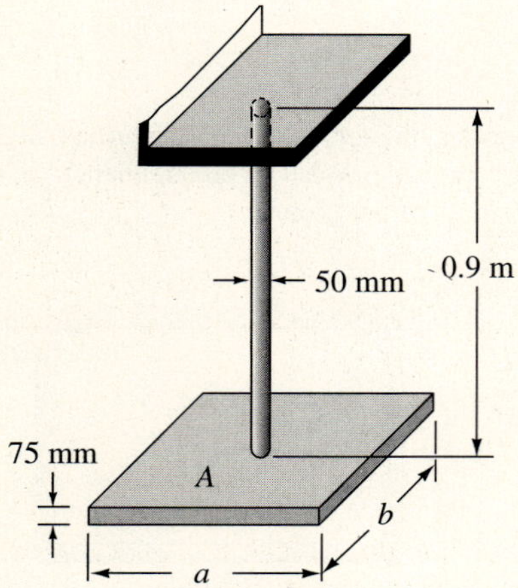

Figure P.19.37.

19.38. A rod of weight W and length L is restrained in the vertical position by two identical springs having spring constant K. A vertical load P acts on top of the rod. What value of P, in terms of W, L, and K, will cause the rod to have a natural frequency of oscillation about A approaching zero for small oscillations? What does this signify physically? [*Hint:* For small θ we may take $\cos\theta = 1$.]

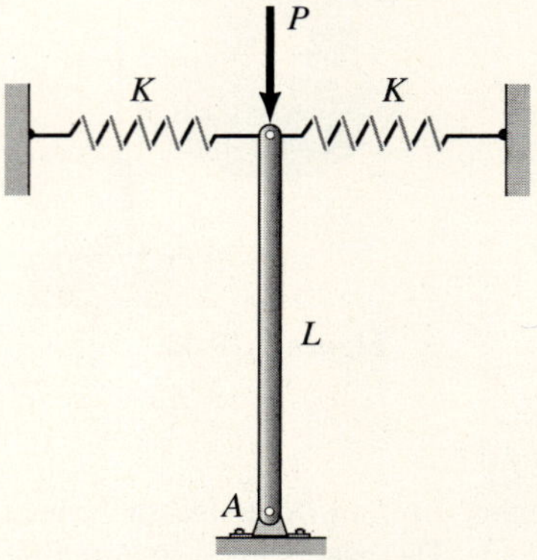

Figure P.19.38.

19.39. A 1 m rod weighing 60 N is maintained in a vertical position by two identical springs having each a spring constant of 50 N/mm. What vertical force P will cause the natural frequency of the rod about A to approach zero value for small oscillations? [*Hint:* For small θ we can say that $\cos\theta = 1$.]

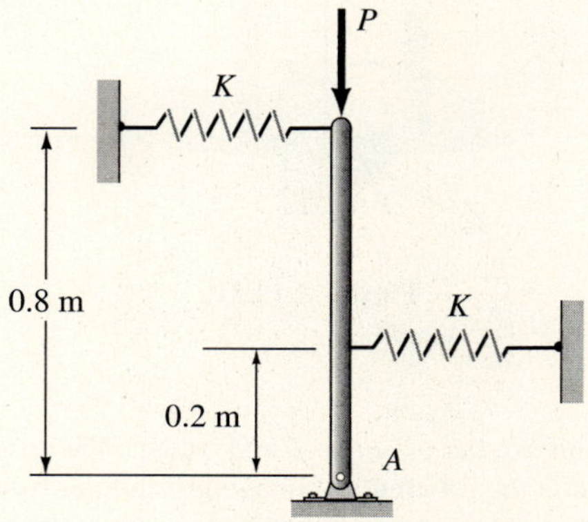

Figure P.19.39.

19.40. What is the natural frequency of torsional vibration for the stepped cylinder? The mass of the cylinder is 45 kg, and the radius of gyration is 0.46 m. The following data also apply:

$$D_1 = 0.3 \text{ m}$$
$$D_2 = 0.6 \text{ m}$$
$$K_1 = 0.875 \text{ N/mm}$$
$$K_2 = 1.8 \text{ N/mm}$$
$$W_A = 178 \text{ N}$$

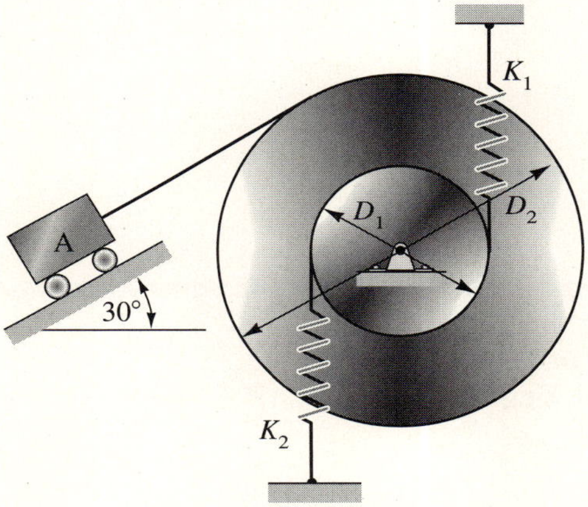

Figure P.19.40.

19.41. A disc A weighs 445 N and has a radius of gyration of 0.45 m about its axis of symmetry. Note that the center of mass does not coincide with the geometric center. What are the amplitude of oscillation and frequency of oscillation if, at the instant that the center of mass is directly below B, the disc is rotating at a speed of 10 mrad/s counterclockwise?

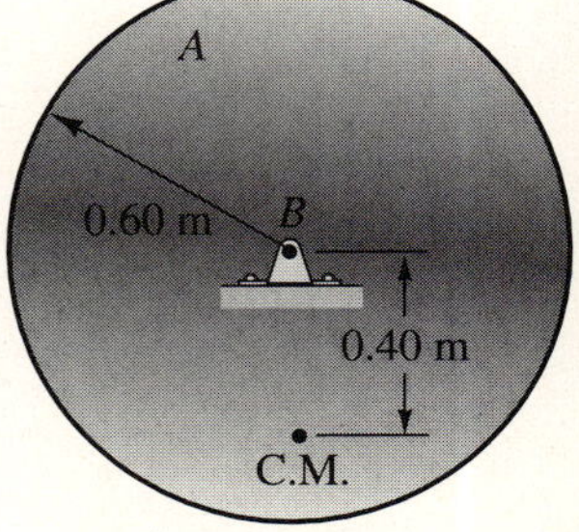

Figure P.19.41.

19.42. A disc B is suspended by a flexible wire. The tension in the top wire is 4.45 kN. If the disc is observed to have a period of lateral oscillation of 0.2 s for very small amplitude and a period of torsional oscillation of 5 s, what is the radius of gyration of the disc about its geometric axis? The torsional spring constant for each of the wires is 1.47 N m/rad.

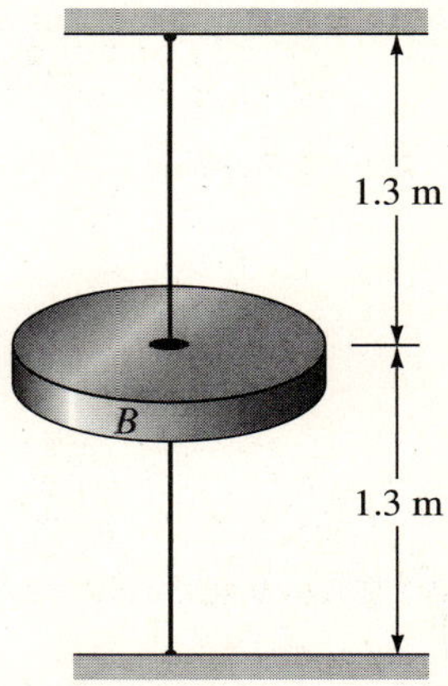

Figure P.19.42.

19.43. Two discs are forced together such that, at the point of contact, a normal force of 225 N is transmitted from one disc to the other. Disc A weighs 900 N and has a radius of gyration of 0.4 m about C, whereas disc B weighs 225 N and has a radius of gyration about D of 0.3 m. What is the natural frequency of oscillation for the system, if disc A is rotated 10° counterclockwise and then released? The center of mass of B coincides with the geometric center.

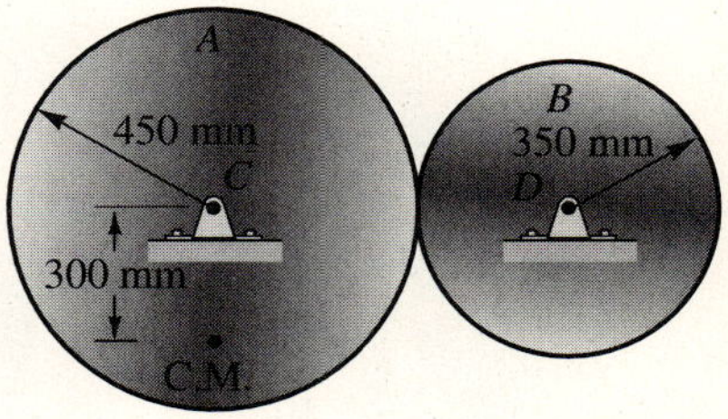

Figure P.19.43.

19.44. In Problem 19.43, find the minimum coefficient of friction for no slipping between the discs. From Problem 19.43, $\omega_n = 3.86$ rad/s.

*19.4 Examples of Other Free-Oscillating Motions

In the previous sections, we examined the rectilinear translation of a rigid body under the action of a linear restoring force as well as the pure rotation of a rigid body under the action of a linear restoring torque. In this section, we shall first examine a body with one degree of freedom undergoing *plane motion* governed by a differential equation of motion of the form given in the previous section. The dependent variable for such a case varies harmonically with time, and we have a *vibratory plane motion*. Consider the following example.

Example 19.5

Shown in Fig. 19.13(a) on an inclined plane is a uniform cylinder maintained in a position of equilibrium by a linear spring having a spring constant K. If the cylinder rolls without slipping, what is the equation of motion when it is disturbed from its equilibrium position?

We have here a case of plane motion about a configuration of equilibrium. Using xyz as a *stationary* reference, we shall measure the displacement x of the center of mass from the equilibrium position and accordingly shall need to consider only those forces and torques developed as the cylinder departs from this position. Accordingly, we have for **Newton's law** for the mass center [see Fig. 19.13(b)]:

$$-f - Kx = M\ddot{x} \qquad \text{(a)}$$

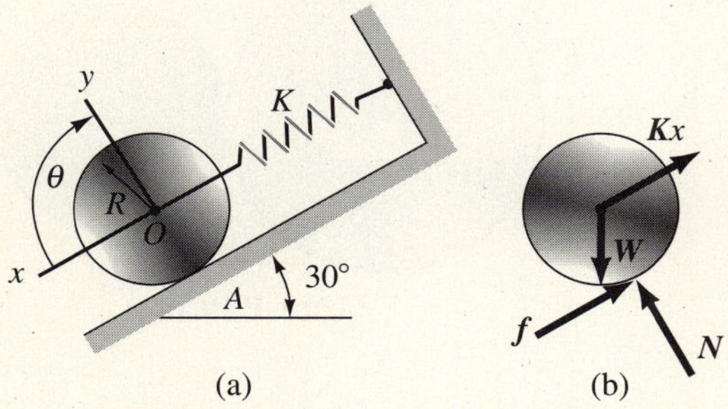

Figure 19.13. Plane-motion vibration.

Example 19.5 (Continued)

Now employ the **angular-momentum** equation about the geometric axis of the cylinder at O. Using θ to measure the rotation of the cylinder about this axis from the equilibrium configuration, we get

$$-fR = \tfrac{1}{2} MR^2 \ddot{\theta} \qquad \text{(b)}$$

Noting from **kinematics** that $\ddot{x} = -R\ddot{\theta}$ as a result of the no-slipping condition, we have for Eq. (b):

$$fR = \frac{1}{2} MR^2 \left(\frac{\ddot{x}}{R}\right)$$

Therefore,

$$f = \tfrac{1}{2} M\ddot{x} \qquad \text{(c)}$$

Substituting for f in Eq. (a) using this result, we have

$$M\ddot{x} = -\tfrac{1}{2} M\ddot{x} - Kx$$

Therefore,

$$\ddot{x} + \frac{2}{3}\frac{K}{M} x = 0 \qquad \text{(d)}$$

We could also have arrived at the differential equation above by noting that we have instantaneous pure rotation about the line of contact A as a result of the no-slipping condition. Thus, the **angular-momentum** equation can be used as follows about the point of contact on the cylinder:

$$\left(\tfrac{1}{2} MR^2 + MR^2\right)\ddot{\theta} = KxR \qquad \text{(e)}$$

Noting as before that $\ddot{\theta} = -\ddot{x}/R$, we get

$$-\frac{3}{2} MR^2 \frac{\ddot{x}}{R} = KxR$$

Therefore, as before

$$\ddot{x} + \frac{2}{3}\frac{K}{M} x = 0 \qquad \text{(d)}$$

Example 19.5 (Continued)

We may solve the differential equation to give us

$$x = x_0 \cos \sqrt{\frac{2}{3}\frac{K}{M}}t + \frac{\dot{x}_0}{\sqrt{\frac{2}{3}K/M}} \sin \sqrt{\frac{2}{3}\frac{K}{M}}t \tag{g}$$

where x_0 and $\dot{x}_0$ are the initial position and speed of the center of mass, respectively. Since $\theta = -x/R$ (we have here only one degree of freedom[4] as a result of the no-slipping condition), we have for θ from Eq. (g):

$$\theta = \frac{-x_0}{R} \cos \sqrt{\frac{2}{3}\frac{K}{M}}t - \frac{\dot{x}_0}{R\sqrt{\frac{2}{3}K/M}} \sin \sqrt{\frac{2}{3}\frac{K}{M}}t \tag{h}$$

[4]See footnote 3 on page 975.

*19.5 Energy Methods

Up to now, the procedure has been primarily to work with *Newton's law* or the *angular-momentum equation* in reaching the differential equation of interest. There is an alternative approach to the handling of free vibration problems that may be very useful in dealing with simple systems and in setting up approximate calculations for more complex systems. Suppose we know for a one-degree-of-freedom system that only linear restoring forces and torques do work during possible motions of the system. Then, the agents developing such forces are *conservative* force agents and may be considered to store potential energy. You will recall from Section 13.3 that the *total mechanical energy* for such systems is *conserved*. Thus, we have

$$\text{PE} + \text{KE} = \text{constant} \tag{19.16}$$

Also, we know from our present study that the system must oscillate harmonically when disturbed and then allowed to move freely with only the linear restoring agents doing work. Thus, if κ is the independent coordinate measured from the static-equilibrium configuration, we have

$$\kappa = A\sin(\omega_n t + \beta) \tag{19.17}$$

Hence,

$$\dot{\kappa} = A\omega_n \cos(\omega_n t + \beta) \tag{19.18}$$

Now at the instant when $\kappa = 0$, we are at the static-equilibrium position and the potential energy of the system is a minimum. Since the total mechanical energy must be conserved at all times once such a motion is under way, it is also clear that the kinetic energy must be at a maximum at that instant. If we take the lowest potential energy as zero, then we have for the total mechanical energy simply the maximum kinetic energy. Also, when the body is undergoing a change in direction of its motion at the outer extreme position, the kinetic energy is zero instantaneously, and accordingly the potential energy must be a maximum and equal to the total mechanical energy of the system. Thus, we can equate the maximum potential energy with the maximum kinetic energy.

$$(\text{KE})_{\text{max}} = (\text{PE})_{\text{max}} \tag{19.19}$$

In computing the $(\text{KE})_{\text{max}}$, we will involve $(\dot{\kappa})_{\text{max}}$ and hence $A\omega_n$, whereas for the $(\text{PE})_{\text{max}}$ we will involve $(\kappa)_{\text{max}}$ and hence A. In this way we can set up quickly an equation for ω_n, the natural frequency of the system. For example, if we have the simple linear spring-mass system of Fig. 19.1, we can say:

$$(\text{PE}) = \tfrac{1}{2}Kx^2$$

Therefore,

$$(\text{PE})_{\text{max}} = \tfrac{1}{2}K(x_{\text{max}})^2 = \tfrac{1}{2}KA^2$$

where we have made use of our knowledge that $x = A\sin(\omega_n t + \beta)$. And, noting that $\dot{x} = A\omega_n \cos(\omega_n t + \beta)$, we have

$$(\text{KE})_{\text{max}} = \tfrac{1}{2}M(\dot{x}_{\text{max}})^2 = \tfrac{1}{2}M(A\omega_n)^2$$

Now, equating these expressions, we get

$$\tfrac{1}{2}KA^2 = \tfrac{1}{2}M(A\omega_n)^2$$

Therefore,

$$\omega_n = \sqrt{\frac{K}{M}}$$

which is the expected result. We next illustrate this approach in a more complex problem.

Example 19.6

A cylinder of radius r and weight W rolls without slipping along a circular path of radius R as shown in Fig. 19.14. Compute the natural frequency of oscillation for small oscillation.

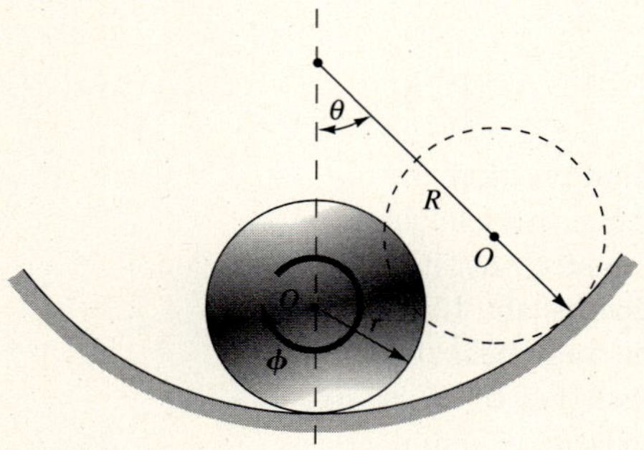

Figure 19.14. Cylinder rolls without slipping.

This system has one degree of freedom. We can use ϕ, the angle of rotation of the cylinder about its axis of symmetry, as the independent coordinate, or we may use θ as shown in the diagram. To relate these variables for no slipping we may conclude, on observing the motion of point O, that for small rotation:

$$(R - r)\theta = r\phi$$

Therefore,

$$\theta = \frac{r}{R - r}\phi \tag{a}$$

The only force that does work during the possible motions of the system is the force of gravity W. The torque developed by W about the point of contact for a given θ is easily determined after examining Fig. 19.15 to be

$$\text{torque} = Wr\sin\theta = Wr\sin\left(\frac{r}{R - r}\phi\right) \tag{b}$$

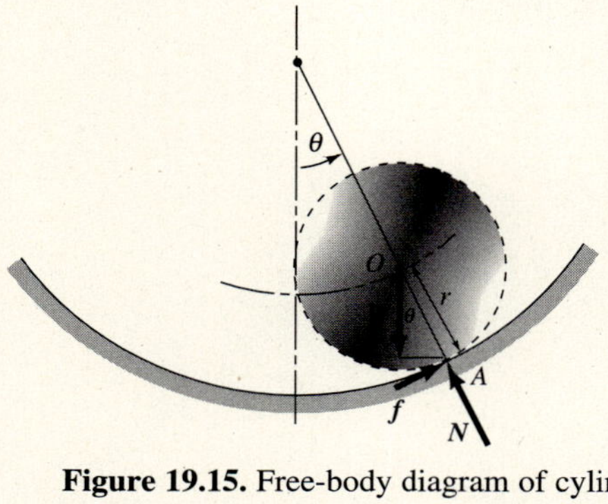

Figure 19.15. Free-body diagram of cylinder.

Example 19.6 (Continued)

This is a restoring torque, and because we limit ourselves to *small oscillations* it becomes $W[r^2\phi/(R-r)]$, which is clearly a linear restoring torque. Because the force doing work on the cylinder is *conservative*, and because it results in a *linear restoring torque*, we can employ the energy formulation of this section.

The motion may be considered to be given as follows:

$$\theta = C\sin(\omega_n t + \beta) \tag{c}$$

or, using Eq. (a),

$$\phi = \frac{R-r}{r}C\sin(\omega_n t + \beta) \tag{d}$$

Expressing the maximum potential and kinetic energies and using C for θ_{max} and the lowest position of O as the datum, we have:

$$\begin{aligned}(\text{PE})_{\text{max}} &= W(R-r)(1-\cos\dot{\theta}_{\text{max}})\\ &= W(R-r)(1-\cos C)\end{aligned} \tag{e}$$

$$\begin{aligned}(\text{KE})_{\text{max}} &= \frac{1}{2}\frac{W}{g}(R-r)^2\dot{\theta}^2_{\text{max}} + \frac{1}{4}\frac{W}{g}r^2\dot{\phi}^2_{\text{max}}\\ &= \frac{1}{2}\frac{W}{g}(R-r)^2(C\omega_n)^2 + \frac{1}{4}\frac{W}{g}r^2\left(\frac{R-r}{r}C\omega_n\right)^2\end{aligned} \tag{f}$$

We have used Eq. (d) in the last expression of Eq. (f). Expanding $\cos C$ in a power series and retaining the first two terms $(1-C^2/2)$, we then get, on equating the right sides of the above equations:

$$W(R-r)\frac{C^2}{2} = \frac{W}{2g}\left[(R-r)^2 + \frac{(R-r)^2}{2}\right]\omega_n^2 C^2$$

Therefore,

$$\omega_n = \sqrt{\frac{g}{\frac{3}{2}(R-r)}} \tag{g}$$

PROBLEMS

19.45. A cylinder of diameter 1 m is shown. The center of mass of the cylinder is 0.3 m from the geometric center, and the radius of gyration is 0.6 m at the center of mass. What is the natural frequency of oscillation for small vibrations without slipping? The cylinder weighs 220 N. Work the problem by two methods.

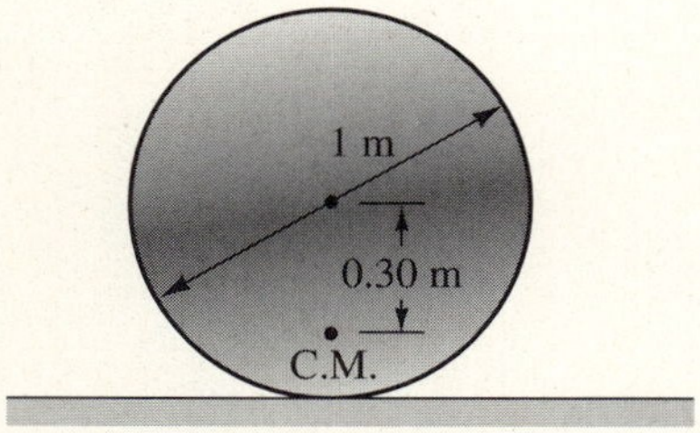

Figure P.19.45.

19.46. A stepped cylinder is maintained along the incline by a spring having a spring constant K. What is the formulation for the natural frequency of oscillation for the system? What is the maximum friction force? Take the weight of the cylinder as W and the radius of gyration about the geometric centerline O as k. The initial conditions are $\theta = \theta_0$ at $t = 0$, and $\dot{\theta} = \dot{\theta}_0$ at $t = 0$. There is no slipping.

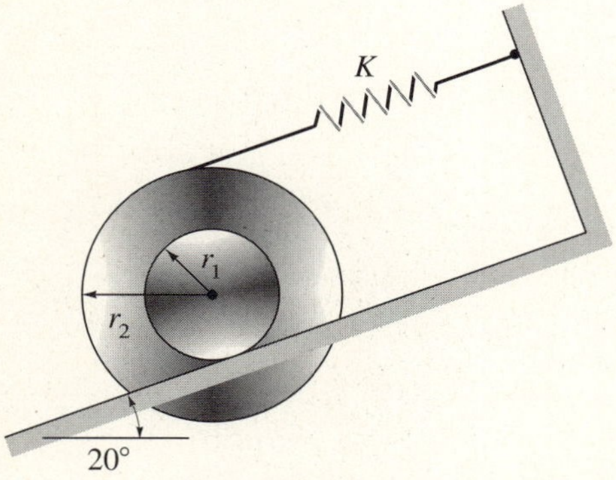

Figure P.19.46.

19.47. Two masses are attached to a light rod. The rod rides on a frictionless horizontal rail. $M_1 = 45$ kg and $M_2 = 14$ kg. What is the natural frequency of oscillation of the system if a small impulsive torque is applied to the system when it is in a rest configuration? Consider the masses as particles. [*Hint:* Consider motion about the center of mass of the system.]

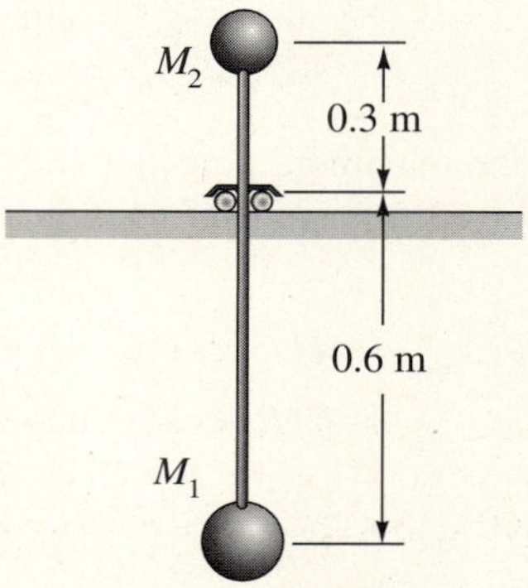

Figure P.19.47.

19.48. Work Problem 19.30 by energy methods.

19.49. Work Problem 19.13 by energy methods.

19.50. Work Problem 19.40 by energy methods.

19.51. A *manometer* used for measuring pressures is shown. If the mercury has a length L in the tube, what is the formulation for the natural frequency of movement of the mercury?

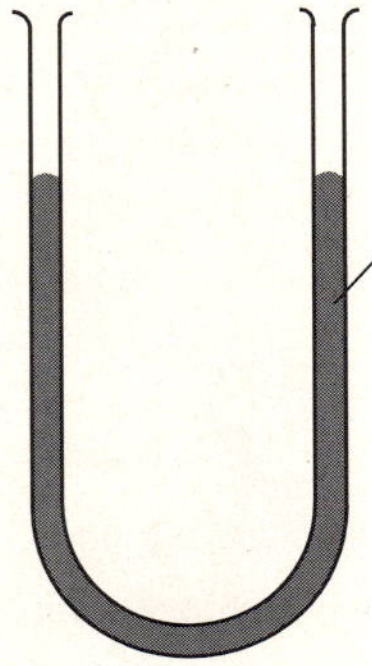

Figure P.19.51.

19.52. An *inclined manometer* is often used for more accurate pressure measurements. If the mercury in the tube has a length L, what is the natural frequency of oscillation of the mercury in the tube?

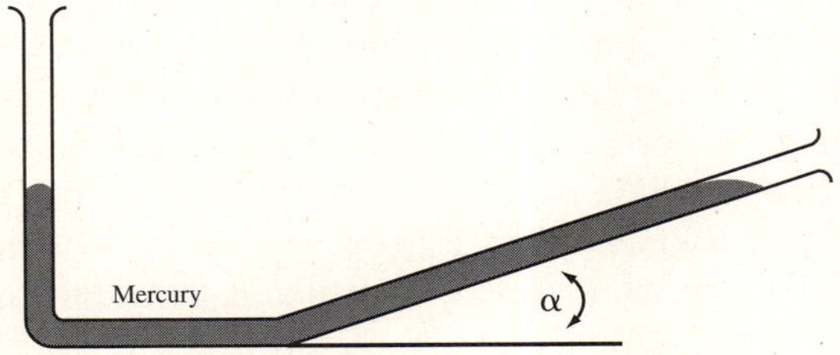

Figure P.19.52.

19.53. A *differential* manometer is used for measuring high pressures. What is the natural frequency of oscillation of the mercury?

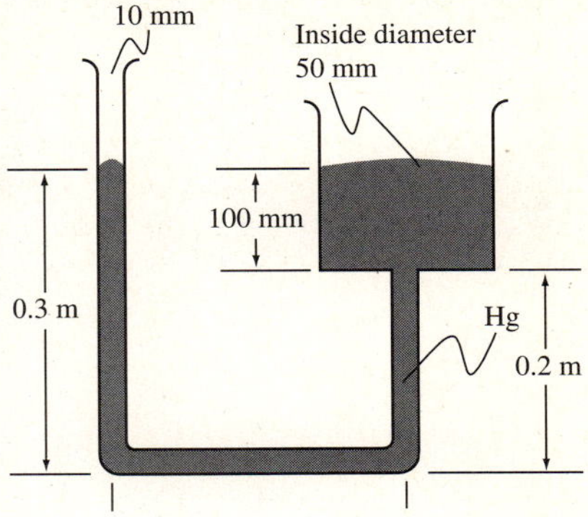

Figure P.19.53.

19.54. A stepped cylinder rides on a circular path. For small oscillations, what is the natural frequency? Take the radius of gyration about the geometric axis O as k and the weight of the cylinder as W.

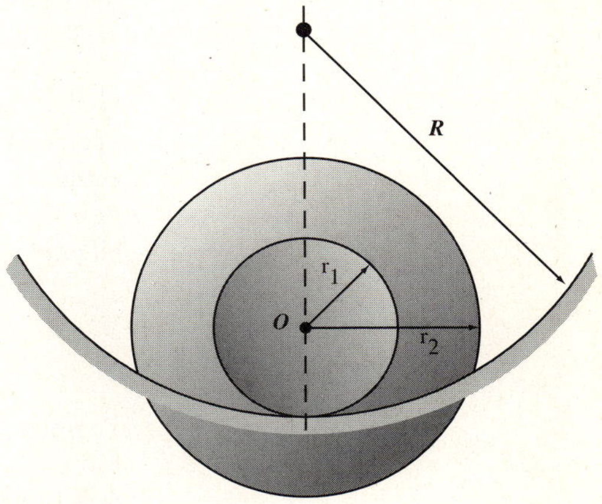

Figure P.19.54.

19.6 Linear Restoring Force and a Force Varying Sinusoidally with Time

We shall now consider the case of a sinusoidal force acting on a spring–mass system (Fig. 19.16) shown with stationary reference xy. The sinusoidal force has a frequency of ω (not to be confused with ω_n, the natural frequency) and an amplitude of F_0. At time $t = 0$, the mass will be assumed to have some known velocity and position, and we shall investigate the ensuing motion.

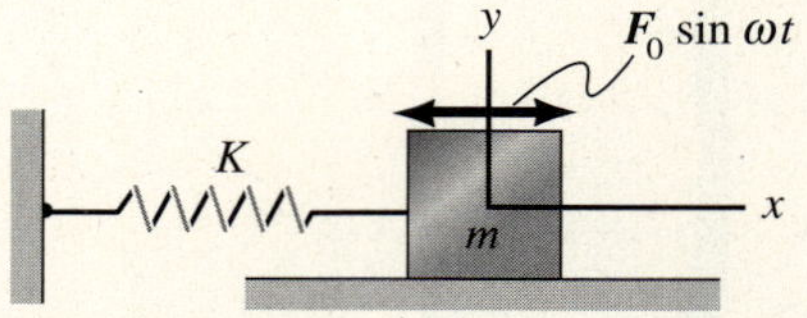

Figure 19.16. Spring–mass system with harmonic disturbance.

Measuring the position x from the unextended position of the spring, we have for *Newton's law*:

$$m\frac{d^2x}{dt^2} = -Kx + F_0 \sin \omega t \qquad (19.20)$$

Rearranging the equation so that the dependent variable and its derivatives are on the left-hand side and dividing through by m, we get the standard form:

$$\frac{d^2x}{dt^2} + \frac{K}{m}x = \frac{F_0}{m}\sin \omega t \qquad (19.21)$$

If the right-hand side is zero the equation is termed *homogeneous*. This was the equation studied in Section 19.5. If any function of t or constant appears on the right side, as in the case above, the equation is *nonhomogeneous*.

The general solution of a nonhomogeneous differential equation of this type is found by getting the general solution of the corresponding homogeneous equation and then finding a *particular solution* that satisfies the full equation. The sum of these solutions, then, is the general solution of the equation. Often, the solution for the homogeneous equation is termed the *complementary solution*.

In this case, we have already ascertained the complementary solution:

$$x_c = C_1 \sin\sqrt{\frac{K}{m}}t + C_2 \cos\sqrt{\frac{K}{m}}t \qquad (19.22)$$

To get a particular solution x_p, we can see by inspection that a function of the form $x_p = C_3 \sin \omega t$ will give a solution if the constant C_3 is chosen properly. Substituting this function into Eq. 19.21, we thus have

$$-C_3\omega^2 \sin \omega t + \frac{K}{m}C_3 \sin \omega t = \frac{F_0}{m}\cos \omega t$$

Clearly, the value of C_3 must be

$$C_3 = \frac{F_0/m}{K/m - \omega^2} \tag{19.23}$$

We can now express the general solution of the differential equation at hand:

$$x = C_1 \sin\sqrt{\frac{K}{m}}t + C_2 \cos\sqrt{\frac{K}{m}}t + \frac{F_0/m}{K/m - \omega^2}\sin\omega t \tag{19.24}$$

Note that there are two arbitrary constants which are determined from the initial conditions of the problem. Do not use the results of Eq. 19.6 for these constants, because we must now include the particular solution in ascertaining the constants. When $t = 0$, $x = x_0$ and $\dot{x} = \dot{x}_0$. We apply these conditions to Eq. 19.24:

$$\begin{aligned} x_0 &= C_2 \\ \dot{x}_0 &= C_1\sqrt{\frac{K}{m}} + \frac{F_0/m}{K/m - \omega^2}\omega \end{aligned} \tag{19.25}$$

Solving for the constants, we get

$$\begin{aligned} C_2 &= x_0 \\ C_1 &= \frac{\dot{x}_0}{\sqrt{K/m}} - \frac{\omega F_0/m}{(K/m - \omega^2)\sqrt{K/m}} \end{aligned} \tag{19.26}$$

Returning to Eq. 19.24, notice that we have the superposition of two harmonic motions—one with a frequency equal to $\sqrt{K/m}$, the natural frequency ω_n of the system, and the other with a frequency ω of the "driving function" (i.e., the nonhomogeneous part of the equation). The frequencies ω and ω_n are not the same in the general case. The phasor representation then leads us to the fact that since the rotating vectors have different angular speeds, the resulting motion cannot be represented by a single phasor, and hence the motion is not harmonic. The two parts of the motion are termed the *transient* part, corresponding to the complementary solution, and the *steady-state* part, corresponding to the particular solution, having frequencies ω_n and ω, respectively. With the introduction of friction (next section), we shall see that the transient part of the motion dies out while the steady state persists as long as there is a disturbance present.

Let us now consider the steady-state part of the motion in Eq. 19.24. Dividing numerator and denominator by K/m, we have for this motion, which we denote as x_p:

$$x_p = \frac{F_0/K}{1 - (\omega^2 m/K)}\sin\omega t = \frac{F_0/K}{1 - (\omega/\omega_n)^2}\sin\omega t \tag{19.27}$$

It will be useful to study with respect to ω/ω_n the variation of the magnitude of the steady-state amplitude x_p for $F_0/K = 1$, namely

$$\left|\frac{1}{1-(\omega/\omega_n)^2}\right|$$

shown plotted in Fig. 19.17. As the forcing frequency approaches the natural frequency, this term goes to infinity, and thus the amplitude of the forced vibration approaches infinity. This is the condition of *resonance*. Under such circumstances, friction, which we neglect here but which is always present, will limit the amplitude. Also, when very large amplitudes are developed, the properties of the restoring element do not remain linear, so that the theory which predicts infinite amplitudes is inapplicable. Thus, the linear, frictionless formulations cannot yield correct amplitudes at resonance in real problems. The condition of resonance, however, does indicate that large amplitudes are to be expected. Furthermore, these amplitudes can be dangerous, because large force concentrations will be present in parts of the restoring system as well as in the moving body and may result in disastrous failures. It is therefore important in most situations to avoid resonance. If a disturbance corresponding to the natural frequency is present and cannot be eliminated, we may find it necessary to change either the stiffness or the mass of a system in order to avoid resonance.

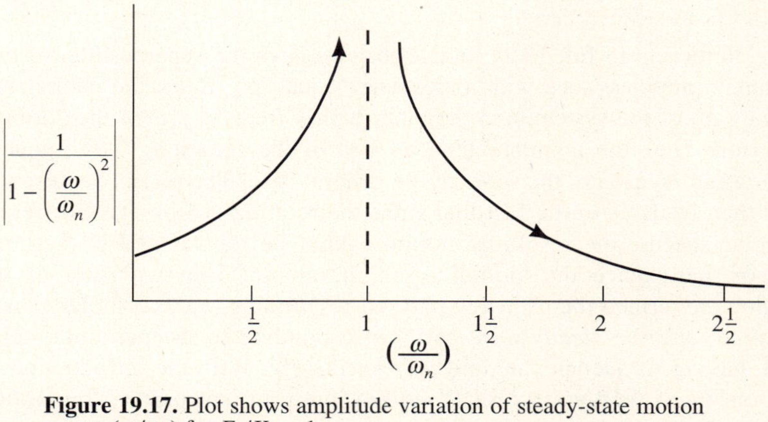

Figure 19.17. Plot shows amplitude variation of steady-state motion versus (ω/ω_n) for $F_0/K = 1$.

From Fig. 19.17 we can conclude that the amplitude will become small as the frequency of the disturbance becomes very high. Also, considering the amplitude C_3 for steady-state motion (Eq. 19.23), we see that below resonance the sign of this expression is positive, and above resonance it is negative, indicating that below resonance the motion is in *phase* with the *disturbance* and above resonance the motion is directly 180° *out of phase* with the *disturbance*.

Example 19.7

A motor mounted on springs is constrained by the rollers to move only in the vertical direction (Fig. 19.18). The assembly weighs 2.6 kN and when placed carefully on the springs causes a deflection of 2.5 mm. Because of an unbalance in the rotor, a disturbance results that is approximately sinusoidal in the vertical direction with a frequency equal to the angular speed of the rotor. The amplitude of this disturbance is 130 N when the motor is rotating at 1720 r/min. What is the steady-state motion of this system under these circumstances if we neglect the mass of the springs, the friction, and the inertia of the rollers?

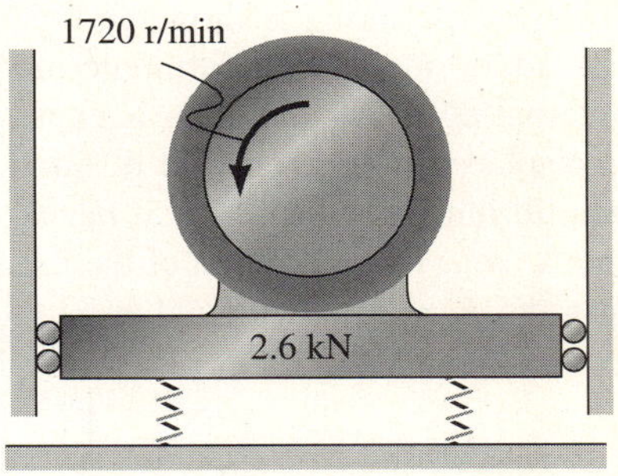

Figure 19.18. Motor with unbalanced rotor.

The spring constant for the system is

$$K = \frac{2600}{2.5} = 1040 \text{ N/mm} = 1.04 \text{ MN/m}$$

and the natural frequency becomes

$$\begin{aligned} \omega_n &= \sqrt{\frac{1.04 \times 10^6}{2600/9.81}} \\ &= 62.6 \text{ rad/s} = 9.97 \text{ Hz} \end{aligned}$$

The steady-state motion is

$$\begin{aligned} x_p &= \frac{F_0/K}{1-(\omega/\omega_n)^2} \sin \omega t \\ &= \frac{130/(1.04 \times 10^6)}{1-[1720/(60)(9.97)]^2} \sin \frac{1720}{60}(2\pi)t \\ &= -1.720 \times 10^{-5} \sin 180.1t \text{ m} \end{aligned}$$

$$x_p = -0.0172 \sin 180.1t \text{ mm}$$

Note that the driving frequency is above the natural frequency. In starting up motors and turbines, we must sometimes go through a natural frequency of the system, and it is wise to get through this zone as quickly as possible to prevent large amplitudes from building up.

Example 19.8

A mass on a spring is shown in Fig. 19.19. The support of the spring at x' is made to move with harmonic motion in the vertical direction by some external agent. This motion is expressed as $a \sin \omega t$. If at $t = 0$ the mass is displaced in a downward position a distance of 25 mm from the static equilibrium position and if it has at this instant a speed downward of 75 mm/s, what is the position of the mass at $t = 5$ s? Take $a = 125$ mm, $\omega = 10$ rad/s, $K = 73$ kN/m, and $m = 15$ kg.

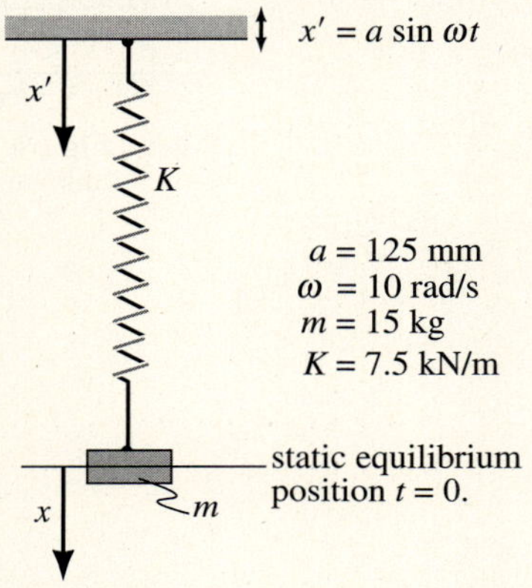

Figure 19.19. Spring–mass system with disturbance.

Let us express **Newton's law** for the mass. Note that the extension of the spring is $x-x'$. Hence,

$$m\frac{d^2x}{dt^2} = -K(x - x')$$

Replacing x' by the known function of time, we get, upon rearranging the terms:

$$\frac{d^2x}{dt^2} + \frac{K}{m}x = \frac{Ka}{m}\sin \omega t$$

This is the same form as Eq. 19.21 for the case where the disturbance is exerted on the mass directly. The solution, then, is

$$x = C_1 \sin\sqrt{\frac{K}{m}}t + C_2 \cos\sqrt{\frac{K}{m}}t + \frac{a}{1 - (\omega/\sqrt{K/m})^2}\sin \omega t$$

Putting in the numerical values of $\sqrt{K/m}$, etc., we have

$$x = C_1 \sin 22.4t + C_2 \cos 22.4t + 157 \sin 10t \text{ mm.}$$

Example 19.8 (Continued)

Now impose the initial conditions to get

$$25 = C_2$$
$$75 = 22.4C_1 + (157)(10)$$

Therefore,

$$C_1 = -66.74$$

The motion, then, is given as

$$x = -66.74 \sin 22.4t + 25 \cos 22.4t + 157 \sin 10t \text{ mm.}$$

When $t = 5$ s, the position of the mass relative to the lower datum is given as

$$(x)_5 = -66.74 \sin(22.4)(5) + 25 \cos(22.4)(5) + 157 \sin 50 = 29.64 \text{ mm.}$$

You may approximate the setup of this problem profitably with an elastic band supporting a small body as shown in Fig. 19.20. By oscillating the free end of the band with varying frequency from low frequency to high frequency, you can demonstrate the rapid change of phase between the disturbance and the excited motion as you pass through resonance. Thus, at low frequencies both motions will be in phase and at frequencies well above resonance the motion will be close to being 180° out of phase. Without friction this change, according to the mathematics, is discontinuous, but with the presence of friction (i.e., in a real case) there is actually a smooth, although sometimes rapid, transition between both extremes.

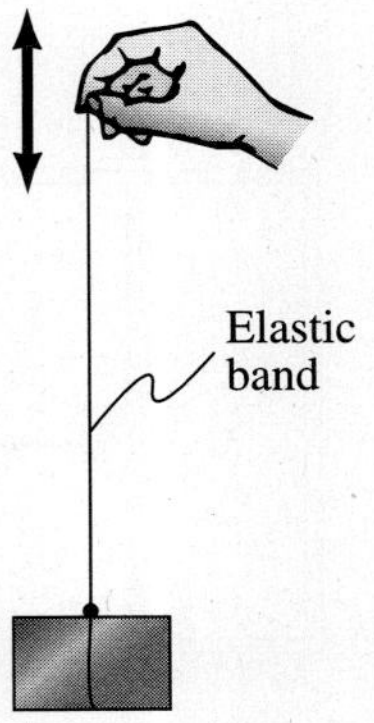

Figure 19.20. Simple resonance and change of phase demonstration.

PROBLEMS

19.55. A mass is held by three springs. Assume that the rolling friction on the floor is negligible, as are the inertial effects of the rollers. The spring constants are:

$$K_1 = 5.4 \text{ N/mm}$$

$$K_2 = 3.6 \text{ N/mm}$$

$$K_3 = 1.8 \text{ N/mm}$$

A sinusoidal force having an amplitude of 22 N and a frequency of $10/\pi$ Hz acts on the body in the direction of the springs. What is the steady-state amplitude of the motion of the body?

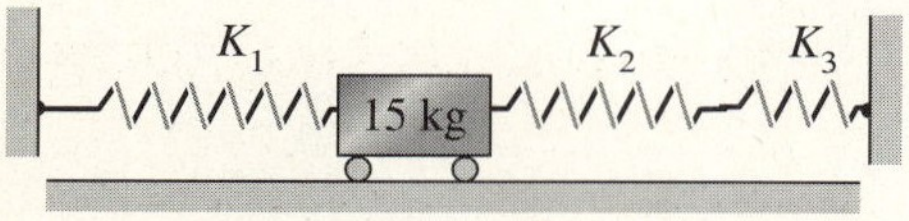

Figure P.19.55.

19.56. In Problem 19.55, the initial conditions are:

(a) The initial position of the body is 75 mm to the right of the static-equilibrium position.

(b) The initial velocity is zero.

(c) At $t = 0$, the sinusoidal disturbing force has a value of 22 N in the positive direction.

Find the position of the body after 3 s.

19.57. A sinusoidal force, with amplitude F of 22 N and frequency $1/2\pi$ Hz, acts on a body having a mass of 22 kg. Meanwhile, the wall moves with a motion given as 8 cos t mm. For a spring constant $K = 8.8$ N/mm, what is the amplitude of the steady-state motion? There is no friction.

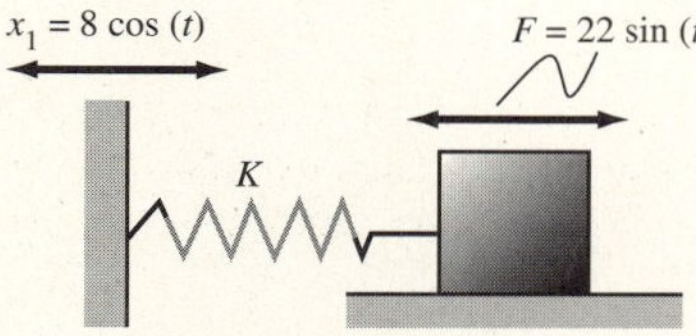

Figure P.19.57.

19.58. In Problem 19.57, suppose that the disturbing force F is 22 sin ($t + \pi/4$). What is the amplitude of the steady-state motion?

19.59. A torque $T = A \sin \omega t$ is applied to the disc. Express the solution for the transient torsional motion and the steady-state torsional motion in terms of arbitrary constants of integration. Take the shear modulus of elasticity of the shaft as G [*Hint:* Recall that K_t for a shaft is GJ/L.]

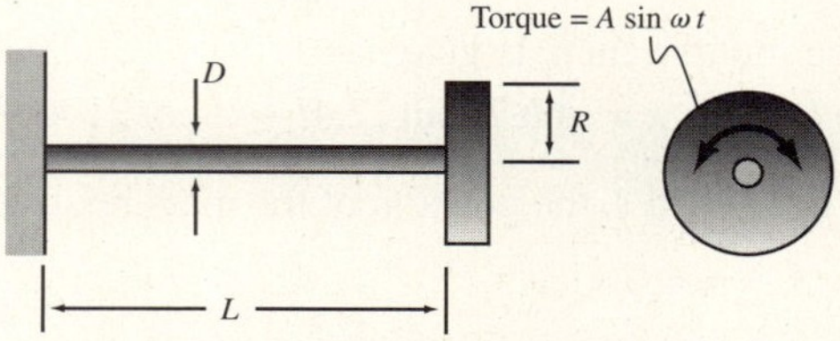

Figure P.19.59.

19.60. A *vibrograph* is a device for measuring the amplitude of vibration in a given direction. The apparatus is bolted to the machine to be tested. A seismic mass M in the vibrograph rides along a rod CD under constraint of a linear spring of spring constant K. If the machine being tested has a harmonic motion x of frequency ω in the direction of C–D, then M will have a steady-state oscillatory frequency also of frequency ω. The motion of M relative to the vibrograph is given as x' and is recorded on the rotating drum. Show that the amplitude of motion of the machine is

$$\left| \frac{(\omega/\omega_n)^2 - 1}{(\omega/\omega_n)^2} \right|$$

times the amplitude of the recorded motion x', where $\omega_n = \sqrt{K/M}$.

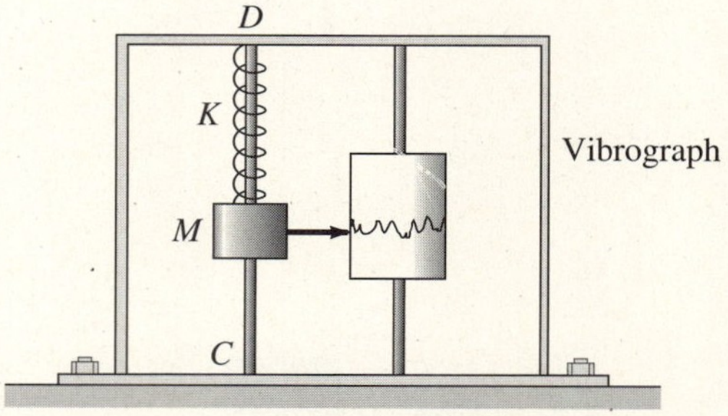

Figure P.19.60.

19.61. A vibrograph is attached rigidly to a diesel engine for which we want to know the vibration amplitude. If the seismic spring-mass system has a natural frequency of 10 Hz, and if the seismic mass vibrates relative to the vibrograph with an amplitude of 1.27 mm when the diesel is turning over at 1000 r/min, what is the amplitude of vibration of the diesel in the direction of the vibrograph? The seismic mass weighs 4.5 N. See Problem 19.60 before doing this problem.

19.62. Explain how you could devise an instrument to measure torsional vibrations of a shaft in a manner analogous to the way the vibrograph measures linear vibrations of a machine. Such instruments are in wide use and are called *torsiographs*. What would be the relation of the amplitude of oscillations as picked up by your apparatus to that of the shaft being measured? See Problem 19.60 before doing this problem.

19.63. A trailer of weight W moves over a washboard road at a constant speed V to the right. The road is approximated by a sinusoid of amplitude A and wavelength L. If the wheel B is small, the center of the wheel will have a motion x closely resembling the aforementioned sinusoid. If the trailer is connected to the wheel through a linear spring of stiffness K, formulate the steady-state equation of motion x' for the trailer. List all assumptions. What speed causes resonance?

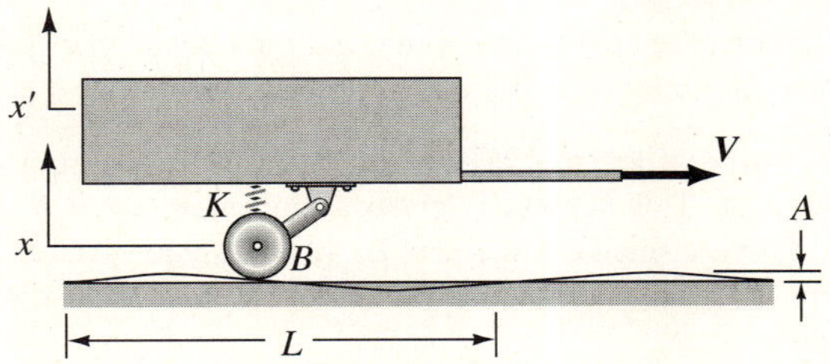

Figure P.19.63.

19.64. In Problem 19.63, compute the amplitude of motion of the trailer for the following data:

$$W = 5.34 \text{ kN}$$

$$V = 44 \text{ m/s}$$

$$K = 43.8 \text{ N/mm}$$

$$L = 10 \text{ m}$$

$$A = 100 \text{ mm}$$

What is the resonance speed V_{res} for this case? From Problem 19.63, we have

$$x'_p = \frac{A}{\left|1 - (2\pi V/L)^2 (W/gK)\right|} \sin \frac{2\pi Vt}{L}$$

19.65. A cantilever beam of length L has an electric motor A weighing 100 N fastened to the end. The tip of the cantilever beam descends 12 mm when the motor is attached. If the *center of mass* of the armature of the motor is a distance 2 mm from the axis of rotation of the motor, what is the amplitude of vibration of the motor when it is rotating at 1750 r/min? The armature weighs 40 N. Neglect the mass of the beam.

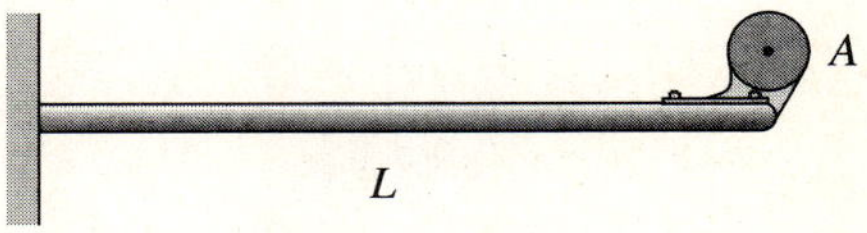

Figure P.19.65.

19.66. Suppose that a 2 N block is glued to the top of the motor in Problem 19.65, where the maximum strength of the bond is 0.5 N. At what minimum angular speed ω of the motor will the block fly off?

19.67. An important reason for mounting rotating and reciprocating machinery on springs is to decrease the transmission of vibration to the foundation supporting the machine. Show that the amplitude of force transmitted to the ground, F_{TR}, for such cases is

$$F_0 \left| \frac{1}{1 - (\omega/\omega_n)^2} \right|$$

where F_0 is the disturbing force from the machine. The factor $|1/[1 - (\omega/\omega_n)^2]|$ is called the *relative transmission factor*. Show that, unless the springs are soft, $(\omega_n < \omega/\sqrt{2})$, the use of springs actually increases the transmission of vibratory forces to the foundation.

19.68. In Example 19.7, what is the amplitude of the force transmitted to the foundation? What must K of the spring system be to decrease the amplitude by one-half? See Problem 19.67.

19.69. A machine weighing W N contains a reciprocating mass of weight w N having a vertical motion relative to the machine given approximately as $x' = A \sin \omega t$. The machine is mounted on springs having a total spring constant K. This machine is guided so that it can move only in the vertical direction. What is the differential equation of motion for this machine? What is the formulation for the amplitude of the machine for steady-state operation?

19.70. A mass M of 0.5 kg is suspended from a stiff rod AB via a spring whose spring constant K is 100 N/m. The end of rod AB is given a vertical sinusoidal motion $\delta_A = 2 \sin 14t$ mm, with t in seconds. What is the maximum force on the rod at C long after the motion has started?

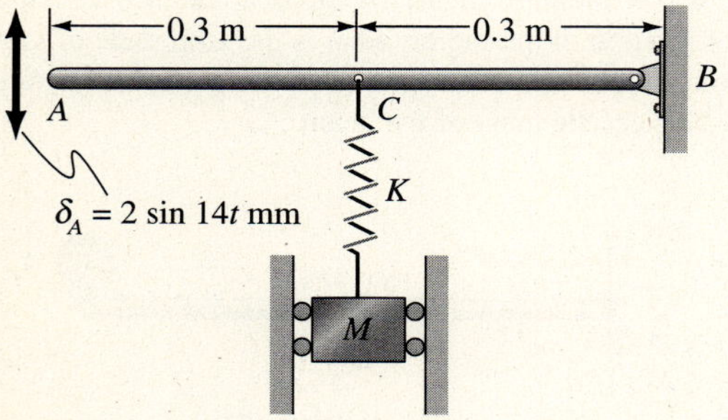

Figure P.19.70.

19.71. In Problem 19.70, what range of frequencies of the motion δ_A must be excluded to keep the maximum force at C less than 7 N? Consider only steady-state motion. [*Hint:* Note that below resonance disturbance and motion are in phase whereas above resonance they are 180° out of phase. Therefore, x_p in Eq. 19.27 will be positive below resonance and negative above resonance.]

19.72. A bob B of weight W is suspended from a vehicle A which is made to have a motion $x_A = \delta \sin \omega t$. If δ is very small, what should ω be so that bob B has an amplitude of motion equal to 1.5δ?

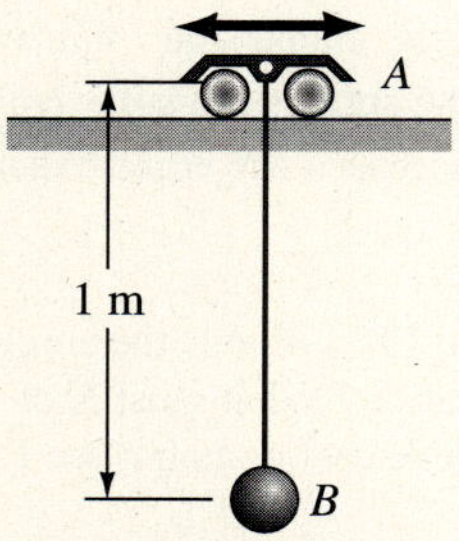

Figure P.19.72.

19.73. Two spheres each of mass $M = 2$ kg are welded to a light rod that is pinned at B. A second light rod AC is welded to the first rod. At A we apply a disturbance $F_o \sin \omega t$. At the other end C, there is a restraining spring which is unstretched when AC is horizontal. If the amplitude of steady-state rotation of the system is to be kept below 20 mrad, what ranges of frequencies ω are permitted? The following data apply:

$$l = 300 \text{ mm}$$
$$K = 7.0 \text{ N/mm}$$
$$F_o = 10 \text{ N}$$
$$a = 100 \text{ mm}$$

See the hint in Problem 19.71.

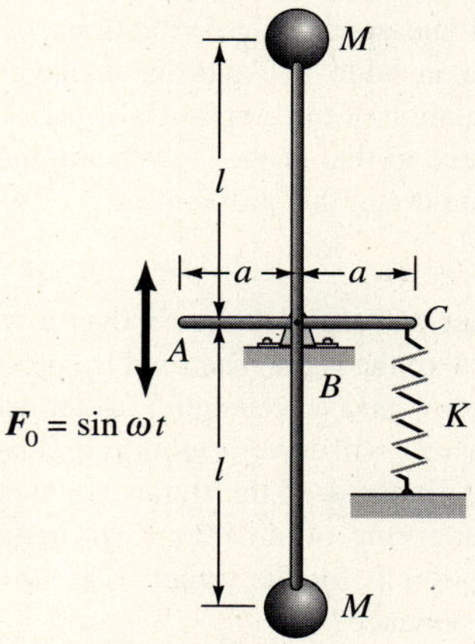

Figure P.19.73.

19.74. In Problem 19.73, what is the angle of rotation of the system 10 s after the application of the sinusoidal load? The system is stationary at time $t = 0$. Take $\omega = 13$ rad/s.

19.75 A rod of length L and weight W is suspended from a light support at A. This support is given a movement $x_A = \delta \sin \omega t$, where δ is very small compared to L. At what frequency, ω, should A be moved if the amplitude of motion of tip, B, is to be 1.5δ?

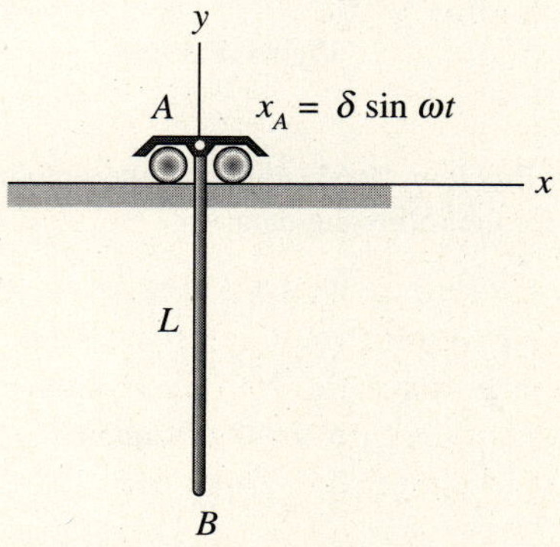

Figure P.19.75.

19.7 Linear Restoring Force with Viscous Damping

We shall now consider the case in which a special type of friction is present. In the chapters on statics, you will recall, we considered coulombic or dry friction for the cases of sliding and impending motion. This force was proportional to the normal force at the interface of contact and dependent on the material of the bodies. At this time, we shall consider the case of bodies separated from each other by a thin film of fluid. The frictional force (called a *damping* force) is independent of the material of the bodies but depends on the nature of the fluid and is proportional for a given fluid to the relative velocity of the two bodies separated by the film. Thus,

$$f = -c\left(\frac{dx}{dt}\right)_{\text{rel}} \tag{19.28}$$

where c is called the *coefficient of damping*. The minus sign indicates that the frictional force opposes the motion (i.e, the friction force must always have a sign opposite to that of the relative velocity).

In Fig. 19.21 is shown the spring-mass model with damping present. We shall investigate possible motions consistent with a set of given initial conditions. The differential equation of motion is

$$m\frac{d^2x}{dt^2} = -Kx - c\frac{dx}{dt}$$

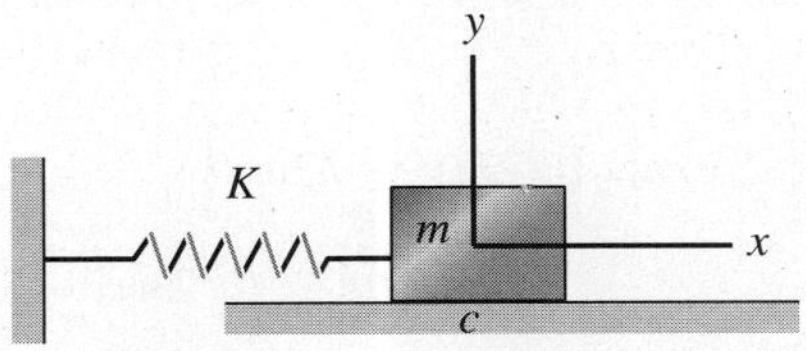

Figure 19.21. Spring–mass system with damping.

In standard form, we get

$$\frac{d^2x}{dt^2} + \frac{c}{m}\frac{dx}{dt} + \frac{K}{m}x = 0 \tag{19.29}$$

This is a homogeneous, second-order, differential equation with constant coefficients. We shall expect two independent functions with two arbitrary constants to form the general solution to this equation. Because of the presence of the first derivative in the equation, we cannot use sines or cosines for trial solutions, since the first derivative changes their form and prevents a cancellation of the time function. Instead, we use e^{pt} where p is determined so as to satisfy the equation. Thus, let

$$x = C_1e^{pt}$$

Substituting, we get

$$C_1 p^2 e^{pt} + \frac{c}{m} C_1 p e^{pt} + \frac{K}{m} C_1 e^{pt} = 0$$

Canceling out $C_1 e^{pt}$, we get

$$p^2 + \frac{c}{m} p + \frac{K}{m} = 0$$

Solving for p, we write

$$p = \frac{-c/m \pm \sqrt{(c/m)^2 - 4K/m}}{2} = -\frac{c}{2m} \pm \sqrt{\left(\frac{c}{2m}\right)^2 - \frac{K}{m}} \tag{19.30}$$

It will be helpful to consider three cases here.

Case A

$$\frac{c}{2m} > \sqrt{\frac{K}{m}}$$

Here the value p is *real*. Using both possible values of p and employing C_1 and C_2 as arbitrary constants, we get

$$x = C_1 \exp\left\{\left[-c/2m + \sqrt{(c/2m)^2 - K/m}\right]t\right\} + C_2 \exp\left\{\left[-(c/2m) - \sqrt{(c/2m)^2 - K/m}\right]t\right\} \tag{19.31}$$

Rearranging terms in Eq. 19.31 we get the following standard form of solution:

$$x = \exp[-(c/2m)t]\left\{C_1 \exp\left[\sqrt{(c/2m)^2 - K/m}\, t\right] + C_2 \exp\left[-\sqrt{(c/2m)^2 - K/m}\, t\right]\right\} \tag{19.32}$$

Since $c/m > \sqrt{(c/2m)^2 - K/m}$, we see from Eq. 19.32 that the first exponential dominates and so as the time t increases, the motion can only be that of an exponential of *decreasing* amplitude. Thus, there can be no oscillation. The motion is illustrated in Fig. 19.22 and is called *overdamped* motion.

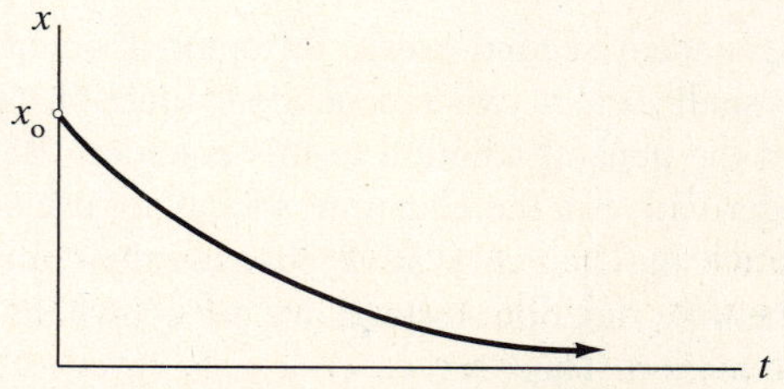

Figure 19.22. Overdamped motion.

Case B

$$\frac{c}{2m} < \sqrt{\frac{K}{m}}$$

This means that we have a negative quantity under the root in Eq. 19.30. Extracting $\sqrt{-1} = i$, we can then write p as follows:

$$p = -\frac{c}{2m} \pm i\sqrt{\frac{K}{m} - \left(\frac{c}{2m}\right)^2}$$

The solution then becomes

$$x = \exp[-(c/2m)t]\left\{C_1 \exp\left[i\sqrt{K/m - (c/2m)^2}\, t\right] + C_2 \exp\left[-i\sqrt{K/m - (c/2m)^2}\, t\right]\right\} \quad (19.33)$$

From complex-number theory, we know that $e^{i\theta}$ may be replaced by $\cos\theta + i\sin\theta$ and thus the equation above can be put in the form

$$x = \exp[-(c/2m)t]\left\{C_1\left[\cos\sqrt{\frac{K}{m} - \left(\frac{c}{2m}\right)^2}\, t + i\sin\sqrt{\frac{K}{m} - \left(\frac{c}{2m}\right)^2}\, t\right] + C_2\left[\cos\sqrt{\frac{K}{m} - \left(\frac{c}{2m}\right)^2}\, t - i\sin\sqrt{\frac{K}{m} - \left(\frac{c}{2m}\right)^2}\, t\right]\right\} \quad (19.34)$$

Collecting terms and replacing sums and differences of arbitrary constants including i by other arbitrary constants, we get the result:

$$x = \exp[-(c/2m)t]\left[C_3 \cos\sqrt{\frac{K}{m} - \left(\frac{c}{2m}\right)^2}\, t + C_4 \sin\sqrt{\frac{K}{m} - \left(\frac{c}{2m}\right)^2}\, t\right] \quad (19.35)$$

The quantity in brackets represents a harmonic motion which has a frequency less than the free undamped natural frequency of the system. The exponential term to the left of the brackets, then, serves to decrease continually the amplitude of this motion. A plot of the displacement against time for this case is illustrated in Fig. 19.23, where the upper dashed envelope corresponds in form to the exponential function $e^{-(c/2m)t}$. We call this motion *underdamped* motion.

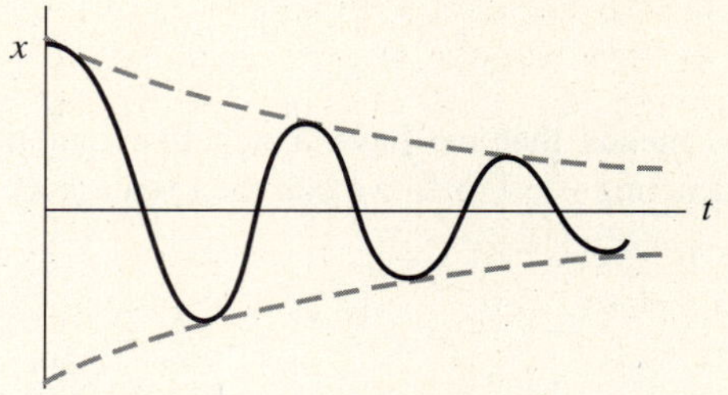

Figure 19.23. Underdamped motion.

Case C

$$\frac{c}{2m} = \sqrt{\frac{K}{m}}$$

Since this is the dividing line between the overdamped case and one in which oscillation is possible, the motion is termed a *critically damped motion*. We have here *identical* roots for p given as

$$p = -\frac{c}{2m} \tag{19.36}$$

and accordingly for such a case the general solution to Eq. 19.29 according to the theory of differential equation is then

$$x = (C_1 + C_2 t)e^{-(c/2m)t} \tag{19.37}$$

First we see from this equation that we do *not* have an oscillatory motion. Also, you will recall from the calculus that as t goes to infinity an exponential of the form e^{-At}, with A a positive constant, goes to zero faster than Ct goes to infinity. Accordingly, Fig. 19.22 can be used to picture the plot of x versus t for this case.

The damping constant for this case is called the *critical* damping constant and is denoted as c_{cr}. The value of c_{cr} clearly is

$$c_{cr} = 2\sqrt{Km} \tag{19.38}$$

It should be clear that, for a damping constant less than c_{cr}, we will have underdamped motion while for a damping constant greater than c_{cr} we will have overdamped motion.

In all the preceding cases for damped free vibration, the remaining step for a complete evaluation of the solution is to compute the arbitrary constants from the initial conditions of the particular problem. Note that in discussing damped motion we shall consider the "natural frequency" of the system to be that of the corresponding *undamped* case and shall refer to the actual frequency of the motion as the frequency of free, damped motion.

Example 19.9

Springs and dashpots are used in packaging delicate equipment in crating so that during transit the equipment will be protected from shocks. In Fig. 19.24, we have shown a piece of equipment whose weight W is 500 N. It is supported in a crate by one spring and two dashpots (or shock absorbers). The value of K for the spring is 30 N/mm and the coefficient of damping, c, is 1 N s/mm for each dashpot. The crate is held above a rigid floor at a height h of 150 mm. It is then released and allowed to hit the floor in a plastic impact. What is the maximum deflection of W relative to the crate?

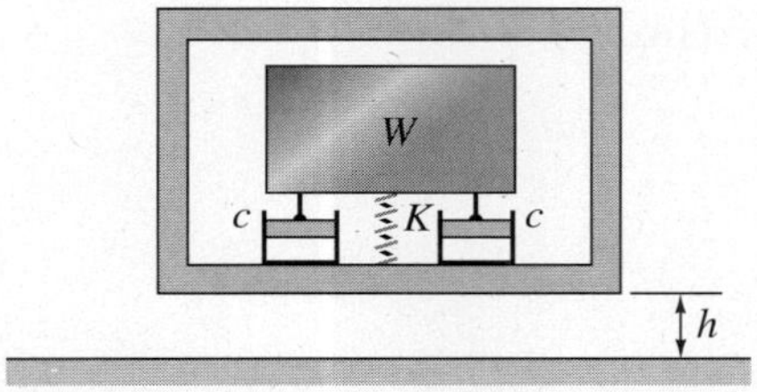

Figure 19.24. Packaging to reduce breakage.

As a first step, we compute the *critical damping* to find what regime we are in.

$$\begin{aligned} c_{cr} &= 2\sqrt{Km} = 2\sqrt{(30)(1000)(500)/g} \\ &= 2473 \text{ N s/m} \end{aligned} \tag{a}$$

The total damping coefficient for our case is

$$c_{total} = (2)(1)(1000) = 2000 \text{ N s/m}$$

We are therefore *underdamped*. The motion is then given as follows:

$$x = e^{-(c/2m)t}\left[C_3 \cos\sqrt{\frac{K}{m} - \left(\frac{c}{2m}\right)^2}\,t + C_4 \sin\sqrt{\frac{K}{m} - \left(\frac{c}{2m}\right)^2}\,t\right] \tag{b}$$

Note that

$$\frac{c}{2m} = \frac{2000}{(2)(500)/9.81} = 19.62 \text{ s}$$

$$\frac{K}{m} = \frac{(30)(1000)}{500/9.81} = 589 \text{ s}^{-2}$$

Example 19.9 (Continued)

Hence,

$$x = e^{-19.62t}(C_3 \cos 14.27t + C_4 \sin 14.27t) \qquad \text{(c)}$$

When $t = 0$, at the instant of impact, take

$$x = 0 \quad \text{and} \quad \dot{x} = \sqrt{2gh} = \sqrt{(2)(9.81)(0.15)} = 1.716 \text{ m/s}$$

The first condition renders $C_3 = 0$. For the second condition, note first that

$$\dot{x} = e^{-19.62t}[C_4(14.27)14.27t] - 19.62e^{-19.62t}(C_4 \sin 14.27t)$$

For the second condition ($\dot{x} = 1.716$ at $t = 0$) we get

$$1.716 = C_4(14.27)$$

Therefore,

$$C_4 = 0.1202$$

Thus, we have for x:

$$x = 0.1202e^{-19.62t} \sin 14.27t \qquad \text{(d)}$$
$$\dot{x} = e^{-19.62t}(1.716 \cos 14.27t - 2.358 \sin 14.27t) \qquad \text{(e)}$$

Set $\dot{x} = 0$ and solve for t in order to get the maximum deflection of W.

$$1.716 \cos 14.27t - 2.358 \sin 14.27t = 0$$

Therefore,

$$\tan 14.27t = 0.7274$$

The smallest t satisfying the equation above is

$$t = 0.0441 \text{ s}$$

The value of x for this time is from Eq. (d):

$$x = 0.1202e^{-(19.62)(0.0441)} \sin[14.27(0.0441)]$$

$$x = 0.0298 \text{ m} = 29.8 \text{ mm}$$

Hence, W moves a maximum distance of 29.8 mm downward after impact.

Example 19.10

A block W of 200 N (see Fig. 19.25) moves on a film of oil which is 0.1 mm in thickness under the block. The area of the bottom surface of the block is 20×10^3 mm^2. The spring constant K is 2 N/m. If the weight is pulled in the x direction and released, what is the nature of the motion?

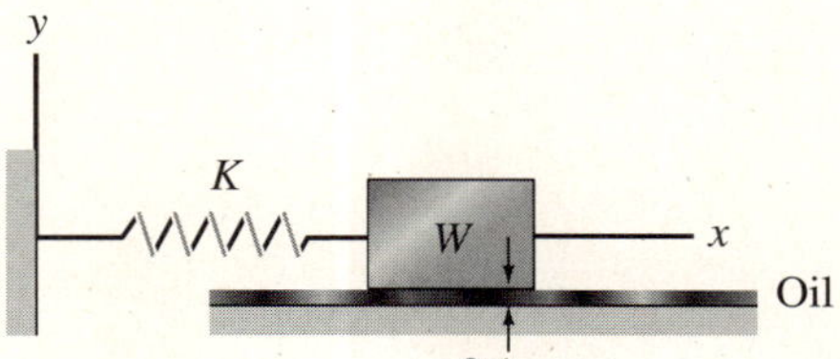

Figure 19.25. Spring–mass on film of oil.

You may have learned in physics that friction force per unit area (i.e., shear stress) on the block W from the oil is given by **Newton's viscosity law** as:

$$\tau = \mu\left(\frac{\partial V}{\partial y}\right)_{\text{block}} \tag{a}$$

where τ is the shear stress (force per unit area), μ is the *coefficient of viscosity* (not to be confused with the coefficient of friction), and $\partial V/\partial y$ is the slope of the velocity profile at the block surface (see Fig. 19.26). Now the oil will stick to the surfaces of the block W and the ground surface. And so

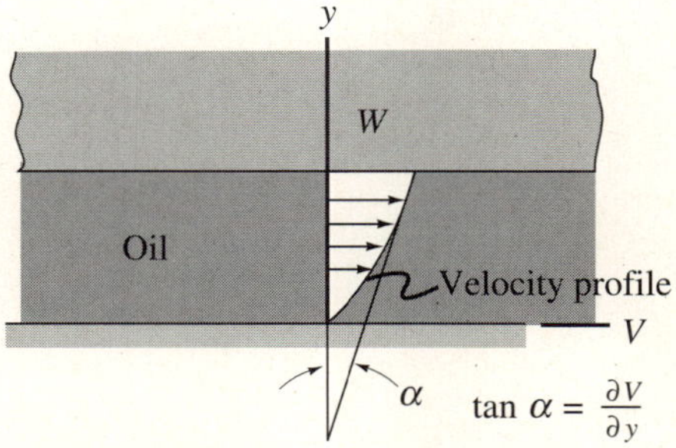

Figure 19.26. Slope of velocity profile at bottom of W.

Example 19.10 (Continued)

we can approximate the velocity profile as shown in Fig. 19.27, where we have used a straight-line profile connecting zero velocity at the bottom and velocity $\dot{x}$ of the block W at the top. Such a procedure gives good results when the film of oil is thin as in the present case. The desired slope $(\partial V/\partial y)_{\text{block}}$ is then approximated as

$$\left(\frac{\partial V}{\partial y}\right)_{\text{block}} = \frac{\dot{x}}{0.1 \times 10^{-3}} \qquad \text{(b)}$$

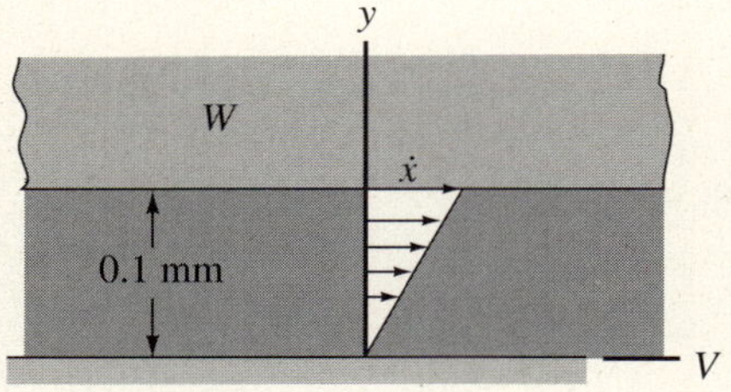

Figure 19.27. Approximate velocity profile.

The coefficient of viscosity can be found in handbooks. For our case, let us say that $\mu = 8$ mPa s.

It is now an easy matter to compute the coefficient of damping c. Thus, the friction force is

$$f = \tau A = -\left[(8 \times 10^{-3})\left(\frac{\dot{x}}{0.1 \times 10^{-3}}\right)\right](20 \times 10^3/10^6) = -1.600\dot{x} \text{ N} \qquad \text{(c)}$$

Thus, $c = 1.600$. The critical damping for the problem is

$$c_{cr} = 2\sqrt{Km} = 2\sqrt{(2)\left(\frac{200}{g}\right)} = 12.77$$

Thus, motion clearly will be *underdamped*. The frequency of oscillation is then

$$\omega = \sqrt{\frac{K}{m} - \left(\frac{c}{2m}\right)^2}$$

$$= \sqrt{\frac{2}{200/g} - \left[\frac{1.600}{(2)(200)/g}\right]^2}$$

$$\omega = 0.311 \text{ rad/s} = 49.5 \text{ mHz}$$

Similar problems involving viscous friction of lubricants and oils are given in the homework section.

*19.8 Linear Restoring Force, Viscous Damping, and a Harmonic Disturbance

In the spring-mass problem shown in Fig. 19.28 we include driving function $F_0 \cos \omega t$ along with viscous damping. The differential equation in the standard form then becomes

$$\frac{d^2x}{dt^2} + \frac{c}{m}\frac{dx}{dt} + \frac{K}{m}x = \frac{F_0}{m}\cos \omega t \tag{19.39}$$

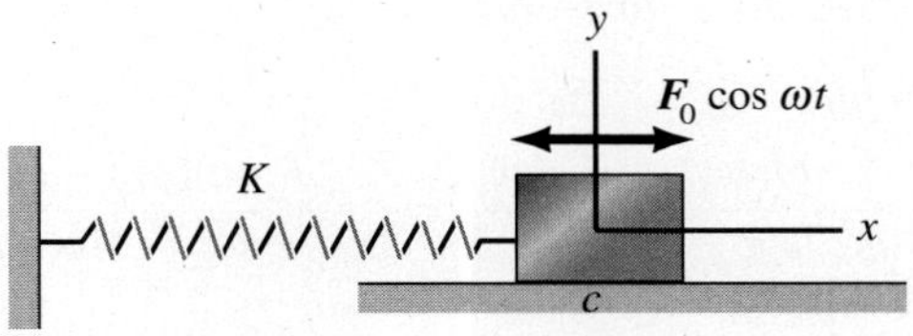

Figure 19.28. Spring–mass system with damping.

Equation 19.39 is a nonhomogeneous equation. The general solution will be the homogeneous solution worked out in Section 19.7, plus any particular solution of Eq. 19.39.

Because there is a first derivative on the left side of the equation, we cannot expect a particular solution of the form $D \cos \omega t$ to satisfy the differential equation. Instead, from the method of *undetermined coefficients* we shall try the following:

$$x_p = D \sin \omega t + E \cos \omega t \tag{19.40}$$

The constants D and E are to be adjusted to facilitate a solution. Substituting into the differential equation, we write

$$-D\omega^2 \sin \omega t - E\omega^2 \cos \omega t + \frac{c}{m}\omega D \cos \omega t - \frac{c}{m}\omega E \sin \omega t$$
$$+ \frac{K}{m} D \sin \omega t + \frac{K}{m} E \cos \omega t = \frac{F_0}{m}\cos \omega t$$

Collecting the terms, we have

$$\left(-D\omega^2 - \frac{c}{m}\omega E + \frac{K}{m}D\right)\sin \omega t + \left(-\frac{F_0}{m} - E\omega^2 + \frac{c}{m}\omega D + \frac{K}{m}E\right)\cos \omega t = 0$$

We set each coefficient of the time functions equal to zero and thus get two simultaneous equations in the unknowns E and D:

$$-D\omega^2 - \frac{\omega c}{m}E + \frac{K}{m}D = 0$$
$$-\frac{F_0}{m} - E\omega^2 + \frac{\omega c}{m}D + \frac{K}{m}E = 0$$

Rearranging and replacing K/m by ω_n^2, we get

$$D\left(\omega^2 - \omega_n^2\right) + E\left(\frac{\omega c}{m}\right) = 0$$

$$D\left(-\frac{\omega c}{m}\right) + E\left(\omega^2 - \omega_n^2\right) = -\frac{F_0}{m}$$

Using Cramer's rule, we see that the constants D and E become

$$D = \frac{\begin{vmatrix} 0 & \omega c/m \\ -F_0/m & \omega^2 - \omega_n^2 \end{vmatrix}}{\begin{vmatrix} \omega^2 - \omega_n^2 & \omega c/m \\ -\omega c/m & \omega^2 - \omega_n^2 \end{vmatrix}} = \frac{(F_0/m)(\omega c/m)}{(\omega^2 - \omega_n^2)^2 + (\omega c/m)^2}$$

$$E = \frac{\begin{vmatrix} \omega^2 - \omega_n^2 & 0 \\ -\omega c/m & -F_0/m \end{vmatrix}}{\begin{vmatrix} \omega^2 - \omega_n^2 & \omega c/m \\ -\omega c/m & \omega^2 - \omega_n^2 \end{vmatrix}} = \frac{(F_0/m)(\omega_n^2 - \omega^2)}{(\omega^2 - \omega_n^2)^2 + (\omega c/m)^2}$$

The entire solution can then be given as

$$x = x_c + \frac{(F_0/m)(\omega_n^2 - \omega^2)}{(\omega^2 - \omega_n^2)^2 + (\omega c/m)^2}\cos\omega t + \frac{F_0\omega c/m^2}{(\omega^2 - \omega_n^2)^2 + (\omega c/m)^2}\sin\omega t \qquad (19.41)$$

The constants of integration are present in the complementary solution x_c and are determined by the initial condition to which the *entire* solution given above is subject.

The complementary solution here is a *transient* in the true sense of the word, because it dies out in the manner explained in Section 19.7. The particular solution is a harmonic motion with the same frequency as the disturbance. Only the amplitude of this motion is affected by the damping present. Note that, mathematically, the amplitude of the steady-state motion cannot become infinite with damping present unless F_0 becomes infinite.

We now write the steady-state solution in the following way:

$$x_p = \frac{(F_0/m)(\omega_n^2 - \omega^2)}{(\omega^2 - \omega_n^2)^2 + (\omega c/m)^2}\cos\omega t + \frac{(F_0/m)(\omega c/m)}{(\omega^2 - \omega_n^2)^2 + (\omega c/m)^2}\sin\omega t \qquad (19.42)$$

We can represent this formulation in a phasor diagram as shown in Fig. 19.29. It should be clear that we can give x_p in the following form:

$$x_p = A\cos(\omega t - \alpha) \qquad (19.43)$$

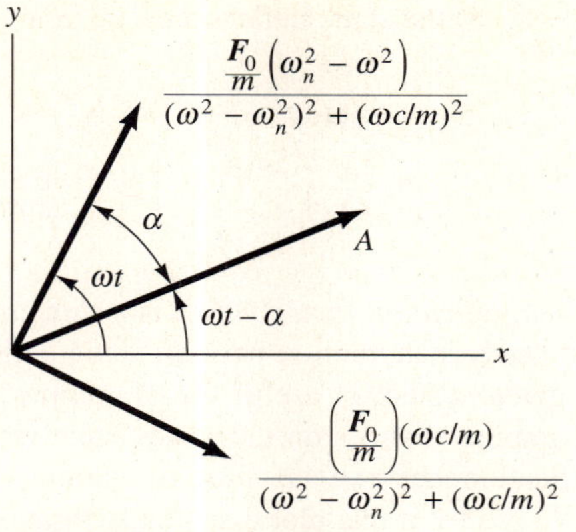

Figure 19.29. Phasor diagram.

where the amplitude A is given as

$$A = \left\{\left[\frac{(F_0/m)(\omega^2 - \omega_n^2)}{(\omega^2 - \omega_n^2)^2 + (\omega c/m)^2}\right]^2 + \left[\frac{(F_0/m)(\omega c/m)}{(\omega^2 - \omega_n^2)^2 + (\omega c/m)^2}\right]^2\right\}^{1/2}$$

$$= \frac{F_0}{m}\frac{\sqrt{(\omega^2 - \omega_n^2)^2 + (\omega c/m)^2}}{(\omega^2 - \omega_n^2)^2 + (\omega c/m)^2}$$

$$= \frac{F_0 / m}{\sqrt{(\omega^2 - \omega_n^2)^2 + (\omega c/m)^2}}$$

$$= \frac{F_0}{\sqrt{(m\omega^2 - K)^2 + (\omega c)^2}} \tag{19.44}$$

and where α, the phase angle, is given as

$$\alpha = \tan^{-1}\left[\frac{(F_0/m)(\omega c/m)}{(\omega^2 - \omega_n^2)^2 + (\omega c/m)^2} \cdot \frac{(\omega^2 - \omega_n^2)^2 + (\omega c/m)^2}{(F_0/ m)(\omega_n^2 - \omega^2)}\right]$$

$$= \tan^{-1}\frac{\omega c}{K - m\omega^2} \tag{19.45}$$

We may express the amplitude A in yet another form by dividing numerator and denominator in Eq. 19.44 by K and by recalling from Eq. 19.38 that the ratio $2\sqrt{Km}/c_{cr}$ is unity. Thus, we get

$$A = \frac{F_0/K}{\sqrt{\left[\left(\frac{\omega}{\omega_n}\right)^2 - 1\right]^2 + \frac{1}{K^2}\left(\frac{2\sqrt{Km}}{c_{cr}}\right)^2(\omega c)^2}}$$

$$= \frac{\delta_{st}}{\sqrt{\left[\left(\frac{\omega}{\omega_n}\right)^2 - 1\right]^2 + \left[2\left(\frac{c}{c_{cr}}\right)\left(\frac{\omega}{\omega_n}\right)\right]^2}} \tag{19.46}$$

where $F_0/K = \delta_{st}$, is the static deflection. The term

$$= \frac{1}{\sqrt{\left[\left(\frac{\omega}{\omega_n}\right)^2 - 1\right]^2 + \left[2\left(\frac{c}{c_{cr}}\right)\left(\frac{\omega}{\omega_n}\right)\right]^2}}$$

is called the *magnification factor* which is a dimensionless factor giving the amplitude of steady-state motion per unit static deflection. Accordingly, this factor for a given system is useful for examining the effects of frequency changes or damping changes on the steady-state vibration amplitude. A plot of the magnification factor versus ω/ω_n for various values of c/c_{cr} is shown in Fig. 19.30. We see from this plot that small vibrations result when ω is kept far from ω_n. Additionally, note that maximum amplitude does not occur at resonance but actually at frequencies somewhat below resonance. Only when the damping goes to zero does the maximum amplitude occur at resonance. However, for light damping we can usually consider that when $\omega/\omega_n = 1$, we have an amplitude very close to the maximum amplitude possible for the system.

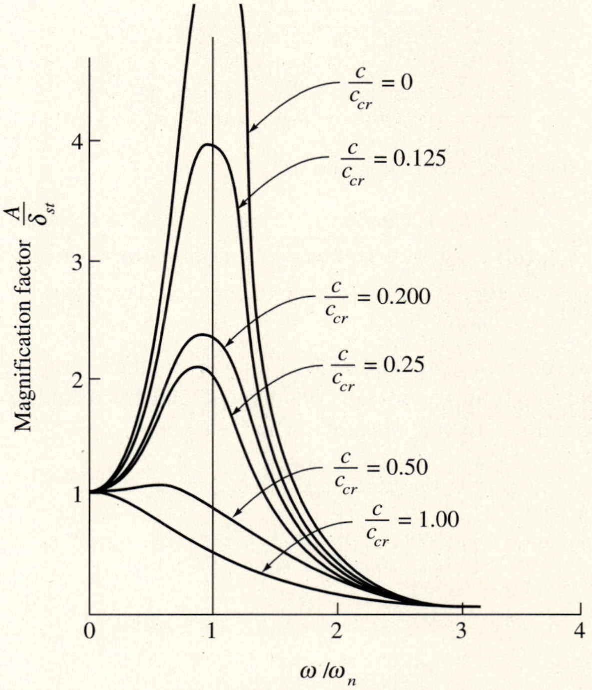

Figure 19.30. Magnification factor plot.

Example 19.11

A *vibrating table* is a machine that can be given harmonic oscillatory motion over a range of amplitudes and frequencies. It is used as a test apparatus for imposing a desired sinusoidal motion on a device.

In Fig. 19.31 is shown a vibrating table with a device bolted to it. The device has in it a body B of mass 7.3 kg supported by two springs each of stiffness equal to 5.4 kN/m and a dashpot having a damping constant c equal to 87.6 N s/m. If the table has been adjusted for a vertical motion x' given as 25 sin $40t$ mm with t in seconds, compute:

1. The steady-state amplitude of motion for body B.
2. The maximum number of g's acceleration that body B is subjected to for steady-state motion.
3. The maximum force that body B exerts on the vibrating table during steady-state motion.

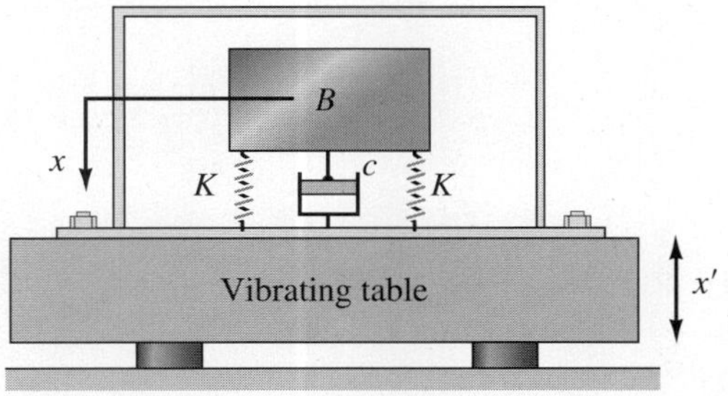

Figure 19.31. A device on a vibrating table.

Measuring the vertical position of body B from the static-equilibrium position with coordinate x, we have from **Newton's law**:

$$M\ddot{x} + c(\dot{x} - \dot{x}') + K(x - x') = 0 \quad \text{(a)}$$

Using $P \sin \omega t$ to represent x' for now, we have

$$\ddot{x} + \frac{c}{M}\dot{x} + \frac{K}{M}x = \frac{cP\omega}{M}\cos \omega t + \frac{KP}{M}\sin \omega t \quad \text{(b)}$$

Letting $cP\omega = F_1$ and $KP = F_2$, we have

$$\ddot{x} + \frac{c}{M}\dot{x} + \frac{K}{M}x = \frac{F_1}{M}\cos \omega t + \frac{F_2}{M}\sin \omega t \quad \text{(c)}$$

Example 19.11 (Continued)

Using a phasor-diagram representation, we can combine the forcing functions into one expression as follows:

$$\frac{F_1}{M}\cos\omega t + \frac{F_2}{M}\sin\omega t = \frac{\sqrt{F_1^2 + F_2^2}}{M}\cos(\omega t - \alpha)$$
$$= \frac{R}{M}\cos(\omega t - \alpha)$$

where $\alpha = \tan^{-1}(F_2/F_1)$ and where $R = \sqrt{F_1^2 + F_2^2}$. Thus, we have

$$\ddot{x} + \frac{c}{M}\dot{x} + \frac{K}{M} = \frac{R}{M}\cos(\omega t - \alpha) \qquad \text{(d)}$$

Except for the phase angle α, Eq. (d) is identical in form to Eq. 19.39. Clearly, if we are interested only in the steady-state amplitude, the phase angle α is of no consequence. Hence setting $\alpha = 0$, we can use the result given by Eq. 19.44 with R taking the place of F_0. Thus, for the amplitude A of mass B, we have

$$A = \frac{R}{\sqrt{(M\omega^2 - K)^2 + (\omega c)^2}} \qquad \text{(e)}$$

The following numerical values apply:

$$M = 73 \text{ kg}, \quad K = 10.8\times 10^3 \text{ N/m}, \quad \omega = 40 \text{ rad/s},$$
$$c = 87.6 \text{ N s/m}, \quad P = 0.025 \text{ m}$$
$$R = \sqrt{F_1^2 + F_2^2} = \left[(\omega cP)^2 + (KP)^2\right]^{1/2}$$
$$= \left\{[(40)(87.6)(0.025)]^2 + \left[(10.8\times 10^3)(0.025)\right]^2\right\}^{1/2} = 283.8$$

Hence, we have for A:

$$A = \frac{283.8}{\sqrt{(11.68\times 10^3 - 10.8\times 10^3)^2 + [(40)(87.6)]^2}} \qquad \text{(f)}$$
$$= 78.6 \text{ mm}$$

To get the maximum steady-state acceleration[5] for body B we compute $|(\ddot{x}_p)|_{max}$. Thus, we have from Eq. 19.43:

[5]We wish to remind you that the *steady-state* solution corresponds to the *particular* solution $\ddot{x}_p$.

Example 19.11 (Continued)

$$\ddot{x}_p = -A\omega^2 \cos(\omega t - \alpha)$$

$$\left|(\ddot{x}_p)\right|_{\max} = \omega^2 A$$

$$= (1600)(0.0786) = 125.8 \text{ m/s}^2 \qquad \text{(g)}$$

$$= \frac{125.8}{9.87} g = \boxed{12.82g}$$

The maximum force transmitted to the body by the springs and dashpot during steady-state motion is established clearly when the body B has its greatest acceleration in the upward direction. We have for the maximum force F_B on noting that $\dot{x}_p = 0$ when $\ddot{x}_p$ is maximum:

$$F_B = W_B + M_B(\ddot{x}_p)_{\max}$$

$$= (7.3)(9.81) + (7.3)(125.8) = 990 \text{ N} \qquad \text{(h)}$$

If there were no spring-dashpot system between B and the vibratory table, the maximum force transmitted to B would be

$$F_B = W_B + \frac{M_B}{g}(\ddot{x}')_{\max}$$

$$= (7.3)(9.81) + (7.3)\left[(0.025)(40)^2\right] = 364 \text{ N} \qquad \text{(i)}$$

We see from Eq. (f) that the amplitude of the induced motion on B is three times what it would be if there were no spring-damping system present to separate B from the table. And from Eqs. (h) and (i) we see that the presence of the spring-damping system has resulted in a considerable *increase* in force acting on body B. Now the use of springs and dashpots for suspending or packaging equipment is generally for the purpose of reducing—not increasing—the amplitude of forces acting on the suspended body. The reason for the increase in these quantities for the disturbing frequency of 40 rad/s is the fact that the natural frequency of the system is 37.8 rad/s, thus putting us just above resonance. To protect the body B for disturbances of 40 rad/s, we must use considerably softer springs.

As an exercise at the end of the section you will be asked in Problem 19.90 to compute K for permitting only a maximum of 12 mm amplitude of vibration for this problem.

*19.9 Oscillatory Systems with Multi-Degrees of Freedom

We shall concern ourselves here with a very simple system that has two degrees of freedom, and we shall be able to generalize from this simple case. In the system of masses shown in Fig. 19.32, the masses are equal, as are the spring constants of the outer springs. We neglect friction, windage, etc. How can we describe the motion of the masses subsequent to any imposed set of initial conditions?

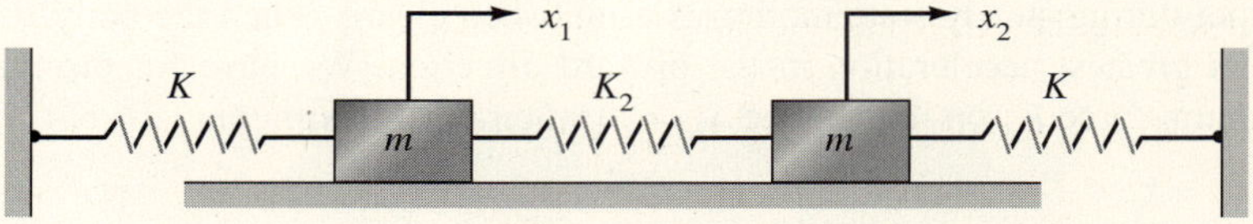

x_1, x_2 measured from equilibrium configuration

Figure 19.32. Two-degree-of-freedom system.

We first express *Newton's law* for each mass. To do this, imagine the masses at any position x_1, x_2 measured from the equilibrium configuration, and then compute the forces. Assume for convenience that $x_1 > x_2$. The spring K_2 is in compression for this supposition, and hence it produces a negative force on the mass at x_1 and a positive force on the mass at x_2. The equations of motion then are:

$$\frac{d^2x_1}{dt^2} = -Kx_1 - K_2(x_1 - x_2) \qquad (19.47a)$$

$$\frac{d^2x_2}{dt^2} = -Kx_2 + K_2(x_1 - x_2) \qquad (19.47b)$$

If you imagine that the masses are at any other nontrivial position, you will still arrive at the above equations.

Because both dependent variables appear in both differential equations, they are termed *simultaneous* differential equations. We rearrange the equations to the following standard form:

$$\frac{d^2x_1}{dt^2} + \frac{K}{m}x_1 + \frac{K_2}{m}(x_1 - x_2) = 0 \qquad (19.48a)$$

$$\frac{d^2x_2}{dt^2} + \frac{K}{m}x_2 - \frac{K_2}{m}(x_1 - x_2) = 0 \qquad (19.48b)$$

Finding a solution is equivalent to finding two functions of time $x_1(t)$ and $x_2(t)$, which when substituted into Eqs. 19.48 (a) and (b) reduce each equation to an identity. Only second derivatives and zeroth derivatives appear in these equations, and we would thus expect that sine or cosine functions of time would yield a possible solution. And since both $x_1$and x_2 appear in the same equation, these time functions must be of the same form in order to allow a cancellation of the time function. A trial solution, therefore, might be:

$$x_1 = C_1 \sin(pt + \alpha) \tag{19.49a}$$

$$x_2 = C_2 \sin(pt + \alpha) \tag{19.49b}$$

where C_1, C_2, α, and p are as yet undetermined. Substituting into Eq. 19.48 and canceling out the time function, we get:

$$-C_1 p^2 + \frac{K}{m} C_1 + \frac{K_2}{m}(C_1 - C_2) = 0 \tag{19.50a}$$

$$-C_2 p^2 + \frac{K}{m} C_2 - \frac{K_2}{m}(C_1 - C_2) = 0 \tag{19.50b}$$

Rearranging the above equations, we write:

$$\left(-p^2 + \frac{K}{m} + \frac{K_2}{m}\right)C_1 - \frac{K_2}{m} C_2 = 0 \tag{19.51a}$$

$$-\frac{K_2}{m} C_1 + \left(-p^2 + \frac{K}{m} + \frac{K_2}{m}\right)C_2 = 0 \tag{19.51b}$$

One way of ensuring the satisfaction of this equation is to have $C_1 = 0$ and $C_2 = 0$. This means, from Eqs. 19.49 (a) and (b), that x_1 and x_2 are always zero, which corresponds to the static equilibrium position. While this is a valid solution, since this static equilibrium is a possible motion, the result is trivial. We now ask: Is there a means of satisfying these equations without setting C_1 and C_2 equal to zero?

To answer this, solve for $C_1$and C_2 in terms of the coefficients, as if they were unknowns in the above equations. Using Cramer's rule, we then have:

$$C_1 = \frac{\begin{vmatrix} 0 & -K_2/m \\ 0 & -p^2 + K/m + K_2/m \end{vmatrix}}{\begin{vmatrix} -p^2 + K/m + K_2/m & -K_2/m \\ -K_2/m & -p^2 + K/m + K_2/m \end{vmatrix}}$$

$$C_2 = \frac{\begin{vmatrix} -p^2 + K/m + K_2/m & 0 \\ -K_2/m & 0 \end{vmatrix}}{\begin{vmatrix} -p^2 + K/m + K_2/m & -K_2/m \\ -K_2/m & -p^2 + K/m + K_2/m \end{vmatrix}} \tag{19.52}$$

Notice that the determinant in the numerator is in each case zero. If the denominator is other than zero, we must have the trivial solution $C_1 = C_2 = 0$, the significance of which we have just discussed. A *necessary* condition for a *nontrivial* solution is that the denominator also be zero, for then we get the indeterminate form 0/0 for C_1 and C_2. Clearly, C_1 and C_2 can then have possible values other than zero, and so the required condition for a nontrivial solution is:

$$\begin{vmatrix} -p^2 + K/m + K_2/m & -K_2/m \\ -K_2/m & -p^2 + K/m + K_2/m \end{vmatrix} = 0 \tag{19.53}$$

Carrying out this determinant multiplication, we get:

$$\left(-p^2 + \frac{K}{m} + \frac{K_2}{m}\right)^2 = \left(\frac{K_2}{m}\right)^2 \tag{19.54}$$

Taking the roots of both sides, we have:

$$-p^2 + \frac{K}{m} + \frac{K_2}{m} = \pm\frac{K_2}{m} \tag{19.55}$$

Two values of p^2 satisfy the necessary condition we have imposed. If we use the positive roots, the values of p are:

$$p_1 = \sqrt{\frac{K}{m}}$$
$$p_2 = \sqrt{\frac{K}{m} + \frac{2K_2}{m}} \tag{19.56}$$

where p_1 and p_2 are found for the plus and minus cases, respectively, of the right side of Eq. 19.55.

Let us now return to Eqs. 19.51 (a) and (b) to ascertain what further restrictions we may have to impose to ensure a solution, because these equations form the criterion for acceptance of a set of functions as solutions. Employing $\sqrt{K/m}$ for p in Eq. 19.51 (a), we have:

$$\left(-\frac{K}{m} + \frac{K}{m} + \frac{K_2}{m}\right)C_1 - \left(\frac{K_2}{m}\right)C_2 = 0 \tag{19.57}$$

From this equation we see that when we use this value of p it is necessary that $C_1 = C_2$ to satisfy the equation. The same conclusions can be reached by employing Eq. 19.51(b). We can now state a permissible solution to the differential equation. Using A as the amplitude in place of $C_1 = C_2$, we have:

$$x_1 = A\sin\left(\sqrt{\frac{K}{m}}t + \alpha\right) \tag{19.58a}$$

$$x_2 = A\sin\left(\sqrt{\frac{K}{m}}t + \alpha\right) \tag{19.58b}$$

If we examine the second value of p, we find that for this value it is required that $C_1 = -C_2$. Thus if we use B for C_1 and use β as the arbitrary value in the sine function, another possible solution is:

$$x_1 = B\sin\left(\sqrt{\frac{K}{m} + \frac{2K_2}{m}}t + \beta\right) \tag{19.59a}$$

$$x_2 = -B\sin\left(\sqrt{\frac{K}{m} + \frac{2K_2}{m}}t + \beta\right) \tag{19.59b}$$

Let us consider each of these solutions. In the first case, the motions of both masses are in phase with each other, have the same amplitude, and thus move together with simple harmonic motion with a natural frequency $\sqrt{K/m}$. For this motion, the center spring is not extended or compressed, and, since the mass of the spring has been neglected, it has no effect on this motion. This explains why the natural frequency has such a simple formulation.

The second possible independent solution is one in which the amplitudes are equal for both masses but the masses are 180° out of phase. Each mass oscillates harmonically with a natural frequency greater than the preceding motion. Since the masses move in opposite directions in the manner described, the center of the middle spring must be stationary for this motion. It is as if each mass were vibrating under the action of a spring of constant K and the action of half the length of a spring with a spring constant K_2 (Fig. 19.33), which explains why the natural frequency for this motion is $\sqrt{(K + 2K_2)/m}$. (It will be left for you to demonstrate in an exercise that halving the length of the spring doubles the spring constant.)

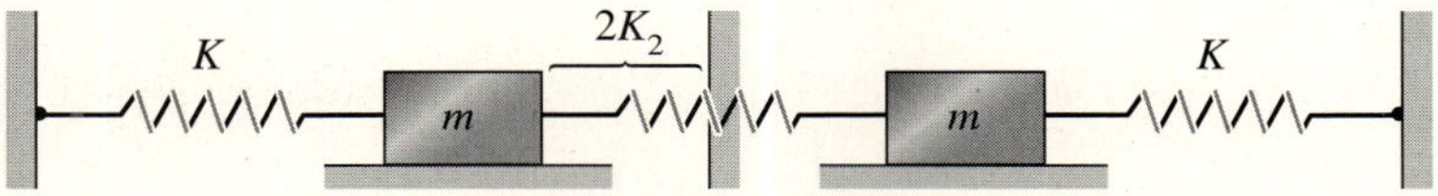

Figure 19.33. Bodies 180° out of phase.

Each of these motions as given by Eqs. 19.58 and 19.59 is called a natural *mode*. The first mode refers to the motion of lower natural frequency, and the second mode identifies the one with the higher natural frequency. It is known from differential equations that the general solution is the sum of the two solutions presented:

$$\begin{aligned} x_1 &= \left[A\sin\left(\sqrt{\frac{K}{m}}t + \alpha\right) \right] + \left[B\sin\left(\sqrt{\frac{K}{m} + \frac{2K_2}{m}}t + \beta\right) \right] \\ x_1 &= \left[A\sin\left(\sqrt{\frac{K}{m}}t + \alpha\right) \right] + \left[-B\sin\left(\sqrt{\frac{K}{m} + \frac{2K_2}{m}}t + \beta\right) \right] \end{aligned} \qquad (19.60)$$

first mode of motion — second mode of motion

Four constants are yet to be determined: A, B, α, and β. These are the constants of integration and are determined by the initial conditions of the motion—that is, the velocity and position of each mass at time $t = 0$.

From this discussion we can make the following conclusions. The general motion of the system under study is the superposition of two modes of motion of harmonic nature that have distinct natural frequencies with amplitudes and phase angles that are evaluated to fit the initial conditions. Thus the basic modes are the "building blocks" of the general free motion.

If the masses, as well as the springs, were unequal, the analysis would still produce two natural frequencies and mode shapes, but these would neither be as simple as the special case we have worked out nor, perhaps, as intuitively obvious.

As we discussed in the first paragraph of this section, two natural frequencies correspond to the two degrees of freedom. In the general case of n degrees of freedom, there will be n natural frequencies, and the general free vibrations will be the superposition of n modes of motion that have proper amplitudes and are phased together in such a way that they satisfy $2n$ initial conditions.

PROBLEMS

19.76. A body of weight W N is suspended between two springs. Two identical dashpots are shown. Each dashpot resists motion of the block at the rate of c N s/m. What is the equation of motion for the block? What is c for critical damping?

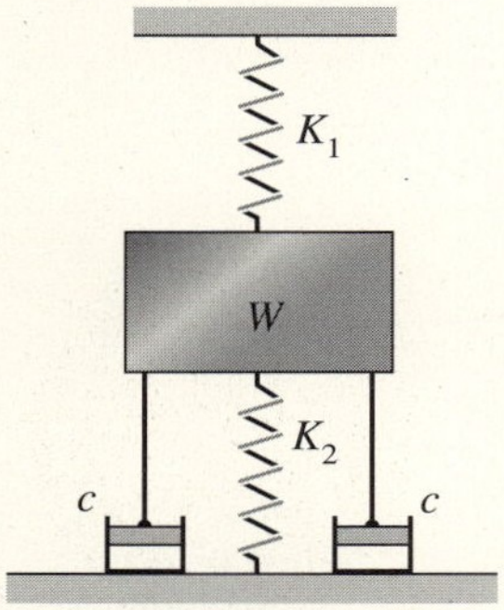

Figure P.19.76.

19.77 In Problem 19.76, the following data apply:

$$W = 445\text{ N}$$
$$K_1 = 8.8\text{ N/mm}$$
$$K_2 = 14.0\text{ N/mm}$$
$$c = 825\text{ N s/m}$$

Is the system underdamped, overdamped, or critically damped? If the weight W is released 150 mm above its static-equilibrium configuration, what are the speed and position of the block after 0.1 s? What force is transmitted to the foundation at that instant?

19.78. The damping constant c for the body is 7.3 N s/m. If, at its equilibrium position, the body is suddenly given a velocity of 3 m/s to the right, what will the frequency of its motion be? What is the position of the mass at $t = 5$ s?

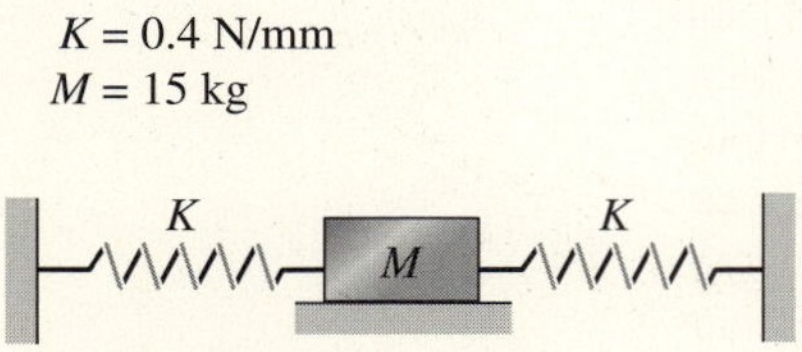

Figure P.19.78.

19.79. The damping in Problem 19.78 is increased so that it is twice the critical damping. If the mass is released from a position 75 mm to the right of equilibrium, how far from the equilibrium position is it in 5 s? Theoretically, does it ever reach the equilibrium position?

19.80. A plot of a free damped vibration is shown. What should the constant C_3 be in Eq. 19.35 for this motion? Show that ln (x_1/x_2), where x_1 and x_2 are two succeeding peaks, can be given as $(c/4\text{m})\tau$. The expression ln (x_1/x_2) is called the *logarithmic decrement* and is used in vibration work.

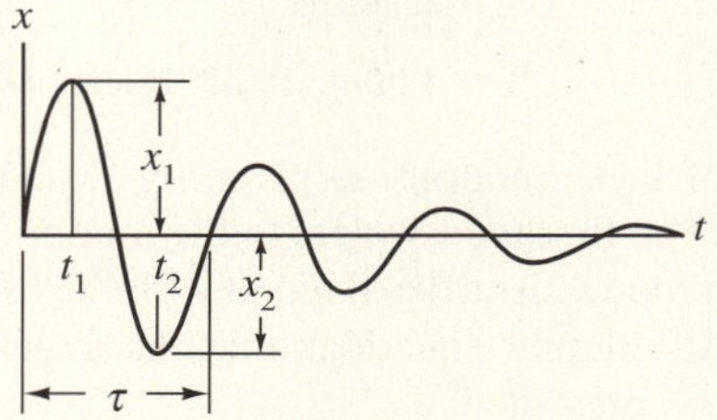

Figure P.19.80.

19.81. A rod of length 2.25 m and weight 200 N is shown in the static-equilibrium position supported by a spring of stiffness $K = 14$ N/mm. The rod is connected to a dashpot having a damping force c of 69 N s/m. If an impulsive torque gives the rod an angular speed clockwise of 0.5 rad/s at the position shown, what is the position of point A at $t = 0.2$ s?

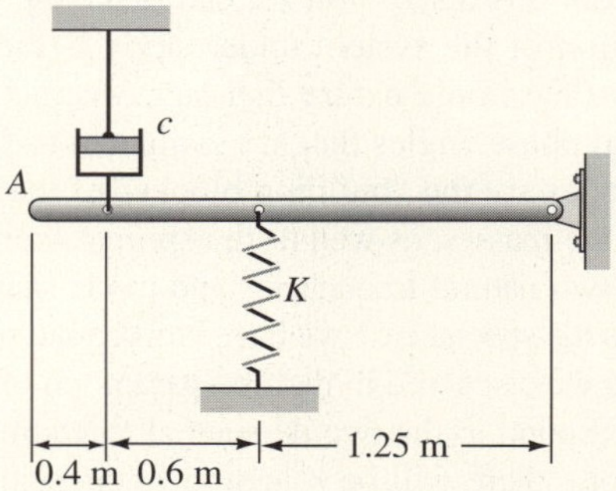

Figure P.19.81.

19.82. A spherical ball of weight 134 N is welded to a vertical light rod which in turn is welded at B to a horizontal rod. A spring of stiffness $K = 8.8$ N/mm and a damper c having a value 179 N s/m are connected to the horizontal rod. If A is displaced 75 mm to the right, how long does it take for it to return to its vertical configuration?

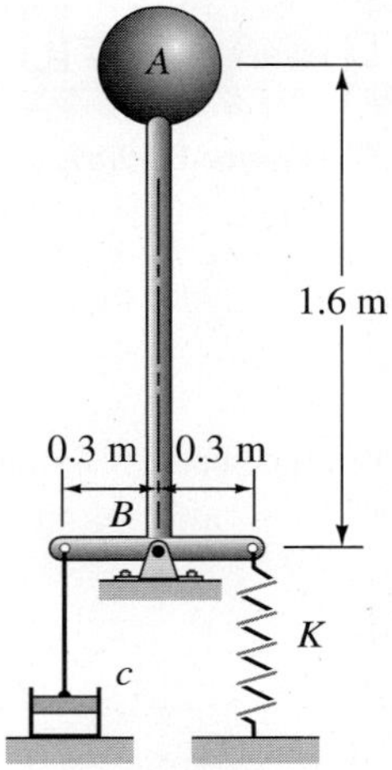

Figure P.19.82.

19.83. Cylinder A of weight 200 N slides down a vertical cylindrical chute. A film of oil of thickness 0.1 mm separates the cylinder from the chute. If the air pressure before and behind the cylinder is maintained at the same value of 110 kPa, what is the maximum velocity that the cylinder can attain by gravity? The coefficient of viscosity of the oil is 8 mPa s.

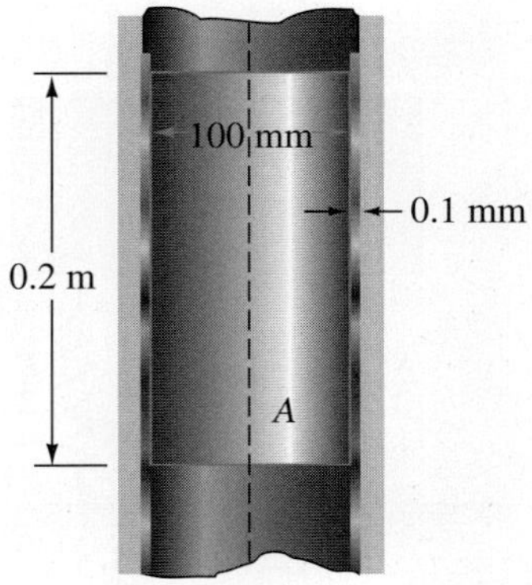

Figure P.19.83.

19.84. A disc A with a mass of 5 kg is constrained during rotation about its axis by a torsional spring having a constant K_T equal to 20 μN m/rad. The disc is in a journal having a diameter 2 mm larger than the disc. Oil having a viscosity 8.5 mPa s fills the outer space between disc and journal. If we assume a linear profile for the oil film, what is the frequency of oscillation of the disc if it is rotated from its equilibrium position and then released? The diameter of the disc is 40 mm and its length is 30 mm. The oil acts only on the disc's outer periphery.

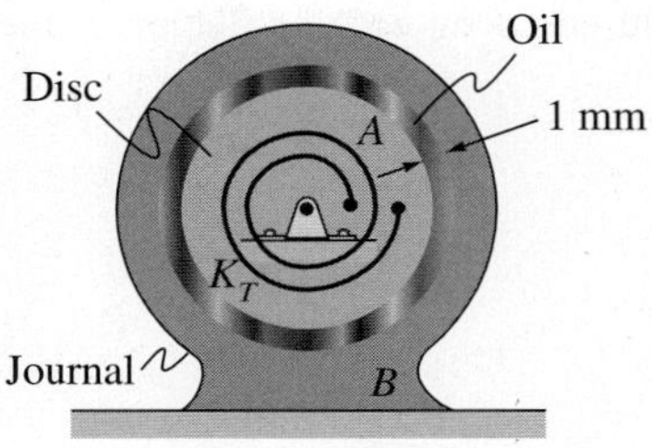

Figure P.19.84.

19.85. A block W weighing 60 N is released from rest at a configuration 100 mm above its equilibrium position. It rides on a film of oil whose thickness is 0.1 mm and whose coefficient of viscosity is 9.5 mPa s. If K is 50 N/m, how far down the incline will the block move? The block is 0.20 m on each edge.

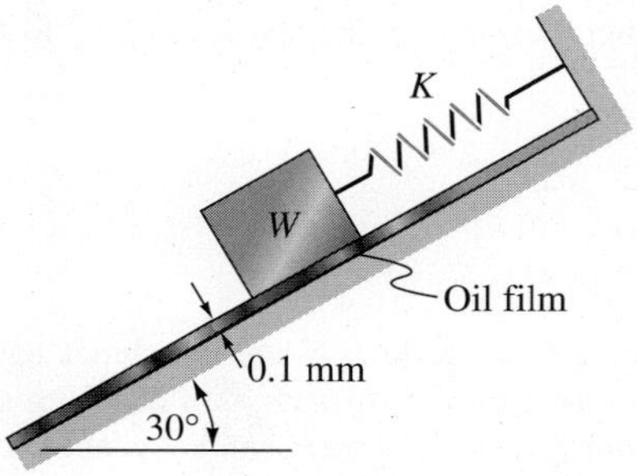

Figure P.19.85.

***19.86.** In Problem 19.85, set up two simultaneous equations to determine the spring constant K so that, after W is released, W comes back to its equilibrium position with no oscillation. As a short project, solve for K using a computer.

19.87. A disc B of diameter 100 mm rotates in a stationary housing filled with oil of viscosity 6 mPa s. The disc and its shaft have a mass of 30 g and a radius of gyration of 20 mm. The shaft and disc connect to a device that supplies a linear restoring torque of 5 N mm/rad. Use a linear velocity profile for the oil and find how long each oscillation of the disc about its axis takes.

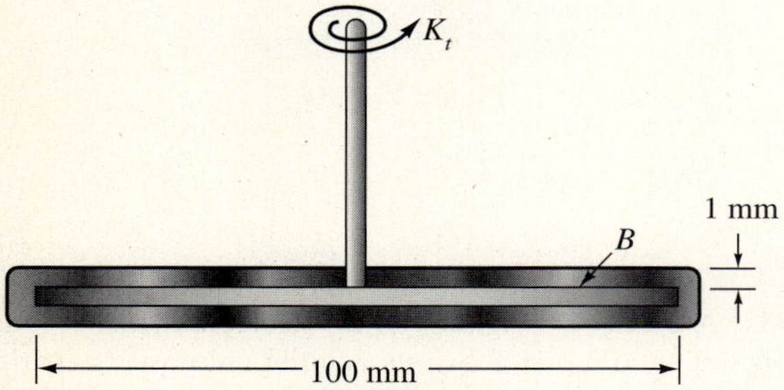

Figure P.19.87.

19.88. Examine the case of the spring-mass system with viscous damping for a sinusoidal forcing function given as $F_0 \sin \omega t$. Go through the steps in the text leading up to Eq. 19.41 for this case.

19.89. A force $F = 35 \sin 2t$ N acts on a block having a weight of 285 N. A spring having stiffness K of 550 N/m and a dashpot having a damping factor c of 68 N s/m are connected to the body. What is the amplitude of steady-state motion for the body, and what is the maximum force transmitted to the wall?

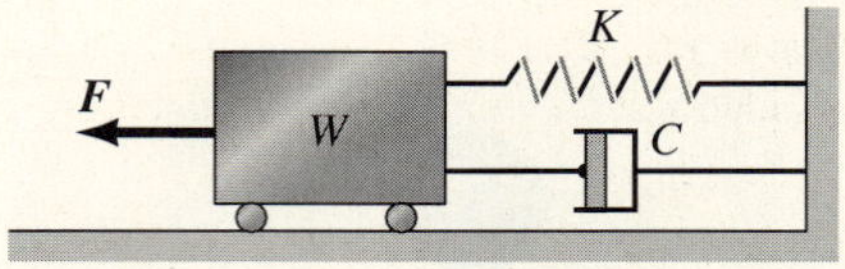

Figure P.19.89.

19.90 In Example 19.11, compute K for an amplitude of steady-state vibration of 12 mm.

19.91. A block of mass M rests on two springs, the total spring constant of which is K. Also, there is a dashpot of constant c. A small sphere of mass m is attached to M and is made to rotate at a speed of ω. The distance from the center of rotation to the sphere is r. Derive the equation of motion for the block first by considering the motions of M and m separately as single masses. Show that you could reach the same equation of motion by lumping the masses M and m into one body of mass $(M + m)$ on which a sinusoidal disturbance equal to $mr\omega^2 \sin \omega t$ (from the rotating sphere) is applied.

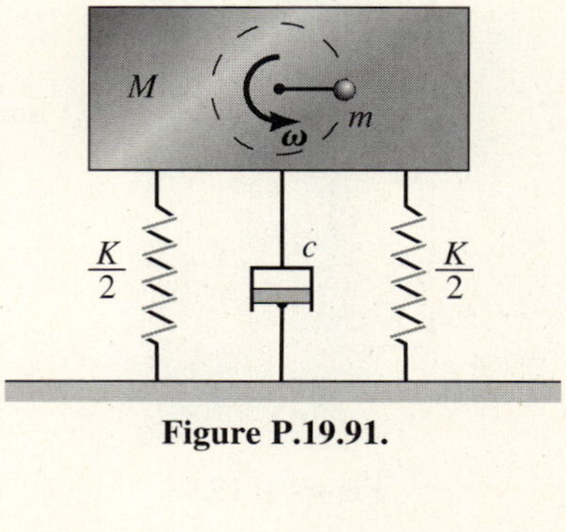

Figure P.19.91.

19.92. A platform weighing 222 N deflects the spring 50 mm when placed carefully on the spring. A motor weighing 22 N is then clamped on top of the platform and rotates an eccentric mass m which weighs 1 N. The mass m is displaced 150 mm from the axis of rotation and rotates at an angular speed of 28 rad/s. The viscous damping present causes a resistance to the motion of the platform of 275 N m/s. What is the steady-state amplitude of the motion of the platform? See Problem 19.91 before doing this problem.

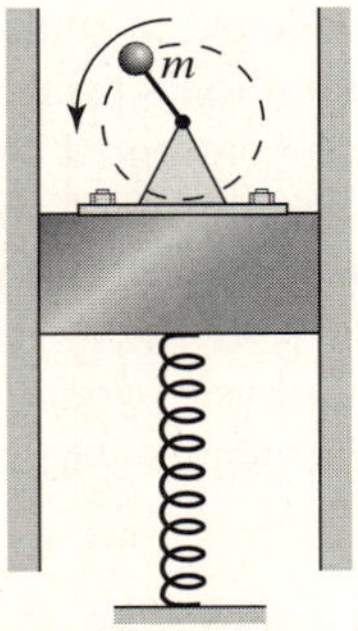

Figure P.19.92.

19.93. A body weighing 143 N is connected by a light rod to a spring of stiffness K equal to 2.6 N/mm and to a dashpot having a damping factor c. Point B has a given motion x' of 30.5 sin t mm with t in seconds. If the center of A is to have an amplitude of steady-state motion of 20 mm, what must c be?

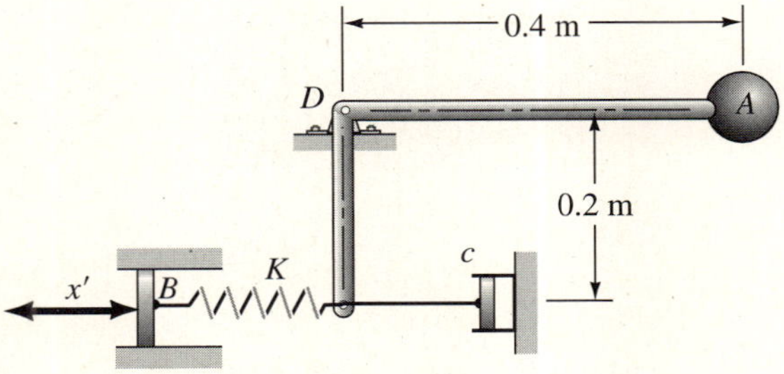

Figure P.19.93.

***19.94.** In Problem 19.15, if we include the inertial effects of rod BC, how many degrees of freedom are there? If rod BC weighs 22 N set up the differential equations of motion for the system.

***19.95** Two bodies of equal mass, M = 15 kg, are attached to walls by springs having equal spring constants K_1 = 0.8 N/mm and are connected to each other by a spring having a spring constant K_2 = 0.9 N/mm. If the mass on the left is released from a position $(x_1)_0$ = 75 mm at t = 0 with zero velocity and the mass at the right is stationary at x_2 = 0 at this instant, what is the position of each mass at the time t = 5 s? The coordinates x_1 and x_2 are measured from the static-equilibrium positions of the body.

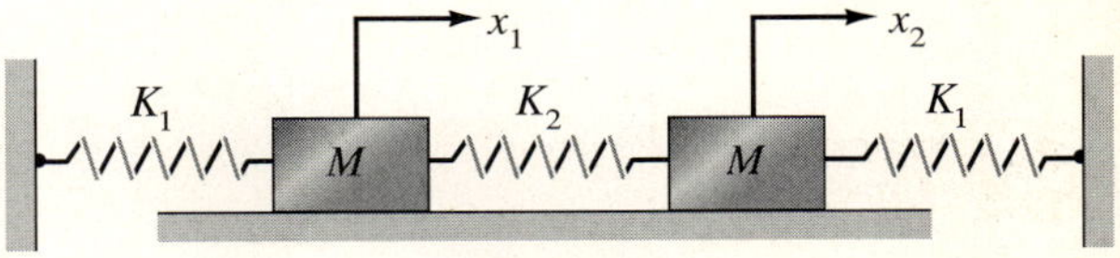

Figure P.19.95.

***19.96.** Let K_2 in Fig. P.19.95 be very small compared to K_1. Assume one mass has been released at t = 0 from a position displaced from equilibrium with zero velocity, while the other mass is released from the initial stationary position at that instant with zero velocity. Show that one mass will have a maximum velocity while the other will have a minimum velocity and that there will be a continual transfer of kinetic energy from one mass to the other at a frequency equal to the beat frequency of the natural frequencies of the system.

[*Hint*: Study the phasors $A \cos \sqrt{\frac{K_1}{M}}t$ and $A \cos \sqrt{\frac{K_1}{M} + \frac{2K_2}{M}}t$.]

19.10 Closure

This introductory study of vibration brings to a close the present study of particle and rigid-body mechanics. As you progress to the study of deformable media in your courses in solid and fluid mechanics you will find that particle mechanics and, to a lesser extent rigid-body mechanics, will form cornerstones for these disciplines. And in your studies involving the design of machines and the performance of vehicles you will find rigid-body mechanics indispensable.

It should be realized, however, that we have by no means said the last word on particle and rigid-body mechanics. More advanced studies will emphasize the variational approach introduced in statics. With the use of the calculus of variations, such topics as Hamilton's principle, Lagrange's equation,[6] and Hamilton-Jacobi theory will be presented and you will then see a greater unity between mechanics and other areas of physics such as electromagnetic theory and wave mechanics. Also the special theory of relativity will most surely be considered.

Finally, in your studies of modern physics you will come to more fully understand the limitations of classical mechanics when you are introduced to quantum mechanics.

[6]Some of these topics are covered in the author's text written with C.L. Dym, "Energy and Finite Elements Methods in Structural Mechanics" Taylor and Francis.

PROBLEMS

19.97. If $K_1 = 2K_2 = 1.8K_3$, what should K_3 be for a period of free vibration of 0.2 s? The mass M is 3 kg.

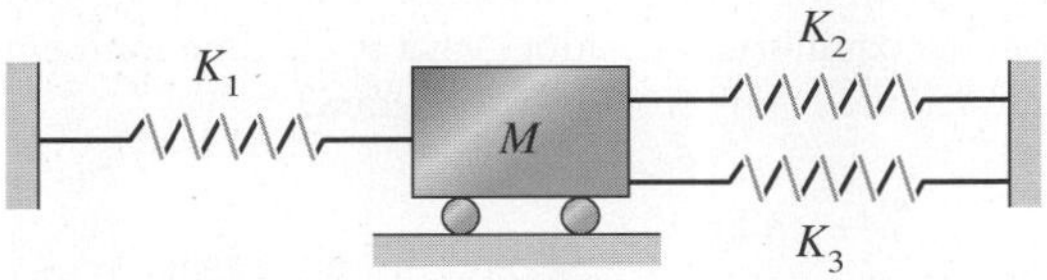

Figure P.19.97.

19.98. In Problem 19.55, determine the natural frequency of the system. If the mass is deflected 50 mm and then released, determine the displacement from equilibrium after 3 s. Finally, determine the *total* distance traveled during this time.

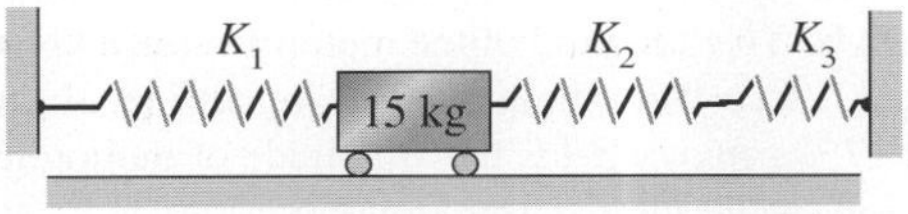

Figure P.19.98.

19.99. Find the natural frequency of motion of body A for small rotation of rod BD when we neglect the inertial effects of rod BD. The spring constant K_2 is 0.9 N/mm and the spring constant K_1 is 1.8 N/mm. The weight of block A is 178 N. Neglect friction everywhere. Rod BD weighs 44 N.

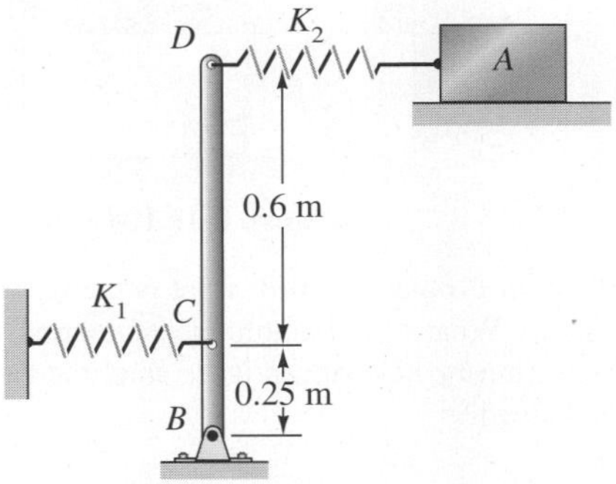

Figure P.19.99.

19.100. What is the equivalent spring constant for small oscillations about the shaft AB? Neglect all mass except the block at B, which is 45 kg. The shear modulus of elasticity G for the shaft is 100 GN/m^2. What is the natural frequency of the system for torsional oscillation of small amplitude?

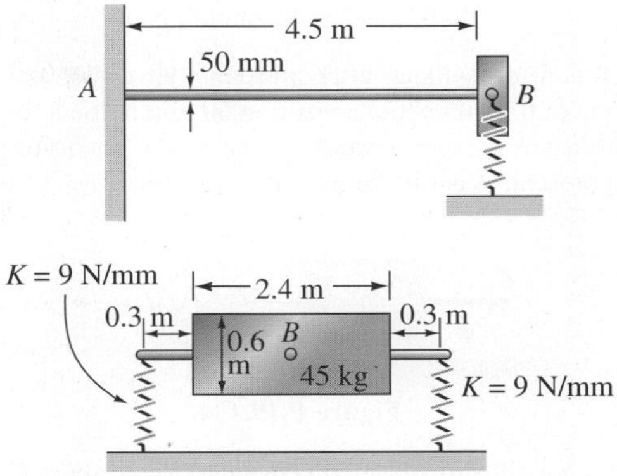

Figure P.19.100.

19.101. What is the radius of gyration of the speedboat about a vertical axis going through the center of mass, if it is noted that the boat will swing about this vertical axis one time per second? The mass of the boat is 500 kg.

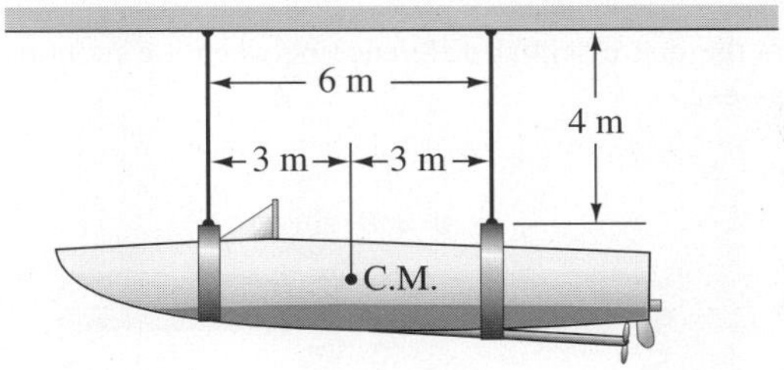

Figure P.19.101.

19.102. A rod of length L and mass M is suspended from a frictionless roller. If a small impulsive torque is applied to the rod when it is in a state of rest, what is the natural frequency of oscillation about this state of rest?

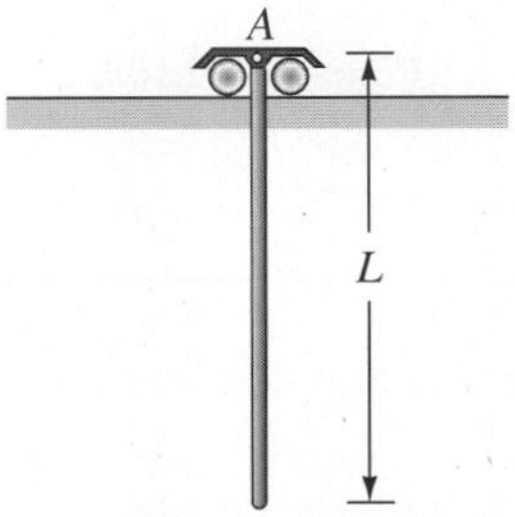

Figure P.19.102.

*19.103. Solve Problem 19.12 by energy methods.

19.104. A block is acted on by a force F given as

$$F = 90 + 22 \sin 80t \text{ N}$$

and is found to oscillate, after transients have died out, with an amplitude of 0.5 mm about a position 50 mm to the left of the static-equilibrium position corresponding to the condition when no force is present. What is the weight of the body?

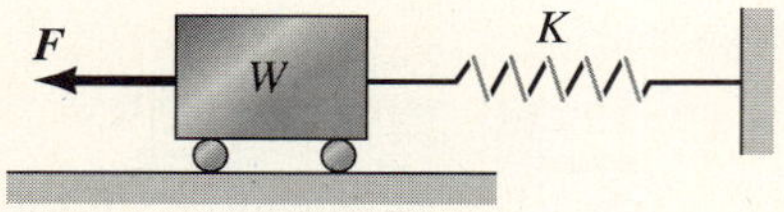

Figure P.19.104.

19.105. A light stiff rod ACB is made to rotate about C such that θ varies sinusoidally with $\omega = 20$ rad/s. What should the amplitude of rotation θ_0 be to cause a steady-state amplitude of vibration of M to be 20 mm? The following data apply:

$$M = 3 \text{ kg}$$
$$K = 2.5 \text{ N/mm}$$
$$l = 1 \text{ m}$$

What is the one essential difference between the motions of the two masses?

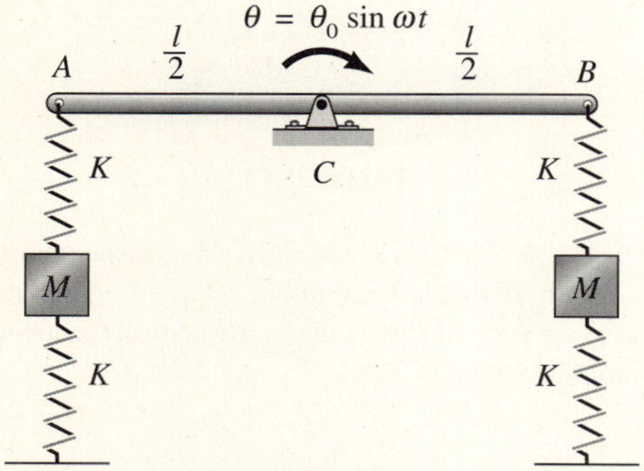

Figure P.19.105.

19.106. In Problem 19.105, what is the angle θ after 5 s? System is stationary at the outset and AB is horizontal. Take $\omega = 39$ rad/s and θ_0 as 20 mrad.

19.107. A body rests on a conveyor moving with a speed of 1.5 m/s. If the damping constant is 30 N s/m, determine the equilibrium force in the spring. If the body is displaced 75 mm to the left from the equilibrium position, what is the time for the mass to pass through the equilibrium position again?

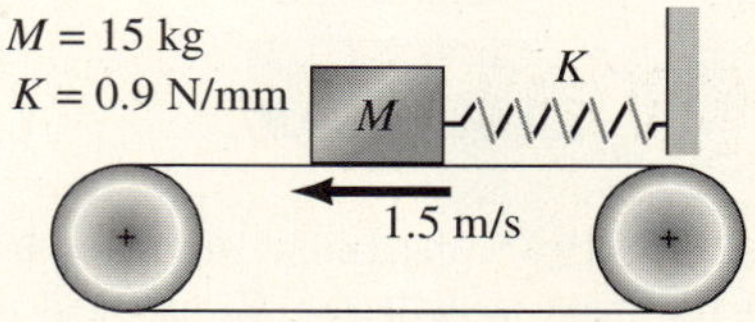

Figure P.19.107.

*19.108. A motor is mounted on two springs of stiffness 8.8 N/mm each and a dashpot having a coefficient c of 96 N s/m. The motor weighs 222 N. The armature of the motor weighs 89 N with a center of mass 5 mm from the geometric centerline. If the machine rotates at 1750 r/min, what is the amplitude of motion in the vertical direction of the motor? Determine the maximum force transmitted to the ground.

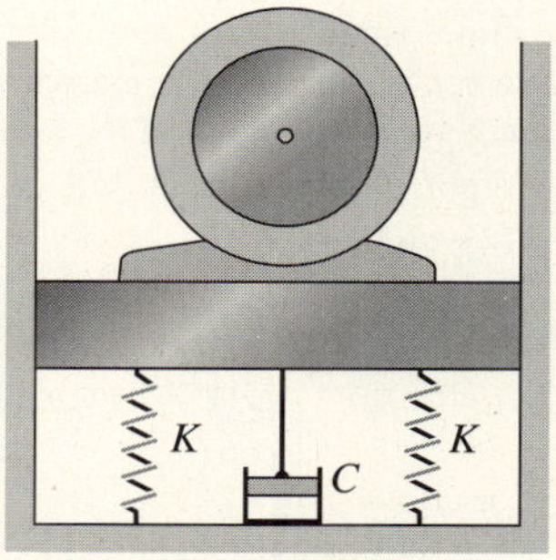

Figure P.19.108.

*19.109. In Problem 19.108, what is the resonant condition for the system? What is the amplitude of motion for this case? To what value must c be changed if the amplitude at this motor speed is to be halved?

APPENDIX I

Integration Formulas

1. $\int \frac{x\,dx}{a+bx} = \frac{1}{b^2}[a + bx - a\ln(a+bx)]$

2. $\int \frac{dx}{a^2 - x^2} = \frac{1}{2a}\ln\left(\frac{a+x}{a-x}\right)$

3. $\int \sqrt{x^2 \pm a^2}\,dx = \frac{1}{2}\left[x\sqrt{x^2 \pm a^2} \pm a^2 \ln\left(x + \sqrt{x^2 \pm a^2}\right)\right]$

4. $\int \sqrt{a^2 - x^2}\,dx = \frac{1}{2}\left(x\sqrt{a^2 - x^2} + a^2 \sin^{-1}\frac{x}{a}\right)$

5. $\int x\sqrt{a^2 - x^2}\,dx = -\frac{1}{3}\sqrt{(a^2 - x^2)^3}$

6. $\int x\sqrt{a + bx}\,dx = -\frac{2(2a - 3bx)\sqrt{(a+bx)^3}}{15b^2}$

7. $\int x^2\sqrt{a^2 - x^2}\,dx = -\frac{x}{4}\sqrt{(a^2 - x^2)^3} + \frac{a^2}{8}\left(x\sqrt{a^2 - x^2} + a^2 \sin^{-1}\frac{x}{a}\right)$

8. $\int x^2\sqrt{a^2 \pm x^2}\,dx = \frac{x}{4}\sqrt{(x^2 \pm a^2)^3} \mp \frac{a^2}{8}x\sqrt{x^2 \pm a^2} - \frac{a^4}{8}\ln\left(x + \sqrt{x^2 \pm a^2}\right)$

9. $\int \frac{dx}{\sqrt{a^2 - x^2}} = \sin^{-1}\frac{x}{a}$

10. $\int \frac{dx}{\sqrt{x^2 + a^2}} = \ln\left(x + \sqrt{x^2 + a^2}\right) = \sinh^{-1}\frac{x}{a}$

11. $\int x^m e^{ax}\,dx = \frac{x^m e^{ax}}{a} - \frac{m}{a}\int x^{m-1}e^{ax}\,dx$

12. $\int x^m \ln x\,dx = x^{m+1}\left(\frac{\ln x}{m+1} - \frac{1}{(m+1)^2}\right)$

13. $\int \sin^2\theta\,d\theta = \frac{1}{2}\theta - \frac{1}{4}\sin 2\theta$

14. $\int \cos^2 \theta \, d\theta = \frac{1}{2}\theta + \frac{1}{4}\sin 2\theta$

15. $\int \sin^3 \theta \, d\theta = -\frac{1}{3}\cos \theta(\sin^2 \theta + 2)$

16. $\int \cos^m \theta \sin \theta \, d\theta = -\dfrac{\cos^{m+1} \theta}{m+1}$

17. $\int \sin^m \theta \cos \theta \, d\theta = \dfrac{\sin^{m+1} \theta}{m+1}$

18. $\int \sin^m \theta \, d\theta = -\dfrac{\sin^{m-1} \theta \cos \theta}{m} + \dfrac{m-1}{m}\int \sin^{m-2} \theta \, d\theta$

19. $\int \theta^2 \sin \theta \, d\theta = 2\theta \sin \theta - (\theta^2 - 2)\cos \theta$

20. $\int \theta^2 \cos \theta \, d\theta = 2\theta \cos \theta + (\theta^2 - 2)\sin \theta$

21. $\int \theta \sin^2 \theta \, d\theta = \frac{1}{4}\left[\sin \theta(\sin \theta - 2\theta \cos \theta) + \theta^2\right]$

22. $\int \sin m\theta \cos m\theta \, d\theta = -\dfrac{1}{4m}\cos 2m\theta$

23. $\int \dfrac{d\theta}{(a + b\cos \theta)^2}$

$$= \frac{1}{(a^2 - b^2)}\left(\frac{-b\sin \theta}{a + b\cos \theta} + \frac{2a}{\sqrt{a^2 - b^2}}\tan^{-1}\frac{\sqrt{a^2 - b^2}\tan\dfrac{\theta}{2}}{a + b}\right)$$

24. $\int \theta \sin \theta \, d\theta = \sin \theta - \theta \cos \theta$

25. $\int \theta \cos \theta \, d\theta = \cos \theta + \theta \sin \theta$

APPENDIX II

Computation of Principal Moments of Inertia

We now turn to the problem of computing the principal moments of inertia and the directions of the principal axes for the case where we do not have planes of symmetry. It is unfortunate that a careful study of this important calculation is beyond the level of this text. However, we shall present enough material to permit the computation of the principal moments of inertia and the directions of their respective axes.

The procedure that we shall outline is that of extremizing the mass moment of inertia at a point where the inertia-tensor components are known for a reference xyz. This will be done by varying the direction cosines l, m, and n of an axis k so as to extremize I_{kk} as given by Eq. 9.13. We accordingly set the differential of I_{kk} equal to zero as follows:

$$\begin{aligned} dI_{kk} = {} & 2lI_{xx}\,dl + 2mI_{yy}\,dm + 2nI_{zz}\,dn \\ & -2lI_{xy}\,dm - 2mI_{xy}\,dl - 2lI_{xz}\,dn \\ & -2nI_{xz}\,dl - 2mI_{yz}\,dn - 2nI_{yz}\,dm = 0 \end{aligned} \tag{II.1}$$

Collecting terms and canceling the factor 2, we get

$$\begin{aligned} (lI_{xx} - mI_{xy} - nI_{zz})dl + (-lI_{xy} + mI_{yy} - nI_{yz})dm \\ + (-lI_{xz} - mI_{yz} + nI_{zz})dn = 0 \end{aligned} \tag{II.2}$$

If the differentials dl, dm, and dn were *independent* we could set their respective coefficients equal to zero to satisfy the equation. However, they are not independent because the equation

$$l^2 + m^2 + n^2 = 1 \tag{II.3}$$

must at all times be satisfied. Accordingly, the differentials of the direction cosines must be related as follows[1]:

$$l\,dl + m\,dm + n\,dn = 0 \tag{II.4}$$

[1]We are thus extremizing I_{kk} in the presence of a constraining equation.

We can of course consider any two differentials as independent. The third is then established in accordance with the equation above.

We shall now introduce the *Lagrange multiplier* λ to facilitate the extremizing process. This constant is an arbitrary constant at this stage of the calculation. Multiplying Eq. II.4 by λ and subtracting Eq. II.4 from Eq. II.2 we get when collecting terms:

$$\left[(I_{xx} - \lambda)l - I_{xy}m - I_{xz}n\right]dl + \left[-I_{xy}l + (I_{yy} - \lambda)m - I_{yz}n\right]dm + \left[-I_{xz}l - I_{yz}m + (I_{zz} - \lambda)n\right]dn = 0 \quad \text{(II.5)}$$

Let us next consider that m and n are independent variables and consider the value of λ so chosen that the coefficient of dl is zero. That is,

$$(I_{xx} - \lambda)l - I_{xy}m - I_{xz}n = 0 \quad \text{(II.6)}$$

With the first term Eq. II.5 disposed of in this way, we are left with differentials dm and dn, which are independent. Accordingly, we can set their respective coefficients equal to zero in order to satisfy the equation. Hence, we have in addition to Eq. II.6 the following equations:

$$\begin{aligned} -I_{xy}l + (I_{yy} - \lambda)m - I_{yz}n &= 0 \\ -I_{xz}l - I_{yz}m + (I_{zz} - \lambda)n &= 0 \end{aligned} \quad \text{(II.7)}$$

A necessary condition for the solution of a set of direction cosines l, m, and n, from Eqs. II.6 and II.7, which does not violate Eq. II.3[2] is that the determinant of the coefficients of these variables be zero. Thus:

$$\begin{vmatrix} (I_{xx} - \lambda) & -I_{xy} & -I_{xz} \\ -I_{xy} & (I_{yy} - \lambda) & -I_{yz} \\ -I_{xz} & -I_{yz} & (I_{zz} - \lambda) \end{vmatrix} = 0 \quad \text{(II.8)}$$

This results in a cubic equation for which we can show there are three real roots for λ. Substituting these roots into any two of Eqs. II.6 and II.7 plus Eq. II.3, we can determine three direction cosines for each root. These are the direction cosines for the principal axes measured relative to xyz. We could get the principal moments of inertia next by substituting a set of these direction cosines into Eq. 9.13 and solving for I_{kk}. However, that is not necessary, since it can be shown that the three Lagrange multipliers *are* the principal moments of inertia.

[2]This precludes the possibility of a trivial solution $l = m = n = 0$.

APPENDIX III (REPEAT OF CHAPTER 9)

Second Moments and Product Second Moments of Area[1]

9.1 Introduction

In this chapter, we shall consider certain measures of mass distribution relative to a reference. These quantities are vital for the study of the dynamics of rigid bodies. Because these quantities are so closely related to second moments and products of area, we shall consider them at this early stage rather than wait for dynamics. We shall also discuss the fact that these measures of mass distribution—the second moments of inertia of mass and the products of inertia of mass—are components of what we call a second-order tensor. Recognizing this fact early will make more simple and understandable your future studies of stress and strain, since these quantities also happen to be components of second-order tensors.

9.2 Formal Definition of Inertia Quantities

We shall now formally define a set of quantities that give information about the distribution of mass of a body relative to a Cartesian reference. For this purpose, a body of mass M and a reference xyz are presented in Fig. 9.1. This reference and the body may have any motion whatever relative to each other. The ensuing discussion then holds for the instantaneous orientation shown at time t. We shall consider that the body is composed of a continuum of particles, each of which has a mass given by $\rho\, dv$. We now present the following definitions:

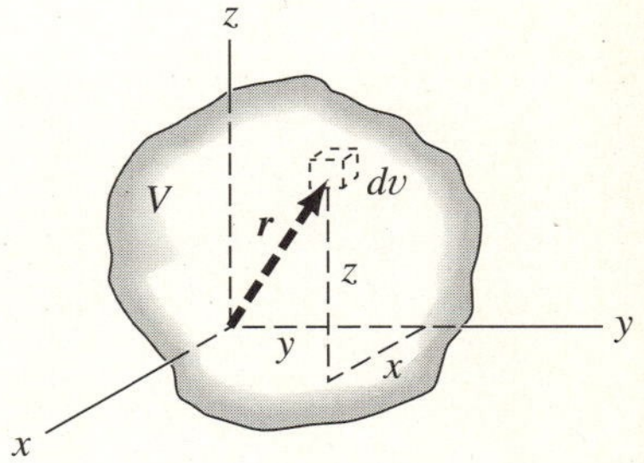

Figure 9.1. Body and reference at time t.

[1]This chapter may be covered at a later stage when studying dynamics. In that case, it should be covered directly after Chapter 15.

$$I_{xx} = \iiint_V (y^2 + z^2)\rho \, dv \tag{9.1a}$$

$$I_{yy} = \iiint_V (x^2 + z^2)\rho \, dv \tag{9.1b}$$

$$I_{zz} = \iiint_V (x^2 + y^2)\rho \, dv \tag{9.1c}$$

$$I_{xy} = \iiint_V xy\, \rho \, dv \tag{9.1d}$$

$$I_{xz} = \iiint_V xz\, \rho \, dv \tag{9.1e}$$

$$I_{yz} = \iiint_V yz\, \rho \, dv \tag{9.1f}$$

The terms I_{xx}, I_{yy}, and I_{zz} in the set above are called the *mass moments of inertia* of the body about the x,y, and z axes, respectively.[2] Note that in each such case we are integrating the mass elements $\rho\, dv$, times the *perpendicular distance squared* from the mass elements to the coordinate axis about which we are computing the moment of inertia. Thus, if we look along the x axis toward the origin in Fig. 9.1, we would have the view shown in Fig. 9.2. The quantity $y^2 + z^2$ used in Eq. 9.1a for I_{xx} is clearly d^2, the perpendicular distance squared from dv to the x axis (now seen as a dot). Each of the terms with mixed indices is called the *mass product of inertia* about the pair of axes given by the indices. Clearly, from the definition of the product of inertia, we could reverse indices and thereby form three additional products of inertia for a reference. The additional three quantities formed in this way, however, are equal to the corresponding quantities of the original set. That is,

$$I_{xy} = I_{yx}, \quad I_{xz} = I_{zx}, \quad I_{yz} = I_{zy}$$

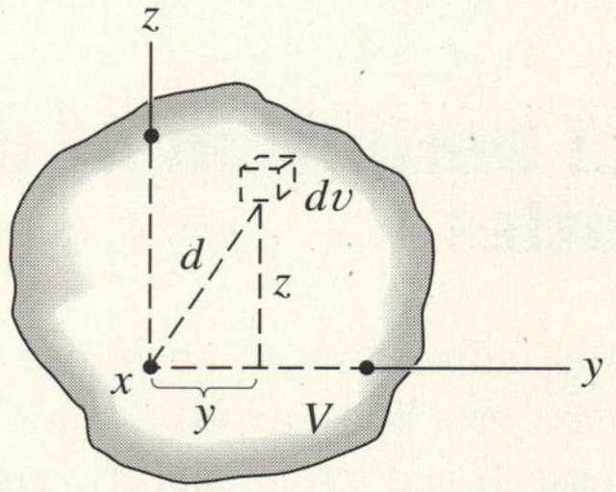

Figure 9.2. View of body along x axis.

We now have nine inertia terms at a point for a given reference at this point. The values of the set of six independent quantities will, for a given body,

[2]We use the same notation as was used for second moments and product second moments of area, which are also sometimes called moments and products of inertia. This is standard practice in mechanics. There need be no confusion in using these quantities if we keep the context of discussions clearly in mind.

depend on the *position* and *inclination* of the reference relative to the body. You should also understand that the reference may be established anywhere in space and *need not* be situated in the rigid body of interest. Thus there will be nine inertia terms for reference xyz at point O outside the body (Fig. 9.3) computed using Eqs. 9.1, where the domain of integration is the volume V of the body. As will be explained later, the nine moments and products of inertia are components of the inertia tensor.

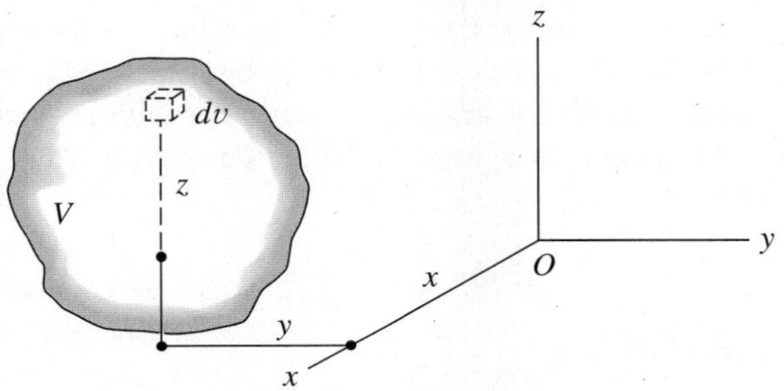

Figure 9.3. Origin of xyz outside body.

It will be convenient, when referring to the nine moments and products of inertia for reference xyz at a point, to list them in a matrix array, as follows:

$$I_{ij} = \begin{pmatrix} I_{xx} & I_{xy} & I_{xz} \\ I_{yx} & I_{yy} & I_{yz} \\ I_{zx} & I_{zy} & I_{zz} \end{pmatrix}$$

Notice that the first subscript gives the row and the second subscript gives the column in the array. Furthermore, the left-to-right downward diagonal in the array is composed of mass moment of inertia terms while the products of inertia, oriented at mirror-image positions about this diagonal, are equal. For this reason we say that the array is *symmetric*.

We shall now show that the sum of the mass moments of inertia for a set of orthogonal axes is independent of the orientation of the axes and depends only on the position of the origin. Examine the sum of such a set of terms:

$$I_{xx} + I_{yy} + I_{zz} = \iiint_V (y^2 + z^2)\rho\, dv + \iiint_V (x^2 + z^2)\rho\, dv + \iiint_V (x^2 + y^2)\rho\, dv$$

Combining the integrals and rearranging, we get

$$I_{xx} + I_{yy} + I_{zz} = \iiint_V 2(x^2 + y^2 + z^2)\rho\, dv = \iiint_V 2|\mathbf{r}|^2 \rho\, dv \qquad (9.2)$$

But the magnitude of the position vector from the origin to a particle is *independent* of the inclination of the reference at the origin. Thus, *the sum of the moments of inertia at a point in space for a given body clearly is an invariant with respect to rotation of axes.*

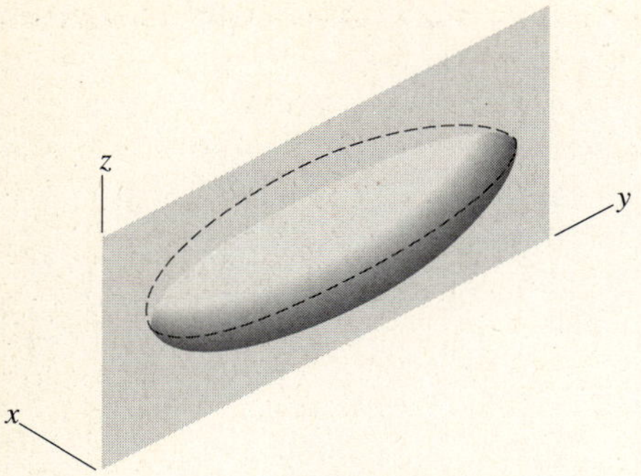

Figure 9.4. *zy* is plane of symmetry.

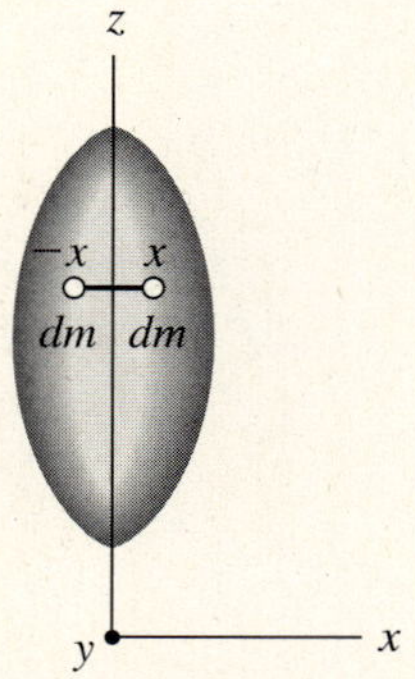

Figure 9.5. View along *y* axis.

Clearly, on inspection of the equations 9.1, it is clear that the moments of inertia must always be positive, while the products of inertia may be positive or negative. Of interest is the case where one of the coordinate planes is a *plane of symmetry* for the mass distribution of the body. Such a plane is the *zy* plane shown in Fig. 9.4 cutting a body into two parts, which, by definition of symmetry, are mirror images of each other. For the computation of I_{xz}, each half will give a contribution of the same magnitude but of opposite sign. We can most readily see that this is so by looking along the *y* axis toward the origin. The plane of symmetry then appears as a line coinciding with the *z* axis (see Fig. 9.5). We can consider the body to be composed of pairs of mass elements *dm* which are mirror images of each other with respect to position and shape about the plane of symmetry. The product of inertia I_{xz} for such a pair is then

$$xz\,dm - xz\,dm = 0$$

Thus, we can conclude that

$$I_{xz} = \underbrace{\int xz\,dm}_{\substack{\text{right}\\\text{domain}}} - \underbrace{\int xz\,dm}_{\substack{\text{left}\\\text{domain}}} = 0$$

This conclusion is also true for I_{xy}. We can say that $I_{xy} = I_{xz} = 0$. But on consulting Fig. 9.4, you should be able to readily decide that the term I_{zy} will have a positive value. Note that those products of inertia having *x* as an index are zero and that the *x* coordinate axis is normal to the plane of symmetry. Thus, we can conclude that *if two axes form a plane of symmetry for the mass distribution of a body, the products of inertia having as an index the coordinate that is normal to the plane of symmetry will be zero.*

Consider next a body of *revolution*. Take the *z* axis to coincide with the axis of symmetry. It is easy to conclude for the origin *O* of *xyz* anywhere along the axis of symmetry that

$$I_{xz} = I_{yz} = I_{xy} = 0$$
$$I_{xx} = I_{yy} = \text{constant}$$

for all possible *xy* axes formed by rotating about the *z* axis at *O*. Can you justify these conclusions?

Finally, we define *radii of gyration* in a manner analogous to that used for second moments of area in Chapter 8. Thus:

$$I_{xx} = k_x^2 M$$
$$I_{yy} = k_y^2 M$$
$$I_{zz} = k_z^2 M$$

where k_x, k_y, and k_z are the radii of gyration and *M* is the total mass.

Example 9.1

Find the nine components of the inertia tensor of a rectangular body of uniform density ρ about point O for a reference xyz coincident with the edges of the block as shown in Fig. 9.6.

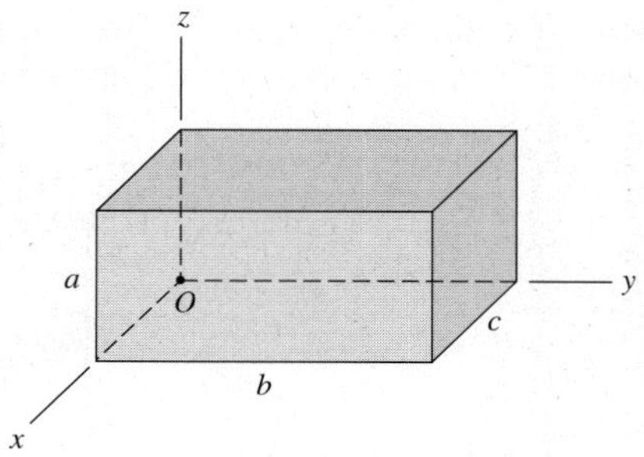

Figure 9.6. Find I_{ij} at O.

We first compute I_{xx}. Using volume elements $dv = dx\,dy\,dz$, we get on using simple multiple integration:

$$\begin{aligned} I_{xx} &= \int_0^a\int_0^b\int_0^c (y^2 + z^2)\rho\,dx\,dy\,dz \\ &= \int_0^a\int_0^b (y^2 + z^2)c\rho\,dy\,dz = \int_0^a\left(\frac{b^3}{3} + z^2 b\right)c\rho\,dz \qquad \text{(a)} \\ &= \left(\frac{ab^3c}{3} + \frac{a^3bc}{3}\right)\rho = \frac{\rho V}{3}(b^2 + a^2) \end{aligned}$$

where V is the volume of the body. Note that the x axis about which we are computing the moment of inertia I_{xx} is *normal* to the plane having sides of length a and b, i.e., along the z and y axes. Similarly:

$$I_{yy} = \frac{\rho V}{3}(c^2 + a^2) \qquad \text{(b)}$$

$$I_{zz} = \frac{\rho V}{3}(b^2 + c^2) \qquad \text{(c)}$$

We next compute I_{xy}.

$$\begin{aligned} I_{xy} &= \int_0^a\int_0^b\int_0^c xy\,\rho\,dx\,dy\,dz = \int_0^a\int_0^b \frac{c^2}{2}\,y\rho\,dy\,dz \\ &= \int_0^a \frac{c^2b^2}{4}\rho\,dz = \frac{ac^2b^2}{4}\rho = \frac{\rho V}{4}cb \qquad \text{(d)} \end{aligned}$$

Note for I_{xy}, we use the lengths of the sides along the x and y axes.

$$I_{xz} = \frac{\rho V}{4}ac \qquad \text{(e)}$$

$$I_{yz} = \frac{\rho V}{4}ab \qquad \text{(f)}$$

We accordingly have, for the inertia tensor:

$$I_{ij} = \begin{pmatrix} \frac{\rho V}{3}(b^2 + a^2) & \frac{\rho V}{4}cb & \frac{\rho V}{4}ac \\ \frac{\rho V}{4}cb & \frac{\rho V}{3}(c^2 + a^2) & \frac{\rho V}{4}ab \\ \frac{\rho V}{4}ac & \frac{\rho V}{4}ab & \frac{\rho V}{3}(b^2 + c^2) \end{pmatrix} \qquad \text{(g)}$$

Example 9.2

Compute the components of the inertia tensor at the center of a solid sphere of uniform density ρ as shown in Fig. 9.7.

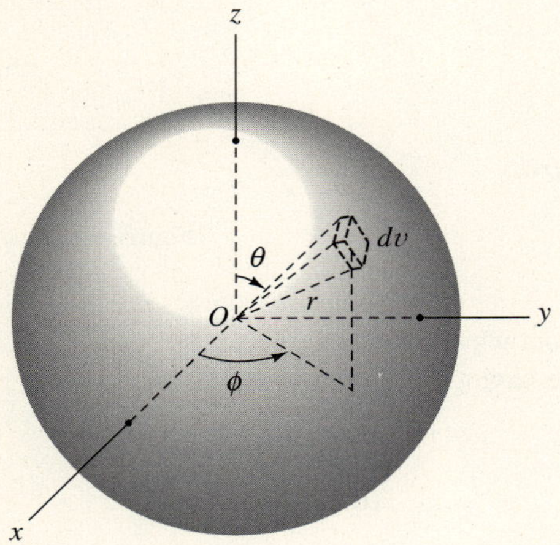

Figure 9.7. Find I_{ij} at O.

We shall first compute I_{yy}. Using spherical coordinates, we have[3]

$$
\begin{aligned}
I_{yy} &= \iiint\limits_V (x^2 + z^2)\rho\, dv \\
&= \int_0^R \int_0^{2\pi} \int_0^{\pi} \left[(r \sin\theta \cos\phi)^2 + (r\cos\theta)^2 \right] \rho \left(r^2 \sin\theta\, d\theta\, d\phi\, dr \right) \\
&= \int_0^R \int_0^{2\pi} \int_0^{\pi} \left(r^4 \sin^3\theta \cos^2\phi \right) \rho\, d\theta\, d\phi\, dr \\
&\qquad + \int_0^R \int_0^{2\pi} \int_0^{\pi} \left(r^4 \cos^2\theta \sin\theta \right) \rho\, d\theta\, d\phi\, dr \\
&= \rho \int_0^R \int_0^{2\pi} \left(r^4 \cos^2\phi \right) \left(\int_0^{\pi} \sin^3\theta\, d\theta \right) d\phi\, dr \\
&\qquad + \rho \int_0^R \int_0^{2\pi} r^4 \left(\int_0^{\pi} \cos^2\theta \sin\theta\, d\theta \right) d\phi\, dr
\end{aligned}
$$

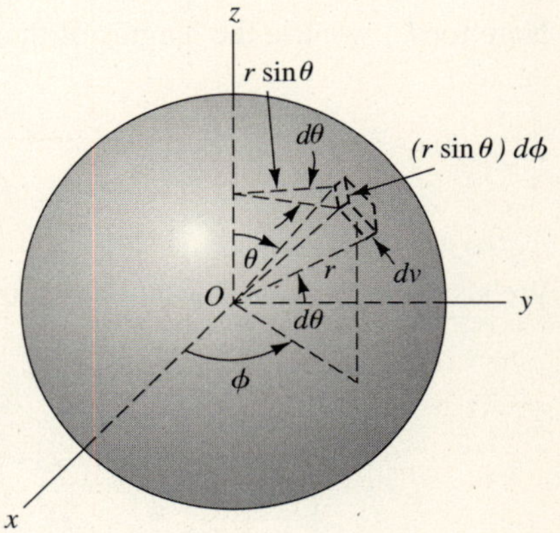

Figure 9.8. $dv = (r \sin\theta\, d\phi)(dr)(r\, d\theta) = r^2 \sin\theta\, d\theta\, d\phi\, dr$.

[3]For those unfamiliar with spherical coordinates, we have shown in Fig. 9.8 a more detailed study of the volume element used. The volume dv is simply the product of the three edges of the element shown in the diagram.

Example 9.2 (Continued)

With the aid of integration formulas from Appendix I, we have

$$I_{yy} = \rho \int_0^R \int_0^{2\pi} r^4 \cos^2 \phi \left[-\tfrac{1}{3} \cos \theta (\sin^2 \theta + 2) \right]_0^{\pi} d\phi \, dr$$
$$+ \rho \int_0^R \int_0^{2\pi} r^4 \left(-\frac{\cos^3 \theta}{3} \right) \Big|_0^{\pi} d\phi \, dr$$
$$= \rho \int_0^R \int_0^{2\pi} r^4 \cos^2 \phi \tfrac{4}{3} \, d\phi \, dr + \rho \int_0^R \int_0^{2\pi} (r^4) \tfrac{2}{3} \, d\phi \, dr$$

Integrating next with respect to ϕ, we get

$$I_{yy} = \rho \int_0^R (r^4)\left(\tfrac{4}{3}\right)(\pi) \, dr + \rho \int_0^R (r^4)\left(\tfrac{2}{3}\right)(2\pi) \, dr$$

Finally, we get

$$I_{yy} = \rho \frac{R^5}{5} \frac{4}{3} \pi + \rho \frac{R^5}{5} \frac{4}{3} \pi$$
$$\therefore I_{yy} = \frac{8}{15} \rho \pi R^5$$

But

$$M = \rho \tfrac{4}{3} \pi R^3$$

Hence,

$$I_{yy} = \tfrac{2}{5} MR^2$$

Because of the point symmetry about point O, we can also say that

$$I_{xx} = I_{zz} = \tfrac{2}{5} MR^2$$

Because the coordinate planes are all planes of symmetry for the mass distribution, the products of inertia are zero. Thus, the inertia tensor can be given as

$$I_{ij} = \begin{pmatrix} \tfrac{2}{5} MR^2 & 0 & 0 \\ 0 & \tfrac{2}{5} MR^2 & 0 \\ 0 & 0 & \tfrac{2}{5} MR^2 \end{pmatrix}$$

9.3 Relation Between Mass-Inertia Terms and Second Moment of Area Terms

We now relate the second moment and product second moment of area studied in Chapter 8 with the inertia tensor. To do this, consider a plate of constant thickness t and uniform density ρ (Fig. 9.9) A reference is selected so that the xy plane is in the midplane of this plate. The components of the inertia tensor are rewritten for convenience as

$$\begin{aligned} I_{xx} &= \rho \iiint_V (y^2 + z^2)\, dv, & I_{xy} &= \rho \iiint_V xy\, dv \\ I_{yy} &= \rho \iiint_V (x^2 + z^2)\, dv, & I_{xz} &= \rho \iiint_V xz\, dv \\ I_{zz} &= \rho \iiint_V (x^2 + y^2)\, dv, & I_{yz} &= \rho \iiint_V yz\, dv \end{aligned} \tag{9.3}$$

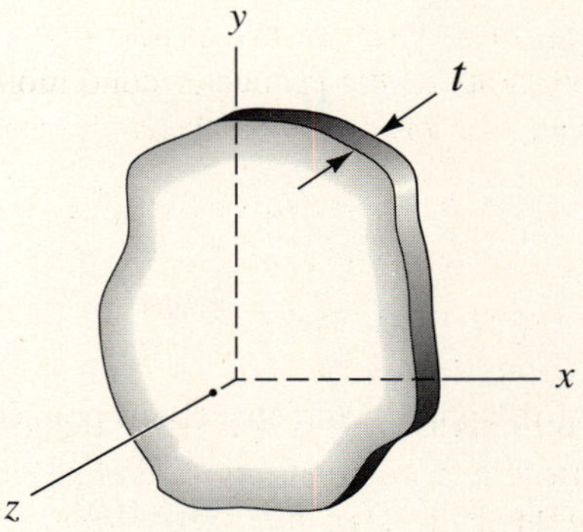

Figure 9.9. Plate of thickness t

Now consider that the thickness t is *small* compared to the lateral dimensions of the plate. This means that z is restricted to a range of values having a small magnitude. As a result, we can make two simplifications in the equations above. First, we shall set z equal to zero whenever it appears on the right side of the equations above. Second, we shall express dv as

$$dv = t\, dA$$

where dA is an area element on the *surface* of the plate, as shown in Fig. 9.10. Equations 9.3 then become

$$I_{xx} = \rho t \iint_A y^2 \, dA, \qquad I_{xy} = \rho t \iint_A xy \, dA$$
$$I_{yy} = \rho t \iint_A x^2 \, dA, \qquad I_{xz} = 0$$
$$I_{zz} = \rho t \iint_A (x^2 + y^2) \, dA, \qquad I_{yz} = 0$$

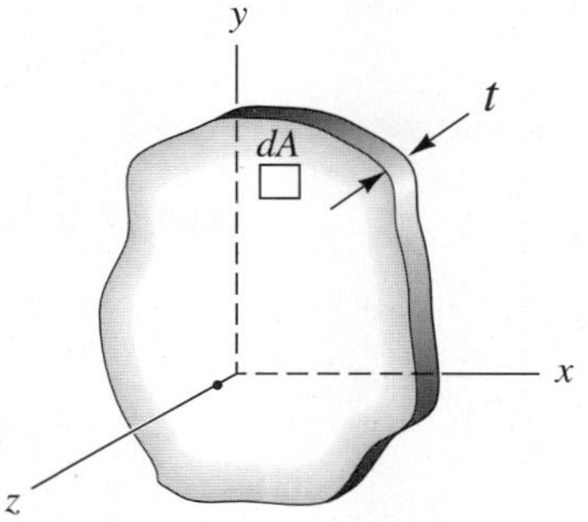

Figure 9.10. Use volume elements $t\,dA$.

Notice, now, that the integrals on the right sides of the equations above are second moments and product second moments of *area* as presented in Chapter 8. Denoting mass-moment and product of inertia terms with a subscript M and second moment and product second moments of area terms with a subscript A, we can then say for the nonzero expressions:

$$(I_{xx})_M = \rho t (I_{xx})_A$$
$$(I_{yy})_M = \rho t (I_{yy})_A$$
$$(I_{zz})_M = \rho t (J)_A$$
$$(I_{xy})_M = \rho t (I_{xy})_A$$

Thus, for a thin plate with a constant product ρt throughout, we can compute the inertia tensor components for reference xyz (see Fig. 9.9) by using the second moments and product second moments of area of the surface of the plate relative to axes xy.

It is important to point out that ρt is the *mass per unit area* of the plate. Imagine next that t goes to zero and simultaneously ρ goes to infinity at rates such that the product ρt becomes unity in the limit. One might think of the resulting body to be a *plane area*. By this approach, we have thus formed a plane area from a plate and in this way we can think of a plane area as a special mass. This explains why we use the same notation for mass moments and products of inertia as we use for second moments and product second moments of area. However the units clearly will be different. We now examine a plate problem.

Example 9.3

Determine the inertia tensor components for the thin plate (Fig. 9.11) relative to the indicated axes xyz. The weight of the plate is 2 mN/mm^2. For the top edge, $y = 2\sqrt{x}$ with x and y in millimetres.

It is clear that for ρt we have, remembering that this product represents mass per unit area:

$$\rho t = \frac{2\times 10^{-3}}{9.81} = 0.204 \text{ g/mm}^2 \qquad \text{(a)}$$

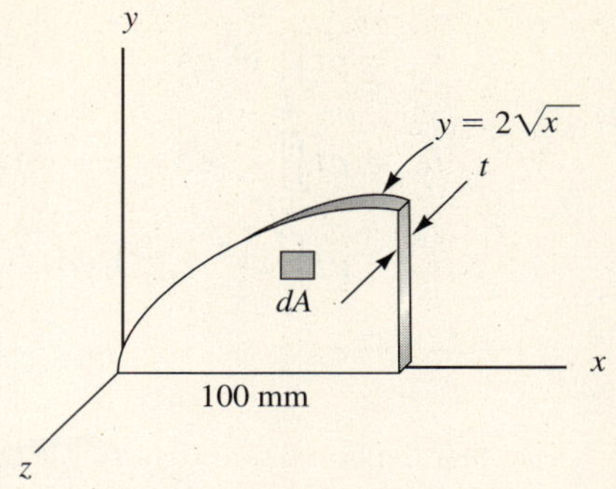

Figure 9.11. Plate of thickness t.

We now examine the second moments and product of area for the surface of the plate about axes xy. Thus,[4]

$$(I_{xx})_A = \int_0^{100}\int_{y=0}^{y=2\sqrt{x}} y^2\,dy\,dx$$

$$= \int_0^{100} \frac{y^3}{3}\Big|_0^{2\sqrt{x}} dx = \int_0^{100} \frac{8}{3}x^{3/2}dx$$

$$= \frac{8}{3}\frac{x^{5/2}}{\frac{5}{2}}\Big|_0^{100} = \left(\tfrac{8}{3}\right)\left(\tfrac{2}{5}\right)\left(100^{5/2}\right)$$

$$= 106.7\times 10^3 \text{ mm}^4$$

$$(I_{yy})_A = \int_0^{100}\int_{y=0}^{y=2\sqrt{x}} x^2\,dy\,dx$$

$$= \int_0^{100} x^2 y\Big|_0^{2\sqrt{x}} dx = \int_0^{100} x^2\left(2\sqrt{x}\right)dx$$

$$= 2\frac{x^{7/2}}{\frac{7}{2}}\Big|_0^{100} = 2\left(\tfrac{2}{7}\right)\left(100^{7/2}\right)$$

$$= 5.71\times 10^6 \text{ mm}^4$$

$$(I_{xy})_A = \int_0^{100}\int_{y=0}^{y=2\sqrt{x}} xy\,dy\,dx$$

$$= \int_0^{100} x\frac{y^2}{2}\Big|_0^{2\sqrt{x}} dx = \int_0^{100} 2x^2\,dx$$

$$= 2\left(\frac{100^3}{3}\right) = 667\times 10^3 \text{ mm}^4$$

[4]Note we have multiple integration where one of the boundaries is variable. The procedure to follow should be evident from the example.

Example 9.3 (Continued)

Using Eq. (a), we can then say for the nonzero inertia tensor components:

$$(I_{xx})_M = (0.204 \times 10^{-3})(106.7 \times 10^3) = 21.76 \text{ kg mm}^2$$
$$(I_{yy})_M = (0.204 \times 10^{-3})(5.71 \times 10^6) = 1.165 \text{ Mg mm}^2$$
$$(I_{xy})_M = (0.204 \times 10^{-3})(667 \times 10^3) = 136.1 \text{ kg mm}^2$$

Note that the nonzero inertia tensor components for a reference xyz on a plate (see Fig. 9.9) are *proportional* through ρt to the corresponding area-inertia terms for the plate surface. This means that all the formulations of Chapter 8 apply to the aforementioned nonzero inertia tensor components. Thus, on rotating the axes about the z axis we may use the transformation equations of Chapter 8. Consequently, the concept of *principal axes* in the midplane of the plate at a point applies. For such axes, the product of inertia is zero. One such axis then gives the maximum moment of inertia for all axes in the midplane at the point, the other the minimum moment of inertia. We have presented such problems at the end of this section.

What about principal axes for the inertia tensor at a point in a general three-dimensional body? Those students who have time to study Section 9.7 will learn that there are *three principal axes* at a point in the general case. These axes are *mutually orthogonal* and the *products of inertia are all zero* for such a set of axes at a point.[5] Furthermore, one of the axes will have a maximum moment of inertia, another axis will have a minimum moment of inertia, while the third axis will have an intermediate value. The sum of these three inertia terms must have a value that is common for all sets of axes at the point.

If, perchance, a set of axes xyz at a point is such that xy and xz form *two planes of symmetry* for the mass distribution of the body, then, as we learned earlier, since the z axis and the y axis are normal to the planes of symmetry, $I_{xy} = I_{xz} = I_{yz} = 0$. Thus, all products of inertia are zero. This would also be true for *any* two sets of axes of xyz forming two planes of symmetry. Clearly, axes forming two planes of symmetry must be *principal axes*. This information will suffice in most instances when we have to identify principal axes. On the other hand, consider the case where there is only *one plane of symmetry* for the mass distribution of a body at some point A. Let the xy plane at A form this plane of symmetry. Then, clearly, the products of inertia between the z axis that is normal to the plane of symmetry xy and *any axis* in the xy plane at A must be zero, as pointed out earlier. Obviously, the z axis must be a principal axis. The other two principal axes must be in the plane of symmetry, but generally cannot be located by inspection.

[5]The third principal axis for a plate at a point in the midplane is the z axis normal to the plate. Note that $(I_{zz})_M$ must always equal $(I_{xx})_M + (I_{yy})_M$. Why?

PROBLEMS

9.1. A uniform homogeneous slender rod of mass M is shown. Compute I_{xx} and $I_{x'x'}$.

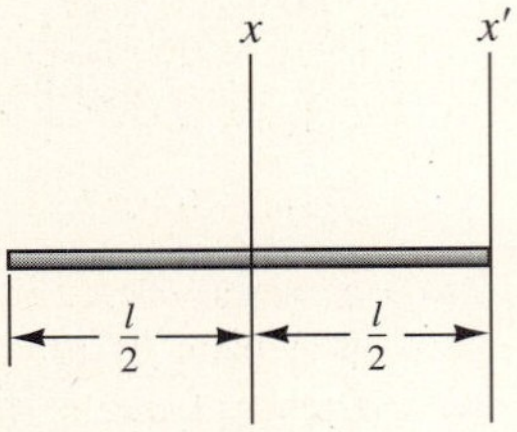

Figure P.9.1.

9.2. Find I_{xx} and $I_{x'x'}$ for the thin rod of Problem 9.1 for the case where the mass per unit length at the left end is 7.5 kg/m and increases linearly so that at the right end it is 12 kg/m. The rod is 6 m in length.

9.3. Compute I_{xy} for the thin homogeneous hoop of mass M.

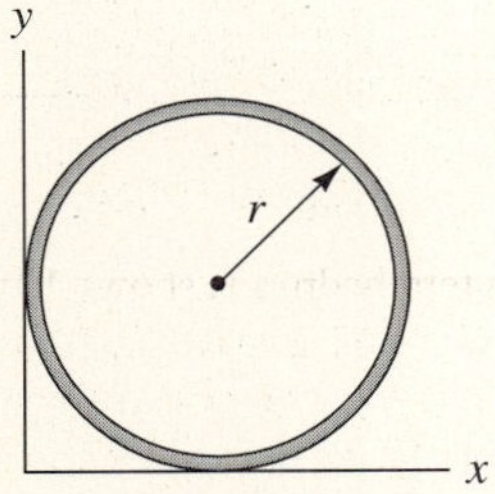

Figure P.9.3.

9.4. Compute I_{xx}, I_{yy}, I_{zz}, and I_{xy} for the homogeneous rectangular parallelepiped.

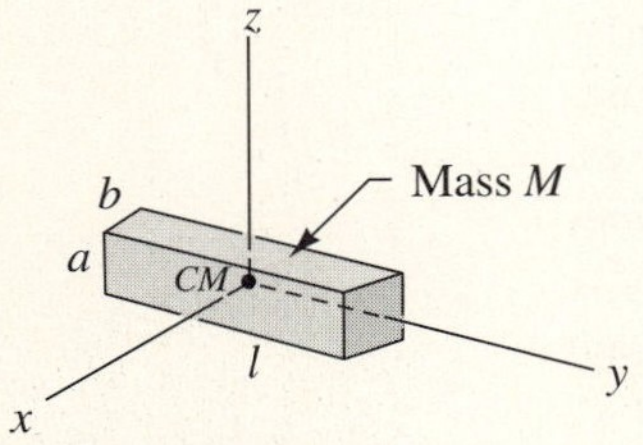

Figure P.9.4.

9.5. A wire having the shape of a parabola is shown. The curve is in the yz plane. If the weight of the wire is 0.3 N/m, what are I_{yy} and I_{xz}? [*Hint:* Replace ds along the wire by $\sqrt{(dy/dz)^2 + 1}\, dz$.]

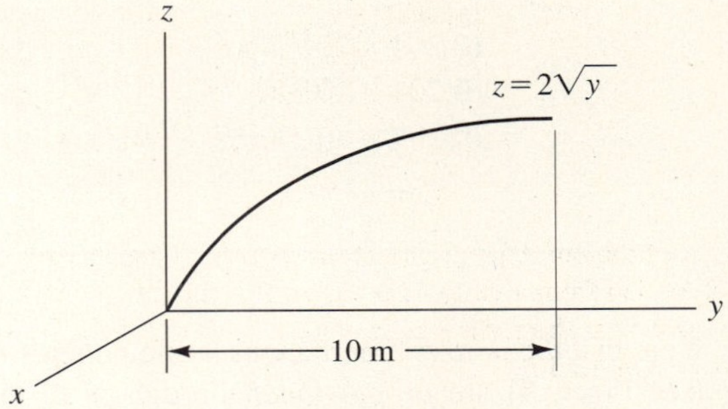

Figure P.9.5.

9.6. Compute the moment of inertia, I_{BB}, for the half-cylinder shown. The body is homogeneous and has a mass M.

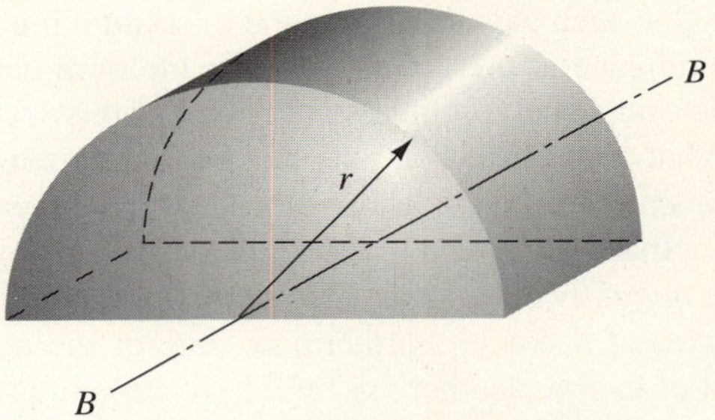

Figure P.9.6.

9.7. Find I_{zz} and I_{xx} for the homogeneous right circular cylinder of mass M.

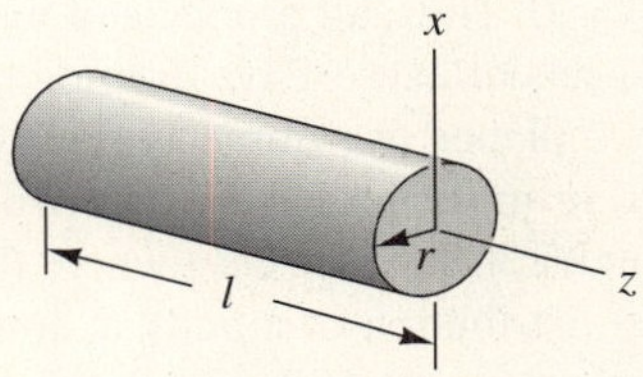

Figure P.9.7.

9.8. For the cylinder in Problem 9.7, the density increases linearly in the z direction from a value of 1 Mg/m^3 at the left end to a value of 1.8 Mg/m^3 at the right end. Take r = 30 mm and l = 150 mm. Find I_{xx} and I_{zz}.

9.9. Show that I_{zz} for the homogeneous right circular cone is $\frac{3}{10}MR^2$.

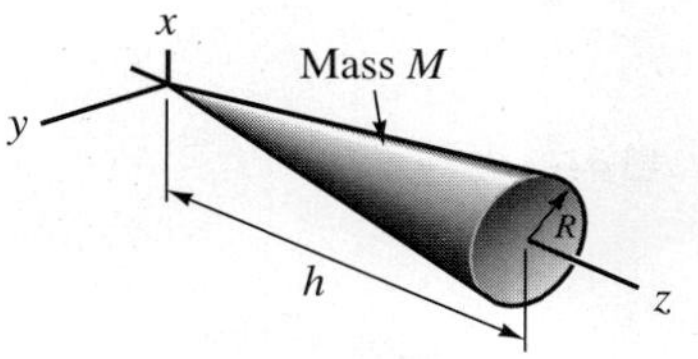

Figure P.9.9.

9.10. In Problem 9.9, the density increases as the square of z in the z direction from a value of 0.2 g/mm^3 at the left end to a value of 0.4 g/mm^3 at the right end. If $r = 20$ mm and the cone is 100 mm in length, find I_{zz}.

9.11. A body of revolution is shown. The radial distance r of the boundary from the x axis is given as $r = 0.2x^2$ m. What is I_{xx} for a uniform density of 1.6 Mg/m^3?

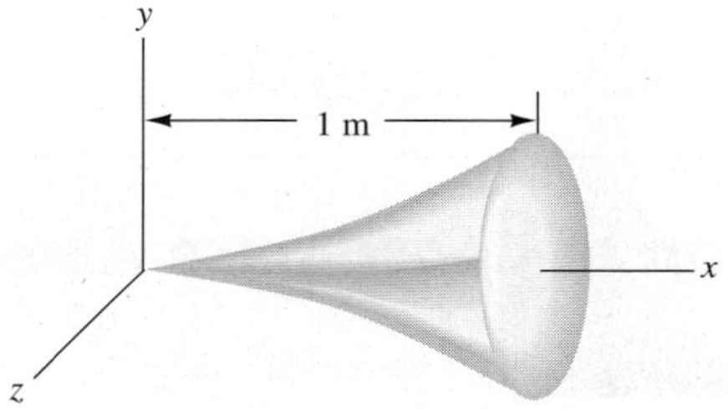

Figure P.9.11.

9.12. A thick hemispherical shell is shown with an inside radius of 40 mm and an outside radius of 60 mm. If the density ρ is 7 Mg/m^3, what is I_{yy}?

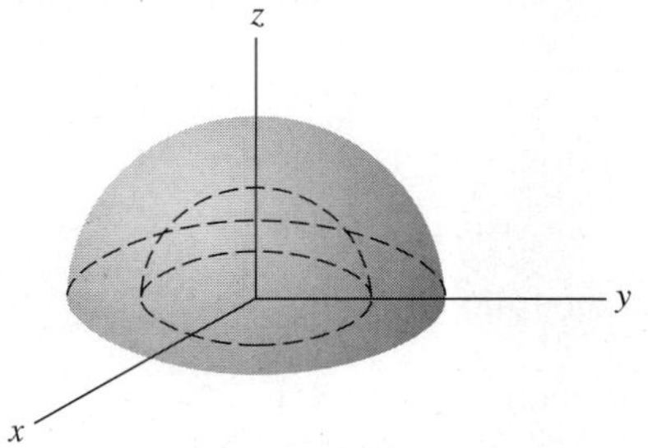

Figure P.9.12.

9.13. Find the mass moment of inertia I_{xx} for a very thin plate forming a quarter-sector of a circle. The plate weighs 0.4 N. What is the second moment of area about the x axis? What is the product of inertia? Axes are in the midplane of the plate.

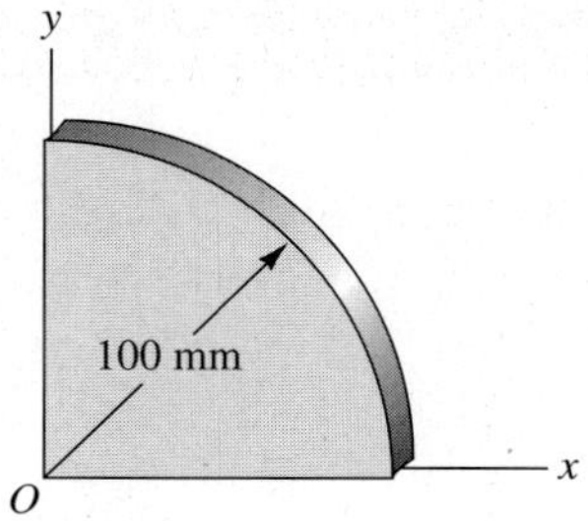

Figure P.9.13.

9.14. Find the second moment of area about the x axis for the front surface of a very thin plate. If the weight of the plate is 2 mN, find the mass moments of inertia about the x and y axes. What is the mass product of inertia I_{xy}?

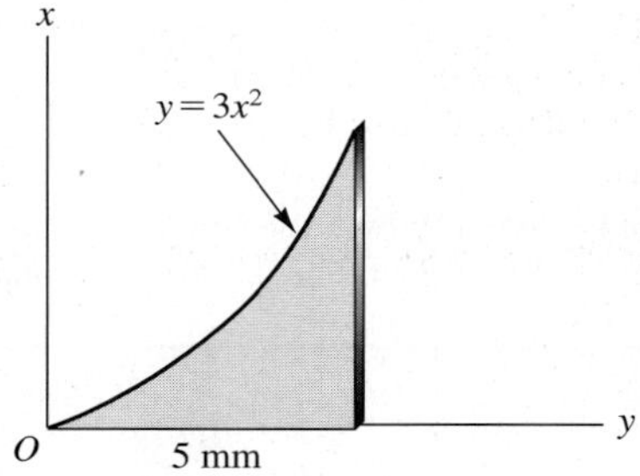

Figure P.9.14.

***9.15.** A uniform tetrahedron is shown having sides of length a, b, and c, respectively, and a mass M. Show that $I_{yz} = \frac{1}{20}Mac$. (*Suggestion:* Let z run from zero to surface ABC. Let x run from zero to line AB. Finally, let y run from zero to B. Note that the equation of a plane surface is $z = \alpha x + \beta y + \gamma$, where α, β, and γ are constants. The mass of the tetrahedron is $\rho abc/6$. It will be simplest in expanding $(1 - x/b - y/c)^2$ to proceed in the form $[(1 - y/c) - (x/b)]^2$, keeping $(1 - y/c)$ intact. In the last integration replace y by $[-c(1 - y/c) + c]$, etc.).

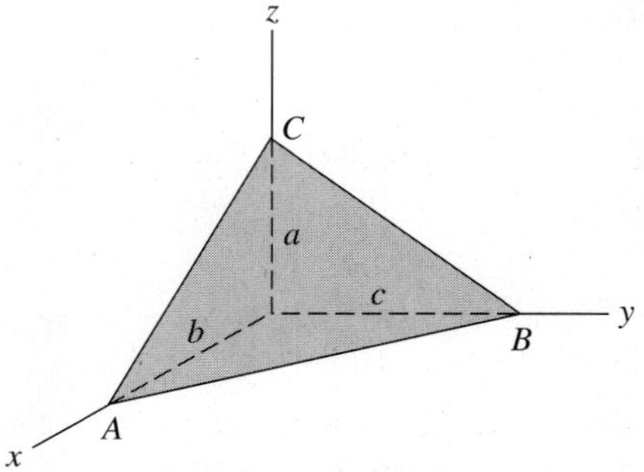

Figure P.9.15.

9.16. In Problem 9.13, find the three principal mass moments of inertia at O. Use the following results from problem 9.13

$$(I_{xx})_M = 101.9 \text{ kg mm}^2$$
$$(I_{xy})_M = 64.9 \text{ kg mm}^2$$

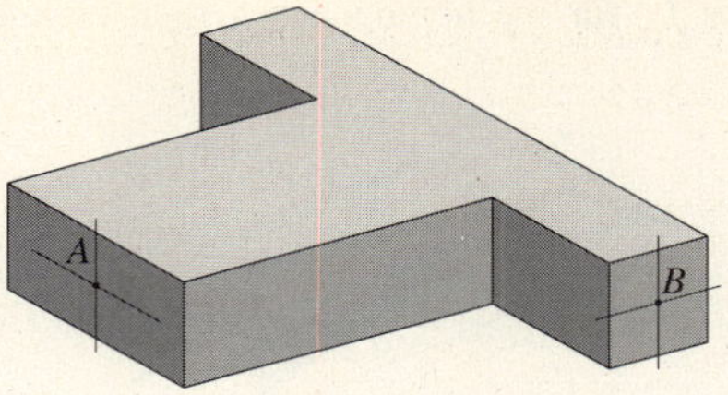

Figure P.9.18.

9.17. In Problem 9.14, compute the values of the three principal mass moments of inertia at O. From Problem 9.14 we have the results

$$(I_{xx})_M = 205 \text{ kg mm}^2$$
$$(I_{yy})_M = 3.82 \text{ kg mm}^2$$
$$(I_{xy})_M = 23.9 \text{ kg mm}^2$$

9.18. Can you identify by inspection any of the principal axes of inertia at A? At B? Explain. The density of the material is uniform.

9.19. By inspection, identify as many principal axes as you can for mass moments of inertia at positions A, B, and C. Explain your choices. The mass density of the material is uniform throughout.

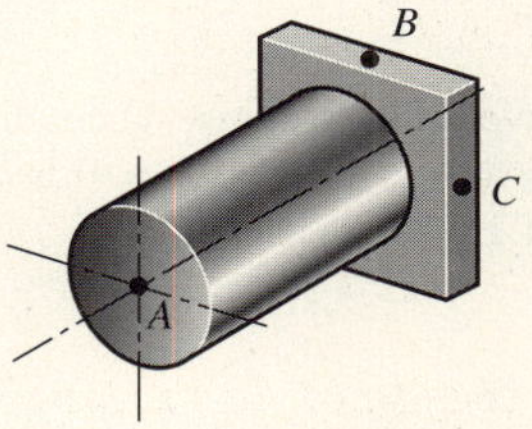

Figure P.9.19.

9.4 Translation of Coordinate Axes

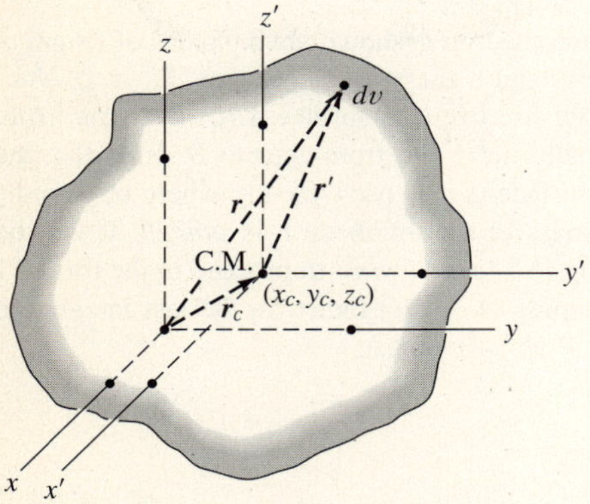

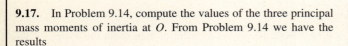

Figure 9.12. xyz translated from $x'y'z'$ at C.M.

In this section, we will compute mass moment and product of inertia quantities for a reference xyz that is displaced under a translation (no rotation) from a reference $x'y'z'$ at the center of mass (Fig. 9.12) for which the inertia terms are presumed known. Let us first compute the mass moment of inertia I_{zz}.

Observing Fig. 9.12, we see that

$$\boldsymbol{r} = \boldsymbol{r}_c + \boldsymbol{r}'$$

Hence,

$$x = x_c + x'$$
$$y = y_c + y'$$
$$z = z_c + z'$$

We can now formulate I_{zz} in the following way:

$$I_{zz} = \iiint_V (x^2 + y^2)\rho\, dv = \iiint_V \left[\left(x_c + x'\right)^2 + \left(y_c + y'\right)^2\right]\rho\, dv \tag{9.4}$$

Carrying out the squares and rearranging, we have

$$I_{zz} = \iiint_V (x_c^2 + y_c^2)\rho\, dv + 2\iiint_V x_c x'\rho\, dv$$
$$+2\iiint_V y_c y'\rho\, dv + \iiint_V (x'^2 + y'^2)\rho\, dv \tag{9.5}$$

Note that the quantities bearing the subscript c are constant for the integration and can be extracted from under the integral sign. Thus,

$$I_{zz} = M(x_c^2 + y_c^2) + 2x_c \iiint\limits_V x'\, dm + 2y_c \iiint\limits_V y'\, dm + \iiint\limits_V (x'^2 + y'^2)\rho\, dv \quad (9.6)$$

where $\rho\, dv$ has been replaced in some terms by dm, and the integration $\iiint\limits_V \rho\, dv$ in the first integral has been evaluated as M, the total mass of the body. The origin of the primed reference being at the center of mass requires of the first moments of mass that $\iiint x'\, dm = \iiint y'\, dm = \iiint z'\, dm = 0$. The middle two terms accordingly drop out of the expression above, and we recognize the last expression to be $I_{z'z'}$. Thus, the desired relation is

$$I_{zz} = I_{z'z'} + M(x_c^2 + y_c^2) \quad (9.7)$$

By observing the body in Fig. 9.12 along the z and z' axes (i.e., from directly above), we get a view as is shown in Fig 9.13. From this diagram, we can see that $y_c^2 + x_c^2 = d^2$, where d is the perpendicular distance between the z' axis through the center of mass and the z axis about which we are taking moments of inertia. We may then give the result above as

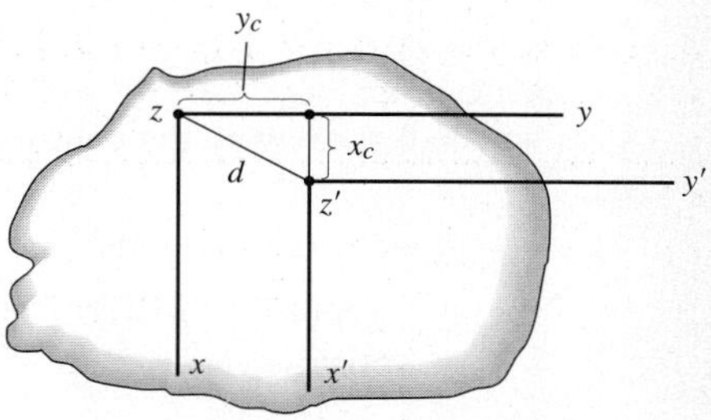

Figure 9.13. View along z direction (from above).

$$I_{zz} = I_{z'z'} + Md^2 \quad (9.8)$$

Let us generalize from the previous statement.

The moment of inertia of a body about any axis equals the moment of inertia of the body about a parallel axis that goes through the center of mass, plus the total mass times the perpendicular distance between the axes squared.

We leave it to you to show that for products of inertia a similar relation can be reached. For I_{xy}, for example, we have

$$I_{xy} = I_{x'y'} + Mx_c y_c \quad (9.9)$$

Here, we must take care to put in the proper signs of x_c and y_c as measured *from* the xyz reference. Equations 9.8 and 9.9 comprise the well-known *parallel-axis theorems* analogous to those formed in Chapter 8 for areas. You can use them to advantage for bodies composed of simple familiar shapes, as we now illustrate.

Example 9.4

Find I_{xx} and I_{xy} for the body shown in Fig. 9.14. Take ρ as constant for the body. Use the formulations for moments and products of inertia at the center of mass as given on the inside front cover page.

We shall consider first a solid rectangular prism having the outer dimensions given in Fig. 9.14, and we shall then subtract the contribution of the cylinder and the rectangular block that have been cut away. Thus, we have, for the overall rectangular block which we consider as body 1,

$$\begin{aligned}(I_{xx})_1 &= (I_{xx})_c + Md^2 = \tfrac{1}{12}M(a^2+b^2)+Md^2\\ &= \tfrac{1}{12}[(\rho)(6)(2.4)(4.5)](2.4^2+4.5^2)+[(\rho)(6)(2.4)(4.5)](1.2^2+2.25^2) \quad \text{(a)}\\ &= 561.816\rho\end{aligned}$$

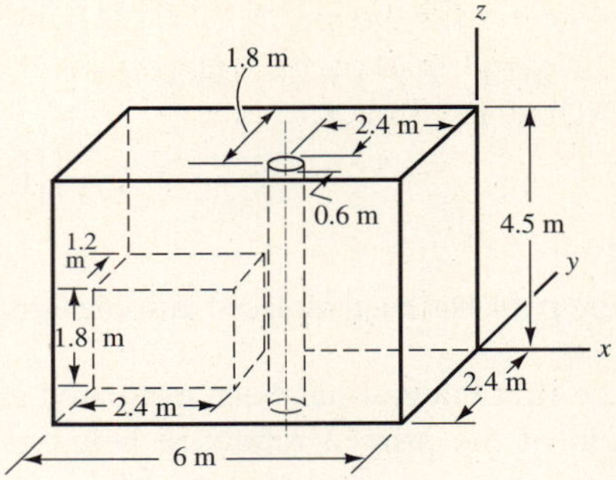

Figure 9.14. Find I_{xx} and I_{xy}.

From this, we shall take away the contribution of the cylinder, which we denote as body 2. Using the formulas from the inside front cover page,

$$\begin{aligned}(I_{xx})_2 &= \tfrac{1}{12}M(3r^2+h^2)+Md^2\\ &= \tfrac{1}{12}\left[\rho\pi(0.3)^2(4.5)\right]\left[3(0.3)^2+4.5^2\right]+\left[\rho\pi(0.3)^2(4.5)\right]\left[1.8^2+2.25^2\right] \quad \text{(b)}\\ &= 12.74\rho\end{aligned}$$

Also, we shall take away the contribution of the rectangular cutout (body 3):

$$\begin{aligned}(I_{xx})_3 &= \tfrac{1}{12}M(a^2+b^2)+Md^2\\ &= \tfrac{1}{12}[\rho(2.4)(1.8)(1.2)](1.2^2+1.8^2)+[\rho(2.4)(1.8)(1.2)](0.6^2+0.9^2) \quad \text{(c)}\\ &= 8.087\rho\end{aligned}$$

The quantity I_{xx} for the body with the rectangular and cylindrical cavities is then

$$I_{xx} = (561.816 - 12.74 - 8.087)\rho$$

$$I_{xx} = 541\rho \quad \text{(d)}$$

We follow the same procedure to obtain I_{xy}. Thus, for the block as a whole, we have

$$(I_{xy})_1 = (I_{xy})_c + Mx_cy_c$$

At the center of mass of the block, both the $(x')_1$ and $(y')_1$ axes are normal to planes of symmetry. Accordingly, $(I_{xy})_c = 0$. Hence,

$$\begin{aligned}(I_{xy})_1 &= 0 + [\rho(6)(2.4)(4.5)](-1.2)(-3)\\ &= 233.28\rho \quad \text{(e)}\end{aligned}$$

For the cylinder, we note that both the $(x')_2$ and $(y')_2$ axes at the center of mass are normal to planes of symmetry. Hence, we can say that

$$\begin{aligned}(I_{xy})_2 &= 0 + [\rho(\pi)(0.3^2)(4.5)](-2.4)(-1.8)\\ &= 5.50\rho \quad \text{(f)}\end{aligned}$$

Example 9.4 (Continued)

Finally, for the small cutout rectangular parallelepiped, we note that the $(x')_3$ and $(y')_3$ axes at the center of mass are perpendicular to planes of symmetry. Hence, we have

$$\begin{aligned}(I_{xy})_3 &= 0 + [\rho(2.4)(1.8)(1.2)](-0.6)(-4.8)\\ &= 14.93\rho \qquad \text{(g)}\end{aligned}$$

The quantity I_{xy} for the body with the rectangular and cylindrical cavities is then

$$I_{xy} = (233.28 - 5.50 - 14.93)\rho = \boxed{212.9\rho} \qquad \text{(h)}$$

*9.5 Transformation Properties of the Inertia Terms

Let us assume that the six independent inertia terms are known at the origin of a given reference. What is the mass moment of inertia for an axis going through the origin of the reference and having the direction cosines l, m, and n relative to the axes of this reference? The axis about which we are interested in obtaining the mass moment of inertia is designated as kk in Fig. 9.15.

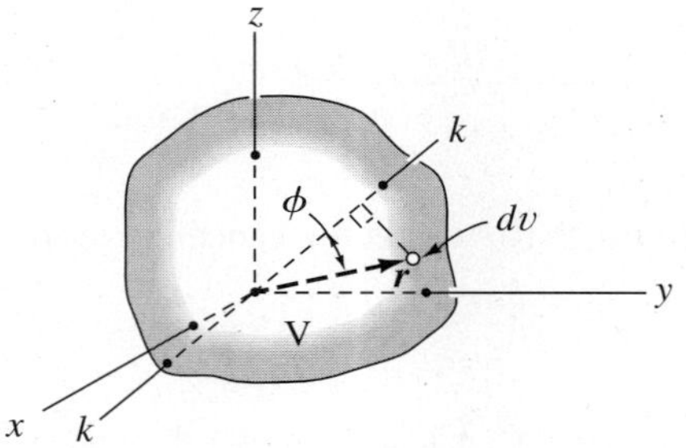

Figure 9.15. Find I_{kk}.

From previous conclusions, we can say that

$$I_{kk} = \iiint_V [|\boldsymbol{r}|(\sin\phi)]^2 \rho\, dv \qquad (9.10)$$

where ϕ is the angle between kk and $\boldsymbol{r}$. We shall now put $\sin^2 \phi$ into a more useful form by considering the right triangle formed by the position vector $\boldsymbol{r}$ and the axis kk. This triangle is shown enlarged in Fig. 9.16. The side a of the triangle has a magnitude that can be given by the dot product of $\boldsymbol{r}$ and the unit vector $\boldsymbol{\epsilon}_k$ along kk. Thus,

$$a = \boldsymbol{r} \cdot \boldsymbol{\epsilon}_k = (x\boldsymbol{i} + y\boldsymbol{j} + z\boldsymbol{k}) \cdot (l\boldsymbol{i} + m\boldsymbol{j} + n\boldsymbol{k}) \tag{9.11}$$

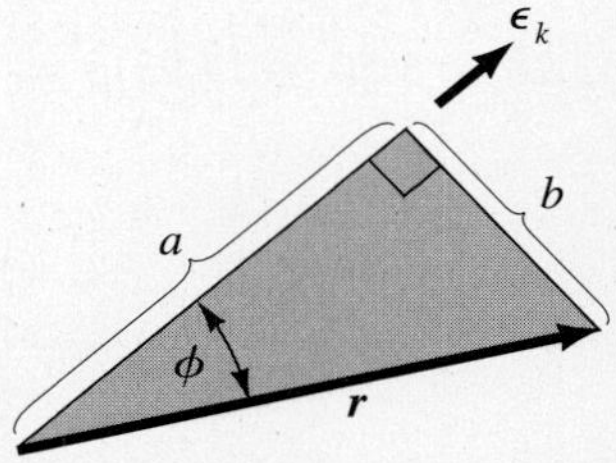

Figure 9.16. Right triangle formed by $\boldsymbol{r}$ and kk.

Hence

$$a = lx + my + nz$$

Using the Pythagorean theorem, we can now give side b as

$$b^2 = |\boldsymbol{r}|^2 - a^2 = (x^2 + y^2 + z^2) \\ - (l^2x^2 + m^2y^2 + n^2z^2 + 2lmxy + 2lnxz + 2mnyz)$$

The term $\sin^2 \phi$ may next be given as

$$\sin^2 \phi = \frac{b^2}{r^2} = \frac{(x^2 + y^2 + z^2) - (l^2x^2 + m^2y^2 + n^2z^2 + 2lmxy + 2lnxz + 2mnyz)}{x^2 + y^2 + z^2} \tag{9.12}$$

Substituting back into Eq. 9.10, we get, on canceling terms,

$$I_{kk} = \iiint_V \big[(x^2 + y^2 + z^2) \\ - (l^2x^2 + m^2y^2 + n^2z^2 + 2lmxy + 2lnxz + 2mnyz)\big]\rho\, dv$$

Since $l^2 + m^2 + n^2 = 1$, we can multiply the first bracketed expression in the integral by this sum:

$$I_{kk} = \iiint_V [(x^2 + y^2 + z^2)(l^2 + m^2 + n^2) \\ - (l^2x^2 + m^2y^2 + n^2z^2 + 2lmxy + 2lnxz + 2mnyz)]\rho\, dv$$

Carrying out the multiplication and collecting terms, we get the relation

$$I_{kk} = l^2 \iiint_V (y^2 + z^2)\rho\, dv + m^2 \iiint_V (x^2 + z^2)\rho\, dv + n^2 \iiint_V (x^2 + y^2)\rho\, dv$$
$$- 2lm \iiint_V (xy)\rho\, dv - 2ln \iiint_V (xz)\rho\, dv - 2mn \iiint_V (yz)\rho\, dv$$

Referring back to the definitions presented by Eqs. 9.1, we reach the desired transformation equation:

$$I_{kk} = l^2 I_{xx} + m^2 I_{yy} + n^2 I_{zz} - 2lm I_{xy} - 2ln I_{xz} - 2mn I_{yz} \qquad (9.13)$$

We next put this in a more useful form of the kind you will see in later courses in mechanics. Note first that l is the direction cosine between the k axis and the x axis. It is common practice to identify this cosine as a_{kx} instead of l. Note that the subscripts identify the axes involved. Similarly, $m = a_{ky}$ and $n = a_{kz}$. We can now express Eq. 9.13 in a form similar to a matrix array as follows on noting that $I_{xy} = I_{yx}$, etc.

$$\begin{aligned} I_{kk} = \; & I_{xx}a_{kx}^2 - I_{xy}a_{kx}a_{ky} - I_{xz}a_{kx}a_{kz} \\ & - I_{yx}a_{ky}a_{kx} + I_{yy}a_{ky}^2 - I_{yz}a_{ky}a_{kz} \\ & - I_{zx}a_{kz}a_{kx} - I_{zy}a_{kz}a_{ky} + I_{zz}a_{kz}^2 \end{aligned} \qquad (9.14)$$

This format is easily written by first writing the matrix array of I's on the right side and then inserting the a's remembering to insert minus signs for off-diagonal terms.

Let us next compute the product of inertia for a pair of mutually perpendicular axes, Ok and Oq, as shown in Fig. 9.17. The direction cosines of Ok we shall take as l, m, and n, whereas the direction cosines of Oq we shall take as l', m', and n'. Since the axes are at right angles to each other, we know that

$$\boldsymbol{\epsilon}_k \cdot \boldsymbol{\epsilon}_q = 0$$

Therefore,

$$ll' + mm' + nn' = 0 \qquad (9.15)$$

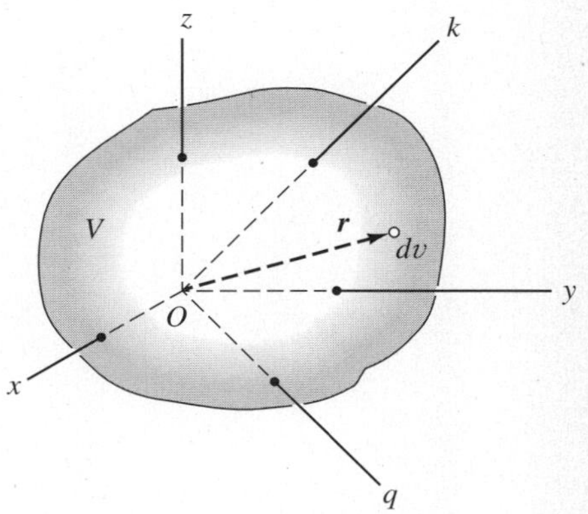

Figure 9.17. Find I_{kq}.

Noting that the coordinates of the mass element $\rho\, dv$ along the axes Ok and Oq are $\boldsymbol{r} \cdot \boldsymbol{\epsilon}_k$ and $\boldsymbol{r} \cdot \boldsymbol{\epsilon}_q$, respectively, we have, for I_{kq}:

$$I_{kq} = \iiint_V (\boldsymbol{r} \cdot \boldsymbol{\epsilon}_k)(\boldsymbol{r} \cdot \boldsymbol{\epsilon}_q)\rho\, dv$$

Using xyz components of $\boldsymbol{r}$ and the unit vectors, we have

$$I_{kq} = \iiint_V [(x\boldsymbol{i} + y\boldsymbol{j} + z\boldsymbol{k}) \cdot (l\boldsymbol{i} + m\boldsymbol{j} + n\boldsymbol{k})]$$
$$\times [(x\boldsymbol{i} + y\boldsymbol{j} + z\boldsymbol{k}) \cdot (l'\boldsymbol{i} + m'\boldsymbol{j} + n'\boldsymbol{k})]\rho\, dv \qquad (9.16)$$

Carrying out the dot products in the integrand above, we get the following result:

$$I_{kq} = \iiint_V (xl + ym + zn)(xl' + ym' + zn')\rho\, dv$$

Hence,

$$I_{kq} = \iiint_V (x^2 ll' + y^2 mm' + z^2 nn' + xylm' + xzln' + yxml' + yzmn' + zxnl' + zynm')\rho\, dv \tag{9.17}$$

Noting from Eq. 9.15 that $(ll' + mm' + nn')$ is zero, we may for convenience add the term, $(-x^2 - y^2 - z^2)(ll' + mm' + nn')$, to the integrand in the equation above. After canceling some terms, we have

$$I_{kq} = \iiint_V (-x^2 mm' - x^2 nn' - y^2 ll' - y^2 nn' - z^2 ll' - z^2 mm' + xylm' + xzln' + yxml' + yzmn' + zxnl' + zynm')\rho\, dv$$

Collecting terms and bringing the direction cosines outside the integrations, we get

$$\begin{aligned} I_{kq} = {} & -ll'\iiint_V (y^2 + z^2)\rho\, dv - mm'\iiint_V (x^2 + z^2)\rho\, dv \\ & - nn'\iiint_V (y^2 + x^2)\rho\, dv + (lm' + ml')\iiint_V xy\rho\, dv \\ & + (ln' + nl')\iiint_V xz\rho\, dv + (mn' + nm')\iiint_V yz\rho\, dv \end{aligned} \tag{9.18}$$

Noting the definitions in Eq. 9.1, we can state the desired transformation:

$$I_{kq} = -ll'I_{xx} - mm'I_{yy} - nn'I_{zz} + (lm' + ml')I_{xy} + (ln' + nl')I_{xz} + (mn' + nm')I_{yz} \tag{9.19}$$

We can now rewrite the previous equation in a more useful and simple form using a's as direction cosines. Thus, noting that $l' = a_{qx}$, etc., we proceed as in Eq. 9.14 to obtain

$$\begin{aligned} -I_{kq} = {} & I_{xx}a_{kx}a_{qx} - I_{xy}a_{kx}a_{qy} - I_{xz}a_{kx}a_{qz} \\ & - I_{yx}a_{ky}a_{qx} + I_{yy}a_{ky}a_{qy} - I_{yz}a_{ky}a_{qz} \\ & - I_{zx}a_{kz}a_{qx} - I_{zy}a_{kz}a_{qy} + I_{zz}a_{kz}a_{qz} \end{aligned} \tag{9.20}$$

Again you will note that the right side can easily be set forth by first putting down the matrix array of I_{ij} and then inserting the a's with easily determined subscripts while remembering to insert minus signs for off-diagonal terms.

Example 9.5

Find $I_{z'z'}$ and $I_{x'z'}$ for the solid cylinder shown in Fig. 9.18. The reference $x'y'z'$ is found by rotating about the y axis an amount 30°, as shown in the diagram. The mass of the cylinder is 100 kg.

It is simplest to first get the inertia tensor components for reference xyz. Thus, using formulas from the inside front cover page we have

$$I_{zz} = \frac{1}{2}Mr^2 = \frac{1}{2}(100)\left(\frac{1.3}{2}\right)^2 = 21.13 \text{ kg m}^2$$

$$I_{xx} = I_{yy} = \frac{1}{12}M(3r^2 + h^2)$$

$$= \frac{1}{12}(100)\left[(3)\left(\frac{1.3}{2}\right)^2 + 3^2\right]$$

$$= 85.56 \text{ kg m}^2$$

Noting that the xyz coordinate planes are planes of symmetry, we can conclude that

$$I_{xz} = I_{yx} = I_{yz} = 0$$

Next, evaluate the direction cosines of the z' and the x' axes relative to xyz. Thus,

For z' axis:

$$a_{z'x} = \cos 60° = 0.5$$
$$a_{z'y} = \cos 90° = 0$$
$$a_{z'z} = \cos 30° = 0.866$$

For x' axis:

$$a_{x'x} = \cos 30° = 0.866$$
$$a_{x'y} = \cos 90° = 0$$
$$a_{x'z} = \cos 120° = -0.5$$

First, we employ Eq. 9.14 to get $I_{z'z'}$.

$$I_{z'z'} = (85.56)(0.5)^2 + (21.13)(0.866)^2$$

$$= \boxed{37.24 \text{ kg m}^2}$$

Finally, we employ Eq. 9.20 to get $I_{x'z'}$.

$$-I_{x'z'} = (85.56)(0.5)(0.866) + (21.13)(0.866)(-0.5)$$

Therefore,

$$\boxed{I_{x'z'} = -27.90 \text{ kg m}^2}$$

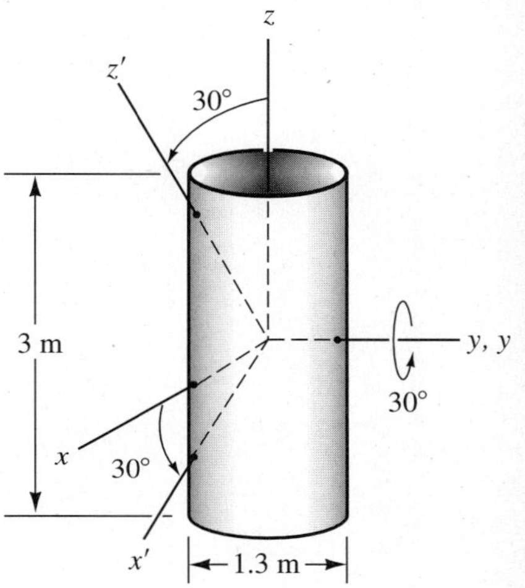

Figure 9.18. Find $I_{z'z'}$ and $I_{x'z'}$.

*9.6 Looking Ahead: Tensors

By making the axis Ok in Fig. 9.17 an x' axis at O and using the direction cosines for this axis ($a_{x'x}$, $a_{x'y}$, $a_{x'z}$), we can formulate $I_{x'x'}$ from Eq. 9.14. By a similar procedure, we can consider axis Ok to be a y' axis at O and we can formulate $I_{y'y'}$ using for this axis direction cosines $a_{y'x}$, $a_{y'y}$, $a_{y'z}$ from Eq. 9.14. Similarly, for $I_{z'z'}$. We can thus get the mass moments of inertia for reference x', y', z' at O rotated arbitrarily relative to xyz. Also by considering the Ok and Oq axes to be x' and y' axes, respectively, with $a_{x'x}$, $a_{x'y}$, and $a_{x'z}$ as direction cosines for the x' axis and $a_{y'x}$, $a_{y'y}$, and $a_{y'z}$ as direction cosines for the y' axis, we can evaluate $I_{x'y'}$ at O using Eq. 9.20. This approach can similarly be followed to find $I_{x'z'}$ and $I_{y'z'}$. Thus, employing Eqs. 9.14 and 9.20 as parent equations, we can develop equations for computing the nine inertia quantities for a reference $x'y'z'$ rotated arbitrarily relative to xyz at O in terms of the nine known inertia quantities for reference xyz at O. Thus, once the nine inertia quantities are known for one reference at some point, they can be determined for *any* reference at that point. We say that the inertia terms *transform* from one set of components for xyz at some point O to another set of components for $x'y'z'$ at point O by means of certain transformations formed from Eqs. 9.14 and 9.20.

We now define a symmetric,[6] *second-order tensor as a set of nine components*

$$\begin{pmatrix} A_{xx} & A_{xy} & A_{xz} \\ A_{yx} & A_{yy} & A_{yz} \\ A_{zx} & A_{zy} & A_{zz} \end{pmatrix}$$

which transforms with a rotation of axes according to the following parent equations. For the diagonal terms,

$$\begin{aligned} A_{kk} = {} & A_{xx}a_{kx}^2 + A_{xy}a_{kx}a_{ky} + A_{xz}a_{kx}a_{kz} \\ & + A_{yx}a_{ky}a_{kx} + A_{yy}a_{ky}^2 + A_{yz}a_{ky}a_{kz} \\ & + A_{zx}a_{kz}a_{kx} + A_{zy}a_{kz}a_{ky} + A_{zz}a_{kz}^2 \end{aligned} \tag{9.21}$$

For the off-diagonal terms,

$$\begin{aligned} A_{kq} = {} & A_{xx}a_{kx}a_{qx} + A_{xy}a_{kx}a_{qy} + A_{xz}a_{kx}a_{kz} \\ & + A_{yx}a_{ky}a_{qx} + A_{yy}a_{ky}a_{qy} + A_{yz}a_{ky}a_{kz} \\ & + A_{zx}a_{kz}a_{qx} + A_{zy}a_{kz}a_{qy} + A_{zz}a_{kz}a_{qz} \end{aligned} \tag{9.22}$$

On comparing Eqs. 9.21 and 9.22, respectively, with Eqs. 9.14 and 9.20, we can conclude that the array of terms

$$I_{ij} = \begin{pmatrix} I_{xx} & -I_{xy} & -I_{xz} \\ -I_{yx} & I_{yy} & -I_{yz} \\ -I_{zx} & -I_{zy} & I_{zz} \end{pmatrix} \tag{9.23}$$

is a second-order tensor.

[6]The word "symmetric" refers to the condition $A_{12} = A_{21}$, etc., that is required if the transformation equation is to have the form given. We can have nonsymmetric second-order tensors, but since they are less common in engineering work, we shall not concern ourselves here with such possibilities.

You will learn that because of the common transformation law identifying certain quantities as tensors, there will be extremely important common characteristics for these quantities which set them apart from other quantities. Thus, in order to learn these common characteristics in an efficient way and to understand them better, we become involved with tensors as an entity in the engineering sciences, physics, and applied mathematics. You will soon be confronted with the stress and strain tensors in your courses in strength of materials.

To explore this point further, we have shown an infinitesimal rectangular parallelepiped extracted from a solid under load. On three orthogonal faces we have shown nine force intensities (i.e., forces per unit area). Those with repeated indices are called *normal stresses* while those with different pairs of indices are called *shear stresses*. You will learn, that knowing nine such stresses, you can readily find three stresses, one normal and two orthogonal shear stresses, on *any* interface at any orientation inside the rectangular parallelepiped. To find such stresses on an interface knowing the stresses shown in Fig. 9.19, we have the *same transformation equations* given by Eqs. 9.21 and 9.22. Thus stress is a *second-order tensor*.

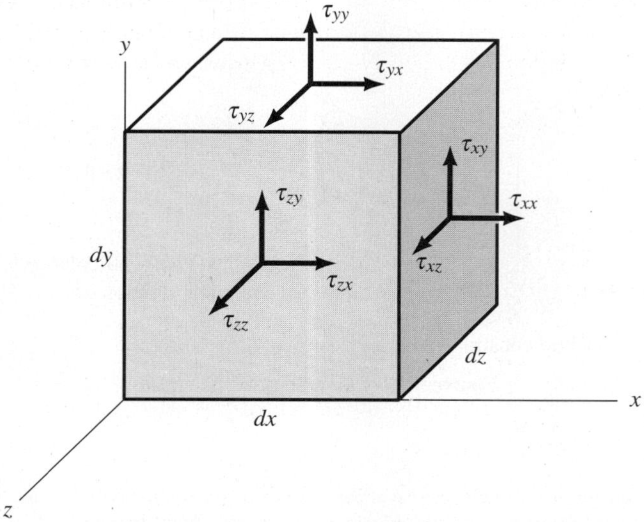

Figure 9.19. Nine stresses on three orthogonal interfaces at a point.

A *two-dimensional* simplification of τ_{ij} involving the quantities τ_{xx}, τ_{yy}, and τ_{xy} (= τ_{yx}) as the only nonzero stresses is called *plane stress*. This occurs in a thin plate loaded in the plane of symmetry as shown in Fig. 9.20. Plane stress is the direct analog of *second moments and products of area*, which is a two-dimensional simplification of the inertia tensor. Clearly, plane stress and second moments and products of area have the same transformation equations, which are Eqs. 8.26 through 8.28 with τ_{xx}, τ_{yy}, τ_{xy}, $\tau_{x'y'}$ replacing I_{xx}, I_{yy}, $-I_{xy}$, $-I_{x'y'}$ respectively.

In solid mechanics, you will also learn that there are nine terms ϵ_{ij} that describe deformation at a point. Thus consider the undeformed an infinitesimal

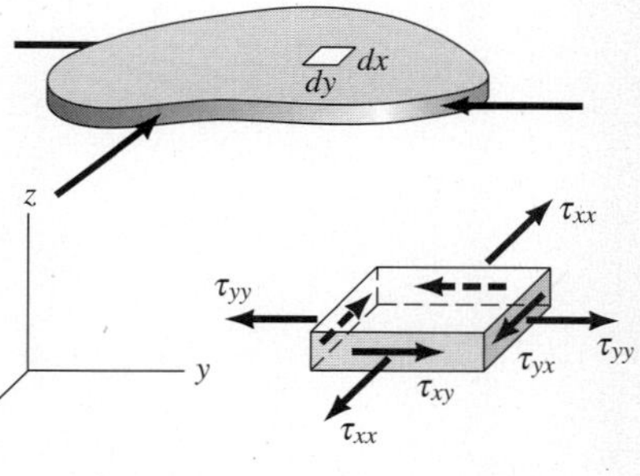

Figure 9.20. The case of plane stress.

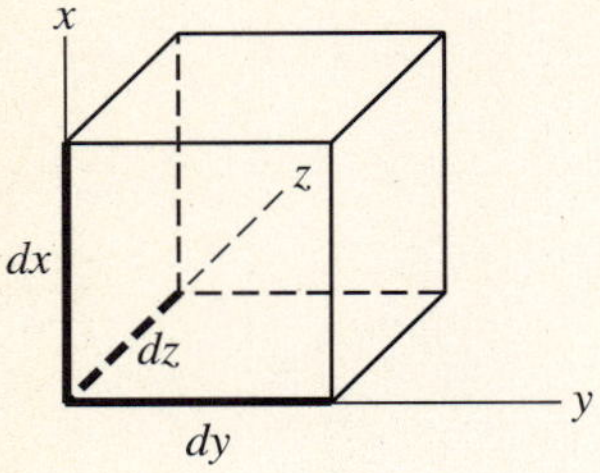

Figure 9.21. An infinitesimal rectangular parallelepiped with three edges highlighted.

rectangular parallelepiped in Fig. 9.21. When there is a deformation there are *normal strains* ϵ_{xx}, ϵ_{yy}, ϵ_{zz} along the direction of the darkened edges which give the changes of length per unit original length of these edges. Furthermore, when there is a deformation, there are six *shear strains* $\epsilon_{xy} = \epsilon_{yx}$, $\epsilon_{xz} = \epsilon_{zx}$, $\epsilon_{yz} = \epsilon_{zy}$ that give the change in angle in radians from that of the right angles of the three darkened edges. Knowing these quantities, we can find any other strains in the rectangular parallelepiped. These other strains can be found by using transformation Eqs. 9.21 and 9.22 and so *strain* is also a *second-order tensor.*

The two-dimensional simplification of ϵ_{ij} involving the quantities, ϵ_{xx}, ϵ_{yy}, and ϵ_{xy} $(= \epsilon_{yx})$ as the only nonzero strains is called *plane strain* and represents the strains in a prismatic body constrained at the ends with loading normal to the centerline in which the loading does not vary with z (see Fig. 9.22). Also, the prismatic body must not be subject to bending. Plane strain is an analogous mathematically to plane stress and second moments and products of area. All three are two-dimensional simplifications of second-order symmetric tensors and have the *same transformation equations* as well as other mathematical properties. Finally, in electromagnetic theory and nuclear physics, you will be introduced to the quadruple tensor.[7]

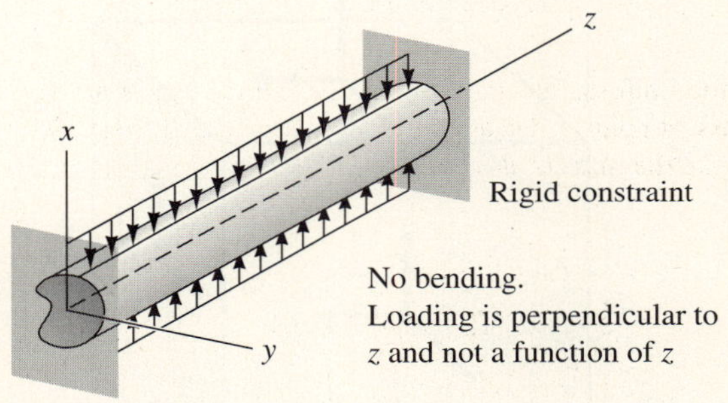

Figure 9.22. Example of plane strain.

[7]Vectors may be defined in terms of the way components of the vector for a new reference are related to the components of the old reference at a point. Thus, for any direction n, we have for component A_n:

$$A_n = A_x a_{nx} + A_y a_{ny} + A_z a_{nz} \tag{a}$$

Using Eq. (a), we can find components of vector $\boldsymbol{A}$ with respect to $x'y'z'$ rotated arbitrarily relative to xyz. Thus, all vectors must transform in accordance with Eq. (a) on rotation of the reference. Obviously, the vector, as seen from this point of view, is a special, simple case of the second-order tensor. We say, accordingly, that vectors are *first-order tensors*.

As for scalars, there is clearly no change in value when there is a rotation of axes at a point. Thus,

$$T(x', y', z') = T(x, y, z) \tag{b}$$

for $x'y'z'$ rotated relative to xyz. Scalars are a special form of tensor when considered from a transformation point of view. In fact, they are called *zero-order tensors*.

PROBLEMS

In the following problems, use the formulas for moments and products of inertia at the mass center to be found in the inside front cover page.

9.20. What are the moments and products of inertia for the xyz and $x'y'z'$ axes for the cylinder?

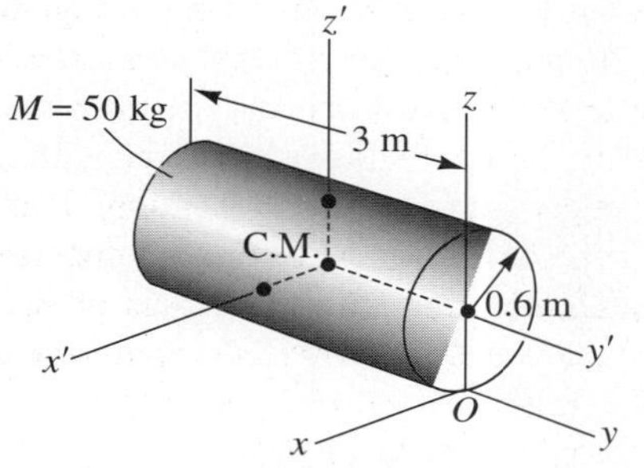

Figure P.9.20.

9.21. For the uniform block, compute the inertia tensor at the center of mass, at point a, and at point b for axes parallel to the xyz reference. Take the mass of the body as M kg.

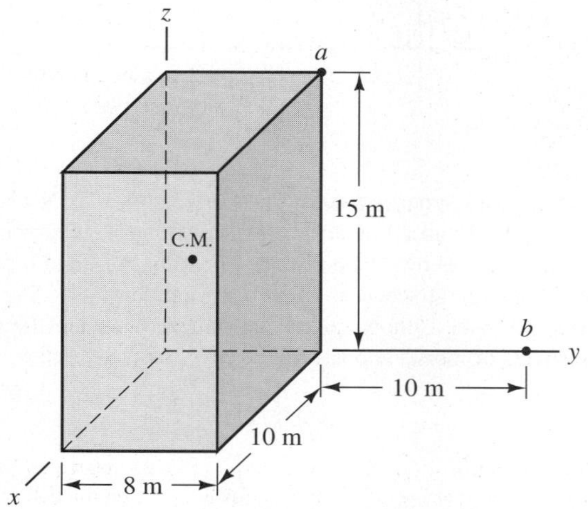

Figure P.9.21.

9.22. Determine $I_{xx} + I_{yy} + I_{zz}$ as a function of x, y, and z for all points in space for the uniform rectangular parallelepiped. Note that xyz has its origin at the center of mass and is parallel to the sides.

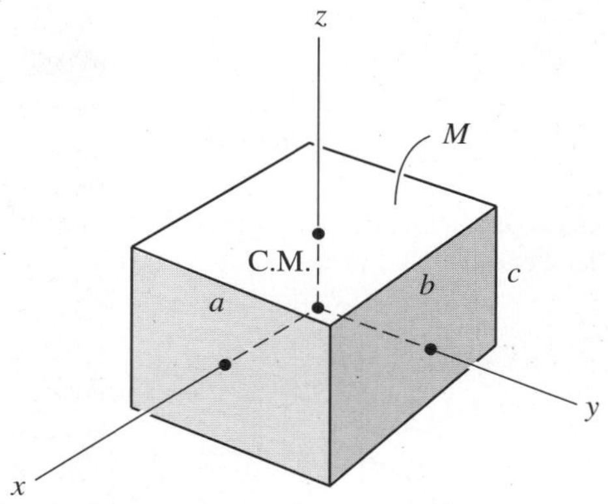

Figure P.9.22.

9.23. A thin plate weighing 100 N has the following mass moments of inertia at mass center O:

$$I_{xx} = 15 \text{ kg m}^2$$
$$I_{yy} = 13 \text{ kg m}^2$$
$$I_{xy} = -10 \text{ kg m}^2$$

What are the moments of inertia $I_{x'x'}$, $I_{y'y'}$, and $I_{z'z'}$ at point P having the position vector:

$$\boldsymbol{r} = 0.5\boldsymbol{i} + 0.2\boldsymbol{j} + 0.6\boldsymbol{k} \text{ m}$$

Also determine $I_{x'z'}$ at P.

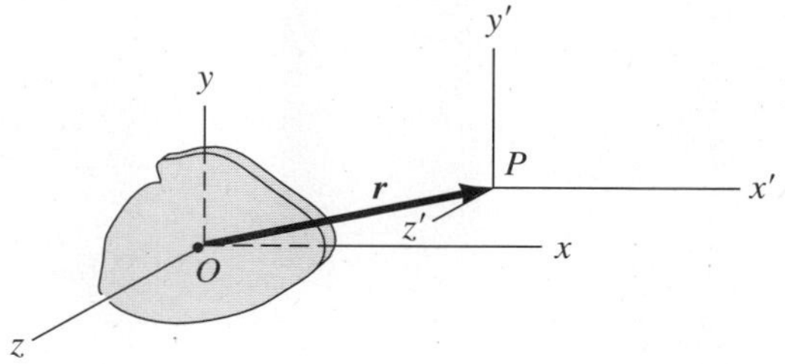

Figure P.9.23.

9.24. A crate with its contents weighs 20 kN and has its center of mass at

$$r_c = 1.3\boldsymbol{i} + 3\boldsymbol{j} + 0.8\boldsymbol{k} \text{ m}$$

It is known that at corner *A*,

$$I_{x'x'} = 5.5 \text{ Mg m}^2$$
$$I_{x'y'} = -1.5 \text{ Mg m}^2$$

for primed axes parallel to *xyz*. At point *B*, find $I_{x''x''}$ and $I_{x''y''}$ for double-primed axes parallel to *xyz*.

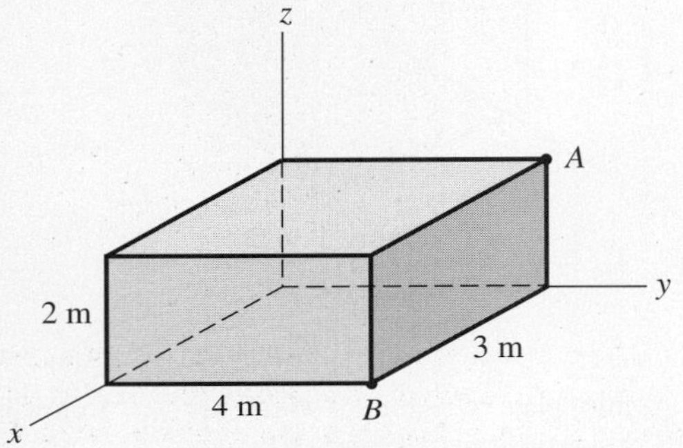

Figure P.9.24.

9.25 A cylindrical crate and its contents weigh 500 N. The center of mass is at

$$r_c = 0.6\boldsymbol{i} + 0.7\boldsymbol{j} + 2\boldsymbol{k} \text{ m}$$

It is known that at *A*,

$$(I_{yy})_A = 85 \text{ kg m}^2$$
$$(I_{yz})_A = -22 \text{ kg m}^2$$

Find I_{yy} and I_{zy} at *B*.

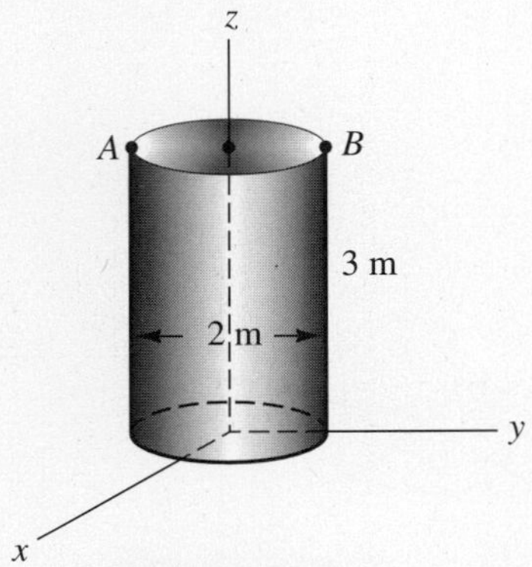

Figure P.9.25.

9.26. A block having a uniform density of 5 Mg/m^3 has a hole of diameter 40 mm cut out. What are the principal moments of inertia at point *A* at the centroid of the right face of the block?

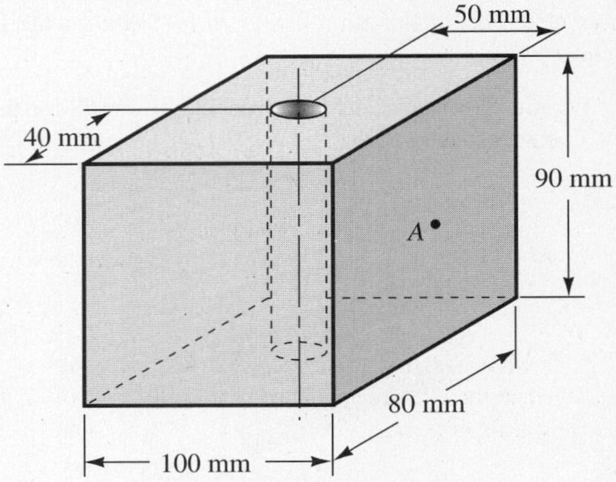

Figure P.9.26.

9.27. Find maximum and minimum moments of inertia at point *A*. The block weighs 20 N and the cone weighs 14 N.

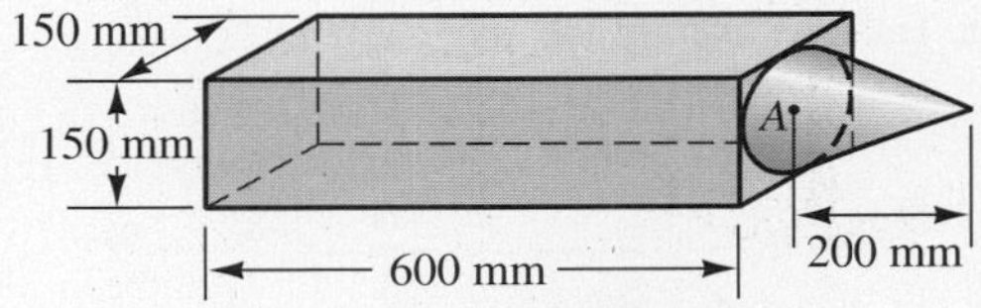

Figure P.9.27.

9.28. Solid spheres *C* and *D* each weighing 25 N and having radius of 50 mm are attached to a thin solid rod weighing 30 N. Also, solid spheres *E* and *G* each weighing 20 N and having radii of 30 mm are attached to a thin rod weighing 20 N. The rods are attached to be orthogonal to each other. What are the principal moments of inertia at point *A*?

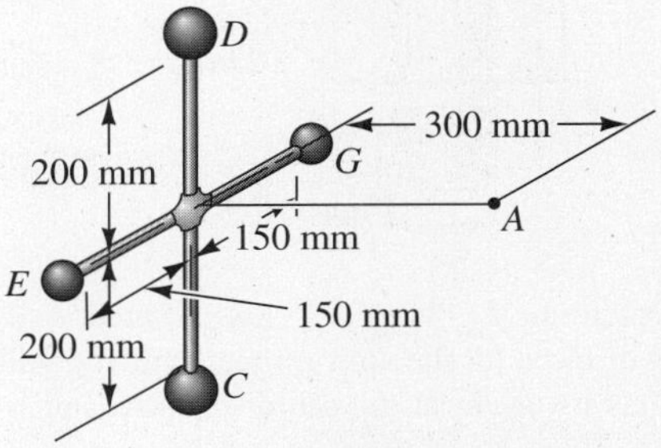

Figure P.9.28.

9.29. A cylinder is shown having a conical cavity oriented along the axis *A–A* and a cylindrical cavity oriented normal to *A–A*. If the density of the material is 7.2 Mg/m^3, what is I_{AA}?

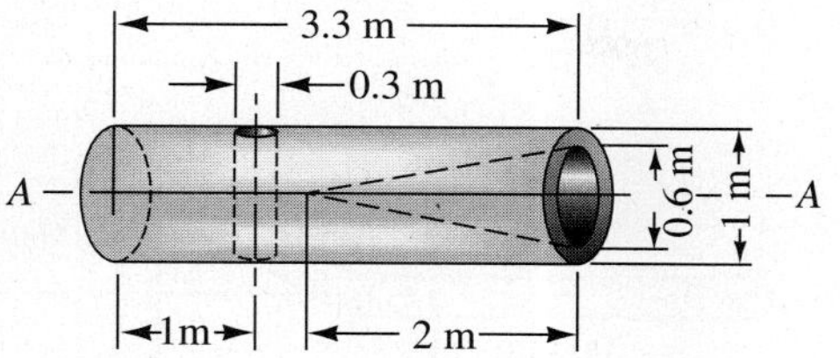

Figure P.9.29.

9.30. A flywheel is made of steel having a density of 7.85 Mg/m^3. What is the moment of inertia about its geometric axis? What is the radius of gyration?

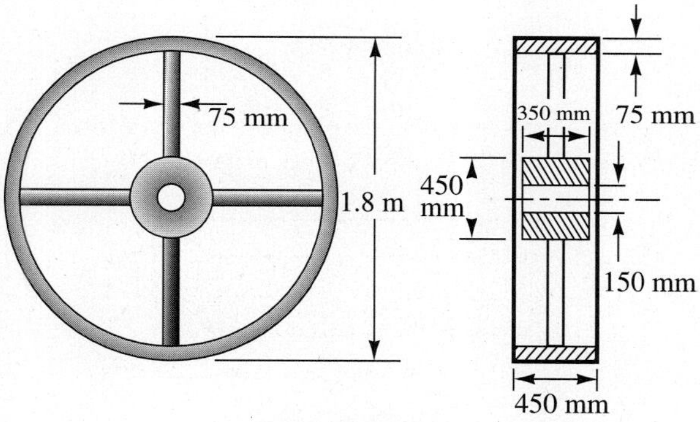

Figure P.9.30.

9.31. Compute I_{yy} and I_{xy} for the right circular cylinder, which has a mass of 50 kg, and the square rod, which has a mass of 10 kg, when the two are joined together so that the rod is radial to the cylinder. The *x* axis lies along the bottom of the square rod.

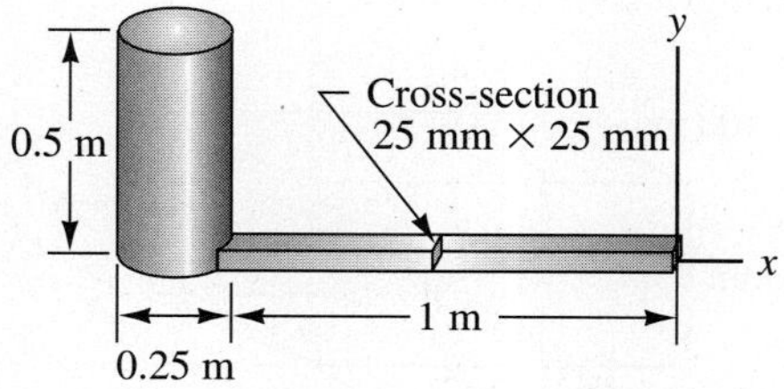

Figure P.9.31.

9.32. Compute the moments and products of inertia for the *xy* axes. The density is 7.85 Mg/m^3 throughout.

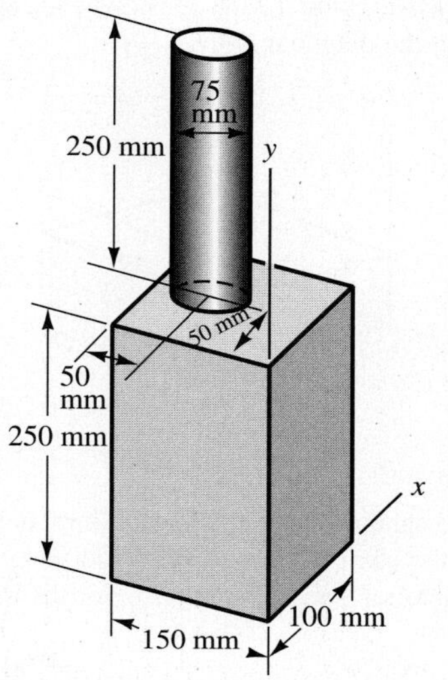

Figure P.9.32.

9.33. A disc *A* is mounted on a shaft such that its normal is oriented 10° from the centerline of the shaft. The disc has a diameter of 600 mm, is 25 mm in thickness, and a mass of 50 kg. Compute the moment of inertia of the disc about the centerline of the shaft.

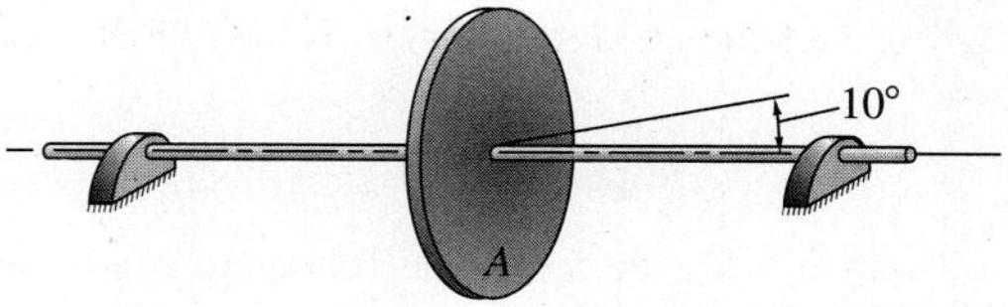

Figure P.9.33.

9.34. A gear *B* having a mass of 25 kg rotates about axis *C–C*. If the rod *A* has a mass distribution of 7.5 kg/m, compute the moment of inertia of *A* and *B* about the axis *C–C*.

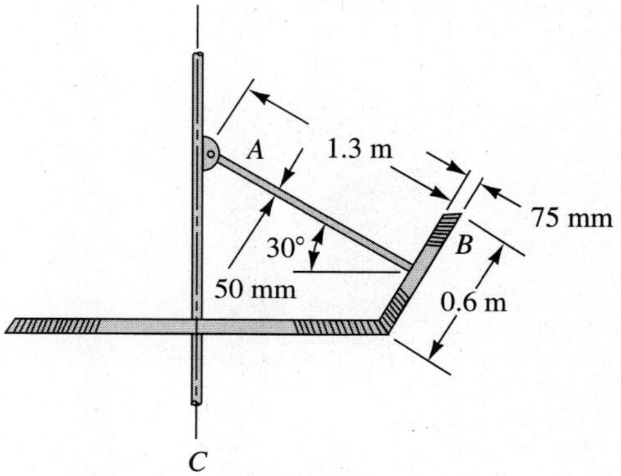

Figure P.9.34.

9.35. A block weighing 100 N is shown. Compute the moment of inertia about the diagonal D–D.

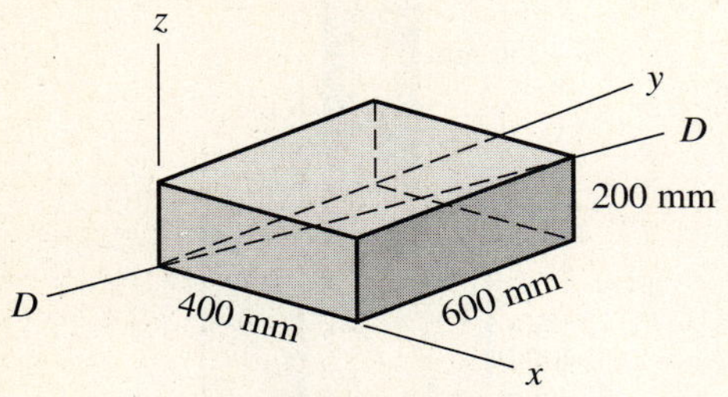

Figure P.9.35.

9.36. A solid sphere A of diameter 300 mm and weight 450 N is connected to the shaft B–B by a solid rod weighing 30 N/m and having a diameter of 1 in. Compute $I_{z'z'}$ for the rod and ball.

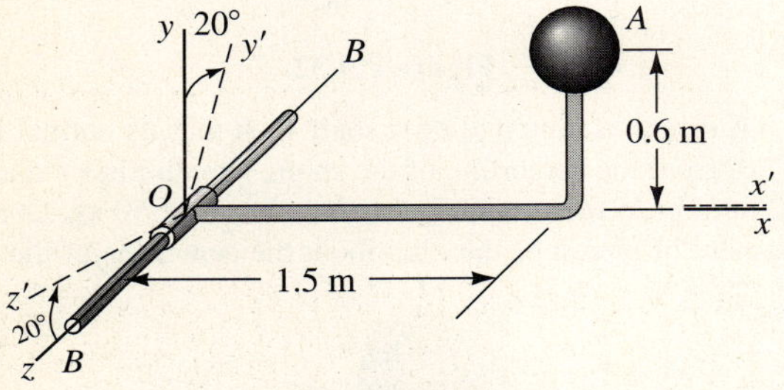

Figure P.9.36.

9.37. In Problem 9.13, we found the following results for the thin plate:

$$I_{xx} = I_{yy} = 101.9 \text{ mg m}^2$$
$$I_{xy} = 64.9 \text{ mg m}^2$$

Find all components for the inertia tensor for reference $x'y'z'$. Axes $x'y'$ lie in the midplane of the plate.

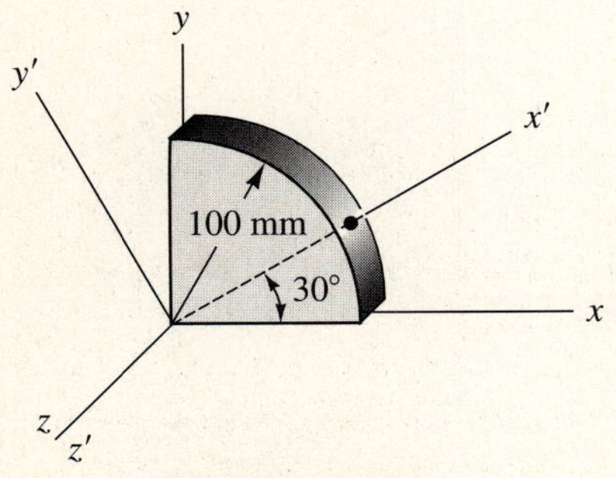

Figure P.9.37.

9.38. A bent rod weighs 0.1 N/mm. What is I_{nn} for

$$\boldsymbol{\epsilon}_n = 0.3\boldsymbol{i} + 0.45\boldsymbol{j} + 0.841\boldsymbol{k}?$$

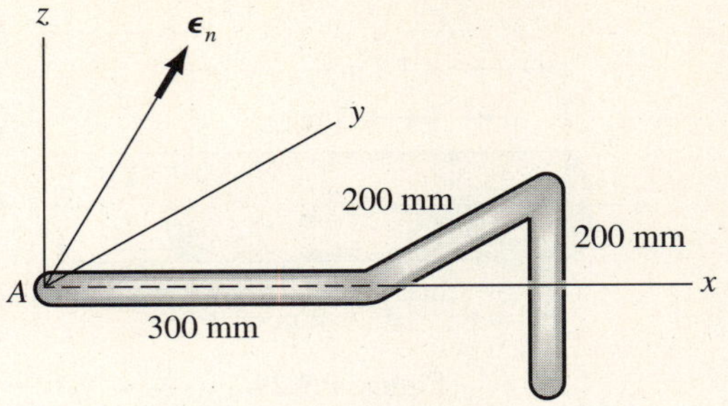

Figure P.9.38.

9.39. Evaluate the matrix of direction cosines for the primed axes relative to the unprimed axes.

$$a_{ij} = \begin{pmatrix} a_{x'x} & a_{x'y} & a_{x'z} \\ a_{y'x} & a_{y'y} & a_{y'z} \\ a_{z'x} & a_{z'y} & a_{z'z} \end{pmatrix}.$$

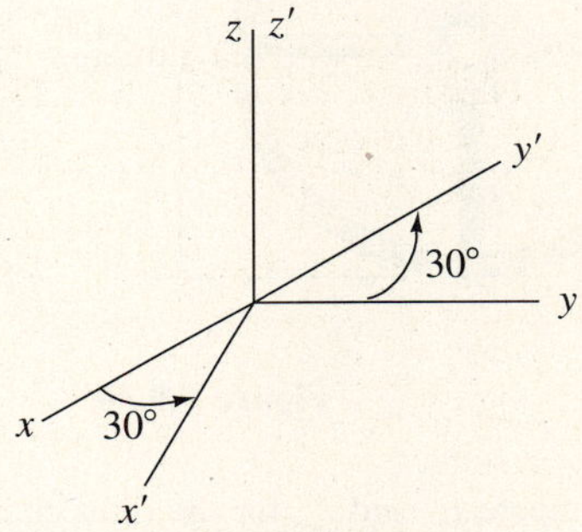

Figure P.9.39.

9.40. The block is uniform in density and weighs 10 N. Find $I_{y'z'}$.

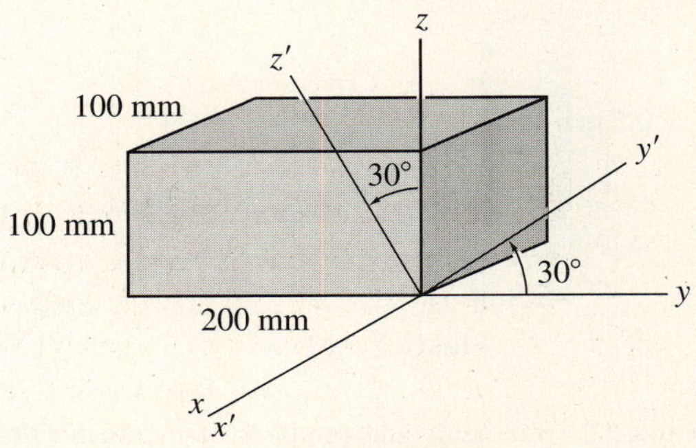

Figure P.9.40.

***9.41.** A thin rod of length 300 mm and weight 12 N is oriented relative to $x'y'z'$ such that

$$\boldsymbol{\epsilon}_n = 0.4\boldsymbol{i}' + 0.3\boldsymbol{j}' + 0.866\boldsymbol{k}'$$

What is $I_{x'y'}$?

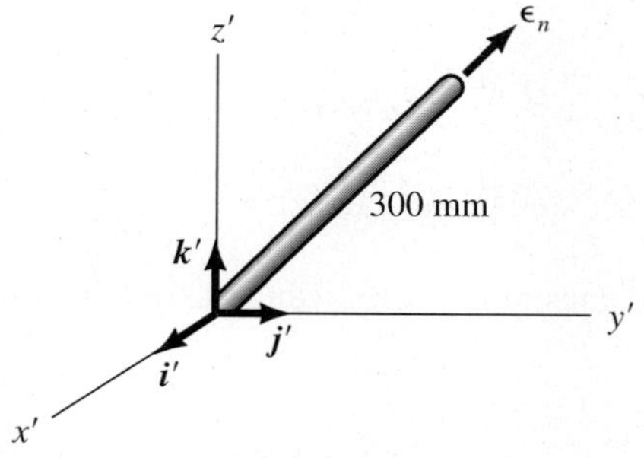

Figure P.9.41.

***9.42.** Show that the transformation equation for the inertia tensor components at a point when there is a rotation of axes (i.e., Eqs. 9.14 and 9.20) can be given as follows:

$$I_{kq} = \sum_j \sum_i a_{ki} a_{qj} I_{ij}$$

where k can be x', y', or z' and q can be x', y', or z', and where i and j go from x to y to z. The equation above is a compact definition of *second-order tensors*. Remember that in the inertia tensor you must have a minus sign in front of each product of inertia term (i.e., $-I_{xy}$, $-I_{yz}$, etc.). [*Hint:* Let $i = x$; then sum over j; then let $i = y$ and sum again over j; etc.]

***9.43.** In Problem 9.42, express the transformation equation to get $I_{y'z'}$ in terms of the inertia tensor components for reference xyz having the same origin as $x'y'z'$.

*9.7 The Inertia Ellipsoid and Principal Moments of Inertia

Equation 9.14 gives the moment of inertia of a body about an axis k in terms of the direction cosines of that axis measured from an orthogonal reference with an origin O on the axis, and in terms of six independent inertia quantities for this reference. We wish to explore the nature of the variation of I_{kk} at a point O in space as the direction of k is changed. (The k axis and the body are shown in Fig. 9.23, which we shall call the physical diagram.) To do this, we will employ a geometric representation of moment of inertia at a point that is developed in the following manner. Along the axis k, we lay off as a distance the quantity OA given by the relation

$$OA = \frac{d}{\sqrt{I_{kk}/M}} \tag{9.24}$$

where d is an arbitrary constant that has a dimension of length that will render OA dimensionless, as the reader can verify. The term $\sqrt{I_{kk}/M}$ is the *radius of gyration* and was presented earlier. To avoid confusion, this operation is shown in another diagram, called the inertia diagram (Fig. 9.24), where the new ξ, η, and ζ axes are *parallel* to the x, y, and z axes of the physical diagram. Considering all possible directions of k, we observe that some surface will be formed about the point O', and this surface is related to the shape of the body through Eq. 9.14. We can express the equation of this surface quite

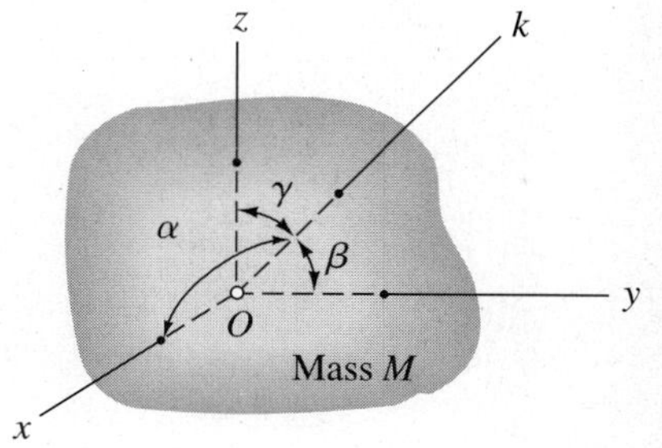

Figure 9.23. Physical diagram.

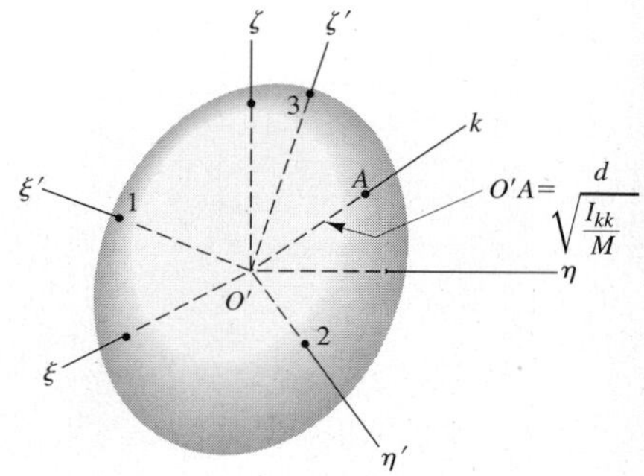

Figure 9.24. Inertia diagram.

readily. Suppose that we call ξ, η, and ζ the coordinates of point A. Since $O'A$ is parallel to the line k and thus has the direction cosines a_{kx}, a_{ky}, and a_{kz} that are associated with this line, we can say that

$$\begin{aligned} a_{kx} &= \frac{\xi}{O'A} = \frac{\xi}{d\sqrt{M/I_{kk}}} \\ a_{ky} &= \frac{\eta}{O'A} = \frac{\eta}{d\sqrt{M/I_{kk}}} \\ a_{kz} &= \frac{\zeta}{O'A} = \frac{\zeta}{d\sqrt{M/I_{kk}}} \end{aligned} \tag{9.25}$$

Now replace the direction cosines in Eq. 9.13, using the relations above:

$$\begin{aligned} I_{kk} = {} & \frac{\xi^2}{Md^2/I_{kk}} I_{xx} + \frac{\eta^2}{Md^2/I_{kk}} I_{yy} + \frac{\zeta^2}{Md^2/I_{kk}} I_{zz} \\ & + 2\frac{\xi\eta}{Md^2/I_{kk}}(-I_{xy}) + 2\frac{\xi\zeta}{Md^2/I_{kk}}(-I_{xz}) + 2\frac{\eta\zeta}{Md^2/I_{kk}}(-I_{yz}) \end{aligned} \tag{9.26}$$

We can see that I_{kk} cancels out of the preceding equation, leaving an equation involving the coordinates ξ, η, and ζ of the surface and the inertia terms of the body itself. Rearranging the terms, we then have

$$\begin{aligned} & \frac{\xi^2}{Md^2/I_{xx}} + \frac{\eta^2}{Md^2/I_{yy}} + \frac{\zeta^2}{Md^2/I_{zz}} \\ & \quad + \frac{2\xi\eta}{Md^2}(-I_{xy}) + \frac{2\xi\zeta}{Md^2}(-I_{xz}) + \frac{2\eta\zeta}{Md^2}(-I_{yz}) = 1 \end{aligned} \tag{9.27}$$

Considering analytic geometry, we know that the surface is that of an ellipsoid (see Fig. 9.24), and is thus called the *ellipsoid of inertia*. The distance squared from O' to any point A on the ellipsoid is inversely proportional to the moment of inertia (see Eq. 9.24) about an axis in the body at O having the same direction as $O'A$. We can conclude that the inertia tensor for any point of a body can be represented geometrically by such a second-order surface, and this surface may be thought of as analogous to the arrow used to represent a vector graphically. The size, shape, and inclination of the ellipsoid will vary for each point in space for a given body. (Since all second-order tensors may be represented by second-order surfaces, you will, if you study elasticity, also encounter the ellipsoids of stress and strain.)[8]

An ellipsoid has three orthogonal axes of symmetry, which have a common point at the center, O' (see Fig. 9.24). In the figure, these axes are shown as $O'1$, $O'2$, and $O'3$. We pointed out that the shape and inclination of the ellipsoid of inertia depend on the mass distribution of the body about the *origin* of the xyz reference, and they have nothing to do with the choice of *the orientation of the xyz* (and hence the $\xi\eta\zeta$) reference at the point. We can

[8]See I.H. Shames, *Mechanics of Deformable Solids*, Prentice-Hall, Inc., Englewood Cliffs, N.J., 1964, Chap. 2. Also, Krieger Publishing Co., N.Y., 1979.

therefore imagine that the xyz reference (and hence the $\xi\eta\zeta$ reference) can be chosen to have directions that coincide with the aforementioned symmetric axes, $O'1$, $O'2$, and $O'3$. If we call such references $x'y'z'$ and $\xi'\eta'\zeta'$, respectively, we know from analytic geometry that Eq. 9.27 becomes

$$\frac{(\xi')^2}{Md^2/I_{x'x'}} + \frac{(\eta')^2}{Md^2/I_{y'y'}} + \frac{(\zeta')^2}{Md^2/I_{z'z'}} = 1 \tag{9.28}$$

where ξ', η', and ζ' are coordinates of the ellipsoidal surface relative to the new reference, and $I_{x'x'}$, $I_{y'y'}$, and $I_{z'z'}$ are mass moments of inertia of the body about the new axes. We can now draw several important conclusions from this geometrical construction and the accompanying equations. One of the symmetrical axes of the ellipsoid above is the longest distance from the origin to the surface of the ellipsoid, and another axis is the smallest distance from the origin to the ellipsoidal surface. Examining the definition in Eq. 9.24, we must conclude that the minimum moment of inertia for the point O must correspond to the axis having the maximum length, and the maximum moment of inertia must correspond to the axis having the minimum length. The third axis has an intermediate value that makes the sum of the moment of inertia terms equal to the sum of the moment of inertia terms for all orthogonal axes at point O, in accordance with Eq. 9.2. In addition, Eq. 9.28 leads us to conclude that $I_{x'y'} = I_{y'z'} = I_{x'z'} = 0$. That is, the products of inertia of the mass about these axes must be zero. Clearly, these axes are the *principal axes* of inertia at the point O.

Since the preceding operations could be carried out at any point in space for the body, we can conclude that:

At each point there is a set of principal axes having the extreme values of moments of inertia for that point and having zero products of inertia.[9] *The orientation of these axes will vary continuously from point to point throughout space for the given body.*

All symmetric second-order tensor quantities have the properties discussed above for the inertia tensor. By transforming from the original reference to the principal reference, we change the inertia tensor representation from

$$\begin{pmatrix} I_{xx} & (-I_{xy}) & (-I_{xz}) \\ (-I_{yx}) & I_{yy} & (-I_{yz}) \\ (-I_{zx}) & (-I_{zy}) & I_{zz} \end{pmatrix} \text{ to } \begin{pmatrix} I_{x'x'} & 0 & 0 \\ 0 & I_{y'y'} & 0 \\ 0 & 0 & I_{z'z'} \end{pmatrix} \tag{9.29}$$

In mathematical parlance, we have "diagonalized" the tensor by the preceding operations.

[9]A general procedure for computing principal moments of inertia is set forth in Appendix II.

9.8 Closure

In this chapter, we first introduced the nine components comprising the inertia tensor. Next, we considered the case of the very thin flat plate in which the *xy* axes form the midplane of the plate. We found that the mass moments and products of inertia terms $(I_{xx})_M$, $(I_{yy})_M$, and $(I_{xy})_M$ for the plate are proportional, respectively, to $(I_{xx})_A$, $(I_{yy})_A$, and $(I_{xy})_A$, the second moments and product of area of the plate surface. As a result, we could set forth the concept of principal axes for the inertia tensor as an extension of the work in Chapter 8. Thus, we pointed out that for these axes the products of inertia will be zero. Furthermore, one principal axis corresponds to the maximum moment of inertia at the point while another of the principal axes corresponds to the minimum moment of inertia at the point. We pointed out that for bodies with two orthogonal planes of symmetry, the principal axes at any point on the line of intersection of the planes of symmetry must be along this line of intersection and normal to this line in the planes of symmetry.

Those readers who studied the starred sections from Section 9.5 onward will have found proofs of the extensions set forth earlier about principal axes from Chapter 8. Even more important is the disclosure that the inertia tensor components change their values when the axes are rotated at a point in exactly the same way as many other physical quantities having nine components. Such quantities are called second-order tensors. Because of the common transformation equation for such quantities, they have many important identical properties, such as principal axes. In your course in strength of materials you should learn that stress and strain are second-order tensors and hence have principal axes.[10] Additionally, you will find that a two-dimensional stress distribution called *plane stress* is related to the stress tensor exactly as the moments and products of area are related to the inertia tensor. The same situation exists with strain. Consequently, there are similar mathematical formulations for plane stress and the corresponding case for strain (plane strain). Thus, by taking the extra time to consider the mathematical considerations of Sections 9.5 through 9.7, you will find unity between Chapter 9 and some very important aspects of strength of materials to be studied later in your program.

In Chapter 10, we shall introduce another approach to studying equilibrium beyond what we have used thus far. This approach is valuable for certain important classes of statics problems and at the same time forms the groundwork for a number of advanced techniques that many students will study later in their programs.

[10]See I.H. Shames, *Introduction to Solid Mechanics*, 2nd ed., Prentice-Hall, Inc., Englewood Cliffs, New Jersey, 1989.

PROBLEMS

9.44. Find I_{zz} for the body of revolution having uniform density of 0.2 kg/mm^3. The radial distance out from the z axis to the surface is given as

$$r^2 = -4\,z\ \text{mm}^2$$

where z is in millimetres. [*Hint:* Make use of the formula for the moment of inertia about the axis of a disc, $\frac{1}{2}Mr^2$.]

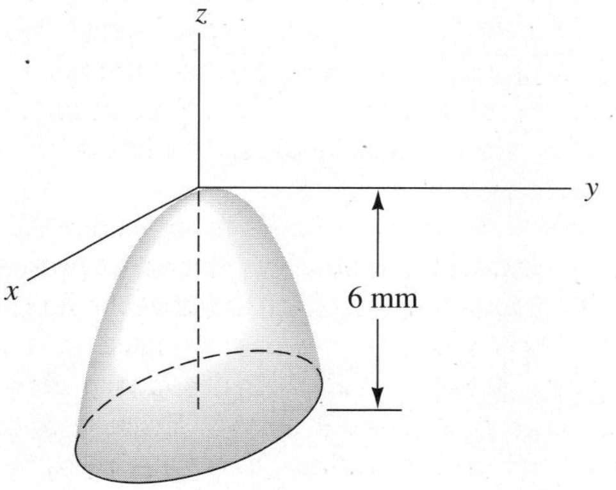

Figure P.9.44.

9.45. In Problem 9.44, determine I_{zz} without using the disc formula but using multiple integration instead.

9.46. What are the inertia tensor components for the thin plate about axes xyz? The plate weighs 2 N.

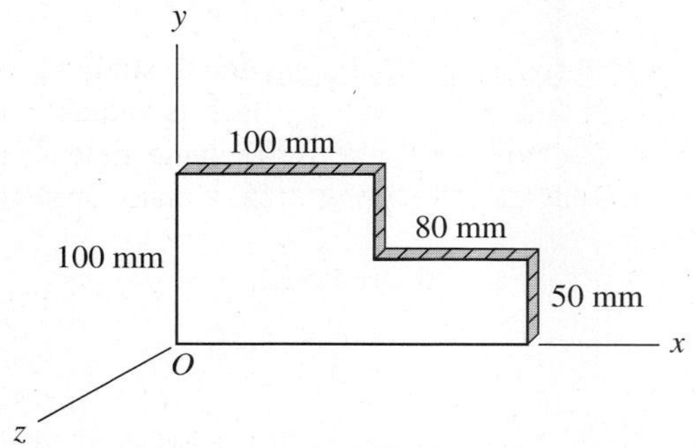

Figure P.9.46.

9.47. In Problem 9.46, what are the principal axes and the principal moments of inertia for the inertia tensor at O?

9.48. What are the principal mass moments of inertia at point O? Block A weighs 15 N. Rod B weighs 6 N and solid sphere C weighs 10 N. The density in each body is uniform. The diameter of the sphere is 50 mm.

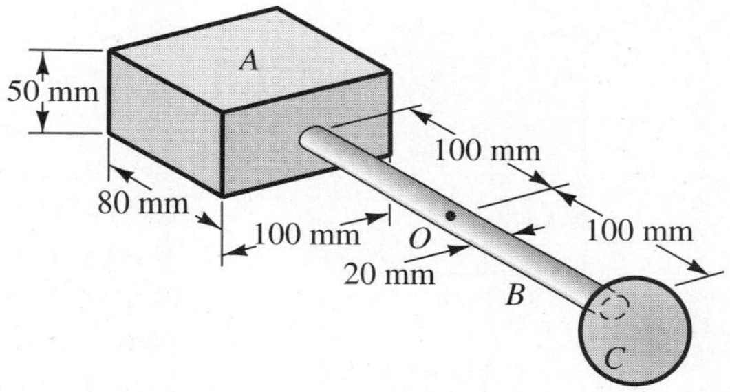

Figure P.9.48.

9.49. The block has a density of 15 kg/m.^{3} Find the moment of inertia about axis AB.

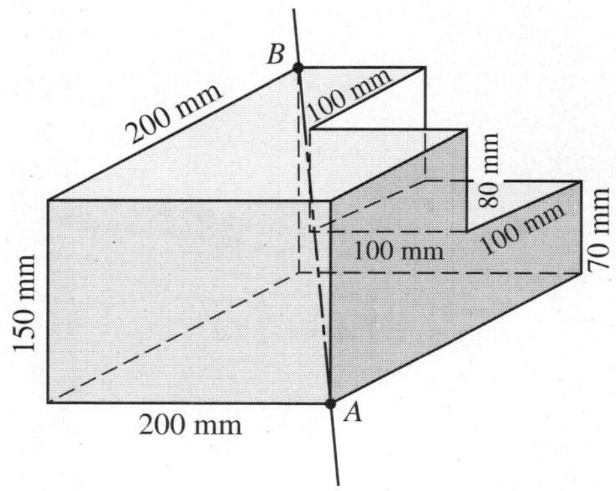

Figure P.9.49.

9.50. A crate and its contents weighs 10 kN. The center of mass of the crate and its contents is at

$$r_c = 0.4\boldsymbol{i} + 0.3\boldsymbol{j} + 0.6\boldsymbol{k} \text{ m}$$

If at A we know that

$$I_{yy} = 800 \text{ kg m}^2$$
$$I_{yz} = 500 \text{ kg m}^2$$

find I_{yy} and I_{yz} at B.

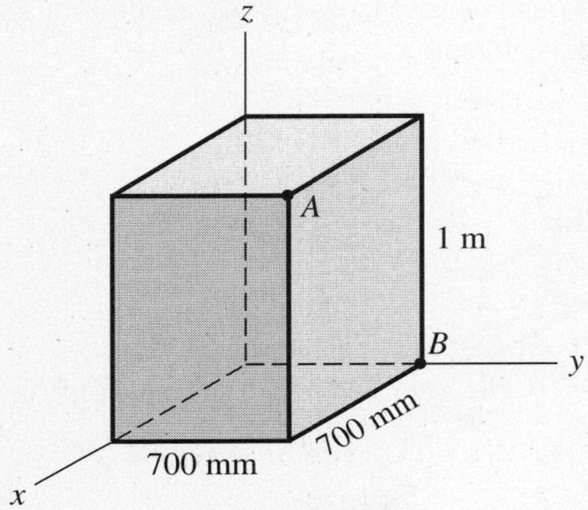

Figure P.9.50.

9.51. A semicylinder weighs 50 N. What are the principal moments of inertia at O? What is the product of inertia $I_{y'z'}$? What conclusion can you draw about the direction of principal axes at O?

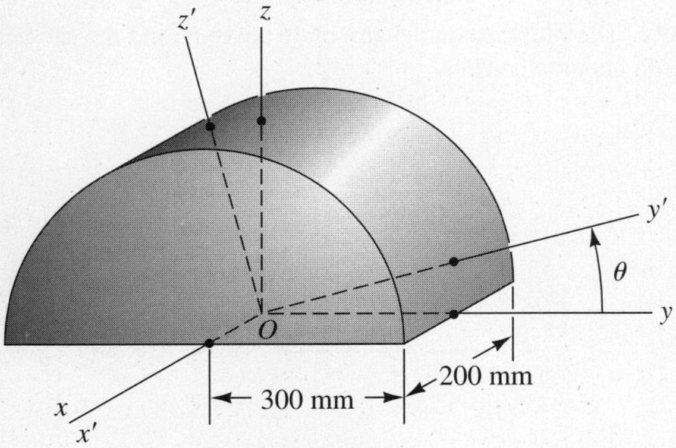

Figure P.9.51.

9.52. Find I_{yy} and I_{yz}. The diameter of A is 0.3 m. B is the center of the right face of the block. Take $\rho = \rho_0$ kg/m^3.

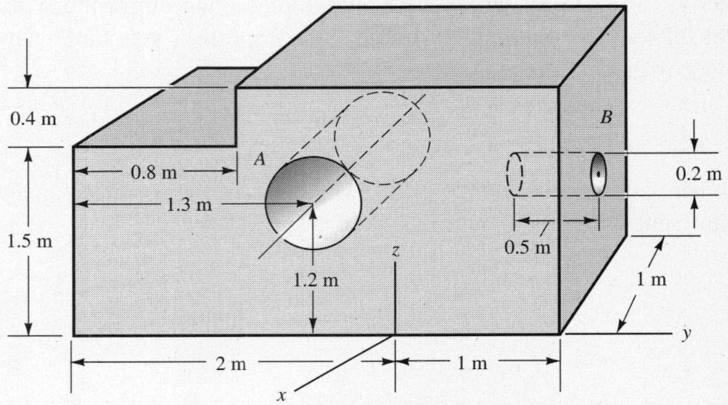

Figure P.9.52.

9.53. A body is composed of two adjoining blocks. Both blocks have a uniform mass density ρ equal to 10 kg/m^3.

(a) Find the mass moments of inertia, I_{xx} and I_{zz}

(b) Find the product of inertia I_{xy}.

(c) Is the product of inertia $I_{yz} = 0$ (yes or no)? Why?

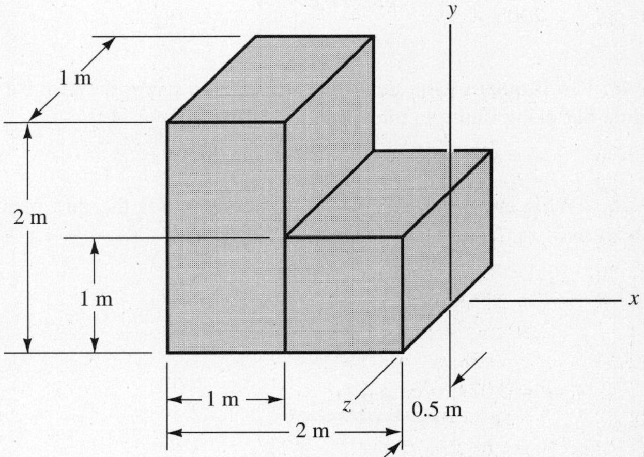

Figure P.9.53.

ANSWERS

11.2. $\dot{x} = 4.59$ m/s $\dot{y} = 1.605$ m/s.

11.4. $\boldsymbol{a} = 0.6765\boldsymbol{i} - 1.2075\boldsymbol{j} - 0.3177\boldsymbol{k}$ m/s^2.

11.6. $a(3) = 12$ m/s^2 $d = 60.94$ m.

11.8. 1.2032 m/s 0.4056 m/s^2.

11.10. $\vec{V} = 2\boldsymbol{i} - 1.296\boldsymbol{j}$ m/s.
$a = 0.6\boldsymbol{i} + 9.21\boldsymbol{j}$ m/s^2.
$a_n = 8.05$ m/s^2.

11.12.
$$\boldsymbol{V} = V_n\boldsymbol{i} + \left[-V_A + V_C \sin\alpha + \frac{V_{c^2}\left(\cos^2\alpha\right)(t)}{\left\{h^2 + \left[V_c(\cos\alpha)(t)\right]^2\right\}^{\frac{1}{2}}} \right]\boldsymbol{j}.$$

11.14. $x^4 + 4.048(y - x) = 0$ $d = 1.594$ m.

11.16. $2.76 < d < 7.66$ m.

11.18. $\delta = 9.148$ m.

11.20. $d = 26.42$ m.

11.22. 1.23 m.

11.24. $\bar{x} = 92.5$ km $\bar{y} = 8.56$ km.

11.26. $d = 280$ m $\Delta\tau = 2.04$ s.

11.28.
$$400 \cos\alpha\left[\sin\beta - \frac{\cos\beta}{4}\right] = -21.72.$$
$$12\,000 = [400 \cos\alpha \cos\beta + 1.178]\left(\frac{400 \sin\alpha}{4.905}\right).$$

11.30. $\boldsymbol{a}_1 = 12.5\boldsymbol{j}$ m/s^2 $\boldsymbol{a}_2 = 21.65\boldsymbol{i} - 12.5\boldsymbol{j}$ m/s^2 $\boldsymbol{a}_3 = 0$.

11.32. $a_n = 0.481$ m/s^2 $V = 20.93$ m/s.

11.34. $\omega = 17.27$ r/min.

11.36. $\boldsymbol{\epsilon}_n = 0.995\boldsymbol{i} - 0.0995\boldsymbol{j}$ $R = 101.5$ m.

11.38. $|\boldsymbol{a}| = 3.176$ m/s^2.

11.40. $V = 247.6$ m/s.

11.42. 2.30 m/s^2 14.84 m/s^2.

11.44. $x = -3.58$ m $y = 0.89$ m.

11.46. $\boldsymbol{a} = 0.02\boldsymbol{\epsilon}_t = 0.01504\boldsymbol{\epsilon}_n$ m/s^2.

11.48. $\dot{\boldsymbol{a}} = 5\boldsymbol{\epsilon}_l + 6\boldsymbol{\epsilon}_{lt}$ m/s^3.

11.50. $\dot{\theta} = 1.063$ rad/s.
$\ddot{\theta} = 0.275$ rad/s^2.

11.52. $\Omega = 53.3$ rad/s $\dot{\Omega} = 72$ rad/s2·

11.54. $\boldsymbol{V} = 21.5\boldsymbol{\epsilon}_{-} + 358\boldsymbol{\epsilon}_\theta$ mm/s.
$\boldsymbol{a} = -517\boldsymbol{\epsilon}_{-} + 127.4\boldsymbol{\epsilon}_\theta$ mm/s^2.

11.56. $\boldsymbol{V} = -3\boldsymbol{\epsilon}_{-} + 3\boldsymbol{\epsilon}_\theta$ m/s.
$\boldsymbol{a} = -16.60\boldsymbol{\epsilon}_{\bar{r}} - 28.8\boldsymbol{\epsilon}_\theta$ m/s^2.

11.58. $\boldsymbol{V}(2) = 0.7524\boldsymbol{\epsilon}_{\bar{r}} + 1.139\boldsymbol{\epsilon}_\theta + 24\boldsymbol{\epsilon}_z$ m/s.
$\boldsymbol{a}\,(2) = -1.064\boldsymbol{\epsilon}_{\bar{r}} + 2.364\boldsymbol{\epsilon}_\theta + 12\boldsymbol{\epsilon}_z$ m/s^2.

11.60. $\boldsymbol{r} = 11.66\boldsymbol{\epsilon}_{\bar{r}} + (t^3 + 10)\,\boldsymbol{\epsilon}_z$ m.
$\boldsymbol{V} = 11.66\boldsymbol{\epsilon}_{\bar{r}} + 3t^2\boldsymbol{\epsilon}_z$ m/s.
$\boldsymbol{a} = 6t\,\boldsymbol{\epsilon}_z$ m/s^2.

11.62. $\boldsymbol{V} = 1.061\boldsymbol{\epsilon}_{\bar{r}} + 0.04242\boldsymbol{\epsilon}_\theta + 1.061\boldsymbol{\epsilon}_z$ m/s.
$\boldsymbol{a} = -5.268\boldsymbol{\epsilon}_{\bar{r}} + 0.4329\boldsymbol{\epsilon}_\theta + 5.345\boldsymbol{\epsilon}_z$ m/s.

11.64. $\boldsymbol{a} = 1200\boldsymbol{\epsilon}_{\bar{r}} + 900\boldsymbol{\epsilon}_\theta + 5.73\boldsymbol{\epsilon}_z$ mm/s^2.

11.66. $\boldsymbol{V} = 1.500\boldsymbol{\epsilon}_{-} + 0.600\boldsymbol{\epsilon}_\theta + 2.60\boldsymbol{\epsilon}_z$ m/s.
$\boldsymbol{a} = -1.200\boldsymbol{\epsilon}_{\bar{r}} + 6\boldsymbol{\epsilon}_\theta - 9.81\boldsymbol{\epsilon}_z$ m/s^2.

11.68. $\boldsymbol{V} = -3.183 \times 10^{-3}\boldsymbol{\epsilon}_{\bar{r}} + 3.680\boldsymbol{\epsilon}_\theta$ m/s.
$\boldsymbol{a} = -23.18\boldsymbol{\epsilon}_{\bar{r}} + 0.330\boldsymbol{\epsilon}_\theta$ m/s^2.

11.70. $\bar{r} = 0.1146 + 9.192\,z^2$ m.
$\dot{Z} = 2$ mm/s.
$\dot{Y} = 7.346$ mm/s.

11.72. Particle 1 $2.397\boldsymbol{i} - 0.397\boldsymbol{j}$ m/s.
Particle 2 $0.397\boldsymbol{i} + 2.603\boldsymbol{j}$ m/s, etc.

11.74. $\boldsymbol{F} = -10.68 \times 10^3 + 7.36\boldsymbol{j}$ N.

11.76. $\omega = 25.3$ rad/s $\Omega = 521.4$ rad/s.

11.78. $\boldsymbol{V}_{xyz} = -3\boldsymbol{j} - 3.6\boldsymbol{k}$ m/s.
$\boldsymbol{a}_{xyz} = 1.5\boldsymbol{j} - 10.41\boldsymbol{k}$ m/s^2.

11.80. Hit at 3.946 m to right of centre.

11.82. Hit at 62.7 m from bow.

11.84. $\boldsymbol{F} = -65.2\boldsymbol{k}$ mN.

11.86. $\alpha = 18.55°$.

11.88. 6.75 s $\beta = 80.60°$.

11.90. $\boldsymbol{F} = -5.73\boldsymbol{i} + 46.1\boldsymbol{j}$ N.

11.92. $\boldsymbol{V}_{xyz} = 86\boldsymbol{i} + 19\boldsymbol{j} - 99.22\boldsymbol{k}$ m/s.
$\boldsymbol{a}_{xyz} = -1094\boldsymbol{i} + 4.4\boldsymbol{j} - 8.9\boldsymbol{k}$ m/s^2.

11.94. $\boldsymbol{F} = 163.1\boldsymbol{i} + 244.6\boldsymbol{j} + 10.09 \times 10^3\boldsymbol{k}$ N.

11.96. $\boldsymbol{V} = 136\boldsymbol{i} + 80\boldsymbol{j} + 5\boldsymbol{k}$ m/s.
$\boldsymbol{r} = 1523\boldsymbol{i} + 1004\boldsymbol{j} + 106\boldsymbol{k}$ m.

11.98. Hit at midpoint of freighter.

11.100. $y = 34.19$ km.

11.102. $\boldsymbol{V} = 1.742\boldsymbol{i} + 0.836\boldsymbol{j}$ m/s.
$\boldsymbol{a} = 6.338\boldsymbol{i} + 10.322\boldsymbol{j}$ m/s.

11.104. 39.55 m/s.

11.106. $A = 635.8$ m^2.

11.108. 2.55×10^{-3} g.

11.100. $\boldsymbol{V} = 0.3\boldsymbol{\epsilon}_{\bar{r}} + 6\boldsymbol{\epsilon}_\theta + 0.6\boldsymbol{\epsilon}_z$ m/s.
$\boldsymbol{a} = -6\boldsymbol{\epsilon}_{\bar{r}} + 1.2\boldsymbol{\epsilon}_\theta$ m/s^2 $\boldsymbol{a} \bullet \boldsymbol{\epsilon} = -4.56$ m/s^2.

11.112. $a = 2.19$ m/s^2.

11.114. $\bar{x} = 17.49$ m.
$\bar{y} = 15.29$ m.
$\dot{x} = 10$ m/s
$\dot{y} = 0.1623$ m/s.

11.116. $d = 982.2$ m $\delta = 0.46$ m.

11.118. $\alpha = 33.38°$.

11.120. $\boldsymbol{F}_{max} = 19.62\boldsymbol{i} - 29.43\boldsymbol{j} + 304\boldsymbol{k}$ mN.

11.122. $\boldsymbol{V} = 149\boldsymbol{i} + 200\boldsymbol{j} + 60\boldsymbol{k}$ m/s.
Tensile force on rod is 45 N.

11.124. $\boldsymbol{\epsilon}_0 = 0.0879\boldsymbol{i} + 0.0439\boldsymbol{j} + 0.1758\boldsymbol{k}$.

12.2. $\boldsymbol{r} = 10\boldsymbol{i} + 8\boldsymbol{j} + 7\boldsymbol{k}$ m.

12.4. 29.36 m 2.446 s.

12.6. 6.26 s.

12.10. $d = 10.21$ mm.

12.12. 148.6 m.

12.14. 10.37 m.

12.16. 62.90 m/s.

12.18. $V_A = 3.92$ m/s $V_B = 1.91$ m/s.

12.20. 4 m/s.

12.22. 435 m/s 16.49 km.

12.24. 29.73 m/s 19.41 m.

12.26. 372.7 N 4.12 m/s^2.

12.28. 4.81 m distance = 4.81 m.

12.30. 77.53 m/s^2.

12.32. $d = 13.86$ mm $l = 20$ mm.

12.34. 22.4 m/s $d = 108.1$ m.

12.36. $A = 45.76$ m^2.

12.38. $d = 745.5$ mm $V = -0.845$ m/s.

12.40. $t = 0.522$ s.

12.42. $t = 8.32$ h.

12.44. $V = 0.488$ m/s.

12.46. $t = 2.10$ s $h = 27.93$ m.

12.48. $h = 23.54$ km.

12.50. $F = 10$ kN.

12.54. $\dot{\theta} = 31.53$ r/min.

12.56. $d = g/\omega^2$.

12.58. 350 N.

12.60. $\omega = 9.77$ rad/s.

12.62. $\boldsymbol{\epsilon}_n = -0.594\boldsymbol{i} + 0.396\boldsymbol{j} + 0.701\boldsymbol{k}$.

12.66. $V_E = 9.98$ m/s $V_C = 7.05$ m/s.

12.68. $\ddot{\theta} = 1.131$ rad/s^2 $T = 216$ mN.

12.70. $\Delta t = 2.11$ h.

12.74. $V_{max} = 4.193$ km/s $\tau = 5.36$ h.

12.76. $V = 86.92$ m/s.

12.78. $\tau = 548$ h.

12.80. 216.6 m/s.

12.82. $h = 105$ km.

12.86. $V_r = 1.773$ km/s $\epsilon = 0.336$.

12.90. $\omega = 4.43$ rad/s.

12.92. $a_t = -8.45$ m/s^2.

12.94. $$(F_n)_{\text{TOTAL}} \frac{110}{g}\left[\frac{(1+a^4\sinh^2 ax)^{3/2}}{a^3\cosh ax}\right]^{-1} \quad (9)$$
$$+110\cos[\tan^{-1}(a^2\sinh ax)].$$

12.96. From S = 15.66 mm to 138.0 mm.

12.100. $(\boldsymbol{V}_1 - \boldsymbol{V}_C)_{at\ t=2} = -146\boldsymbol{i} + 4.64 \times 10^3\boldsymbol{j} + 25.3\boldsymbol{k}$ m/s.
$(\boldsymbol{V}_2 - \boldsymbol{V}_C)_{at\ t=2} = -174\boldsymbol{i} + 4.64 \times 10^3\boldsymbol{j} + 96\boldsymbol{k}$ m/s.

12.102. $F = -30.822$ kN.

12.104. 2.01 m.

12.106. 0.1171 m/s.

12.108. $\Delta t = 47.9$ s.

12.112. 10.73 Mm 1.131 h 10.85 km/s.

12.114. $(\Delta V)_1 = 314.2$ m/s $(\Delta V)_2 = 302.2$ m/s.

12.116. $F = 199.5$ N.

12.118. $\Delta V = 113.3$ m/s.

12.120. 1.036 kN.

12.122. $-Mg + K(r_0 - r) = M(\ddot{r} - r\dot{\theta}^2)$
$g\theta = r\ddot{\theta} + 2\dot{r}\dot{\theta}$.

12.124. $\boldsymbol{a}_c = 0.0853\,(30t + 75)\boldsymbol{i} - 50\boldsymbol{k}$ m/s^2.
$(x_c) - (x_c)_0 = 706.7$ m.
$(y_c) - (y_c)_0 = 4.01$ m.
$(z_c) - (z_c)_0 = -204.0$ m.

12.126. $V = 8.3$ mm/s.

12.128. $V = 0.157$ m/s.

12.130. 0.90 m/s.

12.132. $V = 4.706$ m/s.

12.134. 11.58 m.

12.136. $\dot{s}_B = 0.309$ m/s.

12.138. 0.5934 g.

13.2. $V = 10.39$ m/s.

13.4. $V = 3.98$ m/s.

13.6. $V = 1.79$ m/s.

13.8. $F = 15.46$ kN.

13.10. -45.2 m/s^2.

13.12. $V = 2.54$ m/s.

13.14. $KE = 1.78$ kJ.

13.16. $F = 8.50$ kN.

13.18. 0.211 m/s.

13.20. $\omega = 8.74$ rad/s.

13.22. $d = 1.835$ m.

13.24. $d = 0.312$ m.

13.26. $V = 18.75$ m/s.

13.28. $W = 121.0$ kN.

13.30. 94 mm.

13.32. 16.59 kN.

13.34. Power = $14 \times 10^3\, a_0 V + \kappa V^3$.
for $a_0 = 0$ Power = κV^3.

13.36. $\Delta t = 7.5$ s.

13.38. 8.40 kN.

13.40. $V = 1.02$ m/s.

13.42. $\delta = 106.1$ mm.

13.46. $d = -5z\cos x + xy + 2y^2z + \text{const.}$

13.48. $V = 1.938$ m/s.

13.50. $V = 0.389$ m/s.

13.52. $\dot{V} = 33.33$ m/s².
13.54. $h = 8.26$ m.
13.56. $V = 10.39$ m/s.
13.58. $\delta = 477$ mm.
13.60. $V = 16.42$ m/s.
13.62. $V = 2.853$ m/s.
13.64. $\delta = 2.398$ mm.
13.66. $d = 149.12$ m $F_{AB} = 58.955$ kN $F_{CB} = 29.477$ kN.
13.68. $V = \left[\frac{q}{L}(L^2 - a^2)\right]^{\frac{1}{2}}$.
13.70. $(\mathcal{W}_K)_{\text{internal}} = -\frac{1}{2}\frac{\mathcal{W}_1}{g}V_0^{\,2} + 5.7(\mathcal{W}_1 + \mathcal{W}_2)$.
13.72. $KE = 61.02$ J.
13.74. $V = 2.05$ m/s.
13.78. $KE = 0.1549$ J Error = 1.937%.
13.80. $V = 6.138$ m/s.
13.82. $d = 1.835$ m.
13.84. $F = 9$ N $f = 3$ N.
13.86. $(V_C)_f = 6.26$ m/s $(V_C)_{f=0} = 7.67$ m/s.
13.88. $F_C = 115.35$ N.
13.90. $\omega = 6.83$ rad/s.
13.92. $d = 3.823$ m $\begin{cases} f = 0 \text{ horizontal surface} \\ f = 83.35 \text{ N on incline.} \end{cases}$
13.94. $V = 1.414$ m/s $f_{\text{total}} = 915$ N.
13.96. $\omega = 3.17$ rad/s.
13.98. $V_A = 3.655$ m/s $a = 1.308$ m/s².
13.100. $\omega = 0.499$ rad/s.
13.102. 0.83 kW Power Input = 3.93 kW.
13.104. 409 W.
13.106. $\theta = 43.7°$.
13.110. 3.98 MW.
13.112. 11.7 MW.
13.114. $V = 1.716$ m/s 58.5 N.
13.116. $\omega = 11.90$ rad/s $f = 130.0$ N.
13.118. 48.41 m/s.
13.120. $\mathcal{W} = 7.99$ kg $F_n = 161.9$ N.
13.122. $V = 9.17$ m/s $f = 85.7$ N.
13.124. $V =$ zero.
13.126. $V = 0.4937$ m/s $f = 186.1$ N.

14.2. $\boldsymbol{V} = 333\boldsymbol{i} + 400\boldsymbol{j} + 4 \times 10^3\boldsymbol{k}$ m/s.
14.4. $V = 53.96$ m/s to left.
14.6. $t = 1.631$ s $\Delta t = 0.7136$ s.
14.8. $\boldsymbol{F}_{AV} = 12.8\boldsymbol{i} - 14.40\boldsymbol{j} + 48.0\boldsymbol{k}$ N.
14.10. $\boldsymbol{I} = -4.17\boldsymbol{i} - 1.68\boldsymbol{j} + 0.42\boldsymbol{k}$ kg m/s.
14.12. $t = 0.646$ s.
14.14. $t = 7.55$ s $t = 13.21$ s $F = 22.71$ kN.
14.16. 1.452 kW.
14.18. $T_2 = 105.8$ N.
14.20. $T_x = 341.5$ Ns.
14.22. $\Delta V = 2.78$ m/s.
14.24. $V = 0.402$ m/s.
14.28. $V_A = 5.183$ m/s $V_B = 10.37$ m/s.
14.30. 4.905 m/s.
14.32. $F_\omega = 4.88$ kN.
14.34. $(V_B)_f = 4.10$ m/s ΔKE = 18.15 J.
14.36. $(V_A)_f = -1.130$ m/s.
14.38. $\delta = 144$ mm.
14.42. $\theta = 12.15°$.
14.44. $\delta = 0.750$ m.
14.46. 4.648 m/s $\delta = 0.36$ m.
14.48. $\delta = 0.7990$ m.
14.50. 4.8 m/s.
14.52. $(\boldsymbol{V}_A)_f = -0.566\boldsymbol{i} + 4.245\boldsymbol{j}$.
$(\boldsymbol{V}_B)_f = 2.404\boldsymbol{j}$.
14.56. $\Delta = 51.42$ mm $\delta = 59$ mm.
14.58. 50.5°.
14.60. Drag = 3.53 mnV^2.
14.62. Drag = 0.140 mnV^2.
14.64. $F_x = \frac{4}{3}\frac{\pi SR^2}{c}$ 3.53 N.
14.66. 10.36 km/s.
14.68. $d = 0.205$ m.
14.70. $\omega_2 = 8.312$ rad/s.
14.72. $\omega_2 = 5.191$ rad/s.
14.74. 1.813 km/s.
14.76. 1.748 km/s.
14.78. $h = 183$ km.
14.80. $(\Delta V)_1 = 315.3$ m/s.
$(\Delta V)_2 = 301.4$ m/s.
14.82. 6346 km 9390 km.
14.84. $r_{\max} = 53.395$ Mm.
14.86. $r_A = 14.29$ Mm $\Delta V = 3.254$ km/s.
14.88. $V_0 = 205.8$ m/s.
$V_r = 229.4$ m/s.
14.90. $\boldsymbol{P} = 9.914\boldsymbol{i} - 0.211\boldsymbol{j} + 2.881\boldsymbol{k}$ kg m/s.
$\boldsymbol{H}_0 = 2.884\boldsymbol{i} - 1.922\boldsymbol{j} - 37.997\boldsymbol{k}$ kg m²/s.
$\boldsymbol{H}_a = -8.121\boldsymbol{i} - 24.750\boldsymbol{j} - 1.801\boldsymbol{k}$ kg m²/s.
14.92. $\dot{\omega} = -20.0$ rad/s².
14.94. $\dot{\omega} = \frac{T - 8s_2Vm\omega}{4m(s_1^{\,2} + s_2^{\,2})}$.
14.98. $\dot{\omega} = 593$ Hz = 3725 rad/s.
14.100. 21.0 m/s².
14.102. $H = 42.52$ Gg m²/s.
14.104. $\dot{\omega} = 37.5$ mrad/s² $\boldsymbol{V}_C = 0.1749\boldsymbol{i} + 0.262\boldsymbol{j}$ m/s².
14.106. $\dot{\omega} = 2.04$ rad/s².
14.108. $V = 1.281$ m/s.

14.110. $(V_2)_t = 5.769$ km/s.
$(V_2)_r = 9.366$ km/s.
$\Delta V = 13.793 - 7.715 = 6.078$ km/s.
14.112. $x = 4.464$ m.
14.114. $V_f = 13.43$ m/s.
14.116. $x = -20.566$ km.
14.118. $V = 0.821$ m/s.
14.120. $\delta = 0.671$ m $\quad V_B = 53.7$ m/s.
14.122. $\theta_A = 5.47°$ $\quad \theta_B = 29.9°$.
14.124. $\boldsymbol{V}_f = 54.05\boldsymbol{i} + 56.95\boldsymbol{j} - 8.74\boldsymbol{k}$ m/s.
14.126. $(V_A)_f = 28.57$ m/s $\quad (V_B)_f = 33.07$ m/s.
14.128. $H_0 = 9.87$ kg m²/s $\quad P_0 = 0$.
14.130. $H_0 = 6871$ kg m/s².
14.132. $h = 3$ Mm $\quad r_{max} = 9.371$ Mm.
14.134. $F = 1.8$ kN.
14.136. $V = 4.953$ m/s $\quad \Delta KE = 3.40$ J.
14.138. $E = 58.96$ Mev $\quad l_B = -0.325$ $\quad m_B = 0.18746$
$n_B = -0.97417$.

15.4. $-10.66\boldsymbol{i} + 7.31\boldsymbol{j} - 35.18\boldsymbol{k}$ m/s.
15.6. $-521\boldsymbol{i} + 108.7\boldsymbol{j} + 65.2\boldsymbol{k}$ m/s.
15.8. $\dot{\boldsymbol{\rho}} = 10\boldsymbol{i} + 5\boldsymbol{j}$ mm/s.
$\ddot{\boldsymbol{\rho}} = -1.250\boldsymbol{k}$ m/s².
15.10. $\boldsymbol{\omega} = 0.940\boldsymbol{i} + 0.342\boldsymbol{j} + 1.8\boldsymbol{k}$ rad/s.
$\dot{\boldsymbol{\omega}} = -0.616\boldsymbol{i} + 1.692\boldsymbol{j}$ rad/s².
15.12. $\boldsymbol{\omega} = -0.4\boldsymbol{i} + 169.0\boldsymbol{j} - 169.0\boldsymbol{k}$ rad/s.
$\dot{\boldsymbol{\omega}} = -0.2\boldsymbol{i} - 68.32\boldsymbol{j} - 68.32\boldsymbol{k}$ rad/s².
15.14. $\dot{\boldsymbol{\omega}} = 3.14\boldsymbol{i} + 1.356\boldsymbol{j}$ krad/s².
$\ddot{\boldsymbol{\omega}} = 407\boldsymbol{j} + 14.2\boldsymbol{k}$ krad/s³.
15.16. $\boldsymbol{\omega} = -121.6\boldsymbol{j}$ rad/s.
$\dot{\boldsymbol{\omega}} = 3.823\boldsymbol{i}$ krad/s².
15.18. $\dot{\boldsymbol{\omega}} = 0.12\boldsymbol{j}$ rad/s².
$\ddot{\boldsymbol{\omega}} = -24\boldsymbol{i}$ mrad/s³.
15.20. $\boldsymbol{V}_A = 0.5\boldsymbol{i} + 0.1212\boldsymbol{j}$ m/s.
$\omega = 0.4788$ rad/s.
15.22. $\dot{\boldsymbol{\omega}} = 0.18\boldsymbol{i} - 0.1\boldsymbol{j} + 0.2\boldsymbol{k}$ rad/s².
15.24. $\dot{\boldsymbol{V}} = -20.5\boldsymbol{i} - 2.14\boldsymbol{j} + 17.10\boldsymbol{k}$ m/s².
15.26. $\boldsymbol{V} = -243\boldsymbol{i} + 207\boldsymbol{j} + 25\boldsymbol{k}$ m/s.
15.28. $\omega = 56.7$ rad/s $\quad \boldsymbol{V}_B = 29\boldsymbol{j} - 12.03\boldsymbol{k}$ m/s.
15.30. $V_A = -0.6$ m/s.
15.32. $\omega = 4$ rad/s $\quad V_D = -6\boldsymbol{i} - 19\boldsymbol{j}$ m/s.
15.34. $\omega_C = -10$ rad/s.
15.36. $\boldsymbol{V}_C = -0.548\boldsymbol{i} - 0.373\boldsymbol{j}$ m/s.
$\boldsymbol{a}_C = -43.4\boldsymbol{i} - 63.7\boldsymbol{j}$ m/s².
15.38. $\omega_{AB} = -9.33$ rad/s $\quad \boldsymbol{V}_B = -12.38\boldsymbol{j} - 7.149\boldsymbol{k}$ m/s.
15.40. $\omega_{BA} = -54.80$ rad/s $\quad V_A = 23.97$ m/s.
15.44. $V_y = 14.66$ m/s $\quad \omega = -5$ rad/s.
15.46. $\boldsymbol{V}_B = 11.95\boldsymbol{i} - 4.949\boldsymbol{j}$ m/s.
$\boldsymbol{a}_B = -57.77\boldsymbol{i} - 57.77\boldsymbol{j}$ m/s².
$\dot{\theta} = 2.625$ rad/s.

15.48. $\omega_{AB} = -0.609$ rad/s $\quad \dot{\omega}_{AB} = 14.71$ rad/s².
15.50. $\boldsymbol{V}_B = 5.03\boldsymbol{i} + 2\boldsymbol{j} - 3\boldsymbol{k}$ m/s.
15.52. $\boldsymbol{V}_D = -0.15\boldsymbol{i} - 4.410\boldsymbol{j} - 0.813\boldsymbol{k}$ m/s.
$\boldsymbol{a}_D = -0.375\boldsymbol{j} + 3\boldsymbol{k}$ m/s².
15.54. $\boldsymbol{\omega}_{AB} = -17.6\boldsymbol{k}$ mrad/s.
$\dot{\boldsymbol{\omega}} = -0.434\boldsymbol{k}$ mrad/s².
15.56. $\omega_A = 3.643$ rad/s.
$\dot{\omega}_A = -6.45$ rad/s².
15.58. $\omega = 1.273$ rad/s $\quad V_C = 2.163$ m/s.
15.60. $V = 1.156$ m/s.
15.62. $V_A = 5.478$ m/s $\quad a_A = -929.6$ m/s².
15.64. $\omega_H = 3.93$ rad/s $\quad \dot{\omega}_H = 20.7$ rad/s².
15.66. $\omega_C = 5.00$ rad/s.
15.68. $\omega_{AB} = -5$ rad/s $\quad \dot{\omega}_{AB} = -360.8$ rad/s².
15.70. $\boldsymbol{V}_{XYZ} = 8.86\boldsymbol{i} + 5\boldsymbol{j}$ m/s.
15.72. $\boldsymbol{V}_{XYZ} = -5.196\boldsymbol{i} - 3\boldsymbol{j} - 15\boldsymbol{k}$ m/s.
15.74. $\boldsymbol{V}_{XYZ} = 1.30\boldsymbol{i} + 1.8\boldsymbol{j} + 0.130\boldsymbol{k}$ m/s.
15.76. $\omega_1 = 15.40$ rad/s $\quad \omega_2 = -31.37$ rad/s.
15.78. $\omega_1 = 1.985$ rad/s.
15.80. $\boldsymbol{V}_{XYZ} = -2.04\boldsymbol{i} + 4.92\boldsymbol{j} - 2.66\boldsymbol{k}$ m/s.
15.82. $\boldsymbol{V}_{\perp barrel} = -0.908\boldsymbol{i} - 0.5253\boldsymbol{j} - 0.6\boldsymbol{k}$ m/s.
15.84. $\boldsymbol{V}_{XYZ} = 4.00\boldsymbol{i}$ m/s.
15.86. $V_{XYZ} = 11.37$ m/s rel. to ground.
$V_{xyz} = 2.94$ m/s rel. to rod.
15.88. $V_{xyz} = 3.756$ m/s $\quad V_{XYZ} = 7.165$ m/s.
15.90. $\omega_D = 10.8$ rad/s.
15.92. $\boldsymbol{V}_{XYZ} = 1.5\boldsymbol{i} - 1.5\boldsymbol{j} - 6\boldsymbol{k}$ m/s.
$\boldsymbol{a}_{XYZ} = -30\boldsymbol{i} - 90\boldsymbol{j} - 10.5\boldsymbol{k}$ m/s².
15.94. $\boldsymbol{a}_{XYZ} = -780\boldsymbol{i} - 52\boldsymbol{j} - 520\boldsymbol{k}$ m/s².
15.96. $\boldsymbol{a}_{XYZ} = -161.9\boldsymbol{k}$ m/s².
15.98. $\boldsymbol{a}_{XYZ} = -5.08\boldsymbol{i} + 2.314\boldsymbol{j} + 2.70\boldsymbol{k}$ m/s².
15.100. $\boldsymbol{a}_{XYZ} = -1.950\boldsymbol{i} - 18.69\boldsymbol{j} - 1.207\boldsymbol{k}$ m/s².
$|\boldsymbol{a}_{XYZ}| = 1.919$ g's.
15.102. $\boldsymbol{a}_{XYZ} = -25.6\boldsymbol{j} - 57.6\boldsymbol{k}$ m/s².
15.106. $\boldsymbol{a}_{\perp gun\ barrel} = -0.403\boldsymbol{i} - 0.233\boldsymbol{j} - 1.647\boldsymbol{k}$.
15.108. $\boldsymbol{V}_{XYZ} = -1.113\boldsymbol{i} + 0.1202\boldsymbol{j}$ m/s.
$\boldsymbol{a}_{XYZ} = 0.0969\boldsymbol{i} - 0.1859\boldsymbol{j}$ m/s².
15.110. $\boldsymbol{a}_{XYZ} = -0.361\boldsymbol{i} - 0.771\boldsymbol{j}$ m/s².
$|\boldsymbol{a}_{XYZ}| = 0.087$ g's.
15.112. $\boldsymbol{a}_{XYZ} = -73.22\boldsymbol{i} - 112.63\boldsymbol{j} + 67.3\boldsymbol{k}$ m/s².
15.114. $\boldsymbol{a}_{XYZ} = -0.4\boldsymbol{j} + (-15\boldsymbol{j}) + 2(5\boldsymbol{i} + 20\boldsymbol{k}) \times (0.2\boldsymbol{j}) + (0.02\boldsymbol{i} - 10\boldsymbol{j}) \times (-0.01\boldsymbol{j}) + (0.5\boldsymbol{i} + 20\boldsymbol{k}) \times [(0.5\boldsymbol{i} + 20\boldsymbol{j}) \times (-0.01\boldsymbol{j})]$ m/s².
15.116. $\boldsymbol{F} = -0.5097\boldsymbol{i} + 0.1346\boldsymbol{j} - 0.1702\boldsymbol{k}$ N.
15.118. $\boldsymbol{a}_{XYZ} = -15.50\boldsymbol{i} + 0.405\boldsymbol{j} - 7.80\boldsymbol{k}$ m/s².
$(\boldsymbol{a}_A)_{XYZ} = -1.05\boldsymbol{i} + 0.665\boldsymbol{j} - 2.2\boldsymbol{k}$ m/s².
15.122. $\boldsymbol{F} = 60\boldsymbol{j}$ N.
15.124. Axial force = 30.7 N compression
Bending moment = 27 Nm.
15.126. $F_{axial} = 1.791$ N.
15.128. –37.69 m/s² $\quad$ 3.20 N in plus x direction.

15.130. $\dot{\boldsymbol{\omega}} = -4\boldsymbol{i} + 2\boldsymbol{j} + 30\boldsymbol{k}$ rad/s².
15.132. $\omega_C = 2.07$ rad/s $\quad \dot{\omega}_C = -1.799$ rad/s².
15.134. $\boldsymbol{a}_{XYZ} = -100\boldsymbol{i} + 5\boldsymbol{j} - 100\boldsymbol{k}$ m/s².
15.136. 48.66 m/s².
15.138. $\omega_B = -0.5077$ rad/s
$\dot{\omega}_B = -1.0115$ rad/s².
15.140. $\boldsymbol{V}_{XYZ} = 6\boldsymbol{i} - 9\boldsymbol{j}$ m/s.
$\boldsymbol{a}_{XYZ} = -6\boldsymbol{i} + 1.5\boldsymbol{j} - 150\boldsymbol{k}$ m/s².
15.142. $\boldsymbol{V}_B = 1.500\boldsymbol{i} - 1.500\boldsymbol{j}$ m/s.
15.144. $V_C = 10$ m/s $\quad \omega_{CD} = 20$ rad/s $\quad \omega_{BC} = 0$.
15.146. $\boldsymbol{V}_{XYZ} = 2\boldsymbol{j}$ m/s
$\boldsymbol{a}_{XYZ} = -13.33\boldsymbol{k}$ m/s².

16.2. 2.12 Nm.
16.4. 16.70 m down.
16.6. $F = 3.28$ N.
16.8. $P = 678$ N.
16.10. $a_B = -3.27$ m/s².
16.12. $t = 0.411$ s.
16.14. $\ddot{\theta}_A = -0.8657$ rad/s².
16.16. $F_{AB} = 163.4$ N.
16.18. 552 rev.
16.20. $\sigma_{rr} = 480.5$ N/nm².
16.22. $F_x = 387.6$ N $\quad F_y = 749.4$ N.
16.24. $B_y = C_y = 2.244$ kN $\quad B_z = C_z = 5.27$ N
16.26. $A_y = B_y = 14.02$ kN
16.28. $\dot{\theta}_A = -0.9583$ rad/s.
16.30. $\ddot{\theta} = \dfrac{2y}{L}$
16.32. $\ddot{\theta} = -49.97$ rad/s².
16.34. $a_0 = -3.70$ m/s² $\quad \ddot{\theta} = -1.052$ rad/s².
16.36. –2.08 m.
16.38. $\ddot{\theta} = -14.96$ rad/s².
16.40. 28.9 rad/s².
16.42. 1.442 m/s².
16.44. 32.2 mm.
16.46. $P = 5.88$ kN.
16.48. 3.43 m/s².
16.50. $f = 0.283W$.
16.52. $f = 24.53$ N $\quad N = 85.0$ N.
16.54. 0.405 m/s².
16.56. $T = 335.9$ N.
16.58. $\boldsymbol{a} = \boldsymbol{0}$.
16.60. $\boldsymbol{a} = -2.512\boldsymbol{i} + 6.26\boldsymbol{j}$ m/s².
16.64. 13.08 m/s $\quad \dot{\theta} = 10.90$ rad/s.
16.66. $\ddot{X} = 49.4$ mm/s² $\quad \ddot{Y} = 0 \quad \ddot{\theta} = 0.8429$ rad/s².
16.68. $V_0 = 7.732$ m/s.
16.70. $P = 8.75$ N.
16.72. $BD = 49.5$ N $\quad AC = 37.0$ N.
16.74. $\dot{X} = -3.343$ m/s.
$(\dot{X})_{.3} = -3.294$ m/s.
16.76. $\ddot{\theta} = 19.62$ rad/s² $\quad \ddot{Y} = -4.905$ m/s².
16.78. $\ddot{Y}_A = -2.18$ m/s² $\quad \ddot{Y}_C = 2.18$ m/s².
16.80. $T_B = 246$ N $\quad T_A = 201$ N.
16.82. $\dot{\omega} = 7.57$ rad/s² $\quad \ddot{X} = -2.27$ m/s².
16.84. $\ddot{\theta}_{BC} = -6.43$ rad/s² $\quad \ddot{\theta}_{AB} = 4.29$ rad/s².
16.86. $\ddot{\theta}_{AB} = 6.54$ rad/s².
16.88. $\boldsymbol{a}_A = 1.1308\boldsymbol{i} + 0.6529\boldsymbol{j}$ m/s²
$\dot{\omega} = 0.17125$ rad/s².
$\boldsymbol{a}_B = 0.52549\boldsymbol{i} + 0.04752\boldsymbol{j}$ m/s².
16.90. 0.222 rev.
16.92. $A_x = -32$ N $\quad A_y = -212.4$ N
$B_x = -176.6$ N $\quad B_y = -26.40$ N
16.94. $B_x = 0 \quad A_x = 0$
$B_y = 0$–235 N $\quad A_y = -254$ N
$A_z = -50$ N.
16.96. $\dot{\omega} = 6.36$ rad/s² $\quad B_x = 33.95$ N $\quad B_y = 0$
$A_x = 28.9$ lb $\quad A_y = 0$.
16.98. $B_x = 25$ N $\quad B_y = 60.2$ N
$A_x = 25$ N $\quad A_y = -60.2$ N.
16.100. $B_x = 200$ N $\quad B_y = -21.93$ N
$A_x = 200$ N $\quad A_y = -21.93$ N.
16.102. $M_x = -9.171$ Nm $\quad V_x = -48.9$ N
$M_y = 20.9$ Nm $\quad V_y = 8.66$ N
$M_z = -0.3415$ Nm.
16.104. $r_3 = 0.378$ m $\quad \theta_3 = 249.4°$
$r_4 = 0.327$ m $\quad \theta_4 = 192.2°$.
16.106. $y_D = -16$ mm $\quad z_D = 0$
$y_C = 136$ mm $\quad z_C = 0$.
16.108. $\theta_A = 270°$ $\quad L_A = 318.2$ mm.
16.110. $\theta_B = 223.1°$ $\quad r_B = 1.715$ m
$\theta_4 = 254.3°$ $\quad r_4 = 2.191$ m.
16.112. $T = 768.9$ N.
16.114. $t = 9.174$ s.
16.116. 16.90 m/s².
16.118. –41 rad/s².
16.120. 287 N.
16.122. $D_x = 6.96$ N $\quad D_y = 31.23$ N.
16.124. $\ddot{\theta}_A = -3.26$ rad/s².
16.126. $\ddot{\theta} = 58.0$ mrad/s²
$X = 5$ m $\quad Y = 5$ m.
16.128. $\boldsymbol{a}_D = 4.06\boldsymbol{i} - 7.20\boldsymbol{j}$ m/s².
16.130. $N_A = 85.46$ N $\quad N_B = 191.58$ N
$F_A = 17.09$ N $\quad F_B = 38.32$ N.
16.132. $\ddot{\theta} = -18.34$ rad/s².
16.134. $A_x = 0 \quad A_y = -105.1$ N
$A_z = 490.5$ N.
16.136. $\theta_C = 226.2°$ $\quad r_C = 0.256$ m
$\theta_D = 266.8°$ $\quad r_D = 0.416$ m.

16.138. $A_x = -192.7$ N $A_y = 2.50$ kN
$M_A = 5.648$ kNm.

17.2. 6.12 J.
17.4. 24.5 J.
17.6. 45.8 J 4.954 J.
17.8. $KE = 27.46$ J.
17.10. $\omega = \left[\dfrac{8W_2ag}{3W_1r^2 + 2W_2(r-a)^2}\right]^{\frac{1}{2}}$.
17.13. $V_B = -0.575$ m/s.
17.14. $\omega = 2.79$ rad/s.
17.18. $V_D = 5.40$ m/s.
17.20. $\omega_A = 19.08$ rad/s $\omega_B = 9.54$ rad/s.
17.22. $V_C = 3.65$ m/s.
17.24. $\omega_z = 2.87$ rad/s.
17.26. $\Delta = 0.255$ m.
17.30. $\dot{\theta}_H = 35.30$ rad/s.
17.32. $V = 2.354$ m/s $f = 79.4$ N.
17.34. $V = 4.60$ m/s $f = 7.48$ kN.
17.36. 25.3 rad/s.
17.38. $V_A = 4.1$ m/s.
17.40. $(H_0)_x = 0$ $(H_0)_y = 16.2\pi$ kg m/s
$(H_0)_z = 99.96\pi$ kg m/s.
17.42. $(H_c)_x = 1.941$ kg m²/s $(H_c)_y = 0.510$ kg m²/s
$(H_c)_z = 0$.
17.44. $h = 1.016$ m.
17.46. $T_1 = 1.13$ kN $P = 678.1$ N.
17.48. $L = 9.38$ m.
17.50. $\omega = 2.732$ r/min.
17.52. $V_A = -8.38$ m/s.
17.54. $\omega = 9.09$ rad/s.
17.56. $t = 46.7$ ms.
17.58. $y_{max} = 1.431$ km 116.1°.
17.62. $\omega_A = -108.7$ rad/s.
17.64. $\omega = 9.90$ rad/s.
17.66. $V_A = -11.04$ m/s.
17.68. $(w_2)_f = -165.2$ rad/s $(\omega_1)_f = 24.22$ rad/s.
17.70. $\omega_f = 7.64$ rad/s.
17.72. $\boldsymbol{V}_B = 1.542\boldsymbol{i} - 1.028\boldsymbol{j}$ m/s.
17.74. $f = 15.05$ Hz.
17.76. -5.92 m/s $\omega_f = 70.14$ rad/s.
17.78. $[(V_C)_y]_f = -12.39$ m/s
$[(V_C)_x]_f = 12.85$ m/s.
17.80. 22.50 J.
17.82. $d = 0.1246$ m.
17.84. $V = 2.23$ m/s.
17.86. $(H_C)_x = -589.3$ kg m²/s
$(H_C)_y = -1000.6$ kg m²/s
$(H_C)_z = -588.6$ kg m²/s.
17.88. 2.32 m/s.
17.90. $\omega = 4.11$ rad/s.
17.92. $\boldsymbol{V}_E = -0.3117\boldsymbol{i} - 0.434\boldsymbol{j}$ m/s.
17.94. $\omega = 14.78$ rad/s.

18.2. $\dot{\omega}_x = 12.19$ rad/s²
$\dot{\omega}_y = 7.31$ rad/s²
$\dot{\omega}_z = 30$ mrad/s²
18.4. -1.20 Nm.
18.6. $149.8\boldsymbol{i}$ Nm.
18.8. $\boldsymbol{a} = -122.6\boldsymbol{i} - 5\boldsymbol{j} + 22.9\boldsymbol{k}$ m/s².
18.10. Shear force = 147.2 N
Tensile force = 112.5 kN
Moment = -2.934 kNm.
18.14. $\boldsymbol{M} = 8.35\boldsymbol{i} - 0.586\boldsymbol{j} + 3.59\boldsymbol{k}$ Nm.
18.16. $\boldsymbol{M} = -148.8\boldsymbol{i} + 300\boldsymbol{j} - 108.6\boldsymbol{k}$ Nm.
18.18. Moment about the centre of mass of *CD* due to motion of *CD*: $M_{CD} = -3.18 \times 10^{-9}$ nNm.
Moment about the centre of mass of *EG* due to motion of *EG*: $M_{EG} = -1.147 \times 10^{-9}$ nNm.
18.20. $\boldsymbol{M} = -2.462\boldsymbol{i} + 1.095\boldsymbol{j} - 2.190\boldsymbol{k}$ kNm.
18.22. $\boldsymbol{M} = 1.95\boldsymbol{i}$ kNm.
18.24. $T_x = 3.72$ Nm.
$T_y = -5.24$ nNm.
$T_z = -20.95$ Nm.
18.26. $\boldsymbol{M} = 149.4\boldsymbol{i}$ Nm.
18.28. $\boldsymbol{M} = -2.39 \times 10^3 \sin 0.419t\boldsymbol{i} + 42.9 \times 10^3 \cos 0.419t\boldsymbol{j}$ Nm.
18.30. $\boldsymbol{M} = 8.35\boldsymbol{i} - 0.586\boldsymbol{j} + 3.59\boldsymbol{k}$ Nm.
18.32. $\dot{\psi} = 2$ rad/s.
$\dot{\theta} = 0$
$\dot{\phi} = -13.33$ rad/s.
18.34. 1.744 krad/s.
18.36. $T_x = -33.07$ Nm.
18.38. $\dot{\psi} = 2.83$ mrad/s.
Result is an angular acceleration in the direction of the torque equal to 48.8 mrad/s².
18.42. $\dot{\psi} = -5.43$ rad/s;
5.82 rad/s.
18.44. 2.414 Nm/rad.
18.46. 0.504 Nm
from the posts in a direction normal to the plate.
18.48. $\boldsymbol{H} = 40.5\boldsymbol{i} + 116.6\boldsymbol{k}$ kg m²/s.

18.50. $\dfrac{\dot{\psi}}{\sin\alpha} = \dfrac{\dot{\phi}}{\sin(\theta-\alpha)}$

$\omega^2 = \dot{\psi}^2 + \dot{\phi}^2 + 2\dot{\phi}\dot{\psi}\cos\theta$

$\dot{\phi} = \dfrac{I_{xx} - I_{zz}}{I_{zz}}\dot{\psi}\cos\theta$

18.52. $\dot{\psi} = 1.313$ rad/s
$\dot{\phi} = 2$ rad/s
$\theta = 12.9°$.

18.54. $\dot{\psi} = 3.26$ rad/s.

18.56 $\theta = 13.98°$
$\dot{\phi} = -44.2$ mrad/s
$\dot{\psi} = 0.1264$ rad/s.

18.58. $\theta = 128.3°$
$\dot{\phi} = -1.194$ rad/s.

18.60. $\boldsymbol{M} = -29.96\boldsymbol{i} + 3210\boldsymbol{j} + 0.412\boldsymbol{k}$ Nm.

18.61. $\boldsymbol{K} = -0.1262\boldsymbol{i} + 0.1433\boldsymbol{j} + 9.17\boldsymbol{k}$ N
$\boldsymbol{H} = -0.700\boldsymbol{i} + 0.1433\boldsymbol{j}$ N.

18.62. 15.69 rad/s.

18.63. 2.31 krad/s or 65.4 mrad/s.

18.64. $R = 60.9$ m.

18.65. $\theta = 47.3°$
$\dot{\phi} = -0.1283$ rad/s
$\dot{\psi} = 0.525$ rad/s.

18.66. $\omega_x = 0.299$ rad/s
$\omega_y = 0.190$ rad/s
$\omega_z = 0.423$ rad/s.

18.67. $\dot{\psi} = 0.2064$ rad/s
Z in direction of z axis.

18.68. $\dot{\phi} = 1.905$ rad/s
$\theta = 71.87°$.

19.4. $K_2 = 3.90$ Nm.

19.6. $\omega_n = \sqrt{\dfrac{2\mu g}{L}}$ rad/s.

19.8. 0.438 Hz.

19.10 0.557 Hz.

19.12. 19.81 rad/s; 8.34 N/mm.

19.14. 29.5 rad/s.

19.16. $\ddot{r} + \left(\dfrac{K}{m} - \omega^2\right)r = \dfrac{Kr_0}{m}$

$\omega_n = \sqrt{\dfrac{K}{m} - \omega^2}$

19.18. 15.49 rad/s 84.85 rad/s.

19.20 1.469 s.

19.22. $\ddot{x} + 0.326x + 12.30 \times 10^{-3}x^2 + 155.0 \times 10^{-6}x^3 = 0$
$\omega = 90.9$ mHz.

19.24. $(K_t)_{eq} = 24.3$ μNm/rad
$\omega_n = 304$ rad/s
$\theta = -7.42°$.

19.26. $\omega_n = 1$ Hz.

19.28. 7.88 Hz
$\theta = -0.342°$.

19.30. $\omega_n = \sqrt{\dfrac{2Kr^2}{R^2[(M/2)-(W_1/g)]}}$

19.32. $\dfrac{1}{12}\dfrac{W}{g}L^2\ddot{\theta} + \dfrac{Wd^2}{4l}\theta = 0$

$\omega_n = \sqrt{\dfrac{3d^2g}{lL^2}}$ rad/s.

19.34 218.4 rad/s.

19.36. 1.377°.

19.38. $P = [(2KL - W/2)]$.

19.40. 2.32 rad/s.

19.42. $k = 0.518$ m.

19.44. $\mu = 0.087$.

19.46. $\omega_n = \sqrt{\dfrac{K(r_1+r_2)^2}{(W/g)(k^2+r_1^2)}}$

$f_{max} = \dfrac{[I_{00}\omega^2 - K(r_1+r_2)(r_2)]A}{r_1}$.

19.48. $\omega_n = \sqrt{\dfrac{2Kr^2}{\frac{1}{2}MR^2 + (W_1/g)R^2}}$.

19.50. 2.32 rad/s.

19.52. $\omega_n = \sqrt{\dfrac{(3+\sin\alpha)g}{L}}$.

19.54. $\omega_n = \sqrt{\dfrac{g}{(R-r_1)[1+(k^2/r_1^2)]}}$.

19.56. $x = 1.65$ mm.

19.58. 9.92 mm.

19.64. 110.5 mm
$V = 14.28$ m/s.

19.66. $\omega = 15.95$ rad/s.

19.68. 17.89 N $K = 553$ kN/m.

19.70. 9.805 N.

19.72. 1.808 rad/s.

19.74. −11.28°.

19.76. $m\ddot{x} + 2c\dot{x} + (K_1 + K_2)x = 0$

$c_{cr} = \sqrt{\dfrac{w}{g}(K_1 + K_2)}$

19.78. $x(5) = -113.7$ mm $f \cong 7.30$ rad/s.
19.82. $t = 0.321$ s.
19.84. $\omega = 22.5$ mHz.

19.86.
$$e^{-0.311\tau}\left[\left(-0.05-\frac{0.0155}{\alpha}\right)e^{-\alpha\tau}+\left(-0.05-\frac{0.0155}{\alpha}\right)e^{-\alpha\tau}\right]=0$$
$$e^{-0.311\tau}[(-0.05\alpha-0.01555)e^{\alpha\tau}-(-0.05\alpha-0.01555)e^{-\alpha\tau}]$$
$$-0.311e^{-0.311\tau}\left[\left(-0.05-\frac{0.01555}{\alpha}\right)e^{\alpha\tau}+\left(-0.05-\frac{0.01555}{\alpha}\right)e^{\alpha\tau}\right]=0$$

19.90. 1.45 kN/m.
19.92. 0.705 mm.
19.94. 2 degrees of freedom $\ddot{x} + 456x - 235\theta = 0$
$\ddot{\theta} = 223x + 334.4_\theta = 0.$
19.97. $K_3 = 799$ N/m.
19.98. 49.8 mm; 2.003 m.
19.99. 2.35 rad/s.
19.100. 48.6 rad/s.
19.101. $k = 0.748$ m.
19.102. $\omega_n = \sqrt{\dfrac{6g}{L}}$
19.103. $\omega = 19.80$ rad/s.
19.104. $W = 64.$ 6N.
19.105. $\theta_0 = 60.8$ mrad.
19.106. 3.014 mrad.
19.107. 0.221 sec.
19.108. $A = 1.454$ mm
$(F_{tr})_{max} = 337$ N.
19.109. 23.6 rad/s
11.15 mm
$c = 192.0$ Ns/m.

9.2. 175.5 kg m^2 621 kg m^2.
9.4. $I_{yy} = \frac{1}{12}M(a^2 + b^2)$ $I_{zz} = \frac{1}{12}M(b^2 + l^2)$
$I_{xx} = \frac{1}{12}M(a^2 + l^2)$.
9.6. $\frac{1}{2}Mr^2$.
9.8. 26.7 g m^2
3.95 g m^2.
9.10. 1.723 g m^2.
9.12. 3.959 g m^2.
9.14. $I_{xx} = 100.4 \times 10^3$ mm^4.
$I_{yy} = 1.875 \times 10^3$ mm^4
$I_{xy} = 11.72 \times 10^3$ mm^4
$(I_{xx})_M = 205$ kg mm^2
$(I_{yy})_M = 3.82$ kg mm^2
$(I_{xy})_M = 23.9$ kg mm^2.
9.16. 37.0 kg mm^2 166.8 kg mm^2 203.8 kg mm^2.
9.20. $I_{y'y'} = 9$ kg m^2 $I_{z'z'} = I_{x'x'} = 42$ kg m^2
$I_{xx} = 172.5$ kg m^2 $I_{yy} = 27$ kg m^2
$I_{zz} = 154.5$ kg m^2.
9.22. $I_{xx} + I_{yy} + I_{zz} = \frac{M}{6}(a^2 + b^2 + c^2)$
9.24. $I_{x''x''} = 3.870$ Mg m^2
$I_{y''y''} = 4.615$ Mg m^2.
9.26. 12.58 g m^2 12.39 g m^2 3.91 g m^2.
9.28. 1.650 kg m^2 1.438 kg m^2 0.513 kg m^2.
9.30. $k = 762$ mm $I = 1.106$ Mg m^2.
9.32. $I_{xx} = 2.429$ kg m^2 $I_{yy} = 433$ g m^2
$I_{xy} = 347$ g m^2
9.34. $I_{cc} = 37.81$ kg m^2.
9.36. 124.53 kg m^2.
9.38. 0.524 kg m^2.
9.40. −6.96 g m^2.
9.44. 362 kg mm^2.
9.46. $(I_{xx})_M = 0.534$ g m^2
$(I_{yy})_M = 1.659$ g m^2
$(I_{zz})_M = 2.19$ g m^2
$(I_{xy})_M = 0.568$ g m^2
$(I_{xz})_M = (I_{yz})_M = 0.$
9.48. 49.3 g m^2 1.87 g m^2 50.3 g m^2.
9.50. 1.075 g m^2 92.35 kg m^2.
9.52. $I_{yy} = 7.58\rho_0$ kg m^2 $I_{yz} = -1.789\rho_0$ kg m^2.

Index